BIOLOGY OF PLANTS

SECOND EDITION

BIOLOGY OF PLANTS

SECOND EDITION

PETER H. RAVEN

Missouri Botanical Garden and
Washington University, St. Louis

RAY F. EVERT

University of Wisconsin, Madison

HELENA CURTIS

WORTH PUBLISHERS, INC.

BIOLOGY OF PLANTS, Second Edition

MANUFACTURED IN THE UNITED STATES OF AMERICA

LIBRARY OF CONGRESS CATALOG CARD NUMBER 75-42980

SECOND PRINTING NOVEMBER 1976

ISBN 0-87901-054-1

DESIGNED BY MALCOLM GREAR DESIGNERS

COMPOSITION AND STRUCTURES BY NEW ENGLAND TYPOGRAPHIC SERVICE

ILLUSTRATIONS BY JOHN KYRK AND ELLEN MARIE DUDLEY

PICTURE EDITOR: ANNE FELDMAN

COVER: *Farm Garden with Sunflowers* BY GUSTAV KLIMT, ÖSTERREICHISCHE GALERIE, VIENNA, AUSTRIA

WORTH PUBLISHERS, INC.

444 PARK AVENUE SOUTH

NEW YORK, NEW YORK 10016

PREFACE TO THE SECOND EDITION

Since the first edition of BIOLOGY OF PLANTS appeared in 1970, we have received from teachers and students many constructive comments on the book's strengths and shortcomings, and some sound advice on how it might be improved. Ray Evert, one of those who taught from BIOLOGY OF PLANTS and had excellent ideas for making it better, has now joined us as coauthor. The substantial improvements in this edition owe a great deal to his talents as a scholar and teacher.

Over the past six years there have been major advances in our understanding of virtually every discipline of plant biology. The second edition takes this recent information into account and attempts to provide a thoroughly up-to-date introduction to the world of plants and those organisms traditionally associated with them. In addition, we believe that this time we have provided a more balanced treatment of all the major disciplines of plant biology.

Although we have reorganized and largely rewritten the text, we have tried to retain the best features of the first edition. Again the interrelationships of evolution, diversity, and ecology are emphasized throughout, and the dual themes of evolution and ecology still pervade the book.

We recognize that there is no "best" way to present plant biology, either in lecture or in laboratory. The new sequence of topics adopted for this edition reflects, in part, the present organization of the course in general botany that one of the authors has given over the past sixteen years. We have tried to make each section of the book as independent as possible, so that topics do not have to be presented in lockstep with our preferred sequence.

The section on plant anatomy, now Section 5, has been rewritten entirely, emphasizing the continuity of the plant body. In this new edition the coverage of plant anatomy has been expanded from two to five profusely illustrated chapters. All but a few of the illustrations in this section were prepared especially for BIOLOGY OF PLANTS. Throughout the book the abundant drawings, photographs, and electron micrographs will not only help students to understand the narrative but should also provide valuable resources for laboratory work.

We have expanded the section on the diversity of living organisms, also strengthening its evolutionary theme and its focus on interrelationships. More common examples are used, particularly in the chapters on algae and fungi.

There is so much new material, so many clarified presentations, that we cannot list them all here. Throughout the many months of rewriting we were most concerned with helping students really to understand what goes on and with conveying the excitement of current research and past observation. We hope that our enthusiasm throughout the book and the greater attention given to the inherently more difficult topics will foster curiosity and understanding in our readers.

PETER H. RAVEN
RAY F. EVERT
January 1976 HELENA CURTIS

ACKNOWLEDGMENTS TO THE FIRST EDITION

This book could not have been written without the help, at every stage, of experts in various areas of botanical research and teaching. We wish to acknowledge especially the assistance of Winslow R. Briggs of Harvard University, who helped in particular with the chapters dealing with plant physiology, and who also reviewed, criticized, and made valuable suggestions concerning other parts of the manuscript. Peter M. Ray of Stanford University was the principal reviewer of Chapter 25 on water movement in plants, and R. Paul Levine and Daniel I. Arnon were advisors for the chapters on photosynthesis.

Others who reviewed various parts of the manuscript and to whom we wish to express our thanks are: Constantine J. Alexopoulos, University of Texas; Lewis E. Anderson, Duke University; Daniel I. Axelrod, University of California, Davis; Harlan P. Banks, Cornell University; Charles E. Barr, State University of New York, Brockport; R. G. S. Bidwell, Case Western Reserve University; David W. Bierhorst, University of Massachusetts; W. Dwight Billings, Duke University; Olov Björkman, Carnegie Institution, Stanford; Walter F. Bodmer, Stanford University; Harold C. Bold, University of Texas; Allan M. Campbell, Stanford University; Elizabeth G. Cutter, University of California, Davis; Andrew J. Darlington, Imperial Cancer Research Fund, London; Theodore Delevoryas, Yale University; E. Lloyd Dunn, Stanford University; Paul R. Ehrlich, Stanford University; Ursula Goodenough, Harvard University; R. H. Hill, Cambridge University; Richard W. Holm, Stanford University; Albert W. Johnson, San Diego State College; George Johnson, Stanford University; Clifford B. Knight, East Carolina College; Bastiaan J. D. Meeuse, University of Washington, Seattle; Harold A. Mooney, Stanford University; Robert Ornduff, University of California, Berkeley; Johannes Van Overbeek, Texas A & M University; Marcus M. Rhoades, Indiana University; Frank B. Salisbury, Utah State University; W. B. Schofield, University of British Columbia; G. Ledyard Stebbins, University of California, Davis; William C. Steere, Director, New York Botanical Garden; Richard A. White, Duke University; Richard H. Wilson, University of Texas.

One of the authors would like to express his appreciation to his wife, Tamra Engelhorn Raven, for her assistance and encouragement.

By introducing young people not only to what is known about botany and ecology but particularly to the many unsolved problems in these fields, we hope to enlist new talents and new enthusiasms in working toward the solutions on which all of our futures depend.

April 1970

PETER H. RAVEN
HELENA CURTIS

ACKNOWLEDGMENTS TO THE SECOND EDITION

Special thanks and acknowledgment are due to Susan E. Eichhorn, of the University of Wisconsin, Madison, who supervised much of the organization and typing of the manuscript, as well as most of the preparation of illustrations for the second edition. Special recognition and thanks are also due Ellen M. Dudley for her many superb drawings and Damian S. Neuberger, University of Wisconsin, Madison, for his outstanding photographic work.

We are grateful to Barbara G. Pickard of Washington University, St. Louis, for her meticulous and detailed suggestions on Sections 2, 6, and 7, and to George B. Johnson of the same institution for his help with Chapter 3 and Section 3. William F. Whittingham, University of Wisconsin, Madison, was the principal reviewer of Chapter 11 on fungi, and he reviewed, criticized, and made valuable suggestions on other parts of the manuscript as well. William J. Woelkerling, University of Wisconsin, Madison, was the principal reviewer of Chapter 12 on the algae and slime molds.

Among the others who have reviewed parts of the manuscript or supplied specific information for this edition and to whom we wish to express our thanks are Paul J. Allen, University of Wisconsin, Madison; Daniel I. Axelrod, University of California, Davis; Harlan P. Banks, Cornell University; Wayne M. Becker, University of Wisconsin, Madison; E. S. Beneke, Michigan State University; H. Weston Blaser, University of Washington; Winslow Briggs, Carnegie Institute of Washington; Allan M. Campbell, Stanford University; Rita R. Colwell, University of Maryland; William L. and Chicita F. Culberson, Duke University; Marshall R. Crosby, Missouri Botanical Garden; Dennis J. Donaher, Wayne State University; Eva F. Esterman, San Francisco State University; D. H. Franck and G. C. Gerloff, University of Wisconsin, Madison; Martin Gibbs, Brandeis University; George Gould, City College of San Francisco; Mason E. Hale, Jr., Smithsonian Institution; C. E. Henrickson, University of Kentucky; Paul Horgen, University of Toronto; Robert E. Hungate, University of California, Davis; Carl S. Keener, Pennsylvania State University, University Park; T. T. Kozlowski, University of Wisconsin, Madison; Norma J. Lang, University of California, Davis; Thomas McInnis, Jr., Clemson University; Louis Odell, Towson State College; Elmer Palmatier, University of Rhode Island; Richard W. Pippen, Western Michigan University; David Porter, University of Georgia; David H. Rembert, Jr., University of South Carolina; Thomas L. Rost, University of California, Davis; Roy and Barbara Saigo, University of Wisconsin, Eau Claire; J. W. Schopf, University of California, Los Angeles; Roger Y. Stanier, Institut Pasteur, Paris; Robert Thornton, University of California, Davis; Winifred Trakimas, SUNY Agricultural and Technical College, Farmingdale; John Troughton, Carnegie Institute of Washington; Joseph E. Varner, Washington University, St. Louis; Dean P. Whittier, Vanderbilt University; Clyde Willson, University of California, Berkeley; Roy H. Wishard, El Camino College; Z. Wochok, University of Alabama, University; and Ann Wylie, University of Otago.

Others—all of the University of Wisconsin, Madison—who contributed to the preparation of the second edition and to whom we wish to express our thanks are Gene Coffman and David M. Selk, photographers; Mary E. Lauder and Ellen Thom, typists; Herbert M. Clarke and Roland R. Dute, for plant identifications; and Geoffrey A. Kuter, for cultures of various organisms.

The second author would like to express his gratitude to his wife and children for their patience and encouragement during preparation of the second edition.

PETER H. RAVEN
RAY F. EVERT
HELENA CURTIS

January 1976

CONTENTS IN BRIEF

TABLE OF CONTENTS

BIOLOGY OF PLANTS

SECOND EDITION

INTRODUCTION

When a particle of light strikes a molecule of chlorophyll, an electron is jolted out of the molecule and raised to a higher energy level. Within a fraction of a second, it returns to its previous energy state. All life on this planet is dependent upon the energy momentarily gained by the electron. The process by which some of the energy given up by the electron in returning to its original energy level is converted into chemical energy—energy in a form usable by living systems—is known as photosynthesis. Photosynthesis is the vital link between the physical and the biological world, or, as Nobel laureate Albert Szent-Györgyi said, more poetically: "What drives life is a little electric current, kept up by the sunshine."

Only a few types of organisms—the green plants, the algae, and some bacteria—possess a type of chlorophyll that can, when embedded in the membranes of a living cell, carry out this energy conversion. However, once the energy is trapped in chemical form, it becomes available as an energy source to all other organisms, including man. Thus we are totally dependent upon photosynthesis, a process which our laboratories are just beginning to understand and to which green plants are so exquisitely adapted.

THE EVOLUTION OF PLANT LIFE

Like all other living organisms, plants have a long evolutionary history. The planet earth itself is some 4.7 billion years old, an accretion of dust and gases swirling in orbit around the star that is our sun. The earliest known fossils are about 3.2 billion years old and consist of small, relatively simple cells.

As events are reconstructed, these first cells were formed by a series of chance events. The raw materials—carbon, oxygen, hydrogen, and nitrogen—were present in the gases of the early atmosphere. (These four elements make up 98 percent of the tissue of all living things today.)

I-1
A cluster of cells of the freshwater green alga Cosmarium botrytis. *Algae are photosynthetic organisms that are either single-celled or have relatively simple multicellular structures (compared with plants) and are adapted to life in water. They are the subject of Chapter 12.*

Through the thin atmosphere, the rays of the sun beat down on the harsh, bare surface of the young earth, bombarding it with light, heat, and ultraviolet radiation. Water vapor cooled in the upper atmosphere, fell on the crust of the earth as rain, and steamed up again, driven by the sun's heat. Violent rainstorms, accompanied by lightning, released electrical energy. Radioactive substances in the earth's crust emitted their energy, and molten rock and boiling water erupted from beneath the earth's surface. The energy in this vast crucible broke apart the simple gases of the atmosphere and reformed them into more complicated molecules.

According to present hypotheses, the compounds that were formed in the atmosphere tended to be washed out by the driving rains and to collect in the oceans, which grew larger as the earth cooled. As a consequence, the ocean became an increasingly rich mixture of organic molecules. Some organic molecules have a tendency to aggregate in groups; in the primitive ocean, these groups probably took the form of droplets, similar to the droplets formed by oil in water. Such droplets of organic molecules appear to have been the forerunners of primitive cells, the first forms of life.

These organic molecules also served, according to present theories, as the source of energy for the earliest forms of life. The primitive cells or cell-like structures were able to use these compounds, which were abundant in the "primordial soup," to satisfy their energy requirements.

Such cells are known as *heterotrophs,* a category of organisms that today includes all living things classified as animals or fungi and many of the one-celled organisms, the bacteria and the protists. *Hetero* comes from the Greek word meaning "other," and *troph* comes from *trophos,* "one that feeds." A heterotrophic organism is one that is dependent upon others—that is, upon an outside source of organic molecules—for its energy.

As the primitive heterotrophs increased in number, they began to use up the complex molecules on which their existence depended—and which had taken millions of years to accumulate. Organic molecules in free solution (not inside a cell) became more and more scarce. Competition began. Under the pressure of this competition, cells that could make efficient use of the limited energy sources now available were more likely to survive than cells that could not. In the course of time, by the long, slow process of weeding out the less fit, cells evolved that were able to make their own energy-rich molecules out of simple nonorganic materials. Such organisms are called *autotrophs,* "self-feeders." Without the evolution of autotrophs, life on earth would soon have come to an end.

The most successful of the autotrophs were those that evolved a system for making direct use of the sun's energy—the process of photosynthesis.

The earliest photosynthetic organisms, although simple in comparison to modern plants, were much more complex than the primitive heterotrophs. To capture and use the sun's energy required first, a complex pigment system that could catch and hold the energy of a ray of light and, linked to this system, a way of fixing the energy in an organic molecule.

Thus the flow of energy in the biosphere came to assume its modern form: radiant energy channeled through photosynthetic autotrophs to all other forms of life.

Photosynthesis and The Coming of Oxygen

As photosynthetic organisms increased in number, they changed the face of the planet. This biological revolution came about because one of the most efficient strategies of photosynthesis—the one employed by nearly all living autotrophs—involves splitting the water molecule (H_2O) and releasing its oxygen. Thus, as a consequence of photosynthesis, the amount of oxygen gas (O_2) in the atmosphere increased. This had two important consequences. First, some of the oxygen molecules in the outer layer of atmosphere were converted to ozone (O_3) molecules. When there is a sufficient quantity of ozone molecules in the atmosphere, they filter the ultraviolet rays, highly destructive to living organisms, from the sunlight that reaches the earth. By about 450 million years ago, organisms, protected by the ozone layer, could survive in the surface layers of water and on the land.

Second, the increase in free oxygen opened the way to a much more efficient utilization of the energy-rich molecules formed by photosynthesis. As we shall see in Chapter 5, respiration* yields far more energy than can be extracted by any anaerobic (oxygenless) process. Until the atmosphere became aerobic, the only cells that evolved were *prokaryotic*—simple cells that lacked nuclear envelopes and that did not have their genetic material organized into complex chromosomes. Today, the surviving prokaryotes are the bacteria and the blue-green algae. According to the fossil record, the increase of relatively abundant free oxygen was accompanied by the first appearance of *eukaryotic* cells—cells with complex chromosomes, nuclear envelopes, and membrane-bounded organelles. All living systems are composed of eukaryotic cells, except for the bacteria, the blue-green algae, and the viruses (which actually are in a "twilight zone" somewhere between the living and the nonliving world).

* It should be noted that respiration has two meanings in biology. One is the breathing in of oxygen and breathing out of carbon dioxide; this is also the ordinary, nontechnical meaning of the word. The second meaning of respiration is the oxidation of food molecules by cells—that is, the breaking down of energy-rich, carbon-containing molecules and their use by the cell as an energy source. This process, sometimes qualified as cellular respiration, is what we are concerned with here.

(a)

(b)

I–2
A modern heterotroph and a photosynthetic autotroph. (a) *A fungus,* Pholiota squarrosoides, *growing on an old cottonwood log in southeastern Alaska.* Pholiota, *which, like other fungi, absorbs its food in an organic form—often from other organisms—is heterotrophic. Fungi are the subject of Chapter 11.* (b) *Dutchman's breeches* (Dicentra cucullaria), *one of the first plants to flower in spring in the deciduous woods of eastern North America. Like most vascular plants, Dutchman's breeches is rooted in the soil; photosynthesis takes place chiefly in the deeply divided leaves of this autotrophic organism. The flowers are produced in well-lighted conditions before the leaves appear on the surrounding trees. The underground portions of the plant live for many years and spread to produce new plants vegetatively under the thick cover of decaying leaves and other organic material on the forest floor.*

The Sea and The Shore

Early in evolutionary history, the principal photosynthetic organisms were microscopic cells floating on the surface of the sunlit waters. Energy abounded as did carbon, hydrogen, and oxygen, but as the cellular colonies multiplied, they quickly depleted the mineral resources of the open ocean. (It is this shortage of essential minerals that is the limiting factor in any modern plans to harvest the seas.) As a consequence, life began to drift toward the shores, where the waters were rich in nitrates and minerals carried down from the mountains by rivers and streams and scraped from the coasts by the ceaseless waves.

The rocky coast presented a much more complicated environment than the open sea and, in response to these evolutionary pressures, living organisms became more complex in structure and more diversified. Some 650 million years ago organisms evolved in which many cells, connected by strands of protoplasm, were linked together to form an integrated, multicellular body. In these primitive organisms we see the beginnings of the modern plants and animals.

On the turbulent shore, those photosynthetic organisms that were multicellular were better able to maintain their position against the action of the waves, and, in meeting the challenge of the rocky coast, new forms developed. Typically, these organisms evolved relatively

I–3
Pinus ponderosa *in Yosemite, California, is an example of a vascular plant.*

strong walls for support and specialized structures to anchor their bodies to the rocky surfaces. As these multicellular organisms increased in size, they were confronted with the problem of how to supply food to the dimly lit portions of their bodies where photosynthesis was not taking place. As a consequence of these new pressures, specialized food-conducting tissues evolved that, extending down the center of their bodies, connected the photosynthesizing parts with the lower, nonphotosynthesizing structures.

The Transition to Land

The body of the familiar plant can best be understood in terms of its long history and, in particular, in terms of the evolutionary pressures involved in the transition to land. The requirements of a photosynthetic organism are relatively simple: light, water, carbon dioxide for photosynthesis, oxygen for respiration, and a few minerals. On the land, light is abundant, as are oxygen and carbon dioxide, both of which circulate more freely than in the water, and the soil is generally rich in minerals. The critical factor is water. Land animals, generally speaking, are mobile and able to seek out water just as they seek out food. Fungi, though immobile, remain largely below the surface of the soil or within whatever damp organic material they feed upon. Plants utilize an alternative evolutionary strategy. Roots anchor the plant in the ground and collect the water required for maintenance of the plant body and for photosynthesis. A continuous stream of water moves into the root hairs, up through the roots and stems, and then out through the leaves. All of the above-ground portions of the plant that are ultimately concerned with photosynthesis are covered with a waxy *cuticle* that retards water loss. However, the cuticle also prevents the necessary exchange of gases between the plant and the surrounding air. The solution to this dilemma is found in specialized openings called *stomata* (singular, stoma), which open and close in response to environmental and physiological signals, thus helping the plant maintain a balance between its water losses and its oxygen and carbon dioxide requirements.

In younger plants and those with a seasonal life span (annuals) the stem is also a photosynthetic organ. In longer-lived plants (perennials), the stem may become thickened and woody and covered with cork, which also retards water loss. In both cases, the stem serves both to support the chief photosynthetic organs, holding them to the light, and to house the plant's intricate and efficient vascular system. The vascular system has two major components: the *xylem,* through which water passes upward through the plant body, and the *phloem* (pronounced flow-em), through which food manufactured in the leaves and other photosynthetic parts of the plant is transported throughout the plant body. It is this efficient conducting system that has given the main group of modern plants, the vascular plants, their name.

Perhaps also as a consequence of their immobility, vascular plants, unlike animals, continue to grow throughout their life spans. All plant growth originates in localized regions of perpetually embryonic tissues; these regions are called *meristems.* Meristems located at the tips of all roots and shoots are called *apical meristems.* These are involved with the extension of the plant body. Thus the roots move continually toward new sources of water, and the photosynthetic regions are continually expanded and extended toward the light. This type of growth that originates from apical meristems is known as *primary growth.*

The type of growth that results in the thickening of stems, branches, and roots is known as *secondary growth.* It originates in a second kind of meristematic tissue, the *vascular cambium.*

Thus, in sum, the vascular plant is characterized by a root system that serves to anchor the plant in the ground and to collect water and minerals from the soil; a stem or trunk that raises the photosynthetic parts of the plant body towards its energy source, the sun; the highly specialized photosynthetic organs, the leaves; all of which—root, stem, and leaves—are interconnected by a complicated and efficient system for the transport of food and water. All of these characteristics are adaptations to a photosynthetic existence on land.

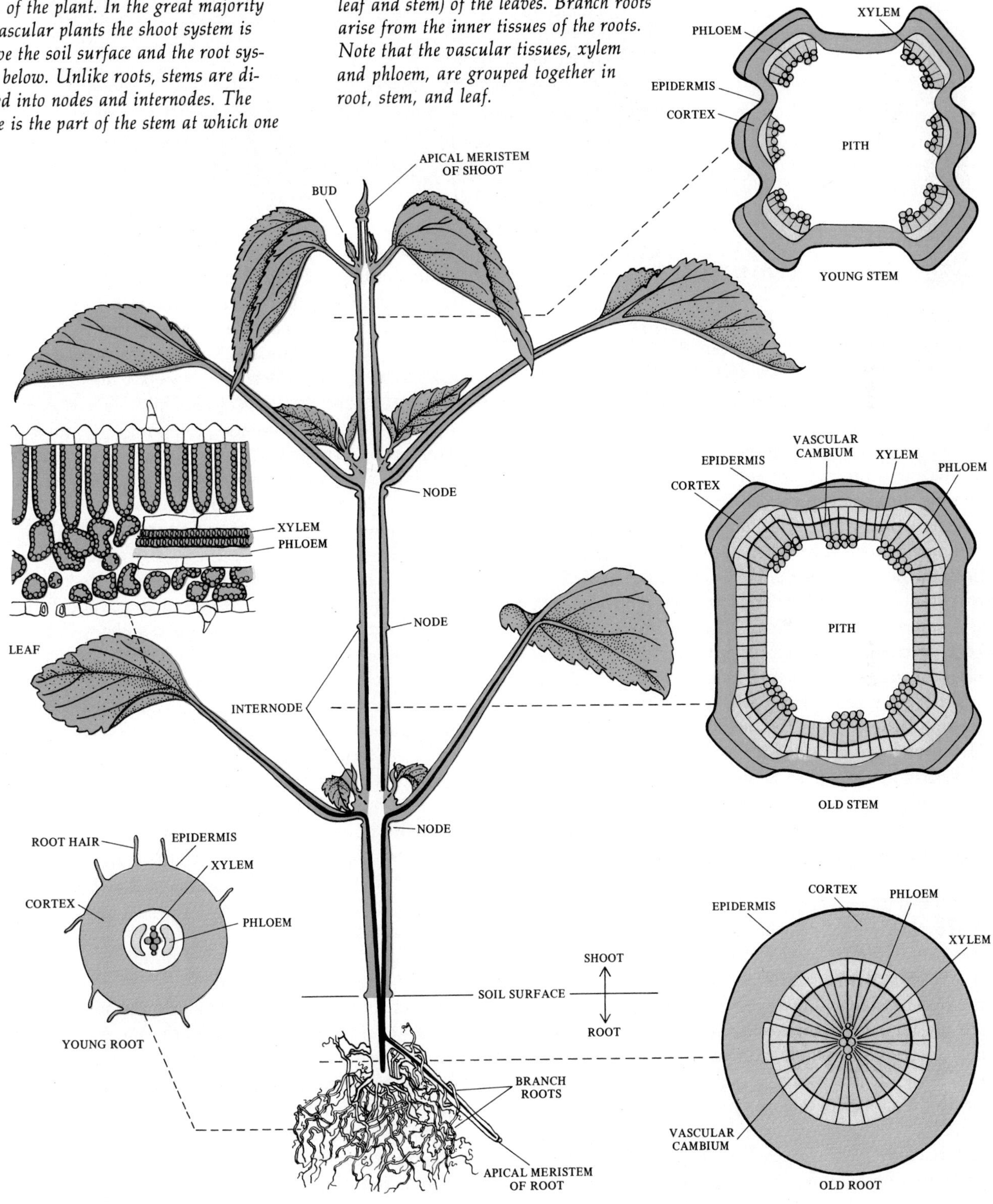

I-4

Diagram of a Salvia *plant, showing the principal organs and tissues of the modern vascular plant body. The organs—root, stem, and leaf—are composed of tissues, groups of cells with distinct structures and functions. Collectively, the roots make up the root system, and the stems and leaves together, the shoot system, of the plant. In the great majority of vascular plants the shoot system is above the soil surface and the root system below. Unlike roots, stems are divided into nodes and internodes. The node is the part of the stem at which one or more leaves are attached, and the internode is the part of the stem between two successive nodes. (In* Salvia, *two leaves are opposite each other at each node. Each pair is at right angles to the pair of leaves at the node above or below it.) Buds (embryonic shoots) commonly arise in the axils (upper angle between leaf and stem) of the leaves. Branch roots arise from the inner tissues of the roots. Note that the vascular tissues, xylem and phloem, are grouped together in root, stem, and leaf.*

I–5
The prairies of the central United States are among the richest agricultural areas in the world, capable of retaining their productivity indefinitely if properly managed. This uncultivated prairie is dotted with wild flowers.

THE EVOLUTION OF COMMUNITIES

The invasion of the land by the plants changed the face of the continents. Looking down from an airplane on one of our country's great expanses of desert or on the peaks of the Sierra Nevada, one can begin to imagine what the world looked like before the coming of the plants. Yet even in these inhospitable regions, the traveler who goes by land will find green plants of an astonishing variety punctuating the expanses of rock and sand. And in those parts of the world where the climate is more temperate and the rains more favorable, the plant communities dominate the land and determine its character. In fact, to a large extent, they *are* the land. Rain forest, meadow, woods, prairie, tundra—each of these words brings to mind the portrait of a landscape. And the main features of the landscape are the plants—enclosing us in a dark green cathedral in our imaginary rain forest, carpeting the ground beneath our feet with wild flowers in a meadow, moving in great golden waves as far as the eye can see across our imaginary prairie. Only when we have sketched these biomes in broad strokes, that is, in terms of trees and shrubs and grasses, can we fill in the other features of our landscape—a deer, an antelope, a rabbit, a wolf.

How do vast plant communities, such as those we can see on a continental scale, come into being? We can trace to some extent the evolution of the different kinds of plants and animals that populate them. Even with accumulating knowledge, however, we have only begun to glimpse the far more complex pattern of development, through time, of the whole system of organisms that make up these various communities. Such communities, along with the nonliving environment of which they are a part, are known as ecological systems, or *ecosystems.* We shall be discussing ecosystems in greater detail in the final chapters of this book. For now, however, it is sufficient to regard an ecosystem as forming a sort of corporate entity, made up of transient individuals. Some of these individuals, the larger trees, live as long as several thousand years; others, the microorganisms, live only a few hours or even minutes. Yet the ecosystem as a whole tends to be remarkably stable; once in balance, it will not change for centuries. Your grandchild will someday perhaps walk along a woodland path once followed by your greatgrandparents, and where they saw a pine tree, a mulberry bush, a meadow mouse, wild blueberries, or a towhee, this child, if this woodland still exists, will see these same kinds of plants and animals and in the same numbers.

Although many of the organisms in an ecosystem are competing for resources, the system as a whole functions as an integrated unit. The death of a solitary cell floating on the ocean's surface is likely to involve the dissipation of its stored energy and the breakdown of its chemical constituents. In an ecosystem, virtually every living thing, down to the smallest bacterial cell or fungal spore, provides a food source for some other living organism. In this way, the energy captured by green plants is transferred in a highly regulated way through a number of different types of organisms before it is dissipated. Moreover, interactions among the organisms themselves, and between the organisms and the nonliving environment, produce an orderly cycling of elements such as nitrogen and phosphorus. Energy itself must be constantly added to the ecosystem, but the elements, as we shall see in future chapters, are cycled through the organisms, returned to the soil, decomposed by soil bacteria and fungi, and recycled. These transfers of energy and this cycling of elements involve complicated sequences of events, and in these sequences each group of organisms has its own particular and highly specific place. As a consequence, it is impossible to change a single element in an ecosystem without the risk of destroying the carefully developed balance on which its stability depends.

I-6
The four seasons in a deciduous forest in Illinois. In such forests, characteristic of much of the North Temperate zone, the trees produce their leaves early in spring and begin to manufacture food; they lose them again in the autumn and enter an essentially dormant condition, thus passing the unfavorable growing conditions of winter. Food manufactured in the leaves is carried throughout the plant and deep into the earth to reach the farthest roots. At the same time, water and minerals are carried in a continuous stream up through the roots, stems, and leaves. Most of the water is lost from the leaf as water vapor through the same specialized pores in the leaves, the stomata, through which carbon dioxide enters. Many herbs grow under the trees, and a number of these flower very early in the spring, before the leaves of the trees have reached full size and shade the forest floor. Most of the trees shed their pollen in large quantities in spring, and it is carried by the wind, sometimes reaching the flowers of other trees of the same species.

PLANTS AND MAN

Man is a relative newcomer to the world of living things. If we were to measure the entire history of the earth on a 24-hour time scale, starting at midnight, cells would appear in the warm seas at about dawn and then we would pass through the day and the twilight. The first multicellular organisms would not be present until well after dark, and man's earliest appearance (about one million years ago) would be at less than ½ minute before the day's end. Yet man, more than any other animal—indeed almost as much as the plants that invaded the land—has changed the surface of the planet, shaping the biosphere according to his needs, his ambitions, or his follies.

At the close of the Pleistocene epoch, some 10,000 years ago, man was already the most widely distributed land mammal. Humans, who then numbered about 5 million, were hunter-gatherers, living in small nomadic bands. The stage was set for the first major advance that allowed rapid expansion of the human population: the development of agriculture.

Origins of Agriculture

The reasons for the change to the agricultural way of life are not clear. One factor seems to have been changes in the climate. The most recent of the glaciations began to retreat about 18,000 years ago, withdrawing slowly for about 6000 years. As the glaciers retreated, the plains of northern Europe and of North America, once cold grasslands or steppes, gave way to forest. The great herbivores that roamed these steppes retreated northward and eventually vanished; the woolly mammoth was last seen in Siberia about 12,000 years ago. While some animals became extinct, men adapted, as they had during the entire period of violent climatic changes that have marked their evolutionary history. With the migratory animals gone, man shifted his attention to smaller game, such as deer, which tended to be resident in the same area throughout the year. Fishing became an important part of the economy at this time also; ponds and lakes were filling, and streams were rushing with waters from the melting glaciers. Hunters made canoes and paddles and seines and other fishing equipment. The hunting and gathering of small animals—instead of large migratory herbivores—undoubtedly resulted in a less nomadic existence for the hunters and formed a prelude to the agricultural revolution. (However it should be noted that similar fluctuations had occurred previously in the history of man without leading to a cultural revolution.)

The Transition

The earliest traces of agriculture are found in an area in the Near East generally known as the Fertile Crescent. Here were the raw materials required by an agricultural economy: cereals, which are grasses with seeds capable

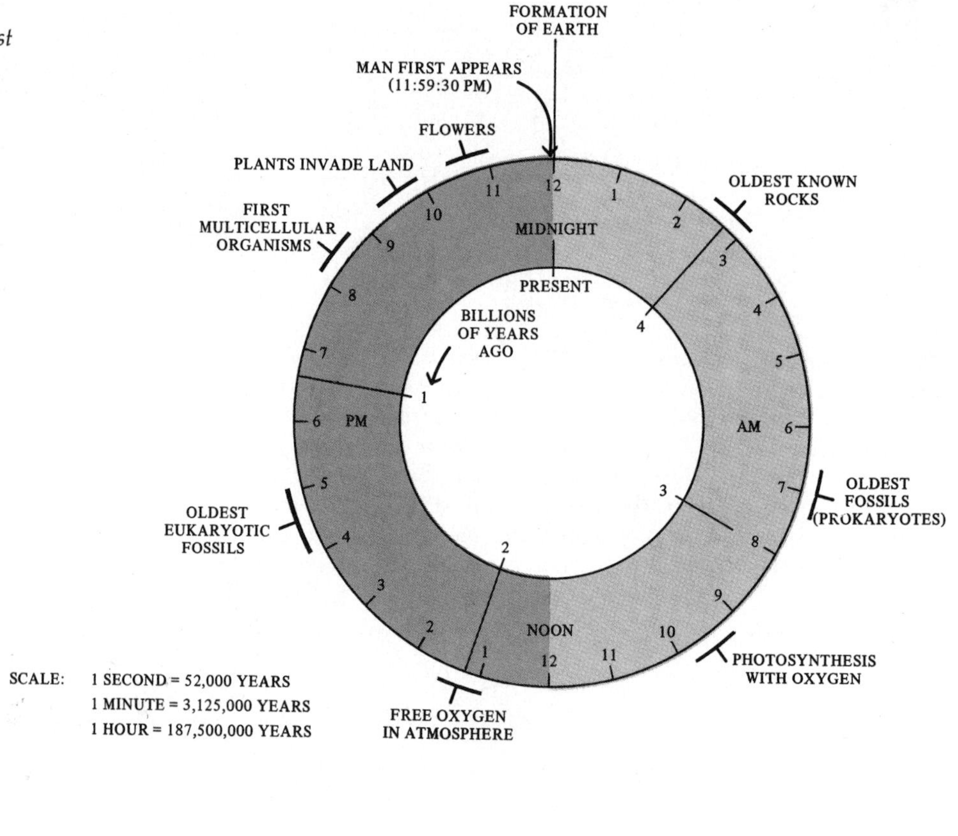

I-7
The clockface of biological time. Life first appears relatively early in the earth's history, before 7:00 A.M. on a 24-hour scale. The first multicellular organisms do not appear until the twilight of that 24-hour day, and man himself is a late arrival—at about 20 seconds to midnight.

of being stored for long periods without serious deterioration, and herbivorous herd animals, which can be readily domesticated. The grasses in the Fertile Crescent were wild wheats and barley; they still grow wild in these foothills. The animals were wild sheep and goats.

Although we do not know exactly when or where it first happened, the first deliberate planting of seeds can be seen as the logical end of a simple series of events. The wild cereals are weeds, ecologically speaking; that is, they grow readily on open or disturbed areas, patches of bare land where there are few other plants to compete with them. Also, early man, judging from his campsites, was not very tidy. It is easy to envision how seeds might be spilled or discarded with garbage on open land around man's habitations. And the presence of the archaeological remains of what are clearly permanent dwellings indicates that men were staying in one place long enough to recognize and reap their accidental harvest. From this it would have been an easy step to the deliberate saving of seeds and tending of crops.

Because men selected the seeds that they gathered and planted, they soon produced changes in the wild strains. In wild wheats, for instance, the stalk (rachis) on which the flower clusters grow and on which, eventually, the seeds develop, becomes brittle when the seeds mature and breaks off, scattering the seed. Among these wild plants, occasional mutants can be found in which the rachis is not brittle. It is these plants, at a disadvantage in the wild, that are more likely to be harvested by men and, as a consequence, more likely to be planted. Thus, just as men came to be dependent upon their crop plants, the plants they grew for food—such as cereals that could not seed themselves—came to be dependent upon man.

About 11,000 years ago, new cultures appeared around the Fertile Crescent. They were characterized by implements associated with the harvesting and processing of grains, such as flint sickle blades, grinding stones, or stone mortars and pestles.

By 8100 years ago, agricultural communities were established in eastern Europe; by 7000 years ago (about 5000 B.C.), agriculture had spread to the western Mediterranean and up the Danube into central Europe, and by 4000 B.C., to Britain. During this same period, agriculture originated separately in Central and South America, and perhaps slightly later in the Far East.

From the Old World sites, we have fossil imprints of cultivated wheat and barley, remains of domesticated goats, sheep, and cattle, and pottery vessels, stone bowls, and mortars. The farmers of the New World grew corn, pumpkins, squash, gourds, and cotton. The potato, the sweet potato, the peanut, and the tomato are also examples of New World crops. Today, of course, the vast majority of all the foods we eat is the result of deliberate cultivation.

I-8
Primitive methods of wheat production in Lebanon.

The Consequences of the Agricultural Revolution

The change to agriculture had profound consequences. Populations were no longer nomadic. Thus, they could store food not only in silos and granaries, but in the form of domesticated animals. In addition to food stores, other possessions could be accumulated to an extent far beyond that previously possible. Even land could be owned and accumulated and passed on by inheritance. Thus, the world became divided into semipermanent groups of haves and have-nots, as it is today.

Because the efforts of a few could produce enough food for everyone, the communities became diversified. People became tradesmen, artisans, bankers, scholars, poets, all the rich mixture of which a modern community is composed. And these people could live much more densely than ever before. For hunting and food-gathering economies, 5 square kilometers, on the average, are required to provide enough for one family to eat.

One immediate and direct consequence of the agricultural revolution was an increase in populations. A striking characteristic of hunting groups is that they vigorously limit their numbers. A woman on the move cannot carry more than one infant along with her household baggage, minimal though that may be. When simple means of birth control—often just abstention—are not effective, she resorts to abortion or, more probably, infanticide. In addition, there is a high natural mortality, particularly among the very young, the very old, the ill, the disabled, and women at childbirth. As a result, populations dependent upon hunting tend to remain small.

Once families became sedentary, there was no longer the same urgent need to limit the number of births, and probably there was also a decrease in the mortality rates.

The Population Explosion

About 25,000 years ago there were perhaps 3 million people. By the close of the Pleistocene epoch, some 10,000 years ago, the human population probably numbered a little more than 5 million, spread over the entire world. By 4000 B.C., about 6000 years ago, the population had increased enormously, to more than 86 million, and by the time of Christ, it is estimated, there were 133 million people. In other words, the population increased more than 25 times between 10,000 and 2000 years ago.

By 1650, the world population had reached 500 million, many people living in urban centers, and the development of science, technology, and industrialization had begun, bringing about further profound changes in the life of man and his relationship to nature.

By 1976, there were more than 4 billion people on our planet (Table I-1). This is an almost incomprehensibly large figure; moreover, the rate of increase of this enormous population is also unprecedented. The birth rate in the United States has decreased drastically since 1972, and the population could stabilize during the next century. For the world as a whole, however, the population is growing at about 2.2 percent per year. This means that about 175 people are added to the world population every minute, about 250,000 each day, and 90 million every year. If this rate of increase is sustained, in place of the 4 billion people living in the year 1975, there will be 7 billion people on earth by the year 2000. Over a century, a 2.2 percent growth rate leads to an eightfold increase in population.

At the time of the World Food Conference held in Rome in November 1974, it was estimated that at least 460 million people were suffering from hunger and severe malnutrition, whereas another 1.5 billion were con-

Table I-1 *A demographic summary of peoples of the world. (Taken from Environmental Fund 1975)*

	1975					1958–1963	1965
AREA	POPULATION ESTIMATES (MILLIONS)	GROWTH RATE (%)	BIRTH RATE (PER 1000)	DEATH RATE (PER 1000)	POPULATION UNDER 15 (%)	GROWTH RATE (%)	BIRTH RATE (PER 1000)
Africa	420.1	2.8	47.0	21.0	44	2.3	46
Asia	2407.4	2.5	39.0	14.0	>40	1.8	38
Europe	474.2	0.8	15.4	10.4	26	0.9	19
Latin America	327.6	2.9	38.0	11.0	43	-	-
North America*	242.4	1.0	15.0	9.1	27	1.6	22
Oceania	20.9	2.1	23.0	9.7	33	2.1	27
World	4146.9	2.2	35.0	13.0	37	2.0	36

* Canada and the United States. At least half the growth shown derives from immigration.

sidered undernourished. The less developed countries of the world, with populations growing much more rapidly than the average, had an immediate need at that time for an additional 8 to 10 million metric tons of grain per year. Unless production is increased drastically, the United Nations has estimated that the annual deficit could reach 85 to 100 million metric tons of grain by 1985. The high cost of producing fertilizer due to the increasing world energy shortage is contributing severely to the problem. Consider the fact that the more than 5 million tractors in the United States alone require 30 billion liters (8 billion gallons) of fuel, the equivalent of the energy content in the food produced. As fossil fuels become more scarce and more expensive, the costs of food will continue to increase. For Americans, who spend on the average less than 20 percent of their personal income on food, this already occasions serious concern. For those in developing nations who may spend 80 to 90 percent of their income on food, it can be a death sentence.

Although there are gains to be made in bringing additional land under cultivation, the most promising approach seems to lie in the development of existing crops grown on presently cultivated land, in terms not only of their yield, but also of their protein content. Agricultural technology, including methods of irrigation, can also be improved considerably. These are among the most vital areas of concern to the plant scientist. The effort to stimulate agriculture by the development of new crop plants—especially grains—has been called the Green Revolution.

Enormous progress is being made. The production of wheat in Mexico has quadrupled since 1950. Between 1968 and 1972, India and Pakistan doubled their wheat production. China, the most populous nation in the world, has become agriculturally self-sustaining, largely as the result of adopting these new strains. Techniques of breeding, fertilizing, and irrigation are being applied to rice and other crops in developing countries throughout the world. A new manmade hybrid, *Triticale,* one of the most promising products of this program, is described on page 162. Among the important areas of current research are the improvement of photosynthetic efficiency and the fixation of atmospheric nitrogen; both will be discussed in later chapters.

Despite its acknowledged success, this massive effort has come under criticism in recent years. One reason is the increasing cost of fertilizer; these new grains require intensive cultivation. Because the large landowners are able to afford the investment in fertilizer and farming equipment that the small-scale farmers cannot, these new agricultural developments are seen as accelerating the consolidation of farm lands into a few large holdings by the very wealthy. Bad weather—both droughts and floods—diminished yields in 1972, 1973, and 1974. Most serious of all, although food production is still outstrip-

(a)

(b)

I-9
(a) *Norman Borlaug, who was awarded the Nobel Peace Prize in 1970. Borlaug is the leader of a research project sponsored by The Rockefeller Foundation under which new strains of wheat have been developed in Mexico. Widely planted, these new strains have changed the status of Mexico from that of a wheat importer when the program began in 1944 to that of an exporter by 1964.* (b) *Field workers weeding a rice plot at the International Rice Research Institute in the Philippines. The most recent strains are highly disease-resistant; however, they will grow only on irrigated land, which forms only about 30 percent of the cropland in Asia.*

I-10
Modern wheat production.

ping population growth, the Green Revolution, at its most productive, will not be able to keep pace for long with the rapid growth of the world's population.

Finally, there is a more fundamental though more elusive reason for the dissatisfaction with the Green Revolution. When it was first introduced, it appeared to many to be an almost magical solution to problems so enormous and distressing that they had seemed insoluble. It is now clear, however, that poverty and famine and the unrest and violence they may bring will not be solved by a "technological fix." The Green Revolution must of course go forward. At the same time, we must recognize that the broader solutions are social, political, and ethical, involving not only the growth of food but its distribution, not only the limiting of populations but the raising of living standards of these populations to tolerable levels.

This introduction has ranged from the beginnings of life on this planet to the evolution of land plants and of plant communities to the development of agriculture and of modern society, with its most pressing current problem, the unprecedented growth of the human population. These broad topics are of interest to many people other than botanists. As we turn to Chapter 1, in which our attention narrows to a cell so small it cannot be seen by the unaided eye, it is well to keep in the back of our minds these broader concerns. A basic knowledge of plant biology is useful in its own right and essential in many fields of endeavor, but it is also increasingly relevant to some of society's most crucial problems and to the difficult decisions that will face us in choosing among the proposals for diminishing them. Thus this book is dedicated not only to the botanists of the future, whether teachers or researchers, but also to the informed citizens, scientists and laymen alike, in whose hands such decisions lie.

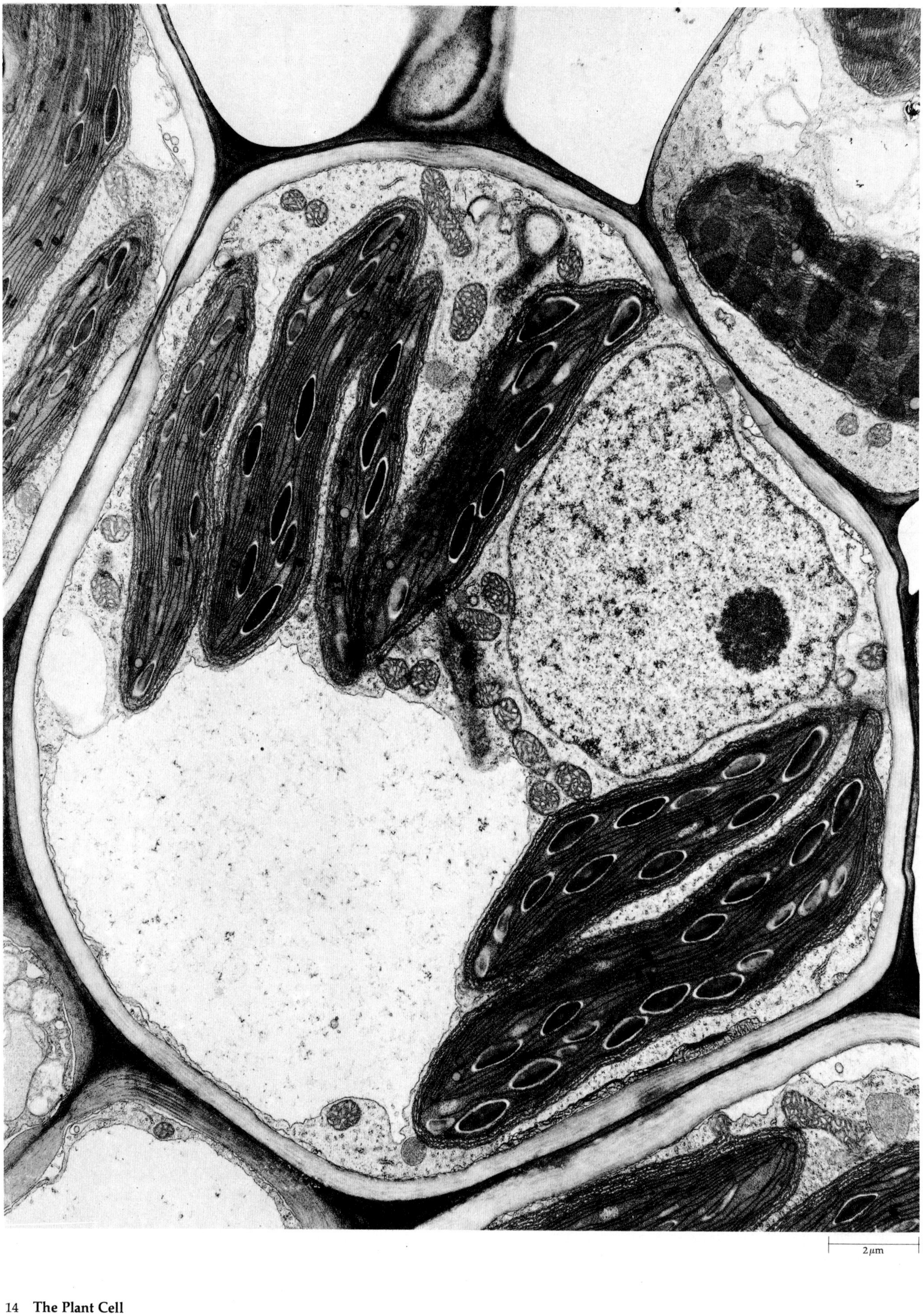
2μm

CHAPTER 1

Introduction to the Cell

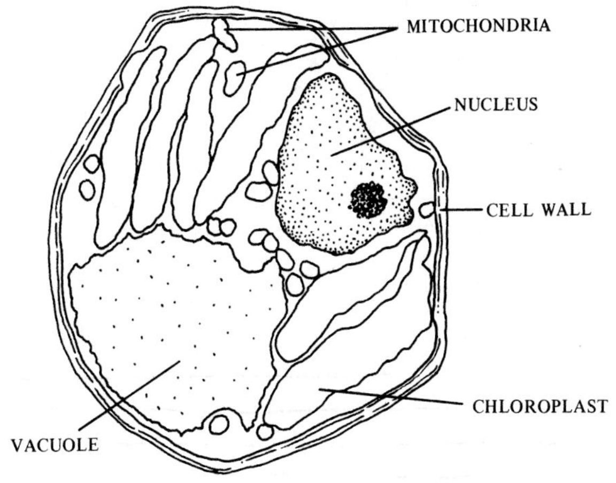

1–1
At left is a cross section of a plant cell enlarged about 9000 times its actual size by an electron microscope. The labeled drawing above identifies some of the cell's structures and organelles, those specialized cell parts that resemble and function as organs. Throughout this chapter, we will name and point out specific cell components in micrograph captions, even though it may be a few pages before the text describes their functions.

This particular cell is a bundle sheath cell from the leaf of a corn (Zea mays) *plant. Cells like this one are part of the veins of the leaf, and completely surround the water- and food-conducting tissues (xylem and phloem) of the veins. The nucleus—the site of the cell's genetic information—can be seen at the upper right. Several chloroplasts—organelles concerned with photosynthesis—and numerous mitochondria—organelles involved in the conversion of food to usable energy—can also be seen in this bundle sheath cell. The large, relatively clear area at the lower left is a vacuole, the principal ingredient of which is water.*

Cells are the structural and functional units of life. The smallest living organisms are composed of single cells. The largest are made up of billions of cells, each of which still lives a partly independent existence. The realization that all organisms are composed of cells was one of the most important conceptual advances in the history of biology because it provided a unifying theme for the study of all living things. When studied at the cellular level, even the most diverse organisms are remarkably similar to one another, both in their physical organization and in their biochemical properties. The *cell theory* (which is, of course, not a theory at all but the observation of a fact) was formulated early in the nineteenth century, well before the presentation of Darwin's theory of evolution, but these two great unifying concepts are, in fact, closely related. In the similarities among cells, we catch a glimpse of a long evolutionary history that links modern organisms, including plants and ourselves, with the first cellular units that took shape billions of years ago.

There are many, many different kinds of cells. Within our own bodies are more than 100 different and distinct cell types. In a teaspoon of pond water, you are likely to find several different kinds of one-celled organisms, and in a whole pond, there are probably several hundred clearly different kinds. Plants are composed of cells superficially quite different from those of our own body, and insects have many cells of kinds not found either in plants or in vertebrates. Thus, the first remarkable fact about cells is their diversity.

The second, even more remarkable fact is their similarity. Every cell is a self-contained and at least partially self-sufficient unit, bounded by an outer membrane, the plasma membrane or plasmalemma (often called simply the cell membrane), that controls the passage of materials in and out of the cell and so makes it possible for the cell to differ biochemically and structurally from its surroundings. As we shall see, this membrane has the same basic structure throughout the living world. Within this membrane is the cytoplasm which, in most cells, includes a variety of formed bodies as well as various dissolved and suspended molecules. Every cell also has at least one chromosome. In all types of cells, it is the DNA of the chromosome that encodes the genetic characteristics (see Chapter 3), and this *genetic code* is very much the same from bacterium to oak tree to man. (In evolutionary terms, this means that the genetic code was probably in existence in its present form more than 3 billion years ago.)

Considered functionally, the similarities among cells are even more striking. Let us look at the four most important functions of cells.

1. Cells have the ability to take in energy and to convert it from one form to another. The original source of energy for living systems on this planet is the sun. The chlorophyll-containing cells of green plants and of algae transform this radiant energy to chemical energy by the process of photosynthesis. The nonphotosynthetic cells of plants, animals, and other organisms derive their energy by the oxidation of the products of photosynthesis. All life in this biosphere depends on the carefully controlled flow of energy from the sun through living cells back into the environment.
2. Cells use this energy to build and maintain their structure. Under the influence of the environment, an inanimate object—such as a mountain—will gradually fall apart, erode, and finally vanish. Living things, however, in the course of their interactions with the environment, not only are able to maintain themselves but also to grow and to evolve. (This relationship between energy and order is one of the most interesting principles of modern biology.)
3. In order to maintain and build more structure, the cell must take in substances ("building materials") from the outside. About 90 elements are found naturally on the earth's surface and in its atmosphere. Of these, living organisms utilize only 24, and 99 percent of the bulk of living tissue is made up of 6 of these. Also, the elements that are used are often found within the cell in proportions very different from those in which they exist outside. In short, although cells constantly exchange materials with the environment, they maintain a relatively stable internal environment that is quite different from that of their surroundings.
4. Cells reproduce themselves, passing on to their daughter cells all the capacities that they themselves possess.

ORIGIN OF THE CELL THEORY

In the seventeenth century, Robert Hooke, using a microscope of his own construction, noticed that cork and other plant tissues are made up of small cavities separated by walls. He called these cavities "cells," meaning "little rooms." The word did not take on its present meaning, however, for more than 150 years.

In 1838, Matthias Schleiden, a German botanist, came to the conclusion that all plant tissues are organized in the form of cells. In the following year, zoologist Theodor Schwann extended Schleiden's observation to animal tissues and proposed a cellular basis for all life. The cell theory is of tremendous and central importance to biology because it emphasizes the basic sameness of all living systems and so brings an underlying unity to widely varied studies involving many different kinds of organisms.

In 1858, the cell theory took on an even broader significance when the great pathologist Rudolf Virchow generalized that cells can arise only from preexisting cells: "Where a cell exists, there must have been a preexisting cell, just as the animal arises only from an animal and the plant only from a plant. . . . Throughout the whole series of living forms, whether entire animal or plant organisms or their component parts, there rules an eternal law of continuous development."

In the broad perspective of evolution, Virchow's concept takes on an even larger significance. There is an unbroken continuity between modern cells—and the organisms that they compose—and the primitive cells that first appeared on Earth more than 3 billion years ago.

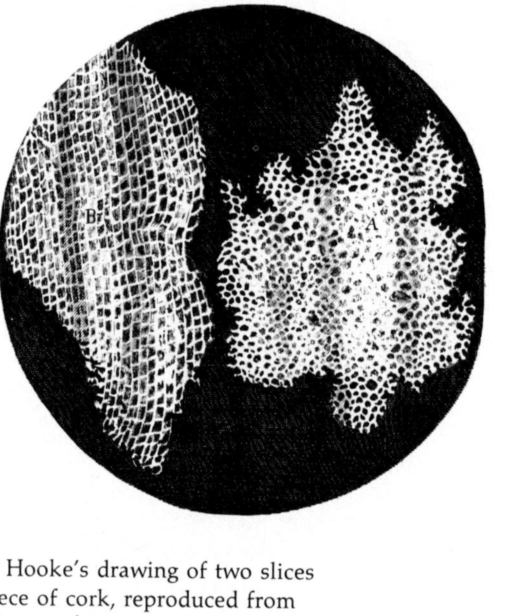

Robert Hooke's drawing of two slices of a piece of cork, reproduced from his *Micrographia* (1665). Hooke was the first to use the word "cells" to describe these tiny compartments into which living organisms are organized.

VIEWING THE MICROSCOPIC WORLD

Most cells can be seen only with the aid of a microscope. The units of measurement generally used for describing cells are micrometers and nanometers. (See Table 1–1 below.) Unaided, the human eye has a resolving power of about 1/10 millimeter, or 100 micrometers. This means that if you look at two lines that are less than 100 micrometers apart, they merge into a single line. Similarly, two dots less than 100 micrometers apart look like a single blurry dot. To separate structures closer than this, optical instruments such as microscopes are used. The best light microscope has a resolving power of 0.2 micrometer or about 200 nanometers and so improves on the naked eye about 500 times. It is theoretically impossible to build a light microscope that will do better than this.

Notice that resolving power and magnification are two different things; if you take a picture through the best light microscope of two lines that are less than 0.2 micrometer, or 200 nanometers, apart, you can enlarge that photograph indefinitely, but the two lines will still blur together. By using more powerful lenses, you can increase magnification, but this will not improve resolution.

THE ELECTRON MICROSCOPE

With the electron microscope, resolving power has been increased almost 400 times over that provided by the light microscope. This is achieved by using "illumination" consisting of electron beams instead of light rays. Under the very best conditions, electron microscopy currently affords a resolving power of about 0.5 nanometers, roughly 200,000 times greater than that of the human eye. (A hydrogen atom is about 0.1 nanometer in diameter.)

In a conventional electron microscope, a beam of electrons transmitted through the specimen leaves its imprint on a screen below. Areas in the specimen that permit the transmission of more electrons—"electron-transparent regions"—show up light in color, and "electron-dense areas" are dark.

The conventional electron microscope has one great disadvantage. Electrons have a very small mass and must travel in a vacuum; electron beams can only pass through specimens that are exceedingly thin. To prepare them for the electron microscope, specimens must therefore be killed and embedded in hard materials so that they can be sliced by special cutting instruments. This means, of course, that the high resolving powers of the electron microscope can be applied only to tissues that are no longer alive. Also, it is sometimes difficult to determine what changes may be produced in the material in the course of its preparation.

A few kinds of cells, almost all viruses, and many of the structures within cells can be seen only with the electron microscope.

THE SCANNING ELECTRON MICROSCOPE

In a scanning electron microscope, the electrons whose imprints are recorded come from the surface of the specimen. The electron beam is focused into a fine probe which is used to scan the specimen; complete scanning takes a few seconds to a few minutes. As a result of the electron bombardment from the probe, the specimen emits low-energy secondary electrons. Variations in the surface of the specimen alter the number of secondary electrons emitted. Holes and fissures appear dark, whereas knobs and ridges are light. Electrons scattered from the surface plus secondary electrons are collected, amplified, and transmitted to a screen which is scanned in synchrony with the electron probe. Figure 2–12 on page 54 was prepared with a scanning electron microscope, as were several others in this text.

Table 1–1 *Measurements Used in Microscopy*

1 centimeter (cm)	=	1/100 meter = 0.4 inch
1 millimeter (mm)	=	1/1,000 meter = 1/10 cm
1 micrometer (μm)*	=	1/1,000,000 meter = 1/10,000 cm
1 nanometer (nm)	=	1/1,000,000,000 meter = 1/10,000,000 cm
1 angstrom (Å)	=	1/10,000,000,000 meter = 1/100,000,000 cm
		or
1 meter	=	10^2 cm = 10^3 mm = 10^6 μm = 10^9 nm = 10^{10} Å

* Micrometers were formerly known as microns, indicated by the Greek letter μ, which corresponds to our m and is pronounced "mew."

PROKARYOTES VS. EUKARYOTES

As we noted earlier, there are many different types of cells with striking diversities and similarities. Within these patterns of diversity and similarity, it is possible to identify two fundamentally distinct types of cellular organization: prokaryotic and eukaryotic. We shall examine prokaryotic cells first, because they are simpler and, in the evolutionary sense, more ancient. However, eukaryotic cells, a category that includes all plant cells, will be the chief focus of our attention in this and succeeding chapters.

The terms *prokaryote* and *eukaryote* are derived from the Greek word *karyon*, meaning "nut" or "kernel (nucleus)." Prokaryotes are "before the nucleus," and eukaryotes are "with a true nucleus." Modern prokaryotes include the bacteria and the blue-green algae. (A fuller description of this group can be found in Chapter 10.)

PROKARYOTIC CELLS

The bacterium *Escherichia coli* is a good example of a modern prokaryote. These bacteria, which are very common, are inhabitants, usually harmless, of the human digestive tract. Because they are frequently used in modern studies of biochemistry and genetics, *E. coli* may be the most thoroughly understood of all living systems.

As can be seen in Figure 1–2, *E. coli* is quite simple in its general structure. The organisms are rod-shaped and are about 2 micrometers (2000 nanometers) long and 1 micrometer in diameter. Each consists of a cell wall, a plasma membrane, and the cytoplasm. The DNA, which is a single long molecule within the cytoplasm, contains the genetic information of the cell. (The way in which this information is coded into the molecule will be described in Chapter 3.) In the bacterial cell, the DNA is in an extremely condensed form; if the molecule were fully extended, it would be about 700 times longer than the entire cell. It is also in the form of a circle—that is, there are no ends to the molecule.

Often *E. coli* cells contain two molecules of DNA, and sometimes even four. This is because the DNA replicates before the cell divides and, in a fast-growing culture, DNA replication may get well ahead of cell division.

The rest of the cytoplasm, as seen in the electron microscope, usually appears quite dense, owing largely to the presence of numerous granules of stored food. The only structures with definite size and shape in the cytoplasm are the *ribosomes*. Ribosomes are made up of approximately 65 percent RNA and 35 percent protein. They are small bodies, only about 20 nanometers in diameter. Each one is composed of two almost spherical subunits, one larger than the other. Ribosomes are the sites at which amino acids are linked together to form proteins. (These materials will be fully described in Chapter 2.)

The plasma membrane of *E. coli* (and of every other living cell) is only 7.5 to 9.0 nanometers thick. Because the membrane is so thin, obviously it can consist of only a few layers of molecules. Chemical analysis shows that it is composed of approximately 60 percent protein and 40 percent lipid. Under the electron microscope, this membrane appears to have a three-ply structure, with two electron-dense layers enclosing a less dense one.

The plasma membrane of *E. coli* performs two complex and critical functions: the determination of what passes into and out of the cell and the provision of a work surface upon which most of the biochemical reactions that take place in the cell occur. In *E. coli* and other prokaryotes, the plasma membrane often contains many folds and convolutions (mesosomes) that extend into the interior of the cell and greatly increase the area of working surface within the cell. In addition, certain prokaryotes, particularly the blue-green algae, have complex systems of internal membranes.

Outside the plasma membrane is the *cell wall*, which in *E. coli* is about 10 nanometers thick. The cell wall is a complex structure made up of lipopolysaccharides (complex molecules in which lipids are combined with polysaccharides and with proteins). The wall is rigid and gives the cell its shape, as well as protecting it from the environment. Outside the cell wall of *E. coli* there is a thin sheath of slime; extending beyond that are short hairlike structures (7.5 to 10 nanometers in diameter) called *pili*.

Many bacteria (although not *E. coli*) have *flagella* (Figure 1–3). Bacterial flagella are long, slender appendages that terminate in the cytoplasm. They move bacteria

1–2
Escherichia coli, *example of a modern prokaryote. The DNA occurs in the less dense material in the center of the cell. The most dense small bodies in the cytoplasm are ribosomes. The scale marker at the bottom right corner of the micrograph provides a reference for size. These bacteria have been magnified 24,500 times actual size.* [$(24{,}500)\ (0.5\mu m) \cong 1.2cm \cong \frac{1}{2}$ *inch, the length of the scale marker.*]

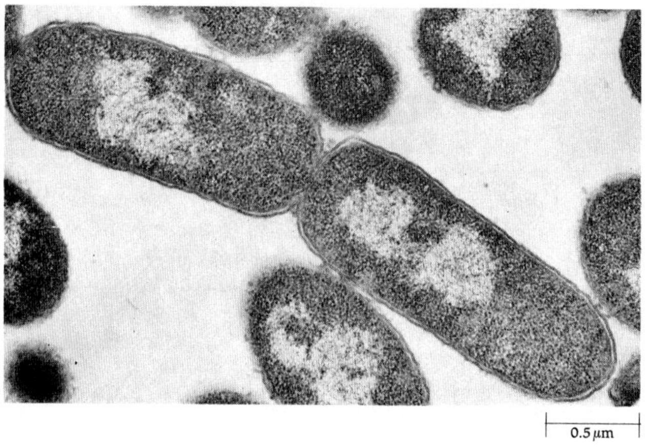

1–3
A bacterial cell (Proteus mirabilis) *with numerous pili and six long flagella.*

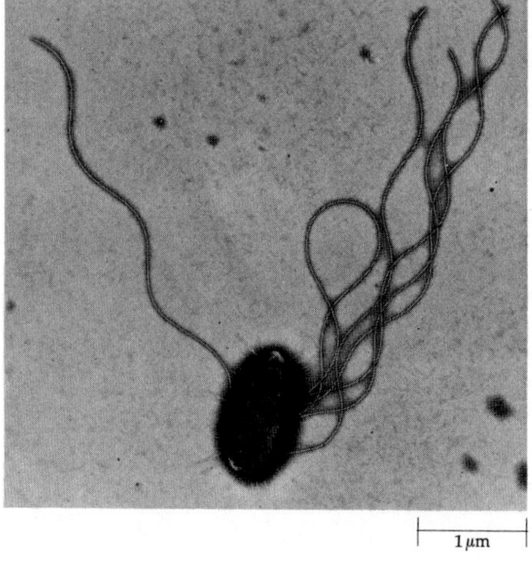

through the liquid medium in which the cells live. Bacterial cells such as *E. coli* that do not have flagella are not motile.

EUKARYOTIC CELLS

All cells of plants and animals, of algae (except the blue-greens), and of fungi are eukaryotic. As you can see in Figure 1–4, the eukaryotic cell is structurally much more complex than the prokaryotic cell. It is also much larger. Individuals of *E. coli* are only about 2 micrometers long and 1 micrometer in diameter. The unicellular green alga shown in Figure 1–4 is about 30 micrometers in diameter.

Another example of a eukaryotic cell is shown in Figure 1–5. This is a parenchyma cell, which is a relatively unspecialized plant cell and, unlike some other cells of the mature plant body, is alive at maturity.

Like the prokaryotic cell we examined previously, the parenchyma cell, like all eukaryotic cells, is bounded by a plasma membrane. Like all plant cells (but unlike animal cells), it is enclosed in a semirigid cell wall, which is

1–4
Chlamydomonas, *example of a modern eukaryote. The cell contains a membrane-bounded nucleus and several types of complex organelles. Primitive eukaryotes are believed to have resembled this one-celled alga.*

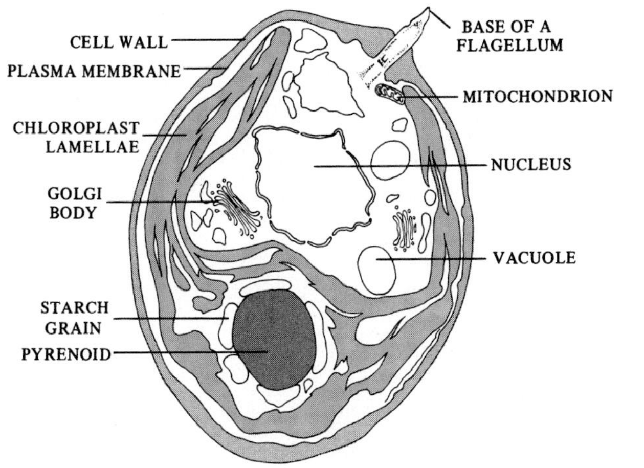

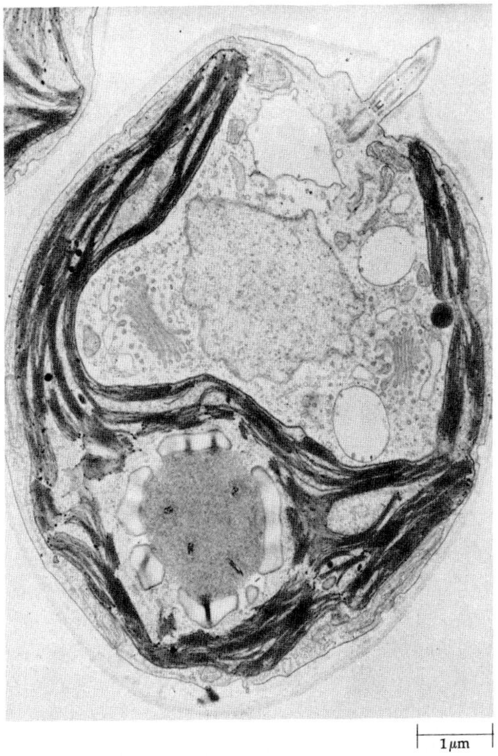

1–5
Transverse section of a parenchyma cell from a vein of a corn (Zea mays) *leaf, showing many of the cellular components characteristic of plant cells.*

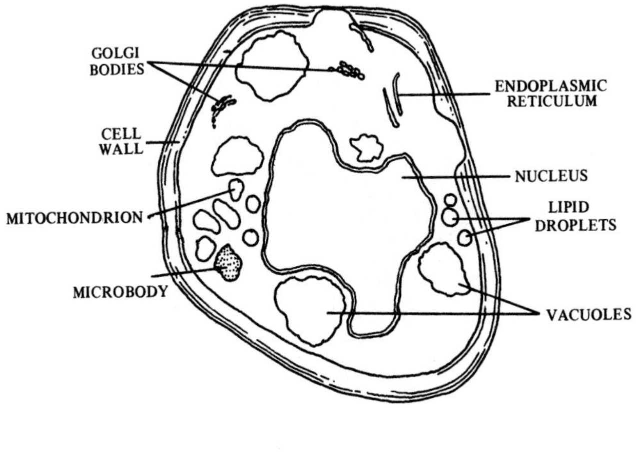

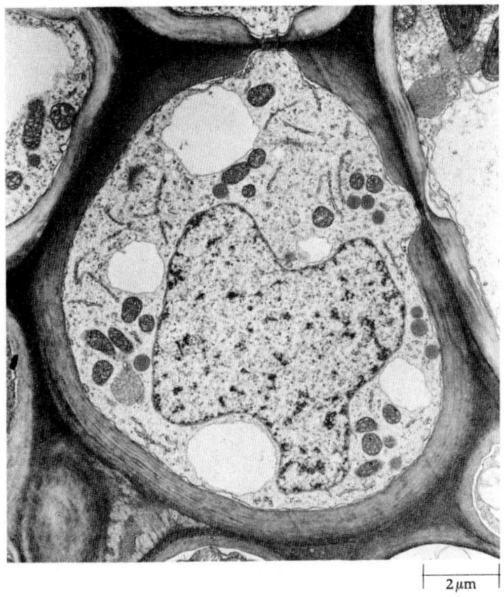

outside the plasma membrane; the living unit inside the wall is known as the *protoplast*, and the contents of the cell is sometimes known as the *protoplasm*. Protoplasm consists of *cytoplasm* and *nucleus*. Like the cytoplasm of prokaryotes, the cytoplasm of the eukaryotic cell includes ribosomes (although the ribosomes of the eukaryotic cell are larger). Unlike the prokaryotic cell, however, the eukaryotic cell is characterized by numerous systems of membranes and organelles, which are remarkably similar from cell to cell. Some organelles, such as mitochondria, are found in virtually all cells; others are found only in particular types of cells. Chloroplasts, for example, are found only in photosynthetic cells, but they are present in cells as diverse as those of a leaf or a single-celled alga.

The portion of the cytoplasm in which the nucleus, membrane systems, and various organelles are suspended is known as the cytoplasmic ground substance. In a living plant cell, such as that of the pond weed *Elodea* (see Figure 1–9), the ground substance is constantly in motion and you can observe the organelles as well as various substances suspended in the ground substance being swept along passively in orderly fashion in the moving currents. This movement is known as *cytoplasmic streaming*, or *cyclosis*, and it continues so long as the cell is alive. How cyclosis takes place remains a mystery, although many experiments have been conducted on it. It undoubtedly facilitates the exchange of materials within the cell and between the cell and the environment, although it is not known whether or not this is its primary function. In organisms such as the slime molds (Chapter 12), cytoplasmic streaming apparently plays a role in cell movement, although here again the mechanism is not understood.

All eukaryotic cells, at some stage of their development, contain a nucleus. The nucleus is bounded by a double membrane called the nuclear envelope. It contains chromatin, which in nondividing nuclei is often difficult to distinguish from the *nucleoplasm*, or the ground substance, of the nucleus. The structure of the nucleus and the process by which it divides (mitosis) will be considered in detail in the final pages of this chapter.

JOURNEY INTO A LEAF

GRASS PLANT, life size.
This is a common grass (Hordeum glaucum), *the sort that grows in fields and abandoned lots everywhere. This particular plant is broken off at the roots. Magnified 10 times, the circled portion of the stem is shown in the drawing below.*

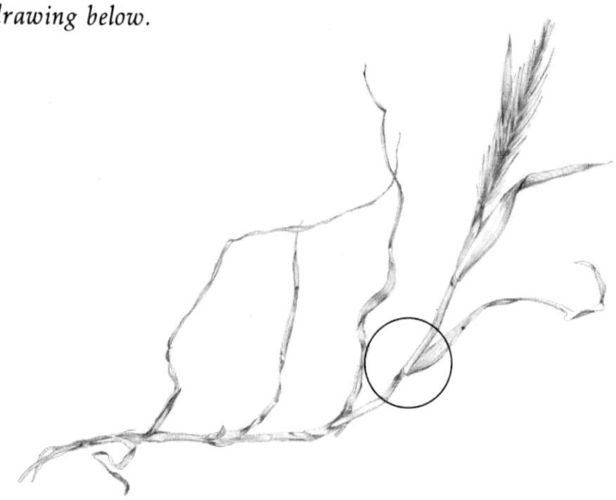

STEM, magnified 10 times.
The leaves form a sheath around the stem, and their veins run parallel to one another.

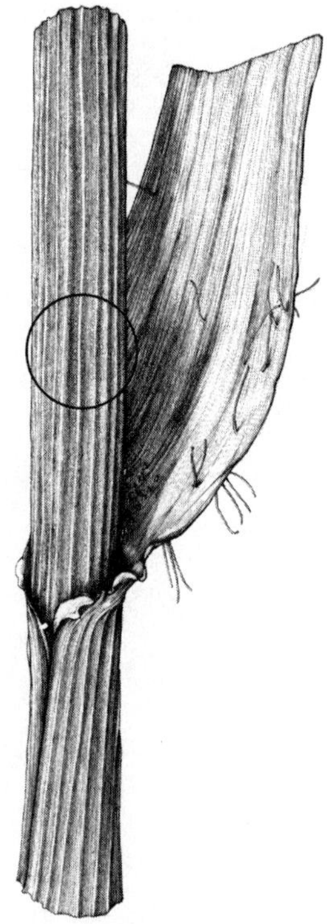

LEAF SURFACE, magnified 100 times.
Now the individual cells of the leaf become visible. Near the bottom, a few surface epidermal cells have been removed to reveal the internal cells, which are shown at the right, now magnified 1000 times.

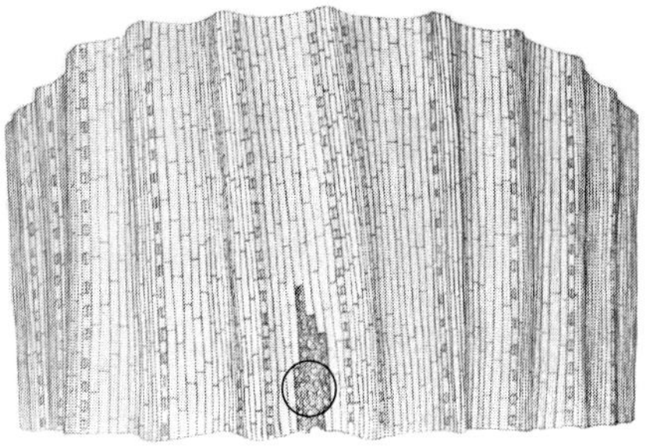

LEAF INTERIOR, as seen with the light microscope, magnified 1000 times.
These are the cells of the mesophyll that manufacture high-energy compounds photosynthetically. In addition to the numerous functional components within the cell, there are extensive water-filled vacuoles. The cells are surrounded by air spaces.

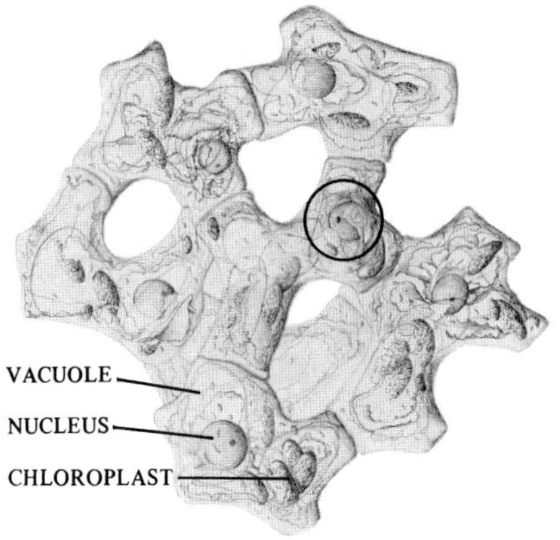

CELL INTERIOR, as seen with the electron microscope, magnified 10,000 times.
This portion of one cell contains a representative sampling of the components in a typical plant cell. The functions of each labelled structure will be described in the following pages.

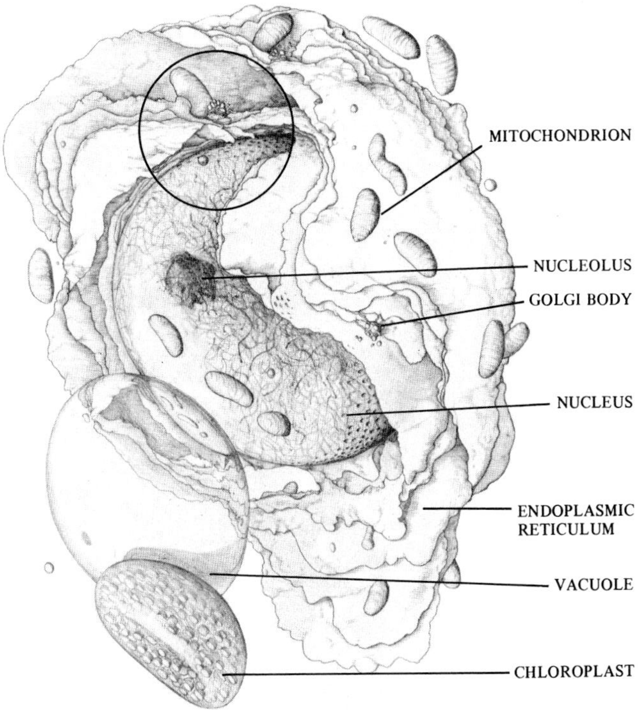

CELL INTERIOR, as seen with the electron microscope, magnified 50,000 times.
The chromatin material within the nucleus contains strands of DNA in which the cell's information is stored; these chromatin strands seem to attach to the nuclear membrane at the periphery of the pores. The Golgi body is packaging and transporting newly manufactured molecules.

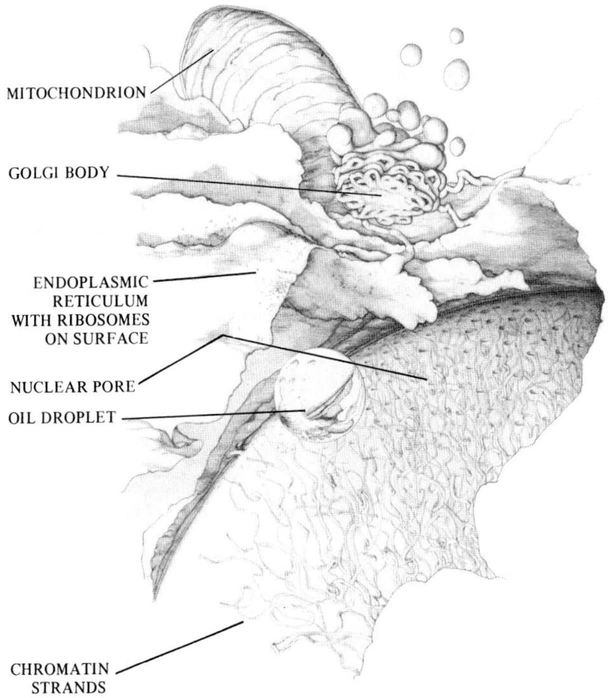

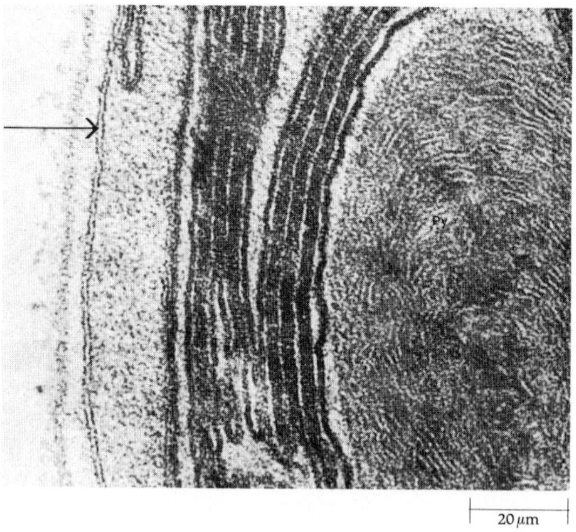

1–6
Cellular membranes under high magnification appear to have a three-ply, sandwichlike structure, as seen in this plasma membrane (pointed out by the arrow) of the unicellular green alga Chlorella pyrenoidosa.

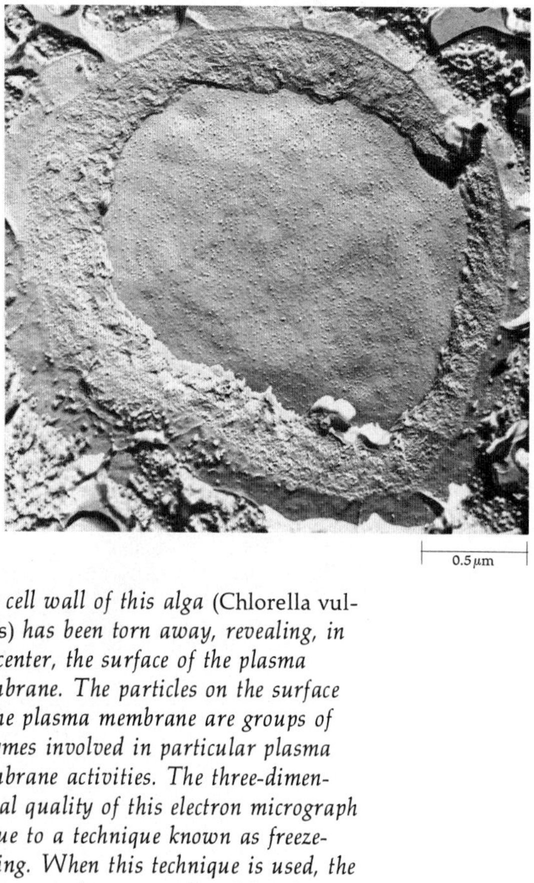

1–7
The cell wall of this alga (Chlorella vulgaris) *has been torn away, revealing, in the center, the surface of the plasma membrane. The particles on the surface of the plasma membrane are groups of enzymes involved in particular plasma membrane activities. The three-dimensional quality of this electron micrograph is due to a technique known as freeze-etching. When this technique is used, the specimen is frozen rapidly. After the water is removed from the surface, a replica is made of the specimen. It is the replica that is magnified and photographed.*

Membranes and Membrane Systems

As we noted at the beginning of the chapter, the protoplasts of both prokaryotic and eukaryotic cells are delimited by a plasma membrane. In addition to this outer membrane, eukaryotic cells characteristically contain a number of membrane systems and membrane-bounded vacuoles and organelles. Wherever membranes are found within a living cell (eukaryotic or prokaryotic), they have a characteristic three-ply structure (Figure 1–6) The interpretation of this structure in molecular terms will be discussed in the next chapter.

The plasma membrane may be smooth or irregular in outline. Sometimes deep invaginations of the plasma membrane penetrate the cytoplasm. Increasing evidence indicates that portions of these invaginations are pinched off in the cytoplasm. Such membrane-bounded vesicles then disappear in the cytoplasm, and their contents apparently are incorporated into the cell. This process, which has long been recognized in animal cells, is known as *pinocytosis*. A reverse pinocytosis also occurs in plant cells. In this process, membrane-bounded vesicles, which are formed within the cell, migrate to the plasma membrane, fuse with it, and then discharge their contents into the region of the wall.

The Tonoplast and Vacuoles

Vacuoles are membrane-bounded regions within the cell that are filled with liquid, called *cell sap*, rather than with protoplasm. They are surrounded by the *vacuolar membrane* or *tonoplast*. The cell sap contains ions and small organic molecules, often in far larger quantities than are found in the protoplasm, as well as some enzymes and other large molecules.

The immature plant cell typically contains large numbers of small vacuoles, which increase in size and coalesce as the cell enlarges (Figure 1–8). In the mature

1–8
In higher plants, as the cells mature, the vacuoles scattered through the cytoplasm enlarge by taking up water and coalesce. Most of the growth of the cell results from enlargement of the vacuole, and in a mature cell, the vacuole often occupies a major part of the cell volume.

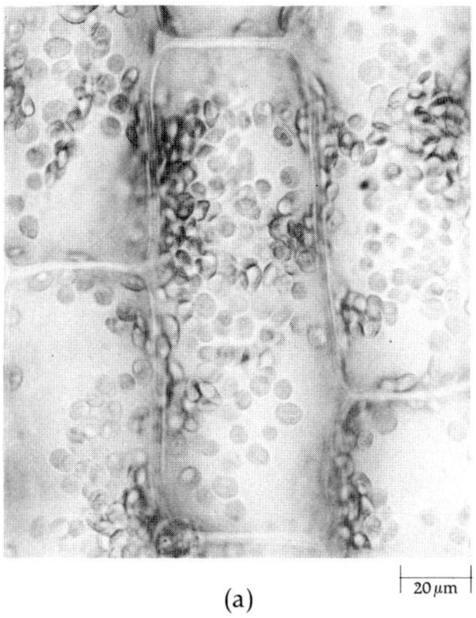

(a)

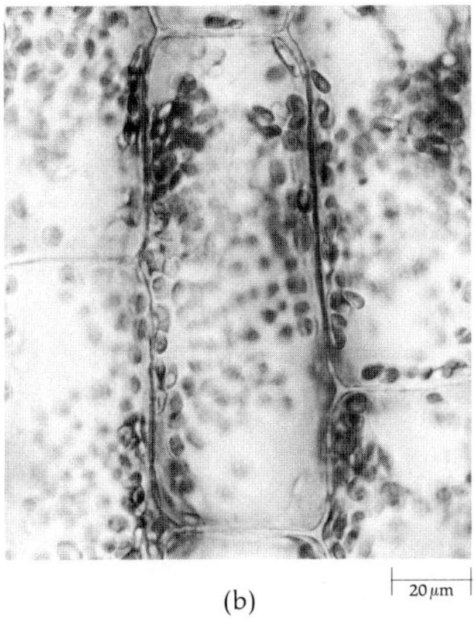

(b)

1–9
Elodea cell. (a) *Upper surface of cell.* (b) *Middle of cell. The numerous disk-shaped structures are chloroplasts located in the cytoplasm along the wall. In surface view* (a), *the disk-shaped chloroplasts appear circular in outline. In* (b) *the chloroplasts, which have their broad surfaces facing the surface of the wall, appear elongate. Notice the absence of chloroplasts in the center of the cell—that is, within the vacuole—in* (b).

cell, as much as 90 percent of the volume may be taken up by the vacuole, with the protoplasm consisting of a thin peripheral layer closely pressed against the cell wall (Figure 1–9). (Not all young plant cells, however, contain only small vacuoles; the cells of the vascular cambium, for example, may contain very large ones.)

The principal component of the cell sap is water, and other components vary with the type of plant and its physiological state. Vacuoles typically contain mineral salts and sugars, and some contain dissolved proteins. The tonoplast plays an important role in the active transport of certain ions into the vacuole and the retention of them there. Thus ions may accumulate in the cell sap in concentrations far in excess of those in the surrounding cytoplasm. Sometimes the concentration of a particular material in the vacuole is sufficiently great for it to form crystals (Figures 1–10 and 1–11). Vacuoles are usually slightly acidic. Some of them, like the vacuoles in citrus fruits, are very acidic—hence the tart, sour taste of the fruit. If such vacuoles are broken, the escaping fluid damages the surrounding protoplasm.

The vacuole is also a site of pigment deposition. The blues, violets, purples, dark reds, and scarlets of plant cells are usually caused by a group of pigments known as the anthocyanins (see page 380). Unlike most other plant pigments, the anthocyanins are readily soluble in water, and they are dissolved within the cell sap. They are responsible for the red and blue colors of many vegetables (radishes, turnips, cabbages), fruits (grapes, plums, cherries), and a host of flowers (cornflowers, geraniums, delphiniums, roses, and peonies). Sometimes they are so brilliant that they mask the chlorophyll in the leaves, as in the red maple. Anthocyanins are the pigments responsible for the brilliant red colors of some leaves in the fall. Some of these pigments form in response to cold weather and bright sunlight, whereas others, masked by chlorophyll in green leaves, are visible only when the chloroplasts die and bleach.

1–10
Druses, compound crystals composed of calcium oxalate, in parenchyma cells of the Begonia *stem, as seen in ordinary* (a) *and polarized* (b) *light.*

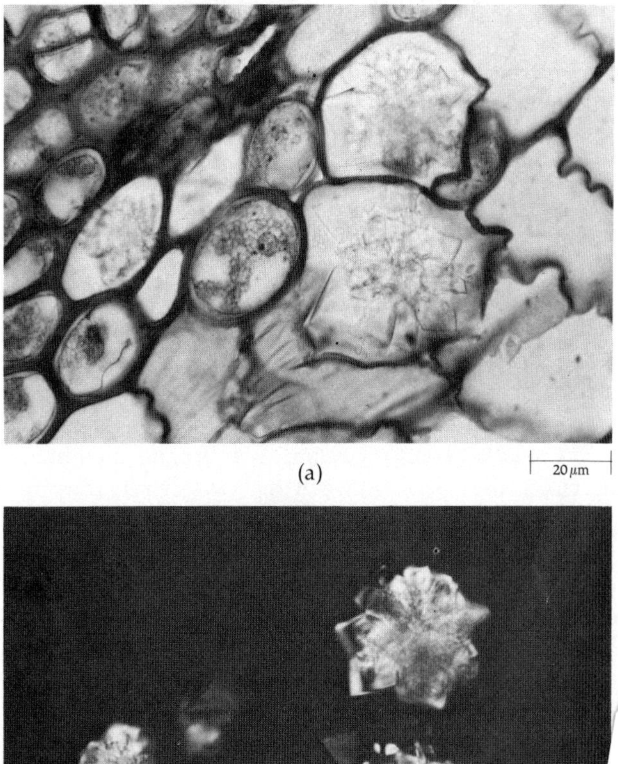

(a)

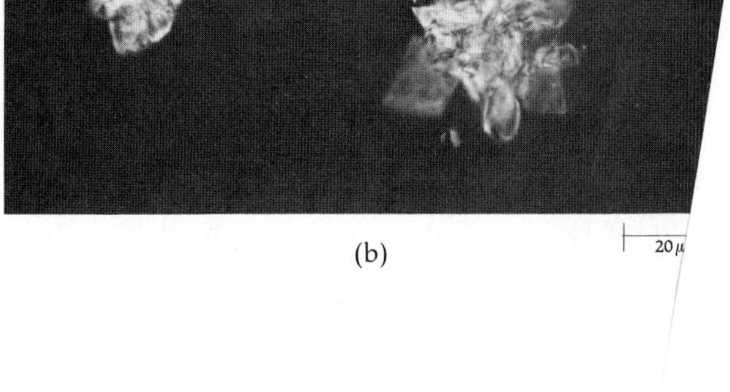

(b)

1–11
A bundle of raphides, needlelike crystals of calcium oxalate, in leaf cell of the snake plant (Sansevieria), *as seen in ordinary* (a) *and polarized* (b) *light.*

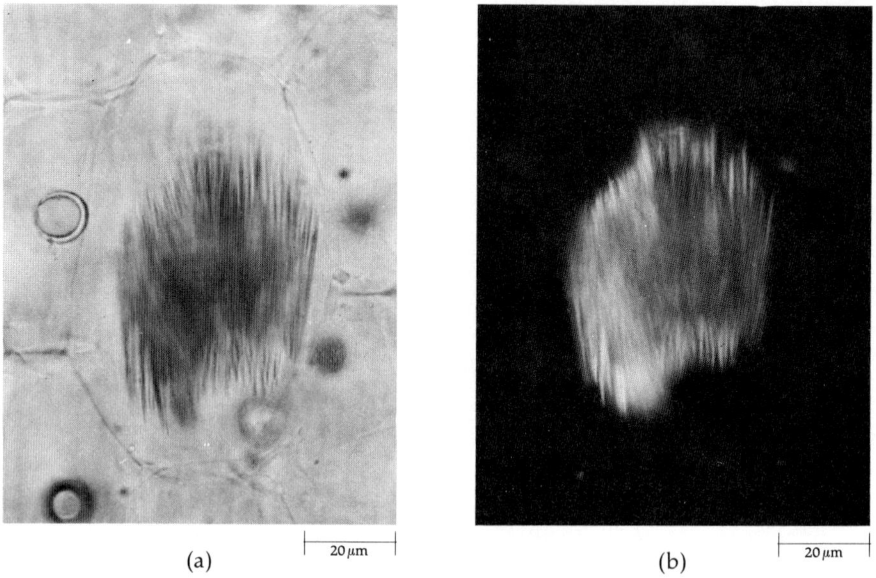

Apparently the vacuole, in addition to being involved with the accumulation of ions and molecules of various types and in the water balance of the cell, is also involved with the breakdown of macromolecules and the recycling of their components within the cell. Vacuoles are thus comparable in function with the organelles, known as lysosomes, that occur in animal cells.

Endoplasmic Reticulum

With its folds and branched, interconnecting system of membranes, the *endoplasmic reticulum* structures the cytoplasm and divides it into numerous compartments. Between and within these compartments and on the surface of the endoplasmic reticulum, many of the vital processes of the cell take place.

The endoplasmic reticulum is a double membrane structure, the space between the two membranes appearing transparent in electron micrographs (Figure 1–6). The membrane-bounded cavities vary considerably in size and shape in different types of cells and under different physiological conditions. In some cells, the network consists of fine tubules 50 to 100 nanometers in diameter. In others, the cavities may be much larger, forming flattened sacs, called *cisternae.* Both forms of endoplasmic reticulum may be found within a given cell.

The endoplasmic reticulum appears to function as a communications system through the cell. In some electron micrographs, it can be seen to be continuous with the outer membrane of the nuclear envelope. In fact, these two structures together seem to form a single membrane system. When the nuclear envelope breaks down at cell division, its fragments are similar to portions of endoplasmic reticulum. It is easy to visualize the endoplasmic reticulum as a system for channeling materials—for example, proteins and lipids—to different parts of the cell, which, as we are beginning to see, is a very complex "factory" involved in a great number of different and simultaneous activities.

Ribosomes, the sites of protein synthesis in the cell, are abundant in the cytoplasm of metabolically active cells and may occur free in the cytoplasm or be attached to the endoplasmic reticulum. They commonly occur at both places. Endoplasmic reticulum with ribosomes attached to it is called rough endoplasmic reticulum and that lacking ribosomes is called smooth endoplasmic reticulum. Ribosomes actively involved in protein synthesis occur in clusters or aggregates called *polyribosomes,* or *polysomes* (Figure 1–12). Cells that are synthesizing proteins in large quantities often contain extensive systems of polysome-bearing endoplasmic reticulum. Indeed, protein synthesis undoubtedly constitutes one of the principal roles of the endoplasmic reticulum in both plant and animal cells.

The Golgi Apparatus

Golgi bodies, or *dictyosomes,* are groups of flat, disk-shaped sacs or cisternae, which are often branched into a complex series of tubules at their margins (Figure 1–13). Around the edges of the disks can usually be seen numerous, approximately spherical vesicles which are formed and pinched off at the margins of the disks. The term *Golgi apparatus* is used to refer collectively to all of the Golgi bodies of a given cell.

Most authorities now agree that Golgi bodies function as collecting and packaging centers. Evidently the poly-

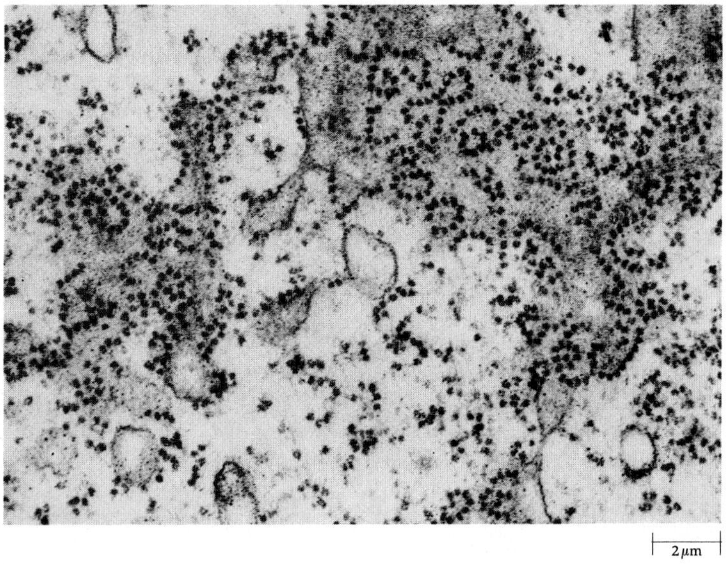

1–12
Groups of ribosomes (polyribosomes) on the surface of endoplasmic reticulum. The endoplasmic reticulum is a network of membranes that fills the cytoplasm of the eukaryotic cell, dividing it into channels and compartments and providing surfaces on which chemical reactions take place. Ribosomes are the sites at which amino acids are assembled into proteins. This electron micrograph shows a portion of an epidermal cell of a radish root.

saccharides synthesized by many plant cells collect in the vesicles, which then migrate to and fuse with the plasma membrane. The vesicles then discharge their contents to the cell exterior, and their polysaccharide contents become part of the cell wall. The enzymes responsible for the syntheses of these polysaccharides are carried in the Golgi bodies.

Golgi bodies seem always to be found in association with the basal bodies of flagella and cilia (pages 26–28), as well as with centrioles (page 42), in cells that contain these structures. This association suggests strongly that the Golgi bodies may also produce or deliver the enzymes involved in synthesizing the structural proteins (fibrous proteins) that make up the flagella, cilia, or spindle fibers (page 42). In at least some plant cells, much of the plasma membrane seems to be formed from the membranes bounding Golgi vesicles.

Microtubules

Microtubules, which are found in all cells, are long, thin structures about 24 nanometers in diameter and variable in length. They are polymers—large molecules composed of many identical subunits. Each subunit is a protein molecule called tubulin. The outer, electron-dense walls of microtubules are about 5 nanometers thick, and they surround an inner, electron-transparent center about 12 nanometers in diameter (Figure 1–14). In these outer walls, the tubulin appears to be arranged in protofilaments, linear arrays of subunits parallel to the long axis of the tubule. Most investigators agree that there are 13 protofilaments per tubule and that the protofilaments are composed of globular subunits of tubulin 4 to 4.5 nanometers in diameter.

1–13
The Golgi body consists of a group of flat, membranous sacs associated with vesicles that apparently bud off from the sacs. It serves as a packaging center for the cell and is concerned with secretory activities in both plant and animal cells.

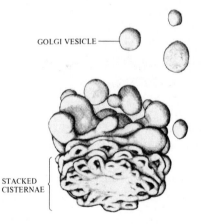

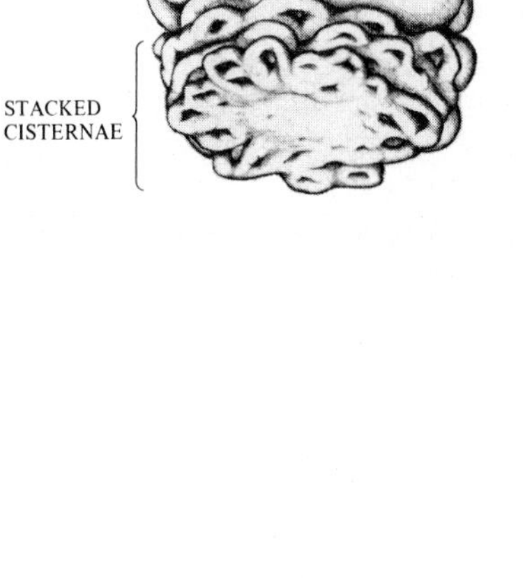

1-14
(a) *Oblique section of a cell wall showing elongate microtubules beneath it. The dark dots in the wall are plasmodesmata* (pages 35-36). (b) *This section has been cut in a plane perpendicular to the microtubules, which can be seen in cross section just inside the wall. The microtubules are separated from the wall by the plasma membrane.*

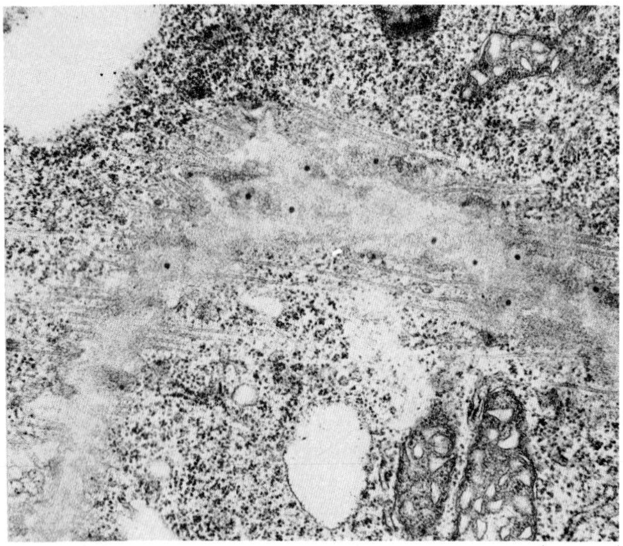

(a)

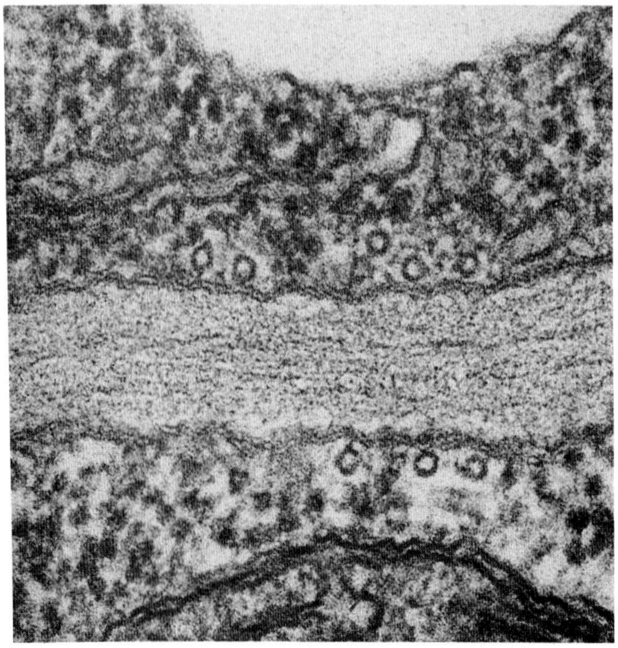

(b)

Microtubules occur just inside the plasma membrane of nondividing cells and are believed to be involved in the orderly growth of the cell wall, especially through their control of the alignment of the cellulose microfibrils which are deposited by the cytoplasm in adding to the cell wall. Microtubules also serve to direct Golgi vesicles toward the developing wall and to orient other cytoplasmic components, such as the nucleus, mitochondria, plastids, and lipid droplets, within the cell. Microtubules are present in the spindle fibers which form in dividing cells and apparently play a role in cell-plate formation, which takes place between cells that are dividing. In addition, we shall see that microtubules are important components of flagella and cilia and apparently are involved in the movement of those structures.

Flagella and Cilia

Flagella and cilia (singular: flagellum and cilium) are hairlike structures that project from the free surfaces of many different types of plant and animal cells. They are relatively constant in diameter (about 0.2 micrometer), but they vary in length from about 2 to 150 micrometers. By convention, the ones that are longer or are present alone or in small numbers are usually referred to as flagella, whereas the shorter ones or those occurring in greater numbers are cilia. There is no definite distinction between them, however, and we shall use the term *flagellum* to refer to both.

In algae, fungi, protozoa, and very small animals, flagella are locomotor organs, propelling the microorganism through the water. In plants, these organelles are found only in the sex cells (gametes) and then only in plants that have motile gametes, such as *Ginkgo*, the maidenhair tree. Some flagella, called tinsel flagella, bear one or two rows of minute, lateral appendages, whereas others, termed whiplash flagella, lack such processes (Figure 1-15).

One of the most intriguing of the discoveries made possible by the electron microscope is the internal structure of flagella. Each flagellum has a precise internal organization (Figure 1-16). An outer ring of nine pairs of microtubules surrounds two additional microtubules in the center. This basic pattern of organization is found in all flagella except those of bacteria.

Isolated flagella can beat all by themselves, so the beat must be explainable in terms of the structure of the flagellum itself. One suggestion is that the microtubules are contractile, so that each pair can contract individually. Alternatively, some investigators believe that the motion is caused by a creeping, tractor-fashion, of the outside tubules along the center pair. Recent experiments have shown that abrupt changes (fluxes) in calcium ion concentration across the plasma membrane, similar to those involved in the nervous systems of higher animals, control the movement of cilia and flagella.

1–15
(a) *The two types of flagella.* (b) *Whiplash flagellum and a portion of a tinsel flagellum on the zoospore of an Oomycete,* Pythium infestans. (c) *Enlargement of a partly disintegrated tinsel flagellum showing the strands of which it is composed.*

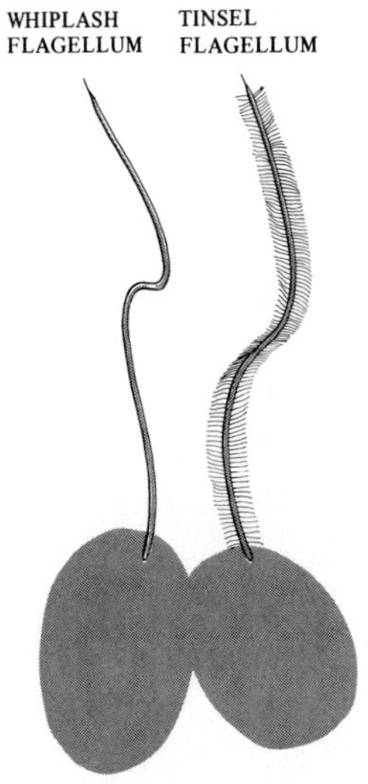

(a)

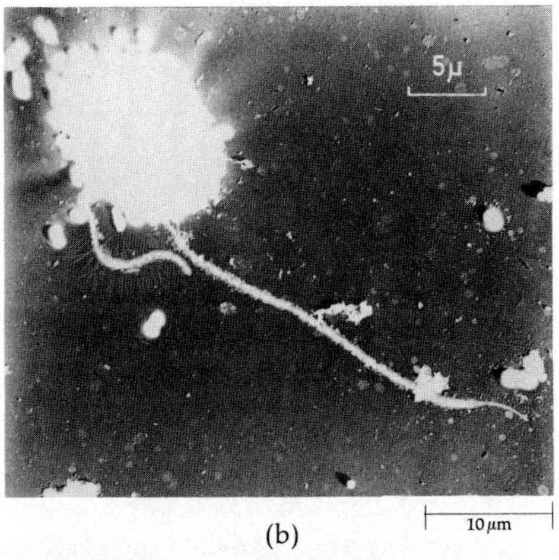

(b)

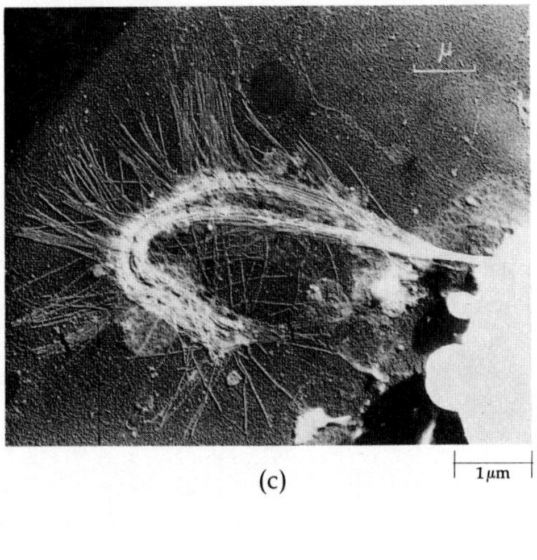

(c)

1–16
Structure of the flagellum. Two central microtubules are encircled by nine pairs of microtubules. The arms are believed to be enzymes involved in the energy reactions that produce movement. Flagella and cilia are found in nearly all major groups of organisms, and all of those found in eukaryotes have this same basic 9-plus-2 structure.

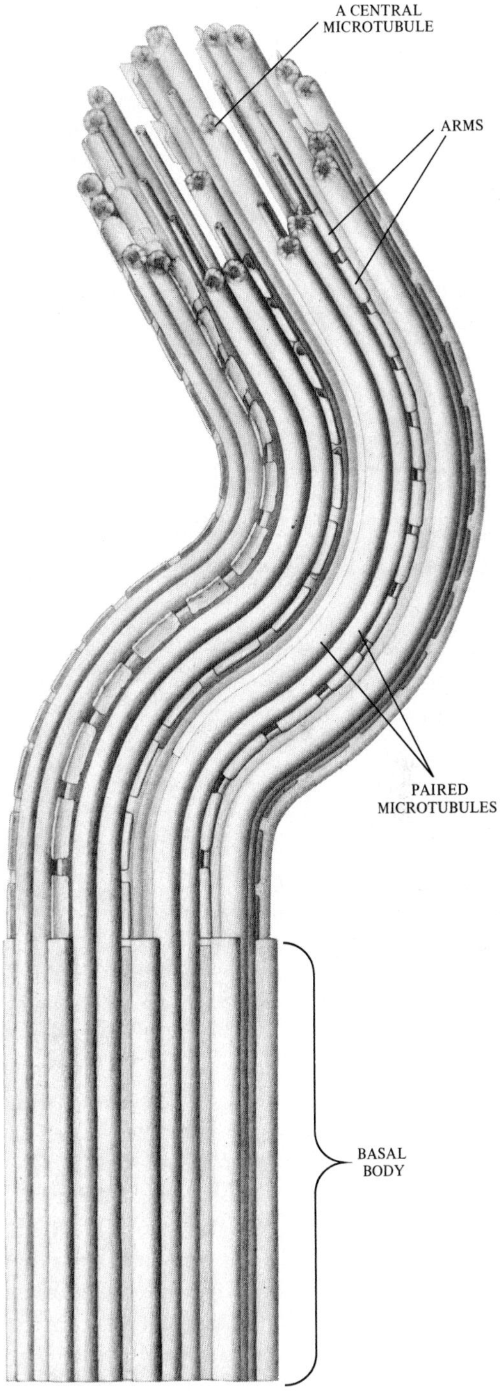

1–17
(a) *Longitudinal section of a flagellum of* Chlamydomonas reinhardtii. *Notice that the membrane surrounding the flagellum is continuous with the plasma membrane. A mitochondrion can be seen near the basal bodies. The larger body to the right is a portion of a chloroplast.*
(b) *Three basal bodies from* Chlamydomonas reinhardtii. *The one in the center has been sectioned transversely. At the time of cell division, the flagella are lost and two new basal bodies and new flagella are formed. Note that the basal body differs from the flagellum in structure; the basal body has a circle of nine triplet microtubules and no central microtubules. This same structure is found in centrioles.*

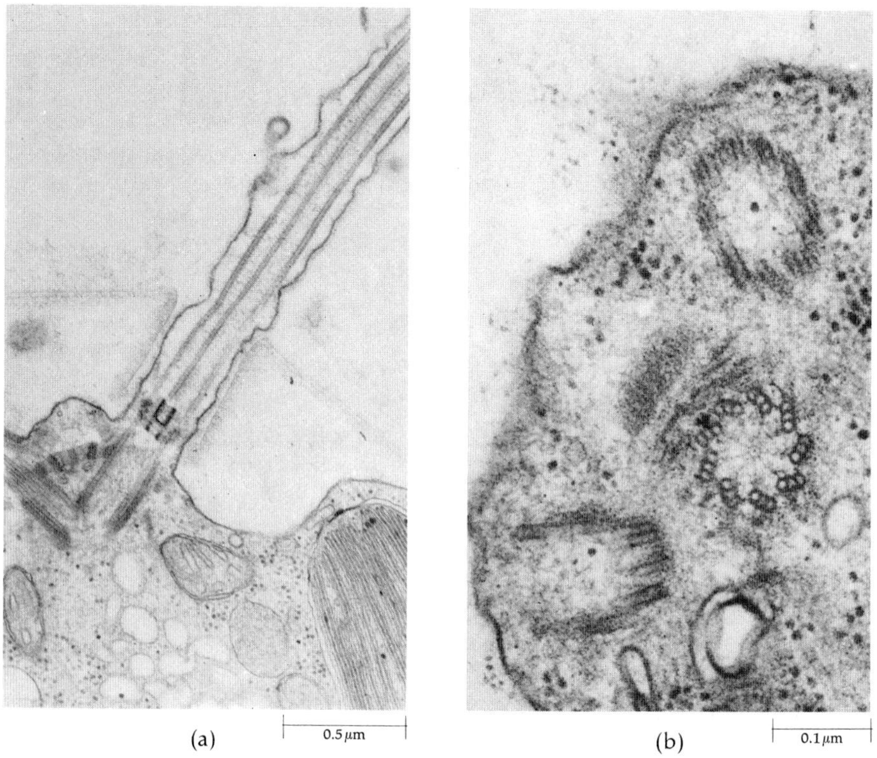

Flagella grow out of cylinder-shaped organelles in the cytoplasm known as *basal bodies* (Figure 1–17). Basal bodies have an internal structure that somewhat resembles the structure of the flagellum except that, in the basal body, the outer tubules occur in triplets rather than in pairs, and the two central tubules are absent.

Plastids

Together with vacuoles and cell walls, *plastids* are characteristic components of plant cells. Each plastid is delimited by an envelope consisting of two three-ply membranes, and internally the plastid is differentiated into a system of membranes and a more or less homogeneous ground substance, the *stroma*.

Mature plastids commonly are classified on the basis of the kinds of pigments they contain. *Chloroplasts* contain chlorophylls and carotenoid pigments. In higher plants, chloroplasts usually are disk-shaped and between 4 and 6 micrometers in diameter. A single mesophyll ("middle of leaf") cell may contain 40 to 50 chloroplasts, a square millimeter of leaf some 500,000. The chloroplasts usually occur in the cytoplasm with their broad surfaces parallel to the cell wall, as shown in Figure 1–9. It has been reported that chloroplasts are able to move independently within the cytoplasm, enabling them to orient their surfaces in relation to the light.

The internal structure of the chloroplast is complex (Figures 1–18 and 1–19). The stroma is traversed by an elaborate system of membranes in the form of flattened sacs called *lamellae* or *thylakoids*. Each lamella or thylakoid is similar to the plastid envelope in that it consists of two membranes. All of the thylakoids are believed to constitute an interconnected system. Chloroplasts are characterized by the presence of *grana* (singular: granum)—stacks of disklike thylakoids resembling a stack of coins. The lamellae of the various grana are connected with each other by lamellae (the stroma or intergrana lamellae) traversing the stroma. The chlorophylls and carotenoid pigments are found within the thylakoid membranes.

Chloroplasts often contain small lipid droplets and starch grains as well. The starch grains are temporary storage products and accumulate only when the plant is actively photosynthesizing. They may be lacking in the chloroplasts of plants kept in the dark for as little as 24 hours, but often reappear after the plant has been in the light for only 3 or 4 hours.

The interior surfaces of chloroplast lamellae contain numerous granules called quantasomes (Figure 1–20), structures once believed to represent the basic morphological units involved in the light reactions of photosynthesis. However, it is now uncertain whether the quantasome granule is a functional unit. Chloroplasts also contain small ribosomes similar to those of prokaryotes, and strands of DNA can often be seen in the stroma-free regions.

Chloroplasts do not occur in the blue-green algae, which are prokaryotic organisms similar to bacteria in

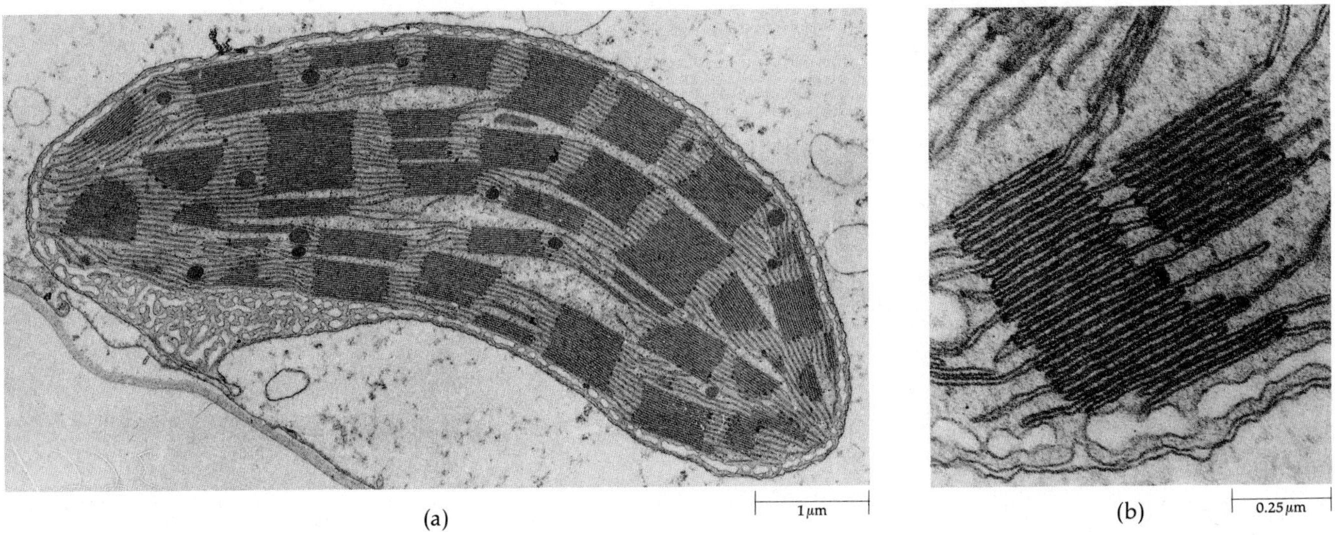

(a) (b)

1–18
(a) *A chloroplast of a corn* (Zea mays) *leaf.* (b) *Detail showing grana composed of stacks of disklike thylakoids.*

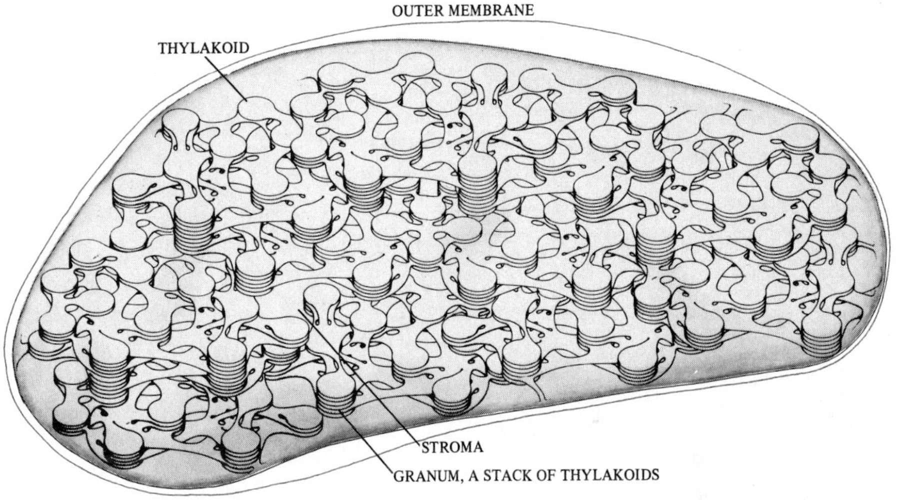

1–19
The inner structure of a chloroplast. A thylakoid (or lamella) is composed of a pair of membranes fused to form a closed disk. A granum is made up of thylakoids stacked one on the other. All the chlorophyll in the chloroplast is associated with these membranous structures which, as you can see, are believed to form a single interconnected system.

1–20
Part of the membrane has been torn away from the surface of this lamella. The particles that are revealed on the inner surface have been termed quantasomes. The relation of this cobblestone-like structure to the process of photosynthesis is not yet known.

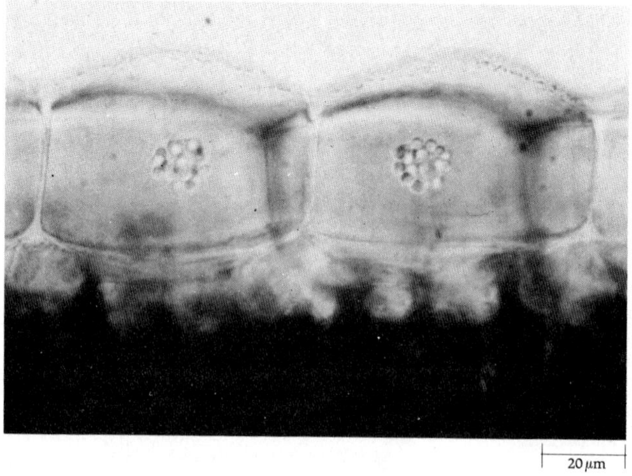

1–21
Leucoplasts clustered around the nucleus in epidermal cells of a Zebrina *leaf. The dense appearance of these cells is due to the presence of anthocyanin in their vacuoles.*

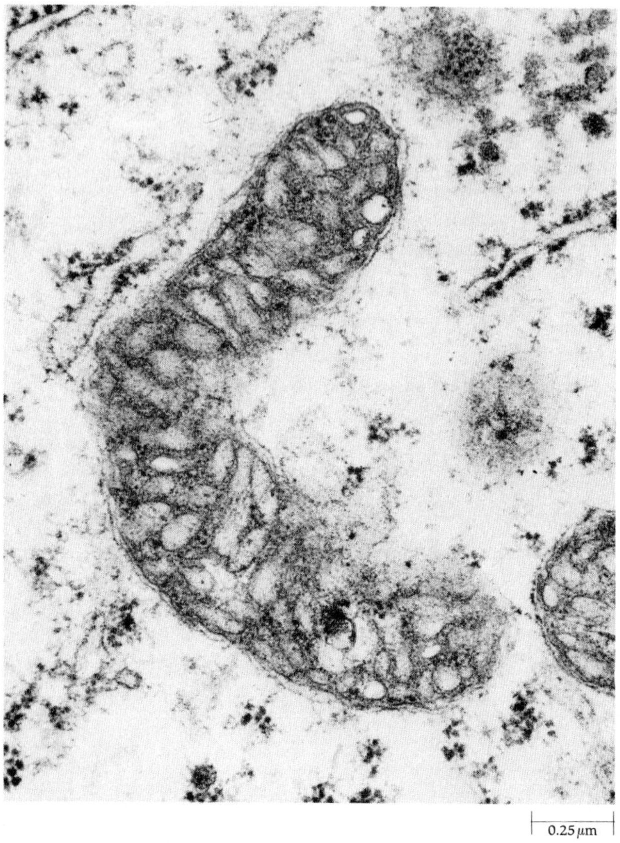

1–22
Mitochondrion of a cell from a leaf of corn (Zea mays). *A single cell may contain as many as 700 mitochondria.*

their structure. These algae contain a chlorophyll of the type found in higher plants (chlorophyll *a*), but it is organized in a network of lamellae running continuously through much of the cytoplasm (see Figure 1–4). Chloroplasts, with their highly ordered structure, are found only in eukaryotic organisms.

Two other kinds of plastids commonly found in the cells of higher plants are *chromoplasts* and *leucoplasts.* As the name implies, chromoplasts are colored plastids. Of variable shape, chromoplasts synthesize and retain the carotenoid pigments, which are yellow, orange, or red in color. They may develop from previously existing green chloroplasts by a transformation in which the chlorophyll and internal membrane structure of the chloroplast disappear and masses of carotenoids accumulate.

Leucoplasts (Figure 1–21) are nonpigmented plastids. Some synthesize starch, but others are thought to be capable of forming a variety of substances, including oils and proteins. Upon exposure to light, leucoplasts may develop into chloroplasts.

The various kinds of mature plastids—chloroplasts, chromoplasts, and leucoplasts—develop from small, colorless bodies known as *proplastids.* Proplastids become fully green and differentiated into chloroplasts only in the presence of light. This is why seedlings grown in the dark are pale in color. With their double-membrane envelope and short internal lamellae, proplastids often are difficult to distinguish from mitochondria. Proplastids divide by fission and are apportioned among daughter cells in regions of active cell division. In unicellular algae and in some plants, such as liverworts and ferns, mature chloroplasts apparently divide by fission and are distributed to the daughter cells, growing to full size after division.

Mitochondria

Like chloroplasts, mitochondria are bounded by two membranes and contain many internal membranes (Figures 1–22 and 1–23) which greatly increase the surface area available to enzymes and the reactions associated with them. Mitochondria are generally smaller than plastids, measuring about half a micrometer in diameter and usually 1 to 4 micrometers in length, although some may be much longer and there is apparently great variability in shape. Most are just visible under the light microscope.

Mitochondria are the sites of breakdown (oxidation) of organic molecules to release energy and the conversion of this energy to molecules of ATP, the chief chemical energy source for all cells. (These processes are discussed further in Chapter 5.) In time-lapse studies, mitochondria can be seen to be in constant motion, turning and twisting and moving from one part of the cell to another. They also appear to coalesce and to divide. They tend to congregate where energy is required. In cells in which the plasma membrane is very active in transporting materials in or out of the cell, they often can be found arrayed along the membrane surface. In single-celled algae that move by means of flagella, mitochondria are typically clustered at the bases of the flagella.

The outer membrane of the mitochondrial envelope serves to enclose the organelle, whereas the inner one is convoluted into folds, pleats, or shelflike projections known as cristae. The greater the energy requirements of a cell, the more cristae its mitochondria are likely to contain. The cristae are surrounded by a liquid stroma containing proteins, RNA, strands of DNA, ribosomes, and various solutes. Mitochondria can be separated undamaged from the rest of the cell and purified. As a consequence, it has been possible to identify the enzyme systems associated with them and to make quite precise correlations between mitochondrial structures and functions.

Bacterial cells do not have mitochondria. In fact, the entire bacterial cell is not much larger than a mitochondrion. But *Escherichia coli* and many other bacteria are capable of respiration. The enzyme systems that carry out respiration in *E. coli* form a part of the membrane surface of the bacterial cell, in much the same way that the enzyme systems that carry out respiration in mitochondria form a part of its internal membranes. This fact has become part of the evidence for current theories about the evolutionary origin of mitochondria and chloroplasts.

1–23
A mitochondrion. The inner membrane folds inward, forming cristae, which greatly increase the working surface of the organelle.

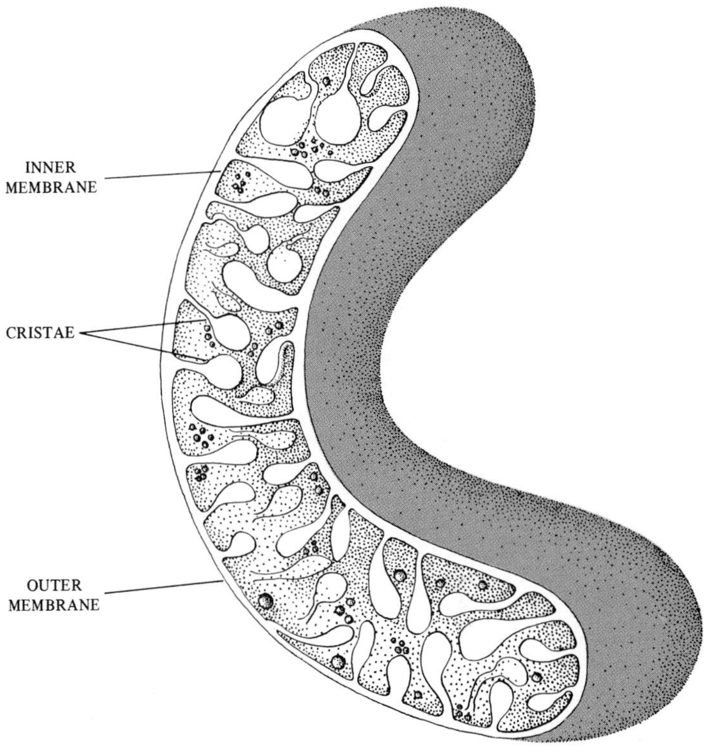

Microbodies

Unlike plastids and mitochondria, which are bounded by two membranes, microbodies are spherical organelles bounded by a single membrane. They range in diameter from 0.5 to 1.5 micrometers. Microbodies have a granular interior, sometimes with a crystalline, proteinaceous inclusion (Figure 1–24). Microbodies generally are asso-

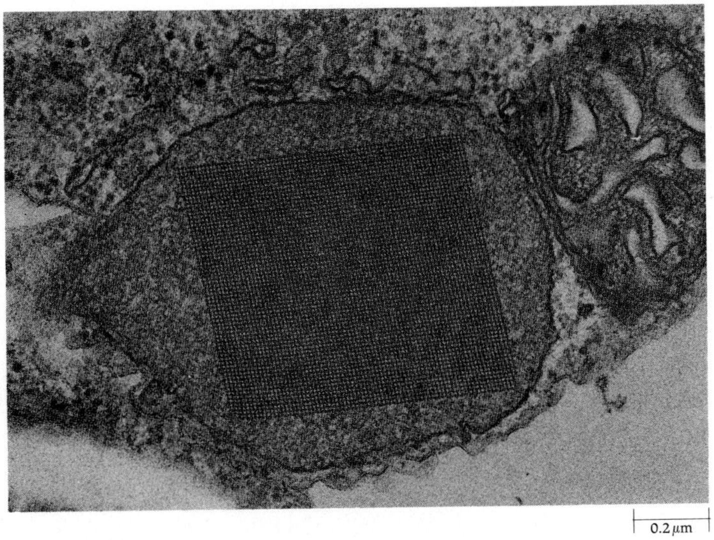

1–24
A microbody with a large crystalline inclusion, from a leaf of a tobacco plant (Nicotiana tabacum). *Note part of mitochondrion, at right.*

THE EVOLUTIONARY ORIGIN OF MITOCHONDRIA AND CHLOROPLASTS

Mitochondria and chloroplasts share a number of unusual properties. Both are bounded by a double membrane and have a complex internal membrane structure. Both have a capacity for growth and division that is relatively independent of the rest of the cell, and, as we shall see in Chapters 5 and 6, both go about the production of ATP, the molecules that provide chemical energy for all cells, in similar ways.

In addition to these striking similarities, it has been found in recent years that both mitochondria and chloroplasts contain characteristic forms of DNA and also RNA, as well as ribosomes. In some of these organelles, it has been demonstrated that the DNA is present in the form of a closed circle, like that of bacteria.

The genetic role of this DNA is currently being investigated in several laboratories. Isolated chloroplasts have the capacity to synthesize RNA, which, as we shall see in Chapter 3, is usually carried out only under the direction of chromosomal DNA. The ability to form chloroplasts and the pigments associated with them is largely controlled by chromosomal DNA interacting in some poorly understood fashion with chloroplast DNA. Thus chloroplasts cannot be formed in the absence of chloroplast DNA.

Chloroplast ribosomes resemble bacterial ribosomes in several ways. For example, the ribosomes of both prokaryotes and chloroplasts are only about two-thirds as large as the ribosomes found in the cytoplasm and on the endoplasmic reticulum of the eukaryotic cell. The synthesis of protein in the ribosomes of mitochondria, chloroplasts, and bacteria is inhibited by the antibiotic cyclohexamide, a substance that has no effect on the ribosomes of eukaryotic cells.

On the basis of the accumulating evidence, it seems probable that mitochondria and chloroplasts originated as free-living prokaryotes that found shelter within larger heterotrophic cells. These larger cells were the forerunners of the eukaryotes. The smaller cells, which contained (and still contain) all the mechanisms necessary to trap and convert energy from their surroundings, donated these useful capacities to the larger ones. Cells with these respiratory assistants had a clear advantage over their contemporaries, and undoubtedly soon multiplied at their expense. All modern eukaryotes contain mitochondria, and all autotrophic eukaryotes contain chloroplasts; both seem to have been acquired by independent symbiotic events. (Symbiosis is the close association between two or more different organisms that may be, but is not necessarily, beneficial to each.) The larger and more complex cells of eukaryotes seem to protect their symbiotic organelles from environmental extremes. As a consequence, eukaryotes were able to invade the land and the acidic waters, where the prokaryotic blue-green algae are absent but the eukaryotic green algae abound.

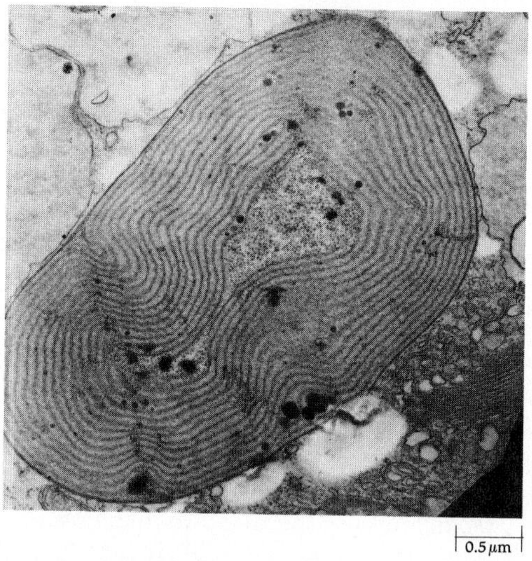

A blue-green algal cell (*Glaucocystis nostochinearum*) contained within the larger cell of a green alga. Chloroplasts are believed to have had their evolutionary origin in events such as these.

ciated with one or two cisternae of rough endoplasmic reticulum, which usually are smooth on the surfaces facing the microbodies.

Some microbodies, called peroxisomes, play an important role in glycolic acid metabolism associated with photosynthesis (see page 118). Others, called glyoxysomes, contain enzymes necessary for the conversion of fats into carbohydrates during germination in many seeds.

Spherosomes

Spherosomes are spherical structures that impart a granular appearance to the cytoplasm of a plant cell when viewed with the light microscope. They range in diameter from 0.5 to 2 micrometers, falling within the size range of mitochondria. Many structures identified as spherosomes contain mostly lipids and apparently are centers of lipid synthesis and accumulation (Figure 1–5). In many preparations these oil globules, or lipid droplets, are bounded by a single membrane, believed by some investigators to have its origin in the endoplasmic reticulum. Whether other structures of similar size and appearance should also be called spherosomes is a matter of current discussion.

The Cell Wall

If one were to select one feature that above all others characterizes plant cells and distinguishes them from animal cells, that feature would be the presence outside the living portion of the cell of a nonliving semirigid or rigid cell wall containing cellulose. Its presence is the basis of many of the characteristics of plants as organisms. The cell wall limits the size of the cell and prevents

1–25
(a) Diagram of plant cell walls. The cells are separated from one another by the middle lamella. On each side of the middle lamella is the primary wall of an individual cell. In plant cells with secondary walls, this wall is deposited inside the primary wall. Some plant cells are dead at maturity. (b) Electron micrograph of two adjacent cell walls of vessel elements, the cells that form the tubes through which water is conducted. You can see the middle lamella, the primary walls, and the layered secondary walls. The cells are from the wood of a black locust (Robinia pseudo-acacia) *tree. The protoplasts are no longer present.*

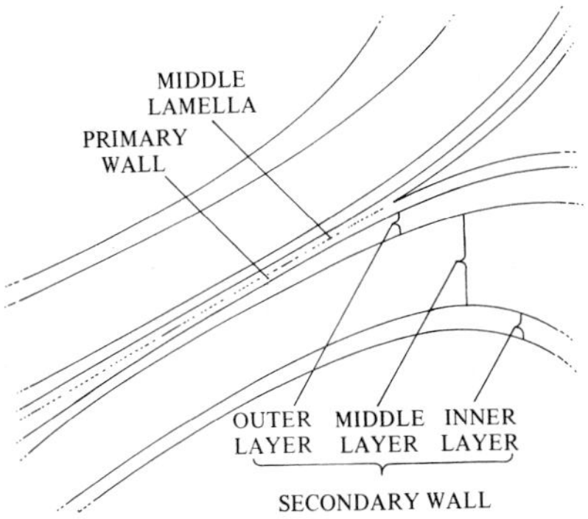

(a)

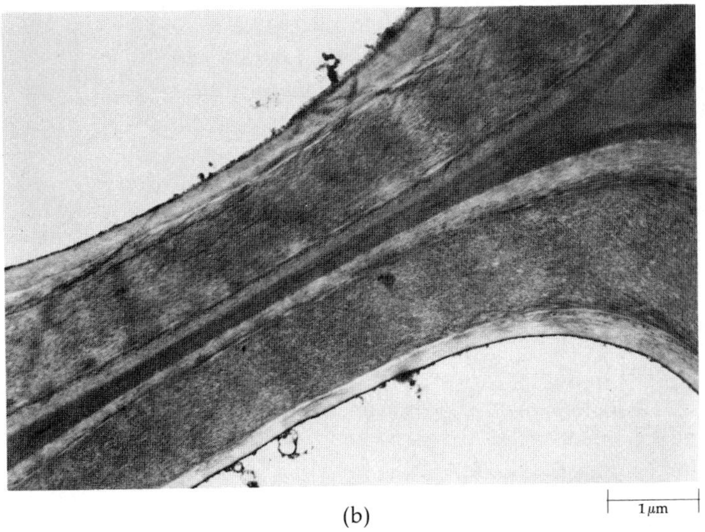

(b)

it from being ruptured from undue enlargement of the protoplast resulting from the intake of water by the vacuole. Cells that have relatively thin walls depend on turgor pressure (see page 522) to keep them rigid, whereas those with thick walls provide mechanical support to the plant part containing them.

Plant cell walls vary greatly in thickness, depending partly on the role the cells play in the structure of the plant and partly on the age of the individual cell. Developmental studies, coupled with the use of the electron microscope, polarized light, and x-rays, indicate that there are two layers in all plant cell walls: the *intercellular substance,* or *middle lamella,* and the *primary wall.* In addition, many cells deposit another wall layer, the *secondary wall.* The middle lamella occurs between the primary walls of adjacent cells, and the secondary wall, if present, is laid down by the protoplast of the cell on the inner surface of the primary wall (Figure 1–25).

Chemical Composition of Cell Walls

The most characteristic component of plant cell walls is cellulose, which forms much of the structural framework of the wall. Cellulose is made of repeating molecules of glucose attached end to end (see Chapter 2). These long, thin molecules are united in microfibrils, each of which may contain as many as 2000 molecules. These microfibrils (Figure 1–26), which measure 25 to 30 nanometers in width each, wind together to form fine threads, or fibrils, and the fibrils, in turn, may coil around one another like strands in a cable. Each "cable," or macrofibril,

1–26
Surface of the cell wall of the algal cell Chaetomorpha, *showing cellulose microfibrils, each of which is made up of hundreds of molecules of cellulose.*

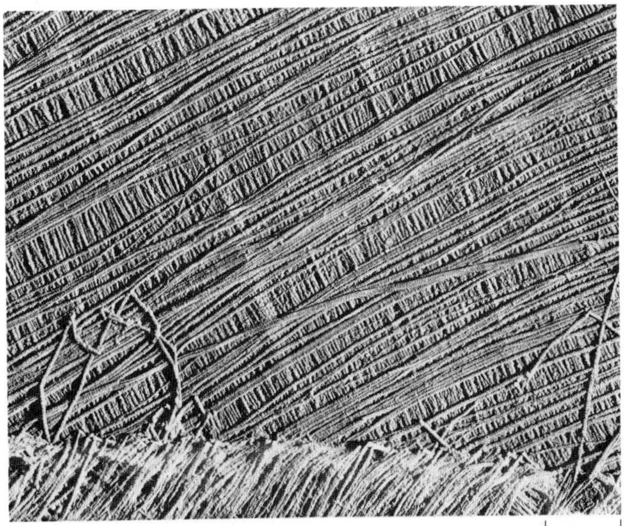

contains about 500,000 cellulose molecules and measures up to 0.5 micrometers in width. Individual macrofibrils may reach 4 micrometers in length. Cellulose wound in this fashion is as strong as an equivalent thickness of steel.

Because of the arrangement of the cellulose molecules into orderly three-dimensional lattices, cellulose has crystalline properties. These crystalline properties can easily be demonstrated with plane-polarized light (light vibrating in one plane only). If a crystalline substance such as cellulose is viewed with a light microscope between two crossed polaroids (crystalline prisms that can polarize light)—one polaroid placed between the light source and microscope and the other between the ocular and the eye—it will appear bright in a dark field. Because of its crystalline properties, the cellulose changes the plane of polarization of the light reaching the polaroid above the ocular, thus allowing some light to reach the eye (Figure 1–27). Cellulose and other substances affecting light in this manner are said to be optically anisotropic. Substances that do not affect light this way are said to be optically isotropic.

The cellulose fibrils of the wall exist in a cross-linked matrix of large, noncellulosic molecules. Some of these are polysaccharides called *hemicelluloses.* Others are the *pectic substances,* which are closely related to the hemicelluloses chemically. There are three principal kinds of pectic compounds: pectic acids, pectin, and protopectin. Pectic acid, which is soluble in water and may be present in the protoplast, is probably the precursor of pectin and protopectin. Pectin, an important component of primary walls, is sold commercially as a gelling agent in fruit jellies. Protopectin, in association with calcium or magnesium, functions as the cement that holds plant cells together.

Lignin is one of the most important constituents of the secondary wall and after cellulose is the most abundant plant polymer. Physically, lignin is rigid, and it serves to add rigidity to the wall. In the soil, lignin decomposes very slowly, and very few kinds of microorganisms are able to break it down. Lignin may be present in all three wall layers. Characteristically, lignification begins in the middle lamella, then spreads to the primary wall, and finally to the secondary wall.

Gums, mucilages, and lipids, especially waxes, characterize the cell walls of particular kinds of plants, occurring in the walls of those cells that perform specific functions in the plant. For example, cutin is a common waxy substance found in the outer cell walls of the surface, or epidermal, cells of higher plants. It greatly restricts water loss through these cell walls.

The Middle Lamella

The *middle lamella* is composed mainly of pectic substances. Lacking cellulose, it is optically isotropic. Frequently, it is difficult to distinguish the middle lamella from the primary wall, especially in cells that develop thick secondary walls. In lignified walls, the middle lamella is the first part of the wall to become lignified.

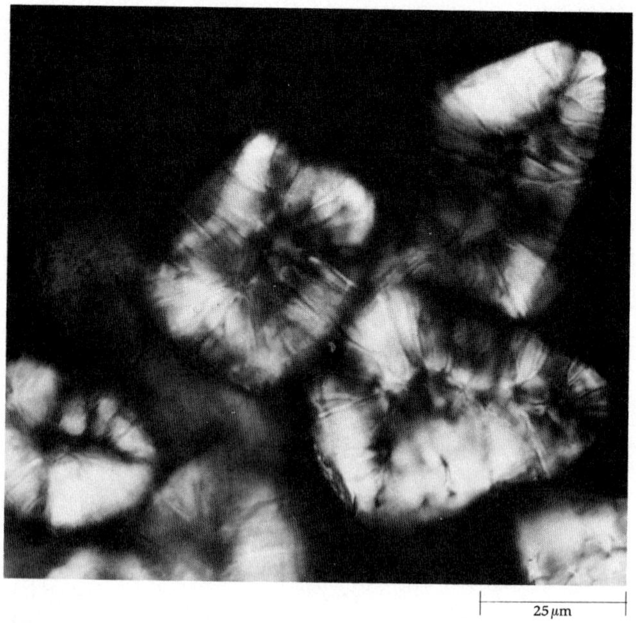

1–27
Stone cells (sclereids) from the flesh of the fruit of a pear (Pyrus communis) *fruit seen in polarized light. The stone cells have very thick secondary walls.*

The Primary Wall

The wall layer deposited before and during growth of the cell is called the *primary wall.* Many cells have only a primary wall. In addition to cellulose, primary walls contain hemicelluloses and some pectin. They also contain hydroxyproline-rich proteins and may become lignified. Because of the presence of cellulose, the primary wall is optically anisotropic. The pectic component imparts plastic properties to the wall, which make it possible for the primary wall to be stretched or extended.

Initially, the cellulose microfibrils of the primary wall form an irregular mesh or network with a predominantly transverse orientation. As the wall increases in surface area, the orientation of the outer microfibrils becomes more nearly longitudinal, or parallel to the long axis of the cell. The initial orientation of the microfibrils is determined by the orientation of microtubules located next to the plasma membrane within the cytoplasm.

Actively dividing cells commonly contain only primary walls, as do most mature cells involved with such metabolic processes as photosynthesis, respiration, and secretion. Cells such as these—that is, cells with primary walls and living protoplasts—are able to dedifferentiate (lose their specialized cellular form), divide, and redifferentiate (regain their form). Consequently, it is largely cells with only primary walls that are involved in wound healing and regeneration in the plant.

Usually, primary cell walls are not of uniform thickness throughout, but have thin areas called *primary pit-fields* (Figure 1–28). Cytoplasmic "threads" (plasmodesmata) connecting the living protoplasts of adjacent cells are commonly aggregated in the primary pit-fields.

The Secondary Wall

As we mentioned previously, many plant cells have only a primary wall, but in others a secondary wall is deposited by the protoplast inside the primary wall. This occurs mostly after the cell has ceased to grow and the primary wall is no longer enlarging in area. It is partly because of this that the secondary wall is structurally distinct from the primary wall. Secondary walls are particularly important in specialized cells that have functions such as conduction and strengthening; in these cells, the protoplast often dies after it has laid down the secondary wall. Cellulose is more abundant in secondary walls than in primary walls, and the secondary wall is therefore rigid and not readily stretched. The other polysaccharides are correspondingly rare or absent. Similarly, proteins, which may be relatively abundant in primary cell walls, are absent in secondary cell walls.

Frequently, three distinct layers, designated S_1, S_2, and S_3, for outer, middle, and inner layer, respectively, can be distinguished in the secondary wall (Figure 1–25). The layers differ from one another in the orientation of their cellulose fibrils. Within any one layer the fibrils are essentially parallel to one another. The laminated structure of such secondary walls, like that seen in plywood, greatly increases the strength of the wall. The cellulose fibrils are laid down in a denser pattern, with the matrix of other polysaccharides more limited than in the primary wall, and this is one of the reasons that the secondary wall increases the rigidity of the cell. Lignin is common in cells that form secondary walls and is always found, as we shall see in Chapter 19, in such wood cells as fibers, tracheids, and vessel members. Secondary walls are strongly anisotropic.

In some cell types—especially those concerned with conducting and strengthening—the secondary cell wall is laid down over only a portion of the primary wall. It may take the form of rings, spirals, or a network, these patterns giving the cells in which they occur an unusual and often bizarre appearance. Their formation is preceded by the appearance of dense, oriented bands of microtubules in the cytoplasm.

When the secondary cell wall is deposited, it is not laid down over the primary pit-fields of the primary wall. As a consequence, characteristic depressions, or *pits*, are formed in the secondary wall (Figure 1–28). Pits are also formed in some instances in areas where there are no primary pit-fields.

A pit in a cell wall usually occurs opposite a pit in the wall of the cell with which it is in contact. The middle lamella and two primary walls between the two pits are called the *pit membrane*. The two opposite pits plus the membrane constitute a *pit-pair*. Two principal types of pits are found in cells that have secondary walls: *simple* and *bordered*. In bordered pits, the secondary wall arches over the *pit cavity*. In simple pits, there is no overarching.

1–28
Primary pit-fields, pits, and plasmodesmata. (a) *Parenchyma cell with primary walls and primary pit-fields, thin areas in the walls. As shown here, plasmodesmata commonly traverse the wall at the primary pit-fields.* (b) *Cells with secondary walls and numerous simple pits.* (c) *A simple pit-pair.* (d) *A bordered pit-pair.*

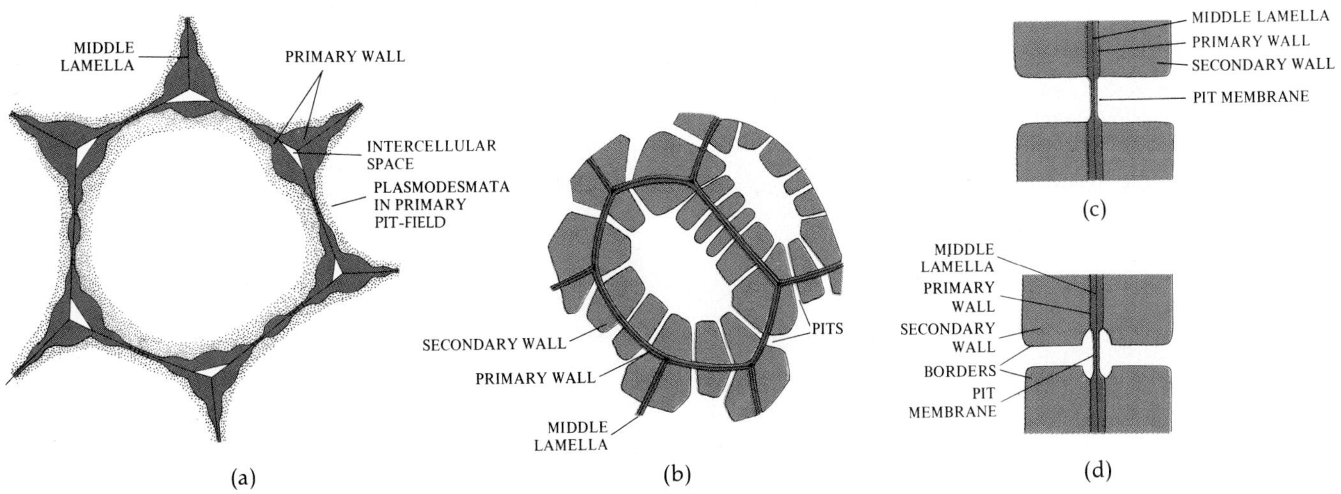

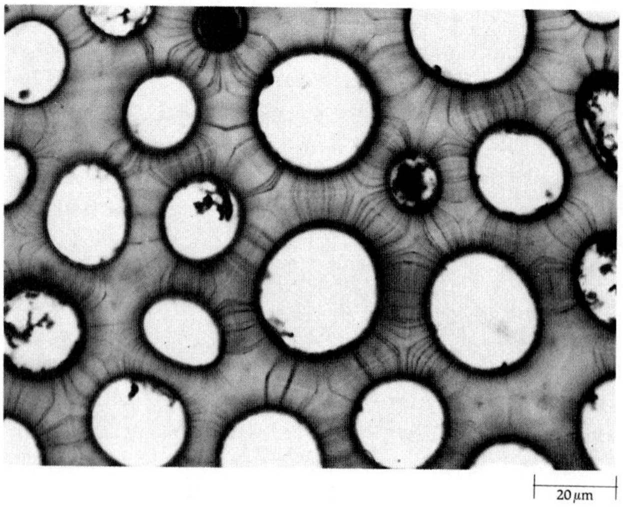

1–29
Light micrograph of plasmodesmata in the thick primary walls of persimmon (Diospyros) *endosperm, the nutritive tissue within the seed.*

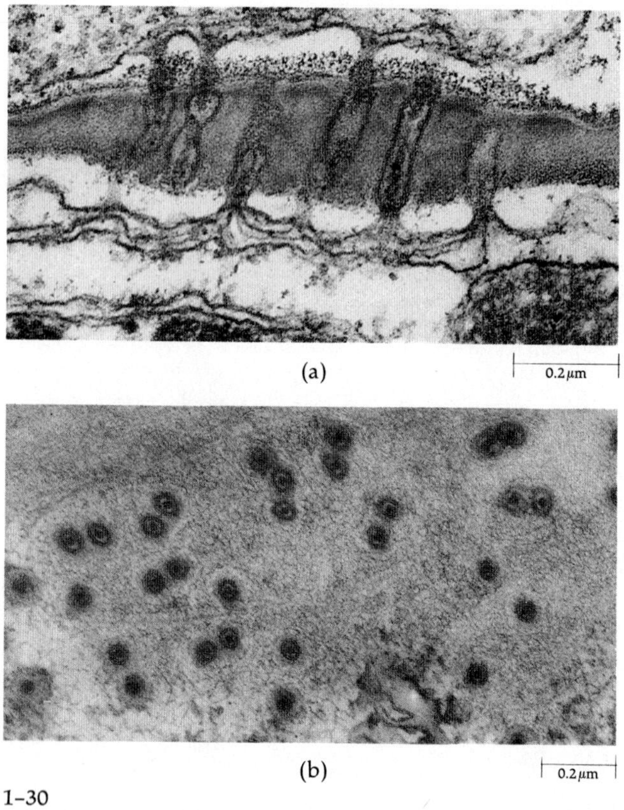

1–30
Electron micrograph of (a) *Plasmodesmata, in longitudinal view, connecting the protoplasts of two barley* (Hordeum) *leaf cells.* (b) *Transverse view of plasmodesmata in parenchyma cell walls of the phloem of an elm* (Ulmus) *tree.*

Plasmodesmata

The protoplasts of adjacent plant cells are connected with one another by fine strands of cytoplasm known as *plasmodesmata* (singular: plasmodesma). Although such structures long have been seen with the light microscope, they were difficult to interpret (Figure 1–29); it was not until they were observed with the electron microscope that their nature was confirmed (Figure 1–30).

Plasmodesmata may occur throughout the cell wall or be aggregated in primary pit-fields or the membranes between pit-pairs. With the electron microscope, plasmodesmata appear as channels lined by plasma membrane and traversed by a tubule of endoplasmic reticulum approximately 4 nanometers in diameter. The plasmodesmata serve to unify the individual cells of the plant body into an integrated organism.

The Nucleus

The nucleus is often the most prominent structure within the cytoplasm of eukaryotic cells. The nucleus performs two important functions: (1) It controls the ongoing activities of the cell. As we shall see in Chapter 3, it exerts this control by determining which protein molecules are produced by the cell and when. (2) It stores the genetic information, passing it on to the daughter cells in the course of cell division.

In eukaryotic cells, the nucleus is delimited from the cytoplasm by a double membrane (two three-ply membranes) called the *nuclear envelope*. The nuclear envelope, as seen in the electron microscope, contains a large number of circular pores 30 to 100 nanometers in diameter (Figure 1–31), the inner and outer membranes being joined around each pore to form the margin of the pore. The pores are not merely holes in the envelope; each has a complicated structure.

As mentioned previously, at many places the outer membrane of the envelope is continuous with the endoplasmic reticulum.

If the cell is treated by special staining techniques, thin strands and grains of *chromatin* can be seen within the nuclear boundaries in the nucleoplasm. Chromatin is made up of DNA combined with proteins. During the process of cell division the chromatin becomes progressively more condensed until it takes the form of *chromosomes*. As in prokaryotic organisms, the hereditary information is carried in molecules of DNA. The content of DNA per cell is much higher in eukaryotic organisms than in prokaryotes. In prokaryotes, the circular molecules of DNA are essentially free in the cytoplasm. In eukaryotes, they are organized into much larger units that involve proteins in addition to DNA. Exactly how chromosomes are organized is not known, but we do know that their fundamental units consist of long DNA-protein fibers. The theory most favored at present is

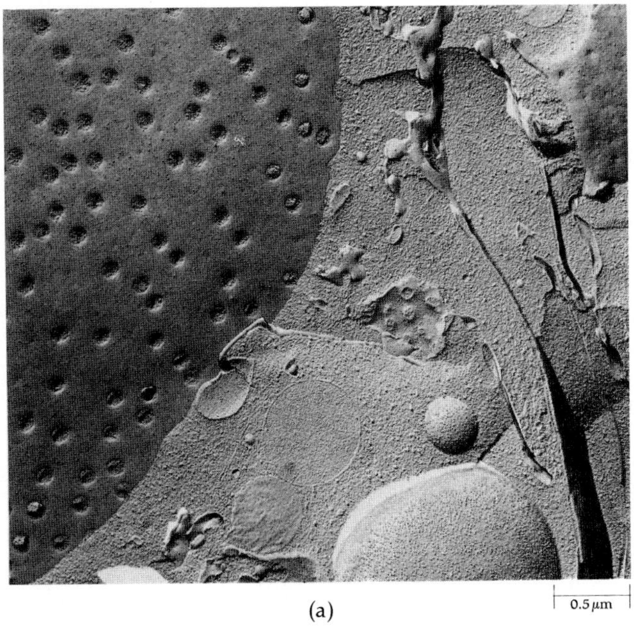

(a)

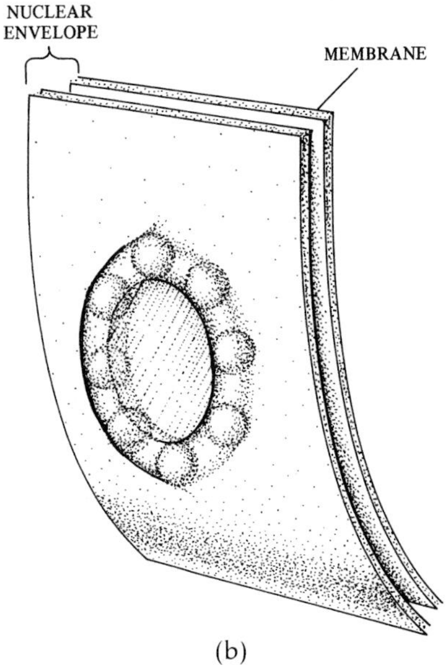

(b)

1–31
(a) *The nuclear pores are clearly revealed in this freeze-etch preparation of the surface of the nuclear membrane that fills the left half of the micrograph. Fragments of endoplasmic reticulum can be seen near the nucleus. The cell is from the root tip of an onion.* (b) *Diagram of a nuclear pore. The function of the eight globular structures encircling the pore is unknown.*

that the eukaryotic chromosome consists of one very long DNA-protein fiber which has a regular repeating structure that resembles beads on a string (Figure 1–32). The "beads" are DNA complexed with histone proteins and the "string" consists of short stretches of DNA between clusters.

Different organisms vary in the number of chromosomes present in their cells. *Haplopappus gracilis,* a desert annual, has 4; cabbage, 20; a frog, 26; a sunflower, 34; bread wheat, 42; man, 46; a goldfish, 94; and one fern of the genus *Ophioglossum,* about 1250.

The number of chromosomes usually present in the cells of higher plants and animals is known as the *diploid* number, and this number is generally constant in all cells. When reproductive cells, spores or gametes, are formed, the number of chromosomes is reduced to one-half the diploid number. Such cells are known as *haploid* cells. When male and female gametes come together, they fuse to form a single cell, the *zygote,* which has the diploid chromosome number. For instance, in man, the sperm cell and the egg cell, both of which are haploid, each contain 23 chromosomes. The chromosome number of the zygote is 46. Meiosis, the process in which the chromosome number of the organism is reduced to the haploid number, will be described in Chapter 7.

Some plant cells—certain root cells, for example—have more than twice the haploid number of chromosomes. Such cells are known as *polyploid cells,* and as we shall see in Chapter 8, some entire organisms are also polyploid. Haploid cells are conventionally designated as 1*n,* and diploid cells as 2*n.* Polyploid cells may be 3*n,* 4*n,* or some other multiple of *n.* (The *n* stands for the number of chromosomes in the sex cells of the particular organism.)

Often the only structures discernible within the nuclei with the light microscope are the approximately spherical structures known as *nucleoli* (singular: nucleolus).

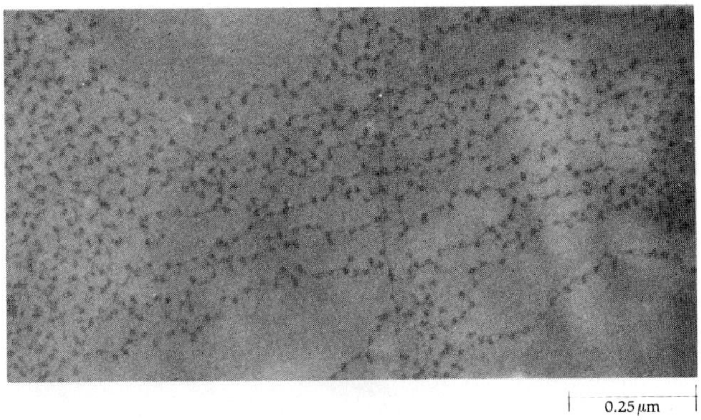

1–32
DNA-protein fibers streaming out of a chicken red blood cell nucleus. The distance between beads is about 14 nanometers, and the diameter of each bead is about 6.9 nanometers.

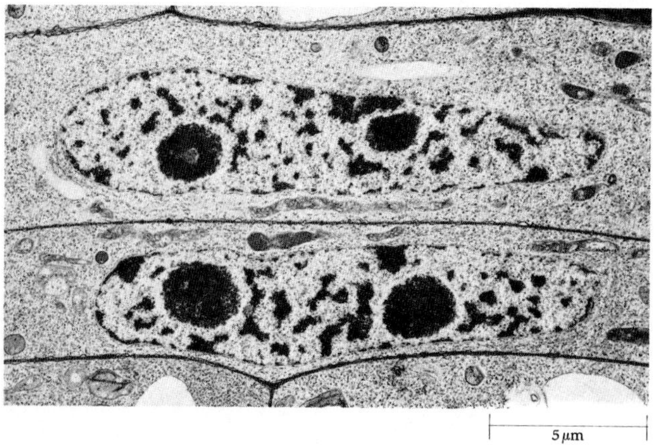

1–33
Parts of two immature cells from the vein of the corn (Zea mays) *leaf. Each nucleus contains two nucleoli.*

Nucleoli are present in interphase (nondividing) nuclei (Figure 1–33). They disappear during early stages of nuclear division and reappear during final stages of division. Normally there is one nucleolus to each haploid set of chromosomes. However, the nucleoli often coalesce and appear as one large structure in the interphase nucleus. Consisting mostly of protein, the nucleoli also contain about 5 percent RNA.

Table 1–2 summarizes the differences among the principal types of cells.

CELL DIVISION

In one-celled organisms, cells grow to a certain size by assimilating materials from the environment and synthesizing these materials into new structural and functional molecules. When such a cell reaches a certain size, it divides. The two daughter cells, each about half the size of the original mother cell, then begin growing again. In a one-celled organism, cell division may occur every day or even every few hours, producing a succession of organisms that are potentially immortal. In many-celled organisms, cell division is the means by which the organism grows. In all of these instances, the new cells produced are structurally and functionally similar both to the parent cell and to one another.

Cell division in eukaryotes consists of two overlapping stages, *mitosis* and *cytokinesis*. In mitosis, the nuclear membrane breaks down; the chromosomes, which have previously been duplicated, are divided equally; and two new nuclei are formed. In cytokinesis, the cytoplasm of the parent cell is divided into two parts, each containing one of the nuclei.

Table 1–2 *Comparison of Prokaryotic, Plant, and Animal Cells*

	PROKARYOTE	PLANT	ANIMAL
Cell membrane	Present	Present	Present
Cell wall	Present (noncellulose polysaccharide plus protein)	Present (cellulose)	Absent
Nucleus	No nuclear envelope	Surrounded by nuclear envelope	Surrounded by nuclear envelope
Chromosomes	Single, containing only DNA	Multiple, containing DNA and protein	Multiple, containing DNA and protein
Endoplasmic reticulum	Absent	Usually present	Usually present
Mitochondria	Absent	Present	Present
Plastids	Absent	Present in many cell types; chloroplasts in photosynthetic cells	Absent
Ribosomes	Present (smaller)	Present	Present
Golgi bodies	Absent	Present	Present
Lysosomes	Absent	Usually absent (as known in animal cells)	Often present
Vacuoles	Absent	Usually large single vacuole in mature cell	Small or absent
9 + 2 cilia or flagella	Absent	Absent (in higher plants)*	Often present
Centrioles	Absent	Absent (in higher plants)*	Present

* Except for a few, which have motile sperm.

At the end of cell division, the two new daughter cells will always contain exactly the same complement of genetic material. They will also usually be about the same size and have about the same numbers of the different organelles. The precise division of the genetic material is the result of mitosis; the division of the cytoplasm is the result of cytokinesis.

THE CELL CYCLE

Dividing cells pass through a regular sequence of events, known as the cell cycle. Completion of the cycle requires varying periods of time, depending on both the type of cell and external factors, such as temperature or available nutrients. Whether it lasts an hour or a day, however, the relative amount of time spent at each phase is about the same.

The S (synthesis) phase of the cell cycle is the period during which the genetic material (DNA) is duplicated. G (gap) phases precede and follow the S phase. The G_1 period occurs after mitosis and precedes the S phase; the G_2 period follows the S phase and occurs before mitosis. The G and S phases together are referred to as *interphase*.

The G_1 phase, between mitosis and chromosome synthesis, is principally a period of growth of the cytoplasmic material, including all the various organelles. Also, during this G_1 period, according to current hypothesis, substances are synthesized that either inhibit or stimulate the S phase and the rest of the cycle, thus determining whether or not cell division will occur. During the G_2 phase, structures involved directly with mitosis, such as the spindle fibers, are synthesized.

Some cells pass through successive cell cycles repeatedly. This group includes the one-celled organisms and certain cells in growth centers. Some specialized cells lose their capacity to replicate once they are mature. A third group of cells, such as those that form callus after a wound, retains the capacity to divide but does so only under special circumstances.

Mitosis

Mitosis, or nuclear division, is a continuous process that is conventionally divided into four major phases: *prophase*, *metaphase*, *anaphase*, and *telophase*.

At the start of mitosis, the chromosomes appear as long, slender threads. They become shorter and shorter and move to the center of the cell when they are maximally contracted. They then split longitudinally into two identical halves, which appear to be pulled to opposite poles of the cell by spindle fibers. The two groups of chromosomes are genetically equivalent. At the end of mitosis, the contracted chromosomes relax again, and they are reconstituted into the nuclei of the two daughter cells. (The major phases of mitosis are shown in Figure 1-36.)

1-34
The cell cycle. Dividing cells go through four principal phases, including the phases of mitosis and the S phase, during which the chromosomes are duplicated. Separating mitosis and the S phase are two G phases. The first of these (G_1) is a period of general growth and replication of cytoplasmic organelles. During the second (G_2), structures directly associated with mitosis, such as the spindle fibers, are synthesized. After the G_2 phase comes mitosis, which is, in turn, divided into four phases. Mitosis actually occupies only 5 to 10 percent of the cell cycle.

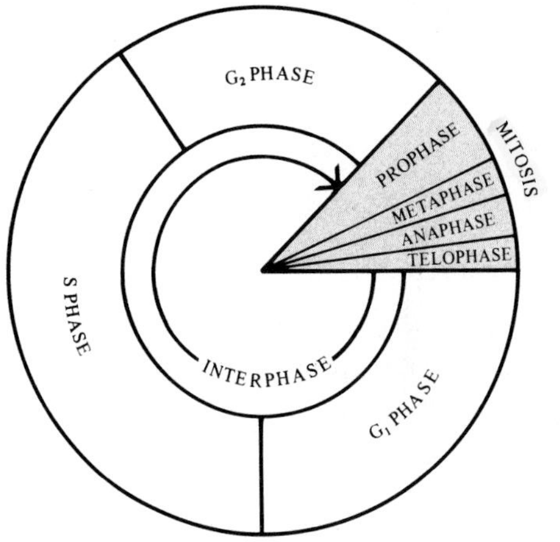

THE CELL LIFE CYCLE

1-35
Diagram of a chromosome at the beginning of mitosis. The chromosomal material has replicated during interphase so that each chromosome now consists of two identical parts, called chromatids. The chromatids are joined together at their centromeres.

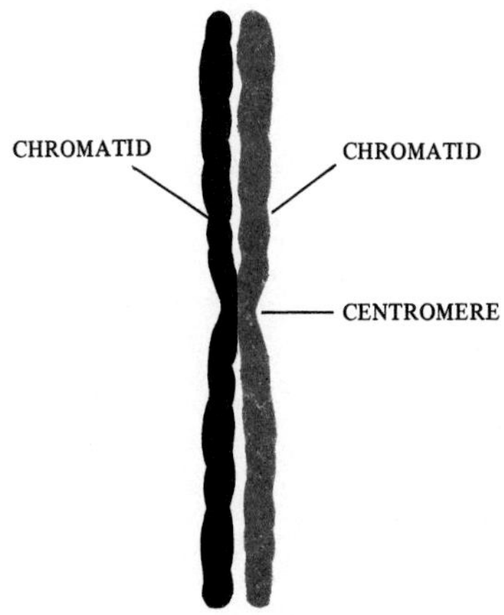

1-36
Mitosis, a diagrammatic representation with two pairs of chromosomes. During prophase, the chromosomes become visible as long double threads (pairs of chromatids); the threads shorten and thicken; finally, the nucleolus and nuclear membrane disappear.

The appearance of the spindle marks the beginning of metaphase, during which the chromosomes migrate to the equatorial plane of the spindle. At full metaphase (shown here) the centromeres of the chromosomes lie in the plane.

Anaphase begins as the centromeres divide and separate, providing each of the sister chromatids, which now become chromosomes, with a centromere. As shown here, the daughter chromosomes then move to opposite poles of the spindle.

Telophase—more or less the reverse of prophase—begins when the daughter chromosomes have completed their migration.

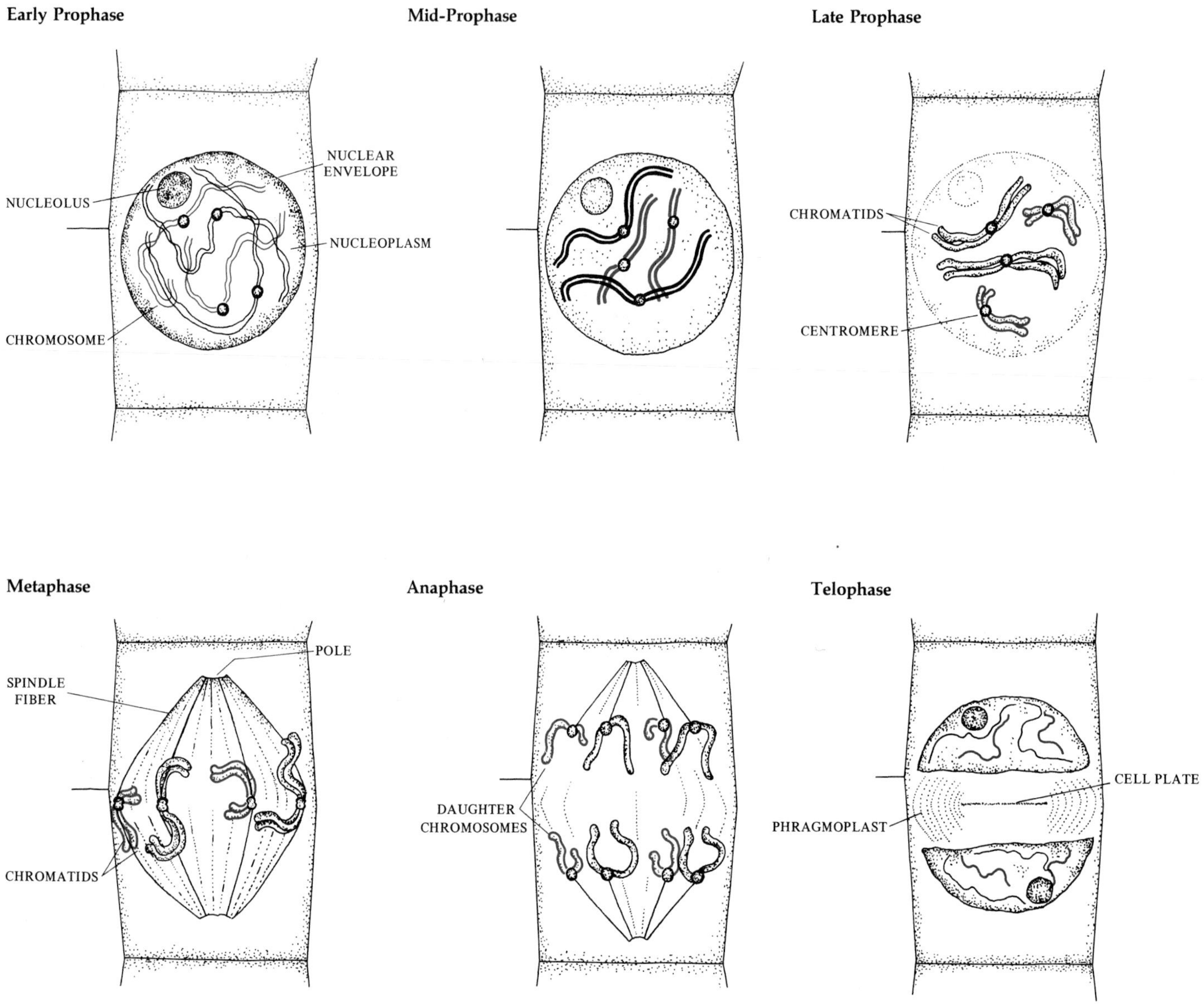

Interphase is the period between mitotic divisions. During interphase, the genetic material is duplicated in preparation for mitosis.

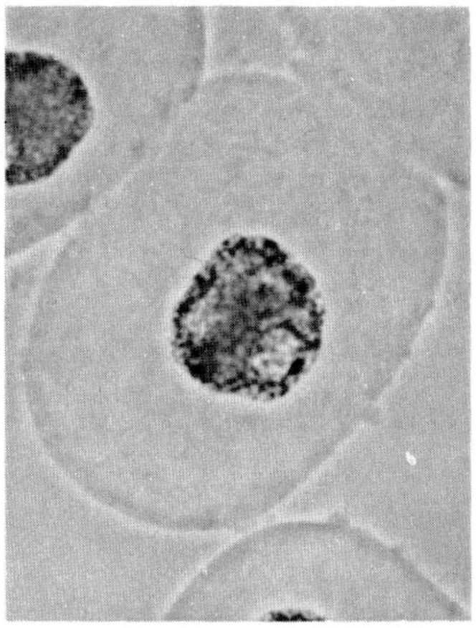

Early prophase. The genetic material is becoming condensed and the individual chromosomes are becoming visible.

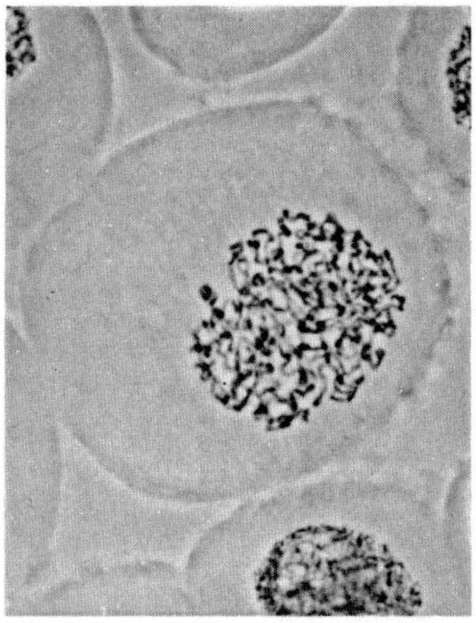

Late prophase. The chromosomes, or pairs of chromatids, have further condensed and are becoming more distinct.

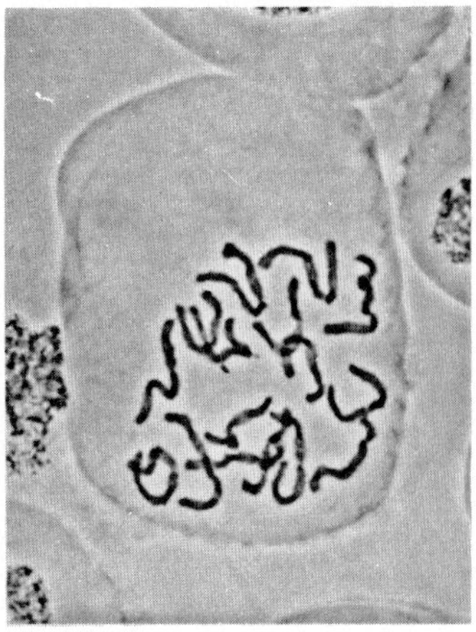

Metaphase. The chromosomes are becoming arranged in the equatorial plane of the cell. They appear to be guided by spindle fibers (not visible in these preparations) attached to the centromeres.

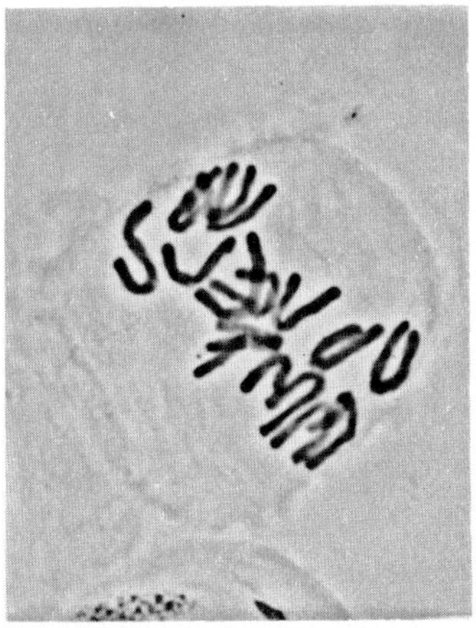

Early anaphase. The centromeres have begun to divide. Some of the sister chromatids are still held together at their tips.

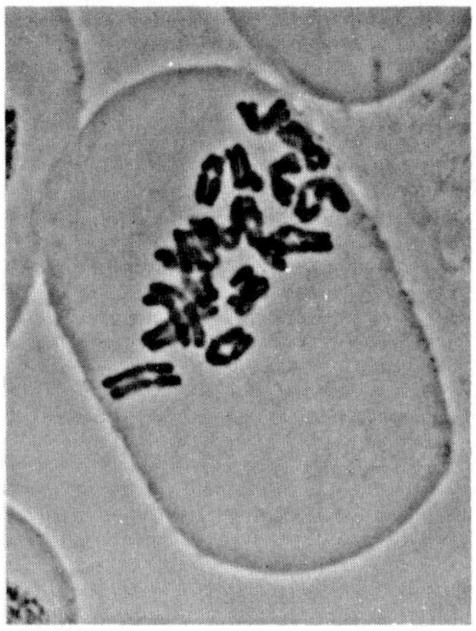

Late anaphase. The chromatids (new chromosomes) are now completely separated and are moving apart.

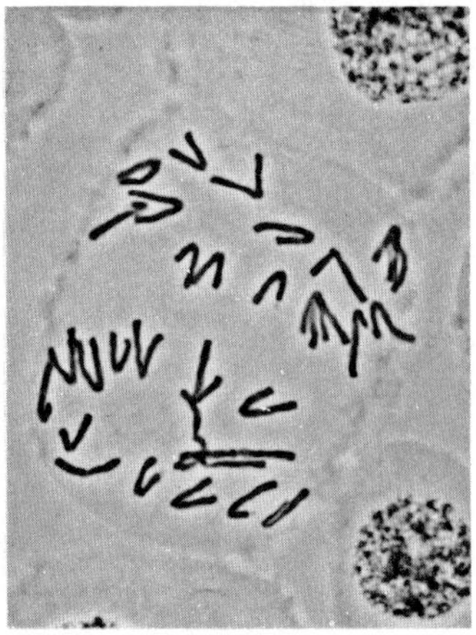

Early telophase. The daughter chromosomes have reached the opposite poles of the dividing cell.

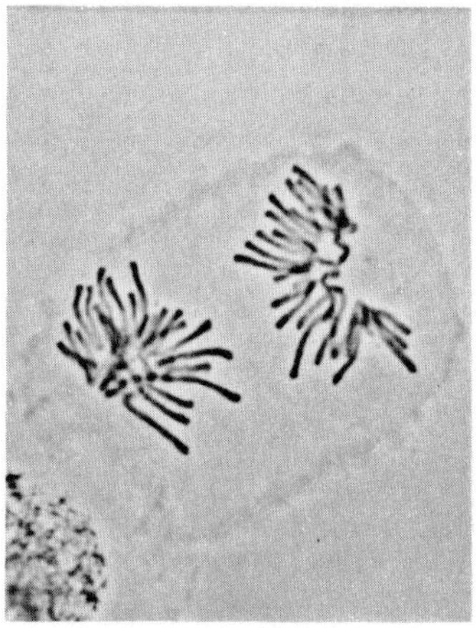

Late telophase. Mitosis is completed and the chromosomes are once more becoming diffuse.

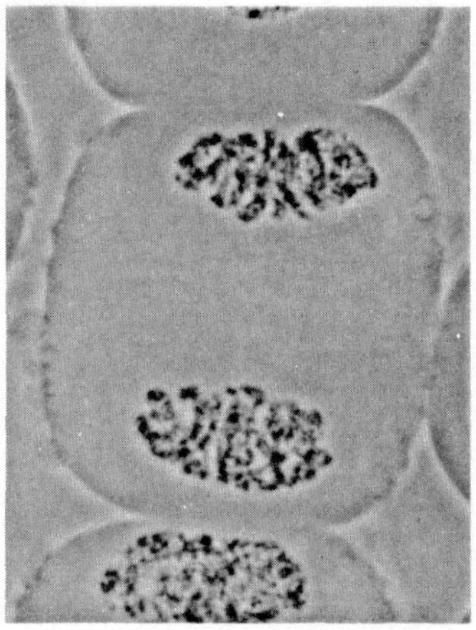

Interphase. A new cell wall is forming, completing the separation of the two daughter cells.

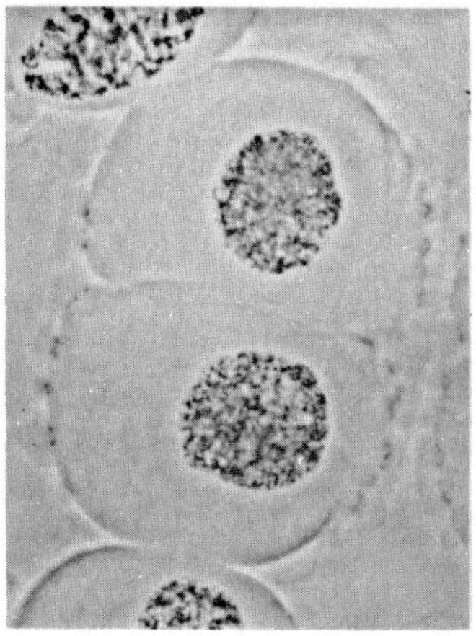

Before mitosis begins, the nucleus must assume the proper position within the cell for division to take place. Recent evidence indicates that microtubules may play a role in positioning of the nucleus. In many cells, microtubules form a ringlike band that outlines the equatorial plane of the future mitotic spindle.

Prophase

The beginning of mitosis is signaled by the chromosomes gradually becoming visible. They are first seen as elongated threads, which shorten and thicken. During this period, which is known as prophase, the chromosomes become more condensed, probably as a result of losing water, and also undergo contractions. As they become more distinct, it is possible to recognize that each chromosome is composed of two longitudinal halves. Each of the halves is called a *chromatid*, and the chromatids are joined in a narrowed area known as the *centromere* (also called the *kinetochore*), where the chromatids are not contracted (Figure 1–35). The centromere, which has its own characteristic location on each chromosome, divides the chromosome into two arms of varying length.

As prophase progresses, the nucleoli grow smaller and finally disappear. Shortly afterward, in most cell types, the nuclear envelope appears to break down, putting the contracted chromosomes into direct contact with the cytoplasm. The breakdown of the nuclear envelope marks the end of prophase (Figure 1–37).

Near the nucleus in many cells it is possible to detect a pair of organelles, the *centrioles*. Centrioles are identical to the basal bodies of flagella and cilia, and, like them, apparently have the ability to organize some of the protein molecules into long, slender microtubules similar to those that make up the 9-plus-2 structure of flagella. *Spindle fibers* contain bundles of microtubules, each about 20 nanometers in diameter (Figure 1–38). During the mitotic prophase, the two members of a pair of centrioles migrate to the opposite poles of the cell and then apparently organize the formation of the spindle fibers. There are two sets of fibers, each radiating from an opposite pole. Together they form a three-dimensional structure that is spindle-shaped—that is, it is thickest in the equatorial plane and tapering to a point at each end.

In motile cells, the same organelles often function as centrioles and then as basal bodies in organizing flagella

1–37
A late prophase nucleus of a parenchyma cell in the horsetail (Equisetum) *stem. The nuclear envelope is in the process of breaking down, marking the end of prophase.*

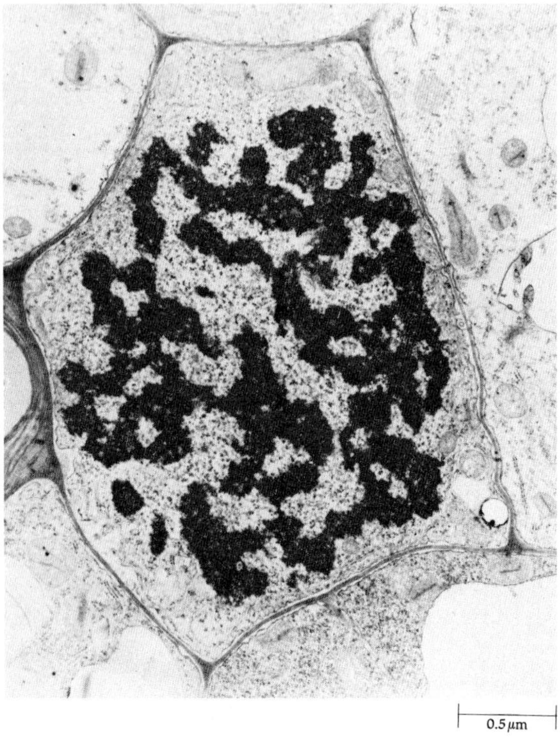

1–38
Some microtubules—components of the spindle fibers—in this dividing cell can be seen leading directly into the dense chromosomal material; others, like the one indicated by the arrow, pass continuously from pole to pole.

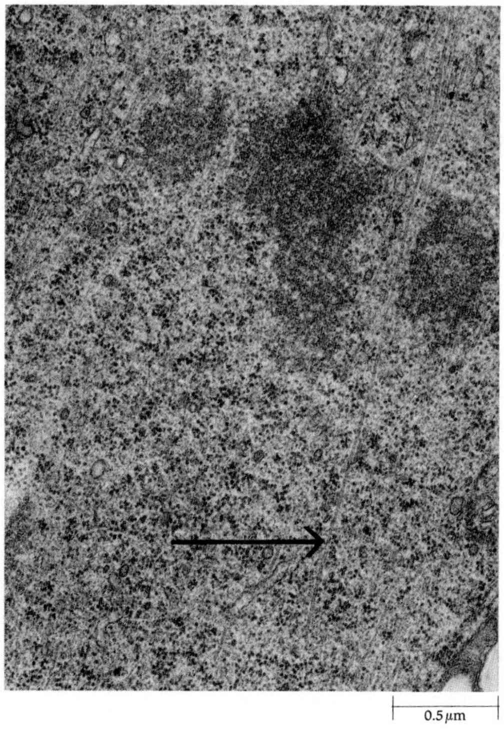

or cilia. Nonmotile cells of plants lack centrioles or similar structures, but in these cells the organization of spindle fibers proceeds just as it does in cells that do possess these organelles. Recently it has been shown that the polymerization of tubulin into microtubules can be initiated by the centromeres themselves. In special cases, where nonmotile land plants produce motile cells, as in the formation of sperm cells in a fern, centrioles appear in the cells, organize spindle fibers in the last mitotic division leading to the production of the sperm cells, and then function as basal bodies, organizing flagella or cilia.

Metaphase

Metaphase begins with the appearance of the spindle. During metaphase, the chromosomes, still double, become arranged so that each centromere lies on the equatorial plane of the spindle. Each chromosome appears to be attached to the spindle fibers by its centromere. Some of the spindle fibers pass from one pole to the other and have no chromosome attached to them.

When the chromosomes have all moved to the equatorial region, the cell has reached full metaphase. The chromosomes are now in position to separate.

Anaphase

During anaphase, the chromatids separate from one another; they are now called daughter chromosomes. First, the centromere divides and the two daughter chromosomes move away from one another toward opposite poles. Their centromeres, which still appear to be attached to the spindle fibers, move first, and the arms drag behind. The two daughter chromosomes pull apart, with the tips of the longer arms separating last.

The spindle fibers attached to the chromosomes shorten and cause the chromatids to divide and the daughter chromosomes to separate. The fibers appear to move, but, in fact, the microtubules are apparently continuously formed at one end of the spindle fiber and disassembled at the other. In the process, it appears as if the spindle fibers were tugging the daughter chromosomes toward the poles by their centromeres.

At the end of the anaphase, the two identical sets of chromosomes have been separated and moved to opposite poles.

1–39
Photomicrograph of dividing cells in the tip of an onion root. By comparing these to the drawings on the preceding pages, you should be able to identify the various phases of mitosis.

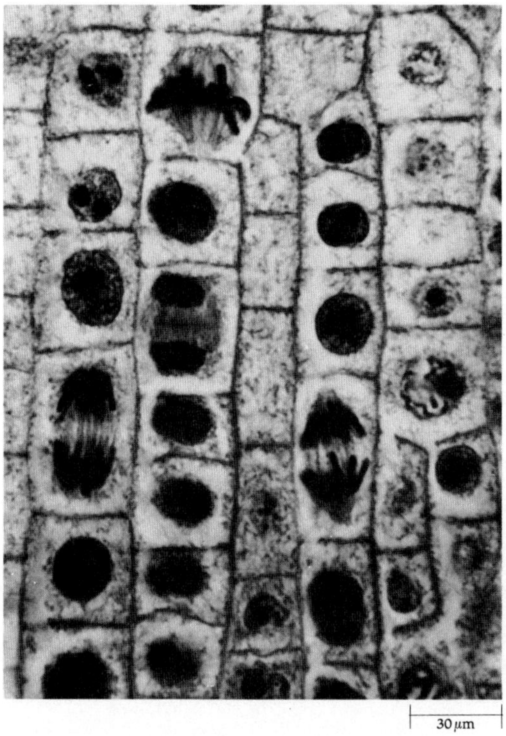

Telophase

During telophase, the separation of the two identical sets of chromosomes is made final as nuclear envelopes are organized around each of them. The membranes of these nuclear envelopes are derived from rough endoplasmic reticulum. The spindle apparatus disappears. Nucleoli also re-form at this time. They are attached to special areas known as nucleolar organizers, which are located on particular chromosomes. A cell may have two, four, or more nucleoli, but these often fuse into a single spherical mass if they come into contact. In the course of telophase, the chromosomes become increasingly indistinct, elongating to become slender threads again. When these processes are completed and the chromosomes have once more disappeared from view, mitosis is over and the two daughter nuclei have entered interphase.

During the process of mitosis, two daughter nuclei are produced that are identical to one another and to the nucleus that divided to produce them. This is important, for the nucleus, as we shall see in more detail in Chapter 3, is the control center of the cell. It contains coded instructions that specify the production of proteins, many of which mediate cellular processes by acting as enzymes and some of which serve directly as structural elements in the cell. This hereditary blueprint must be passed on exactly to the daughter cells, and its exact duplication is ensured in eukaryotic organisms by the organization of chromosomes and their division in the process of mitosis.

The duration of mitosis varies with the tissue and the

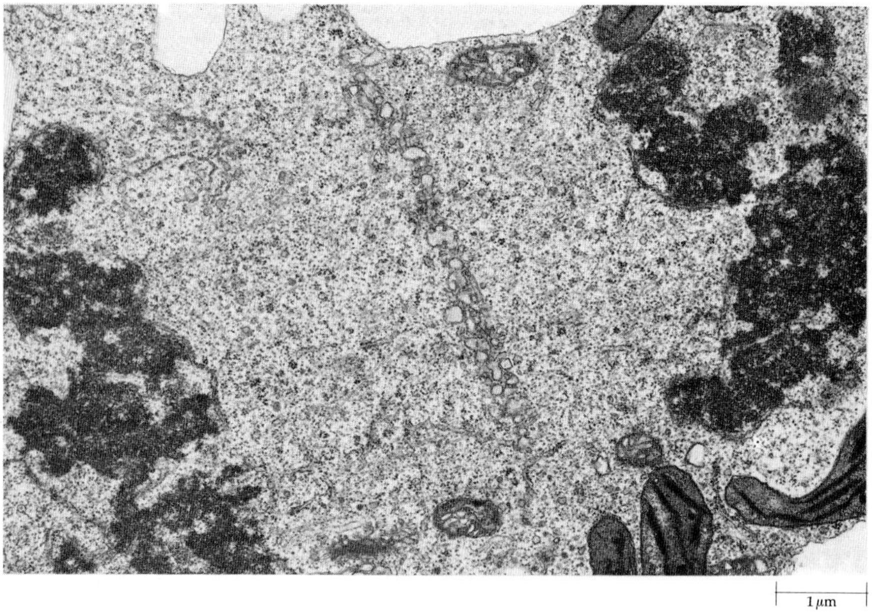

1-40
The chromosomes are separated and the cell plate is beginning to form, completing the separation of the two daughter cells.

organism, or species, involved. However, prophase is the longest phase and anaphase the shortest. In a root tip, the relative lengths of time for each of the four phases may be: prophase, 1 to 2 hours; metaphase, 5 to 15 minutes; anaphase, 2 to 10 minutes; and telophase, 10 to 30 minutes.

Cytokinesis

As we noted previously, cytokinesis is the process by which the cytoplasm is divided. In most organisms, cells divide by ingrowth of the cell wall, if present, and constriction of the plasma membrane, a process that cuts through the spindle fibers. In all plants (bryophytes and vascular plants) and a few algae, cell division takes place by the formation of a *cell plate* (Figure 1-40).

At early telophase, a spindle-shaped system of fibrils called the *phragmoplast* arises between the two daughter nuclei. The fibrils of the phragmoplast, like those of the mitotic spindle, are composed of microtubules. As seen with the light microscope, small droplets appear across the equatorial plane of the phragmoplast and gradually fuse, forming the cell plate, which grows outward until it reaches the wall of the dividing cell—completing the separation of the two daughter cells. With the electron microscope, the fusing droplets can be seen to be vesicles derived from Golgi bodies. The vesicles presumably contain pectic substances that form the middle lamella. Apparently plasmodesmata are formed at this time, as segments of endoplasmic reticulum are "caught" between fusing vesicular contents.

Following formation of the middle lamella, each protoplast deposits a primary wall next to the middle lamella. In addition, each daughter cell deposits a new wall layer around the entire protoplast in continuity with the wall at the cell plate. Enlargement of the daughter cells stretches and ruptures the original wall of the parent cell.

SUMMARY

All living matter is composed of cells. Cells are extremely varied in their appearance, ranging in structure and function from independent, single-celled organisms to the highly specialized interdependent cell types found in complex multicellular plants and animals. Cells are remarkably similar, however, in their basic structure. All cells are bounded by an outer membrane (known variously as the plasma membrane, plasmalemma, or cell membrane). The contents of the living cell consist of cytoplasm and nucleus. The cytoplasm includes molecules suspended and in solution and specialized structures, the organelles. The nucleus contains the hereditary information in the form of DNA.

Cells are also similar in their functions. The functions shared by cells are: (1) the capacity to extract energy from the environment and change it from one form to another; (2) the capacity to use this energy to build more organic molecules and so to maintain themselves and grow; (3) the capacity to selectively exchange materials with the environment; (4) the capacity to reproduce.

Cells are of two general types: prokaryotic and eukaryotic. *Escherichia coli,* a bacterium, is an example of a prokaryote. This organism consists of cytoplasm bounded by a membrane, the plasma membrane, and a cell wall. Membranes control the passage of substances in and out of cells and also provide surfaces for enzymatic activity. The cytoplasm of *E. coli* contains granules of stored food materials and a coiled, circular molecule of DNA, which codes the genetic information.

Also found in the cytoplasm of *E. coli* (and in both the nucleus and the cytoplasm of eukaryotes) are ribosomes, small particles formed of two almost spherical subunits, one larger than the other. They are the sites of the manufacture of protein.

The unicellular green alga *Chlamydomonas* is an example of a eukaryote. It consists of a wall and the protoplast. The protoplast includes the cytoplasm and the nucleus. Cytoplasm is constantly in motion, a phenomenon known as cytoplasmic streaming or cyclosis.

The cytoplasm of eukaryotic cells, like that of the *E. coli* cell, is delimited from the cell wall by a plasma membrane. Cell membranes in general have a three-ply structure and are about 8 nanometers thick. In eukaryotes, membranes not only control the passage of materials in and out of the cell but in and out of organelles and from one cellular compartment to another.

A membrane—the vacuolar membrane or tonoplast—separates the cytoplasm from the vacuoles. Vacuoles are highly characteristic of plant cells and are filled with cell sap, an aqueous solution of materials including a variety of sugars, mineral salts, and other substances. Young cells commonly contain numerous small vacuoles, which enlarge and coalesce as the cell matures. The enlargement of the vacuoles is the major cause of increase in size of the cell.

The cytoplasm contains an extensive system of membranes, the endoplasmic reticulum, a double-membrane structure. The endoplasmic reticulum has numerous connections with the nuclear envelope and often has numerous ribosomes attached to it. Ribosomes are also found free (not attached to endoplasmic reticulum) and, in eukaryotes, are present in both cytoplasm and nucleus.

Golgi bodies, or dictyosomes, are groups of flat, disk-shaped sacs (membranes), which bud off numerous vesicles. The Golgi bodies apparently serve as collection and packaging centers for complex carbohydrates and other substances, which are transported to the surface of the cell in the vesicles. The vesicles also serve as a source of plasma membrane material.

Intimate structural and functional associations exist among the various membranes and membrane systems of the cells.

Microtubules are elongate structures of variable length, polymers of the protein tubulin. They play a role in mitosis, cell-plate formation, the orderly growth of the cell wall, and the movement of flagella.

Flagella are hairlike organelles that project from the surface of some plant cells, serving as locomotor organelles. All flagella of eukaryotic cells have the same highly characteristic 9-plus-2 structure. Flagella grow out of basal bodies, minute cylinder-shaped structures that also have a highly characteristic arrangement of fibrils, which occur in a circle of nine triplets.

Together with vacuoles and cell walls, plastids are characteristic components of plant cells. Each plastid is bounded by an envelope consisting of two membranes. Mature plastids are classified on the basis of the kinds of pigments they contain: Chloroplasts contain chlorophylls and carotenoid pigments, chromoplasts contain carotenoid pigments, and leucoplasts are nonpigmented. Chloroplasts contain grana—stacks of disklike sacs called lamellae or thylakoids—in which the pigments are located. The various kinds of plastids develop from small, colorless bodies known as proplastids.

Like chloroplasts, mitochondria are bounded by two membranes, the inner one of which is folded to form an extensive inner membrane system. This membrane system increases the surface area available to enzymes and the reactions associated with them. Mitochondria are the principal sites of oxidation of organic energy-yielding molecules.

Microbodies and spherosomes are small organelles bounded by a single unit membrane. They contain groups of enzymes that have particular functions in the cell.

The cell wall is a major distinguishing feature of the plant cell. It determines the structure of the cell, the texture of plant tissues, and many important characteristics that distinguish plants as organisms. Cellulose is present in the cell wall of higher plants as rigid fibrils composed of a large number of cellulose molecules. The cellulose fibrils of the wall exist in a cross-link matrix of large, noncellulosic molecules, such as hemicelluloses and pectin. Lignin may also be present in all three wall layers, but is especially characteristic of secondary cell walls.

Almost all plant cells contain a primary wall, which is thin and generally plastic. Microtubules, which are found in the cytoplasm, often lying parallel to the cell wall, are thought to guide the laying down of cell-wall substance for the primary wall. Cellulose molecules and perhaps also the enzymes involved in putting them together seem to be carried to the cell wall in vesicles that bud off the Golgi apparatus.

Outside the primary wall is a middle lamella, composed largely of noncellulosic polysaccharides, which binds the primary wall of one cell to that of adjacent cells. The middle lamella is organized from the first elements of the cell plate that is laid down between dividing cells. In addition to a primary wall, some cells have a secondary wall, which is not readily stretched and provides mechanical support. Cells with secondary walls often die after the wall is laid down, leaving the cell to serve as a conduit for water or as a rigid supporting structure.

The integration of the plant body takes place by

means of the plasmodesmata, which pass from cell to cell. In the primary wall, plasmodesmata often are aggregated in thin areas known as primary pit-fields, and in living cells with secondary walls, they pass from cell to cell through the pit membranes of pits, areas where no secondary wall material is deposited.

The nucleus, which is the control center of the cell, is often the most prominent structure within the protoplast. It is surrounded by a nuclear envelope which is composed of two membranes. Within the envelope is the chromatin, which during nuclear division is organized into chromosomes. The chromatin and chromosomes consist of DNA and proteins.

Dividing cells pass through a cell cycle consisting of a G_1 phase of general growth and replication of organelles; an S phase, during which chromosomes are duplicated; a G_2 phase of synthesis of structure directly associated with mitosis; and mitosis. The G_1, S, and G_2 phases are collectively known as interphase.

When the cell is in interphase, the chromosomes are visible only as thin strands and grains, the chromatin. During mitosis, the chromosomal material condenses, and each chromosome can be seen to consist of two longitudinal halves, the chromatids, held together at the centromere (kinetochore). The centromere divides, and each chromatid—now called a daughter chromosome—moves to one of the daughter cells. In this way, the genetic material is equally distributed between the two new cells. Cell-plate formation begins during the telophase of mitosis, and this process results in the separation of the two daughter cells.

Centrioles, which are structurally identical to basal bodies, play a role in the mitosis of most animal and some plant cells, marking the poles of the spindles and apparently serving to organize the spindle fibers. Centrioles are usually found only in plant cells that either have flagella or are in the process of developing them, such as sperm cells. In such cells, centrioles may serve both in mitosis and as basal bodies in organizing the flagella.

Mitosis is generally followed by cytokinesis, the division of the cytoplasm. In many organisms, cytokinesis results from constrictions in the cytoplasm between the two nuclei. In plants and certain algae, the cytoplasm is divided by the coalescing of vesicles containing pectic substance and derived from the Golgi apparatus to form a cell plate. Once the cytoplasm is divided, the protoplasts lay down new cell walls.

CHAPTER 2

The Molecular Composition of Cells

Table 2-1 *Atomic Composition of Three Representative Species of Organisms** (Percentage by Weight)

ELEMENT	MAN	ALFALFA	BACTERIA
C	19.37	11.34	12.14
H	9.31	8.72	9.94
N	5.14	0.825	3.04
O	62.81	77.90	73.68
P	0.63	0.71	0.60
S	0.64	0.10	0.32
CHNOPS total:	97.90	99.60	99.72

* After Morowitz, H. J.: *Energy Flow in Biology*, Academic Press, New York, 1968.

As we noted in the previous chapter, despite their variety cells are remarkably similar in their basic structures. If we examine them at the molecular level, the similarities are even more striking.

The earth and its atmosphere contain about 90 naturally occurring different types of atoms, or elements. (The basic chemical principles that describe the behavior of these elements are reviewed in Appendix A.) Of all these available types of atoms, only a relatively few were selected in the course of evolution to form the complex, highly organized material of the cells of living organisms. In fact, as you can see in Table 2-1, about 99 percent (by weight) of living matter is composed of only six elements. These elements are found in cells chiefly in organic compounds and, in particular, in water, which makes up more than half of all living matter.

WATER

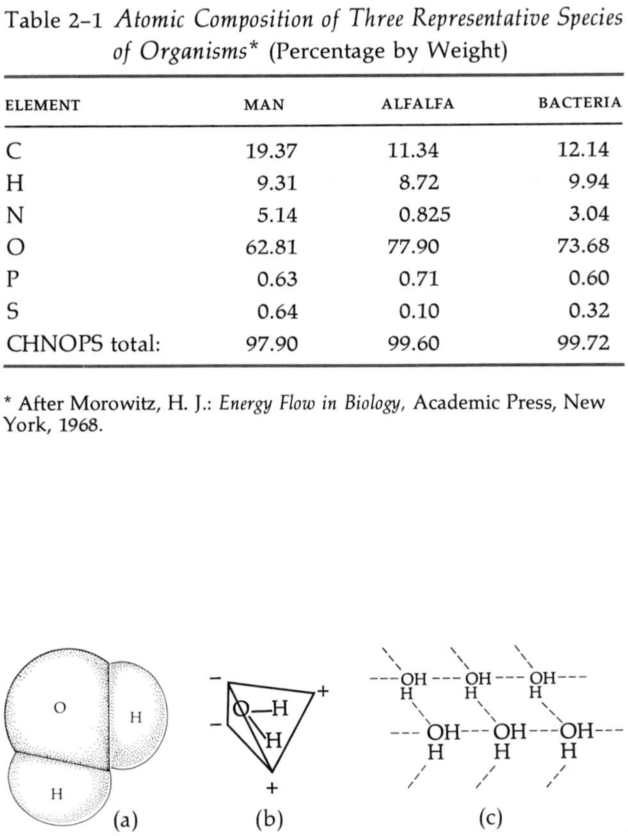

2-1
The water molecule is trilobed as shown in the space-filling model (a). *However, the single electrons of the hydrogen atom are shared with the oxygen atom and are located nearer the latter. As a consequence, the molecule has two slight negative charges and two slight positive charges, which make it more nearly tetrahedral* (b). *Thus each water molecule can bond to four others. The dashed lines indicate hydrogen bonds* (c).

Water is composed of a number of very small molecules held together by the mutual attraction of positively charged and negatively charged atoms. The water molecule itself has no net positive or negative charge. As you can see in Figure 2-1, the oxygen atom and the two hydrogen atoms form a triangle. The two hydrogen electrons are more strongly attracted by the oxygen atom than by the hydrogen nuclei, and so they tend to be located near the oxygen atom. As a result, the two hydrogen atoms carry local positive charges and the oxygen atom carries two local negative charges, although the molecule as a whole is electrically neutral.

When the positively charged hydrogen atom of the water molecule comes next to an atom carrying a sufficiently strong electronegative charge—such as the oxygen atom in another water molecule—the force of the attraction forms a bond between them, which is known as a hydrogen bond. Any single hydrogen bond has only an exceedingly short lifetime (about 10^{-11} second). All together, however, they have considerable strength. It is

2–2
Because of the polarity of water molecules, water can serve as a solvent for ions and other polar molecules. This diagram shows table salt (NaCl) dissolving in water as the water molecules cluster around the individual ions, separating them from each other.

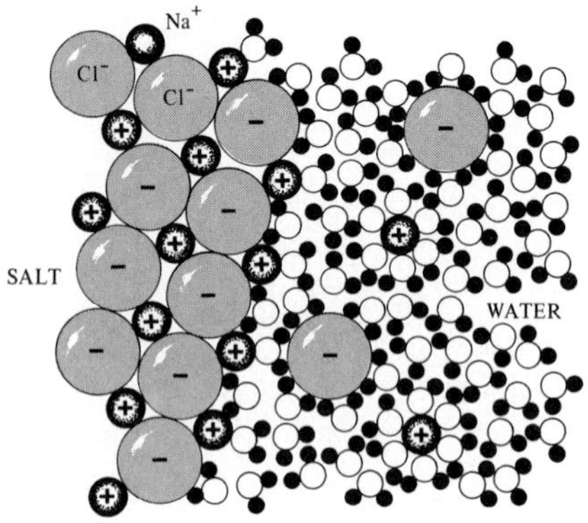

Table 2–2 *Some Important Classes of Organic Compounds*

ORGANIC COMPOUNDS	FUNCTIONS	COMPONENTS	COMPOSITION
Carbohydrates	Energy source, structural material, building blocks for other molecules	Simple sugars	Carbon, hydrogen, and oxygen
Fats (Lipids)	Energy storage, structural material	Fatty acids, glycerol	Carbon, hydrogen, and oxygen
Proteins	Structural materials, catalysts	Amino acids	Carbon, hydrogen, oxygen, nitrogen, and sulfur
Nucleic acids	Patterns for protein synthesis	Nitrogenous bases, sugars, and phosphates (nucleotides)	Carbon, hydrogen, oxygen, nitrogen, and phosphorus

because of these multiple, shifting hydrogen bonds that water shows strong intermolecular attractions and yet is not very viscous, an arrangement that is responsible for many of its unusual properties.

Many substances within living systems are found in solution. (A solution is a uniform mixture of two or more substances.) The polarity of the water molecules is responsible for water's role as a solvent. The charged water molecules tend to pull apart molecules such as NaCl (table salt) into their constituent ions. Then, as shown in Figure 2–2, the water molecules cluster around and so segregate the charged ions. Many of the molecules important in living systems also bear areas of positive and negative charge and so attract water molecules in the same way. Such molecules are said to be hydrophilic ("water-loving"). Molecules that lack local positive or negative charges and so do not attract water molecules are said to be hydrophobic ("water-fearing"). Because water molecules are attracted to one another, they tend to exclude these noncharged molecules, which then, as a result, associate with one another in what is known as hydrophobic interactions. The association of lipids in cell membranes (page 54) is an example of such an interaction.

More about the properties of water will be found in Chapter 25.

ORGANIC COMPOUNDS

Organic compounds, by definition, are compounds that contain both carbon and hydrogen, so the term includes, for example, many of the new synthetic fibers. In this book, however, we shall use "organic compounds" to refer to the relatively complex carbon-containing molecules found in living systems. Just as living materials are composed for the most part of relatively few types of atoms, these atoms are arranged to form a relatively few organic compounds that, in various combinations, make up most of the dry weight of living organisms. The four principal types of organic compounds are carbohydrates, lipids, proteins, and nucleic acids (Table 2–2).

2–3
Some biologically important monosaccharides. Five-carbon sugars (pentoses) and six-carbon sugars (hexoses) can exist in both chain and ring forms, as shown in (a). *The ring form is the form usually assumed in water. By convention, carbon atoms are not labeled in the ring forms. However, the position occupied by each carbon atom has a number. The numbering of the carbon atoms in a glucose molecule is shown in* (b).

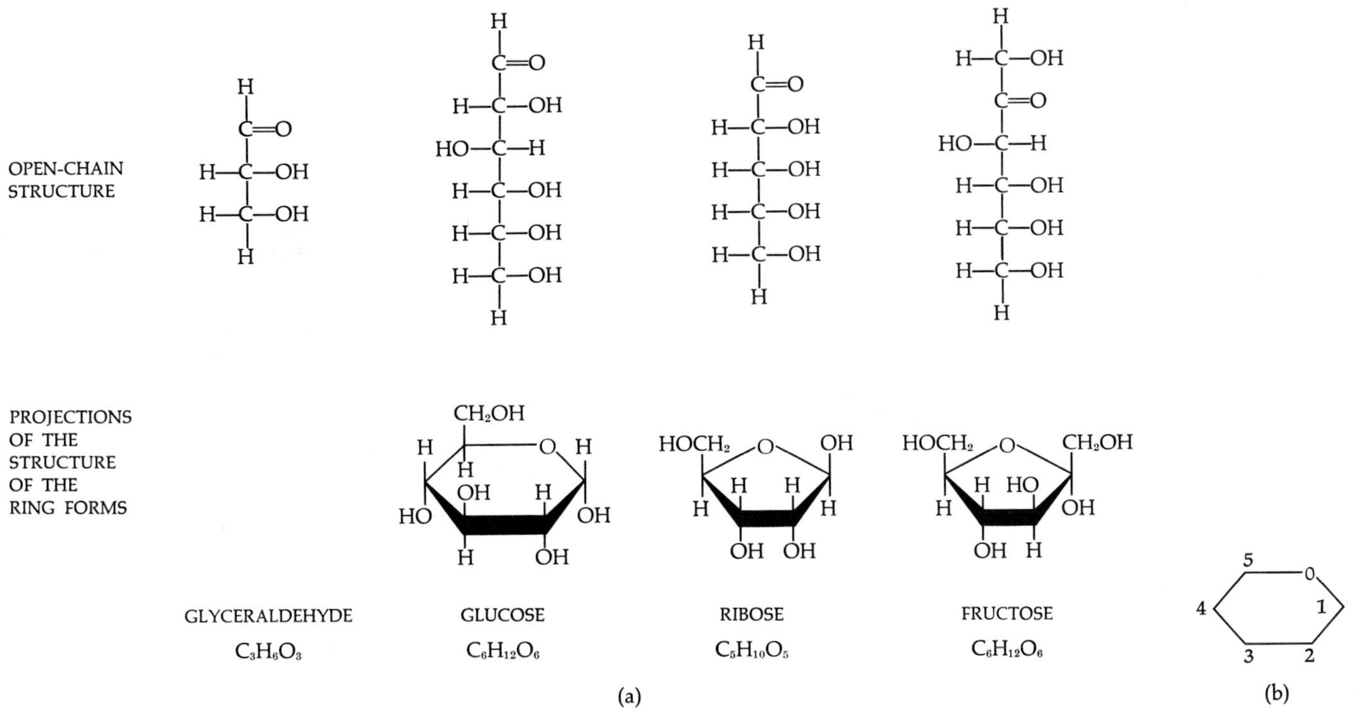

CARBOHYDRATES

Carbohydrates—sugars, starches, and related substances—are compounds that contain carbon combined with hydrogen and oxygen in the proportion of one carbon atom to two hydrogen atoms to one oxygen atom (CH_2O). The simplest carbohydrates are the *monosaccharides* (single sugars), or simple sugars, made up of a chain of carbon atoms to which H and O atoms are attached. Examples of some common monosaccharides are shown in Figure 2–3. As the figure indicates, the five- and six-carbon sugars can also exist in ring form; in fact, they are normally found in this form when they are dissolved in water. Two monosaccharides combine to form a *disaccharide*. As you can see in Figure 2–4, the union is accomplished by the removal of a molecule of water from the pair of monosaccharide molecules. Such joined molecules can be broken apart by *hydrolysis*, the addition of a molecule of water at each linkage, to form monosaccharide units again. Hydrolysis is an exergonic, or "downhill" reaction; the bonding energy of its prod-

2–4
The monosaccharides glucose and fructose combine to make the disaccharide sucrose. Sucrose is the form in which sugar is generally transported in plants. Bonds between monosaccharides are formed by the removal of a molecule of water. Formation of sucrose requires an energy input by the cell of 5.5 kilo calories per mole.

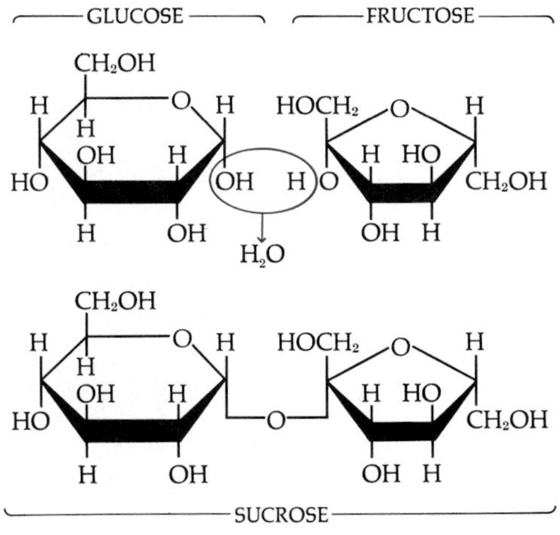

ucts is less than that of the original molecule. Conversely, linking two monosaccharides into a disaccharide requires an input of energy.

The monosaccharide glucose is the form in which sugar is most often transported through animal systems. A combination of two glucose units forms the disaccharide maltose (malt sugar), which is abundant in germinating grains; a combination of glucose and fructose forms sucrose (cane sugar), which is the form in which most sugar in plants is transported. Sucrose is our common table sugar. Lactose (milk sugar) is a disaccharide composed of glucose and galactose.

The ultimate source of sugar in virtually all cells is photosynthesis. In the process of photosynthesis, energy from the sun is converted to the chemical bond energy of the sugar molecule. When the sugar molecule is broken down again, the energy of the chemical bonds is released (Figure 2–5). The breakdown of 1 mole (see Appendix A) of glucose yields 686,000 calories. (A calorie is the amount of heat necessary to raise the temperature of 1 gram of water 1°C.)

Polysaccharides

In the cell, monosaccharides are linked together in long chains to form *polysaccharides* (many sugars). Polysaccharides, because of their large size, are insoluble and cannot pass through plasma membranes; they therefore constitute a storage form for sugar. *Starch*, which is built up of many glucose molecules (glucose "residues"), is the chief storage polysaccharide in higher plants, and glycogen is the common storage form for sugar in higher animals and in fungi (Figure 2–6). These carbohydrates must be hydrolyzed to monosaccharides before they can be used as energy sources or transported through living systems.

Polysaccharides are also important structural compounds. In plants, the principal structural polysaccharide is cellulose (Figure 2–7). Although cellulose is made up of the same building materials as starch, the arrangement of its long-chain molecules makes it rigid, and so its biological role is extremely different. Also, because the bonds linking the glucose units in cellulose are different from those in starch, cellulose is not readily hydrolyzed. As we mentioned in Chapter 1, in addition to cellulose, plant cell walls may contain a variety of other polysaccharides (Figure 2–8), pectic compounds (Figure 2–9), and lignin (Figure 2–10).

Once monosaccharides are incorporated into the plant cell wall in the form of cellulose, they are no longer available to the plant as an energy source. In fact, only some fungi, bacteria, and protozoa, and a very few higher animals (silverfish, for example) possess enzyme systems capable of breaking down cellulose. Other or-

2–5
Energy changes taking place during chemical reactions can be measured very precisely by means of a calorimeter. In the calorimeter, heat energy released by the reaction is transferred to the water jacket surrounding the flask and measured in calories.

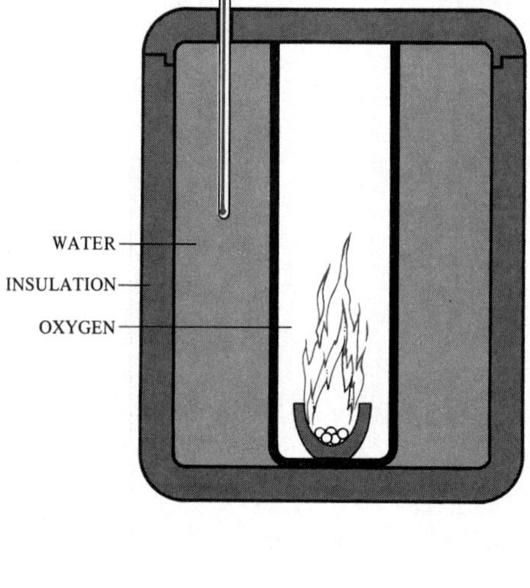

RADIOCARBON DATING

All organic materials contain carbon, and all of this carbon previously existed in the form of carbon dioxide and made its way into the living world by means of photosynthesis. Most of the carbon atoms present in carbon dioxide are carbon 12 (^{12}C), but a certain fixed proportion are carbon 14 (^{14}C), a radioactive isotope of carbon. Carbon 14 is produced as a result of bombardment by high-energy particles from outer space and occurs in small amounts as heavy carbon dioxide. Plant cells use carbon dioxide to make glucose and other organic molecules, accepting $^{14}CO_2$ as readily as $^{12}CO_2$. All animals are directly or indirectly dependent upon these molecules for food; thus a fixed proportion of carbon atoms in the tissues of all living things is radioactive carbon 14. After death, no more carbon is ingested, so the proportions shift, with the radioactive carbon 14 decaying slowly and the carbon 12 remaining the same. Carbon 14 has a half-life of 5730 years, so a fossil that old should contain just half the carbon 14 of a living plant. By measuring the ratio of ^{14}C to ^{12}C in a fossil or even in a man-made structure of wood or some other once-living material, the objects can be dated quite accurately. Carbon 14 dating is particularly useful for studying archeological remains, most of which lie within its time span. This dating method depends on the assumption that the proportion of ^{14}C to ^{12}C has remained constant in the atmosphere within the time span under study.

2–6
In plants, accumulated sugars are stored in the form of starch. Starch is composed of two different types of polysaccharides, amylose (a) *and amylopectin* (b). *A single molecule of amylose contains 1000 or more glucose units in a long, unbranched chain which winds to form a helix* (c). *Starch molecules, perhaps because of their helical nature, tend to cluster into granules. In* (d), *an electron micrograph of a portion of a tomato leaf cell, the large pale objects are starch grains and the dark layered organelles enclosing them are chloroplasts. A molecule of amylopectin is made up of about 48 to 60 glucose units arranged in shorter, branched chains.* (e) *and* (f) *show starch grains in amyloplasts (starch-forming plastids) of the potato tuber.* (e) *was photographed with ordinary light and* (f) *with polarized light.*

The common storage form for sugar in higher animals, fungi, and blue-green algae is glycogen, which resembles amylopectin in general structure except that each molecule contains only 16 to 24 glucose units.

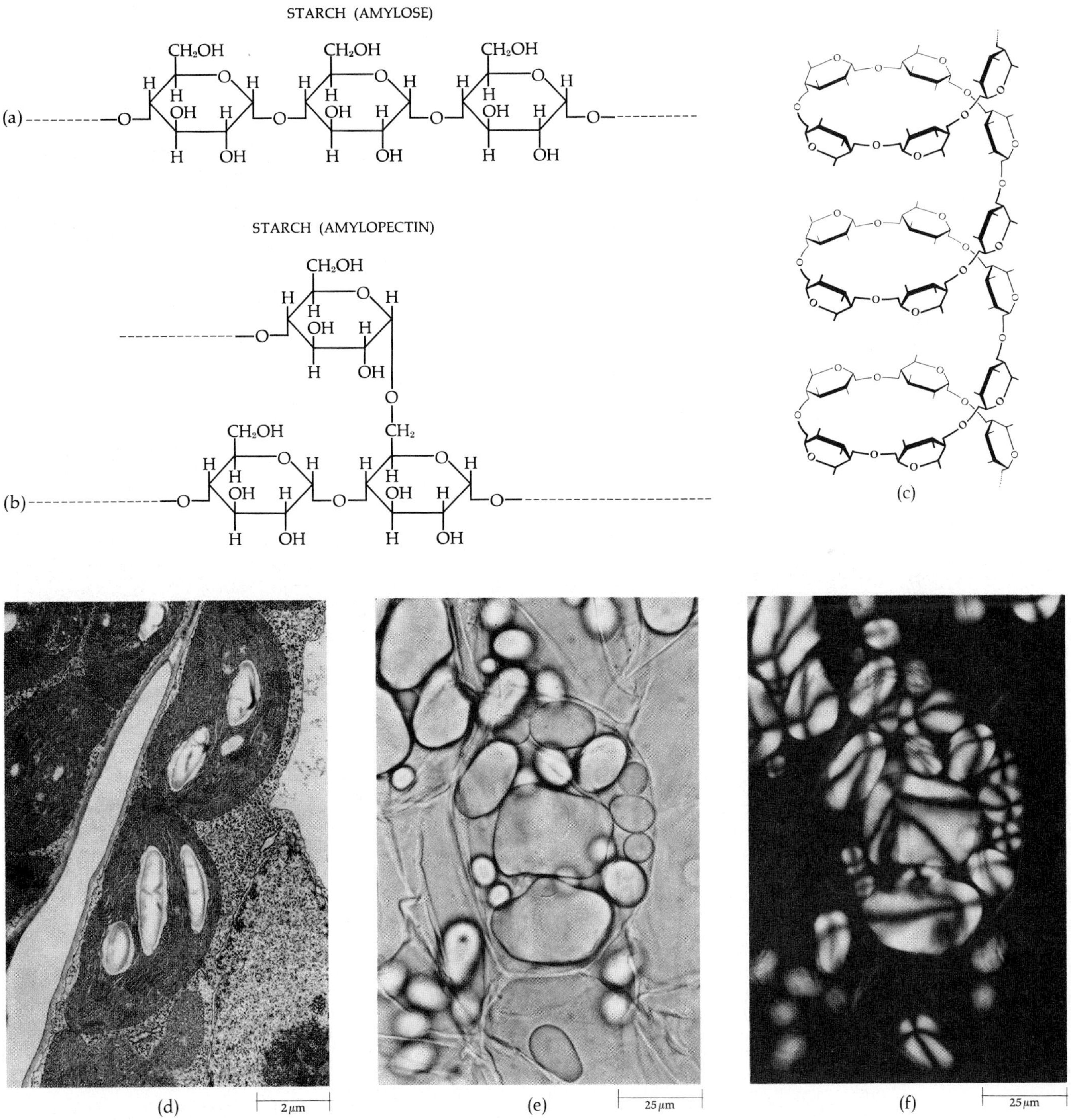

2–7
(a) *In living systems, alpha and beta forms of the ring-structured glucose molecule are in equilibrium. The molecules pass through the straight-chain form to get from one ring structure to another.* (b) *Cellulose consists entirely of beta-glucose units. The OH groups (in color), which project from both sides of the chain, form hydrogen bonds with neighboring OH groups, resulting in the formation of bundles of cross-linked parallel chains* (c). *In the starch molecule* (d), *composed of alpha-glucose units, most of the OH groups capable of forming cross-linkages project into the interior of the coiled molecule* (e).

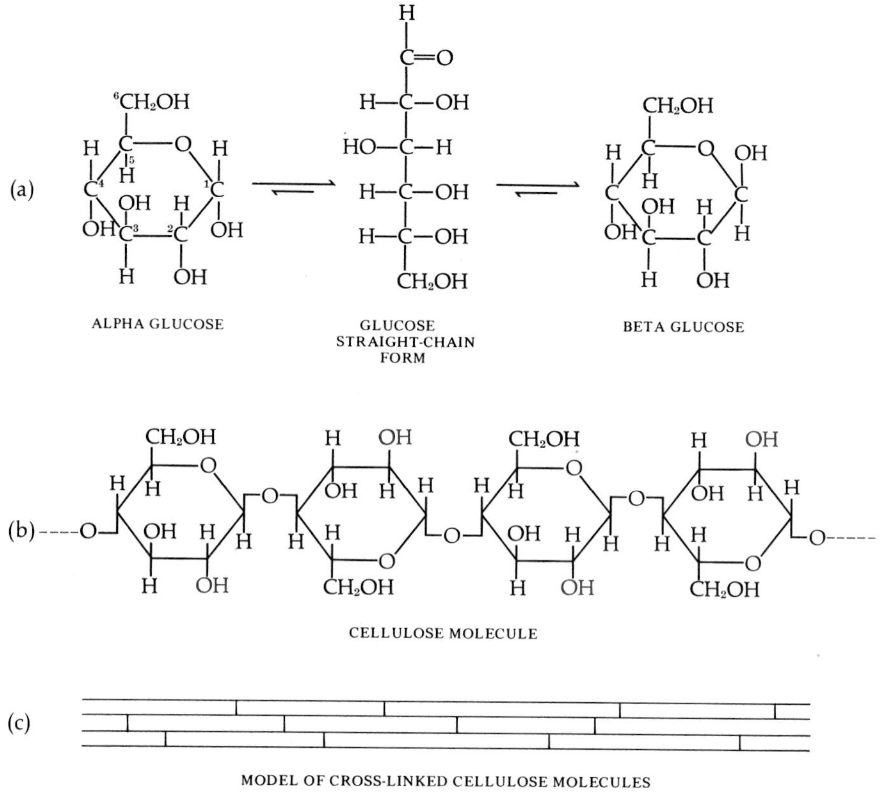

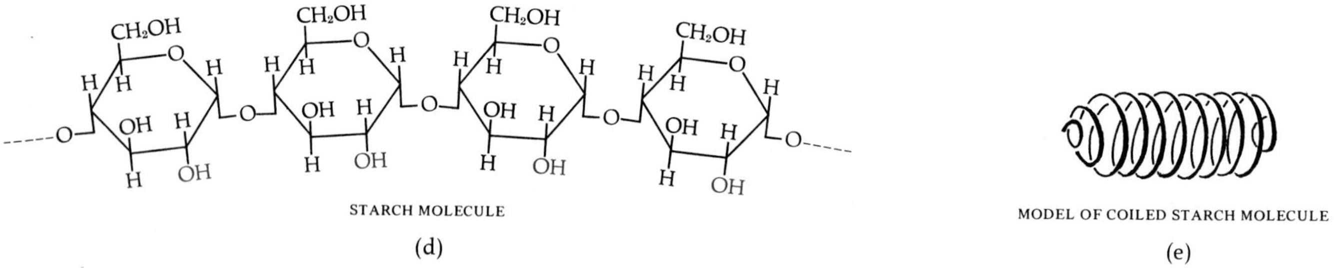

2–8
Building blocks of some polysaccharides other than cellulose that are found commonly in the cell walls of higher plants. These polysaccharides are called, respectively, galactans, mannans, xylans, and arabans, and all are widespread in the cell walls of higher plants. All four of these structures are drawn in the alpha-configuration.

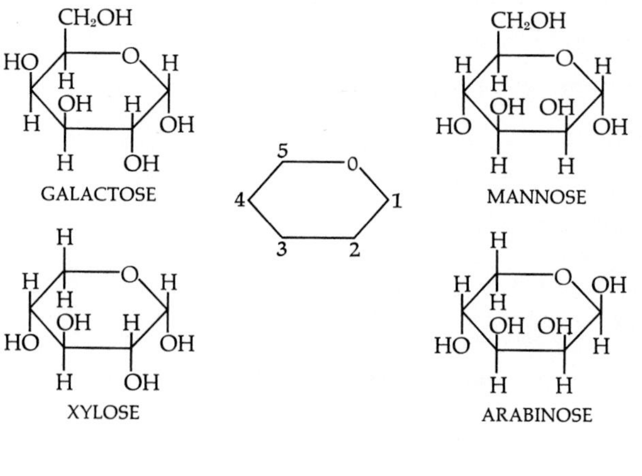

2-9
Pectic compounds are built up of residues of alpha-galacturonic acid (a), *which is a derivative of glucose. Pectic acid is the particular pectic compound shown in* (b). *Calcium and magnesium salts of pectic acid make up most of the middle lamella that binds adjacent plant cells together. In pectin and protopectin, different proportions of the hydrogen ions (shown in color) are replaced with methyl (CH_3—) groups. Protopectin is a common constituent of the cell wall and less soluble than pectin, which is commonly found dissolved in plant juices.*

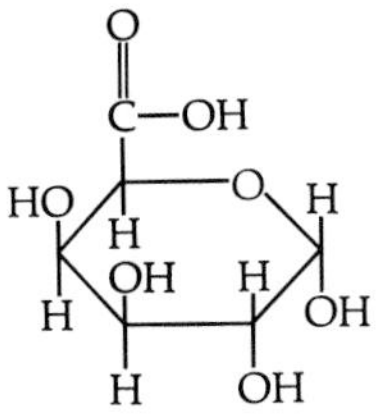
GALACTURONIC ACID
(a)

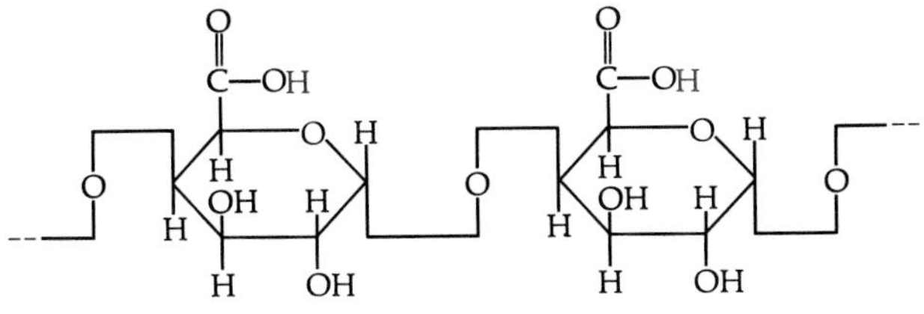
PECTIC ACID
(b)

2-10
One of the repeating units of lignin. These units are apparently put together in various ways to form the highly branched and complex molecules that make up lignin. The R and R′ groups vary in the massive branched molecules.

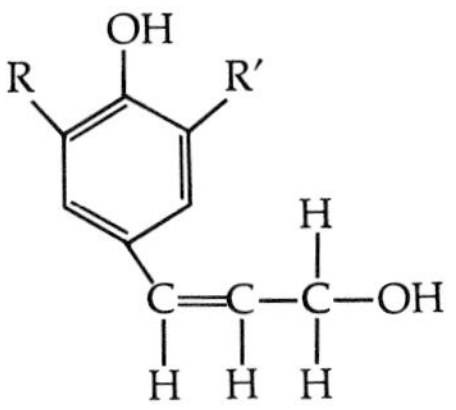

2-11
A fat molecule consists of three fatty acids joined to a glycerol molecule. As with the polysaccharides, these bonds are formed by the removal of a molecule of water. A fatty acid such as stearic acid, in which each carbon atom holds as many hydrogen atoms as possible, is known as saturated. A fatty acid such as oleic acid, in which any carbon atom holds only one instead of two hydrogens, is unsaturated. Unsaturated fatty acids are more common in plants than in animals.

ganisms, such as ruminants, termites, and cockroaches, are able to utilize cellulose as a source of energy only because of the microorganisms that inhabit their digestive tracts.

Another important structural polysaccharide is chitin, which is the principal structural component of fungal cell walls and also of the exoskeletons of insects and crustaceans.

LIPIDS

Lipids are fatty or oily substances. They have two principal distinguishing characteristics: (1) They are generally insoluble in water (although soluble in other lipids), and (2) they contain a large proportion of carbon–hydrogen bonds and, as a consequence, release a larger amount of energy in oxidation than other organic compounds. Fats, on the average, yield about twice as many calories as an equivalent amount of carbohydrates. These two characteristics determine their roles as structural materials and as energy reserves.

Fats

Fat is the principal form in which lipids are used for storage. Cells synthesize fats from sugars. A fat consists of three fatty acids joined to a glycerol molecule (Figure 2-11). Fatty acids are long hydrocarbon chains that carry a terminal carboxyl group, giving them the characteristics of a weak acid. The glycerol forms a link with the carboxyl group, releasing a molecule of water. (Like the polysaccharides and the proteins, fats are broken down by hydrolysis.) The glycerol thus serves as a binder or carrier for the fatty acids. The physical nature of the fat is determined by the chain lengths of the fatty acids and by whether the acids are saturated or unsaturated. In saturated fatty acids, all the carbon atoms hold as many hydrogen atoms as possible. Unsaturated fatty acids contain carbon atoms joined by double bonds; such carbon atoms are able to form additional bonds with other atoms (hence the term unsaturated). Unsaturated fats,

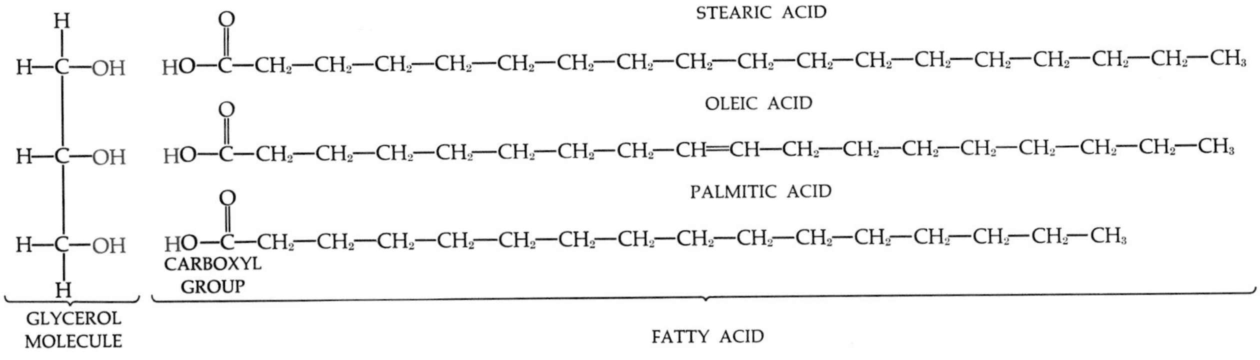

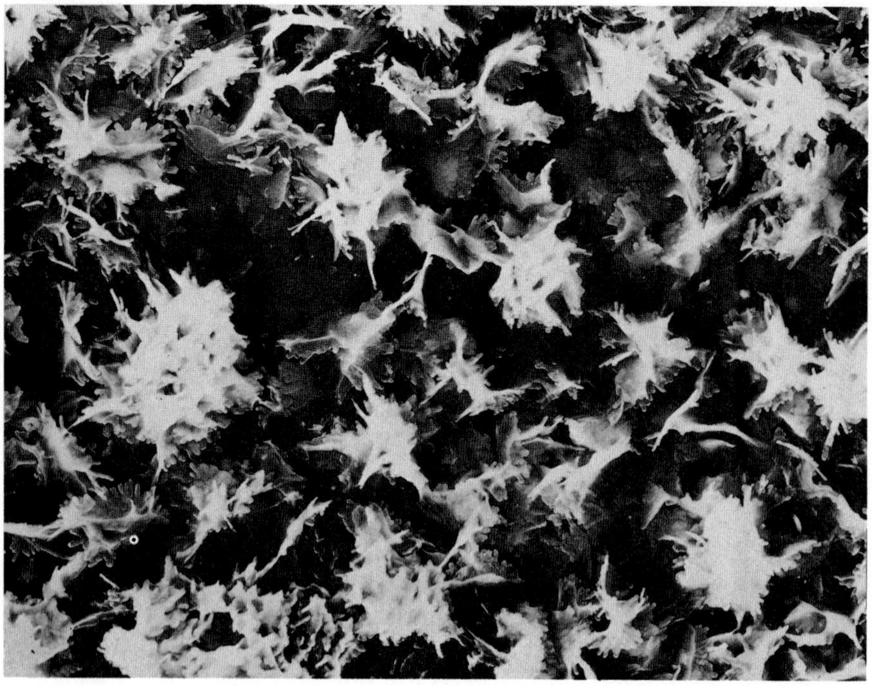

2–12
The upper surface of the leaf of Eucalyptus cloeziana *showing deposits of wax. Beneath these deposits is the cuticle, a wax-containing layer covering the outer walls of the epidermal cells. Biosynthesis of waxes, which protect exposed plant surfaces from water loss, is a property of all groups of vascular plants.*

which tend to be oily liquids, are more common in plants than in animals; examples are olive oil, peanut oil, and corn oil. Animal fats, such as lard, contain saturated fatty acids and usually have higher melting temperatures.

Waxes

Waxes are also made of fatty acids. They are formed by the union of long-chain alcohols (instead of glycerol) and fatty acid molecules. Plant waxes are important constituents of *cutin*, which covers the epidermis of leaves, stems, and fruits, and of *suberin*, the waterproofing material of the walls of cork cells (Figure 2–12).

Phospholipids

Closely related to the fats are the phospholipids—various compounds in which glycerol is attached to only two fatty acids (Figure 2–13), with the third space occupied by a molecule containing phosphorus. Phospholipids are very important in cellular structure, particularly in the membranes of cells.

The phosphate end of the phospholipid molecule is soluble in water, whereas the fatty acids are not. If phospholipids are added to water, they tend to form a film along its surface, with their polar heads under the water and the insoluble fatty acid chains protruding above the surface (Figure 2–14a). In the watery interior of the cell, phospholipids tend to align themselves in rows, with the insoluble fatty acids oriented toward one another and the phosphate ends directed outward (Figure 2–14b). Such configurations are important in the structure of cell membranes. The inner electron-transparent layer of the three-ply membrane characteristic of all cells consists of a double row of lipid molecules.

PROTEINS

Proteins, like polysaccharides, are *polymers*—large molecules made up of a number of similar molecular subunits, the *monomers*. In proteins, the molecular subunits are nitrogen-containing molecules known as *amino acids*. Only 20 different kinds of amino acids are generally found in proteins, but because protein molecules are large and complex, often containing several hundred amino acids, the number of different amino acid sequences and, therefore, the possible variety of protein molecules, is enormous. A single cell of the bacterium *Escherichia coli* has 600 to 800 different kinds of proteins at any one time, and the cell of a plant or animal will probably have several times that number.

Proteins are broadly classified into two types, fibrous and globular.

1. *Fibrous proteins* are the simpler of the two kinds of proteins. They contribute to the structural framework of the animal body and help to determine it, in much the same way that cellulose serves this function for plants. They form such structures as tendons, ligaments, and cartilage of the animal

2–13
A phospholipid molecule consists of two fatty acids linked to a glycerol molecule, as in a fat, and a phosphate group (indicated by colored screen) linked to the glycerol's third carbon. It also usually contains an additional chemical group, indicated by the letter R. The fatty acid tails are nonpolar (lacking positive and negative charges) and therefore insoluble in water; the phosphate and R groups are soluble.

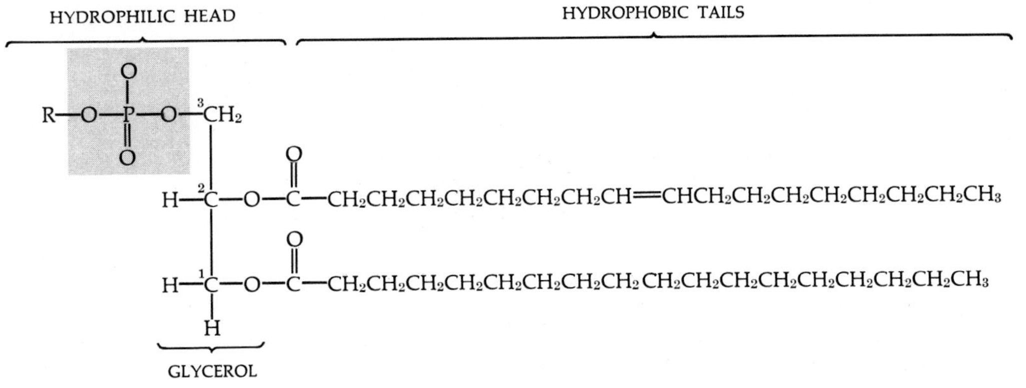

body. Fibrous proteins are often also called structural proteins.

2. *Globular proteins* have more subtle and complicated functions. They play important dynamic or metabolic roles in living systems. For example, enzymes, which are mostly globular proteins, serve as specific catalysts in the various chemical reactions of the body. The specialized proteins of cell membranes are globular.

In plants, the largest concentration of proteins is found in seeds, in which as much as 40 percent of dry weight may be protein. These proteins are not, strictly speaking, either fibrous or globular; they appear to function as storage forms of amino acids that will be needed by the developing plant embryo.

Amino Acids

Figure 2–15a shows the basic structure of an amino acid. It consists of an amino group and a carboxyl group bonded to a carbon atom, the so-called alpha-carbon. The R stands for the rest of the molecule, which varies in structure. It is this R group that determines the identity of any particular amino acid. Figure 2–15b shows the full structure of the 20 different kinds of amino acids found in proteins.

The amino acids are grouped according to their electrical charges. These charges are important in determining the properties of the various amino acids and, in particular, of the proteins formed from combinations of them.

2–14
The glycerol-phosphate combination is soluble in water, or hydrophilic, whereas the fatty acids are insoluble, or hydrophobic. As a consequence, when placed in water, the molecules tend to form a film on the water surface with their hydrophilic heads beneath the surface and their hydrophobic tails projecting above it (a). *In the cell, they tend to align themselves in rows with their soluble heads pointing outward* (b).

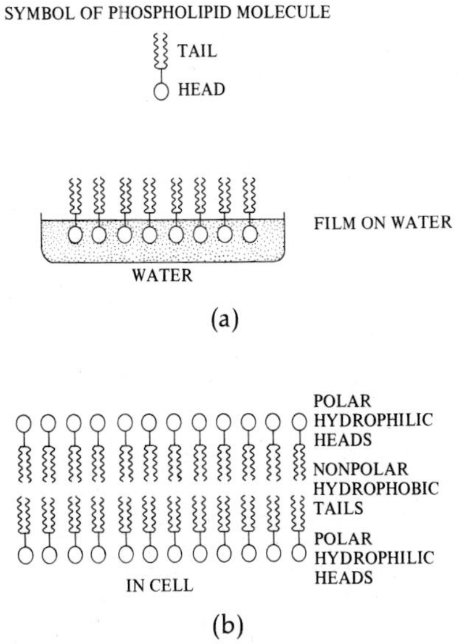

2–15

(a) *Every amino acid contains an amino group (NH_2) and a carboxyl group (COOH) bonded to a central carbon atom. A hydrogen atom and a side group are also bonded to the same carbon atom. This basic structure is the same in all amino acids. The R stands for the side group, which is different in each kind of amino acid.*

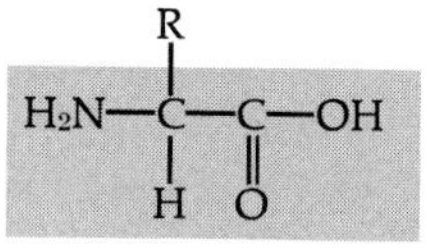

(b) *The 20 kinds of amino acids found in proteins. As you can see, their basic structures are the same, but they differ in their side groups. Because of differences in their side groups, amino acids may be nonpolar (with no difference in charge between one part of the molecule and another), polar but with the two charges balancing one another out so that the amino acid as a whole is uncharged, negatively charged (acidic), or positively charged (basic). The nonpolar molecules are not soluble in water, whereas the charged and polar molecules are.*

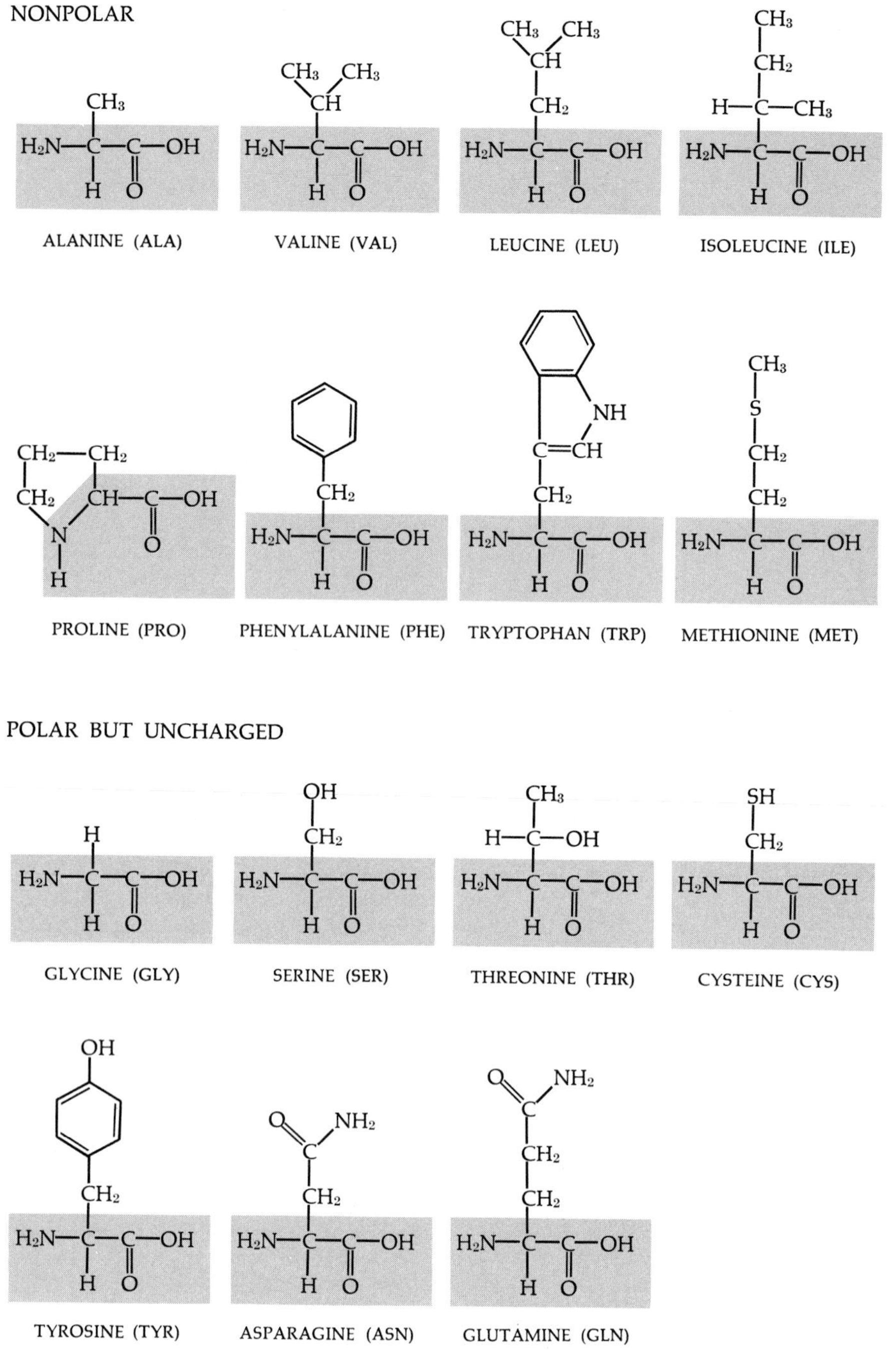

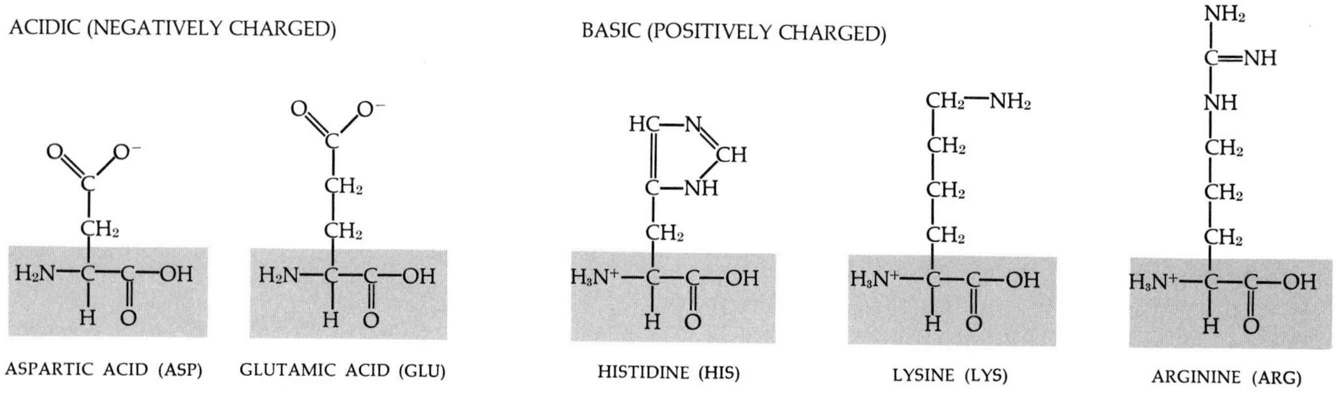

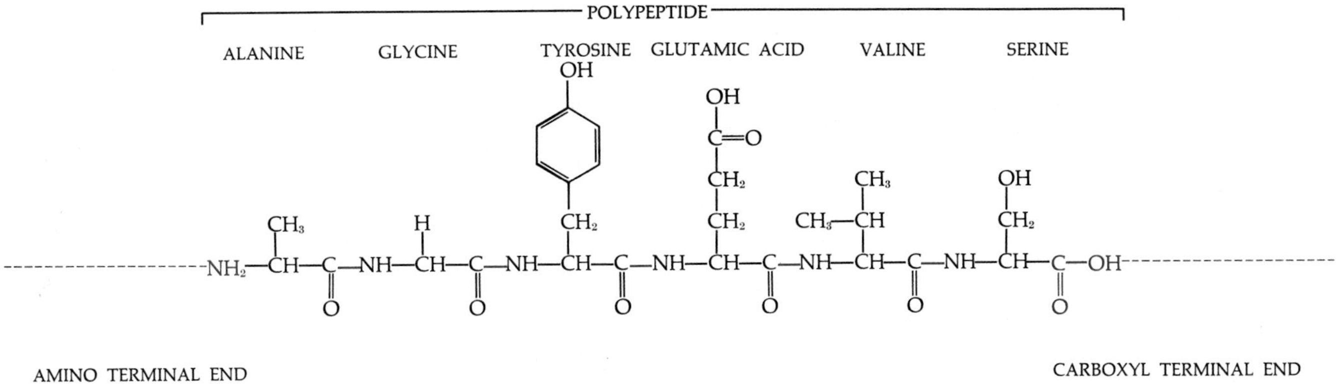

2–16
The links between amino acid residues are known as peptide bonds. As is the case with the bonds between sugar residues, peptide bonds are formed by the removal of a molecule of water. The bonds, shown here in color, always form between the carboxyl (COOH) group of one amino acid and the amino (NH_2) group of the next. As a consequence, the basic structure of a protein is always a long, unbranched molecule. This linear arrangement of the amino acids is known as the primary structure of the protein.

Polypeptides

A chain of amino acids is known as a *polypeptide*. As in the case of polysaccharides, a molecule of water is removed to create each link in the chain. The amino group of one amino acid always links to the carboxyl group of the next; this bond is known as the *peptide bond* (Figure 2–16). Amino acids joined by peptide bonds are often referred to as amino acid residues. Because of the way in which the amino acids are bonded together, there is always a free amino group at one end of the chain and a free carboxyl group at the other. The end with the amino group is called the N terminal and the one with the carboxyl group the C terminal.

Amino acids are also linked together by disulfide bonds, which are covalent bonds formed between two sulfur atoms of two cysteine residues. A disulfide bond may link two cysteine residues in the same polypeptide chain or in two different chains.

Proteins are large polypeptides. They have molecular weights ranging from 10^4 (10,000) to more than 10^5 (100,000)—in comparison with water, which has a molecular weight of 18, and glucose, which has a molecular weight of 180.* (The average weight of an amino acid residue is about 120; if you remember this figure, you can rapidly calculate the approximate number of amino acids in a protein on the basis of its total molecular weight.)

Levels of Protein Organization

The sequence of amino acids in the polypeptide chain is referred to as the *primary structure* of the protein. These chains often coil in a helix *(secondary structure)*, and the helix may be folded to produce a globular protein molecule *(tertiary structure)*. Finally, two or more proteins may combine to form the *quaternary structure*.

* The way in which molecular weights are determined is described in Appendix A.

Primary Structure

The primary structure of a protein is simply the linear sequence of amino acids in the polypeptide chain. Every different kind of protein has a different primary structure. The primary structure of one protein, the enzyme lysozyme, is shown in Figure 2–17.

When biochemists first began to demonstrate the complexity of the primary structure of proteins, speculative thinkers in the field of biology were quick to see that the amino acids, the number of which was so provocatively close to the number of letters in our own alphabet, could be arranged in a variety of different ways and that these different arrangements might account for both the great diversity and the great specificity of biochemical reactions within the cell.

Secondary Structure

In the cell, polypeptide chains do not lie out flat, as they usually appear in diagrams; they spontaneously assume regular coiled structures in three dimensions. This coiled arrangement is the secondary structure of a protein. The most common secondary structure is the alpha helix, which resembles a spiral staircase (Figure 2–18). The alpha helix is very uniform in its geometry, with a turn of the helix occurring every 3.6 amino acids. The helical structure of polypeptide chains, as Linus Pauling demonstrated, is maintained by hydrogen bonds across successive turns of the spiral, with the hydrogen atom in the amino group of one amino acid bonded to an oxygen atom of the carboxyl group of an amino acid in the next coil.

2–17
The primary structure of the enzyme lysozyme is shown in this two-dimensional model. Lysozyme is a protein containing 129 amino acid subunits, commonly called residues. These residues form a polypeptide chain that is cross-linked at four places by disulfide (—S—S—) bonds. (From The Three-Dimensional Structure of an Enzyme Molecule *by David C. Phillips. Copyright © 1966 by Scientific American, Inc. All rights reserved.)*

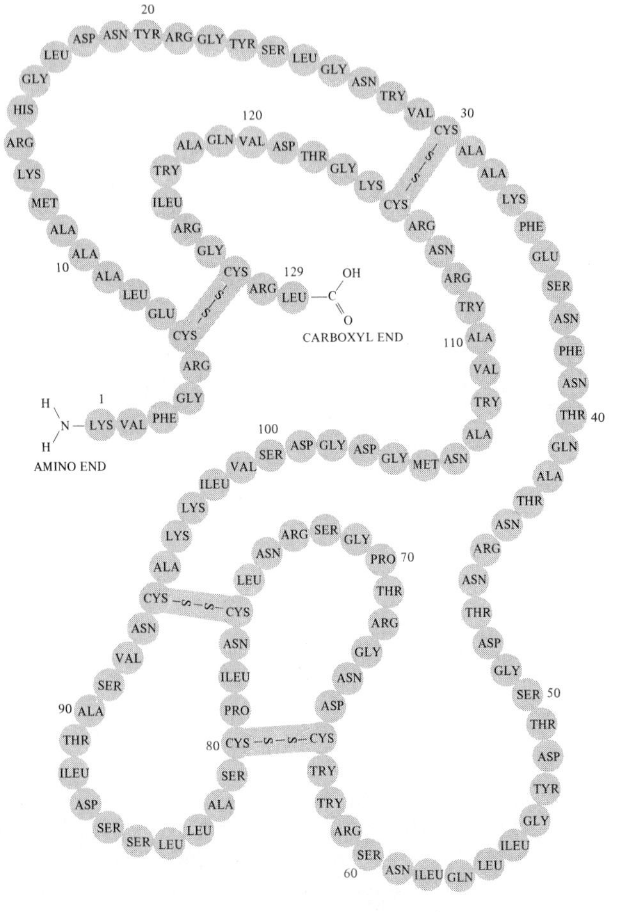

2–18
The alpha helix, the most common secondary structure of proteins. The configuration is very regular in its geometry. In this figure, only the carbon and nitrogen atoms of the "backbone" of the amino acid residues are shown. A turn occurs every 3.6 residues. The helix is held in shape by hydrogen bonds that form between the amino acids, linking oxygen and hydrogen atoms at regular intervals.

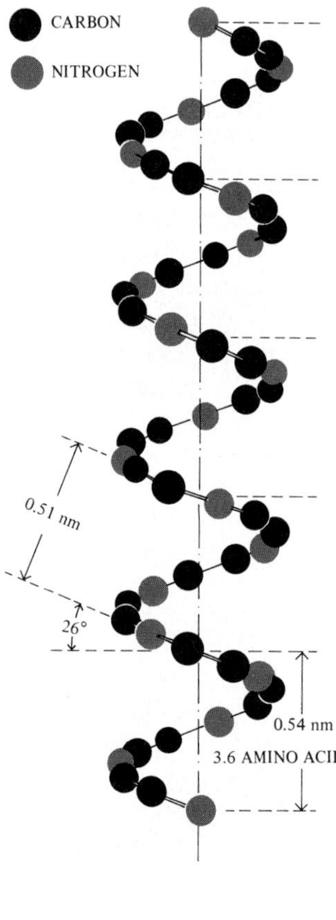

Tertiary and Quaternary Structure

Polypeptide chains may also fold up to form globular structures, which are the tertiary structure of protein. Figure 2–19 shows the folding of the main chain in the enzyme lysozyme. Most biologically active proteins, such as enzymes and hormones, are globular.

The tertiary structure is determined by the primary structure and is maintained largely by hydrogen bonds and by interactions among the various amino acid residues based on their various charges (see Figure 2–15) and between the amino acid residues and water molecules. These bonds are relatively weak and can be broken quite easily by physical or chemical changes in the environment, such as heat or increased acidity. This breakdown is called denaturation. When proteins are

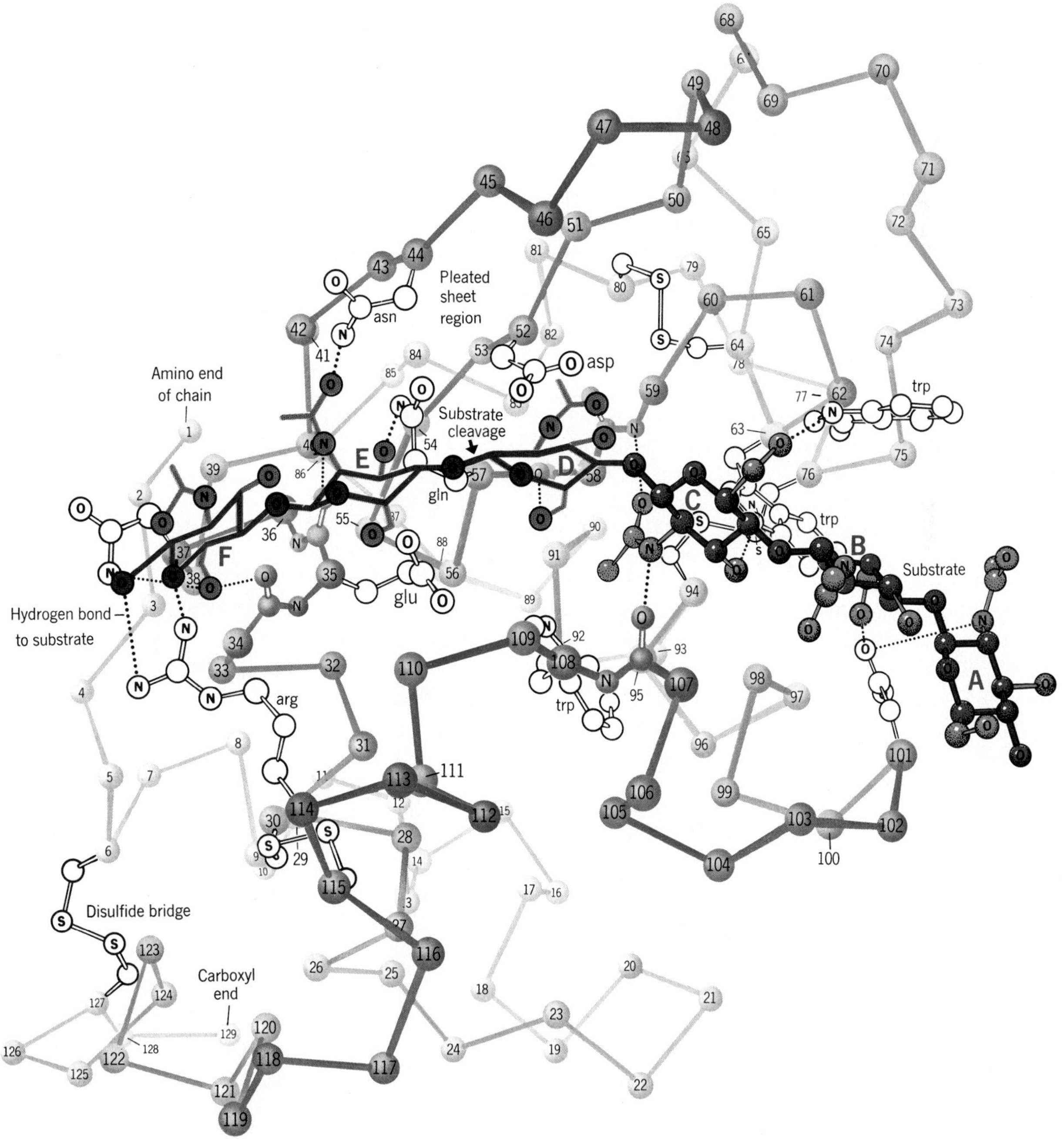

2–19
From this drawing you can gain a rough idea of the complex nature of the precise folding and coiling that make up the tertiary structure of a polypeptide. Again, it is the main chain of the enzyme lysozyme that is shown. The crevice that forms the active site (see following page) runs horizontally across the molecule. The substrate is shown in a darker color. (From the Structure and Action of Proteins *by R. E. Dickerson and I. Geis. W. A. Benjamin, Menlo Park, Calif., Publisher. Copyright 1969 by Dickerson and Geis.)*

denatured, the polypeptide chains are unfolded, causing a loss of the biological activity of the protein. Organisms cannot live at extremely high temperatures because their enzymes and other proteins become unstable and nonfunctional.

The quaternary structure results from specific interactions between two or more polypeptide chains. These chains may or may not be identical. Many, probably most, enzymes consist of two or more polypeptide chains.

Enzymes

Enzymes are globular proteins that act as catalysts. Catalysts are substances that accelerate the rate of a chemical reaction, remaining unchanged in the process. Because they remain unaltered, catalysts may be used over and over again and so they are typically effective in very small amounts.

In the laboratory, the rates of chemical reactions are usually accelerated by the application of heat, which increases the force and frequency of collisions among molecules. But in a cell in nature, hundreds of different reactions are going on at the same time, and heat would speed up all these reactions indiscriminately. Moreover, heat would melt the lipids, denature the proteins, and have other generally destructive effects on the cell. Because of catalytic enzymes, cells are able to carry out chemical reactions at great speed and at comparatively low temperatures. (If enzymes were not present, the reactions would take place anyway, but at a rate so slow that their effects would be negligible.) Enzymes speed the rate of reaction by bringing various molecules in alignment with one another (see Figure 2–21). A single enzyme molecule may catalyze as many as several thousand reactions per second.

The chemical, or chemicals, upon which an enzyme acts is known as its *substrate*. (In the reaction diagrammed in Figure 2–21, sucrose is the substrate.) The site on the surface of the globular enzyme molecule into which the substrate fits is the *active site*; usually there is only one active site in each enzyme molecule. Only a few amino acid residues are involved in any particular active site. Some of these may be adjacent to one another in the primary structure, but often they are brought into proximity to one another by the intricate foldings of the chains involved in the tertiary structure. In the enzyme-substrate complex, the substrate fits the active site as a key fits into a lock. The remarkable specificity of enzymes is due to this geometrical relationship of chemical groups. An enzyme will only accept a molecule that has a complementary fit.

Enzymes are generally named by adding the suffix *-ase* to the root of the name of the substrate. Amylase cat-

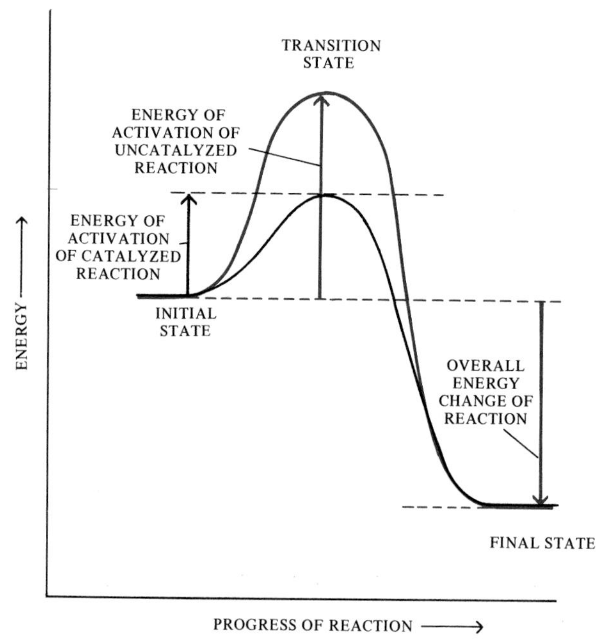

2–20
Chemical reactions require an input of energy to get them started—analogous to the spark which lights a fire or the push which sends the boulder on its downhill course. An uncatalyzed reaction requires more activation ("input") energy than a catalyzed one, such as an enzymatic reaction. Note, however, that the overall energy change from the initial state to the final state is the same.

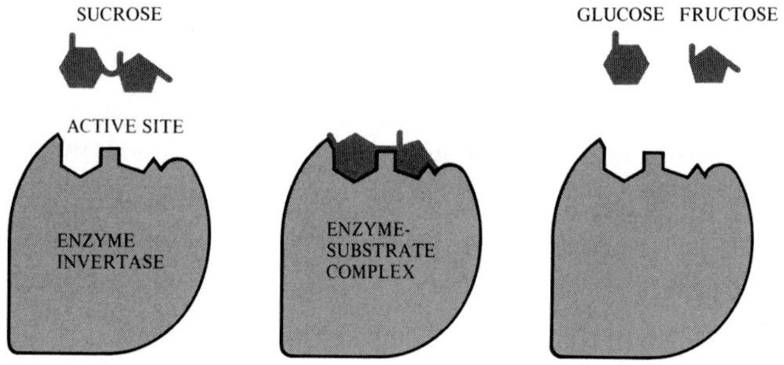

2–21
The enzyme sucrase (invertase) hydrolyzes sucrose to yield a molecule of glucose and a molecule of fructose. As emphasized in the diagram, the closeness of the fit of the enzyme and its substrate determines the biological activity of the enzyme.

alyzes the hydrolysis of amylose (starch), and sucrase catalyzes the hydrolysis of sucrose into glucose and fructose. Over 1000 enzymes are now known, each of them capable of catalyzing some specific chemical reaction.

Protein Shape and Biological Control

The chemical forces holding a protein into its particular tertiary structure are very weak ones. The binding of a substrate molecule to an enzyme often results in changes in the enzyme's three-dimensional shape, the change helping to promote the catalytic process.

The ability of proteins to bend flexibly from one shape to another when binding external compounds is a fundamental property of enzymes and explains much of what we know about how enzyme activities are regulated and coordinated within the cell. Some enzymes are active only when their conformation is changed by the binding to the enzyme of a small activating regulatory molecule (Figure 2–22). Much of the metabolism in the cell is coordinated by means of such regulatory signals. For example, in the human body, ingested glucose is normally converted to glucose-6-phosphate, which either can be oxidized immediately to produce energy in the form of molecules of ATP or stored in the form of glycogen, a chain of sugar molecules. Which pathway is chosen is determined by the concentration of regulator molecules in the blood and their consequent effects upon the enzymes involved in these pathways.

Proteins, Lipids, and Cellular Membranes

Lipids make up roughly half, by weight, of cellular membranes. The other 50 percent of the membrane is protein. The three-ply structure characteristic of all cellular membranes is sometimes described as a "butter sandwich," with proteins and the phosphate ends of phospholipid molecules comprising the outer, electron dense layers—the bread—and the insoluble fatty acid chains of the phospholipids the butter. A few of the protein molecules extend across the lipid bilayer com-

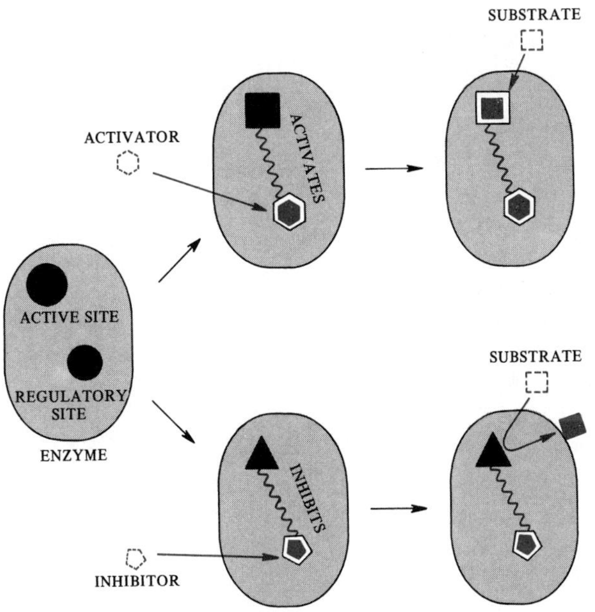

2–22
Simplified scheme for illustrating the mechanism by which enzyme activity is controlled by the induced conformational changes brought about by small regulatory molecules that are not themselves involved in the primary enzymatic reaction. The complex polypeptide chain that comprises the actual enzyme is represented in this convention by the light gray area; the black circles symbolize the shapes of the enzyme's active site and regulatory site in the absence of any bound molecules. The binding of the activator molecule (colored hexagon) *induces a conformational change at the regulatory site* (symbolized by the change from a circle to a hexagon) *and also alters the enzyme's structure* (symbolized by the colored zigzag line), *which in turn changes the shape of the active site* (symbolized by the change from a circle to a square). *As a result the substrate* (colored square) *can now bind more easily at the active site. In the reverse process the inhibitor molecule* (colored pentagon) *induces a different type of conformational change in the regulatory site* (symbolized by the change from a circle to a pentagon), *which is transmitted through the enzyme's structure, so that the resulting alteration at the active site* (symbolized by the change from a circle to a triangle) *makes the active site repel the substrate. (From* Protein Shape and Biological Control *by Daniel E. Koshland Jr. Copyright © 1973 by Scientific American, Inc. All rights reserved.)*

pletely, either singly or in pairs (see Figure 2–23), others lie at or near either membrane surface, and others penetrate the membrane a short distance. The lipid bilayer is about 4.5 nanometers thick.

The differences in membranes from cell to cell and from organism to organism have to do mainly with the characteristics of the numerous proteins incorporated in them. These include many enzymes that control the reactions taking place on the "working surface" of the membranes. Recently, electron micrographs have been prepared showing structures that have been interpreted as assemblies of enzymes engaged in particular functions. Protein molecules are also involved with the transport of substances through membranes, which will be discussed in Chapter 26. Certain components are apt to be concentrated in different areas of the membrane, depending on the position of the cell in the tissue and the functions in which it is involved. Most of the protein molecules of the membrane are associated with the inner membrane layer.

Because of the lipid component of the membrane, it is largely fluid, with the consistency of a light oil. Proteins that extend through this layer are immobilized, as are the lipids bound to them; but when such constraints do not prevent it, both lipid and protein molecules are free to move around within this oily film in response to physiological conditions.

NUCLEOTIDES AND THEIR DERIVATIVES

Nucleotides are other major building blocks used by living systems. They are complex molecules. Each is made up of three subunits: (1) a phosphate group, (2) a five-carbon (pentose) sugar, and (3) a nitrogen base, so-called because its ring structure contains nitrogen as well as carbon. Nucleotides contain two types of nitrogen bases: *pyrimidines,* which have a single ring, and *purines,* which have two fused rings (Figure 2–24).

Nucleic Acids

Nucleic acids are polymers of nucleotides. They are of two types, DNA and RNA. Their structure and their role in the cell will be discussed in the next chapter. DNA is the molecule in which the genetic information is stored. RNA serves as a translator and transmitter of this genetic information. It is through RNA that the DNA dictates the structure of proteins and thus the structure and function of the cell.

The three pyrimidines found in nucleic acids are thymine, cytosine, and uracil. DNA contains thymine and cytosine; RNA contains cytosine and uracil. The two purines are adenine and guanine; RNA and DNA both contain these purines. DNA and RNA also contain slightly different types of sugar, deoxyribose and ribose, respectively. "Deoxy" means "minus one oxygen," and as you can see in Figure 2–24, this is the only difference between these two pentose sugars.

The way in which these molecules are put together, the significance of their arrangement, and how this information was discovered are the subjects of the next chapter.

Energy-Exchange Molecules

Nucleotides and compounds derived from them serve a variety of functions within the cell. One nucleotide derivative of prime importance is adenosine triphosphate, or ATP. Molecules of ATP are present in all living cells and participate in many biochemical reactions. As you can see in Figure 2–25, ATP is composed of adenine, plus a five-carbon sugar (ribose), plus three phosphates.

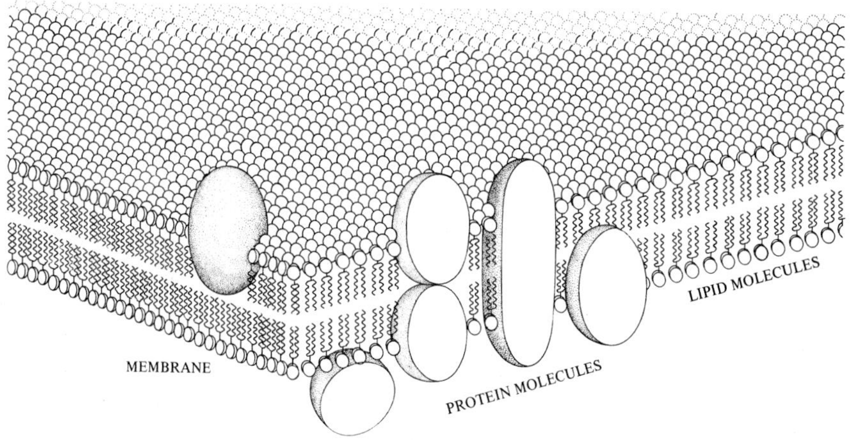

2–23
The arrangement of phospholipids and proteins characteristic of cellular membranes, with some kinds of protein near the surface, especially the inner surface, of the membranes, and others extending through it. (From A Dynamic Model of Cell Membranes *by Roderick A. Capaldi. Copyright © 1974 by Scientific American, Inc. All rights reserved.)*

2–24
The building blocks of RNA and DNA. Each nucleotide building block contains a phosphate group, a sugar, and a purine or pyrimidine. The purines in both groups are the same, but one type of DNA nucleotide contains thymine whereas its RNA counterpart contains uracil. The only other difference between the two is the presence of one more oxygen atom on the sugar (ribose) component of the RNA. As you will see in Chapter 3, however, the biological roles of the two are profoundly different.

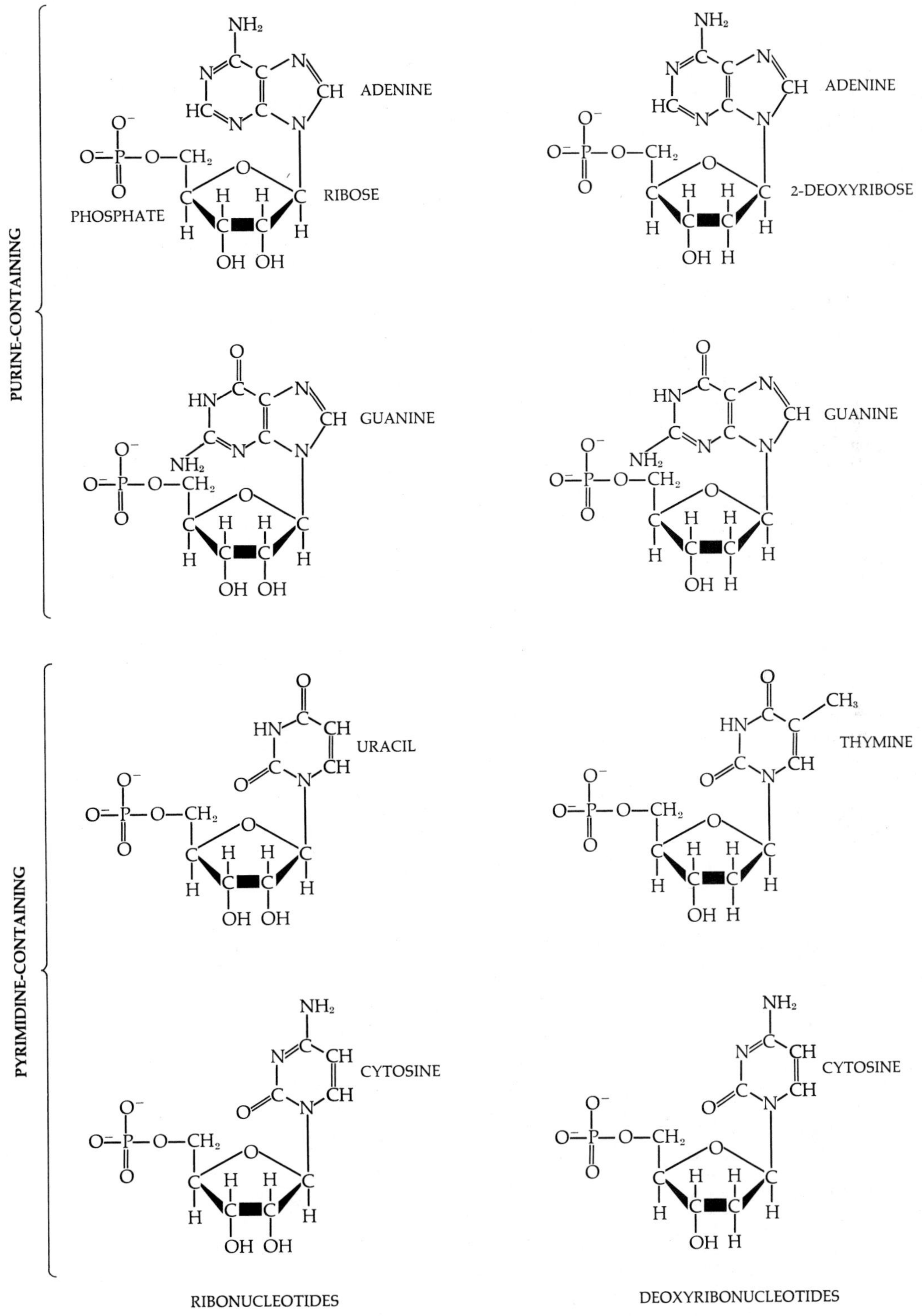

2-25
(a) *Adenine, a nitrogen base.* (b) *Ribose, a five-carbon sugar.* (c) *A phosphate group.* (d) *ATP, adenine plus ribose plus three phosphate groups. The wavy line* ~ *linking each of the last two phosphates to the molecule indicates a high-energy bond.*

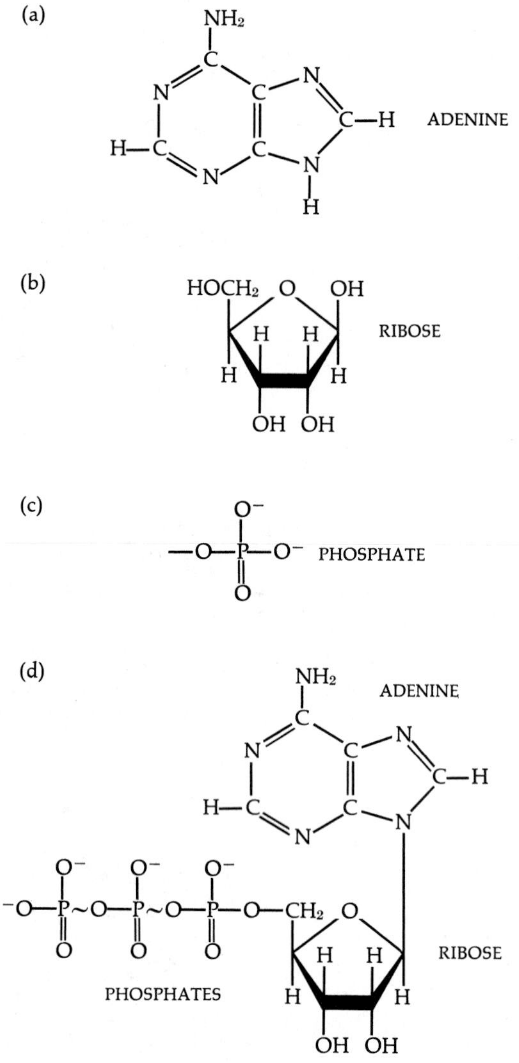

The third phosphate can be removed from ATP by hydrolysis, leaving ADP (adenosine diphosphate) and a phosphate:

$$\text{ATP} + H_2O \longrightarrow \text{ADP} + \text{phosphate}$$

In the course of this reaction, about 7000 calories per mole of ATP are released, a relatively large amount of chemical energy. Removal of the second phosphate produces AMP (adenosine monophosphate) and releases a slightly larger amount of chemical energy. To indicate that relatively large amounts of chemical energy are involved, the bonds linking these two phosphates to the rest of the molecule are often called high-energy bonds, symbolized by ~. This term is somewhat misleading, however, because the energy released during this reaction does not arise entirely from the bond. The products of these reactions, ADP or AMP and phosphate, contain 7000 calories less than the reactants, ATP or ADP and H_2O. This difference in energy between reactants and products is due only in part to bond energy. It is also a result of the internal structure of the ATP or ADP molecules. The phosphate groups each carry negative charges and so tend to repel each other.

In most reactions that take place within a cell, the terminal phosphate groups of ATP or ADP are not simply removed but transferred to another molecule. The amounts of energy released in such transfers are of a magnitude well suited to the step-by-step activities of living cells. (A corollary to this is that the activities of living systems have evolved in such a way as to take advantage of the energy made available by hydrolysis of the high-energy bonds of ATP and ADP.)

The ADP (or AMP) molecule is subsequently recharged—regaining its phosphate group(s) and again becoming ATP—with the energy obtained from the oxidation of a carbon-containing compound. (In photosynthetic cells, as we shall see, the energy of sunlight is also used directly to form ATP from ADP.) We shall discuss the formation of ATP at greater length in Chapter 4.

Let us look at a simple example of an energy exchange involving ATP. As we mentioned previously, sucrose is formed from the monosaccharides glucose and fructose; this is an energy-requiring (endergonic) reaction. The energy for the reaction can be supplied by coupling the synthesis of sucrose to the removal of a phosphate group from the ATP molecule.

First, the terminal phosphate group of ATP is transferred to the glucose molecule.

$$\text{ATP} + \text{glucose} \longrightarrow \text{glucose phosphate} + \text{ADP}$$

In this reaction, some of the 7000 calories available from the hydrolysis of ATP are conserved by the transfer of the phosphate group to the glucose molecule, which thus becomes "energized."

Next glucose phosphate reacts with fructose to form sucrose:

Glucose phosphate + fructose ⟶ sucrose + phosphate

In this second step, the phosphate group is released from the glucose and most of the energy made available by its release (energy originally derived from the ATP) is used to form the bond between glucose and fructose. The free phosphate formed is then available, with an input of energy, to recharge an ADP molecule to ATP.

As we noted in Figure 2–4, the formation of sucrose requires 5.5 kilocalories per mole. The conversion of ATP to ADP + phosphate releases 7 kilocalories per mole. Thus most of the energy in the phosphate bond has been effectively utilized. The ATP–ADP molecule serves as a universal energy carrier, shuttling between energy-releasing reactions, such as the breakdown of glucose, and energy-requiring ones.

Nucleotides play important roles in a variety of other energy-exchange molecules. One such compound is NAD, which contains a nucleotide of adenine plus a phosphate (Figure 4–8). The widespread use of the purine–sugar–phosphate combination in energy transfers may have to do with the fact that these comparatively large charged molecules do not pass through membranes and thus cannot "escape" in and out of cells or membrane-bounded organelles.

SUMMARY

Living matter is composed of only a few of the naturally occurring elements. The bulk of living matter is water, which is a collection of small molecules held together by hydrogen bonds. Most of the rest of living material is composed of organic compounds, which are carbohydrates, lipids, proteins, and nucleic acids.

Carbohydrates serve as a primary source of chemical energy for living systems and as important structural elements in cells. The simplest carbohydrates are the monosaccharides (single sugars) such as glucose and fructose. Monosaccharides can be combined to form disaccharides (two sugars) such as sucrose and polysaccharides (chains of many submolecules of sugar) such as starch and cellulose. These molecules usually can be broken apart again by hydrolysis.

Lipids are another source of energy and also of structural materials for cells. Compounds in this group, which includes fats, waxes, and phospholipids, are generally insoluble in water.

Proteins are very large molecules composed of long chains of amino acids known as polypeptides. There are 20 different amino acids in proteins, and from these an enormous number of different protein molecules is built. Fibrous proteins serve as structural elements, and globular proteins play important dynamic and metabolic roles. The principal levels of protein organization are: (1) primary structure, the amino acid sequence; (2) secondary structure, the coiling or spiraling of the polypeptide chain; and (3) tertiary structure, the folding of the coiled chain into various shapes.

Because of enzymes, which are globular proteins, cells are able to catalyze reactions at high speed and at relatively low temperatures. The specificity of enzymes is due to the lock-and-key fit of the active site on the enzyme with the substrate molecule.

Nucleotides are composed of three submolecules: a nitrogen base, a phosphate group, and a five-carbon sugar. Nucleotides containing deoxyribose sugar form DNA; those containing ribose sugar form RNA. Nucleic acids consist of nucleotides linked together in long chains.

Sugars, fats, and other energy-storing molecules are broken down by the cell and used to form ATP from ADP. ATP supplies the energy for most of the energy-requiring activities of the cells. The ATP molecule consists of a nitrogen base, adenine; a ribose sugar; and three phosphate groups. Two of the groups are linked to the molecule by high-energy bonds—bonds that release a relatively large amount of energy when the terminal phosphate is transferred to another compound. ATP participates as an energy carrier in most series of reactions that take place in living systems.

CHAPTER 3

The Chemistry of Heredity

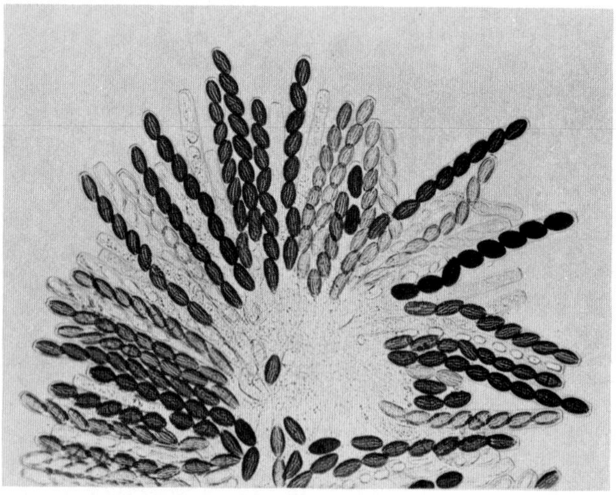

3–1
In order to test their hypothesis that genes act by directing the formation of specific enzymes, Beadle and Tatum chose the bread mold Neurospora crassa *for their experiments. Like many other fungi,* Neurospora *is composed of many filaments (called hyphae) packed closely together. A mass of hyphae is called a mycelium. When two fungal mycelia encounter one another, portions of their mycelia may, if they are of different mating strains, fuse and form a "zygote" from which fruiting bodies are produced. Within the fruiting bodies are spore cases called asci (singular, ascus), and within the asci are sexual spores which, in turn, may germinate to produce new mycelia. Here you can see the spores aligned within the asci.*

Ever since man first started to look at the world around him, he has puzzled and wondered about heredity. Why is it that the offspring of all living things—whether dandelions, dogs, aardvarks, or oak trees—always resemble their parents, never some other species? Why does a child have his mother's eyes or his father's chin or, even more puzzling, his grandfather's nose? These questions are recorded in the writings of the Greeks, long before the birth of Christ, and were probably old even then.

Such questions have always been important ones. Throughout history, biological inheritance has been a major factor in determining the distribution of wealth, power, land, and royal privileges. And from the point of view of the study of biology, self-replication is at the heart of any definition we may have of life.

Some say that the twentieth century will be remembered as the time in history when man first reached the moon. Others predict that it will be best known as the age in which man discovered the nature of DNA and unraveled the secret of heredity. Both so clearly rank among the great adventures of the human spirit that perhaps even time will not be able to choose between them.

One Gene, One Enzyme

In order to tell the story of molecular genetics—the name given to this area of biological research—it is necessary to go back to a series of crucial experiments performed in the 1940s by George Beadle and Edward L. Tatum, who later shared a Nobel prize for this work. By this time, it was known that chromosomes consisted of DNA and protein and that the hereditary factors, known as genes, were carried on the chromosomes. But no one knew what genes were, chemically speaking, or how they exerted their effects. By this time also, the tremendous importance of enzymes in cellular activities had become known, and biological interest was centered on proteins in general and enzymes in particular. Beadle and Tatum formulated an hypothesis: A gene acts by directing the formation of a specific enzyme, which, in turn, regulates the rate of (catalyzes) a specific chemical reaction—one gene, one enzyme. The problem lay in how to prove this hypothesis, how to take an inherited trait and prove that it was the result of the action of an enzyme.

Their solution was an ingenious one; they turned the problem around. Instead of selecting a genetic characteristic and working out its chemistry, they decided to begin with known biochemical reactions—reactions controlled by enzymes—and see how genetic changes (mutations) affected these reactions.

First, they searched for and found an organism that could grow in a medium in which every single chemical ingredient was known: the red bread mold *Neurospora crassa*. *Neurospora* requires only a carbon source (for which it can use any one of a number of simple sugars), one vitamin (biotin), and a few salts. All the amino acids, other vitamins, polysaccharides, and various substances essential for its growth and functioning the mold can make for itself. On the other hand, if a substance—a particular amino acid, for example—is provided in the medium, *Neurospora* will use it "as is" rather than synthesize its own. The synthesis of an amino acid or a vitamin requires a series of reactions that are each mediated by a particular enzyme. If, as a result of mutation, *Neurospora* were to lose the ability to synthesize one of these enzymes, it could no longer grow in the simple minimal medium, but it could survive on one supplemented with the amino acid it was not capable of making itself. This provided a simple test of whether or not an hereditary change had occurred (Figures 3–1 and 3–2).

Beadle and Tatum exposed the spores of *Neurospora* to x-irradiation to increase the mutation rate. They then permitted the spores to develop in a supplemented medium. Those that grew were crossed with wild-type *Neurospora*. ("Wild type" is a term frequently used in genetics to designate the common strain of an organism, the one usually found in nature.) Spores were collected from the hybrid and allowed to grow out in the supplemented medium, and then portions of the mold were tested for growth in the minimal medium. The loss of ability to grow in the minimal medium suggested that an enzyme or enzymes had been lost. By patient trial and error, Beadle and Tatum were able to pinpoint such

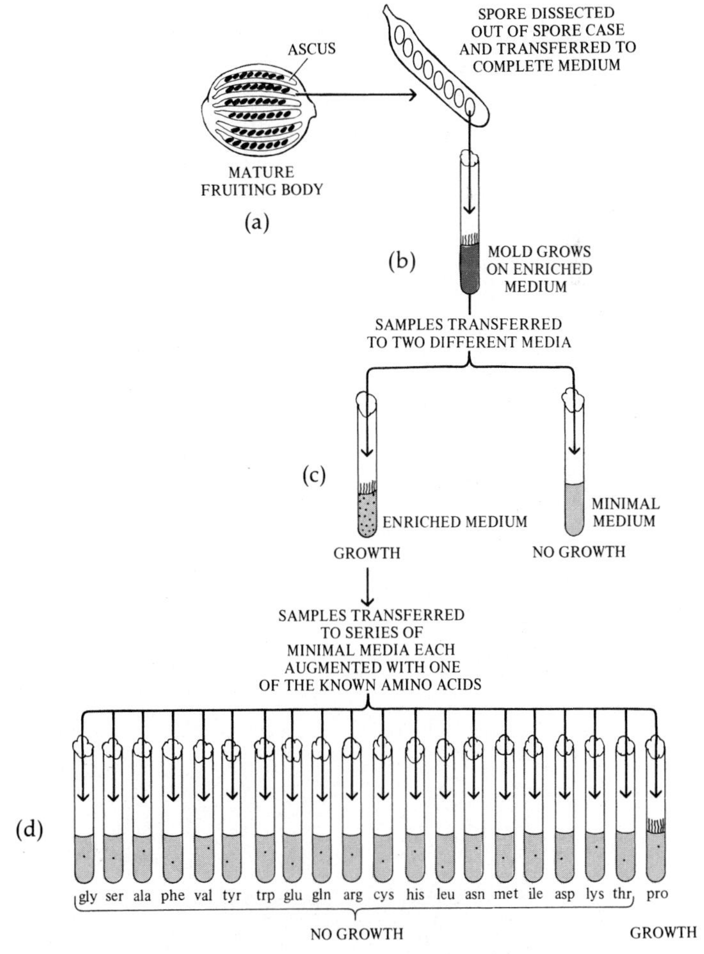

3–2
How Beadle and Tatum tested the mutants of Neurospora. *By these experiments, they were able to show that a change in a single gene results in a change in a single enzyme.* (a) *Asci are removed from fruiting bodies of* Neurospora *and the sexual spores dissected out.* (b) *Each spore is transferred to an enriched medium, containing all* Neurospora *normally needs for growth plus supplementary amino acids.* (c) *A fragment of the mycelium is tested for growth in the minimal medium. If no growth is observed on the minimal medium, it may mean that a mutation has occurred that renders this mutant incapable of making a particular amino acid, and so tests are continued.* (d) *Subcultures of mycelia that grow on the enriched medium but not on the minimal one are tested for their ability to grow in minimal media supplemented with only one of the amino acids. As in the example shown here, a mold that has lost its capacity to synthesize the amino acid proline is unable to survive in a medium that lacks that amino acid. Further tests are then made to discover, in each case, which enzymatic step has been impaired.*

3–3
Three enzymatic reactions involved in the biosynthesis of the amino acid arginine. To test their hypothesis that each enzyme is produced by a separate gene, Beadle and Tatum sought mutant strains of Neurospora *that could not carry out one particular biochemical step, such as the production of citrulline from ornithine, or arginine from citrulline. Such mutants could be analyzed by observing whether they could or could not grow in a particular medium. For example, a mold that could not perform step 2 could satisfy its arginine requirements with arginine or citrulline, but not with ornithine.*

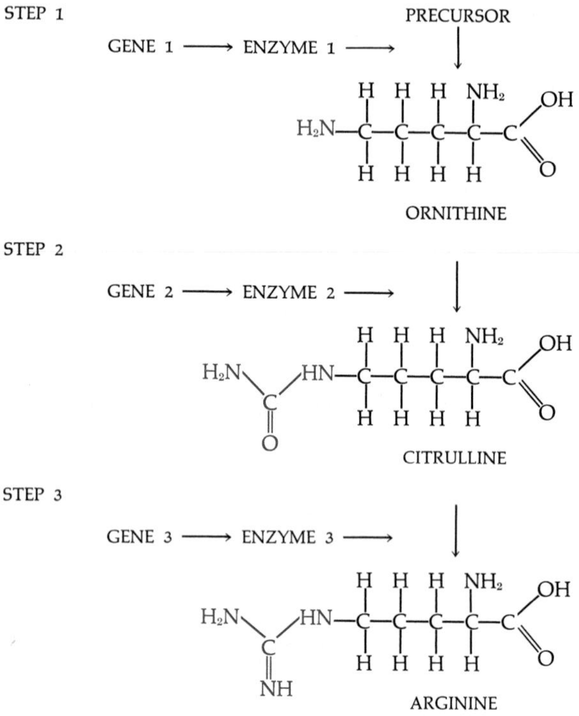

losses. For example, a strain that had lost the ability to make the amino acid tryptophan could grow in the minimal medium plus only tryptophan but could not, under any circumstances, grow without the tryptophan; one that could not make lysine could grow in the minimal medium plus lysine, and so forth. Many such strains were produced that had lost specific biosynthetic abilities as the result of mutation.

It soon became possible to identify the exact enzyme involved. For instance, three different mutants were found that could not synthesize the amino acid arginine. By crossbreeding experiments, these mutations were shown to involve different genes. Beadle and Tatum hypothesized that the mutations each involved a different one of a series of enzymes required to make the amino acid. (See Figure 3–3.) This hypothesis could be proved in two ways:

1. If an arginine-requiring strain regained the ability to make arginine when citrulline was added to the medium, it was clear that the missing enzyme was not enzyme 3. On the other hand, if the strain could make arginine when ornithine was added, it was clear that enzymes 2 and 3 were present but enzyme 1 or some prior enzyme was absent.
2. The mutant molds could be broken up and analyzed chemically. In this way, cellular extracts from the mutants could be shown to be lacking in the activity associated with the particular enzyme—and present in the wild-type *Neurospora.*

At the time these experiments were first performed, enzymatic assembly lines had been worked out for very few organisms. Since that time, however, it has been found that a great variety of cells, including bacteria, yeast, and even the cells that make up the tissues of the human body, are very similar in their enzyme systems and in their stepwise synthesis of the various basic cellular nutrients.

One Gene, One Protein

The Beadle-Tatum hypothesis that a particular gene was responsible for a particular enzyme was quick to gain acceptance. That all enzymes are proteins had already been demonstrated in the 1930s. Not all proteins are enzymes, however. Some are hormones, substances made in one part of an organism and transported to another part of that organism (by the bloodstream, for example), where they have a specific effect. Others are structural proteins, such as tubulin. These proteins, too, are under gene control.

This expansion of the original concept did not modify it in principle: "One gene, one enzyme," as the theory was first abbreviated, was simply amended to the less memorable but more precise "one gene, one polypeptide chain." In other words, enzymes and other protein molecules are the direct products of genes. And these proteins were seen as making up a sort of language—"the language of life"—that spelled out the directions for all the many activities of the cell.

THE CHEMISTRY OF THE GENE: DNA VS. PROTEIN

Chromosomes, you will recall, are complexes of DNA and protein. The question, then, was which of the two carries the genetic information. At that time, many prominent investigators, particularly those who had been studying proteins, believed that the genes themselves were proteins, that the chromosomes contained master models of all the proteins that would be required by the cell, and that enzymes and other proteins active in cellular life were copied from these master models. This was a logical hypothesis, but as it turned out, it was wrong.

There was also accumulating evidence for the role of DNA as the genetic material, based on a variety of different types of data: First, as revealed by specific stains, DNA is present in the chromosomes of all cells and is found there almost exclusively. Second, in general, the body cells of any plant or animal contain twice as much DNA as is found in the sex cells. Third, as shown in Table 3–1, the proportions of purines and pyrimidines are the same in all cells of a given species but vary from one species to another. Such variations among different types of organisms would be essential in a chemical that spelled out the "language of life." (Do you notice anything else interesting about the proportions of purines and pyrimidines?) Fourth, DNA isolated from bacterial cells can act as a transforming factor, endowing other bacterial cells with new genetic characteristics (see Figure 3–4). Fifth, in the course of the infection of bacterial cells by certain bacterial viruses (bacteriophages), DNA and only DNA enters the cell and acts to direct the formation of new viruses (see Figure 3–5).

Despite this evidence, it was not until Watson and Crick made their now historic discovery of the structure of DNA that its genetic role came to be generally accepted.

Table 3–1 *Composition of DNA in Several Species**

	MOLES PER 100 GRAM-ATOMS, PERCENT			
SOURCE	ADENINE	GUANINE	CYTOSINE	THYMINE
Man	30.4	19.6	19.9	30.1
Ox	29.0	21.2	21.2	28.7
Salmon sperm	29.7	20.8	20.4	29.1
Wheat germ	28.1	21.8	22.7	27.4
Escherichia coli	26.0	24.9	25.2	23.9
Sheep liver	29.3	20.7	20.8	29.2

* After Chargaff, Erwin, *Essays on Nucleic Acids,* 1963.

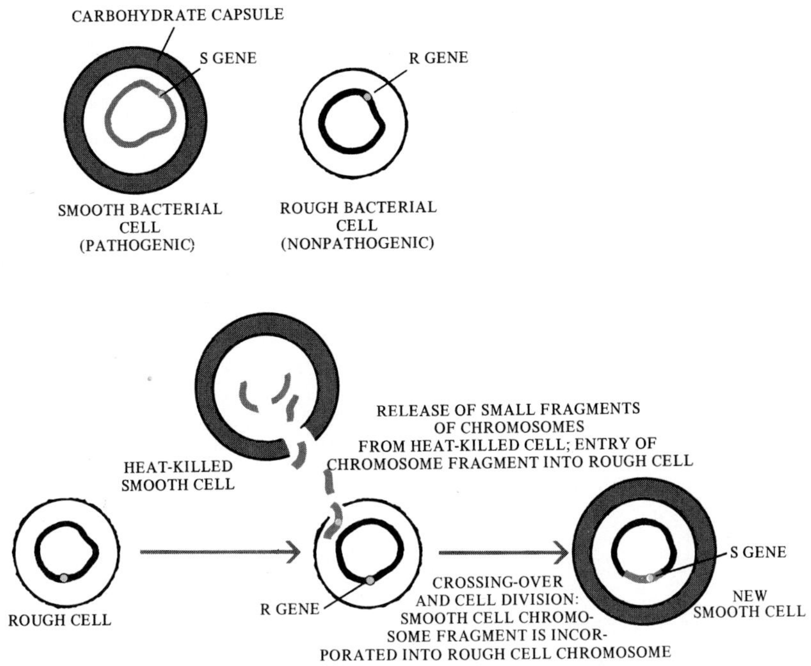

3–4
Identification of the transforming factor, a crucial experiment in establishing the role of DNA. Smooth-coated pneumococcal cells are pathogenic (disease-causing). Rough-coated pneumococci are not. These traits are hereditary; progeny of smooth-coated cells will be smooth-coated, and those of rough-coated cells, rough-coated. If smooth-coated cells are killed and the cellular remains added to a culture of rough-coated cells, some of the rough-coated cells become smooth-coated and produce smooth-coated progeny. This phenomenon, known as transformation, was first observed in 1928. In 1944, it was proved that the "something" that changes the genetic makeup of the rough-coated cells—the transforming factor—is DNA. The diagram shows the modern interpretation of transformation.

3-5

Summary of a series of important early experiments in the history of molecular biology. Bacteriophages (or phages, from the Greek phagein, *to eat), viruses that infect bacterial cells, consist of protein and DNA. By growing virus-infected bacterial cells in media containing radioactive sulfur (^{35}S) or radioactive phosphorus (^{32}P), phages were produced in which the proteins and the DNA were labeled separately. (DNA has no sulfur, and proteins have no phosphorus.) The labeled phages were then permitted to infect cells growing in a nonradioactive medium. These ingenious experiments revealed that only the DNA of the phage enters the cell and that, therefore, it must be the DNA that carries the hereditary information necessary for the production of new, complete phage particles.*

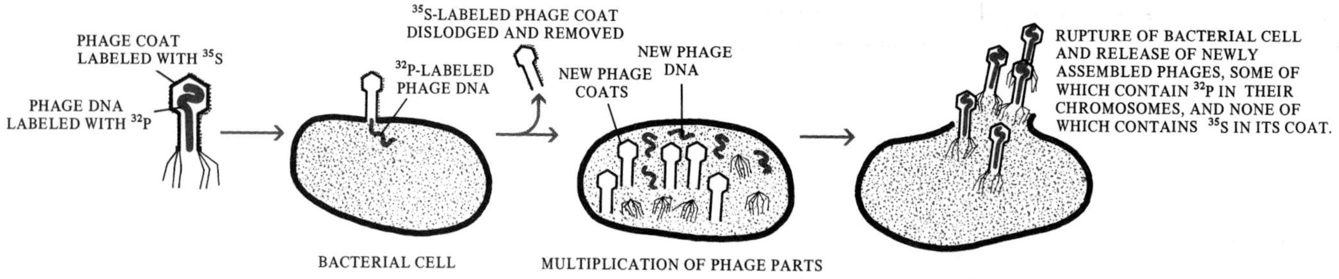

THE NATURE OF DNA

In 1951, when James D. Watson went to England on a fellowship, he arranged to work with Francis Crick of the Cavendish Laboratory. Watson and Crick were among those scientists who were convinced that DNA, not protein, was the genetic material and so, in Watson's words, "the most golden of all molecules." They set out to determine how the molecule was put together.

Watson and Crick had two types of information with which to work. First, if DNA *were* the genetic material, it would have to meet at least four requirements:

1. It must carry genetic information from cell to cell and from generation to generation. Further, it must carry a great deal of information. Consider how many instructions must be contained in the set of genes that directs, for example, the development of an elephant, a tree, or even a bacterium.
2. It must copy itself, for the chromosome does this before every cell division. Moreover, it must do this with great precision; from mutation-rate data we know that a human gene is, on the average, copied for millions of years without a mistake.
3. On the other hand, it must sometimes change or mutate. When a gene changes, that is, when "a mistake is made," the "mistake" must be copied rather than what was originally there. This is a most important property, perhaps the unique attribute of living things, for without the capacity to replicate "errors," there could be no evolution by natural selection.
4. It must have some mechanism for "reading out" the stored information and translating it into action in the living individual.

Watson and Crick were well aware that only if the DNA molecule could be shown to have the size, the configuration, and the complexity required to code the tremendous store of information needed by living things and to make exact copies of this code, would it be possible for it to be the genetic material.

The second type of information came from previous biochemical studies of the molecule. Among these data, some of which were contradictory, were the following:

1. The DNA molecule was very large and also long and thin.
2. The three components (nitrogenous base, sugar, and phosphate) were arranged in nucleotides, as shown previously on page 63.
3. According to x-ray diffraction studies made by Rosalind Franklin and Maurice Wilkins at King's College, London (see Figure 3-6), the long molecule of DNA was made up of regularly repeating units that appeared to be arranged in a spiral or, more accurately, a helix.
4. According to Linus Pauling, the structure of DNA might be similar to that of protein. Pauling had shown in 1950 that the secondary structure of some proteins is helical (see page 58), and that this structure is maintained by hydrogen bonds between the amino acids.
5. Chargaff's analyses had shown, as you may have noticed in Table 3-1, that the ratio of nucleotides containing adenine to those containing thymine is 1 to 1 and so is the ratio of nucleotides containing guanine to those containing cytosine.

Watson and Crick did not do experiments in the usual sense, but rather undertook to examine all the facts about DNA and assemble them into a meaningful whole. They worked with the accumulated data, with the measurements provided by Wilkins' and Franklin's x-ray photographs, and with a model made of scrap tin, which they attempted to put together in such a way that it would meet all the physical and chemical requirements (Figure 3-7).

The most important question for them was whether or not the chemical structure of DNA would actually reflect its biological function. "In pessimistic moods," Watson recalled in a review of these investigations, "we often worried that the correct structure might be dull—that is, that it would suggest absolutely nothing." It turned out, in fact, to be unbelievably "interesting."

The Double Helix

By piecing together the various data, Watson and Crick were able to deduce that DNA does not have a single-stranded helix structure, as do proteins, but is a huge entwined double helix.

The banister of a spiral staircase forms a single helix. If you take a ladder and twist it into the shape of a helix, keeping the rungs perpendicular, this will form a crude model of a double helix. The two railings are made up of sugar and phosphate molecules, alternating. The perpendicular rungs of the ladder are formed by the nitrogenous bases—adenine (A), thymine (T), guanine (G), and cytosine (C)—one base for each sugar-phosphate, with two bases forming each rung. The paired bases meet across the helix and are joined together by hydrogen bonds, the relatively weak, omnipresent chemical bonds that Pauling had demonstrated in his studies of the structure of proteins. (See Figure 3-8.)

The distance between the two sides, or railings, according to Wilkins' measurements, is 2 nanometers. Two purines in combination would take up more than 2 nanometers, and two pyrimidines would not reach all the way across. But if a purine paired in each case with a pyrimidine, there would be a perfect fit. The paired bases, the rungs of the ladder, would therefore always be purine-pyrimidine combinations.

Watson and Crick noticed first that the nucleotides along any one chain of the double helix could be assembled in any order: ATGCGTACATTGCCA, and so on. Because a DNA molecule may be several hundred nucleotides long, there is a possibility for great variety of arrangements. The number of paired bases ranges from

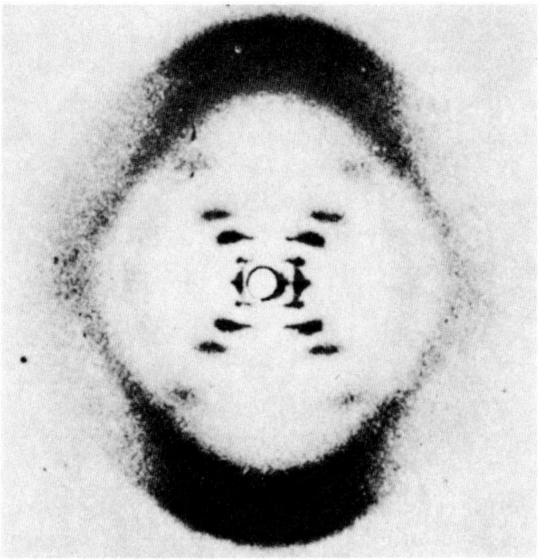

3-6
X-ray diffraction photograph of DNA in the B form taken by Rosalind Franklin late in 1952. The patterns evident in such photographs showed that DNA is made up of regularly repeating units, suggested that these units were arranged in the form of a helix, and made it possible to calculate distances between the atoms. (From The Double Helix *by James D. Watson. Copyright © 1968 by James D. Watson. Reprinted by permission of Atheneum Publishers.)*

3-7
Watson (left) and Crick with their tin model of DNA.

3-8
A representation of the Watson-Crick concept of the DNA molecule. At the left the molecule is drawn in side view as though enclosed in an imaginary transparent cylinder with a central axis (dotted lines). The nitrogenous base pairs (heavy lines) are flat molecules occupying the central area in the cylinder (rectangles in cross sections at right). The base pairs represented by thymine (T) and adenine (A) at level A and cytosine (C) and guanine (G) at level B are linked by hydrogen bonds represented as broken lines. The bases are stacked one above the other at intervals of 0.34 nm and rotated 36 degrees with each step. As a result of this rotation, the successive side views of the bases appear as lines of different lengths according to whether the rectangular areas enclosing them are seen in side or end view. The backbone of the molecule consists of two chains, each composed of deoxyribose sugar molecules (S), each linked to a phosphate group (P) above and to one below to form a continuous chain projecting into the cylinder. The oxygen of the sugar molecule is shown as a small circle at the apex of the pentagon. The sugar molecules on the opposite sides of each base pair are seen to be oriented in opposite directions, thus producing the antiparallel orientation of the backbones. In this side view, the molecular composition of the backbones is shown only over small sections of the molecule, because the twisting of the backbones would otherwise make the drawing confusing. Instead, the continuity of the backbones is represented by ribbons drawn on the surface of the cylinder to represent the projection of their paths. These ribbons are separated by about 120 degrees from each other around the circumference of the cylinder. (The Etkin model, from BioScience, *1973.)*

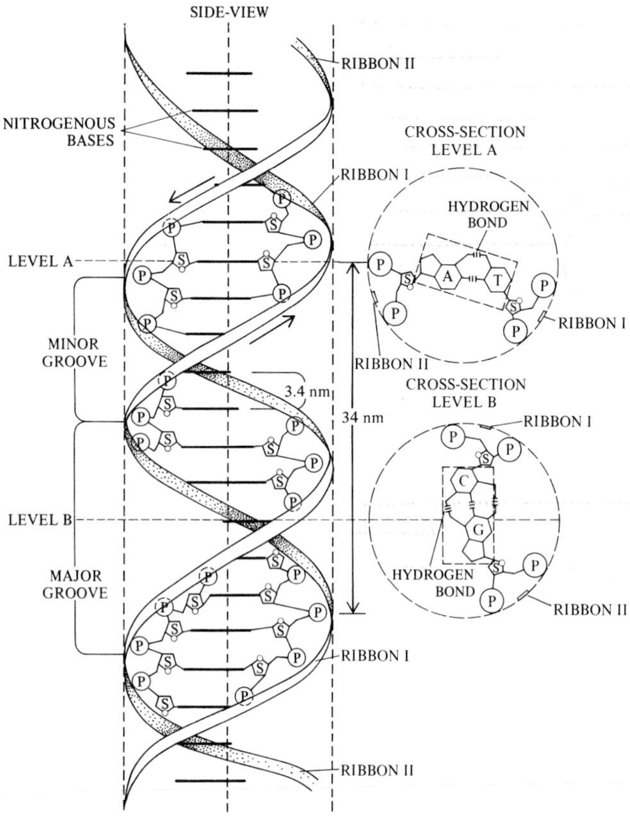

about 5000 for the simplest known virus up to an estimated 5 billion in the 46 chromosomes of man. The DNA from a single human cell—which, if extended in a single thread, would be about 1.5 meters long—contains information equivalent to some 600,000 printed pages averaging 500 words each, or a library of about a thousand books. In short, the DNA molecule could, indeed, carry the necessary information.

The Molecule That Copies Itself

The most exciting discovery came, however, when they set out to construct the matching strand. They encountered an interesting and important restriction. Not only could purines not pair with purines and pyrimidines not pair with pyrimidines, but because of the configurations of the molecules, adenine could pair only with thymine and guanine only with cytosine. This restriction could perhaps have been predicted from Chargaff's data. Thus, one strand of DNA dictates very precisely the structure of the other strand. When the molecule reproduces itself, it simply "unzips" down the middle, the bases breaking apart at the hydrogen bonds. The two strands separate, and new strands form along each old one, using the raw materials in the cell. If a T is present on the old strand, only an A can fit into place on the new strand; a G will pair only with a C, and so on. In this way, each strand forms a copy of the original partner strand, and two exact replicas of the molecule are produced. (See Figures 3-9 and 3-10.) The age-old question of how he-

3–9
Watson and Crick hypothesized that old DNA strands are retained intact and that new ones form along each of the old ones when DNA replicates itself. This hypothesis, known as semiconservative replication, *was strongly supported by the experiments of M. Meselson and F. Stahl. In these experiments,* Escherichia coli *cells were grown in a medium containing "heavy" nitrogen (^{15}N) until they accumulated a "heavy" DNA. The bacteria were then transferred to a medium in which only "light" nitrogen (^{14}N) was available. The first generation grown in light nitrogen had DNA molecules that contained half light nitrogen (the newly-made strands) and half heavy nitrogen (the old strands). The second generation had two types of DNA molecules, one composed entirely of light nitrogen, and one made up of half light and half heavy, each present in equal amounts. In subsequent generations, the proportion of light to heavy increased, but some DNA that was half heavy was still present, showing that the DNA strands containing heavy nitrogen were being retained intact and that, therefore, the molecule of DNA in each daughter strand must consist of one "new" and one "old" strand.*

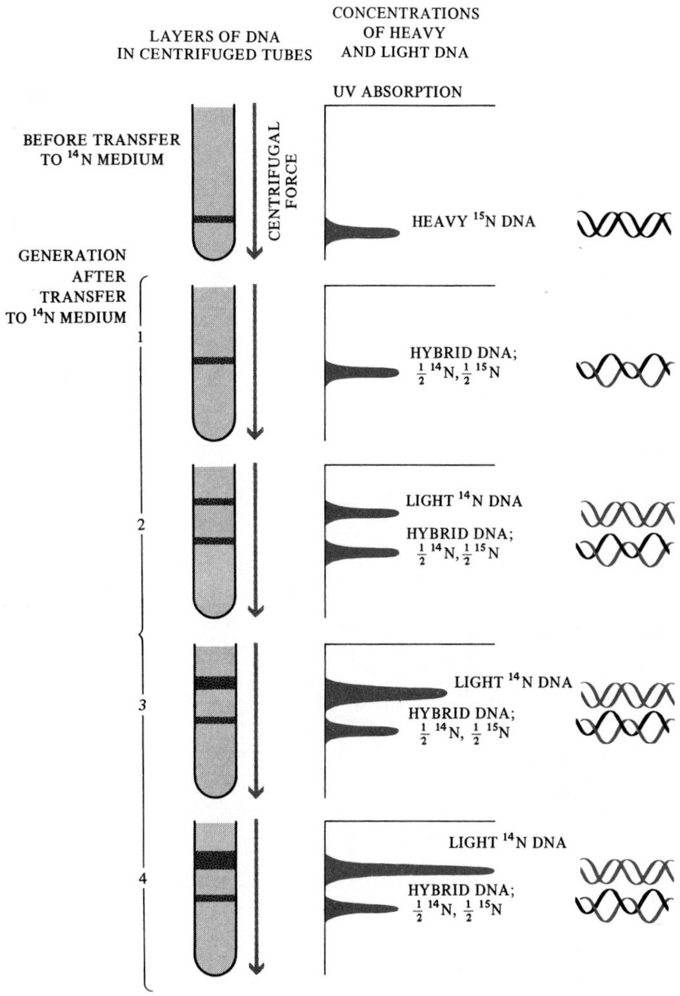

3–10
DNA replicating. The bacterium, E. coli, *was placed in medium containing thymine labeled with tritium (^3H) and was allowed to incorporate this radioactive thymine into its DNA for two generations. Then its DNA was extracted and placed on a photographic plate which recorded the position of the H^3 atoms in the DNA (a technique known as autoradiography). We see here the circular chromosome of one bacterial cell undergoing the act of duplication in which the two strands of the double helix separate and have new partners synthesized along them.*

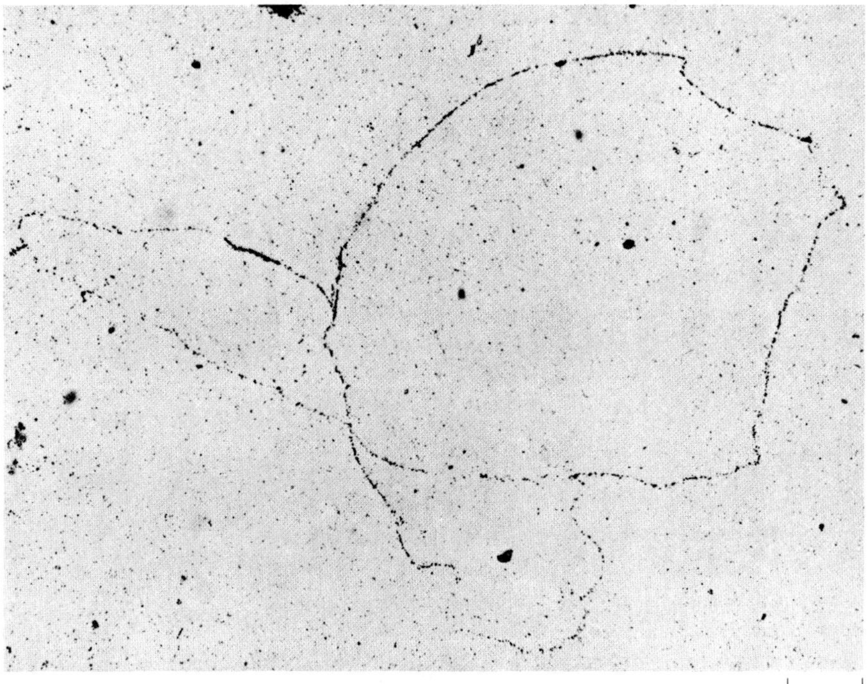

reditary information is duplicated and passed on, duplicated and passed on, for generation after generation had, in principle, been answered by the discovery of the structure of DNA.

In what is considered one of the great understatements of all time, Watson and Crick wrote, in their original brief published report: "It has not escaped our notice that the specific pairing we have postulated immediately suggests a possible copying mechanism for the genetic material."

HOW DO GENES WORK?

Watson and Crick disclosed the chemical nature of the gene and suggested the way in which it duplicated itself. The still unanswered question was how the information stored in the giant DNA molecule could influence structure or function. How, for instance, does DNA change a harmless pneumococcus to a virulent one, determine the shape of a leaf or the odor of a flower, or explain why your eyes are the same color as your mother's?

When it was believed that genes were proteins, it was necessary only to assume that the chromosomes contained master models of all the proteins that would be required by the cell and that enzymes and structural proteins were copied from these master models as needed. When DNA was accepted as the chemical basis of the gene, this simple concept had to be abandoned. If the proteins, with their 20 amino acids, were the language of life, to extend the metaphor of the 1940s, the DNA molecule, with its four nitrogenous bases, could be envisioned as a sort of code for this language. So the term *genetic code* came into being.

As it turned out, the idea of a genetic code was useful not only as a dramatic metaphor but also as a working analogy. Biochemists approached the problem using the methods of cryptographers. There are 20 biologically important amino acids in proteins, and there are four different nucleotides in DNA. If each nucleotide coded one amino acid, there would be a maximum number of only four amino acids. If all possible arrangements of nucleotide pairs were used (4^2, or 16), it still would not be quite enough. Therefore, following the code analogy, at least three nucleotides must specify each amino acid. This provides for 4^3, or 64, possible combinations. This postulate was verified several years later.

Translating the Code—DNA to Protein

The steps between DNA and protein, it was discovered, involve the sister molecule of DNA, ribonucleic acid (RNA). It had long been suspected that RNA was involved in protein synthesis, because cells that were synthesizing large amounts of protein invariably contained large amounts of RNA.

RNA differs from DNA in a few significant respects. The sugar, or ribose, component of the molecule contains one more atom of oxygen, and in place of thymine, RNA has another pyrimidine, uracil (U). (See page 63.) Also, the RNA molecule is found only rarely in a fully double-stranded form, which means that its properties and activities are quite different from those of DNA.

There are three types of RNA: *messenger RNA* (mRNA), *transfer RNA* (tRNA), and *ribosomal RNA.*

Messenger RNA is a large molecule, ranging in size from a few hundred nucleotides to about 10,000. It is formed along one strand of the DNA helix by the same base-pair copying principle that applies to the formation of a new strand of DNA (Figure 3–11). Uracil "fits" in the place of thymine. As previously hypothesized, a sequence of three bases specifies one amino acid; such a sequence is known as a *codon* (Figure 3–11). Once formed, the long single-stranded molecule attaches by one end to a ribosome.

3–11
Protein biosynthesis starts with the formation of a strand of mRNA along one of the strands of DNA. The mRNA thus formed is a complementary copy of the DNA strand, with uracil in place of thymine.

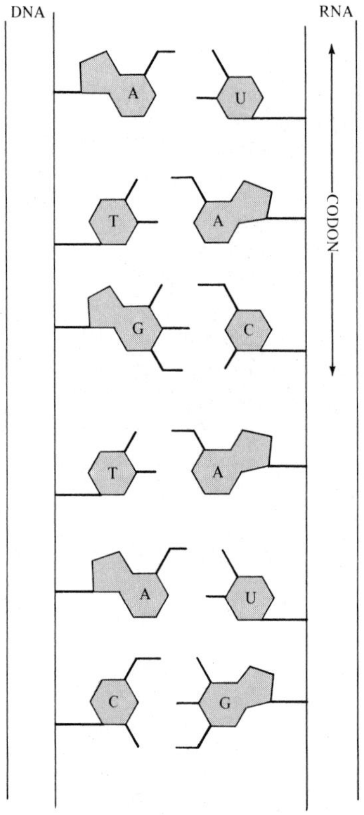

Transfer RNA has sometimes been called the dictionary of the language of life because it translates DNA language into protein language. There are many different types of tRNA, apparently one specific type for each of the codons of the genetic code. Each tRNA molecule consists of a strand of 77 nucleotides forming a long single strand that folds back on itself (Figure 3–12). As can be seen in Figure 3–13, part of the specificity of a particular tRNA molecule resides in the nature of its *anticodon* triplet sequence, which determines when it will recognize a particular triplet codon on the mRNA. Equally as important to the tRNA molecule's specificity is its ability to bind a particular amino acid (that one coded by the anticodon). This specificity is achieved with the aid of specific enzymes, known as activating enzymes, which recognize both a particular amino acid and a particular tRNA. These activating enzymes are thus key elements in the translation of the genetic message. They determine which amino acid will be associated with which tRNA (and thus with which triplet anticodon).

The base sequences of a number of different tRNA molecules have been discovered so far. Each type of tRNA differs in its nucleotide sequence, but all of them have the same number of bases and all appear to have a similar shape.

At the point at which the mRNA molecule touches the ribosome, the tRNA molecule, with its captive amino acid, zeros into position. As the ribosome moves along the mRNA strand, the next tRNA molecule moves into place with its amino acid in tow. At this point, the first tRNA molecule detaches itself from the mRNA molecule. The energy in the bond that holds the tRNA molecule to the amino acid is now utilized, in participation with a specific enzyme, to forge the peptide link between the two amino acids, and the tRNA, released, becomes available once more. These molecules apparently can be used over and over again.

3–12
Two-dimensional (left) and three-dimensional (right) structure of the tRNA molecule from yeast responsible for the transfer of the amino acid phenylalanine. The three-dimensional structure was worked out by a group from the Massachussetts Institute of Technology and announced in early 1974. This tRNA molecule contains 77 nucleotides. Fourteen of the 77 bases are not the typical A, G, U, and C but are closely related structures. The significance of these atypical bases is not yet known, but, by preventing base pairing, they may play a role in regulating the tertiary structure of the molecule and therefore its functioning. Regions of base pairing are indicated by dots. Each portion of the tRNA molecule appears to have a unique function: The acceptor end is where the amino acid is bound; the TΨC loop is the same in every tRNA and probably governs binding to ribosomes; the DHU loop region, in contrast, differs from tRNA to tRNA and may be involved in determining which amino acid is enzymatically added to the acceptor end; and the anticodon loop contains the three bases complementary to the three-base codon of mRNA, in this case G-A-A.

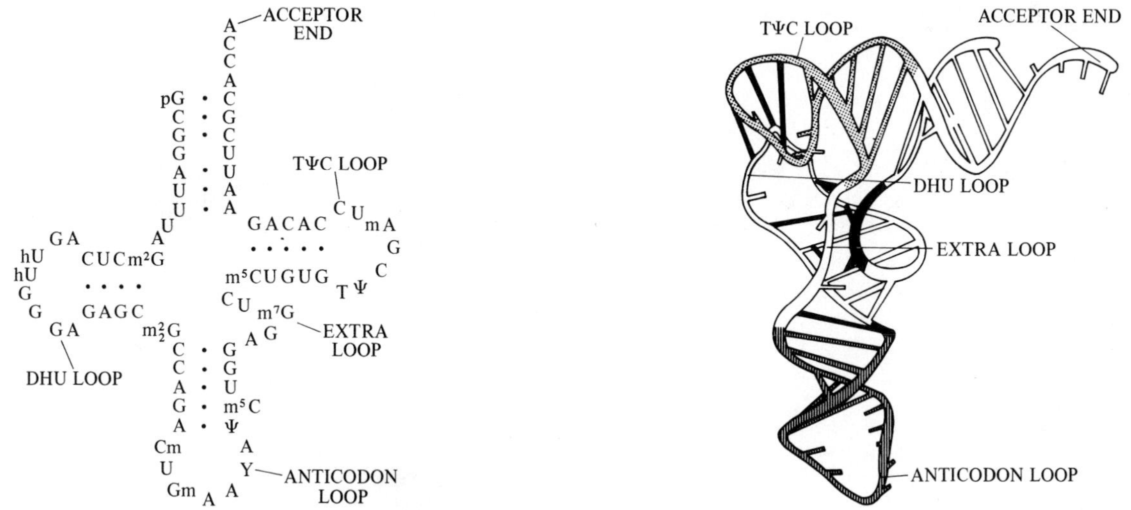

3–13
How a protein is made. As many as 64 different kinds of tRNA molecules are formed on the DNA in the nucleus of the cell. These molecules are so structured that each can be attached (by a special enzyme) at one end to a specific amino acid. Each carries somewhere in the molecule an anticodon that fits an mRNA codon for that particular amino acid. The process of protein biosynthesis begins when an mRNA strand is formed on the DNA template in the nucleus and travels to the cytoplasm. A ribosome attaches to the strand, and at the point of attachment, the matching tRNA molecule, with its amino acid, plugs in momentarily to the codon in the mRNA. As the ribosome moves along the mRNA strand, a tRNA linked to its particular amino acid fits into place and the first tRNA molecule is released, leaving behind its amino acid. As the process continues, the amino acids are brought into line one by one, following the exact order laid down by the DNA code, and are formed into a protein chain, which may be anywhere from fifty to hundreds of amino acids long.

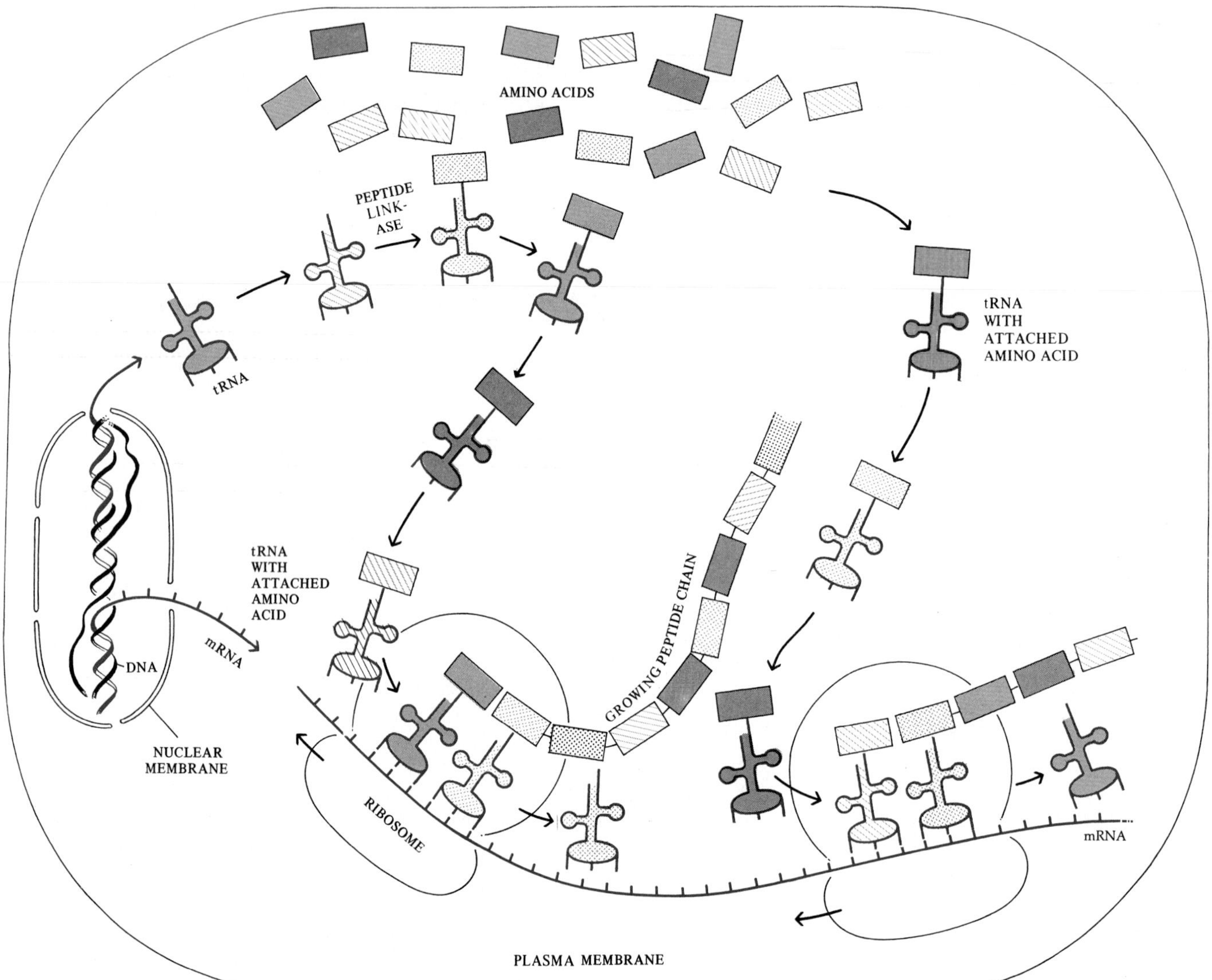

3–14
A bacterial gene in action. In the micrograph, you can see molecules of RNA polymerase, the enzyme that regulates the transcription of RNA from DNA. The one at the far right is approximately at the point where transcription begins. Several different mRNA strands (shown in color in the diagram) are being formed simultaneously. The longest one, at the left, was the first one synthesized. As each mRNA strand peels off the DNA molecule (active chromosome segment), ribosomes attach to the RNA, translating it into protein. The protein molecules are not visible in this micrograph of a bacterial gene.

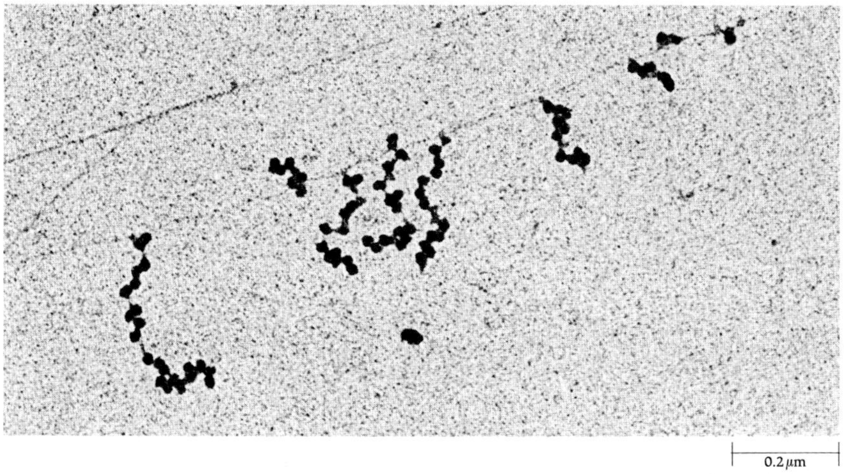

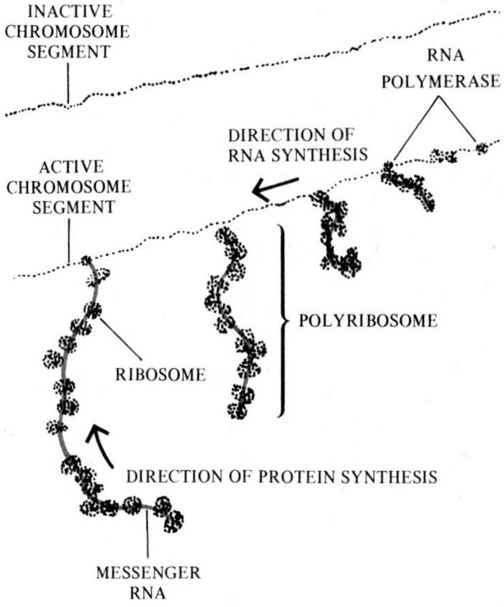

Messenger RNA appears to have a much briefer life, at least in *Escherichia coli.* It is usually "read" by several ribosomes simultaneously, thereby producing several polypeptide chains in a matter of seconds. Then it may be read by another group of ribosomes once or twice more; after that it is destroyed. The average lifetime of an mRNA molecule in *E. coli* is 2 minutes, although its lifetime in other types of cells may be longer.

Although the great bulk of cellular RNA is to be found in the ribosomes, the role of ribosomal RNA in protein synthesis is not fully understood. Like mRNA and tRNA, ribosomal RNA is formed along one strand of the DNA helix; and like tRNA but unlike mRNA, it is long-lived. The ribosome, a compact mass of RNA and protein, consists of two approximately spherical particles, the larger of which contains a special groove or notch into which the tRNA molecule fits. The way in which one tRNA molecule is ejected from this binding site to make way for the next and the way in which the ribosome moves along the mRNA molecule are not fully understood.

In summary, the mRNA is a long, single-stranded molecule that is formed along the DNA helix and that carries the DNA message to the ribosomes, the point of assembly for proteins. The sequence of codons on the mRNA molecule determines the order in which the smaller molecules of tRNA line up, one after the other, along the mRNA molecule. Because each tRNA molecule is attached to a specific amino acid, the order in which the tRNA molecules line up along the long mRNA strand determines precisely the order in which the amino acids are linked together to form proteins (Figure 3–14).

Breaking the Code

The studies that elucidated the role of RNA in protein synthesis were carried out, using the mRNA, tRNA, ribosomes, DNA, and enzyme systems of *E. coli* cells. This raised a question that interested Marshall Nirenberg of the United States Public Health Service. Could the *E. coli* system translate mRNA from other cells? Nirenberg took fractions containing mRNA from a number of different cell sources and added them to the *E. coli* system. In every case, new protein was formed. Even RNA from tobacco mosaic virus, which naturally multiplies only in the cells of the tobacco plant, could be "read" as mRNA by the ribosomes and tRNA molecules of the bacterial cells. As far as these and all subsequent experiments have shown, the code is a universal language common to all living things. In other words, it is probably some 3.5 billion years old!

Leaving others to ponder the evolutionary implications of his finding, Nirenberg raced ahead to give it practical application. If *E. coli* could read a foreign message and translate it, perhaps it could read a totally synthetic message, one dictated by the scientists themselves. A means for synthesizing RNA was available. Severo Ochoa of New York University had found an enzyme (RNA polymerase) that, when added to a test tube containing ribonucleotides and ATP, would link the ribonucleotides together in a long strand of RNA. The trouble with the method, from Nirenberg's point of view, was that there was no way to control the order in which an assortment of ribonucleotides would be assembled. For Nirenberg's purposes, the order was of the utmost importance. He wanted to know the exact contents of any message that he dictated.

A simple solution for this seemingly perplexing problem suddenly presented itself—an RNA molecule that consisted of only one ribonucleotide repeated over and over again, the ribonucleotide uracil. Nirenberg prepared 20 different test tubes, each of which contained cellular extracts of *E. coli* with ribosomes, tRNA, ATP, the necessary enzymes, and amino acids. In each test tube, one of the amino acids, and only one, carried a radioactive label. The synthetic poly-U, as it was called, was added to each test tube. In 19 of the test tubes, nothing detectable occurred, but in one, the one to which radioactive phenylalanine had been added, the investigators were able to detect newly formed, radioactive polypeptide chains. When the polypeptide was analyzed, it was found to consist only of phenylalanines, one after another. Nirenberg had dictated the message "uracil . . . uracil . . . uracil . . . ," and a clear answer had come back, "phenylalanine . . . phenylalanine . . . phenylalanine. . . ."

Within the year following Nirenberg's discovery, tentative codes were worked out by Nirenberg, Ochoa, and others for all the amino acids, using synthetic mRNA. A synthetic polynucleotide made up entirely of adenine (poly-A), for instance, makes a peptide chain composed entirely of lysine. If two parts of uracil are combined with one part of guanine, the peptide that is dictated will be composed largely of valine, so it was presumed that the code for valine is GUU or UUG or UGU.

The Order of the Bases

How is it possible to tell the order of the bases in a codon? A, G, and C, for instance, can be arranged in nine different combinations. And there is no way yet available to isolate a portion of the giant DNA molecule and compare it directly to the protein it produces.

Two ingenious methods have been used to solve this problem; one was devised by Nirenberg and the other by H. G. Khorana, now at MIT in Cambridge.

Nirenberg found that, by adding nucleotides one at a time, it is possible to synthesize triplets—that is, codons—in which the bases appear in known, predetermined order. This single triplet, although useless for the biochemical functions of the cell, will serve to bind the appropriate transfer RNA, with its amino acid, to a ribosome. Unbound tRNA will slip right through a cellulose filter, but tRNA attached to a ribosome will be caught in the filter. Nirenberg prepared various combinations of

3–15
Khorana's method for "breaking" the genetic code involved a variety of synthetic RNAs with repeating sequences of bases. The synthetic nucleotide chains, containing two or three different bases in groups of two, three, or four, were used to produce polypeptide chains. When introduced into cell-free systems containing the machinery for protein synthesis, the base sequences are read off as triplets (as shown in the center) and produce the amino acid sequences shown on the right. Notice that because the starting point of the sequence is not clearly defined under these experimental conditions, the GAC combination can be read as GAC . . . GAC . . . GAC . . . or ACG . . . ACG . . . ACG . . . or CGA . . . CGA . . . CGA. . . . Three different polypeptide chains containing one repeating amino acid result. This offered a direct proof of the hypothesis, formulated 10 years earlier, that the codon is a triplet. (The actual combinations shown here are not necessarily those used in the original experiments.)

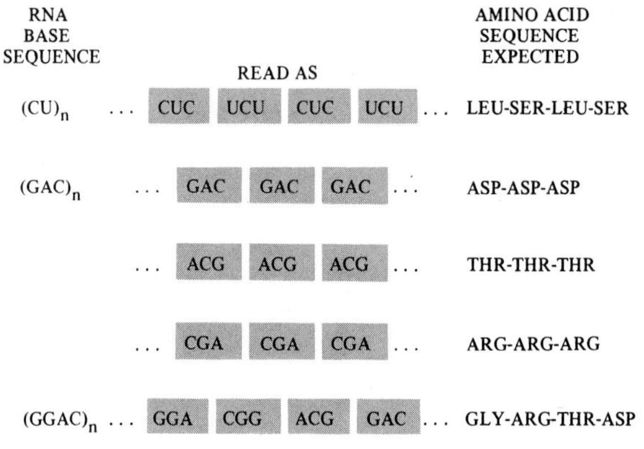

RNA BASE SEQUENCE
READ AS
AMINO ACID SEQUENCE EXPECTED
(CU)n
. . . CUC UCU CUC UCU . . .
LEU-SER-LEU-SER
(GAC)n
. . . GAC GAC GAC . . .
ASP-ASP-ASP
. . . ACG ACG ACG . . .
THR-THR-THR
. . . CGA CGA CGA . . .
ARG-ARG-ARG
(GGAC)n
. . . GGA CGG ACG GAC . . .
GLY-ARG-THR-ASP

synthetic nucleotide triplets, tRNA, amino acids with radioactive labels (to facilitate their detection), and cell-free ribosome preparations. These combinations, after a brief period of exposure to one another, were passed through the cellulose filter. Then any radioactive material present in the filter was removed and analyzed. Knowing the triplet codon that he had put in the test tube, Nirenberg had only to identify the amino acid, bound by its tRNA to the ribosome, for an exact translation of his code triplet.

In an alternative approach, Khorana and his co-workers devised a method for the synthesis of chains of DNA or RNA in which two or three nucleotides could be repeated over and over again in a known sequence. Thus, he could make deoxyribose strands of TCTCTCTCTC, which would form a double helix with strands of AGAGAGAGAG, and strands of TGTGTGTGTG, which would form a double helix with ACACACACAC. Then, using RNA polymerase (the enzyme that makes mRNA on the DNA template), he could obtain messenger strands of AGAGAGAGAG, UCUCUCUCUC, ACACACACAC, and UGUGUGUG-UG.

Each of these RNA chains, when used as a messenger in the cell-free system, produced polypeptide chains of alternating amino acids. Poly-AG produced arginine and glutamic acid over and over again; poly-UC, serine and leucine; poly-AC, threonine and histidine; and poly-UG, cysteine and valine. This is, of course, what you would expect from a triplet code, since the messenger would be read AGA . . . GAG . . . AGA. . . . This is the clearest direct proof that the mRNA is read sequentially (that is, one nucleotide after another) and also that the codon consists of an uneven number of nucleotides. Khorana later synthesized nucleotide chains consisting of known sequences of three and more nucleotides (Figure 3–15), and eventually an entire gene that replicated itself.

New techniques are now making it possible to isolate and purify the stretch of DNA coding for single specific genes. The few genes that have been sequenced to date exhibit a high degree of internal structure (see Figure 3–17), suggesting that the three-dimensional structure of a gene may be important in determining how that gene's primary structure is translated.

Biological Implications

All but three of the 64 trinucleotides have now been identified in terms of a particular amino acid (Figure 3–16). These three are stop signals which cause the chain to terminate.

Because 61 combinations code 20 amino acids, you can see that there are a number of "synonyms" among the codons. Most often, these synonyms differ only in the third nucleotide. These third-nucleotide differences re-

3–16
The genetic code, consisting of 64 triplet combinations (codons) and their corresponding amino acids (see page 56). Since 61 triplets code 20 amino acids, there are "synonyms," as many as six for leucine, for example. Most of the synonyms, as you can see, differ only in the third nucleotide. Of the 64 codons, only 61 specify particular amino acids. The other three codons are stop signals which cause the chain to terminate. The code is shown here as it would appear in the mRNA molecule. How would you determine the corresponding DNA codes?

FIRST LETTER	SECOND LETTER: U		C		A		G		THIRD LETTER
U	UUU	phe	UCU		UAU	tyr	UGU	cys	U
	UUC		UCC	ser	UAC		UGC		C
	UUA	leu	UCA		UAA	stop	UGA	stop	A
	UUG		UCG		UAG	stop	UGG	trp	G
C	CUU		CCU		CAU	his	CGU		U
	CUC	leu	CCC	pro	CAC		CGC	arg	C
	CUA		CCA		CAA	gln	CGA		A
	CUG		CCG		CAG		CGG		G
A	AUU		ACU		AAU	asn	AGU	ser	U
	AUC	ile	ACC	thr	AAC		AGC		C
	AUA		ACA		AAA	lys	AGA	arg	A
	AUG	met	ACG		AAG		AGG		G
G	GUU		GCU		GAU	asp	GGU		U
	GUC	val	GCC	ala	GAC		GGC	gly	C
	GUA		GCA		GAA	glu	GGA		A
	GUG		GCG		GAG		GGG		G

3-17
Entire structure of a gene from MS2 virus, an RNA virus, as deduced by a Belgian team headed by Walter Fiers and reported in 1972. When the gene is not coding, it folds up into the flowerlike configuration shown here. The degeneracy in the genetic code, the fact that several RNA triplet combinations code for the same amino acid, seems to have been used to maximize base-pairing possibilities in the nucleic acid.

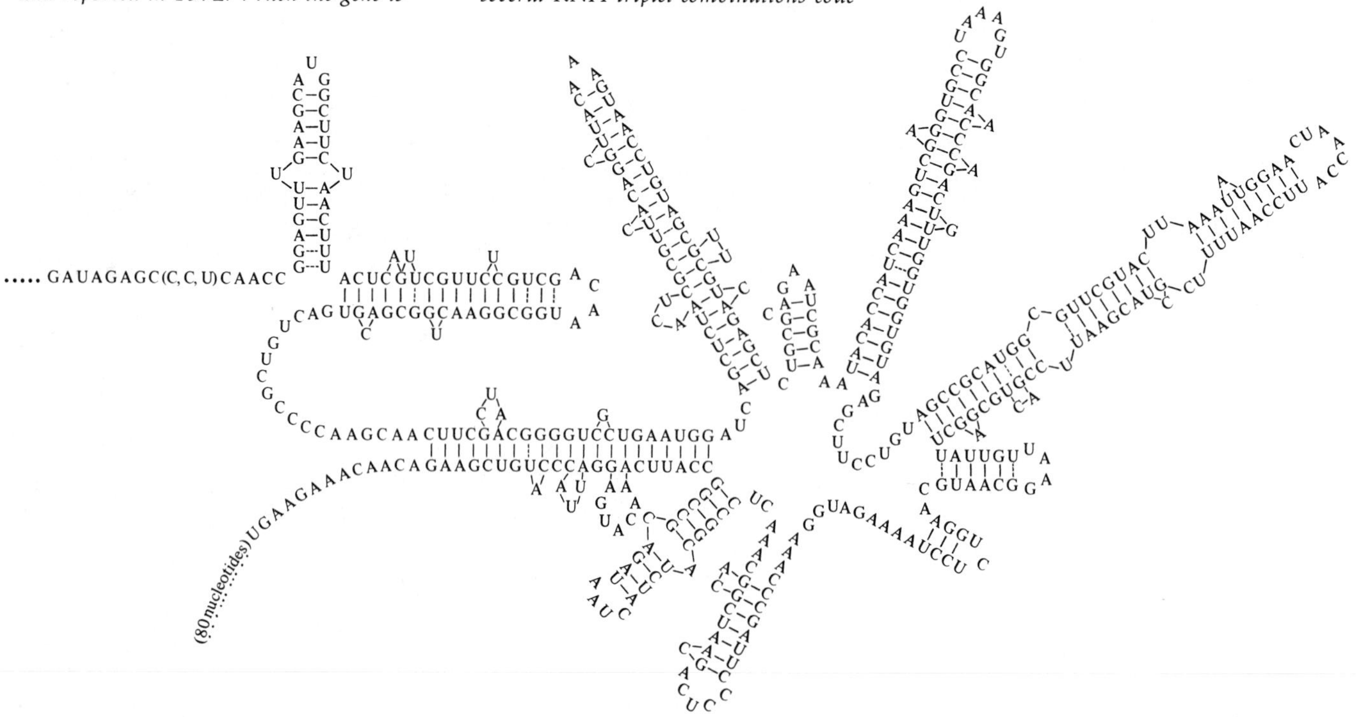

flect the manner in which the tRNA anticodon binds to the mRNA codon: Only the first two tRNA bases bind their complementary mRNA bases tightly, the third tRNA base being permitted to "wobble" somewhat, so that several possible mRNA bases might fit well enough to complete the triplet. The genetic code is called degenerate because several mRNA code words may specify the same amino acid. This degeneracy, however, may have important biological implications (Figure 3-17).

REGULATING GENE FUNCTIONS

The chromosomes of *Escherichia coli* contain enough DNA to code between 2000 and 4000 different proteins. Yet biochemists estimate that only about 600 to 800 different kinds of enzymes and other proteins are present in the cell at any particular time. Furthermore, these proteins are found in varying amounts, with certain much-used enzymes in large concentrations and others barely discernible. As far as it is possible to determine, normal bacterial cells exercise very close control over the amounts of each type of protein.

For example, cells of *E. coli* supplied with the disaccharide lactose as a carbon and energy source need the enzyme beta-galactosidase to split the disaccharide into glucose and galactose before they can use it (Figure 3-18). In cells growing on lactose, approximately 3000 molecules of beta-galactosidase are present in every normal *E. coli* cell. This represents about 3 percent of all the protein in the cell. In the absence of lactose, however, there is an average of only one molecule of this enzyme per cell. When the galactosidase is needed, it is produced from new mRNA.

Mutants of *E. coli* have been found that produce the enzyme even in the absence of lactose. These mutants are at a disadvantage compared with cells with a balanced protein synthesis, because by making an enzyme in the absence of its substrate they are using their energies and resources uneconomically. Substances such as lactose that increase the amount of a particular enzyme produced by a cell are known as *inducers,* and the enzymes whose rate of production they influence are known as inducible enzymes.

Some substances act to repress enzyme production. For example, *E. coli* can make all its own amino acids from a carbon source and ammonia (NH_3). If a particular amino acid—histidine, for example—is present in the medium, the cell will then stop making all of the enzymes associated with the biosynthesis of histidine. Such enzymes are known as repressible enzymes. In bacterial cells, mRNA molecules are broken down very soon after they are produced. Therefore, control over mRNA production directly controls the rate of enzyme synthesis.

3-18
Splitting of lactose to galactose and glucose requires the enzyme beta-glactosidase. Beta-galactosidase is an inducible enzyme; that is, its production is regulated by an inducer—in this case, lactose.

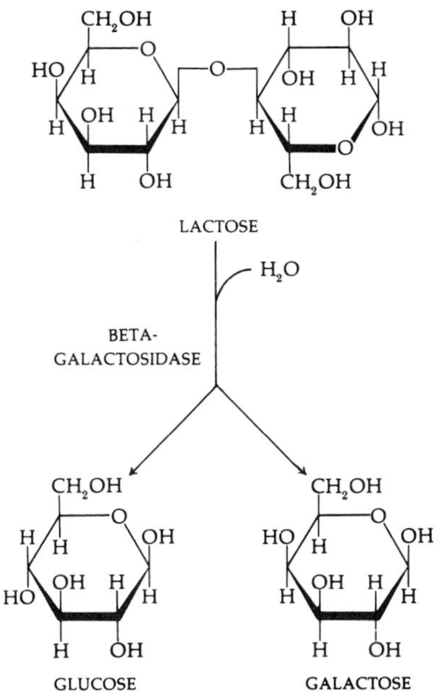

3-19
According to the operon theory of Jacob and Monod, the synthesis of proteins may be regulated by interactions involving either a repressor and inducer or a repressor and corepressor. (a) *In the case of inducible enzymes, such as beta-galactosidase, the repressor molecule is active until it combines with the inducer (in this case, lactose). It is then inactivated, and so the genes in the operon are no longer repressed.* (b) *In the case of repressible enzymes, the repressor is not active until it combines with the corepressor. Thus, in the absence of the corepressor, the genes in the operon are active.*

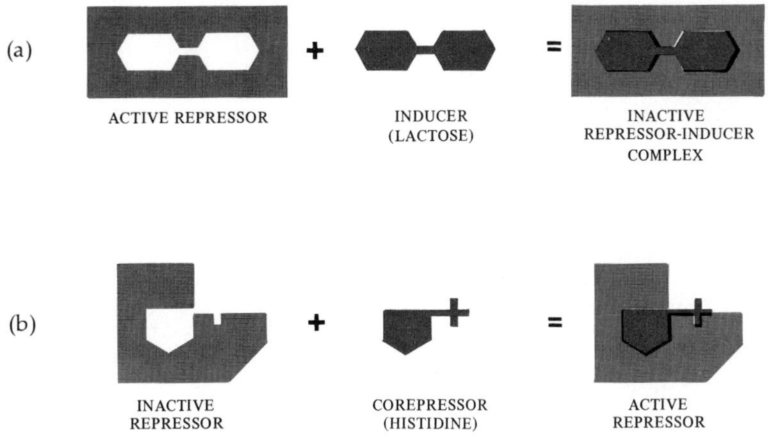

The Operon

To explain how bacterial cells regulate enzyme biosynthesis, François Jacob and Jacques Monod, in the work for which they received the Nobel prize in 1965, developed the hypothesis of the *operon.* An operon is a group of related genes all aligned together along a single segment of DNA. The operon consists of three different types of genes: the *promoter,* located at the site at which formation of the mRNA begins; the *operator,* an on-off switch which acts to stop the production of mRNA under certain conditions; and one or more *structural genes,* the genes that actually code for enzymes or other proteins. In the beta-galactosidase system, the three structural genes code for a messenger RNA that dictates the structure of three enzymes, beta-galactosidase, and two other enzymes involved in lactose metabolism. The operon genes are adjacent to one another and are transcribed consecutively, one after the other along a single strand of DNA, forming a single mRNA molecule. The mRNA molecules produced are active for only a very short time, after which they are broken down by specific enzymes.

The activity of the operon is controlled by yet another gene, the *regulator,* which is not necessarily adjacent to the operon. The regulator codes for a protein, known as the *repressor,* which apparently binds to the DNA at the site of the operator gene. Because the operator gene is located between the promoter and the structural genes, this blocks the production of mRNA.

The repressor is controlled by another "signal" compound. In the case of inducible enzymes, this compound is the inducer. The inducer binds with the repressor molecule and changes its shape so that it can no longer attach itself to the operator gene of the operon. In the absence of the repressor, mRNA molecules are formed along the structural genes, and from these molecules proteins are produced. When the supply of the inducer is exhausted, the regulator once again assumes control, and mRNA production and protein formation cease. In the beta-galactosidase system, the inducer is lactose or a closely related compound derived from lactose. The lactose or related compound binds with the repressor and inactivates it, thus permitting enzyme biosynthesis to proceed (Figures 3-19a and 3-20a). In the case of repressible enzymes, the "signal" compound is a corepressor. The repressor is active only when bound with the corepressor (Figures 3-19b and 3-20b).

The presence of regulators and operons has now been demonstrated in higher organisms, such as corn and the evening primrose. In these plants, mutations have been detected that result in the unrestrained production of groups of enzymes similar to the unrestrained production of beta-galactosidase in *E. coli* that have mutations in the operator genes.

3–20

The operon. An operon is a group of genes forming a functional unit. There are usually several structural genes, which code for different proteins, often a group of enzymes that work sequentially in a particular enzyme sequence. The transcription can begin only at the site marked "promoter." (a) *In operons activated by inducers, the regulator gene codes for a protein that represses transcription of mRNA from the structural genes. The regulator, which need not be adjacent to the operon on the chromosome, directs production of a repressor protein. This repressor acts upon the operator gene, apparently by binding to the DNA at this site and blocking the formation of mRNA. In order to start transcription again, another compound, the inducer, is needed. The inducer counteracts the effects of the repressor, probably by binding to it and changing its shape. The repressor can no longer bind to the operator site, so synthesis of mRNA proceeds.* (b) *In operons regulated by the corepressors, transcription continues until the repressor and corepressor combined bind to the DNA of the operon. The repressor alone cannot halt transcription.*

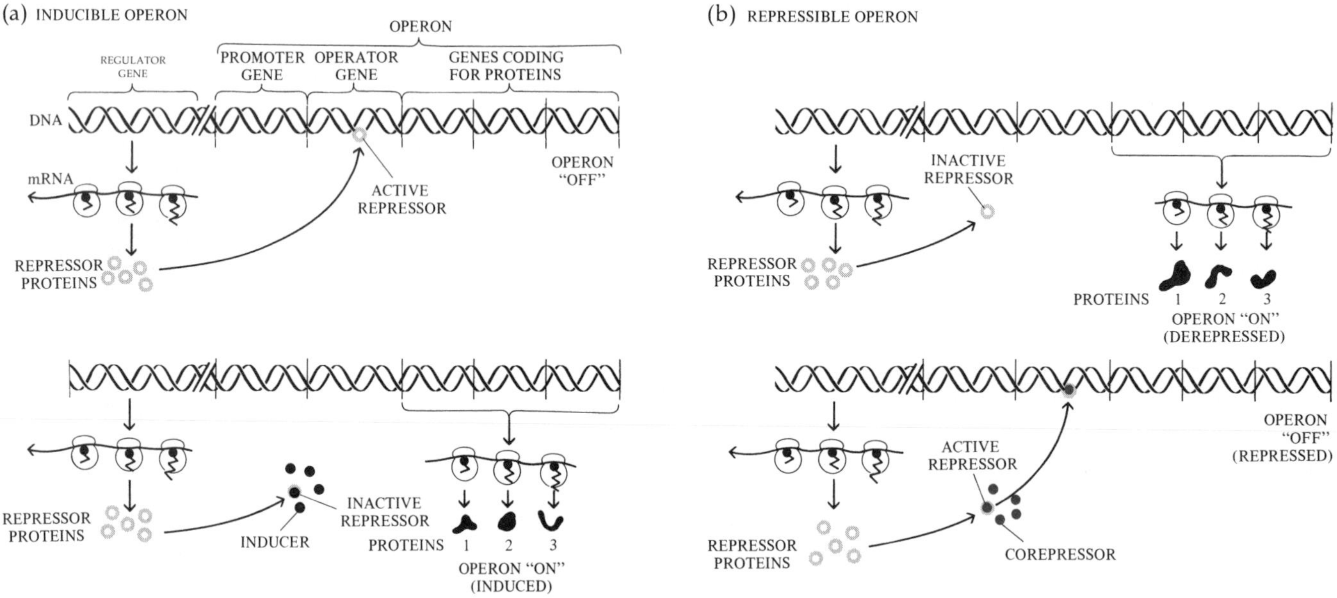

3–21

A repressor protein (the white spherical form in the central lefthand portion of the micrograph) attached to the lactose operon.

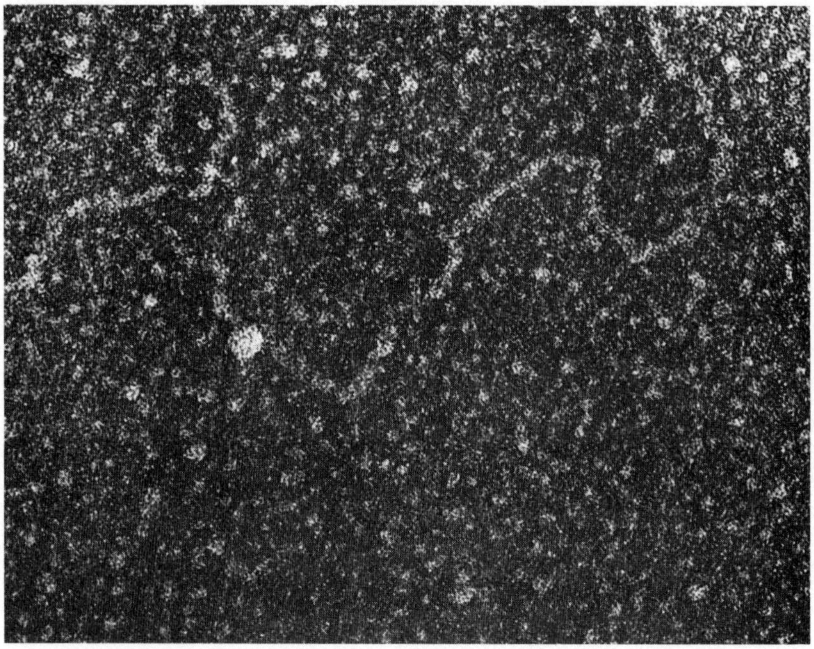

Regulation by Feedback Inhibition

In addition to the sorts of mechanisms just discussed, which alter the functions of cells by altering the types of enzymes they contain, there are a number of different kinds of control systems that work by affecting the enzymes themselves. The simplest is direct feedback inhibition.

When the product of a particular biosynthetic pathway accumulates, this product may combine with an early (usually the first) enzyme in the pathway, rendering that enzyme inoperative. As long as the final product of the pathway is present in high concentration, no additional molecules can pass along this particular pathway, because the first step is effectively blocked. When the concentration of the final product is lowered as the product molecules present in the cell are utilized, the enzyme governing the first step in the pathway escapes from its combined form, and the pathway begins to function again. In this way, the cell can regulate with great precision the quantities of various molecules.

Fine adjustment control systems such as this end-product feedback inhibition are examples of a property of proteins known as *allosteric* ("other shape") *interactions*. The bonds determining the tertiary structure of a protein are weak, and the binding of a particular molecule (an allosteric effector) at one site on the protein may so affect the weak interactions determining its shape that the conformation of the protein is changed (Figure 3–22). When a protein molecule undergoes a change in shape of this sort, a second site, perhaps an enzyme active site, which may be on quite a different part of the protein molecule, can be affected.

In considering allosteric interactions, it is important to recall that enzymes are, in general, bound in fixed positions on membranes within the cell. It is now clear that for many biosynthetic pathways, all the enzymes concerned with the entire process are bound at one site on a membrane. A molecule that is being passed along a biosynthetic pathway, such as that involved in the synthesis of fatty acids, may not exist free in the cytoplasm. Rather it remains at a fixed site, bound to enzymes that are concerned with the synthetic pathway, changing in structure as it is affected by one enzyme after another.

The importance of such a cellular strategy is clear. Only so many molecules can exist simultaneously in a cell, and it is more efficient not to have many intermediate molecules in a given biosynthetic pathway present at any one time if only the final product is utilized by the cell. Concentrating the enzymes involved in a particular pathway into a multi-enzyme particle, in which the molecule being changed by the action of these enzymes remains in position in the particle until it is completely converted into the final product, results in a "solid state" system in which intermediates never exist free and unbound in the cytoplasm.

3–22
An allosteric ("other shape") effector, represented by the dumbbell-shaped structure, can bind to an enzyme and, by disrupting the weak bonds determining its tertiary structure, alter the active site. The altered enzyme cannot interact with its normal substrate but may be able to work upon a second substrate.

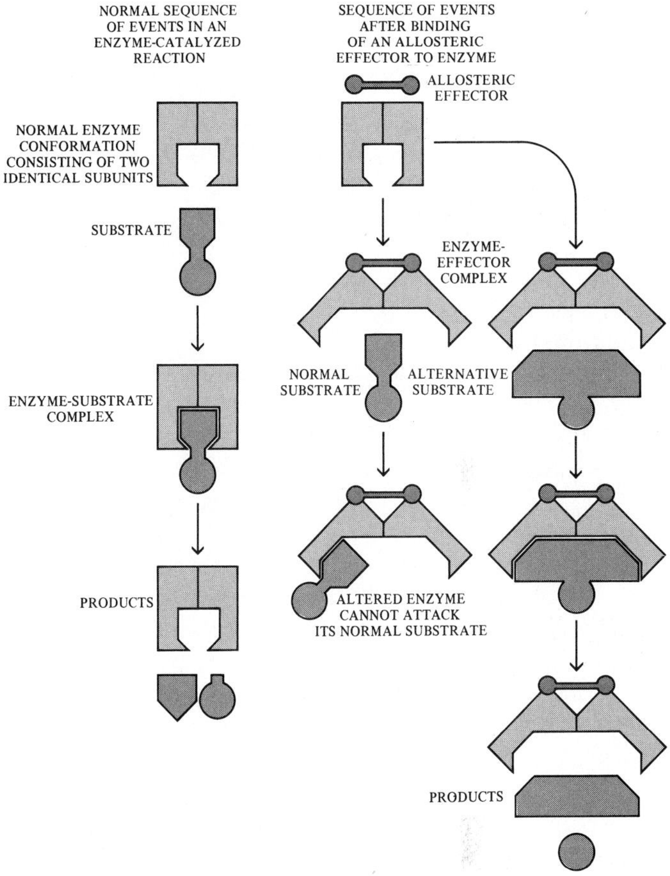

Important examples of such multi-enzyme complexes within the cell are those that contain the enzymes concerned with ATP formation in mitochondria and chloroplasts. One of the important features of such multi-enzyme complexes is that an allosteric change in one of the enzymes, apparently usually the first one in the system, will turn off the entire system and thus regulate the concentration of all intermediates of the system in the cell.

THE GENETIC CONTROL OF DIFFERENTIATION

Differentiation is the developmental process by which a relatively unspecialized cell or tissue undergoes a progressive change to become more specialized in function or structure.

Joachim Hämmerling, working in the early 1930s, studied the role of the cytoplasm in differentiation by taking advantage of some unusual properties of the marine alga *Acetabularia.* The vegetative body of *Acetabularia* consists of a single huge cell 2.5 to 5 centimeters in height. Individuals have a cap, a stalk, and a "foot," all of which are differentiated portions of the single cell. If the cap is removed, the cell will rapidly regenerate a new one. Different species of *Acetabularia* have different kinds of caps.

Hämmerling found, first of all, that the cell of *Acetabularia* is able to synthesize protein for up to 2 months after the nucleus is removed, which meant that much of the control of protein synthesis resided in the cytoplasm. He then set out to determine to what extent control *was* exerted by the nucleus. Taking a nucleus from an individual of *A. mediterranea,* a species that has a round cap, he transplanted it into the cell of an individual of *A. crenulata,* a species that has a jagged one. He wanted to find out what sort of cap this compound cell, with the cytoplasm of *A. crenulata* and nucleus of *A. mediterranea,* would produce.

The answer was unexpectedly clearcut: The compound cell produced the cap characteristic of *A. crenulata,* indicating that the control of differentiation resided entirely in the cytoplasm. Hämmerling was not satisfied with these results, however. Removing the cap, he waited for the cell to regenerate a second one. This second cap was intermediate between those of the two species. When he removed it, the cell regenerated a cap entirely typical of *A. mediterranea.* The cap-shaping substances that had been produced in the cytoplasm under the influence of the original *A. crenulata* nucleus had been exhausted, and the nucleus of *A. mediterranea* was in control. (See Figure 3–23.)

In a second series of experiments, Hämmerling put two nuclei into a single cell. He found that a cell that contained one nucleus from *A. mediterranea* and one from *A. crenulata* would produce intermediate caps indefinitely.

Subsequent work has demonstrated that the cap-determining substances are probably messenger RNAs, transcribed by nuclear DNA and concentrated at the cap end of the cell.

Many of the factors that are involved in differentiation are directly involved in the control of the nuclear DNA ⟶ mRNA ⟶ protein sequence that we have outlined in this chapter. Control may be exercised at each level in the process. In view of this, it is not surprising that a wide variety of nonspecific agents such as heat and cellular pH can alter cellular differentiation. As we shall see in Chapter 23, there is evidence that some plant hormones also work by regulating the production of mRNA and so coordinating the steps in the development of the organism.

Control of Multicellular Differentiation

We have already discussed the role of induction and repression in controlling the production of specific mRNA molecules within a cell. In eukaryotic organisms, other mechanisms are available to control mRNA production. These are related to the structure of eukaryotic chromosomes, which, as you will remember, consist of DNA bound with proteins.

To appreciate these strategies, however, it will first be necessary to review the nature of multicellular differentiation. In *Escherichia coli,* each cell divides to produce

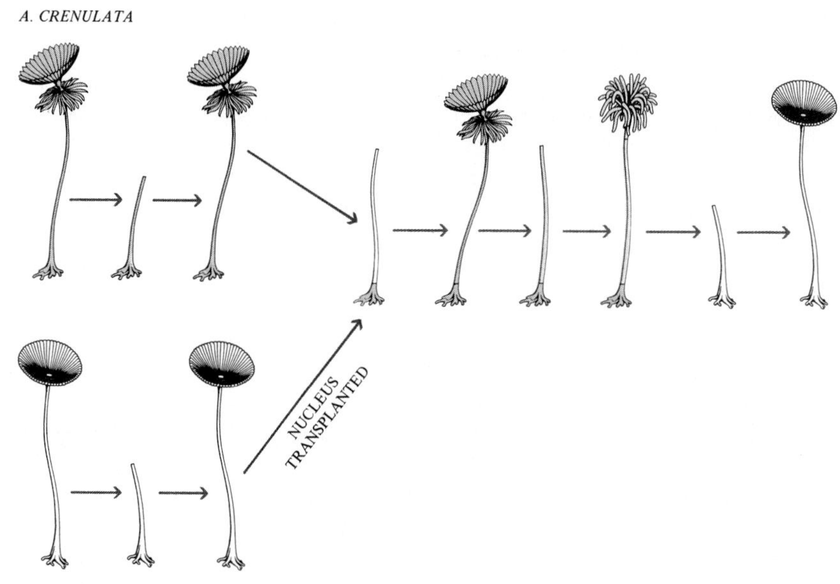

3–23
The control that the nucleus exercises over cell structures was illustrated by an experiment performed with the large, one-celled alga Acetabularia. *Two species were used.* A mediterranea *and* A. crenulata. *The two species differ in the shape of their caps. In either species, if the cap is removed, the cell will grow another one just like it. If the cap and nucleus are removed from* A. crenulata *and a nucleus from* A. mediterranea *is grafted on to the cell, as shown, the new cap that forms will resemble that of* A. crenulata. *If that cap is removed, another cap will grow that is intermediate between the two species. If that cap is removed, the regenerated cap will be identical to the* A. mediterranea *cap.*

another that exactly resembles it in structure and function. In higher organisms, groups of cells that *differ* from one another both in their structure and in their function arise in the course of cell division. But genes control structure and function. Does this mean that the cells of higher organisms, even though they all arise from the same cell, somehow contain different genetic information?

Equality of Genetic Information in All Cells

At the close of the nineteenth century, the German embryologist August Weismann proposed that, as cells divide in the forming embryo, the nuclear material present in the zygote is parceled out unequally to the various daughter cells, so that the determinants for one type of activity go to only one group of cells. In the early 1900s, the famous embryologist Hans Spemann devised an experiment to test this hypothesis. Spemann tied off one portion of a salamander egg and let the other half of the egg, which contained the nucleus, divide a number of times, forming 16 to 32 cells. Then he let the nucleus from one of these cells slip through to the unnucleated half. This half then went ahead and developed into an embryo that was a Siamese twin to the first. In short, the nucleus was still able to provide the genetic material for the development of a complete embryo. It had not lost any of its "determinants."

More recent experiments indicate that every cell of an organism is totipotent; that is, it contains the entire quota of genetic material present in the fertilized egg. This concept of the totipotent cell is in accord, of course, with the observation that all diploid cells from a particular type of organism contain the same quantities of DNA. In a plant, the nectar-secreting cell, the photosynthetic cell of the leaf, and the storage cell of the root would all appear to have exactly the same genetic material. The difference must lie in which genes are active and which inactive. Some genes must be inactive for many cellular generations. For instance, in the ferns, centrioles and flagella are seen only in the male gamete. Yet the information for making these structures is carried in all the cells of the organism. It is estimated that at least 80 percent of the genes in a cell are inactive at any given time.

Mechanisms for Gene Activation

As we mentioned previously, the DNA of eukaryotic cells is always closely associated with proteins. A large proportion of this protein is in the form of histones, a group of small, negatively charged polypeptide molecules that are bound to the DNA. There is some evidence that this binding with histones inactivates the DNA and that only when a particular gene is somehow released from the histones can mRNA be synthesized.

The hypothesis that genes are turned on and off sequentially (though not necessarily by histones) is confirmed by some observations in insects. In some flies, certain cells contain giant chromosomes; these chromosomes are the result of the repeated replication of the nuclear material without any accompanying separation of the strands. Because the chromosomes are so large, they can readily be seen even at interphase. Many years ago, cytologists noticed enlarged regions in these chromosomes, which they called "chromosome puffs." Chromosome puffs, which are areas where the coherence of the chromosome filaments is loosened, are believed to represent areas where particular genes are active. This is supported by the fact that the puffs can often be shown to be the sites of RNA synthesis. In insects treated with molting hormone, which triggers a whole series of new metabolic events inside the cell, two new puffs are produced within a matter of hours after the treatment.

As we mentioned in Chapter 1, certain cell lines contain more than the diploid number of chromosomes. These "extra" chromosomes may represent a strategy for increasing the production of proteins. In roots, for example, many cells are polyploid. For instance, in corn *(Zea mays)*, most of the differentiated cells in the root have either twice or four times as high a DNA content as normal diploid cells, and certain specialized cells concerned with the conduction of water (vessel elements) have up to 16 times the diploid amount before they mature. It has not yet been possible to connect the amount of DNA in such cells with the sort of differentiation the cells undergo, but future work in the area appears promising.

Cytoplasmic Control of Differentiation

In many organisms, various cytoplasmic particles also play a direct role in cellular differentiation. These include not only organelles such as plastids and mitochondria, which contain their own DNA complements, but also other sorts of particles that affect differentiation in various ways. If they are unequally partitioned between the daughter cells at mitosis, the resulting cell lines may have very different fates.

In a similar way, microchemical gradients within cells, that is, minute changes in the quantity of a substance present as one moves across the cell, play a major role in differentiation. For example, in the brown alga *Fucus*, a gradient of stored insoluble food particles is apparently set up in the fertilized egg by the action of gravity. This gradient in turn determines the position of the spindle apparatus in the first division, setting up two distinctive cell lines from the very first cell division in the life of the individual. Similar observations have been made in the embryology of animals. Such a simple strategy as the unequal division of a cell may also be of fundamental importance in parceling out different kinds of elements

3–24
Genes in action. The genes shown here are nucleolar genes engaged in the synthesis of ribosomal RNA. Approximately 100 RNA molecules are being synthesized simultaneously on each active gene. As you can see, the molecules at the beginning of the gene are shorter than the more nearly completed molecules at the end of the gene segment. (The arrows indicate the direction in which transcription is proceeding.) Each RNA fibril is attached to a spherical granule about 12.5 nm in diameter located at the DNA axis. These are almost certainly RNA polymerase molecules involved in transcription. Each gene is separated from the others by a segment of inactive DNA.

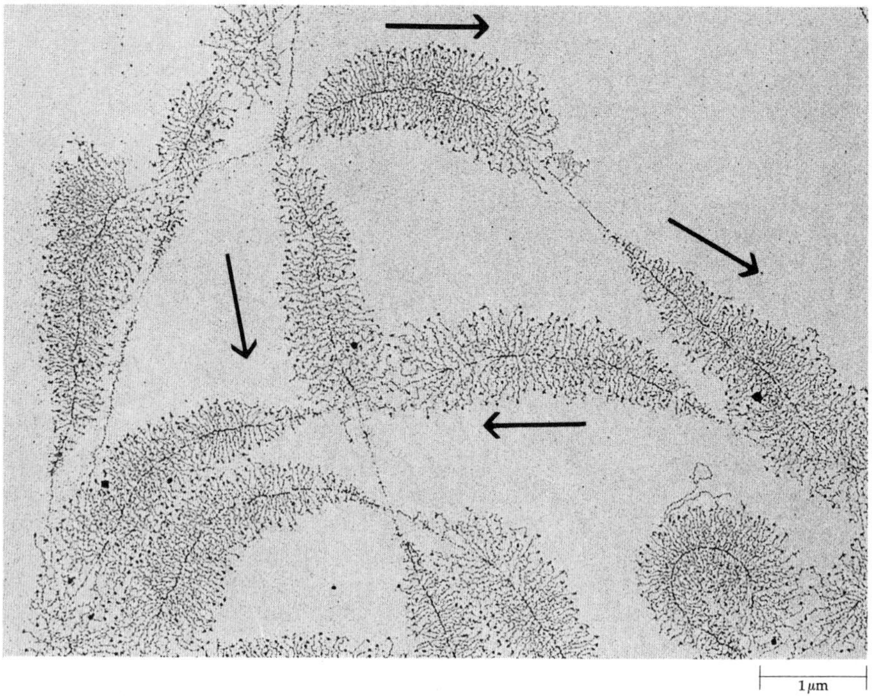

present in the cytoplasm and in determining the fate of the resulting cell lines.

In the cells of higher organisms, where many substances are diffusing at different rates and in different directions, and tissues of various kinds are often packed together in close proximity, comparable but more complex effects must be very common. Not only are gradients set up within individual cells and tissues that lead to an extremely subtle control of developmental processes, but also the differentiation of any one cell may be determined largely by its position in the body of the developing plant or animal. Some of the ways in which hormonal and other factors interact in the development of higher plants will be discussed in Section 6.

The characteristics of an entire cell are thus determined by processes of many different kinds that interact with one another and are controlled by numerous feedback mechanisms which ensure a high degree of cellular consistency and self-regulation but which allow differentiation to take place when appropriate. In these complex interactions, we catch a glimpse of the true meaning of the cell theory and of the importance of the cell as a crucially important level of biological integration.

SUMMARY

Evidence for the role of DNA in heredity includes the following: (1) DNA occurs almost entirely in the chromosome. (2) In general, all the cells of an organism contain equal amounts of DNA except spores, gametes, and other haploid cells, which contain half as much. (3) The proportions of purines and pyrimidines are different in different species. (4) DNA can alter hereditary traits in bacterial cells. (5) In infection with some viruses, DNA is the only viral component that enters the infected cell, where it directs the formation of new viruses.

The function of DNA as the repository of genetic information received general scientific acceptance in the 1950s after the work of Watson and Crick had revealed the structure of the DNA molecule. The key features of this structure are: (1) The molecule is sufficiently complex and varied to carry a tremendous load of information; and (2) the molecule can be self-replicating because of the base-pairing principle by which adenine can pair only with thymine and guanine only with cytosine.

The way in which the order of bases in the DNA molecule specifies the order of amino acids in a protein is now known. Each series of three nucleotides along a DNA strand is the codon for a particular amino acid. The information in the codon is transferred from the DNA by means of a long, single strand of messenger RNA (mRNA). The mRNA forms along one of the strands of DNA, following the principles of base pairing first suggested by Watson and Crick, and therefore is complementary to it.

The mRNA strand attaches to a ribosome, a macromolecular complex composed of RNA and protein. At the point at which the strand of mRNA is in contact with the protein, small molecules of transfer RNA (tRNA), which serve as adaptors between the nucleic acid and the proteins, are bound temporarily to the mRNA strand.

This bonding is believed to take place by the same base-pairing principle that holds together the two strands of the double helix of DNA and that attracts the forming mRNA to the strand of DNA. Each tRNA molecule carries the specific amino acid called for by the mRNA codon into which the tRNA plugs. Thus, following the sequence dictated by the DNA, the amino acid units are brought into line one by one and are formed into a polypeptide chain.

In a brilliant experiment involving a synthetic RNA, poly-U, Nirenberg was able to "break" the DNA/RNA code. Subsequent work by Nirenberg, Khorana, and others succeeded in identifying the remaining codons and in proving that the messenger RNA is read sequentially and that the codon consists of three nucleotides.

The biosynthesis of protein in the cell is closely regulated. One method of regulation is the operon system proposed by Jacob and Monod: The operon consists of a group of genes turned on and off simultaneously by a chemical substance produced by a regulator gene. This system, which was discovered in *Escherichia coli,* appears also to operate in cells of higher organisms.

Another method for the regulation of protein biosynthesis is direct feedback inhibition. In feedback inhibition, the product of a particular biosynthetic pathway combines with the first enzyme (or some other enzyme early in the sequence) and inhibits its action. Such combinations are known as allosteric interactions.

The nucleus exerts control over cytoplasmic differentiation, as revealed in studies of the one-celled marine alga *Acetabularia.* The way in which this control is exerted is presumably by the production of specific mRNAs. In a multicellular organism, each cell has the same genetic makeup. Differences among the cells in structure and function are determined by activation and inactivation of particular genes or groups of genes. Some evidence suggests that genes may be activated when they are released from their close association with histones.

SUGGESTIONS FOR FURTHER READING

BAKER, J. J., and G. E. ALLEN: *Matter, Energy and Life: An Introduction for Biology Students,* 2nd ed., Addison-Wesley Publishing Co., Inc., Reading, Mass., 1970.*

A book for students who have had no previous chemistry or physics, which deals with such topics as the structure of matter, the formation of molecules, the course and mechanism of chemical reactions, as well as with the chemistry of living systems.

BONNER, J. and J. E. VARNER: *Plant Biochemistry,* 3rd ed., Academic Press, Inc., New York, 1976.

The best modern account of this rapidly expanding field, this is a collection of articles by leading specialists that includes much structural interpretation.

BROWN, WALTER V., and ELDRIDGE M. BERTKE: *Textbook of Cytology,* 2nd ed., The C. V. Mosby Co., St. Louis, 1974.

A well-illustrated account of all species of cellular structure, emphasizing chromosomes.

DE ROBERTIS, E. D. P., W. W. NOWINSKI, and F. A. SAEZ: *Cell Biology,* 5th ed., W. B. Saunders Co., Philadelphia, 1970.

A thorough discussion of cell form, function, and genetics.

GIESE, ARTHUR C.: *Cell Physiology,* 4th ed., W. B. Saunders Co., Philadelphia, 1973.

A description of the major problems in cell physiology; it is intended for the more advanced student.

GOODENOUGH, URSULA, and R. LEVINE: *Genetics,* Holt, Rinehart and Winston, Inc., New York, 1974.

A thoroughly up-to-date account of molecular genetics.

HAYES, W.: *The Genetics of Bacteria and Their Viruses,* 2nd ed., Halsted Press, New York, 1969.

The fundamental reference in the field. This edition is vastly improved and indeed a superb account of the genetics of these organisms.

INGRAM, V. M.: *The Biosynthesis of Macromolecules,* 2nd ed., W. A. Benjamin, Inc., Menlo Park, Calif., 1970.*

A modern account of what is known about the biosynthesis of proteins and other large molecules that play a fundamental role in cellular organization.

JENSEN, W. A., and R. B. PARK: *Cell Ultrastructure,* Wadsworth Publishing Co., Inc., Belmont, Calif., 1967.*

A collection of electron micrographs of a wide range of cells and organisms. Intended to supplement biology textbooks used in introductory courses.

LEDBETTER, MYRON C., and KEITH R. PORTER: *Introduction to the Fine Structure of Plant Cells,* Springer-Verlag, Berlin and New York, 1970.

An outstanding collection of plant electron micrographs, perhaps the most technically accomplished of its kind available in a collected form.

LEHNINGER, ALBERT L.: *Biochemistry,* 2nd ed., Worth Publishers, Inc., New York, 1975.

The best biologically oriented biochemistry text available.

* Available in paperback.

LOEWY, A. G., and P. SIEKEVITZ: *Cell Structure and Function*, 2nd ed., Holt, Rinehart and Winston, Inc., New York, 1969.

An outstanding elementary text on cell structure and function.

MCELROY, W. D.: *Cell Physiology and Biochemistry*, 3rd ed., Prentice-Hall, Inc., Englewood Cliffs, N.J., 1971.*

A brief introduction to the basic principles of cell physiology and biochemistry.

O'BRIEN, T. P., and MARGARET E. MCCULLY: *Plant Structure and Development. A Pictorial and Physiological Approach*, The Macmillan Company, London, 1969.*

A beautifully illustrated atlas of cellular structure and development in the higher plants.

ROBARDS, A. W. (Ed.): *Dynamic Aspects of Plant Ultrastructure*, McGraw-Hill Book Company, New York, 1974.

An outstanding series of reviews of the cellular components discussed in this section, and of the specialized conducting cells considered in Section 5, by leading specialists in each field.

THOMAS, LEWIS: *The Lives of a Cell: Notes of a Biology Watcher*, Viking Press, New York, 1974.

Thomas, a physician and medical researcher, reveals the extent to which science can tune our intellectual antennae, broaden our perceptions, and extend our appreciation of ourselves and of the world around us. Anyone who wants to refute the contention that science destroys human values need look no further than these short, sensitive essays.

TROUGHTON, JOHN, and F. B. SAMPSON: *Plants: A Scanning Electron Microscope Survey*, John Wiley & Sons, Inc., New York, 1974.*

A scanning electron microscope study of some anatomical features in plants and the relationship of these features to physiological processes.

WATSON, J. D.: *The Double Helix*, Atheneum Publishers, New York, 1968.*

"Making out" in molecular biology. A lively account of how to become a Nobel laureate.

———: *Molecular Biology of the Gene*, 2nd ed., W. A. Benjamin, Inc., Menlo Park, Calif., 1970.

For the student who wants to go more deeply into the subject of modern genetics, this book is generally agreed to be outstanding.

WHITE, E. H.: *Chemical Background for the Biological Sciences*, 2nd ed., Prentice-Hall, Inc., Englewood Cliffs, N.J., 1970.*

A brief account of the principles of chemistry that are essential for an understanding of biological processes.

WHITEHOUSE, H. L.: *Towards an Understanding of the Mechanism of Heredity*, 3rd ed., St. Martin's Press, Inc., New York, 1973.*

A most interesting book, which discusses the field in terms of its hypotheses and how they were tested; a series of "case histories."

* Available in paperback.

SECTION 2

Photosynthesis and Respiration: Harvesting the Sun

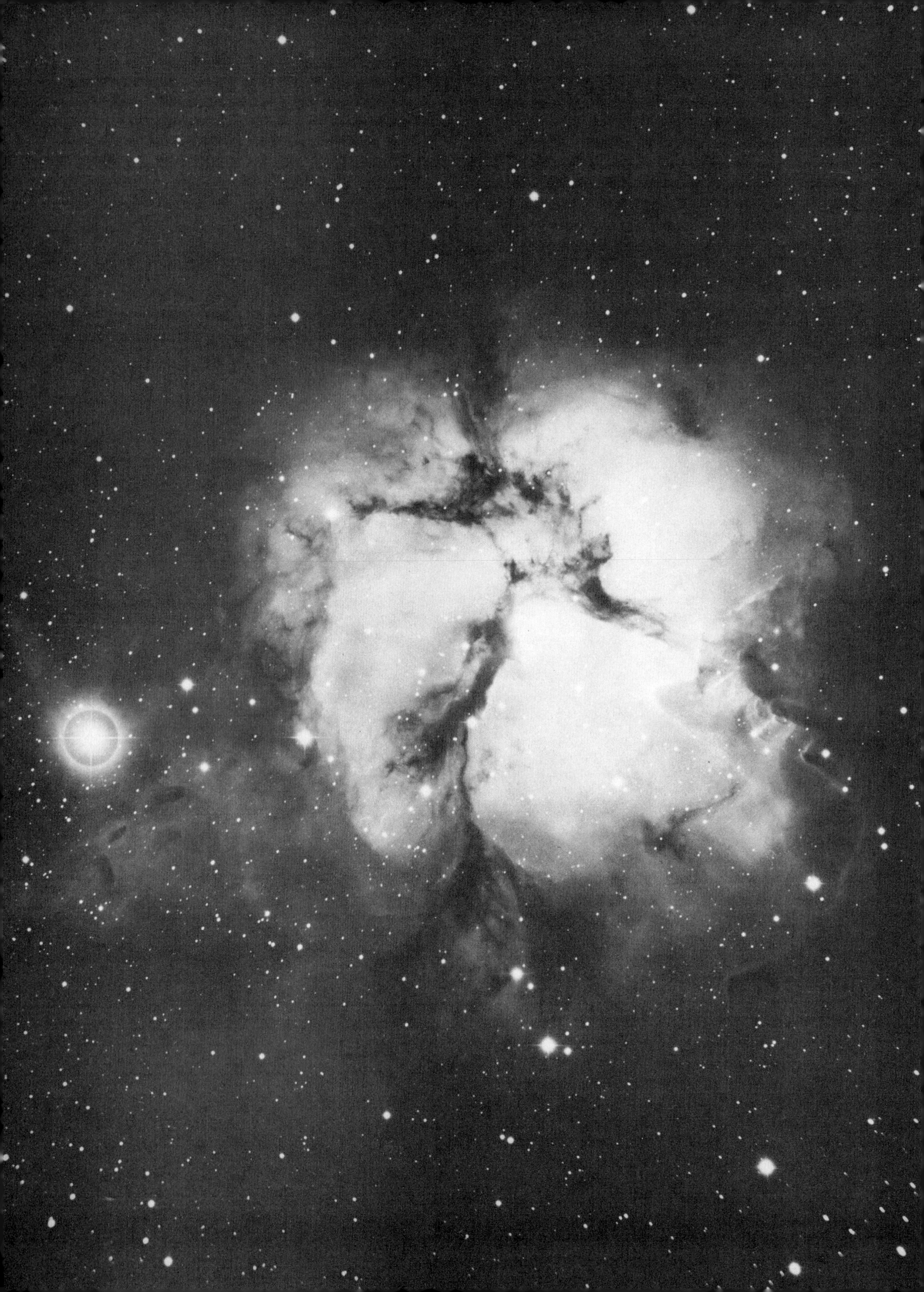

CHAPTER 4

The Flow of Energy

Life here on earth depends on the flow of energy from the thermonuclear reaction taking place at the heart of the sun. The amount of energy delivered by the sun is 13×10^{23} calories per year—the number 13 followed by 23 zeros. It is a difficult quantity to imagine. For example, the amount of energy striking the earth every day is the equivalent of 1 million Hiroshima-sized atom bombs. About one-third of this energy is immediately reflected back as light (as it is from the moon). Much of it is absorbed by the earth and converted to heat. Some of this absorbed heat energy serves to evaporate the waters of the ocean, producing the clouds that, in turn, produce the rainfall on the land. Solar energy, in combination with other factors, is also responsible for the movements of air and of water that help set patterns of climate.

Green plants absorb a small fraction of this light energy and transform it to chemical energy. This process of photosynthesis is the vital link between these vast energy resources of the sun and the energy needs of the living world.

AN HISTORICAL PERSPECTIVE

The role of plants in the economy of nature was not recognized until comparatively recently. Aristotle and the other Greeks, observing that the life processes of animals were dependent upon the food they ate, thought that plants derived their food from the soil.

A little over 300 years ago, the Belgian physician Jan Baptista van Helmont (Figure 4–2) offered the first experimental evidence that soil alone does not nourish the plant. Van Helmont grew a small willow tree in an earthenware pot for 5 years, adding only water to the pot. At the end of 5 years, the willow had increased in weight by 74.4 kilograms, whereas the earth had decreased in weight by only 57 grams. On the basis of these results, van Helmont concluded that all the substance of the plant was produced from the water and none from the soil.

4–1
Birth of a star. Our sun, like all stars, began as a swirling cloud of hydrogen gas in interstellar space.

4–2
"I took an earthen vessel, in which I put 200 pounds of earth that had been dried in a furnace, which I moistened with rainwater, and I implanted therein the trunk or stem of a willow tree, weighing five pounds. And at length, five years being finished, the tree sprung from thence did weigh 169 pounds and about three ounces. When there was need, I always moistened the earthen vessel with rainwater or distilled water, and the vessel was large and implanted in the earth. Lest the dust that flew about should be co-mingled with the earth, I covered the lip or mouth of the vessel with an iron plate covered with tin and easily passable with many holes. I computed not the weight of the leaves that fell off in the four autumns. At length, I again dried the earth of the vessel, and there was found the same 200 pounds, wanting about two ounces. Therefore 164 pounds of wood, bark and roots arose out of water only." Jan Baptista van Helmont.

4–3
Joseph Priestley—". . . no vegetable grows in vain. . . ."

This experiment is of general interest to those concerned with tracing the history of science because it is one of the first carefully designed biological experiments ever reported. As we shall see, van Helmont's conclusions were too broad.

Priestley's "Planetary Ventilation"

Fire, one of the "elements" of the ancient Greeks, intrigued the early alchemists and the Renaissance chemists who were their successors. And the question of what happens when an element is consumed by fire was similarly a focus of study and debate among the chemists of the eighteenth century.

Like his contemporaries, Joseph Priestley (1733–1804) believed that the flames leaping away from a burning object, such as a candle, represented something escaping from it. This "something," the inflammable principle, was called phlogiston. Air was needed to burn the candle because it was air that carried the phlogiston away. When the air became too saturated with phlogiston, combustion could no longer continue. And an intriguing additional observation was made about phlogiston: Phlogisticated air not only could not support burning, it could not support animal life. The phlogiston theory was the first theory in chemistry that began to deal with the unity of chemical processes.

On August 17, 1771, Priestley "put a sprig of mint into air in which a wax candle had burned out and found that, on the 27th of the same month, another candle could be burned in this same air." Priestley believed, as he reported, that he had "accidentally hit upon a method of restoring air that has been injured by the burning of candles." The "restorative which nature employs for this purpose," he stated, is "vegetation." Priestley extended his observations and soon showed that air "restored" by vegetation was not "at all inconvenient to a mouse." Priestley's experiments offered the first logical explanation of how the air remained "pure" and able to support life despite the burning of countless fires and the breathing of many animals. When he was presented with a medal for his discovery, the citation read in part: "For these discoveries we are assured that no vegetable grows in vain . . . but cleanses and purifies our atmosphere."

Here we see the germination of the concept of a vital relationship between plants and animals.

Only in the Light

Priestley's reports that plants dephlogisticate the air were of great interest to his fellow chemists, but they soon attracted criticism because the experiments could not readily be confirmed. In fact, when Priestley tried to do the experiments again himself, he did not get the same results. (It is possible that he moved his equipment

4–4
Lavoisier conducting an experiment on respiration. This drawing was made by Mme. Lavoisier, who included herself in the scene. Lavoisier was guillotined on May 8, 1794, during the French Revolution. The judge presiding over the case is reported to have said, "The Republic has no need of savants."

to a dark corner of his laboratory.) It was a Dutch physician, Jan Ingenhousz (1730–1799), who was finally able to confirm Priestley's work with an important addition. He found that dephlogistication takes place only in sunlight. Plants at night or in the shade, he reported, "contaminate the air which surrounds them, throwing out an air hurtful to animals." He also observed that only the green parts of plants restore the air and that "the sun by itself has no power to mend air without the concurrence of plants."

In 1782, Pastor Jean Senebier of Geneva showed that the amount of restored air produced by a plant is equal to the amount of phlogisticated air available to it. Slowly the different parts of the puzzle were being assembled.

Lavoisier and the Chemical Revolution

While Ingenhousz was performing his experiments on plants, Antoine Lavoisier (1743–1794), working in his private laboratory, was carrying out the experiments that put chemistry on an essentially modern basis (Figure 4–4). Lavoisier, through his work, helped to establish the law* of *conservation of matter.* This principle is one of the foundations of modern science and can be stated very simply: *Although matter may be transformed from one form to another, it is never lost.* We assume the law of conservation of matter every time we write a chemical equation.

By the 1780s, Lavoisier was prepared to make a direct attack on the powerful phlogiston theory. Most of his arguments were based on extremely careful measurements. Although his actual measurements were inaccurate by present-day standards, he established the principles of measurement that brought about a chemical revolution. He showed that a metal gained weight when it was heated, a fact difficult to reconcile with the theory that it had lost phlogiston. He then proved that the gain in weight resulted from a combination with what he called "eminently respirable air." (He later came to call this air oxygen.) He showed that when charcoal was heated, it combined with oxygen to produce a gas that was a compound of carbon and oxygen. This compound we now know as carbon dioxide.

Among his many discoveries, those that are most directly connected with the story we are following here concern animal respiration. Working with the mathematician P. S. Laplace (1749–1827), Lavoisier confined a guinea pig in oxygen for 10 hours and measured the carbon dioxide produced. He also measured the amount of oxygen used by a man active and at rest. In these experiments, he was able to show that the combustion of carbon compounds with oxygen to form carbon dioxide and water is the true source of animal heat and that during physical work, oxygen consumption increases. "Respiration is merely a slow combustion of carbon and hydrogen, which is similar in every respect to that which occurs in a lighted lamp or candle, and, from this point of

* Note that the word "law" in science has quite a different meaning than when it is applied to human social systems. A scientific law cannot be broken; if it is, it simply ceases to exist!

view, animals that breathe are really combustible bodies which burn and are consumed."

In modern language:

$$(CH_2O) + O_2 \rightarrow CO_2 + H_2O + \text{heat}$$

The work of Ingenhousz spanned the prematurely terminated career of Lavoisier. Quick to adopt Lavoisier's ideas about gases, Ingenhousz hypothesized that the plant was not just exchanging "good air" for "bad" and so making the world habitable for animal life. In the sunshine, he suggested, a plant absorbs the carbon from carbon dioxide, "throwing out at that time the oxygen alone, and keeping the carbon to itself as nourishment."

Nicholas Theodore de Saussure (1767–1845) applied Lavoisier's principles of quantitative measurements and showed that equal volumes of CO_2 and O_2 are exchanged during photosynthesis and that the plant does indeed retain the carbon. He also showed that more weight was gained by the plant during photosynthesis than could be accounted for by the carbon taken in as carbon dioxide. In other words, the carbon in the dry matter of plants came from carbon dioxide, but, equally important, the rest of the dry matter, with the exception of minerals from the soil, came from water.

Thus, all the components were identified—carbon dioxide, water, and light—and it became possible to write the overall photosynthetic equation:

$$CO_2 + H_2O + \text{light} \rightarrow (CH_2O) + O_2$$

If you compare this with the previous equation, you can see that photosynthesis is apparently respiration (or combustion) in reverse. It was not until the concept of energy became rigorously formulated, however, that the full impact of this idea was felt.

VAN NIEL'S HYPOTHESIS

In 1796, Ingenhousz had suggested that in photosynthesis carbon dioxide is split to yield carbon and oxygen, with the oxygen released as gas. Subsequently, the proportion of carbon, hydrogen, and oxygen atoms in sugars and starches was found to be about one atom of carbon per molecule of water (CH_2O), as the word "carbohydrate" indicates. Thus, in the overall reaction for photosynthesis—

$$CO_2 + H_2O + \text{light energy} \rightarrow (CH_2O) + O_2$$

—it was generally assumed that the carbohydrate came from a combination of the carbon and water molecules and that the oxygen was released from the carbon dioxide molecule. This entirely reasonable hypothesis was widely accepted. But, as it turned out, it was quite wrong.

The investigator who upset this long-held theory was C. B. van Niel of Stanford University. Van Niel, then a graduate student, was investigating photosynthesis in different types of photosynthetic bacteria. In their photosynthetic reactions, bacteria reduce carbon to carbohydrates, but they do not release oxygen. Among the types of bacteria van Niel was studying were the purple sulfur bacteria, which require hydrogen sulfide for photosynthesis. In the course of photosynthesis, globules of sulfur (S) are excreted or accumulated inside the bacterial cells. In these bacteria, van Niel found that this reaction takes place during photosynthesis:

$$CO_2 + 2H_2S \xrightarrow{\text{light}} (CH_2O) + H_2O + 2S$$

This finding was simple enough and did not attract much attention until van Niel made a bold extrapolation. He proposed that the generalized equation for photosynthesis is

$$CO_2 + 2H_2A \xrightarrow{\text{light}} (CH_2O) + H_2O + 2A$$

In this equation, H_2A stands for some oxidizable substance, such as hydrogen sulfide, free hydrogen, or any one of several other compounds used by photosynthetic bacteria—or water. In the photosynthetic algae and green plants, H_2A is water. In short, van Niel proposed that it was the water that was split by photosynthesis, not the carbon dioxide.

This brilliant speculation, first proposed in the early 1930s, was not proved until many years later when, eventually, isotopic oxygen (^{18}O) made it possible to trace the oxygen from water to free oxygen:

$$CO_2 + 2H_2{}^{18}O \xrightarrow{\text{light}} (CH_2O) + H_2O + {}^{18}O_2$$

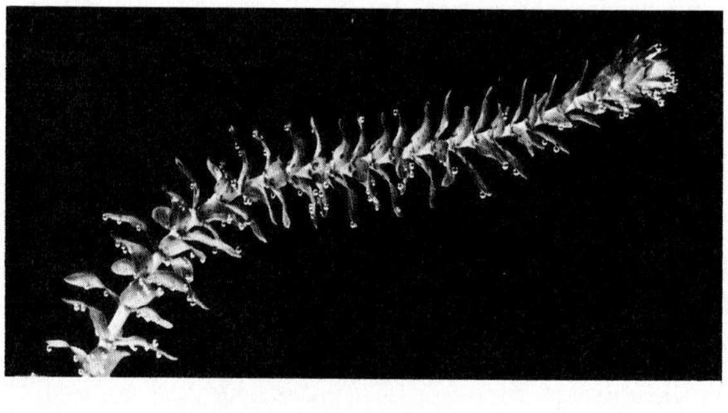

The bubbles on the leaves of this pondweed, *Elodea*, growing under water, are bubbles of oxygen, one of the products of photosynthesis. Van Niel was the first to hypothesize that the oxygen produced in photosynthesis came from the lysis of water rather than the splitting of carbon dioxide.

LIFE AND ENERGY: BIOENERGETICS

The law of the conservation of matter, set forth by Lavoisier, prepared the way for an even broader concept, the law of *conservation of energy.* This, too, is a simple law. It states: *Energy can be changed from one form to another, but it cannot be created or destroyed.* The total energy of any closed system remains constant, regardless of the physical or chemical changes it may undergo. Light is a form of energy, as is electricity. Light can be changed to electrical energy, and electrical energy can be changed to light (for example, by letting it flow through the tungsten wire in a light bulb). Energy can be stored in chemical bonds. When we burn carbohydrates, such as wood or paper, much of this energy is released in the form of heat. Motion is another kind of energy. A gasoline engine, for example, generates energy in the form of motion. In order to do this, it must use up energy in another form—the chemical energy in the gasoline. By burning the gasoline as fuel, the engine changes the chemical energy to heat and then changes the heat to motion.

Energy can be stored in various other forms. A boulder on a hill is an example of stored energy. The energy required to push the boulder up the hill is "stored" in the boulder as potential energy. When the boulder rolls downhill, the energy is released as kinetic (motion) energy.

Some useful energy is always dissipated as heat; this is why it is impossible to design a perpetual motion machine. When a boulder rolls downhill, the potential energy it received on being brought up the hill is converted to kinetic energy and also to heat energy due to friction. In a gasoline engine, about 75 percent of the energy originally present in the fuel is dissipated to the surroundings in the form of heat. Similarly, an electric motor converts only 25 to 50 percent of the electric energy into mechanical energy, the rest going into heat. (Heat is simply the random motion of atoms or molecules. The higher the heat, the greater the motion.)

Heat, then, is the common channel of energy "loss." It is also one of the most readily measurable forms of energy. For these reasons, the study of energy changes is usually referred to as *thermodynamics.* The law of conservation of energy is the *first law of thermodynamics.*

When we say that energy is "lost" as heat, what we actually mean is that it is no longer available to do work, because the heat is evenly distributed. Energy stored in a stick of dynamite can do work; that part of the energy that is released as heat at the time of the explosion is no longer available for work. The energy in the boulder at the top of the hill is also available to do work. If it is difficult to visualize a rolling boulder as having the capacity to do work, it is probably because boulders are not conventionally harnessed for this purpose. Substitute water, for which the first law of thermodynamics applies equally well. Suppose that, instead of the boulder, we transport water to the top of our imaginary hill and let it rush down again. We know from experience that the water could be used to turn a series of paddle wheels and that machines powered by such wheels could be used to grind corn or do other work useful in human terms.

This brings us to the *second law of thermodynamics.* Stated in its simplest form, this law says that *in all natural processes, energy in a form to do work is eventually converted to heat energy, which is dissipated out into the surroundings.* Another way of expressing this same law is to say that in any system, order tends to decrease and disorder to increase. A third, to return to our previous analogy, is that boulders do not roll uphill. Two further examples are shown in Figure 4–5. This second law, unlike the first one, shows the *direction* that natural processes take. As a consequence, it is sometimes known as "time's arrow."

Now let us look at these two laws with regard to biological systems. The cell is a specialist in energy exchange, interconverting electrical energy, light energy,

4–5
Illustrations of the second law of thermodynamics showing the increase of disorder, or randomness, in two physical systems. Such flows of energy never reverse spontaneously.

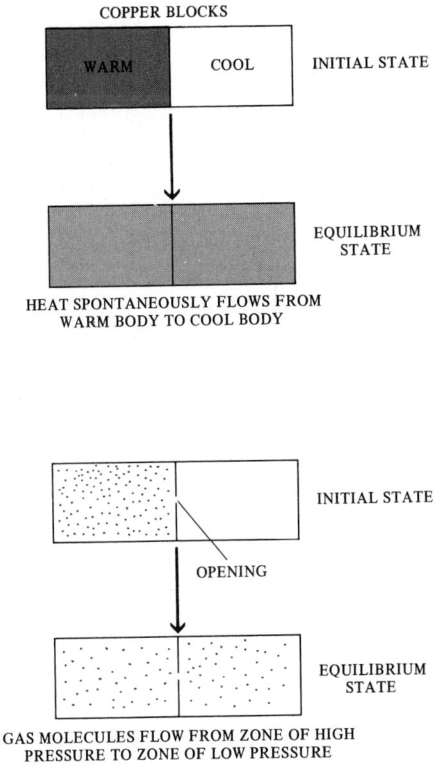

kinetic energy, and chemical energy and also, as we shall see, shifting the energy from one type of chemical bond to other forms that are more convenient for long-term storage or for immediate use. Most of the attributes that we would select as characteristic of living things are forms of energy exchange. They constantly take in useful energy from outside sources and release it into the environment in a less useful form. The ultimate outside source is, of course, the sun. Living organisms can continue to operate at this vast perpetual deficit only because of the tremendous stores of energy that enter the living world from this outside source.

In the words of Julius Robert Mayer, a nineteenth-century physicist who was one of the founders of the science of thermodynamics, photosynthesis is the process by which "the fleeting sun rays are fixed and skillfully stored for future use."

4–6
When the boulder is pushed uphill, kinetic energy is converted to potential energy and heat. When the boulder rolls downhill, potential energy is converted to kinetic energy and heat (which is also a form of energy).

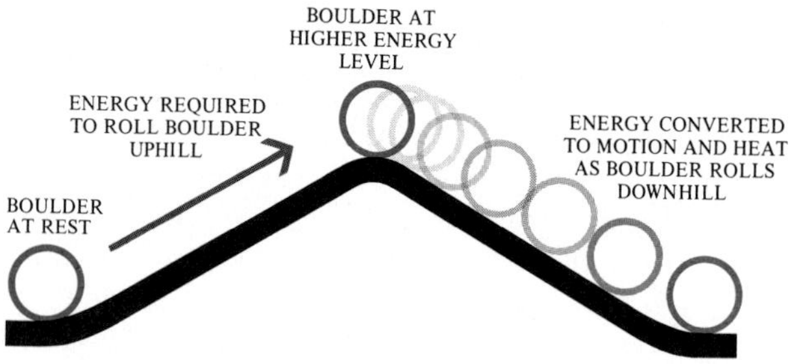

Energy and the Atom

In order to understand the general principles of energy conversion in living systems, it is helpful to review briefly the structure of atoms. Atoms, as you will recall, are made up of a central nucleus that contains protons (positively charged particles) and neutrons and an outer cloud of electrons, which are negatively charged particles moving in orbitals around the nucleus. Electrons occur at certain fixed distances from the nucleus, arranged in electron shells, which have different energy levels. Energy conversions in chemical systems involve the movement of electrons from one energy level to another or the transfer of electrons from one molecule to another.

In order to explain how the movement of such a minute particle can involve energy and why its distance from the nucleus determines its energy level, let us return to our analogy of the boulder.

A boulder sitting still on flat ground has no energy. If you push it up a hill, it gains potential energy. So long as it sits on the peak of the hill, it neither gains nor loses energy. If it is permitted to roll down the hill, some of the potential energy is converted into kinetic energy, and some to heat (Figure 4–6). Similarly, a spaceship poised on the launching pad has no energy. The chemical energy of rocket fuel is required to blast it away from earth, but no input energy is required for its return flight.

The electron is like the rocket in that an input of energy can raise it to a higher level, farther away from the nucleus. So long as it remains in this higher level, it possesses potential energy. When it returns to a lower level, its potential energy is released. The fall of electrons from higher to lower levels is the source of energy for all biological work. An atom or group of atoms in which one or more electrons have been pushed to a higher energy level is said to be in an *excited* state.

The initial process involved in the energy conversion of living systems is the absorption of light by the chlorophyll of green plants. By this process, the chlorophyll molecule is excited—that is, one of its electrons is elevated to a higher energy level. The electron only remains at this higher level for a fraction of a second. Then it either returns to its previous energy level or escapes from the molecule. In an isolated chlorophyll molecule, the electron returns to its ground state and its energy is released as heat or light (fluorescence). In the green plant cell, however, chlorophyll molecules are packed tightly together with other molecules and the dislodged electron is therefore passed from one molecule to another in such a way that the energy released in its fall becomes available for the work of the cell.

Oxidation and Reduction

The loss of an electron is known as *oxidation*, and the gain of an electron is *reduction*. Oxidation and reduction take place simultaneously, because an electron that is lost by one atom is accepted by another. For example, the transfer of an electron from chlorophyll to another molecule is an oxidation-reduction reaction in which the excited chlorophyll acts as the electron donor, or reducing agent (Figure 4–7).

Electron loss is traditionally referred to as oxidation because oxygen is very commonly the electron acceptor. Oxygen has an extremely strong affinity for electrons, and it is abundant in the modern atmosphere. As a consequence, the transferred electron is often (although not necessarily always) accepted by an oxygen atom. Often an electron travels in company with a proton—in short, as part of a hydrogen atom. In that case, oxidation results in the removal of the hydrogen ion (proton) and its electron, and reduction results in the transfer of both a hydrogen ion and an electron to another substance. The

4-7
When an atom (or a molecule) loses an electron, it is said to be oxidized. When it gains an electron, it is reduced.

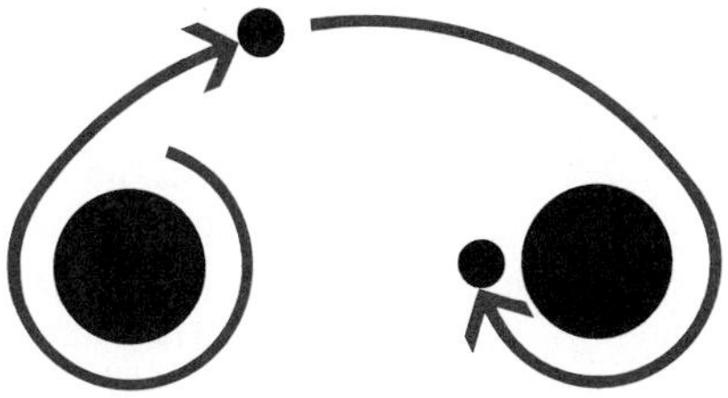

reduction of oxygen thus results in the formation of water. The reduction of carbon dioxide—which occurs in photosynthesis—is the formation of carbohydrate (CH_2O) from carbon dioxide, hydrogen ions, and electrons. The oxidation of carbohydrate—which occurs in respiration—yields carbon dioxide and water and releases the energy put into the molecule during photosynthesis. In eukaryotes, photosynthesis takes place in the chloroplasts, and respiration is completed in the mitochondria.

As we noted previously, oxidation and reduction always take place simultaneously. However, if a reaction results in a net increase in chemical bond energy (at the expense of some other kind of energy such as heat or light), it is often referred to as a reduction process. Conversely, if there is a net decrease in chemical bond energy (with a release of energy as heat or light), it is referred to as an oxidation process.

Energy "Loss"

In the course of rapid oxidation, such as occurs when a piece of paper or a stick of wood is burned, most of the energy in the chemical bonds is lost as heat. In order to define exactly what we mean when we speak of energy "loss," we need to look once again at the particles of which matter is composed.

Atoms and molecules are in constant motion. In a gas, the molecules are relatively far apart and move at high speeds. The rate of motion of the molecules is directly related to their temperature. If you heat the gas, the molecules move faster. If you cool the gas, the molecular motion slows down. If you slow it down sufficiently, the attractions between molecules begin to predominate and the gas condenses into a liquid. As the temperature of a liquid is lowered, the kinetic activity of the molecules decreases further. This decrease in activity permits the molecules to pack a little more closely together, until the liquid freezes to a solid. In a solid state, each molecule is confined to a definite small space between neighboring molecules. The molecules still continue to vibrate within this space, however, although the motion becomes slower and slower as the temperature decreases. Molecular movement does not stop until absolute zero, which is −273°C (−460°F).

Thus heat is one form of energy, the kinetic energy of molecules, and when we say that energy has been lost as heat, we mean that the energy has been converted to this new form, in which it is no longer available to do cellular work.

Energy Conversion in Living Systems

In living systems, only about half of the energy released from a carbohydrate molecule in the course of oxidation is converted to heat energy. The rest is retained as chemical energy, which can be used by the cell to carry out vital activities such as the biosynthesis of proteins and other compounds, the transport of materials across cell membranes, and movements of the cell or of its structures. The reason that so little of the energy is converted to heat, as compared with the heat produced by combustion, for example, is that the electrons are not transferred rapidly from the carbohydrate molecule to oxygen. To return to our analogy, such a rapid transfer would be like letting the boulder rush down a steep hill. In living systems, the electrons that have been "pumped up" to a high level by light energy during photosynthesis are passed "downhill" during respiration from one electron-accepting molecule to the next, each with a stronger attraction for the electron than the one before it, until the electrons finally reach oxygen, which has a very strong attraction for the negatively charged particles. Energy is released at particular steps during the downhill transfer and captured in a form that can be used by the cell.

Molecules Involved in Electron Transfer

The ability of the living cell to regulate the flow of energy depends on the presence within the cell of a number of highly specialized molecules involved in electron transfer. One such molecule is NAD (nicotinamide adenine dinucleotide), the structure of which is shown in Figure 4-8. NAD exists within the cell for only one purpose: It accepts electrons (and hydrogen atoms) from molecules that are being oxidized or gives up electrons (and hydrogen atoms) to molecules that are being reduced. The oxidation and reduction reactions are carried out by special enzymes known as hydrogenases, and NAD and similar compounds are sometimes referred to as coenzymes because their existence is essential to the particular enzyme in carrying out its functions.

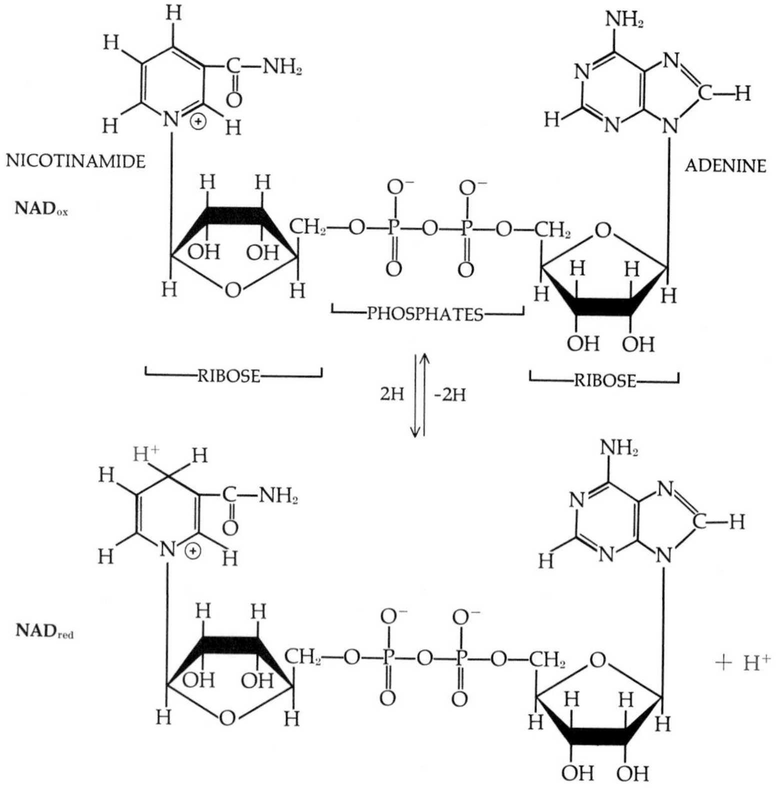

4–8
NAD (nicotinamide adenine dinucleotide) is made up of a submolecule of nicotinamide attached to a ribose sugar and a phosphate, which are attached in turn to a nucleotide of adenine. When NAD_{ox} becomes reduced, two hydrogen atoms (two protons and two electrons) are transferred to the nicotinamide ring. One of them is released as a hydrogen ion (H^+). Thus NAD_{ox} accepts a hydrogen ion (a proton) and two electrons to become NAD_{red}.

There are other molecules also involved in electron transfer that resemble NAD in structure, including NADP (which differs from NAD only in having a third phosphate group) and FP (flavoproteins).

All of these molecules are constructed around a nucleotide, as is ATP, which suggests that all may have had a common ancestor. However, this common biochemical ancestor must have existed millions and in fact billions of years ago, because these electron-transfer molecules are found not only throughout the eukaryotic world but in prokaryotes as well. These molecules all play a role in the reactions that will be described in the next two chapters.

Another type of molecule involved in the transfer of electrons is a *cytochrome* (Figure 4–9). Cytochromes contain an atom of iron held in a nitrogen-containing ring; chlorophyll and hemoglobin have similar components, with chlorophyll containing a magnesium atom rather than iron, and hemoglobin, like the cytochromes, containing iron. Each cytochrome also is surrounded by a protein made up of about 100 amino acids. Cytochromes differ in their protein chains and also in the energy levels at which they hold electrons. Characteristically, they function as part of an electron transport chain in which a cytochrome will accept an electron from a compound that holds it at a higher level and pass it to a cytochrome that holds it at a slightly lower level, which then passes it "downhill" to yet another. Such electron transport chains are found, as you will see in Chapters 5 and 6, both in mitochondria and in chloroplasts.

4–9
Cytochromes are molecules in which an atom of iron is held in a nitrogen-containing (porphyrin) ring. They are involved in electron transfer. It is the iron that actually combines with the electrons, each iron atom accepting an electron as it is reduced from Fe^{3+} to Fe^{2+}, that is, from the ferric to the ferrous form.

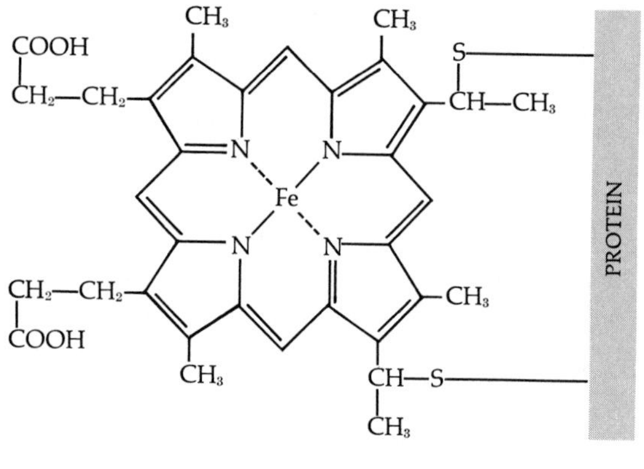

ATP: A Universal Currency

Now that we have glimpsed the complex machinery, it is appropriate to ask what this machinery is designed to produce. The answer is simple: ATP. As we noted in Chapter 2, ATP is aptly described as the universal currency of the cell. Even the very reactions that produce ATP require ATP to drive them forward. ATP is involved in the biosynthesis from small molecules of the many different types of large molecules required by the cell for its structures and functions. ATP is also involved in the active transport of materials across cell membranes. In cells that move by means of flagella, the chemical energy of ATP is converted to kinetic energy. In some organisms, the chemical energy is even converted back to light energy—which, of course, was its original source.

SUMMARY

In the process of photosynthesis, energy from the sun is stored in the chemical bonds of carbohydrate molecules. In the process of respiration, these carbohydrate molecules are oxidized and some of the energy released during these reactions is used to produce ATP.

The energetics of living systems are governed by two physical principles, known as the laws of thermodynamics. The first law states that energy can be changed from one form to another but that it cannot be created or destroyed. The second law states that in any closed system the useful energy—the energy available to do work—declines.

Living cells are specialists in energy conversions. The ultimate source of energy for living systems is light. The energy of light, absorbed by chlorophyll, raises electrons to higher energy levels. Electrons raised to a higher energy level may be passed from one atom or molecule to another. The loss of an electron is oxidation, and the gain of an electron is reduction. Oxidation and reduction take place simultaneously. Photosynthesis involves the reduction of carbon dioxide to carbohydrate. The oxidation of carbohydrates involves the reduction of oxygen to form water. When carbohydrate is oxidized rapidly, much of the energy stored in the molecule during photosynthesis is "lost" as heat. In cells, carbohydrates are oxidized in a series of small separate steps that enable the cell to conserve almost half of the energy of the carbohydrate molecule in the form of ATP.

Cells are able to capture and convert the energy of the falling electron because of the existence of highly specialized molecules including electron acceptors, such as NAD (nicotinamide adenine dinucleotide) and cytochromes.

CHAPTER 5

Respiration

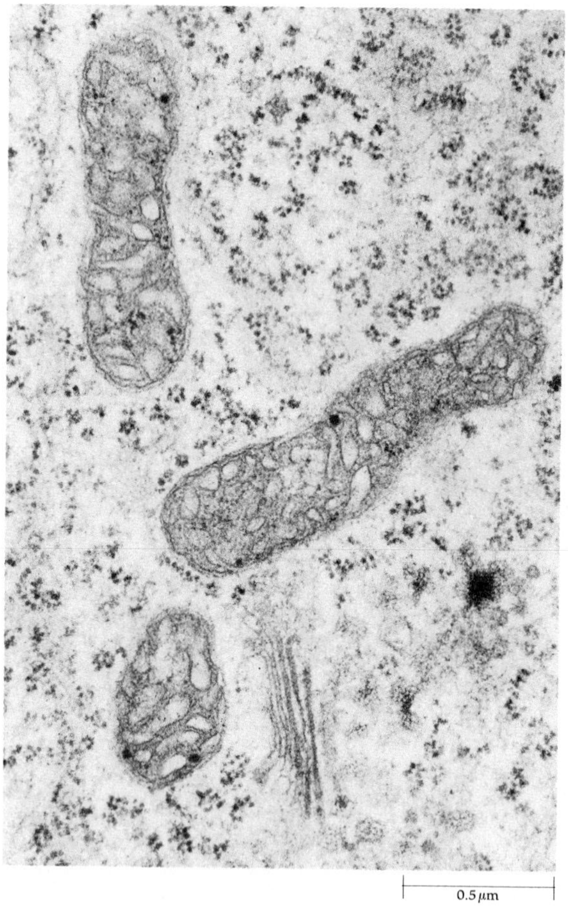

5-1
Mitochondria from leaf cell of Zea mays. *Mitochondria are the sites of cellular respiration by which chemical energy is transferred from carbon-containing compounds to ATP. Most of the ATP is produced on the surfaces of the cristae by enzymes that form a part of the structure of these membranes.*

Respiration is the means by which the energy of carbohydrates is transferred to ATP, the universal energy-carrier molecule (page 64) and so made available for the immediate energy requirements of the cell. In the following pages, we shall show in some detail how a cell breaks down carbohydrates and uses the released energy to recharge ADP to ATP, thereby storing energy in the terminal "high-energy" phosphate bond of that molecule. We describe the process in detail, not because we believe it should be memorized but because it provides an excellent illustration both of chemical principles described in previous chapters and also of the way in which cells perform biochemical work.

As we mentioned in Chapter 2, energy-yielding carbohydrate molecules are generally found in plants as sucrose or as starch. A necessary preliminary to the respiratory sequence is the hydrolysis of these transport or storage molecules to monosaccharides (single sugars). Respiration itself is generally considered to begin with glucose, which, you will recall, is a building block of both sucrose and starch. Glucose is also the transport molecule for carbohydrates in animal systems, in which respiration is the chief source of chemical energy.

The glucose molecule is broken down gradually in three distinct stages: *glycolysis,* the *Krebs cycle,* and the *electron transport chain.* In glycolysis, the six-carbon glucose molecule is broken down to two three-carbon molecules of pyruvic acid. In the second and third stages, the Krebs cycle and the electron transport chain, the pyruvic acid molecules are broken down to carbon dioxide and water. The overall equation, in a simplified form, is

$$C_6H_{12}O_6 + 6O_2 \rightarrow 6CO_2 + 6H_2O + \text{energy}$$

carbohydrate + oxygen → carbon dioxide + water + energy

As the glucose molecule is oxidized, some of the energy that was packed into it by the reactions that took place in the chloroplast is extracted in a series of small, discrete steps and packaged in the form of ATP.

In accord with the second law of thermodynamics, some of this chemical energy is dissipated as heat energy. In birds, mammals, and to some extent, lower vertebrates, the heat generated by cellular respiration is conserved by various mechanisms so that the body temperature of the organism generally remains above that of the ambient air. In plants, respiration is generally so slow that any heat produced has virtually no effect on body temperature. However, during the rapid growth associated with flowering, some plants, such as *Philodendron* and the eastern skunk cabbage, have been found to have temperatures as much as 20°C above air temperatures.

GLYCOLYSIS

Glycolysis, which takes place in the ground substance of the cytoplasm, involves three important events. First, the energy-rich glucose molecule is broken down into two simpler, energy-poorer molecules of pyruvic acid. The energy change can be expressed in terms of calories. One mole of glucose contains 686,000 calories; that is, if a mole of glucose is burned completely, 686,000 calories will be released as heat. Two moles of pyruvic acid have a total energy content of about 547,000 calories. Therefore, the energy released during glycolysis is about 139,000 calories per mole of glucose. The equation for the overall reaction is:

$$\underset{\text{glucose}}{C_6H_{12}O_6} \rightarrow \underset{\text{pyruvic acid}}{2CH_3COCOOH} + \underset{\text{protons}}{4H} + \underset{\text{electrons}}{4e}$$

As you can see, the glucose is oxidized in this reaction—that is, it gives up electrons—although no free oxygen is involved.

If this reaction took place all at once, all of the energy released would be dissipated. However, the breakdown of glucose is accomplished by a series of separate reactions, each of which is catalyzed by a different enzyme.

The chemical modification that occurs at each step in the series is very small. The molecule is broken apart, a little at a time. In this way, only a small amount of energy is released each time a chemical bond is broken. If the energy of the molecule were liberated all at once, it would produce heat, as Lavoisier recognized. A sudden burst of heat would be of little use to the cell and could be destructive. By controlling the release of energy, the cell loses only a portion of it as heat and is able to conserve or recover much of it in the form of useful chemical energy.

The second important feature of glycolysis is *phosphorylation*, the process by which ATP is formed from ADP by the addition of inorganic phosphate (P_i). The phosphorylation that takes place during glycolysis is known as *substrate phosphorylation*. Four molecules of ATP are formed by substrate phosphorylation. These four ATPs are not net gain, however, because two molecules of ATP are put into the reaction, one at the first (Figure 5–2) and one at the third enzymatic step. The two ingoing ATPs provide the energy for this rearrangement and for "priming" the subsequent steps in the reaction. So the useful energy conserved from glycolysis consists of two ATPs.

The difference between ADP and ATP in terms of useful energy is approximately 7000 calories per mole. Therefore, in the oxidation of glucose to pyruvic acid, 14,000 calories are captured in the form of ATP.

In the third important event of glycolysis, the protons (hydrogen ions) and electrons given up by glucose during oxidation are accepted by the electron-acceptor molecule, nicotinamide adenine dinucleotide (NAD), described in the previous chapter, and NAD_{ox}, accepting two electrons, becomes NAD_{red}.

Two molecules of NAD_{red} are formed from each molecule of glucose during glycolysis. (We shall set these aside for the moment in our energy calculation, because they do not reach their ultimate destination until the respiratory cycle is completed.) Therefore, in order to write the complete equation of glycolysis, we must take into account not only glucose and pyruvic acid but also NAD_{ox}, NAD_{red}, ATP, the inorganic phosphate P_i, and ADP:

$$C_6H_{12}O_6 + 2ADP + 2P_i + 2NAD_{ox} \rightarrow 2C_3H_4O_3 + 2ATP + 2NAD_{red}$$

Thus, the end products of the glycolysis of one molecule of glucose are two molecules of pyruvic acid, two molecules of ATP, and two molecules of NAD_{red}.

5–2
The first step of the nine steps of glycolysis described on the following pages involves the transfer of a "high-energy" phosphate group from ATP to glucose. By this step, energy is put into the reaction. Like all the steps in glycolysis, this one is catalyzed by a specific enzyme.

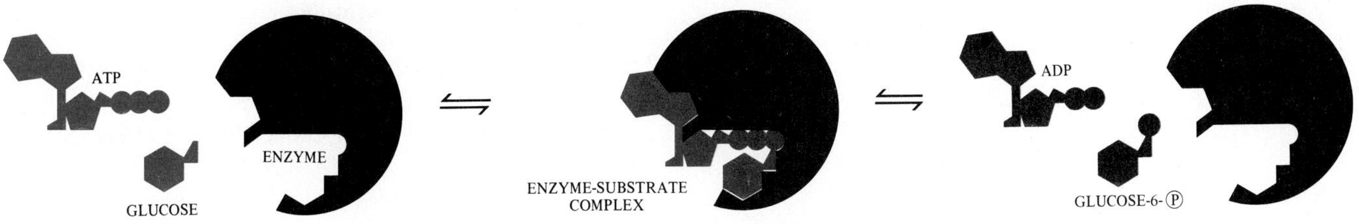

5-3
The steps of glycolysis.

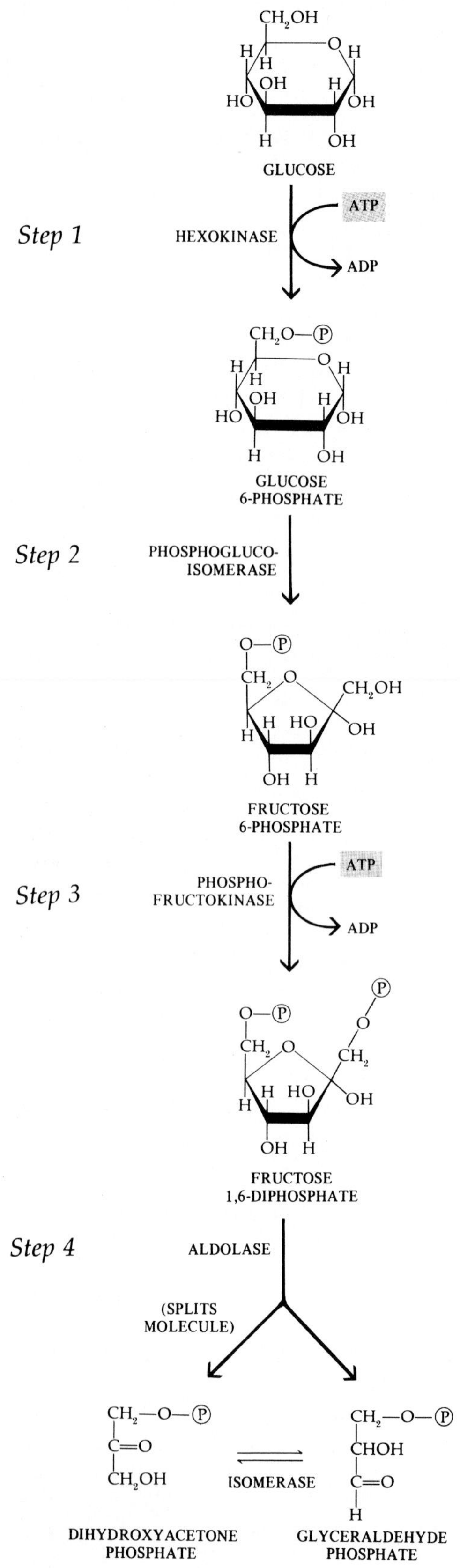

The Steps of Glycolysis

Do not try to memorize these steps, but follow them closely. Note especially the formation of ATP from ADP and of NAD_{red} from NAD_{ox}. These represent the cell's net gain from this energy transaction. (See Figure 5-3.)

Step 1: The first steps in glycolysis require an input of energy. This activation energy is supplied by the hydrolysis of ATP to ADP. The terminal phosphate group is transferred from an ATP molecule to the glucose molecule, to make glucose-6-phosphate. (This is also the first step, you may recall, in the biosynthesis of sucrose.) The combining of ATP with glucose to produce glucose-6-phosphate and ADP is an energy-yielding reaction, as we saw previously (page 64). Some of the energy released from the ATP is conserved in the chemical bond linking the phosphate to the sugar molecule. This reaction is catalyzed by a specific enzyme (hexokinase), and each of the reactions that follows is similarly regulated by a specific enzyme.

Step 2: The molecule is reorganized, again with the help of a particular enzyme. The six-sided ring characteristic of glucose becomes a five-sided fructose ring. As shown in Figure 2-3, glucose and fructose both have the same number of atoms—$C_6H_{12}O_6$—and differ only in the arrangements of these atoms. This reaction can proceed in either direction; it is pushed forward by the accumulation of glucose-6-phosphate and the disappearance of fructose-6-phosphate as the latter enters step 3.

Step 3: This step, which is similar to step 1, results in the attachment of a phosphate to the first carbon of the fructose molecule, producing fructose-1,6-diphosphate, that is, fructose with phosphates in the 1 and 6 positions. Note that in the course of the reactions thus far, two molecules of ATP have been converted to ADP and no energy has been recovered.

Step 4: The molecule is split into two three-carbon molecules. The two are interconvertible. However, because the glyceraldehyde phosphate is used up in subsequent reactions, all of the dihydroxyacetone phosphate is eventually converted to glyceraldehyde phosphate. Thus, all subsequent steps must be counted twice to account for the fate of one glucose molecule. With the completion of step 4, the preparatory reactions that require an input of ATP energy are complete.

Step 5: Glyceraldehyde phosphate molecules are oxidized—that is, hydrogen atoms with their electrons are removed—and NAD_{ox} becomes NAD_{red}. This is the first reaction from which the cell gains energy.

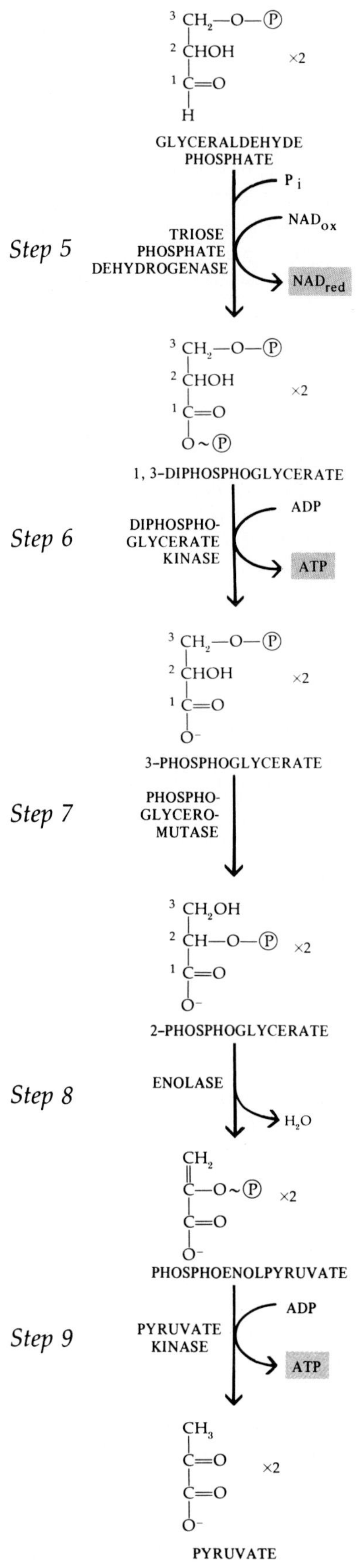

Energy from this oxidation reaction is also used to attach phosphate groups to what is now the 1 position of each of the glyceraldehyde molecules. (The designation P_i indicates inorganic phosphate available as a phosphate ion in solution in the cytoplasm.) Note that a high-energy bond is formed.

Step 6: The high-energy phosphate is released from the diphosphoglycerate molecule and used to recharge a molecule of ADP (a total of two molecules of ATP per molecule of glucose). This is a highly exergonic reaction and so pulls all the previous reactions forward.

Step 7: The remaining phosphate group is transferred from the 3 position to the 2 position.

Step 8: In this step, a molecule of water is removed from the three-carbon compound, and as a consequence of this internal rearrangement of the molecule, a high-energy phosphate bond is formed.

Step 9: The high-energy phosphate is transferred to a molecule of ADP, forming another molecule of ATP (again, a total of two molecules of ATP per molecule of glucose). This is also a highly exergonic reaction, and so the sequence runs downhill with accelerating force as it ends.

In summary, the complete sequence begins with one molecule of glucose. Energy is put into the sequence at steps 1 and 3 by the transfer of a phosphate group from an ATP molecule—one at each step—to the sugar molecule. The six-carbon molecule splits at step 4, and from this point onward, the sequence is exergonic. At step 5, a molecule of NAD_{ox} takes energy from the system and becomes NAD_{red}. At steps 6 and 9, molecules of ADP take energy from the system, form additional phosphate bonds, and become ATP. Thus one glucose molecule, using energy from the phosphate bonds of two ATP molecules to initiate the glycolytic reaction, produces two NAD_{red} from two NAD_{ox} molecules and four ATP from four ADP molecules.

$$\text{Glucose} + 2\text{ATP} + 4\text{ADP} + 2\text{NAD}_{ox} \longrightarrow$$
$$2\text{ pyruvate} + 2\text{ADP} + 4\text{ATP} + 2\text{NAD}_{red}$$

Phosphates and water (formed by the splitting of ATP) also participate in the reaction.

Two molecules of pyruvate* remain, and these two molecules still contain a large amount of the energy stored in the original glucose molecule. This series of reactions is carried out by virtually all living cells—from prokaryotes to the eukaryotic cells of plants and man.

* Pyruvic acid dissociates, producing pyruvate and a hydrogen ion (proton). The two forms (pyruvic acid and pyruvate) exist in dynamic equilibrium, and the two terms are used interchangeably.

THE MITOCHONDRION

The next stages of cellular respiration, the Krebs cycle and the electron transport chain, both require oxygen, unlike glycolysis, and in eukaryotic cells they take place in the specialized organelle known as the mitochondrion (see Figure 5-1). As shown in Figure 1-23, mitochondria are surrounded by two membranes. The outer one is smooth, and the inner one folds inward. The folds are called cristae. The more active a cell, the more numerous are both its mitochondria and the cristae within them. Within the inner compartment, surrounding the cristae, is a dense solution containing enzymes, coenzymes, water, phosphates, and other molecules involved in respiration. The mitochondrion is a self-contained chemical plant. The outer membrane lets most small molecules in or out freely, but the inner one permits the passage only of certain molecules, such as pyruvate and ATP, and restrains the passage of others. The enzymes of the Krebs cycle are in solution in the inner compartment. The enzymes and other components of the electron transport chain are built into the surfaces of the cristae. In the mitochondria, pyruvic acid from glycolysis is oxidized to carbon dioxide and water, completing the breakdown of the glucose molecule.

THE KREBS CYCLE

Before entering the Krebs cycle, each of the three-carbon pyruvic acid molecules is oxidized. The third carbon (the one in the carboxyl group) is removed, forming carbon dioxide. In the course of this exergonic reaction, a molecule of NAD_{red} is produced from NAD_{ox}. The glucose molecule has now been oxidized to two acetyl (CH_3CO) groups. These groups are momentarily accepted by a coenzyme known as coenzyme A, which is a large molecule, a portion of which is a nucleotide and a portion of which is a vitamin, pantothenic acid, one of the B-complex vitamins. The combination of the acetyl group and CoA is known simply as acetyl CoA (Figure 5-4).

Fats and amino acids can also be converted to acetyl CoA and enter the respiratory sequence at this point. A fat molecule is first hydrolyzed to glycerol and three fatty acids. Then successive two-carbon groups are removed, beginning at the carboxyl end. A molecule such as palmitic acid (Figure 2-11), which contains 16 carbon atoms, yields eight molecules of acetyl CoA. Amino acids also can be converted, by various reactions, to acetyl CoA.

In the Krebs cycle (Figure 5-5), the two-carbon acetyl group is combined with a four-carbon compound (oxaloacetic acid) to produce a six-carbon compound (citric acid). In the course of the cycle, two of the six carbons are oxidized to CO_2, and oxaloacetic acid is regenerated—thus making this series literally a cycle. Each turn around the cycle uses up one acetyl group and regenerates a molecule of oxaloacetic acid, which is then ready to begin the sequence again. In the course of these steps, some of the energy released by the oxidation of the carbon atoms is used to convert ADP to ATP (one molecule per cycle), some is used to produce NAD_{red} from NAD_{ox} (three molecules per cycle), and some is used to reduce a second electron carrier, a flavoprotein,

5-4
The three-carbon pyruvic acid molecule is oxidized to the two-carbon acetyl group that is combined with coenzyme A to form acetyl CoA. The oxidation of the pyruvic acid molecule is coupled to the reduction of NAD_{ox} Acetyl CoA enters the Krebs cycle.

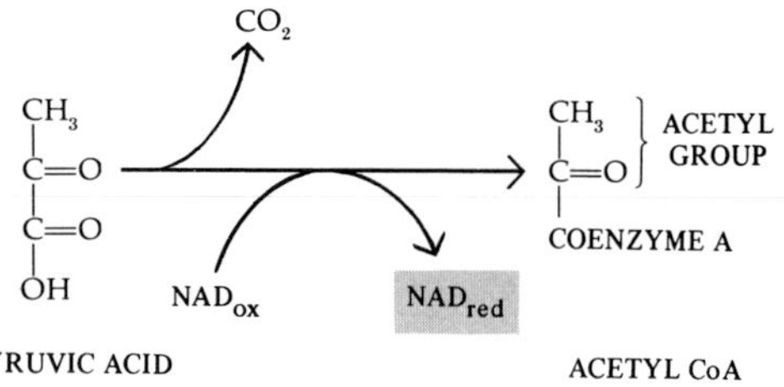

5-5
Summary of the Krebs cycle. In the course of the cycle, the carbons donated by the acetyl group are oxidized to carbon dioxide and the hydrogen atoms are passed to electron carriers. As in glycolysis, a specific enzyme is involved at each step. One molecule of ATP, three molecules of NAD_{red}, and one molecule of FP_{red} represent the energy yield of the cycle.

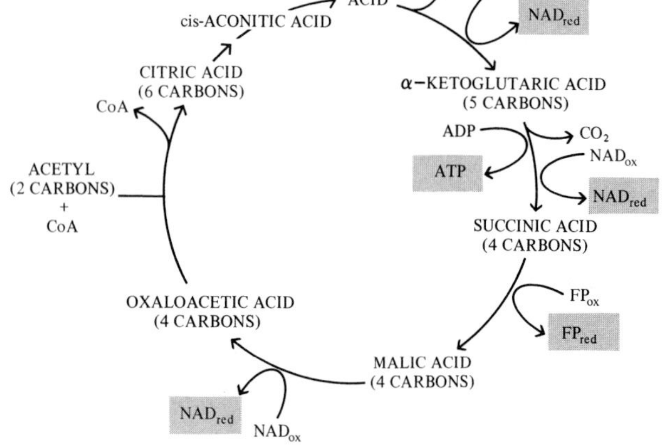

THE ANAEROBIC PATHWAY

In eukaryotic cells (and most prokaryotes also), pyruvic acid usually follows the pathway we describe, ending up completely oxidized as carbon dioxide and water.

However, if oxygen is not present, or is present only in low concentrations, the pyruvic acid molecule may continue to be broken down by oxygenless (anaerobic) processes, ending up either as lactic acid (in some bacteria, fungi, and animal cells) or as ethyl alcohol (in yeast and most plant cells).

Yeast cells, for example, can live without oxygen. Under such anaerobic conditions, they convert glucose to pyruvic acid by the glycolytic sequence shown in Figure 5–3, just as do other cells. The pyruvic acid so formed is then converted to ethyl alcohol (ethanol):

CH₃C COOH → CO₂
PYRUVIC ACID (FROM GLYCOLYSIS) → CH₃—C—H ACETALDEHYDE → CH₃CH₂OH ETHANOL
NAD_red (FROM GLYCOLYSIS) ↓ NAD_ox

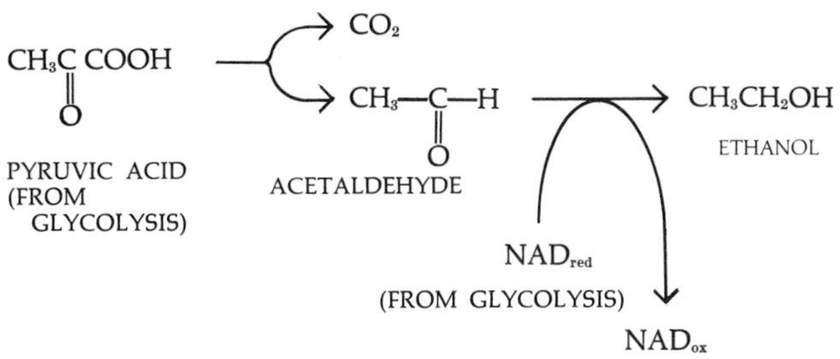

When the glucose-filled juices of grapes and other fruits are extracted and stored in airtight kegs, the yeast cells, present as a bloom on the skin of the fruit, turn the fruit juice to wine by converting glucose into ethanol. This process is called fermentation. Yeast, like all living things, has a limited tolerance for alcohol, and when a certain concentration (about 12 percent) is reached, the yeast cells die and fermentation ceases.

The fermentation of glucose to alcohol is carried on by some fungi other than yeasts and by some bacteria. Higher plants deprived of oxygen carry on alcoholic fermentation in the same way. Under natural conditions, this is most likely to occur in roots when soil is waterlogged. The roots of most species are injured or killed by being deprived of oxygen for several days.

The reduction of pyruvic acid to alcohol is an energy-requiring reaction and is accompanied by the oxidation of NAD_{red} (produced in the earlier stages of glycolysis) to NAD_{ox}.

The complete, balanced equation for the fermentation of glucose to ethyl alcohol is

$$C_6H_{12}O_6 + 2ADP + 2P_i \rightarrow$$
GLUCOSE
$$2CH_3CH_2OH + 2CO_2 + 2ATP$$
ETHYL ALCOHOL

Two molecules of ATP are generated from ADP and P_i in the course of fermentation. During this process, about 7 percent of the total available energy of the glucose molecule—about 52,000 calories—is released, with about 93 percent remaining in the two alcohol molecules. And, of the 52,000 calories released during fermentation, only 14,000 are trapped and stored in the two molecules of ATP. So, in terms of energy yield, anaerobic fermentation is relatively inefficient.

The fact that the glycolytic sequence does not require oxygen suggests that this series of reactions evolved early, before free oxygen was available, and that the primitive heterotrophs used glycolysis, or something very much like it, to extract energy from the organic compounds that they absorbed from their watery surroundings. Only a few types of organisms that depend solely on fermentation still exist on our planet. One of these, as an example, is the bacterium (Clostridium tetani) that causes tetanus. This dangerous and sometimes fatal disease develops in puncture wounds, such as those made by nails, where oxygen does not reach the damaged tissue and the bacteria flourish.

BIOLUMINESCENCE

Luminescent fungi (Mycena lux-coeli) photographed by their own light. In most organisms, the energy transferred to ATP is used in cellular work. In some, however, the chemical energy is sometimes reconverted to light energy. Bioluminescence is probably an accidental by-product of energy exchanges in most luminescent organisms, such as the fungi shown here. In some, however, such as fireflies, in which the flashes serve as mating signals, bioluminescence has come to serve a useful function.

abbreviated FP (one molecule of FP_{red} from FP_{ox} per cycle).

$$\text{oxaloacetic acid} + \text{acetyl CoA} + \text{ADP} + 3\text{NAD}_{ox} + \text{FP}_{ox} \longrightarrow \text{oxaloacetic acid} + 2\text{CO}_2 + \text{CoA} + \text{ATP} + 3\text{NAD}_{red} + \text{FP}_{red}$$

ELECTRON TRANSPORT

The glucose molecule is now completely oxidized. Some of its energy has been used to produce ATP from ADP. Most of it, however, still remains in electrons removed from the carbon atoms as they were oxidized and passed to the electron carriers NAD and FP. These electrons are at a high energy level. In the course of the electron transport chain, they are passed "downhill" and the energy released is used to form ATP molecules from ADP. This process is known as *oxidative phosphorylation.*

The electron carriers of the electron transport chain of the mitochondria differ from NAD and FP in their chemical structure. Most of them belong to the class of compounds known as cytochromes (see Chapter 4, page 98). Each cytochrome differs in its protein chain and also in the energy level at which it holds the electrons. They thus work in sequence, somewhat like a series of waterwheels. Water, as it flows from a point of higher water potential to one of lower water potential, releases energy which can be harnessed by waterwheels to do work. As electrons flow along the electron transport chain from a higher to a lower energy level, the cytochromes harness the released energy and use it to convert ADP to ATP. At the end of the chain, the electrons are accepted by oxygen and combine with protons (hydrogen ions) to produce water. Each time one pair of electrons passes from NAD_{red} to oxygen, three molecules of ATP are formed from ADP and phosphate. Each time a pair of electrons passes from FP_{red}, which holds them at a slightly lower energy level than NAD_{red}, two molecules of ATP are formed. (See Figure 5-6.)

We are now in a position to see how much of the energy originally present in the glucose molecule has been recovered in the form of ATP.

Glycolysis yielded two molecules of ATP directly and two molecules of NAD_{red}, for a total net gain of eight molecules of ATP.

The conversion of pyruvic acid to acetyl CoA yields

5-6

Summary of respiration. Glucose is first broken down to pyruvic acid, with a yield of two ATP molecules and the reduction (dashed arrows) of two NAD molecules. Pyruvic acid is oxidized to acetyl CoA, and one molecule of NAD is reduced (note that this and subsequent reactions occur twice for each glucose molecule; this electron passage is indicated by solid arrows). In the Krebs cycle, the acetyl group is oxidized and the electron acceptors, NAD and FP, are reduced. NAD and FP then transfer their electrons to the series of cytochromes that make up the electron-transport chain. As these cytochromes pass the electrons "downhill," the energy released is used to make ATP from ADP.

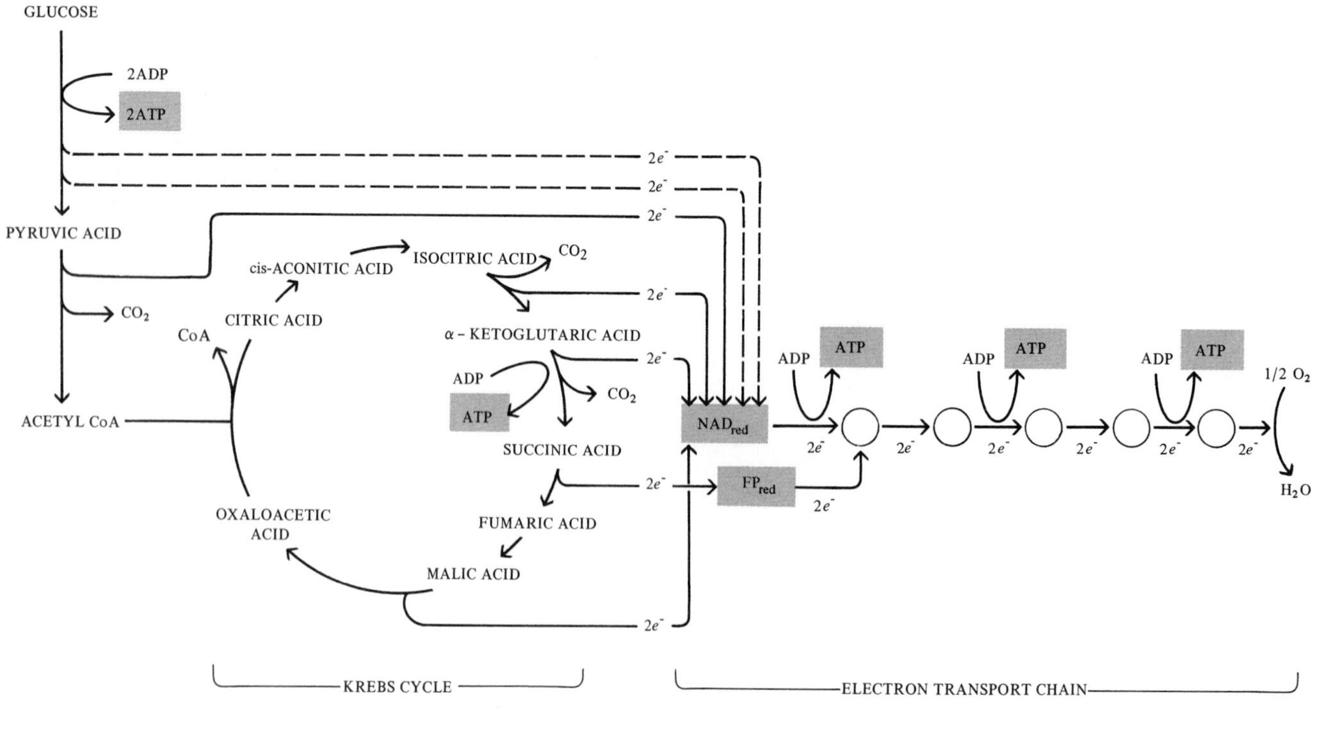

5-7
Energy changes in the oxidation of glucose. The complete respiratory sequence (glucose + $6O_2 \longrightarrow 6CO_2 + 6H_2O$) proceeds with an energy drop of 686 kilocalories. Of this, about 40 percent (266 kilocalories) is conserved in 38 ATP molecules. In anaerobic respiration (glucose $\longrightarrow$ ethyl alcohol + carbon dioxide), by contrast, only two ATP molecules are produced, representing only about 2 percent of the available energy of glucose.

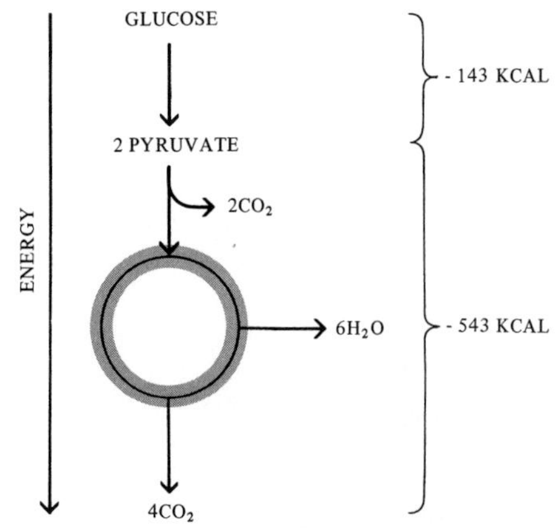

two molecules of NAD_{red} for each molecule of glucose and so produces six molecules of ATP.

The Krebs cycle yields, for each molecule of glucose, two molecules of ATP, six of NAD_{red}, and two of FP_{red}, or a total of 24 ATPs.

As the balance sheet, Table 5-1, shows, the complete yield from a single molecule of glucose is 38 molecules of ATP (Figure 5-7). Note that all but two of the 38 molecules of ATP have come from reactions taking place in the mitochondrion, and all but four involve the oxidation of NAD_{red} or FP_{red} along the electron transport chain.

The total difference in free energy between the reactants (glucose and oxygen) and the products (carbon dioxide and water) is 686 kilocalories. Almost 40 percent of this, about 266 (7 × 38) kilocalories, has been captured in the "high-energy" bonds of the 38 ATP molecules.

Table 5-1 *Summary of Energy Yield from One Molecule of Glucose*

Glycolysis:		2 ATP		
	2 NAD_{red} ⟶	6 ATP	⟶	8 ATP
Pyruvate ⟶ acetyl CoA:	1 NAD_{red} ⟶	3 ATP	(× 2) ⟶	6 ATP
Krebs cycle:		1 ATP		
	3 NAD_{red} ⟶	9 ATP	(× 2) ⟶	24 ATP
	1 FP ⟶	2 ATP		

SUMMARY

The oxidation of glucose, the process of respiration, is a chief source of energy in most cells. As the glucose is broken down in a series of small enzymatic steps, the energy in the molecule is packaged in the form of high-energy bonds in molecules of ATP.

The first phase in the breakdown of glucose is glycolysis, in which the six-carbon glucose molecule is split into two three-carbon molecules of pyruvate, and two new molecules of ATP and two of NAD_{red} are formed. This reaction takes place in the ground substance of the cytoplasm of the cell.

In the course of respiration, the three-carbon pyruvate molecules are broken down to two-carbon acetyl groups, which then enter the Krebs cycle. In the Krebs cycle, the two-carbon acetyl group is broken apart in a series of reactions to carbon dioxide. In the course of the oxidation of each acetyl group, four electron acceptors (three NAD and one FP) are reduced, and another molecule of ATP is formed.

The final stage of breakdown of the fuel molecule is the electron transport chain, which involves a series of electron carriers and enzymes embedded in the inner membranes of the mitochondrion. Along this series of electron carriers, the high-energy electrons accepted by NAD_{red} and FP_{red} during the Krebs cycle pass downhill to oxygen. Each time a pair of electrons passes down the electron-transport chain, ATP molecules are formed from ADP and phosphate. In the course of the breakdown of the glucose molecule, 38 molecules of ATP are formed, most of them in the mitochondrion.

CHAPTER 6

Photosynthesis

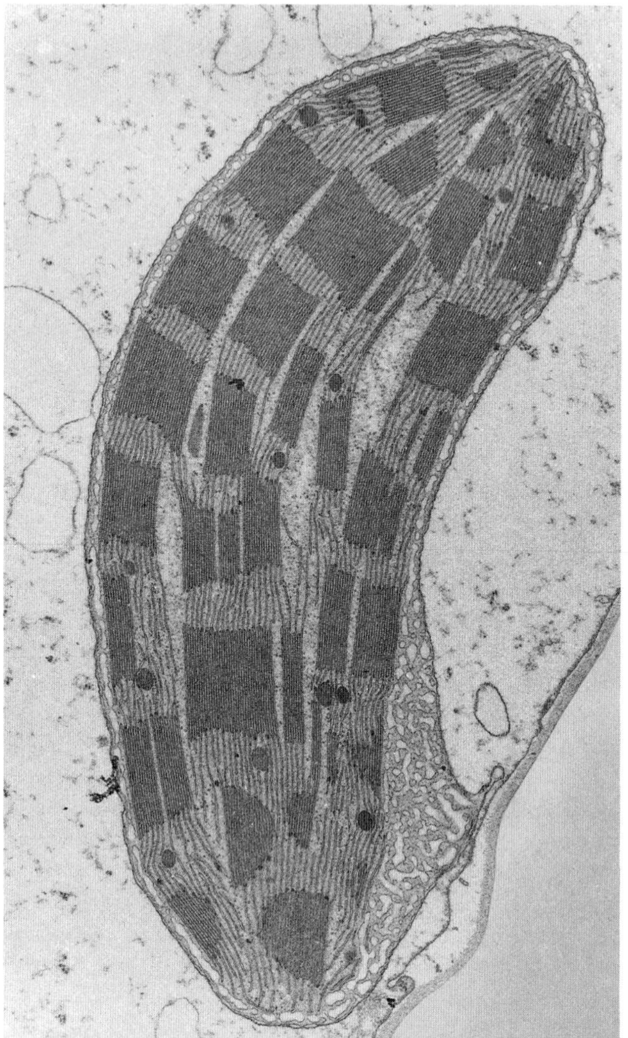

6–1
The chloroplast is the site of photosynthesis in all plants. The light-capturing reactions take place in the lamellae, or thylakoids, where the chlorophylls and other pigments are found. The series of reactions by which this energy is transferred to carbon-containing compounds takes place in the stroma, the material surrounding the photosynthetic membranes.

In the previous chapter, we described the breakdown of carbohydrates to yield the energy required for the many different kinds of activities carried out by living systems. In the pages that follow, we shall complete the circle by describing the way in which the energy coming from the sun in the form of light is captured and converted to chemical energy.

This process—photosynthesis—is the route by which virtually all energy enters our biosphere. Without this flow of energy from the sun, channeled through the chloroplasts of eukaryotic cells, the pace of life on this planet would swiftly diminish and then, following the inexorable second law of thermodynamics, cease altogether.

THE LIGHT REACTIONS

The Role of Pigments

The first step in the conversion of light energy to chemical energy is the absorption of light. A *pigment* is any substance that absorbs light. Some pigments absorb all wavelengths of light and so appear black. Some absorb only certain wavelengths and so transmit or reflect the wavelengths they do not absorb. Chlorophyll, the pigment that makes leaves green, absorbs light principally in the violet and blue wavelengths and also in the red; because it reflects green light, it appears green. Different pigments absorb light energy at different wavelengths. The absorption pattern of a pigment is known as the *absorption spectrum* of that substance (Figure 6–2).

When pigments absorb light, electrons are boosted to a higher energy level with three possible consequences: (1) The energy may be converted to heat. (2) It may be remitted as light energy: When the energy is remitted almost instantaneously, the phenomenon is known as fluorescence; when it is remitted later, it is known as phosphorescence. The light emitted is always at a lower wavelength than the light absorbed, another example of the second law of thermodynamics. (3) The energy may be captured in a chemical bond, as it is in photosynthesis.

An action spectrum defines the relative effectiveness (per incident number of photons, particles of light energy) of different wavelengths of light for light-requiring processes, such as photosynthesis, flowering, and phototropism (the bending of a plant toward light) (Figure 6–3). Similarity between the absorption spectrum of a pigment and the action spectrum of a process is considered as evidence that that particular pigment is responsible for that particular process.

Photosynthetic Pigments

Pigments that participate in photosynthesis include the chlorophylls, the carotenoids, and the phycobilins.

There are several different kinds of chlorophyll, which differ from one another only in details of their molecular structures. Chlorophyll *a* occurs in all photosynthetic eukaryotes and in the prokaryotic blue-green algae, and it is considered to be essential for photosynthesis of the type carried out by organisms of these groups. In the vascular plants, bryophytes, green algae, and euglenoid algae, chlorophyll *b* is also found. Chlorophyll *b* is an accessory pigment and, like the other accessory pigments, broadens the spectrum of light absorption in photosynthesis. When a molecule of chlorophyll *b* absorbs light, the excited molecule transfers its energy to a molecule of chlorophyll *a*, which then proceeds to transform it into chemical energy during the course of photo-

6–2
The absorption spectrum of a pigment is measured by a spectrophotometer. This device directs a beam of light of each wavelength at the object to be analyzed and records what percentage of light of each wavelength is absorbed by the pigment sample as compared with a reference sample. Because the mirror is lightly (half) silvered, half of the light is reflected and half is transmitted. The photoelectric cell is connected to an electronic device that automatically records the percentage absorption at each wavelength.

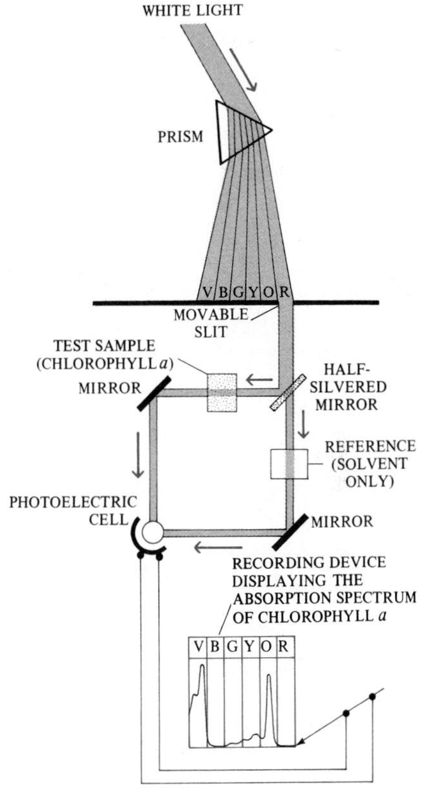

6–3
Results of an experiment performed in 1882 by T. W. Englemann revealing the action spectrum of photosynthesis in a filamentous alga. Like more recent investigators, Englemann used the rate of oxygen production to measure the rate of photosynthesis. Unlike his successors, however, he lacked sensitive devices for detecting oxygen. As his oxygen indicator, he chose bacteria that are attracted by oxygen. In place of the mirror and diaphragm usually used to illuminate objects under view in his microscope, he substituted a "microspectral apparatus" which, as its name implies, produced a tiny spectrum of colors that it projected upon the slide under the microscope. Then he arranged a filament of algal cells parallel to the spread of the spectrum. The oxygen-seeking bacteria congregated mostly in the areas where the violet and red wavelengths fell upon the algal filament. As you can see, the action spectrum for photosynthesis Englemann revealed in this experiment paralleled the absorption spectrum of chlorophyll. He therefore concluded that photosynthesis depends on the light absorbed by chlorophyll. This is an example of the sort of experiment that scientists refer to as "elegant." An elegant experiment is not only brilliant but also very simple in design and conclusive in its results.

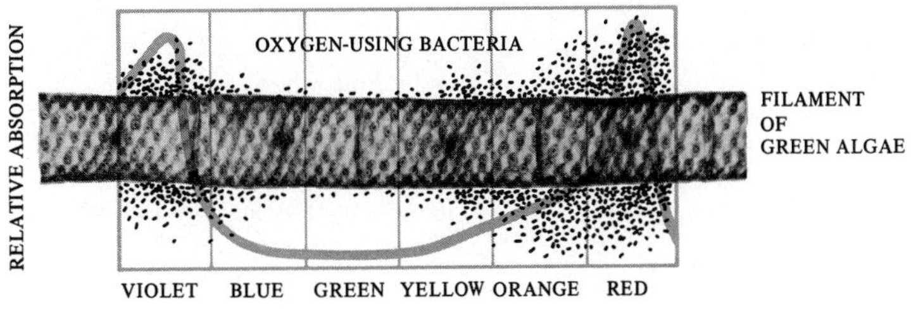

6-4
(a) Chlorophyll a *is a large molecule with a central core of magnesium held in a porphyrin ring. Attached to the ring is a long, insoluble carbon–hydrogen chain, which serves to anchor the molecule in the internal membranes of the chloroplast. Chlorophyll* b *differs from chlorophyll* a *in having a CHO group in place of the encircled CH_3 group. Alternating single and double bonds, known as conjugated bonds, such as those in the porphyrin ring of chlorophylls, are common among pigments. (b) The estimated absorption spectra of chlorophyll* a *and chlorophyll* b. *(Prepared by Govindjee.)*

(a)

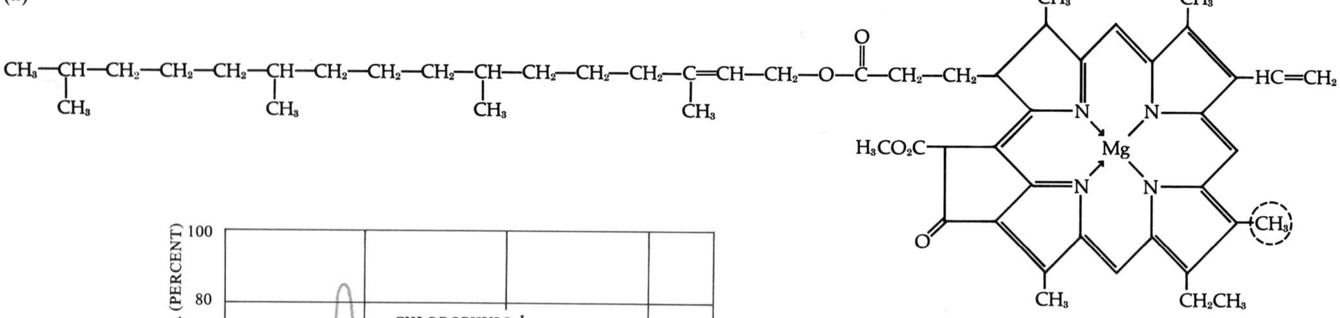

(b)

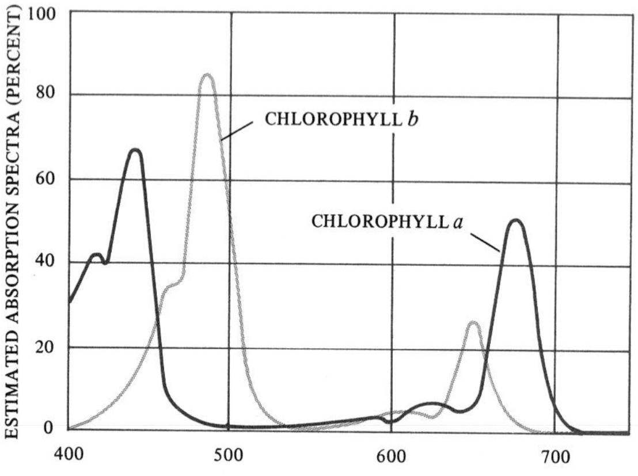

synthesis. Because chlorophyll *b* absorbs light of different wavelengths from chlorophyll *a* (Figure 6–4), it extends the range of light that can be used for photosynthesis. In the leaves of green plants, chlorophyll *b* generally constitutes about one-fourth of the total chlorophyll content. Chlorophyll *c* takes the place of chlorophyll *b* in other groups of plants.* Two other classes of pigments are involved in the capture of light energy in photosynthesis—the *carotenoids* and the *phycobilins*. Carotenoids are red, orange, or yellow fat-soluble pigments found in all chloroplasts and also, in association with chlorophyll *a*, in the prokaryotic blue-green algae. Like the chlorophylls, the carotenoid pigments of chloroplasts are embedded in the thylakoid membranes. Carotenoids that do not contain oxygen are called *carotenes*, and those that do are called *carotenols* (Figure 6–5). In the green leaf, the color of the carotenoids is masked by the much more abundant chlorophylls. In some tissues, such as those of a ripe red tomato or the petals of a yellow flower, the carotenoids predominate. Not being water-soluble, carotenoids are not found free in the cytoplasm, but, like the chlorophylls, are bound to proteins within the plastids.

The third major class of accessory pigments, the phycobilins, are found in the blue-green algae, in the chloroplasts of the red algae, and in a few other groups of eukaryotic algae. Unlike the carotenoids, the phycobilins are water-soluble. The structures of two phycobilins are shown in Figure 10–25 (page 197).

Pigment Systems

If chlorophyll molecules are isolated in a test tube and light is permitted to strike them, they fluoresce, and none of the light is converted to any form of energy useful to living systems. However, chlorophyll molecules embedded with accessory molecules in the fatty membrane of a thylakoid can trap this light energy in such a way that it can be converted to chemical energy. In the chloroplast, chlorophyll and other molecules are packed into what are known as photosynthetic units. Each unit contains from 250 to 400 molecules of pigment which can absorb light and pass it on to a reactive molecule. The energy absorbed by the reactive molecule is sufficient to force an electron completely away from the molecule, like a spaceship going out of orbit and heading for the moon. The reactive molecule is thus oxidized and minus an electron, a situation sometimes referred to as

* Photosynthetic bacterial cells contain either bacteriochlorophyll (in purple bacteria) or chlorobium chlorophyll (in the green sulfur bacteria). These molecules are illustrated in Figure 10–19. Chlorophyll *d* is a substance which may or may not occur naturally in red algae; see page 265.

6-5
(a) Related carotenoids. Cleavage of the beta-carotene molecule at the point shown yields two molecules of vitamin A. Oxidation of vitamin A yields retinene, the pigment involved in vision. In the carotenoids, the conjugated bonds are located in the carbon chains. Chloroplasts contain both carotenes and carotenols. Zeaxanthin, a carotenoid, is the pigment responsible for the yellow color of corn kernels. (b) *The estimated absorption spectrum of carotenoids in the chloroplast. (Prepared by Govindjee.)*

(a)

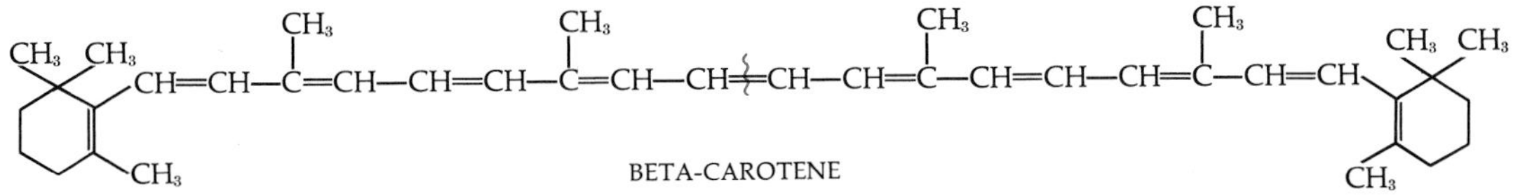

BETA-CAROTENE

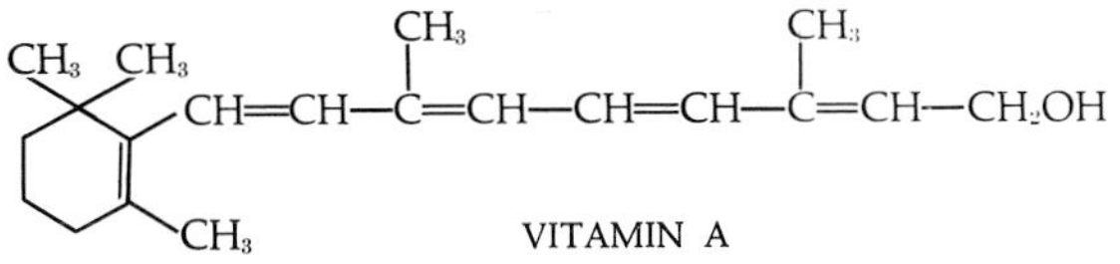

VITAMIN A

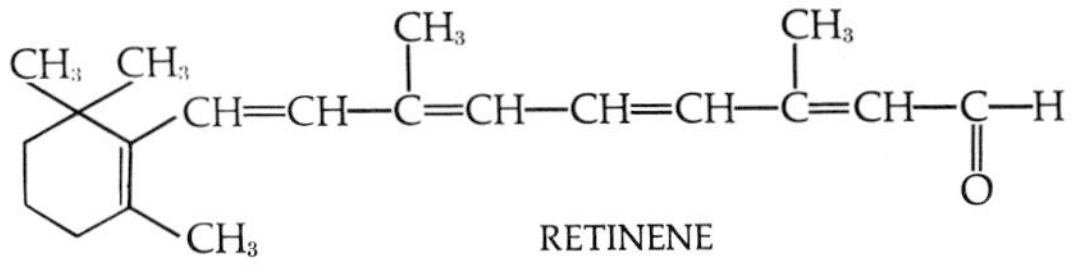

RETINENE

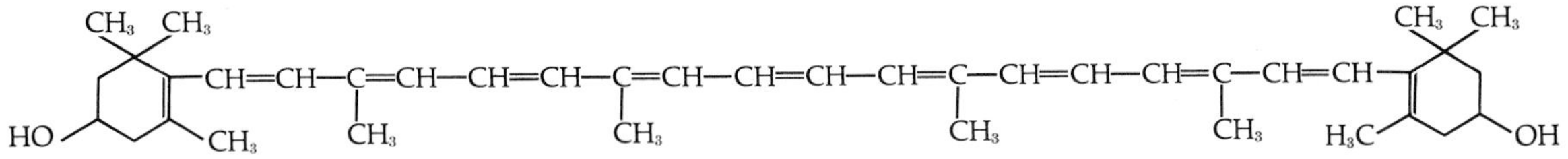

ZEAXANTHIN

having an electron "hole." (Remember the "hole"; it is important in the interpretation of subsequent events.)

In the leaves of plants, there are two different kinds of pigment systems, known simply as Pigment System I and Pigment System II. Pigment System I contains a larger proportion of chlorophyll *a* to chlorophyll *b* than Pigment System II. Carotenoids of different types are present in both systems, with System I containing mostly carotenes and System II mostly carotenols. These pigment systems can be separated from one another by centrifuging (which separates particles of different weights).

In Pigment System I, the reactive molecule is a form of chlorophyll *a* known as P_{700} because one of the peaks of its absorption spectrum is at 700 nanometers, slightly to the right of the usual chlorophyll *a* peak. When P_{700} is oxidized, it bleaches, which is how it was detected. No one has managed to detect P_{700} in isolation. It therefore seems that P_{700} is not an unusual kind of chlorophyll but rather chlorophyll *a* that has unusual properties because of its arrangement in the membrane and its position in relation to other molecules. Pigment System II also contains a specialized chlorophyll molecule that is capable of passing an electron on to an electron acceptor—probably also a form of chlorophyll *a*. Its reaction peak is at 680 nanometers. Thus, each pigment system forms part of a different photochemical system.

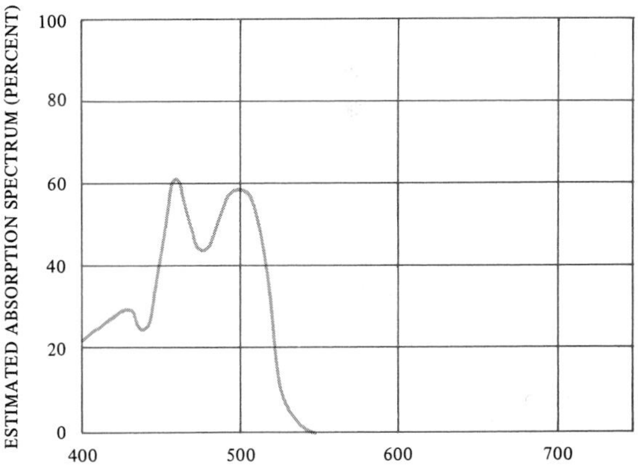

(b)

When white light passes through a prism, it is sorted into a spectrum of different colors. This separation takes place because each color has slightly different wavelengths.

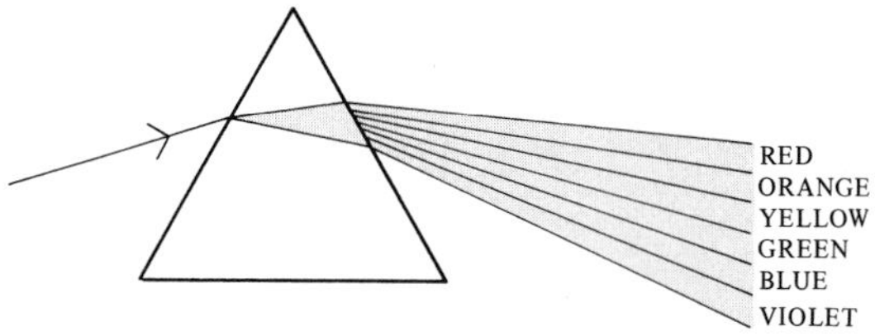

Almost 300 years ago, the English physicist Sir Isaac Newton (1642–1727) separated visible light into a spectrum of colors by letting it pass through a prism. By this experiment, Newton showed that white light is actually made up of a number of different colors, ranging from violet at one end of the spectrum to red at the other. Their separation is possible because light of different colors is bent at different angles in passing through the prism. Newton believed that light was a stream of particles (or, as he termed them, "corpuscles") because, in part, of its tendency to travel in a straight line.

In the nineteenth century, through the genius of James Clerk Maxwell (1831–1879), it came to be known that what we experience as light is in truth a very small part of a vast continuous spectrum of radiation, the electromagnetic spectrum. As Maxwell showed, all the radiations included in this spectrum travel in waves. The wavelengths—that is, the distances from one peak to the next—vary from those of x-rays, which are measured in nanometers, to those of low-frequency radio waves, which are measured in kilometers. The shorter the wavelength, the greater the energy. Within the spectrum of visible light, red light has the longest wavelength, violet the shortest. Another feature that these radiations have in common is that, in a vacuum, they all travel at the same speed—300,000 kilometers per second.

By 1900 it had become clear, however, that the wave theory of light was not adequate. The key observation, a very simple one, was made in 1888: When a zinc plate is exposed to ultraviolet light, it acquires a positive charge. The metal, it was soon deduced, becomes positively charged because the radiation energy dislodges electrons, forcing them out of the metal atoms. Subsequently it was discovered that this photoelectric effect, as it is known, can be produced in all metals. Every metal has a wavelength critical for the effect; the radiation (visible or invisible) must be of that wavelength or a shorter (more energetic) wavelength for the effect to occur. (The hypothesis that electrons orbit atoms at particular energy levels, as formulated by Bohr and others, is based on these observations.)

With some metals, such as sodium, potassium, and selenium, the critical wavelength is within the spectrum of visible light, and as a consequence, visible light striking the metal can set up a moving stream of electrons (such a stream is, of course, an electric current). The electric eyes that open doors for you at supermarkets or airline terminals, exposure meters, and television cameras all operate on this principle of turning light energy into electrical energy.

WAVE OR PARTICLE?

Now here is the problem. The wave theory of light would lead you to predict that the brighter the light, the greater the force with which the electrons would be dislodged. But as we have already seen, whether or not light can eject the electrons of a particular metal depends not on the brightness of the light but on its wavelength. A very weak beam of the critical wavelength or a shorter wavelength is effective, whereas a stronger (brighter) beam of a longer wavelength is not. Furthermore, as was shown in 1902, increasing the brightness of the light increases the number of electrons dislodged, but not the velocity at which they are ejected from the metal. To increase the velocity, one must use a shorter wavelength of light. Nor is it necessary for energy to be accumulated in the metal. With even a dim beam of a critical wavelength, an electron may be emitted the instant the light hits the metal.

To explain such phenomena, the particle theory of light was proposed by Albert Einstein in 1905. According to this theory, light is composed of particles of energy called photons. The energy of a photon is not the same for all kinds of light but is, in fact, inversely proportional to the wavelength—the longer the wavelength, the lower the energy. Photons of violet light, for example, have almost twice the energy of photons of red light, the longest visible wavelength.

The wave theory of light permits physicists to describe certain aspects of its behavior mathematically, whereas the photon theory permits another set of mathematical calculations and predictions. These two models are no longer regarded as opposed to one another; rather, they are complementary, in the sense that both are required for a complete description of the phenomenon we know as light.

The coexistence of these two theories further illustrates the scientific method. If a scientist defines light as a wave and measures it as such, it behaves like a wave. Similarly, if one defines it as a particle, one gets information about light as a particle. As Einstein once said, "It is the theory which determines what we can observe."

THE FITNESS OF LIGHT

Light, as Maxwell showed, is only a tiny band in a continuous spectrum. From the physicist's point of view, the difference between light and darkness—so dramatic to the human eye—is only a few nanometers of wavelength. There is no qualitative difference at all marking the borders of the light spectrum. Why does this particular small group of radiations, rather than some other, bathe our world in radiance, make the leaves grow and the flowers burst forth, cause the mating of fireflies and palolo worms, and, when reflecting off the surface of the moon, excite the imaginations of poets and lovers? Why is it that this tiny portion of the electromagnetic spectrum is respon-

Visible light represents only a very small portion of the vast electromagnetic spectrum.

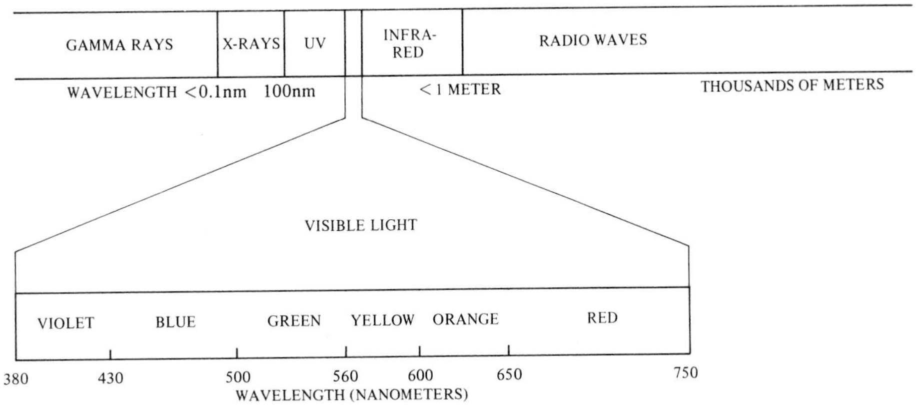

sible for vision, for phototropism (movement toward light), for photoperiodism (the changes that take place in an organism with the changing length of day and night in the changing seasons), and also for photosynthesis, on which all life depends? Is it an amazing coincidence that all these biological activities depend on these same wavelengths?

George Wald of Harvard, one of the greatest living experts on the subject of light and life, says no. He thinks that if life exists elsewhere in the universe, it is probably dependent upon this same fragment of the vast spectrum. Wald bases this conjecture on two points. First, living things, as we have seen, are composed of large, complicated molecules held in special configurations and relationships to one another by hydrogen bonds and other weak bonds. Radiation of even slightly higher energies than the energy of violet light breaks these bonds and so disrupts the structure and function of the molecules. Radiations with wavelengths below 200 nanometers drive electrons out of atoms; hence they are called ionizing radiations. The energy of light of wavelengths longer than those of the visible band is largely absorbed by water, which makes up the great bulk of all living things. When such light does reach organic molecules, its lower energy causes them to increase their motion (increasing heat) but does not trigger changes in their structure. Only those radiations within the range of visible light have the property of exciting molecules—that is, of raising electrons from one energy level to another—and so of producing biological changes.

The second reason that the visible band above all others of the electromagnetic spectrum was "chosen" by living things is that it, above all, is what is available. Most of the radiation reaching the earth from the sun is within this range. Higher-energy wavelengths are screened out by the oxygen and ozone high in the atmosphere. Much infrared radiation is screened out by water vapor and carbon dioxide before it reaches the earth's surface.

This is an example of what has been termed "the fitness of the environment"; the suitability of the environment for life and the suitability of life for the physical world are exquisitely interrelated. If they were not, life could not, of course, exist.

(a)

(b)

The spectrum of visible light is slightly different for the bee than for the human. The bee's vision is triggered by light in the ultraviolet portion of the spectrum. Hence a flower such as that of *Calylophus lavandulifolius*, which appears solid yellow to the human eye *(a)*, has distinctive markings for the bee *(b)*.

Model of Light Reactions

Figure 6–6 shows the present model of how the two photochemical systems work together in photosynthesis. Light energy first enters Pigment System II, where it is responsible for two important events: (1) Electrons are boosted uphill from a reactive molecule, and (2) water is split into protons, electrons, and oxygen gas. As the electrons are removed, they are passed uphill at increasingly higher energy levels to a primary electron acceptor. (The electron acceptors in these reactions, some of which are cytochromes (page 98), some of which are other electron transfer molecules whose identities are known, and some of which are unidentified, are all indicated by circles.) The electrons then pass downhill along an electron transport chain similar to the electron transport chain in the mitochondria. In the course of this passage, ATP is formed from ADP. This process is known as *photophosphorylation*, and is comparable to the oxidative phosphorylation process in cellular respiration. The photochemical systems probably evolved separately (with Photochemical System I coming first), and Photochemical System I, as we shall see, can probably operate separately. In general, however, they work together.

In Photosystem I, light energy boosts electrons from P_{700} to an electron carrier from which they are passed downhill to the electron acceptor molecule. The gain from these two steps in photosynthesis is represented by the ATP molecules (the exact number is not certain) and the $NADP_{red}$. Oxygen is a by-product. The $NADP_{red}$ is the chief source of the reducing power used to reduce carbon to organic compounds.

Why are there two photochemical systems rather than one? The answer seems to lie in the problem of the electron "holes." Suppose for a moment that System I were isolated from System II. There would be no way to fill the "holes" left in the P_{700} molecule in System I. This function is now carried out by the electrons boosted up from System II. And what fills the holes in System II? They are filled by electrons from the water molecules. When the electrons pass from the water molecule to the chlorophyll molecule (which has a greater electron affinity), the water molecule falls apart, releasing oxygen gas. Thus, all the pieces fit together.

As we mentioned previously, there is also some evidence that System I can work independently. When this

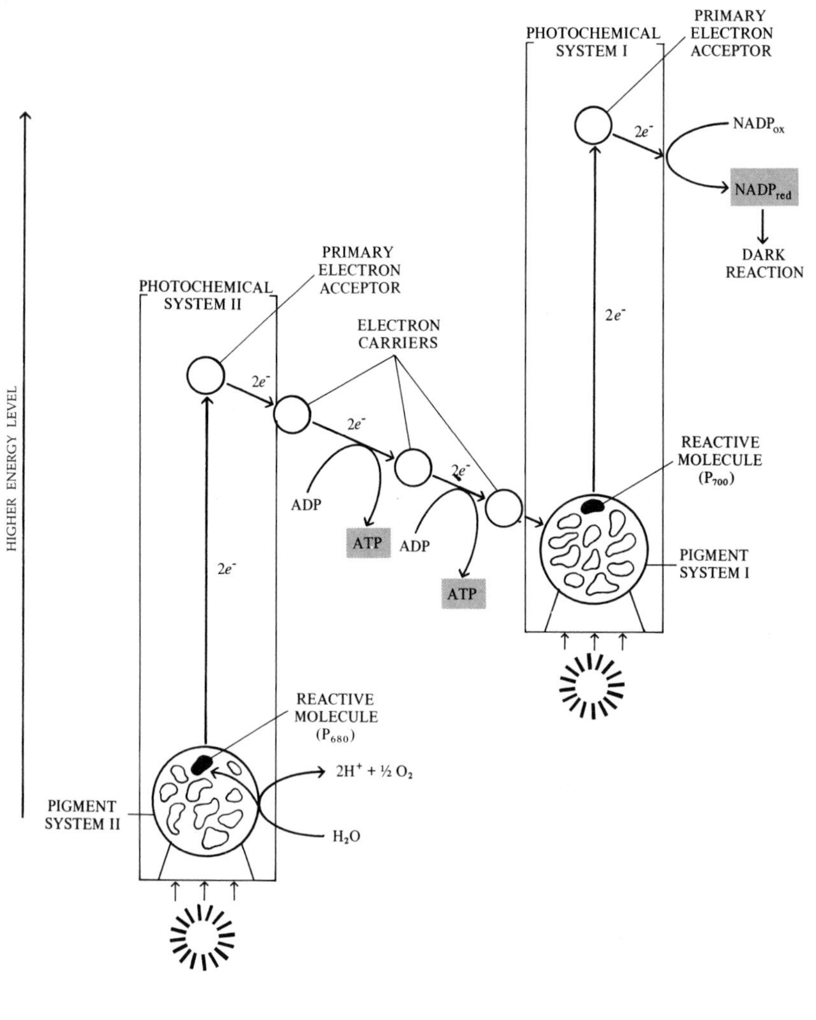

6–6
Light energy trapped in the reactive molecule of Pigment System II boosts electrons uphill from chlorophyll to a primary electron acceptor. These electrons are replaced by electrons pulled away from water molecules, which then fall apart, producing protons and oxygen gas. The electrons are passed from the electron acceptor along an electron transport chain to a lower energy level, the reaction center of Pigment System I. As they pass along the electron transport chain, some of their energy is packaged in the form of ATP. Light energy absorbed by Pigment System I boosts the electrons to another primary electron acceptor. From this acceptor, they are passed via other electron carriers to $NADP_{ox}$ *to form* $NADP_{red}$*. The electrons removed from Pigment System I are replaced by those from Pigment System II. ATP and* $NADP_{red}$ *represent the net gain from the light reactions.*

occurs, no $NADP_{red}$ is formed. In this process, called *cyclic electron flow*, electrons are boosted from P_{700} to an electron acceptor and from there pass downhill through a series of cytochromes back into the pigment system (Figure 6–7). ATP is produced in the course of the passage. It is believed that the most primitive photosynthetic mechanisms did, indeed, work in this way. These early photosynthetic mechanisms probably produced ATP; the fact that ATP is an energy carrier in every known kind of living system testifies to its very early appearance in the history of evolution. These early mechanisms did not give off oxygen, however, and it is not known if they reduced carbon to sugar.

Summary of Light Reactions

The reactions that we have just described are the "light reactions" of photosynthesis. In the course of these reactions, as we saw, light energy is converted to electrical energy—the flow of electrons—and the electrical energy is converted to chemical energy stored in the bonds of $NADP_{red}$ and ATP.

THE DARK REACTIONS

In the second stage of photosynthesis, the energy generated by the light reactions is used to incorporate carbon into organic molecules. These reactions take place in the stroma of the chloroplasts and do not need light, and so they are often referred to as the "dark reactions" (although they normally take place in the light). The dark reactions involve a cycle called the *Calvin cycle*. It is analogous to the Krebs cycle (page 104) in that, in each turn of the cycle, the starting product is again regenerated. The dark reactions convert the chemical energy produced by the light reactions to forms more suitable for storage and transport and also build the basic carbon structures from which all of the other organic molecules of living systems are produced. This last process is known as *carbon fixation*.

The Calvin Cycle

In the Calvin cycle (named after its discoverer, Melvin Calvin of the University of California, Berkeley), the starting (and ending) compound is a five-carbon sugar with two phosphates attached, ribulose-1,5-diphosphate (RuDP). The cycle begins when carbon dioxide enters the cycle and is incorporated into RuDP, which then splits to form two molecules of 3-phosphoglycerate (PGA) (Figure 6–8). The enzyme responsible for this reaction is RuDP carboxylase.

The complete cycle is diagrammed in Figure 6–9. As in the Krebs cycle, each step is regulated by a specific enzyme. At each full turn of the cycle, a molecule of carbon dioxide enters the cycle, is reduced, and a molecule of

6–7
Cyclic electron flow bypasses Photochemical System II and requires only Photochemical System I. ATP is produced from ADP, but oxygen is not released and $NADP_{ox}$ is not reduced.

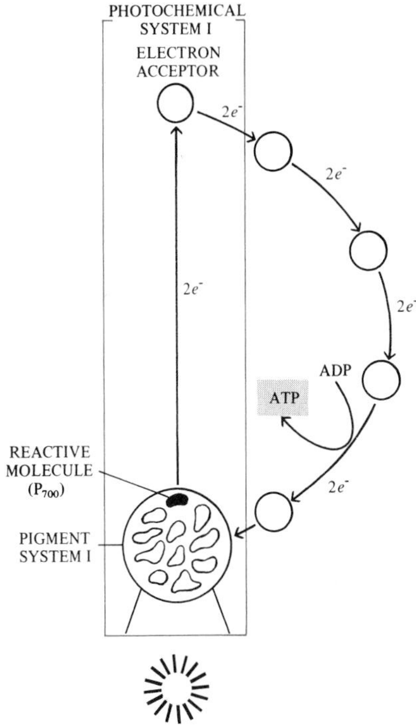

6–8
Calvin and his collaborators briefly exposed photosynthesizing algae to radioactive carbon dioxide ($^{14}CO_2$). They found that the radioactive carbon is first incorporated into ribulose-1,5-diphosphate (RuDP). The -1,5- indicates the position of the carbons to which the phosphates are attached. The RuDP then immediately splits to form two molecules of 3-phosphoglycerate (PGA). The radioactive carbon atom, indicated in color, next appears in one of the two molecules of PGA. This is the first step in the Calvin cycle.

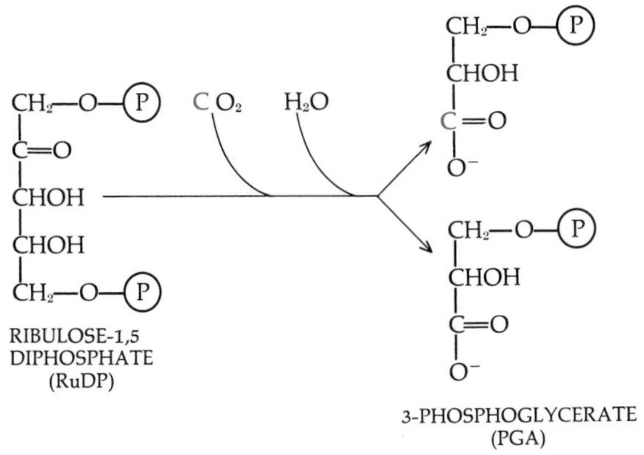

6-9
Summary of the Calvin cycle. Three molecules of ribulose-1,5-diphosphate (RuDP), a five-carbon compound, are combined with three molecules of carbon dioxide, yielding six molecules of 3-phosphoglycerate, a three-carbon compound. These are converted to six molecules of glyceraldehyde phosphate. Five of these three-carbon molecules are combined and rearranged to form three five-carbon molecules of RuDP. The "extra" molecule of glyceraldehyde phosphate represents the net gain from the Calvin cycle. The energy that "drives" the Calvin cycle is in the form of ATP and $NADP_{red}$, produced by the light reactions.

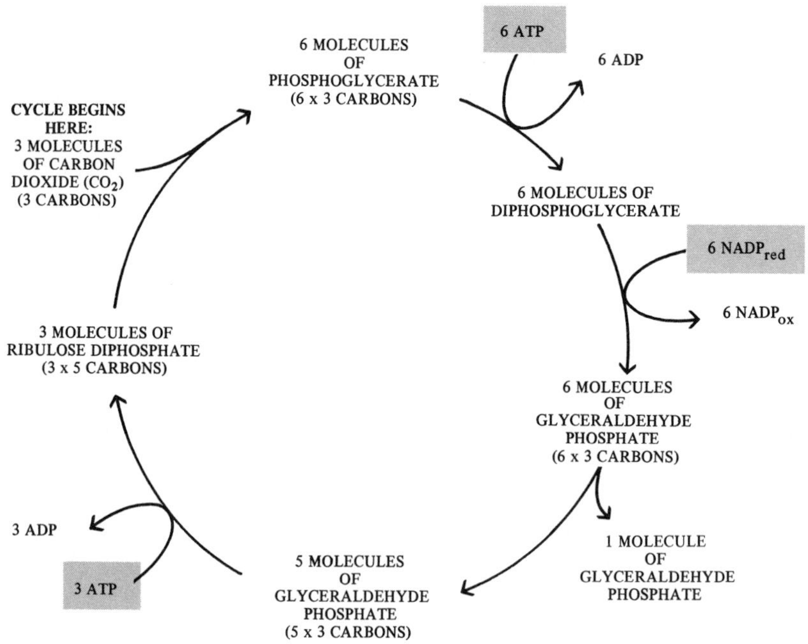

RuDP is regenerated. Six revolutions of the cycle, with the introduction of six atoms of carbon, are necessary to produce a six-carbon sugar, such as glucose. The overall equation is

$$6RuDP + 6CO_2 + 18ATP + 12NADP_{red} \longrightarrow 6RuDP + glucose + 18P_i + 18ADP + 12NADP_{ox}$$

As you can see, the immediate product of the Calvin cycle is glyceraldehyde phosphate. This same sugar-phosphate is formed when the fructose diphosphate molecule is split at the fourth step in glycolysis (page 102). These same steps, using the energy of the phosphate bond, can be reversed to form glucose from glyceraldehyde phosphate.

The Four-Carbon Photosynthetic Pathway

The mechanism of carbon dioxide fixation discovered by Calvin and his co-workers, and discussed above, is not the only one used by green plants to fix carbon dioxide. In the early 1960s, Hugo Kortschak and his colleagues in the laboratory of the Hawaiian Sugar Planters' Association found evidence that the first photosynthetic product in sugarcane was not the three-carbon compound PGA but a four-carbon compound. Shortly afterward, two Australian plant physiologists, M. D. Hatch and C. R. Slack, confirmed the findings of the Hawaiian group and reported that the first photosynthetic product in this newly discovered carbon-fixing mechanism was the four-carbon acid oxaloacetic acid (Figure 6-10). Oxaloacetic acid is formed when carbon dioxide is added to the three-carbon compound phosphoenol pyruvate (PEP), a reaction mediated by the enzyme phosphoenol pyruvate carboxylase (PEP carboxylase). Because the products of the PEP system are four-carbon compounds, plants utilizing this pathway are called C_4, or four-carbon, plants. This distinguishes them from the C_3, or three-carbon, plants in which the immediate product of photosynthesis is the three-carbon compound PGA.

The products of the PEP system, malic and aspartic acids, are not substitutes for PGA. After they are formed, the malic and aspartic acids are broken down enzymatically to yield carbon dioxide and pyruvic acid. The CO_2 is transferred to ribulose-1,5-diphosphate (RuDP) of the Calvin cycle, and the pyruvic acid reacts with ATP to form more molecules of PEP. Thus, it is only in the initial steps of carbon fixation that C_4 photosynthesis differs from the more familiar C_3 photosynthesis.

Anatomy of C_4 Plants

Leaf anatomy in C_4 plants differs from that in C_3 plants. In C_3 plants, chloroplasts of similar appearance are distributed throughout the mesophyll tissue of the leaf, and the Calvin cycle takes place in each cell, beginning with the fixation of atmospheric carbon dioxide by carboxylation of RuDP to form PGA. In C_4 plants, certain cells of the leaf fix most of the atmospheric CO_2. In corn, for example, the veins are surrounded by large bundle sheath cells that contain chloroplasts with nu-

6-10
Carbon dioxide fixation by the Hatch-Slack pathway.

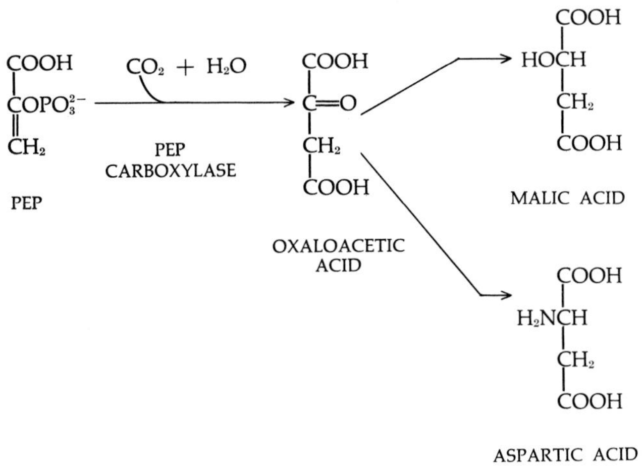

merous starch grains and poorly developed grana. By contrast, the mesophyll cell chloroplasts commonly lack starch grains and contain well-developed grana (see Figure 1-1). Only the mesophyll cells—which lack the Calvin cycle—fix atmospheric carbon dioxide by the PEP system. The malic and aspartic acids formed in the mesophyll cells are then transported to the bundle sheath cells to provide CO_2 to the Calvin cycle located there. Thus, the biochemical events of photosynthesis are compartmentalized anatomically in the corn leaf.

Efficiency of C_4 Plants

One might ask why the C_4 plants should have evolved such a seemingly cumbersome method of providing carbon dioxide to the Calvin cycle. It should be noted that C_4 plants apparently evolved primarily in the tropics and are especially well adapted to high extremes of light, temperature, and dryness. It has been demonstrated that under such adverse environmental conditions, C_4 plants are superior utilizers of available carbon dioxide compared with C_3 plants. This is due to the fact that the enzyme PEP carboxylase has a much greater affinity for carbon dioxide at low concentrations than the enzyme RuDP carboxylase of the Calvin cycle. The C_4 plant's two-cycle system of photosynthesis allows the plant to absorb CO_2 efficiently even at low concentrations and then to feed it into the Calvin cycle, which provides the plant with photosynthetic end products essential for growth. The temperature optima for C_4 photosynthesis are much higher than those for C_3 photosynthesis; and C_4 plants flourish even at temperatures that would eventually be lethal to many C_3 species.

A striking illustration of differing growth patterns of C_4 plants is found in our lawns, which, in the cooler parts of the country at least, consist mainly of C_3 grasses such as Kentucky bluegrass (*Poa pratensis*) and creeping bent (*Agrostis tenuis*). Crabgrass (*Digitaria sanguinalis*), which all too often overwhelms these dark green, fine-leaved grasses with patches of its yellowish green, broader leaves in summer, is an annual and a C_4 grass that grows much more rapidly than the temperate C_3 grasses mentioned above at high temperatures. Corn (*Zea mays*) and sugar cane (*Saccharum officinale*), as well as sorghum (*Sorghum vulgare*) and rice (*Oryza sativa*), are C_4 grasses of tropical orientation, whereas wheat (*Triticum*), rye (*Secale cereale*), and oats (*Avena sativa*) are temperate grasses with C_3 metabolism.

The list of plants known to utilize the four-carbon pathway has grown to over 100 genera, including both monocotyledons and dicotyledons (see page 342). The pathway undoubtedly has arisen many times independently in the course of evolution. Sugarcane, corn, and sorghum are among the best-known C_4 plants. At least a dozen genera have both C_3 and C_4 species. An example of such a genus is *Atriplex*, the saltbushes: *Atriplex patula* is a C_3 plant and *A. rosea* is a C_4 plant.

CAM Photosynthesis

A third pathway of CO_2 fixation, crassulacean acid metabolism (CAM) photosynthesis, has evolved independently in many succulent plants, including cacti and stonecrops. It also involves four-carbon molecules in the fixation of CO_2. In these plants, malic and isocitric acids accumulate in the leaves in darkness and are converted back to CO_2 in the light. This is evidently advantageous in the conditions of high light intensity and water stress under which most succulent plants live. They are largely dependent upon night-time accumulation of carbon for their photosynthesis, because their stomata are closed during the day, retarding water loss. By accumulating organic acids in darkness, much more CO_2 can be fixed in a 24-hour cycle than would otherwise be possible, given the pattern of stomatal opening and closing.

Both C_4 and CAM photosynthesis selectively fix the ^{13}C isotope of carbon, which constitutes about 1 percent of the carbon in nature, producing characteristic $^{13}C/^{12}C$ ratios in plant material formed by the different pathways. This relationship has pointed the way to some significant studies of preserved and fossil plant material.

Photorespiration

Another reason for the comparative inefficiency of C_3 plants has been traced to a process known as *photorespira-*

6-11
Reactions catalyzed by ribulose-1,5-diphosphate (RuDP) carboxylase. Reaction 1 is favored by high CO_2 and low oxygen concentrations. Reaction 2 has been proposed to occur in the presence of low CO_2 and high oxygen concentrations (normal atmospheric conditions).

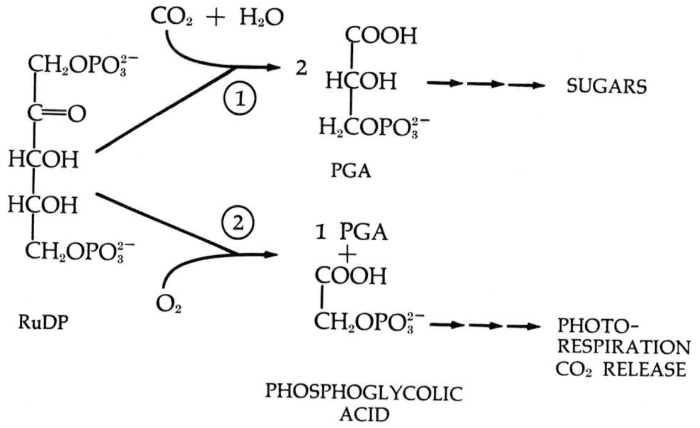

tion. Like the respiratory processes that take place in the mitochondria, photorespiration involves the oxidation of carbohydrates; however, unlike these processes, photorespiration yields neither ATP nor $NADP_{red.}$

Photorespiration is a light-requiring process that takes place in microbodies called peroxisomes.

The primary source of the carbon dioxide produced during photorespiration is believed to be a two-carbon acid, glycolic acid. This two-carbon compound is produced, it is hypothesized, by C_3 plants in the presence of low CO_2 and high O_2 concentrations (see Figure 6-11). Under these conditions, which are normal atmospheric conditions, as much as 50 percent of the photosynthetically fixed carbon may be reoxidized to CO_2 during photorespiration. Thus, photorespiration appears to decrease plant productivity directly. Interestingly, C_4 plants exhibit much less photorespiration than C_3 plants—another reason for their high efficiency as photosynthesizing organisms. Photorespiration appears to be one of the main factors decreasing net productivity in C_3 plants, and efforts are being made in some crop plants to breed for low rates of photorespiration and thus increase yield. The place of photorespiration in the plant's overall economy is still not properly understood, however, and there may be limiting factors built in with respect to other reactions.

SUMMARY

Photosynthesis takes place in two stages, a light-dependent stage, the "light reactions," and a light-independent stage, the "dark reactions." In the first step in photosynthesis, light is absorbed by pigment systems. This absorbed light energy is transformed into chemical energy.

The pigments involved in photosynthesis are chlorophyll *a* and *b* and carotenoids. Light absorbed by pigments boosts their electrons to a higher energy level. Because of the way the pigments are packed into membranes, they are able to transfer these electrons to trapping molecules, which are probably chlorophyll *a*. Two pigment systems have been identified in thylakoids.

In the currently accepted model of the light reactions in photosynthesis, light energy first strikes Pigment System II, which contains several hundred molecules of chlorophyll *a* and chlorophyll *b*. Electrons are passed uphill to an electron acceptor from the reactive molecule. The electron holes are filled from the lysis and oxidation of water. The electrons then pass downhill to Pigment System I along an electron transport chain, in the course of which ATP is generated (photophosphorylation). Light absorbed in Pigment System I results in the ejection of an electron from the reactive molecule chlorophyll P_{700}. The resulting electron holes left in the reactive molecule are filled by the electrons from Pigment System II. The electrons are accepted by the electron carrier molecule, $NADP_{ox}$. The energy yield from the light reactions is represented by molecules of $NADP_{red}$ and the ATP formed by photophosphorylation.

In the dark reactions, which take place in the stroma, the $NADP_{red}$ and ATP produced in the light reactions are used to reduce carbon dioxide to organic carbon. This is accomplished by means of the Calvin cycle. In the Calvin cycle, a molecule of carbon dioxide is combined with the starting material, a five-carbon sugar called ribulose-1,5-diphosphate (RuDP). At each turn of the cycle, one carbon atom enters the cycle. Three turns of the cycle produce a three-carbon molecule, glyceraldehyde phosphate. Two molecules of glyceraldehyde phosphate (six turns of the cycle) can combine to form a glucose molecule. At each turn of the cycle, RuDP is regenerated. The glyceraldehyde can also be used as starting material for other organic compounds needed by the cell.

In so-called C_4 plants, the carbon dioxide is accepted initially by the three-carbon compound phosphoenol pyruvate (PEP) to yield the four-carbon compound oxaloacetic acid. Oxaloacetic acid is then rapidly converted into either malic or aspartic acid, which transfers the CO_2 to RuDP of the Calvin cycle. At low CO_2 concentrations, C_4 plants are more efficient utilizers of CO_2 than C_3 plants.

SUGGESTIONS FOR FURTHER READING

ARNON, D. I.: "The Light Reactions of Photosynthesis," *Proc. Nat. Acad. Sci. U.S.* 68: 2883–2892, 1971.

An excellent review of the two major aspects of photosynthesis. For the ambitious student who desires to read what scientists write about science. See also Bassham, 1971.

BASSHAM, J. A.: "Photosynthetic Carbon Metabolism," *Proc. Nat. Acad. Sci. U.S.* 68: 2877–2882, 1971.

An excellent review of photosynthesis. See also Arnon, 1971.

BJÖRKMAN, O., and J. BERRY: "High-Efficiency Photosynthesis," *Scientific American* 229(4): 80–93, 1973.

A discussion of the superior photosynthetic performance of four-carbon plants and their potential role in agriculture.

CONANT, JAMES BRYANT (Ed.): *Harvard Case Histories in Experimental Science,* vol. 2, Harvard University Press, Cambridge, Mass., 1964.

Case #5, Plants and the Atmosphere, *edited by Leonard K. Nash, describes the early work on photosynthesis presented often in the words of the investigators themselves. The narrative illuminates the historical context in which the discoveries were made.*

GABRIEL, MORDECAI L., and SEYMOUR FOGEL: *Great Experiments in Biology,* Prentice-Hall, Inc., Englewood Cliffs, N.J., 1955.*

Many of the fundamental discoveries of biology as seen firsthand through the eyes of their discoverers. The examples are well chosen, and their value is greatly enhanced by accompanying explanatory notes and chronological tables of key developments in various areas of biology.

HATCH, M. C., C. B. OSMOND, and R. O. SLATYER (Eds.): *Photosynthesis and Photorespiration,* Wiley-Interscience, New York, 1971.

A collection of outstanding review articles on all aspects of photosynthesis and photorespiration, resulting from a meeting held in Canberra, Australia, in 1970.

KREBS, H. A.: "The History of the Tricarboxylic Acid Cycle," *Perspectives in Biology and Medicine* 14:154–170, 1970.

An historical development of a major metabolic pathway, seen in the perspective of the Nobel laureate who contributed more than any other person to the elucidation of the cycle that bears his name.

LEHNINGER, ALBERT L.: *Bioenergetics: The Molecular Basis of Biological Energy Transformations,* 2nd ed., W. A. Benjamin, Inc., Menlo Park, Calif., 1971.*

A solid but readily understandable account of the flow of energy in cells, including a treatment of both mitochondrial and chloroplast function.

LEVINE, R. P.: "The Mechanism of Photosynthesis," *Scientific American* 221(6):58–70, 1969.

A good account of the two photochemical systems and how they provide the electrons, protons, and energy-rich molecules needed to convert carbon dioxide and water into food.

MCELROY, WILLIAM D.: *Cell Physiology and Biochemistry,* 3rd ed., Prentice-Hall, Inc., Englewood Cliffs, N.J., 1971.*

A good short introduction to cell physiology, especially useful in its descriptions of exchanges and uses of energy within cells.

RABINOWITCH, E. B., and GOVINDJEE: *Photosynthesis,* John Wiley & Sons, Inc., New York, 1969.

A comprehensive and readable account of the entire field, now slightly out of date, but still the most useful overall treatment of the subject.

SAN PIETRO, A., F. A. GREER, and T. J. ARMY: *Harvesting the Sun,* Academic Press, Inc., New York, 1967.

A very readable account, which sheds much light on photosynthesis and the factors that affect it.

ZELITCH, ISRAEL: *Photosynthesis, Photorespiration and Plant Productivity,* Academic Press, New York, 1971.

A modern treatment of the process of photosynthesis within cells, leaves, and plant communities.

* Available in paperback.

SECTION 3 Genetics and Evolution

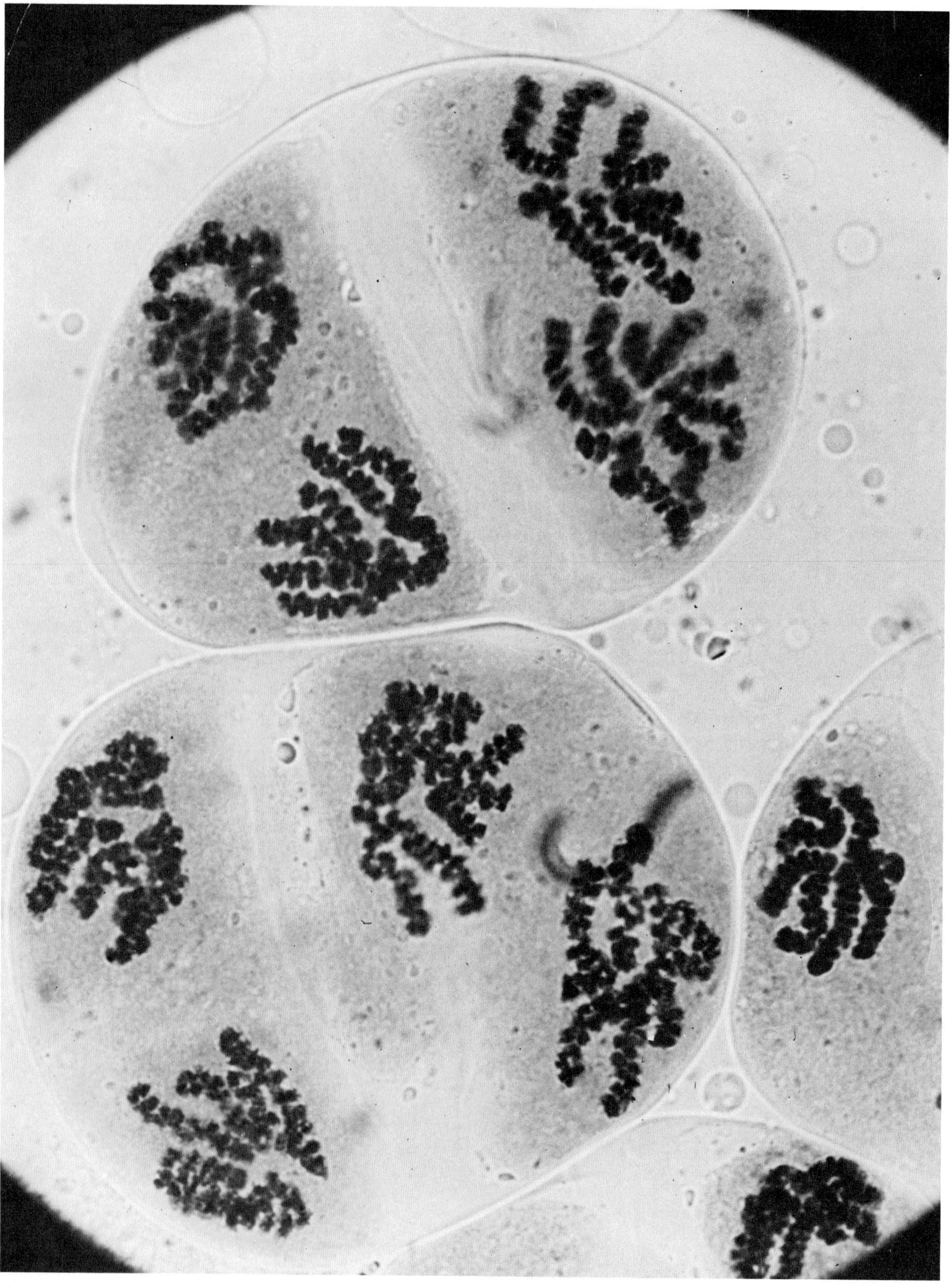

CHAPTER 7

The Genetics of Diploid Organisms

7–1
Gregor Mendel (1822–1884). Mendel carried out his studies in the garden of a monastery in Brno in what is now Czechoslovakia. His work, which was not understood and therefore largely ignored in his lifetime, was rediscovered in the early 1900s.

In Chapter 3, we described the mechanism of heredity in molecular terms; much of this work has been done with prokaryotes. In this chapter, we are going to focus on the genetics of eukaryotic organisms. The branch of genetics discussed here is generally referred to as Mendelian genetics in recognition of the work of Gregor Mendel (Figure 7–1). We have deferred our discussion of Mendelian genetics until this time because of its relevance to evolutionary theory, which we shall also be discussing in this section. Darwin's monumental work on evolution was accomplished in ignorance of Mendel's studies. However, the study of evolution in the twentieth century is almost as dependent upon Mendel and his followers as upon the author of *On the Origin of Species.**

EUKARYOTES VS. PROKARYOTES

One of the major differences between eukaryotes and prokaryotes is that most eukaryotes undergo sexual reproduction, a process not present in prokaryotic organisms. Sexual reproduction involves a regular alternation between *meiosis,* nuclear division in which the number of chromosomes is reduced by half in the formation of haploid cells, and *syngamy,* the reestablishment of the diploid number of chromosomes during the process of fertilization, in which two haploid gametes fuse to form a diploid zygote (Figure 7–2).

Although some eukaryotic organisms do not reproduce sexually, it is evident in most cases that they once did but eventually lost this capacity in the course of their evolutionary history. Later in this chapter, we shall consider why sexual reproduction and diploidy are virtually universal among eukaryotes.

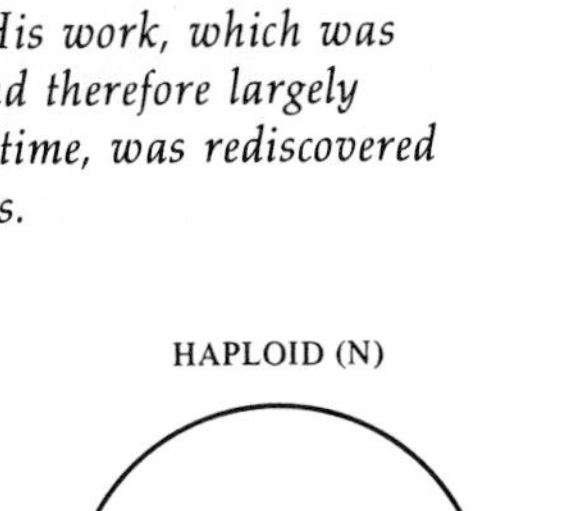

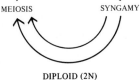

7–2
Sexual reproduction involves a regular alternation between meiosis and syngamy.

* The complete title of Charles Darwin's treatise is *On the Origin of Species by Means of Natural Selection, or the Preservation of Favoured Races In the Struggle for Life.*

All of the organisms with which we shall be dealing in this chapter are diploid for a major portion of their life cycle. Diploid organisms have two sets of chromosomes, one derived from their male parent and another derived from their female parent. It is the interaction of products derived from genes on each of these sets of chromosomes that determines the genetic characteristics of the adult diploid plant or animal.

In Chapter 10, we shall consider the very different problems involved in the genetics of prokaryotes, all of which are haploid, and in Chapter 11, we shall discuss some genetic peculiarities of the fungi. Here, however, we shall be concerned mainly with diploid eukaryotic organisms, primarily the vascular plants.

MEIOSIS

Meiosis takes place only in specialized diploid cells and only at particular times in the life cycle of a given organism. It results in the production of haploid cells, either *gametes* or *spores*. A gamete is a cell that fuses with another gamete to produce a diploid zygote. The zygote may then divide meiotically to give rise to four haploid unicellular organisms or mitotically to form a multicellular diploid organism. The multicellular diploid organism eventually produces haploid spores by meiosis. A spore is a cell that develops into an organism without uniting with another cell. Spores often divide mitotically to produce organisms that are entirely haploid and that eventually give rise to gametes by mitosis.

In understanding meiosis, it is important to remember that the chromosomes of eukaryotes are complex structures involving DNA and protein molecules, perhaps arranged in a long, greatly contorted, continuous fiber.

First Meiotic Division

Meiosis consists of two successive nuclear divisions, and so Roman numerals are used to discriminate between them. In prophase I (prophase of the first meiotic division), the chromosomes, present in the diploid number, first become visible as long, slender threads. As in mitosis (page 40), the chromosomes have duplicated during interphase, so that as prophase I begins each chromosome actually consists of two identical chromatids attached to each other at the centromere. Nevertheless, at this early stage of meiosis each chromosome appears to be single longitudinally rather than double. Before the chromatids become apparent, the chromosomes pair up. The pairing is very precise, beginning at one or more sites along the lengths of the chromosomes and proceeding in a zipperlike fashion. The members of a pair of chromosomes are known as *homologues;* each homologue is derived from a different parent and is made up of two identical chromatids (the duplicates produced during interphase). Thus the homologous pair consists of four chromatids. (As we shall see on page 134, alleles, the alternate forms of the same functional genes in diploid organisms, occupy the same position (locus) along homologous chromosomes. It is at these loci that pairing takes place.) Because of this pairing, meiosis cannot occur in haploid cells, in which homologues are not present. The pairing process itself is called *synapsis,* and associated pairs of homologous chromosomes are called *bivalents.*

During the course of prophase I, the paired threads become more and more tightly contracted, and the chro-

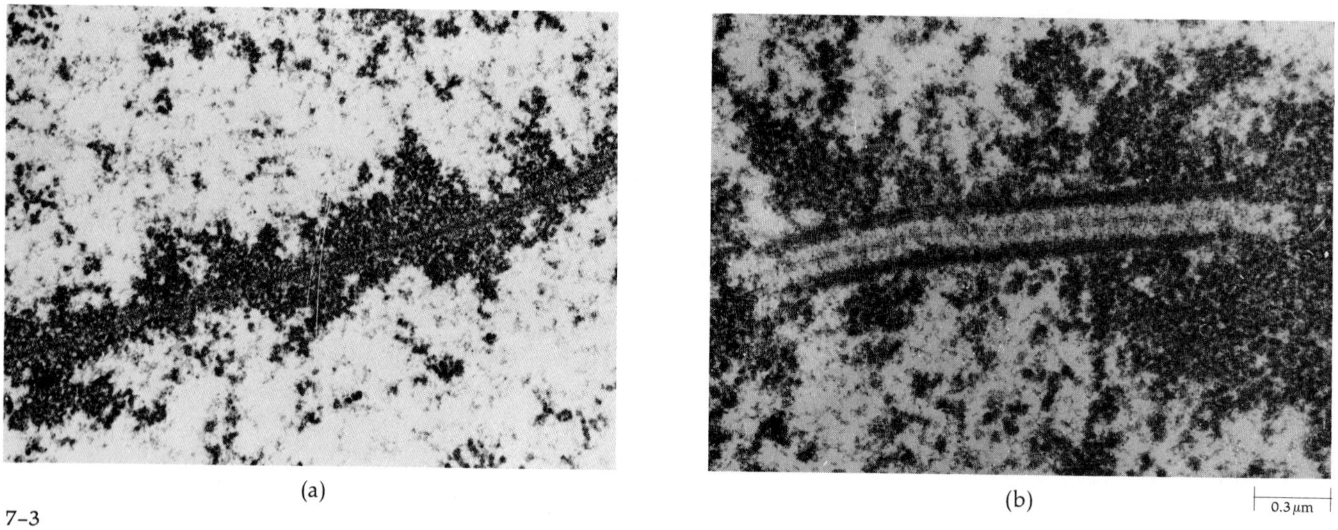

(a) (b)

7-3
(a) *Portion of a chromosome of* Lilium *early in prophase I, prior to pairing. Note the dense axial core. This core, which consists mainly of proteins, may arrange the genetic material of the chromosome in preparation for pairing and genetic exchange.* (b) *Synaptinemal complex in a bivalent of* Lilium. *The chromatin is loosely arranged around the complex.*

mosomes shorten and thicken. Using the electron microscope, it is possible to identify a densely staining axial core, consisting mainly of proteins, in each chromosome (Figure 7–3a). During mid-prophase, the axial cores of a pair of homologous chromosomes approach each other to within 0.1 micrometer to form what is known as a synaptinemal complex (Figure 7–3b).

In favorable material, it can now be seen that each axial core is double, in other words that each bivalent is, as we have noted, made up of four chromatids, two per chromosome. In at least one point on each bivalent, during the time when the synaptinemal complex exists, two homologous chromatids break and exchange segments, thus forming complete chromatids but with exchanged segments. This event is known as *crossing-over*. The visible evidence for this is the crossing, called a *chiasma* (plural: chiasmata) of the chromatids (Figure 7–4).

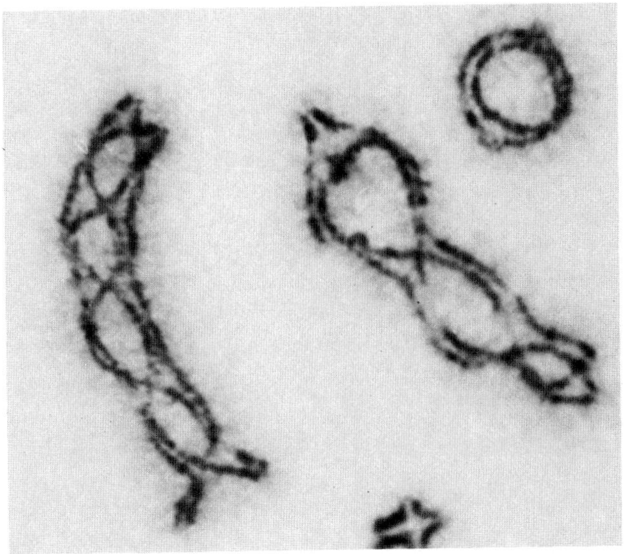

7–4
Chiasmata in chromosomes of a grasshopper, Chorthippus parallelus.

As prophase I proceeds, the synaptinemal complex ceases to exist. Eventually, the nuclear envelope breaks down. The nucleolus usually disappears as RNA synthesis is temporarily suspended. Finally, the homologous chromosomes appear to repulse one another. Their chromatids are held together, however, at the chiasmata. These chromatids tear apart very slowly. As they do separate, some of the chiasmata terminalize, or move toward the end of the chromosome arm. There may be one or more chiasmata in each arm of the chromosome, or only one in the entire bivalent; depending on how many there are, the appearance of the bivalent varies widely. Some of the possibilities can be seen in Figure 7–4.

In metaphase I, the spindle apparatus becomes conspicuous (Figure 7–5). The paired chromosomes, still joined together at their chiasmata, move to the equatorial plane of the cell and their centromeres become attached to spindle fibers. Unlike mitotic metaphase, in which the centromeres line up on the equatorial plane, during meiotic metaphase I the centromeres of the paired chromosomes line up on either side of the equatorial plane.

Anaphase I begins when the homologous chromosomes separate entirely and begin to move toward the poles. Notice again the contrast with mitosis. In mitotic anaphase, the centromeres divide and the identical chromatids are separated. In meiotic anaphase I, the centromeres do not divide and the chromatids remain together; it is the homologues that are separated. (Because of the exchanges of chromatid segments resulting from crossing-over, however, the chromatids are not identical, as they were at the onset of meiosis.)

In telophase I, the coiling of the chromosomes relaxes and the chromosomes tend to become elongated and once again indistinct. Nuclear envelopes are reorganized out of the endoplasmic reticulum as telophase changes gradually to interphase, which may be short or absent depending on the organism involved. The nucleolus reappears. In many plants and animals, the nuclei enter promptly into prophase II of the second meiotic division.

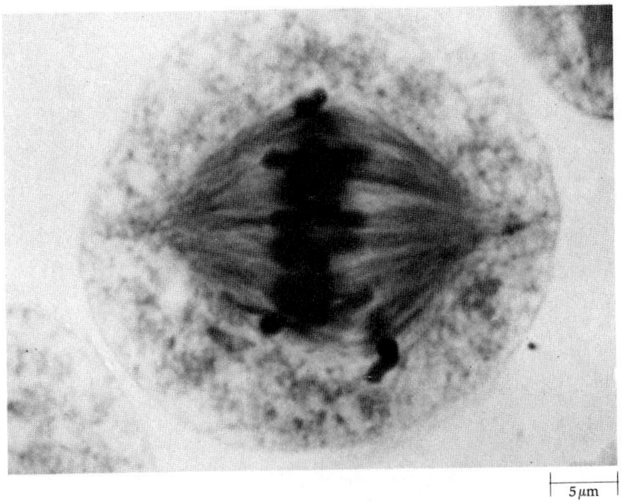

7–5
The spindle in a pollen mother cell of wheat (Triticum aestivum) *during metaphase I of meiosis.*

7–6
Meiosis, a diagrammatic representation with two pairs of chromosomes. Not all stages are shown. Prophase I. *The chromosomes become visible as elongated threads, homologous chromosomes come together in pairs, the paired chromosomes coil round one another, and the paired chromosomes—held together at the chiasmata—become very short.* Metaphase I. *The paired chromosomes come to lie with their centromeres evenly distributed on either side of the equatorial plane of the spindle.* Anaphase I. *The paired chromosomes separate and move to opposite poles. The second meiotic division essentially follows the course of an ordinary mitosis.* Metaphase II. *The chromosomes are lined up at the equatorial plane with their centromeres lying on the plane.* Anaphase II. *The centromeres divide and the chromatids separate and move toward opposite poles of the spindle.* Telophase II. *The chromosomes have completed their migration. Four new nuclei, each with the haploid number of chromosomes, are formed.*

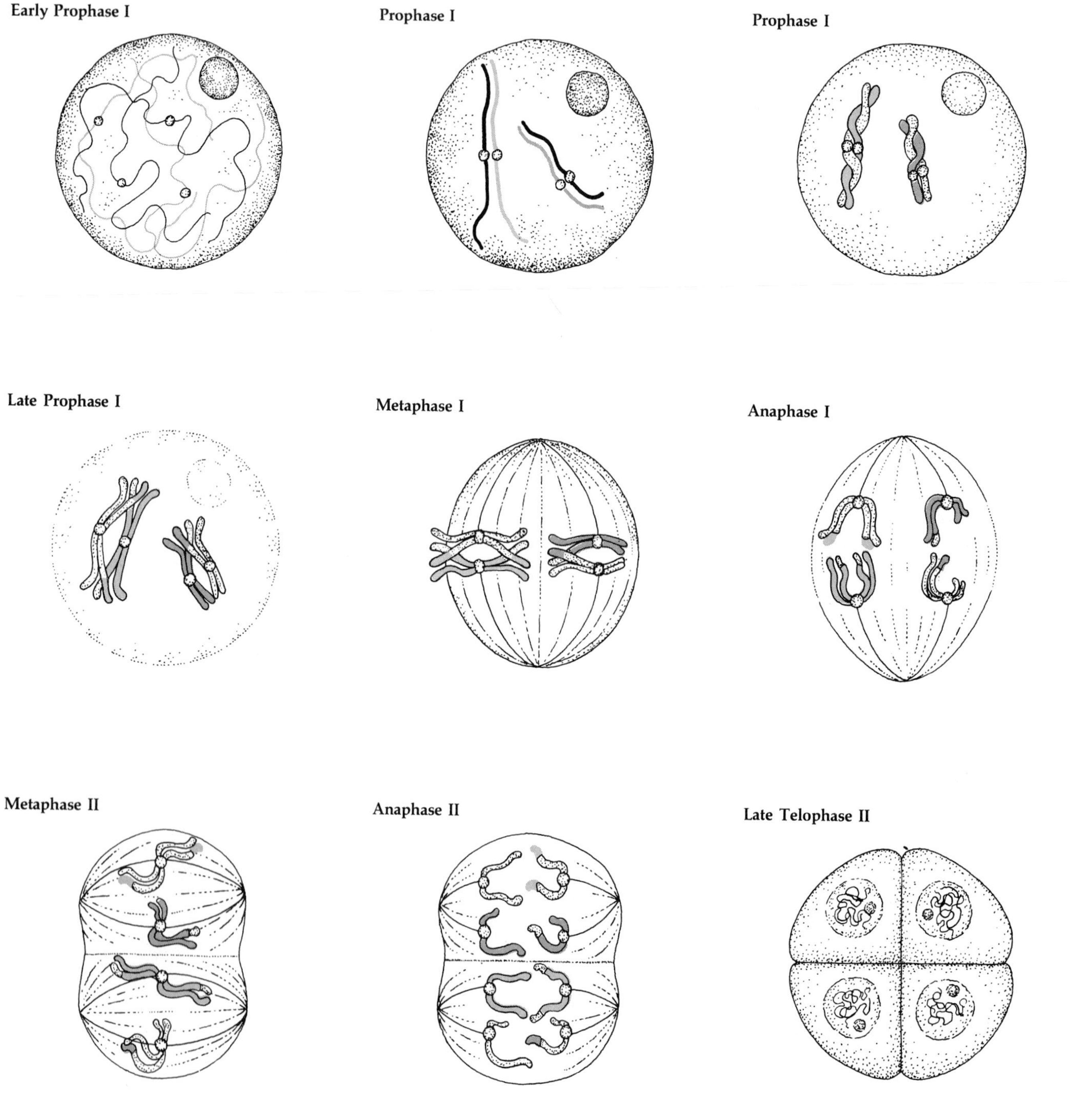

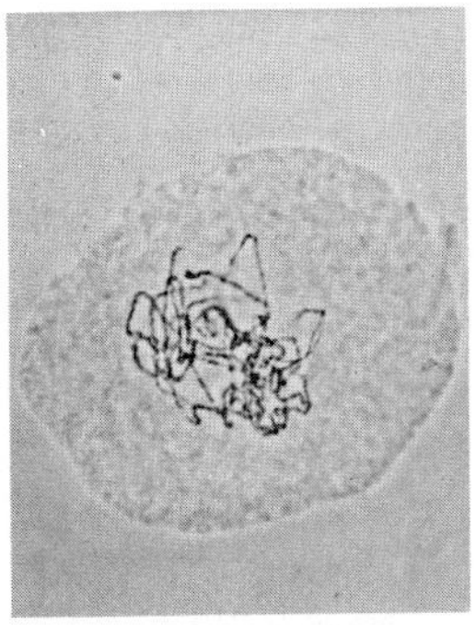

Early Prophase I. The chromosomes appear as threads. Each thread is actually double-stranded, composed of two identical chromatids.

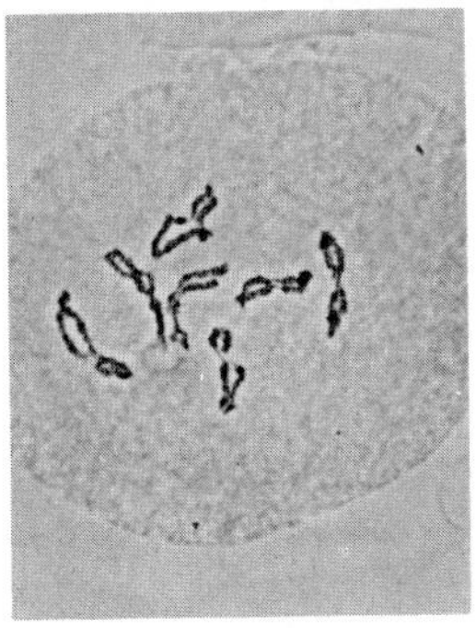

Prophase I. Homologous chromosomes pair. This is a crucial point of difference between meiosis and mitosis. Chiasmata are visible.

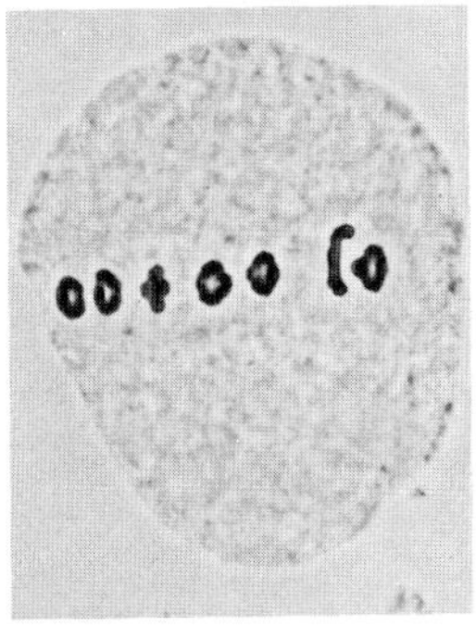

Metaphase I. Bivalents are lined up on the equatorial plane of the cell.

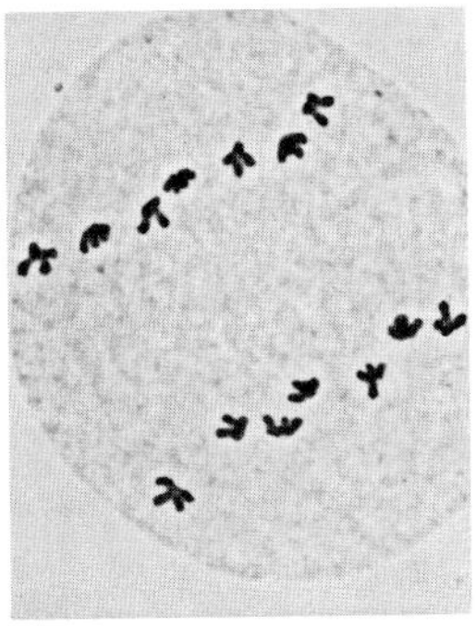

Anaphase I. Homologous chromosomes have separated from each other and are moving toward opposite poles of the cell.

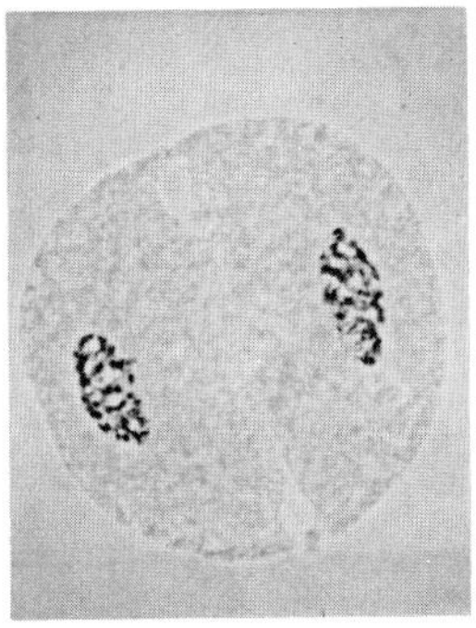

Late Telophase I. Chromosomes are regrouped at each pole and the cell is dividing to form two haploid cells.

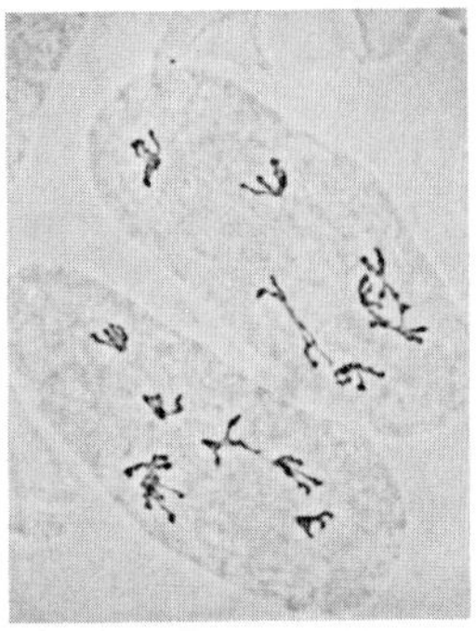

Prophase II. The chromosomes are reappearing. Each still consists of two chromatids, but, because of crossing-over, the chromatids are no longer identical to each other.

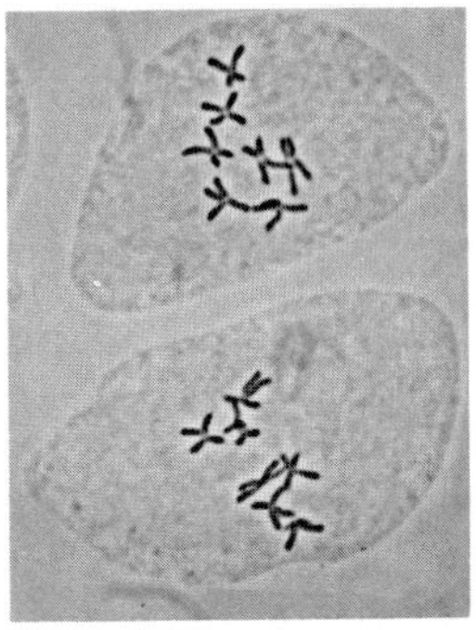

Metaphase II. The chromosomes are lined up on the equatorial plane.

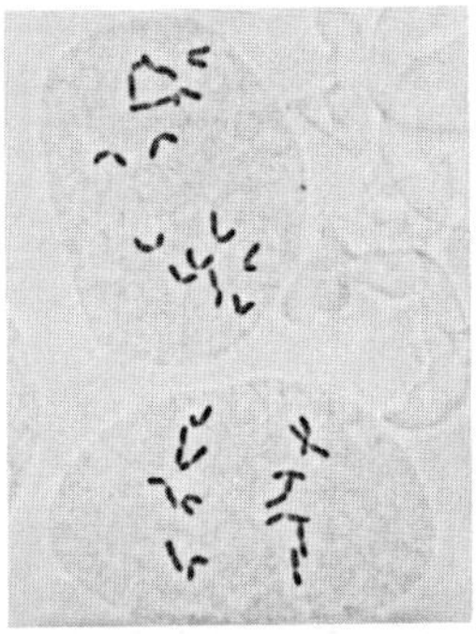

Anaphase II. The centromere of each chromosome has divided and the chromatids, now chromosomes, are moving toward opposite poles.

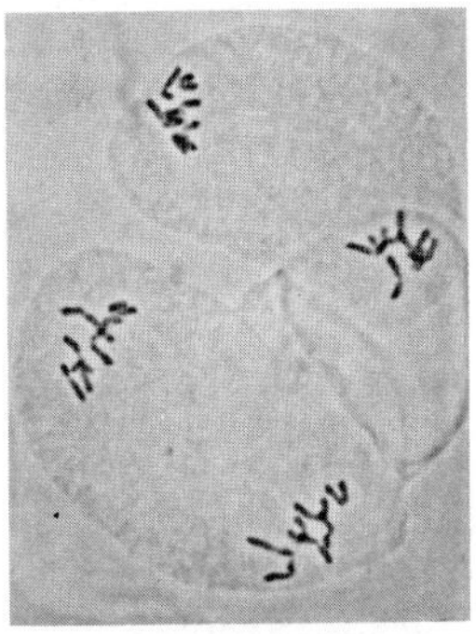

Telophase II. The chromosomes have completely separated and new cell walls are forming.

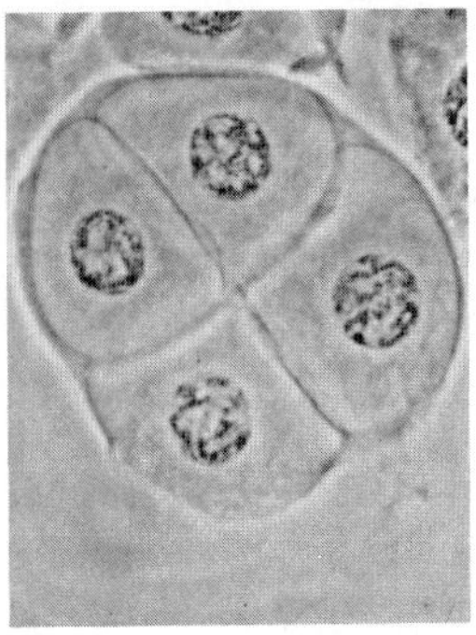

Tetrad. New plasma membranes and cell walls form. The four cells will become differentiated as pollen grains.

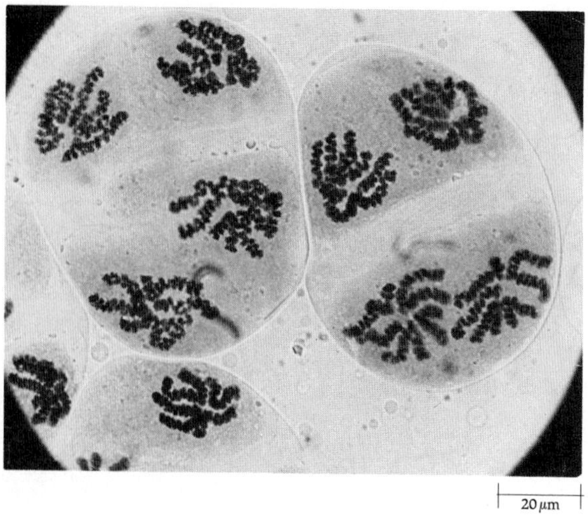

7-7
Contracting in chromosomes of Trillium erectum *(wake-robin). This cell is in second meiotic anaphase, and the chromosomes are almost completely separated.*

Second Meiotic Division

At the beginning of the second meiotic division, the chromatids are still attached by their centromeres. This division proceeds just like mitosis, with the nuclear envelope becoming disorganized and the nucleolus disappearing at the end of prophase II. At metaphase II, the spindle apparatus becomes obvious, and the chromosomes line up with their centromeres on the equatorial plane. At anaphase II, the centromeres divide and are pulled apart, and the newly separated chromosomes move to opposite poles. At telophase II, new nuclear envelopes and nucleoli are organized, and the contracted chromosomes relax as they fade into an interphase nucleus.

The Consequences of Meiosis

The end result of meiosis is that the genetic material present in the diploid nucleus has been replicated only once (before the onset of meiosis) but has been divided up twice. Therefore, each cell has only half as many chromosomes as the original diploid nucleus. More important are the genetic consequences of the process. At meiotic metaphase I, the orientation of the bivalents is random. In other words, the chromosomes derived from one parent are randomly divided between the two new nuclei. (In addition, due to crossing-over, each chromosome may contain segments derived from either parent.) If the original diploid cell had two pairs of homologous chromosomes, $n = 2$, there are 4 possible ways in which the chromosome pairs could line up with respect to one another. If $n = 3$, there are 8 possible distributions; in $n = 4$, there are 16; and so forth. The general formula, as you may be able to see by now, is 2^n. In man, with $n = 23$, the number of possible combinations is 2^{23}, or 8,388,608.

As the number of chromosomes increases, the chance of reconstituting the same set that was present in the original haploid nucleus becomes increasingly small. Quite apart from this, the existence of at least one chiasma in each bivalent makes it almost impossible that any cell produced by meiosis could be genetically the same as any that fused to give rise to the diploid line of cells that was undergoing meiosis.

Meiosis differs from mitosis, described in Chapter 1, in three fundamental ways (Figure 7-8):

1. Two nuclear divisions are involved, producing a total of four nuclei or cells.
2. Each of the four nuclei or cells contains half the number of chromosomes present in the original nucleus from which it was produced.
3. The nuclei or cells produced by meiosis contain wholly new combinations of chromosomes. In contrast with mitosis, in which cells with chromosome complements *identical* to those of the mother cell (and so to each other) are produced, in meiosis, daughter cells or nuclei *different from* the parent type are produced.

The genetic consequences of the behavior of chromosomes in meiosis are profound. Because of meiosis and syngamy, populations of diploid organisms in nature are far from uniform, consisting of individuals that differ from one another in many characteristics.

ASEXUAL REPRODUCTION: AN ALTERNATIVE STRATEGY

Asexual reproduction (also known as vegetative reproduction, or apomixis) results in progeny that are identical to their single parent. In eukaryotes, there is a wide variety of means of asexual reproduction, ranging from the development of an unfertilized egg cell to division in almost equal parts of the parent organism. In all such cases, however, the new eukaryotes are the product of mitosis and are therefore genetically identical to the parent.

Vegetative reproduction is common in higher plants and is accomplished in many different ways (Essay, page 130). Often plants reproduce both sexually and asexually, thus hedging their evolutionary bets (Figure 7-9), but many species reproduce only asexually. Even among these, however, it is clear that their ancestors were capable of sexual reproduction and that, therefore, vegetative reproduction represents an alternative—a "choice" made in response to evolutionary pressures for extreme uniformity. This choice, if rigidly adhered to, severely restricts the ability of the population to adapt to differing conditions.

7–8
A comparison of mitosis and meiosis.

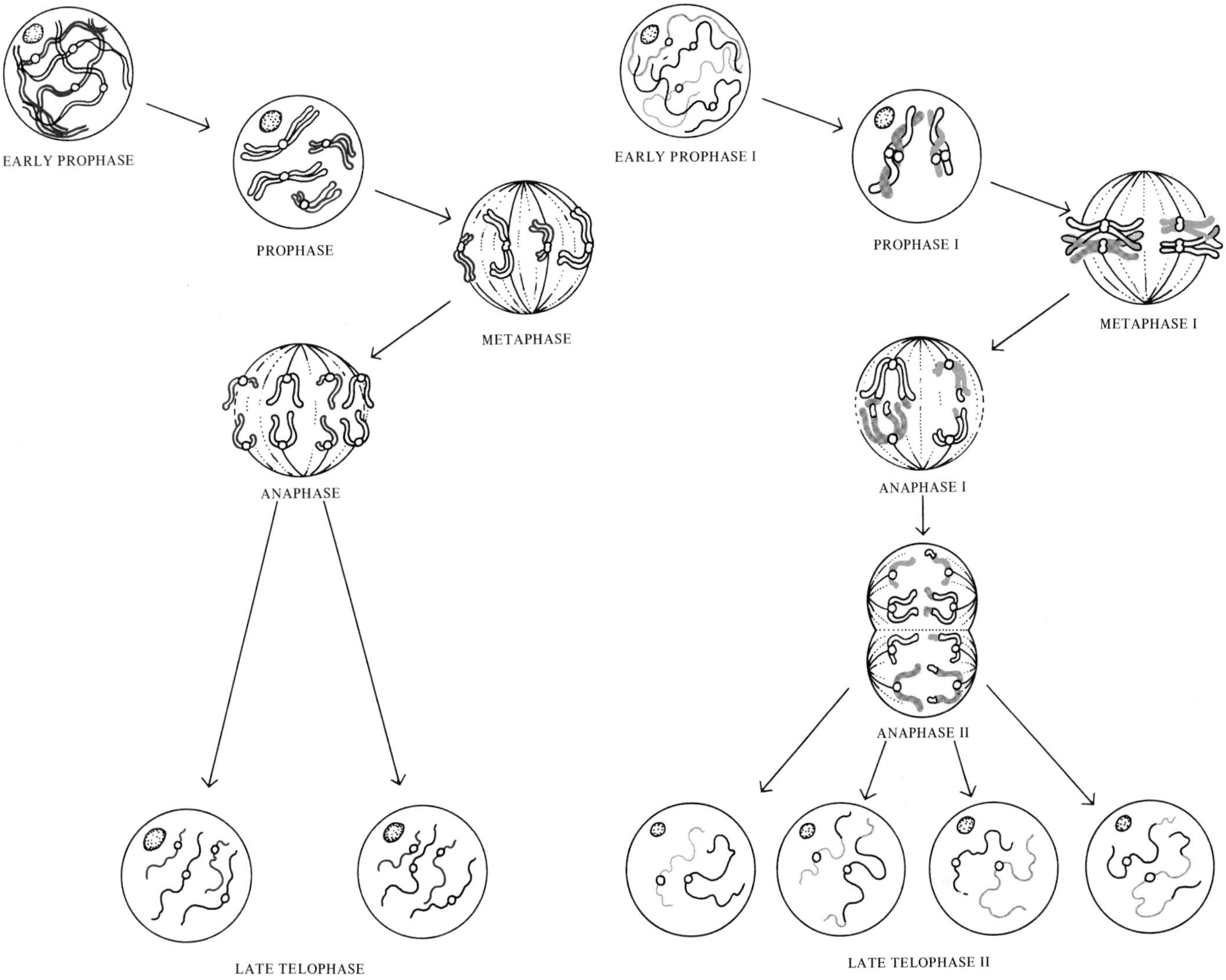

7–9
Violets reproduce both sexually and asexually. The larger flowers are cross-pollinated by insects, and the seeds may be blown or otherwise carried some distance from the parent plant. The smaller flowers, closer to the ground, are self-pollinated and never open. Seeds from these flowers drop close to the parent plant and produce plants that are genetically similar to the parent and so, presumably, apt to grow successfully near the parent. Creeping underground stems (rhizomes) also eventually produce a new seris of genetically identical plants right next to the parent.

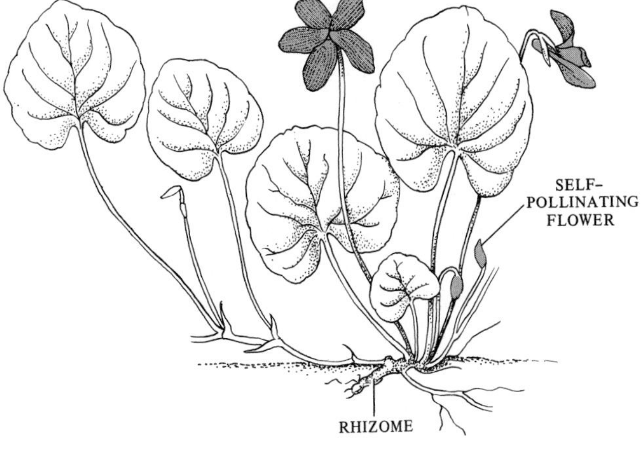

VEGETATIVE REPRODUCTION: SOME WAYS AND MEANS

The forms of vegetative reproduction in plants are many and varied. Some plants reproduce by means of runners, or stolons, long slender stems that grow along the surface of the soil. In strawberry (Fragaria × ananassa), for instance, leaves, flowers, and roots are produced at every other node of the runner. Just beyond the second node, the tip of the runner turns up and becomes thickened. This thickened portion first produces adventitious roots and then a new shoot, which continues the runner.

Rhizomes, or underground stems, are also important reproductive structures, particularly in grasses, in which the rhizomes produce stems bearing leaves and flowers. During growth, the underground stems or rhizomes are capable of invading areas adjacent to the parent plant. Each new node can give rise to a new plant, so that the species is not dependent solely upon seeds for survival. The noxious character of many weeds results from this type of growth pattern. Many garden plants, such as irises, are propagated almost entirely from rhizomes. Corms, bulbs, and tubers—all different kinds of underground stems (see pages 456–457)—are structures specialized for storage and reproduction from which new individuals arise. The roots of some plants—for example, cherry, apple, raspberry, and blackberry—produce "suckers" or sprouts which give rise to new plants. When the root of a dandelion is injured, as by a grazing animal or a spade, each root fragment gives rise to another entire plant.

In some few species, even the leaves are reproductive. One example is the genus Kalanchoë. Kalanchoë daigremontiana is familiar to many people as the "maternity plant" or "mother of thousands," so-called because numerous plantlets arise from meristematic tissue located in notches along the margins of the leaves. The maternity plant is propagated primarily from these small plants. Another is the walking fern, Asplenium rhizophyllum, in which young plants form where the ends of the attenuated leaf tips touch the gound.

Many crop plants and ornamentals are reproduced vegetatively; asexual reproduction ensures the preservation of the most desirable combinations of characteristics.

Potatoes are propagated artificially from segments of the tuber, each with one or more eyes. It is the eyes of the "seed pieces" that give rise to the new plant. Propagation of sweet potatoes generally is accomplished with whole small roots. New shoots arise from adventitious buds. Bananas are propagated from seeds. However, commercial varieties of banana do not produce seeds and are propagated by suckers that develop from buds on the corm. In other kinds of seed plants, including many kinds of citrus, some grasses, and also dandelions, the embryos in their seeds may be produced asexually from the parent plant. Such seeds give rise to individuals that are genetically identical to their parents and therefore provide another instance of asexual reproduction.

The strawberry *(Fragaria × ananassa)* is propagated by runners, long, slender stems that grow along the soil surface, every second node producing roots and then a new shoot. These plants also form flowers and reproduce sexually.

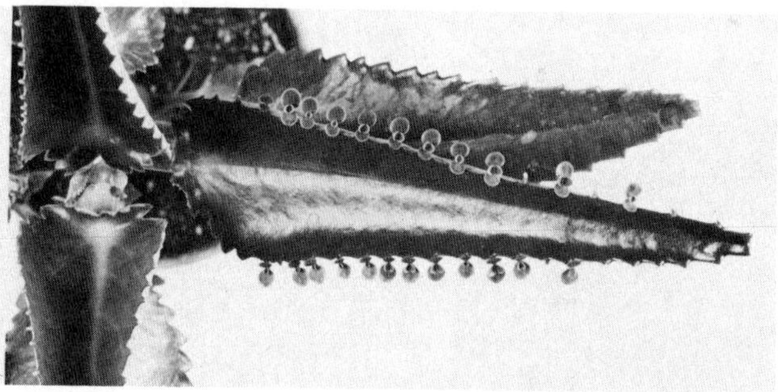

Kalanchoë daigremontiana, the maternity plant. Notice the small plants that have arisen in the notches along the margins of the leaf. When mature, they drop to the soil and take root.

Walking fern, *Asplenium rhizophyllum*, making its way over rock ledges at Keys Ferry, West Virginia. Young plants, visible in the photograph, form where the slender leaf tips touch the ground. In this way, the fern is capable of forming large colonies of genetically identical plants.

SEXUAL REPRODUCTION AND DIPLOIDY

All prokaryotic organisms are haploid. They have several means of achieving genetic recombination, however. In all of these, which are described in more detail in Chapter 10, a portion of a bacterial chromosome is transferred from one cell to another, but there is no mechanism comparable to meiosis by which it can then be transmitted regularly along with the chromosome of the cell that it joins. The only method for regular transmission of the new genetic information found among prokaryotes is the incorporation of the new fragment into the bacterial chromosome, a system that is neither flexible nor readily repeatable.

The Evolution of Sexual Reproduction

Sexual reproduction has a high selective advantage. As we have seen, it occurs only in eukaryotic organisms and involves a regular alternation betwen meiosis and syngamy. One of its most significant features is that it is the mechanism that produces variability in natural populations and, to a certain extent, helps to maintain it. As such, it provides the basis for understanding the process of evolution. In theory, sexual reproduction is unnecessary if an organism is particularly well in tune with its environment. What is needed in such a situation is the accurate reproduction of a particular "winning combination," genetically speaking. In fact, however, natural populations have to keep adjusting to a constantly changing environment, and those that are able to invade new environments in competition with others will have an advantage. Recently, genetic recombination of one kind or another has been found in organisms in which no such processes were believed to exist, such as the coliform bacteria, the blue-green algae, and some Fungi Imperfecti. These discoveries suggest that genetic recombination is more advantageous than had been previously thought and that its advantages extend even to relatively simple, fast-breeding organisms.

Another measure of the evolutionary advantage of sexual reproduction is provided by the amount of energy and other resources that it requires. These include, in higher plants, not only the production of gametes, often in great excess, but also the development of flowers and various other devices that enhance the possibilities—which still are often tenuous—of fertilization of these gametes.

The advantages of sexual reproduction were summarized in 1932 by Nobel laureate Hermann J. Müller:

> There is no basic biological reason why reproduction, variation and evolution cannot go on indefinitely without sexuality or sex; therefore sex is not, in an absolute sense, a necessity, it is a "luxury." It is, however, highly desirable and useful, and so it becomes necessary in a relativistic sense, when our competitor-species are also endowed with sex, for sexless beings, although often at a temporary advantage, can not keep up the pace set by sexual beings in the evolutionary race, and when readjustments are called for, they must eventually lose out.

The Evolution of Diploidy

Eukaryotic organisms may have arisen as long as 1.8 billion years ago, judging from fossils recently discovered in Canada. They were probably haploid and asexual at first. The oldest convincing evidence of sexuality comes from sporelike structures about 900 million years old (Figure 7–10). Once sexual reproduction was established among the unicellular eukaryotes, the stage was set for the evolution of diploidy. There is no direct evidence when the diploid condition evolved, but most, if not all of the multicellular animals and algae that appear in the fossil record 650–700 million years ago were diploid. It seems likely that diploidy arose when two haploid cells combined to form a diploid zygote. The zygote then presumably divided immediately by meiosis, thus restoring the haploid condition. In organisms with this very simple sort of life cycle, the zygote is the only diploid cell.

By "accident"—an accident that took place in a number of separate evolutionary lines—the zygote divided mitotically instead of meiotically and, as a consequence, produced an organism composed of a number of diploid cells. In a few groups of organisms, the haploid stage (and sexuality) was lost completely, as in *Euglena* and most of the amoebas. In the majority of organisms, however, meiosis was not completely suppressed but only delayed. In the animals, this delayed meiosis results

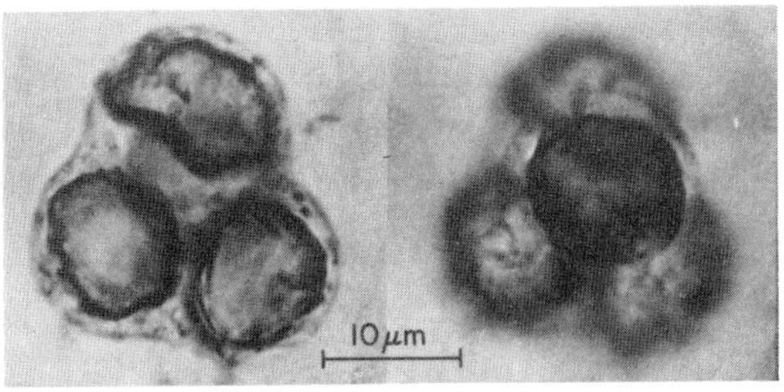

7–10
Spores from the Late Precambrian Bitter Springs chert of central Australia, about 900 million years old and interpreted as products of meiosis. These fossils are the oldest evidence suggestive of special reproduction, involving alternation of generations, eukaryotic sexuality, genetic recombination, and all of the associated phenomena.

in the production of gametes—eggs and sperms. These gametes then fuse, immediately restoring the diploid state. Thus, in the animals, gametes are the only haploid cells.

In plants, meiosis results not in the production of gametes but of spores. Spores are cells that divide directly by mitosis to produce a multicellular haploid organism, in contrast to gametes which can develop only following fusion with another gamete. Haploid organisms of this sort which appear in alternation with diploid forms are found in the plants, in the algae, and in two genera of fungi closely related to one another; this is the phenomenon known as alternation of generations. Among the plants, the haploid, or gamete-producing, generation is called the *gametophyte* ("gamete-plant"), and the diploid, or spore-producing, generation, the *sporophyte* ("spore-plant").

In some modern algae—most of the red algae, many of the green algae, a few of the brown algae—the haploid and diploid forms are the same in external appearance. Such life cycles, in which *isomorphic* alternation of generations is said to occur, are believed to have occurred first in the course of evolution. As we shall see when we discuss polyploidy (which is the occurrence of three, four, or more times the haploid number of chromosomes), an increase in the number of sets of chromosomes does not necessarily change the external appearance of the plant body.

In some groups, however, mutations occurred which, although present in both the haploid and diploid generations, are expressed in only one. In this way, the gametophyte and sporophyte became notably different from one another, and *heteromorphic* alternation of generations originated. Such life cycles are characteristic of plants and brown algae. In some algae, the gametophyte and sporophyte, although markedly different, are equally large and complex. Among the mosses and liverworts, the gametophyte is nutritionally independent from and usually larger than the sporophyte; among the vascular plants, the sporophyte is much larger and more complex than the gametophyte, which is nutritionally dependent upon it.

As we mentioned previously, diploidy permits the storage of more genetic information and perhaps its more subtle expression in the course of development. This may be the reason that the sporophyte is the large, complex, and nutritionally independent generation in vascular plants. Among the bryophytes, the gametophyte is nutritionally independent but the sporophyte is structurally more complex. In those algae that have a heteromorphic alternation of generations, the sporophyte is the large complex and nutritionally independent generation, just as in the vascular plants, except for one genus (*Cutleria*) of brown algae.

Among the vascular plants, which are now the dominant plants, one of the clearest evolutionary trends is

7–11
Evolution of genetic systems. Each of these circles represents a different type of life cycle. The most primitive eukaryotes were undoubtedly haploid for most of their life cycle, as indicated by the single line. In this type of life cycle (lower left), meiosis (indicated by the four spheres) takes place right after fertilization (designated by the arrow). The other life cycles differ from the haploid one in (1) the point at which fertilization takes place and (2) the proportion of the life cycle spent in the diploid state (indicated by the double line). The groups refer to modern organisms with the particular types of life cycles.

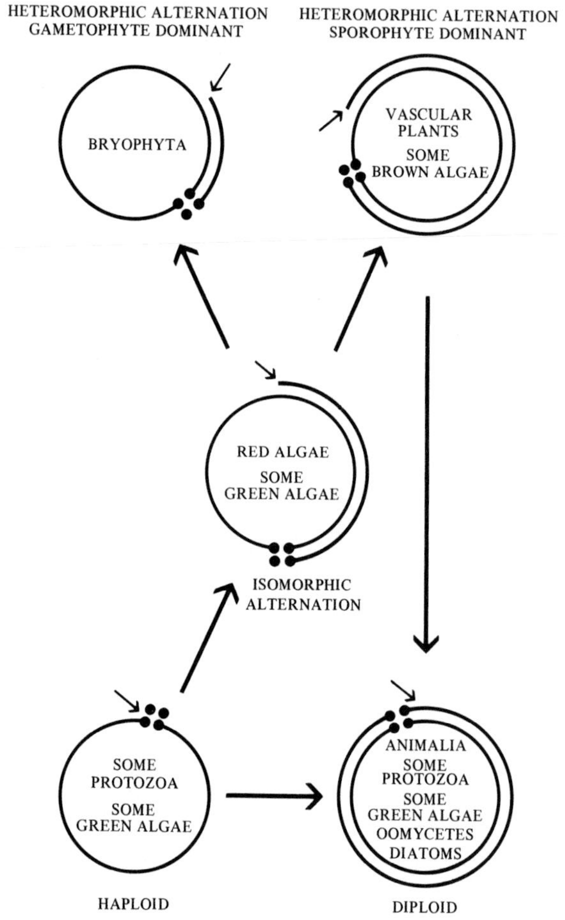

7–12
Diagrams of the principal types of life cycles. In these diagrams, the diploid phase of the cycle takes place beneath the horizontal line, the haploid phase above. The four arrows signify the products of meiosis; the single arrow represents the fertilized egg.

In gametic meiosis (a), *the gametes are formed by meiosis in a diploid individual and fuse to form a diploid zygote that divides to produce another diploid individual. This type of life cycle is characteristic of most animals and some aquatic fungi* (Oomycetes). *It is also the type of life cycle exhibited by the brown alga* Fucus.

In zygotic meiosis (b), *the zygote divides by meiosis to form four haploid cells that divide by mitosis to produce more haploid cells or a multicellular individual that eventually gives rise to gametes by differentiation. This type of life cycle is found in* Chlamydomonas *and a number of other algae.*

In sporic meiosis (c), *the diploid individual produces haploid spores as a result of meiosis. These spores do not function as gametes but undergo mitotic division. This gives rise to multicellular individuals that eventually produce gametes that fuse to form zygotes. The zygotes in turn differentiate into diploid individuals. Such a life cycle, known as alternation of generations, is characteristic of the plants and many algae. A similar sort of life cycle is found in the aquatic fungus* Allomyces *and one closely related genus.*

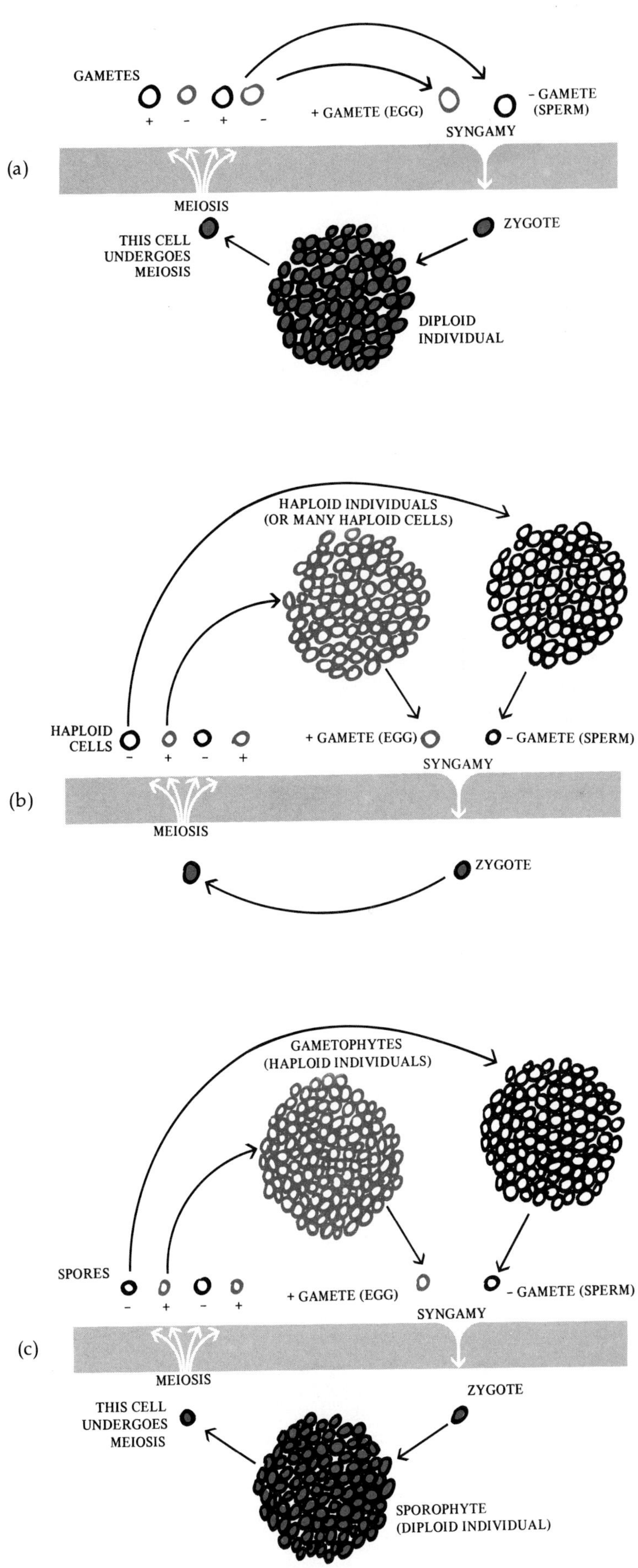

7-13
In a flower, pollen develops in the anther and the egg cells in the ovule. Pollination occurs when pollen grains, trapped on the stigma, germinate and grow down to the ovule. Pollination in most species involves the pollen from one plant (often carried by an insect) being caught on the stigma of another plant (cross pollination). In the pea flower, the stigma and anthers are completely enclosed by the petals. Because the pea flower, unlike most, does not open until after fertilization has taken place, the plant normally self-pollinates. In his crossbreeding experiments, Mendel pried open the bud before the pollen matured and removed the anthers with forceps. Then he pollinated the flower artificially by dusting the stigma with pollen collected from other plants. The fertilized eggs develops into an embryo within the ovule, and ovule and embryo form the pea (the seed).

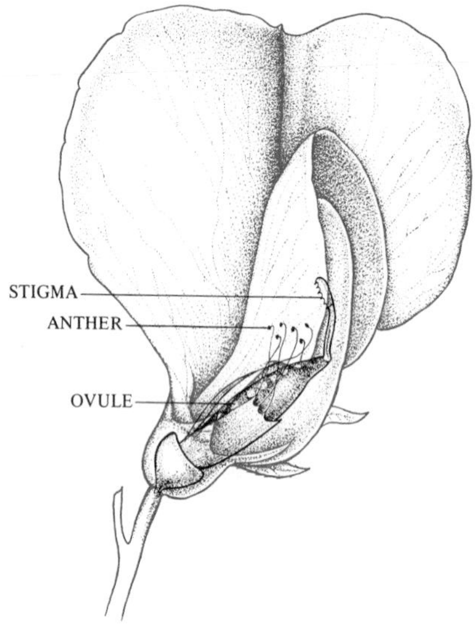

Table 7-1 *Mendel's Pea-Plant Experiment*

TRAIT	DOMINANT	RECESSIVE	F_2 GENERATION DOMINANT	RECESSIVE	TOTAL
Seed form	Smooth	Wrinkled	5,474	1,850	7,324
Seed color	Yellow	Green	6,022	2,001	8,023
Flower position	Axial	Terminal	651	207	858
Flower color	Red	White	705	224	929
Pod form	Inflated	Constricted	882	299	1,181
Pod color	Green	Yellow	428	152	580
Stem length	Tall	Dwarf	787	277	1,064

the increasing dominance of the sporophyte and the increasing suppression of the gametophyte which now, among the flowering plants, consists only (in the female gametophyte) of a fragile body of seven cells completely dependent upon the sporophyte or (in the male gametophyte) a three-celled pollen grain.

GENES AND DIPLOID ORGANISMS

Characteristics are determined in diploid organisms by the interactions between the members of a pair of genes, or *alleles.* The ways in which such alleles interact in producing particular characteristics were first revealed by Mendel. For his studies, Mendel chose cultivated varieties of the garden pea *(Pisum sativum)* and considered well-defined contrasting traits, such as differences in flower color or seed shape. He made large numbers of experimental crosses (Figure 7-13) and, perhaps most important of all, he studied the offspring not just of the first generation but of subsequent generations and their crosses. Table 7-1 lists the seven traits that Mendel used in his experiments.

When Mendel crossed plants with these contrasting traits, he found that in every case, one of the alternate traits could not be seen in the first generation (F_1). All the progeny of the cross between yellow-seeded plants and green-seeded plants were as yellow-seeded as the yellow-seeded parent. The trait for yellow-seededness and the other traits that were seen in the F_1 generation Mendel called *dominant.* The traits that did not appear in the first generation he called *recessive.* When plants of the F_1 generation were allowed to self-pollinate, the recessive trait reappeared in the F_2 generation, as you can see by studying Table 7-1, in ratios of approximately 3 to 1 (Figure 7-14).

These results can be understood easily in terms of meiosis. Consider a cross between a white-flowered plant and a red-flowered plant. The allele for white flower color, which is a recessive trait, is indicated by the lowercase letter *w.** The contrasting allele for red flower color, which is a dominant trait, is indicated by the capital letter *W.* In the strains of garden pea with which Mendel worked, white-flowered individuals had the genetic constitution, or *genotype, ww.* Red-flowered individuals had the genotype *WW.* Individuals such as these, which have two identical alleles at a particular site, or *locus,* on their homologous chromosomes, are said to be *homozygous.* When plants with these contrasting traits are crossed, every individual in the F_1 generation will receive a *W* allele from the red-flowered parent and a

* In standard genetic terminology the symbol of the gene is conventionally the first letter of the word used to describe the characteristic controlled by the deviant form of the gene. The dominant allele is indicated by a capital letter, and the recessive by a lowercase letter, which is how *W* comes to stand for red.

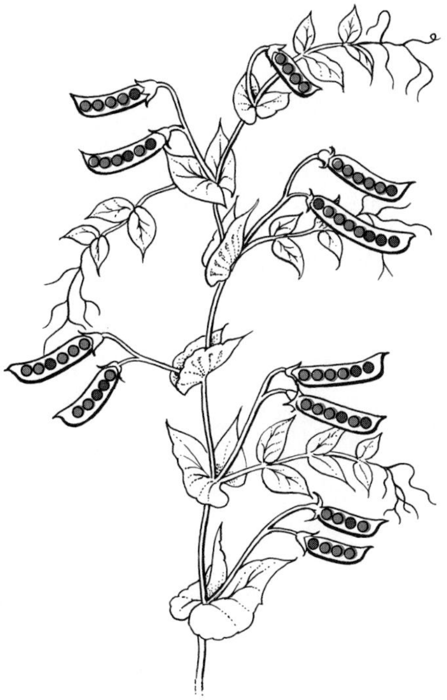

7–14
Garden pea plant with seeds (peas) in the approximate ratio of three yellow seeds for each green seed, as observed by Mendel.

w allele from the white-flowered parent, and thus have the genotype *Ww*. Such an individual is said to be *heterozygous* for this gene.

In the course of meiosis, a heterozygous individual will form two kinds of gametes, *W* and *w*. These will be present in equal proportions and, as indicated in Figure 7–15a, will recombine to form $\frac{1}{4}$*WW* individuals, $\frac{1}{4}$*ww* individuals, and $\frac{1}{2}$*Ww* individuals. In terms of their *phenotype*, or appearance, the heterozygous *Ww* individuals are red-flowered and are thus indistinguishable from the homozygous *WW* individuals. The action of the allele from the red-flowered parent is sufficient to mask the action of the allele from the white-flowered parent. This, then, is the basis for the 3 to 1 ratios that Mendel observed.

How can you tell whether the genotype of a plant with red flowers is *WW* or *Ww*? As you can see in Figure 7–15b, you can tell by crossing it with a white-flowered plant and counting the progeny of the cross. This experiment, which was also performed by Mendel, is known as a testcross. It is also known as a backcross, because it is often, though not necessarily, performed by crossing a member of the F_1 generation with the recessive parent.

The Principle of Segregation

The principle established by these experiments is the *principle of segregation*, sometimes known as Mendel's first law. According to this principle, hereditary characteristics are determined by discrete factors (now called genes) that appear in pairs, one of each pair inherited from each

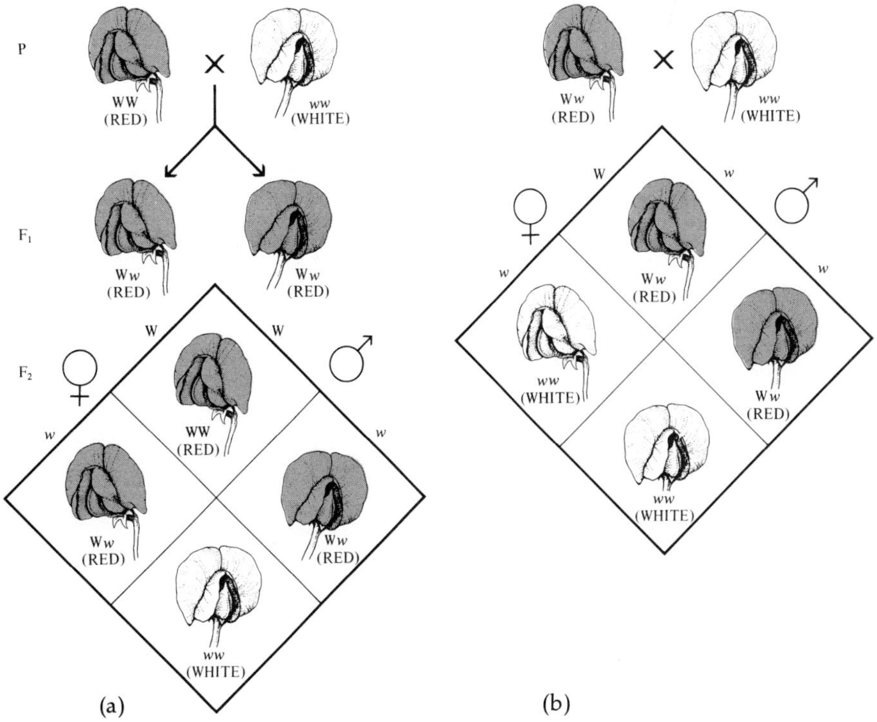

7–15
(a) *A cross between a pea plant with two dominant genes for red flowers* (WW) *and one with two recessive genes for white flowers* (ww). *The phenotype of the offspring in the* F_1 *generation is red, but note that the genotype is* Ww. *The* F_2 *generation is shown by a Punnett square. The* W *allele, being dominant, determines the phenotype. Only when the offspring receives a* w *allele from each parent and the genotype is* ww *does the recessive trait (white) appear. The ratio of dominant to recessive phenotypes is thus always expected to be 3 to 1.*
(b) *Testcross between a pea plant with red flowers and one with white. Although the red-flowering plant is phenotypically identical to the* WW *plant shown in* (a), *results of the testcross reveal that it is heterozygous for this gene.*

parent. Each time the adult offspring produces sex cells, the pairs of factors are separated, or segregated, one member of the pair going to one sex cell, or gamete, and the other member going to another gamete. This concept of a discrete factor explained how a characteristic could persist from generation to generation without being blended with other characteristics and how it could seemingly disappear and then reappear in a later generation. This principle of segregation was, as we shall see, of the utmost importance for evolutionary theory.

Incomplete Dominance

In the examples we have just considered, the action of the dominant allele, when it was present, concealed the existence of the other, recessive allele. Dominant and recessive characteristics are not always so clear-cut, however. In cases of incomplete dominance, the heterozygote is intermediate between the two homozygotes because the action of one allele does not completely mask the action of the other.

For instance, in the snapdragon, a cross between a red-flowered plant and a white-flowered plant produces a pink-flowered one. When the F_1 generation is crossbred, the traits sort themselves out again, the result being one red-flowered (homozygous) plant to two pink-flowered (heterozygous) plants to one white-flowered (homozygous) plant (Figure 7–16a). Thus cases of incomplete dominance conform to Mendel's principle of segregation.

We can explain the results of the snapdragon experiments in biochemical terms. One of the alleles apparently codes for the production of an enzyme necessary for the production of red pigment. If the alternative allele is present in the homozygous condition, it either reduces or halts production of this pigment. This may be because (1) the alternative gene produces no functional enzyme at all; (2) the gene produces a form of the enzyme with reduced efficiency; or (3) the gene produces an enzyme that works to block pigment production. In the first two cases, the second allele will probably be recessive or have a reduced effect in the heterozygote, whereas in the third case, it may act as a dominant gene, blocking pigment production entirely. Looking at heterozygosity in another way, the first allele might be able to direct the production of enough pigment to produce the effect of full color in the heterozygote in cases 1 and 2; the first allele would then be said to be dominant, and the second, recessive.

Independent Assortment

Genes are located on chromosomes. When we consider inheritance patterns involving more than one gene, we find that certain differences in the patterns depend on whether the genes are located on the same or different

7–16
Incomplete dominance. (a) *When a snapdragon with red flowers is crossed with one with white flowers, the resulting progeny* (Ww) *all have pink flowers. When the pink-flowering plants are crossed, the results show that although the gene products blend in the heterozygote, the genes themselves remain discrete and segregate according to the Mendelian ratio.* (b) *In the lima bean* (Phaseolus lunatus), *the gene for spotting* (S) *affects the pigment distribution of the seed coat. The top row of beans is* SS, *the middle row* Ss, *and the bottom row* ss. *As you can see,* S *is incompletely dominant.*

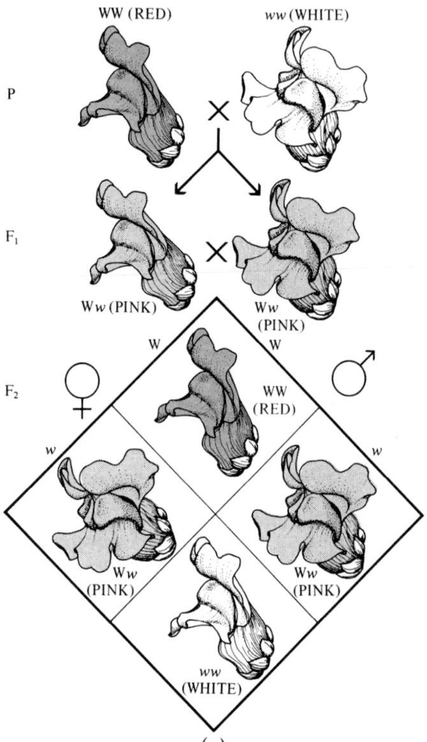

(a)

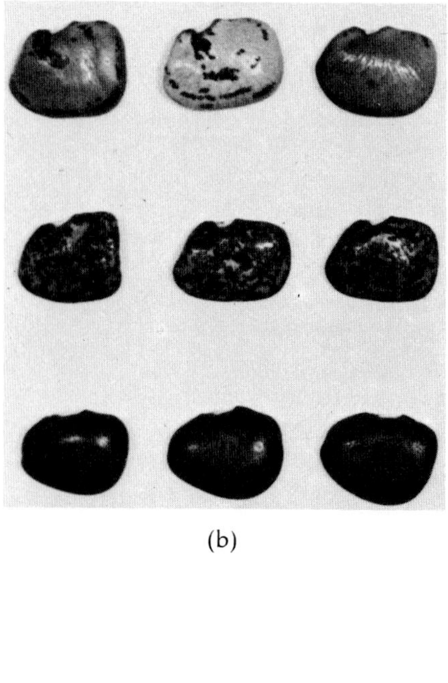

(b)

chromosomes. We shall consider first the relatively simple situation in which the genes are located on different chromosomes, and then the more complicated one in which they are located on the same chromosome.

Mendel himself studied hybrids that involved two pairs of contrasting characteristics. For example, he crossed strains of garden peas in which one parent plant had seeds that were smooth and yellow and the other had seeds that were wrinkled and green. The smooth and yellow traits, you will recall, are both dominant, and the wrinkled and green traits are recessive. As you would expect, all the plants of the F_1 generation had seeds that were smooth and yellow. When the F_1 seeds were planted and the flowers allowed to self-pollinate, 556 plants were produced. Of these, 315 showed the two dominant characteristics, smooth and yellow, and 32 combined the recessive traits, green and wrinkled. All the rest of the seeds were unlike either parent; 101 were wrinkled and yellow, and 108 were smooth and green. Totally new combinations of characteristics had appeared.

Figure 7-17 shows diagrammatically the basis for such results. As you can see, in a cross involving two pairs of dominant and recessive alleles, the ratio of distribution will be 9:3:3:1. The 9 represents the proportion of the F_2 progeny that will show both dominant traits, 1 is the proportion that will show both recessive traits, and 3 and 3 are the proportions of the two possible combinations of dominants and recessives. In the example we have shown, one parent carries both dominant traits and the other parent carries both recessive traits. Suppose that each of the parents carried one recessive and one dominant trait? Would the results be the same? If you are not sure of the answer, try making a diagram of the possibilities.

From these experiments, Mendel formulated his second law, the principle of independent assortment. It states that the inheritance of a pair of factors for one trait is independent of the simultaneous inheritance of factors for other traits, such factors "assorting independently," as though there were no other factors present.

Linkage

When Mendel was performing his experiments, chromosomes had not yet even been visualized, much less recognized as the genetic material. Knowing that genes are located on chromosomes, you can readily see that if two different genes controlling characteristics in which we are interested are located on the same chromosome pair, they will not, in general, segregate independently, and a 9:3:3:1 ratio will not be obtained in the F_2 generation. (Either by good fortune or by selecting the traits he discussed, Mendel did not encounter this phenomenon of <u>*linkage*</u>, which he probably would not have been able to understand. The pea has seven pairs of chromosomes,

7-17
Independent assortment. In one of Mendel's experiments he crossed a plant having round (RR) *and yellow* (YY) *peas with a plant having wrinkled* (rr) *and green* (yy) *peas. (The letters* R *and* Y *are used because round and yellow are the less common forms in nature.) The* F_1 *generation was all round and yellow. In the* F_2 *generation, however, as shown in the Punnett square, the recessive traits reappeared. Furthermore, they appeared in new combinations. As you can see, in a cross such as this involving two pairs of alleles or different chromosomes, the expected ratio in the* F_2 *generation is 9:3:3:1, with 9 representing the two dominant traits, 1 representing the two recessive traits, and 3 and 3 representing the alternative combinations of dominant and recessive.*

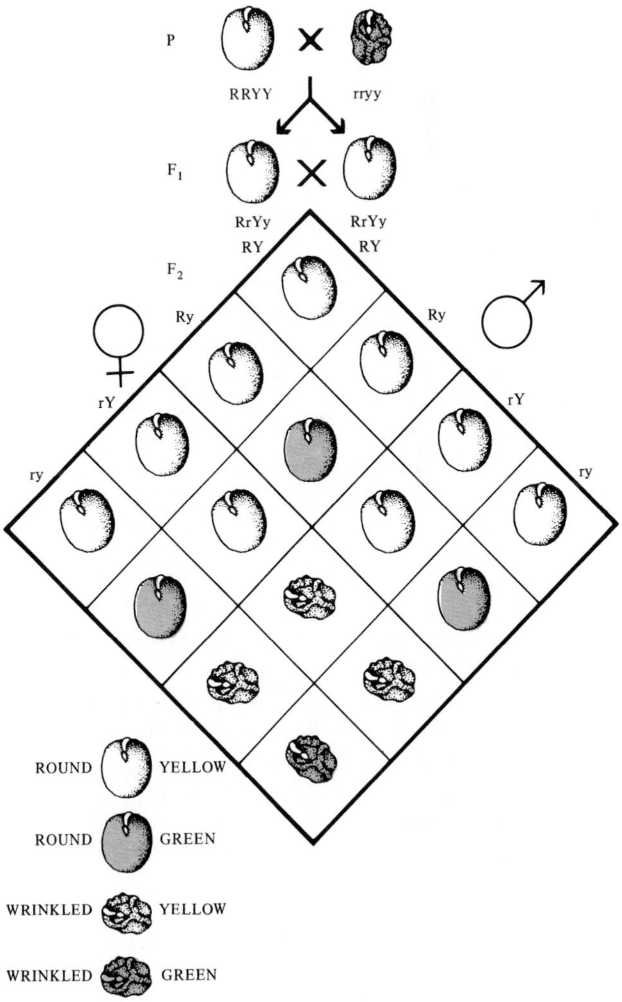

and the seven major traits studied by Mendel are either located on separate chromosomes or so far apart on the same chromosome as to segregate independently.)

Linkage and the related phenomenon of crossing-over were first understood as a result of the work of T. H. Morgan and his group with the fruit fly *Drosophila* in the early 1900s. *Drosophila* (the name means "lover of dew") is a particularly useful material for genetic studies. Colonies are easy to breed and maintain; a new generation of flies can be produced every two weeks; each female lays hundreds of eggs at a time; and the most common species has only four pairs of chromosomes.

Most individuals of *Drosophila* have gray bodies and long wings. These organisms, which show the common characteristic features of the population, are known as "wild types"; characteristics of the wild type are usually dominant. In Morgan's laboratories, two mutant types arose that had black bodies and short, or vestigial, wings. Those characteristics were recessive. The recessive characteristics were linked, and so were the dominants. Thus, individuals tended to be gray with long wings or black with vestigial wings.

Morgan crossed the wild types with mutant types bearing both recessive mutations in a homozygous state. As expected, all the F_1 offspring were phenotypically gray-bodied and long-winged.

A testcross was then carried out between the F_1 and its double-recessive parent, with the expectation of finding that the progeny divided between gray and long and black and vestigial. Large numbers of replicate experiments gave the following unexpected results:

Gray and long	41.5%
Black and vestigial	41.5%
Gray and vestigial	8.5%
Black and long	8.5%

The explanation of these figures lies in the fact (1) that the two genes were on one chromosome and their two alleles on the homologous chromosome, and (2) that sometimes genes could be exchanged between homologous chromosomes. It is now known that crossing-over, the actual breakage and rejoining of chromosomes that results in the appearance of chiasmata, takes place in prophase I of meiosis.

7–18
Hugo de Vries, standing next to Amorphophallus titanum, *a member of the same family as the calla lily. The plant, a native of the Sumatran jungles, has one of the most massive inflorescences of any of the angiosperms. This picture was taken in the arboretum of the Agricultural College at Wageningen, Holland, in 1932.*

MUTATIONS

The studies we have just described on the assortment of genes depend on the existence of differences between the alleles of a particular gene. What is the source of such differences? The first answer to this question was given by the Dutch scientist Hugo de Vries (Figure 7–18).

In studying the inheritance of characteristics in an evening primrose *(Oenothera erythrosepala)*, de Vries found that although the patterns of heredity were generally orderly and predictable, occasionally a characteristic appeared that had not been observed previously in either parental line. De Vries hypothesized that this new characteristic was the phenotypic expression of a change in a gene. Moreover, according to his hypothesis, the changed gene would then be passed along just as the other genes were. De Vries spoke of this hereditary change as a *mutation* and of the organism carrying it as a *mutant*. De Vries reported this work in 1901.

Ironically, only 2 of about 2000 changes in the evening primrose observed by de Vries were actually mutations. The others were due to new genetic combinations or the presence of extra chromosomes rather than actual abrupt changes in any particular gene. However, although most of de Vries' examples were invalid, his definition of a mutant and his recognition of the role that mutation plays in producing variation are still essentially accurate.

It is now known that a mutation can involve as little as one change in a single nucleotide pair or it may involve larger changes, such as deletions of a portion of the chromosome; repetitions of some of the genetic material; translocations, in which a piece of one chromosome becomes attached to another; or inversions, which occur when a segment of chromosome breaks loose, is rotated 180°, and becomes reattached by the "wrong" ends. Mutator genes have recently been discovered that increase the rate of mutation in other genes, sometimes in quite specific ways. Although most mutations are harmful, the capacity to mutate is extremely important because it is this capacity that allows the species to vary, to adapt to changing conditions, and so to evolve. Mutations per locus per cell division in eukaryotes occur spontaneously at a rate of about 5×10^{-6} (one mutant cell per 200,000).

In diploid plants, every gene is normally present in duplicate. When a mutation occurs in a predominantly haploid organism, such as *Neurospora* or one of the prokaryotes, it is immediately exposed to the environment. If favorable, it is selected for; if unfavorable, as is usually the case, it is quickly eliminated from the population. In diploid organisms, the situation is very different. Here each chromosome is present in replicate, and a mutation on one of the replicates, even if it would be unfavorable in a double dose, may have much less effect or even be advantageous when present in a single dose. Thus it may persist in the population. The mutant gene may eventually alter its function, or the selective forces on the population may change in such a way that its effects become advantageous.

PLEIOTROPY

In contrast with the examples given previously, genes normally affect more than one characteristic of an organism. Their products are proteins, either enzymes or structural proteins, and these products are usually involved in multiple interactions in the course of development of the organism. When the science of genetics was relatively new, genes were described conveniently in terms of their most obvious phenotypic effects, that is, long-winged, purple-petaled, hairy, and so forth. One of the first investigators to show how misleading this useful terminology can be was Theodosius Dobzhansky, who worked with Morgan's group at Columbia University. Dobzhansky arbitrarily selected the females of 12 strains of *Drosophila*, each with a single different mutation that changed a specific characteristic, such as eye or body color or wing shape. He then examined the shape of the spermatheca, the organ in which the male's sperms are stored. Ten out of twelve of the mutant strains showed a variation from normal in the size and shape of this particular organ, although observation had indicated that each mutant fly differed from the type commonly found in nature in only a single obvious characteristic. In short, a single mutation could have widespread and unpredictable effects.

Sometime later, another investigator, Hans Gruneberg, approaching the problem from the other direction, studied a whole complex of inherited deformities in the rat, including thickened ribs, a narrowing of the tracheal passage, a loss of elasticity of the lungs, hypertrophy of the heart, blocked nostrils, a blunt snout, and, needless to say, a greatly increased mortality. He was able to demonstrate that all these changes were caused by a single mutation, that is, a mutation involving only one gene. This particular gene produces a protein involved in the formation of cartilage, and because cartilage is one of the most common structural substances of an animal's body, the widespread effects of a mutation of such a gene are not difficult to understand. The capacity of a single gene to affect many characteristics is known as *pleiotropy.*

GENE INTERACTIONS IN DIPLOID ORGANISMS

Not only does a single gene often affect more than one phenotypic characteristic, but conversely there are many characteristics that are controlled by the combined effects of several genes. We shall now discuss some examples of this under two broadly overlapping headings, epistasis and polygenic inheritance.

Epistasis

Early in the present century, the British geneticist William Bateson and his group obtained some surprising results that at first seemed impossible to explain. They crossed two pure-breeding white-flowered varieties of sweet pea *(Lathyrus odoratus)* and found that the progeny all had purple petals! When these F_1 plants were allowed to self-pollinate, they found that of 651 plants that had flowered in the F_2 generation, 382 had purple petals and 269 were white. At first, these figures may seem meaningless, but if you examine them closely, you will see that they fit a 9:7 ratio. This is essentially a 9:3:3:1 ratio in which only the plants (9/16) that show the effects of *two* dominant genes had flowers with purple petals. A situation of this sort is described as *epistasis,* a form of gene interaction whereby one gene interferes with the phenotypic expression of another nonallelic gene (or genes) in such a way that the phenotype is determined effectively by the former and not the latter when both genes occur together in the same genotype.

A similar situation was seen in a cross between a pink-flowered and a white-flowered strain of salvia *(Salvia horminum).* The F_1 plants all had purple flowers. In the F_2 generation, 255 had purple flowers, 92 pink, and 114 had white. These results are harder to explain. Suppose, however, that you take the total, which is 471, and break it down into the expected results of the 9:3:3:1 ratio. The figures would be 259.3, 86.4, 86.4, and 28.9. Thus you can see that, as in the previous example, only the plants with both dominants can make the purple anthocyanin pigment. If a plant has only one dominant, it either makes pink anthocyanin or none, depending on which dominant is present. If both recessives are present, the plant cannot produce anthocyanin. Can you think of an explanation? (Remember the studies with *Neurospora* described in Chapter 3.)

Although it is in part a matter of definitions, epistasis is a much more widespread and important phenomenon than is implied by the formal definition given above. The more we know about heredity, the clearer it is that no gene ever acts in isolation; its effects are always modified by the internal environment, which is, of course, the end product of the interaction of thousands of genes. This is not surprising when we remember that genes act by the production of proteins, which can produce their effects only in the context of the cell. All such interactions between nonallelic genes in the expression of the phenotype can be considered examples of epistasis.

Polygenic Inheritance

Figure 7–19 plots the distribution of ear length in a variety of corn plants, but if it were relabeled, a curve of this shape could just as well represent the distribution of weight among a random assortment of pinto beans, of height among United States males, or of variations in the number of ventral bristles among a population of *Drosophila.* Environmental factors, such as rain in the cornfield, might affect the actual measurements involved in preparing such a graph, but they rarely affect the shape of the curve.

A curve of this sort is known as a bell-shaped, or "normal," curve. The pattern of variation is said to be *continuous.* It indicates that populations cannot be divided into a series of sharply contrasting forms.

The first experiment illustrating how many genes can interact to produce a continuous pattern of variation in plants was carried out by the Swedish scientist H. Nilson-Ehle, using wheat. Table 7–2 shows the phenotypic effects of various combinations of four genes controlling the color intensity of wheat kernels. Skin color in man follows a similar pattern of inheritance.

Such quantitative characteristics are determined by the combined effects of many pairs of genes. Some of the effects of these genes are additive, and some may work in opposite directions. Both epistasis and pleiotropy are important in producing the overall phenotypic effect, which is the result of such complex interactions that it is very difficult or impossible to isolate the effect of any one gene on the characteristic. This situation is the prevalent one in the determination of most characteristics.

7–19
Distribution of ear length of the Black Mexican variety of corn (Zea mays). *This is an example of a characteristic that is determined by the interaction of a number of genes. Such characteristics show continuous variations; if these variations are plotted as a curve, the curve is bell-shaped, as shown here, with the mean, or average, falling in the center of the curve.*

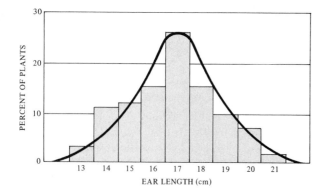

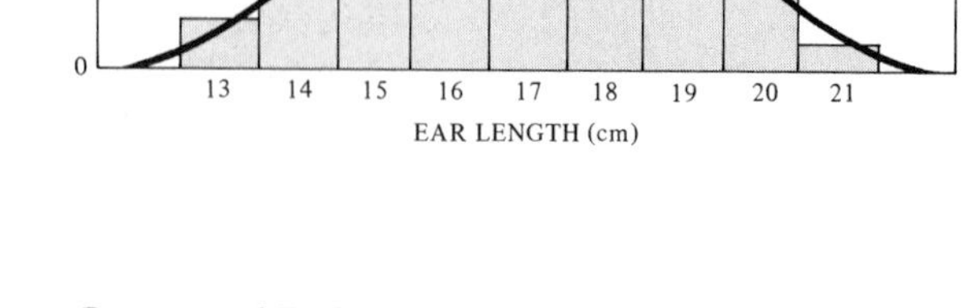

SUMMARY

Sexual reproduction involves two events: The reduction of chromosome number by meiosis and the reestablishment of the diploid number at syngamy. It is characteristic of eukaryotes, although it has been lost in some modern forms.

Meiosis results in the production of either gametes or spores. A gamete is a haploid cell that fuses with another gamete producing a diploid zygote. A spore is a cell that, without fusing with another cell, develops into a mature haploid organism.

Meiosis involves two nuclear divisions and results in a total of four cells (or nuclei), each of which has the haploid number of chromosomes.

In the first meiotic division, the homologous chromosomes pair lengthwise. The chromosomes are double, consisting of two chromatids, and chiasmata form between homologous chromatids. These chiasmata are the result of crossing-over, the exchange of chromosome segments between homologous chromosomes. The bivalents line up at the equatorial plane in a random manner (but with the centromeres of the paired chromosomes on either side of the plane), so that the chromosomes that came from the female parent and those that came from the male parent are completely re-

Table 7-2 *The Genetic Control of Color in Wheat Kernels*

Parents:		$R_1R_1R_2R_2 \times r_1r_1r_2r_2$ (dark red) (white)			
F_1:		$R_1r_1R_2r_2$ (medium red)			
F_2:		Genotype		Phenotype	
1		$R_1R_1R_2R_2$	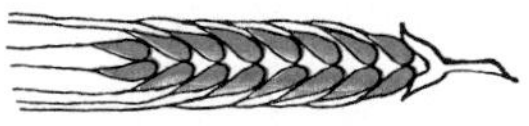	Dark red	15 red to 1 white
2	4	$R_1R_1R_2r_2$	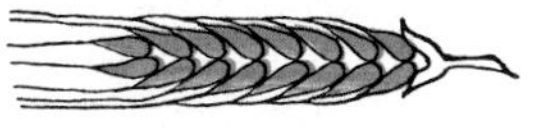	Medium-dark red	
2		$R_1r_1R_2R_2$		Medium-dark red	
4	6	$R_1r_1R_2r_2$	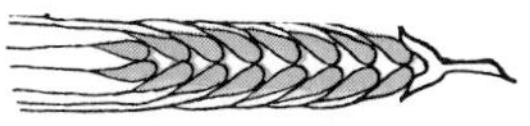	Medium red	
1		$R_1R_1r_2r_2$		Medium red	
1		$r_1r_1R_2R_2$		Medium red	
2	4	$R_1r_1r_2r_2$		Light red	
2		$r_1r_1R_2r_2$		Light red	
1		$r_1r_1r_2r_2$	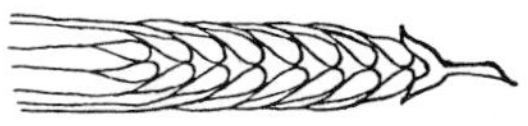	White	

assorted during anaphase I. This reassortment, together with crossing-over, ensures that each of the products of meiosis differs from the parental set of chromosomes and from each other. In this way, meiosis permits the expression of the variability that is stored in the diploid genotype.

In the second meiotic division, the chromosomes divide as in mitosis.

Sexual reproduction results in the production of progeny dissimilar from either parent. Asexual, or vegetative, reproduction results in progeny genetically identical to the parent, thus decreasing variability in a population. In most populations, sexual reproduction confers a significant advantage, because it permits adaptation to changing environmental conditions.

Diploidy evolved subsequent to the process of sexual reproduction. In primitive eukaryotes and all but two genera of fungi, the zygote formed by syngamy divides immediately by meiosis. From these, diploid life cycles were derived on a large number of different occasions, when the zygote divided by mitosis. If the haploid cells produced by meiosis function immediately as gametes, the result is the type of life cycle found in the animals and in certain algae and fungi. If they divide by mitosis, as in many algae and all plants, and two genera of fungi, they are considered spores, and the diploid generation that gave rise to them is the sporophyte. The haploid generation to which they give rise is the gametophyte and eventually it produces gametes by mitosis. If the gametophyte and sporophyte in a particular life cycle are approximately equal in size and complexity, the alternation of generations is said to be isomorphic; if they differ widely, the alternation is heteromorphic.

Diploidy permits the storage of genetic variability. The genetic makeup of an organism is known as its genotype; its appearance, or outward characteristics, is its phenotype. In diploid organisms, a category that includes most familiar plants and animals, every gene is present twice. (These gene pairs are known as alleles.) As a consequence, mutations are more difficult to detect than in haploid organisms. The phenotype in diploids is determined by the interaction of the two alleles present at the same locus on homologous chromosomes. Both alleles may be alike (a homozygous condition), or they may be different (heterozygous).

Although both alleles are present in the genotype, only one may be detected in the phenotype. The gene that is expressed in the phenotype is the dominant gene.

The one that is concealed in the phenotype is the recessive gene. When two organisms that are each heterozygous for a given pair of alleles are crossed, the ratio of dominant to recessive in the phenotype is 3:1. If the action of one allele is insufficient to mask the action of its alternative form, the heterozygotes are distinguishable (incomplete dominance), and the ratio of phenotypes is 1:2:1.

Differences between alleles are the results of mutations, which are changes in genes or groups of genes. Most mutations are disadvantageous, and so are rapidly eliminated in haploid organisms. In diploid organisms, however, they may persist in the population in a masked form and so add to the store of genetic variability.

Genes that are located on different chromosomes segregate independently of one another, whereas those that are located on the same chromosome do not. Genes located on the same chromosome are said to be in the same linkage group. In meiosis, genes do not always remain in the same linkage groups because of crossing-over. Genes act by producing proteins, and it is therefore not surprising that they usually affect not one but many characteristics of the organism. This multiple effect is known as pleiotropy.

A given characteristic is usually controlled by the interaction of more than one gene. Epistasis is the control of the expression of one gene pair by other, nonallelic genes. When multiple genes affect a single trait, inheritance is said to be polygenic, and the pattern of variation for the characteristic will be continuous. Continuous variations follow a distribution that can be represented graphically by a bell-shaped curve. Most quantitative characteristics of organisms, such as weight and height, are polygenic.

CHAPTER 8

The Evolution of Plant Diversity

8–1
Darwin and the H.M.S. Beagle. *The* Beagle, *on which Darwin sailed for 5 years, is shown here in the Straits of Magellan.*

In 1831, as a young man of 22, Charles Darwin set forth on a 5-year voyage as ship's naturalist on the British navy ship *H.M.S. Beagle* (Figure 8–1). The book he wrote about the journey, *The Voyage of the Beagle,* not only is a classic of natural history but also provides insight into those experiences that led directly to Darwin's later proposal of his theory of evolution by natural selection.

In the year that Darwin's historic voyage took place, most scientists—and nonscientists as well—still believed in the theory of "special creation." According to this concept, all the kinds or species (species simply means kind) of organisms had come into existence or been created in their present form. Some scientists, such as Jean Baptiste de Lamarck (1744–1829), had challenged special creation, but their arguments were not sufficiently convincing to shake these tenets, which were not just a scientific hypothesis but were firmly embedded in all of Western culture.

Darwin was able to bring about this great intellectual revolution simply because the mass of evidence he presented was so overwhelming and convincing that there was no longer room for reasonable scientific doubt. Particularly important in the genesis of Darwin's ideas were his experiences during a stay of some 5 weeks in the Galapagos Islands, an archipelago that lies in equatorial waters several hundred miles off the west coast of South America (Figure 8–2). Here he made two particularly important observations: First, the plants and animals found on the islands, although distinctive, were similar to those on the nearby South American mainland. If each kind of plant and animal was created separately and was unchangeable, as was then thought, why did they not resemble, for example, the plants and animals of Africa instead of those of South America? Or, indeed, why were they not unique forms, found nowhere else on earth? Second, people familiar with the islands pointed out the variations that occurred from island to island in such forms as the familiar giant tortoises *(galápagos).* Sailors who took these tortoises on board and kept them as convenient sources of fresh meat on their sea voyages were able to tell which island a particular tortoise came

8-2
The Galapagos are a small cluster of volcanic islands some 950 kilometers off the coast of Ecuador. Since they came into existence more than a million years ago, they have been colonized from time to time by accidental plant and animal voyagers swept by wind or water from the mainland. Some few of these plants and animals have managed to survive, reproduce, and adapt themselves to these bleak islands. The Galapagos are far enough apart so that the intervening sea forms a natural barrier to many organisms. As a consequence, slightly different species have evolved on neighboring islands. Were the living things on each island the product of a separate "special creation"? This question continued to haunt Darwin for many years after the voyage of the Beagle *and led eventually to his formulation of the theory of evolution by natural selection.*

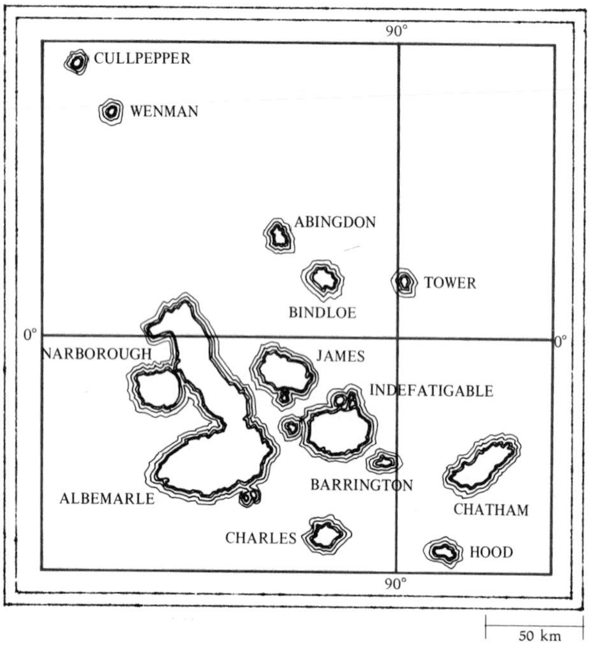

from simply by its appearance. Why, if the Galapagos tortoises had been specially created, did they not all look alike?

Darwin began to wonder if all the tortoises and other strange plants and animals of the Galapagos might not have been derived at different times from the mainland of South America. Once they reached this remote archipelago, they might have spread slowly from island to island, changing bit by bit in relation to local conditions and eventually becoming distinctive races that could be easily recognized by any observer.

In 1838, Darwin read a book that had been published in 1798 by the Reverend Thomas Malthus, a clergyman who sounded an early warning about the explosive growth of the human population. Malthus pointed out that human population growth was so rapid that it would not only soon outstrip the possible supply of food, but would leave "standing room only" on earth. Darwin saw that Malthus' reasoning was theoretically correct, not only for the human population but also for all other populations of plants and animals. For example, a single breeding pair of elephants, the slowest breeding of all animals, could produce some 19 million individuals in 750 years, were all their progeny to survive. Despite this, the number of elephants remains relatively constant, and where there were two individuals 750 years ago, there will, in general, be only two at the present time. But what determines *which* two elephants, out of the possible 19 million, will be the surviving progeny?

Darwin termed the process by which the two survivors were chosen *natural selection*. He used this term to contrast with *artificial* selection, the process by which breeders of domestic plants and animals deliberately change the characters of the races in which they are interested. They do this by controlling which individuals are allowed to breed and which are not. Darwin saw that organisms are also variable in nature. Those individuals that have favorable characteristics are more apt to breed than those that lack them, and this process can be counted on, in the long run, to produce slow but steady changes in the characteristics of populations. Such a process might reasonably be expected to require a long period of time, and it is no coincidence that the theories of the geologist Charles Lyell, who postulated that the earth was much older than it had previously been thought to be, had a profound influence on Darwin; Darwin needed such an earth as a stage upon which to view the unfolding of the diversity of living things. The process of natural selection soon became known as "survival of the fittest," which is an apt phrase but one that must be employed cautiously, as we shall learn in this chapter.

In artificial selection, breeders can normally concentrate their efforts on one or a few characteristics of interest (although there are certain very definite limits, as we shall discuss shortly). In natural selection, however, the entire organism must be fit in terms of the total environment in which it lives.

THE BEHAVIOR OF GENES IN POPULATIONS

The Hardy-Weinberg Law

In the nineteenth century, when most biologists believed in some sort of blending inheritance, it was difficult to understand why rare characteristics did not simply become so diluted that, for all intents and purposes, they disappeared. Darwin could not solve this problem in terms of the knowledge of genetics of his day.

After the rediscovery of Mendel's work, the question reappeared, framed in more modern terms: Why did the dominant alleles not eventually drive out the recessive ones, with a consequent loss of variability for the population as a whole? The answer, although it was not immediately obvious, lay in a proper understanding of the particulate nature of the gene; this particulate nature was pointed out in 1908 simultaneously by G. H. Hardy, an English mathematician, and G. Weinberg, a German physician.

The *Hardy-Weinberg law*, as it is now known, states that in a large population in which there is random mating, in the absence of forces that change the proportions of genes (these will be discussed below), the original proportions of dominant alleles to recessive alleles will be retained from generation to generation.

For example, consider the alleles of a single gene, gene *A*. Let us make up an artificial population, so that half of the individuals are homozygous *AA* and the other half homozygous *aa*. In Figure 8-3 we show by means of Punnett squares that the proportion of *AA*'s (or *aa*'s, or *Aa*'s) in the third (or fourth or fifth) generation will, on the average, be the same as it was in the second generation.

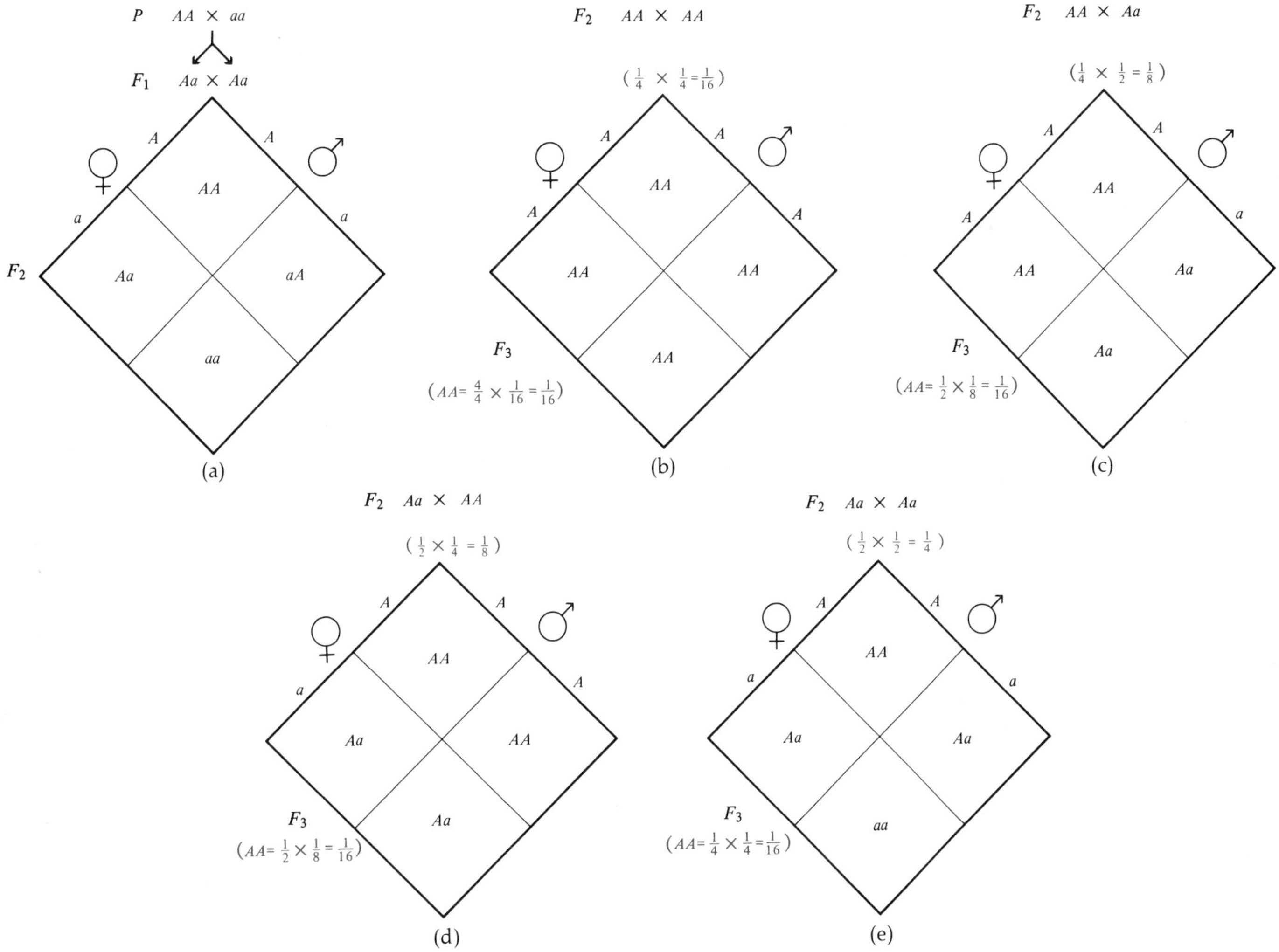

8-3
(a) *When two populations, one homozygous for a dominant gene* (AA) *and one homozygous for its recessive allele* (aa), *are interbred, all of the first filial generation* (F_1) *resemble the dominant parents, even though they are heterozygous* (Aa). *When this generation interbreeds, however, the typical Mendelian 3:1 ratio will become manifest in the* F_2 *generation, which, on the average, will be* $\frac{1}{4}$ AA, $\frac{1}{2}$ Aa, *and* $\frac{1}{4}$ aa. *The probability of producing AAs in the third* (F_3) *generation as a result of random breeding is the following:* (b) *If AAs crossbreed, the progeny will all be AA. The chances of this happening are* $\frac{1}{4} \times \frac{1}{4}$ *or* $\frac{1}{16}$. (c) *If AAs cross with Aas, the chances of producing AAs are* $\frac{1}{16}$ *because only half of the offspring are AAs.* (d) *If Aas cross with AAs, the chances of AAs are, again,* $\frac{1}{16}$. (e) *If Aas cross with Aas, the chances of producing AAs are* $\frac{1}{16}$ *because only a fourth of the offspring are AAs. Thus, in the third generation, AAs, on the average, will number* $\frac{1}{16} + \frac{1}{16} + \frac{1}{16} + \frac{1}{16}$ *or* $\frac{1}{4}$*—the same as in the second generation, and as in the fourth, fifth, sixth, and so on. In short, sexual recombination alone does not change the proportions of different alleles.*

For studies in population genetics, the Hardy-Weinberg principle is usually stated in algebraic terms, with the fractions we use in Figure 8-3 expressed as decimals. In the case of a gene for which there are two alleles in the gene pool, the frequency (represented by the symbol p) of the dominant allele plus the frequency (q) of the recessive allele must together equal the frequency of the whole, or 1; $p + q = 1$. (The frequency of an allele is simply the proportion of that allele in a gene pool in relation to all alleles of the same gene.) This is equivalent to saying that if there are only two alleles, A and a, of a particular gene and if half (0.50) of the alleles in the gene pool are A, the other half (0.50) have to be a. Similarly, if 99 out of 100 (0.99) are A, 1 (0.01) is a.

How then do we find the relative proportions of individuals who are AA, Aa, and aa? These proportions can be calculated by multiplying the frequency of A (male) by that of A (female) [A^2], A (male) times a (female) [Aa], a (male) times A (female) [Aa], and a (male) times a (female) [a^2]. We can express these multiplications in algebraic terms as $p^2 + 2pq + q^2$—which equals, as you may remember from the binomial theorem, $(p + q)^2$. If $p + q = 1$, then $(p + q)^2 = 1$. So we can see that if half (0.50) of the gene pool is A and half a, the proportion of AA will be 0.25, the proportion of Aa will be 0.50, and the proportion of aa will be 0.25, which is exactly what Mendel said—although in slightly different terms. [$A^2 + 2Aa + a^2 = (0.50)^2 + 2(0.50)(0.50) + (0.50)^2 = (0.25) + 2(0.25) + (0.25) = 0.25 + 0.50 + 0.25 = 1$.]

What happens in subsequent generations? As can be seen in Figure 8-4, the genotype frequencies now remain constant indefinitely. The gene frequencies similarly remain constant at $p(A) = 0.5$, $q(a) = 0.5$.

8-4
Possible combinations of gametes in a population consisting of AA, Aa, aa individuals, illustrating the principle expressed in the Hardy-Weinberg equilibrium.

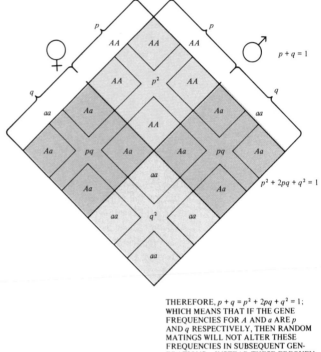

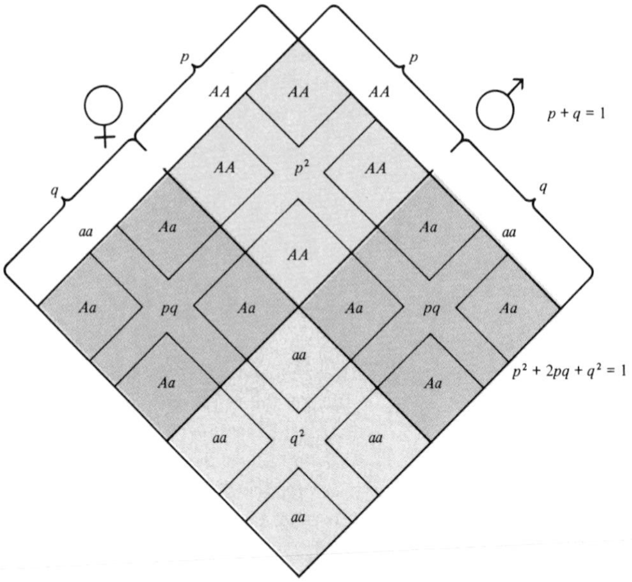

THEREFORE, $p + q = p^2 + 2pq + q^2 = 1$; WHICH MEANS THAT IF THE GENE FREQUENCIES FOR A AND a ARE p AND q RESPECTIVELY, THEN RANDOM MATINGS WILL NOT ALTER THESE FREQUENCIES IN SUBSEQUENT GENERATIONS; INSTEAD THESE FREQUENCIES WILL REMAIN CONSTANT FROM GENERATION TO GENERATION.

Exceptions to the Hardy-Weinberg Law

We know, however, that populations are not in Hardy-Weinberg equilibrium. If they were, no evolution could occur. Four factors operate to produce deviations from this equilibrium.

Population Size

As we mentioned previously, the Hardy-Weinberg law operates only when there is a large population. In a small population, the random loss of individual genotypes—for example, by a failure to breed—can lead to the elimination of one of the alleles from the population.

Migration

Further distortion is caused by migration into or out of a particular population. If individuals with particular genetic characteristics leave a population or enter it in a proportion different from that in which they are already represented in the population, the gene and genotype frequencies will obviously change.

Mutation Rate

If a particular gene is mutating to an allelic form at a rate higher than that at which the reverse mutation is occurring, the population cannot remain in equilibrium for the gene in question.

Selection

The major factor causing exceptions to the Hardy-Weinberg law is *selection*. Selection provides virtually the whole basis for our current understanding of evolutionary change. Mutations are, of course, the basis for variability, but changes in populations take place as a result of selection.

Selection is the nonrandom reproduction of genotypes. In any variable population, some individuals will leave more progeny than others. As a result, certain genes become more frequent in populations, and others become rarer. Although we sometimes think of selection as a creative force, it is important to remember that it is merely a description of events that have already taken place. When we find that the proportion of a certain gene is higher in a given generation than it was in the preceding generation, we say, in the absence of an alternative explanation, that selection has occurred. Selection does not *cause* the changes that occur. However, by continually eliminating certain genetic types from a population under a given set of conditions, it *channels* the variations in characteristics caused by mutation and recombination. In this way, selection results in the production of individuals that can survive better under given conditions than those that were eliminated. In unusual or marginal environments, genotypes that were poorly represented in the main part of the range of a species may be greatly increased by the action of selection. Selection is strictly a mechanistic process; the direction of it must necessarily change as conditions change.

In general, there are three basic types of selection operating within populations: disruptive, stabilizing, and directional (Figure 8-5). *Disruptive selection* is said to have occurred when two or more extreme types in a population increase at the expense of the intermediate forms. *Stabilizing selection*, a process that presumably goes on at all times in all populations, is the continual elimination of extreme individuals. For any quantitative characteristic, some individuals must be produced in each generation that are so extreme that they are at a disadvantage.

8-5
The three different types of selection are indicated by the shift from the black to the colored bell-shaped curves.

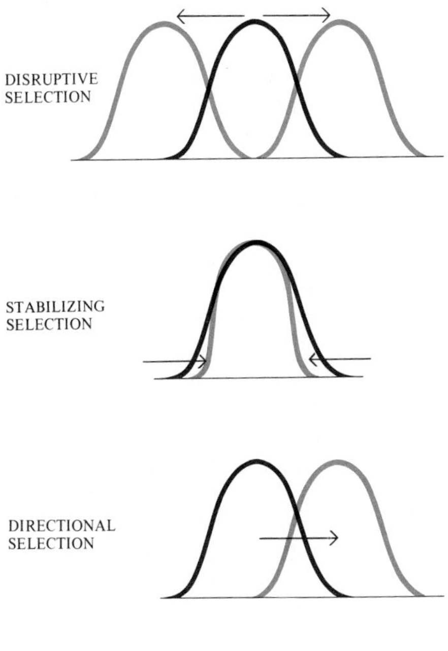

How much variability is allowed will depend on the characteristic in question and on the nature of the environment, but it is difficult to imagine any environment so permissive that it would allow the survival of all individuals, no matter how extreme. The third type of selection, directional selection, is of greatest interest to students of evolution, and it is the one that will be given the most consideration here. We say that *directional selection* has occurred when the mean value of the population shifts in favor of some particular characteristic and when there is no known explanation for this shift.

The nature of directional selection has a greal deal to do with a problem that has puzzled biologists for more than a century. How has variability been maintained in natural populations despite the operation of natural selection during the entire period in which living organisms have been present on earth? One might think that all the variability would have long since been eliminated and that the organisms that exist at present would be superbly adapted to their individual roles and essentially unvariable. There are a number of answers to this question. A major one has to do with the nature of selection against a recessive gene.

In the case of a deleterious recessive gene, even one that causes the death of individuals homozygous for it, the lower the frequency of the gene in the population, the less it will be acted on by natural selection. This is because the proportion of recessive genes that occur in homozygotes decreases precipitously as the frequency of the gene decreases. Where q equals the frequency of gene a, the distribution of the recessive gene will show the pattern outlined below.

	Genotype frequencies			
q	AA	Aa	aa	Percentage of total a in heterozygotes Aa
0.9	0.01	0.18	0.81	10
0.1	0.81	0.18	0.01	90
0.01	0.9801	0.0198	0.0001	99

In short, the lower the frequency of a recessive gene, the less it is exposed to the action of selection in a homozygous form. As the gene frequency drops, the progress toward removal of the gene from the population slows down accordingly. This fact has particular significance for those who favor eugenic improvement of the human race by selectively preventing individuals who carry undesirable recessive traits in homozygous form from breeding. Quite apart from the extraordinary difficulties involved in deciding which traits are undesirable enough to allow this procedure and to make it socially acceptable, there are serious difficulties inherent in the process of selection. If a particular recessive gene is present at a frequency of 0.01 in the human population, so that homozygous individuals make up 0.0001 of the population (about 1 such individual out of each 10,000 "normals"), it will take 100 generations, or about 2500 years, to lower the gene frequency to 0.005 and to reduce the number of defective individuals to 1 in 40,000. And even this much reduction assumes total prevention of breeding by all homozygous individuals, no additional mutations to increase the frequency of the gene, and a human society that would remain stable and maintain the prevention of breeding by such individuals as an important goal for 2500 years. In contrast, dominant genes that are lethal or prevent breeding are eliminated from the population in one generation.

CHANGES IN POPULATIONS

Response to Selection: Genetic Factors

The response of a population to selection is affected by many of the principles of genetics discussed in the preceding chapter. In general, only the phenotype is being selected, and it is the relationship between the phenotype and the environment that determines the reproductive success of an organism. As nearly all characteristics of interest in natural populations are determined by the interaction of many genes, similar individuals can have very different genetic properties. When some feature, such as tallness, is strongly selected for, there is, in general, an accumulation of genes that contribute to this feature and an elimination of those that work in the opposite direction.

But we must never think of selection for a polygenic characteristic as a simple accumulation of one set of genes, with an elimination of their alleles. In our consideration of gene interactions (page 139), we saw that the action of genes is complicated. The expression of a gene takes place only through its production of a protein. Proteins are extremely diverse and act in various ways in mediating different reactions in organisms and in affecting their structure. It is, therefore, not surprising that the action of every gene is affected by the action of other genes. Further, enzymes and other proteins do not normally, or perhaps ever, work in only one highly specific way in the organism. They produce various biochemical effects that may affect the phenotype in many ways (pleiotropy). Gene interactions are of importance in determining the course of selection in a population. Some of the effects of a gene may be beneficial to the organism, or contribute to the characteristic being selected for. Others may be harmful, or work against the characteristic being selected for. As a consequence of gene interactions, we can measure the phenotypic effects of a particular gene only in a particular genetic context. As selection for particular genes changes the representation of the other genes in the population, the effects of the selected genes become gradually changed. Thus the value of the genes in determining a particular characteristic, or the fitness of the organism as a whole, also changes.

Another feature of importance in determining the nature of selective change is the necessity of producing an organism that "works." Strong selection for one feature may be prohibited because it would lead to the accumulation of so many undesirable side effects that the organism would no longer be able to survive, or might lose its fertility. The results of the experiment shown in Figure 8-6 suggest that, as a result of selection, the population depicted reached a new internal genetic balance that allowed the production of fertile, otherwise normal in-

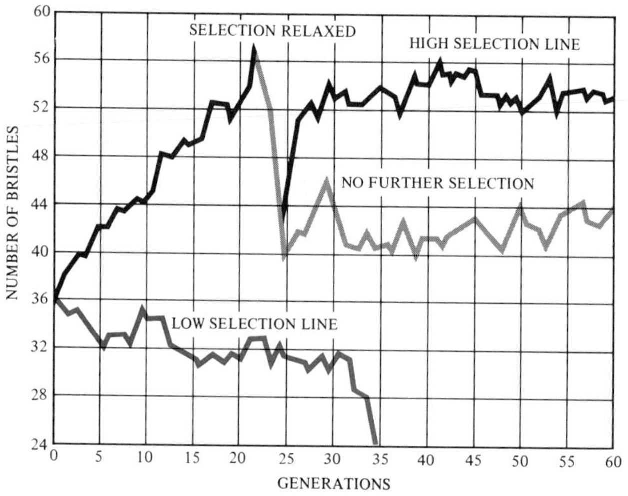

8-6
Selection for bristle number in Drosophila melanogaster. *In the parental stock, the average number of bristles on the ventral surface of the abdomen was 36. One breeding group was selected for an increase in bristle number (high selection line) and one for a decrease in bristle number (low selection line). The high selection line reached 56 bristles in 21 generations, but soon the stock began to become sterile. Selection was discontinued after generation 21 and then resumed at generation 25. This time the previous high bristle number was regained without loss of fertility. The low selection line, which was not permitted to achieve a new internal genetic balance, died out because of sterility. Selection for a single "neutral" gene, such as bristle number, had widespread effects on the genetic construction of the organisms.*

dividuals with 42, instead of 36, abdominal bristles. This balance, which can be looked upon as a measure of the resistance of the population to genetic change, has been termed genetic homeostasis by I. Michael Lerner of the University of California. Homeostasis in physics is a sort of dynamic equilibrium, a resistance to change. Genetic homeostasis tends to keep the population as a whole producing a high proportion of individuals that are well adapted in terms of their environment.

Response to Selection: Phenotypic Factors

Other kinds of limits to selection are imposed by the need for an individual organism to meet conflicting environmental demands. For example, the long tail of the male peacock, with its brilliant coloration, is highly correlated with the mating success of the bird, and it is clearly advantageous in this respect for the bird to have the largest and most showy feathers possible. On the other hand, the bird must also escape from its predators, and here such tail feathers might be a disadvantage. Natural selection strikes some sort of balance between the two selectional forces, and peacocks with tail feathers that are either too long and colorful or too short and drab probably do not make as great a reproductive contribution to subsequent generations as those in which the tails are "just right."

Similarly, in harsh, alpine environments, plants must grow and photosynthesize rapidly if they are to store enough carbohydrate to survive the long, severe winters. Therefore they must respond to the first sign of favorable conditions. But if their response is so finely tuned that they respond even to a temporary thaw in midwinter, they will be eliminated from the population. Desert plants, the seeds of which must germinate whenever sufficient water is available but must not germinate when too little water is available, are in a similar position. As will be discussed on page 507, some have solved this problem by the production of germination inhibitors in their seed coats: Only when enough rain has fallen to leach away these substances will the seeds germinate.

An interesting evolutionary strategy among tropical woody members of the pea family (Fabaceae) has been detected by Daniel Janzen of the University of Michigan. In Central America, Janzen found two kinds of species: Those that have numerous small seeds and those that have rather few, large seeds (Figure 8–7). The plants can spend a certain amount of their food reserve each year on the production of seeds, and they can therefore produce either a large number of small seeds or a small number of large seeds. Why do some of the species follow one path and the others the other path?

Large seeds provide a more abundant food supply for the germinating seedlings. Because each parent tree or shrub will be replaced, on the average, by only one other tree or shrub, it would seem most efficient for it to produce a small number of large seeds with an abundant stored reserve of food. But this is not always possible in the tropics, owing to the very heavy predation on legume seeds by insects, especially the seed beetles (Bruchidae). The adult bruchids lay their eggs on the fruits, and their larvae complete their development within the seeds. Bruchids are often so abundant that they virtually destroy the seed crop of a particular legume in a given year, except for the few seeds removed by a dispersal agent, such as a bird, before the bruchids can find them.

Janzen investigated bruchid infestation among large-seeded and small-seeded groups. To his surprise, he found that the small-seeded species were heavily attacked by bruchids, so that very few viable seeds remained, whereas nearly all the large-seeded species were untouched. Subsequently he found that these large seeds, but not the small ones, contained chemical substances that apparently protected them from the beetles.

Summing up these results, it seems clear that, in an evolutionary sense, the legumes had two options. Either they could produce a very large number of small seeds, with a relatively limited food supply for the germinating seedlings but with a better chance of some of the seeds

8–7
Seeds of Central American legumes. Each of the seeds is derived from a different species of tree or shrub belonging to this plant family. All of the seeds in the lower group are presumably poisonous, and none is attacked by seed-eating insects. Those in the upper group are edible, and heavily attacked by seed-eating beetles of the family Bruchidae, which complete their larval development within the seeds. Exit holes may be seen in several of the seeds of the upper group.

escaping the attacks of the beetles, or they could produce a smaller number of large seeds, with a good food supply for the germinating seedlings. However, they could take this second option only if the seeds were protected in some way from the beetles. This is an example of the sort of evolutionary compromise often found in nature.

In the light of Janzen's reasonable explanation of the situation in tropical legumes, it is interesting to note that on oceanic islands the seeds and fruits of plants are often larger than the mainland relatives of these plants (Figure 8–8). The kinds of pests that attack the plants on the mainland are often absent on the islands, and it seems likely that this provides the opportunity for the plants to produce larger fruits and seeds, which provide a better start for their seedlings. This isolation from pests, like the biochemical defenses of the large-seeded legumes of the American tropics, is a factor that allows large seed size, selected for on other grounds, to prosper more than it could otherwise.

8–8
Island species, isolated from their natural enemies, are often able to produce larger fruits and seeds. Shown here are two related species, one found in Hawaii and one on the Japanese mainland.

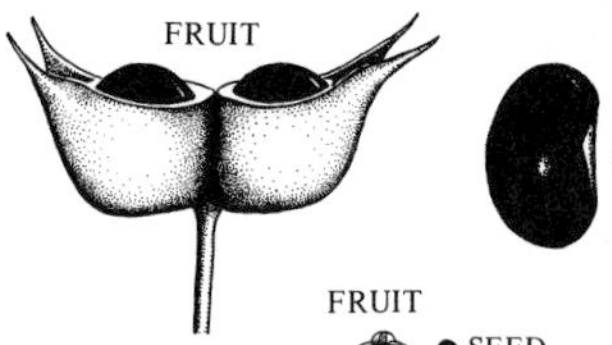

Changes in Natural Populations

In recent years, it has become possible to study changes that have occurred in natural populations. Changes in climate and natural catastrophes have been a constant feature of the world since the beginning, and the populations of organisms present at earlier times responded to these changes in ways similar to those we can observe at present. The nature and rate of some of these changes have been such that we can speak about them as "evolution in action." It is not surprising that many of the relatively rapid changes in natural populations that have been observed over the last century or so have resulted from the activities of man, an ecological dominant without parallel in the history of the world.

Some of the most spectacular changes have been observed in populations of microorganisms, which are haploid. Their mutations are expressed immediately in the phenotype, and selection for or against them proceeds at once. All haploid organisms have a short generation time, which is probably essential for successful changes in populations of microorganisms, because they do not have the store of variability for recombination that is found in diploid organisms.

In plants, strong selective forces have been observed to produce rapid changes in natural populations. A pasture in southern Maryland had been divided many years earlier into two portions—one grazed and one ungrazed. The plants in the grazed part of the pasture were much

GENETIC DIVERSITY AND THE GREEN REVOLUTION

Until recently, the major food plants of the world, like the wild ancestors from which they arose, were genetically very diverse. In fact, since the beginnings of agriculture, more than 10,000 years ago, huge reserves of variability have accumulated in the important crop plants by the processes of mutation, hybridization, artificial and unconscious selection, and adaptation to a wide range of conditions. Thus for crops such as wheat, potatoes, and corn, there are literally thousands of known strains. This genetic variability provides an important safety factor. If one strain were to prove unusually susceptible to a particular pathogen, for example, others could be found that were genetically resistant from which new varieties could be bred.

As a result of the Green Revolution (page 11), however, crop plants have become more and more uniform and hence more vulnerable to destruction. In 1970, for example, the southern corn leaf blight fungus, Helminthosporium maydis, *destroyed approximately 15 percent of the United States corn crop—a loss of approximately a billion dollars. These losses were apparently related to the appearance of a new race of the fungus that is highly virulent for corn of a type that is used extensively in hybrid seed production.*

Such dangers increase as the improved strains being developed continue to replace the many distinct types that existed before. And time is growing short. Indeed, it is already becoming difficult to locate seeds of many of these previous strains.

It is now generally agreed, by both critics and supporters of the Green Revolution, that aggressive research programs are needed in connection with our crop plants, first to monitor the appearance of new strains of plant pathogens and second to maintain the genetic diversity of these crops. Varieties threatened with extinction must be sought out throughout the world and preserved in suitable gene banks. One such genetic reservoir has been established by the United States Department of Agriculture at Fort Collins, Colorado. If such programs are not expanded and maintained, survival of the human race may once again be threatened by its own technological successes.

lower than those in the ungrazed part, and it was first thought that this might be because of the direct action of the grazing. This was tested by digging up some of the plants in the grazed part and planting them in good soil. It was assumed that if they were low merely because they were grazed, they would become tall, like the plants in the ungrazed part of the pasture, under the changed conditions. Some did, but plants of white clover *(Trifolium repens),* Kentucky bluegrass *(Poa pratensis),* and orchardgrass *(Dactylis glomerata)* remained low in stature, indicating that the populations had been rapidly modified genetically by the selective force of grazing. Only those genotypes that produced low-growing individuals were able to set seeds under these conditions.

In Wales, there are a number of abandoned lead mines. The tailings and dumps around these mines are rich in lead and nearly bare of plants. One species of grass, *Agrostis tenuis,* colonizes the mine soil, which contains up to 1 percent lead and 0.03 percent zinc. In one experiment, samples of plants were taken from mine tailings and from a nearby pasture and grown together both in normal soil and in lead-mine soil. In normal soil, the lead-mine plants were definitely slower-growing and smaller than the pasture plants. On the lead-mine soil, the mine sample was normal, but the pasture sample did not grow at all. Half of the pasture plants were dead in three months and had misshapen roots that were rarely more than 2 millimeters long. But a few of the pasture plants (3 out of a sample of 60) showed some resistance to the lead-rich soil. These were doubtless similar to the plants originally selected in the development of the lead-resistant race. The mine was no more than 100 years old, and the resistant race had been developed in this very short period of time.

THE DIVERGENCE OF POPULATIONS

The world is physically and biotically complex, and individual kinds of plants almost always have discontinuous distributions. Their habitats are always discontinuous, whether they be lakes, streams, the tops of mountains, a certain soil, shaded spots in the woods, or whatever. Given this, the movement of genes between the isolated populations is more or less limited, and the populations are correspondingly free to respond to the selective demands of their local situations in different ways. Even though normal gene flow (the movement of genes from one population to another, as a consequence of the immigration of individuals from a different population) may often provide a source of variability for a distant population, it cannot counteract the effects of local selection. The distances necessary to isolate populations effectively vary with the dispersibility of the plants in question, but are often quite small. For several kinds of insect-pollinated plants in temperate regions, a gap of 15 meters may effectively isolate two populations. Rarely will more than 1 percent of the pollen that reaches a given individual come from this far away. Even in plants in which the pollen is spread by wind, very little falls more than 50 meters from the parent plant under normal circumstances, and its chance of reaching a receptive stigma is correspondingly decreased beyond this distance. This is not to say, of course, that two pine trees separated by 50 meters are out of genetic contact, but rather that each will be affected more by the demands of its local environment in producing progeny.

Any two separated populations diverge from one another because they respond to different selective forces. If they regain contact they may merge, or the differences that have accumulated between them may lead to a degree of genetic isolation. Genetic isolation, or reproductive isolation, can come about in a variety of ways, as we shall discuss later in this chapter.

Ecotypic and Clinal Variation

Developmental plasticity is the tendency of individual, genetically identical organisms to differ greatly among themselves because of the action of the environment. Plasticity is much higher in plants than it is among animals. The open system of growth characteristic of plants can more easily be modified in various ways that produce striking differences between the individuals. Environmental factors, as every gardener knows, can cause profound changes in the phenotype of various species of plants. Leaves that develop in the shade are thinner, broader, have a greater volume of airspace, thinner palisade tissue, and fewer stomata than leaves of the same plant that develop in the sun. Day length—the length of time per 24 hours that a plant is exposed to the light—can also affect leaf form. In *Kalanchoë,* for example, plants grown on short days (8 hours of light) had small succulent leaves with smooth edges, whereas plants grown on long (16-hour) days had large thin leaves with notched edges. In view of such observations, it is not surprising that a number of workers proposed, until about 1930, that much of the variability of plants observed in nature was directly caused by the environment and did not have a genetic basis.

Are the differences between races of plants inhabiting different habitats genetically or environmentally controlled? The first definitive answer to this question was provided by the Swedish botanist Göte Turesson, who transplanted races of a number of plant species in southern Sweden into his experimental gardens. In the great majority of the 31 species he investigated experimentally, the differences he observed in nature were genetically controlled; in a very few, environmental modification predominated. Differences in plant stature, time of flowering, color of leaves, and the like were

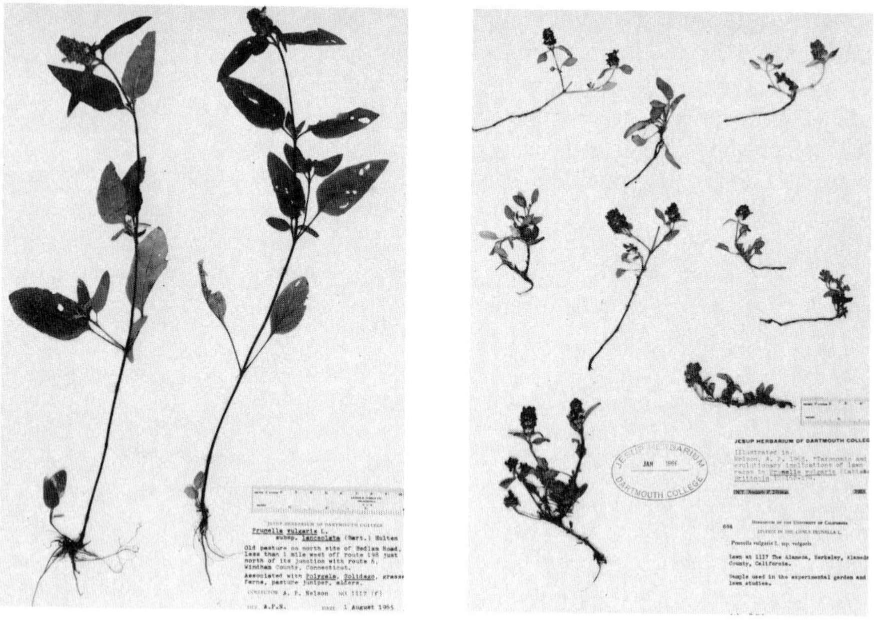

(a) (b)

8-9
Prunella vulgaris is a common herb, belonging to the mint family, which occurs in woods, meadows, and lawns throughout North America. Most populations consist of erect plants such as those shown in (a)*, which were found in an abandoned pasture in Connecticut. Populations in lawns, however, always consist of prostrate plants such as those shown in* (b)*; erect plants cannot survive in lawns. When lawn plants are grown in an experimental garden, some remain prostrate whereas others become erect. The prostrate habit is genetically determined in the first group, environmentally in the second.*

usually genetically controlled. Turesson termed such genetic races, differentiated with respect to particular habitats, *ecotypes*, a term that is still used today.

Of particular interest in the study of ecotypes is the work of Jens Clausen, David Keck, and William Hiesey, which was carried out in California under the auspices of the Carnegie Institution of Washington. These workers dealt experimentally with a number of species of plants of the western United States. They established transplant stations at Stanford, near sea level; in the central Sierra Nevada at Mather, at about 1400 meters elevation; and at Timberline, at 3050 meters elevation (Figure 8-10). Working mainly with plants that could be propagated asexually, so that genetically identical individuals could be grown at all three locations, they expanded certain of Turesson's experiments.

The western United States is a region of environmental contrasts, and it is not surprising that many plant species have developed sharply defined ecotypes in this region. One species studied by the Carnegie Institution group was the perennial herb *Potentilla glandulosa.* A close relative of the strawberry *(Fragaria),* it ranges through a wide variety of climatic zones in California, and native populations occur near each of the three experimental stations mentioned above. When *P. glandulosa* from the different locations were grown side by side in the experimental garden, a number of ecotypic differences became apparent. There appeared to be four major ecotypes. Their morphological characteristics are strongly correlated with the physiological ones critical for survival.

For example, the Coastal Range ecotype consists of plants that grow actively both winter and summer when cultivated at Stanford. They also survive at Mather, where they are subjected to about 5 months of cold winter. Here they become winter-dormant but store enough food in the remainder of the year to carry them through this unfavorable season. At Timberline, plants of this ecotype almost invariably die the first winter; the short growing season at this high elevation does not permit their survival. Other plants that occur in the Coastal Range produce ecotypes with identical physiological responses. Indeed, it is often true that different species of plants growing together are more similar to one another physiologically than they are to other populations of their respective species.

The physiological and morphological characteristics of ecotypes usually have a very complex genetic basis, involving dozens or, in some cases, perhaps even hundreds, of loci. Sharply defined ecotypes are characteristic of regions where the breaks between adjacent habitats are themselves sharply defined. On the other hand, if the environment changes gradually, with no clear break, then populations of plants may do likewise. Patterns of this sort, called *clines*, are often encountered in marine organisms, where the temperature often changes gradually with latitude. They are also characteristic of areas such as the eastern United States, where rainfall gradients may extend over thousands of kilometers. When populations of plants are sampled along a cline, the differences are often proportional to the distance between them. On a small scale, populations of plants vary in similar ways—either gradually or abruptly, depending on the nature of the local environment.

8–10
Transplant gardens of the Carnegie Institution. (a) *Stanford, near sea level;* (b) *Mather, in the central Sierra Nevada at about 1400 meters elevation;* and (c) *Timberline, at 3050 meters elevation.*

Physiological Differentiation

One of the most active fields of research in botany at present concerns the physiological basis for ecotypic differentiation. For example, Scandinavian strains of the goldenrod *Solidago virgaurea* from shaded and exposed habitats show differences in the photosynthetic response to light intensity during growth. The plants from shaded environments grow rapidly under low light intensity, whereas under high light intensity the growth rate is markedly retarded. In contrast, plants from exposed situations grow rapidly under high light intensity. In the cattail *Typha latifolia,* the enzyme malate dehydrogenase is much more resistant to high-temperature inactivation when obtained from plants native to hot climates.

In one experiment, arctic and alpine populations of the widespread herb *Oxyria digyna* were studied, using strains from an enormous latitudinal gradient extending from Thule, Greenland, and Point Barrow, Alaska, to the mountains of California and Colorado. It was found that leaf chlorophyll content increased with latitude, that plants of northern populations had higher respiration rates at all temperatures than plants from farther south, and that high-elevation plants near the southern limits of the species attained photosynthetic light saturation at a higher light intensity than low-elevation plants from farther north. The existence of *Oxyria* over such a wide area and range of ecological conditions is made possible, in part, by the differences in metabolic potential among its constituent populations.

Reproductive Isolation

The genetic system of a population of plants responds to selection as an integrated unit. For this reason, plants from widely different populations may be unable to form hybrids (the offspring of genetically dissimilar parents) or, if they do, the hybrids may not be fertile. In general, the greater the difference in appearance of two populations, the less likely it is that they can hybridize, although there are important exceptions to this generalization.

When two populations become reproductively isolated—unable to form fertile hybrids with one another—they have no effect on each other's subsequent evolution, at least in a genetic sense. Such isolation is therefore one of the most crucial steps in the evolutionary divergence of populations.

Reproductive isolation has often been used as a basis for "defining" the category *species.* However, the very different looking, ecologically distinct "species" in some groups of plants—particularly in long-lived groups such as trees and shrubs—often can hybridize. On the other hand, the "species" in herbaceous and short-lived groups are generally sterile when crossed, and individual populations (varieties) within such "species" are also often sterile. Thus, it is impossible to use a single criterion to define a species in plants. (The question of what constitutes a species will be discussed further in Chapter 9.)

Populations of annual groups of plants presumably change much more rapidly than populations of long-lived groups. Not only do they have much shorter life cycles, but they depend in each new year on establishment from seed, a factor that heightens the effects of natural selection and consequently causes them to diverge from one another rapidly.

There are a number of factors in addition to hybrid sterility that tend to keep populations of plants distinct when they grow together. Some prevent the formation of the hybrid in the first place. For example, two kinds of plants that are capable of forming fertile hybrids may occur in different habitats. In the eastern United States, the scarlet oak *(Quercus coccinea)* occurs together with the black oak *(Q. velutina)* over thousands of square miles; yet hybrids are rare. In general, scarlet oak is found in relatively moist, low areas with acidic soil, and black oak is found in drier, well-drained situations. Only where the habitat is badly disturbed does hybridization become frequent, and here the distinction between the two sorts of habitats becomes blurred.

Among the other mechanisms that prevent the formation of hybrids between species that occur together are seasonal differences in the time of flowering. If two species do not flower together, they will not hybridize in nature even when they grow side by side. The sorts of photoperiodic mechanisms discussed in Chapter 24 provide abundant mechanisms for differentiation of this kind. Again, two species that occur together may differ in their pollination systems, a point that will be discussed again in Chapter 17. If they are visited by and pollinated by different kinds of insects, they will hybridize only when these insects "make mistakes" and visit the "wrong" flower.

Clusters of Species

The evolutionary processes we have been discussing lead to the production of groups of related species in many different geographic areas. The clusters of species on the Galapagos archipelago are famous because of their role in the development of the theory of evolution.

Differentiation on islands is particularly striking where, in the absence of competition, organisms seem much more free to produce bizarre forms than on the mainland. Apparently islands, like lakes, provide favorable situations for the kinds of major evolutionary changes that occur when new genera and families arise. In such localities plants and animals often change much more rapidly than on the mainland, assuming forms never encountered elsewhere. A good example is afforded by the genus *Cyanea,* a member of the family Lobeliaceae that is confined entirely to the Hawaiian Islands. There are at least 60 species of this genus, and they differ strikingly among themselves in almost every characteristic (Figure 8–11). However, clusters of species differentiated to a less spectacular degree are characteristic of all flourishing groups of organisms and are the inevitable result of the evolutionary patterns we have been discussing. Convergent evolution may likewise take place under similar circumstances on different continents or in different areas (Figure 8–12).

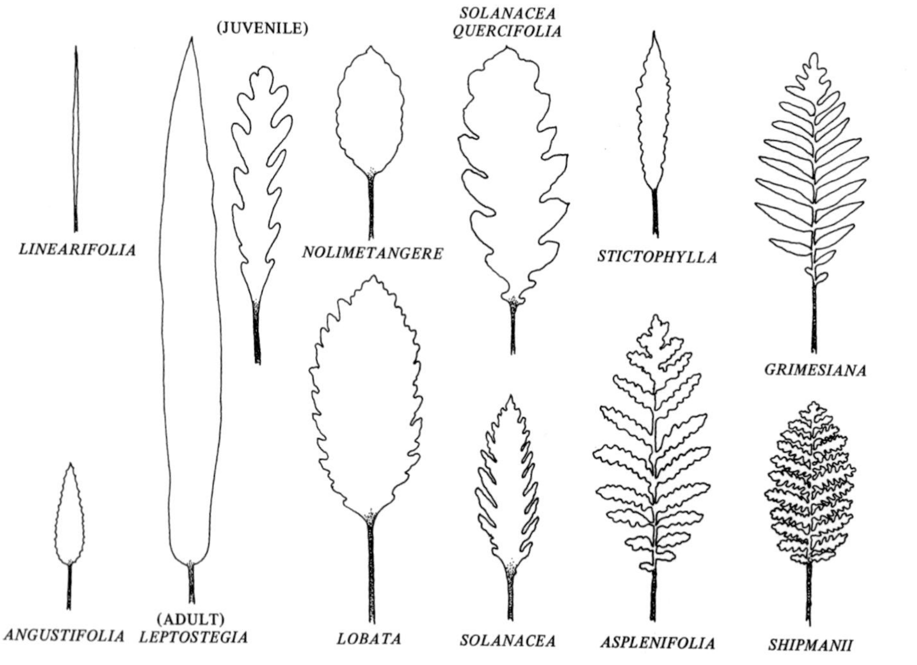

8–11
The species of Cyanea *(a genus found only in the Hawaiian islands) differ widely in leaf size and shape. These differences are often connected with their diverse growth forms and leaf outlines.* Cyanea linearifolia, *for instance, is usually found in dry, sunny locations. Plants with fernlike leaves like those in the right half of the picture are found in shady locations where thin, wide-bladed leaves are more efficient in light-gathering. They probably evolved from species such as* C. lobata *(center).*

8–12
Convergent evolution can produce similarities among totally unrelated organisms. Adaptation to a desert environment has resulted in the similarities of form shown in three species of different plant families: euphorbia, cactus, and milkweed.

THE EVOLUTIONARY ROLE OF HYBRIDIZATION

Even if species hybridize rarely in nature, these hybrids may be important because of the way in which they recombine the parental characteristics. The environment changes repeatedly, and individuals of hybrid origin may often present genetic combinations better suited to the new environment than either parent. Or they may be able to colonize some habitat where neither parent could grow. When the habitats of the parental species are next to one another and sharply distinct, there may be little opportunity for the establishment of hybrids; but where the habitats intergrade or are disturbed, the situation may be very different. Here the recombination of genetic material shared by two species has a greater potential for producing well-adapted offspring than does change within a single population. The degree of variability available when two species are hybridizing is much greater than that available to either of the species by itself, and the chances of putting together individuals suited to the new habitat are correspondingly greater (Figure 8–13). Some of the ways in which this occurs in plants will now be discussed.

The Establishment of Hybrid Populations

The term introgressive hybridization was coined by Edgar Anderson of the Missouri Botanical Garden to describe the following situation. When two species are hybridizing and the hybrids are rare, these hybrids will most frequently cross not with other hybrids but with

8–13
Plane trees (Platanus) *offer examples of well-differentiated populations that have retained the ability to hybridize. Present-day species of this ancient genus have been isolated from one another for at least 50 million years in widely scattered localities around the Northern Hemisphere. One of these, the oriental plane* (P. orientalis), *occurs from the eastern Mediterranean region east to the Himalayas. This handsome tree has been widely cultivated in southern Europe since Roman times, but it cannot be grown in northern Europe away from the moderating influence of the Atlantic. After the discovery of the New World, the eastern North American* P. occidentalis *was brought into cultivation in the colder portions of northern Europe, where it flourished. About 1670, the two very different trees were being grown together in the botanical garden at the University of Oxford, where they hybridized to produce the intermediate and fully fertile London plane,* P. × hybrida. *This fertile hybrid, which is capable of growing in regions that have cold winters, is much more vigorous than either parent. It is now grown as a street tree throughout the temperate regions of the world, because it has a high tolerance for urban pollution.*

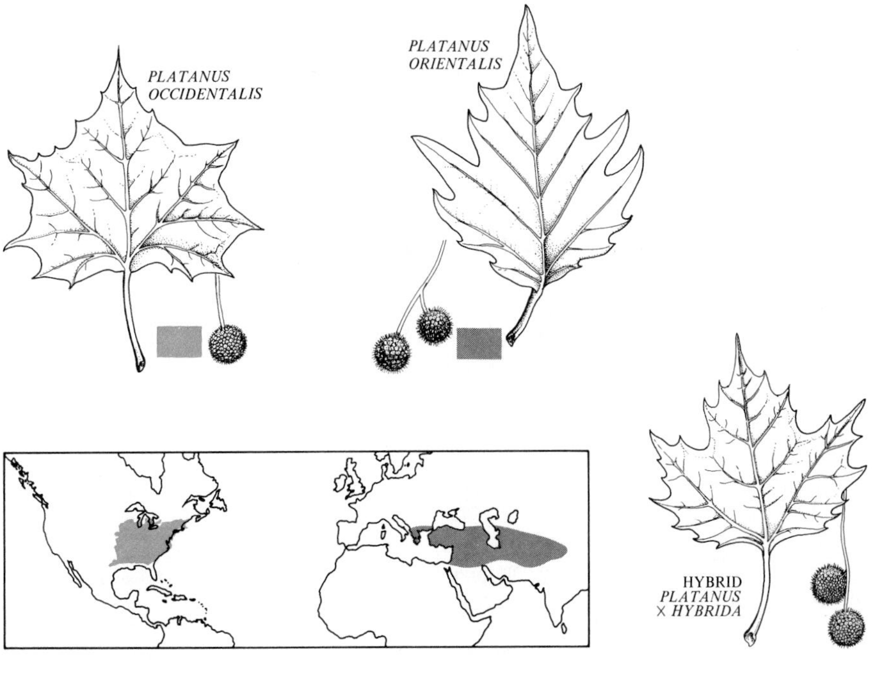

NEW HYBRIDS FROM SOMATIC CELLS

In 1969, a team of Japanese scientists reported that single mesophyll cells isolated enzymatically from tobacco leaves were able to divide and could be grown in laboratory cultures, much like bacteria. In 1970 and 1971, members of this same group showed that these naked protoplasts, grown in a suitable medium, would regenerate cell walls and continue to divide and, when transferred to other media, colonies of these cells could differentiate shoots and roots and eventually form whole plants.

In the meantime, several groups of botanists, working independently, developed methods for fusing plant protoplasts from the same or different species, thus producing cells with the chromosomes of different parents. In 1972, investigators from the Brookhaven National Laboratory in Upton, New York combined the various techniques, fusing two somatic (body) cells of different tobacco species and inducing them to develop into hybrid plants. The isolation of the hybrid protoplasts from the paternal and nonhybrid protoplasts was facilitated by the fact that the hybrid protoplasts were able to grow on culture media that would not support either of the parent strains.

Thus far—at least at this writing—the only hybrid plant to have been cultivated successfully from hybrid somatic cells is that of tobacco. Moreover, the hybrid was formed between two species of tobacco (Nicotiana glauca *and* N. langsdorffii) *which earlier had been hybridized successfully by sexual means. (The production of a hybrid whose characteristics were already well known was a necessary part of the experimental technique. Without this information it would have been difficult, maybe impossible, to identify the hybrid protoplasts in a mixed population of protoplasts.) However, this initial success has, of course, raised the hopes of producing entirely new superplants—organisms that might be capable, for instance, of simultaneously carrying out C_4 photosynthesis and fixing nitrogen—that is, converting atmospheric nitrogen into stable, biologically assimilable compounds. New techniques are being developed to induce the fusion of protoplasts from different genera. So far, fusion of protoplasts from plants as distantly related as corn* (Zea mays) *and soybean* (Glycine max) *has been achieved, but such hybrid cells have not produced new plants.*

Even while these studies are going forward, the isolated protoplasts are being used in a variety of other ways. For instance, it has been found that they can be used in rapid screening tests to determine resistance to particular pathogens or to detect nutritional requirements. In this way, a scientist can work with 10^7 or 10^8 individual cells in a test tube rather than with 10^3 or 10^4 plants on an hectare of land. Also, haploid protoplasts (isolated from anther cultures) can be grown in the same way as the diploid cells and can even be made to form adult plants. These haploid organisms are especially useful in the search for valuable mutations because the mutations cannot be masked, as they are in diploid forms.

Finally, and most speculative of all, there is the possibility of introducing foreign genetic material—selected DNA molecules—into the protoplasts. Such DNA might be of plant or bacterial origin. One way to introduce it would be through the use of viruses (see page 190). Although it will certainly be some time before practical results are achieved by any of these methods, the basic techniques—the isolation of protoplasts and their regeneration into mature plants—are clearly well at hand.

one of the parental species. As a result of repeated backcrossing with rare hybrid individuals, one or both of the parental species may eventually be modified. If the modified individuals are better able to survive than the unmodified ones in part of the range of the species, introgression may be favored selectively. However, the barrier to hybridization between the two original species may not be affected by the introgression that is changing one or both of them, and they may still be free to evolve along separate pathways. Each can remain well adapted to its respective habitat and, by virtue of its increased variability, may also be able to exist in new habitats.

In cases of introgressive hybridization, the hybridizing species retain their distinctness and the new populations are clearly part of one or both of the hybridizing species. In nonintrogressive instances, hybridization may weaken or diminish the distinctness of two clearly defined species by the production of a series of intermediate populations, which may become stabilized if they play a different ecological role from their parents. In the production of such populations, there is tremendous potential for recombining genes from the parents into new, highly adapted populations. As in the case of introgression, the combination of the two gives the populations collectively tremendous potential for response to change.

As you can see in the example in Figure 8–14, it is often difficult to demonstrate with certainty that a particular intermediate population has originated following hybridization between two other species. Nevertheless, evidence is rapidly accumulating that this is an important evolutionary mechanism, which operates in group after group of plants. In some genera, recombination of genetic material between species and the production of new entities seem to be the chief mode for developing new species and responding to a changing environment. These are primarily woody genera, such as the mountain lilacs *Ceanothus* (Figure 8–15) and the manzanitas *Arctostaphylos.* Clusters of species that interact in this way tend to be most common on islands or in highly diverse, rapidly changing areas such as California and its neighboring states.

Both introgressive hybridization and the stabilization of hybrid populations depend for their success on the fertility of the hybrids. Even if the hybrids are sterile, however, they may propagate themselves either by apomixis or polyploidy.

Apomixis and Hybridization

Sterile hybrids may reproduce themselves by a variety of vegetative means (see page 130). Most flexible are systems in which apomixis is prevalent but in which occasional hybridization occurs to produce variable and novel combinations of genes.

An outstanding example of such a system is the ex-

8–14
Richard Straw of California State University, Los Angeles, has hypothesized that hybridization accounts for the origin of the flowering plant Penstemon spectabilis *in the mountains of southern California.* Penstemon grinnellii *has broad, two-lipped, very pale blue flowers that are pollinated mainly by large bees, such as carpenter bees* (Xylocopa). *Another species,* P. centranthifolius, *has long slender red flowers that are visited mostly by hummingbirds. Their suspected hybrid derivative,* P. spectabilis, *is ecologically and morphologically intermediate betweeen the two, with rose-purple flowers. It is pollinated by unusual pollen-gathering wasps of the genus* Pseudomasaris, *which visit neither of the parental species.*

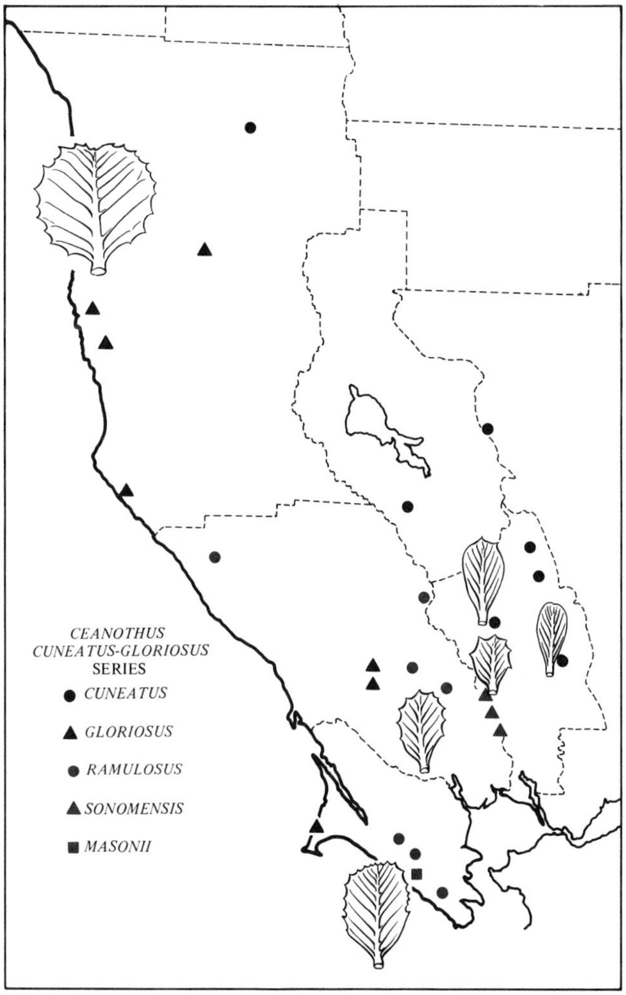

(a)

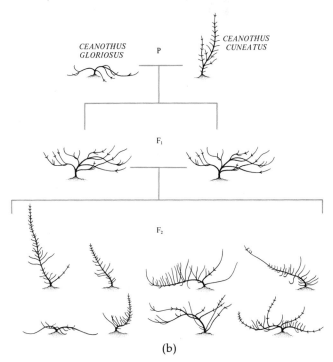

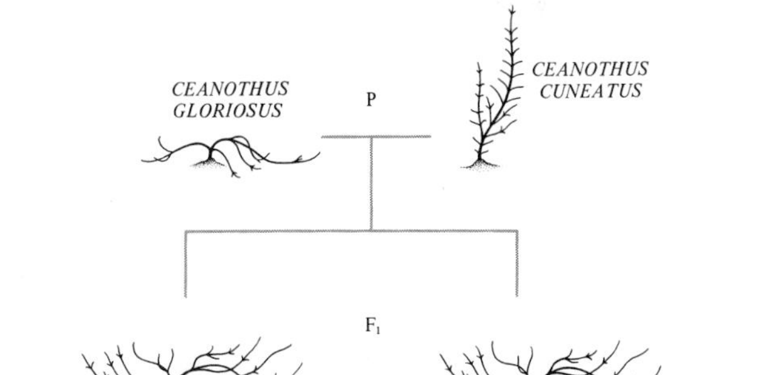

(b)

8–15
The establishment of hybrid populations is an important evolutionary mechanism in many groups of woody plants such as Ceanothus. (a) *Map showing portion of central California, a region that is geologically complex. Two relatively widespread and distinct species, the coastal* C. gloriosus *and the interior* C. cuneatus *(which ranges far to the east and out of the area), have produced three series of hybrid populations that are more or less variable but stabilized and presumably grow better than either parent in the areas where they occur. Leaves from representative populations are shown on the map.* (b) *Segregation of the extremes in appearance in the progeny of an artificial cross between the two species.*

tremely variable Kentucky bluegrass, *Poa pratensis,* which in one form or another occurs all around the Northern Hemisphere. Occasional outcrossing with a whole series of related species has produced hundreds of apomictic races, each well adapted to the ecological characteristics of the region where it grows. Thus we see a flexible system in which new genotypes are produced constantly and the best are preserved by apomixis. In such a system, it makes no difference whether the well-adapted individuals are sterile or not, for their success does not depend on sexual reproduction. Further, apomictically reproducing individuals are particularly successful in arctic regions because pollination by insects is difficult under these conditions and because certain narrowly defined genotypes may be more successful under the rigorous conditions of the Arctic than variable populations of sexually reproducing organisms.

The hundreds of species of hawthorn, *Crataegus,* and blackberries, *Rubus,* found in the eastern United States are apomictic derivatives of complexes in which occasional hybrids occur, as in *Poa pratensis.* In all these cases, the destruction of natural habitats by man on a wholesale scale has made possible the establishment of many new genotypes that would have had no place in the primeval forests of the region.

8–16
When two diploid species cross to form a diploid hybrid, the chromosomes of the hybrid may not be able to pair at meiosis. If so, gametes cannot be formed. If the nonfertile hybrid becomes polyploid (4n), *each chromosome will have a pairing partner at meiosis and viable gametes can be produced.*

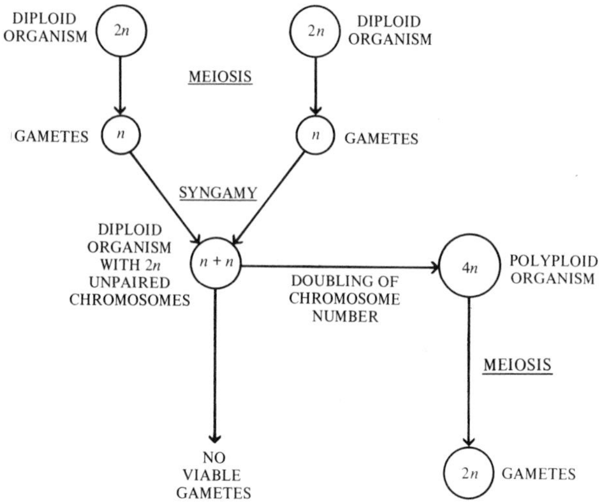

Polyploidy

Cells or individuals with more than two sets of chromosomes are called polyploids (see page 37). Polyploid cells arise at a low frequency as the result of a "mistake" in mitosis so that the chromosomes divide but the cell does not. If such cells divide by further mitosis so that they give rise to a new individual, either sexually or asexually, that individual will have twice the number of chromosomes of its parent. Polyploid individuals can be produced by the use of a drug called colchicine, which inhibits the formation of the cell plate at mitosis through its disruptive effect on microtubule formation.

When two species cross, their hybrid is often sterile because the chromosomes of the species involved are unable to pair properly with one another and so to undergo normal meiosis. If polyploidy occurs in such a hybrid, each chromosome from both parents will be present in duplicate and may be able to pair with its duplicate chromosome. In this way, meiosis will be normal and fertility restored. This process is shown diagrammatically in Figure 8–16.

It is the latter sort of polyploid that is of particular evolutionary importance. Indeed, the latest estimate is that approximately half of the 235,000 kinds of flowering plants have had a polyploid origin.

Like all hybrids, polyploids of hybrid origin are most apt to become established if they are genetically programmed to be successful in some habitat different from that of the two parents. As an evolutionary mechanism, polyploidy appears to function chiefly in stabilizing hybrid derivatives. Each gene is represented at least four times instead of twice, and individuals that exhibit recessive traits are correspondingly rare.

One of the earliest well-documented cases of polyploidy involved hybrids between the radish *Raphanus sativus* and the cabbage *Brassica oleracea.* Both of these species have 18 chromosomes in their somatic cells and regularly form nine pairs of chromosomes at meiotic metaphase I. The hybrid between them, which was obtained with some difficulty, had 18 unpaired chromosomes in meiosis and was completely sterile. In the polyploid that appeared spontaneously among plants of this nature, there were 36 chromosomes in the vegetative cells, and 18 pairs were formed regularly in meiosis. In other words, the hybrid had *all* the chromosomes of the radish and the cabbage in its cells, and these functioned normally to give the polyploid a relatively high fertility.

A number of polyploids have originated as weeds in habitats associated with the activities of man, and sometimes they have been spectacularly successful. One of the best known is a salt marsh grass of the genus *Spartina* (Figure 8–17). A native species of this genus, *S. maritima,* occurs in marshes along the coast of Europe and also Africa. A second species, *S. alterniflora,* was accidentally introduced into Britain from eastern North America in about 1800, and it spread to form large but local colonies.

(a) (b) (c) (d)

8–17
The native British species of salt marsh grass, Spartina maritima, *has 60 chromosomes, visible here in a cell in meiotic anaphase I* (a). *A North American species,* S. alterniflora, *with 62 chromosomes (31 bivalents in meiotic metaphase I)* (b) *was collected from the shores of Southampton Water in 1839, but the date of its introduction into Britain is unknown. Although the two species have never been successfully crossed artificially, specimens of their sterile hybrid,* S. × townsendii, *were first collected from the shore of Southampton Water in 1870. A vigorous polyploid,* S. anglica, *arose spontaneously from this sterile hybrid and was first collected in 1892. The polyploid, which has 122 chromosomes, as shown in meiotic anaphase I* (c), *is now extending the salt marshes of Great Britain* (d).

GENETIC VARIABILITY AND POLYPLOIDY

As at all other levels of morphology, plants exhibit a great deal of genetic variability at the molecular level. Typically if one were to examine an "average" locus within a collection of diploid plants, 10 to 15 percent of the individuals examined would prove to be heterozygous, having different alleles on their two chromosomes. Within the past decade it has become possible to perform just such analyses directly, using the technique of zone electrophoresis. A plant homogenate is placed upon a gel or other supporting medium and subjected to high voltage. Individual proteins migrate in the field, each with a characteristic rate depending on its particular charge, size, and shape. The power of the technique is that it permits examination of discrete gene products, as specific substrates may be employed in staining for a specific enzyme. Thus a genetic variant at a particular enzyme locus may be directly detectable as exhibiting a band of altered mobility when the gel is stained with its specific substrate.

The reasons why plants (and animals) maintain such high levels of variability (polymorphism) is not yet clear, although current investigations suggest that the heterozygous individuals, possessing alternative forms of the enzyme with different biochemical characteristics, may be better able to cope with diverse environmental stresses. In plants, the prisoners of their immediate microhabitat, such physiological buffering may be of immense adaptive value. The power of such an adaptive strategy becomes even more apparent when one considers the implications of polyploidy in plants. Most polymorphic enzymes have several identical subunits—dimeric molecules have two identical subunits, tetramers have four, and so on. Consider a diploid plant with two alleles, α *and* β, *for a dimeric enzyme. A heterozygous individual has not two, but three, alternative dimeric molecules* ($\alpha\alpha$, $\alpha\beta$, *and* $\beta\beta$) *in its cells, rendering it far more versatile biochemically than a homozygote, which can have only one form* ($\alpha\alpha$ *or* $\beta\beta$). *If the enzyme is a tetramer (four subunits), then the heterozygous individual has available no less than five alternative assemblages* ($\alpha\alpha\alpha\alpha$, $\alpha\alpha\alpha\beta$, $\alpha\alpha\beta\beta$, $\alpha\beta\beta\beta$, $\beta\beta\beta\beta$), *each with its own biochemical characteristics. In a polyploid plant, typically a tetraploid, an alteration in any one of the four alternative genes leads in each such individual to a wide array of alternative subunit assemblages. Unlike the polymorphic population, in which only some individuals are heterozygous, here* all *progeny possess the biochemical versatility conferred by genetic variability. Tetraploid plants typically exhibit a far greater range of edaphic (soil) habitat tolerance than do their diploid progenitors, often conferring dramatic adaptive superiority. The generalized biochemical flexibility conferred by increases in ploidy when they occur in a genetically variable species seems one of the best explanations for the widespread occurrence and evolutionary success of ploidy in plants.*

Much remains to be learned of the nature of the adaptive forces operating to maintain this immense reservoir of variability. Increasingly, efforts are being made to study the physiological consequences of the variability by studying enzyme activities in vivo. *This is one of the most active and exciting areas of current botanical research, demanding investigations at many levels of biological organization.*

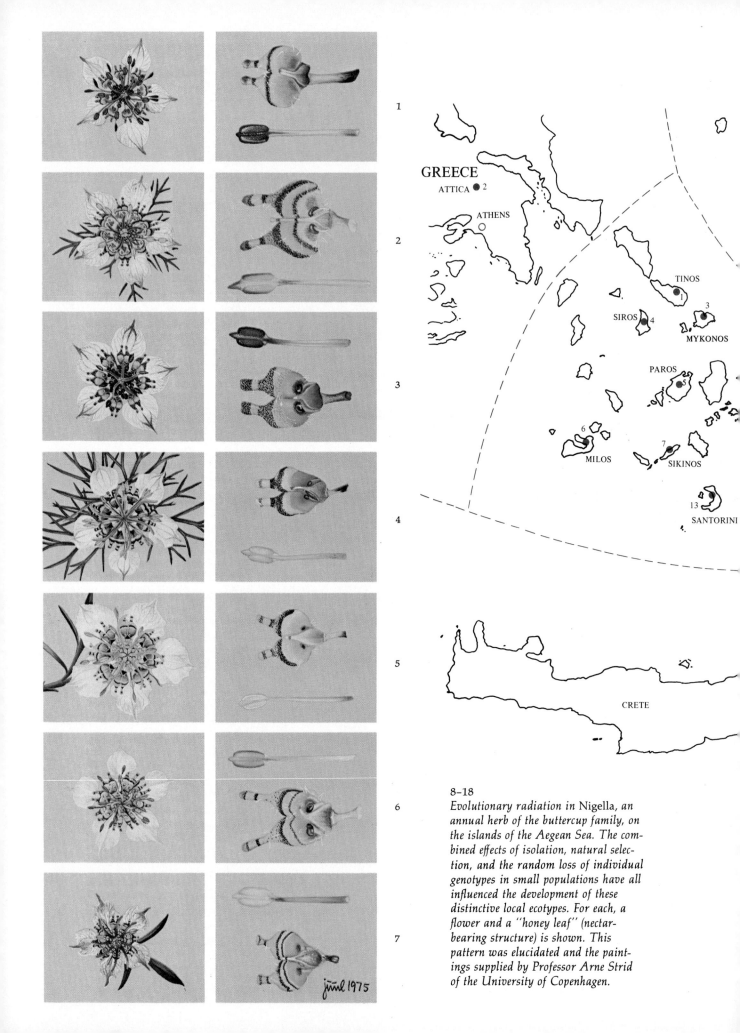

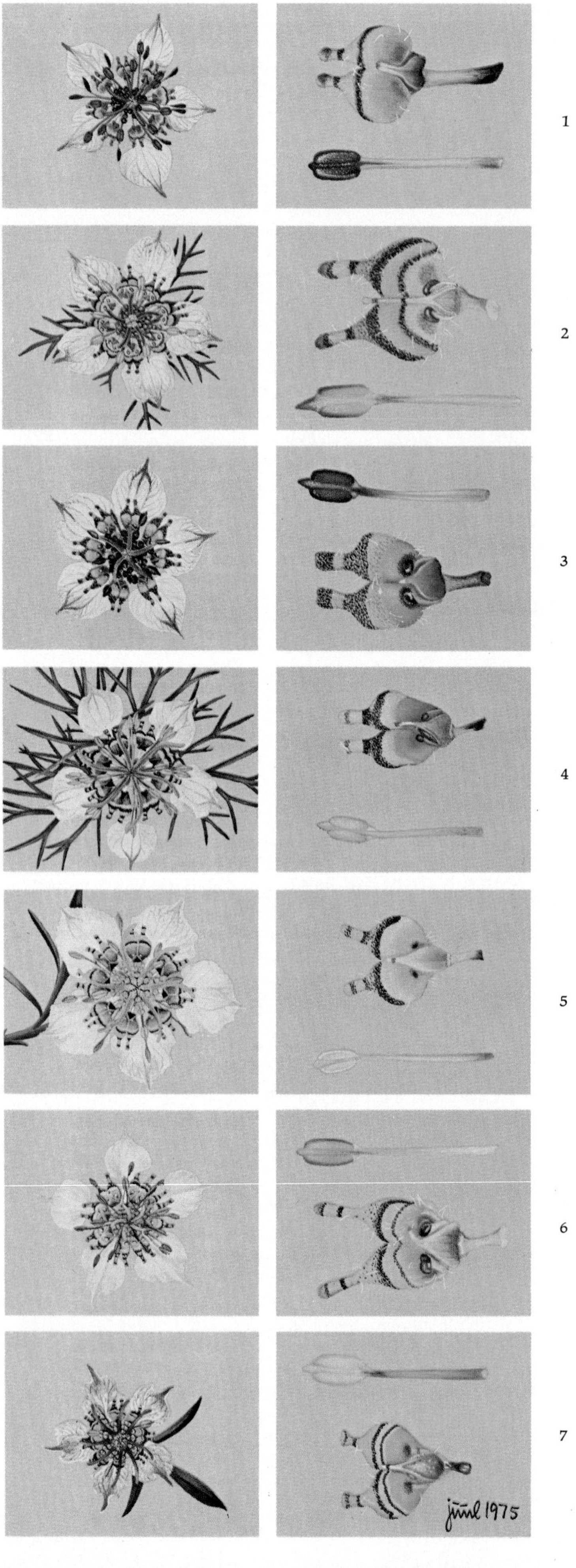

8–18
Evolutionary radiation in Nigella, *an annual herb of the buttercup family, on the islands of the Aegean Sea. The combined effects of isolation, natural selection, and the random loss of individual genotypes in small populations have all influenced the development of these distinctive local ecotypes. For each, a flower and a "honey leaf" (nectar-bearing structure) is shown. This pattern was elucidated and the paintings supplied by Professor Arne Strid of the University of Copenhagen.*

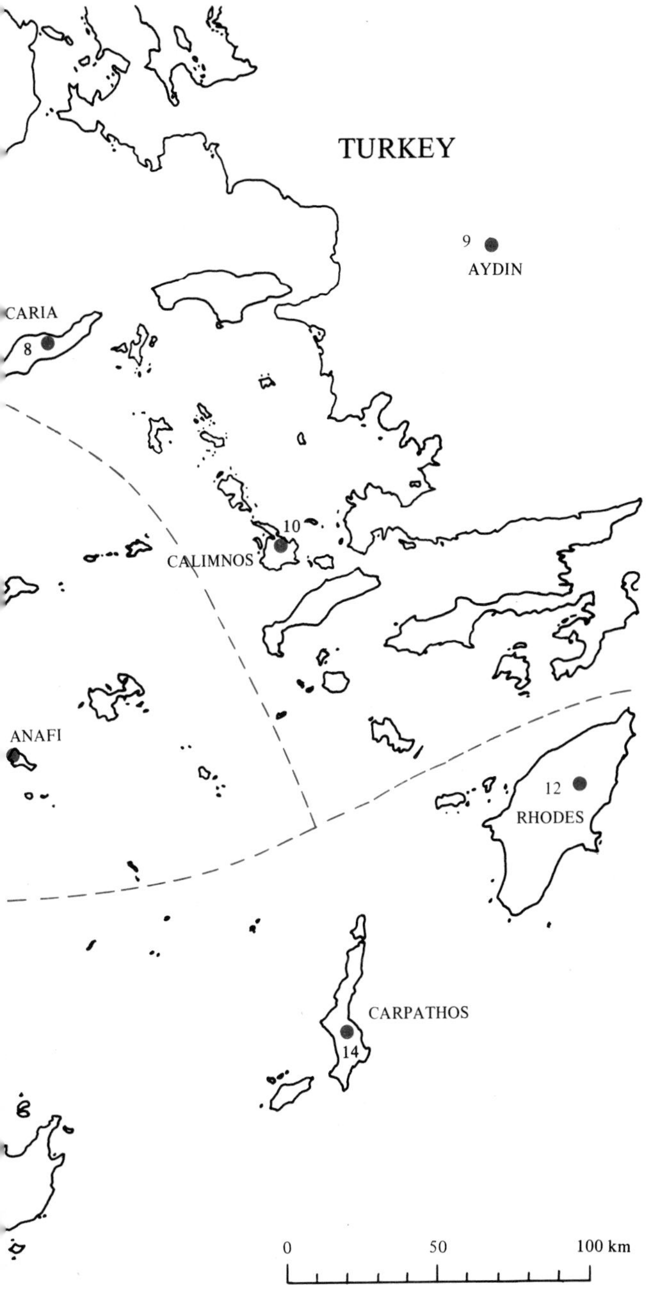

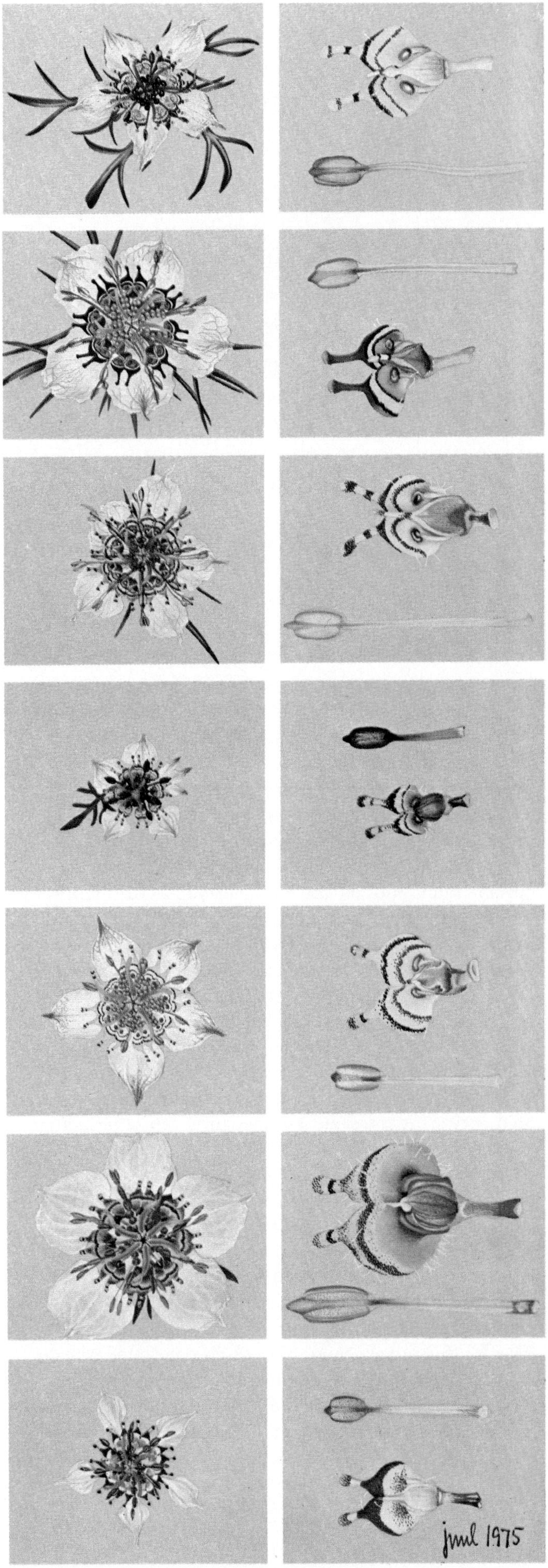

8

9

10

11

12

13

14

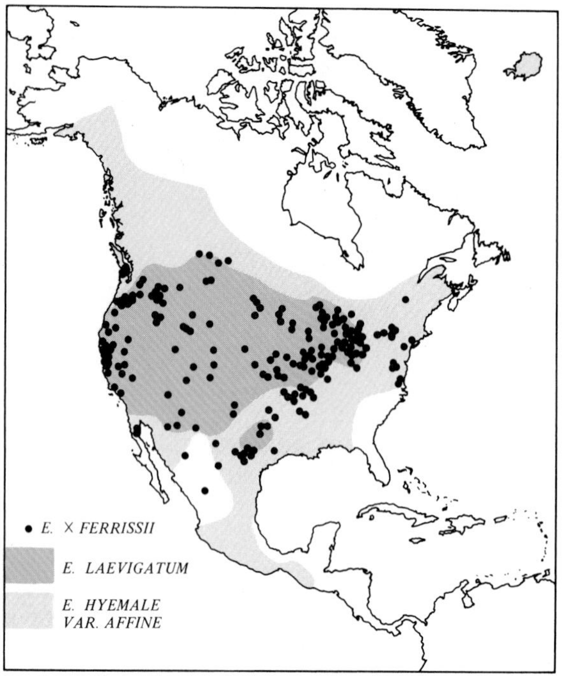

8–19
In the horsetails, genus Equisetum, *one of the most abundant and vigorous of the kinds found in North America,* E. × ferrissii, *is a completely sterile hybrid between* E. laevigatum *and* E. hyemale *(variety* affine*). Horsetails propagate readily from small fragments of underground stems, and the hybrid maintains itself over its wide range through vegetative propagation.*

The native *S. maritima* is low in stature, but *S. alterniflora* is much larger, frequently growing to 0.5 meter tall and occasionally to 1 meter. In the vicinity of the harbor at Southampton, both the native and the introduced species existed side by side throughout the nineteenth century, and by 1870, a sterile hybrid between them was collected that reproduced vigorously by rhizomes. Of the two parental species, *S. maritima* has a somatic chromosome number of 60 and *S. alterniflora* has 62; the hybrid, owing perhaps to some minor meiotic misdivision, also has 62. This sterile hybrid, which was named *S. × townsendii,* persists to the present. About 1890, a vigorous seed-producing polyploid was derived naturally from this hybrid, and it spread rapidly all around the coasts of Great Britain and northwestern France. This polyploid, *S. anglica,* with 122 chromosomes, is often planted to bind mudflats, and such use has contributed to its very rapid spread. At present, it is evidently replacing both parental species.

One of the most important polyploid crops is wheat. The most commonly cultivated species is bread wheat, *Triticum aestivum,* with 42 chromosomes; millions of hectares are planted with this crop. Bread wheat was derived at least 7000 to 8000 years ago following the spontaneous production of a sterile hybrid between a cultivated wheat with 28 chromosomes and a wild weedy grass of the same genus. The 28-chromosome wheat had originated following the hybridization and production of a sterile hybrid between two wild diploid species, each with 14 chromosomes, in the Near East. The grass, which has 14 chromosomes, grew in the fields with the 28-chromosome wheat, as it still does. Their hybridization probably took place in central Europe. The original hybrid had 21 chromosomes and, of course, a disturbed meiosis; however, its 42-chromosome derivative was highly fertile and had desirable characteristics that led to its selection by the early farmers of Europe. Wheats with 28 chromosomes are also still cultivated and are the chief grain used in the production of pasta—such as macaroni and spaghetti—because of the agglutinating properties of their proteins.

In both instances, polyploidy followed hybridization and the production of a sterile hybrid. As so often happens in cases of hybrid origin, the characteristics that allowed the survival of the polyploid were not obvious in the weeds that hybridized with the cultivated wheats; the additional genetic material allowed the formation of new, more favorable combinations of characteristics in the polyploids.

Recent studies indicate that new strains produced by artificial hybridization may further improve agricultural performance. Particularly promising is *Triticale,* a group of man-made hybrids between wheat *(Triticum)* and rye *(Secale).* Some of these hybrids combine the high yield of wheat with the ruggedness of rye and are more resistant to rust than wheat is, a characteristic that could be particularly important in the subtropical and tropical highland portions of the world, where rust is the chief factor limiting wheat production. The most important strain of *Triticale* has 42 chromosomes and was derived by chromosome doubling following hybridization of 28-chromosome wheat with rye (14 chromosomes). By

1974, it was being grown on more than 400,000 hectares (a million acres) in 53 countries, producing yields higher than wheat in many.

In nature, polyploids are selected by the environment, and not directly by man. Events similar to those we have reviewed in the history of wheat must have taken place well over 100,000 times just to account for the present representation of polyploids in the flora of the world. Among them are many of our most important crops—not only wheat, but also cotton, tobacco, sugarcane, bananas, potatoes, and safflower, just to mention a few. To this list can be added many of our most attractive garden flowers, such as chrysanthemums, pansies, and daylilies. Polyploidy is a major evolutionary mechanism.

SUMMARY

Natural selection is the process by which, under environmental pressure, organisms with more favorable characteristics leave more surviving progeny. Natural selection can take place because of genetically determined variations in natural populations.

The simple genetic situation is defined by the Hardy-Weinberg law, which states that in a large population in which there is random mating, and in the absence of forces that change the proportions of alleles, these proportions will remain constant from generation to generation.

Three principal factors that can cause changes in the proportions of genes and deviations from the Hardy-Weinberg equilibrium are differential migration, mutation, and selection. Of these, selection is the most consequential. It is defined as the differential reproduction of genotypes, some being favored over others. Stabilizing selection acts continuously to cull unfit individuals from the population. Disruptive selection acts in certain instances to split a population into two or more segments with different characteristics. Directional selection occurs when the mean value of the population shifts for some particular feature.

Recessive genes are inaccessible to selection in diploid organisms because the less frequent they become, the higher the proportion of genes that is locked up in heterozygotes, with no negative or positive value. For example, if the frequency of a given recessive gene in a human population were 0.01, it would take about 2500 years to reduce it to half this value, even under the most favorable conditions.

Response to selection is complex for a number of reasons. Only the phenotype is accessible to selection, and it can be based in many instances on a wide variety of genotypes. Because of epistasis and pleiotropy, single genes cannot be selected in isolation; selection affects the whole genotype. Because of polygenic inheritance and linkage of genes, the situation is even more complex. Because selection for any characteristic will affect the entire genotype, there are definite limits to selective progress. Genetic homeostasis is defined as a measure of the resistance to change in a population.

A number of instances of very rapid evolutionary change have been measured in natural populations, and the observed rates are more than adequate to account for the evolution of the world's organisms. For example, populations of pasture plants and of the grass *Agrostis tenuis* have also been observed to change rapidly in response to changes in environmental requirements.

Populations adjust to particular environments and form sharply defined units, or ecotypes, if the lines between the environments are sharply drawn. If environments gradually intergrade, populations of plants may form clines with respect to the characteristics under consideration. Even though environmental modification is common in plants, most races have a genetic basis. The physiological basis of differentiation is just beginning to be understood and is a fertile field for investigation.

While populations are changing, they also diverge in those factors allowing them to intercross successfully. After a period of isolation, two populations may be more or less incompatible or may produce sterile hybrids. Related species of relatively long-lived plants, such as trees and shrubs, are less apt to be reproductively isolated than related species of annuals and other short-lived plants. In the latter, genetic barriers often exist also within the species. The difference reflects the rate of response to a given selective force, as well as the fact that if populations of short-lived plants are not separated by some sort of barrier, they will tend to merge. Populations of trees and shrubs, on the other hand, may survive almost indefinitely, with no clear-cut barrier to hybridization, because the rate of replacement is much slower.

Introgressive hybridization is a term used to describe situations in which rare hybrids between two species backcross to their parents. If this happens repeatedly, the characteristics of one or both of the hybridizing species may be modified, even though the barrier between them remains sharply defined. Such modified individuals and populations may have an enhanced survival value under certain environmental conditions.

Hybrid populations derived from two species are common in plants and predominate in most genera of woody plants. This is particularly true in environments with relatively few species, such as oceanic islands, where adjustment to environmental changes may be especially critical, and in regions with sharply defined environmental breaks and a rapidly shifting climate, such as California.

Even if the hybrids between two species are sterile, they may be propagated by vegetative reproduction (apomixis) or become fertile following doubling in chromosome number (polyploidy).

SUGGESTIONS FOR FURTHER READING

ANDERSON, EDGAR: *Plants, Man and Life*, University of California Press, Berkeley, 1968.*

A fascinating account of genetic evolutionary studies of plants that reflects the personality of one of the greatest contemporary students of plants.

CARLQUIST, SHERWIN: *Island Biology*, Columbia University Press, New York, 1974.

This book presents a great deal of fascinating information about the often bizarre evolutionary pathways that have been followed by island plants and animals.

CLAUSEN, JENS: *Stages in the Evolution of Plant Species*, Cornell University Press, Ithaca, N.Y., 1951.

A very useful summary of trends in plant evolution by one of the principal workers in the Carnegie Institution group.

COLINVAUX, PAUL: *Introduction to Ecology*, John Wiley & Sons, Inc., New York, 1973.

A modern synthetic text in ecology written from an evolutionary point of view, and illustrating the interplay between these fields.

CROW, JAMES F.: *Genetic Notes*, Burgess Publishing Co., Minneapolis, 1966.*

An excellent and succinct summary of genetics for the student.

DARWIN, CHARLES: *On the Origin of Species by Means of Natural Selection, or the Preservation of Favoured Races in the Struggle for Life*, Doubleday & Co., Inc., Garden City, N.Y., 1960.*

Darwin's "long argument." Every student of biology should, at the very least, browse through this book to catch its special flavor and to begin to understand its extraordinary force.

———: *The Voyage of the Beagle*, Natural History Library, Doubleday & Co., Inc., Garden City, N.Y., 1962.*

Darwin's own chronicle of the expedition on which he made the discoveries and observations that eventually led him to his theory of evolution. This is the best introduction both to the man and to his work.

EHRLICH, PAUL R., RICHARD W. HOLM, and DENNIS R. PARNELL: *The Process of Evolution*, 2nd ed., McGraw-Hill Book Co., New York, 1975.

A concise and thoughtful treatment of modern evolutionary theory that provides a useful review of the entire field.

GOODENOUGH, URSULA, and R. LEVINE: *Genetics*, Holt, Rinehart and Winston, Inc., New York, 1974.

A thoroughly up-to-date account of genetics from a molecular point of view.

GRANT, VERNE: *Plant Speciation*, Columbia University Press, New York, 1971.

This comprehensive treatment of plant evolution provides a thorough introduction to most aspects of the field.

HOCHACHKA, P., and G. SOMERO: *Strategies of Biochemical Adaptation*, W. B. Saunders Co., Philadelphia, 1973.

An incisive and original treatment of metabolic organization in the light of its evolutionary and ecological implications.

LEVINE, LOUIS: *Biology of the Gene*, The C. V. Mosby Co., St. Louis, 1973.

The best short genetics text available.

MURRAY, JAMES: *Genetic Diversity and Natural Selection*, Hafner Publishing Co., New York, 1972.*

An outstanding brief treatment centering on the role of genetic diversity in natural populations.

PETERS, JAMES A. (ed.): *Classic Papers in Genetics*, Prentice-Hall, Inc., Englewood Cliffs, N.J., 1960.*

Includes papers by most of the scientists responsible for the important developments in genetics—Mendel, Sutton, Morgan, Beadle and Tatum, Watson and Crick, Benzer, etc. You should find this book very interesting; the authors are surprisingly readable, and the papers give a feeling of immediacy that no other account can achieve.

SRB, ADRIAN S., KAY D. OWEN, and R. S. EDGAR: *General Genetics*, 2nd ed., W. H. Freeman and Co. Publishers, San Francisco, 1965.

This older text is a particularly useful book for someone who wishes to study classical genetics because the authors provide not only a clear and relatively simple text but also a remarkably good and long list of questions and problems following each chapter.

STEBBINS, G. LEDYARD: *Processes of Organic Evolution*, Prentice-Hall, Inc., Englewood Cliffs, N.J., 1966.*

A brief review of the entire field by one of its outstanding practitioners.

———: *Variation and Evolution in Plants*, Columbia University Press, New York, 1950.

Although now badly out of date, this volume remains the standard work on plant evolution. It is somewhat technical, but it will amply repay careful study.

STRICKBERGER, MONROE W.: *Genetics*, 3rd ed., The Macmillan Co., New York, 1975.

An up-to-date and cohesive account of the science, this book is much broader in coverage than most texts at this level.

* Available in paperback.

SECTION 4 Diversity of Plants

CHAPTER 9

The Classification of Living Things

9–1
Coral hairstreak butterfly (Harkenclenus titus) *visiting an inflorescence of butterflyweed* (Asclepias tuberosa).

As many as 10 million different kinds of living things share our biosphere. Man differs from these other organisms both in the degree of his curiosity and in his power of speech. As a consequence of these two characteristics, he has long sought to inquire about the other creatures of his world and to exchange information about them. As man's knowledge grew, he discovered that in order to find out what is known about an organism or to report new information about it, it is necessary to know the name that other men have given it. Most familiar organisms have been given common names, but even for the simplest of purposes, common names may be inadequate. Sometimes they are misleading, particularly when we are exchanging information with people from different parts of the world. A dogwood or a cowslip in England may or may not be the same as a plant bearing a similar name in North America. A pine in England or the United States is not the same as a pine in Australia. A yam in the southeastern United States is a totally different vegetable from one a few hundred kilometers away in the British West Indies. When different languages are involved, the problems of communication become hopelessly complex. For this reason, biologists refer to organisms by Latin names, officially recognized by international organizations of botanists and zoologists.

The practice of referring to organisms by Latin names began in medieval times, when Latin was the language of scholarship. As the system developed, organisms were first grouped into *genera* (singular: genus) and then identified by descriptive Latin phrase names, known as polynomials. By the end of the seventeenth century, the first word of the polynomial designated the name of the genus or group to which the plant belonged. Thus all willows were identified by polynomials beginning with the word *Salix,* the phrases describing kinds of roses all began with *Rosa,* and those referring to oak trees with *Quercus;* all of these are names of genera.

9–2
Carolus Linnaeus (1707–1778), the inventor of the binomial system of classification. Linnaeus believed that each living thing corresponded more or less closely to some ideal model and that by classifying them, he was revealing the grand pattern of creation.

THE BINOMIAL SYSTEM

A simplification in the system of naming living things was made by the eighteenth-century Swedish professor, physician, and naturalist, Carolus Linnaeus (Figure 9–2), whose ambition was to classify all the known kinds of plants and animals according to their genera. In 1753, he published a two-volume work, *Species Plantarum* ("the kinds of plants"), which contained brief analytical descriptions of every known species of plant, with references to the earlier works about each one. He used polynomial designations for all species, and in many cases changed them so that they would be directly comparable with those referring to other species in the same genus. Although he regarded the polynomials as the proper names for species, he also made an important innovation in the system. In the margin of his book, opposite the "proper" name of each species, Linnaeus entered a single word, which, together with the generic name, formed a convenient "shorthand" designation for the species. In the book, catnip, which had previously been designated as *Nepeta floribus interrupte spicatus pedunculatis* (Nepeta with flowers in an interrupted pedunculate spike), was described under *Nepeta* and "cataria" was put in the margin, making it *Nepeta cataria*, which is its name today.

The convenience of this system was obvious, and Linnaeus and subsequent authors soon replaced all "proper" names with "shorthand" ones. This binomial ("two-name") system is still used today. The earliest binomial name applied to a particular species is accepted as the correct name for that species, and later names, which may accidentally be applied to the same species, are rejected. The rules governing the application of botanical names to plants are embodied in the *International Code of Botanical Nomenclature*, which is revised at successive International Botanical Congresses held every 5 years. The most recent of these met in Leningrad in July 1975.

A species name consists of two parts—the generic name and the specific name. However, a generic name may be written alone when one is referring to the entire group of species comprising that genus. Thus the evening primrose genus is *Oenothera*. The genus *Oenothera* includes some 120 species of North and South America, a few of which are widely cultivated. One of these is the spectacular *Oenothera muelleri*, a plant with large white flowers that are 10 to 12 centimeters in diameter, which is known only from a few mountaintops in the northeastern Mexican state of Nuevo León. Another species is the weedy *Oenothera biennis*, which is widespread throughout eastern North America and has been introduced in many other temperate regions. (Figure 9–3 shows three species of another genus, *Viola*.)

A specific name is meaningless when written alone; for example, *biennis* could refer to any of scores of species in different genera that happen to have this word as part of their name. For this reason, the specific name is always preceded by the name or the initial letter of the genus that includes the species in question, that is, *Oenothera biennis* or *O. biennis*.

Species can be further divided into subspecies or varieties. Subspecies of one species bear an overall resemblance to one another but exhibit one or more important variations. As a result of these subdivisions, although the binomial name is still the basis of classification, the names of some plants and animals consist of three parts. Thus the northeastern race of O.

9–3
Three members of the violet genus, Viola. (a) *Common violet,* Viola papilionacea. (b) *Long-spurred violet,* V. rostrata. (c) *Pansy,* V. tricolor *var.* hortensis. *These photographs indicate the kinds of differences in leaf shape and margin, flower color and size, and other features that distinguish the species of a single genus, even though there is an overall similarity between all three species illustrated here. The pansy is an annual plant, selected from western European progenitors for cultivation in gardens; the other species are perennials. There are about 500 species of the genus* Viola *in all, most of them in temperate regions of the Northern Hemisphere.*

(a)

biennis is *O. biennis* subsp. *biennis,* whereas that occurring from the Great Lakes region southward in the central United States is *O. biennis* subsp. *centralis.* The peach tree is *Prunus persica* var. *persica,* the nectarine is *Prunus persica* var. *nectarina.* Note that the names of varieties are written in italics also and that the first-named subspecies or variety (chronologically speaking) repeats the name of the species.

WHAT IS A SPECIES?

Biologists have held several philosophically different views about the nature of species and the way in which this important taxonomic entity should be defined. The word itself has no special connotation; it simply means "kind" in Latin. There are many groups of particular kinds of organisms in nature, and these are called species. We now know that these organisms are constantly evolving and that the relationships within and among them are also changing. Sets of populations that resemble one another relatively closely and other sets of populations less closely are called species, but the application of this term differs widely from group to group (Figure 9-4).

In higher animals, "species" is often assumed to refer to a group of organisms that can interbreed with one another but not with individuals of another species. This criterion has not proved to be useful in most groups of plants. When we attempt to cross plants taken from different species, the species may not cross, or the hybrids

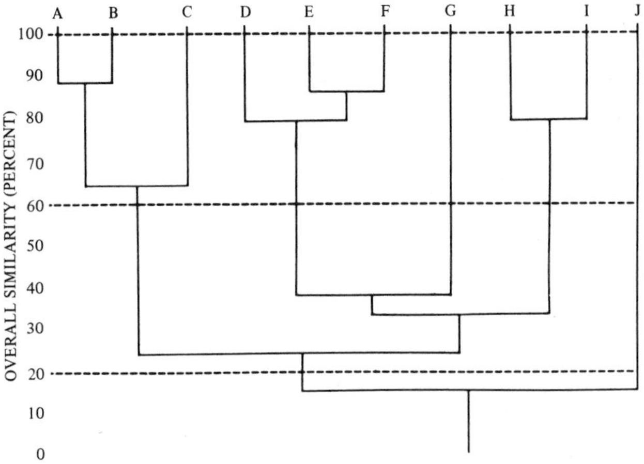

9-4
Overall similarity among ten imaginary species A–J, based on a number of characteristics. The degree of similarity between any two species is indicated by the level at which the lines extending down from those species intersect. No two pairs of species are related at exactly the same level of similarity. Possible ways to recognize taxonomic categories are indicated by the dotted lines; for example, the five lines intersected at the 60 percent level of similarity might be recognized as genera, the two intersected at the 20 percent level as families. It would be equally logical to draw these lines at different levels.

(b)

(c)

9-5
(a) *Diagram of breeding experiments involving populations of the annual plant* Clarkia rhomboidea. *The names of the localities identify the populations that were hybridized. In this experiment, the fertility of the hybrids was measured by the percentage of functional pollen they produced. As you can see, although all of these populations belong to the same species, they form hybrids of widely varying fertility. The populations are separated by geographical gaps of greater or lesser magnitude and would rarely if ever have the opportunity of hybridizing with one another in nature. Work of this nature illustrates clearly why the species is not a "genetic unit."*
(b) *Flower of* Clarkia rhomboidea.

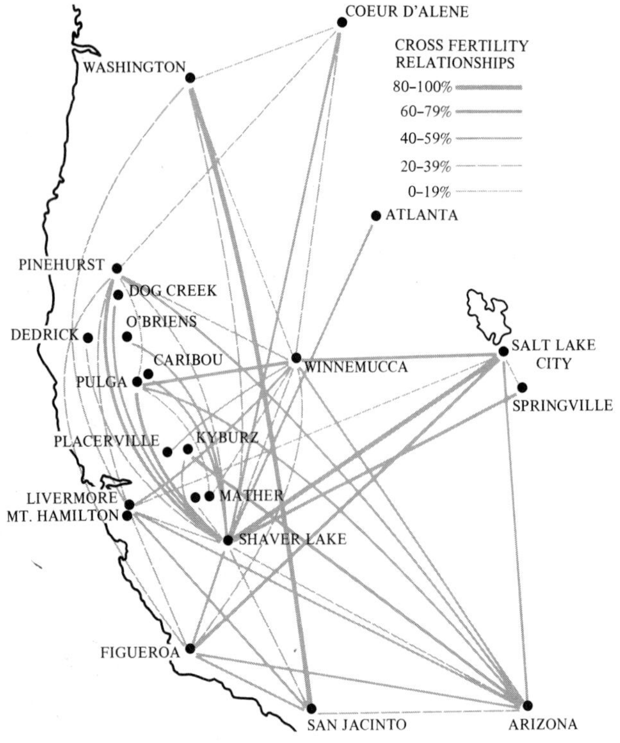

(a)

(b)

may be more or less infertile. This is not always true, however. Among the white oaks, for example, it appears that almost any species can be crossed with almost any other to produce fertile hybrids. Among the herbs, on the other hand, different populations within a single species often produce sterile hybrids when crossed (see Figure 9-5). In a particular group, it is reasonable to say that the greater the differences between the individuals being crossed, the less fertile their hybrids are likely to be, but in going from group to group, it is difficult to correlate the ability to form hybrids with any particular taxonomic status.

Thus the word "species" means different things for different kinds of organisms, which is not surprising considering that evolution in various groups of organisms has followed different paths. For example, genetic recombination is rare among bacteria and is unknown in other large groups (e.g., algae related to *Euglena*), so we should not expect the units we call species in these groups to resemble the units we call species among, for example, the oaks. Species, however, remains a useful term and provides a convenient way to talk about and catalogue organisms.

OTHER TAXONOMIC GROUPS

Linnaeus (and earlier scientists) recognized the plant, animal, and mineral kingdoms, and the *kingdom* is still the major unit used in biological classification. Between the level of genus and the level of kingdom, however, subsequent authors in the nineteenth and twentieth centuries have added a number of categories. Thus, genera are grouped into *families*, families into *orders*, orders into *classes*, and classes into *divisions*.* These categories (also known as *taxa*) may be subdivided or aggregated into a number of less important ones, such as subgenera and superfamilies. By convention, generic and specific names are written in italics, whereas the names of families, orders, classes, and other taxa are not. Sample classifications of corn *(Zea mays)* and the edible mushroom *(Agaricus campestris)* are given in Table 9-1 on page 175.

* "Phylum," which is the equivalent term used in zoological classification, is not recognized as a taxonomic category by the *International Code of Botanical Nomenclature.*

9–6
Monera *is made up of the bacteria and the blue-green algae. These are the only modern prokaryotes.* (a) Lactobacillus acidophilus, *bacteria that sour milk.* (b) *A gelatinous colony of the blue-green alga* Nostoc. (c) Oscillatoria, *a filamentous blue-green alga.*

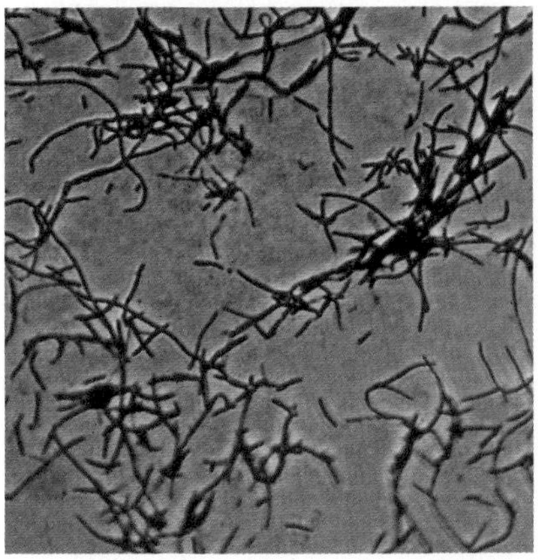

(a) 25 μm

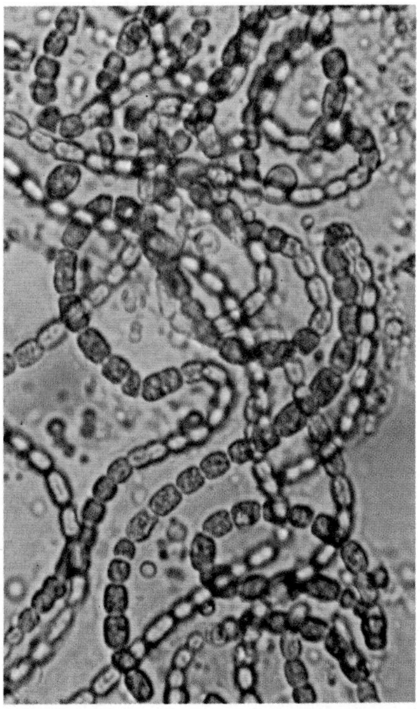

(b) 10 μm

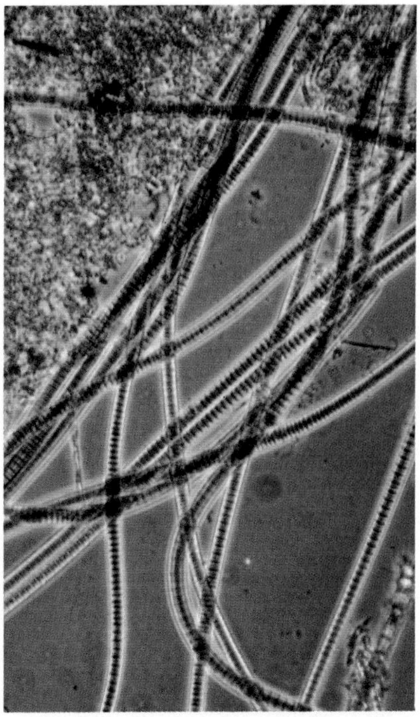

(c) 100 μm

9–7
Protista. (a) Volvox, *a motile colonial green alga (division Chlorophyta).* (b) Ulva, *a green alga (division Chlorophyta), and* Fucus, *a brown alga (division Phaeophyta), on rocks exposed at low tide in southeastern Alaska.* (c) *Rockweed,* Fucus, *a brown alga.*

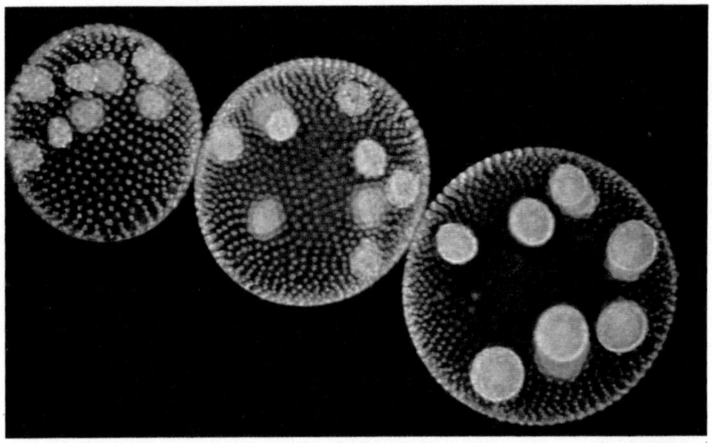

(a) 200 μm

(b)

(c)

9-7 (cont'd.)
Protista. (d) *A multicellular green alga,* Enteromorpha compressa *(division Chlorophyta). Most algae of this genus form tubes of cells two cell layers thick. They tolerate high salinities and often grow in polluted coastal waters.* (e) Ceramium acanthonotum, *a red alga.* (f) *Fruiting bodies of a plasmodial slime mold,* Stemonitis *(division Gymnomycota).* (g) *A chrysophyte diatom in side view, showing the characteristic intricately marked shell.* (h) *Plasmodium of plasmodial slime mold,* Physarum *(division Gymnomycota) under damp polar bark.*

(e)

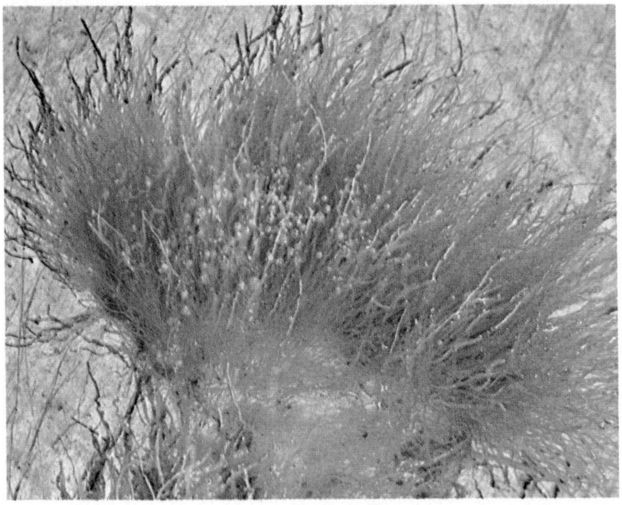

(d)

(f)

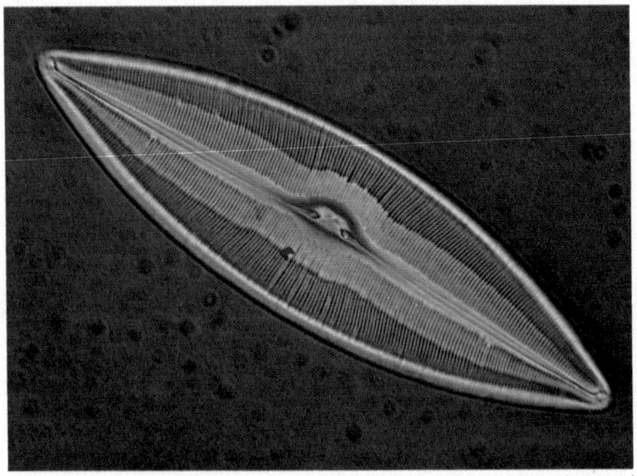

(g)

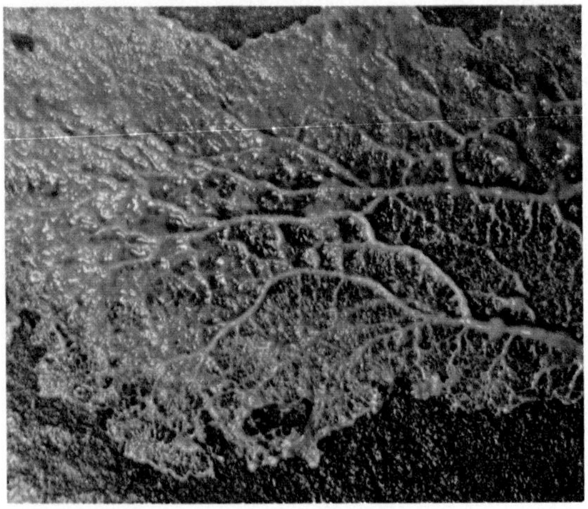

(h)

9-8
Fungi. (a) Tremella, *a jelly fungus.* (b) Cantharellus aurantiacus, *a mushroom.* (c) Cortinarius iodes, *another mushroom.*

(a)

(b)

(c)

9-9
Plants. (a) *A moss,* Climacium dendroides *(division Bryophyta).* (b) *Tree clubmoss,* Lycopodium obscurum *(division Tracheophyta, subdivision Lycophytina).* (c) *A leafy liverwort,* Marchantia polymorpha *(division Bryophyta).*

(a)

(b)

(c)

9–9 (cont'd.)

Plants. (d) *Marsh fern,* Thelypteris palustris *(division Tracheophyta, subdivision Filicophytina).* (e) *Cedar of Lebanon,* Cedrus libani *(division Tracheophyta, subdivision Spermatophytina).* (f) *Wood horsetail,* Equisetum sylvaticum *(division Tracheophyta, subdivision Sphenophytina).* (g) *Goldenrod,* Solidago altissima *(division Tracheophyta, subdivision Spermatophytina).* (h) *Foxtail barley,* Hordeum jubatum *(division Tracheophyta, subdivision Spermatophytina).* (i) *Barrel cactus,* Echinocerous *(division Tracheophyta, subdivision Spermatophytina).* (j) *Purple fringed orchid,* Habenaria fimbriata *(division Tracheophyta, subdivision Spermatophytina).*

(f)

(d)

(e)

(g)

(h)

(i)

(j)

THE MAJOR GROUPS OF ORGANISMS

Before starting our consideration of particular groups of plants, we shall discuss the present system of classification of organisms. Only by doing this can we obtain a sense of plants as part of the entire biological world.

As we have mentioned, in Linnaeus' time, it was thought that there were three kingdoms of "objects": animals, plants, and minerals. Thus living things were thought of as either plant or animal. Animals moved, ate things, and breathed, and the size of their bodies was definitely limited. Plants did not move, eat, or breathe; they seemed to grow indefinitely and branch if injured. Of the two, plants seemed to represent a lower order of life.

As new groups of organisms were discovered, they were classified as either plant or animal. Thus the fungi and bacteria were grouped with the plants, and the protozoa were grouped with the animals. However, taxonomists (scientists specializing in the problems of classification) began to have trouble with forms such as *Chlamydomonas*, a swimming green alga that moves *and* manufactures its own food. Organisms such as *Chlamydomonas* were not clearly either plant or animal, and by the 1930s it was evident that the traditional division of organisms into two kingdoms was little more than a his-

Table 9–1 *Biological Classification. Notice how much you know about an organism when you know its place in the system. The descriptions here do not define the various categories but tell you something about their characteristics.*

CORN

Category	*Name*	*Description*
Kingdom	Plantae	Organisms that have chlorophyll *a* and *b* contained in chloroplasts, and more specialized than the green algae.
Division	Tracheophyta	Vascular plants; plants with a specialized system of xylem and phloem.
Subdivision	Spermatophytina	Seed plants
Class	Angiospermae	Flowering plants; ovules enclosed in an ovary, pollination indirect
Subclass	Monocotyledoneae	Embryo with one cotyledon; flower parts usually in 3s; many scattered vascular bundles in the stem
Order	Commelinales	Monocots with fibrous leaves; characterized by reduction and fusion in flower parts
Family	Poaceae	Hollow-stemmed monocots with reduced greenish flowers; fruit a specialized achene (caryopsis); the grasses
Genus	*Zea*	Robust grasses with separate staminate and carpellate flower clusters; caryopsis fleshy
Species	*Zea mays*	Corn

EDIBLE MUSHROOM

Category	*Name*	*Description*
Kingdom	Fungi	Primarily nonmotile, multinucleate, heterotrophic, absorptive organisms in which cellulose does not predominate in the cell walls
Division	Mycota	Includes all members of kingdom Fungi
Class	Basidiomycetes	Dikaryotic fungi that form a basidium which bears four spores
Order	Agaricales	Fleshy fungi with radiating gills or pores
Family	Agaricaceae	Agaricales with gills
Genus	*Agaricus*	Dark-spored soft fungi with a central stalk and gills free from the stalk
Species	*Agaricus campestris*	The common edible mushroom

9–10
Electron micrographs of a prokaryote, the blue-green alga Anabaena cylindrica *(a), and a eukaryote, the single-celled green alga* Cyanidium caldarium *(b). Note the much greater complexity of the eukaryotic cell. The large, lobed chloroplast of* Cyanidium, *which here appears in two parts, is biochemically similar to and comparable in size with an entire blue-green alga and unlike the chloroplasts of most other similar eukaryotes. It is likely that it is in fact a symbiotic blue-green alga that has assumed the function of a chloroplast. The three-dimensional quality of these electron micrographs are due to a technique known as freeze-etching. When this technique is used, the specimen is frozen rapidly and split apart. A replica is made of the surface of the specimen, which is what is shown here.*

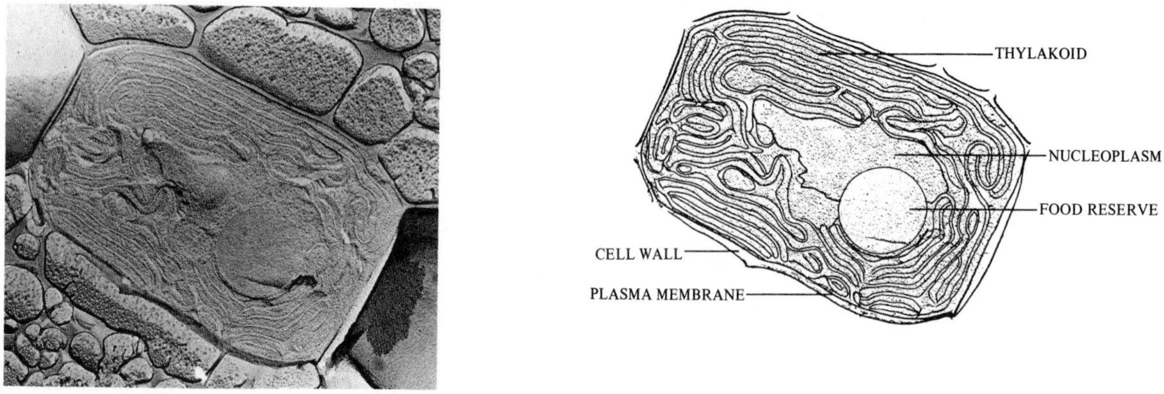

(a)

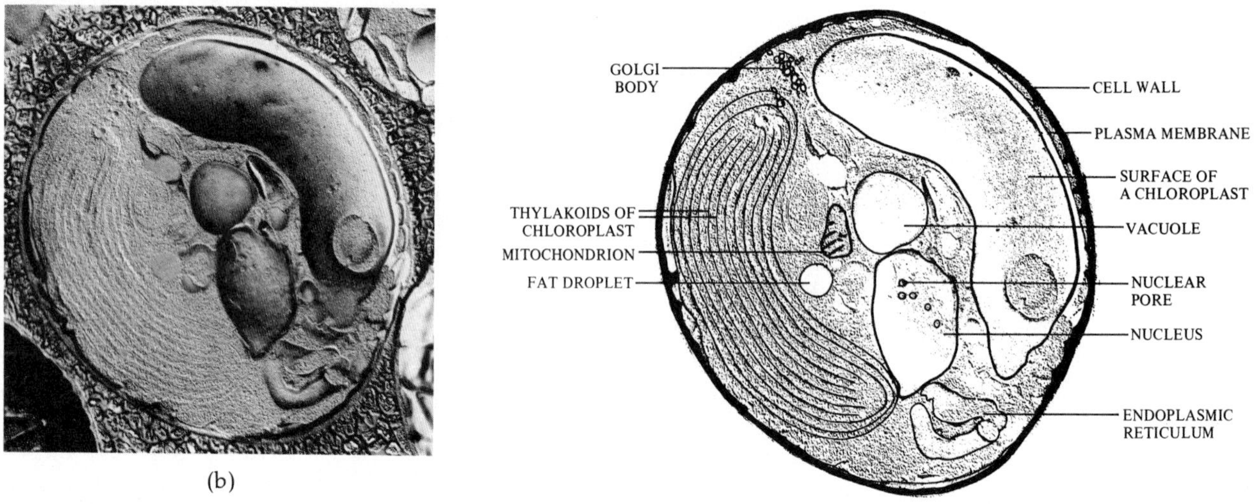

(b)

torical curiosity.

Unfortunately, no completely acceptable alternative has been proposed. The relationships among organisms are complex at all levels, and it is perhaps unreasonable to expect to be able to group all organisms into a series of clearly and precisely defined kingdoms. As a result, the old division into plants and animals is still widely reflected in the organization of college and university science departments, research projects, and textbooks (including this one).

The Prokaryotes

As new principles of overall classification have emerged, it has become evident that the most fundamental division in the living world is the distinction between the prokaryotes and the eukaryotes (Figure 9–10). In the bacteria and blue-green algae, the two groups of prokaryotic organisms, there are no membrane-bound cellular organelles, no microtubules, and no complex 9-plus-2 structure of flagella. The genetic material is borne on a single circular molecule of DNA and is not associated with proteins. Although several mechanisms leading to genetic recombination are known in prokaryotes, genetic recombination is infrequent and accomplished by means other than sexual reproduction. The cell walls of bacteria and blue-green algae contain muramic acid, and in this and a number of other biochemical peculiarities, the prokaryotes are different from all other organisms. These reasons provide ample grounds for recognizing the bacteria and blue-green algae as constituting a separate "kingdom," the kingdom Monera.

The Eukaryotes

All eukaryotes have a definite nucleus that is bounded by a double membrane. Within the nuclear envelope are complex chromosomes in which the DNA is associated with proteins in most groups. These chromosomes segregate regularly by mitosis, which involves the formation of spindle fibers that are organized in all organisms except the higher plants, where they have mostly been lost, by organelles known as centrioles. Organelles exactly similar to centrioles, called basal bodies, organize complex flagella and cilia (with their 9-plus-2 microtubules) in eukaryotes that possess flagella or cilia. Microtubules also occur within the cytoplasm of eukaryotes. Complex organelles such as mitochondria occur in the cells of all eukaryotes; and vacuoles, which are bounded by a single membrane, also occur widely in eukaryotes, especially plants. Both mitochondria and vacuoles are lacking in prokaryotes.

Many eukaryotes also exhibit two important features that are not found in prokaryotes: multicellularity and sexual reproduction. The cells of prokaryotes sometimes remain together in filamentous or even three-dimensional masses following cell division, but there are no protoplasmic connections between the individual cells and hence no overall integration of the whole filament or mass. In plant eukaryotes, the protoplasts of contiguous cells are connected by plasmodesmata, which traverse their walls. In animals, there are no cell walls, and the protoplasts are in more direct contact. Although the different lines of eukaryotes were derived independently from unicellular ancestors, multicellularity is characteristic of the group as a whole and is not found among the prokaryotes.

Sexual reproduction involves a regular alternation between meiosis and syngamy (union of gametes) and occurs only in eukaryotic organisms.

Formal Classification of Living Organisms

Recently it has been suggested that living organisms be divided among five kingdoms: the prokaryotic Monera and four eukaryotic groups (Figure 9–11). Of the eukaryotic groups, the Protista are believed to have given rise to the three remaining eukaryotic groups—the plants, the animals, and the fungi. These mostly multicellular

9–11
One scheme of evolutionary relationships among organisms. The solid lines indicate phylogenetic relationships and the dotted lines indicate the establishment of symbiotic relationships. All eukaryotic organisms were derived from a single line of cells containing mitochondria which had their derivation in symbiotic bacteria. From this single line of cells there developed the diverse assemblage of one-celled organisms known as the Protista. Those members of the Protista that developed symbiotic relationships with photosynthetic prokaryotes gave rise to the several different lines of modern algae. From particular single-celled protista, the fungi, plants, and animals evolved.

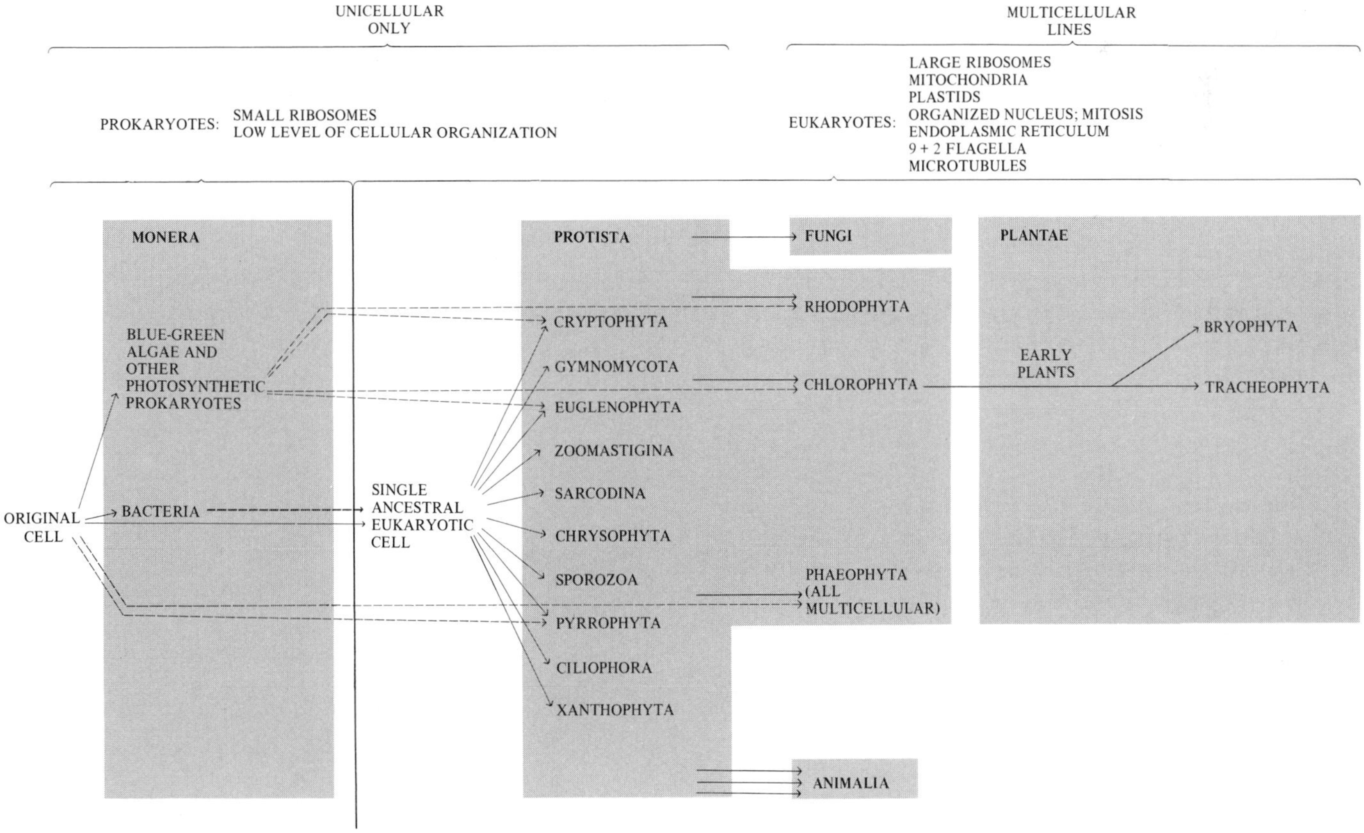

groups differ fundamentally in their mode of nutrition. In general, plants manufacture their food, animals ingest it, and fungi absorb it. For the purposes of this classification, the viruses (see Chapter 10) are considered separately. They have a direct relationship with the bacteria, as we shall see, but differ in organization from all cellular forms of life.

The following is a synopsis of the system of classification used in this book (see Table 9-2).*

Kingdom Monera

The Monera comprise prokaryotic organisms, which lack nuclear envelopes, plastids, mitochondria, and advanced 9-plus-2 flagella. They exhibit solitary unicellular or colonial unicellular organization and lack protoplasmic connections between the cells. The nutritive mode of absorption predominates, but some groups are photosynthetic or chemosynthetic. Reproduction is predominantly by cell division, although genetic recombination occurs in several groups. They are either motile by simple flagella or gliding, or are nonmotile. (The Monera, which consist of the bacteria, including blue-green algae, are discussed in Chapter 10.)

Kingdom Protista

The Protista are here regarded as comprising all organisms traditionally regarded as protozoa (one-celled animals) as well as all eukaryotic algae (Figure 9-7 on page 171). In addition, the slime molds (see page 276), often regarded as relatives of the fungi, are included in the Protista. In sum, Protista constitute a heterogeneous assemblage of unicellular, colonial, and multicellular eukaryotes. Nutritive modes are diverse—photosynthesis, absorption, ingestion, or a combination of these. The reproductive cycles of the Protista are varied, but typically involve both cell division and sexual reproduction. They are motile by 9-plus-2 flagella, are amoeboid, or are nonmotile. The algae, which traditionally are treated as plants, and the slime molds, division Gymnomycota, are discussed in Chapter 12. Other Protista—the heterotrophic divisions known as Protozoa—are not considered in this text.

Kingdom Animalia

The animals are multicellular organisms with wall-less eukaryotic cells lacking plastids and photosynthetic pigments. Nutrition is primarily ingestive with digestion in an internal cavity, but some forms are absorptive, and a number of groups lack an internal digestive cavity. The level of organization and tissue differentiation in higher forms far exceeds that of other kingdoms, with evolution

* Modification of the five kingdom schemes of Whittaker (R. H. Whittaker: *Science*, vol. 163, pp. 150–160, 1969) and Margulis (L. Margulis: *Evolution*, vol. 25, pp. 242–245, 1971).

Table 9-2 *Classification of Major Groups of Living Organisms Traditionally Regarded as Plants and Considered in this Text [See Appendix C for summary descriptions of these groups.]*

	KINGDOM MONERA
PROKARYOTES	Bacteria, including blue-green algae
	KINGDOM PROTISTA
ALGAE	Division Chlorophyta (green algae)
	Division Phaeophyta (brown algae)
	Division Rhodophyta (red algae)
	Division Chrysophyta (diatoms and golden-brown algae)
	Division Xanthophyta (yellow-green algae)
	Division Pyrrophyta (dinoflagellates)
	Division Euglenophyta (euglenoids)
SLIME MOLDS	Division Gymnomycota (slime molds)
	Class: Myxomycetes (true slime molds)
	Class: Acrasiomycetes (cellular slime molds)
	KINGDOM FUNGI
FUNGI	Division Mycota (true fungi)
	Class: Chytridiomycetes (chytrids)
	Class: Oomycetes (water molds)
	Class: Zygomycetes (bread molds)
	Class: Ascomycetes (sac fungi)
	Class: Basidiomycetes (club fungi)
	KINGDOM PLANTAE
BRYOPHYTES	Division Bryophyta (nonvascular plants)
	Class: Musci (mosses)
	Class: Anthocerotae (hornworts)
	Class: Hepaticae (liverworts)
	Division Tracheophyta (vascular plants)
SEEDLESS VASCULAR PLANTS	Subdivision Psilophytina (whisk fern)
	Subdivision Lycophytina (lycophytes, including club mosses)
	Subdivision Sphenophytina (horsetails)
	Subdivision Filicophytina (ferns)
SEED PLANTS	Subdivision Spermatophytina (seed plants)
GYMNOSPERMS	Class: Cycadinae (cycads)
	Class: Ginkgoinae (ginkgo)
	Class: Coniferinae (conifers)
	Class: Gnetinae (vessel-containing gymnosperms)
ANGIOSPERMS	Class: Angiospermae (flowering plants)
	Subclass: Dicotyledoneae (dicots)
	Subclass: Monocotyledoneae (monocots)

in particular of complex sensory and neuromotor systems. The motility of the organism (or, in sessile forms, of its parts) is based on contractile fibrils. Reproduction is predominantly sexual, with haploid stages other than gametes almost nonexistent above the lowest divisions. Animals are not discussed in this book.

Kingdom Fungi

The fungi have traditionally been grouped with plants, but the evidence that they are an independent line is now overwhelming. Aside from the low level of differentiation of their bodies, they have little in common with the algae. Most fungi are nonmotile organisms, and they all are eukaryotic. They do not have plastids or photosynthetic pigments and absorb their nutrients from either dead or living organisms. Complex life cycles are well developed in higher forms. Some fungi produce motile, flagellated spores. Reproductive cycles typically involve both sexual and asexual processes. (See Chapter 11 for a discussion of the fungi.)

Kingdom Plantae

The plants—the bryophytes and vascular plants—include all green organisms more specialized than and derived from the green algae. All are multicellular and composed of vacuolate eukaryotic cells with cellulosic cell walls. Their principal mode of nutrition is photosynthesis, but a few have become heterotrophic. In general, their evolution has occurred in relation to their successful invasion of the land. Structural differentiation has occurred, with trends toward organs of photosynthesis, anchorage, and support. In higher forms such organization has produced specialized photosynthetic, vascular, and covering tissues. Reproduction is primarily sexual, with cycles of alternating haploid and diploid generations, the former being progressively reduced in the more advanced members of the kingdom. The bryophytes are discussed in Chapter 13, and the vascular plants in Chapters 14–17.

SUMMARY

Biologists have developed methods of naming and classifying living things that permit them to designate very precisely the organism with which they are working, an essential factor in scientific communication. The classification of an organism also reveals its relationship to other living things.

Plants are identified scientifically by two names, a binomial. The first word in the binomial designates the name of the genus (plural, genera), and the second word designates the species. The entire binomial is the species name. Species are sometimes subdivided into varieties. Genera are grouped into families, families into orders, orders into classes, and classes into divisions (sometimes called phyla). Divisions are grouped into kingdoms, the kingdom being the largest unit used in classification of the living world. In this text, living organisms are grouped into five kingdoms: (1) kingdom Monera, which includes all the prokaryotes (blue-green algae, bacteria, and related organisms); (2) kingdom Protista, which includes the Protozoa, eukaryotic algae, and slime molds (Protozoa are not discussed in this book); (3) kingdom Animalia, which includes multicellular organisms that are not photosynthetic; (4) kingdom Fungi, the fungi; and (5) kingdom Plantae, which includes the bryophytes and vascular plants, those photosynthesizing organisms that are more advanced than the algae.

SUGGESTIONS FOR FURTHER READING

BELL, PETER R., and C. L. F. WOODCOCK: *The Diversity of Green Plants*, 2nd ed., Addison-Wesley Publishing Co., Inc., Reading, Mass., 1972.*

A concise modern survey of the diversity of green plants.

BOLD, H. C.: *Morphology of Plants*, 3rd ed., Harper & Row, Publishers, Inc., New York, 1973.

A well-illustrated and ample treatment of the diversity of plants.

CORNER, E. J. H.: *The Life of Plants*, Mentor Books, New American Library, Inc., New York, 1968.*

A renowned botanist with a flair for poetic prose describes the evolution of plant life, telling how plants modified their structures and functions to meet the challenge of a new environment as they invaded the shore and spread across the land.

DELEVORYAS, T.: *Plant Diversification*, Holt, Rinehart & Winston, Inc., New York, 1966.*

This brief volume stresses evolutionary trends in presenting a readable account of the groups traditionally considered to be plants.

MARGULIS, LYNN: *Origin of Eukaryotic Cells*, Yale University Press, New Haven and London, 1970.

A fascinating discourse on the origin of eukaryotic cells by serial symbiotic events, beautifully illustrated and well reasoned.

ROSS, HERBERT H.: *Biological Systematics*, Addison-Wesley Publishing Co., Inc., Reading, Mass., 1974.

A nicely balanced brief overview of taxonomy, both plant and animal.

SCAGEL, R. F., et al.: *An Evolutionary Survey of the Plant Kingdom*, 2nd ed., Wadsworth Publishing Co., Inc., Belmont, Calif., 1969.

An exhaustive, thoroughly illustrated review.

* Available in paperback.

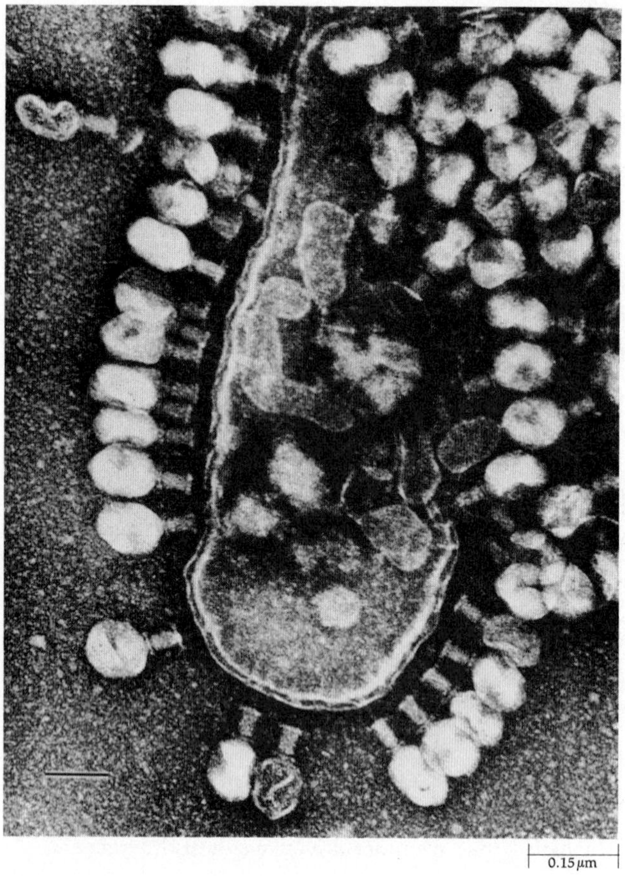

10–1
Bacterial viruses (T2) infecting a cell of Escherichia coli. *Some of the viruses have discharged their DNA into the bacterial cell, which has begun to break down (lyse).*

CHAPTER 10

Bacteria and Blue-Green Algae

The bacteria and blue-green algae are the simplest of all living organisms and probably bear the closest resemblance of any modern organism to the earliest forms of life on earth. They resemble one another in basic architecture, being the two living groups of prokaryotic organisms. Indeed, blue-green algae are best regarded as one of many groups of specialized bacteria and are so treated in this book. Together the bacteria and blue-green algae constitute the kingdom Monera, which, with the viruses, will be the subject of this chapter.

Both bacteria and blue-green algae, being prokaryotes, lack organized nuclei with a nuclear envelope. They do not have complex chromosomes like those of eukaryotic organisms, and they do not reproduce sexually, although some groups have mechanisms that lead to genetic recombination.

Nearly all prokaryotes have a rigid cell wall. Polypeptides are incorporated into this basic wall structure, serving as links between the polysaccharide units. This unusual construction serves to distinguish the prokaryotes from the eukaryotes.

No prokaryote is truly multicellular. Although some bacteria, including most blue-green algae, form filaments or masses of cells, these are connected only because their cell walls fail to separate completely following cell division or because they are held together within a common mucilaginous capsule or sheath. There are no protoplasmic connections, such as plasmodesmata, between the individual cells. Multicellularity, like sexual reproduction, occurs only in eukaryotic organisms.

Prokaryotic organisms lack cellular organelles in their cytoplasm, but they have other structures that play similar roles. Their plasma membranes often have many folds and convolutions extending into the interior of the cell. Such membranes increase the surface area to which enzymes are bound, which facilitates the separation of different enzymatic functions. In the bacteria, at least, the DNA molecules, following replication and separation, become attached to structures extending inward from the plasma membranes. These structures hold the molecules in different regions of the cell and thus play a part in cell division that is comparable to that of the spindle fibers in mitosis. Some photosynthetic bacteria, including blue-green algae, bear their photosynthetic pigments in thylakoids. In other photosynthetic bacteria, the chlorophyll is located in discrete spherical bodies called chromatophores.

Blue-green algae are specialized bacteria that have a particular system of photosynthesis involving chlorophyll *a*. Chlorophyll *a* is found in the chloroplasts of all photosynthetic eukaryotes—further evidence to support the hypothesis presented in Chapter 6 that chloroplasts (and mitochondria) are derived from symbiotic prokaryotes.

THE BACTERIA

Bacteria are the smallest, oldest, and most abundant group of organisms in the world. Most of them are only about 1 micrometer in diameter, with some being only one-tenth of that size, but others range up to 10 (or rarely, even 30 or more) micrometers in length. Bacteria go back at least 3.2 billion years in the fossil record (Figure 10–3). The total weight of all the bacteria in the world exceeds that of all other organisms combined despite their extremely small size.

Bacteria occur in all habitats and, largely because of their great metabolic versatility, can survive in many environments that support no other form of life. They have been found in the icy wastes of Antarctica, the boiling waters of natural hot springs, and even in the dark depths of the ocean. Some are *obligate anaerobes*, that is, organisms that can only live in the absence of air, whereas others, *facultative anaerobes*, can survive without oxygen but grow more vigorously if it is supplied (aerobic respiration yields much more energy than anaerobic, as we noted in Chapter 5).

Living bacteria have recently been found in samples of rock and ice recovered by drilling from depths of up to 430 meters in Antarctica. These bacteria, which are at least 10,000 and possibly a million years old, were lying dormant at temperatures ranging from about −7 to −14°C but immediately resumed normal activity when their temperature was elevated. These findings confirm the fact that bacteria are capable of lying in a state of "suspended animation" for extremely long periods.

Recently, the limits of tolerance of certain bacteria have been investigated in relation to the possible existence of extraterrestrial life on various planets with known atmospheres. For example, the extreme alkalinity of Jupiter's atmosphere has been considered one of the arguments against the chance of life there; however, bacteria from the Livermore Valley of California have

10–2
Despite their great abundance and the obvious effects associated with many of their activities, bacteria are so small that they were not actually seen until the late seventeenth century. At that time, Antony van Leeuwenhoek (1632–1723), a cloth merchant, made the first microscope lenses with sufficient magnification. Leeuwenhoek spent most of his leisure time at microscopy, and his observations include the first visual records we have of protozoa, yeasts, red blood cells, sperm cells, and many other small objects, in addition to the bacteria. His microscopes, as shown in a replica (a), *were very small and had to be held up to the eye. The single lens was fixed in position and the specimen was held on the point of the screw opposite the lens.*
(b) *Leeuwenhoek's drawing, made in 1683, of bacteria found in his mouth.*

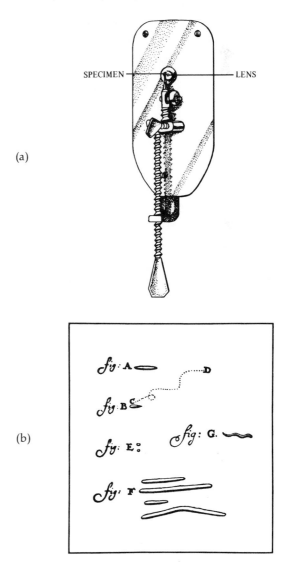

(a)

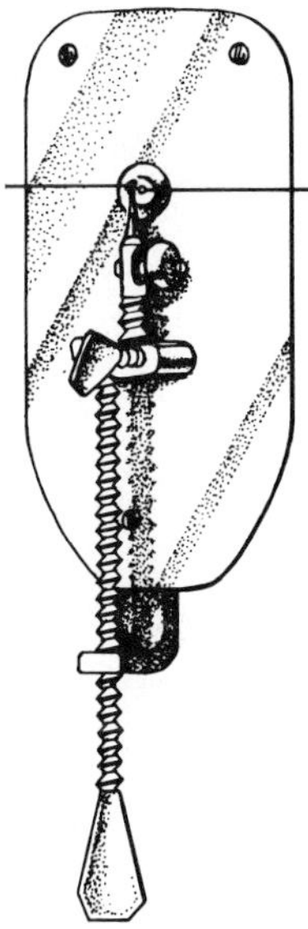

(b)

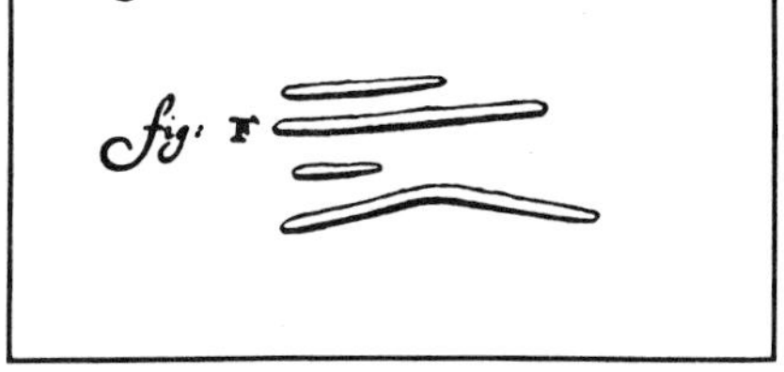

10–3
Alga-like microfossils (Archaeosphaeroides barbertonensis) *in thin sections* (a) *and bacteria-like microfossils* (Eobacterium isolatum) *in surface replicas* (b). *They were found in the same deposit of South African black chert and are about 3.2 billion years old, the oldest fossils now known. Until quite recently, the oldest known fossils were approximately 0.6 billion years old. The dramatic extension backward into time of our knowledge of the origins of life has been made possible by several developments in seemingly unrelated fields. Radiometric dating methods (comparing the proportions of isotopes in minerals included in the rocks) have made possible accurate determinations of age for very old sediments. Increasing geological knowledge has indicated the best places to look for these fossils. Increasingly sophisticated methods have been used for the detection of organic molecules in old rocks. And the electron microscope has been used to find very small fossils in ultrathin sections of ancient rocks.*

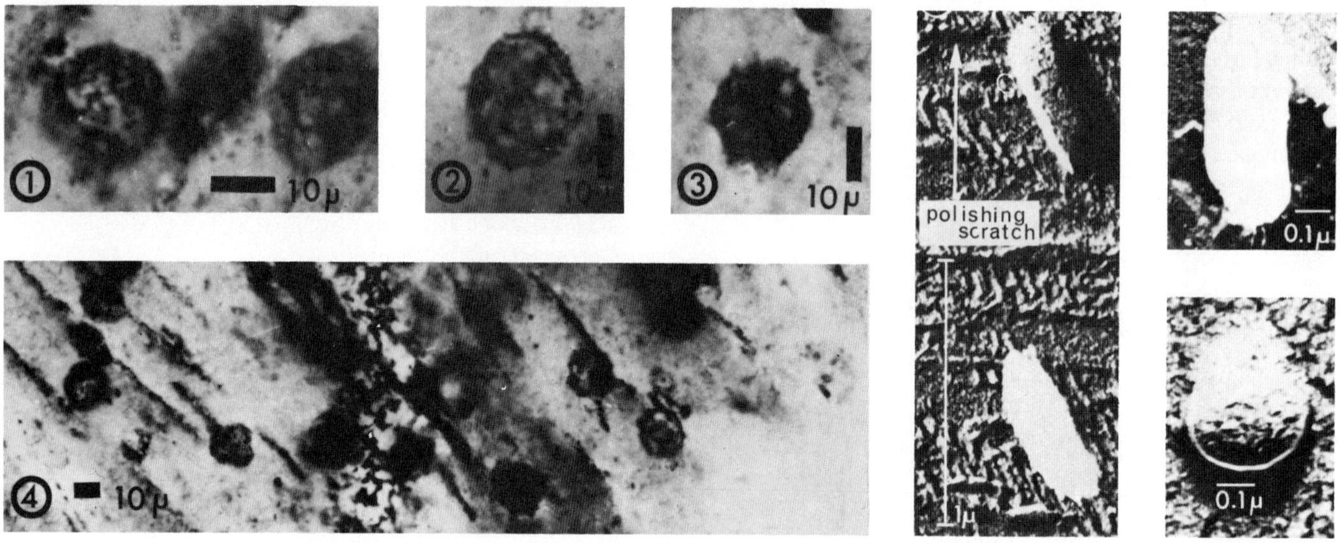

(a) (b)

SENSORY RESPONSES IN BACTERIA

Despite the relative simplicity of bacterial cells, they are able to respond to stimuli, as evidenced by their capacity to locate areas where there are concentrations of oxygen or food molecules. Recently scientists at the University of California, working with a strain of Salmonella, *were able to show that the cells could move up a very slight concentration gradient of a nutrient amino acid (serine). How are the bacteria able to distinguish an area of lower concentration from an area of only very slightly higher concentration? There seem to be only two possible answers to this question. One is that they sense the gradient instantaneously by comparing the concentrations of serine at their "head" with that at their "tail." This possibility has been rejected because the bacteria are so extremely small that the differences in concentration are too minute and too variable. The second hypothesis, supported by preliminary data, is that they are able to compare concentrations at many body lengths apart. In other words, information about the concentration at one time and place is stored by the bacteria and compared with information received at another time and place.*

Some photosynthetic bacteria exhibit phototaxis—moving toward light by differential motion of their flagella. Escherichia coli *and other flagellated bacteria similarly exhibit chemotactic (chemical-sensing) responses that are made possible by changes in the direction of flagellar rotation. When presented simultaneously with both attractant and repellent, they respond to whichever one is present in the more effective concentration, indicating that they probably have a processing mechanism that compares opposing signals from different chemoreceptors, sums these signals up, and then communicates the sum to the flagella. The entire system has been shown to be controlled by proteins in the surface membranes of the cells.*

One of the nonsulfur photosynthetic bacteria, Halobacterium halobium, *which occurs in brine, has patches of a purple pigment in its plasma membrane that is analogous to the visual pigment of the eye. This bacterium is mobile and responds phototactically to violet light, suggesting that its pigment detects light as does the analogous visual pigment.*

These are remarkably complex responses for such small and simply constructed organisms, and they foreshadow the many kinds of tropic responses found among the eukaryotes—that is, responses to external stimuli in which the direction of movement is determined by the direction from which the most intense stimulus comes.

10-4
Kakabekia (a) *a curious organism nearly 2 billion years old and originally thought to have no known living relatives, was discovered in the Gunflint chert of southern Ontario. In 1964, S. M. Siegel of the University of Hawaii was testing the viability of living organisms when subjected to several possible primitive atmospheres. In a soil sample from Wales, exposed to an atmosphere of 25 percent methane, 25 percent ammonia, 10 percent oxygen, and 40 percent nitrogen, several types of organisms remained viable. Among them was an organism* (b) *that appears to be identical with* Kakabekia, *rediscovered after a lapse of 2 billion years. This organism has what might be described as a "fossil" metabolism: It requires ammonia but no oxygen.* Kakabekia, *which grows from 2 micrometers to about 15 micrometers in diameter, is prokaryotic but resembles no known group of organisms.*

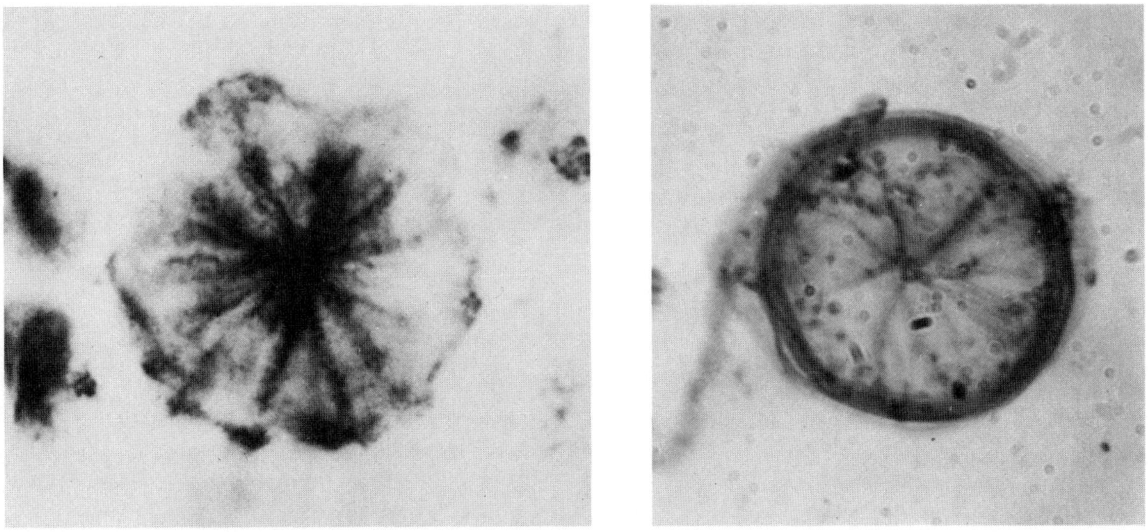

(a) (b)

been found in water with an alkalinity as high as pH 11.5, and they grow and reproduce in solutions of sodium hydroxide that are equally basic. Others, collected near the base of Mt. Shasta, can tolerate not only extreme alkalinity but also the presence of high concentrations of ammonia. The prokaryote *Kakabeckia* (Figure 10-4) was discovered in an investigation of likely conditions for the early earth atmosphere.

Bacteria are most important, however, for the role they play in the world ecosystem. Like the fungi, they are decomposers; without bacteria and fungi, the organic material built up by plants and animals would accumulate, and gradually all organisms would be overwhelmed by the products of their own metabolism. Through the action of the decomposers, materials incorporated into the bodies of once-living organisms are released and made available for successive generations of living things. In a single gram of fertile agricultural soil, there may be 2.5 billion bacteria, 400,000 fungi, 50,000 algae, and 30,000 protozoa.

In addition to their ecological role, bacteria are important in a number of other ways. They are responsible for many of the most serious diseases of man and other animals, including tuberculosis, cholera, anthrax, gonorrhea, diphtheria, and tetanus, and cause a wide range of economically significant diseases of plants. More than 200 species of bacteria are currently recognized as plant pathogens in the United States alone. In recent years, bacteria have become important as the source of a number of important antibiotics, including tyrothricin, bacitracin, subtilin, and polymyxin B. One of the subgroups of bacteria, the actinomycetes, provide still other antibiotics: streptomycin, aureomycin, neomycin, and terramycin. Almost all cheeses are produced as a result of bacterial fermentation of lactose into lactic acid, which coagulates milk proteins. In a similar fashion, bacteria are used commercially in the production of acetic acid and vinegar, various amino acids, and enzymes.

The Structure of Bacteria

In Chapter 1, we considered the colon bacterium *Escherichia coli* as a representative prokaryotic organism. Many of the characteristics of *E. coli* are common to all bacteria and bacteria-like organisms. Bacteria resemble plants in their possession of a rigid cell wall outside the plasma membrane, but unlike most plants they usually have no cellulose in their cell wall. Their cells sometimes have appendages called flagella and pili. Within the cyto-

10-5
Divergent types of bacteria. (a) *Actinomycetes are largely responsible for the "moldy" odor of decaying organic material in soil where they are abundant.* Mycobacterium tuberculosis, *the causative agent of tuberculosis, is an actinomycete, and potato scab, a plant disease, is caused by another genus.* Streptomyces fradiae, *shown here, is the commercial source of the antibiotic neomycin. Actinomycetes are the organisms responsible for nitrogen fixation in nonleguminous plants, forming symbiotic associations with the roots of over a dozen genera.* (b) *Gliding bacteria consist of filamentous forms with relatively large cells that exhibit a "gliding" motility similar to that of the blue-green algae. Waves of contraction cause periodic alterations in the form of the cell and are responsible for its movements.* Beggiatoa, *shown here, is a chemoautotroph. The conspicuous granules in the cells are sulfur, produced by the oxidation of hydrogen sulfide—a process utilized by the bacteria for the production of energy. The gliding bacteria are essentially like colorless blue-green algae, a fact that underscores the fundamental similarity between these prokaryotes.* (c) *The myxobacteria have a pattern of organization similar to that of the slime molds (see page 276). As shown here in these micrographs of* Stigmatella aurantiaca, *the vegetative rods glide together to form swarms that eventually differentiate into "fruiting bodies" into which many bacterial cells are incorporated.* (d) *The spirochaetes are spiral bacteria up to 500 micrometers long, which is an enormous size for bacteria. They move by means of a helical wave moving along the body of the cell, or by a snakelike motion.* Treponema pallidum, *shown here, is the causative agent of syphilis.*

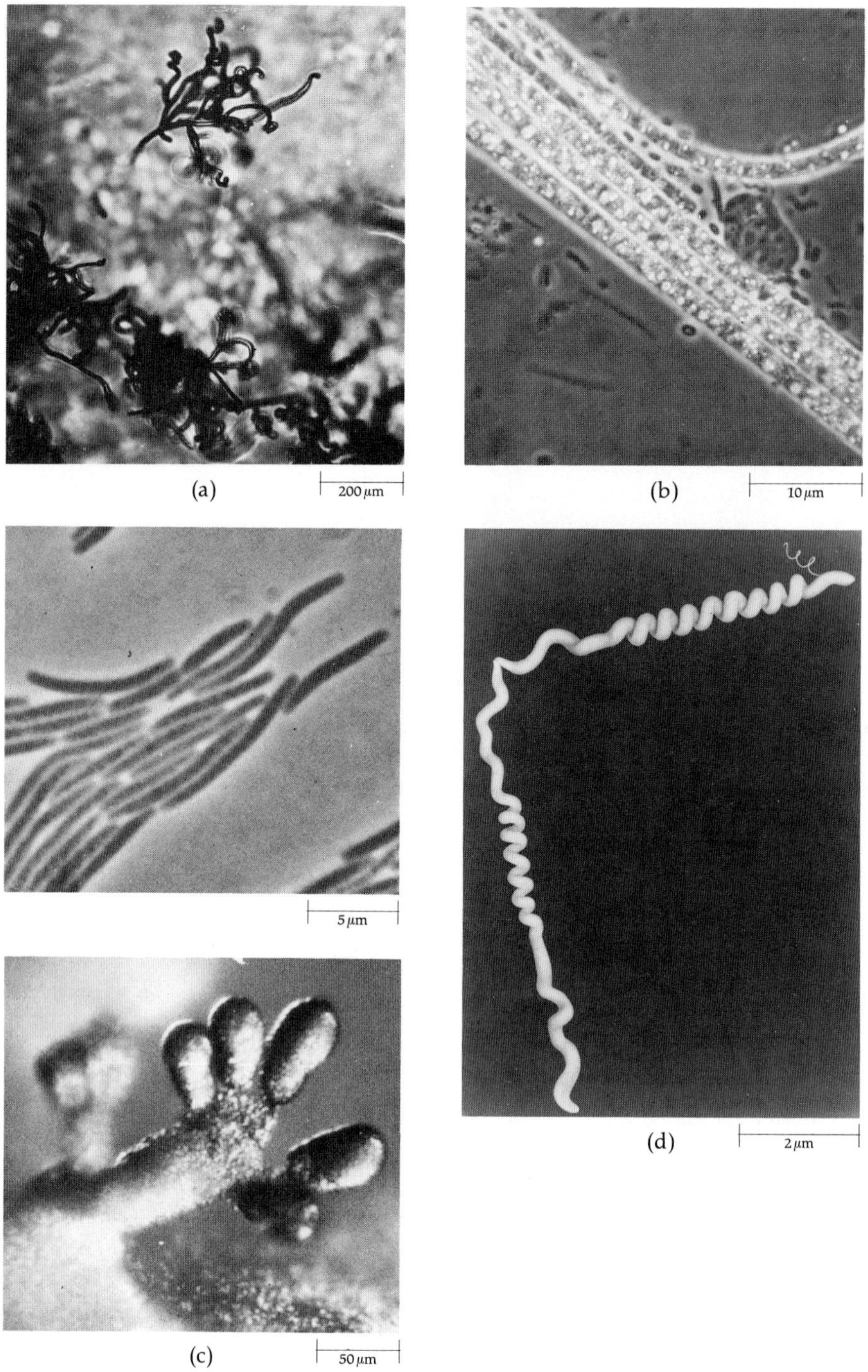

plasm, ribosomes and various granular inclusions, as well as pools of nucleic acid, are found. On the whole, the cells, being prokaryotic, are far less organized and less complex than those of the eukaryotes, although there is a surprising diversity of form and cellular organization (Figure 10-5).

Bacterial Form

Bacteria may be distinguished on the basis of their shapes. Straight, rod-shaped forms are known as <u>*bacilli*</u>, spherical ones are called <u>*cocci*</u>, and long curved ones are called <u>*spirilla*</u> (Figure 10-6). Cocci may stick together in pairs after division (diplococci), they may occur in clusters (staphylococci), or they may form chains (streptococci). The organism that causes pneumonia is a diplococcus, whereas the staphylococci are responsible for many serious infections characterized by boils or abscesses.

Rod-shaped bacilli are more commonly found alone than are the spherical cocci. When they do remain together, they spread out end to end, in filaments, because they always divide transversely. Because these filaments are funguslike in appearance, the combining form *myco-*

10-6
The three major form-groups of bacteria: (a) *bacilli;* (b) *cocci; and* (c) *spirilla. The rod-shaped bacteria include those microorganisms that cause lockjaw* (Clostridium tetani), *diphtheria* (Corynebacterium diphtheriae), *and tuberculosis* (Myobacterium tuberculosis), *as well as the familiar* Escherichia coli. *Among the cocci are* Diplococcus pneumoniae, *the cause of bacterial pneumonia;* Streptococcus lactis, *a common milk-souring agent; and* Nitrosococcus nitrosus, *soil bacteria that oxidize ammonia to nitrites. The spirilli, which are less common, are helically coiled bacteria. Cell shape is a relatively constant feature in most species of bacteria. (Carolina Biological Supply Co.)*

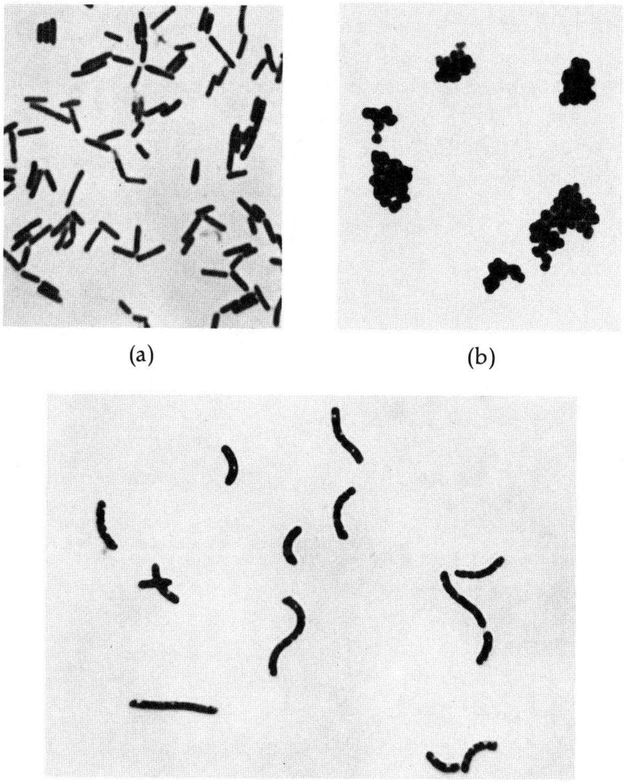

(c)

10-7
(a) *Numerous flagella in* Proteus mirabilis *surround a pair of dividing cells.*
(b) *A cell of* Bdellovibrio bacteriovorus. *These cells, which are abundant in soil and sewage, move by means of their single posterior flagellum. The cell shown here is attached to a host cell of* Erwinia amylovora, *a member of a highly important genus that causes plant diseases. Individuals of* B. bacteriovorus *fasten to other cells by a special attachment site at the apex.*

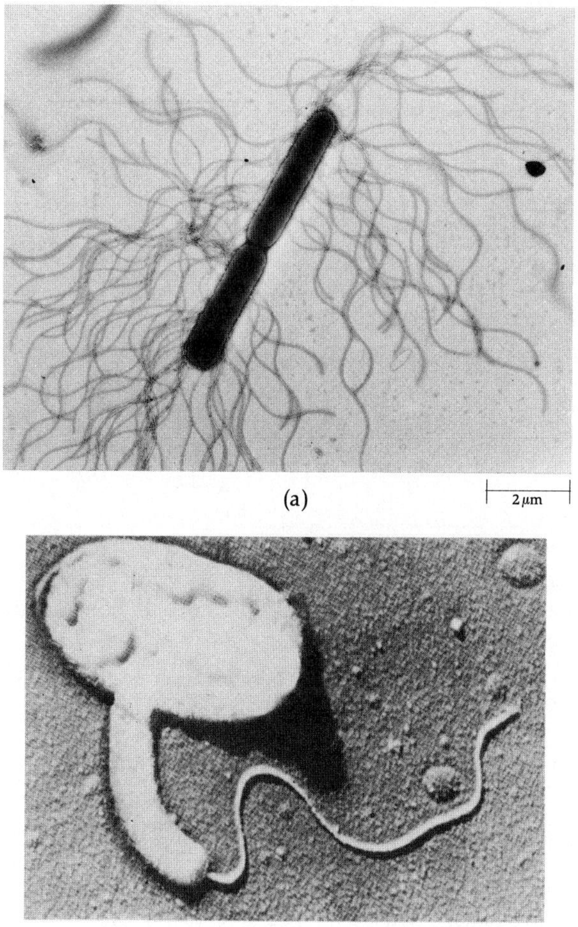

(b)

(from the Greek word for fungus) is often a part of the name of these organisms and is a clue to their appearance. *Mycobacterium tuberculosis,* for example, is a rod-shaped bacterium that forms a filamentous funguslike growth in culture, although not in the host.

Flagella and Pili

Some kinds of bacteria have very slender, rigid, helical flagella that they rotate to swim from place to place.

Bacterial flagella are long (3 to 12 micrometers), slender, and wavy. Because they are only 10 to 20 nanometers in diameter, they are usually too fine to be seen by ordinary microscopic techniques. In some bacteria, they are well distributed over the cell, and in others, they are restricted to one or both ends (Figure 10-7). Bacterial flagella seem to be composed entirely of a single special type of protein—flagellin.

Pili are shorter (up to several micrometers) and straighter than flagella and are only about 7.5 to 10 nanometers in diameter (Figure 10-8). Like flagella, they are formed from bodies in the cytoplasm and they consist entirely of protein, although their proteins are different from those of the flagella and their structure is

10–8
Cell of Escherichia coli, *negatively gram stained, showing the numerous pili.*

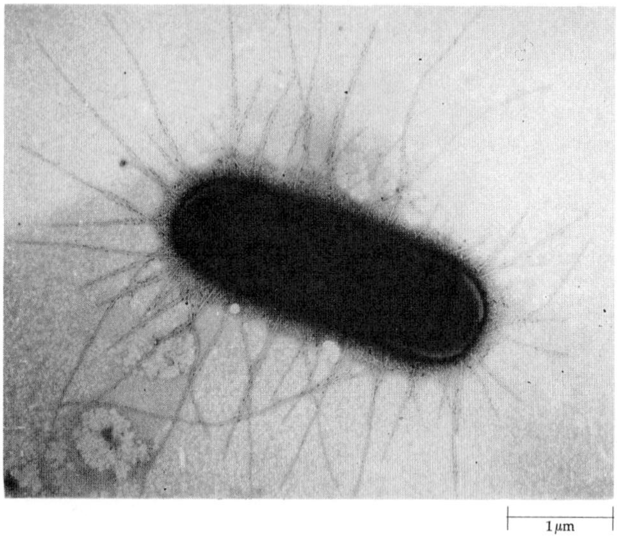

distinct. Special hollow pili are formed on bacterial cells during conjugation; their exact function is not known—they may serve as a bridge for the transfer of DNA, they may pull the conjugating cells together, or they may have some other, unknown function. Ordinary pili evidently help bacteria to find and attach themselves to appropriate surface membranes. A better understanding of the function of bacterial pili may aid in combatting bacterial diseases.

Cell Wall

The bacterial cell wall is complex, and, unlike the cell walls found in most plants, it lacks cellulose (Figure 10–9). Instead, there is a network of molecules of another polysaccharide connected by polypeptide crosslinks. One of these amino sugars, muramic acid, is found in all bacterial cell walls.

This network makes up the glycopeptide, or basal, wall layer, and it is covered in most species by additional layers, with specific large molecules exposed on the cell surface. It is these layers that are responsible for the rigidity of the cell wall. In many bacteria, the glycopeptide layer makes up the basic structure of the cell wall; the walls in these bacteria are 15 to 80 nanometers thick. In other bacteria, large molecules of lipopolysaccharide, a polysaccharide chain with lipids attached to it, are deposited over the glycopeptide layer. Such bacteria have cell walls that are only about 10 nanometers thick.

A Danish microbiologist, Hans Christian Gram, discovered that bacterial cell walls that lack the lipopoly-

10–9
(a) *The structural formulas of the two sugars that alternate to form the polysaccharide backbones of the bacterial cell wall.* (b) *A diagram of the bacterial cell wall structure. The backbones of alternating NAM and NAG are crosslinked by five glycine molecules which attach to short peptide chains of amino acids linked to the NAM molecules. An amino acid in the peptide chain (lysine) forms three peptide bonds, one with the terminal glycine molecule and one with each of the amino acids above and below it.*

These bonds are formed in an unusual way—by the transfer of a peptide bond from one molecule to another. Penicillin blocks this transfer. Because its structure mimics that of a dipeptide, it is able to bind the active site of the specific enzyme involved and thus inactivate it. Without crosslinks, the cell wall cannot hold the cell together. Thus, penicillin acts specifically on growing bacterial cells and not on the cells of an animal host.

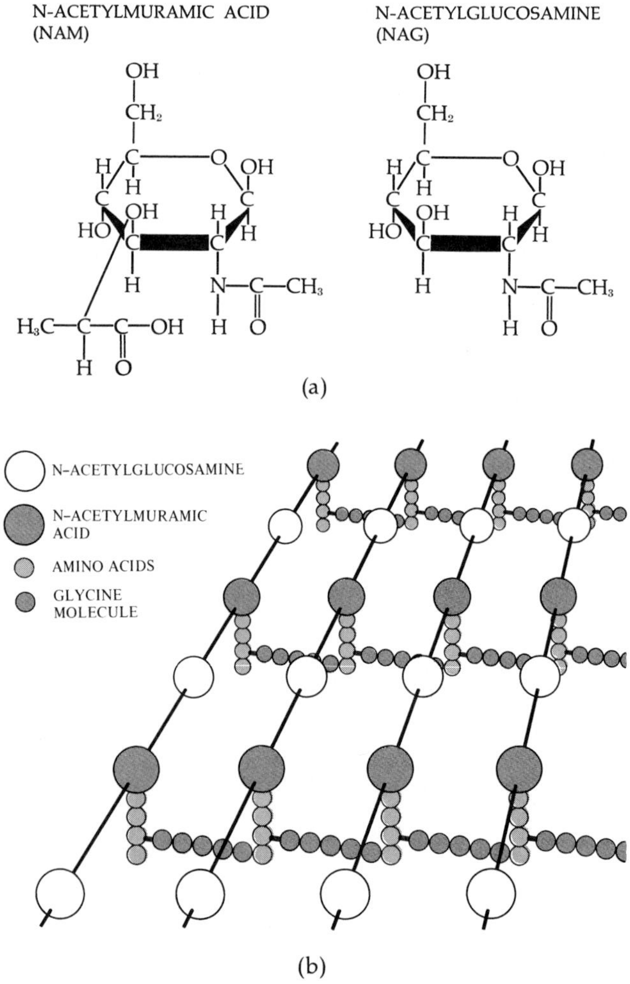

saccharide layer pick up a purple dye—gentian violet—and those in which it is present do not. This staining technique was first used to detect the presence of bacteria in animal tissues: Those bacteria that pick up the stain are known as gram-positive, whereas the others are gram-negative (see Figure 10–8). Gram staining is now widely used as a basis for classifying bacteria, because it reflects a fundamental difference in the architecture of the cell wall. This architecture in turn affects various other characteristics of the bacteria, such as their patterns of resistance to antibiotics. Actinomycin, for example, is a relatively large antibiotic molecule that apparently cannot pass through the fine mesh of the cell wall in gram-negative bacteria. In gram-positive bacteria, it does pass through and disrupts protein synthesis by binding to the DNA double helix, which also makes it toxic to the host cells and limits its wider clinical use.

Outside the cell wall of bacteria is often found a gelatinous layer, the capsule, which is apparently secreted by the bacterial protoplast through the wall.

10–10
Dividing cells of Escherichia coli. *In this preparation, the DNA appears white. The thick wall between the cells will soon split, separating the two daughter cells.*

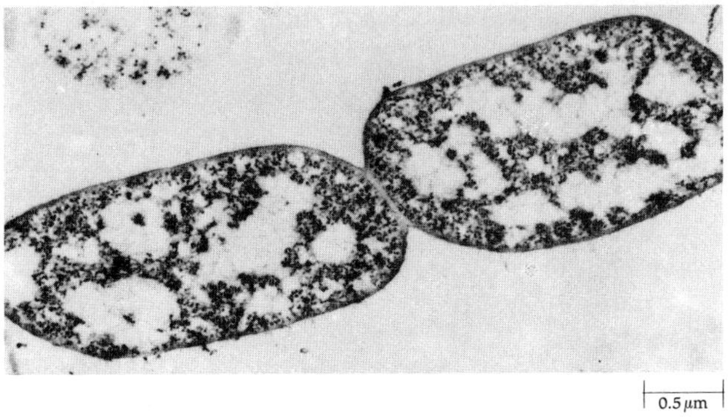

Cytoplasm

As we saw in Chapter 1, the bacterial cytoplasm is bounded by a plasma membrane. Within this membrane, which has many enzymes localized on its inner surface, is cytoplasm with a number of ribosomes and granular inclusions, as well as one or more bodies of chromatin. Bacterial cells usually contain at least two such bodies, because cell division lags behind the division of the genetic material. Each of the bodies consists of a closed loop of double-stranded DNA about 1000 times as long as the cell in which it occurs.

Bacterial Genetics

Cell Division

The chief mode of reproduction in the bacteria is asexual; each cell simply increases in size and eventually splits in two. In the process of fission, the plasma membrane and cell wall grow inward and eventually divide the cell in two. The new wall is thicker than the ordinary cell walls, and it soon splits from the outside toward the center of the old cell, separating the two daughter cells (Figure 10–10). Chains of bacteria are formed when this new cell wall does not divide or when it divides incompletely.

In bacterial cell division, the two daughter molecules of DNA are regularly attached to the plasma membrane at different points. The function of these attachments is similar to that of the spindle fibers in mitosis in that they hold the two molecules of DNA in the different portions of the original cell that will become the daughter cells (Figure 10–11).

10–11
A schematic diagram of attachment of bacterial chromosomes to the plasma membrane, which results in the distribution of one chromosome to each daughter cell. Such attachment serves the same function as that of the spindle fibers during mitosis in eukaryotes.

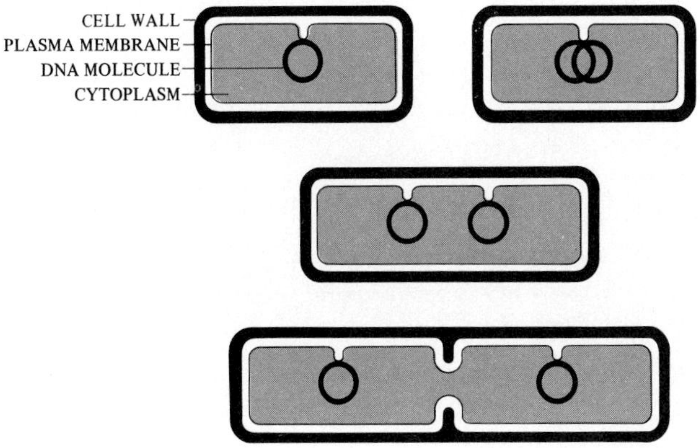

10–12
Light microscopic photo of Bacillus megaterium *dispersed in India ink. Capsules stand out as translucent halos.*

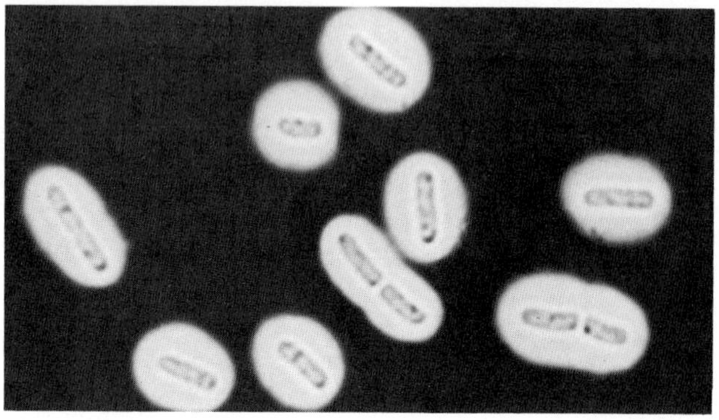

In addition to the normal mode of cell division, a few genera of bacteria, mainly bacilli, can form thick-walled spores within their cells. These spores are much more resistant to heat, to radiation, and to dehydration than the more common form and may remain viable even when boiled in water for as long as 2 hours (Figure 10–13). A bacterial spore is typically about 1 micrometer in diameter and consists essentially of a small cell, often called the core, surrounded by a number of specialized layers that are peculiar to spores. They are formed at a high frequency when a certain nutrient, usually the carbon source, becomes exhausted, provided that other nutrients are present in adequate supply. The formation of the spore is a simple example of differentiation, and it has received much study in recent years; mutants are known in which the process is interrupted at various points.

Genetic Recombination

Although complex sexual reproduction involving alternating meiosis and syngamy is limited to eukaryotic organisms, genetic recombination has been observed to take place in bacteria. In the most general terms, this genetic recombination involves the transfer of a portion of a DNA molecule from one bacterial cell to another. This fragment may simply act in concert with the DNA molecule of the cell which it enters, with both producing messenger RNA, or it may actually become incorporated into the main DNA molecule of the recipient cell, in which case it is passed on to the daughter cells with the rest of the hereditary material. The process of recombination itself is the same whether the fragments are passed from cell to cell by direct contact, are carried from cell to cell by viruses, or enter the cells from solution as "naked" DNA.

10–13
Colony of Clostridium botulinum. *The type of food poisoning (botulism) suffered by consumers of home-canned foods is associated with the survival of the spores of this bacterium, which are extremely resistant to heat. They occasionally persist in food that has not been treated properly.* Clostridium, *which grows only in the absence of oxygen, produces its powerful toxin inside the can or jar. The toxin itself is destroyed by boiling for 15 minutes but can be fatal even in minute quantities. The toxin is the most powerful known, and 1 gram would be enough to kill 14 million people. Botulism is rare in the United States; between 1970 and 1974, there were 121 reported cases and 28 reported deaths.*

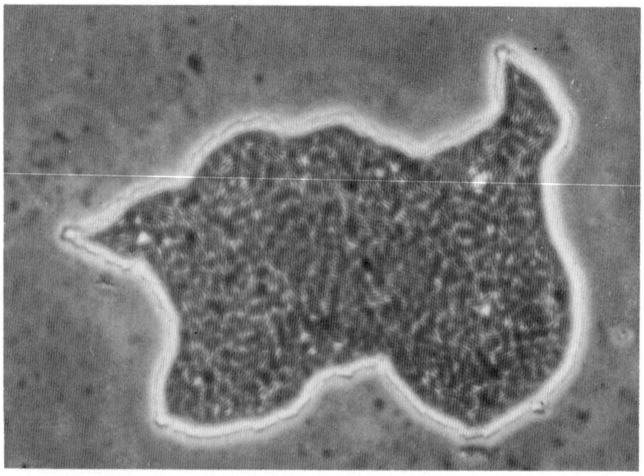

Recombination by Bacterial Conjugation

It is now established that in the cells of many bacteria, in addition to the large circular molecule of DNA known as the bacterial chromosome, there are relatively small fragments of DNA which usually form closed circles and are present in the cytoplasm. Some of these fragments, known as *episomes*, can be integrated into and replicated with the chromosomes.

Some strains of *Escherichia coli*, the organism in which genetic recombination in prokaryotes was first demonstrated, are known as F^+ strains and others as F^- strains. The F^+ strains, which are relatively rare, contain a type of episome called an F factor. When F^+ individuals are mixed with F^- individuals, which do not contain the F factor, the F factors move to the plasma membrane of the F^+ cell. Here they organize the formation of special hollow pili which probably act as conjugation bridges pass-

10–14
Conjugating individuals of Escherichia coli. *The DNA is presumably passing from one cell to the other through a special hollow conjugation pilus.*

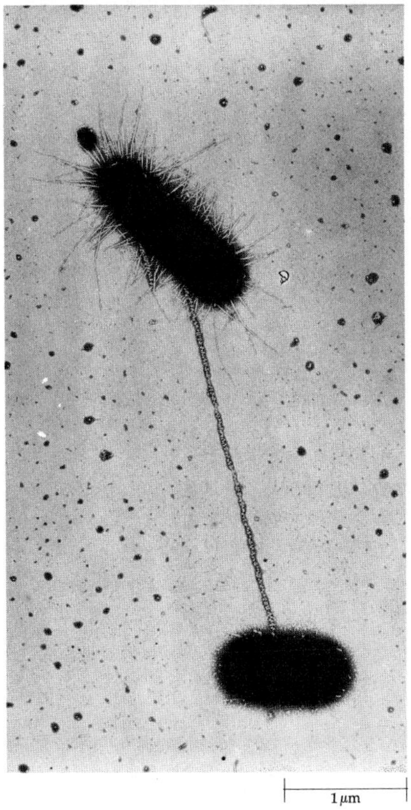

10–15
Diagram of bacterial conjugation. (a) *In the first series, an* F^+ *cell is shown conjugating with an* F^- *cell. The F factor organizes the conjugation pilus, replicates, and may pass through the pilus. The recipient cell then becomes* F^+. (b) *In the second series, an Hfr cell is shown conjugating. In the Hfr strain, the F factor is a part of the chromosome. The chromosome breaks at the point of attachment and a portion of it enters the* F^- *cell. The F factor, which is on the far end of the DNA strand, is seldom transferred to the recipient cell.* (c) *Depending on the length of time the cells are permitted to conjugate, different marker genes (indicated by black dots) enter the recipient cell. By interrupting conjugation at various intervals, it has been possible to make maps of the location of genes on the chromosome of* E. coli.

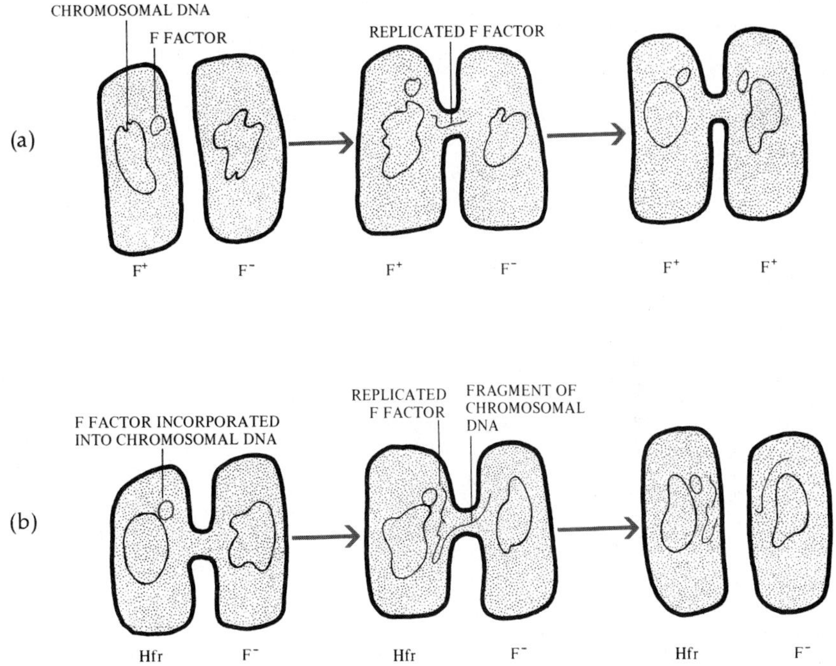

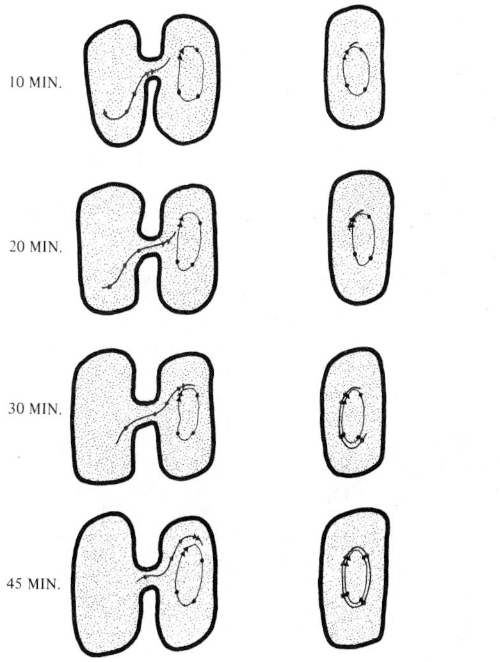

ing from the F^+ to the F^- cells (Figure 10–14). F factors, which are circular molecules of DNA, continuously replicate themselves, passing through the pili into the F^- cells. Soon the entire culture consists of F^+ individuals, all the cells now containing this particular episome.

In such F^+ strains, the F factor is occasionally incorporated into the bacterial chromosome. F^+ individuals in which the F factor is incorporated into the bacterial chromosome give rise to bacterial strains, known as "high-frequency" (Hfr) strains, in which conjugation and transfer of genetic material occur far more frequently. In these strains, the bacterial chromosome, or a portion of it, may pass through the conjugation pilus into an F^- cell. This chromosomal material may then be incorporated into the chromosome of the recipient cell (Figure 10–15).

Other Forms of Bacterial Recombination

In 1952, N. D. Zinder tried to determine whether genetic recombination can occur in the bacterium *Salmonella typhimurium.* Mixing 20 different strains in 79 different combinations, he was able to show recombination in 9 of them. He then maintained two different strains of *Salmonella* in the two arms of a U-tube, with an ultrafine glass filter separating them, and in one arm of the culture he found individuals in which genetic recombination had taken place (Figure 10–16). Because bacteria could not have passed through the filter, Zinder knew that the process he was observing was fundamentally different from bacterial conjugation, which had been demonstrated a few years earlier.

Eventually it was found that one of the strains (A22) was infected by a latent virus (bacteriophage). From time to time these viruses would set up an infection in the bacterial cells, multiply within them, and finally break them down. When the phage particles were assembled before cell lysis, they occasionally picked up small portions of bacterial chromosome and incorporated them within their own protein coats along with phage DNA. These unusual phages then passed through the filter and infected cells of *Salmonella* strain A2. In this way, phages occasionally introduced bacterial DNA derived from strain A22 into the cells of strain A2. The raw material was then present for genetic recombination (Figure 10–17).

The virus here acted merely as a vector—a sort of guided missile carrying the DNA from one strain to another. The transfer of genetic material from one bacterial strain to another by phage infection has become known as *transduction.*

Another kind of genetic recombination in bacteria was described in Chapter 3 (page 69). In *Diplococcus pneumoniae,* the causative agent of pneumonia, there are two types of colonies: rough (R) and smooth (S). The strains that produce smooth colonies readily produce capsules and are virulent. As early as 1928, F. Griffith showed that it was possible to transform a harmless R strain of *Diplo-*

10–16
Apparatus used to demonstrate transduction. In this experiment, bacterial cells of one strain (A22) were placed on one side of the U-shaped tube and cells of another strain (A2) on the other side. A filter was placed between them to prevent conjugation. Genetic traits characteristic of the A22 strain showed up in the A2 strain despite the presence of the filter that prevented the passage of bacteria. It was subsequently found that viruses (phages) latent in the A22 strain of bacterial cells occasionally produce an active infection. Some of the phage particles may carry portions of the DNA of the host cell through the filter to the A2 strain of cells.

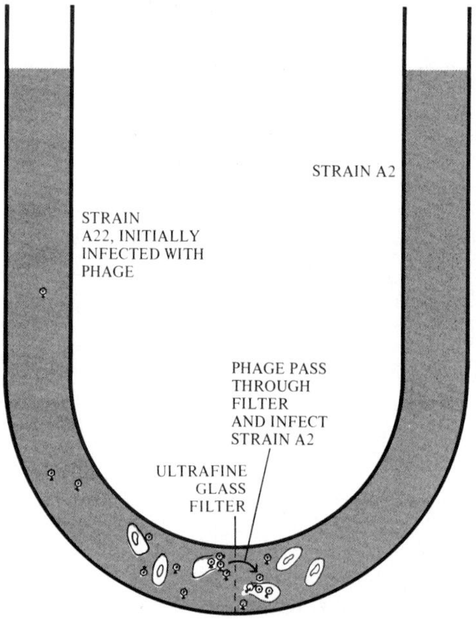

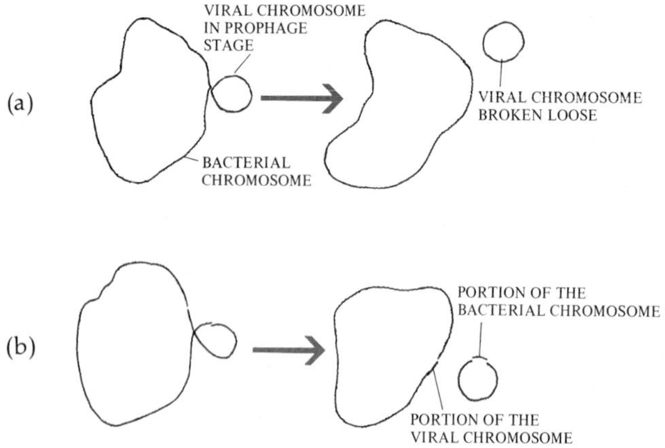

10–17
Some latent (temperate) phages occupy a specific place on the chromosome of the bacterial cell. When the phage sets up an infection, it breaks loose from the bacterial chromosome (a). *At this time it may take a portion of the bacterial DNA with it* (b). *Only one or a few genes are transferred by this process. When the particular type of phage has its specific place on the chromosome, as is often the case, the same gene or genes will be transferred.*

coccus into a virulent S strain by exposing the R strain to heat-killed S strain cells.

As we noted previously, this phenomenon is known as transformation. When bacterial cells are broken down by chemical treatment or by heat, fragments of DNA are released into solution and may pass into other cells. The demonstration that DNA was the genetically active material involved in such transformations was made by Avery, MacLeod, and McCarty in 1944 and was the first direct evidence for the genetic role of DNA. Transformation is now known to occur in many different groups of bacteria.

Factors in Bacterial Evolution

Genetic recombination makes it possible for bacteria to exchange whole blocks of genetic material, and such exchanges are doubtless important in the overall pattern of variation of a bacterial colony and its ability to adjust to different environments. However, conjugation has been demonstrated only in the laboratory and only in a relatively few bacteria. One of the important groups in which it does occur is the family Enterobacteriaceae, a group that includes not only *Escherichia coli,* but also the well-known pathogens *Shigella, Streptomyces,* and *Salmonella.* In these, recombination—often involving bacteria belonging to different genera—has provided a very important source of variation. In this way, resistance to antibiotics is spread rapidly between species of bacteria. It may be carried in nonpathogenic bacteria, such as *E. coli* (normally found in the gut) and then passed rapidly to such pathogenic bacteria as may be introduced.

The major source of variability in most bacteria, and the reason that they are able to adjust to such a wide variety of environmental conditions, is certainly mutation (Figure 10–18). For a given gene, it has been calculated there will be about 1 mutant cell per 10^7 (10 million) individuals. The amount of DNA in *E. coli* is equivalent to approximately 5000 genes. Thus, in a culture of *E. coli,* there is about 1 mutant cell per 2000 individuals; 0.05 percent of the individuals in the culture will have a mutant phenotype in each cell division. In a culture that approaches 10^9 cells—one that has divided about 30 times—the frequency of mutant individuals will be about 30 $\times$ 0.05 percent, as high as 1.5 percent.

Bacteria multiply very rapidly. For *E. coli,* the population may, under optimal conditions, double every 12.5 minutes. The number of mutant individuals produced in such a population, as you can see, is very high. Most organisms have a higher mutation rate than bacteria, but the rapid generation time of bacteria combined with mutations is responsible for their extraordinary adaptability.

Bacterial Metabolism

Heterotrophs

Most bacteria are heterotrophs, organisms that cannot make organic compounds from simple inorganic substances but must obtain them from other organisms. The largest group of heterotrophic bacteria are the saprobes. Saprobes are organisms that obtain their nourishment from dead organic matter. The saprobic bacteria and fungi are responsible for the decay and recycling of organic material in the soil, and many of the characteristic odors associated with soil come from substances produced by bacteria.

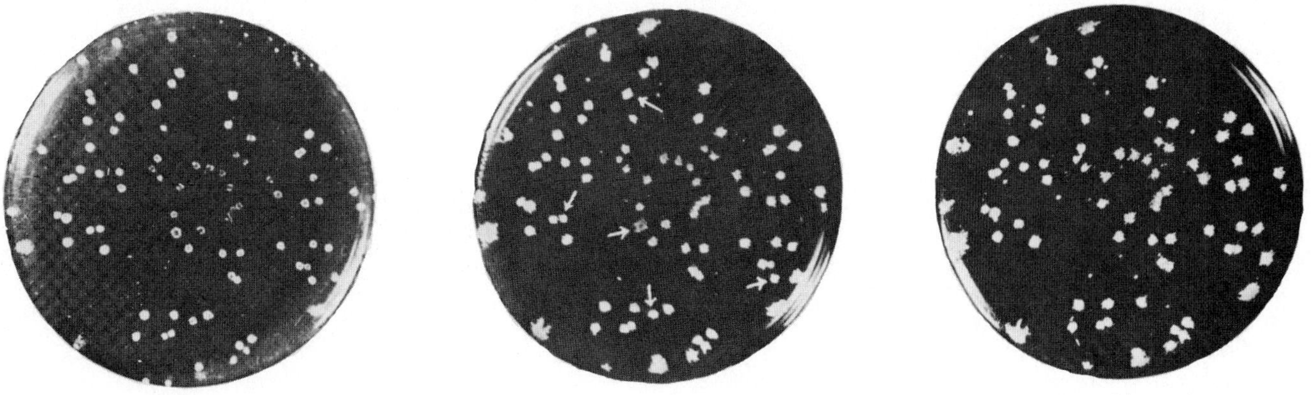

(a) (b) (c)

10–18
Replica plating as a means for detecting the presence of mutations in bacteria. (a) *is the initial plate. Colonies were transferred from* (a) *to* (b) *and* (c) *by means of a velveteen disk that placed every colony in the same position.* (a) *and* (b) *are on a complete medium, whereas a number of growth factors are lacking in* (c). *Under these conditions, the colonies indicated in* (b) *by arrows are unable to grow; they are mutants that were already present but undetected in* (a).

Photosynthetic Bacteria

There are three groups of photosynthetic bacteria, other than the blue-green algae, which will be discussed on page 197. These are the green sulfur bacteria, the purple sulfur bacteria, and the purple nonsulfur bacteria. (The colors of the third group may actually range from purple to red or brown.) Like the green plants, photosynthetic bacteria contain chlorophyll. The chlorophyll found in the green sulfur bacteria, chlorobium chlorophyll, differs chemically in several ways from the chlorophyll *a* of plants. The chlorophyll found in the two groups of purple bacteria is bacteriochlorophyll, which is chemically very similar to chlorophyll *a* and is a pale blue-gray in color (Figure 10–19). The colors of the purple bacteria are due to the presence of several different yellow and red carotenoids, which function as accessory pigments in photosynthesis.

In the photosynthetic sulfur bacteria, the sulfur compounds play the same role in photosynthesis that water does in a green plant. That is,

$$CO_2 + 2H_2S \xrightarrow{\text{light}} (CH_2O) + H_2O + 2S$$

As we saw in Chapter 6, an understanding of the course of photosynthesis in the sulfur bacteria was the key that led Cornelis van Niel to propose the generalized equation for photosynthesis:

$$CO_2 + 2H_2A \xrightarrow{\text{light}} (CH_2O) + H_2O + 2A$$

All bacterial photosynthesis is carried out anaerobically, and it never results in the production of molecular oxygen (O_2).

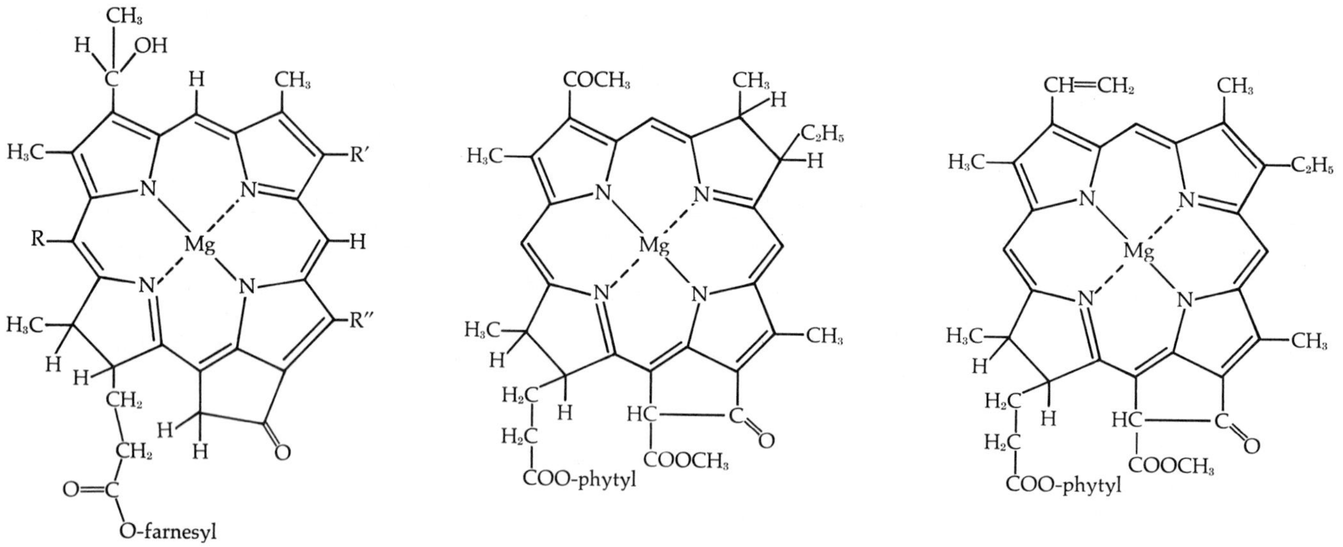

CHLOROBIUM CHLOROPHYLL BACTERIOCHLOROPHYLL CHLOROPHYLL *a*

(a) (b)

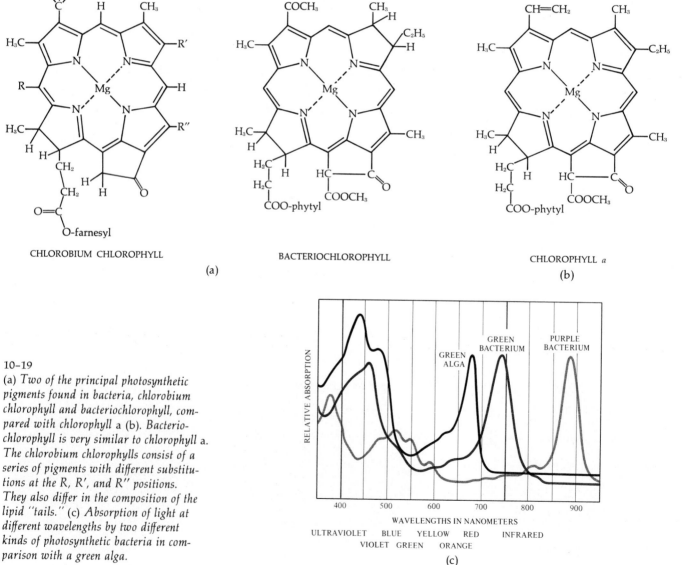

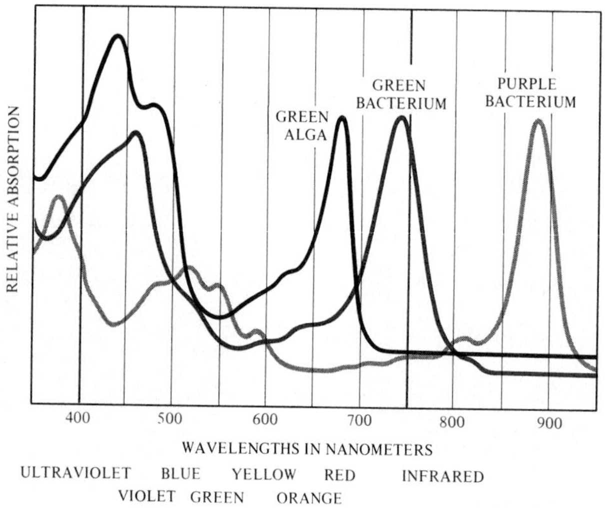

(c)

10–19
(a) *Two of the principal photosynthetic pigments found in bacteria, chlorobium chlorophyll and bacteriochlorophyll, compared with chlorophyll* a (b). *Bacteriochlorophyll is very similar to chlorophyll* a. *The chlorobium chlorophylls consist of a series of pigments with different substitutions at the R, R′, and R″ positions. They also differ in the composition of the lipid "tails."* (c) *Absorption of light at different wavelengths by two different kinds of photosynthetic bacteria in comparison with a green alga.*

In the nonsulfur photosynthetic bacteria, other compounds, including alcohols, fatty acids, and keto acids, serve as hydrogen donors for the photosynthetic reaction. So, as van Niel first deduced, despite its overwhelmingly great importance, both in terms of the world ecosystem and in terms of the numbers of kinds of organisms in which it occurs, green plant photosynthesis can be regarded biochemically as a special case of the generalized photosynthetic reaction.

Because of their requirement for hydrogen sulfide or a similar substrate, the photosynthetic sulfur bacteria are able to grow only in areas that contain large amounts of decaying organic material, distinguishable by their sulfurous odor. In these bacteria elemental sulfur may accumulate in deposits within the cell.

Chemoautotrophic Bacteria

Chemoautotrophic bacteria, unlike photosynthetic ones, require the presence of oxygen and do not utilize the energy of sunlight. The energy used to drive their synthetic reactions is obtained from the oxidation of inorganic molecules such as nitrogen, sulfur, and iron compounds, or from the oxidation of gaseous hydrogen.

Bacterial Ecology

Soil Bacteria

Different groups of microorganisms are involved in specific stages of the decomposing and recycling processes in the soil (Figure 10–20). Many bacteria and fungi break down carbon-containing compounds, releasing CO_2 into the atmosphere. The most important compounds originating from plants are cellulose and lignin, and secondary ones are pectic substances, starch, and sugars. It has been estimated that more than 90 percent of the CO_2 production in the biosphere results from the activity of bacteria and fungi.

Some microorganisms break down proteins into peptides, which are subsequently broken into their constituent amino acids. As we shall see in Chapter 26, many microorganisms have the ability to break down amino acids, with consequent release of ammonium (NH_4^+), a process known as ammonification. The ammonia can be oxidized to nitrite (NO_2^-) by chemoautotrophic bacteria of the genus *Nitrosomonas*, and the nitrites are oxidized to nitrates (NO_3^-) by *Nitrobacter*. These reactions constitute the process of *nitrification*. Both these processes release energy, which is used to reduce carbon dioxide to carbohydrate. Several other kinds of bacteria are capable of reversing the process and changing the nitrates back into nitrites or ammonia.

Denitrification, the conversion of nitrates into nitrogen gas or nitrous oxide, results in the loss of nitrogen from the soil. The reverse of this process, which is extremely important biologically, is *nitrogen fixation*. Several genera of bacteria and blue-green algae, as well as some fungi (yeasts), are capable of nitrogen fixation. Outstanding among them is the symbiotic bacterium *Rhizobium* (see Chapter 26), which forms nodules on the roots of legumes and a few other plants.

Sulfur is made available to plants (which cannot utilize elemental sulfur) by chemoautotrophic bacteria such as *Thiobacillus*, which oxidize elemental sulfur to sulfates:

$$2S + 2H_2O + 3O_2 \rightarrow 2H_2SO_4$$

Sulfates are accumulated by plants and their sulfur is incorporated into proteins. The degradation of proteins, discussed as one aspect of the nitrogen cycle in Chap-

10–20
The complexity of interactions that occur among soil organisms is suggested by this photograph, which shows bacteria growing on an agar plate to which penicillin has been added. Down the center of the plate is a dense colony of Staphylococcus epidermidis, *a strain that is resistant to penicillin because it produces an enzyme called penicillinase that breaks it down. The tiny colonies near the* Staphylococcus *colony are* Neisseria gonorrhoeae, *the causative agent of gonorrhea.* Neisseria *is susceptible to penicillin, but is able to grow in the part of the medium where the* Staphylococcus *has broken the penicillin down. This demonstration suggests the way that natural communities are organized.*

ter 26, liberates amino acids, some of which contain sulfur. A number of bacteria are capable of breaking down these amino acids with the consequent release of hydrogen sulfide (H_2S). Sulfates are also reduced to H_2S by certain soil microorganisms, such as *Desulfovibrio* (Figure 10–21).

10–21
The sulfur cycle. As you can see by an examination of this and other cycles, without the activities of bacteria and fungi, carbon, nitrogen, sulfur, and many other elements would remain locked in the molecules into which they are incorporated. Soil-dwelling heterotrophs break down complex organic molecules and release chemical substances that can then enter into biological cycles. Without the recycling of organic substances, all organisms would soon be overwhelmed by the products of their own metabolism. Man is releasing large quantities of sulfur into the atmosphere by burning fossil fuels at a rate that will equal the emission from natural sources by the year 2000. This is placing a severe strain on the mechanisms of the sulfur cycle as shown in this figure.

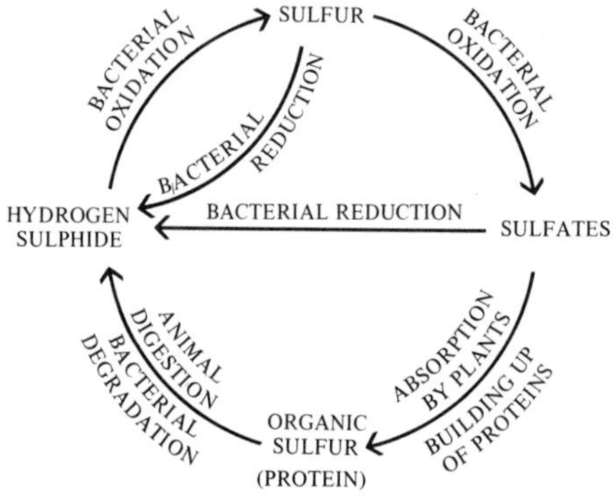

Parasitic and Symbiotic Bacteria

Some heterotrophic bacteria break down organic material still incorporated into the bodies of living organisms. The disease-causing (pathogenic) bacteria belong to this group, as do a number of other nonpathogenic forms. Some of these bacteria have little effect on their hosts, and some are actually beneficial. For example, when all the bacterial inhabitants of the human intestinal tract are removed—as can happen, for example, following prolonged antibiotic therapy—the tissues are much more vulnerable to disease-causing bacteria and fungi. And, as will be discussed in Chapter 26, the relationship between the symbiotic bacteria of the genus *Rhizobium* and the legumes they inhabit is obviously of mutual benefit.

In general, microbes cause diseases when their metabolism results in the production of toxic substances that enter the body of the organisms in which they occur. When the effect of these toxic substances is severe, the disease is considered serious. If death of the host organism results too soon, the relationship will not be as favorable as possible for the parasite, which will then have to find a new host. Most bacterial diseases of man are fatal in only a minority of cases, even if they remain untreated.

Human Diseases

Some of man's diseases are caused by air-borne bacteria. Among the better known of these is diphtheria, which is caused by the bacterium *Corynebacterium diphtheriae.* This organism produces a powerful toxic substance that circulates rapidly throughout the body and causes serious damage to the heart muscle, nervous tissue, and kidneys. Diphtheria is now rare, because most children are immunized against it in infancy. Other serious diseases are caused by air-borne bacteria of the genus *Streptococcus,* which are associated with scarlet fever, rheumatic fever, and other infections. Tuberculosis, caused by *Mycobacterium tuberculosis,* is still a leading cause of death in man, despite improved methods of detection, but the number of reported cases in the United States had decreased to 30,210 by 1974. Other air-borne bacterial diseases include bacterial pneumonia (mostly caused by *Diplococcus pneumoniae*) and whooping cough (caused by *Bordetella pertussis*).

A number of other diseases of bacterial origin are spread in food or water. Thus typhoid fever and paratyphoid are caused by bacteria of the genus *Salmonella,* bacillary dysentery is caused by *Shigella dysenteriae,* and typhus is caused by *Rickettsia prowazekii* (Figure 10–22). Undulant fever, caused by bacteria of the genus *Brucella,* affects both cattle and man and is usually contracted through drinking milk or eating milk products from an infected cow. Pasteurization of milk destroys *Brucella,* and the disease has become extremely rare in many portions of the world.

Bacteria play an enormous role in spoiling food and other stored organic products, and some organisms of this type are pathogenic. Food poisoning by *Clostridium botulinum* is rare but extremely dangerous (see Figure 10–13). *Staphylococcus* food poisoning is fairly common, but it is fortunately much less serious. In recent years, a number of virulent strains of *Staphylococcus* have been associated with serious infections. Many of these are resistant to penicillin, and some produce the enzyme penicillinase, which breaks it down. They frequently come into contact with penicillin-producing fungi and often grow with them in nature, so this resistance confers a natural advantage.

10-22
Typhus has often played a crucial role in human history. At the siege of Granada in 1789, 17,000 Spanish soldiers were killed by typhus, 3000 in combat. In the Thirty Years War, the Napoleonic campaigns, and the Serbian campaign during World War I, typhus was also the decisive factor. More human lives have been taken by rickettsial diseases than any other form of illness except malaria. (a) *Typhus is caused by* Rickettsia prowazekii, *one of the Rickettsiae, a group of very small bacteria-like organisms that are not able to grow on cell-free media. Most are parasites of arthropods, and there are probably millions of undescribed species.* (b) Rickettsia *is spread from rats to man by fleas. It can then be transmitted by body lice and, under crowded conditions, large numbers of people can be infected in a very short time.* (c) *The human louse,* Pediculus humanus, *and* (d) *the rat flea,* Xenopsylla cheopis. (c, Eric V. Gravé)

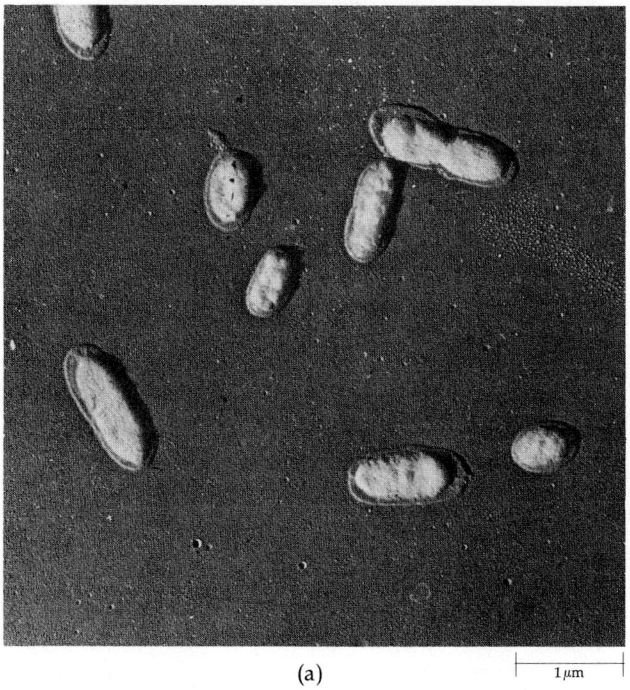
1 μm

(a)

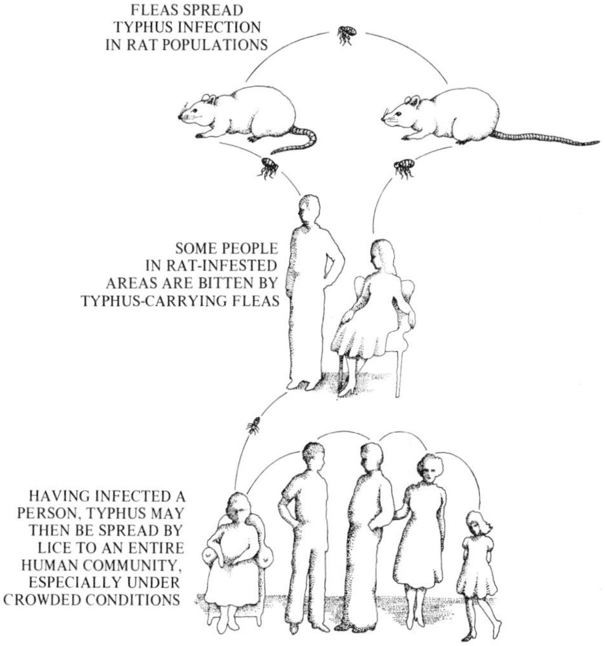
FLEAS SPREAD
TYPHUS INFECTION
IN RAT POPULATIONS
SOME PEOPLE
IN RAT-INFESTED
AREAS ARE BITTEN BY
TYPHUS-CARRYING FLEAS
HAVING INFECTED A
PERSON, TYPHUS MAY
THEN BE SPREAD BY
LICE TO AN ENTIRE
HUMAN COMMUNITY,
ESPECIALLY UNDER
CROWDED CONDITIONS

(b)

Widespread treatment with antibiotics has led to the production of resistant strains in many pathogens, and this has come to constitute a very serious public health problem. The overpromotion of antibiotics by drug companies and their subsequent overprescription by some physicians contribute severely to this problem; antibiotics should never be prescribed for viral diseases, such as colds, for which they are totally ineffective. In addition, the wholesale addition of antibiotics to animal feeds to increase yield has contributed substantially to the reservoir of drug-resistant bacteria in nature. In the Enterobacteriaceae, a family of bacteria that contains many important pathogens, the phenomenon of conjugation, discussed on page 188, leads to the transfer of resistant genes from genus to genus, and to their persistence in populations of nonpathogenic bacteria and ultimate transfer to pathogenic ones when they come into contact. Viruses, too, play a role in the transfer of immunity to drugs among bacteria, partly by transduction and partly by phenomena that will be discussed later in the chapter.

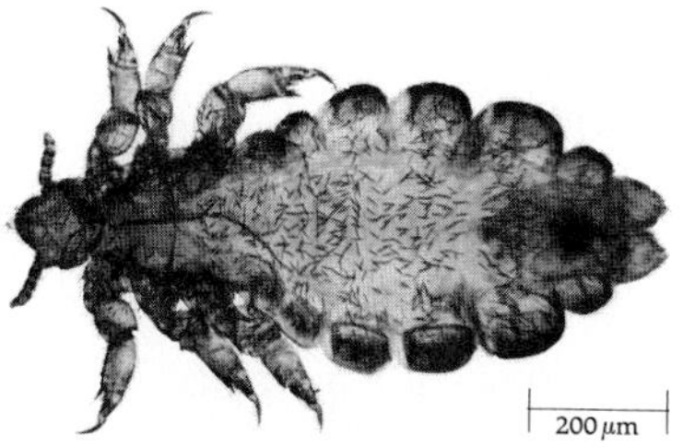
200 μm

(c)

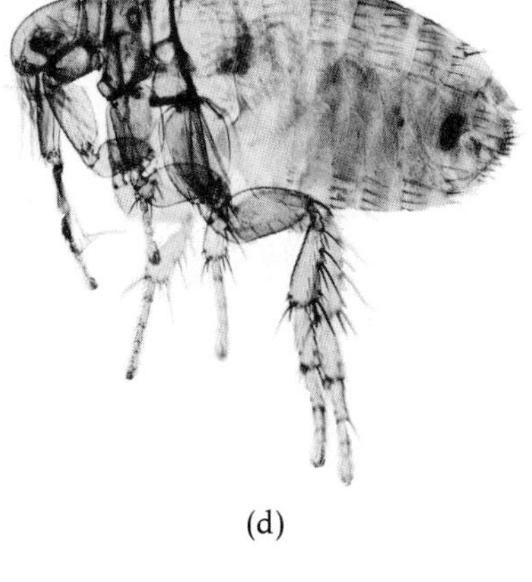

(d)

10-23
(a) *Diagram of a cell of PPLO (pleuropneumonialike organism), also known as* Mycoplasma. *These are essentially very small bacteria, only recently and with difficulty grown in cell-free culture, and include the smallest known cellular organisms, some only 0.3 micrometer in diameter. Long known to cause respiratory diseases in man and other animals, in 1967 PPLO was discovered to play an important role as a plant pathogen. One such plant disease is aster yellows. In* (b), *which shows three Chinese asters,* Callistephus chinensis, *the plant on the left is healthy, that in the center has diseased leaves, and the one on the right is severely infected.* (c) *The six-spotted aster leafhopper,* Macrosteles fascinfrons, *is the most common vector of aster yellow.* (d) *The electron micrograph shows PPLO in a cell from an aster stalk. Cell at right is free from infection.* (e) *PPLO in the salivary gland of* Macrosteles. *More than 50 plant diseases, many involving yellowing and stunting, have been traced to mycoplasmas, which are usually found in the sieve tubes of the phloem. One of the most important is lethal yellowing of coconut palms, which causes the loss of hundreds of thousands of trees annually; another is threatening the lavender perfume industry in southern France.*

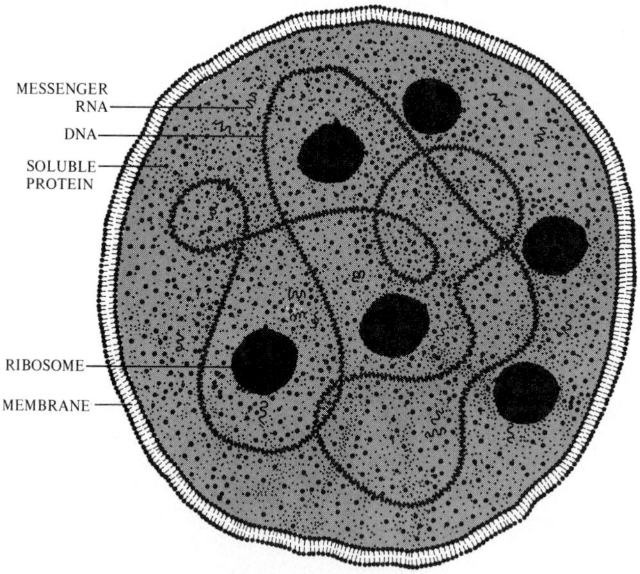

(a)

(b)

(c)

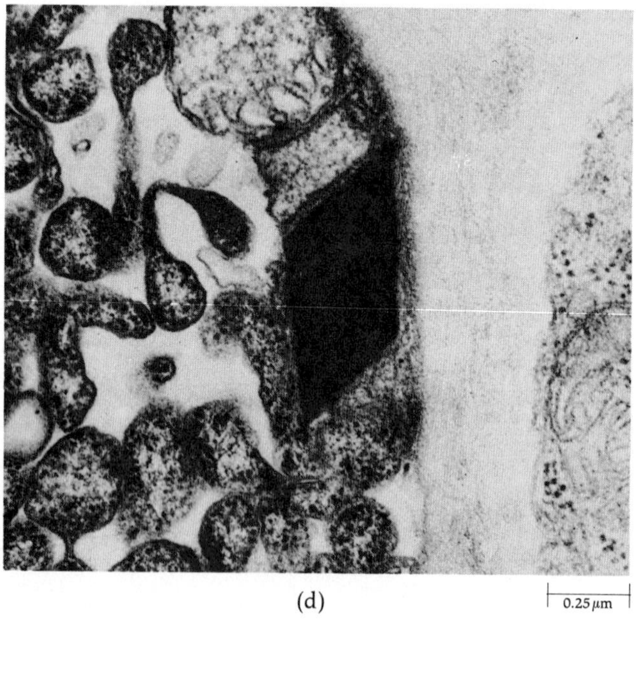

(d)

0.25 μm

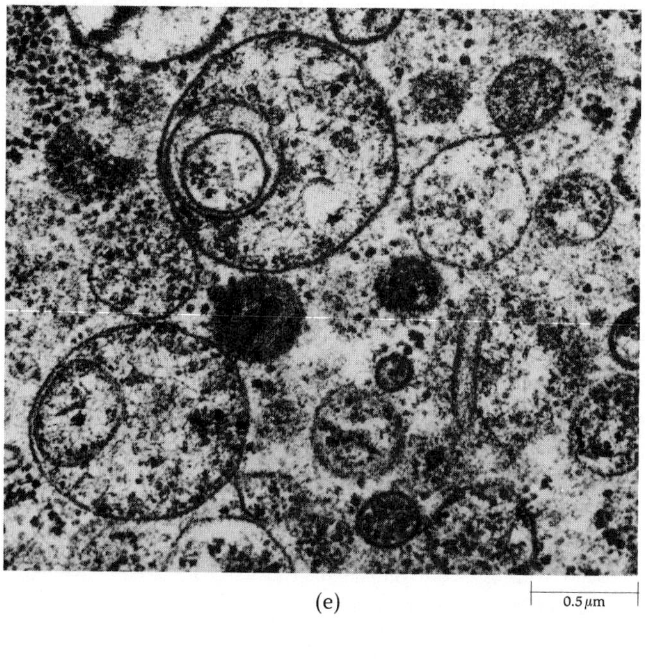

(e)

0.5 μm

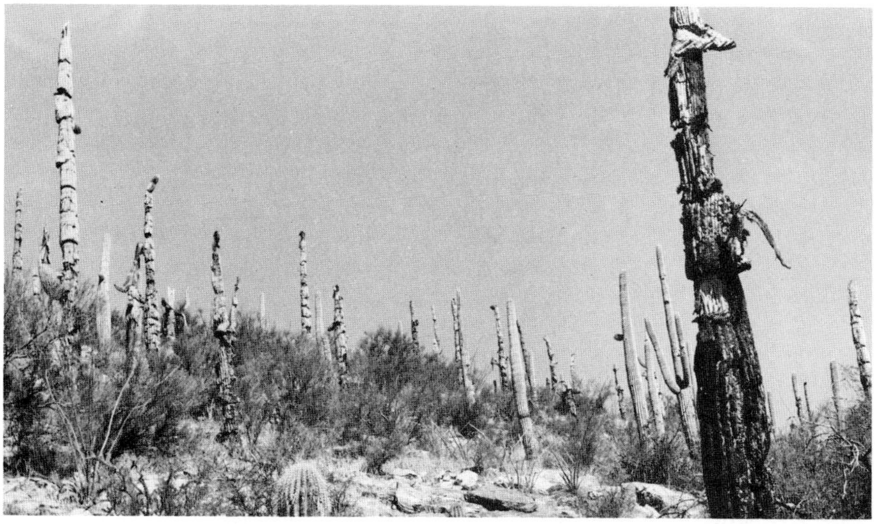

10–24
This population of the giant saguaro cactus (Carnegiea gigantea) *at an elevation of 1000 meters on the south slope of the Santa Catalina Mountains near Tucson, Arizona, was attacked by bacteria after severe freezing conditions in January 1962 weakened the cacti. In this way the bacterium is controlling the population of the giant cactus at the margins of its range.*

Many economically important diseases of plants are also associated with bacteria (Figures 10–23 and 10–24). Some of the more important diseases of plants are called soft rots, blights, or wilts. Sometimes death is associated with an invasion and plugging of the xylem by large numbers of bacteria and the slime they produce. In other instances, the diseases are associated with the production of toxic substances, as in many of the diseases of man and animals.

THE BLUE-GREEN ALGAE

The blue-green algae are photosynthetic prokaryotes organized much like other bacteria. All blue-green algae contain chlorophyll *a*, which is also found in all photosynthetic eukaryotes. They have several kinds of accessory pigments, including several kinds of carotenoids that may also be found in photosynthetic eukaryotes and some of the other bacteria. In addition, the cells of blue-green algae may contain two phycobilins: phycocyanin, a blue pigment, which is always present, and phycoerythrin, a red one, which is sometimes absent (Figure 10–25). Chlorophyll and the accessory pigments do not occur in plastids but are scattered in a membranous system in the peripheral portion of the cell.

The blue-green algae have a cell wall that does not contain cellulose but is made up of the same sort of polysaccharides linked with polypeptides that occur in other bacteria. They also contain lipopolysaccharides, and are thus gram-negative. There seem to be four distinct wall layers, the innermost of which is the rigid glycopeptide layer, just as in other bacteria. The chief carbohydrate storage product of the blue-green algae is a polysaccharide known as cyanophycean starch, which is probably identical with glycogen (see page 50). They

10–25
The two phycobilins known to occur in blue-green algae. These molecules always occur complexed with proteins and assume different properties according to the proteins with which they are associated. These phycobilins also occur in the eukaryotic red algae. [M stands for methyl; E for ethyl; P for propionic acid residue; V for vinyl (—CH=CH$_2$).]

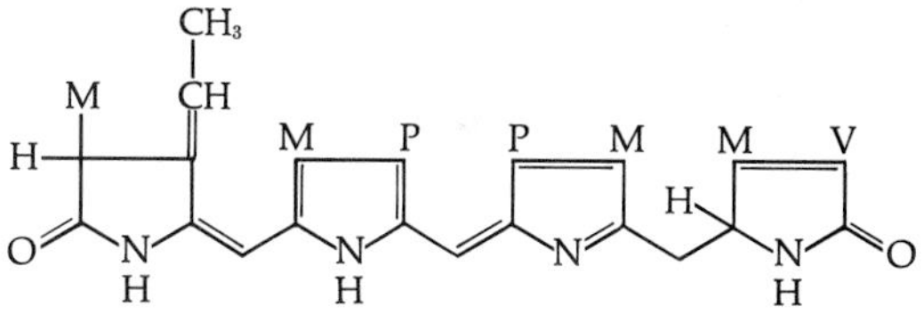

PHYCOERYTHRIN

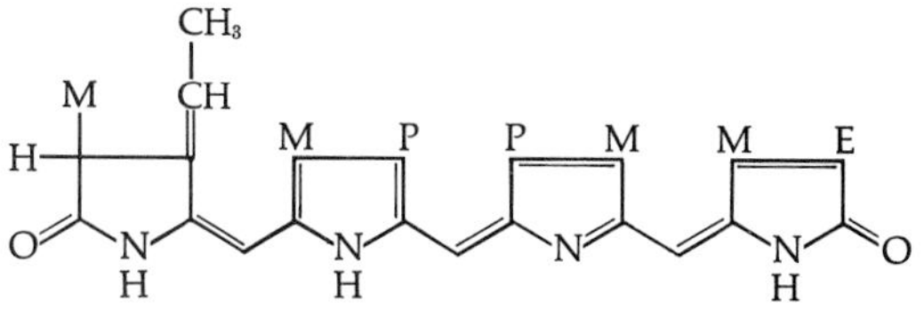

PHYCOCYANIN

also accumulate and store lipids and proteins. The color and overall morphology of the blue-green algae led to their historical grouping with the algae, but in terms of their cellular organization they are best regarded as a specialized group of bacteria, which the American bacteriologist Roger Y. Stanier has termed the Cyanobacteria (from the Greek *kyanos,* blue). The existence of forms such as *Beggiatoa* (see Figure 10–5b), which are intermediate between other bacteria and the blue-greens, strengthens the connection between them.

All blue-green algae are microscopic as individuals, but they often grow in large masses as much as 1 meter or more in length. Some blue-green algae are unicellular (Figure 10–26), others are filamentous, a few form branched filaments, and a very few form plates or irregular colonies (Figure 10–27). Any cell of a blue-green alga may divide, and the resulting subunits may fall away to form new colonies. As in other "multicellular" bacteria, the cells are attached to one another only by their outer walls or gelatinous sheaths, and each cell leads an independent life (Figures 10–28 and 10–29). Not all blue-green algae have a sheath. When present, the mucilaginous sheath or coating is often deeply pigmented, particularly in species that spread up onto the land, and their colors include a light golden yellow, brown, red, emerald green, blue, violet, and blue-black. In addition, the carotenoids and phycobilins modify the color of the cells in which they occur. Despite their name, only about half of the blue-green algae are actually blue-green in color. Indeed, the Red Sea was named because of the dense concentrations or "blooms" of the marine planktonic blue-green alga *Trichodesmium* that frequently occur in it. Similar blooms often occur in polluted fresh water as the result of a concentration of many different species in this group, some of which produce toxins active against fish and mammals.

The cells lack cilia, flagella, or any other type of locomotive organelle, yet some filamentous blue-green algae are capable of motion. This may consist of simple gliding or may be combined with rotation around a longitudinal axis. Short segments that break off from the parent may glide away to a new site at rates as rapid as 10 micrometers per second. Movement may be connected with the extrusion of mucilage through small pores in the cell wall, together with the production of contractile waves in one of the surface layers of the wall. Unicellular blue-green algae may also be capable of intermittent, jerky movements.

Reproduction in unicellular blue-green algae is by cell division (Figure 10–30). Colonial and filamentous forms exhibit several types of fragmentation. Multicellular fragments are termed hormogonia (singular: hormogonium). Some filamentous genera, especially *Nostoc* and *Anabaena,* are capable of forming heterocysts, enlarged cells with a multilayered wall in which the thylakoids are reorganized into a concentric or reticulate pattern. In other circumstances, the vegetative cells of these genera may become transformed into thick-walled spores called *akinetes,* which accumulate proteinaceous cyanophycin granules. The thylakoids in akinetes retain the same arrangement as they have in vegetative cells. Akinetes are highly resistant to adverse environments and have

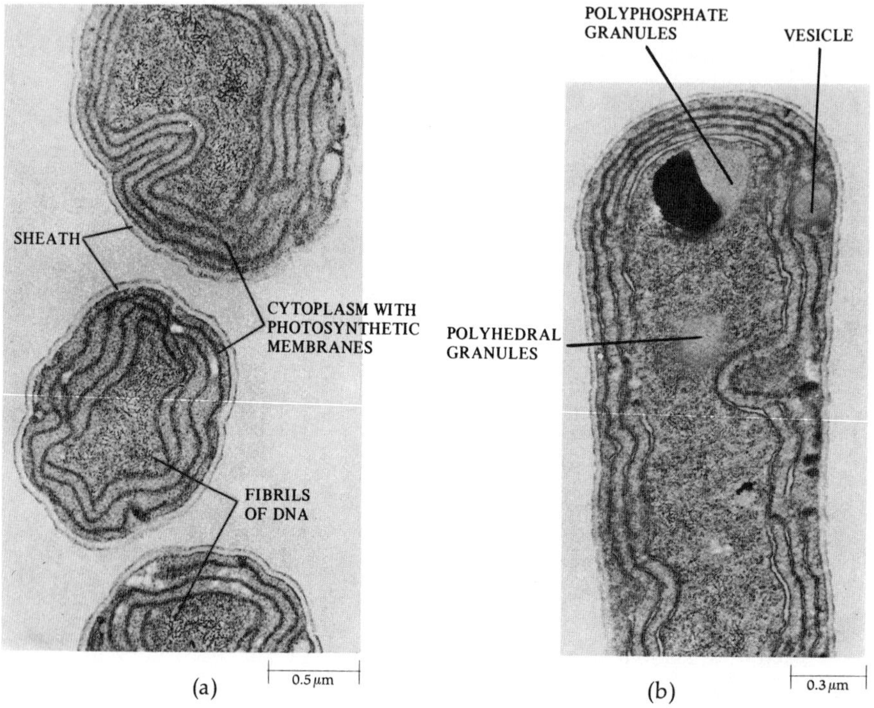

10–26
Synechococcus, *a blue-green alga in which the cells do not adhere to form filaments.* (a) *Electron micrograph of fixed cells showing sheath, cytoplasm with photosynthetic membranes, and central area with fibrils of DNA.* (b) *Electron micrograph of median section showing three types of storage bodies: polyphosphate granules, polyhedral granules, and a vesicle.*

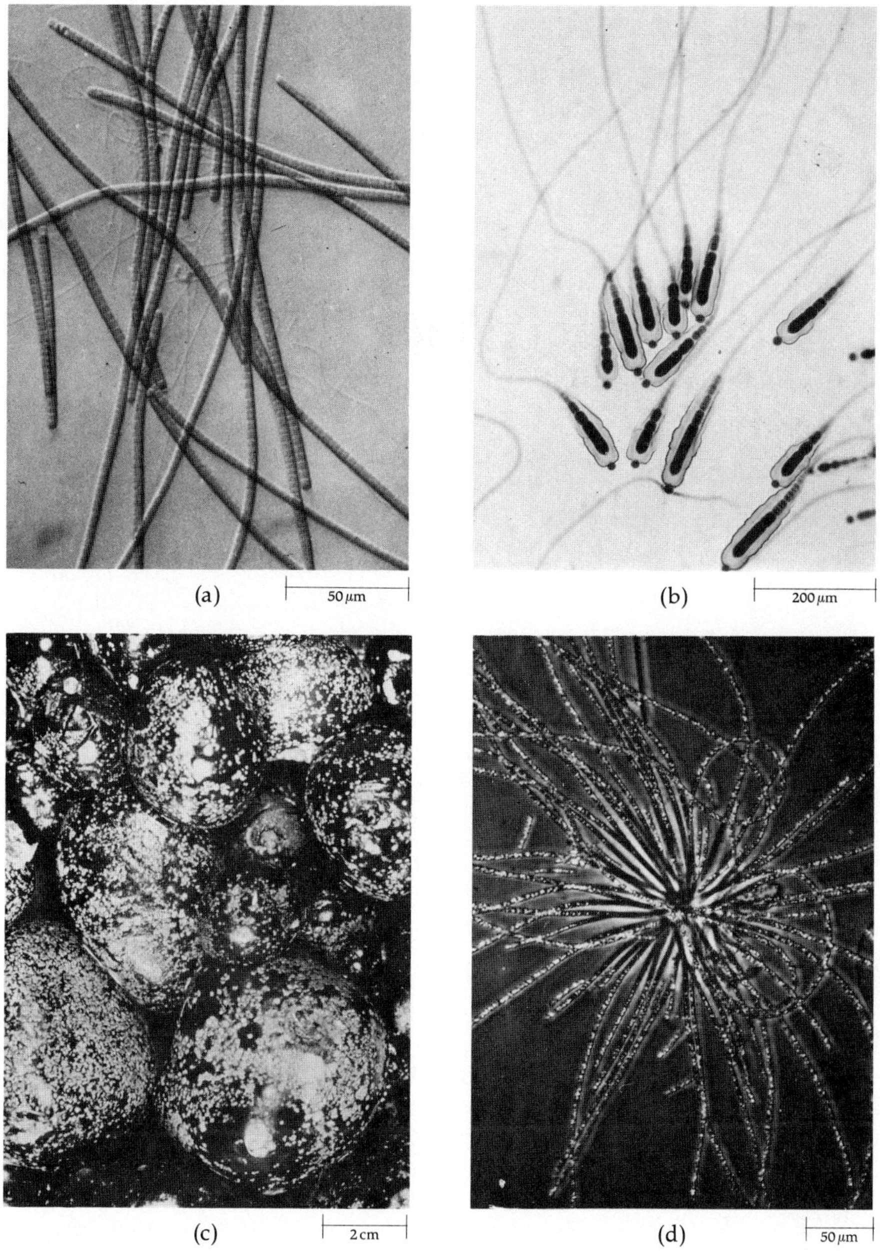

10–27
Four common genera of blue-green algae: (a) Oscillatoria, *a filamentous form, in which the only form of reproduction is the breaking off of portions of the filament;* (b) Gloeotrichia, *a filamentous form with a basal heterocyst. Unlike* Oscillatoria, Gloeotrichia *is capable of forming akinetes, which form just above the heterocysts.* (c) *Gelatinous "balls" of* Nostoc commune, *each containing hundreds of filaments. These "balls" occur frequently in freshwater habitats.* (d) Thiothrix, *a genus that lacks chlorophyll. This species obtains energy by the oxidation of H_2S. The filaments, which are attached to the substrate at the base and so form a characteristic rosette, are filled with sulfur droplets. (a, Eric V. Gravé)*

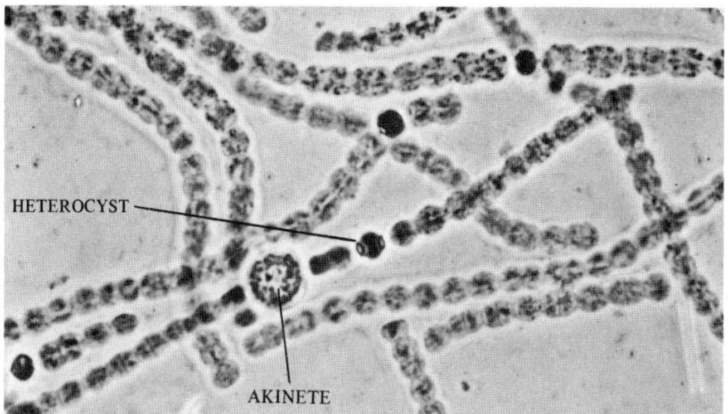

10–28
Anabaena, *a nitrogen-fixing blue-green alga, composed of barrel-shaped cells held in a gelatinous matrix. Like* Gloeotrichia, Anabaena *forms akinetes. Electron micrographs of* Anabaena *can be seen in Figures 10–29 and 10–30.*

10–29
Cell of the blue-green alga Anabaena azollae *showing major features visible with the electron microscope. The gelatinous sheath of this cell has been destroyed in preparing the specimen for electron microscopy. The numerous, very small granules are ribosomes.*

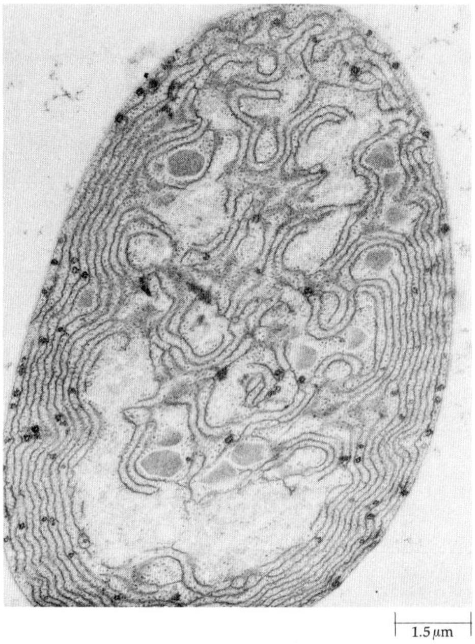

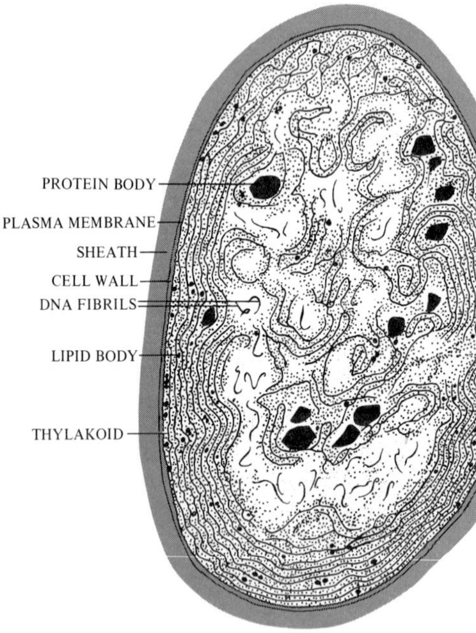

EVOLUTION OF PHOTOSYNTHESIS

The oldest known sedimentary rocks, from Western Greenland, are about 3.76 billion years old. The oldest fossil-bearing ones, however, are from 3 to 3.36 billion years old and occur in South Africa. In these rocks are found bacteria and organic spheroidal bodies that are similar to blue-green algae; they occur in strata about 3.2 billion years old. Here also is found chemical evidence that suggests that photosynthesis was taking place some 3.3 billion years ago.

There are two kinds of chemical evidence. First, modern plants, in the process of photosynthesis, selectively accumulate ^{12}C *in preference to its heavier isotope* ^{13}C*. Thus the ratio of* ^{12}C *to* ^{13}C *is higher in organic material that accumulated as a result of photosynthesis than in organic material formed in other ways. Such enrichment is not found in South African rocks 3.36 billion years old, but appears dramatically in those about 3.3 billion years old and is characteristic of those younger, thus indicating a possible date for the initiation of photosynthesis. The second kind of chemical evidence is that compounds have been found in the rock that are probably breakdown products of the chlorophyll molecule itself.*

Further evidence of the early occurrence of photosynthesis is provided by the accumulations of calcium carbonate similar to those produced by modern blue-green algae in limestone in Rhodesia, some 2.7 billion years old. The blue-green algae are essentially a specialized group of bacteria that carries out photosynthesis utilizing chlorophyll a *and evolving* O_2*, just as do eukaryotic algae and plants. They are the only group of prokaryotes alive today that carries out photosynthesis in this fashion.*

germinated to produce new individuals after as many as 87 years.

Genetic recombination, apparently similar in mechanism to that in other bacteria, occurs in blue-green algae, but its extent remains to be demonstrated. In some instances, transformation by the incorporation of exogenous DNA has been demonstrated.

Distribution and Classification

Owing to their wide geographical ranges and extreme variability of habitats, the taxonomy of the blue-green algae is extraordinarily complicated. Although over 7500 names have been proposed for species of this group, experimental work carried out in the past decade suggests that there may actually be as few as 200 nonsymbiotic distinct species. As an example, *Microcoleus vaginatus* occurs in wet soil or fresh or brackish water from northern Greenland to Antarctica, and from the floor of Death Valley to the top of Pike's Peak. Studies of the organism have revealed that under changing environmental conditions, a single colony may undergo such extreme changes that its members come to resemble

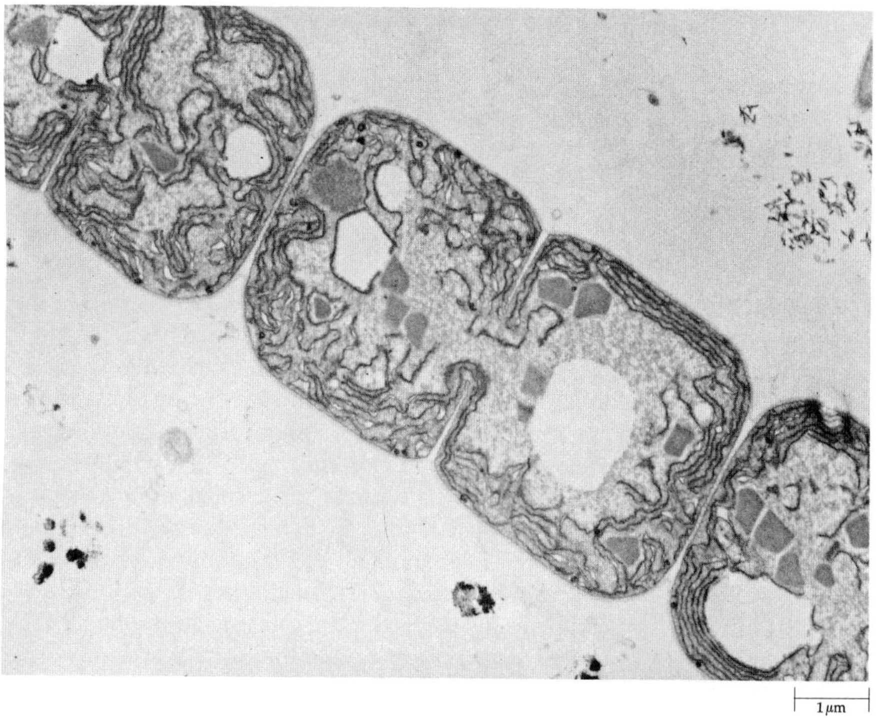

10–30
Cell division in Anabaena. *This electron micrograph shows the cell margins growing inward during the process of division. The sort of cell division shown here is characteristic of all organisms except plants and a few genera of algae, in which a cell plate is formed.*

dozens of different species. It is clear that extensive studies will be necessary before we can begin to understand the extent of the variability in these seemingly simple organisms.

Like other bacteria, the blue-green algae sometimes grow in extremely inhospitable environments, from the near-boiling water of hot springs to the frigid lakes of Antarctica, where they abound. They were the first colonists on the new island of Surtsey, near Iceland, following its eruption and appearance above the sea. On the other hand, blue-green algae are absent in acidic waters, where eukaryotic algae are abundant.

Some blue-green algae occur in marine plankton. Many marine blue-green algae are found in limestone and lime-rich substances such as coralline algae (see page 266) and the shells of mollusks. A number of fresh-water forms, particularly those that occur around hot springs, often deposit thick layers of lime in their colonies. In Yellowstone National Park, the filamentous *Mastigocladus* occurs in hot water at temperatures up to 55°C, and the unicellular *Synechococcus,* up to 73 to 75°C. In soil, including desert soils, they are abundant, with 20,000 to 50,000 individuals per gram being a representative figure.

Many species of blue-green algae can fix nitrogen, and in Southeast Asia, rice can be grown on the same land for years without the addition of fertilizers because of the rich growth of nitrogen-fixing blue-green algae in the rice paddies. Because of these nitrogen-fixing capacities, the blue-greens are able to colonize bare areas of rock and soil. Recent studies have demonstrated that the formation of heterocysts in *Nostoc* and other genera is inhibited by the presence of ammonia or nitrates, but that when these nitrogen-containing substances fall below a threshold, heterocysts begin to appear. At the same time, the activity of an enzyme, nitrogenase, which reduces nitrogen to ammonia, a form in which it can be incorporated by living organisms, becomes evident. When heterocysts are formed, they shut off the production of normal vegetative proteins and produce three new protein components that are probably constituents of nitrogenase. Once a heterocyst has started to produce nitrogenase, its synthesis cannot be shut off by adding ammonium ion to the medium. The spacing of heterocysts and the factors that control their development have been the subject of much study in recent years, and the filamentous blue-green algae *Nostoc* and *Anabaena* seem to provide simple systems for the study of development.

Blue-green algae occur commonly as symbiotes in amoebas, flagellated protozoa, some diatoms, some green algae that lack chlorophyll, other blue-green algae, some higher plants, and coenocytic fungi. When they do occur as symbiotes, they commonly lack a cell wall. Under such circumstances, they are functionally chloroplasts. In fact, in these forms, the blue-green algal cell divides at the same time as the host cell, the process being similar to chloroplast division. Their relationship to the evolution of particular groups of eukaryotic algae will be examined further in Chapter 12.

Blue-green algae are also frequent as the photosynthetic component in lichens and occur in various bryophytes and vascular plants, where they may perform a nitrogen-fixing function. In turn, they may be hosts for fungi (mainly chytrids) and viruses.

THE VIRUSES

The viruses do not fit easily into any of the traditional categories into which living organisms are classified, and the problem of categorizing them is made even more difficult by the fact that there is considerable doubt about whether or not they should be considered living. Microbiology has long been an important subspecialty of botany, however, and the viruses, together with the bacteria, have traditionally been included in the plant kingdom. A more important reason for discussing them here is the fact that they are extremely interesting organisms and represent one of the most valuable tools of modern genetics. They are also of great importance as agents of disease, being the causative agents for smallpox, chickenpox, measles, German measles, mumps, influenza, colds (often complicated by secondary infections by bacteria), infectious hepatitis, yellow fever, polio, and rabies.

Viruses are also responsible for many important diseases of domestic animals and plants. Although many types of virus disease can be prevented by immunization, once established they are relatively difficult to control, as they do not respond to antibiotics.

The Nature of Viruses

The existence of viruses was first recognized when it was found that the causative agents of certain diseases could pass through the porcelain filters commonly used to trap bacteria. In size, they range from about 17 to more than 300 nanometers. Thus viruses are comparable to molecules in size, a hydrogen atom being about 0.1 nanometer in diameter and a large protein molecule being a few hundred nanometers in its greatest dimension. Large viruses are about three times as large (0.3 micrometer) as the smallest cellular organism (0.1 micrometer).

Viruses are parasites that can multiply only within a host cell and are highly specific with regard to the type of cell in which they can multiply. In the host cell, they essentially "take over" the direction of the metabolism, using their own nucleic acids to "command" the host cytoplasm to produce more virus particles. They compete with the genetic material of the host cell, which is similar to their own, in regulating cell functions. Cold viruses multiply in the mucous membranes of the respiratory tract, breaking down tissue and producing the all-too-familiar cold symptoms. Measles viruses and other rash-causing viruses multiply in the cells of the skin. The polio virus—only about 28 nanometers in diameter—multiplies in the intestinal tract and sometimes in the nerve cells. Even bacterial cells have their own set of viral parasites; indeed, one of the techniques for rapid identification of unknown bacteria is to expose them to a spectrum of known bacterial viruses—the bacteriophages—and see which type destroys them.

Plant cells that are completely free of viruses may actually be exceptional, and the implications of this statement for agriculture may be profound. For example, when virus-free strains of rhubarb were developed in Britain by culturing groups of cells from the apical meristems, their yield was 60 to 90 percent greater than those with the "normal" virus infections. Many plants may have chronic infections of viruses, which lower their general vigor but only rarely become acute. A number of cultivated plants with variegated foliage owe this characteristic to a viral infection. It has been suggested that every species of organism, including the prokaryotes, may have at least one specific virus associated with it, in which case there may be literally millions of species of viruses in existence.

Until the 1930s, viruses were considered to be extremely small bacteria. Evidence against this point of view began to accumulate in 1933, when Wendell Stanley prepared from infected plants an extract of a common virus, the tobacco mosaic virus, and purified it. The purified virus precipitated in the form of crystals. Crystallization is one of the chief tests for the presence of a single, uncontaminated chemical compound, and so, clearly, viruses were not composed of the complex variety of organic compounds that characterizes even so small a living thing as a bacterial cell. But when these needlelike crystals were put back into the solution and reapplied to a tobacco leaf, the infection characteristic of tobacco mosaic virus was produced.

The tobacco mosaic virus was subsequently identified as a large nucleoprotein (about 300 nanometers long), that is, a protein in combination with a nucleic acid (Figure 10–31). In the case of tobacco mosaic viruses, as well as some others, the nucleic acid, which is the genetic material, is RNA instead of DNA. In other viruses, DNA serves as the genetic material, again combined with a protein. Viruses are the only organisms that do not contain both DNA and RNA.

The Structure of Viruses

By the use of electron microscopy, the structure of a large number of viruses has now been elucidated. The most common architectural arrangement is an icosahedron, a 20-sided figure, which, as geometricians have calculated, is the most efficient symmetrical arrangement that subunits can be in to form an outer shell with maximum internal capacity. (The geodesic domes of R. Buckminster Fuller are constructed along the same principles.) All viruses previously thought to be spherical have been shown to be icosahedral. Among the icosahedral viruses are those that cause colds (Figure 10–32), polio, chickenpox, fever blisters, human warts, and many types of cancers in mice and other animals.

The bacteriophages used in the genetic studies described in Chapter 3, although only about 100 nano-

10–31
Diagram of tobacco mosaic virus (TMV) representing about half of its total length. This virus has a central core of RNA and a protein coat composed of 2200 identical protein molecules, each containing 158 amino acid residues, folded into a spindle-shaped subunit. At the narrower end of the "spindle" is a groove into which the RNA fits. Before the structure of any virus was known, Crick and Watson predicted that the protein coats of viruses could prove to be made up of a large number of identical subunits. Can you explain the basis of their prediction?

10–32
(a) *Adenovirus, one of the many viruses that cause colds in humans. This virus is an icosahedron. Each of its 20 sides is an equilateral triangle made up of protein subunits. There are 252 subunits in all.* (b) *A model of the adenovirus, made up of 252 tennis balls.*

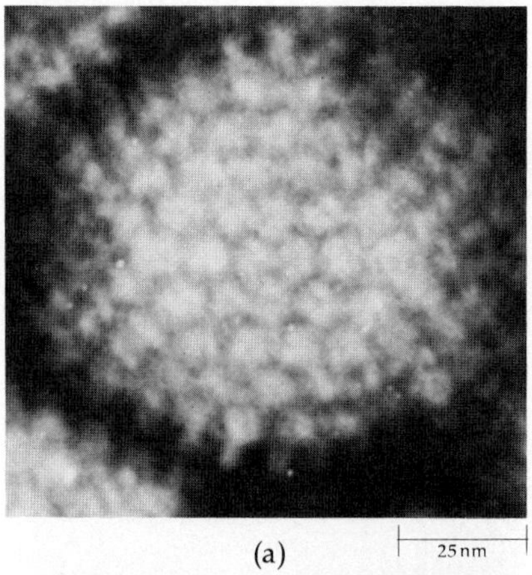

(a)

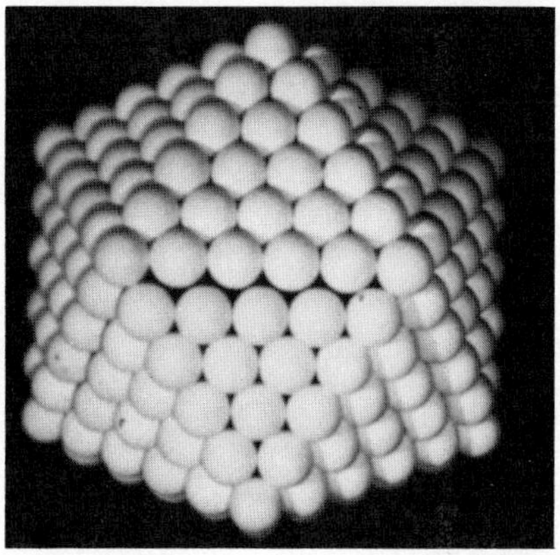

(b)

meters long, are more complex in structure (see Figures 10–1 and 10–33). Each particle is fashioned of at least five separate proteins: the repeating units that make up the hexagonal head, the tail core, the submolecules of the contractile sheath, the base plate of the tail, and the tail fibers. The DNA is coiled within the hexagonal head of the bacteriophage. At least one additional protein is produced; this is the enzyme lysozyme which causes the breakdown (lysis) of the bacterial cell at the end of the infectious cycle. Clear spots (plaques) appear in the bacterial colonies in areas where active lysis by phages is occurring.

Some of the larger viruses, such as those that cause smallpox and influenza, have outer envelopes that appear to be composed of portions of the host plasma membrane carried off by the virus particle on the completion of infection.

10–33
(a) *T4 bacteriophage as shown in an electron micrograph. The T (for type) bacteriophages are a group of tadpole-shaped viruses that attack* Escherichia coli. *They have been essential tools in the study of molecular genetics. More complex than most viruses, the T4 bacteriophage is made up of several different proteins. The head, in which the DNA is enclosed, is composed of repeating identical protein subunits, like the coat of TMV. The tail is a hollow core encased in a contractile sheath containing molecules of ATP that provide the energy for the contraction. At the top of the tail is a base plate, which looks like a six-pointed star. From the six points radiate six long fibers which apparently fold back along the sheath. The tail fibers attach to the bacterial cell wall, drawing the base plate to it. The tail protein then contracts, forcing the hollow tail core through the cell wall and plasma membrane like a microsyringe. The DNA molecule, which is some 650 times longer than the protein head in which it is contained, then passes into the bacterial cell, leaving the protein coat on the outside.* (b) *Model of a T4 bacteriophage.*

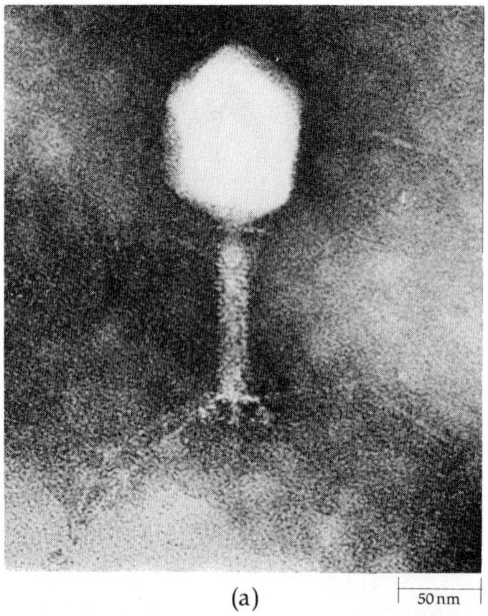

(a)

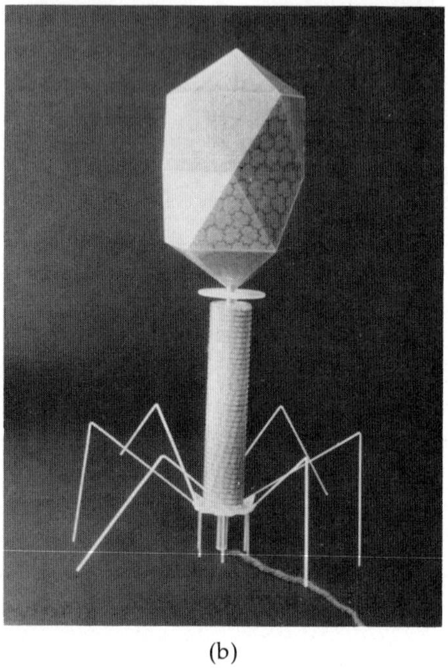
(b)

The Replication of Viruses

The mode of replication of particular viruses depends on their genetic constitution. The protein coat of viruses determines their attachment to host plasma membrane and their entry into the cell. But all viruses shed this coat before they begin to replicate themselves. In some, such as the bacteriophages, it is left outside the host cell; in some, it is shed within; and in others, it is digested by the enzymes of the host cell. When this has been accomplished, one of two things happens.

1. A reaction may occur within the cell that prevents virus multiplication, and the viral DNA may be inserted in a linear fashion into the bacterial chromosome. In such a state, the virus is called a *prophage*. Phages that are capable of existing in prophage form are called *temperate* phages. The viruses involved in transduction are temperate phages (see Figure 10–17). Temperate phages do not destroy their host cells unless they "escape" from the host chromosome, and they are in turn virtually unassailable by the host's immune defense systems. This phenomenon has been demonstrable only in DNA bacteriophages.
2. The virus may multiply. Virus multiplication takes place in three steps. First, the virus nucleic acids direct the host cells to produce new viral enzymes. Viral nucleic acids and structural proteins are then synthesized, each in its appropriate amount. Finally, these materials are assembled into virus particles. These steps generally overlap in time, often involve extensive genetic regulation, and lead to the production of many—often thousands—of new virus particles per cell. When the process of viral multiplication is complete, the particles escape from the host cell, which is generally dead by that time.

In viruses that contain DNA—such as the vaccinia virus, an organism that causes a poxlike disease in cat-

VIRAL INFECTIONS

The influenza virus (A_2, Hong Kong 1/68) mutates rapidly (a). *Changes in its genetic material (which is RNA) result in changes in its protein coat. Because immunity to a virus is, in effect, immunity to the specific proteins of its protective coat, these new viral strains are able to infect previously immune populations. The virus is surrounded by a lipoprotein envelope through which protrude stubby protein spikes. Flu epidemics of influenza strains A and B are shown in* (b). *The length of time between epidemics presumably reflects the time required for a new mutant to become established. Influenza is the only infectious disease that appears periodically in life-threatening global epidemics; in the winter of 1968–1969, more than 51 million cases of Hong Kong flu were reported in the United States alone, with at least 20,000 and perhaps as many as 80,000 deaths which could be attributed to this cause.*

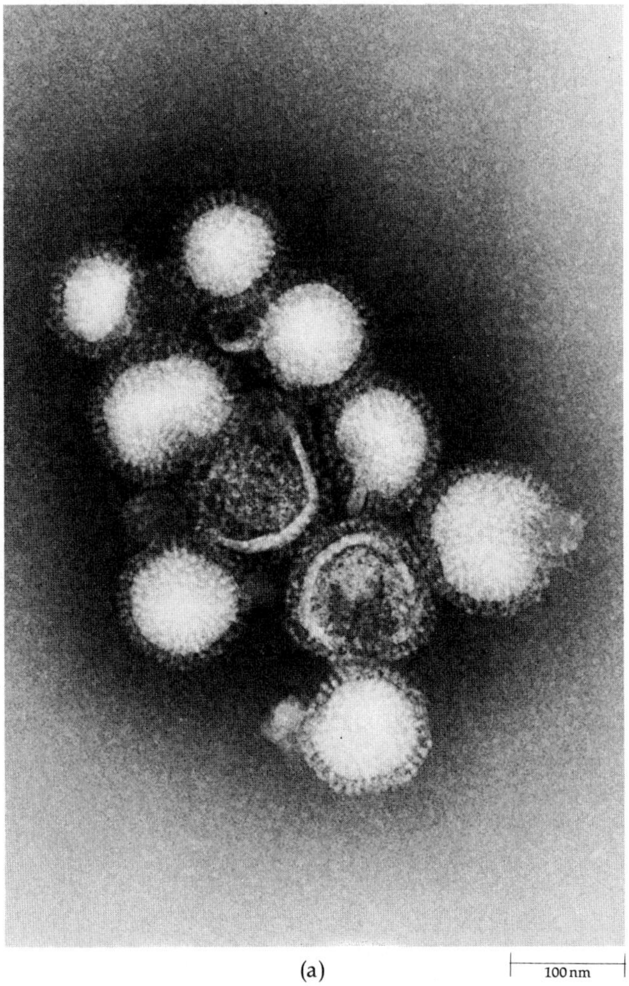

(a)

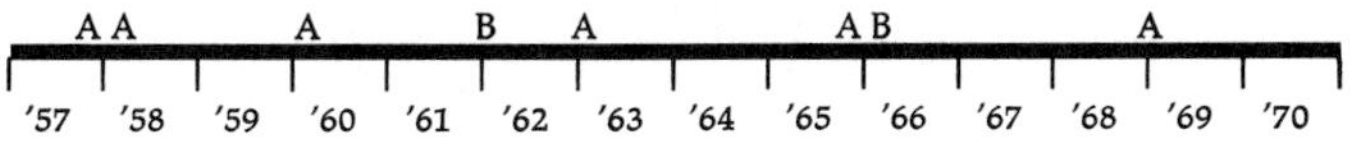

(b)

tle—the viral DNA simply directs the synthesis of a series of different messenger RNA molecules which direct the production of different proteins. The DNA is usually double-stranded, but single-stranded DNA occurs in some very small bacterial viruses.

In most RNA viruses, such as the tobacco mosaic virus, the RNA is single-stranded. This viral RNA replicates itself, presumably by directing the formation of a complementary strand which then serves as the template for new viral RNA molecules. The viral RNA also takes over the ribosomes of the host cell and acts as messenger RNA. In this role, it is responsible for the synthesis of enzymes and virus coat proteins.

Viral activity may profoundly affect the metabolism of the host cell. In the bacterium *Clostridium*, it has been shown for some strains that the production of the lethal toxins associated with botulism takes place only with the active and continued participation of specific bacteriophages. Noninfected bacterial cells do not produce the toxin. Even more surprisingly, infection by other specific bacteriophages causes the same bacterial strain to produce the toxins associated with gas gangrene and many other diseases in animals. The causative organisms associated with botulism and gas gangrene had hitherto been considered to be different species of *Clostridium*, but they are, at least in part, the same species infected by different bacteriophages.

Viruses and Cancer

It is well known that viruses cause cancers and cancer-like growths in many animals and plants (Figure 10–34), and the possibility that viruses are a cause of cancer in humans has been much discussed. Viruses can change from infectious to noninfectious forms, and they mutate readily to produce strains with new sets of characteristics. Even though it has not been possible to isolate cancer-causing viruses from any human cancer, such viruses could be present in very small numbers and still produce the metabolic disorders and consequent rampant cell division characteristic of cancer. They could also be present in prophage form, in which case they would be essentially undetectable. Viruses affect their hosts differently depending on the genetics of the host and its physiological condition, and the presence of one kind of virus in a cell sometimes enhances the effect of another. Physical or chemical injury or x-irradiation can also play a role in activating certain viruses, which suggests that known cancer-causing substances, such as tobacco tar or radioactive materials, may exert their effects by activating a virus. The fact that there are many different forms of cancer, which may have different causes, makes it difficult to prove conclusively whether or not viruses are a cause of human cancer, but this subject is under active study in research centers all over the world.

10-34
(a) *Tumors produced by the wound tumor virus in sweet clover* (Melilotus alba). (b) *The clover leafhopper* (Agallia constricta)—*male, female, and nymph—a vector of wound tumor virus. Electron micrographs of wound tumor virus in* (c) *the host plant, and* (d) *an epidermal cell of the clover leafhopper. The arrows in* (c) *point to virus particles. Viruses are being produced in the honeycomb-like areas at the upper right of* (d). *Individual viruses can be seen in the dark area below. Many other viruses cause diseases in plants, often associated with the phloem, in which various other growth abnormalities occur, often with serious economic loss.*

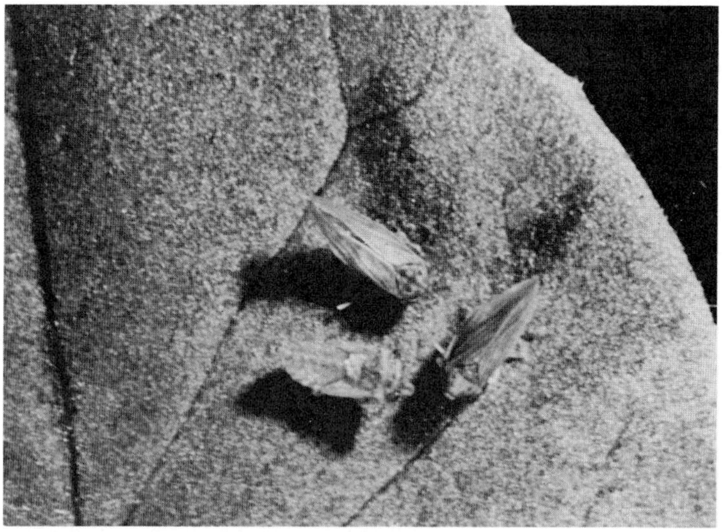

(b)

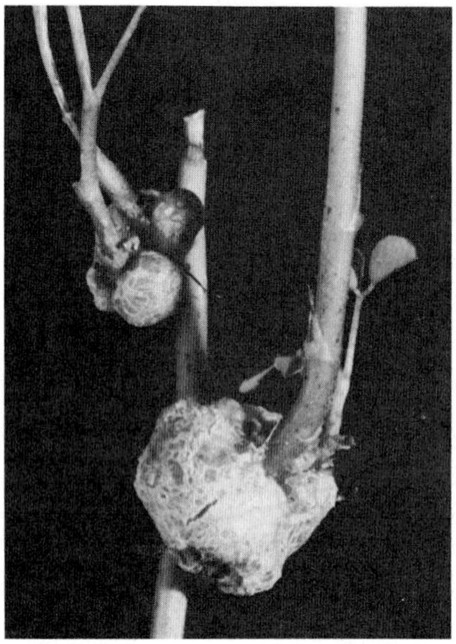

(a)

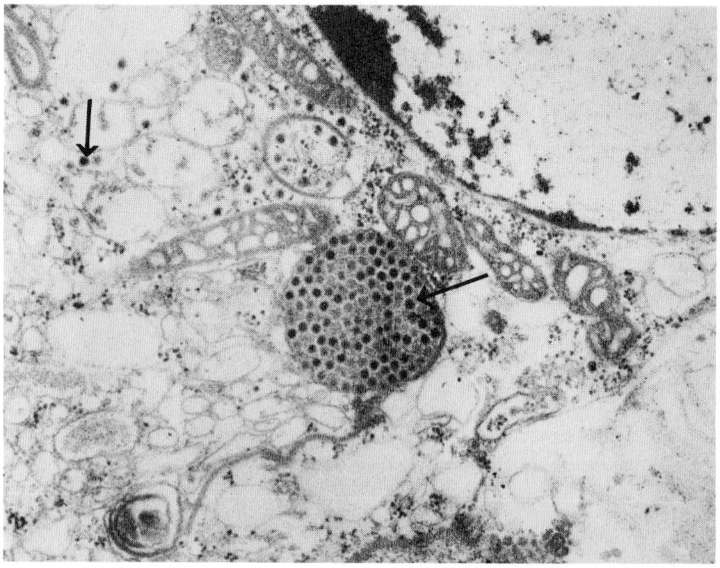

(c)

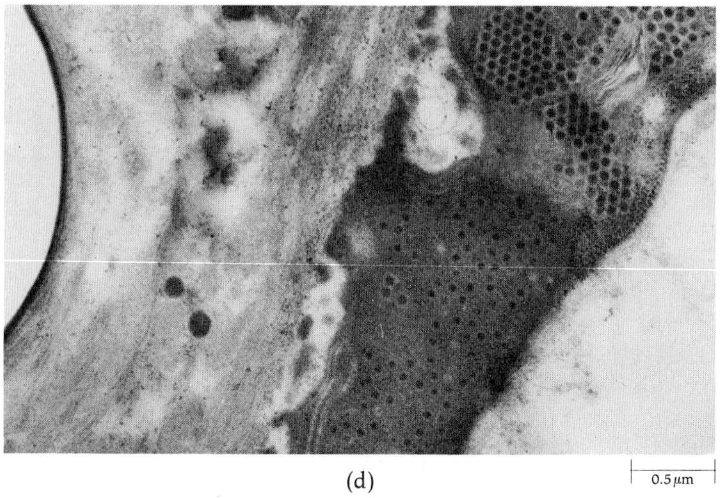

(d)

The Origin of Viruses

Because of the simplicity of viral structure, some workers have suggested that viruses represent the direct descendants of the first self-replicating units from which the earliest cells eventually evolved. This is clearly not so, for viruses exist today only by virtue of their ability to insert themselves into the machinery of their host cells. They compete with the nucleic acids of the host cells and take over the genetic and metabolic activities of the cells in directing the formation of new virus particles. They must have come into being after cells in which the genetic code was already established had evolved.

Viruses are essentially similar to bacterial chromosomes (double-stranded DNA viruses) or molecules of messenger RNA (single- and double-stranded RNA viruses) packaged in a protein overcoat. The phenomenon of transformation, discussed earlier, suggests the likely mode of origin of viruses. Fragments of DNA or RNA occasionally find their way into cells in one way or another. When they do, they may exert an effect upon the genetic processes of the host cell. If they can replicate themselves and spread from cell to cell, they may persist. If they can produce a protective protein wall, they become, in effect, viruses. It is also possible that the largest and most complex viruses, the pox viruses, may have evolved from degenerate bacteria.

Doubtless viruses have evolved on many different occasions over the past 3 billion years. In fact, current evidence seems to suggest that they are still evolving today.

SUMMARY

Bacteria, including blue-green algae, comprise the living prokaryotic organisms. Prokaryotes lack an organized nucleus and cellular organelles and do not reproduce sexually. Their genetic material is incorporated into a single circular molecule of DNA. Prokaryotes have rigid cell walls and are the only organisms in which polypeptides are incorporated into the basic structure of the cell wall. No prokaryote is truly multicellular; although the cells may not divide completely, and thus form filaments or masses, there are no cytoplasmic connections between them.

The bacteria share with the fungi the role of decomposers in the world ecosystem. They are versatile metabolically: most bacteria are heterotrophic, but some are photosynthetic, and others are chemoautotrophic. Some are aerobic, others are obligate anaerobes, and still others are facultative anaerobes. A number of genera play important roles in the cycling of nitrogen, sulfur, and carbon.

Most photosynthetic bacteria, which are anaerobic, utilize hydrogen sulfide or some other hydrogen donor instead of water, and do not liberate oxygen. The chemosynthetic autotrophs derive their energy from the oxidation of inorganic molecules. Those that oxidize proteins and amino acids convert ammonium into nitrates, thus playing an important role in the nitrogen cycle. Many bacteria are important as pathogens.

Bacteria have a rigid cell wall made up of a polysaccharide connected by polypeptide crosslinks. In some groups, large molecules of lipopolysaccharide overlie this basic structure; these cells do not stain with gentian violet and are considered gram-negative, in contrast with the members of the first group, which do stain and are considered gram-positive. Bacterial cells may be spherical (cocci), rod-shaped (bacilli), or spiral (spirilli). They may adhere in groups or filamentous or solid masses if the cell wall does not divide completely, and such forms are given special names. They may have flagella and be motile, or have shorter structures known as pili; the flagella may be located all over the cell or restricted to one or both ends.

Genetic recombination in bacteria involves the transfer of DNA from one cell to another. This may come about by conjugation, by transformation (passive incorporation of fragments of DNA in the medium), or by transduction (injection by bacteriophage). However, mutation, combined with a high reproductive rate, is a much more important source of variability in the prokaryotes than recombination.

The blue-green algae are photosynthetic prokaryotes in which chlorophyll *a* is the photosynthetic pigment just as in the photosynthetic eukaryotes. They also have accessory pigments in the form of carotenoids and phycobilins which aid in photosynthesis. In various combinations, these pigments give the blue-green algae their characteristic colors. Their cell walls resemble those of other bacteria, but they lack pili or flagella. In addition, they are able to adapt to inhospitable environments, and some forms are able to fix nitrogen. A few blue-green algae are anaerobes. They characteristically store their food reserves as glycogen.

Viruses are infectious agents composed of an inner core of nucleic acid, either RNA or DNA, and an outer protective coating of protein. They cannot reproduce themselves outside of living cells. In DNA viruses, the viral DNA competes with the DNA of the host cell in directing its activities. In RNA viruses, the RNA, which is usually single-stranded, acts as messenger RNA in the host cell, becoming associated with the ribosomes and serving as a template for the synthesis of proteins. Each different type of virus has a highly specific architecture, one of the most common arrangements being the icosahedron.

SUGGESTIONS FOR FURTHER READING

AGRIOS, G. N.: *Plant Pathology*, Academic Press, Inc., New York, 1969.

A modern discussion of the field, including many detailed drawings of the disease cycles, which is based on a biochemical approach to the problems of parasitism.

ALEXANDER, M.: *Microbial Ecology*, John Wiley & Sons, Inc., New York, 1971.

A basic text on microbial ecology, straightforward in presentation but not overly simplistic.

BROCK, T. D.: *Biology of Microorganisms*, Prentice-Hall, Inc., Englewood Cliffs, N.J., 1970.

An interesting and often entertaining presentation of microbiology, including algae and protozoa, with the slant toward the whole cell and its ecology.

BUCHANAN, R. E., and N. E. GIBBONS (eds.): *Bergey's Manual of Determinative Bacteriology*, 8th ed., The Williams and Wilkins Co., Baltimore, 1974.

A comprehensive treatise on the identification and classification of microorganisms.

CARR, N. G., and B. A. WHITTON (eds.): *The Biology of Blue-Green Algae*, University of California Press, Berkeley, 1973.

A comprehensive text, with contributions by many experts in the field, which provides an excellent review of the current state of knowledge of these organisms.

DAVIS, B. D., et al.: *Principles of Microbiology and Immunology*, Hoeber Medical Division of Harper & Row, Publishers, Inc., N.Y., 1968.

A comprehensive account of all aspects of the biology of bacteria and viruses, biochemically oriented and written from a medical point of view.

FOGG, G. E., W. D. STEWART, P. FAY, and A. E. WALSBY: *The Blue-Green Algae*, Academic Press, London and New York, 1973.

A useful synthesis of current knowledge of these organisms, including an extensive bibliography.

GOODHEART, CLYDE R.: *An Introduction to Virology*, W. B. Saunders Company, Philadelphia, 1969.

A general account of all aspects of virology, designed as a textbook for advanced undergraduates, and particularly well illustrated.

LURIA, S. E., and J. E. DARNELL, JR.: *General Virology*, 2nd ed., John Wiley & Sons, Inc., New York, 1967.

A comprehensive account of the viruses, perhaps the best general account of the group.

PELCZAR, MICHAEL J., and ROGER D. REID: *Microbiology*, 3rd ed., McGraw-Hill Book Company, New York, 1972.

A standard text that deals with the morphology, biochemistry, and ecological role of prokaryotes, viruses, and some eukaryotes.

SMITH, PAUL F.: *The Biology of Mycoplasmas*, Academic Press, New York and London, 1971.

A comprehensive account of this fascinating group of organisms, which clearly points up their potential utility in biochemical and physiological research.

STANIER, R. Y., M. DOUDOROFF, and E. A. ADELBERG: *The Microbial World*, 2nd ed., Prentice-Hall, Inc., Englewood Cliffs, N.J., 1963.

An introduction to the biology of microorganisms, with special emphasis on the properties of bacteria.

STEVENSON, L. H., and R. R. COLWELL: *Estuarine Microbial Ecology*, University of South Carolina Press, Columbia, 1973.

A collection of papers dealing with the microbial ecology of estuaries and coastal waters.

CHAPTER 11

The Fungi

11–1
Mycelium of a Basidiomycete on a fallen tree trunk.

The fungi are as distinct from the algae, bryophytes, and vascular plants as they are from the animals. They are discussed in this text both because of their great intrinsic interest and because they have been traditionally grouped with the plants. We treat them, however, as a distinct kingdom, the Fungi, which is one of the five main groups of living organisms discussed in Chapter 9.

The fungi, together with the bacteria, are the decomposers of the biosphere (Figure 11–1), and their activities are as necessary to the continued existence of the world we know as are those of the food producers. As we have seen, decomposition by fungi and bacteria releases carbon dioxide into the atmosphere and returns nitrogenous compounds and other materials to the soil where they can be used again by green plants and eventually by animals. It has been estimated that the top 2 decimeters of fertile soil may contain nearly 5 metric tons of fungi and bacteria per hectare.

As decomposers, fungi often come into direct conflict with human interests. A fungus makes no distinction between a rotten tree that has fallen in the forest and a railroad tie; it is just as likely to attack one as the other. Equipped with a powerful arsenal of enzymes which break down organic products, fungi are often nuisances and are sometimes highly destructive. This is especially true in the tropics, because warmth and dampness promote fungal growth; it is estimated that during World War II less than 50 percent of the military supplies sent to tropical areas arrived in usable condition. Fungi attack cloth, paint, cartons, leather, waxes, jet fuel, insulation for cables and wires, photographic film, and even the coating of the lenses of optical equipment—in fact, almost any conceivable substance. Even in the temperate regions, they are the scourge of commercial food producers, growing on bread, fresh fruits, vegetables, meats, and other products.

The economic importance of fungi as commercial pests is enhanced by their ability to grow under a wide range of conditions. Thus some strains of *Cladosporium herbarum,* which attack meat in cold storage, will grow at

−6°C, and one species of *Chaetomium* has an optimum growth at 50°C and can even be grown at 60°C if transferred gradually to that temperature. Fungal spores can often survive at still more extreme temperatures.

The abilities of fungi that make them such important commercial pests may make them commercially valuable in the future. Some fungi have the biologically unusual ability to hydrolyze cellulose to form glucose, which can in turn be used for many industrial processes. *Trichoderma viride,* one of the worst pests in the South Pacific in World War II, is now being cultured by the United States Army, and plans have been made to construct a pilot plant capable of converting 90 metric tons of cellulose from such sources as newspapers and other municipal debris, straw, and peanut shells annually into glucose. Such plants could be operating on a large scale by 1980.

Many fungi have switched from their roles as decomposers to attackers of living organisms (Figure 11–2). Fungi are the most important single cause of plant diseases; well over 5000 species of fungi attack economically valuable crop and garden plants as well as many wild ones. In their attacks on living trees, fungi cause the annual loss of over 20,000 cubic meters of timber in the state of California alone. Other fungi cause serious diseases of man and domestic animals.

Some 100,000 species of fungi have been described, and it is estimated that as many as 200,000 more may be awaiting discovery by mycologists (students of the fungi). There may actually be as many species of fungi as of plants, although far fewer have been described thus far. The fungi have no direct evolutionary connection with the plants and apparently were independently derived from single-celled eukaryotes. They are extremely diverse, both metabolically and structurally, and, in general, the lines of diversity found in the fungi are markedly dissimilar to those found in any other group of organisms.

Traditionally, the slime molds and fungi are considered together. At one time these two groups of organisms, along with the bacteria, were placed in a single division, the Fungi. With the exception of some bacteria, this division consisted entirely of heterotrophic, morphologically simple organisms. With increasing realization of the fundamental differences between prokaryotic and eukaryotic organisms, biologists removed the bacteria from the division Fungi. And, because of their peculiar combination of characteristics, the slime molds have been assigned to the kingdom Protista; consequently, they will be discussed in the same chapter (12) as the algae. Thus the so-called true fungi have been elevated to the rank of kingdom, consisting of a single division, Mycota.

BIOLOGY OF THE FUNGI

All fungi are filamentous or unicellular, and structures such as mushrooms consist simply of a number of filaments packed tightly together. Fungal filaments are known as *hyphae,* and a mass of hyphae is a *mycelium* (Figures 11-1 and 11-3). Growth of the hyphae takes place at their tips, but proteins are synthesized throughout the mycelium and carried to the tips of the hyphae by cytoplasmic streaming, a phenomenon that is very well developed among the fungi. Within 24 hours, a

11–2
Hornet killed by a parasitic fungus.

11–3
(a) *Mycelium of a species of* Aspergillus. *The dark bodies are asexual spore-producing structures. This mold was grown by placing a culture dish containing a simple medium on a New York City windowsill for half an hour.* (b) *Mycelium of the water mold* Achlya ambisexualis, *showing a sporangium releasing zoospores (swimming spores).* (a, Eric V. Gravé)

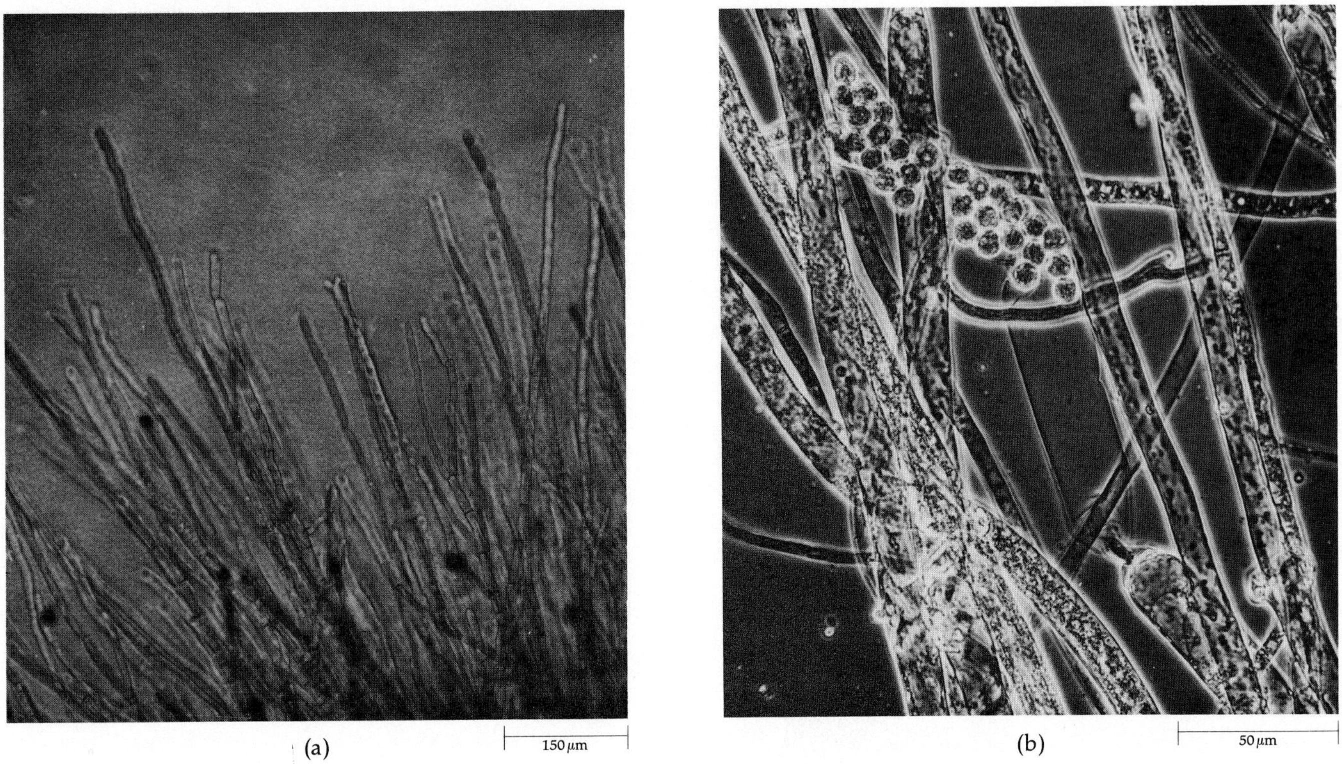

fungus colony may produce more than a kilometer of new mycelium. The words "mycelium" and "mycologist" are derived from the Greek word *myketos,* a fungus.

With their rapid growth and filamentous form, fungi have a very different relationship to the environment from that found in any other group of organisms. The surface-to-volume ratio of fungi is very high, which means that they are in intimate contact with the environment. With a few exceptions, no part of a fungus is more than a few micrometers from the external environment and is separated from it by only a thin cell wall and the plasma membrane. As a consequence, a fungus with an extensive mycelium can have a profound effect on its surroundings, for example, in binding soil together. The hyphae may often fuse, even if they have grown from different spores, and thus increase the intricacy of the network.

The maintaining of this sort of intimate relationship between fungus and environment requires a high metabolic rate. It further requires that all parts of the fungus be metabolically active, and the sorts of quiescent layers of tissue found, for example, in the higher plants are absent in the fungi. Enzymes and other substances secreted by fungi have an immediate effect on the surroundings and are of great importance for the maintenance of the fungus itself.

All fungi are heterotrophic. They obtain their food either as saprobes, organisms that live on dead organic material, or as parasites, organisms that feed on living matter. In either case, the food is ingested by absorption after it has been partially digested by enzymes that are secreted outside the cell wall. Some fungi, including a number of yeasts, can release energy by anaerobic respiration, such as in the production of ethyl alcohol from glucose. Glycogen is the primary storage polysaccharide in fungi, as it is in animals.

Saprobic fungi sometimes have somewhat specialized hyphae known as *rhizoids* that anchor them to the substrate. Parasitic ones often have specialized hyphae called *haustoria* that penetrate the cells of other organisms and absorb nourishment directly from them (Figure 11–4).

All fungi have cell walls and most produce spores of some type. Most fungi are nonmotile, but members of a few classes produce motile flagellated cells (see Figure 11–3b).

11-4
Haustorium of the rust fungus, Melampsora lini, *growing in a mesophyll cell of common flax* (Linum usitatissimum) *leaf.*

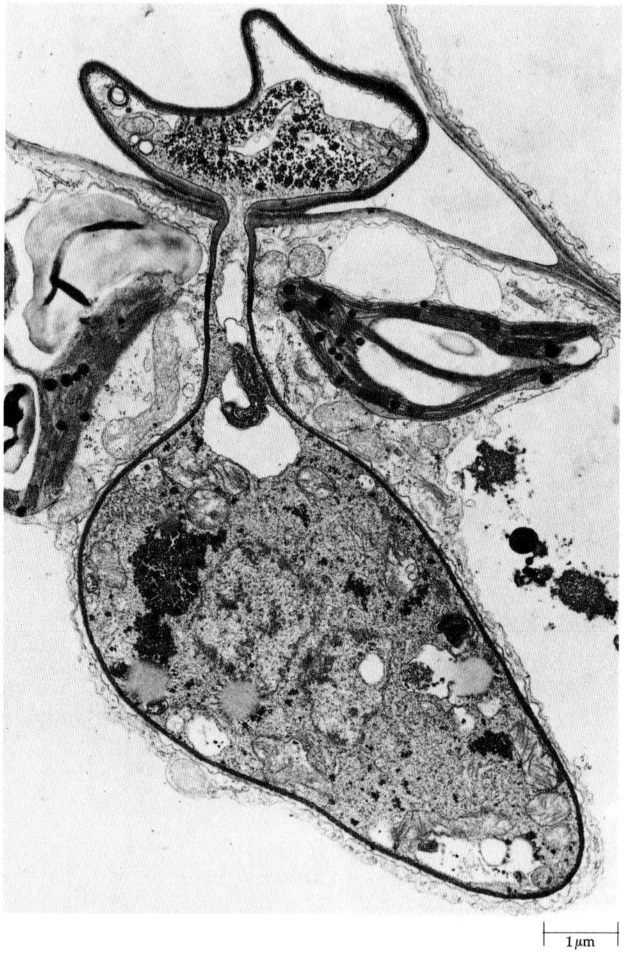

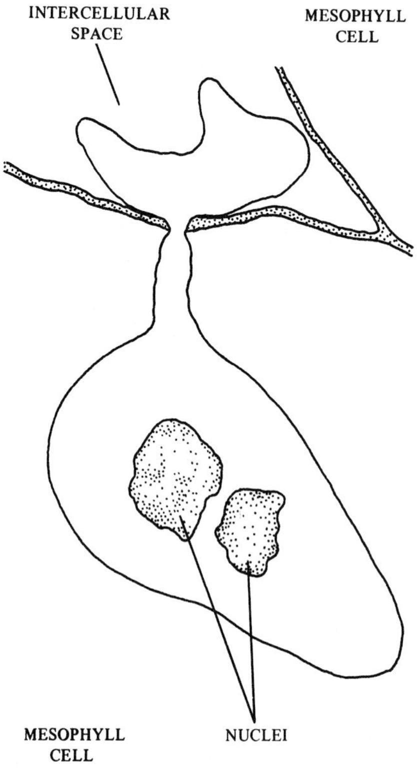

11-5
Structure of chitin, which consists of beta-1,4-linked N*-acetylglucosamine units. A similar linkage is found in cellulose and in the bacterial cell wall, suggesting that it provides a particularly strong, compact polysaccharide. Chitin is characteristic of the cell walls of many fungi and of arthropods.* N*-acetylglucosamine is one of the two principal building blocks in the cells walls of prokaryotes; the other is* N*-acetylmuramic acid.*

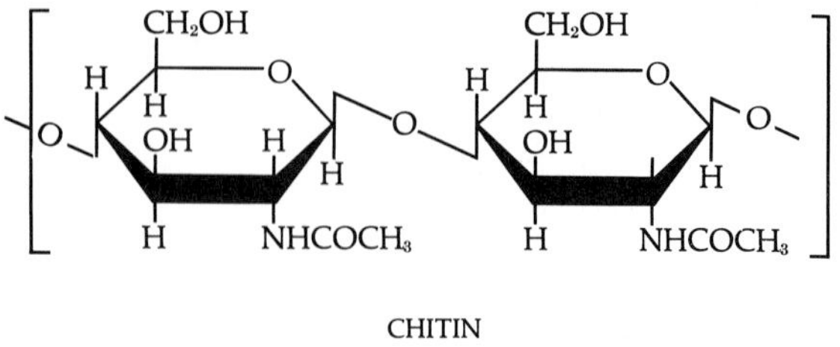

CHITIN

The Fungal Cell Wall

In the plants and in most groups of algae, the cell wall is built on a foundation of cellulose. Other molecules, such as hemicelluloses and pectic substances, are found in the openings between the cellulose microfibrils. In most groups of fungi, the cell wall is composed primarily of another polysaccharide, *chitin* (Figure 11-5), which is the same material found in the hard shells of arthropods. In the Oomycetes, however, the cell walls are composed largely of D-glucose units with a different linkage from that found in cellulose. The Oomycetes probably are not directly related to the other fungi, having been derived independently from protistan ancestors.

THE EVOLUTION OF FUNGI

The first fungi were probably unicellular eukaryotic organisms that probably have no living counterparts. From these were derived *coenocytic* fungi, in which many nuclei are found in a common cytoplasm (coenocytic means "contained in a common vessel").

Several groups of fungi have a coenocytic mycelium (Figure 11-6). All these groups used to be classified together as Phycomycetes, which means algal fungi, because of the similarities that exist between these fungal groups and certain groups of algae. At one time it was suggested that the Phycomycetes were derived from algal progenitors through the loss of chlorophyll. However, today most botanists regard any similarities between algae and the so-called Phycomycetes as examples of evolutionary parallelism. In addition, the old class Phycomycetes has been divided into several classes which are certainly not descended from an immediate common ancestor. Among these classes are the Chytridiomycetes, the Oomycetes, and the Zygomycetes.

In the "higher" fungi, the mycelia are septate—divided by cell walls—but the crosswalls, or septa, are perforated, and in some fungi the cytoplasm and its inclusions stream quite freely along the hyphae (Figure 11-7). The Ascomycetes seem to have been derived from coenocytic forms similar to the Zygomycetes, and the Basidiomycetes from the Ascomycetes. In both Ascomycetes and Basidiomycetes, the cell walls are made up of chitin and glucans. In general, in the course of fungal evolution, the proportion of chitin in the cell walls increased. The earliest fossils that have been definitely identified as fungi are from the Ordovician, 450 to 500 million years ago, but filaments that resemble fungi occur in strata about 900 million years old. Fungi may be as old as any other eukaryote and might even have been derived independently from prokaryotes as much as 2 billion years ago. At any rate, all of the major groups were in existence by the close of the Carboniferous period, some 300 million years ago.

REPRODUCTION

In all fungi, the reproductive structures, both sexual and asexual, are separated from the hyphae by complete septa. These reproductive structures are called *gametangia* if they are directly involved in the production of eggs and sperms, or gametes, and *sporangia* if they are involved in the production of asexual spores. Male gametangia in the fungi are called *antheridia* (but these unicellular structures are quite different from the multicellular antheridia found in the bryophytes and vascular plants), and female gametangia are called *oogonia* or ascogonia. Antheridia and oogonia are found only in a few classes of coenocytic fungi.

In a few primarily aquatic groups of coenocytic fungi, the spores are flagellated and motile. They either have two flagella, one of which is a whiplash flagellum and one a tinsel flagellum (see Figure 1-15, page 27) or only one, usually a whiplash flagellum. In all other groups of fungi, nonmotile spores are the characteristic means of reproduction. Some spores are very small and so can remain suspended in the air for long periods and be carried for great distances. This presumably accounts for the very wide ranges of distribution of certain fungi. Often the sporangia of terrestrial fungi are raised above the mycelium, the spores being easily caught up and transported by air currents, and some have been reco-

11-6
Electron micrograph of a hyphal tip of Aphanomyces euteiches *of the class* Oomycetes, *showing a portion of the coenocytic hypha. Although only a single nucleus appears in this micrograph, others are scattered throughout the hypha.*

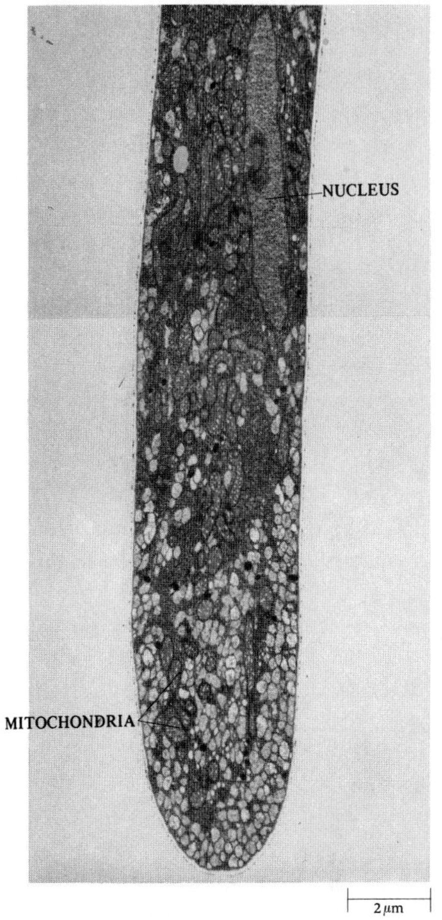

11-7
Neurospora crassa, *an Ascomycete. Electron micrograph of a perforated crosswall with a nucleus in the perforation.*

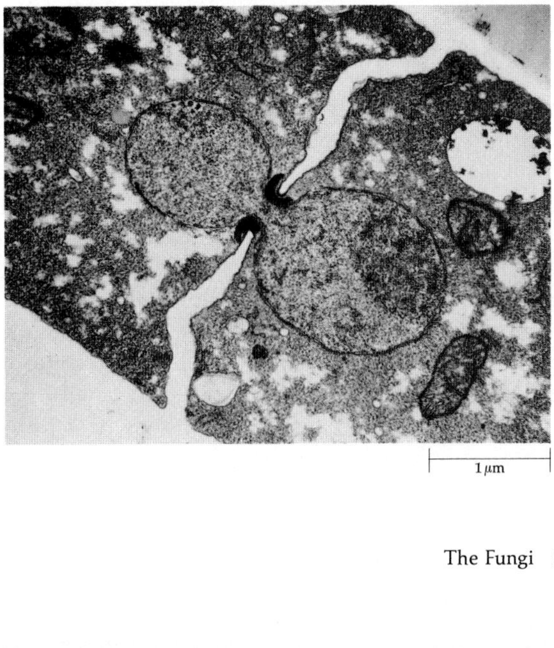

vered nearly a hundred kilometers above the surface of the earth. Fungus spores often are spread by adhering to the bodies of insects and other animals. The bright colors and powdery textures associated with many types of molds are produced by the spores. Often the mycelium is located beneath the surface of the substrate upon which the fungus is growing.

Mitosis and Meiosis

Fungi have a great many biological peculiarities that we are just beginning to understand. One of the most intriguing involves the process of nuclear division. In the fungi, mitosis and meiosis are different from these processes in plants and animals. The nuclear envelope does not dissociate and re-form but is constricted near its midpoint between the two daughter nuclei, and the spindle apparatus is formed within the nuclear envelope. Centrioles are found in those fungi that produce motile spores. Most interesting of all is the recent discovery that the chromosomes of most fungi, like those of dinoflagellates (Chapter 12), lack the histone proteins (page 85) that are characteristic of the chromosomes of all other eukaryotes, or have only small quantities of them. This confirms the view that fungi are not related directly to plants, animals, or other known eukaryotes and therefore deserve status as a separate kingdom. It does appear, however, that Oomycetes have chromosomes similar to those of other eukaryotes in histone content. This is another line of evidence that Oomycetes are not related directly to other fungi. In fact, it is probably more logical to regard them as members of the kingdom Protista than as fungi.

Heterokaryosis and Parasexuality

Heterokaryosis

Among the genetic peculiarities that set the fungi off from other groups of organisms, none is more significant than the phenomenon of *heterokaryosis*, which was discovered by the German mycologist H. Burgeff in 1912. A strain of fungus is heterokaryotic if the nuclei found in a common cytoplasm are genetically different, either because of mutation or because of the fusion of genetically distinct hyphae, which appears to be common in nature. If the nuclei are genetically similar, the strain is *homokaryotic*.

Heterokaryosis is extremely important in the genetics and evolution of the fungi. If the genetically different nuclei in heterokaryotic fungi are segregated, phenotypically distinct hyphae can be created. Thus, even if there are only two different sorts of nuclei in a mycelium, three distinct phenotypes (one the same as the original one, containing both types of nuclei) can be derived from it.

The results of heterokaryosis are somewhat similar to those of the diploid condition in organisms; that is, the appearance and physiological characteristics of a heterokaryotic organism are determined by the interaction of the genetically different nuclei. As recessive mutations accumulate in some of the nuclear lines, they may be compensated for by the genes present in other nuclei (Figure 11–8). Finally, because many of the nuclear lines are unable to exist, or to compete, in a homokaryotic state, heterokaryotic strains are often favored by selection.

11–8
Arg-1 and Arg-10 are mutants of Neurospora, *each of which lacks a different enzyme involved in the production of arginine. Neither will grow on a minimal medium. However, a heterokaryon formed between the two strains grows readily on the minimal medium, because, by combining information contained in the two nuclei, the heterokaryon can synthesize the missing amino acid.*

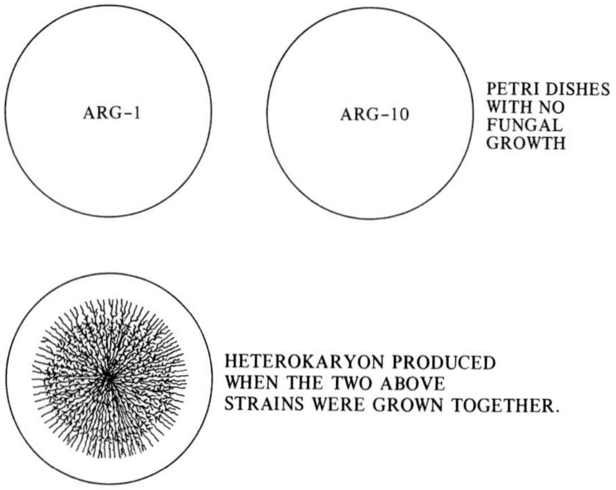

Parasexuality

The *parasexual cycle* in fungi was discovered in 1952 by G. Pontecorvo and J. A. Roper at the University of Glasgow. Working with *Aspergillus nidulans,* Pontecorvo and Roper found that haploid nuclei occasionally fuse in a heterokaryotic mycelium to produce diploid nuclei. Some of these diploid nuclei are heterozygous; in other words, they are produced by the fusion of genetically different nuclei. Pontecorvo estimates that in *Aspergillus nidulans* there is 1 diploid heterozygous nucleus for every 1000 haploid nuclei.

Within the diploid nucleus, the chromosomes may become associated. If this occurs, some crossing-over can take place, although such a phenomenon is infrequent. Haploid nuclei may or may not re-form. If they do re-form, different kinds of haploid nuclei are produced.

These can then take part in new heterokaryotic combinations.

It has recently been demonstrated that the parasexual cycle exists in several other groups of fungi. The significance of this cycle in nature has yet to be assessed completely. However, it appears to be a flexible and commonly occurring system, especially among those fungi that do not reproduce sexually or in which sexual reproduction is infrequent.

MAJOR GROUPS OF FUNGI

Class Chytridiomycetes

With the exception of the Oomycetes, the Chytridiomycetes, with about 750 species, are the only fungi that possess motile cells in their life cycles. They are aquatic, and each motile cell contains a single posterior flagellum of the whiplash type. This is the principal distinguishing feature of the Chytridiomycetes. The absence of chlorophyll and the presence of flagellated motile cells have led some mycologists to suggest a protozoan origin for this group. It is doubtful that they are related directly to other fungi, but additional study will be necessary before this question can be answered.

The Chytridiomycetes include the simplest of fungi, microscopic, unicellular organisms that do not develop a true mycelium. Many species are parasitic on or in algae, other aquatic fungi, and submerged parts and the pollen grains or spores of higher plants. The cell walls of some Chytridiomycetes appear to be composed mainly of chitin; cellulose may also be present. Some species reproduce solely asexually through formation of zoospores, whereas others reproduce both asexually and sexually. *Allomyces* is representative of the class. It forms a true coenocytic mycelium and exhibits an isomorphic alternation of generations (Figure 11-9), essentially similar to the life cycles of the green algae *Cladophora* and *Ulva.*

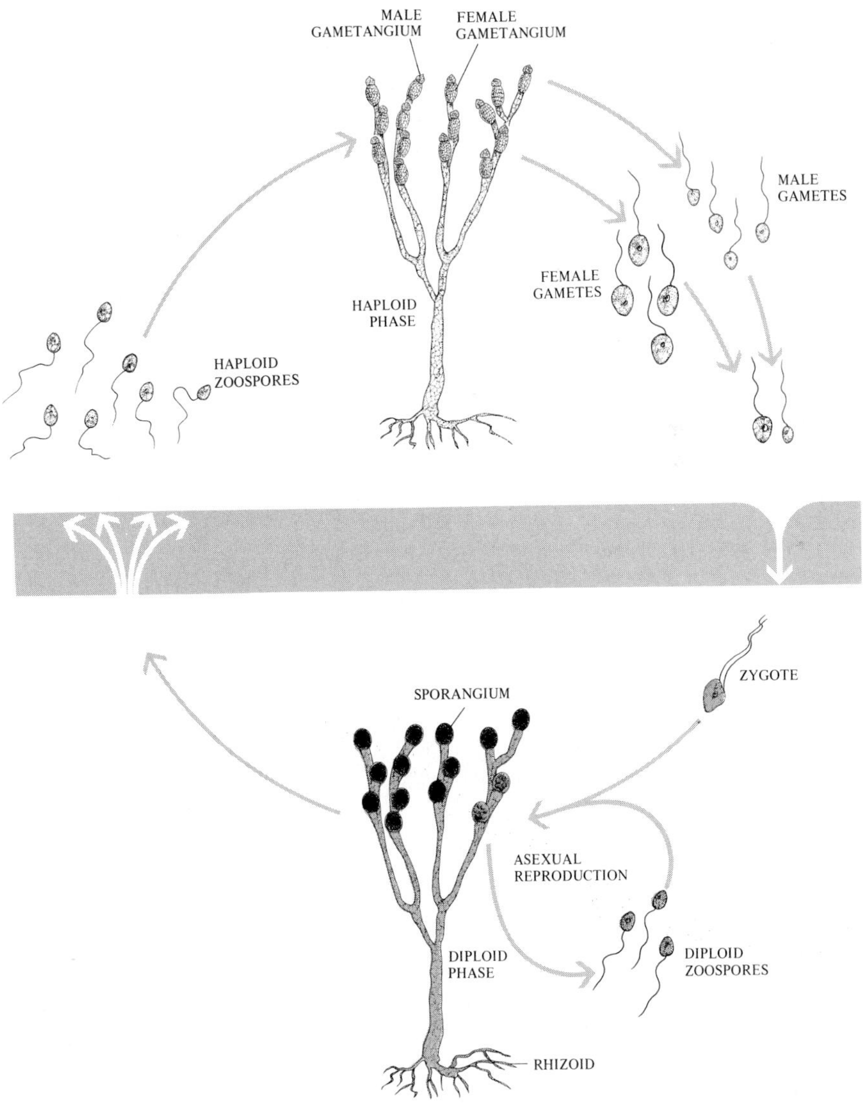

11-9
Some species of Allomyces, *such as* A. macrogynus *(shown here) exhibit an isomorphic alternation of generations. The gametophytes and sporophytes are indistinguishable until they begin to form reproductive organs. The gametophytes produce colorless female gametangia and orange male gametangia, usually close together and in equal numbers. The male gametes, which are about half the size of the females, are attracted by a hormone, sirenin, produced by the female gametes. The zygote loses its flagella, rounds up, and soon germinates to produce a diploid sporophyte. The sporophyte forms two kinds of sporangia: (1) asexual sporangia, colorless mitosporangia that release diploid zoospores, which in turn germinate and repeat the diploid generation; and (2) sexual sporangia, thick-walled, reddish-brown meiosporangia, which can withstand severe environmental conditions. After dormancy, meiosis occurs in the meiosporangia, resulting in the formation of haploid zoospores that develop into gametophytes which at maturity produce gametangia.*

Class Oomycetes

The class Oomycetes, with about 475 known species, is characterized by the presence of zoospores with two flagella—one tinsel and one whiplash (see Figure 1-15, page 27). The organisms in the class range from unicellular forms to greatly branched, coenocytic filamentous mycelia. On the basis of apparent mitotic peculiarities, different chromosomal composition, and their distinctive cell walls, as mentioned earlier, it is virtually certain that the Oomycetes are not directly related to other fungi and would better be placed as one of the many diverse elements in the "catch-all" kingdom Protista. Most species of Oomycetes reproduce by both sexual and asexual means. Sexual reproduction is oogamous (in which one of the gametes, the egg, is large and nonmotile, and the other gamete, the sperm, is smaller and motile) and results in formation of a thick-walled zygote, the *oospore,* which serves as a resting spore.

One large group of the class Oomycetes is aquatic—the so-called water molds. The members of this group are the most abundant in fresh water of all the fungi and are the easiest to isolate from it. They can be cultured on corn meal agar. Most of them are saprobic, but a few are parasitic, including species that cause diseases of fish and fish eggs.

In some water molds, such as Saprolegnia (Figure 11-10), sexual reproduction can occur with male and female sex organs borne on the same individual. Such individuals are said to be *homothallic*. In other water molds, two different individuals (*heterothallic* individuals) are required for sexual reproduction, as in some species of *Achyla*

11-10
Life cycle of Saprolegnia, *a water mold. Present evidence indicates that the mycelium is diploid. Reproduction is mainly asexual. Biflagellated zoospores released from a sporangium swim for a while and then encyst. Each eventually gives rise to a secondary zoospore, which also encysts and then germinates to produce a new mycelium.*

During sexual reproduction, oogonia and antheridia are formed on the somatic hyphae. Meiosis apparently occurs within these structures. The oogonia are enlarged cells in which a number of spherical eggs, or oospheres, are produced. The antheridia develop from the tips of other filaments of the same individual and produce numerous male nuclei. The life cycle of Saprolegnia *is similar to that of the brown alga* Fucus, *with meiosis taking place in gametangia borne on a diploid individual and giving rise directly to gametes. In mating, the antheridia grow toward the oogonia and develop tubular processes called fertilization tubes, which penetrate the oogonia. Male nuclei travel down the fertilization tubes to the female nuclei and fuse with them. Following nuclear fusion, a thick-walled resistant oospore is produced. On germination, the zygote develops into a hypha, which then produces a sporangium, beginning the cycle anew.*

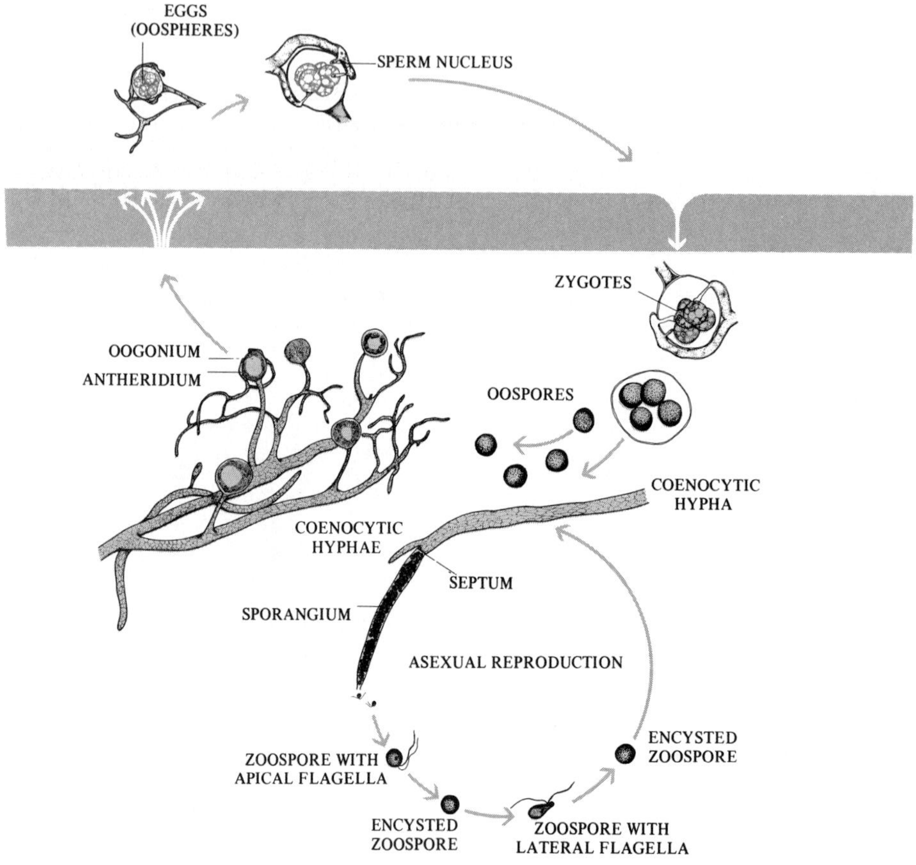

11–11
Achlya ambisexualis, *a water mold that produces male and female sex organs on different individuals.* (a) *Empty sporangium with zoospores encysted at its mouth, a distinctive feature of* Achlya. (b) *Sex organs, showing fertilization tubes extending from the antheridium through the wall of the oogonium to the oospheres.* (c) *Appearance of hyphae before addition of hormone A (antheridiol).* (d) *Antheridial branches produced in response to crystalline hormone A (antheridiol), 2 hours after the addition of the hormone.* (e) *Antheridial branches growing to a plastic particle on which hormone A has been absorbed.*

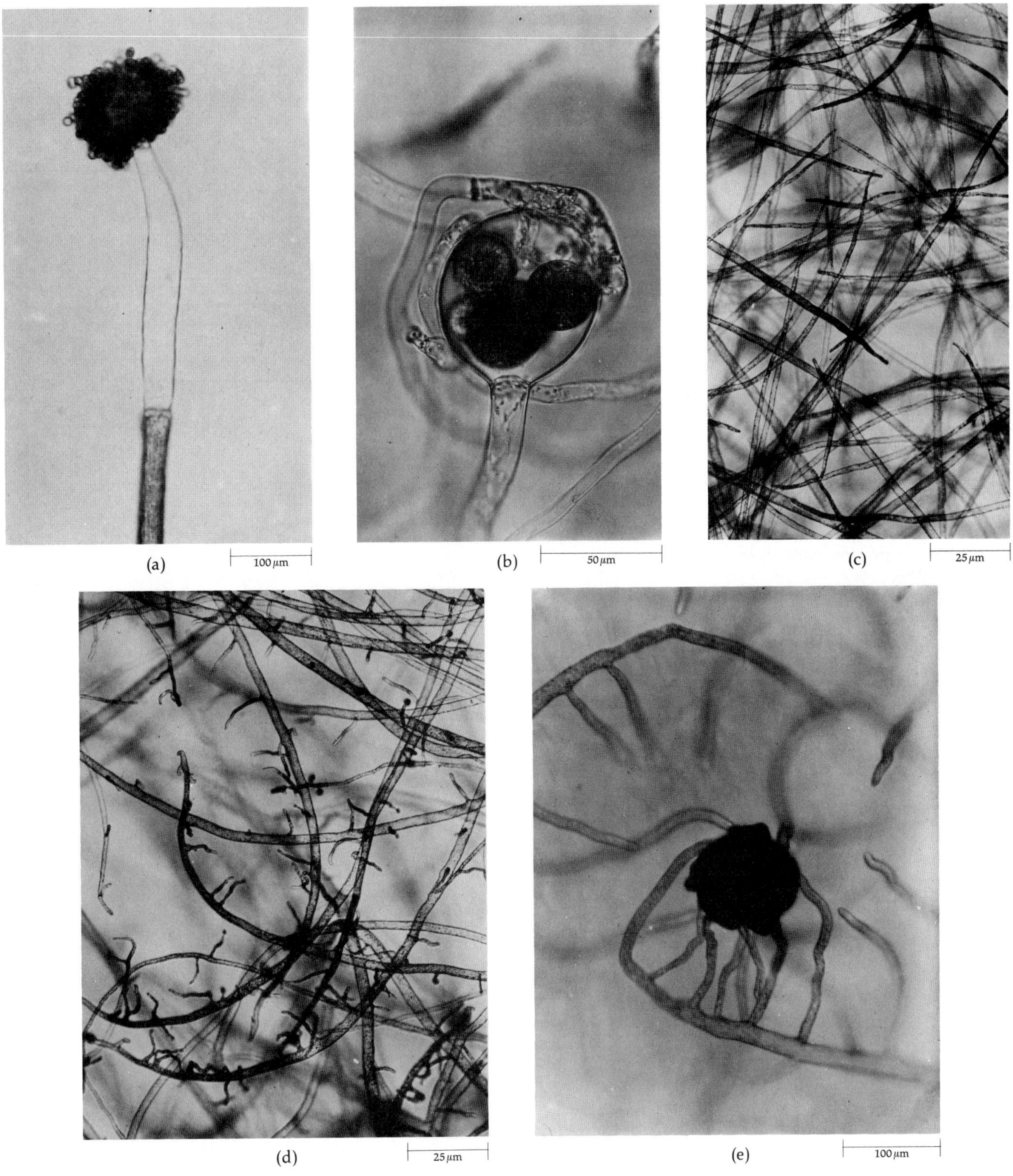

(Figure 11–11). The late John R. Raper of Harvard University published a series of papers describing the hormonal control of sexuality in these two fungi. The female vegetative thallus secretes a substance that induces the initial development of antheridial branches on the male thallus. Raper called this substance hormone A, and it was later characterized by Alma Barksdale and named antheridiol. After the antheridial initials appear, the male secretes another hormone, hormone B, which induces the formation of oogonial initials on the female thallus. With the appearance of young oogonia, the antheridial hyphae are attracted to the female gametangia and distinct, mature antheridia become apparent. Raper attributed this latter response to hormone C, which presumably is secreted by the female, but subsequent research suggested that hormone A can elicit the same developmental reaction. After the antheridia are formed, there is a differentiating response in the female gametangium, presumably hormonally induced, which leads to the maturation of the oogonium and its enclosed female gametes (eggs). Thus, even in the fungi, sexual reproduction may involve a highly coordinated sequence of events which are hormonally induced.

Another group of Oomycetes, the order Peronosporales, is primarily terrestrial, although the organisms still form motile zoospores that swim when free water is available. Among this group are several forms of great economic importance. As C. J. Alexopoulos has said, "At least two of them have had a hand—or should we say a hypha!—in shaping the economic history of an important portion of mankind." The first of these (Figue 11–12) is *Phytophthora infestans* (*Phytophthora* literally means "plant destroyer"), the cause of the late blight of potatoes that produced the great potato famines in Ireland. The famine of 1845–1847, which was caused by this fungus, was responsible for over 1 million deaths from starvation and initiated large-scale emigration from Ireland to the United States; within a decade, the population of Ireland dropped from 8 million to 4 million. Virtually the entire Irish potato crop was wiped out in a single week in the summer of 1846.

Another economically important member of this group is *Plasmopara viticola,* which is the cause of downy mildew of grapes. This fungus was accidentally introduced into France in the late 1870s on American stock that had been imported because of its resistance to other diseases. The mildew soon threatened the entire French wine industry. It was eventually brought under control by a combination of good fortune and skillful observation. French vineyard owners in the vicinity of Medoc

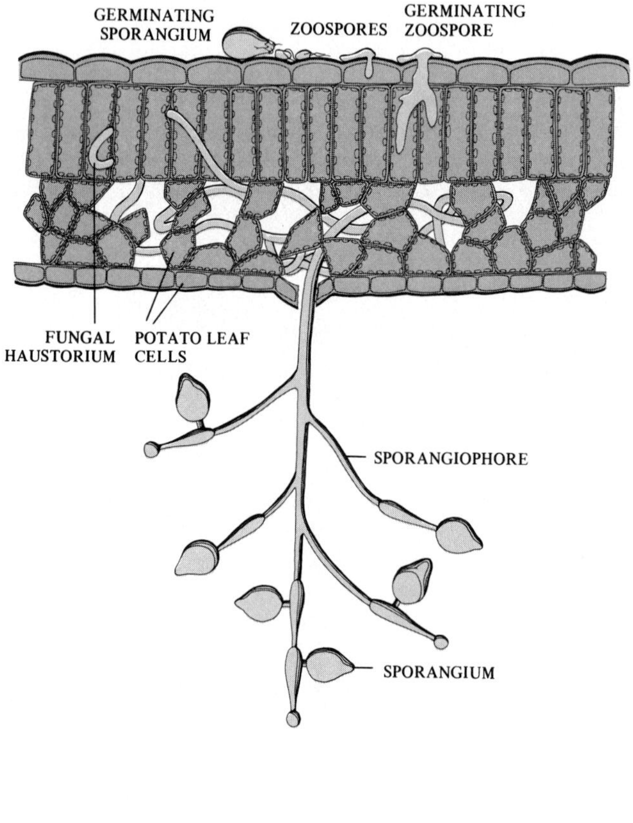

11–12
Phytophthora infestans, *cause of the late blight of the potato. The cells of the potato leaf are shown in gray. In the presence of water, the sporangia either germinate directly through a germ tube or burst to release mobile zoospores which swim to their site of germination.*

customarily put an evil-looking mixture of copper sulfate and lime on vines growing along the roadside to discourage passersby from picking them. A professor from the University of Bordeaux who was studying the problem of the mildew noticed that these plants were free from the symptoms of the disease. After conferring with the vineyard owner, the professor prepared his own mixture of chemicals—the Bordeaux mixture—which was made generally available as a fungicide in 1882. The Bordeaux mixture was the first fungicide for plant diseases.

Class Zygomycetes

The Zygomycetes are terrestrial fungi that live in the soil on decaying plant or animal matter. About 600 species of Zygomycetes have been described. Most are saprobes, but some are parasites of plants, insects, or small soil animals. The term Zygomycetes refers to the chief characteristic of the class, namely, the production of sexual resting spores called *zygospores* (Figure 11–13b). Whereas the Oomycetes are oogamous (reproduction involves a large, nonmotile egg and a small, motile sperm), the Zygomycetes are isogamous (reproduction involves two gametes of the same size). In addition, unlike the Oomycetes, the Zygomycetes produce no flagellated spores at any stage of the life cycle. Asexual reproduction is by means of nonmotile spores.

One of the most common members of this class is *Rhizopus stolonifer*, a black bread mold that forms cottony masses on the surface of moist bread exposed to air. The mycelium of *Rhizopus* is composed of three different types of haploid hyphae. The bulk of the thallus is composed of rapidly growing hyphae that are aseptate (not divided by cross walls into cells or compartments) and multinucleate (coenocytic). From these, arching hyphae called *stolons* grow upward. The stolons form rhizoids where their apices come into contact with the substrate. Sporangia form on the tips of the sporangiophores ("sporangia bearers"), which are erect branches formed directly above the rhizoids. Each sporangium begins as a swelling into which a number of nuclei flow and is eventually cut off from the sporangiophores by the formation of a septum. The protoplasm within is cleaved and a cell wall is formed around each spore. The sporangium becomes black as it matures, giving the mold its characteristic color. Each spore, when liberated, can germinate to produce a new mycelium. The life cycle of *Rhizopus stolonifer* is illustrated in Figure 11–14.

11–13
Rhizopus stolonifer, *the black bread mold that forms cottony masses on the surface of moist bread or other substrates exposed to air.* (a) *Gametangia.* (b) *Zygospore.*

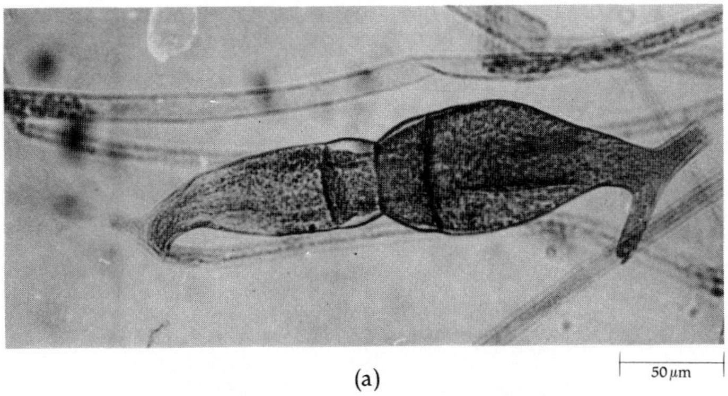

(a)

(b)

11–14
In Rhizopus stolonifer, *sexual reproduction takes place only between different mating strains, which are morphologically indistinguishable. When both mating strains are present, hormones are produced that cause hyphal tips to come together and to develop into gametangia, which become separated from the rest of the thallus by the formation of septa* (Figure 11–13a). *The walls between the two touching gametangia dissolve, and the two multinucleate protoplasts come together. The + and − nuclei may fuse in pairs to form a young zygospore with several diploid nuclei. The zygospore then develops a thick, rough, black coat and becomes dormant, often for several months. Meiosis probably takes place at the time of germination. The zygospore cracks open and produces a sporangium that is similar to the asexually produced sporangium, and the cycle begins again.*

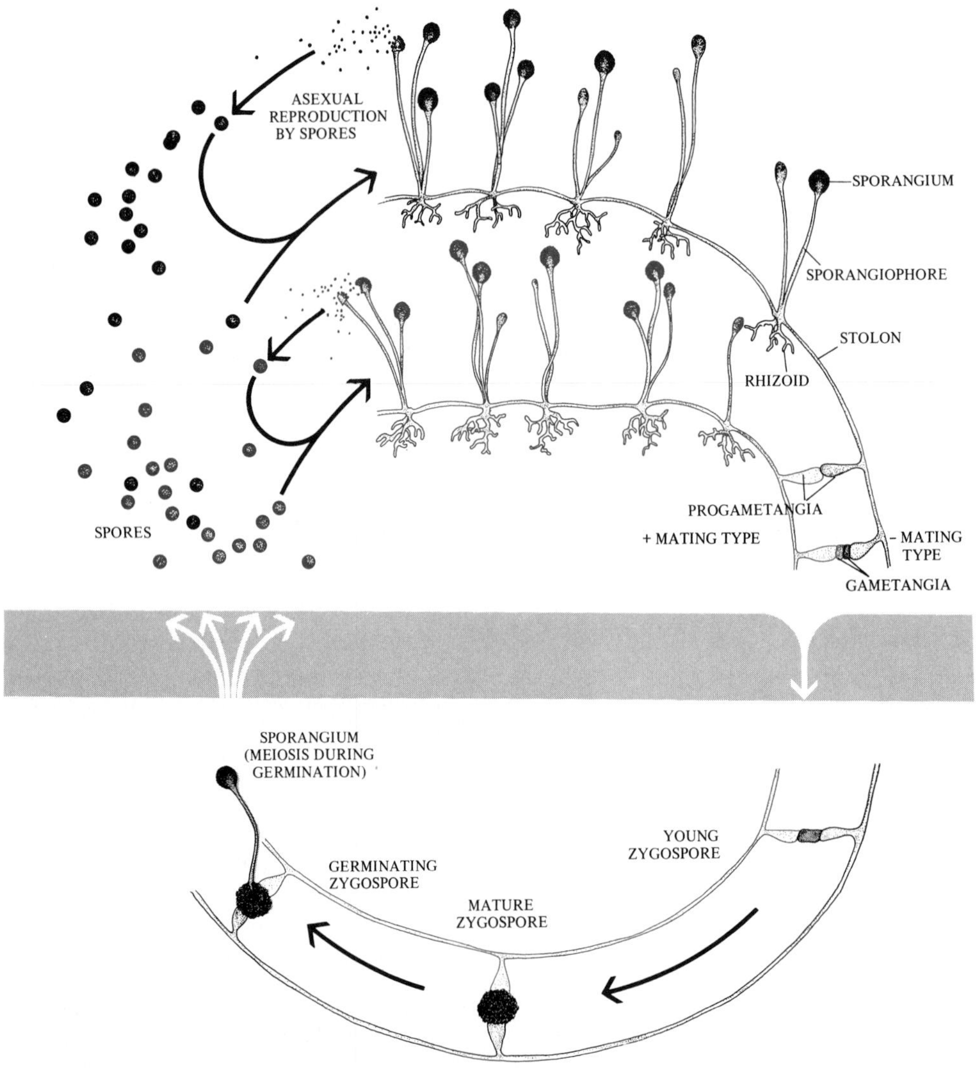

PHOTOTAXIS IN A FUNGUS

In Pilobolus, *a Zygomycete that grows on dung, the sporangia are shot toward the light. The sporangium is oriented toward the light so that all light rays entering the subsporangial swelling converge on a basal photoreceptive area. Light focused elsewhere promotes maximum growth of the sporangiophore on the side away from the light. The high turgor pressure of the sap in the vacuole of the subsporangial swelling splits it and blasts the sporangium off to a distance of 2 meters or more. When it has been fired off, the sporangiophore collapses. The sporangium adheres where it lands, and if this happens to be a blade of grass, it may be eaten by a herbivore. It then passes through the digestive tract of the herbivore unharmed and is deposited in the dung to begin the cycle anew.*
(a) *A dense colony* Pilobolus *growing on horse manure, with the sporangia oriented toward the light.* (b) *The top of a single sporangiophore.* (c) *A sporangium being shot off into space as the subsporangial swelling collapses. The sporangium is trailing a stream of vacuolar sap.* (a,b, Eric V. Gravé)

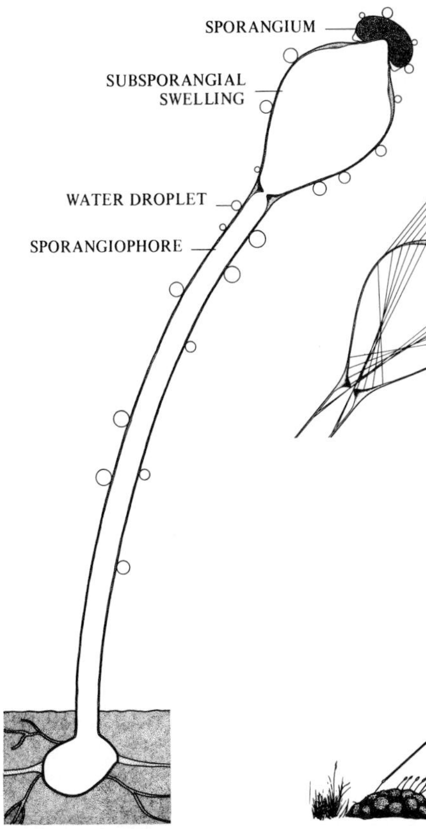

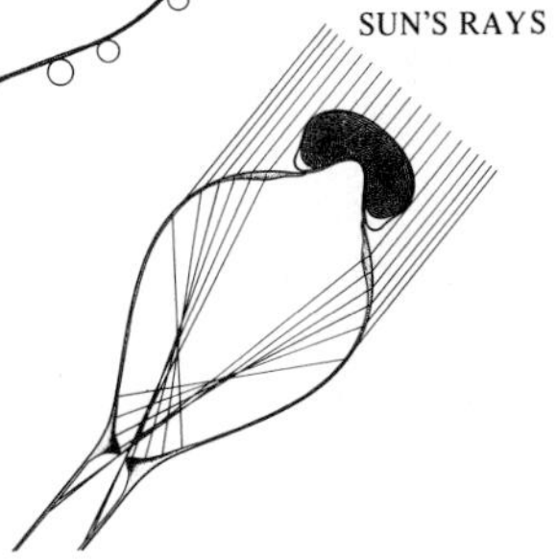

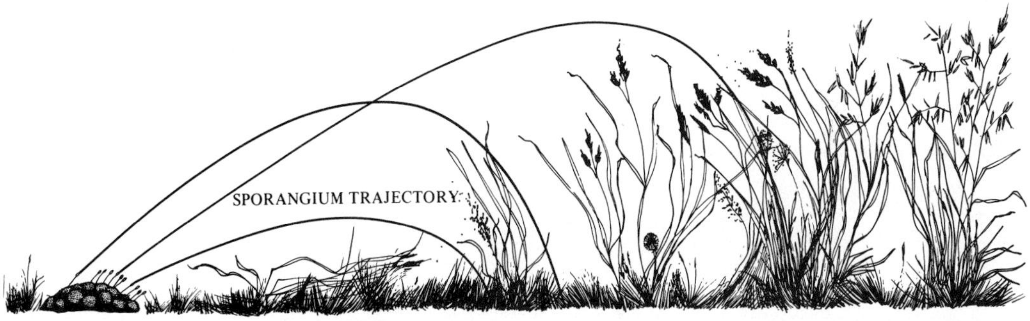

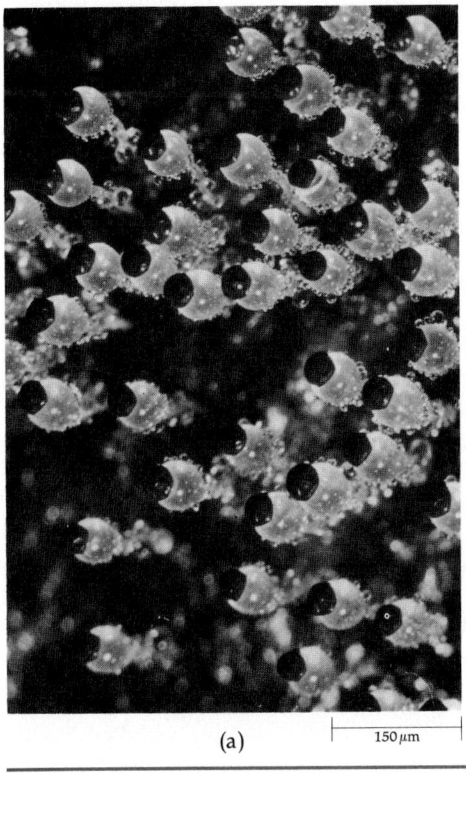

(a)

(b)

(c)

ERGOTISM

Many Ascomycetes are parasitic on higher plants. The disease of plants called ergot is caused by Claviceps purpurea, *a parasite of rye* (Secale cereale) *and other grasses. Although ergot seldom causes serious damage to the crop of rye, it is dangerous because a small amount mixed with rye grains is enough to cause severe illness among domestic animals or among the people who eat bread made with the flour. Ergotism, the toxic condition caused by eating grain infected with ergot, is often accompanied by gangrene, nervous spasms, psychotic delusions, and convulsions. It occurred frequently during the Middle Ages, when it was known as St. Anthony's fire. In one epidemic in 994, more than 40,000 people died. In 1722, ergotism struck down the cavalry of Czar Peter the Great on the eve of battle for the conquest of Turkey and thus changed the course of history. As recently as 1951, there was an outbreak in a small French village in which 30 people became temporarily insane, imagining that they were pursued by demons and snakes; 5 of the villagers died. Ergot, which causes muscles to contract and blood vessels to constrict, is used in medicine. It is also the initial source for the psychedelic drug lysergic acid diethylamide (LSD), the structure of which is shown in* (b). *In* (a) *the purple-black, hard resting structures of* Claviceps *are seen among the spikelets of rye.*

(a)

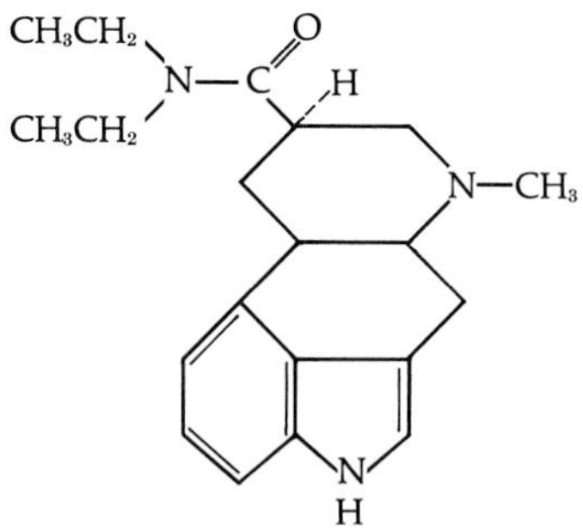

D-LYSERGIC ACID DIETHYLAMIDE

(b)

Class Ascomycetes

The Ascomycetes comprise about 30,000 described species, including a number of familiar and economically important kinds. Most of the blue-green, red, and brown molds that cause food spoilage are Ascomycetes, including the salmon-colored bread mold *Neurospora,* which has played such a notable role in the history of modern genetics. Ascomycetes are the cause of a number of serious plant diseases, including the powdery mildews that attack fruits, the chestnut blight (due to the fungus *Endothia parasitica,* accidentally introduced from northern China), and Dutch elm disease (caused by *Ceratocystis ulmi,* a European fungus). Yeasts are also Ascomycetes, as are the delicious morels and truffles (Figure 11–15). As a whole, the group is relatively poorly known, and thousands more species—some undoubtedly of great economic importance—await scientific description.

Characteristics of the Ascomycetes

Ascomycetes, like most fungi, are filamentous when they are growing. In general, their hyphae are septate (divided by crosswalls) instead of the aseptate hyphae characteristic of the Oomycetes and most Zygomycetes. The crosswalls are perforated, however, and cytoplasm with its included nuclei can move freely through them (Figure 11–7; page 213). The hyphal cells of the vegetative mycelium may be either uninucleate or multinucleate. Some species of Ascomycetes are homothallic, others are heterothallic.

Asexual reproduction in the majority of Ascomycetes is by formation of specialized spores, called *conidia* (from the Greek meaning "fine dust"), which are cut off at the tips of the modified hyphae known as *conidiophores* ("spore bearers"). No flagellated cell appears at any point in the life cycle of these fungi.

Sexual reproduction in Ascomycetes always involves the formation of an *ascus* ("little sac"), a structure that is characteristic of the group and distinguishes the Ascomycetes from all other fungi (Figure 11–16). Ascus formation usually takes place within a complex structure composed of tightly interwoven hyphae, the *ascocarp.* Many ascocarps are macroscopic, and they have been used extensively in the classification of the Ascomycetes. Ascocarps may be open and more or less cup-shaped (apothecia; Figure 11–15b), closed and spherical in shape (cleistothecia; Figure 11–16b), or flask-shaped and with a small pore through which the ascospores escape (perithecia; Figure 11–16c), and the asci are usually borne

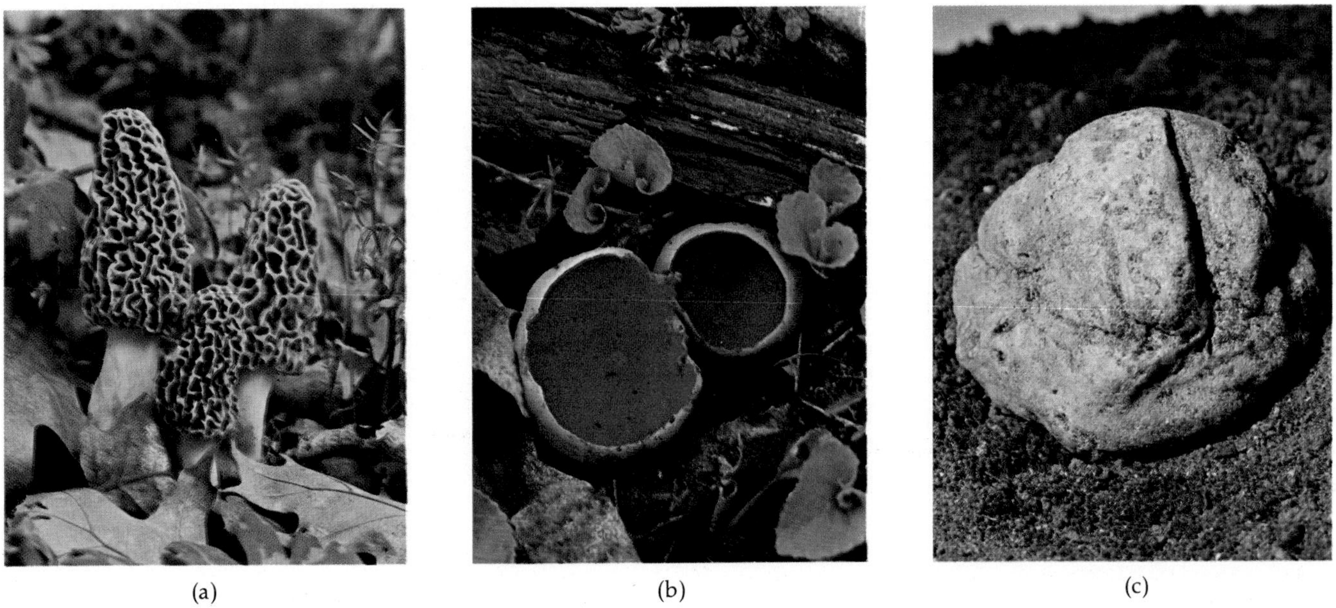

11–15
Representative large Ascomycetes. (a) *Common morel,* Morchella esculenta. *The true morels are among the most edible and choice of the fungi. Mushroom gatherers look for them when the oak leaves are "the size of a mouse's ear."* (b) *Scarlet cup,* Sarcoscypha coccinea, *a beautiful fungus of the woods.* (c) *A truffle,* Tuber melanosporum. *In the truffles, which are highly prized by gourmets, the spore-bearing structures are produced below ground and remain closed, liberating the ascospores only when the ascocarp decays or is broken open by animals. They are mycorrhizal on oaks and are searched for by specially trained dogs and pigs. Neither morels nor truffles have been successfully grown commercially.*

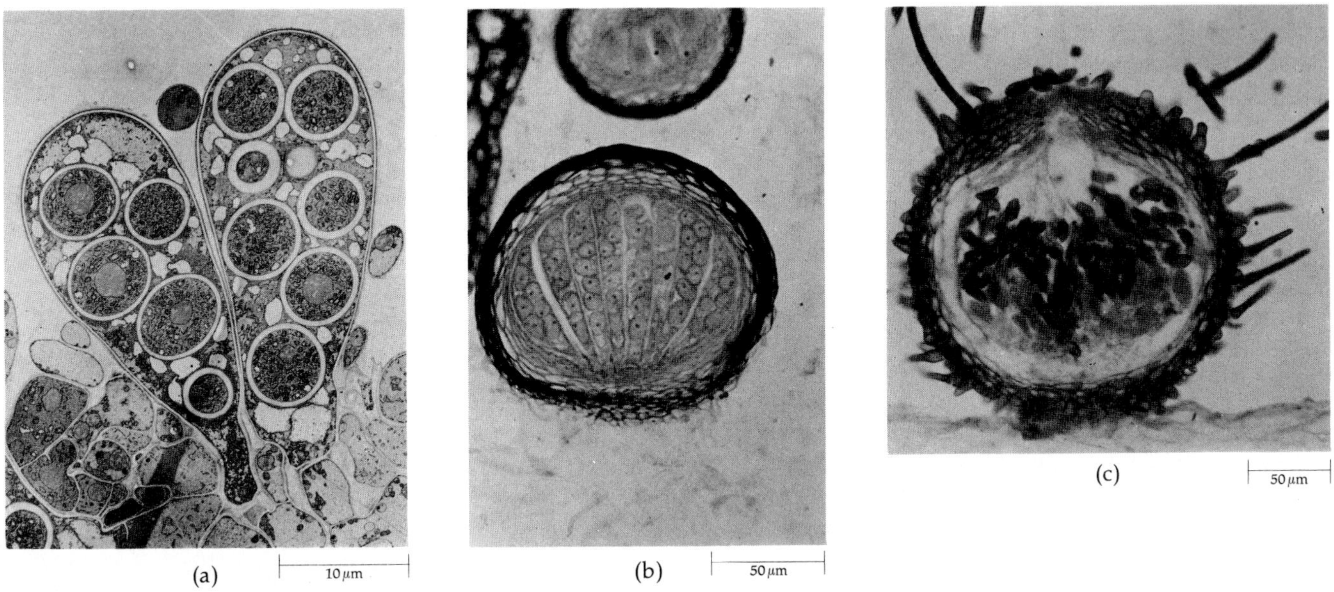

11–16
Asci. (a) *Electron micrograph showing two asci of* Ascodesmis nigricans *in which ascospores are maturing.* (b) *Ascocarp of* Erysiphe aggregata, *showing enclosed asci and ascospores. This completely enclosed type of ascocarp is called a cleistothecium.* (c) *Ascocarp of* Chaetomium erraticum, *showing the enclosed asci and ascospores. This sort of ascocarp, with a small opening, is known as a perithecium. Note the small pore at the top of the perithecium. Asci of* Neurospora *are shown in Figure 3–1 on page 66.*

11-17
Section through hymenial layer of Morchella *showing asci with ascospores. The narrow filaments among the asci are the sterile paraphyses.*

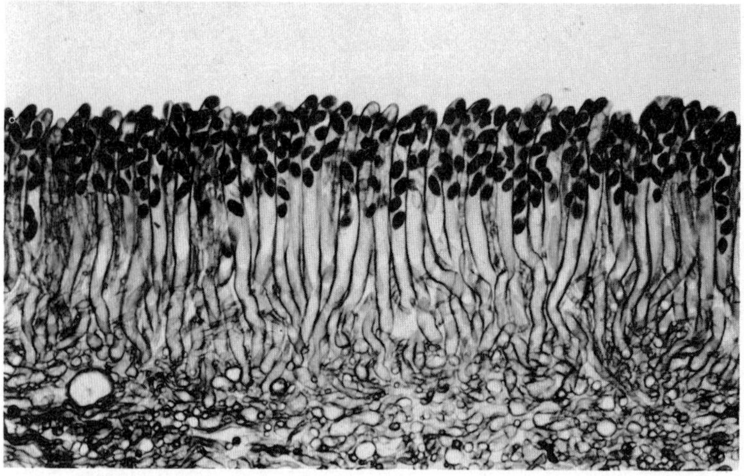

on the inner surface of the ascocarp. This layer of asci is called the *hymenium* or hymenial layer (Figure 11-17). Sterile hairs or *paraphyses* often form part of the hymenium.

Figure 11-18 illustrates the characteristic life cycle of an Ascomycete. The mycelium is initiated with germination of an ascospore, and soon after it begins to form conidiophores which bear conidia. Many crops of generally multinucleate conidia are produced during the growing season, and it is the conidia that are primarily responsible for propagation and dissemination of the fungus.

Ascus formation occurs on the same mycelium that produces conidia and is preceded by the formation of generally multinucleate gametangia called antheridia and ascogonia. The male nuclei of the antheridium pass into the *ascogonium* via an outgrowth, the *trichogyne,* of the ascogonium. *Plasmogamy,* the fusion of the two protoplasts, has now taken place. The male nuclei may then pair with the genetically different female nuclei *but do not fuse with them.* Hyphal filaments now begin to grow out of the ascogonium and elongate into *ascogenous hyphae.* As the ascogenous hyphae develop, pairs of nuclei migrate into them and simultaneous mitotic divisions occur in the hyphae and ascogonium. Cell division in the developing ascogenous hyphae occurs in such a way that the resultant cells are binucleate or *dikaryotic.*

The ascus forms at the tip of a few-celled dikaryotic, ascogenous hypha. In the formation of an ascus, one of the binucleate cells of the dikaryotic hypha grows over to form a hook, or crozier. In this hooked cell, the two nuclei divide in such a way that their spindle fibers are parallel and more or less vertical in orientation. Two of the daughter nuclei are close to one another at the end of the hook; one of the others is near the tip and the other is near the basal septum of the hook. Two septa are then formed; these divide the hook into three cells, the middle one of which becomes the ascus. It is in this middle cell that *karyogamy* takes place, the two parental nuclei fusing to form a diploid nucleus (zygote), the only diploid nucleus in the life cycle of Ascomycetes. Soon after karyogamy the young ascus begins to elongate. The diploid nucleus then undergoes meiosis, which is generally followed by one mitotic division giving a total of four or eight nuclei. These haploid nuclei are then cut off in organized segments of the cytoplasm to form *ascospores.* In most Ascomycetes, the ascus becomes turgid at maturity and finally bursts, squirting its ascospores explosively into the air. Although most species expel the ascospores only about 2 centimeters from the ascus, some may propel them as far as 3 decimeters.

Because of the linear arrangement of the ascospores in the ascus in *Neurospora* (see Figure 3–1, page 66), the individual spores can be dissected out and cultured individually. As a consequence, each instance of recombination and crossing-over that occurs during meiosis can be detected and analyzed, which has made *Neurospora* an important tool in understanding genetic processes.

Unicellular Ascomycetes: The Yeasts

Although some yeasts are Ascomycetes, they are peculiar in several respects. They are predominantly unicellular, without ascocarp formation, and reproduce by the pinching off of small buds (Figure 11-19a) or by fission instead of by asexual spore formation. Sexual reproduction in yeasts takes place when either two cells or two ascospores unite and form a zygote. The zygote nucleus may produce diploid buds or may undergo meiosis to produce four haploid nuclei. There may be subsequent mitotic division. Within the zygote wall, now the ascus, walls are laid down around these nuclei so that eight ascospores are formed, which are liberated when the ascus wall breaks down (Figure 11-19b). The ascospores

11–18
Characteristic life cycle of an Ascomycete. Asexual reproduction takes place by way of specialized spores, known as conidia. Sexual reproduction involves the formation of asci and ascospores.

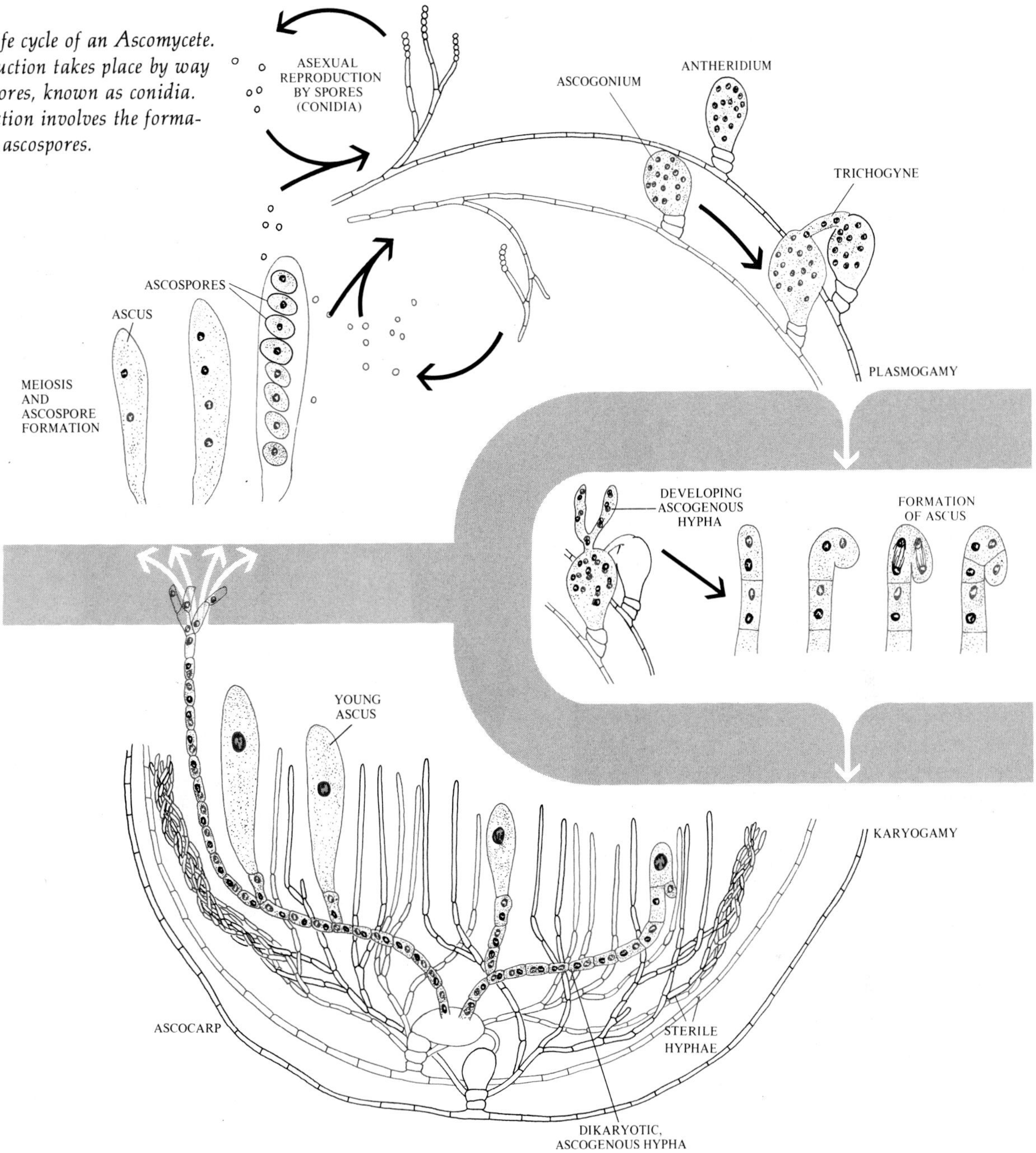

11–19
Yeasts. (a) *Budding cells of* Saccharomyces cerevisiae, *the common bread yeast.* (b) *Asci with ascospores of* Schizosaccharomyces octosporus.

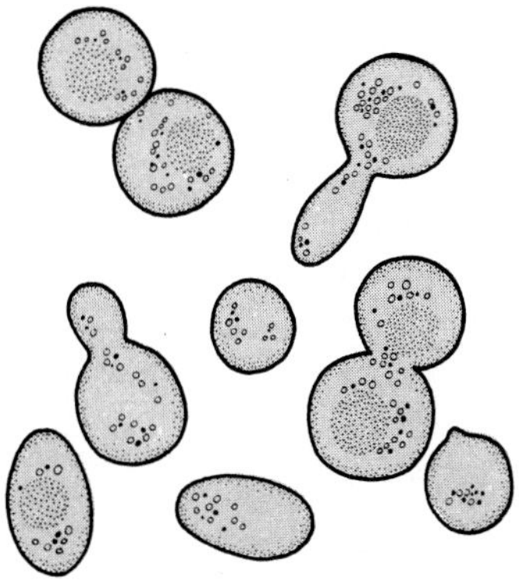
(a)

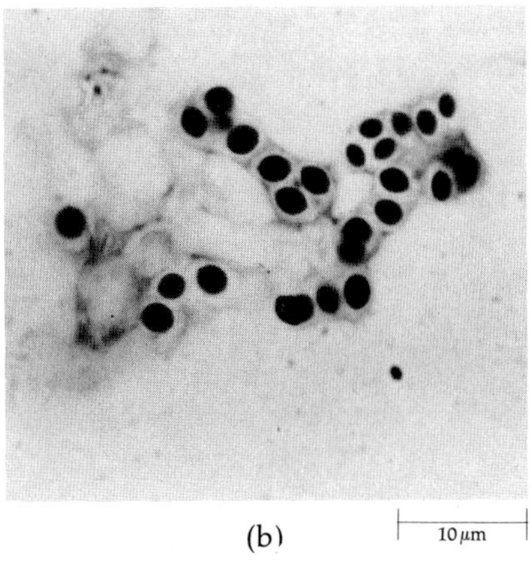

(b)

either bud asexually or fuse with another cell to repeat the sexual process.

The yeasts are important to man because of their ability to ferment carbohydrates, breaking down glucose to produce ethyl alcohol and carbon dioxide in the process. Thus yeasts are utilized by brewers and vintners and also by bakers, alcohol being the important product for the former and carbon dioxide for the latter. Many domestically useful strains of yeast have been developed by selection and breeding; but wild species are also of great importance, as in the fermentation of wine, in which the yeast species that are naturally present are often supplemented, at the present time, by pure cultures. In brewing beer, on the other hand, only pure yeast cultures are used, because the medium is sterilized by heating prior to fermentation. Some of the flavors of wine come directly from the grape, but most arise from the direct action of the yeast. Most of the yeasts important in the production of wines, cider, saké, and beer are strains of a single species, *Saccharomyces cerevisiae,* although other species also play a role. In baking bread, this yeast is virtually the only one used in modern times. Some species of yeast are important as human pathogens, causing such diseases as thrush and cryptococcosis. A number of them, especially *Saccharomyces cerevisiae,* are exceedingly important laboratory organisms for genetic research.

Superficially the yeasts appear to be simple, and perhaps primitive, forms of the Ascomycetes. It seems more probable, however, that their ancestors were the more complex mycelium-forming fungi. There are some 39 genera with 350 recognized species, found in a wide range of terrestrial and aquatic habitats in which a suitable carbon source is available.

The Fungi Imperfecti

The class Fungi Imperfecti or Deuteromycetes is a large group comprising some 25,000 described species of fungi in which the sexual phase (the perfect phase) is unknown (Figure 11-20). The sexual phase may be unknown either because the fungi have not been sufficiently studied or because it apparently has become extinct. Most imperfect fungi are clearly Ascomycetes which apparently reproduce only by means of conidia (Figure 11-21); asci and ascospores—the sexual, or perfect phase—have not been associated with them. A few are Basidiomycetes, as shown by the septa and clamp connections characteristic of that group, which will be discussed on page 232. Parasexual cycles are common and seem to compensate partially for the lack of sexual reproduction in maintaining variability. In general, the classification of the Fungi Imperfecti is based upon the mode of formation of their conidia, which is now being analyzed with the aid of transmission and scanning electron microscopy.

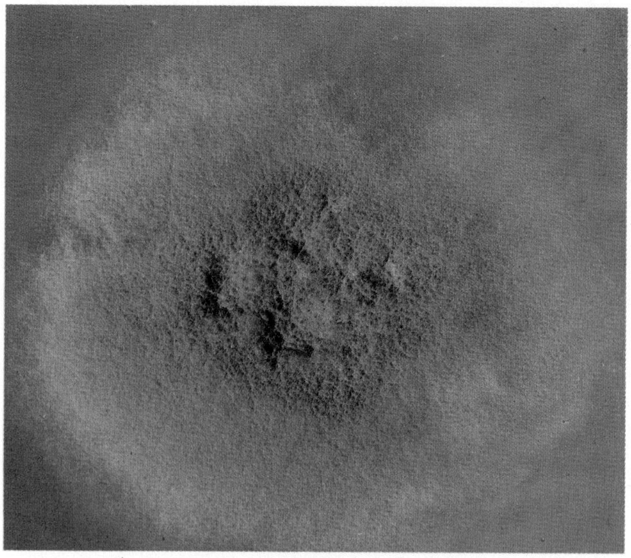

(a)

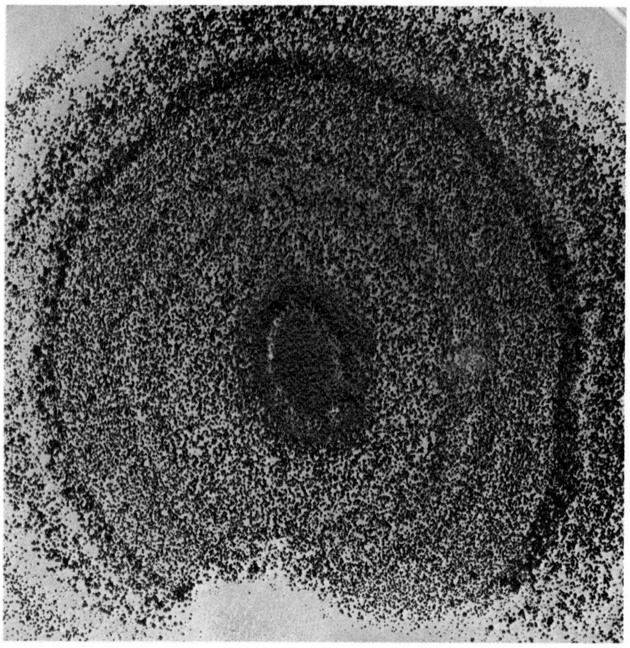

(b)

11-20
Two of the most common genera of Fungi Imperfecti, Penicillium *and* Aspergillus. (a) Penicillium dupontii. (b) Aspergillus niger, *showing the concentric growth pattern produced by successive "pulses" of spore production.*

PREDACEOUS FUNGI

Among the most highly specialized of the fungi are the predaceous fungi that have developed a number of mechanisms for capturing small animals which they use as food. Some secrete a sticky substance on the surface of their hyphae in which passing protozoa, rotifers, small insects, or other animals become glued. More than 50 species of fungi capture small roundworms (nematodes) that abound in the soil. In the presence of a population of roundworms (or even of water in which the worms have been growing), the hyphae of the fungi produce loops which swell rapidly, closing the opening like a noose, when a nematode rubs against its inner surface. Presumably the stimulation of the cell wall increases the amount of osmotically active material in the cell, causing water to enter the cells and increase their turgor with rapidity. (a) *The predaceous imperfect fungus* Arthrobotrys dactyloides *has trapped a nematode. The traps consist of rings, each comprising three cells, which swell rapidly to about three times their original size and garrote the nematode. Once the worm has been trapped, fungal hyphae grow into its body and digest it. When triggered, the ring cells can expand completely in a period of less than a tenth of a second.* (b) *Another nematode-trapping fungus,* Dactylella drechsleri. *This species traps the worms with small adhesive knobs.*

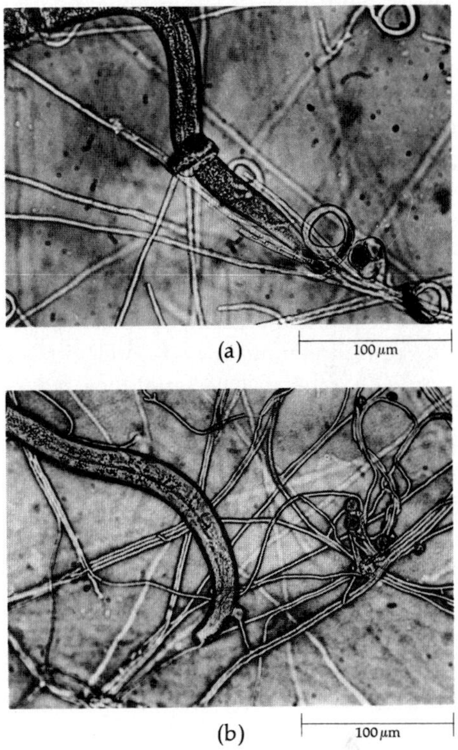

There are many thousands of Fungi Imperfecti, including a number that are of great economic importance. For example, certain members of the genus *Penicillium* give to some types of cheeses the flavor, odor, and character so highly prized by gourmets. One such mold, *P. roquefortii,* was first found in caves near the French village of Roquefort. Legend has it that a peasant boy left his lunch, a fresh piece of mild cheese, in one of these caves and on returning several weeks later found it marbled, tart, and redolent. Only cheeses from the area around these particular caves are permitted to bear the name of Roquefort. Another species of the same genus, *P. camembertii,* gives Camembert cheese its special qualities. In the Orient, soy sauce and soy paste are produced by fermenting soybeans with *Aspergillus tamarii* and other Fungi Imperfecti. *Aspergillus oryzae* is important in the initial steps in brewing saké, the traditional alcoholic beverage of Japan; the yeast *Saccharomyces cerevisiae* is important later in the process. Citric acid is produced commercially in large amounts from colonies of *Aspergillus* grown under very acid conditions. Recently, the enrichment of foods for livestock by fermentation, as by *Aspergillus oryzae,* to increase the protein content, has been investigated at several experimental farms in Europe and America.

Antibiotics are substances produced by one living organism that injure other living organisms (such as bacteria) and so may be therapeutically useful to man. Some

11–21
The characteristic conidiophores of Aspergillus *(tightly clumped) and* Penicillium *(more open branching).*

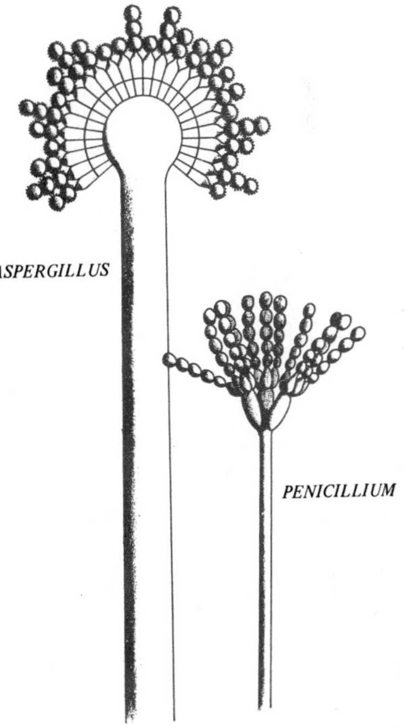

important antibiotics are produced by Fungi Imperfecti. The first antibiotic was discovered by Sir Alexander Fleming, who noted in 1928 that a strain of *Penicillium* that had contaminated a culture of *Staphylococcus* growing on a nutrient agar plate had completely halted the growth of the bacteria. Ten years later Howard Florey and his associates at Oxford purified penicillin and later came to the United States to promote the large-scale production of the drug. The demand during World War II was so great that the production of penicillin was increased from a few million units per month in 1942 to more than 700 billion units in 1945. Penicillin is effective in curing a wide variety of bacterial diseases, including not only pneumonia but also scarlet fever, syphillis, gonorrhea, diphtheria, rheumatic fever, and many others. Many hundreds of antibiotics have now been discovered, and some of these undoubtedly play a significant ecological role in nature also.

Another group of Fungi Imperfecti, the dermatophytes, are the cause of ringworm and of athlete's foot (or ringworm of the foot), diseases which are especially prevalent in the tropics. During World War II, more soldiers had to be sent back from the South Pacific because of skin infections than because of wounds sustained in battle. Another fungus, *Candida albicans*, which is yeastlike under certain conditions, causes thrush and other infections of the mucous membranes. Fungal spores are continually inhaled, and some cause internal diseases, which, especially if they attrack the lungs, may be serious or fatal.

In recent years, fungi that cause diseases in humans, such as *Candida* and *Aspergillus fumigatus*, have become more frequent. The chemicals administered to transplant patients to suppress their normal immune reactions, so that they will be able to accept transplanted organs and tissues, similarly reduce the patients' defenses against fungi and make them much more susceptible to fungal disease. Some chemical treatments, such as those for acute leukemia, also seem to reduce defenses against fungal disease, and therefore more medical attention is now being directed toward such diseases.

The Lichens

The lichens are a large group of Ascomycetes* that can grow only in intimate association with living algal cells. Obtaining nutrition from these algae, they have invaded the harshest habitats, such as bare rock (Figure 11–22), and have diversified into about 25,000 species of varied appearance. One, *Verrucaria serpuloides*, is even a permanently submerged marine lichen. Algae from about 26 different genera, including blue-green algae, are found in combination with these Ascomycetes. The most frequent are the green algae *Trebouxia* and *Trentepohlia* and the blue-green alga *Nostoc;* one of the three is found in about 90 percent of all lichens.

11–22
A mosiac of crustose lichens growing on a bare rock surface.

Lichens are extremely widespread in nature; they occur from arid desert regions to the arctic and grow on bare soil, tree trunks, sunbaked rocks, fence posts, and wind-swept alpine peaks all over the world (Figure 11–23). They are often the first colonists of bare rocky areas. In Antarctica there are more than 350 species of lichens but only two species of vascular plants. Seven of the species of lichens occur in the Queen Maud Mountains, at 86°03′ S latitude. The colors of lichens range from white to black, through shades of red, orange, yellow, and green, and there are many unusual chemical compounds in them. Some of them are so tiny that they are almost invisible; others, like the reindeer mosses, may cover kilometers of land with ankle-deep growth. Except for the reindeer mosses, the lichens are of little direct economic importance to man. However, they have long been the subject of biological investigations because of the intriguing nature of the association between fungus and its included algae. The fungal partner apparently plays the major role in determining the form of the lichen; recently, however, it has been demonstrated that a single fungus with different types of algae can produce morphologically very different individuals traditionally placed in different genera. The algae found in lichens also occur as free-living species, whereas the lichen fungi are generally found in nature only in the lichens.

Some lichens produce special fragments of the thallus called *soredia* composed of fungal hyphae and algae (Fig-

* There are also about a dozen species of Basidiomycetes that form associations with algae, but they are very closely related to free-living groups; they do not form a distinctive group within the lichens, which are otherwise all Ascomycetes.

11–23
Lichens. (a) *A foliose ("leafy") lichen* (Parmelia perforata) *growing on dead branches of a tree in Mississippi.* (b) *Old man's beard* (Usnea *sp.*), *a hanging lichen that often occurs in masses on the limbs of trees.*

(a)

(b)

11–24
Fruticose ("shrubby") lichens. (a) *Goldeye lichen,* Teloschistes chrysophthalmus, (b) *British soldiers,* Cladonia cristatella. *The individuals are about 3 millimeters tall.* (c) Cladonia subtenuis, *widely called reindeer "moss" even though it is actually a lichen.*

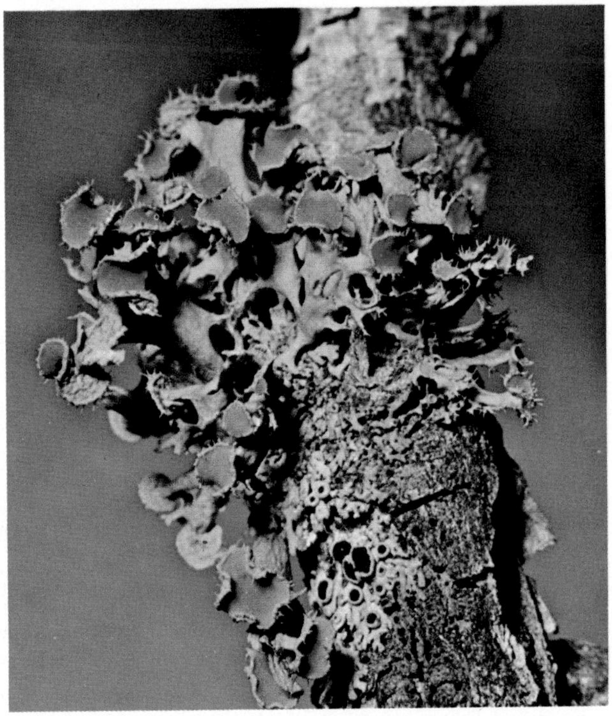

(a)

(b)

(c)

11–25
Cross section of a lichen (Lobaria verrucosa). *The simplest lichens consist of a crust of fungal hyphae entwining scattered colonies of algae. In the higher lichens, however, the hyphae and the algal cells are organized in a thallus with a definite growth form and characteristic internal structure. The lichen shown here has four distinct layers: (1) the upper cortex, a protective surface of heavily gelatinized fungal hyphae; (2) the algal layer, which consists of algal cells and loosely interwoven, thin-walled hyphae; (3) the medulla, which is a thick layer of loosely packed, colorless, weakly gelatinized hyphae. This layer, which makes up about two-thirds of the thickness of the thallus, appears to serve as a storage area, with enlarged fat cells in the fungal hyphae; and (4) the lower cortex, which is thinner than the upper cortex and covered with fine projections that attach it to the substrate. Soredia are fragments of algal cells and fungal hyphae by which the lichens reproduce and spread.*

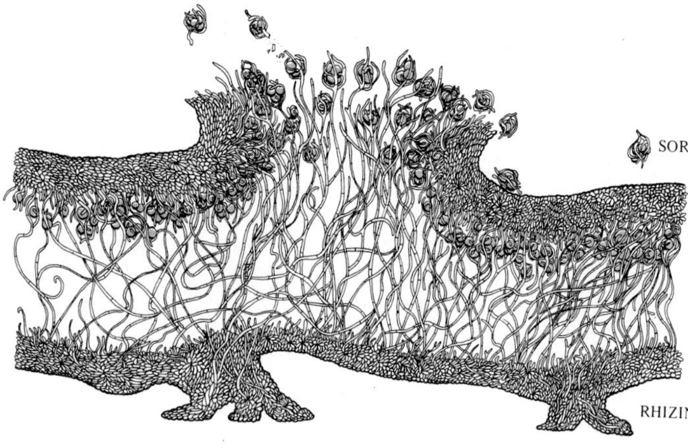

ure 11–25). Lichen fungi also often form ascocarps (Figure 11–26), which are similar in every way to those of nonlichenized fungi except that in lichens they may be durable and produce spores continuously over a number of years.

Biology of the Lichens

Why can the lichens survive under environmental conditions so adverse to any other form of life? At one time, it was thought that the secret of the lichen's success was that the fungal tissue protected the alga from drying out. Actually, one of the chief factors in lichen survival seems to be the fact that they dry out very rapidly. Lichens are frequently very desiccated in nature, with a water content ranging from only 2 to 10 percent of their dry weights. When the lichen dries out, photosynthesis ceases; and in this state of "suspended animation" blazing sunlight or great extremes of heat or cold can be endured. Cessation of photosynthesis depends, at least in large part, on the fact that the upper cortex of the lichen becomes thicker and more opaque when dry, cutting off the passage of light energy. A wet lichen is destroyed by light intensities or temperatures that do not harm a dry lichen.

When a lichen is wetted by rain, it absorbs 3 to 35 times its own weight in water in a very short time. If a dry, brittle lichen thallus is submerged in water, it will become soft and pliable within a few minutes. This is the simple physical process of imbibition—the lichen takes up water in much the same way a blotting paper does—and a dead lichen absorbs about as much water as a live one.

The lichen reaches its maximum vitality, as judged by the rate of photosynthesis, after it has been soaked with water and begun to dry. Its rate of photosynthesis reaches a peak when the water content is 65 to 90 percent of the maximum it can hold; below this level, if the lichen continues to lose water, the rate of photosynthesis decreases. In many environments, the water content of the thallus varies markedly in the course of a day, with most photosynthesis taking place only during a few

11–26
Cross section of a lichen showing the apothecium, a saucer-shaped open ascocarp.

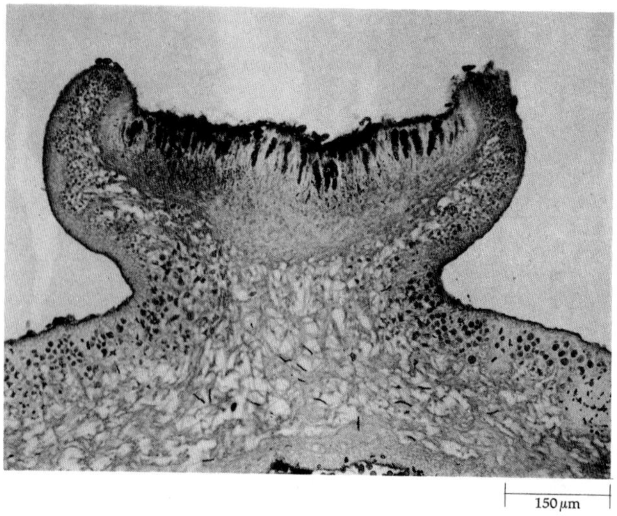

hours, usually in the early morning after wetting by fog or dew. As a consequence, lichens are geared to an extremely slow rate of growth, their radius increasing at a rate of about 0.1 to 10 millimeters a year. Calculated on this basis, some mature lichens may be as much as several thousand years old. They achieve their most luxuriant growth development only on fog-shrouded mountains (Figure 13–1, page 281) and coasts.

Lichens apparently absorb some minerals from their substrate (this is suggested by the fact that particular species are characteristically found on particular kinds of rocks or soil or tree trunks), but most of the elements absorbed by lichens enter the body through the air and in rainfall. Lichens absorb elements from rainwater rapidly and concentrate them within their thallus. Because they have no means of excreting these elements, lichens are particularly susceptible and sensitive to toxic compounds, and lichen growth is a very sensitive indication of the toxic components of polluted air, and especially of sulfur dioxide. Absorption of toxic compounds by lichens causes degradation of the chlorophyll.

In at least one instance, the capacity of lichens to absorb substances from rainfall led to surprising consequences. In studies of fallout following atomic bomb testing, it was found that Alaskan Eskimos and Scandinavian Laplanders had unusually high levels of radioactivity in their bodies. These findings were totally unexpected because it had been calculated that the amount of fallout reaching the ground at the poles was much less than in the temperate regions. The significant factor was the reindeer mosses (actually lichens), which had absorbed the isotopes from the atmosphere and concentrated them. As their name implies, the reindeer mosses are the chief source of food for the reindeer and caribou, which in turn are the chief source of food for the Eskimos and Laplanders, who thus became the repositories of radioactive strontium and cesium.

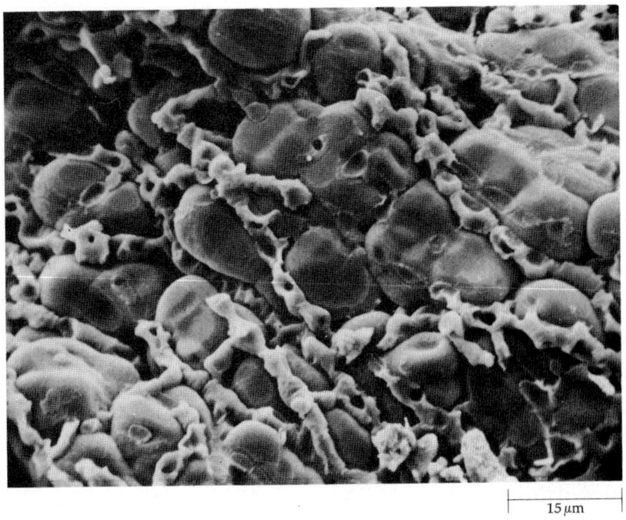

11–27
Scanning electron micrograph of algal layer of the lichen Heppia lutosa, *showing the fungal hyphae growing among the larger algal cells.*

The Nature of the Relationship Between Algae and Fungi

What is the relationship between the two components of a lichen? It is clear that the fungus obtains nutrients from the alga, because the lichen behaves like a photosynthetic organism, dependent only on light, air, and minerals. In fact, the movement of nutrients from the included algae to the fungus has been established by labeling techniques with ^{14}C-labeled glucose. In those lichens that include blue-green algae, such as *Nostoc*, nitrogen fixation by the alga and transfer to the fungus are also important. In the lichen, the fungal hyphae form a close network around the algal cells (Figure 11–27). In most lichens, the haustoria (the specialized hyphae of parasitic fungi) are closely appressed to the algal cells, only rarely penetrating the protoplast. The association with a fungus profoundly affects the nature of the metabolic output from the algal cells (Figure 11–28).

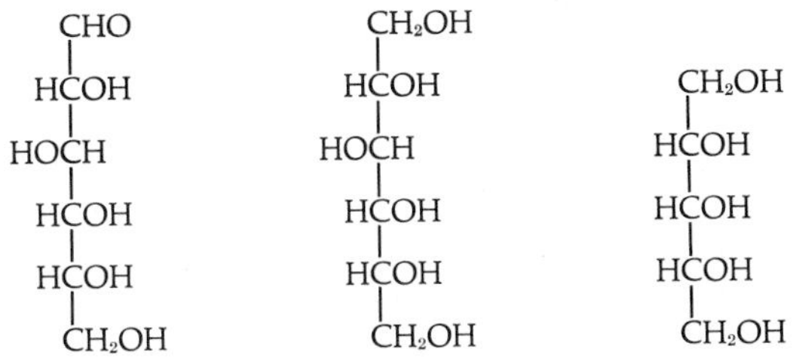

D-GLUCOSE D-SORBITOL D-RIBITOL

11–28
Three compounds produced by lichen algae. Blue-green algae, when they are components of lichens, usually excrete D-glucose. No green alga is known to excrete a sugar under these circumstances. The two alcohols, D-sorbitol and D-ribitol, are excreted in large quantities by different green algae when they are associated with fungi to form lichens. In fungus-free cultures, the metabolic output of the cells is much less and is much more varied.

It is possible to separate the algal and fungal components of a particular lichen and grow them alone in pure culture. Under such conditions, the fungus grows in compact colonies, unlike the lichen thallus. It requires a large number of complex carbohydrates for growth and does not ordinarily form fruiting bodies. Isolated lichen algae grow more rapidly when they are free-living than when they are in the lichen association. Thus it may be most appropriate to think of the lichen partnership as controlled parasitism of the alga by the fungus, rather than symbiosis.

The principal objective of most lichenologists for a number of decades—indeed, ever since separation of the lichen parts was first achieved—was to synthesize a lichen from its composite algae and fungi. This proved astonishingly difficult for reasons that did not become clear until synthesis was finally achieved in the 1960s by Vernon Ahmadjian, then of Clark University. Ahmadjian found that the two forms will establish a lichenlike association only under conditions unfavorable to separate growth, and a natural lichen will separate into its two components under conditions that are favorable to the growth of either partner. In the course of trying to establish "lichenization," he also discovered that the lichen fungi seemed to damage the algal cells and sometimes completely destroyed them.

Contrary to the belief of many earlier lichenologists, who believed that new lichens were established almost exclusively by combinations of algae and fungi from already existing lichens, Ahmadjian proposes that the fungal fruiting bodies so characteristic of the lichens are not useless structures left over from a previous free-living existence but play an important role in lichen production. According to his hypothesis, wind-blown spores from lichen fungi germinate when they come in contact with free-living algal cells, which they then parasitize. Algal cells that can withstand the association survive, and another lichen is produced. If the algal cells cannot survive, both forms perish.

Class Basidiomycetes

The most familiar of all fungi are members of this large class, which includes some 25,000 described species—not only mushrooms, toadstools, stinkhorns, puffballs, and shelf fungi (Figure 11–29), but also those important plant pathogens, the rusts and smuts. The Basidiomycetes are distinguished from all other fungi by the production of *basidiospores,* which are borne outside a club-shaped spore-producing structure, the *basidium.* Many more species undoubtedly await discovery, although the larger Basidiomycetes are the best known of all fungi.

The mycelium of the Basidiomycetes is always septate and in most species passes through three distinct phases—primary, secondary, and tertiary—during the life cycle of the fungus. On germinating, a basidiospore produces the primary mycelium. Initially this mycelium may be multinucleate, but septa are soon formed and the mycelium is divided into monokaryotic (uninucleate) cells. Commonly, the secondary mycelium is produced

11–29
Some representative Basidiomycetes. (a) *Fly agaric,* Amanita muscaria. *Mushrooms at various stages of growth, with one picked to show the gills. This genus of poisonous mushrooms can be recognized by the ring on the stalk and the cup around the base.* (b) *Puffballs,* Lycoperdon ericeterum. *The spores are being discharged from the pore at the top of each puffball and dispersed by the wind.* (c) *Corn smut,* Ustilago maydis. *In the familiar disease of corn* (Zea mays) *caused by this fungus, black, dusty-looking masses of spores are produced in the ears.*

(a)

(b)

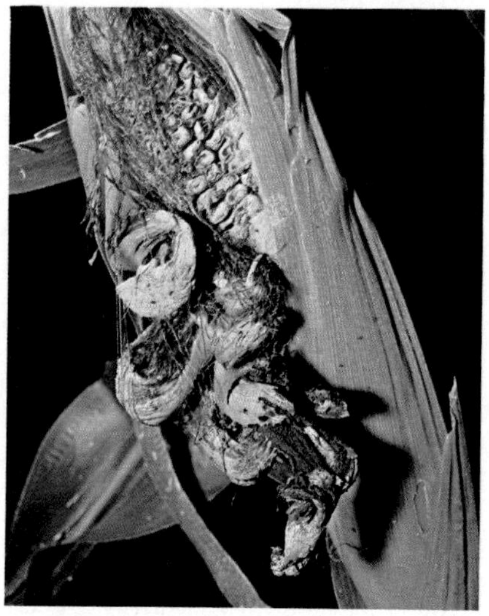
(c)

by fusion of primary hyphae from different mating types—in which case it is heterokaryotic—or it may arise when septa are not formed after nuclear formation—in which case it is homokaryotic. In either case the result is formation of a dikaryotic (binucleate) mycelium, for karyogamy (fusion of gametic nuclei) does not immediately follow plasmogamy (fusion of protoplasts).

The apical cells (those at the apex or tip) of the secondary mycelium usually divide by the formation of clamp connections (Figure 11-30). These clamp connections, which ensure the allocation of one nucleus of each type to the daughter cells, are highly characteristic of the Basidiomycetes. The septa in the secondary mycelia of Basidiomycetes are perforated, but, except in the rusts and smuts, they are lined with a thick, barrel-shaped structure (Figure 11-31). Nuclei may not be able to pass through such septa. In heterokaryotic secondary mycelia in which nuclear migration appears to be taking place, septa are observed that have only simple, enlarged pores, which are similar to those of the Ascomycetes (Figure 11-7).

The tertiary mycelium arises directly from the secondary mycelium and forms the fruiting bodies—*basidiocarps*—of the so-called higher fungi, such as mushrooms and puffballs. Like the secondary mycelium, the tertiary mycelium is dikaryotic. The formation of the basidiocarps may require light.

The Basidiomycetes are divided into two subclasses, the Homobasidiomycetes and the Heterobasidiomycetes.

11-30
(a) *In the Basidiomycetes, dikaryotic hyphae are distinguished by the presence of clamp connections over the septa. These clamp connections are temporary bridges formed during cell division and presumably ensure the proper distribution of the two genetically distinct types of nuclei in the basidiocarp.* (b) *Clamp connections and characteristic septa in a hypha of* Coprinus lagopus.

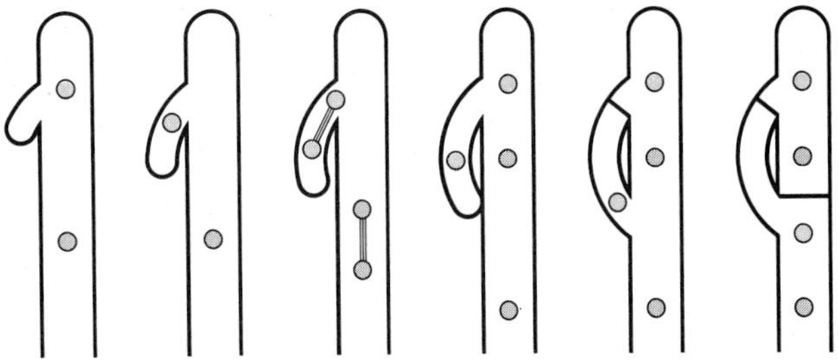

(a)

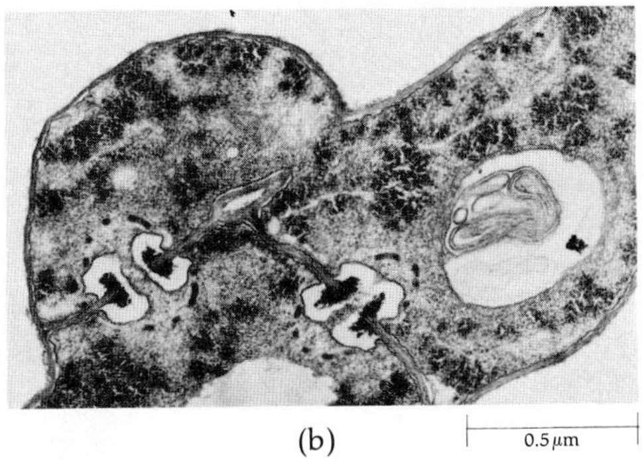

(b)

11-31
Characteristic septum of the secondary mycelia of Basidiomycetes. Shown here is an electron micrograph of a septum in Poria latemarginata, *a common wood-rotting fungus.*

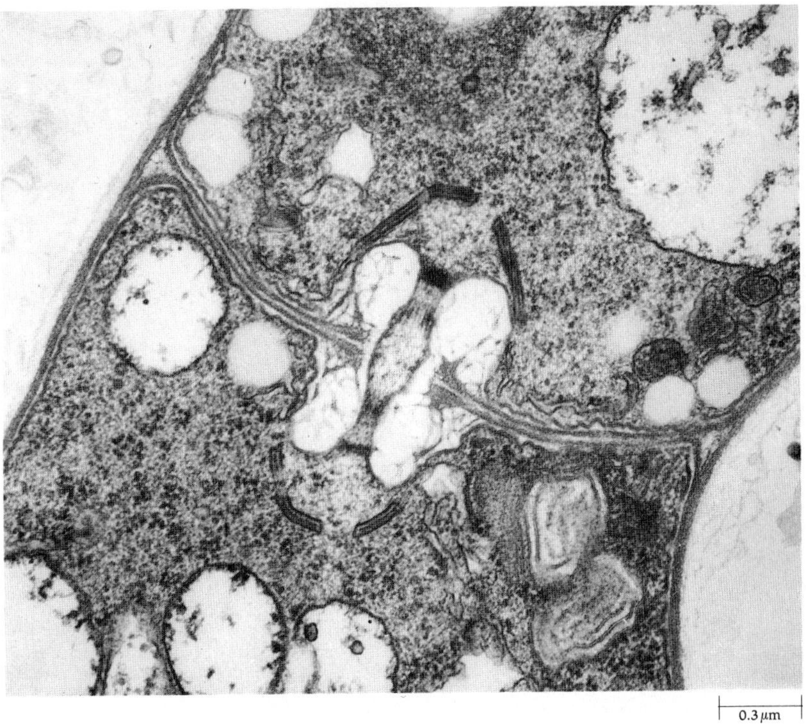

(a) (b) (c) (d)

11-32
Homobasidiomycetes. (a) *A bolete,* Boletus bovinus. *In this group of mushrooms, the gills are replaced by pores.* (b) *Stinkhorn,* Phallus impudicus. *The basidiospores are released in a foul-smelling, sticky mass at the top of the fungus. Flies, such as the individuals of* Mydaea urbana *shown here, visit it for food and spread the spores, which adhere to their legs and bodies in great numbers.* (c) *A shelf fungus,* Polyporus. *Members of this group are responsible for most wood rot.* (d) *A coral fungus,* Clavaria formosa, *in Alaska. The hymenium, the outer sporebearing layer, is borne on all sides of the basidiocarp.*

Subclass Homobasidiomycetes

It is the subclass Homobasidiomycetes to which the mushrooms, shelf fungi, stinkhorns, earthstars, bird's-nest fungi and puffballs belong (see Figures 11-32 and 33). All members of the Homobasidiomycetes produce basidiocarps, which are comparable to the ascocarps of the Ascomycetes. In addition to the presence of basidiocarps, the Homobasidiomycetes are characterized by their club-shaped, aseptate basidia, each of which usually bears four basidiospores on minute, projections, or *sterigmata* (Figure 11-34).

What one recognizes as a mushroom or toadstool is a basidiocarp. (Although "mushroom" is sometimes used to designate the edible forms of basidiocarps and "toadstool" the inedible ones, the dividing line between these two is not clear. In this book, we shall refer to all such forms as mushrooms; this does not mean that they are all edible.) The mushroom consists of a *cap* or *pileus* subtended by a *stalk* or *stipe*. Early in its development—the "button" stage—the mushroom may be covered by a membranous tissue, which ruptures as the mushroom enlarges. In some genera, remnants of this tissue are visible on the upper surface of the cap and, as a cup, or *volva*, at the base of the stipe. The lower surface of the cap consists of radiating strips of tissue called *gills*.

11–33
Two puffballs. (a) *Bird's-nest fungus,* Cyathus striatus. *The lentil-shaped structures in the "nests" contain the basidiospores and are splashed out and thus dispersed by raindrops.* (b) *Earthstar,* Geaster triplex. *In effect, the outer layer of a typical puffball is folded back in this genus.*

(a)

(b)

11–34
Coprinus *sp., a common mushroom.* (a) *Section through gills. The relatively dark margins of the gills constitute the hymenium, or hymenial layer.* (b) *Hymenial layer showing developing basidia and basidiospores.* (c) *Nearly mature basidiospores attached to the basidium by sterigmata.*

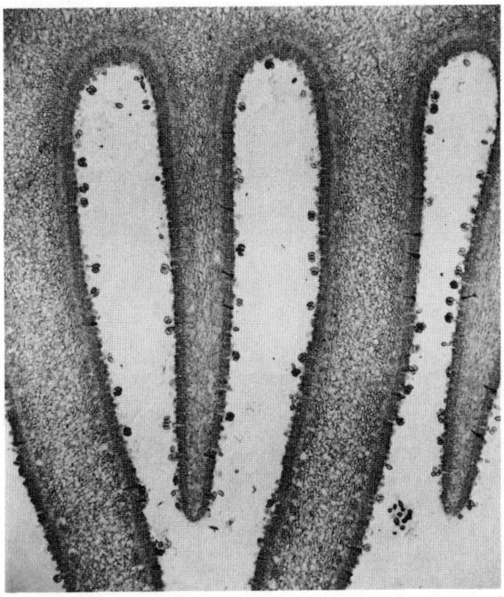

(a) 150 μm

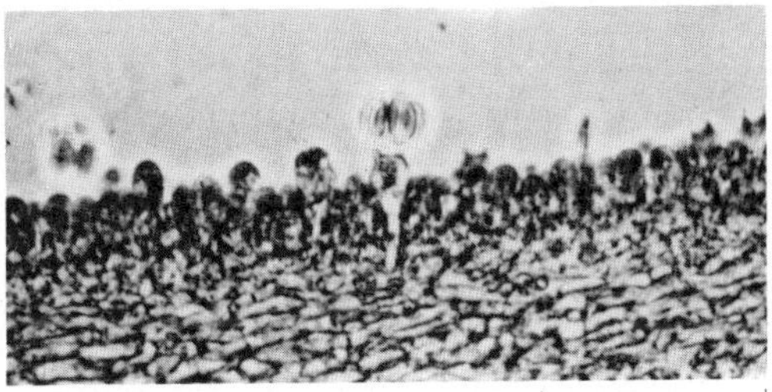

(b) 50 μm

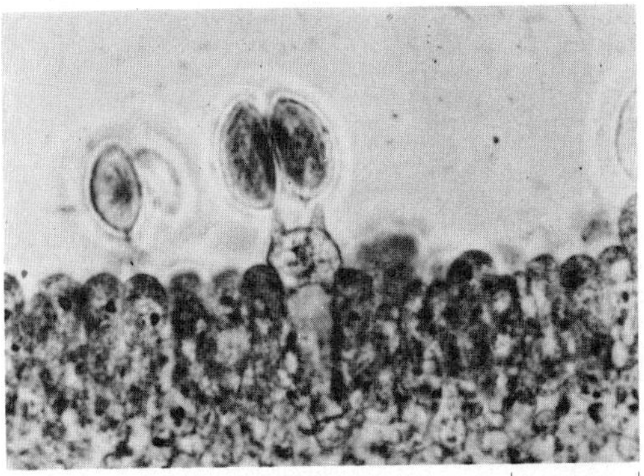

(c) 15 μm

The mycelium from which mushrooms are produced spreads underground, forming a ring, which may grow as large as 30 meters in diameter. In an open area, the mycelium expands evenly in all directions, dying at the center and fruiting at the outer edges, where it grows most actively because this is the area in which there is the most fresh nutritive material. As a consequence, the mushrooms appear in rings, and, as the mycelium grows, the rings become larger and larger in diameter. Such circles of mushrooms are known in European folk legend as "fairy rings" (see Figure 11–35).

The best-known mushrooms belong to the group

11–35
"Fairy ring" formed by the parasol mushroom, Lepiota procera.

(a)

(b)

11–36
Several groups of Indians in southern Mexico and Central America ingest certain Basidiomycetes for their hallucinogenic qualities. The mushrooms figure prominently in their religious ceremonies. One of the most important of the mushrooms is Psilocybe mexicana, *shown here* (a) *growing in a pasture near Huautla de Jiménez, Oaxaca, Mexico.* (b) *The shaman Maria Sabina is ingesting* Psilocybe *in the course of a midnight religious ceremony. Psilocybin* (c) *is the chemical responsible for the colorful visions experienced by those who eat the sacred mushrooms.*

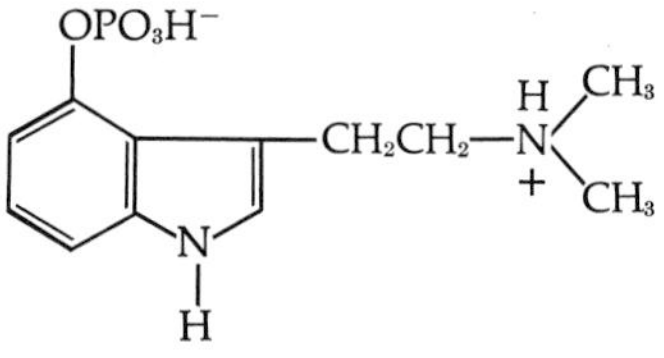

PSILOCYBIN

(c)

known as the gill fungi, which includes *Agaricus campestris*, the common field mushroom. The closely related *A. bisporus* is one of a very few mushrooms that can be cultivated commercially, and some 60 thousand metric tons are produced annually in the United States alone. The gill fungi also include the mushrooms of the genus *Amanita*, which are the most highly poisonous of all mushrooms. Even one bite of the white *Amanita phalloides*, the "destroying angel", can be fatal. Some Basidiomycetes contain chemicals which impart hallucinogenic qualities to them (Figure 11–36).

In nature most Homobasidiomycetes, including mushrooms, reproduce primarily through the formation of basidiospores. A representative life cycle of a mushroom is shown in Figure 11–37. The gills of the mushroom are the sites of basidia and basidiospore formation. The basidium arises on the surface layer or hymenium of the gill as a terminal cell of a dikaryotic hypha. Soon after the young basidium enlarges, karyogamy takes place. This is followed almost immediately by meiosis of the diploid nucleus, resulting in the formation of four haploid nuclei. Each of the four nuclei then migrates into a sterigma, which enlarges at its tip to form a uninucleate basidiospore. The reproductive capacity of a single mushroom is tremendous, with billions of spores being produced by a single basidiocarp. A single puffball may produce several trillion basidiospores.

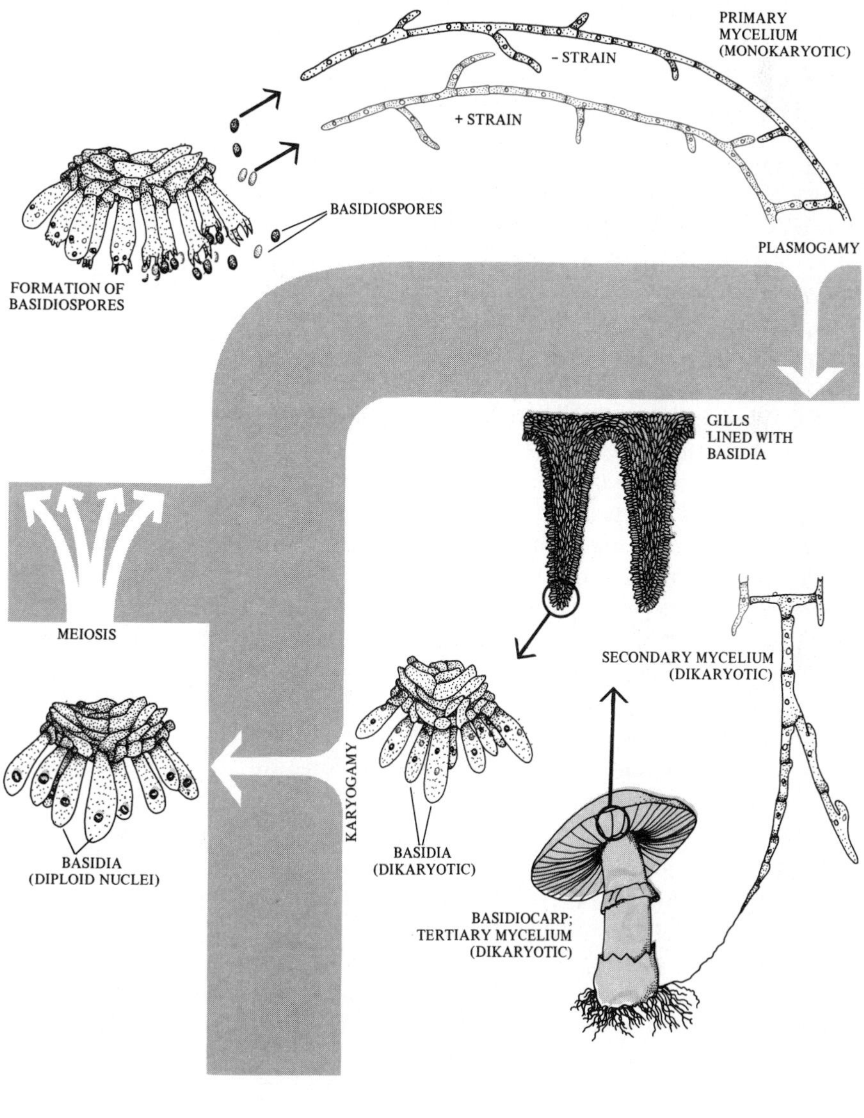

11–37
Life cycle characteristic of the Homobasidiomycetes. The primary mycelia are produced from basidiospores. Secondary dikaryotic mycelia are formed from the primary mycelia. Some secondary mycelia are formed by the fusion of hyphae from different mating types, in which case the mycelia are heterokaryotic. The secondary mycelia divide and differentiate to form the tertiary mycelia that make up the basidiocarp.

Subclass Heterobasidiomycetes

The subclass Heterobasidiomycetes consists of the rusts, smuts, and jelly fungi (Figure 11–38). The jelly fungi are mostly saprobes and like the Homobasidiomycetes, produce basidiocarps. The rusts and smuts are parasitic on vascular plants. They do not form basidiocarps, but produce their spores in clusters, or *sori*. As a group, the Heterobasidiomycetes are characterized by their septate (multicellular) basidia.

The rusts and smuts are of tremendous economic importance, causing millions of dollars worth of damage to crops throughout the world each year.

The life cycles of the rusts may be very complex, and these pathogens are a continual challenge to the plant pathologist whose task it is to keep them under control. Until recently, the rusts were thought to be obligate parasites on vascular plants, but now several species have been maintained in artificial culture. Some smuts are also capable of completing their development under these conditions. A representative smut is shown in Figure 11–29c.

Puccinia graminis, the cause of black stem rust of wheat, will serve to illustrate one of the life cycles of the parasitic Heterobasidiomycetes. It is one of some 7000 species of rusts. Numerous strains or races of *P. graminis* exist and, in addition to wheat, they parasitize other cereal grains such as barley, oats, and rye, and various species of wild grasses. *P. graminis* is a continual source of economic loss for the wheat grower. In one year alone losses in Minnesota, North Dakota, South Dakota, and western Canada amounted to nearly 8 million metric tons. As early as 100 A.D., Pliny described wheat rust as "the greatest pest of the crops." Today plant pathologists combat wheat rust largely by breeding resistant wheat varieties, but mutation and recombination in the rust make any advantage short-lived.

Puccinia graminis is *heteroecious*; that is, it requires two different hosts to complete its life cycle. (*Autoecious* parasites require only one host.) *Puccinia graminis* must spend part of its life cycle on the common barberry, *Berberis vulgaris,* and part on a grass host. One method of attempting to control *P. graminis* has been to eradicate the barberry. In 1755 the crown colony of Massachusetts passed a law ordering "whoever . . . hath any barberry bushes growing in his or their land . . . shall cause the same to be extirpated or destroyed on or before the thirteenth day of June, A.D. 1760."

Infection of the barberry occurs in spring when uninucleate basidiospores infect the barberry leaves. Flask-shaped *spermagonia* (Figure 11–40a and b) are formed primarily on the upper surfaces of the leaves. The form of *P. graminis* that grows on barberry consists of separate + and − strains, so that the basidiospores and the spermagonia derived from them are either + or −. Each spermagonium contains two types of hyphae, one that produces chains of small cells called *spermatia* and the

11–38
A jelly fungus, one representative of the characteristic groups of Heterobasidiomycetes, growing on a dead limb in the Amazon forest of Brazil.

so-called *receptive hyphae*. The spermatia are discharged from an opening in the spermagonium. If a + spermatium of one spermagonium comes in contact with a − receptive hypha of another spermagonium, or vice versa, plasmogamy takes place and dikaryotic hyphae are produced. This brings about dikaryotization of *aecial initials*, which extend downward from the spermagonium. *Aecia* are then formed primarily on the lower surface of the leaf, where they produce dikaryotic *aeciospores* (Figure 11–40a). The aeciospores then infect the wheat.

The first external manifestation of infection on the wheat is the appearance of rust-colored, linear streaks on the leaves and stems (the red stage). These are *uredinia* with unicellular, dikaryotic *uredospores* (Figure 11–40c). Uredospores are produced throughout the summer and reinfect the wheat. In late summer and early fall the red-colored sori gradually become darkened in color and become *telia* (Figure 11–40d) with two-celled, dikaryotic *teliospores* (the black stage). The teliospores are overwintering spores, which infect neither the wheat nor the barberry. Some time prior to germination in spring, the two haploid nuclei fuse with one another to form a diploid nucleus. With the onset of germination, meiosis takes place. A short, cylindrical basidium emerges from each of the teliospore cells, into which the four haploid nuclei migrate. Septations are laid down between the nuclei, which then migrate into the sterigmata and develop into basidiospores. The cycle is now completed.

11-39
Life cycle of Puccinia graminis, *a rust.*

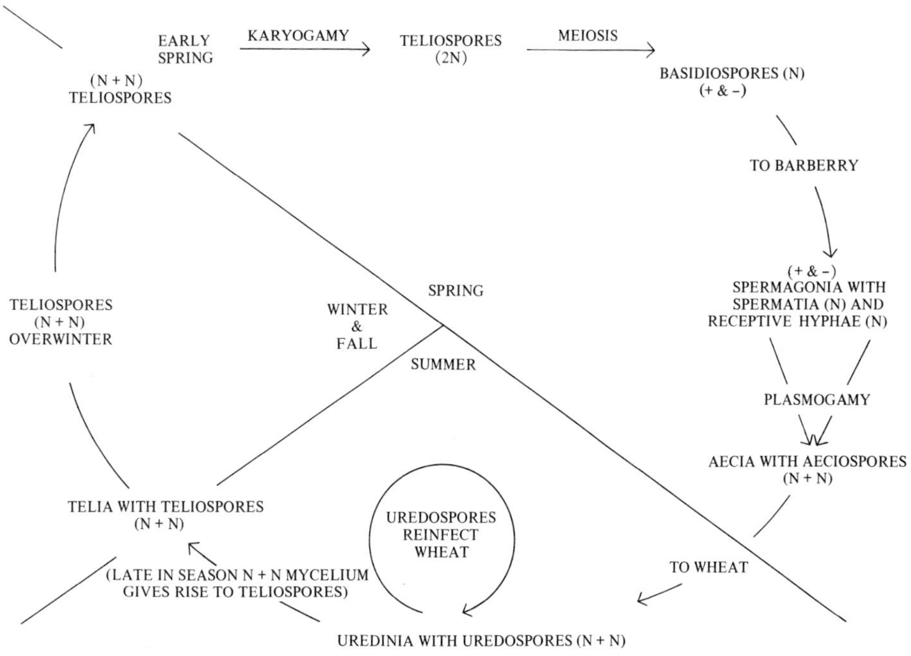

11-40
Some stages in the life cycle of Puccinia graminis, *the cause of black stem rust of wheat.* (a) *Transverse section of infected barberry leaf showing a spermagonium on the upper surface and two aecia on the lower surface. Each aecium contains many chains of aeciospores.* (b) *A spermagonium on the upper surface of a barberry leaf. Note hyphae projecting through the opening in the spermagonium and spermatia-bearing hyphae lining its wall.* (c) *Section of uredinium, with dikaryotic uredospores, on stem of wheat.* (d) *Section of telium, with the two-celled, dikaryotic teliospores, on stem of wheat.*

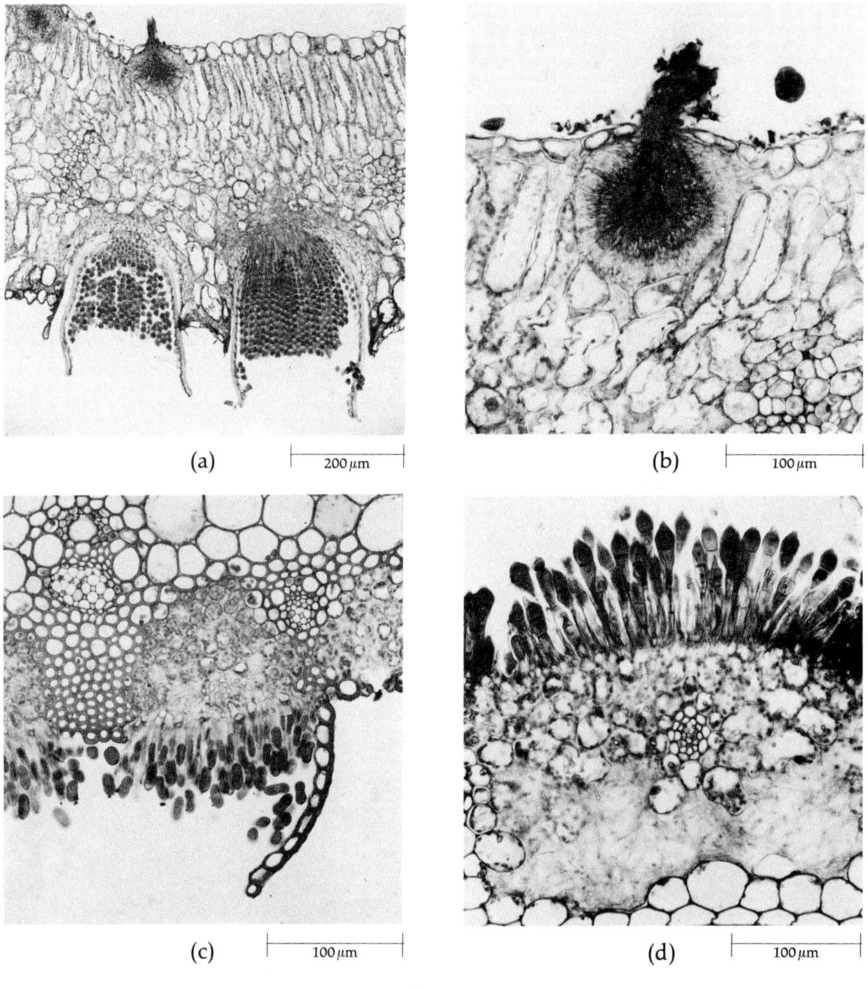

"FUNGUS-ROOTS": MYCORRHIZAE

In recent years, it has been found that a particular kind of association between fungi and higher plants plays a crucial role in mineral nutrition (Figure 11–41). If seedlings of many forest trees are grown in nutrient solutions and then transplanted to prairie and other grassland soils, they fail to grow and eventually die from malnutrition despite the fact that soil analysis shows that there are abundant nutrients in the soil (Figure 11–42). If a small amount (0.1 percent by volume) of forest soil containing fungi is added to the soil around the roots of the seedlings, however, they will grow promptly and normally. The restoration of normal growth is caused by the functioning of *mycorrhizae* ("fungus-roots"), which are intimate and mutually beneficial symbiotic associations between roots and fungi.

Mycorrhizae are thought to occur in well over 90 percent of all families of higher plants. Only a few families are not known to form such associations, among them the mustard family, Brassicaceae, and the knotweed family, Polygonaceae.

Many plants seem to grow normally if they are well supplied with essential elements, such as phosphorus, even if mycorrhizae are lacking, but grow weakly or not at all if they lack mycorrhizae when the essential elements are present in limiting quantities. In view of this, it seems possible that the heavy applications of fertilizer often necessary in agricultural soils might not be required if the fungal and microbial elements normally in these soils were present and in balance. In an analogous relationship, marsh and aquatic plants often lack a fungal associate when growing in water, but the same species require a fungal partner to be able to maintain themselves on drier land.

The importance of mycorrhizae first became apparent in connection with early efforts to grow those most showy of all greenhouse plants, the orchids. Orchids have microscopic seeds that lack endosperm at maturity, although they will readily germinate to develop a few-celled fleshy pad of tissue called a protocorm. Protocorms may persist for as long as 2 years, but they will not grow further unless they are infected by the appropriate fungus.

In mycorrhizal associations in general, only the cortex of the root is invaded by the fungus, which sometimes forms a sheath of hyphae, or a fungal mantle, around the root. Neither the apical meristem nor the vascular cylinder is penetrated by the fungus. Roots with mycorrhizae usually lack root hairs, the role of water and mineral uptake from the soil evidently being assumed by the fungus.

There are two major types of mycorrhizae: *ectomycorrhizae* and *endomycorrhizae*. Most trees of temperate regions, such as pines, oaks, and willows, form ectomycorrhizae, associations in which the roots are enveloped by a hyphal mantle. Within the root, the hyphae of the fungus are mostly confined to the spaces between the

11–41
Network of roots and fungi lifted off a decomposing leaf in the Amazon forest. Current research indicates that in tropical soils, the chief reservoir of minerals is not in the soils themselves but in the soil fungi. Minerals are passed directly from the fungi to living root cells. In this way, little mineral matter leaks into the soil where it can be leached away.

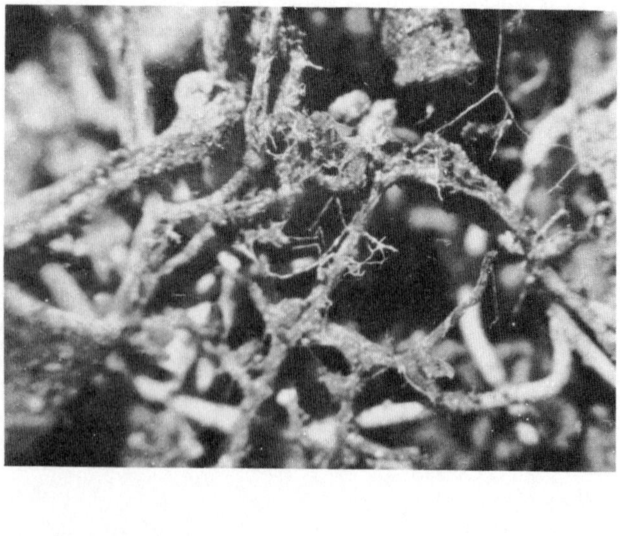

11–42
Mycorrhizae and tree nutrition. Nine-month old seedlings of white pine (Pinus strobus) *were raised for 2 months in a sterile nutrient solution and then transplanted to prairie soil. The seedlings on the left were transplanted directly. The seedlings on the right were placed for 2 weeks in forest soil before being transplanted to the prairie.*

cortical cells (Figure 11-43). Hormones secreted by the fungus cause the root to branch. This, along with the hyphal mantle, imparts a characteristic branched and swollen appearance to the ectomycorrhizae. The fungal components are mainly Basidiomycetes, although Ascomycetes are also common. Probably a majority of mushrooms occur exclusively in mycorrhizal associations, which results in their consistent occurrence with certain kinds of trees and therefore determines their distributions.

Endomycorrhizae are by far the more common type of mycorrhizae, occurring in the majority of flowering plants, including the orchids we have already mentioned, and in several conifers (e.g., *Sequoia* and *Juniperus*). The fungal component is probably always a Zygomycete. The hyphae penetrate the cortical cells where they form coils, swellings, or minute branches (Figure 11-44). Usually the hyphae are eventually digested by the cortical cells. Overall, the most important role of the fungus is the disintegration of soil materials and the absorption and transport of the released soil nutrients to the plant. Orchids, tomatoes, apples, strawberries, and grasses are common flowering plants with endomycorrhizae.

The exact relationship between roots and fungi is not known. Apparently the roots secrete sugars, amino acids, and possibly some other organic substances. Although the evidence is just now accumulating, it seems that the chief role of the fungi in such relationships is in the conversion of the minerals in the soil and in decaying material into available form. For example, if the mycorrhizal fungi in nursery soils are killed by fumigation, many plants are unable to utilize phosphorus, even when it is present in the form of easily soluble compounds. In nature, the growth of seedlings often seems to be limited by the amount of phosphorus available in the seed, until a mycorrhizal association can be set up. These symbiotic fungi are so essential to forest trees that one might consider them as a part of the tree's root system rather than as independent inhabitants of the soil. The distribution of certain flowering plants seems to be controlled by the physiological tolerances of their mycorrhizal fungi. For example, members of the heather family, Ericaceae, cannot tolerate alkaline conditions and are well known for their mycorrhizal associations; it may well be their associated fungi and not the heathers themselves that fail to grow under alkaline conditions.

Indeed, a study of the fossils of early vascular plants, to be discussed in more detail in Chapter 14, has revealed that endomycorrhizae were as frequent as they are in modern vascular plants. This has led K. A. Pirozynzki and D. W. Malloch of the Canada Department of Agriculture to make the very interesting suggestion that the evolution of mycorrhizal associations may have been the critical step allowing the colonization of the land. Given the relatively sterile soils available at the time of the first colonization of the land by plants, the role of my-

11-43
Transverse section of ectomycorrhizal rootlet of Pinus. *The hyphae of the fungus are mostly confined to the spaces between the cortical cells.*

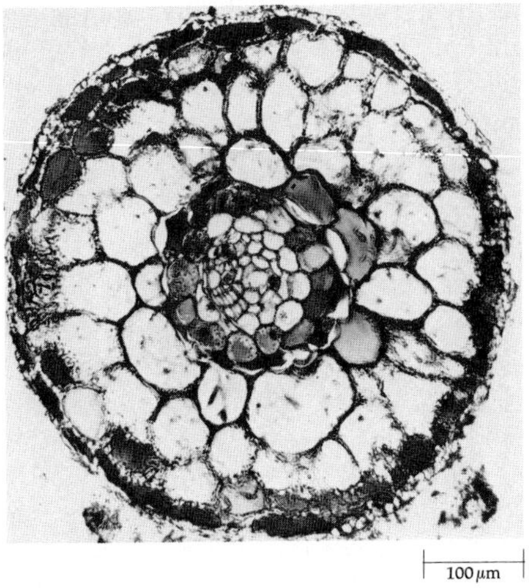

11-44
Fungal hyphae in cortical cells of endomycorrhiza of coral-root orchid (Corallorhiza).

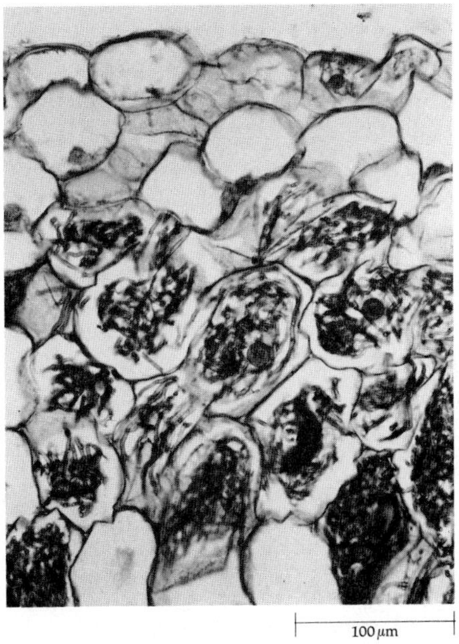

corrhizal fungi (undoubtedly Oomycetes) may have been of crucial significance, particularly in converting ammonia to nitrates and in facilitating the uptake of phosphorus. Thus it may not have been a single organism but rather a symbiotic association of organisms, comparable to a lichen, that initially invaded the land. Just as lichens inhabit sites too extreme for either partner in isolation, so may have the ancestors of the first vascular plants.

SUMMARY

The fungi, division Mycota, together with the bacteria, are the decomposers of the biosphere, breaking down organic products and restoring carbon, nitrogen, and other components to the soil and air. Some 100,000 species are known. They are rapidly growing nonphotosynthetic organisms that characteristically form filaments (known as hyphae). In most, the filaments are highly branched, forming a mycelium. Because of their rapid growth and filamentous form, they have an unusually intimate association with their environment. The hyphae may be septate or aseptate.

Fungi reproduce by means of spores. Among the genetic peculiarities of the groups are the phenomena of heterokaryosis and parasexuality. A heterokaryotic fungus is one in which the nuclei found in a common cytoplasm are genetically different. Parasexuality involves the fusion of haploid nuclei to form diploid nuclei, crossing-over between associated chromosomes, and their ultimate separation to re-form the haploid nuclei. It provides an alternative means for genetic recombination and is found especially among those fungi that do not reproduce sexually.

Glycogen is the primary storage polysaccharide of the fungi. The primary component of most fungal cell walls is chitin. Most fungi are saprobes; that is, they live on organic materials from dead plants and animals. Some, however, are parasites and absorb nourishment from living cells. Rhizoids are the specialized hyphae found in some saprobic fungi that serve to anchor them to the substrate. Parasitic fungi often have specialized hyphae (haustoria) that penetrate the cells of other organisms.

In addition to their role as decomposers, the fungi are economically important to man as destroyers of foodstuffs and other organic materials. The group also includes the yeasts, *Penicillium* and other producers of antibiotics, the cheese molds, and the highly prized morels, truffles, and mushrooms.

The simplest of fungi belong to the class Chytridiomycetes. The Oomycetes and Chytridiomycetes are the only fungi that possess motile cells in their life cycles, and they may not be directly related to the others.

The members of the classes Oomycetes and Zygomycetes have coenocytic mycelia, and their spores are generally formed in sporangia, which are relatively simple saclike structures in which the whole protoplasm becomes converted into spores.

The Oomycetes include the water molds and the order Peronosporales, which includes the organisms that cause potato blight *(Phytophthora infestans)* and downy mildew of grapes *(Plasmopara viticola). Rhizopus stolonifer,* the common black bread mold, is a member of the Zygomycetes. In the Oomycetes the cell wall is composed primarily of cellulose, whereas in Zygomycetes, as in most fungi, it is composed primarily of chitin.

The class Ascomycetes has some 30,000 described species, more than any other group of fungi. Their distinguishing characteristic is the ascus, a saclike structure in which the meiotic spores, known as ascospores, are formed. In the characteristic life cycle, male and female gametangia fuse to produce a filament that is dikaryotic (containing two paired haploid nuclei). The ascus forms at the tip of a dikaryotic hypha. Asexual reproduction may also take place by spore formation; typically the Ascomycetes form a type of asexual spore known as a conidium. The yeasts are single-celled Ascomycetes in which asexual reproduction takes place by fission or budding, and sexual reproduction by the formation of asci not enclosed in fruiting bodies.

The group known as Fungi Imperfecti includes many thousands of species with no known sexual cycle. It includes the mold from which penicillin is extracted *(Penicillium chrysogenum)* and also a number of human pathogens. A parasexual cycle has recently been discovered among members of this group. The cycle involves cellular fusion, nuclear fusion, and reduction in chromosome number; it differs from the sexual cycle in that these phenomena occur randomly rather than during specified times in the life cycle. Although parasexuality has been observed in other fungi, it is more common among this group. Most Fungi Imperfecti are Ascomycetes and a few are Basidiomycetes.

Lichens are Ascomycetes that are obligate parasites on the algae they include. The combination is morphologically and physiologically different from either organism as it exists separately. The ability of the lichen to survive under adverse environmental conditions is related to its ability to withstand dessication and remain dormant when dry.

The class Basidiomycetes includes the most advanced of the fungi. They are also the most familiar and include mushrooms, toadstools, and some important plant pathogens. Their distinguishing characteristic is the production of basidia. The basidium, like the ascus, is the structure on which the meiotic spores (basidiospores) are formed. They are typically four in number and differ from ascospores in that they are formed outside of the basidium. Like the ascus, the basidium is produced at the tip of a dikaryotic hypha. Fusion of the nuclei is followed by meiosis to produce the basidiospores. In the higher Basidiomycetes—the mushrooms and toadstools—the basidia are incorporated into complex fruiting bodies.

Many of the higher Basidiomycetes do not form asexual spores.

Symbiotic associations with fungi, called mycorrhizae, characterize more than 90 percent of all vascular plants. The fungi grow in the cortex of the plants and in some cases also form a dense, feltlike hyphal covering over the roots. Such associations are important in controlling mineral cycling in most soils and may also have been important in the invasion of the land by the algae that became the ancestors of the vascular plants.

SUGGESTIONS FOR FURTHER READING

AHMADJIAN, V.: *The Lichen Symbiosis,* Blaisdell Publishing Co., Waltham, Mass., 1967.

In this small but fascinating book, Ahmadjian presents recent work on the nature of the relationship between the fungal and algal components of a lichen.

AHMADJIAN, VERNON, and MASON E. HALE (eds.): *The Lichens,* Academic Press, Inc., New York, 1973.

A comprehensive account of the lichens, including contributions by 23 authors, on all aspects of the biology of the group.

AINSWORTH, G. C., and A. S. SUSSMAN (eds.): *The Fungi—An Advanced Treatise* (4 volumes), Academic Press, Inc., New York, 1965–1973.

This huge work contains modern summaries of all aspects of fungal biology contributed by experts in each field. It indicates clearly how much remains to be learned about these fascinating and economically important organisms.

ALEXOPOULOS, C. J.: *Introductory Mycology,* 2nd ed., John Wiley & Sons, Inc., New York, 1962.

A brief, crisply written review for students who want to know more about the fungi.

BURNETT, J. H.: *Fundamentals of Mycology,* St. Martin's Press, Inc., New York, 1968.

A modern and thorough account of the fungi that is physiologically and ecologically oriented.

CHRISTENSEN, CLYDE M.: *The Molds and Man. An Introduction to the Fungi,* 3rd ed., University of Minnesota Press, Minneapolis, 1965.*

A highly readable and informative account of the fungi and especially of their complex and important interrelationships with man.

HALE, M. E., JR.: *Lichen Handbook,* Smithsonian Institution, Washington, D.C., 1961.

Not only does this well-written small book provide a guide to the lichens of eastern North America, but it also summarizes their biology in a concise fashion.

HALE, M. E., JR.: *The Biology of Lichens,* 2nd ed., Edward Arnold (Publishers) Ltd., London, 1974.

An outstanding, concise summary of all aspects of the morphology, physiology, systematics, ecology, and economic uses and applications of the lichens.

HARLEY, J. L.: *The Biology of Mycorrhiza,* 2nd ed., Plant Science Monographs, Leonard Hill, London, 1969.

The standard work in the field, readable but an invaluable reference also.

HUDSON, H. J.: *Fungal Saprophytism,* Edward Arnold (Publishers) Ltd., London, Studies in Biology No. 32, 1972.*

A concise and interesting treatment of many aspects of fungal ecology.

INGOLD, C. T.: *Spore Discharge in Land Plants,* Clarendon Press, Oxford, 1939.

A marvelous little book that has been read with pleasure by students and members of the general public for the past three decades.

LARGE, E. C.: *The Advance of the Fungi,* Dover Publications, Inc., New York, 1962.*

A fascinating popular account of the closely interwoven histories of fungi and man, first published in 1940.

SMITH, A. H.: *The Mushroom Hunter's Field Guide,* 2nd ed., The University of Michigan Press, Ann Arbor, Mich., 1963.

A clear, concise, well-illustrated guide to edible mushrooms, enlivened with good advice and pertinent anecdotes.

* Available in paperback.

CHAPTER 12

The Algae and the Slime Molds*

12–1
Algae on the rocks at low tide along the coast of Maine.

The open sea, the shore, and the land are the three great life zones that make up our biosphere. The algae play a role in these first two ancient dwelling places, the sea and the shore, comparable to the role of the land plants in the far younger but more familiar terrestrial world (Figure 12–1).

In the open sea, minute photosynthesizing cells and animals associate together as free-floating *plankton* (from *planktos*, the Greek word for wanderer). The planktonic algae, or *phytoplankton*, are the beginning of the food chain for all the animals that live in the deep waters. They are usually composed of single cells—some quite simple in appearance, others with intricate and delicately detailed forms and sometimes joined into colonies or filaments (Figure 12–2). Planktonic animals, or *zooplankton*, include an abundance of microscopic species, particularly members of the crustacea, such as tiny shrimp, copepods, and their relatives. In addition, large numbers of the larval, or immature, forms of nonplanktonic species are present. Many animal groups are represented, either as adult or larval forms. The animal plankton feed on the phytoplankton (and some on each other); small fish and some larger ones, as well as some of the great whales, feed on the phyto- and zooplankton; and the still larger fish feed on the smaller ones. In this way, the "great meadow of the sea," as it is sometimes called, can be likened to the meadows of the land, serving as the source of animal nourishment.

Along the rocky shore can be found the larger, more complex algae, typically arranged in fairly distinct visible bands or zones in relation to intertidal levels. Their complexity reflects their ability to survive in this difficult life zone, where twice each day they are subject to great fluctuations of humidity, temperature, salinity, and light, in addition to the pounding action of surf and the abrasive action of suspended sand particles churned up by the waves.

* The algae, traditionally considered members of the Plant Kingdom, are treated in this text as Protista, along with the slime molds, traditionally grouped with the fungi.

Anchored offshore beyond the tidelines, algae provide shelter for a rich diversity of microscopic organisms, as well as for larger fish and invertebrate animals that feed on the microorganisms and on each other. These subtidal algal growths may be so dense as to be called "jungles." For example, off the coast of California are kelp-bed jungles composed of giant brown kelps whose broad, 15-meter fronds are buoyed upward on sinuous stalks 30 meters or more from their holdfasts on the bottom. Many large carnivores, including sea otters and tuna, find food and refuge in these kelp beds, which are also harvested directly by man for food and fertilizer.

From apparent oceanic origins, many species of algae became adapted to the "newer" land life zone with its various types of freshwater habitats. Freshwater algae include phytoplankton and macroscopic forms in lakes, ponds, rivers, and streams, as well as—in a less welcome capacity—the forms that inhabit swimming pools and domestic water supplies such as wells and reservoirs. In these aquatic habitats, the algae form the nutritional basis for all the animals, as they do in the seas. Efforts are underway to develop commercial methods for culturing some of these algae to provide protein for man; early yields are approximately two to three times those of conventional crops, especially in sunny tropical regions such as Thailand, Peru, and India. More terrestrial species are found in moist soil (such as sand near a shore or mud); on tree trunks or wooden boards (such as an untreated home deck or patio); bare rocks; and even in the arid soils of the harshest deserts, where they may "come to life" only once or twice a year, after a rainfall. Certain species of algae occupy extremely rigorous environments, such as the near-boiling hot springs at Yellowstone National Park, the highly saline Great Salt Lake of Utah, snowbanks, and glaciers. Even in suspended droplets of water in the atmosphere, certain species may thrive and multiply, falling to earth during storms.

CHARACTERISTICS OF THE ALGAE

Among the eukaryotic algae, a number of distinct lines had evolved by the early Paleozoic era, more than 450 million years ago. The approximately 22,000 described species now in existence are grouped into seven divisions, which are considered to represent some of these evolutionary lines. The term "algae" has been abandoned as a formal term in modern classification, because the various groups are not directly related to one another. Three of these divisions—the Euglenophyta, the Chrysophyta, and the Pyrrophyta—consist almost entirely of unicellular organisms. The other four divisions—Chlorophyta, Rhodophyta, Phaeophyta, and Xanthophyta—include groups that are multicellular.

Nearly all members of these divisions are photosynthetic. In general, they all have a relatively simple construction, which may be a single cell, a filament of cells, a plate of cells, or a solid body more or less comparable to that found in the vascular plants. The cell walls of algae, in general, have a cellulose matrix, with often massive amounts of other polysaccharides that give certain algae a mucilaginous consistency. When algal cells divide, the plasma membranes generally pinch inward from the margin of the cell (furrowing), just as in animals, fungi, and protozoa. Cell plates like those of the higher plants are, however, known in one brown alga and a few green algae. Most algal cells—except those of the red algae—have centrioles (see page 42).

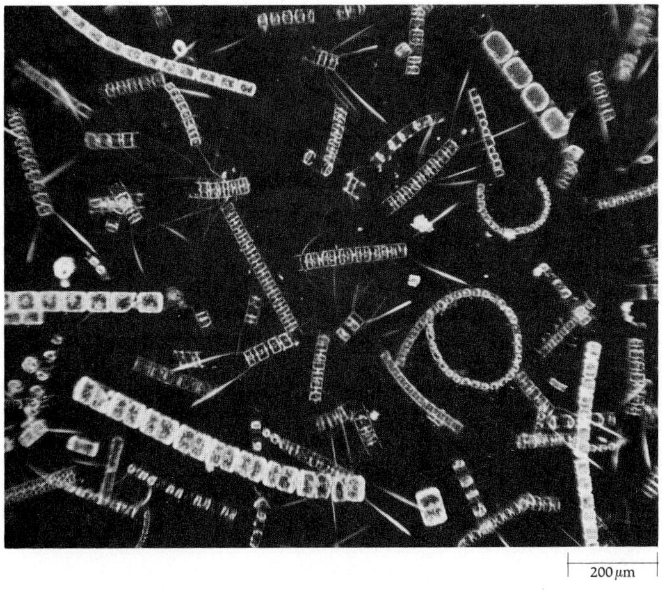

12–2
Phytoplankton, the great meadow of the sea. The organisms shown here represent several different groups of Protista.

Multicellular algae do not have a complex array of tissues like that found in the vascular plants. In certain kelps (brown algae of the division Phaeophyta), however, a central conducting strand consisting of cells that resemble sieve elements is found in the stalk. The reproductive structures of the algae are generally single cells and not multicellular structures with sterile jackets such as are found in the bryophytes and vascular plants.

The divisions of algae differ from one another widely in the nature of their flagella (when present) and in their biochemical characteristics, especially with respect to differences in pigmentation, nature of reserve foods, and cell wall components (Table 12–1). The names of some divisions are derived from the colors of the predominant accessory pigments, which mask the grass-green of the chlorophylls. A wide variety of carotenoids is found in the algae. The xanthophyll fucoxanthin gives the brown algae their characteristic color and name. It is also found in the golden-brown algae and diatoms (division Chrysophyta). The red algae (Rhodophyta) owe their colors to several kinds of phycobilins, accessory pigments that, unlike the carotenoids, are water soluble. In the green

Table 12–1 *Comparative Summary of Characteristics in Seven Divisions of Eukaryotic Algae*

DIVISION	NUMBER OF SPECIES	PHOTOSYNTHETIC PIGMENTS	CARBOHYDRATE FOOD RESERVE	FLAGELLA	CELL WALL COMPONENT	REMARKS
Chlorophyta (green algae)	7000	Chlorophyll *a* and *b*, carotenoids	Starch	2 (or more), apical or subapical, equal, whiplash	Cellulose	Mostly fresh water, but some marine
Phaeophyta (brown algae)	1500	*a*, *c*, carotenoids, including fucoxanthin	Laminarin	2, lateral, forward tinsel, trailing whiplash; in reproductive cells only	Cellulose matrix with alginic acids (polysaccharides)	Almost all marine, flourish in cold ocean waters
Rhodophyta (red algae)	4000	*a*, carotenoids, phycobilins, *d* in some	Floridean starch	None	Cellulose, pectic materials common, xylan in *Porphyra*	Mostly marine, but some fresh water, complex sexual cycle; many species tropical
Chrysophyta (golden-brown algae and diatoms)	6000–10,000	*a*, often *c*, carotenoids, including fucoxanthin	Leucosin	1 or 2, apical, whiplash or tinsel, equal or unequal	Pectic compounds with siliceous material	Mostly marine
Xanthophyta (yellow-green algae)	450	*a*, carotenoids	Leucosin	2 (or more), whiplash or tinsel, unequal	Cellulose, pectic materials; some impregnated with silicon	Mostly fresh water
Pyrrophyta (dinoflagellates)	1100	*a*, *c*, carotenoids	Starch	2, lateral, tinsel	Cellulose	Marine and fresh water; sexual reproduction rare
Euglenophyta (euglenoids)	800	*a*, *b*, carotenoids	Paramylon	1 (to 3), apical, tinsel (one row of hairs)	No cell wall	Mostly fresh water; sexual reproduction unknown

algae the color of the chlorophylls is usually not masked by accessory pigments. However, some types, such as the species of *Chlamydomonas* found on the surface of snow, or *Trentepohlia,* a filamentous alga that grows on branches of vascular plants, develop large amounts of carotenoids as a shield against strong lights and appear red or rust-colored.

A rich diversity of storage products is found in the different divisions of algae, with most having distinctive carbohydrate food reserves, often in addition to lipids. The green algae and dinoflagellates store their carbohydrates as starch, like the bryophytes and vascular plants. In the brown algae, another glucose polymer, laminarin (Figure 12–3), takes the place of starch. The carbohydrate reserves of the red algae are biochemically similar to starch as it occurs in green algae and plants (although it differs from the latter in requiring prolonged boiling for gelatinization), whereas that of the golden-brown and yellow-green algae is very similar to laminarin. One of the most characteristic energy-rich molecules found in the cells of the brown algae is mannitol, a simple alcohol derived from the sugar mannose.

12–3
Laminarin, the principal storage product of brown algae. Like starch, it is made up of glucose residues but unlike starch, there are only about 15 to 30 glucose units per molecule, and their linkage is different.

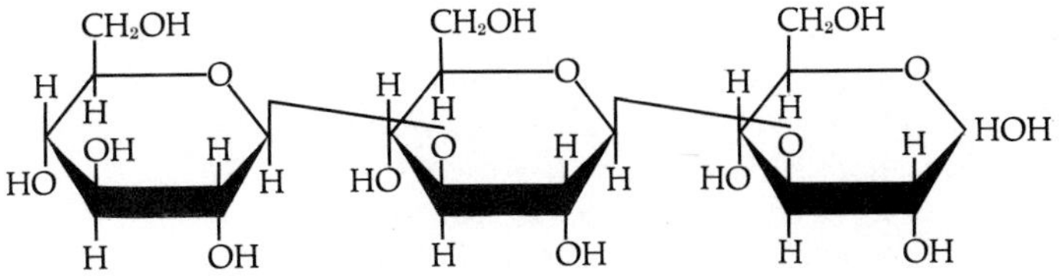

Symbiosis and the Origin of the Chloroplast

In all organisms that produce oxygen as a result of photosynthesis—the blue-green algae, eukaryotic algae, and plants—chlorophyll *a* is involved in the process. All except the blue-green algae have the chlorophyll *a* located in chloroplasts.

As we saw in Chapter 1, chloroplasts seem to have originated when symbiotic prokaryotes, resembling modern blue-green algae, became permanent components of eukaryotic cells. The ease and frequency with which symbiotic relationships involving algae are established are suggested by the wide variety of symbiotic relationships that can be observed today. These modern examples include symbioses with invertebrates, fungi (mostly in lichens), other algae, vertebrates, bryophytes, and vascular plants (in which the algae often grow in cavities in stems, petioles, and leaves). Among invertebrates, for example, some 150 genera belonging to eight different phyla are known to have algal symbionts growing within their cells.

Derivation of the Divisions of Algae

All eukaryotic cells—plant, animal, protistan, or fungal—have endoplasmic reticulum, Golgi bodies, large ribosomes, mitochondria, microtubules, and a complex nucleus. The nuclear apparatus includes a double-membraned nuclear envelope that is continuous with the endoplasmic reticulum, chromosomes that contain DNA in combination with proteins and that undergo mitosis and meiosis with the formation of a spindle apparatus,* and one or more nucleoli. Each of the major lines of eukaryotes, with the exception of the Rhodophyta, includes forms that have flagella with the distinctive 9-plus-2 organization described in Chapter 1. It would thus seem logical to assume that the common ancestor of all living eukaryotes exhibited all these characteristics. If such an organism were living today, it would be classified as a protozoan.

The prokaryotic blue-green algae and the eukaryotic red algae (division Rhodophyta) both have chlorophyll *a,* carotenoids, and phycobilins. Phycobilins are found only in these two groups (and in the division Cryptophyta, a poorly understood and quite distinct group of algae not treated in this text). In blue-green algae these photosynthetic pigments are found throughout the cell, whereas in red algae they are confined to the chloroplasts. The arrangement of the thylakoids in blue-green algae is very similar to that found in chloroplasts. They likewise have similar immunological reactions. The simplest explanation of these facts would seem to be that the chloroplasts of red algae *are,* in effect, blue-green algae that became symbiotic in the cells of some primitive protozoan long ago.

The green algae and the euglenoids (division Euglenophyta) share chlorophylls *a* and *b,* as do the bryophytes and vascular plants. Yet in other respects the green algae and the euglenoids are among the most different of the algal divisions. The green algae are widely accepted as the group ancestral to the bryophytes and vascular plants and are similar to them in biochemistry and cell structure. The euglenoids, on the other hand, are essentially protozoa, with contractile vacuoles and a flexible proteinaceous pellicle in place of a cell wall. On the basis of these relationships, it was once suggested that the chloroplasts

* Except for dinoflagellates and many fungi, in which histone proteins are not combined with the chromosomal DNA.

SYMBIOTIC ALGAE

Green algae, most resembling the genus Chlorella, *are found in many freshwater protozoa, sponges, hydra, and some flatworms. Most of these algae reproduce by simple cell division and are found within the host cell's vacuoles, which divide when the alga divides.*

Another green alga, Platymonas convolutae, *is found mostly in the subepidermal cells of the marine flatworm* Convoluta roscoffensis. *Within the flatworm,* Platymonas *has no cell wall and an irregular shape; its plasma membrane, greatly increased in surface area by fingerlike projections, is more or less in direct contact with the vacuole membrane of the host cell. When removed from the flatworm and cultured,* Platymonas *has a cell wall, four flagella and an eyespot, all lacking when it is symbiotic.*

The most direct relationship between green algae and invertebrates involves certain nudibranchs, which are marine molluscs, and the chloroplasts of some siphonaceous green algae, such as Codium. *The chloroplasts, which are presumably acquired when nudibranchs eat the algae, are found in cells that line their entire respiratory chamber. In the presence of light these chloroplasts carry on photosynthesis so efficiently that individuals of the nudibranch* Placobranchus ocellatus *are reported to evolve oxygen more rapidly than it is consumed.*

Certain dinoflagellates (division Pyrrophyta), which resemble the nonflagellated cells of Chlorella *when they occur as symbionts, inhabit the cells of various marine sponges, coelenterates, molluscs, flatworms, and protozoa. In the giant clams of the family Tridachnidae, the dorsal surface of the inner lobes of the mantle may appear chocolate-brown as a result of the presence of symbiotic algae of this group, which are found in the blood sinuses and probably occur mainly within amoeboid blood cells. Dinoflagellates are also important symbionts in reef-building corals. Coral tissues may contain as many as 30,000 microscopic algae per cubic millimeter. The exact role of these algal symbionts in the coral economy is not known, but tracer studies using radioactive carbon have shown that organic substances pass from the algae to the coral, and it has also been shown that the reef grows much more rapidly when the algae are present. Even though the coral animals are heterotrophic, the reef as a whole is autotrophic.*

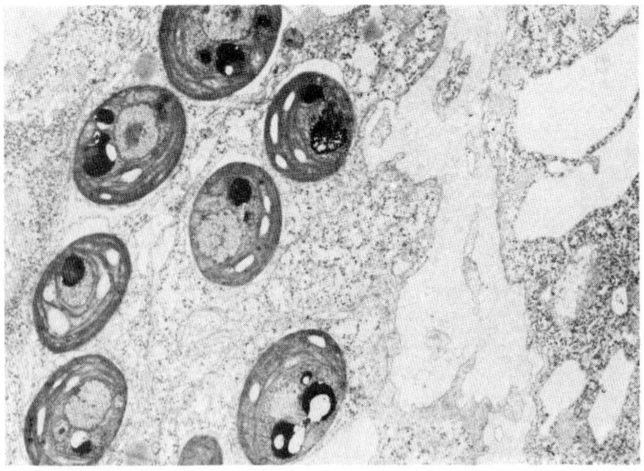

Cells of a species of *Chlorella* growing in a hydra *(Chlorohydra viridissima).* The hydra, a small multicellular animal, provides the algae with protection and shelter. The algae secrete maltose, a source of nourishment for their host.

Nudibranch *(Elysia elsiae).* These animals obtain their chloroplasts by eating siphonous green algae. The chloroplasts divide inside their tissues, making them, by some definitions, plants.

of green algae and euglenoids had a common evolutionary origin by symbiosis; however, the chloroplasts of euglenoids are, in their ultrastructure, more similar to those of the golden-brown algae than to those of the green algae. Dinoflagellates, which appear to have a relatively primitive nuclear apparatus, have relatively advanced chloroplasts. Thus, on the basis of present evidence, it does not seem likely that all chloroplasts originated in a single symbiotic event. A more reasonable explanation would seem to be that multiple symbiotic events involving diverse photosynthetic prokaryotes resulted in the evolution of chloroplasts in different algal lines.

THE GREEN ALGAE: DIVISION CHLOROPHYTA

The green algae are the most diverse of all the algae, both in form and in life history. The group comprises at least 7000 species. Although most green algae are aquatic, they are found in a wide variety of habitats, including the surface of snow, in green patches on tree trunks, and as symbionts in lichens, protozoa, and hydra. Of the aquatic species, a few groups are entirely marine, but the great majority are found in fresh water. Many green algae are microscopic, but some of the marine forms are large; *Codium magnum* of Mexico, for example, sometimes attains a breadth of 25 centimeters and a length of more than 8 meters. This group has a long fossil record, with some simple forms reported from rocks nearly a billion years old.

The Chlorophyta are similar to the bryophytes and vascular plants in several important characteristics. They contain chlorophylls *a* and *b*, store their food as true starch, and have firm cell walls composed, in most genera, of cellulose, with hemicelluloses and pectic substances incorporated into the wall structure. For these reasons, they are believed to be directly related to the evolutionary line from which the bryophytes and vascular plants evolved. Within the Chlorophyta, three fairly distinct lines of progressive evolutionary specialization have been recognized: the volvocine line, the tetrasporine line, and the siphonous line.

The following discussion of diversity of form and reproduction in the Chlorophyta will serve not only to familiarize you with some representative green algae, but also to introduce some basic and important biological phenomena.

Motile Unicellular and Colonial Green Algae

Among the simplest of green algae is the unicellular biflagellate organism *Chlamydomonas*. *Chlamydomonas*, which is one of the most common freshwater green algae in nature, is small (usually less than 25 micrometers long), grass-green, and rounded or pear-shaped (Figure 12–4). It moves very rapidly, with a characteristic darting motion imparted by the beating of the two equal whiplash flagella that protrude from its smaller, anterior pole. Commonly each cell contains a single massive chloroplast. The chloroplast may contain a red pigment body, the *eyespot*, or *stigma*, which may be a shading device associated with a site of light perception, and one or more spherical bodies, the *pyrenoids*. Two or more contractile vacuoles are present near the anterior end of each cell. Pyrenoids are found widely among the algae, but are lacking in the chloroplasts of plants (except the hornworts, page 289). The exact way in which they function is not known, but they seem to be associated with the conversion of sugars to starch for food storage. Starch deposits can be detected surrounding the pyrenoid body. The uninucleate protoplast is surrounded by a thin cellulose wall, inside of which is the plasma membrane.

Under some environmental conditions, cells of *Chlamydomonas* become nonmotile. Cells in this state are usually nonflagellated, and their walls become gelatinous. When conditions change, the flagella may reappear and the cells again become free-swimming. These observations lead us to the conclusion that the evolutionary gaps between flagellated and nonflagellated cells are not necessarily large ones.

Chlamydomonas reproduces both sexually and asexually. During asexual reproduction the nucleus, which is haploid, usually divides twice mitotically, resulting in the production of four daughter cells within the parent cell wall. Each cell then secretes a wall about itself and develops flagella. The original cellulose wall becomes gelatinous, and the daughter cells can then escape from it, although fully formed daughter cells are often retained for

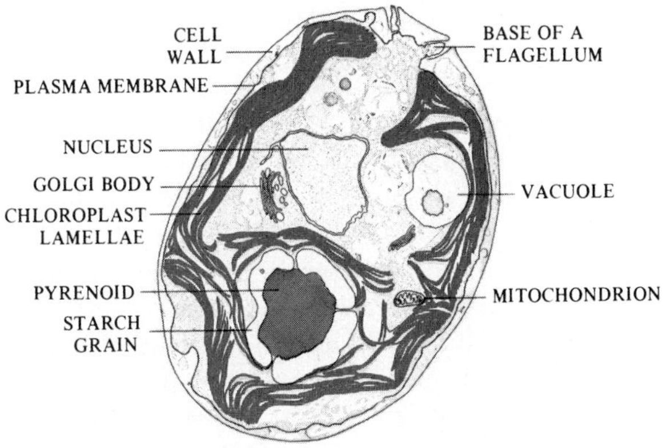

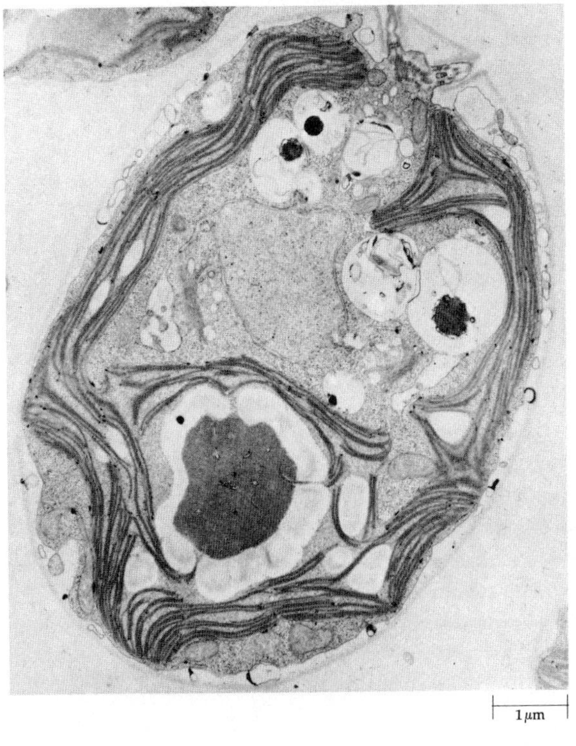

12–4
Electron micrograph of Chlamydomonas, *a unicellular green alga. Only the bases of the flagella can be seen.*

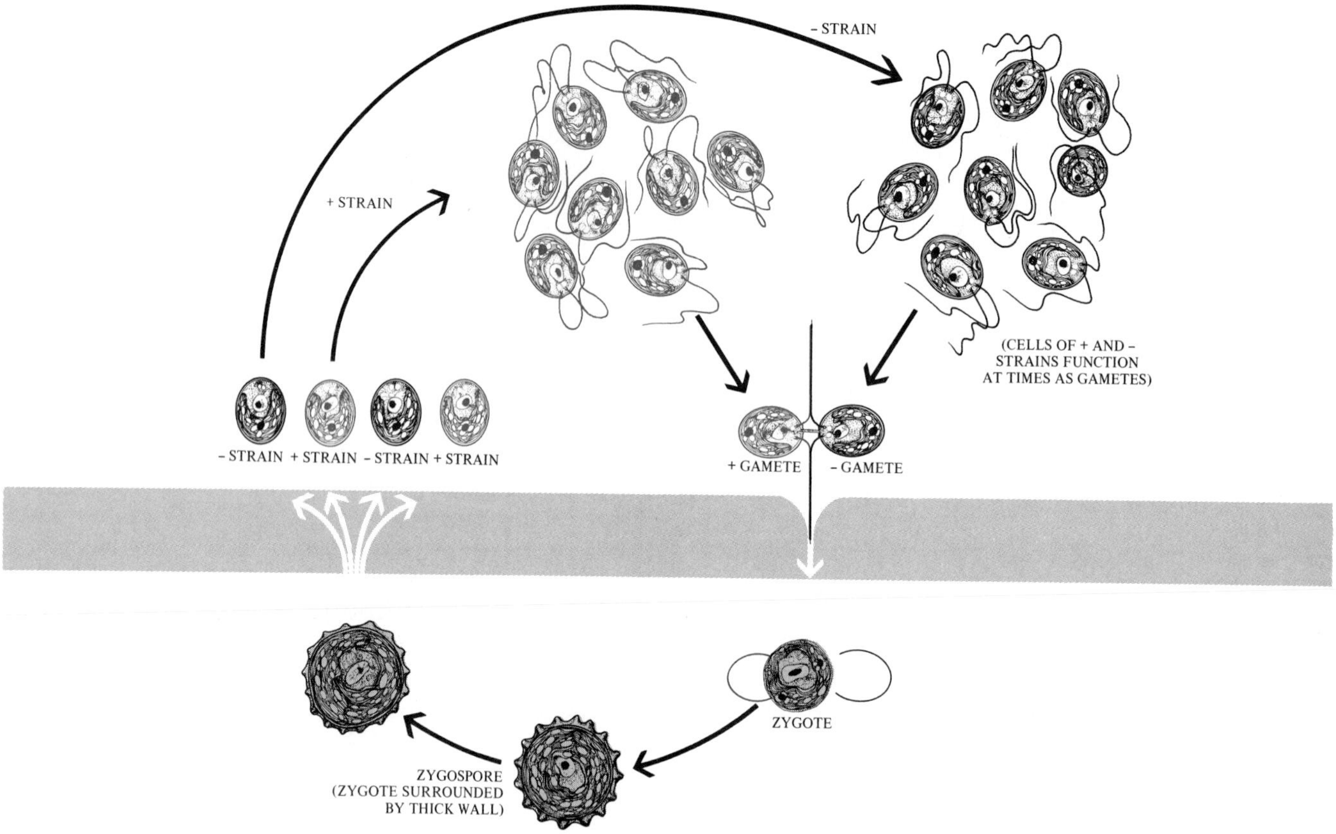

12-5
Life cycle of Chlamydomonas. *Asexual reproduction of the haploid individuals by cell division is the most frequent mode of reproduction. Sexual reproduction, shown here, takes place when individuals of different mating strains come together, cohering at first by their flagella and then by a slender protoplasmic thread, the conjugation tube. The protoplasts of the two cells fuse completely (plasmogamy), followed by the union of their nuclei (karyogamy). A thick wall then is formed around the diploid zygote. After a period of dormancy, meiosis occurs and four haploid individuals emerge.*

some time within the parent cell wall. In ancient flagellates, such aggregations of daughter cells may have been the forerunners of colonial organisms.

Sexual reproduction, which occurs in some species of *Chlamydomonas,* involves the fusion of individuals belonging to different mating strains (Figure 12–5). In some of these species, vegetative cells can function directly as mating cells or gametes. First, a number of individuals become aggregated in clumps. Within these clumps pairs are formed, which stick together first by their flagella and later by a slender protoplasmic thread that connects them at the base of their flagella. As soon as this protoplasmic connection is formed, the flagella become free, and one or both pairs of flagella beat, thus propelling the zygote, the product of sexual union, through the water. The two gametes fuse completely. Soon the four flagella shorten and eventually disappear, and a thick cell wall forms around the diploid zygote. This thick-walled, resistant zygote (zygospore) then undergoes a period of dormancy. Meiosis occurs at the end of the dormant period, resulting in the production of four haploid cells, each of which develops two flagella and a cell wall. These cells can either divide asexually or mate with a cell of another mating strain to produce a new zygote. Thus, *Chlamydomonas* exhibits zygotic meiosis (Figure 7–6), and the haploid phase is the dominant phase in its life history.

In most species of *Chlamydomonas,* the cells of the two mating types, conventionally designated + and −, are identical in size and structure (*isogamy*). In addition to these isogamous species, there are other species in which *anisogamy* (larger female gametes) and *oogamy* (immobile female gametes) occur (Figure 12–6). Thus the single genus *Chlamydomonas* exhibits in its various species the entire range of differences between gametes that occurs in the algae.

12–6
Types of sexual reproduction based on form of gametes. (a) Isogamy. The gametes are equal in size and shape. (b) Anisogamy, one gamete, conventionally termed male, is smaller than the other. (c) Oogamy, in which the female gamete is immobile. Each of these relationships is found in at least one species of Chlamydomonas.

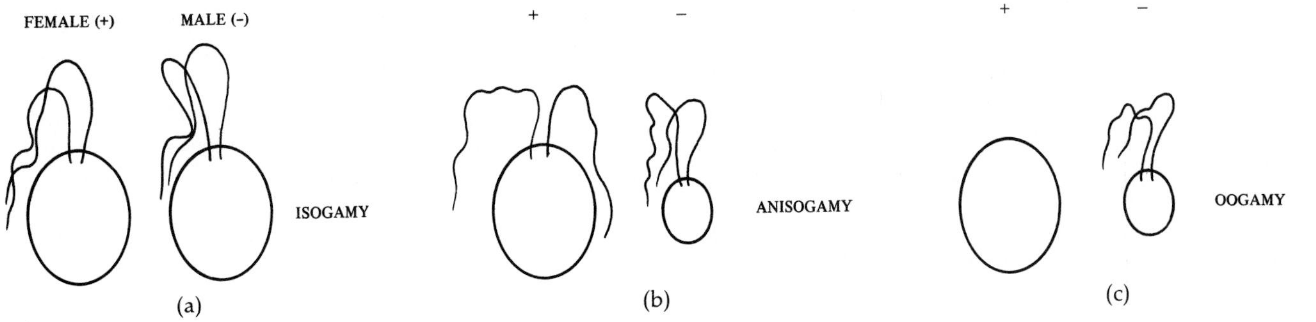

The Volvocine Line

One pathway leading to increased complexity is based on the adhesion of *Chlamydomonas*-like cells in motile colonies propelled by the beating of the flagella of the individual cells (Figure 12–7). This evolutionary line, which is exemplified by the order Volvocales, is known as the volvocine line. In some colonies the cells are connected by cytoplasmic strands that provide for the integration of the whole organism. The simplest member of the volvocine line is *Gonium*. The *Gonium* colony consists of separate cells held together within a matrix. Each colony is made up of 4, 8, 16, or 32 cells (depending on the species) arranged in a slightly curved shield-shaped disk. The flagella of each cell beat separately, pulling the entire colony forward. Each cell in *Gonium* divides to produce an entire new colony.

A closely related colonial organism is *Pandorina*, which forms a tightly packed ovoid or ellipsoid usually consisting of 16 or 32 cells held together within a matrix. The colony is polar, the eyespots being larger in the cells at one end of the colony. Each cell has two flagella, and, because all the flagella point outward, *Pandorina* rolls through the water like a ball. When the cells attain their maximum size, the colony sinks to the bottom and each of the cells divides to form a daughter colony. The latter remain together until all have developed flagella. The parent matrix then breaks open like Pandora's box (which suggested its name), releasing new daughter colonies.

Eudorina (Pleodorina) is also a spherical colony made up of green flagellates, the number of which is 32, 64, or 128 in the most common species. It differs from *Gonium* and *Pandorina* in that some of the cells in the colony, which are smaller than the others and anterior, are incapable of reproducing to form new colonies. Here we see a beginning of specialization of function.

The most spectacular of the volvocine line, and the one from which the group gets it name, is *Volvox*, a hol-

12–7
Several members of the volvocine evolutionary line of green algae drawn to scale. In these algae, cells similar to those of Chlamydomonas *adhere in a gelatinous matrix to form multicellular colonies propelled by the beating of the flagella of the individual cells. Varying degrees of cellular specialization are found in different genera.*

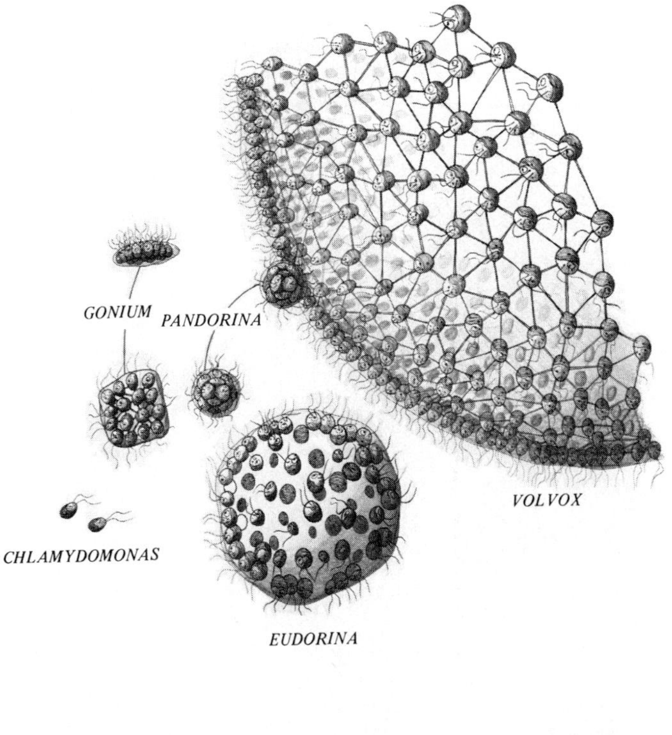

low sphere made up, according to species, of a single layer of 500 to 60,000 tiny biflagellate cells. When *Volvox* whirls through the water, it appears like a spinning universe of individual stars fixed in an invisible firmament.

Volvox exhibits polarity; that is, it has anterior and posterior poles. The flagella of each cell beat in such a way as to spin the entire colony around its axis. *Volvox* orients its anterior end toward the light but moves away from strong light. Most of the cells are vegetative and do not take part in reproduction. For reasons that are not entirely understood, only some of the cells of the posterior hemisphere can form daughter colonies. These cells, which appear identical to others when the colony is young, become larger, greener, and morphologically distinct as the colony matures.

Reproduction in *Volvox* is usually asexual. One of the posterior cells enlarges and then begins to divide. As it divides, the daughter cells are held together by a sticky matrix and by protoplasmic threads. They remain as a little pouch or balloon formed on the inner surface of the original mother colony. During this early stage, all the newly formed cells have their anterior ends toward the center of the daughter sphere. At the point at which the daughter sphere was attached inside the mother colony, there is a small hole. The daughter sphere is attached by protoplasmic strands to neighboring cells. Suddenly, and with remarkable coordination, the young colony turns itself completely inside out, through the hole, like the finger of a glove. Then, after all of the cells point outward, each cell develops flagella and the daughter colony is ready for independent existence. Daughter colonies remain inside the mother colony until the latter breaks apart, their rotational swimming probably rupturing the parent colony (Figure 12–8).

Sexual reproduction in *Volvox* is oogamous. In some species individual colonies produce both eggs and sperms, whereas in others the individual colonies are unisexual—that is, they produce either eggs or sperms. The control of differentiation in *Volvox*, including an analysis of the chemical messengers associated with different modes of reproduction and differentiation, has become an important field of biological investigation.

Increasing specialization in the members of the volvocine line is evident in several ways. First, there is the increase in cell number and size in the colonies. Second, there is an increasing specialization in cell morphology and function. Finally, there is increasingly sexual specialization paralleling that which we have already seen within the genus *Chlamydomonas*. Thus, *Gonium* and *Pandorina* have isogamous reproduction, whereas species of *Volvox* and *Eudorina* are oogamous. Nevertheless, the volvocine line clearly represents an evolutionary "dead end," in that it has not given rise to any more complex group of organisms.

Nonmotile Unicellular and Colonial Green Algae

Chlorella, a nonmotile unicell, was the first alga to be isolated in axenic (pure) culture and has been widely used for experimental studies of photosynthesis. In nature, *Chlorella* is widespread in both fresh and salt water and in soil. Each *Chlorella* cell contains a single cup-shaped chloroplast, with or without a pyrenoid, and a single minute nucleus (see pages 28–30). The only known method of reproduction in *Chlorella* is asexual, each haploid cell dividing mitotically two or three times to give rise to either four or eight nonmotile cells.

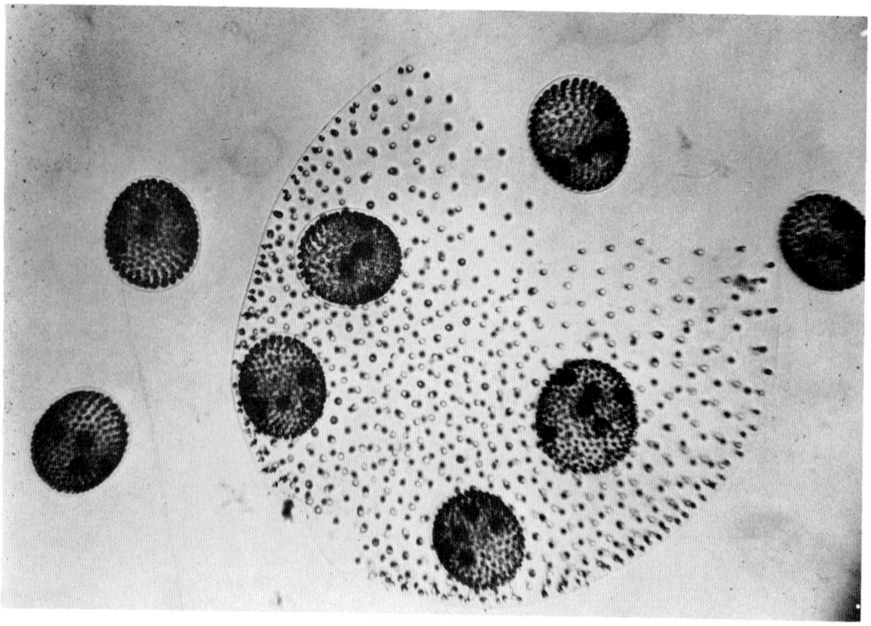

12–8
Volvox aureus. *The parent colonies, which are hollow spheres, each contain several daughter colonies which will eventually be released by the rupture of the parent colony.*

Chlorella is currently under investigation as a potential food source for humans. Pilot farms have been established in the United States, Germany, Japan, and Israel. The Japanese have processed *Chlorella* as a tasteless white powder, rich in vitamins and protein, that can be mixed with flour; it is too costly to market at present, but it could possibly represent a source of food in the future.

Eremosphaera, one of the largest of unicellular green algae, is large enough to be seen with the naked eye (Figure 12–9). It is found in acid water on the bottom of swamps and quiet ponds. Each cell contains many pyrenoid-bearing chloroplasts and a single large nucleus suspended in the center of the cell by numerous radiating strands of cytoplasm. *Eremosphaera* reproduces by both asexual and sexual means. The former is similar to that in *Chlorella.* Sexual reproduction involves the union of biflagellated sperms with nonmotile cells that function as eggs. As in *Chlamydomonas,* meiosis is zygotic.

Hydrodictyon, the "water net," is an example of a nonmotile colony (Figure 12–10). Under favorable conditions it accumulates in massive aggregates in ponds, lakes, and gentle streams. Each colony consists of many cylindrical cells arranged in the form of a large hollow cylinder. Initially uninucleate, each cell eventually becomes multinucleate. At maturity the cell contains a large central vacuole and peripheral cytoplasm in which the nuclei and a large reticulate chloroplast with numerous pyrenoids are located. *Hydrodictyon* reproduces asexually through the formation of uninucleate, biflagellated zoospores. Eventually the zoospores form groups of four to nine, typically six, within the cylindrical parent cell; lose their flagella; and form daughter colonies. Sexual reproduction in *Hydrodictyon* is isogamous, and its meiosis is zygotic.

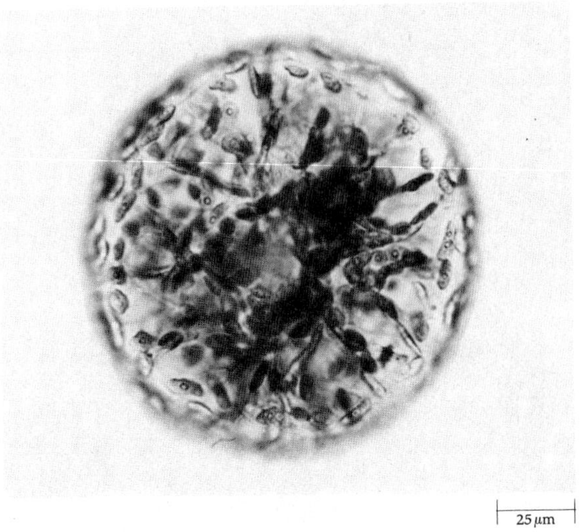

12–9
Eremosphaera. *Numerous chloroplasts can be seen in strands of cytoplasm radiating from the center of the cell.*

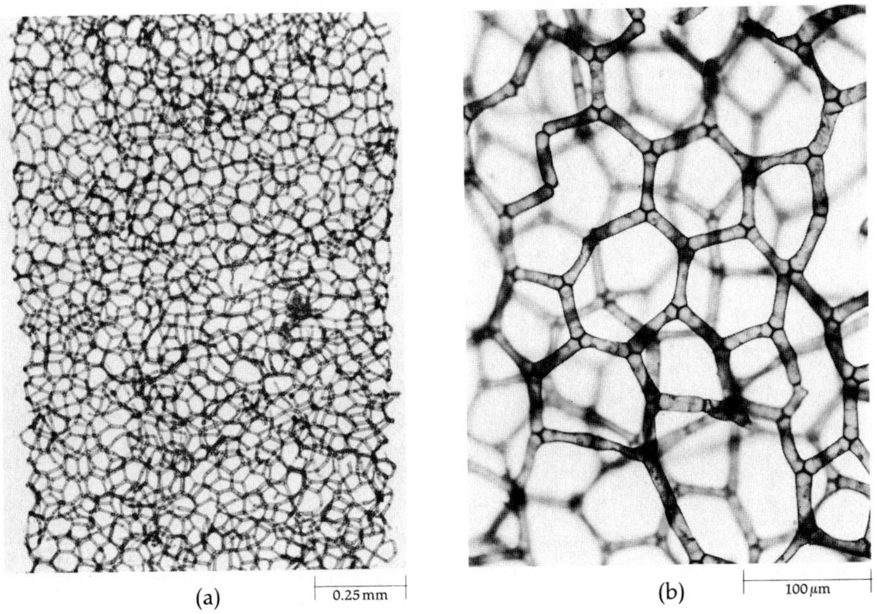

12–10
Hydrodictyon, *a colonial green alga known as the water net.* (a) *Overall view of a portion of a young colony.* (b) *Portion of a young net showing greater detail than* (a).

(a)

(b)

12–11
(a) *A species of* Codium, *abundant along the Atlantic coast.* (b) Valonia, *another siphonous green alga common in tropical waters.*

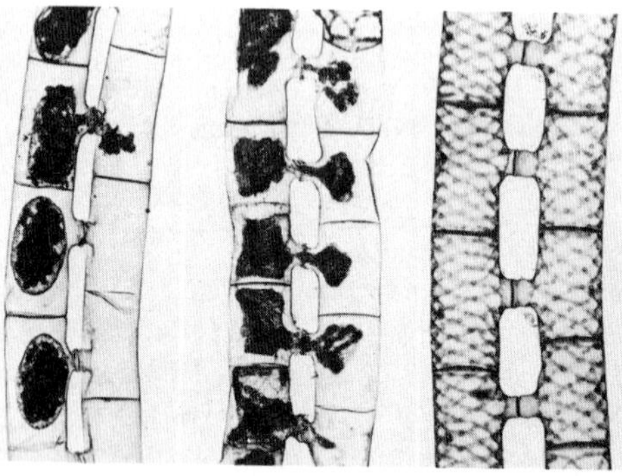

12–12
Sexual reproduction in Spirogyra *follows the formation of conjugation tubes between the cells of adjacent filaments. The contents of the cells of the — strain pass through these tubes into the cells of the + strain. Syngamy takes place within these cells, and the resulting zygote eventually develops a thick, resistant cell wall and is termed a zygospore. The vegetative filaments of* Spirogyra *are haploid, and meiosis occurs during the germination of the zygospores. (Carolina Biological Supply Co.)*

Siphonous Green Algae

The distinctive feature of the siphonous green algae is that their cells are typically multinucleate. In some species, there are essentially no transverse walls or septa dividing cells throughout the entire vegetative phase of the life cycle. Such aseptate, multinucleate cells are coenocytic and result from repeated nuclear division without the formation of cell walls. *Codium magnum,* the largest known green alga, is a member of this line and may reach 8 meters in length. An individual of another species of *Codium* is shown in Figure 12–11a. Only in the reproductive phase of this genus are cells with definite walls cut off. One siphonous green alga, *Valonia,* which is common in tropical waters, has been widely used in physiological experiments requiring large amounts of cell sap and in studies of cell walls. *Valonia,* which reaches the size of a hen's egg, appears to be unicellular, but is actually a large multinucleate vesicle, with separate rhizoids and young branches (Figure 12–11b). One of the best known of siphonous algae is *Acetabularia,* which has been used widely in experiments dealing with the genetic control of differentiation (page 83). With but one exception—*Hydrodictyon*—the siphonous green algae consist entirely of marine forms.

The Siphonous Line

The siphonous line of evolutionary specialization presumably begins with a motile, *Chlamydomonas*-like cell. Lack of coordination between mitosis and cytokinesis results in the development of multinucleate cells. As mentioned, the nonmotile colonial form *Hydrodictyon* belongs to this series. Originally uninucleate, each *Hydrodictyon* cell eventually contains many hundred nuclei. *Codium* and *Valonia* represent advanced members of the siphonous line.

Multicellular Green Algae

One of the most critical evolutionary steps in the history of plants was the transition from the single-celled condition, as in *Chlamydomonas,* to the multicellular condition, in which a series of more or less differentiated cells makes up the mature organism. Various kinds of multicellularity are found among the green algae, depending on the plane in which cell divisions occur. If they are restricted to one plane, division will result in unbranched filaments; if they occur in two planes, branched filaments or a flat sheet may result; and if they occur in three planes, a three-dimensional body will occur. Each of these is represented by particular green algae.

One familiar genus of unbranched green algae is *Spirogyra* (Figure 12–12), which is found as frothy or slimy floating masses in small bodies of water. Each filament is surrounded by a watery sheath that is slimy to the touch. The name *Spirogyra* refers to the helical ar-

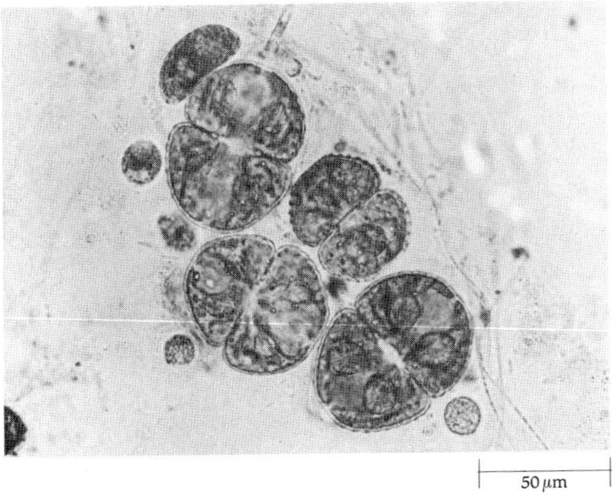

12–13
A single-celled desmid, each cell deeply constricted as is characteristic of the group. These individuals of a species of Cosmarium *are from a green floating mass on top of a rain pool.*

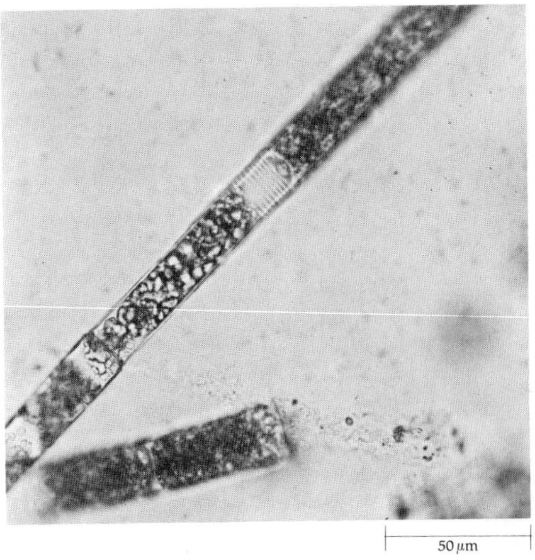

12–14
Oedogonium. *Part of vegetative filament showing annular scars.*

rangement of the one or more ribbonlike chloroplasts found within each uninucleate cell. The chloroplasts contain numerous pyrenoids. Except for fragmentation, no method of asexual reproduction occurs in *Spirogyra.* Morphologically, sexual reproduction in *Spirogyra* must be termed isogamous. However, during sexual reproduction one of the isogametes behaves like a male gamete by migrating across a conjugation tube to fuse with the other isogamete. Meiosis is zygotic.

The desmids (see page 1) are a large group of freshwater green algae related to *Spirogyra* and, like that genus, they lack motile cells. Some are multicellular, but most are unicellular, with a very peculiar cell construction. The cell wall is in two sections with a narrow constriction, the isthmus, between them (Figure 12–13). Some estimates place the number of species of desmids as high as 10,000.

Oedogonium differs from *Spirogyra* in its pattern of growth, which is *intercalary* (that is, between base and apex of the filament). Cells that divide exhibit annular scars that reflect the number of divisions that have taken place (Figure 12–14). The filaments of *Oedogonium* may be free-floating or attached to some object. Some are epiphytic on other algae or aquatic plants. (An epiphyte is an organism that grows nonparasitically on another organism or on some nonliving structure, deriving moisture and nutrients from the air.) Asexual reproduction in *Oedogonium* is by means of zoospore formation, with a single zoospore produced per cell. Each zoospore has a crown of about 120 flagella. Sexual reproduction is oogamous (Figure 12–15). Each male gametangium (antheridium) produces two multiflagellate sperms, each female gametangium (oogonium) only a single egg. Meiosis is zygotic. Both the growth pattern and spores of *Oedogonium* are unique, indicating that it may not be closely related to the other green algae.

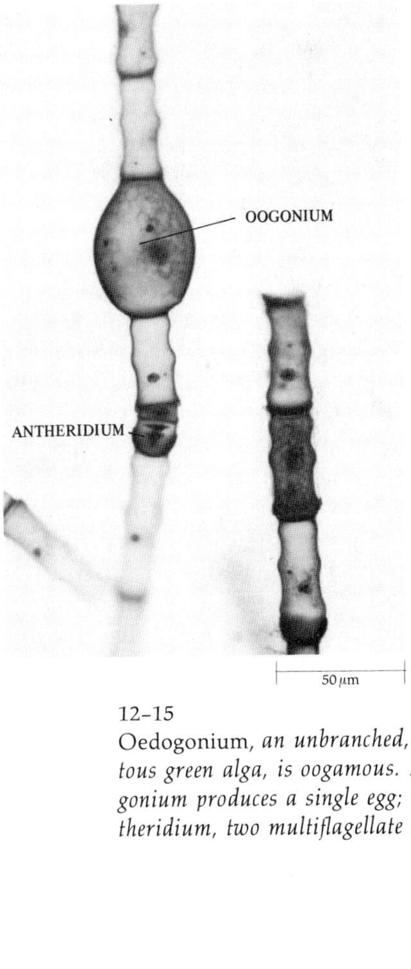

12–15
Oedogonium, *an unbranched, filamentous green alga, is oogamous. Each oogonium produces a single egg; each antheridium, two multiflagellate sperms.*

12-16
Cladophora *is a genus of algae that is widespread in both fresh and salt water. Its filaments are branched.*

12-17
Diagram of life cycle of Cladophora, *which is an alternation of generations. The two generations (gametophyte and sporophyte) are similar in size and morphology. Such alternations of generations are said to be isomorphic.*

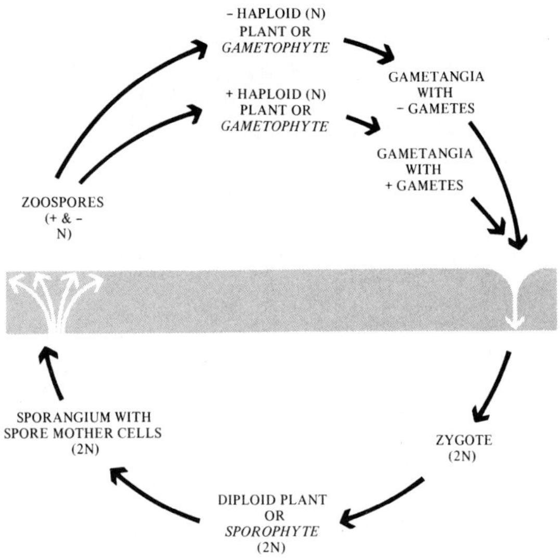

Cladophora is a widespread alga in both fresh and salt water (Figure 12-16). The filaments commonly grow in dense mats, either free-floating or attached to rocks or vegetation. In contrast to the generalized growth of *Spirogyra* and the intercalary growth of *Oedogonium,* that of *Cladophora* is localized near the apices. Branching occurs near the upper portions of relatively young cells. Each cell contains a single peripheral, reticulate chloroplast with many pyrenoids. The center of the cell is filled with a "frothy" cytoplasm containing numerous nuclei.

Alternation of Generations in Green Algae

The life history of *Cladophora* differs significantly from those of the other green algae discussed thus far. In them, meiosis is zygotic and the only diploid cell in the life cycle is the zygote. In *Cladophora,* meiosis is sporic (Figure 7-12, page 133). Two types of individuals exist, one haploid, the other diploid. The diploid individual arises from the zygote by mitosis and produces sporangia with *spore mother cells.* Meiosis occurs within the sporangium, each spore mother cell giving rise to four spores (zoospores in *Cladophora*). The haploid zoospores develop into unisexual (either + or −), haploid filaments. These filaments produce gametangia and gametes (isogametes in *Cladophora*). With the union of gametes, a zygote is produced and the cycle is completed. As mentioned previously (page 215), this type of life cycle, in which diploid, spore-producing individuals (*sporophytes*) alternate with haploid, gamete-producing individuals (*gametophytes*), is called an alternation of generations. In *Cladophora,* the two generations are similar in size and morphology. *Cladophora* thus has an isomorphic alternation of generations (Figure 12-17).

Among the marine green algae is *Ulva,* commonly called sea lettuce (Figure 12-18). Cell division occurs in three planes, but only once in one plane, giving rise to a glistening, flat *thallus* (a simple, relatively little differentiated vegetative body) two cells thick and up to a meter or more long. The thallus is anchored to the substrate by a holdfast produced by protuberances of the basal cells. Each cell of the thallus contains a single nucleus and chloroplast. *Ulva* is anisogamous and has an isomorphic alternation of generations (Figure 12-19) similar to that of *Cladophora.*

12–18
Sea lettuce (Ulva lactuca) *growing in a tidal pool with mussels, limpets, and coralline red algae.*

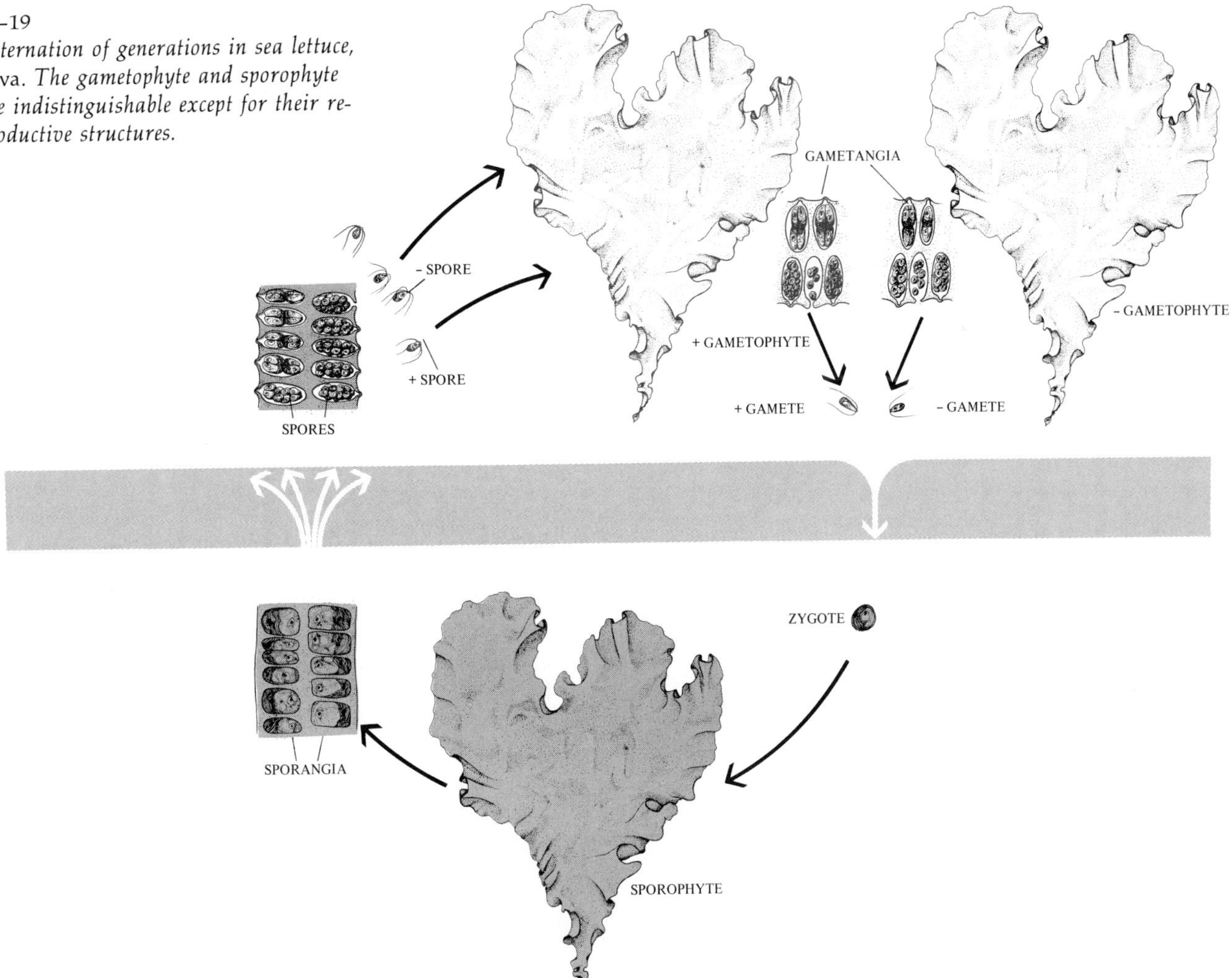

12–19
Alternation of generations in sea lettuce, Ulva. *The gametophyte and sporophyte are indistinguishable except for their reproductive structures.*

12–20
The life cycle of a siphonous green alga, Derbesia, *in which the gametophyte and sporophyte are so dissimilar that they were originally classified in different genera, the gametophyte having been called* Halicystis. *There is a fundamental difference between the cell walls of these two generations: In the gametophyte, they are composed primarily of xylan (with traces of cellulose), whereas in the sporophyte, they are based on mannan (with no cellulose).*

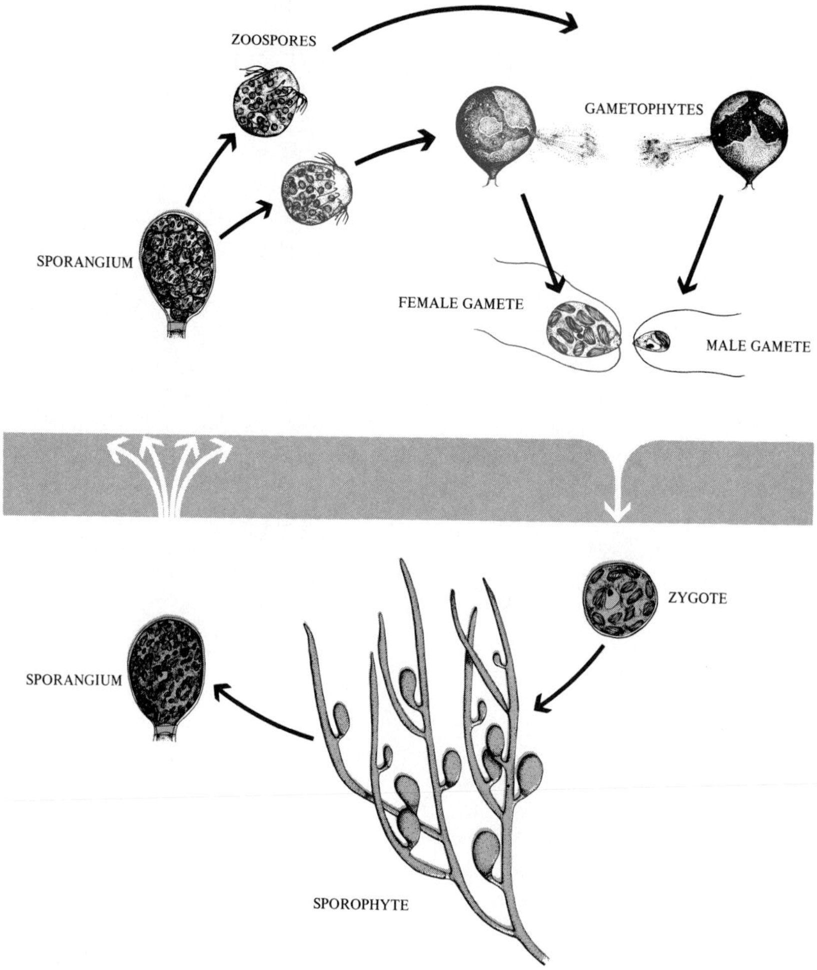

An example of a green alga with a heteromorphic alternation of generations is the siphonous marine *Derbesia (Halicystis).* The gametophyte is a bladder up to 3 centimeters in diameter, the sporophyte a freely branched filament. The two generations are so different in appearance that they were long placed in separate genera (Figure 12–20). *Derbesia* is an excellent organism for classroom use, being easy to culture and handle.

Such green algae as *Chlamydomonas* and *Spirogyra* may be considered to have an alternation of generations in which the sporophyte generation is represented only by the zygote. At the other extreme, *Codium* is diploid, with its gametes being the only haploid cells in the life cycle, as in animals, and so clearly has no alternation of generations.

Ancestors of the Land Plants?

Among the basically filamentous green algae are several genera of special interest in relation to the ancestry of the bryophytes and vascular plants. In most green algae, the cells divide by furrowing, but in a very few, they divide by cell-plate formation like that in the bryophytes and vascular plants. Associated with cell-plate formation is the presence of plasmodesmata, protoplasmic connections through cell walls that are also characteristic of bryophytes and vascular plants. Among the genera of filamentous green algae in which cell-plate formation occurs is *Ulothrix* (Figure 12–21).

Ulothrix, an alga of cold-water streams, ponds, and lakes, has its submerged filaments attached to stones or other objects by means of a special basal cell called a *holdfast* (Figure 12–22). All cells of the filament are essentially similar in appearance and contain a single, ringlike chloroplast, pyrenoid, and nucleus. Asexual reproduction occurs through formation of zoospores with four flagella, sexual reproduction by means of biflagellated isogametes. Meiosis occurs prior to the germination of the zygote, and the *Ulothrix* filament is haploid.

Among the genera closely related to *Ulothrix* is *Fritschiella,* which has a more complex and differentiated thallus. Its branched, filamentous body consists of subterranean rhizoids, a prostrate system near the soil surface, and an erect system with two orders of branches (Figure 12–23). The basal cells often divide in three planes, producing a mass of cells similar to parenchyma tissue, the evolutionary precursor of all other tissues.

Because the cells of these genera do divide by cell

12–21
Ulothrix. *Cell division in this alga is essentially similar to that in the bryophytes and vascular plants. This electron micrograph shows cell-plate formation nearly complete.*

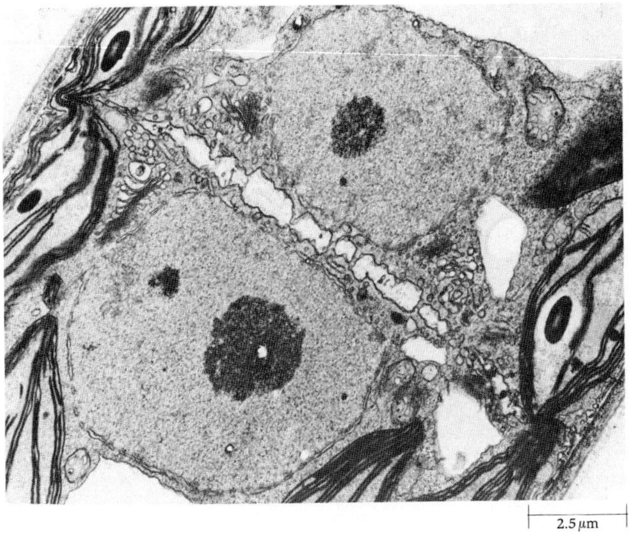

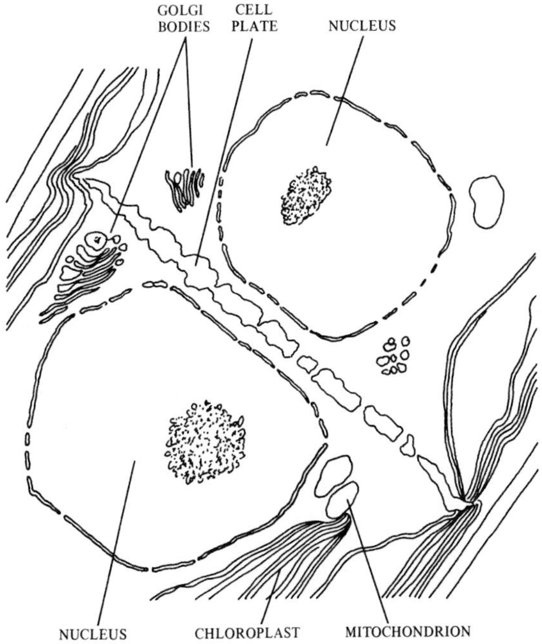

12–22
Ulothrix, *an unbranched, filamentous green alga. The filament on the left with dense contents consists entirely of sporangia in which zoospores are forming. The other filaments are vegetative. A holdfast can be seen on the filament on the right.*

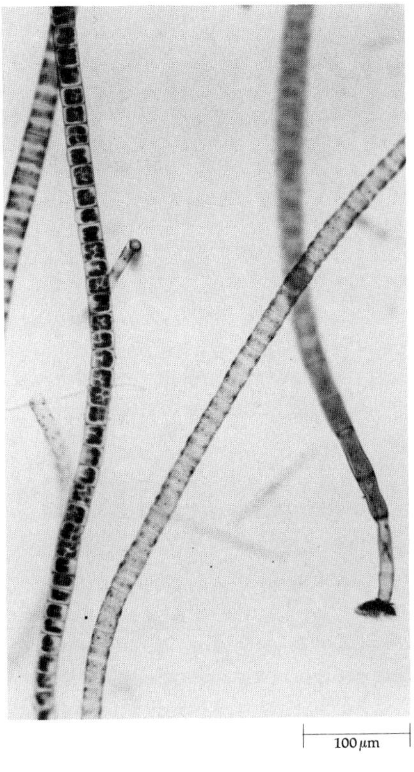

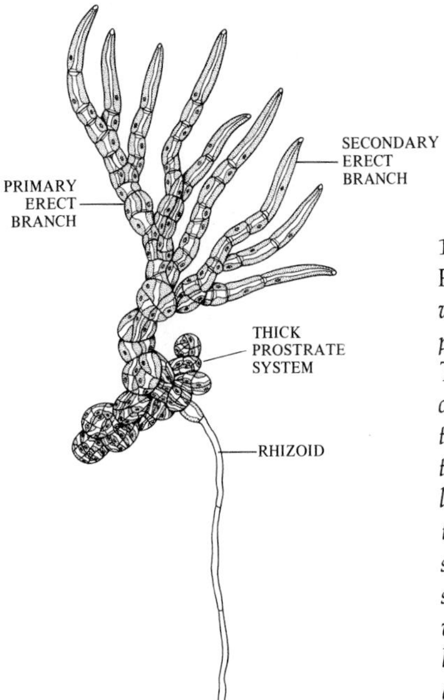

12–23
Fritschiella, *one of the few green algae in which cell divisions take place in three planes, producing a solid vegetative body. These solid bodies, in which the cells are connected by plasmodesmata, resemble those of plants. The degree of differentiation in various tissues is, however, much less complex, and there are differences in the mode of cell division.* Fritschiella *is a semiterrestrial alga, occurring on damp surfaces, such as tree trunks, moist walls, and leaf surfaces. The multicellular rhizoids found near the base of the alga presumably aid it in drought resistance.*

12-24
The most distinctive group of the green algae—so much so that they are sometimes treated as a distinct division—are the stoneworts (class Charophyceae). These algae, which occur in fresh to brackish water, have heavily calcified cell walls. As a consequence, although there are only about 250 living species, the stoneworts are very well represented in the fossil record. In growth pattern, the charophytes are complex, with nodal regions, from which whorls of short branches arise, and coenocytic internodal regions that are naked or more complex. Their sperms are produced in multicellular structures (antheridia) that are more complex than those found in any other group of algae; their eggs are borne in simpler organs (oogonia). The sperms are the only flagellated cells in the life cycle of the charophytes.

The charophytes are the only algae that form multicellular sex organs (gametangia), but they are not comparable to those of the bryophytes and vascular plants, for they lack sterile jacket layers. (a) *Photograph of* Nitella hyalina, *showing the determinate growth pattern and gametangia.* (b) *Segment of* Chara *sp. showing the gametangia. An antheridium is on the left and an oogonium is on the right.*

(a) (b)

plates and are connected by plasmodesmata, and particularly because the complex thallus of genera such as *Fritschiella* is somewhat similar to that of plants, this group of algae has been considered possibly ancestral to the bryophytes and vascular plants. On the other hand, both *Ulothrix* and *Fritschiella* have a specialized mode of cell division, widespread in the green algae, involving a *phycoplast*—a complex system of microtubules all oriented in the plane of cell division. In bryophytes, vascular plants, and certain algae (charophytes (Figure 12-24), *Spirogyra*, and related genera, and a few genera that have been allied with *Ulothrix*, including the very complex *Coleochaete*) the microtubules are oriented perpendicular to the plane of cell division. Thus the search for an ancestor to the bryophytes and vascular plants continues among the multicellular green algae. It is unlikely that any living organism resembles the common ancestor of land plants in detail and, unfortunately, no fossil of a complex green alga has yet been found that is old enough and fits the characteristics of such a hypothetical ancestral form.

The Tetrasporine Line

Beginning with a motile, *Chlamydomonas*-like cell, the tetrasporine line of evolutionary specialization ranges from nonmotile unicells such as *Chlorella* and *Eremosphaera* to more complex multicellular organisms such as *Fritschiella* and *Ulva*, with unbranched filaments (e.g., *Ulothrix*) as intermediate forms. The degree of complexity of the thallus is related to the variation, if any, of the planes of cell division. All members of the tetrasporine line consist of uninucleate cells, and motile cells are produced only during reproduction.

THE BROWN ALGAE: DIVISION PHAEOPHYTA

The brown algae (division Phaeophyta), an almost entirely marine group, comprise most of the conspicuous seaweeds of temperate regions. Although there are only about 1500 species, the brown algae are of considerable interest because they dominate rocky shores throughout

12–25
Brown algae. (a) *Kelp exposed at low spring tide.* Laminaria digitata *fills the foreground with some rockweed,* Fucus serratus, *visible on the rocks at the top corners of the photograph.* (b) *Rockweed* (Fucus vesiculosus) *densely covers many rocky shores that are exposed at low tide. The air-filled bladders on the blades carry them up toward the light. The parts are thick and leathery and moderately drought-resistant. Photosynthetic rates of frequently exposed marine algae are one to seven times as great in air as in water, whereas they are higher in water for those rarely exposed, which accounts in part for the vertical distribution of seaweeds in intertidal areas.*

(a)

(b)

the cooler regions of the world and some, like the kelps, often form extensive beds offshore (Figure 12–25). In clear water, kelps flourish from low-tide level to a depth of 20 to 30 meters. On gently sloping shores, they may extend 5 to 10 kilometers from the coastline. Even in the tropics, where the brown algae are less common, there are the immense floating masses of *Sargassum* (Figure 12–26) in such areas as the vast Sargasso Sea.

The brown algae are also of interest because of their size and often elaborate internal tissue differentiation. The giant kelps *Macrocystis* and *Nereocystis* may have fronds more than 100 meters long under unusually favorable circumstances. Many of the brown algae, such as *Sargassum,* approach the vascular plants in complexity of organization of their vegetative parts. Many kelps exhibit conspicuous differentiation into holdfast, stipe (stalk), and blade regions and have well-defined meristematic regions—areas of sustained cell division—within their bodies. In the kelps, which have regular cell division in three planes, the meristematic regions are generally located within the blade (Figure 12–27), although in

12–26
Sargassum, *a brown alga with a complex pattern of organization of the vegetative parts.*

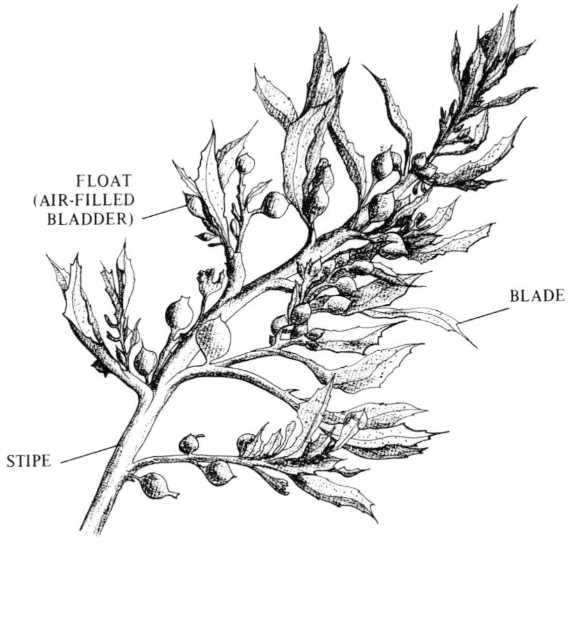

12–27
The leaflike blade of Macrocystis *is produced by the activity of a meristematic region at the junction of the stipe and blade from which additions are made to the stipe and to the terminal blade. The latter splits as it expands, and segments form the "leaves."*

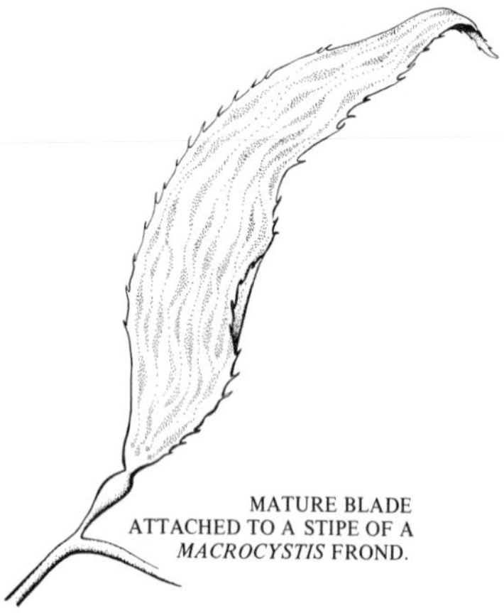

some genera there is a single apical cell, or a group of apical cells, that constitutes the region of active cell division. All cells have centrioles. Growth is rapid and productivity high, particularly among the kelps that are not exposed to the atmosphere; a recent study of three species conducted in Nova Scotia showed that all renewed the tissue in their blades completely between one and five times each year.

In 1974, experimental plantings of kelp on a commercial scale were initiated by the California Institute of Technology near San Clemente Island, off the coast of southern California, to explore its usefulness in producing petroleumlike products for fuel and perhaps also food. *Macrocystis* has been established on a rope raft nearly 3 hectares in size, and its growth will be closely monitored. The raft is anchored to the bottom in some 90 meters of water so that it floats about 12 meters below the surface, the approximate depth at which the kelp normally grows. Individual plants of *Macrocystis* can, under appropriate conditions, grow as much as 0.6 meter per day. If the initial planting is successful, it will be expanded to as much as 40,000 hectares (100,000 acres) by 1985. Proposals are also being considered for introducing *Macrocystis* along the coast of Europe.

The kelps even have strands of elongated conducting cells in the center of the stipe that are similar to phloem. These cells (Figure 12–28) have sieve plates similar to those of the sieve-tube elements of vascular plants, although they must have evolved independently. Marine algae do not need a mechanism for internal water transport, and it is thus not surprising that there are no cells analogous to tracheids or vessel elements in their bodies. They must, however, conduct carbohydrates from the region of synthesis to other, poorly illuminated portions of their bodies far below the surface of the water. Recently, it has been shown experimentally that the

12–28
Kelps do not have a problem in supplying water to the different parts of the body but must conduct carbohydrates from the upper, better illuminated zones where photosynthesis takes place to the darker ones far below the surface. Some seaweeds, such as the giant kelp Macrocystis pyrifera, *have evolved sieve tubes comparable to those of vascular plants.* (a) *A median longitudinal section of a stalk of* Macrocystis, *with sieve tubes on the outside and a loosely organized central region.* (b) *Cross-section showing a sieve plate.*

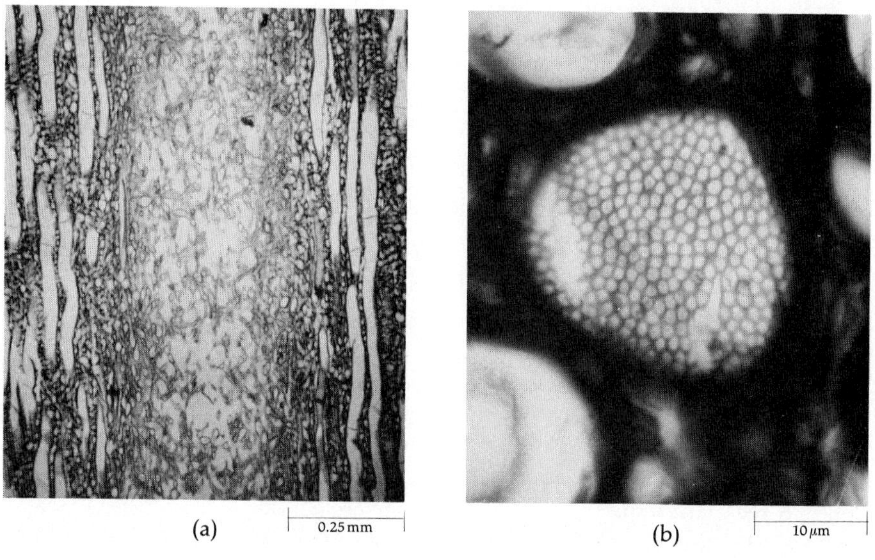

sieve elements in the center of the stipes are the site of this translocation, which takes place in the giant kelp *Macrocystis* at a rate of about 60 centimeters per hour. In many of the kelps that are relatively thick, lateral translocation of carbohydrates from the outer photosynthetic layers to the inner cells also takes place.

Although it has been suggested that the Phaeophyta evolved from motile unicells, neither unicellular nor colonial organisms are found among the brown algae. Various brown algae are filamentous or branched, and some form thin sheets; thus the diversification in this group parallels that in the multicellular green algae.

In addition to chlorophyll *a*, brown algae have chlorophyll *c* as an accessory pigment (Figure 12-29). They also contain various carotenoids, including the abundant xanthophyll fucoxanthin (Figure 12-30), which gives the members of this group their characteristic dark brown or olive-green color. Instead of starch, laminarin is the characteristic carbohydrate storage product (see Figure 12-3), but lipids are also accumulated. Motile cells in this group (zoospores and gametes) have two lateral flagella; the anterior one is of the tinsel variety and the shorter posterior one is a whiplash flagellum. Cell walls in the brown algae are similar to those in most green algae and plants, being based on a matrix of cellulose microfibrils, but there is a well-developed outer layer of mucilaginous compounds. Alginic acid (algin), a gummy substance of considerable importance as a stabilizer, an emulsifier, and a coating for paper, is abundant in the middle lamella and is also found elsewhere in the wall.

The life cycles of most brown algae are essentially similar to those of marine species of *Cladophora* and of *Ulva*; that is, they are alternations of generations. Recently, relatively simple culture methods for marine algae have been developed that have made possible rapid advances in our knowledge of them. Meiosis is sporic, and both generations (gametophyte and sporophyte) are free-living. In some species of brown algae, the alternation is isomorphic, in others it is heteromorphic. The latter type of alternation is exhibited by the common kelp *Laminaria* (Figure 12-31), whereas the filamentous genus *Ectocarpus*, with similar sporophytic and gametophytic generations, illustrates the former. The chemical structure of the molecules that are secreted by

12-29
Chlorophyll c *replaces chlorophyll* b *as one of the accessory pigments in several divisions of algae. Its structure, which was not worked out until the mid-1960s, differs from that of chlorophylls* a *and* b.

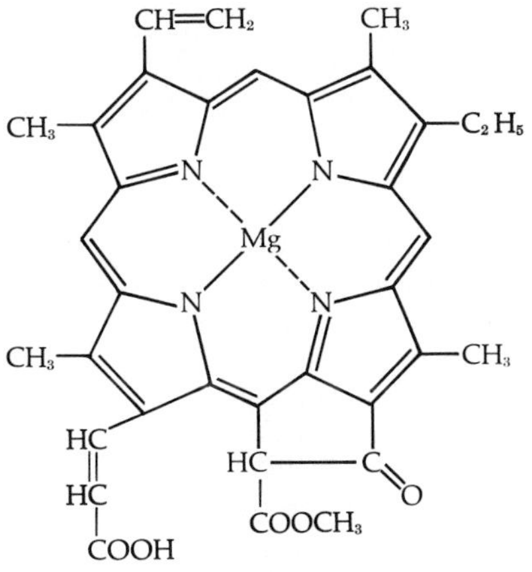

CHLOROPHYLL *c*

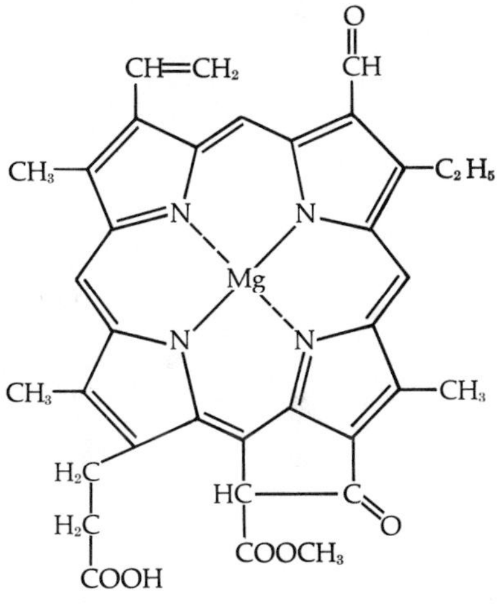

CHLOROPHYLL *b*

12-30
Fucoxanthin, the xanthophyll that gives the characteristic color to brown algae.

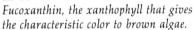

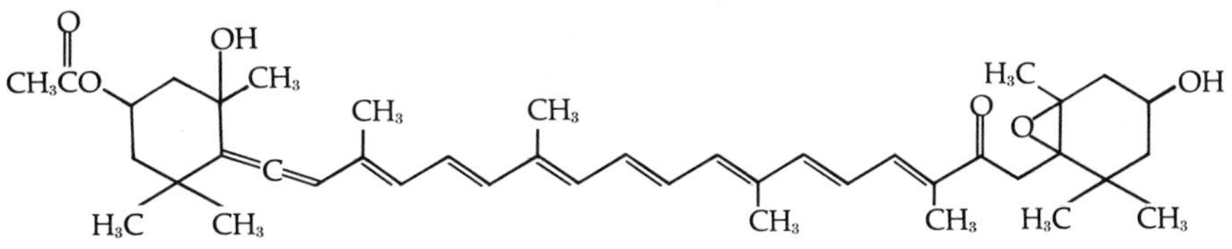

FUCOXANTHIN

12–31

Life cycle of the kelp Laminaria. *Like most of the brown algae,* Laminaria *has an alternation of generations in which the conspicuous generation is the sporophyte. Motile haploid zoospores are produced in the sporangia of* Laminaria *following meiosis. From these zoospores grow the small, filamentous gametophytes, which in turn produce the motile sperms and immotile eggs. In simpler brown algae, the sporophyte and gametophyte are often similar.*

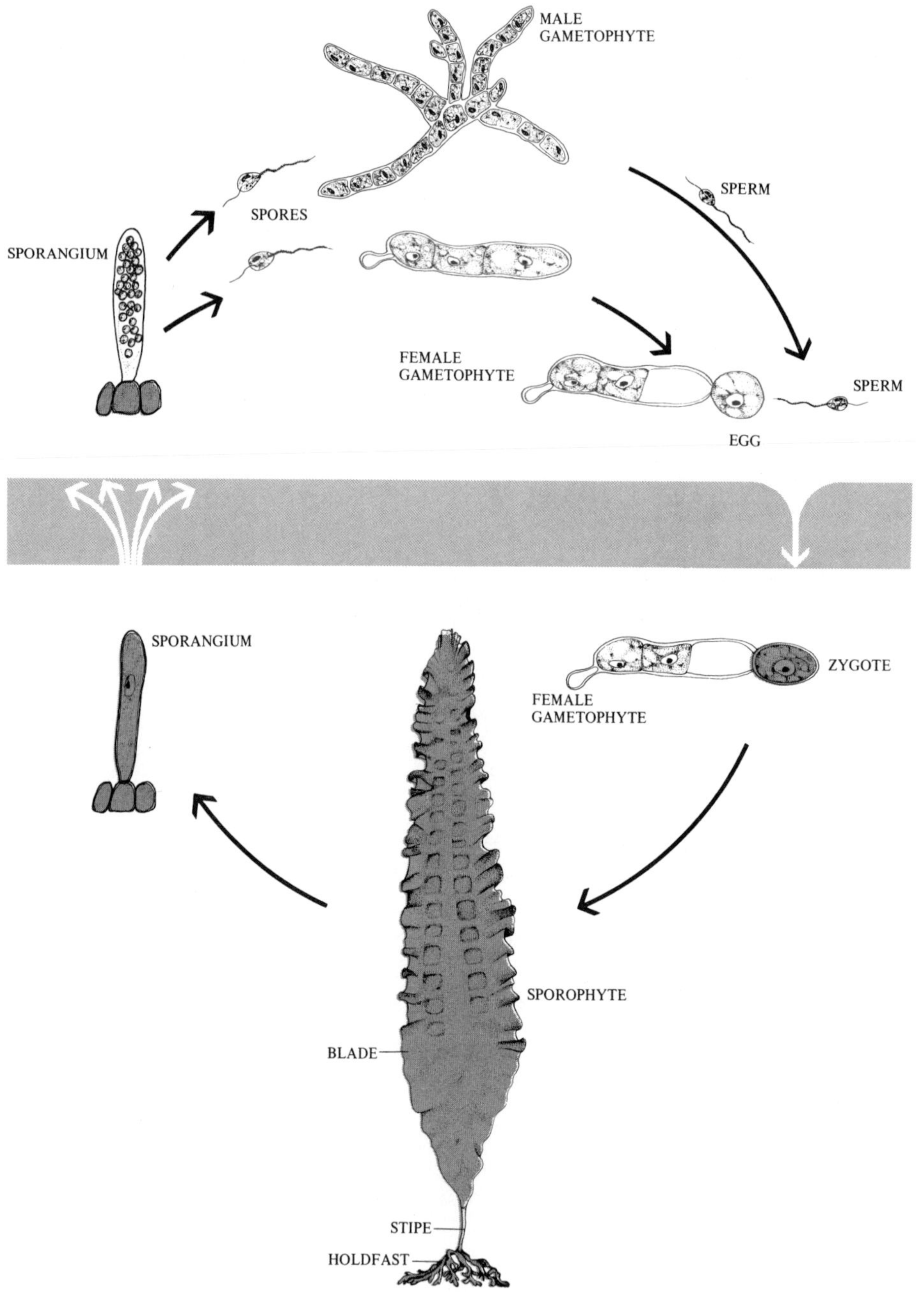

12-32
In Fucus, gametangia are formed in specialized hollow chambers (conceptacles) in fertile areas called receptacles at the tips of the branches of diploid individuals. There are two types of gametangia, oogonia and antheridia. Meiosis is followed immediately by mitosis to give rise to 8 eggs per oogonium and 64 sperms per antheridium. Eventually the eggs and sperms are set free in the water, where fertilization takes place. The life cycle of Fucus has been compared with that in higher animals, which it superficially resembles. Meiosis is gametic, and the zygote grows directly into the new diploid individual.

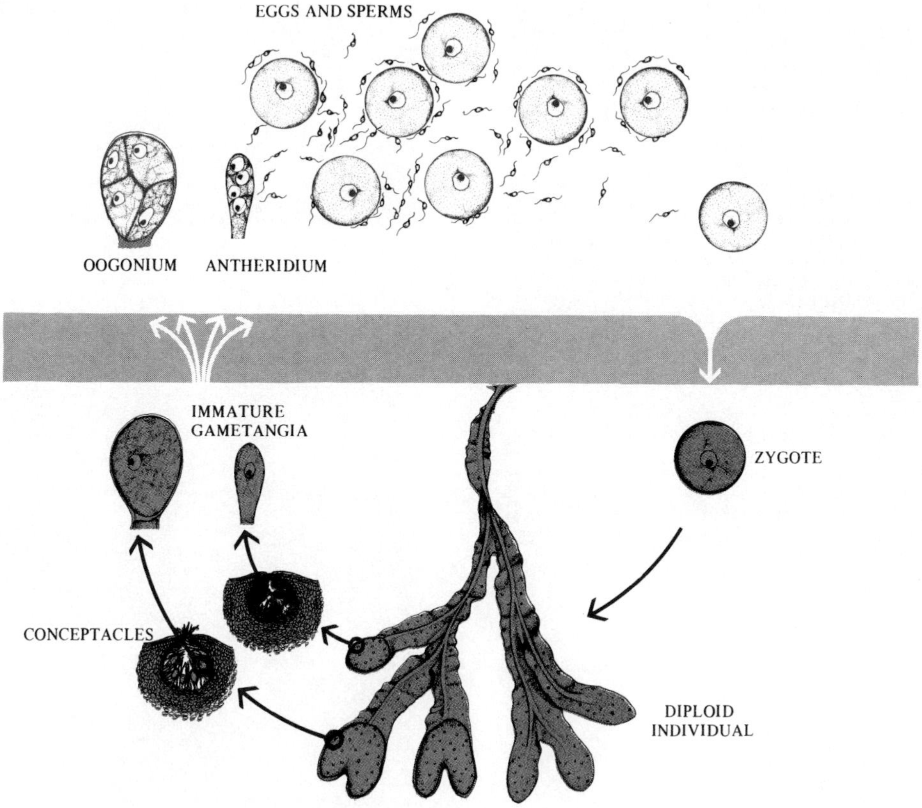

the female gametes and serve to attract the male gametes in several genera of brown algae have now been analyzed chemically and are simple aromatic compounds. *Fucus* serves to illustrate another and very unusual type of life cycle found among the brown algae (Figure 12-32). Here meiosis is gametic, and the zygote grows directly into the new diploid organism (see Figure 7-12, page 133). Zygotic meiosis, which is very common in the division Chlorophyta, has not been observed in the Phaeophyta.

THE RED ALGAE: DIVISION RHODOPHYTA

The red algae have no flagellated cells, are structurally complex (Figure 12-33), and have complex life cycles. In contrast to the green and brown algae, which have centrioles in all cells, the red algae lack centrioles entirely. There are some 4000 species of red algae, and they are particularly abundant in tropical and warm waters, although many are found in the cooler regions of the world. Fewer than 100 species are found in fresh water, but in the sea the number of species is greater than that of all other groups of seaweeds combined.

The water-soluble phycobilins, which mask the color of chlorophyll *a* in the red algae and give them their distinctive color, are accessory pigments and are particularly well suited to the absorption of the green, violet, and blue light that penetrates into deep water (Figure 12-34). The red algae are found at greater depths than any other groups of algae, some having been found at depths of up to 175 meters. Because phycobilins are also found in the blue-green algae, it is likely that the chloroplasts of red algae are derived from ancient symbiotic organisms very similar to modern blue-green algae. Another accessory pigment in some red algae is

(a)

(b)

(c)

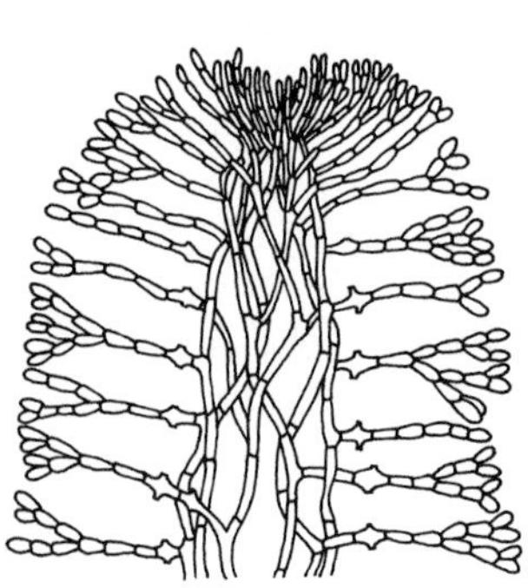

(d)

12–33
Red algae. (a) *In* Pleonosporium dasyoides, *the basically filamentous ground plan of the red algae is clearly evident. It appears that multicellularity in the red algae evolved separately from multicellularity in the green and brown algae, where no similar patterns of organization are found.* (b) *Reef-building coralline alga,* Porolithon craspedium, *from a reef in the Marshall Islands.* (c) *Irish moss* (Chondrus crispus), *an important source of carrageenan and other colloids.* (d) *Detail of structure of one of the simpler red algae,* Cumegloia, *showing its obviously filamentous form.*

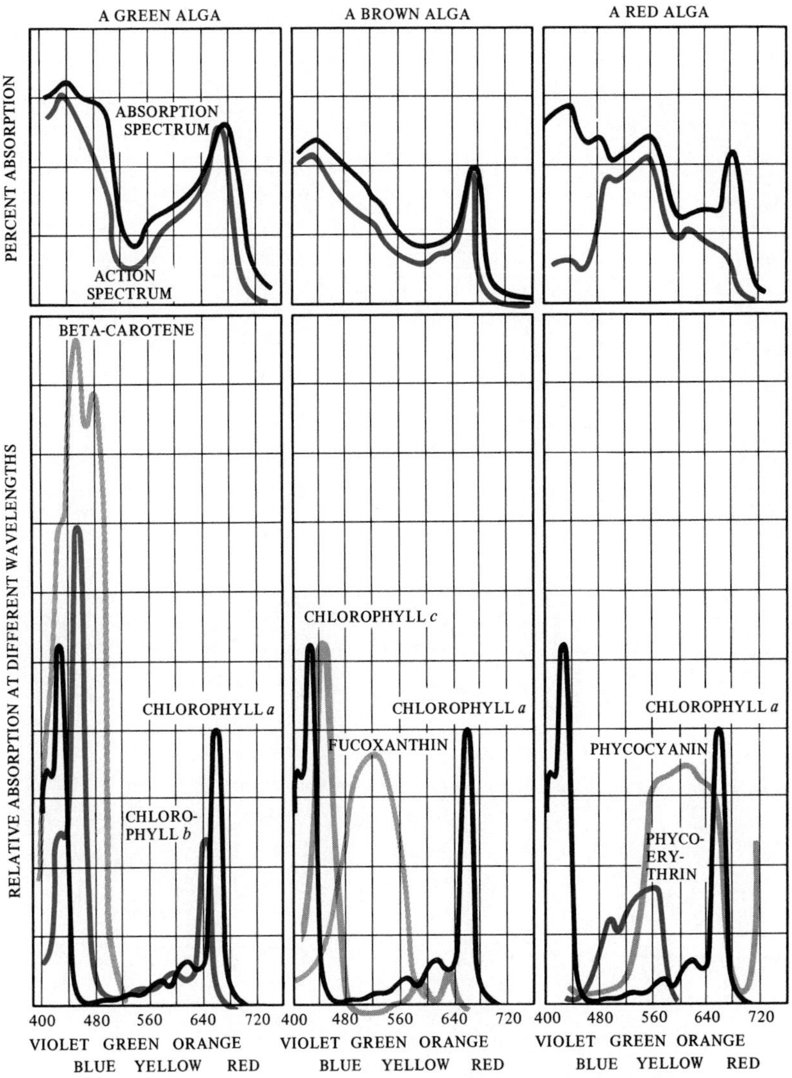

12-34
Absorption and action spectra (see page 108) of a green alga (Ulva taeniata), *a brown alga* (Coilodesme californica), *and a red alga* (Porphyra naiadum). *Below each are shown the absorption spectra for some of the major photosynthetic pigments. As you can see from the action spectra, the red algae are more dependent upon the blue-green wavelengths than are the other two, particularly those wavelengths absorbed by phycocyanin and phycoerythrin. These wavelengths penetrate to deeper water. In red algae found at different depths, the ratio of phycocyanin to phycoerythrin may vary according to the quality of the light that penetrates to these characteristic depths.*

chlorophyll *d* (Figure 12–35). Because chlorophyll *d* has never been detected in the absorption spectra of living red algae, it was once thought to be an oxidation product of chlorophyll *a*. However, it is now considered a genuine pigment. The reserve carbohydrate of the red algae is called florridean starch. The grains of floridean starch usually are found in the cytoplasm on or near the surface of the plastids rather than within them.

The cell walls of red algae include an array of mucilages based on galactose, which give the red algae their characteristic texture; other polysaccharides have also been found in certain red algae.

Red algae usually grow attached to rocks or other algae; there are few floating forms and few that are unicellular or colonial. Most red algae are composed of filaments, although this basic body plan is often difficult to distinguish because the filaments may be packed together very tightly. In many red algae, the cells are in-

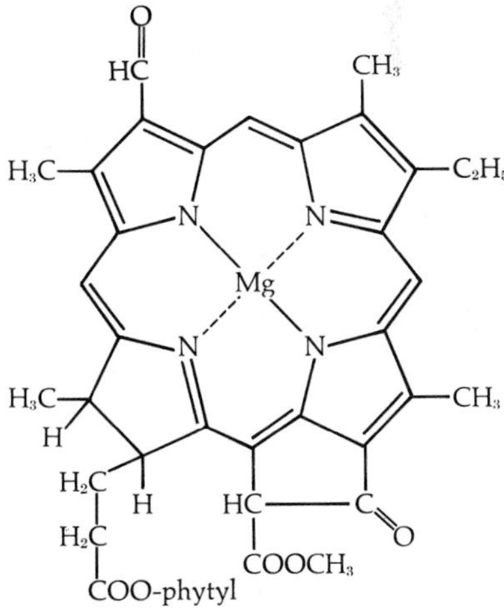

CHLOROPHYLL *d*

12-35
Chlorophyll d has been extracted from red algae, but it is not clear whether or not it is formed during the extraction process.

12-36
Life cycle of Polysiphonia, *a red alga that is widely distributed in marine waters. The gametophytes are derived from haploid tetraspores and are unisexual. The sex organs arise near the tips of the branches. The male sex organs, or spermatangia, occur in dense clusters. Each spermatangium functions directly as a spermatium, or nonmotile sperm. The female sex organ, or carpogonium, develops a long, fingerlike outgrowth, the trichogyne. The enlarged basal portion of the carpogonium contains the nucleus and functions as an egg. Spermatia are carried passively to the trichogyne by water currents. When a spermatium becomes attached to a trichogyne, the walls between spermatium and trichogyne break down at their point of contact. The spermatial nucleus then enters the trichogyne and migrates to the egg nucleus, with which it fuses.*

The series of events following fertilization, which is extremely complicated, results in formation of carposporangia, which produce carpospores. When mature, the carpospores are liberated through an opening in the pericarp. Upon germination, each carpospore gives rise to a tetrasporophyte, which is similar in size and appearance to the gametophytes. The tetrasporophytes produce tetrasporangia, each of which, upon meiosis, gives rise to four haploid tetraspores. The tetraspores develop into gametophytes, and the cycle—an isomorphic alternation of generations—is complete.

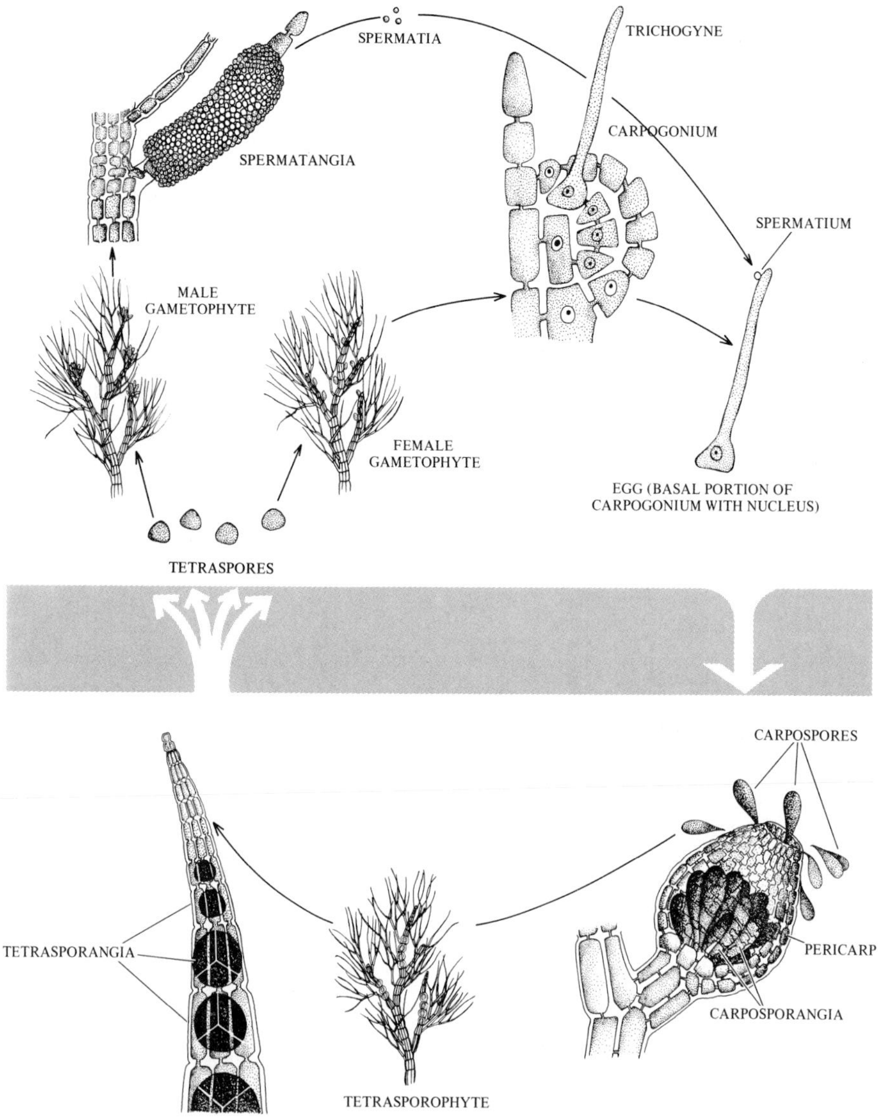

terconnected by prominent plasmodesmata. The simpler families consist of obviously filamentous forms in which cell division takes place mostly in cells within the filament. In the more complex members of this division, the alga body consists of branched filaments that grow principally by the division of apical cells at the tips of the branches. In many families of red algae, these branched filaments are densely interwoven to give the impression of a complex thallus that is superficially comparable to that of the brown algae. The basic life cycle of the red algae is an alternation of generations (Figure 12-36). In most, the gametophyte and sporophyte are isomorphic, but an increasing number of heteromorphic cycles are being discovered.

A number of the red algae become heavily encrusted with calcium carbonate. These coralline algae, as they are called, are extremely important in the building of coral reefs in warm waters, and in such areas they may constitute the principal components of the reef, in some cases almost to the exclusion of the coral animals. Such reefs generally contain about three times as much algae as animal tissue, and the algae control the rate of reef formation under natural conditions. Coralline algae usually form the principal cementing agents in the reefs, and dominate the marine communities on the seaward margins of both reefs and atolls. Inside the lagoons formed by atolls, similarly encrusted siphonaceous green algae may predominate. In the fossil record, calcareous algae (algae that grow on limestone or in soil impregnated with lime) are richly represented from Precambrian time to the present, with marine coralline red algae first appearing in the late Mesozoic era.

THE DIATOMS AND GOLDEN-BROWN ALGAE: DIVISION CHRYSOPHYTA

The diatoms and golden-brown algae (division Chrysophyta) are unicellular organisms that are exceedingly important components of phytoplankton, the microscopic photosynthetic organisms that are suspended in the water. As such, they are a primary source of food for water-dwelling animals, both in marine and freshwater habitats. There are 6000 to 10,000 species in this division.

Like the brown algae, the Chrysophyta have chlorophylls *a* and *c*, the color of which is largely masked by the abundant accessory pigment fucoxanthin. The carbohydrate food reserve of the Chrysophyta is leucosin, which is similar in structure to the laminarin of the brown algae. In Chrysophyta, the products of photosynthesis are normally stored as large oil droplets. In one respect, the Chrysophyta are very different from the brown algae: Their cell walls consist mainly of pectic compounds, which are often impregnated with siliceous materials and thus very rigid. The walls contain no cellulose. On the basis of the existing evidence, it is difficult to decide whether the Chrysophyta are directly related to the brown algae.

The Diatoms: Class Bacillariophyceae

Most species of Chrysophyta belong to this large group of unicellular organisms, which has at least 40,000 valid species (living and extinct) and might have as many as 100,000. There are often tremendous numbers of species of diatoms in small areas. In two small samples of mud from Beaufort, North Carolina, for example, 369 species were identified. Most species of diatoms occur in plankton, but some are bottom dwellers or grow on other algae or plants. They occur both in fresh and salt water.

Diatoms have fine double shells *(frustules)* of polymerized, opaline silica ($SiO_2 \cdot nH_2O$), the two halves (valves) of which fit together, one on top of the other, like a carved pillbox. The delicate markings of these shells, by which the species are identified, have been traditionally used by microscopists to test the excellence of their lenses. Electron microscopy has shown that these fine tracings on the diatom shells are actually composed of a large number of minute, intricately shaped depressions, pores, or passageways that connect the living protoplasm within the shell with the outside environment (Figure 12–37). The most conspicuous features within

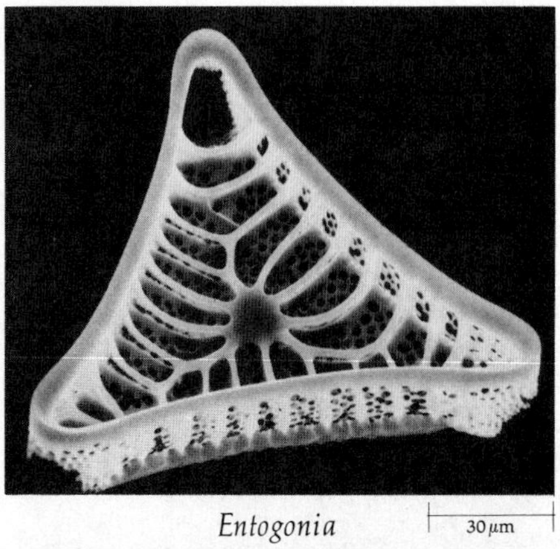

Entogonia

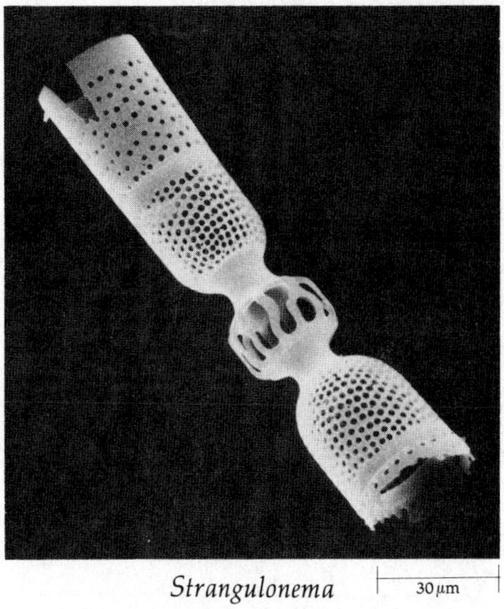

Strangulonema

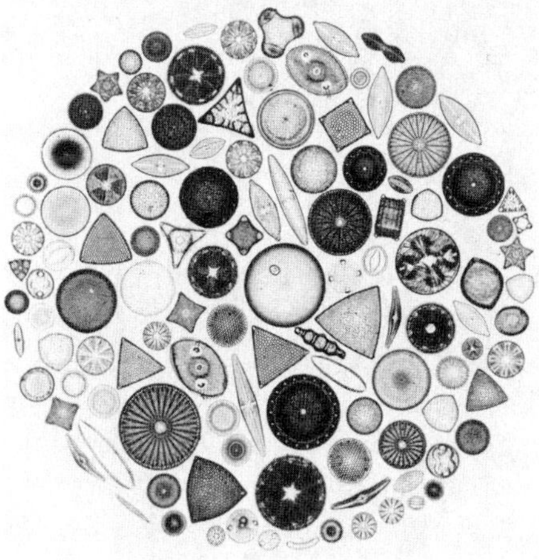

12–37
The siliceous walls of diatoms as seen by the scanning electron microscope (above and center) and the light microscope. Each species has its characteristic pattern of perforations in the walls.

12–38
Reproduction in the diatoms usually takes place by cell division. Each daughter cell receives one of the valves of the previous shell and constructs a second one. The old valve always forms the top lid of the silica box, with the new valve fitting inside it. As a consequence, one cell of each new pair tends to be smaller than the one before it. In some species, the shells are expandable and are enlarged by the growing protoplasm within them. In other species, however, in which the shells are more rigid, cell size is increased at the time of sexual reproduction by the formation of the zygote of an auxospore which, freed from the frustule, expands to the full size characteristic of the species. The walls formed by auxospores are often different from those of vegetative cells, and auxospores have on occasion been described as new genera or species. Once growth is completed, the auxospore secretes new frustules identical to the previous ones in all their intricate markings.

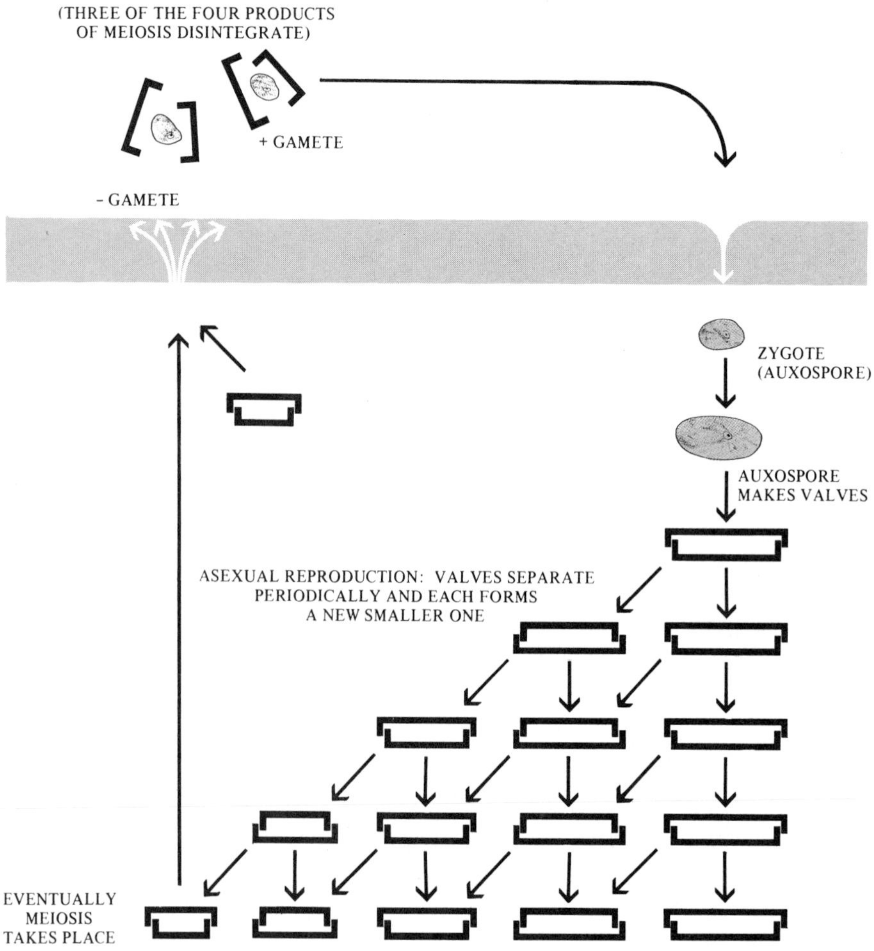

the protoplast of diatoms are the brownish plastids, which contain chlorophyll *a* and *c* as well as fucoxanthin, like the brown algae. Most diatoms reproduce by cell division (Figure 12–38).

On the basis of symmetry, two types of diatoms are recognized: those that are bilaterally symmetrical, the pennate types, and those that are radially symmetrical, the centric types. Centric diatoms are most abundant in marine waters.

Despite their lack of cilia, flagella, or other locomotor organelles, many species of diatoms—but only pennate types—are motile. Their locomotion results from a rigorously controlled secretion that occurs in response to a wide variety of physical and chemical stimuli. All motile diatoms seem to possess a fine groove along each frustule, called the raphe, which is basically a pair of pores connected by a complex slit in the silicate wall of the diatom. Many nonmotile diatoms are attached to one another, with their frustules arranged in long filaments, and the raphe apparently evolved as a locomotor device by modification of the apical pores that secrete the substances that unite these nonmotile diatoms into filaments.

A diatom moves in response to an external stimulus, such as mechanical disturbance, light, heat, or a toxic chemical, by initiating contractions in contractile bundles that lie adjacent to the raphe system. This contraction moves dehydrated crystalline bodies to reservoir areas adjacent to pores in the raphe, and the crystalline bodies are then discharged into the pores. Here they take up water and expand into twisting fibrils. The fibrils move along the raphe until they strike a surface. They immediately adhere to anything they touch, and then contract. If the object to which they adhere is large enough, the diatom moves toward it, depositing a trail of secreted material analogous to the slime trail of a snail. If

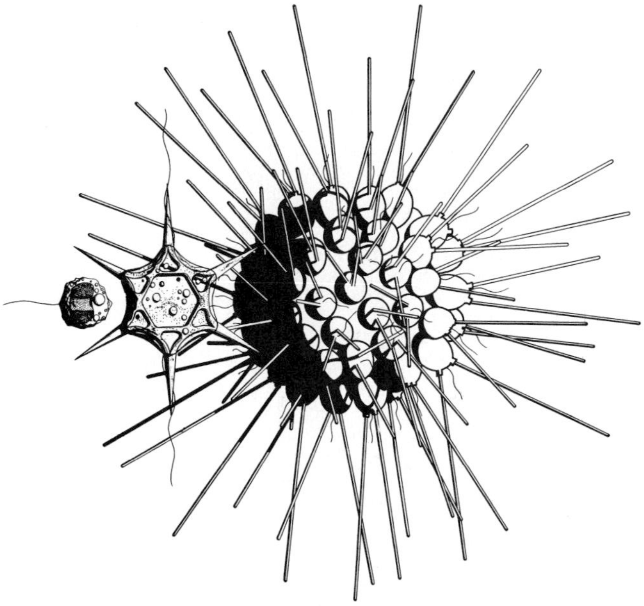

12–39
Diversity in Chrysophyceae. Chrysophaerella *is a colonial organism analogous in some ways to the green alga* Volvox. Distephanus *has conspicuously developed siliceous skeletons, whereas* Chromulina *is naked.*

the object is small, it is transported along the raphe and the diatom remains stationary.

Even motile diatoms are usually at rest. Each can move only a limited distance, for they have a limited number of the required crystalline bodies.

The piled-up silica shells of diatoms, which have collected over millions of years, form the fine, crumbly substance known as diatomaceous earth, which is used as an abrasive in silver polish and for filtering and insulating materials. The paint used on the center lines of roads is impregnated with diatomaceous earth because of its low-angle reflective characteristics. It is estimated that 1 cubic centimeter of diatomaceous earth contains some 4.6 million individuals. In the Santa Maria, California, oilfields there is a subterranean deposit of diatomaceous earth 900 meters thick, and near Lompoc, California, more than 270,000 metric tons of diatomaceous earth are quarried annually for industrial use.

Diatoms first became abundant in the fossil record in the Cretaceous period, some 100 million years ago; many of the species are identical to those still living today, which indicates an astonishing persistence through geological time.

The Golden-Brown Algae: Class Chrysophyceae

The other generally recognized class of Chrysophyta consists of about 1100 species. Until recently, they were thought to be a primarily freshwater group, but within the past decade they have proved to be of extraordinary importance in the marine plankton, particularly the *nannoplankton*. The nannoplankton comprises components of the plankton so small that they pass through an ordinary plankton net, which has mesh openings of 0.040 to 0.076 millimeter. Some of the nannoplankton consist of minute dinoflagellates and diatoms, but representatives of the Chrysophyceae are often abundant. In fact, it is now thought that the Chrysophyceae may be the major food-producing organisms in the ocean.

Some of the golden-brown algae lack cell walls, whereas others have a well-defined wall rich in pectic substances. Many species have superficial or internal siliceous or organic scales or skeletal structures, which may be exceedingly elaborate (Figure 12–39). Many golden-brown algae are motile, having two flagella, whereas others are amoeboid and lack flagella. Except for the presence of chloroplasts, the amoeboid cells are indistinguishable from amoeboid protozoa (phylum Sarcodina), and the two groups may be closely related. Reproduction in the golden-brown algae is largely asexual and involves zoospore formation.

THE YELLOW-GREEN ALGAE: DIVISION XANTHOPHYTA

The Xanthophyta consists of about 450 species, most of which are freshwater organisms. Until recently these organisms were included in the division Chlorophyta, for structurally they resemble many of the green algae. Today some botanists include them with the golden-brown algae and diatoms in the division Chrysophyta, rather than placing them in a separate division.

Unlike the diatoms and golden-brown algae, the yellow-greens lack chlorophyll *c* and fucoxanthin. In addition to chlorophyll *a* and some carotenoid pigments, two

12–40
Vaucheria, *the "water felt," is a coenocytic, filamentous member of the division* Xanthophyta. Vaucheria *is oogamous, producing oogonia and antheridia. The antheridium shown here is empty.*

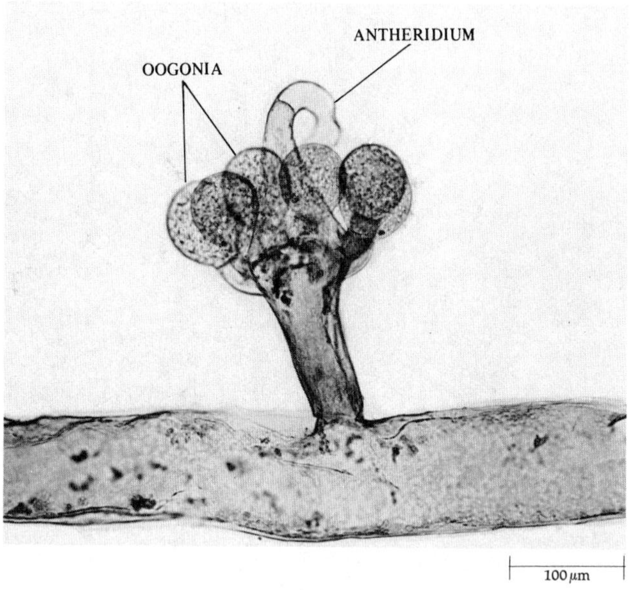

strains of yellow-green algae possess chlorophyll *e*. However, the chlorophyll *e* may be only an alteration form of chlorophyll *c*. The reserve food of the yellow-green algae is leucosin.

Morphologically the Xanthophyta is a very diverse group. It includes unicellular motile, nonmotile, and amoeboid organisms, colonies, and both cellular and coenocytic filamentous forms. The cell wall consists primarily of cellulose and pectin. In some organisms it is impregnated with silicon. The principal method of reproduction is by cell division, although various types of spores may be formed. One of the better-known members of the division is *Vaucheria*, the "water felt." A coenocytic, little-branched filamentous form, *Vaucheria* reproduces both asexually, by formation of large, compound zoospores which appear to be multiflagellate, and sexually. Sexual reproduction in *Vaucheria* is oogamous (Figure 12–40). The normal zoospores of *Vaucheria* and other Xanthophyta resemble those of the brown algae.

THE DINOFLAGELLATES: DIVISION PYRROPHYTA

Most of the Pyrrophyta are unicellular biflagellates. Many are marine, but others occur in fresh water. The principal class, Dinophyceae, includes the dinoflagellates ("spinning flagellates"), of which more than 1000 species are known, many of them of great importance in the plankton. In the dinoflagellates, the flagella beat within two grooves; one of these circles the body like a belt, and the second is perpendicular to the first. The beating of the flagella in their respective grooves causes the organism to spin like a top as it moves. The encircling flagellum is ribbonlike, and both flagella are of the tinsel type.

Many of the dinoflagellates are bizarre in appearance, with stiff cellulose plates forming a wall (theca), which often looks like a strange helmet or part of an ancient coat of armor (Figures 12–41 and 12–42). The plates of the wall are in vesicles inside the plasma membrane, and not outside it, as is the cell wall of most algae. Most dinoflagellates contain chlorophylls *a* and *c*, which are generally masked by carotenoid pigments. Some of the species do not contain chlorophyll and are heterotrophic, but their structure clearly allies them to the other members of the division. Even the autotrophs usually have strong requirements for vitamin B_{12}, like many diatoms, and cannot be regarded as primary producers without qualification. Some are capable of ingesting other cells. It is assumed that the colorless forms are grazers and obtain their nutrition by ingesting other cells or small particulate organic material. Blue-green algae are common symbionts in tropical dinoflagellates.

Dinoflagellates may also play an important role in human affairs. During the winter and spring of 1974, the west coast of Florida was ravaged by its 25th major red tide since 1844. Hundreds of thousands of dead and smelly fish littered the beaches, and millions of tourist dollars were lost. Such red tides are caused by outbreaks or "blooms" of red dinoflagellates of several species, which make the water red or brown in color. The dinoflagellates are ingested not only by fish but also by shellfish, such as mussels, which accumulate them and, depending on the species of dinoflagellate, may become dangerous for human consumption. In the fall of 1972, for example, the lower New England coast experienced its first red tide. Twenty-six people were poisoned by tide-contaminated shellfish, and public confidence was shaken to such an extent that the Massachusetts shellfish industry had only returned to about two-thirds of its former level four years later.

12–41
Pyrrophyta. The armor of some of the species consists of cellulose plates completely enclosed by the plasma membrane. Those species that appear naked have thinner cellulose plates that do not differ fundamentally from those of the armored ones.

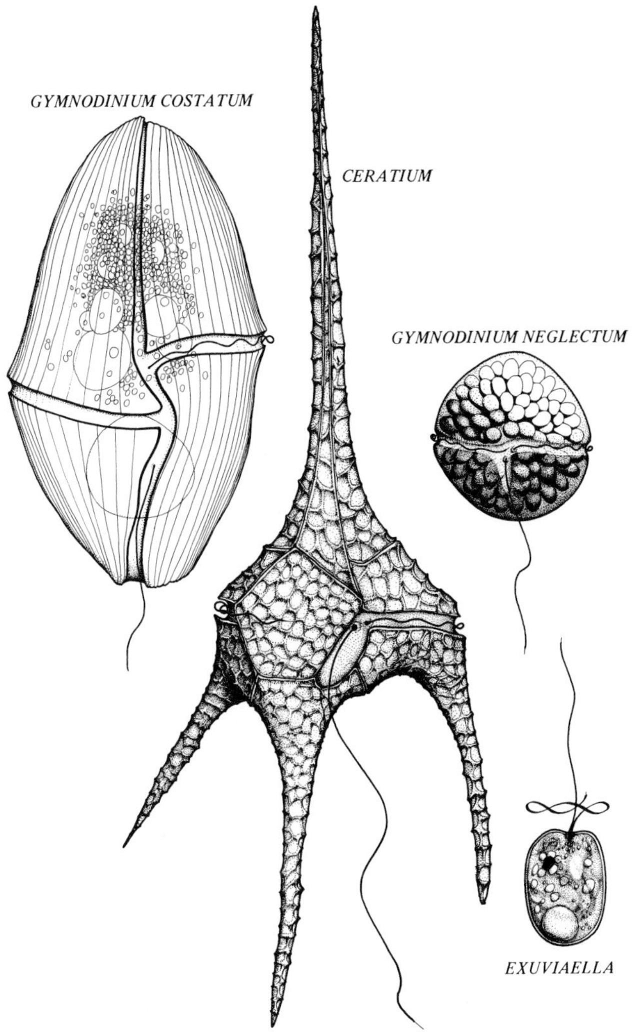

12–42
Dinoflagellates. (a) Ceratium triops, *an armored dinoflagellate.* (b) Noctiluca scintillans, *a bioluminescent marine dinoflagellate. The individual cells are about a millimeter in diameter. This is one of the organisms responsible for the phenomenon of "burning of the waves."*

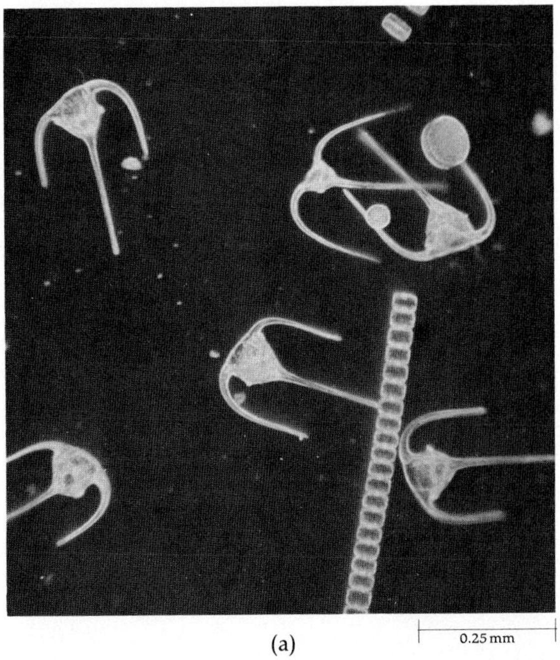

(a)

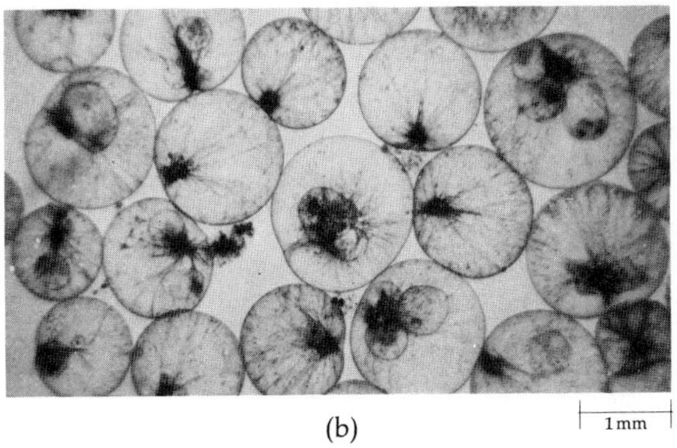

(b)

The poisons produced by some dinoflagellates, such as *Gonyaulax catanella,* are such extraordinarily powerful nerve toxins that they have been studied as possible agents for use in chemical warfare. These toxins are relatively well known. On the other hand, the factors involved in causing the red tides themselves are poorly understood; nutrient and trace metal levels, sewage runoff, ocean salinity and temperature, winds, light, and other factors all seem to play some role. The tides are monitored by checking the level of toxicity of sample shellfish, and currently efforts are underway to track their progress by satellite. Commonly, their occurrence is coupled with a brilliant night-time luminescence of the waves. Many dinoflagellates are bioluminescent, converting chemical energy into flashes of light to produce what ancient seafarers used to call the "burning of the sea at night."

The chief method of reproduction is by longitudinal cell division, with each daughter cell getting one of the flagella and a portion of the theca and then constructing the missing parts in a very intricate sequence. Some nonmotile species form zoospores. In some of these spe-

MITOSIS IN DINOFLAGELLATES

Dinoflagellates have a unique type of mitosis which appears to combine features of both eukaryotes and prokaryotes. Their cellular organization is eukaryotic, but their nuclei retain a number of prokaryotic features. For example, as shown in (a), *the chromosomes of dinoflagellates are always visible and do not contract prior to mitosis. They have no histone associated with them. The nuclear envelope persists during cell division and, apparently, the chromosomes become attached to it, as shown in* (b), *just as the chromosomes of bacteria become attached to the plasma membrane. At the time of mitosis, cytoplasmic channels, as shown in the center of* (b), *invade the dividing nuclei. These channels contain bundles of microtubules, similar to the microtubules of the eukaryotic spindle apparatus. These microtubules, which remain entirely outside the nuclear envelope, are all oriented in one direction and may direct the flow of cytoplasm through the channels and, in this way, regulate the separation of the portions of nuclear envelope with the attached chromosomes. Thus, according to these findings by Donna Kubai and Hans Ris of the University of Wisconsin, the dinoflagellates appear to have evolved their mitotic mechanism separately from other eukaryotic organisms. The organism shown here is* Cryptothecodinium (Gyrodinium) cohnii.

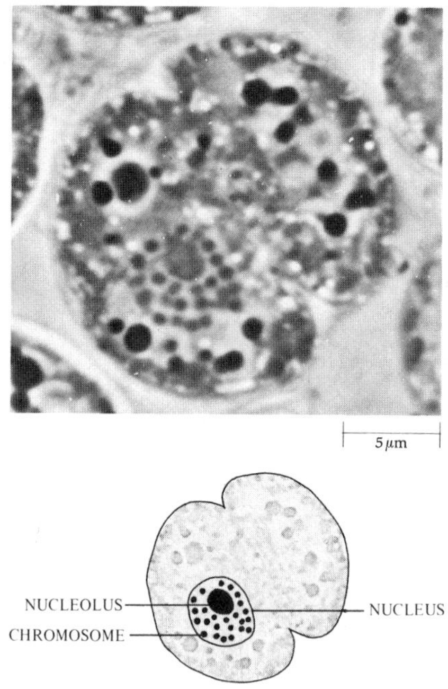

(a)

cies only the zoospore has the typical dinoflagellate structure, whereas the mature forms have cells with no flagella or may even be joined together in filaments. Sexual fusion and genetic recombination are apparently rare in dinoflagellates but have been confirmed in a few species.

THE EUGLENOIDS: DIVISION EUGLENOPHYTA

The euglenoids are protozoa some of which have, in the course of their evolution, acquired chloroplasts that are biochemically similar to, but structurally different from, the chloroplasts of green algae. About a third of the approximately 40 genera of euglenoids have chloroplasts, and these have chlorophylls *a* and *b* together with several carotenoids. Euglenoids store their carbohydrate food reserves in the form of paramylon, a polysaccharide that is not found in any other group of organisms and that, unlike starch, is formed outside the chloroplast.

There are more than 800 species, most of which are found in fresh water, especially water rich in organic material. They range from less than 10 micrometers to more than 500 micrometers (0.5 millimeter) in length and are variable in form. All are unicellular except for the colonial genus *Colacium.* Among the genera that lack chloroplasts, some absorb organic matter, others ingest it.

The euglenoids reproduce by cell division. The parent cell divides longitudinally, forming two new cells that are mirror images of one another. No microtubules or spindle fibers are involved in mitosis, and the nuclear membrane remains intact, with the chromosomes permanently condensed as in dinoflagellates. No sexual reproduction is known in this group.

The division takes its name from *Euglena,* a common genus. Many species of *Euglena* are elongated, as shown in Figure 12–43. The cell is complex and contains numerous small chloroplasts; a nucleus; a long, emergent flagellum with very fine hairs on one side of it; and a short, nonemergent one. The emergent flagellum is usually held in front of the cell like a spinning lasso.

In *Euglena,* the flagella are attached at the base of the flask-shaped opening, the *reservoir,* at the anterior end of the cell. Emptying into the reservoir is the contractile vacuole, which collects excess water from all parts of the cell and discharges it into the reservoir. The cell is delimited by a plasma membrane inside of which are a series of flexible, interlocking proteinaceous strips arranged helically. These strips plus the plasma membrane form a structure termed the *pellicle.* Unlike the stiff walls of plant cells, the flexible pellicle permits *Euglena* to

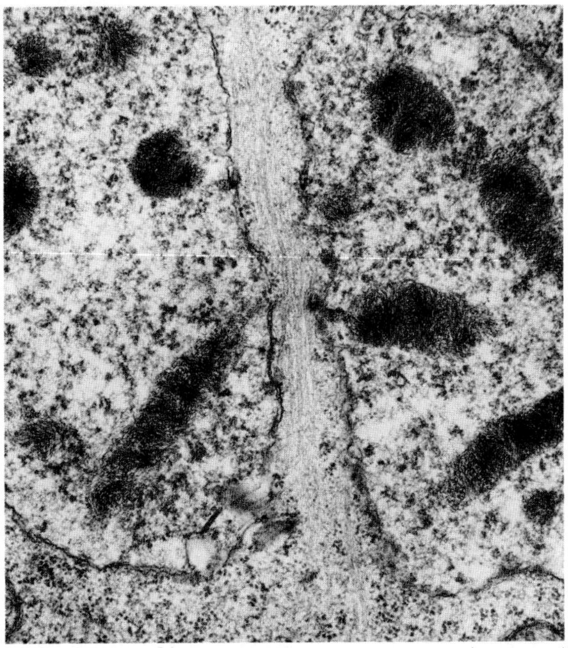

(b) 0.5 μm

12–43
(a) Euglena, *showing two storage bodies of paramylon and the nature of the pellicle.* (b) *The structure of* Euglena, *as interpreted from electron micrographs.*

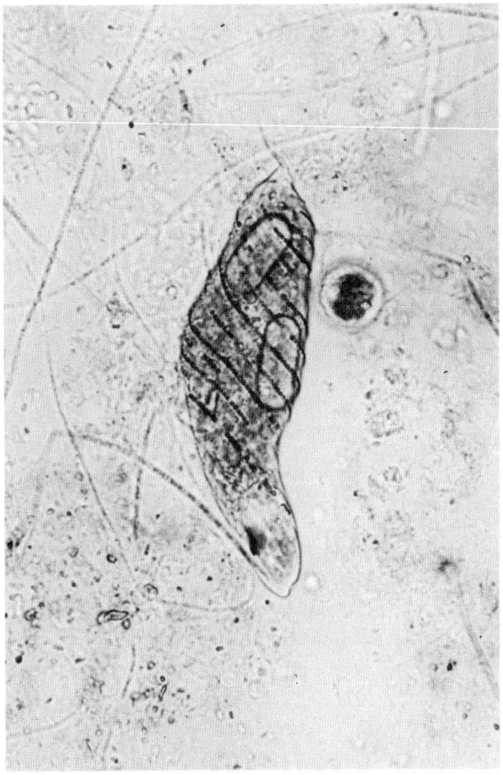

(a) 25 μm

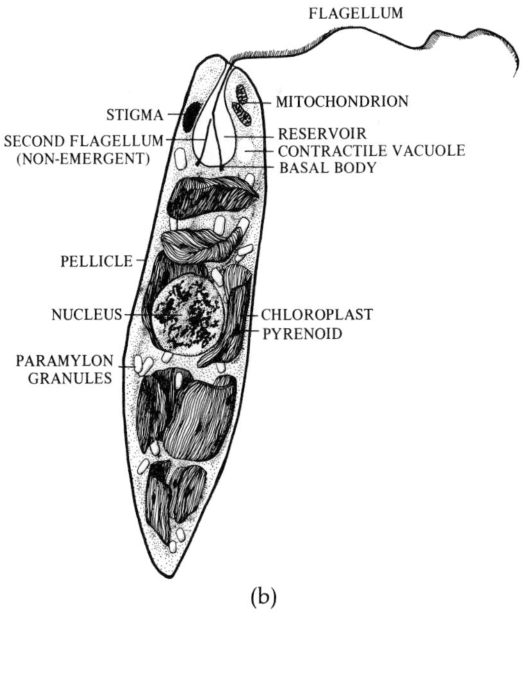

(b)

change its shape, providing an alternative means of locomotion for mud-dwelling forms.

If one leaves a culture of *Euglena* near a sunny window, a clearly visible green cloud will form in the water, and this will move as the light changes, always toward a spot that is bright (but not too bright). If the light is too bright, the individuals of *Euglena* will swim away from it. *Euglena,* like *Chlamydomonas,* is probably able to orient with respect to light because of the presence of a pair of special structures—the stigma and the *photoreceptor,* the latter a swelling at the base of the flagellum. The "shading" of this paraflagellar photoreceptor by the pigmented stigma when the individual euglenoid is in certain orientations with respect to light is important in its directional movement. Only a few nongreen euglenoids have a stigma.

Some species of *Euglena* can survive if they are kept in the dark, where they cannot photosynthesize, as long as they are provided with a carbon source in addition to their other vitamin and mineral requirements. If some strains of *Euglena* are kept in the light, at an appropriate temperature and in a rich medium, the cells may replicate faster than the chloroplasts, producing nonphotosynthetic cells which can survive indefinitely in a suitable medium. It is tempting to speculate that some modern heterotrophs arose from autotrophs in an analogous manner early in evolutionary history.

THE ECONOMIC USES OF ALGAE

People of various parts of the world, especially in the Far East, eat both red and brown algae. One of the red algae, Porphyra *("nori"), which is eaten by many of the inhabitants of the north Pacific Basin, has been cultivated in Japan and China for centuries. The nori industry presently employs more than half a million persons in Japan alone. Various other red algae are eaten on the islands of the Pacific and also on the shores of the north Atlantic. Seaweeds are generally not of high nutritive value, because humans, like most other animals, lack the necessary enzymes to break down the seaweed carbohydrates. Seaweeds do, however, provide necessary salts as well as a number of important vitamins and trace elements and so are valuable supplementary foods. In many north temperate regions, kelp has been harvested for its ash, which is rich in soda and potash and is therefore valuable for industrial processes. Iodine is also produced commercially from kelp. Algae are often harvested and used directly for fertilizer.*

Alginates, derived from beds of kelps such as Macrocystis, *are widely used as thickening agents and colloid stabilizers in the food, textile, cosmetic, pharmaceutical, paper, and welding industries. Off the west coast of the United States,* Macrocystis *beds can be cropped just below the surface several times a year. As mentioned on page 262, attempts are underway to cultivate this giant kelp on a commercial scale.*

One of the most useful direct commercial applications of any alga is the preparation of agar, which is made from a mucilaginous material extracted from the cell walls of a number of genera of red algae. Agar is used to make the capsules that hold vitamins and other drugs, as a dental-impression material, as a base for cosmetics, and as a culture medium for bacteria and other microorganisms. It is also employed as an anti-drying agent in bakery goods, in the preparation of rapid-setting jellies and desserts, and as a temporary preservative for meat and fish in tropical regions. Agar is produced in many parts of the world, but Japan is the principal source. A similar algal colloidal product known as carrageenan is used in preference to agar for the stabilization of emulsions in connection with paints, cosmetics, and dairy products such as ice cream. It has recently been proposed by phycologist Clinton Dawes of the University of South Florida that the common Florida red alga Eucheuma isiforme *be cultivated in tanks to provide carrageenan on a commercial scale.*

A fixed system for the cultivation of *Porphyra.* The nori net is heavily laden with the algae and ready for harvest.

Two fishermen are harvesting *Porphyra* from a floating system with a modern harvesting machine which cuts out the fronds like a lawn mower. The fronds are then pumped into the machine along with sea water, which is discharged through the opening in the front.

THE SLIME MOLDS: DIVISION GYMNOMYCOTA

The slime molds are a group of curious organisms belonging to the kingdom Protista; they are similar to organisms of the phylum Protozoa, especially the amoebas. They have often been grouped with the fungi because they are heterotrophic and because they form sporangia at one stage in their life cycle. Unlike the fungi, however, the Gymnomycota lack a cell wall for most of their life cycle. Also, unlike any plant or fungus, they ingest their food, which consists mainly of bacteria and yeasts. There are two principal classes of Gymnomycota.

The Plasmodial Slime Molds: Class Myxomycetes

During their nonreproductive stages, the Myxomycetes, or "plasmodial" slime molds, are thin streaming masses of protoplasm which creep along in amoeboid fashion (Figure 12–44). This naked mass of protoplasm is called a *plasmodium.* As one of these plasmodia travels, it engulfs bacteria, yeast, fungal spores, and small particles of decayed plant and animal matter, which it digests. It may grow to weigh as much as 20 or 30 grams, and, because slime molds are spread thinly, this amount can cover an area a meter or more in diameter. The plasmodium con-

(a) (b) (c)

12–44
(a) *Plasmodium of a slime mold. Such a plasmodium can pass through a piece of silk or filter paper and come out the other side unchanged. Sporangia of* (b) Arcyria *and* (c) Physarum, *both plasmodial slime molds.*

tains many nuclei but is not divided by cell walls. As the plasmodium grows, the nuclei divide repeatedly and synchronously; that is, all the nuclei in a plasmodium divide at the same time.

The moving plasmodium is typically fan-shaped, with flowing protoplasmic tubules that are thicker at the base of the fan and that spread out, branch, and become thinner. The tubules are composed of slightly solidified protoplasm through which the more liquefied protoplasm flows rapidly. The foremost edge of the plasmodium consists entirely of a thin film of gel separated from the substrate only by a plasma membrane and a slime sheath of unknown chemical composition.

Plasmodial growth continues as long as an adequate food supply and moisture are available. When either of these are in short supply, there is a rapid formation of sporangia. Mounds form on the plasmodium, and each of these develops into a mature sporangium borne at the tip of the stalk. The sporangia are often extremely ornate. Meiosis takes place either inside the developing sporangium during spore formation or afterwards in the young spores. In either case, the mature spores, which have cell walls, are haploid.

The spores are very resistant to environmental conditions; recently some spores that had been kept in the laboratory for more than 60 years were induced to germinate. Thus spore formation in the Myxomycetes apparently serves two functions: (1) survival in adversity and (2) a rearrangement of genetic material by meiosis.

The spores germinate under favorable conditions, and each spore produces one to four flagellated swarm cells. These swarm cells, which are haploid, may have either one or two whiplash flagella. They either may behave as isogametes and fuse in pairs soon after their liberation, or they may lose their flagella and undergo a number of divisions before mating. The zygote, which is diploid, grows by a series of mitotic nuclear divisions to form a new plasmodium.

Most slime molds live in cool, shady, moist places in the woods—on decaying logs, dead leaves, or other damp organic matter. One of the common species *(Physarum cinereum),* however, is sometimes found creeping across city lawns. The plasmodia come in a variety of colors and can be spectacularly beautiful. The function of the pigments is not known with certainty, but they are probably photoreceptors, for only slime molds with pigmented plasmodia require light for fruiting.

The Myxomycetes, of which there are about 450 species, are of little economic importance. They are, however, of scientific interest as a model system of cytoplasmic streaming, because streaming in the Myxomycete plasmodium is the most rapid known and can be seen with only a hand lens. Every minute or so, the streaming movement, which takes place along the conspicuous tubules of the plasmodium, may reverse its course, first coming to a momentary stop and then flowing in the opposite direction. Curiously, it is not clear that the individual rivers of protoplasm, as observed with a lens, flow in the direction of movement of the plasmodium any more often than they flow in the opposite direction, a finding that does nothing to clarify the problem of protoplasmic motion in general.

12–45
Life cycle of a cellular slime mold, Dictyostelium discoideum. (a) *Amoebas in feeding stage. The light gray area in the center of each cell is the nucleus, and the white areas are contractile vacuoles.* (b) *Amoebas aggregating. The direction in which the stream is moving is indicated by the arrow.* (c) *Migrating pseudoplasmodia, each formed of many amoebas. Each "slug" deposits a thick slime sheath which collapses behind it.* (d) *At the end of the migration, the pseudoplasmodium gathers together and begins to rise vertically, differentiating into a stalk and mass of spores suspended in a droplet.*

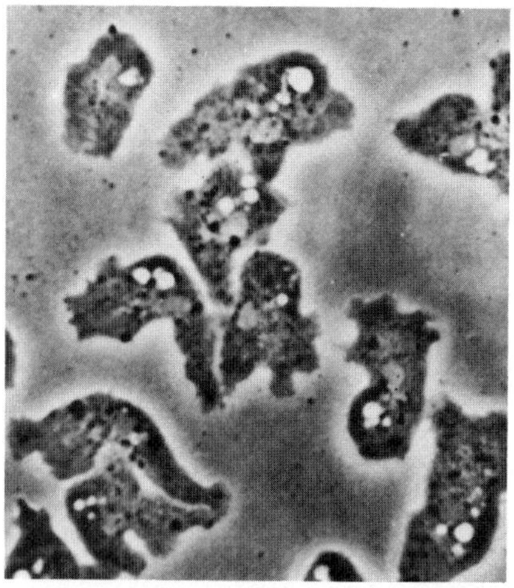

(a) 30 μm

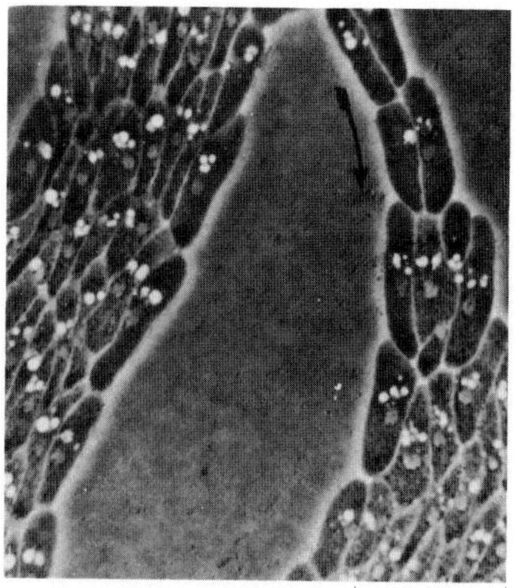

(b) 50 μm

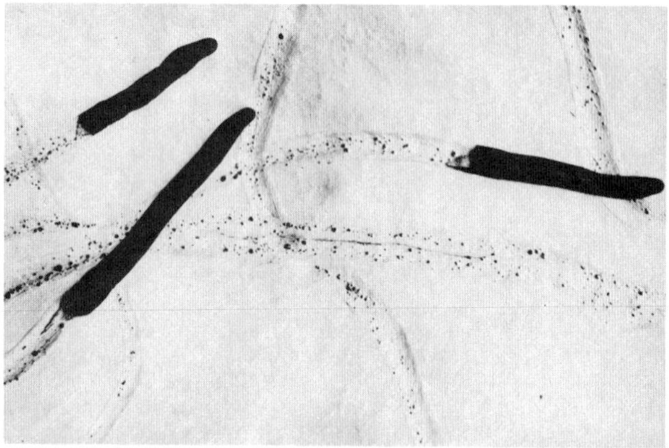

(c) 1 mm

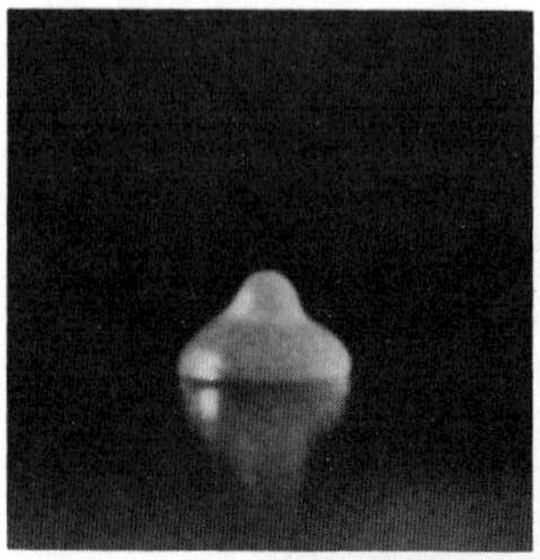

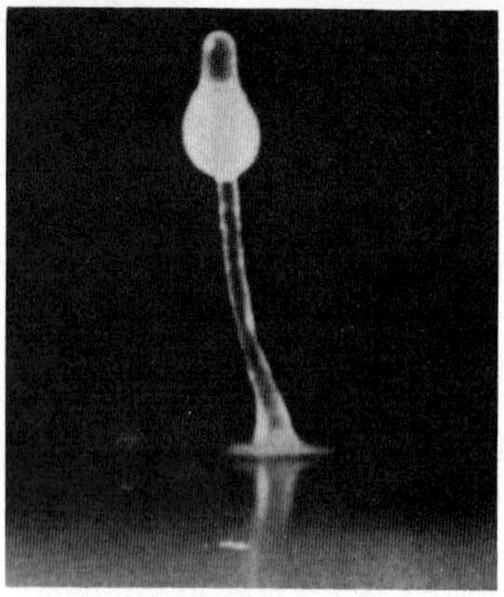

(d) 1 mm

Cellular Slime Molds: Class Acrasiomycetes

The cellular slime molds, like the Myxomycetes, exist for part of their life cycle as amoebalike organisms but differ from the plasmodial slime molds in that the amoebas, on swarming together, retain their identity as individual cells. The cellular slime molds comprise approximately 26 species, grouped into seven genera.

One example of the cellular slime molds is *Dictyostelium discoideum,* which begins life as a group of individual amoebas that are smaller than the common *Amoeba proteus* but not strikingly different in appearance (Figure 12–45). These amoebas grow and divide repeatedly, increasing in number; then quite suddenly they stop growing, swarm together, and adhere to form a many-celled sluglike mobile mass that is capable of moving several centimeters. Movement toward light, especially green light, is characteristic.

The aggregation stage begins when all of the amoebas begin to emit "pulses" of cyclic adenosine monophosphate (AMP) at intervals of approximately 5 minutes (Figure 12–46). An amoeba that receives a signal pulse of cyclic AMP evidently moves toward it, pauses, and then lets off a pulse of its own. The consequence of this is that each amoeba is attracted to its nearest neighbor, because its own signal is automatically delayed until a certain time after the one it has just received. The amoebas form streams, all flowing toward the center of a growing aggregate in a series of waves and sticking together. The amoeba at the center, the pacemaker, is the one that emits pulses of AMP at the fastest rate.

After a period of migration, the cells in the mass begin to differentiate, those in the anterior third of the sluglike mass forming stalk cells, and the remainder forming spores. Evidently the higher concentrations of cyclic AMP at the front of the mass determine the fate of the cells, as isolated amoebas subjected to high concentrations of cyclic AMP differentiate into stalk cells. The

12-46
Cyclic AMP (adenosine monophosphate) is the chemical that attracts the amoebas of the cellular slime molds, causing them to aggregate. In mammals, this same substance acts as a hormone "messenger"; it is triggered by extracellular hormones, such as adrenalin, and acts inside the cells to produce the characteristic hormone effects. Cyclic AMP is formed from ATP. "Cyclic" refers to the fact that the atoms of the phosphate group form a ring.

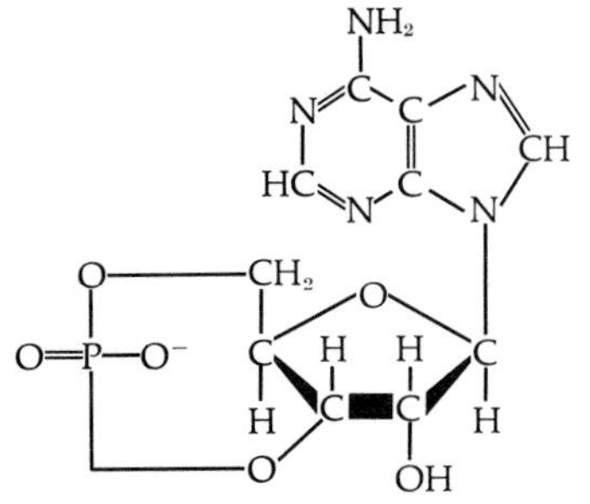

gradient in cyclic AMP seems to develop as the sluglike mass migrates and will arise even if one mass is divided into two, or four, if the resulting masses migrate for a sufficient length of time. As the stalk grows, it is strengthened with fibrils of cellulose, and the mass of amoebas quite abruptly rise up in a fruiting body on the pinnacle of which is a droplet containing hundreds, or, in some instances, thousands of spores. These spores are dispersed, and if they fall on warm damp ground, they germinate. Each releases one small, quite ordinary-looking amoeba, and the entire cycle begins again.

SUMMARY

The algae are a large and diverse group of some 22,000 species of photosynthetic eukaryotic organisms grouped in seven divisions that are viewed as having developed from a series of parallel evolutionary lines. Largely aquatic, the algae are of great importance in marine habitats, playing an ecological role comparable to that of the green plants in land habitats. In the open sea, algae are generally found in the form of plankton. Larger, more complex algae are found along the shore. Many species are found in fresh water, and some occur on land.

The algae, in general, have a relatively simple body composed of one or several cells. Members of five of the seven divisions contain cellulose in their cell walls.

Most algae have unicellular reproductive structures and lack vascular tissues. Despite these general similarities, biochemical and other differences among the groups indicate that they are not closely related. Among the principal criteria for grouping the algae are types of pigments, types of carbohydrate storage products, cell-wall components, and types of (or absence of) flagella.

The various divisions of algae appear to have had their origins in symbiotic relationships between nonphotosynthetic, protozoalike eukaryotic cells and one or more groups of photosynthetic prokaryotes. Members of four of the divisions are multicellular: the Chlorophyta, the Phaeophyta, the Rhodophyta, and the Xanthophyta. All but the second also include unicellular forms. The other three divisions—Chrysophyta, Pyrrophyta, and Euglenophyta—are composed almost entirely of unicellular organisms.

The green algae (division Chlorophyta) are the largest and most diverse group and the one from which the bryophytes and vascular plants probably evolved. It is thought that *Chlamydomonas*, a common unicellular form, resembles a primitive, ancestral green alga. Several evolutionary lines can be traced from a *Chlamydomonas*-like cell.

The Phaeophyta (brown algae) include the largest and most complex of the marine algae. In many types the vegetative body is well differentiated into holdfast, stipe (stalk), and blade. Some approach the vascular plants in complexity of their food-conducting tissues. Members of this division contain chlorophylls *a* and *c* and have abundant quantities of an xanthophyll—fucoxanthin—which gives them their typical olive-green to dark brown color. Their storage product is a polysaccharide, laminarin. The sporophyte is usually larger than the gametophyte.

The Rhodophyta (red algae) are a large group particularly common in warmer marine waters. They nearly always grow attached to a substrate and some grow at great depths (down to 175 meters). They contain phycobilins, which give them their characteristic color.

The Chrysophyta are important components of freshwater and marine phytoplankton. Most of the known species are unicellular organisms known as diatoms. Diatoms are characterized by fine double silicon shells, and abundant fossil records of these shells have been found. The other recognized class of Chrysophyta is the golden-brown algae (Chrysophyceae), a number of which are small cells (1 to 3 micrometers) that are found in large quantities in nannoplankton.

The Xanthophyta (yellow-green algae) consist primarily of unicellular, nonmotile organisms, but include colonial and filamentous forms that parallel the body types found in the class Chlorophyceae of the division Chlorophyta. Hence, the yellow-green algae once were classified among the Chlorophyceae. However, the lack of chlorophyll *b* and of starch in the yellow-greens precludes their being classified with the green algae.

The Pyrrophyta (dinoflagellates) are unicellular biflagellates, many of which are marine. The dinoflagellates are characterized by two flagella which beat in different planes, causing the organism to spin; dinoflagellates often have stiff, bizarrely shaped cellulose walls.

Euglenophyta are a small group of organisms, most of which occur in fresh water and are unicellular. They contain chlorophylls *a* and *b* and store carbohydrates in an unusual starchlike substance, paramylon. The cells lack a cell wall but have a flexible series of protein strips, which make up the pellicle, inside the plasma membrane. The cells are highly differentiated, containing chloroplasts, a contractile vacuole, reservoir, eyespot, and flagella. No sexual cycle is known. They show close affinities to nonphotosynthetic protozoan forms.

The slime molds are amoebalike organisms of the division Gymnomycota of the kingdom Protista. They resemble fungi in that they reproduce by the formation of spores and are heterotrophic. Unlike the fungi, however, they lack a cell wall and they ingest their food.

SUGGESTIONS FOR FURTHER READING

BONEY, A. D.: *A Biology of Marine Algae,* Hutchinson and Co., Ltd., London, 1966.*

An excellent, ecologically and physiologically oriented treatise on marine algae, which accurately summarizes modern information about the group.

BONNER, J. T.: *The Cellular Slime Molds,* 2nd ed., Princeton University Press, Princeton, N.J., 1968.

A record of experimental work with a small but fascinating group of organisms.

CHAPMAN, V. J., and D. J. CHAPMAN: *The Algae,* 2nd ed., Macmillan Company, London and New York, 1973.

A taxonomically oriented text that presents a great deal of information concerning the algae in an organized and useful way and also includes several general chapters on ecology, physiology, and geographical distribution.

DAWSON, E. Y.: *Marine Botany: An Introduction,* Holt, Rinehart and Winston, Inc., New York, 1966.

A short, lively text that covers marine bacteria, fungi, and sea grasses as well as algae.

GRAY, W. D., and C. J. ALEXOPOULOS: *Biology of the Myxomycetes,* Ronald Press Co., New York, 1968.

Well-written book about the biology of the plasmodial slime molds, emphasizing ultrastructural, biochemical, and physiological aspects.

PICKET-HEAPS, J. D.: *Green Algae: Structure, Reproduction and Evolution in Selected Genera,* Sinauer Associates, Sunderland, Mass., 1975.

A beautifully illustrated book providing much insight into the variety of form and function in the cells of the green algae.

ROUND, F. E.: *The Biology of the Algae,* 2nd ed., St. Martin's Press, New York, 1973.

An ecologically and physiologically oriented account of the algae, which introduces a number of interesting aspects of the biology of the group.

SLEIGH, M.: *The Biology of Protozoa,* Edward M. Arnold (Publishers) Ltd., London, 1973.*

A general biology of the protozoa, including chapters on structure, metabolism, reproduction, and ecology, with many excellent illustrations.

* Available in paperback.

CHAPTER 13

Introduction to the Plant Kingdom: The Bryophytes

13–1
Cloud forest at 3650 meters on Mt. Ruwenzori in East Africa with a thick carpet of mosses on the ground. The ephiphytes are Usnea, *a lichen. Inset shows a leafy liverwort on a durable, evergreen palm leaf in this habitat.*

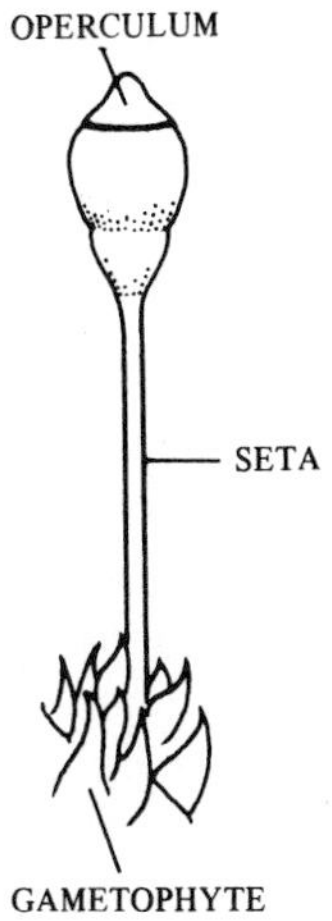

13–2
Urn moss, Physcomitrium turbinatum. *The "urn" is the spore capsule. The protective covering—the calyptra—has fallen from most of the capsules to reveal the lid, or operculum, of the capsule.*

In the preceding chapter, we noted that the bryophytes and the vascular plants are believed to have evolved from some ancient group of green algae. This hypothesis is based on several different lines of evidence. Like the green algae, the bryophytes and vascular plants have chlorophyll *a* as their primary photosynthetic pigment and chlorophyll *b* and carotenoids as accessory pigments. Among the members of all of these groups, starch is the primary carbohydrate food reserve and, with the exception of a few genera of green algae, cellulose is the principal component of their cell walls. Finally, bryophytes and vascular plants both undergo cell-plate formation at cell division, and the only other living organisms that share this characteristic are one brown alga and a few green algae. Because of these similarities, it seems likely that the bryophytes and vascular plants have descended from a remote common ancestor and represent the results of a single invasion of the land by a particular group of green algae.

If this hypothesis is true, the bryophytes and vascular plants must have diverged long ago. The bryophytes first appear in the fossil record in the Devonian period, more than 350 million years ago, and these ancient fossils are quite similar to species living today. Fossil records of vascular plants date from the Silurian period, some 400 million years ago, and so the hypothetical common ancestor of the bryophytes and the vascular plants—a relatively complex green alga—presumably invaded the land somewhat earlier. Presumably the common ancestor of the bryophytes and vascular plants had a well-developed alternation of generations, for this is the type of life cycle exhibited by all bryophytes and vascular plants. Its gametophytes would have borne multicellular gametangia (although today only one group of unique green algae, the stoneworts, exhibit multicellular gametangia), for the gametangia of all bryophytes and vascular plants are multicellular structures. Being an aquatic organism, its sporophytes would have lacked vascular tissue, although conducting strands of elongate, thin-walled cells may have been present. It was probably endomycorrhizal, with an Oomycete partner, as discussed on page 241.

In the course of the transition from water to land, many physiological and morphological adaptations were necessary to prevent drying out of the evolving terrestrial organisms. One of these adaptations was the development of a sterile jacket layer about the sperm- and egg-producing cells of the male and female gametangia, called *antheridia* and *archegonia*, respectively. Similarly, a sterile jacket layer was formed around the spore-producing cells of the *sporangia*.

The onset of the land existence of plants was also correlated with retention of the zygote within the female gametangium and its development there into an embryo, or young sporophyte. Thus, during its early critical stages of development, the young sporophyte is protected by the female gametophyte. In the algae, by contrast, the zygote leads an independent existence.

The aerial parts of most vascular plants are covered with a waxy protective layer, or cuticle, which helps prevent drying out. Most bryophytes apparently lack a cuticle, which in vascular plants is closely correlated with the presence of stomata, specialized pores that function primarily in the regulation of gas exchange. The sporophytes of the hornworts and mosses contain stomata. However, in the hornworts the stomata apparently remain open until late in development, and those of the mosses close only after extreme desiccation (that is, after they are completely dried up).

All bryophytes and vascular plants are oogamous, and all possess a heteromorphic alternation of generations.

13–3
On the facing page are two views of a transverse section of a Polytrichum *stem.* (a) *Overall view of the stem, showing epidermis, cortex, central strand and associated leaves.* (b) *Detailed view of central strand. The thick-walled cells in the center constitute the water-conducting cells, which are surrounded by thin-walled food-conducting cells.*

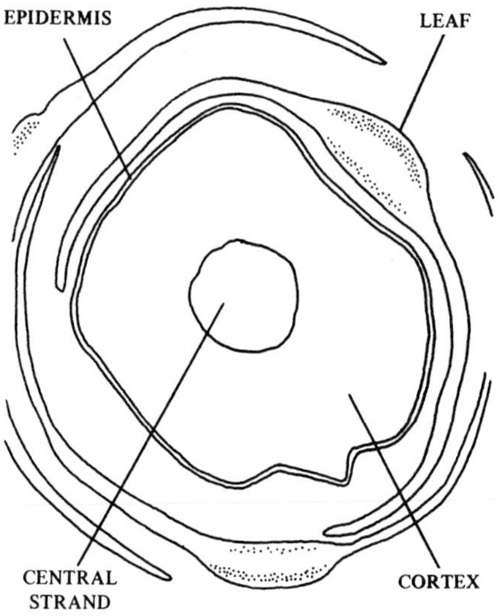

CHARACTERISTICS OF THE BRYOPHYTES

The bryophytes—mosses, liverworts, and hornworts—are relatively small plants, many less than 2 centimeters long, with most less than 20 centimeters long. They are most common in warm and moist areas, especially in the tropics and subtropics, where a variety of species and a luxuriance of individual species can be found. Bryophytes, however, are not confined to such areas. An appreciable number of mosses are found in relatively dry deserts, and many form extensive masses on dry exposed rocks where very high temperatures can be reached. Mosses sometimes dominate the terrain to the exclusion of other plants over large areas of the far north and on rocky slopes above timberline in the mountains. Like the lichens, they are as a group remarkably sensitive to air pollution, especially by sulfur dioxide, and they are often absent or represented by only a few kinds in highly polluted areas. A number of bryophytes (both mosses and liverworts) are aquatic, and some mosses even are found on wave-splashed rocks, although none is truly marine. There are some 24,000 species, more than any other group of plants except the flowering plants; a representative moss is shown in Figure 13–2 on page 281.

Two very important characteristics distinguish the bryophytes from vascular plants. One of these is the absence of the vascular tissues, xylem and phloem, in the bryophytes. Accordingly, all bryophytes, strictly speaking, lack true leaves, true stems, and true roots. Nevertheless, the terms leaf and stem are commonly used when referring to the leaflike and stemlike structures of the gametophytes of leafy liverworts and mosses, and this practice will be followed in this book. It is pertinent to point out that the stems of some mosses contain a central strand of conducting cells that have a number of characteristics in common with xylem and phloem of primitive vascular plants. In the mosses, the water-conducting tissue is surrounded by the food-conducting tissue (Figure 13–3). The presence of these tissues in certain mosses is used as evidence in support of a hypothesis suggesting that bryophytes were derived from primitive vascular plants by regressive evolution. In most bryophytes, the gametophyte is attached to the substrate by means of elongate single cells or filaments of cells called *rhizoids*. The rhizoids generally serve only to anchor the plants, for absorption of water and minerals commonly occurs directly and rapidly throughout the gametophyte. Rootlike structures are lacking.

The second important distinguishing characteristic of the bryophytes is the nature of their alternation of gen-

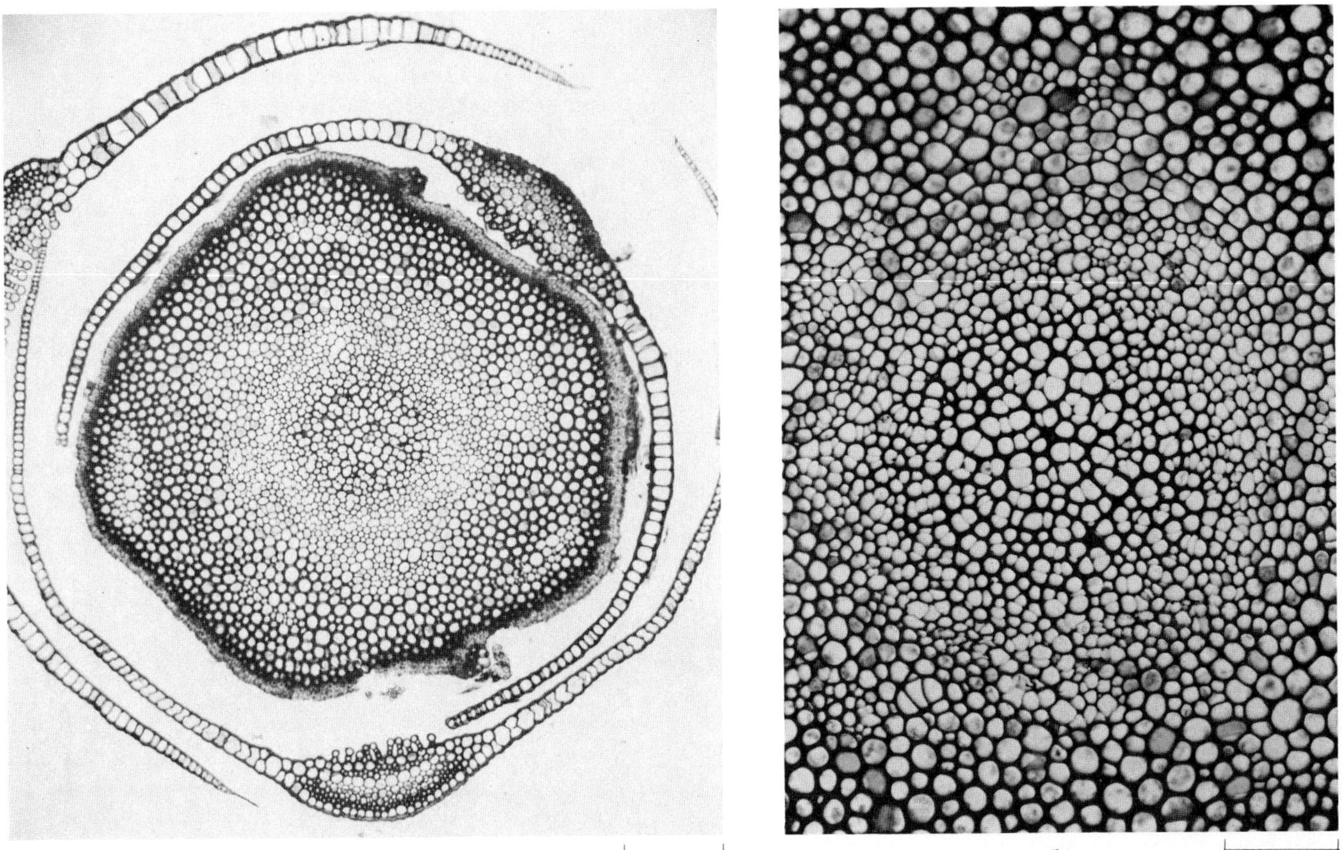

(a) 100 μm (b) 50 μm

erations: The gametophytes are larger and always nutritionally independent, whereas the sporophytes are smaller, varyingly dependent upon, and permanently attached to the gametophytes. In other words, the gametophyte is the conspicuous and dominant generation in the bryophytes. Among algae, life cycles in which the gametophyte is the dominant generation are found only in *Cutleria*, a brown alga, and in a few groups of the green algae, including some that may be related directly to the ancestors of plants. Among vascular plants, the sporophyte is the conspicuous and dominant generation.

The bryophytes have sometimes been referred to as the amphibians of the plant kingdom. In order for fertilization to take place, the biflagellated sperms must swim through water to reach the egg inside an archegonium. The archegonium, which may be stalked, is flask-shaped with a long neck and a basal swollen portion, the *venter*, enclosing a single egg (Figure 13–4a). The central cells of the neck, the *neck canal cells*, disintegrate when the archegonium is mature, leaving a fluid through which the sperms swim to the egg through the neck canal. Malic acid is one of the molecules that seems to attract sperms to the archegonium in bryophytes (and in lower vascular plants as well). The elongate or spherical antheridium is commonly stalked and consists of a one-cell-thick sterile jacket layer surrounding numerous *spermatogenous cells* (Figure 13–4b). Each spermatogenous cell forms a single biflagellated sperm.

The zygote is retained within the venter of the archegonium, where it develops into an embryo (Figure 13–5). For a period, the venter of the archegonium undergoes cell division, keeping pace with growth of the young sporophyte. The enlarged archegonium is called a *calyptra*. At maturity the sporophyte of many bryophytes consists of a *foot*, which remains embedded in the archegonium, a *stalk* or *seta*, and a *capsule* or *sporangium* (Figure 13–5). Generally, the cells of the young and maturing sporophyte contain chlorophyll and carry out photosynthesis, but by the time meiosis occurs in the capsule and the spores are produced, the chlorophyll has usually disappeared. In mosses, the calyptra commonly is lifted upward with the capsule as the seta elongates. Prior to spore dispersal, the protective calyptra falls off and the spores are shed by a variety of mechanisms specialized for dehiscence—the spontaneous bursting open of a structure to discharge its contents.

The division Bryophyta is traditionally divided into three classes: Hepaticae (the liverworts, 9000 species), Anthocerotae (the hornworts, 100 species), and Musci (the mosses, 14,500 species).

13-4
Gametangia of Marchantia. (a) *Portion of an archegoniophore (the stalk supporting the archegonium) showing archegonia in several stages of development.* (b) *Portion of antheridiophore showing developing antheridia.*

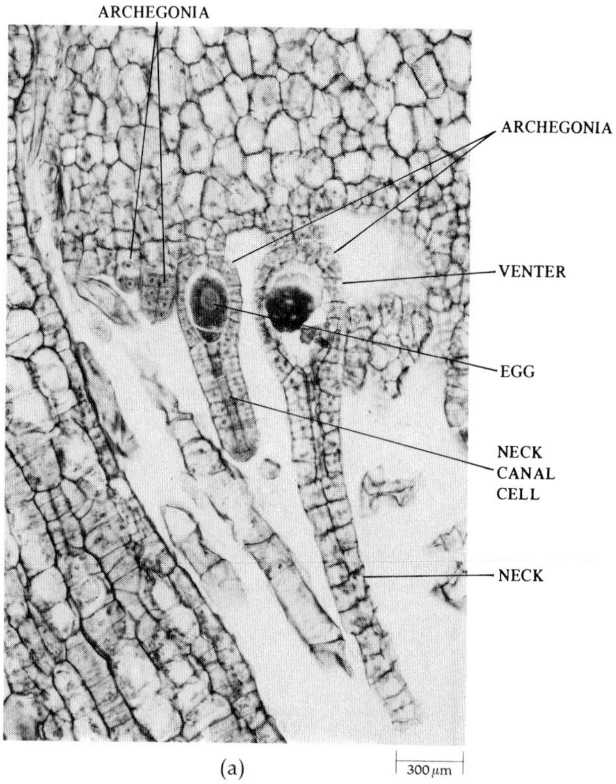

(a)

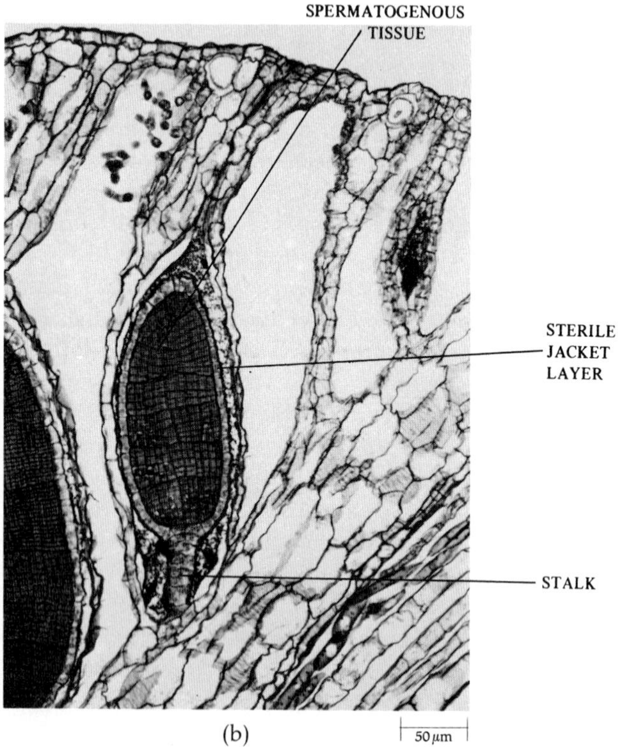

(b)

13-5
Marchantia. *Stages in development of the sporophyte.* (a) *The embryo, or young sporophyte, as an undifferentiated spherical mass of cells within the enlarged venter, or calyptra.* (b) *Foot, seta, and capsule, or sporangium, are now distinct.* (c) *Maturing sporophyte. Note in the micrograph the elaters (spiral, filamentous structures that assist in the dispersion of spores) among spores in capsule.*

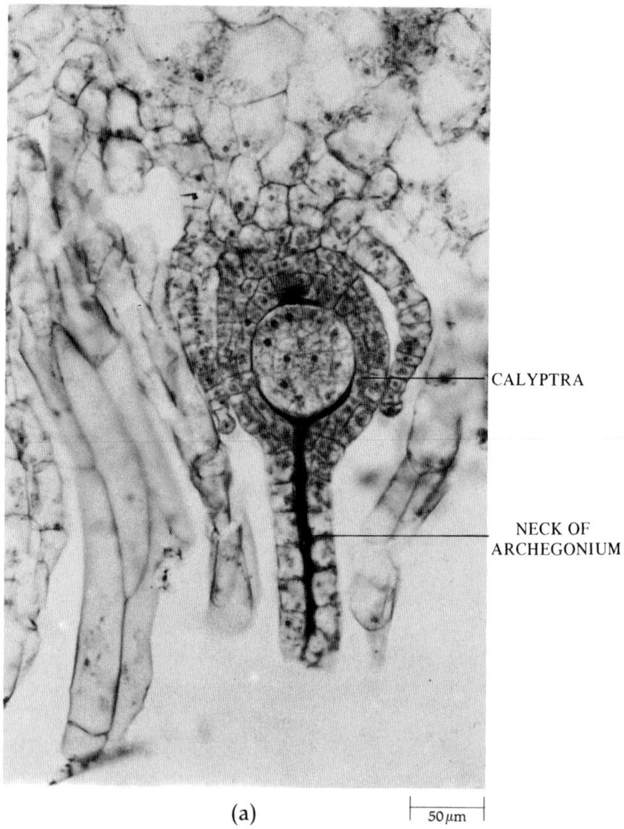

(a)

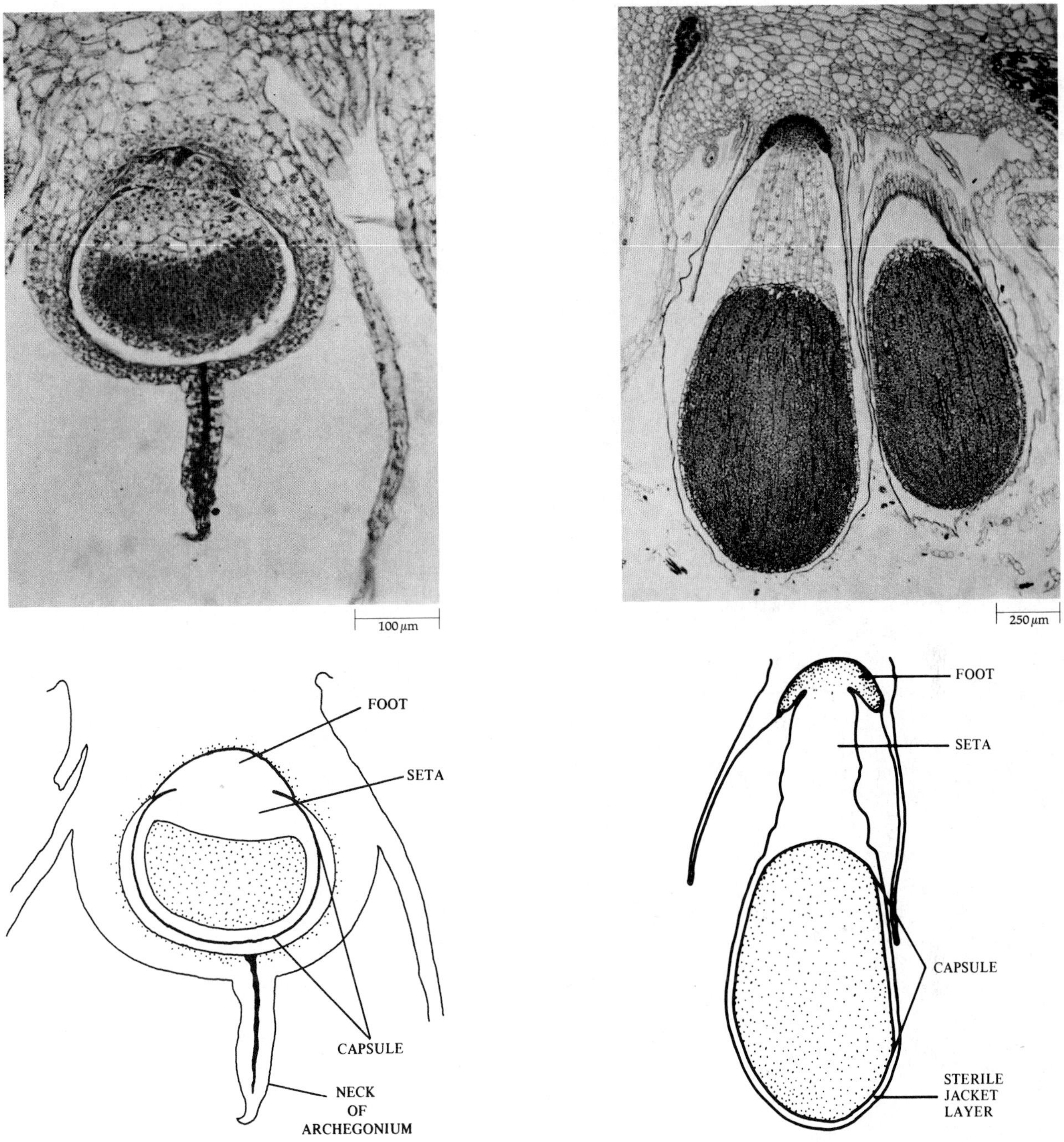

(b) (c)

THE LIVERWORTS: CLASS HEPATICAE

Liverworts are small plants that are generally less conspicuous than mosses. Their name dates from the ninth century, when it was thought, because of the liver-shaped outline of the gametophyte in some genera, that these plants might be useful in treating diseases of the liver—the medieval "Doctrine of Signatures," the theory that the outward appearance of a body signals its special properties. "Wort" simply means "herb" and so appears as part of many plant names.

The gametophytes of some liverworts are flattened, dorsiventral (having distinct upper and lower surfaces) thalli, which grow from an apical meristem. The gametophytes of most species, however, are leafy and grow from a single apical cell, which resembles an inverted pyramid with a base and three sides. Daughter cells are cut off from this single cell. The rhizoids of liverworts are single-celled, unlike those of mosses, which contain several cells each. The gametophytes generally develop directly from spores. The sporophytes of liverworts are, in general, less complex than those of mosses, and their capsules have very different mechanisms for the release of spores.

Thallose Liverworts

Thallose liverworts are nonleafy liverworts that display a wide variety of forms, including members of the order Marchantiales, a familiar and widespread group comprising some 450 species. They can be found on moist, shaded banks and in other suitable habitats such as flowerpots in a cool greenhouse. The thallus is many cells thick—about 30 at the midrib and approximately 10 in the thinner portions—and is sharply differentiated into a thin, chlorophyll-rich upper (dorsal) portion and a thicker, colorless lower (ventral) region (Figure 13-6). The lower surface bears two kinds of rhizoids, as well as rows of scales. The upper surface is divided into raised regions, each of which marks the limits of an underlying air chamber and has a large pore that leads to this chamber. Each chamber has a number of strands of tissue that are rich in chloroplasts, but the walls and floor of the chamber are colorless. There is also some differentiation among the layers of colorless cells below the chamber, particular cells being specialized for storage of starch.

On the basis of sporophyte structure, *Riccia* and *Ricciocarpus* are among the simplest of liverworts. *Ricciocarpus* (Figure 13-7), which is amphibious, is bisexual, that is, both sex organs arise on the same plant. Although some species of *Riccia* are amphibious, most are terrestrial. *Riccia* gametophytes may be unisexual or bisexual. The sporophytes are deeply embedded within the dichotomously branched gametophytes of *Riccia* and *Ricciocarpus* and consist of little more than a sporangium. No special mechanism for spore dispersal occurs in these sporophytes. When the portion of the gametophyte containing mature sporophytes dies and decays, the spores are liberated.

One of the most familiar of liverworts is *Marchantia,* a fairly widespread terrestrial genus, which grows on moist soil and rocks (Figure 13-8). Its dichotomously branched gametophytes are larger than those of *Riccia* and *Ricciocarpus.* Unlike the latter two genera, in which the sex organs are distributed along the upper, or dorsal, surface of the thallus, *Marchantia* has its gametangia restricted to specialized erect structures called gametophores. The gametophytes of *Marchantia* are strictly unisexual, and the male and female gametophytes can readily be identified by their gametophores, which are quite distinct from one another. The antheridia are borne in disk-headed stalks *(antheridiophores)* and the archegonia on "umbrella"-headed stalks *(archegoniophores)* (Figure 13-8).

The sporophytes of *Marchantia* are more highly differentiated than those of *Riccia* and *Ricciocarpus,* consisting of a foot, a short stalk or seta, and a capsule or sporangium (Figure 13-5). In addition to spores, the mature sporangium contains elongate cells, called *elaters,* with spirally arranged hygroscopic (moisture-absorbing) wall thickenings (Figure 13-9). The walls of these cells are sensitive to very slight changes in humidity, and by twisting action they aid in spore dispersal after the capsule dehisces into a number of petal-like segments.

Fragmentation constitutes the principal means of asexual reproduction in the liverworts. Another fairly widespread means of asexual reproduction in the liverworts and mosses is the production of *gemmae,* minute, lens-shaped bodies that can give rise to new plants. The gemmae are produced in special cuplike structures called gemma cups located on the dorsal surface of the gametophyte (Figure 13-10). Gemma cups are produced by *Marchantia,* but not by *Riccia* and *Ricciocarpus.*

13-6
Transverse sections of two thallose liverworts, Marchantia polymorpha (a) *and* Reboulia hemisphaerica (b). *Numerous chlorophyll-bearing cells are evident. Pores permit the exchange of gases in the air-filled chambers that honeycomb the photosynthetic layer, functioning in the same way as the stomata of higher plants. However, unlike stomata, they do not open and close. The photosynthetic layer is thicker in* Reboulia *than in* Marchantia. *In both there are several layers of colorless cells below the photosynthetic ones and various specialized structures that serve to anchor the plant body to the substrate.*

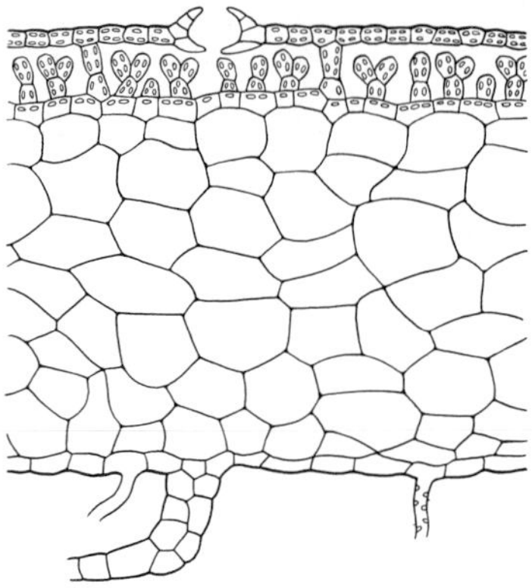
(a)

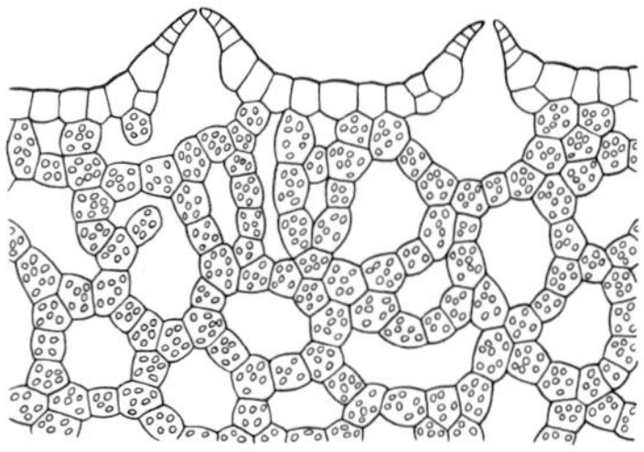
(b)

13-7
Gametophytes of Ricciocarpus natans, *an amphibious liverwort. The system of branching in this organism is dichotomous, that is, the main and subsequent axes fork repeatedly into two branches. The branching is less prominent in the aquatic form* (a) *than in the terrestrial form* (b).

SPORE DISCHARGE IN LIVERWORTS

Three methods of spore discharge in liverworts. Most liverworts are similar to Cephalozia *in their method of spore discharge, whereas some resemble* Marchantia. *The condition found in* Frullania *is less common.*

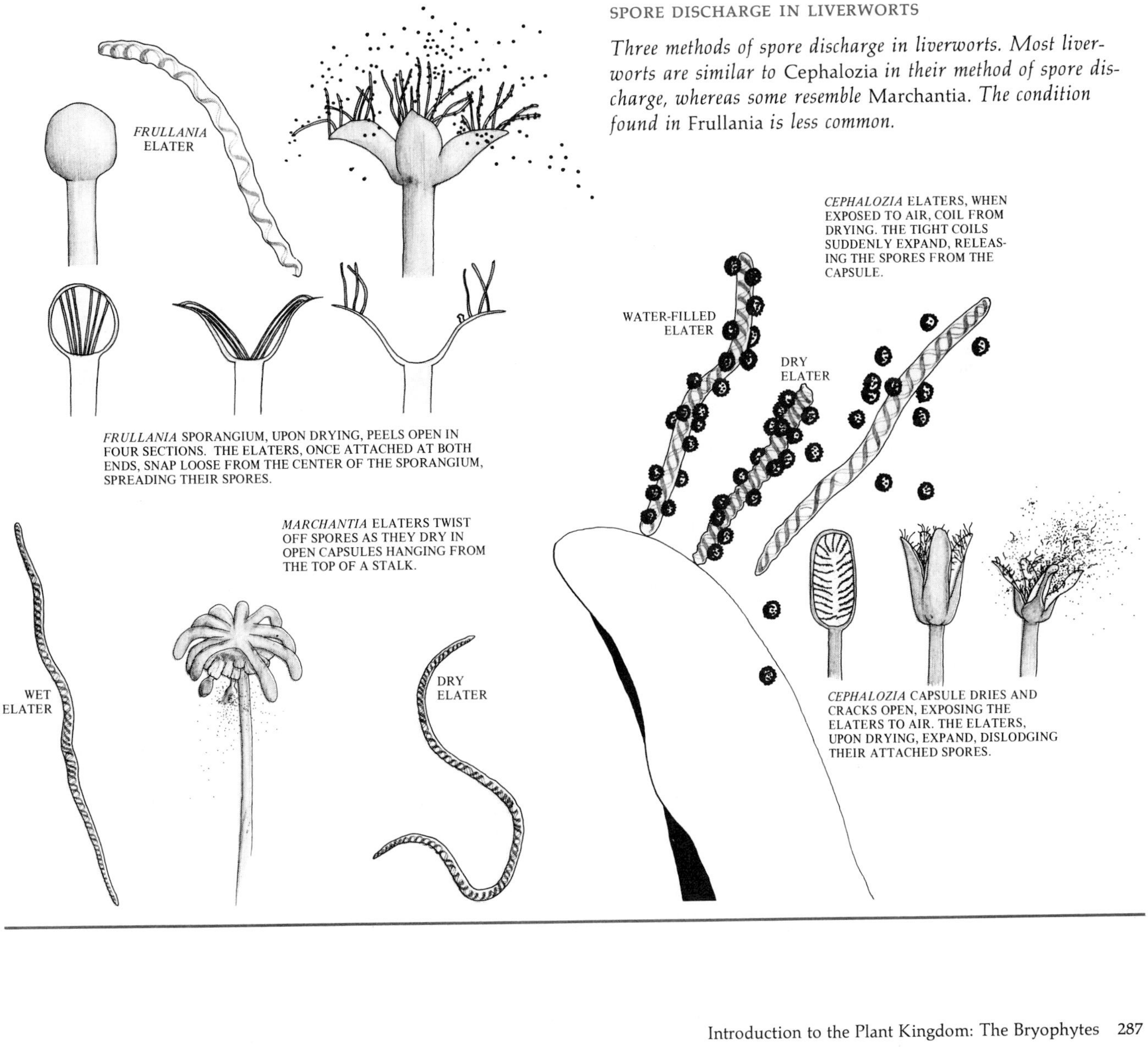

FRULLANIA ELATER
FRULLANIA SPORANGIUM, UPON DRYING, PEELS OPEN IN FOUR SECTIONS. THE ELATERS, ONCE ATTACHED AT BOTH ENDS, SNAP LOOSE FROM THE CENTER OF THE SPORANGIUM, SPREADING THEIR SPORES.
MARCHANTIA ELATERS TWIST OFF SPORES AS THEY DRY IN OPEN CAPSULES HANGING FROM THE TOP OF A STALK.
WET ELATER
DRY ELATER
CEPHALOZIA ELATERS, WHEN EXPOSED TO AIR, COIL FROM DRYING. THE TIGHT COILS SUDDENLY EXPAND, RELEASING THE SPORES FROM THE CAPSULE.
WATER-FILLED ELATER
DRY ELATER
CEPHALOZIA CAPSULE DRIES AND CRACKS OPEN, EXPOSING THE ELATERS TO AIR. THE ELATERS, UPON DRYING, EXPAND, DISLODGING THEIR ATTACHED SPORES.

13–8
The thallose liverwort Marchantia. *The antheridia and archegonia are elevated on specialized stalks above the plant body. Male and female sex organs occur on different plants.*

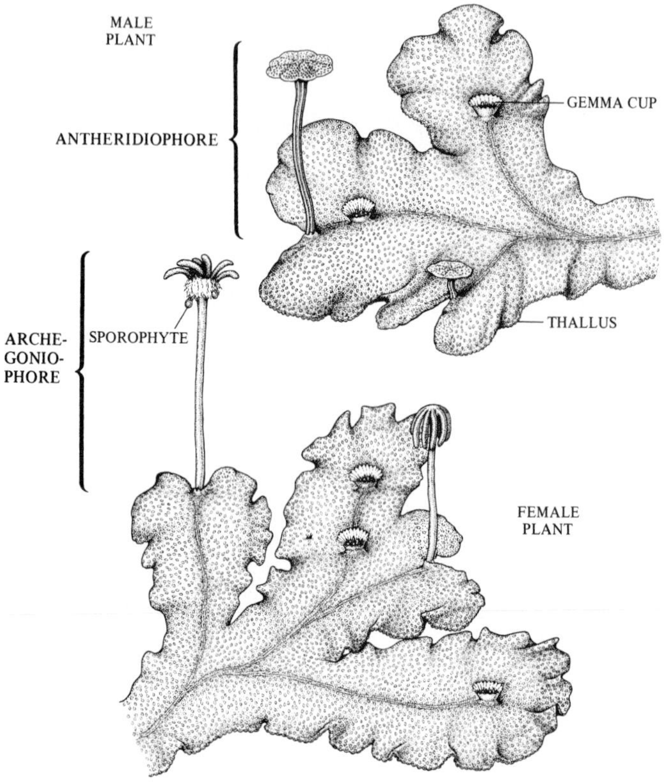

13–9
Mature spores and elaters from a capsule of Marchantia.

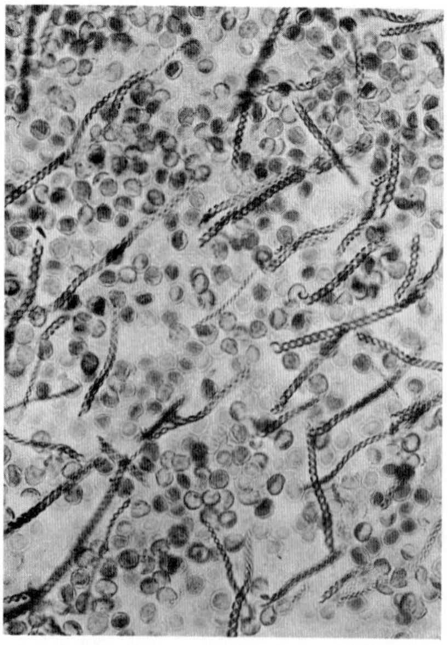

13–10
The liverwort Lunularia cruciata, *with crescent-shaped gemma cups containing gemmae. The gemmae are splashed out by the rain and grow in the vicinity of the parent plant. This liverwort is a common weed of greenhouses, probably because it reproduces so well asexually and so can maintain the genetic stability that is advantageous in such a fixed environment.*

Leafy Liverworts

The leafy liverworts (Figure 13–11) are a diverse group that include more than two-thirds of all known liverworts. The plants are usually well branched and form small mats. Their leaves are often two-lobed, and each grows by means of two distinct apical growing points. The leaves are often arranged in two rows, with a third row of reduced leaves along the lower surface. The leaves of liverworts, like those of mosses, are generally only a single layer of nondifferentiated cells thick. The antheridia and archegonia of the leafy liverworts are characteristically enclosed in a cuplike structure formed by the fusion of two or three leaves.

The leafy liverworts are especially abundant in the tropics and subtropics, in regions of heavy rainfall or high humidity, but they are also present in large numbers in temperate latitudes.

THE HORNWORTS: CLASS ANTHOCEROTAE

The class Anthocerotae consists of five genera, the most familiar of which is *Anthoceros*. *Anthoceros* is worldwide in distribution, generally occurring in moist, shaded habitats. Superficially the gametophyte of *Anthoceros* resembles that of the thallose liverworts (Figure 13–12). However, each cell usually has a single large chloroplast,

13–11
Some examples of leafy liverworts.

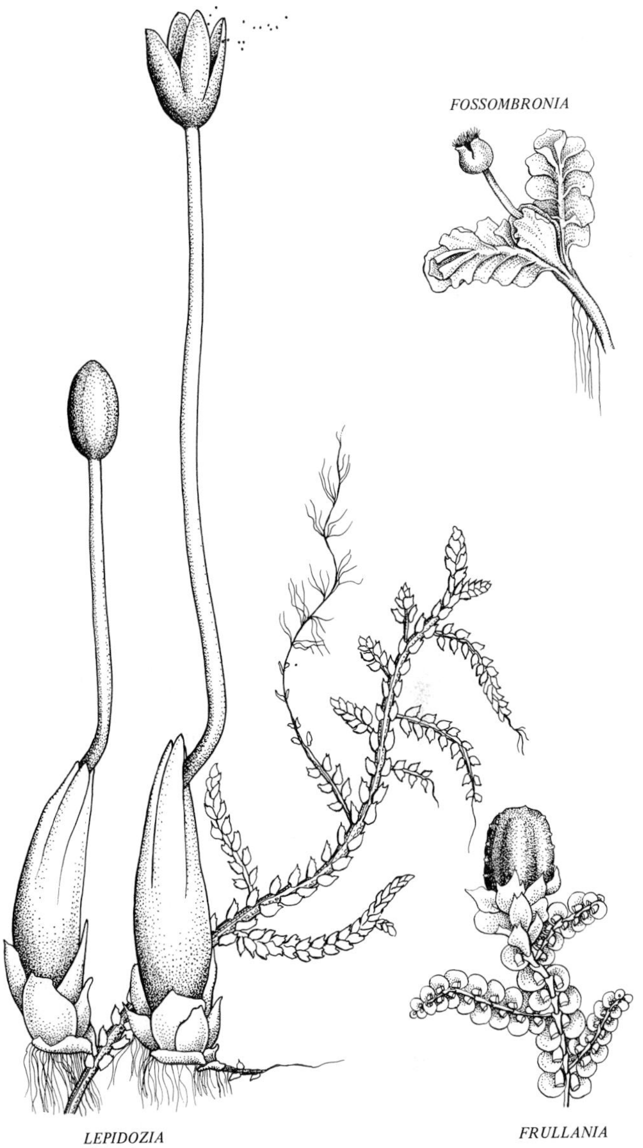

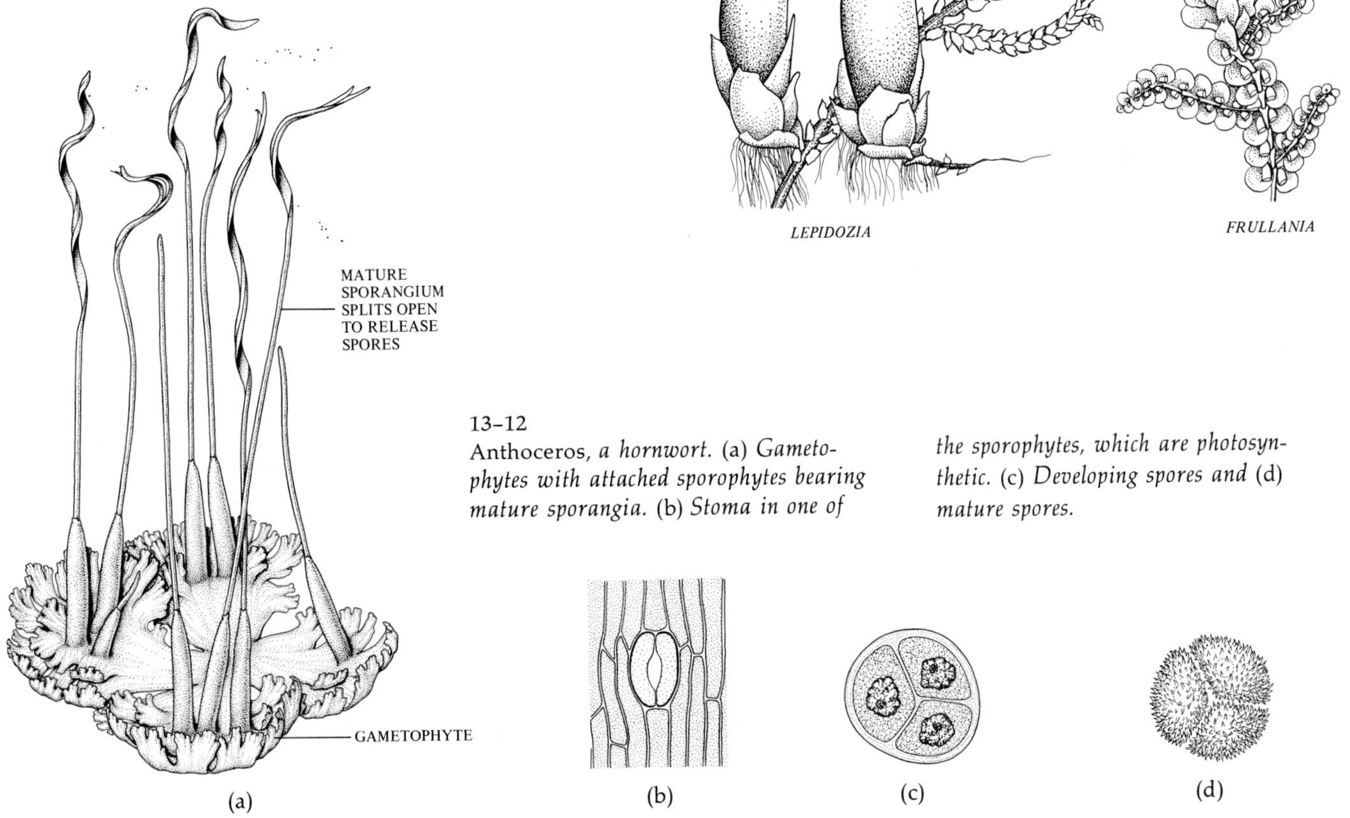

13–12
Anthoceros, *a hornwort.* (a) *Gametophytes with attached sporophytes bearing mature sporangia.* (b) *Stoma in one of the sporophytes, which are photosynthetic.* (c) *Developing spores and* (d) *mature spores.*

rather like those of many algae, in place of the many small, discoid ones found in the cells of other bryophytes and in vascular plants. Each chloroplast contains a pyrenoid, making the resemblance to algae even more striking. The strongly dorsiventrally oriented gametophytes are spherical and often less than 2 centimeters in diameter. They have extensive internal cavities filled with mucilage, not with air, as in *Riccia, Ricciocarpus,* and *Marchantia.* These mucilage-filled cavities are often inhabited by blue-green algae of the genus *Nostoc,* which supply nitrogen to the plants in which they grow.

Some species of *Anthoceros* are unisexual and others are bisexual. The antheridia and archegonia are sunken on the dorsal surface of the gametophyte. Numerous sporophytes may develop on the same gametophyte.

The sporophyte of *Anthoceros* consists of a foot and a long cylindrical sporangium (Figure 13–13). Very early in its development a meristem, or zone of actively dividing cells, develops between foot and sporangium; this meristem is active as long as conditions are favorable for growth. As a result, the sporophyte continues to elongate for a prolonged period of time. Dehiscence of the sporangium begins near its apex and spreads toward its base as the spores mature. The dehiscing sporangium splits longitudinally into two hornlike valves. The sporophyte contains several layers of photosynthetic tissue. Its surface is covered with a cuticle and contains stomata.

At one time, it was speculated that *Anthoceros* was an ancestor of the vascular plants, but in the light of present information, this theory has been largely abandoned.

THE MOSSES: CLASS MUSCI

Many groups of plants contain members that are loosely called "mosses" (reindeer "mosses" are lichens, club "mosses" and Spanish "moss" are vascular plants, and sea "moss" and Irish "moss" are algae), but the true mosses are members of the class Musci, which consists of three subclasses: Bryideae (the "true" mosses), Sphagnideae (the peat mosses), and Andreaeideae (the granite mosses).

True Mosses

The gametophytes of all mosses are represented by two distinct phases: the <u>protonema</u> ("first thread"), which arises directly from a germinating spore, and the leafy gametophyte. In the true mosses the protonema is a uniseriate (the cells occurring in unicellular rows), branching filament, which superficially resembles a filamentous green alga (Figure 13–14). Some branches of the protonema penetrate the substratum and become colorless

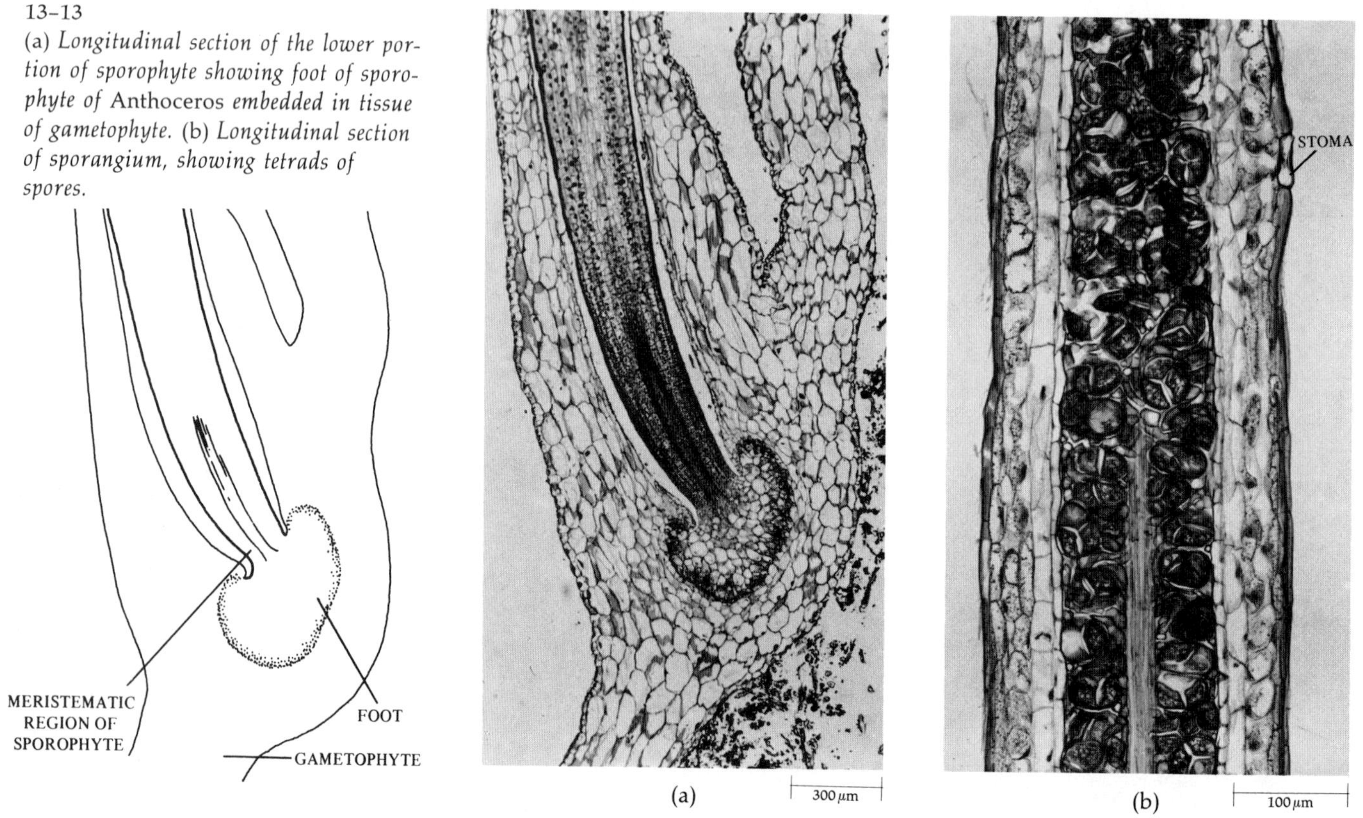

13–13
(a) *Longitudinal section of the lower portion of sporophyte showing foot of sporophyte of* Anthoceros *embedded in tissue of gametophyte.* (b) *Longitudinal section of sporangium, showing tetrads of spores.*

rhizoids, whereas others give rise to minute budlike structures, which develop into the leafy gametophytes (Figure 13–15). Protonemata are likewise found in some liverworts.

In the true mosses, the gametophyte is leafy and usually upright, not dorsiventrally flattened as it is in the leafy liverworts. It grows from an apical initial cell that is similar to that of the leafy liverworts, that is, like an inverted pyramid with three sides. Although three ranks of leaves are produced initially, subsequent twisting of the axis results in displacement of these ranks and the appearance of a spiral leaf arrangement. In some genera (*Fontinalis*, a genus of aquatic mosses, for example) the three-ranked condition of the leaves is obvious in the mature gametophytes.

The gametophytes of mosses range from a few millimeters to 5 decimeters or more in length, and they exhibit varying degrees of differentiation and complexity. All have multicellular rhizoids, and the leaves are normally only one cell layer thick except at the midrib (which is lacking in some genera). In some mosses, such as the common *Polytrichum*, there is often a central strand of elongate cells in the stem which may function in conduction (Figure 13–3), but many other genera lack such specialized tissues.

Two sorts of growth form are common among the gametophytes of mosses (Figure 13–16). In the first, the "cushiony" mosses, the gametophytes are erect and lit-

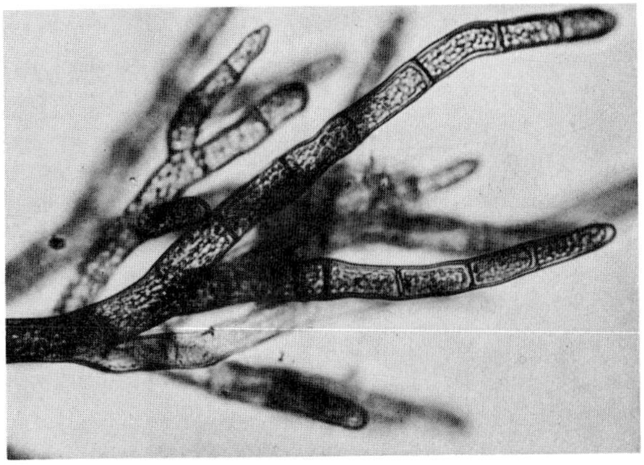

13–14
Protonema of a moss. Protonemata, which are young gametophytes, are characteristic of mosses and some liverworts. They often resemble filamentous green algae.

13–15
Male gametophytes of Polytrichum piliferum, *a moss, showing antheridial heads.*

(a)

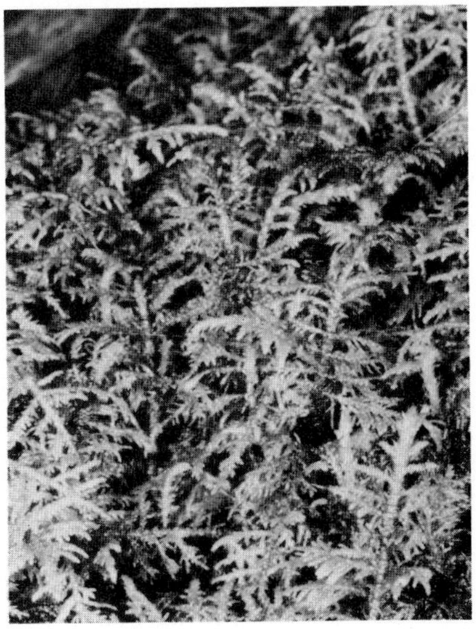

(b)

13–16
Two common growth forms of moss gametophytes. (a) *"Cushiony" form, with gametophytes erect and little branched, in* Polytrichum juniperinum. *Sporophytes, spore capsules atop long, slender seta, can be seen rising above the gametophytes.* (b) *"Feathery" form, with gametophytes creeping and matted, in* Thuidium delicatulum.

tle-branched, usually bearing terminal sporophytes. In the second, the gametophytes are much-branched and "feathery," and the plants are creeping; the sporophytes are borne laterally. This second type of growth pattern is found in those mosses that hang in lush masses from the branches of trees in moist regions. In some mosses the protonema is persistent (withers but remains attached), and the leafy shoots of the gametophyte are minute.

At maturity, gametangia are produced by most leafy gametophytes, either at the tip of the main axis or on a lateral branch (Figure 13–17). In some genera the gametophytes are unisexual, whereas in others both archegonia and antheridia are produced by the same plant.

Sporophytes in the mosses are often small in relation to the gametophytes (Figure 13–18). The capsules are generally elevated on a stalk, which may exceptionally reach 15 to 20 centimeters in length, although some lack a stalk entirely. A short foot at the base of the stalk is embedded in the tissue of the gametophyte. In mosses, the stalk usually elongates early in the development of the sporophyte (in contrast to the situation in liverworts), and the sporophyte is important in photosynthesis. The sporophyte of a moss is therefore much less dependent nutritionally upon the gametophyte than is

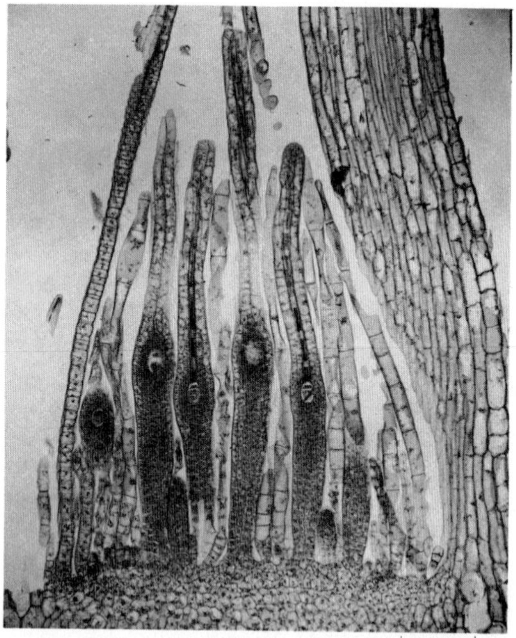

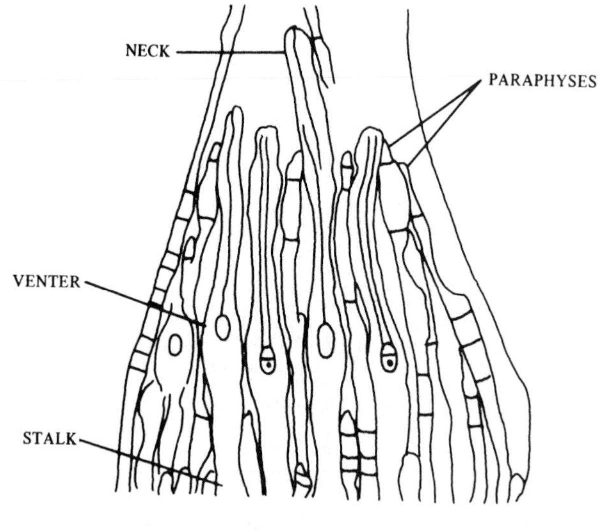

(a) 100 μm

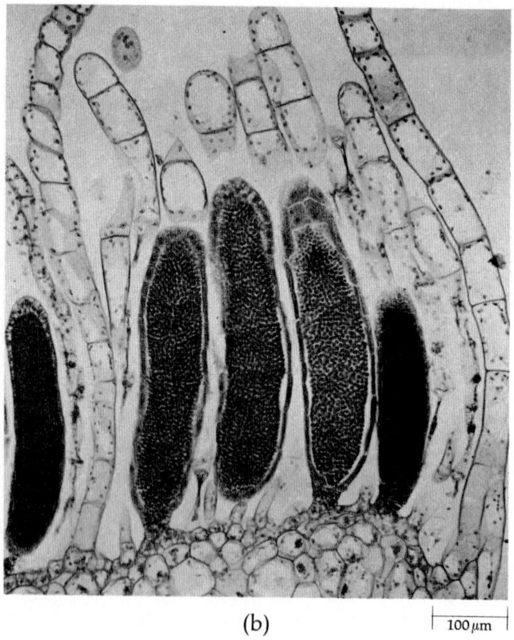

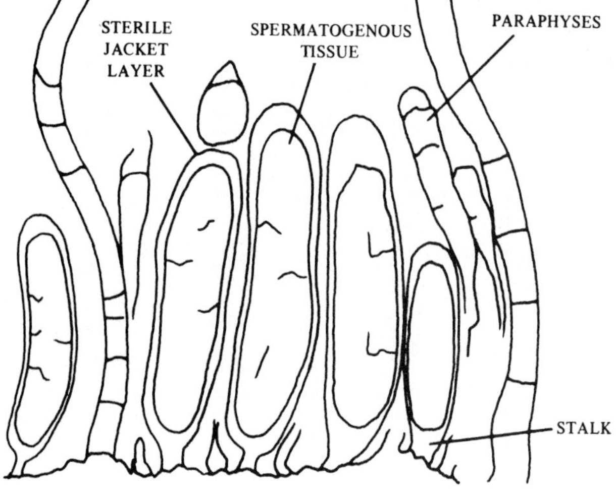

(b) 100 μm

13–17
Gametangia of Mnium, *a moss.* (a) *Longitudinal section through archegonial head showing archegonia surrounded by sterile structures called paraphyses.* (b) *Longitudinal section through antheridial head showing antheridia surrounded by paraphyses.*

13–18
Sporophytes of mosses. (a) Ptilium crista-castrensis, *a feather moss that often carpets the boreal coniferous forests. The female plant on the left bears two capsules elevated on long, twisted setae. Each capsule has a specialized lid, or operculum.* (b) Sphagnum. *This genus is confined chiefly to waterlogged habitats. Many of the capsules shown here have burst.* (c) *Spore-bearing stalks of the hairy moss,* Pogonatum brachyphyllum.

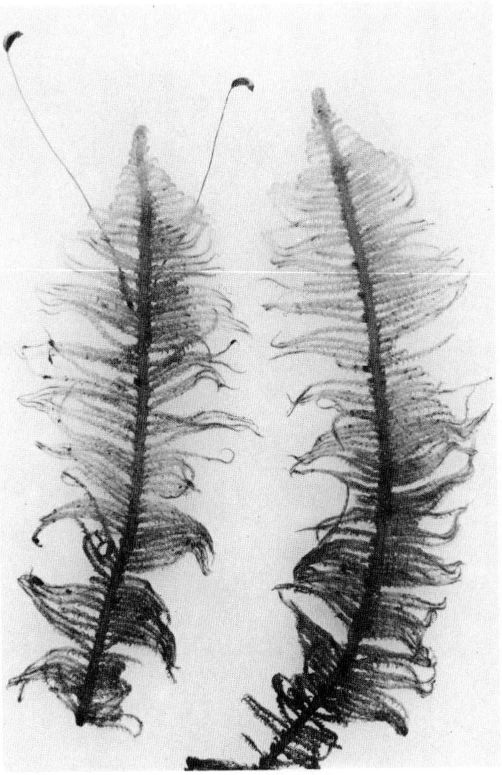

(a)

(b)

(c)

13–19
Three methods of spore discharge in mosses. Brachythecium *(at right) has a peristome ringed with two rows of teeth, which open to release the spores in response to changes in moisture. Most mosses are similar to* Brachythecium *in the way in which their spores are discharged. The mechanisms of spore discharge of* Sphagnum *and* Andreaea *are peculiar to these unusual genera.*

MOIST SPORANGIUM
ANDREAEA SPORANGIUM CONTRACTS AND CRACKS AS IT DRIES, AND THE SPORES FALL OUT.

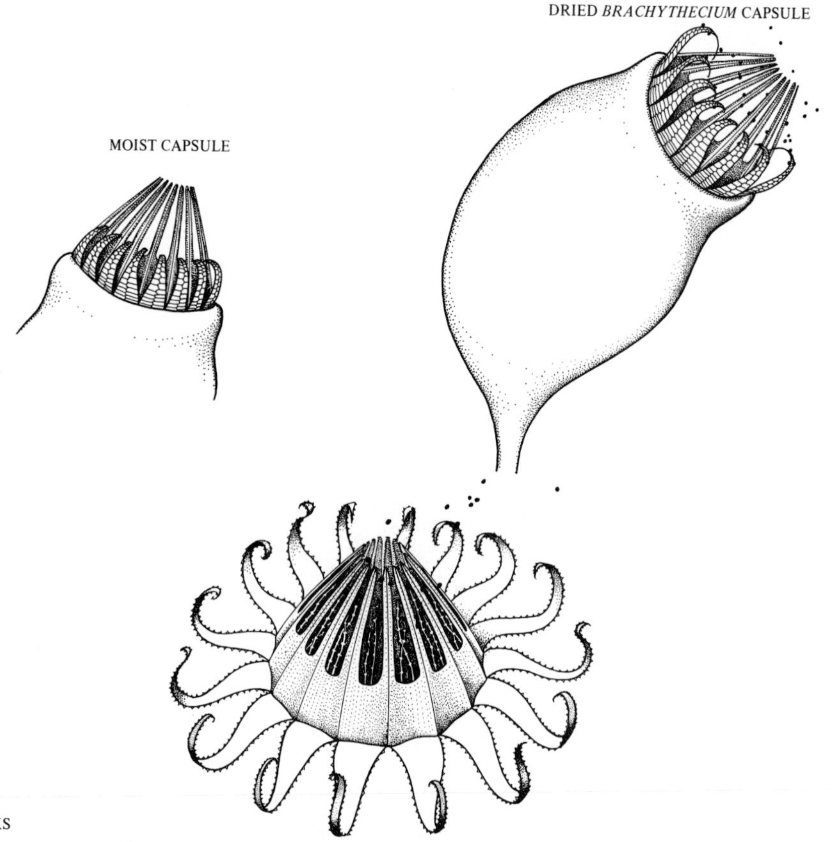
DRIED *BRACHYTHECIUM* CAPSULE
MOIST CAPSULE
DRIED *BRACHYTHECIUM* CAPSULE. THE OUTER SET OF PERISTOMAL TEETH INTERLOCK WITH THE INNER SET UNDER DAMP CONDITIONS. WHEN IT IS DRIER, THE OUTER TEETH PULL AWAY TO ALLOW THE SPORES TO BE DISPERSED IN THE BREEZE.

DRIED SPORANGIUM

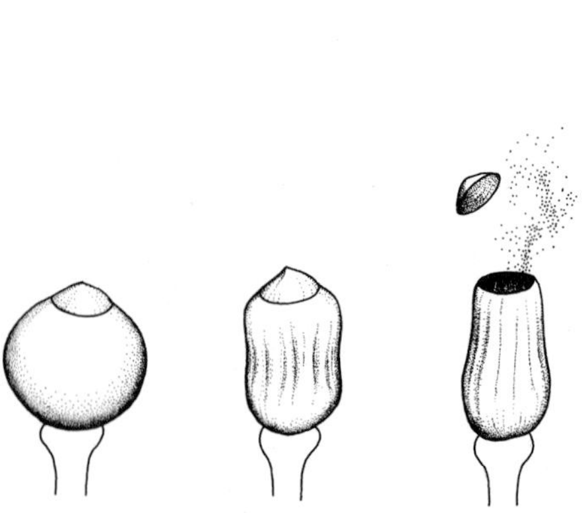
SPHAGNUM CAPSULE SHRINKS AND BURSTS OPEN TO EXPOSE ITS SPORES.

the sporophyte of a liverwort. Stomata, which are lacking on the sporophytes of liverworts, are normally present on the sporophytes of mosses. The sporophytes of mosses have a high degree of internal organization, and a central strand of elongate cells is present in the stalk of many genera.

When the sporophyte of a moss is mature, it gradually loses its ability to photosynthesize and turns yellow, then orange, and then brown. Eventually the lid, or *operculum*, of the capsule, bursts off, revealing an opening that is usually ringed with a series of *peristome* teeth, which regulate spore discharge (Figure 13–19). The peristome is a characteristic of the subclass Bryideae, lacking in the other two subclasses. A representative life cycle of a moss is shown in Figure 13–20.

Asexual reproduction in the true mosses is accomplished largely by fragmentation. Virtually any part of the gametophyte is capable of regeneration, including the sterile parts of the sex organs. Many species produce gemmae, which give rise to new gametophytes.

13–20
In a representative moss life cycle, spores are released from a capsule, which opens when a small lid (operculum) bursts. The spore germinates to form a branched, filamentous protonema, from which a leafy gametophyte develops. Sperms, which are expelled from the mature antheridium, are attracted into the archegonium, where one fuses with the egg cell to produce the zygote. The zygote divides mitotically to form the sporophyte and, at the same time, the venter of the archegonium divides to form the protective calyptra. The sporophyte consists of a capsule, which may be raised on a stalk, also part of the sporophyte, and a foot. Meiosis occurs within the capsule, resulting in the formation of haploid spores.

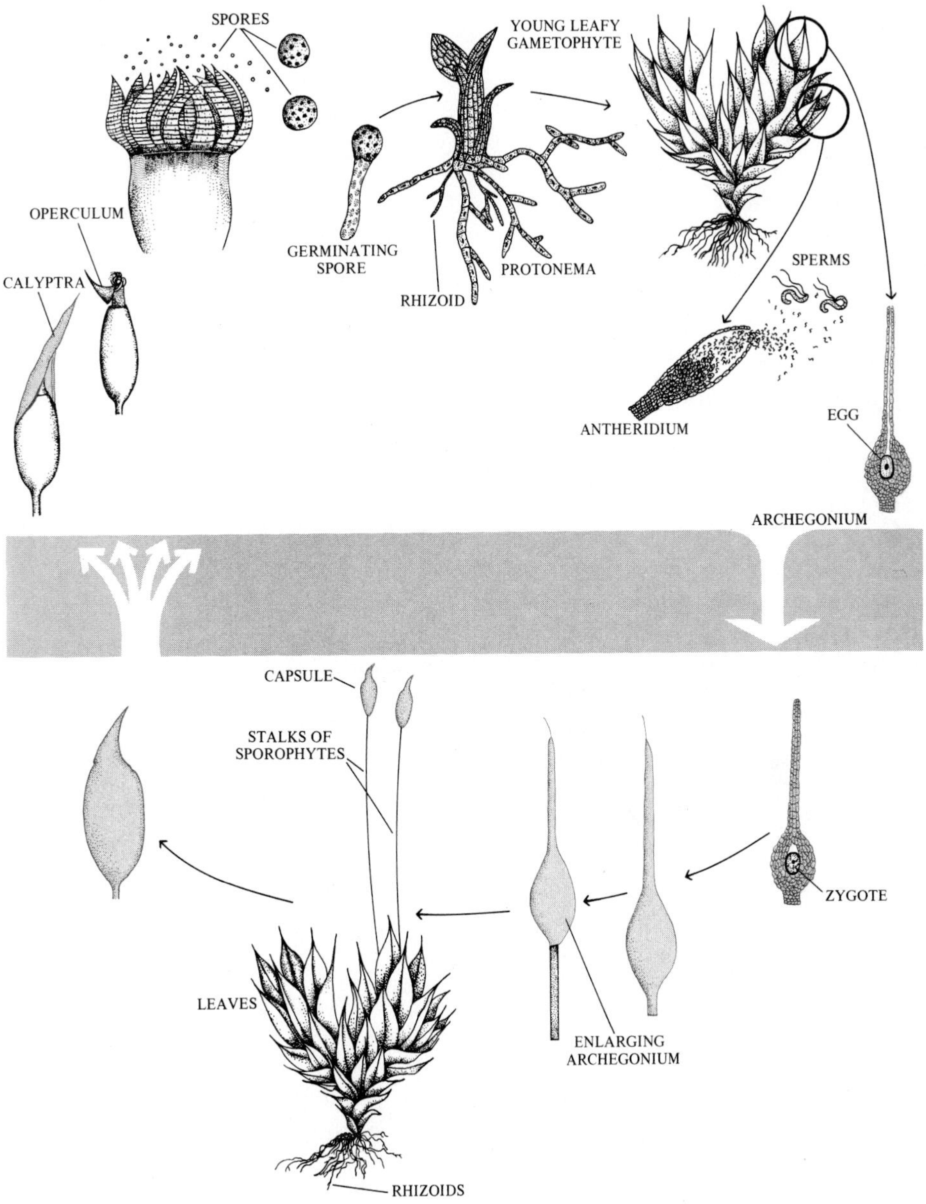

The Peat Mosses

The more than 300 species of peat mosses in the genus *Sphagnum* form a very distinct group that doubtless has been separated from the main line of moss evolution for a long time. The stems of the gametophyte of *Sphagnum* bear clusters of branches, often five at a node, which are more densely tufted near the apex of the stem. The gametophytes of *Sphagnum* arise from a protonema that is platelike instead of filamentous. The leaves in *Sphagnum* lack midribs, and the mature plants lack rhizoids. In the boggy places where they grow, plants of *Sphagnum* are nearly always turgid and thus erect. In the leaves, there is considerable cellular differentiation: There are large dead cells with many pores, which allow for the direct entry of water, and between them there are rows of small, thick-walled living cells with many chloroplasts. Because of the presence of anthocyanins, plants of *Sphagnum* are often reddish or purplish.

The sporophytes of *Sphagnum* (see Figure 13–18b) are also distinctive. The capsules are nearly spherical and are raised on a stalk, the *pseudopodium*, that is part of the gametophyte; the sporophyte itself has a very short stalk. Spore discharge in *Sphagnum* is spectacular. As the capsule matures, its internal tissues shrink, leaving a gas-filled space. Contraction of the maturing capsule results in increased internal pressure, which eventually results in the operculum being blown off, the escaping gas carrying a cloud of spores out with it (Figure 13–19). The most distinctive features of Sphagnideae, as compared with other mosses, are the lack of a peristome and the very peculiar morphology of the gametophyte.

Ecology of Sphagnum

The peat mosses play a very important ecological role in cold and temperate regions all over the world in that they form extensive peat bogs. The mosses contribute to the acidity of their own environment by selective ionic absorption, and the pH in the center of the tufts is often much lower (for example, 4.4 as compared with 6.0) than that of the surrounding soil and water. Because of its superior absorptive qualities and high acidity, peat moss is mixed with soil in gardening to increase the water-holding capacity of the medium. In Ireland and some other northern regions, dried peat is widely used as fuel.

The Granite Mosses

The genus *Andreaea* consists of about 50 species of small, blackish-green or olive-brown tufted rock mosses, which in their own way are as peculiar as *Sphagnum*. Although the gametophyte closely resembles that of the true mosses, it arises from a protonema that is platelike instead of filamentous. Just as in *Sphagnum*, the sporophyte lacks a true stalk, and it is elevated above the leaves on a stalk of gametophytic tissue. The minute capsules of *Andreaea* are marked by four lines of weaker cells along which the capsule splits. The capsule remains intact above and below the dehiscence lines. The resulting four valves are very sensitive to the humidity of the surrounding air, opening widely when dry and closing when moist. This mechanism of spore discharge is different from that of any other moss (Figure 13–19).

SUMMARY

The Bryophyta are an isolated division of plants divided into three classes: the Hepaticae (liverworts), Anthocerotae (hornworts), and Musci (mosses). In their pigments and food reserves, they resemble the green algae and the vascular plants. Some of the hornworts and most of the mosses have stomata on their sporophytes like the vascular plants, and some mosses have a cuticle on their leaves. Although most bryophytes lack specialized vascular strands, and most absorb water directly through the leaves and stems, they are believed to share a remote common ancestor with the vascular plants.

All bryophytes have a well-defined heteromorphic alternation of generations in which the gametophyte is the dominant generation. The male sex organs, antheridia, and female sex organs, archegonia, are multicellular and possess sterile jacket layers. Each archegonium contains a single egg. Numerous sperms are produced by each antheridium. The sperms are coiled, free-swimming, and have two flagella at the anterior end. The sporophyte develops within the archegonium but usually grows out of it, although remaining permanently attached to the gametophyte. In most bryophytes, the sporophyte is differentiated into a foot, a seta (stalk), and a capsule (sporangium). The sporophyte of the hornworts is unique in the presence of a basal meristem, which adds tissue to the sporangium over a prolonged period. The sporophyte of the hornworts is nutritionally quite independent of the gametophyte, as is usually also the case in the mosses before the spores are mature. In liverworts, the sporophyte is nutritionally more dependent, sometimes remaining enclosed by gametophyte tissue at maturity. Upon germinating, the spores of mosses produce a filamentous gametophyte known as a protonema, from which the leafy gametophytes arise; the spores of liverworts and hornworts give rise directly to the gametophyte in its mature form.

SUGGESTIONS FOR FURTHER READING

CONARD, H. S.: *How to Know the Mosses and Liverworts*, Wm. C. Brown Co., Dubuque, Iowa, 1956.*

A good beginner's guide for use in identifying many of the more common bryophytes. It includes a profusely illustrated key and a good glossary.

DOYLE, W. T.: *The Biology of Higher Cryptogams*, The Macmillan Company, New York, 1970.*

A concise account of the biology of the bryophytes and lower vascular plants with an emphasis on developmental and evolutionary relationships.

WATSON, E. V.: *The Structure and Life of Bryophytes*, 3rd ed., Hutchinson & Co., Ltd., London, 1971.

A concise and well-written treatment of the entire group.

* Available in paperback.

CHAPTER 14

The Vascular Plants: Introduction

14–1
Reconstruction of a Carboniferous swamp forest. Most of the tall treelike plants shown here belong to genera that are now extinct. A number of smaller herbaceous plants are related to modern species of Lycopodium *and* Selaginella. *Two cockroaches are visible, one on the large trunk at the left.*

The vascular plants (division Tracheophyta), like all living things, had aquatic ancestors, and the story of their evolution is inseparably linked with the story of their progressive occupation of the land. This evolutionary progression has involved many changes, including the development of roots, which anchor the plants and absorb water and minerals from the soil. Also, as the ancestors of the vascular plants evolved, they began to produce cutin and suberin, waxy substances that protect the plant body from excessive evaporation. Efficient conducting systems also evolved, which, like a system of pipes, conduct water and minerals into the uppermost branches of the tallest trees. Further, there has been a progressive reduction in the gametophytic generation, with the gametophyte becoming more and more protected by and dependent upon the sporophyte. Finally, in a number of lines, seeds evolved structures that protect the embryonic sporophyte from the rigors of a life on land, nourish it, and enable it to withstand unfavorable situations.

As we discussed in Chapter 13, the common ancestor of the bryophytes and vascular plants was probably a relatively complex, multicellular green alga that invaded the land more than 400 million years ago, probably as part of an endomycorrhizal relationship. This green alga presumably had a life cycle in which the sporophyte was the dominant generation.

The vascular plants are the dominant land plants of the biosphere (Figure 14–1). They are members of the division Tracheophyta, which includes five subdivisions with living representatives: the Psilophytina (whisk ferns), the Lycophytina (club mosses, quillworts, and their relatives), the Sphenophytina (horsetails), the Filicophytina (ferns), and the Spermatophytina (seed plants), with about 250,000 living species. In addition, there are three subdivisions of extinct tracheophytes. Other fossils may eventually be found to constitute still other subdivisions. In this chapter, we shall discuss the extinct forms and describe some of the evolutionary advances of the vascular plants as a group. In Chapters 15 and 16, the different subdivisions with living representatives will be discussed in turn.

EARLY VASCULAR PLANTS

Subdivision Rhyniophytina

The earliest known vascular plants belong to the subdivision Rhyniophytina, a group that dates back to the Silurian period, some 400 million years ago. They were seedless plants, consisting of simple, dichotomously branching axes and terminal sporangia. Their plant bodies were not differentiated into roots, stems, or leaves, and their sporangia produced only one type of spore.

The best-known representative of the Rhyniophytina is *Rhynia* (Figure 14–2a). Probably a marsh plant, its leafless, dichotomously branching aerial stems were attached to an underground stem, or rhizome, with tufts of water-absorbing rhizoids. The aerial stems were covered with a cuticle, contained stomata, and served as the principal photosynthetic organs.

The internal structure of *Rhynia* was basically similar to that of many of today's vascular plants. A single layer of superficial cells, the epidermis, surrounded the photosynthetic tissue of the cortex, and the center of the axis consisted of a solid core of xylem surrounded by phloem.

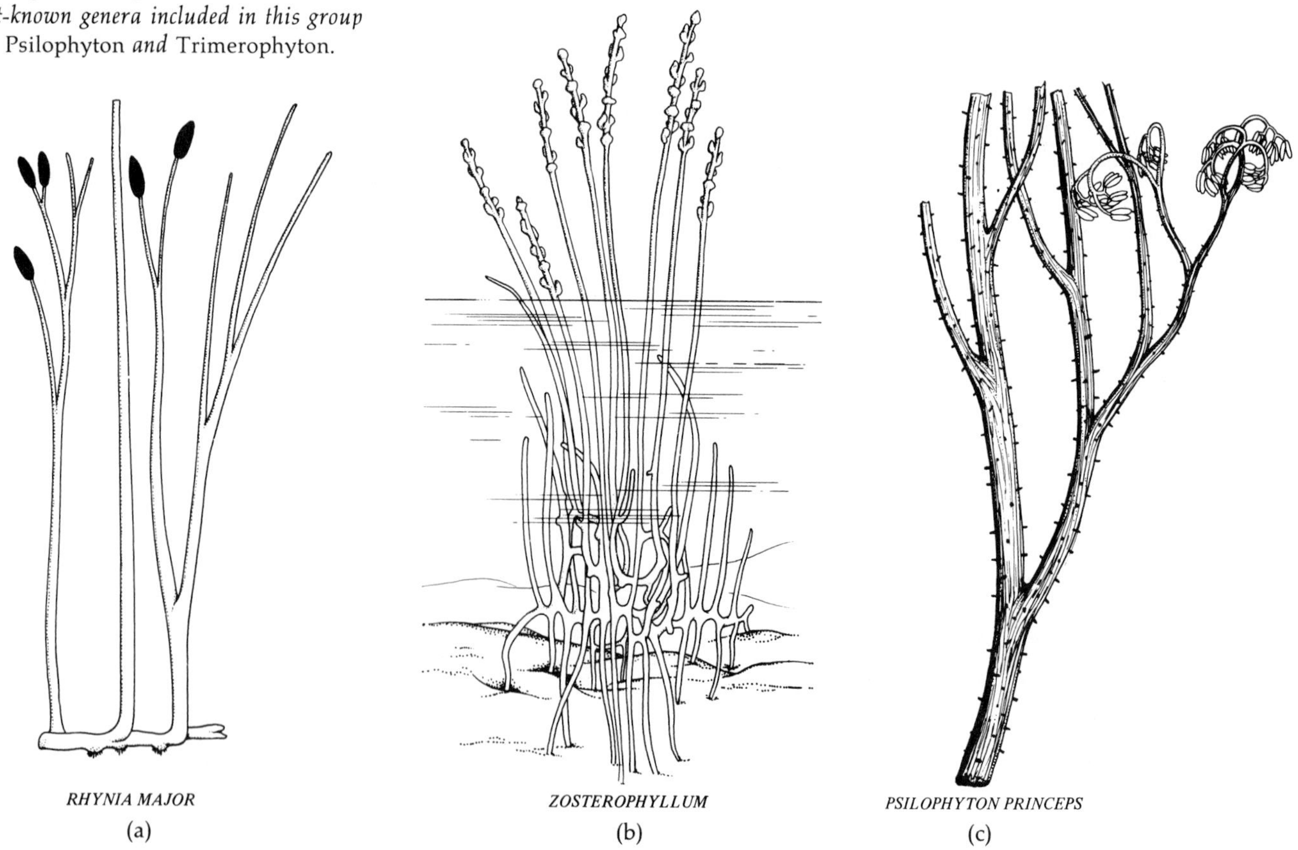

14–2
Early vascular plants. (a) Rhynia major *is a member of the Rhyniophytina, the simplest known vascular plants. The shoot is leafless and dichotomously branched. The sporangia, which are terminal, release the spores by splitting longitudinally.* (b) *In* Zosterophyllum, *the best-known genus of Zosterophyllophytina, the sporangia, which are aggregated into a terminal spike, split along definite slits that form around the outer margin. Zosterophyllophytina are larger than the Rhynophytina, but like the latter, they are mostly dichotomously branched plants. They are either naked, spiny, or toothed. This group may include the ancestors of the Lycophytina.* (c) *The third group of Devonian "psilophytes," Trimerophytina, comprises plants that had a strong central axis with smaller side branches. The side branches are dichotomously branched and have terminal masses of paired sporangia that taper at both ends. The best-known genera included in this group are* Psilophyton *and* Trimerophyton.

Subdivision Zosterophyllophytina

During the Lower and Middle Devonian period, from about 395 to about 350 million years ago, a second group of vascular plants, the Zosterophyllophytina, appeared. Like the Rhyniophytina, these plants were dichotomously branched. It is possible that the group was aquatic. The aerial stems were covered with a cuticle, but only the upper ones contained stomata. In *Zosterophyllum,* the lower branches frequently produced lateral branches that forked into two axes, one that grew upward, the other downward (Figure 14–2b). Unlike the Rhyniophytina, the sporangia of the Zosterophyllophytina were borne laterally on short stalks. Only one type of spore was produced. The internal structure of the Zosterophyllophytina was essentially similar to that of the Rhyniophytina. The Zosterophyllophytina are considered as the likely progenitors of the lycopods (Chapter 15).

Subdivision Trimerophytina

The subdivision Trimerophytina is considered by some paleobotanists to have evolved directly from the Rhyniophytina and to represent the ancestral stock of the ferns, and perhaps the horsetails also. *Psilophyton*-like plants of the Trimerophytina appear in mid-Devonian strata (360 million years old).

In them, the smaller branches dichotomized several times and bore elongate sporangia at their apices (Figure 14–2c). Only one type of spore was produced, and the plants lacked leaves. Except for the presence of a more massive vascular system, the internal structure of the Trimerophytina was essentially similar to that of the Rhyniophytina and Zosterophyllophytina.

Although the Trimerophytina are considered as a distinct group in this book, it has been argued that they should be placed in the Rhyniophytina. Much more study of the Paleozoic groups will be necessary before their relationships can be determined with certainty.

THE STRUCTURE OF VASCULAR PLANTS

From the previous discussion it is clear that the plant bodies of the early vascular plants were dichotomously branched axes that lacked leaves and roots. With evolutionary specialization, morphological and physiological differences arose between various parts of the plant body, bringing about the differentiation of root, stem, and leaf, the organs of the plant (Figure 14–3).

The many different kinds of cells of the plant body are organized into tissues, and the tissues are organized into still larger units called tissue systems. The three tissue systems—*dermal, vascular,* and *fundamental* or *ground*—occur in all organs of the plant, and reveal the basic

14–3
Portions of the three organs of the primary plant body. (a) *Portion of stem, including apical meristem of the shoot (stem and leaves);* (b) *portion of leaf;* (c) *portion of root, including apical meristem. In all three organs, the dermal tissue system is represented by the epidermis and the vascular tissue system by primary xylem and primary phloem. The ground tissue system in the stem is divided into cortex and pith. In the root it is represented by the cortex and in the leaf by palisade and spongy mesophyll tissue. The stem is specialized for support of the leaves and for conduction; the leaf for photosynthesis; the root for absorption and anchorage.*

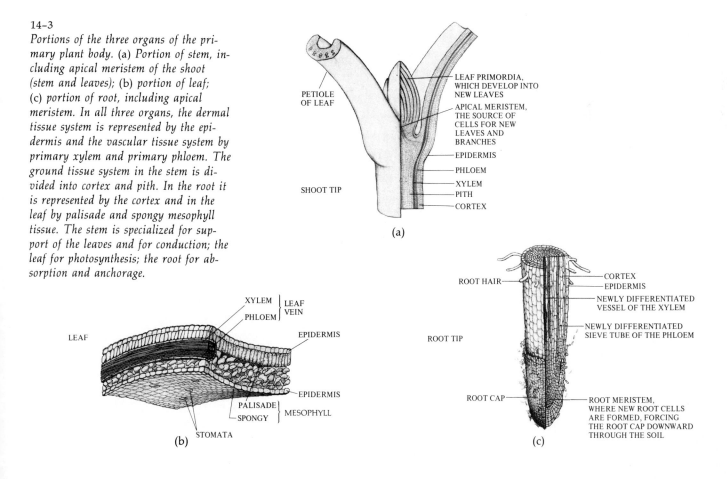

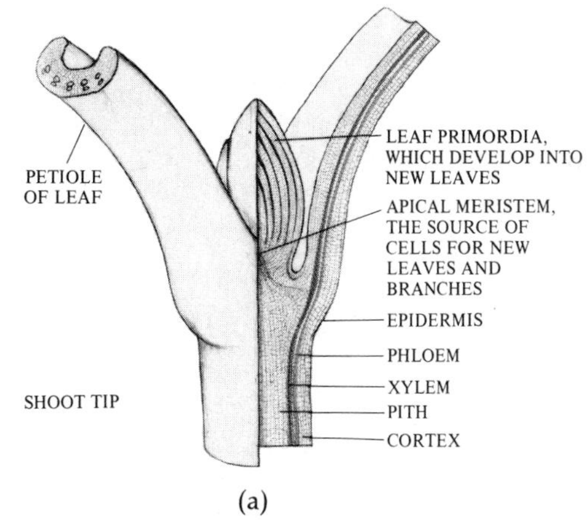

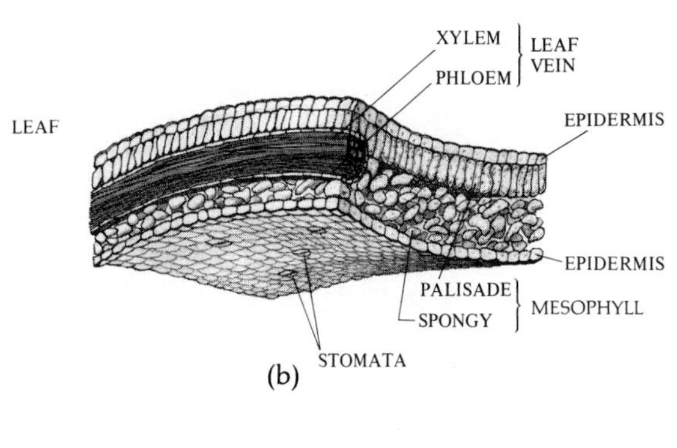

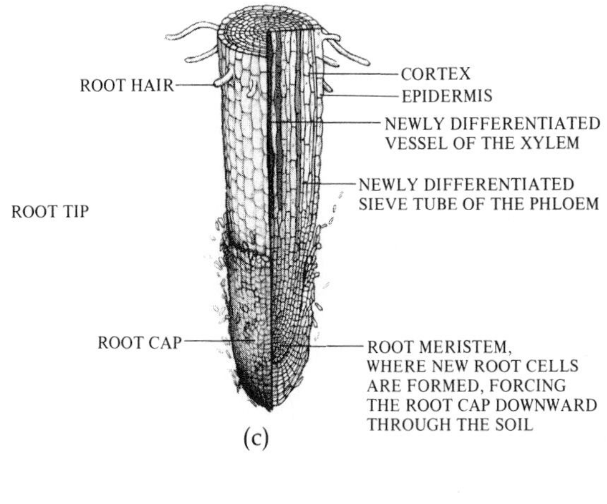

unity of the plant body. The dermal tissue system makes up the outer protective covering of the plant. The vascular tissue system comprises the vascular tissues—xylem and phloem—and is embedded in the ground tissue system (Figure 14–3). In Section 5, we shall learn that the principal differences in the structures of root, stem, and leaf lie primarily in the relative distribution of the vascular and ground tissue systems.

Primary and Secondary Growth

Primary growth may be defined as that growth which takes place relatively close to the tips of roots and stems. It is initiated by the apical meristems and involved primarily with extension of the plant body. Tissues arising during primary growth are called *primary tissues,* and the part of the plant body composed of those tissues is called the *primary plant body* (Figure 14–3). The most primitive of vascular plants consisted entirely of primary tissues, although by itself this does not characterize a primitive plant. Many modern plants are composed entirely of primary tissues.

In addition to primary growth, many seed plants undergo additional growth, which brings about an increase in thickness of the stem and root. Such growth is termed *secondary growth* and involves the activity of a lateral meristem, the *vascular cambium,* which produces the *secondary vascular tissues, secondary xylem,* and *secondary phloem* (see Figure 22–5 on page 462). The production of secondary vascular tissues commonly is supplemented by the activity of the *cork cambium,* which forms a *periderm,* composed mostly of *cork* tissue. The periderm replaces the epidermis as the dermal tissue system of the plant. The secondary vascular tissues and periderm make up the *secondary plant body.* Although only a few living lower vascular plants exhibit secondary growth, secondary growth by a vascular cambium appeared about 360 million years ago in the mid-Devonian among representatives of several unrelated groups.

Steles

The primary vascular tissues—primary xylem and primary phloem—and the pith, if present, collectively constitute what is known as the central cylinder or *stele* of the primary plant body of the stem and root. Two basic types of steles can be recognized, the *protostele* and the *siphonostele.*

The protostele consists of a solid core of vascular tissue, with either the phloem surrounding the xylem or interspersed with it (Figure 14–4). The protostele is regarded as the most primitive type of stele. It is the type of stele found in *Rhynia, Zosterophyllum,* and *Psilophyton.* Protosteles also occur in the Psilophytina, Lycophytina, and in the juvenile stems of other groups. In addition, it is the type of stele found in most roots.

The more advanced siphonostele is characterized by the presence of a central column of ground tissue, the *pith* (Figure 14–4). The vascular tissues are variously arranged around the pith. Siphonosteles occur in the stems of most ferns and seed plants.

Origins of Roots and Leaves

Although the fossil record reveals no information on the origins of roots, as we know them today, it seems reasonable to assume that they evolved from the subterranean portions or rhizomes of the primitive axislike plant body. For the most part, roots are relatively simple structures, which have retained many of the primitive structural characteristics no longer present in the stems of modern plants. This is probably due to the greater uniformity of the underground environment as contrasted with the aerial.

14–4
Longitudinal sections through nodal regions (parts of the stem to which leaves are attached) of stems possessing leaves that are (a) *microphyllous (relatively small with a single strand of vascular tissue) and* (b) *megaphyllous (relatively large, usually with a complex system of veins). With the exception of the unveined leaves of* Equisetum, *microphylls are borne on stems with protosteles, and megaphylls on stems with siphonosteles.* (c) *and* (d) *represent transverse views through the planes represented by the dashed lines in* (a) *and* (b), *respectively Note the presence of pith and leaf gap in* (b) *and* (d), *and their absence in* (a) *and* (c).

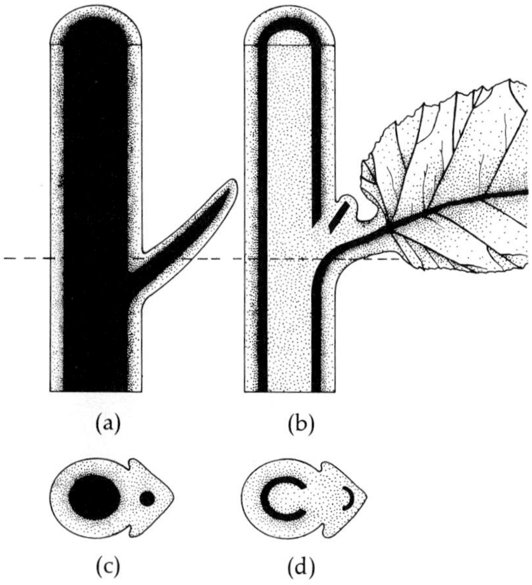

Leaves are the principal lateral appendages of the stem and together with the stem constitute the shoot system of the plant. Two morphologically distinct types of leaves are recognized, *microphylls* and *megaphylls*, and they seem clearly to have evolved in different ways.

Microphylls are relatively small leaves, that contain only a single strand of vascular tissue (Figure 14–4). With the exception of the univeined leaves of *Equisetum*, microphylls are associated with stems possessing protosteles, as in the Psilophytina and Lycophytina. Examination of a microphyllous shoot reveals that the strand of vascular tissue that diverges from the protostele into the leaf, the *leaf trace*, does so without interrupting the pattern of the stele. Although the name microphyll means small leaf, some species of *Isoetes* have fairly long leaves. In addition, certain Carboniferous and Permian members of the Lycophytina had microphylls that reached lengths of 1 meter or more.

14–5
According to one widely accepted theory, microphylls (above) evolved as outgrowths of the main axis of the plant. Megaphylls (below) evolved by fusion of branch systems.

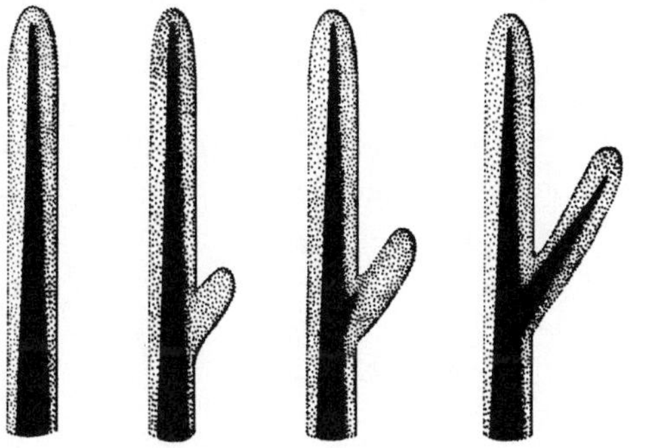

THE EVOLUTION OF MICROPHYLLS

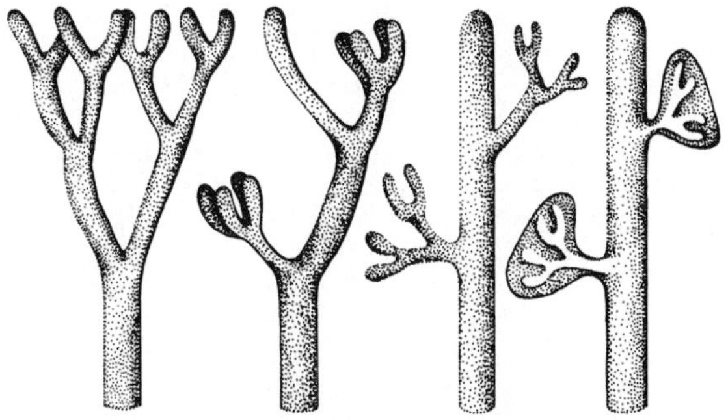

THE EVOLUTION OF MEGAPHYLLS

Microphylls are believed to have evolved as superficial lateral outgrowths of the stem (Figure 14–5). At first these structures were small scalelike or spinelike outgrowths devoid of vascular tissue. Gradually, rudimentary leaf traces developed, which initially extended only to the base of the outgrowth. Finally, the leaf traces extended into the appendage, resulting in formation of the primitive microphyll.

Megaphylls, as the name implies, are commonly large leaves. With few exceptions, megaphylls are associated with stems possessing siphonosteles. The *lamina* or blades of the great majority of megaphylls contain a complex system of veins. In addition, the leaf traces of megaphylls are associated with *leaf gaps* in the stele of the stem. The leaf gap is a region of ground tissue in the stele resulting from the divergence of the leaf trace away from the stele toward the leaf.

Megaphylls are believed to have evolved from entire branch systems by the series of steps illustrated in Figure 14–5. Beginning with a leafless, dichotomously branching axis, unequal branching resulted in more aggressive branches "overtopping" the weaker ones. This was followed with a flattening-out, or "planation," of the subordinated lateral branches. The final step was "webbing," or fusion, of the separate lateral branches to form a primitive lamina.

Although the leaves of *Equisetum* are by definition microphylls, some fossil members of the Sphenophytina had relatively large leaves with dichotomous venation. Possibly the small leaves of *Equisetum* were derived from larger, more complex ones.

It is quite clear that the photosynthetic organs we recognize as leaves evolved in more than one way.

REPRODUCTIVE SYSTEMS

In all vascular plants, the sporophyte is the dominant phase in the life cycle and is larger and structurally much more complex than the gametophyte. Almost all of the preceding remarks in this chapter have been devoted to the sporophyte of vascular plants. We shall now consider some general features of the reproductive structures of the group.

Homospory and Heterospory

During our discussion of early vascular plants, we noted that they produced only one kind of spore. Such vascular plants are said to be *homosporous*. Among living vascular plants, homospory is found in the Psilophytina, Sphenophytina, some of the Lycophytina, and almost all the ferns. Following meiosis, homosporous plants produce only one kind of spore within the sporangium. Upon germination such spores produce bisexual gametophytes, that is, gametophytes that bear both antheridia and archegonia.

Heterospory—the production of two types of spores in two different kinds of sporangia—is found in some of the Lycophytina, in a few ferns, and in all seed plants. Heterospory has arisen many times in unrelated groups of plants during the evolution of vascular plants. It was common at least as early as the Devonian period, more than 350 million years ago. The two types of spores are called *microspores* and *megaspores*, and they are produced in *microsporangia* and *megasporangia*, respectively. Although "micro" implies smallness and "mega" largeness, megaspores are not always larger than microspores, especially in the seed plants. The two types of spores should be defined on the basis of function, not according to relative size. Microspores give rise to male gametophytes (microgametophytes) and megaspores to female gametophytes (megagametophytes). Both types of unisexual gametophytes are much reduced in size as compared with the gametophytes of homosporous vascular plants.

Gametophytes and Gametes

The relatively large gametophytes of homosporous plants are independent of the sporophyte for their nutrition, although the subterranean gametophytes of some species—for example, those of *Lycopodium complanatum* and *L. obscurum*—are clearly heterotrophic, depending upon a mycorrhizal fungus for their existence. Other species of *Lycopodium*, as most ferns, have free-living, photosynthetic gametophytes. In contrast, the gametophytes of heterosporous species are dependent upon stored food derived from the sporophyte for their nutrition.

The evolution of the gametophyte of vascular plants is characterized by a progressive reduction in size and complexity, and the gametophytes of angiosperms are the most greatly reduced of all. The mature megagametophyte of angiosperms commonly consists of only seven cells, one of them an egg cell. When mature, the microgametophyte contains only three cells, and two of them are sperms. Archegonia and antheridia—found in all lower vascular plants—are absent in all angiosperms and in a few gymnosperms (Figure 14-6). In the lower vascular plants, including the ferns, the motile sperms swim through water to the archegonium. These plants must therefore grow in habitats where water is at least occasionally plentiful. In the gymnosperms and angiosperms, entire microgametophytes are carried to the vicinity of the megagametophyte, where they produce special structures, the *pollen tubes,* which bring the sperms near the egg. In this way the seed plants, which avoid the necessity of having free water for their fertilization, can live and reproduce in many areas where other plants cannot.

Evolution of the Seed

One of the most dramatic innovations exhibited by the vascular plants is the evolution of the seed (Figure 14-7). In plants that have seeds, the highly reduced megagametophyte is retained within the megaspore. The

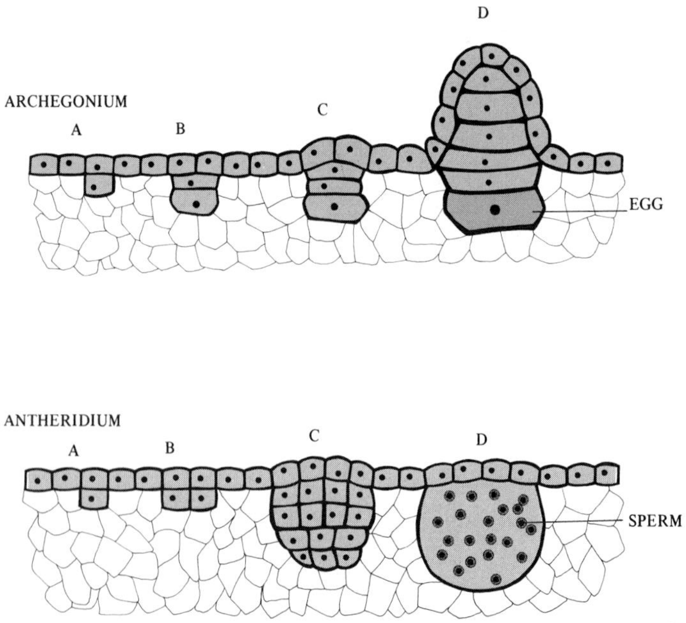

14-6
Stages in the development of multicellular sex organs as seen in the simpler vascular plants. Multicellular gametangia of this sort are characteristic of the lower vascular plants. Archegonia and antheridia are found in most gymnosperms but have been lost in the course of evolution in some gymnosperms and in all angiosperms.

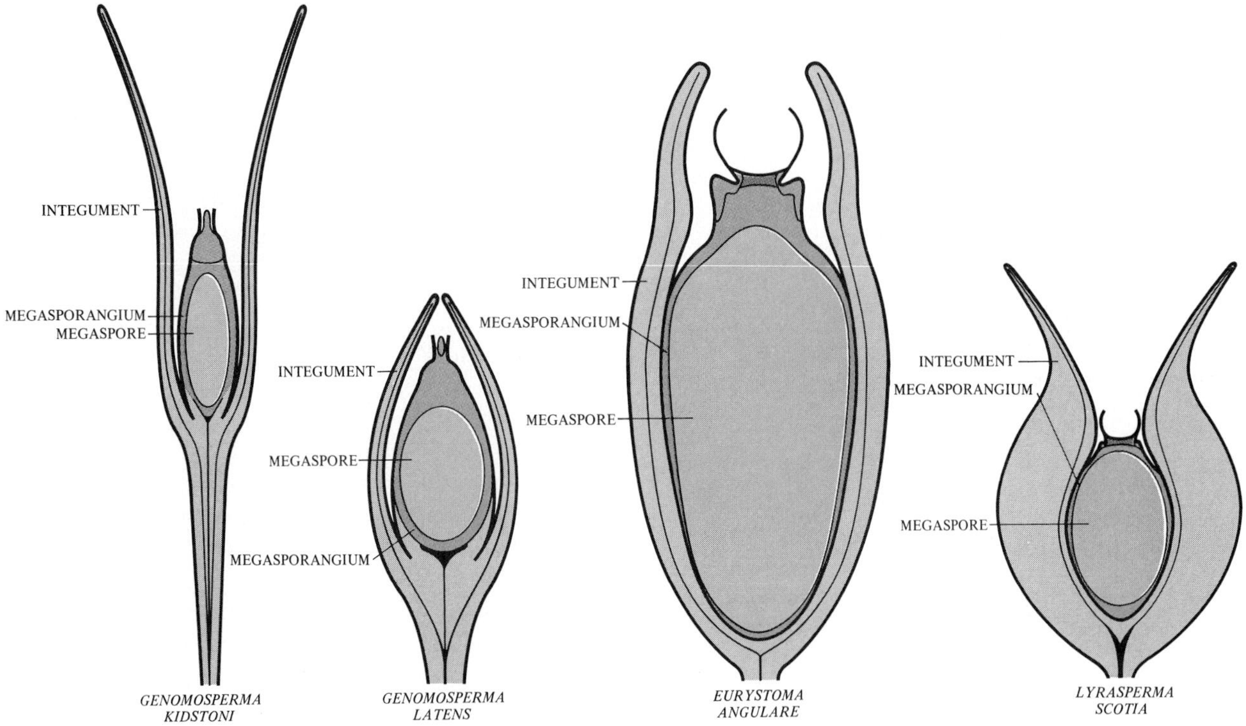

14–7
Seedlike structures in a number of Paleozoic plants showing stages in the evolution of seeds. In Genomosperma *(Greek* genomein, *to become, and* sperma, *seed), eight fingerlike processes arise at the base of the megasporangium. In* G. kidstonii, *these processes are free, in* G. latens, *partly fused together. In other Paleozoic seedlike bodies, these processes are fused partly or completely. The crucial events in the evolution of a seed are (1) the retention of the megagametophyte within the megasporangium of the parent sporophyte, and (2) the enclosure of the megasporangium in some sort of protective coat, the integument.*

megaspore, in turn, is retained within the megasporangium, commonly called the *nucellus* in seed plants. The megasporangia of seed plants, unlike those of other heterosporous plants, are enveloped by one or two layers of tissue, the *integuments.* This entire structure—the megasporangium plus its integument or integuments—is known as the *ovule.*

Following fertilization, the integuments develop into a seed coat, and a seed is formed. In other words, it is the ovule that develops into a seed. In most modern seed plants an embryo, or young sporophyte, develops within the seed before dispersal. In addition, all seeds contain stored food of some kind.

The evolution of the seed is one of the principal factors responsible for the dominance of seed plants in today's flora, because it permits the young sporophyte, protected by the seed coat and provided with a supply of stored food, to remain dormant until conditions are favorable for it to resume growth.

The oldest known seeds are found in late Devonian strata, about 350 million years old (Figure 14–8). During the remainder of the Paleozoic era and the Mesozoic era that followed, a wide variety of seed types evolved.

The angiosperms, which are overwhelmingly the most successful vascular plants at the present time, have not been identified with certainty in the fossil record until the Cretaceous period some 125 million years ago. Thus, they are relative newcomers in the broad picture of vascular plant evolution.

14-8
Reconstruction of a late Devonian plant (Archeosperma arnoldii) *with seedlike structures.* (a) *Branched stalk bearing four cupules in each of which a single megaspore is retained.* (b) *Megaspore. These fossils demonstrate that by about 350 million years ago, what were essentially seeds had evolved in this group of plants.*

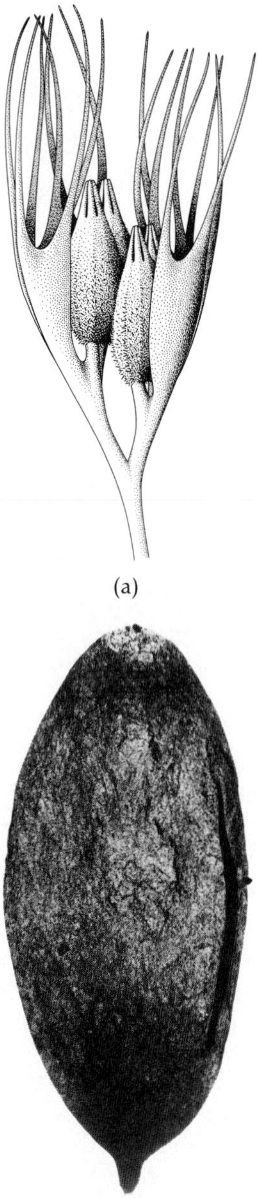

(a)

(b)

SUMMARY

Vascular plants are those plants containing xylem and phloem and exhibiting an alternation of generations in which the sporophyte is the dominant and nutritionally independent phase.

The earliest known vascular plants belong to the subdivision Rhyniophytina and are found in strata of the Silurian period, some 400 million years old. The plant bodies of the Rhyniophytina and their contemporaries were simple, dichotomously branching axes lacking roots and leaves. With evolutionary specialization, morphological and physiological differences arose between various parts of the plant body, bringing about the differentiation of root, stem, and leaf.

The plant bodies of many vascular plants consist entirely of primary tissues. Today, secondary growth is confined largely to the seed plants, although it occurred in several unrelated fossil groups. The primary vascular tissues and associated ground tissues exhibit two basic arrangements: (1) the protostele, which consists of a solid core of vascular tissue; and (2) the siphonostele, which contains a pith.

Roots evolved from the underground portions of the primitive plant body. The leaf originated in more than one way. Microphylls, univeined leaves whose leaf traces are not associated with leaf gaps, evolved as superficial lateral outgrowths of the stem. Megaphylls, leaves with complex venation and leaf traces associated with leaf gaps, evolved from branch systems. Microphyllous shoots are associated with protosteles, megaphyllous shoots with siphonosteles.

Vascular plants are either homosporous or heterosporous. Homosporous plants produce only one type of spore, which gives rise to a bisexual gametophyte. Heterosporous plants give rise to microspores and megaspores, which function as the precursors of male gametophytes and female gametophytes, respectively. The gametophytes of heterosporous plants are much reduced in size, compared with those of homosporous plants. In the history of the vascular plants, heterospory has evolved a number of times. There has been a long continued evolutionary trend toward the reduction in size and complexity of the gametophyte, culminating in the angiosperms. Primitive vascular plants have archegonia and antheridia, but these have been lost in a few gymnosperms and in all angiosperms.

A seed is a structure formed by maturation of an ovule. Following fertilization, an embryo, or young sporophyte, is formed and the integument or integuments of the ovule develop into a seed coat. The oldest known seed plants are known from the late Devonian period, about 350 million years ago.

SUGGESTIONS FOR FURTHER READING

BIERHORST, D. W.: *Morphology of Vascular Plants,* The Macmillan Company, New York, 1971.

An encyclopedic and magnificently illustrated treatise on the morphology of the vascular plants by one of their leading contemporary students.

FOSTER, A. S., and E. M. GIFFORD, JR.: *Comparative Morphology of Vascular Plants,* 2nd ed., W. H. Freeman and Company, San Francisco, 1974.

A well-organized, general account of the vascular plants, containing much summary and interpretative material.

CHAPTER 15

The Seedless Vascular Plants

15–1
Lady fern, Athyrium filix-foemina, *growing along the base of a fallen, moss-covered tree.*

The vascular plants can be divided artificially into two major groups, the seedless vascular plants and the seed plants. In the previous chapter, three entirely fossil subdivisions of seedless vascular plants were discussed. In this chapter, we shall consider the four major subdivisions of seedless vascular plants with living representatives: the Psilophytina, the Lycophytina, the Sphenophytina, and the Filicophytina. The largest group is the last, the ferns, with some 12,000 species throughout the world (Figure 15–1).

SUBDIVISION PSILOPHYTINA

The Psilophytina are represented by two living genera, *Psilotum* and *Tmesipteris. Psilotum* is tropical and subtropical in distribution and can be found growing in the United States in Florida, Louisiana, Arizona, Texas, and Hawaii. *Tmesipteris* is restricted in distribution to Australia, New Caledonia, New Zealand, and other islands of the South Pacific. At one time, *Psilotum* and *Tmesipteris* were included in the same subdivision as *Rhynia* (page 298) and *Psilophyton* (page 299). However, new paleobotanical information has brought about the reclassification of the latter two fossil genera. In addition, it has recently been suggested that *Psilotum* and *Tmesipteris* are, in fact, primitive ferns. Because of the simplicity of their sporophytes, they are placed in the separate subdivision Psilophytina in this book.

Psilotum is unique among living vascular plants in that it lacks both roots and leaves. The sporophyte consists of a dichotomously branching aerial portion with small scalelike outgrowths, and a rhizoid-bearing subterranean portion, or rhizome (Figure 15-2), which is mycorrhizal, the fungi appearing in the cortical cells. The stele of *Psilotum* (Figure 15–3) has been interpreted as a protostele by some morphologists and as a siphonostele by others. The problem lies in the nature of the tissue in the center of the stele—that is, whether it is xylem or sclerified ground tissue of a pith (page 300).

15–2
Whisk fern, Psilotum nudum. (a) *Habit (the characteristic form or appearance of an organism), showing rhizome and dichotomously branching aerial branches.* (b) *Detail of aerial branches, showing sporangia and scale-like outgrowths.*

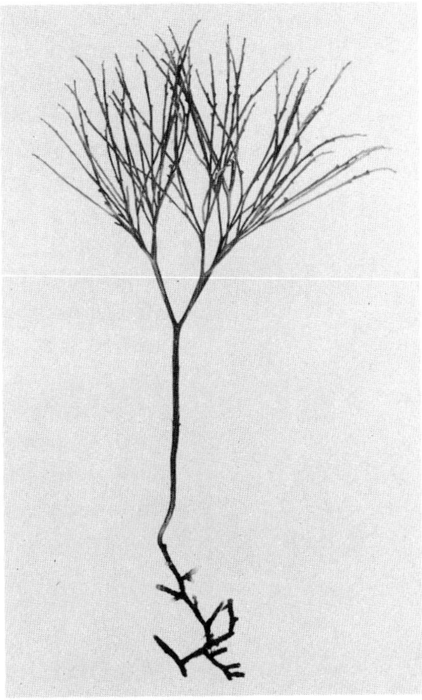

(a)

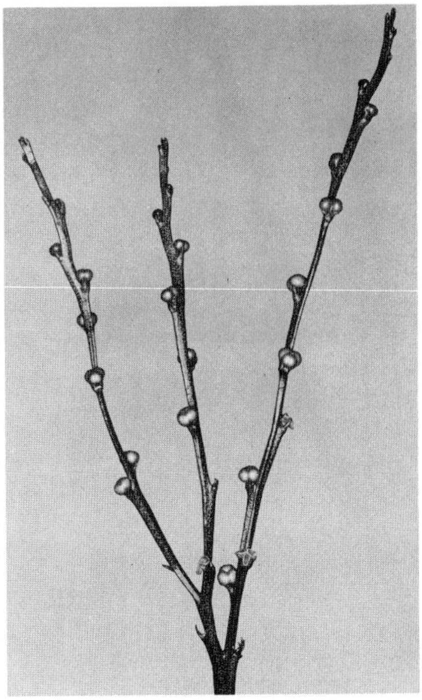

(b)

15–3
Psilotum nudum. (a) *Transverse section of stem, showing mature tissues.* (b) *Detail of protostele, showing xylem and phloem.*

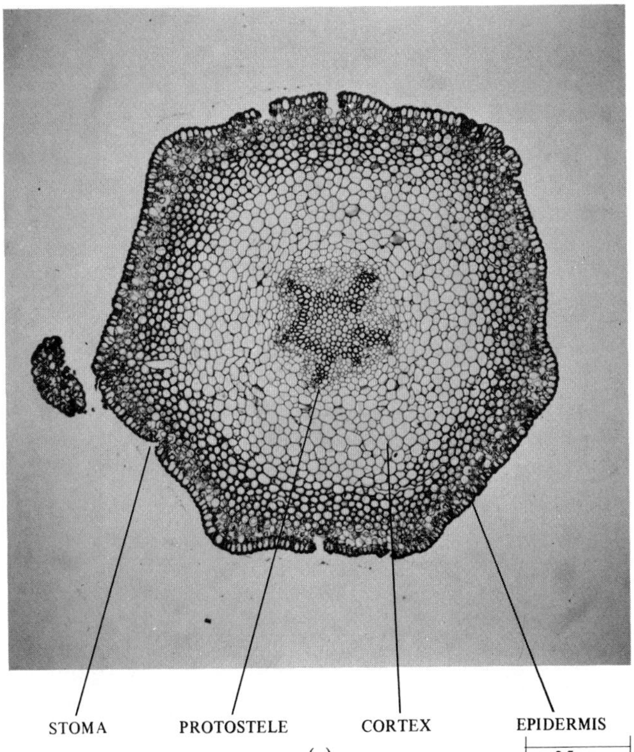

(a)

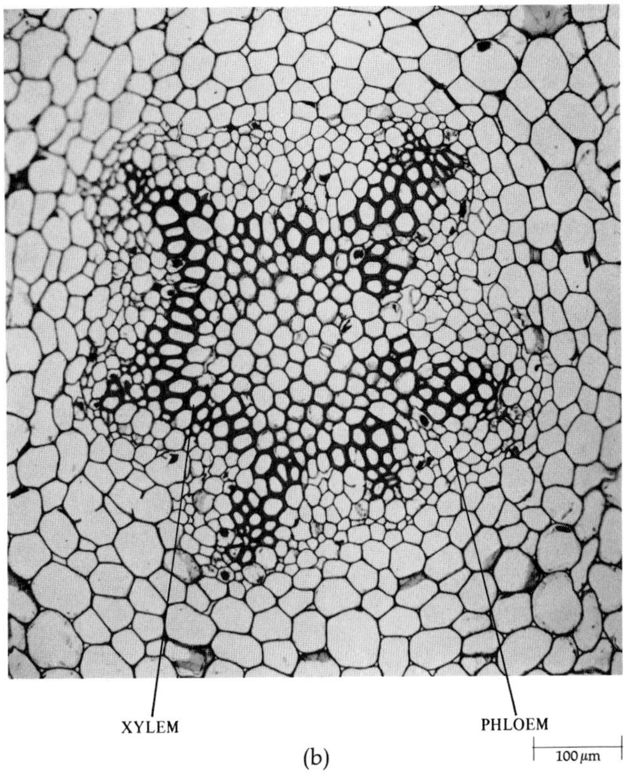

(b)

15-4
Psilotum nudum. *Longitudinal section through portions of two sporangia, showing mature spores. A scalelike appendage subtends the sporangia.*

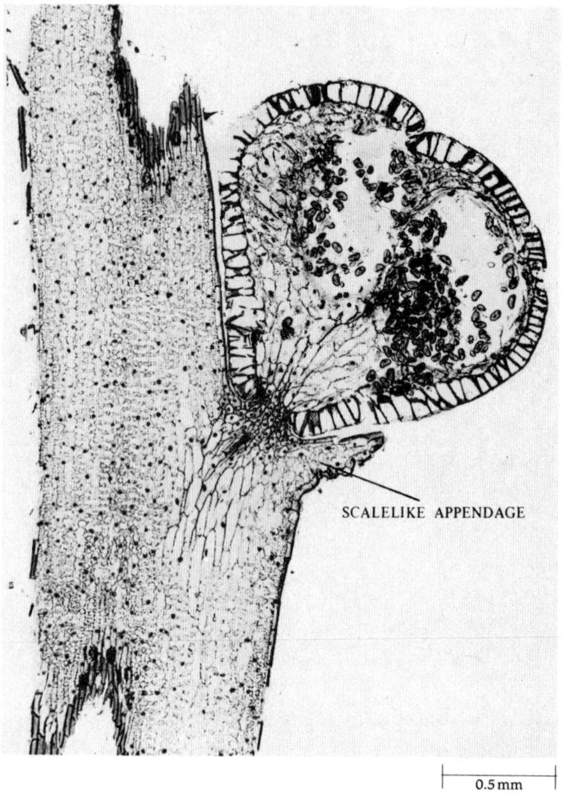

Psilotum is homosporous, the spores being produced in sporangia borne on the ends of short, lateral branches (Figure 15-4). Upon germination, the spores give rise to bisexual gametophytes, which resemble portions of the rhizome (Figure 15-5). Like the rhizome, the subterranean gametophyte contains a symbiotic fungus. In addition, some gametophytes contain vascular tissue. The sperms of *Psilotum* are multiflagellate and require water to swim to the egg. Initially the sporophyte is attached to the gametophyte by a foot. Eventually it becomes detached from the foot, which remains embedded in the gametophyte.

Tmesipteris grows as an epiphyte on tree ferns and other plants (Figure 15-6). The leaflike appendages of *Tmesipteris* are interpreted by some morphologists as flattened branchlets rather than as microphylls. In other respects, *Tmesipteris* is essentially similar to *Psilotum*.

SUBDIVISION LYCOPHYTINA

The five living genera and approximately 1000 living species of Lycophytina are the representatives of an evolutionary line that extends back to the Devonian period. It seems likely that the progenitors of the Lycophytina were *Zosterophyllum*-type plants (see Figure 14-2, page 298). The lycopods early split into two major groups. One group remained herbaceous and is still represented in today's flora. The second (the lepidoden-

15-5
Psilotum nudum. (a) *Gametophyte, which is bisexual—that is, it bears both antheridia and archegonia.* (b) *Detail of gametophyte.*

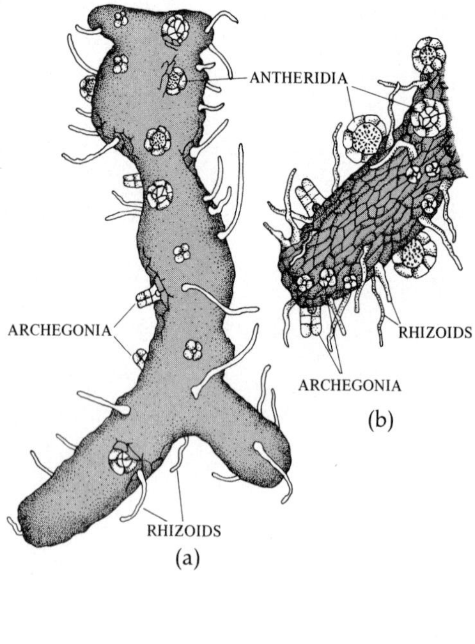

15-6
Tmesipteris tannensis *growing on the trunk of a tree fern in New Zealand.*

drids) became woody and treelike and was among the dominant plants of the coal-forming forests of the Carboniferous period (Figure 14–1, page 297). The lepidodendrids became extinct in the Permian period, about 280 million years ago. Some of them bore seedlike structures analogous to those of modern seed plants.

15–7
Lycopodium obscurum. *Habit, showing rhizome and aerial branches, with microphylls, of sporophyte. Note the adventitious roots arising from rhizome and the three young strobili at tips of stems, above.*

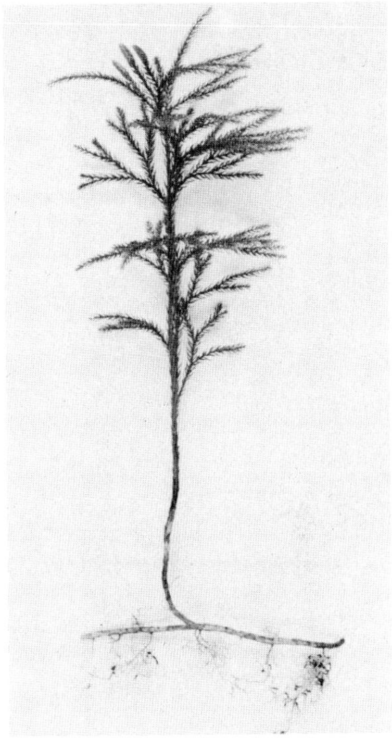

Lycopodium

Perhaps the most familiar living representatives of the Lycophytina are the club mosses, *Lycopodium*. The approximately 200 species of this genus extend from arctic regions into the tropics but rarely form conspicuous elements in any plant community. Most tropical species are epiphytes and are thus rarely seen, but several of the temperate species form mats which may be evident on forest floors. Because they are evergreen, they are most noticeable in winter.

The sporophyte of *Lycopodium* consists of a branching rhizome from which aerial branches and adventitious roots arise (Figure 15–7). The leaves of *Lycopodium*, which are usually arranged in spirals, are microphylls, and the stem and root are protostelic (Figure 15–8).

Lycopodium is homosporous. The sporangia occur singly upon the upper surface of fertile microphylls called *sporophylls*—modified leaf or leaflike organs that bear sporangia (Figure 15–9). In some species, the sporophylls are similar to ordinary microphylls and are interspersed among the sterile microphylls (Figure 15–10a). In others, nonphotosynthetic sporophylls are

THE FOSSIL FUELS

In the process of photosynthesis, plants and algae use the energy of the sun to convert carbon dioxide and water into carbohydrates. These carbohydrates are then oxidized either by the plants themselves, by heterotrophs that eat the plants, by decomposing or, less frequently, by burning (as in a forest fire). In the course of oxidation, carbon dioxide is again formed. The amount of carbon dioxide involved in photosynthesis on an annual basis is about 100 billion metric tons. The amount returned as carbon dioxide as a result of oxidation of these living materials is about the same, differing only by 1 part in 10,000. This very slight imbalance is caused by the burying of organisms in sediment or mud under conditions in which oxygen is excluded and decay is only partial. This accumulation of partially decayed material is known as peat. The peat may eventually become covered with sedimentary rock and so placed under pressure. Depending on time, temperature, and other factors, peat may become compressed into soft or hard coal, petroleum, or natural gas—the so-called fossil fuels.

During certain periods in the earth's history, the rate of fossil fuel formation was greater than at other times. One such period was the Carboniferous, some 300 million years ago (see Figure 14–1). The lands were low, covered by shallow seas or swamps and, in what are now the temperate regions, conditions were favorable for growth year round. The principal large plants were ferns, tree-sized lycophytes and horsetails, and seed ferns, and other primitive gymnosperms; their fossils are found near the large coal seams of this period.

Although man has burned peat and coal for domestic uses for several thousand years, fossil fuels have been mined and consumed on a large scale only since about 1900. The growth of industries, of cities, and of the human population closely parallels the great increase in fuel consumption. Although fossil fuel is still being formed, the rate of formation is so slow that, for all practical purposes, supplies are not renewable and so finite. As the resources become depleted, the cost in energy of extracting the fuel will increase until the process is no longer economically attractive, and at that time, alternative sources will be vigorously sought. Many have been suggested, including the direct use of solar energy, harnessing wind and flowing water (actually indirect uses of solar energy), the surge of the tides, or the heat energy beneath the surface of the earth. The most plausible new energy source is nuclear power which, like the energy of fossil fuels, has its source in reactions that took place long ago in the heart of the sun.

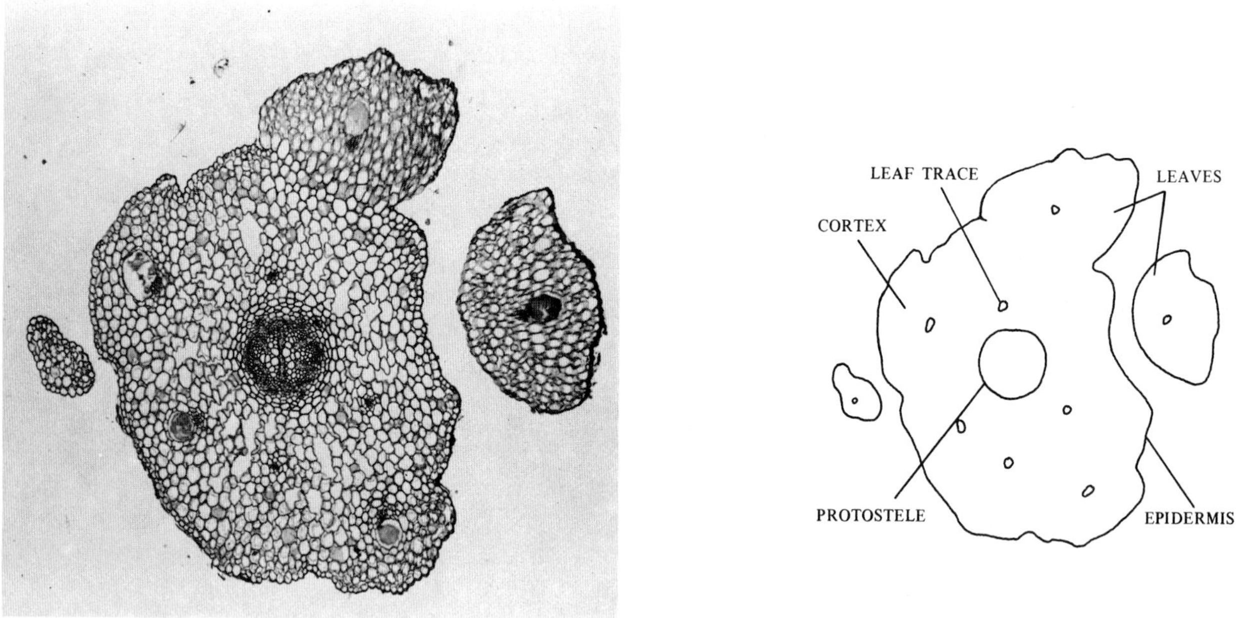

(a) 0.25 mm

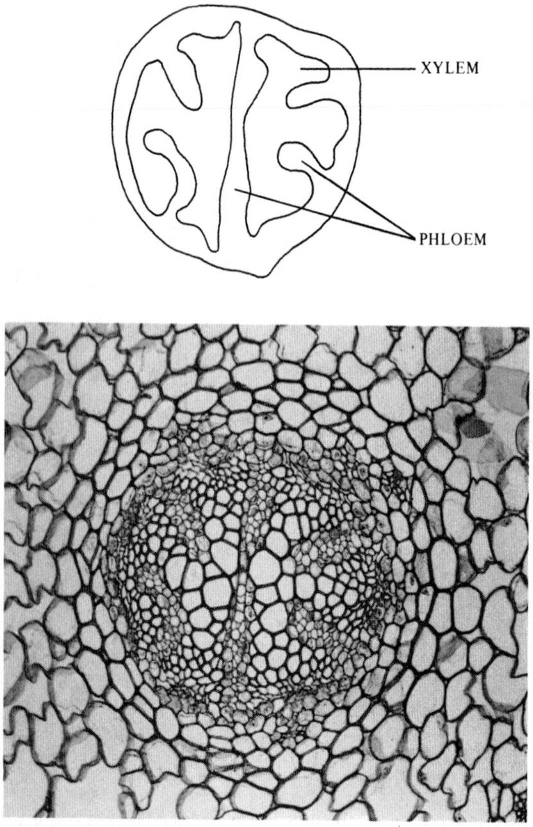

(b) 100 μm

15–8
Lycopodium. (a) *Transverse section of stem showing mature tissues. Leaf trace is a vascular bundle connecting vascular tissues of stem with those of base of leaf.* (b) *Detail of protostele, showing xylem and phloem.*

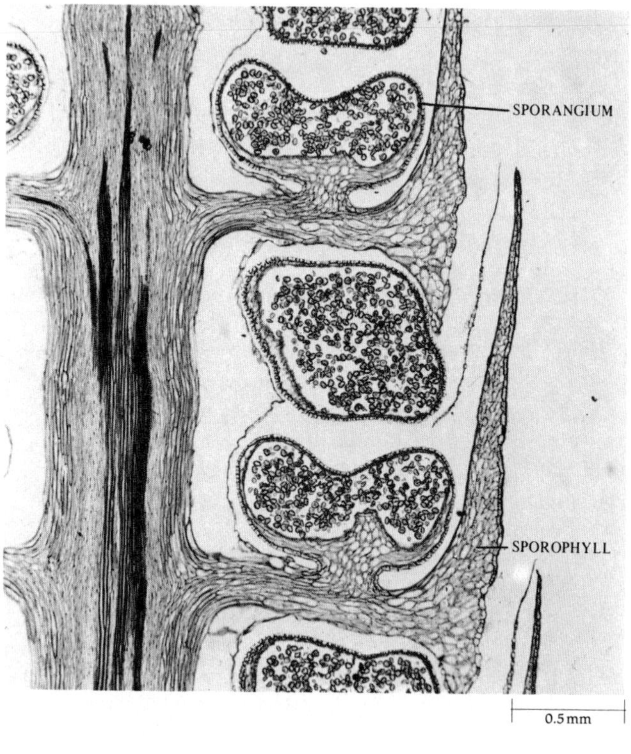

0.5 mm

15–9
Lycopodium. *Longitudinal section showing portion of mature strobilus (cone at the end of an aerial branch which bears sporophylls).* Lycopodium *is homosporous, producing only one type of spore. Each sporophyll bears a single sporangium. Note that all of the spores are of similar size.*

grouped into cones (strobili) at the ends of the aerial branches (Figure 15-10b).

Upon germination, the spores of *Lycopodium* give rise to bisexual gametophytes, which, depending on the species, are either green, irregularly lobed masses or branching, subterranean, and nonphotosynthetic variable structures. Like the gametophytes of *Psilotum* and *Tmesipteris*, those of the species of *Lycopodium* with subterranean gametophytes contain a symbiotic fungus. The development and maturation of archegonia and antheridia in a *Lycopodium* gametophyte requires 6 to 15 years. Some subterranean forms reportedly live for as long as 25 years, and they may even produce a series of sporophytes in successive archegonia as they continue to grow.

Water is required for fertilization, the biflagellated sperms swimming through water to the archegonium and then down its neck. Following fertilization, the zygote develops into an embryo, which grows within the venter of the archegonium. The young sporophyte may remain attached to the gametophyte for a long time (Figure 15-11), but eventually it becomes an independent plant.

The life cycle of *Lycopodium* is summarized in Figure 15—12.

(a)

(b)

15-10
Some species of Lycopodium *have their sporophylls grouped in strobili, whereas others have them interspersed among sterile microphylls.* (a) Lycopodium lucidulum *lacks strobili.* (b) Lycopodium clavatum *with strobili.*

15-11
Lycopodium. *Gametophyte with young sporophyte attached to it.*

15–12
Life cycle of Lycopodium. Lycopodium *is homosporous—that is, the sporophyte produces only one sort of spore as a result of meiosis. The gametophyte developing from this spore then produces both archegonia, each of which contains an egg, and antheridia, each of which produces many biflagellated motile sperms. The gametophytes of some species of* Lycopodium *are subterranean and require the presence of a symbiotic (mycorrhizal) fungus for normal growth. Water is necessary for fertilization to take place, the sperm swimming through it to the archegonium. The zygote is formed and begins its development within the archegonium, as in some seed plants.*

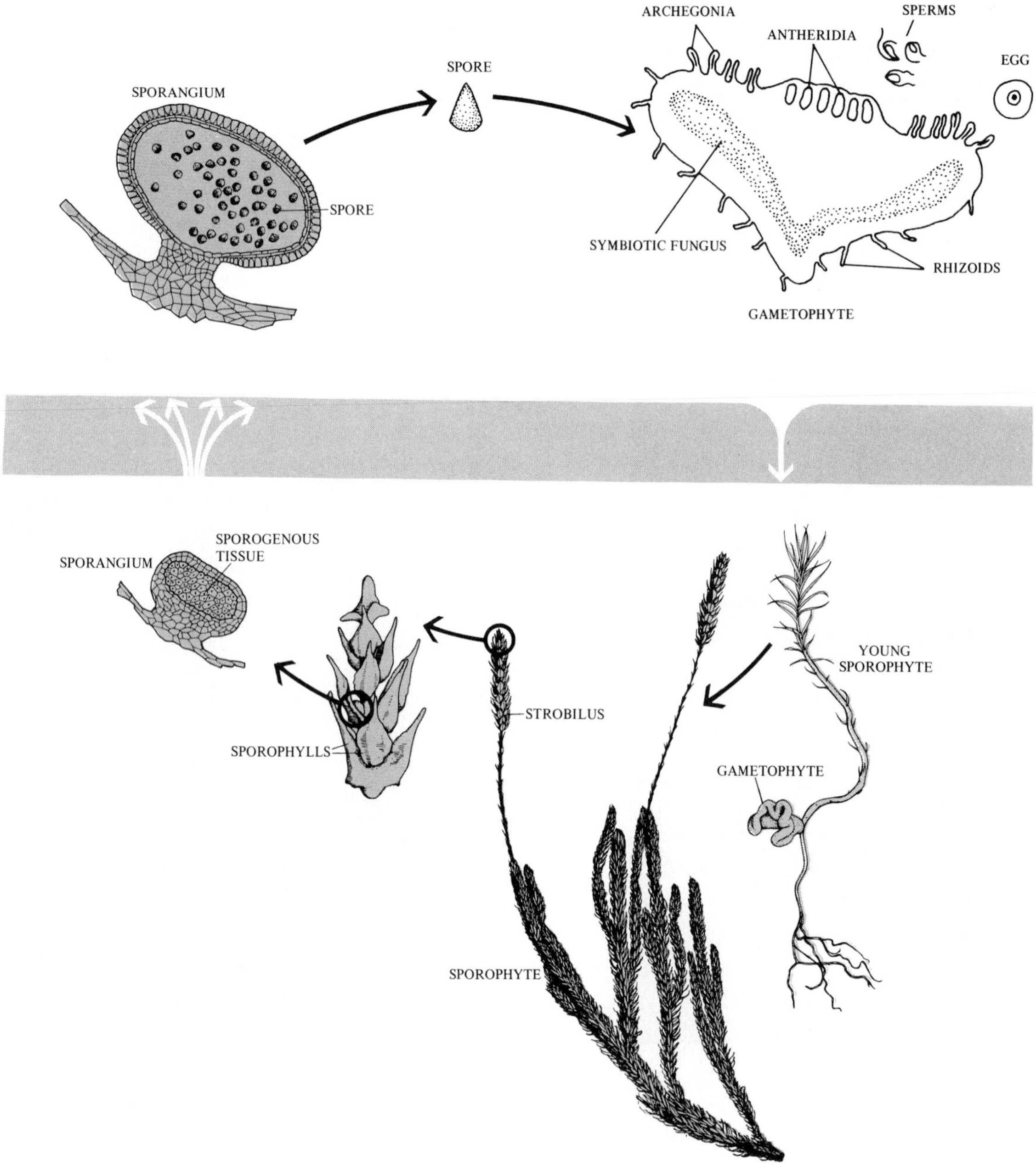

(a)

(b)

15–13
Selaginella. (a) *Habit of* Selaginella kraussiana, *a prostrate, creeping plant. Adventitious roots can be seen arising from stem bearing microphylls.* (b) Selaginella rupestris *with strobili.*

Selaginella

Among the living genera of lycopods, *Selaginella* has the most species, with about 700. These are mainly tropical in distribution. Many grow in moist situations; a few inhabit desert regions but become dormant during the driest part of the year. Among the latter is the so-called resurrection plant, *Selaginella lepidophylla,* which is native to Mexico and ranges north to Texas and New Mexico.

Basically, the herbaceous sporophyte of *Selaginella* is similar to that of *Lycopodium,* bearing microphylls and having sporophylls arranged in strobili (Figure 15–13). Unlike *Lycopodium,* in *Selaginella* a small scalelike outgrowth, called a ligule, develops near the base of the upper surface of each microphyll and sporophyll (Figure 15–14). The stem and root are protostelic (Figure 15–15).

Whereas *Lycopodium* is homosporous, *Selaginella* is heterosporous. This constitutes the most significant difference between the two genera. Each sporophyll bears a single sporangium on its upper surface. Megasporangia are borne by *megasporophylls* and microsporangia by *microsporophylls.* Both kinds of sporangia occur in the same strobilus (Figure 15–14).

Being heterosporous, *Selaginella* produces unisexual gametophytes (Figure 15–16a). The male gametophytes (microgametophytes) develop from microspores. Four microspores are produced by meiosis from each microspore mother cell. At maturity the male gametophyte consists of a single prothallial or vegetative cell and an antheridium, which gives rise to many biflagellated sperms. The male gametophyte develops within the microspore and lacks chlorophyll. The microspore wall must rupture for the sperms to be liberated.

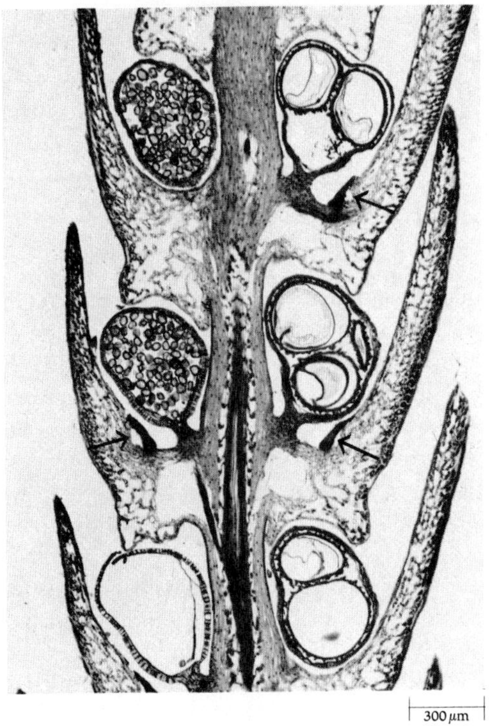

15–14
Selaginella. *Portion of strobilus, showing microsporangia with microspores on left and megasporangia with megaspores on right. Ligule (arrows) can be seen at the base of some sporophylls.*

15–15
Selaginella. (a) *Transverse section of stem, showing mature tissues. The protostele is suspended in the middle of the stem by elongate cortical cells (endodermal cells), called trabeculae.* (b) *Detail of protostele, showing xylem and phloem.*

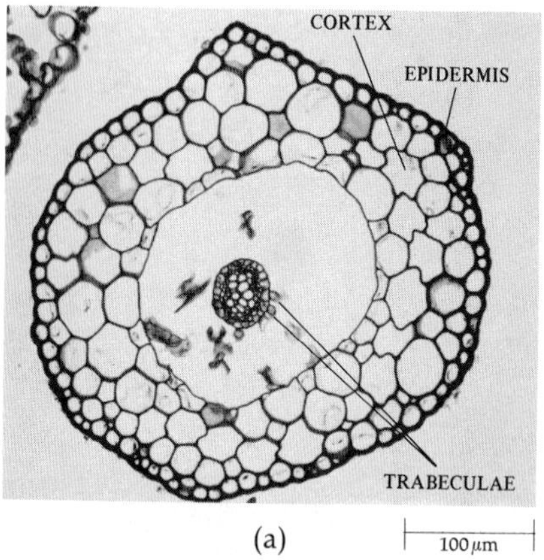

(a)

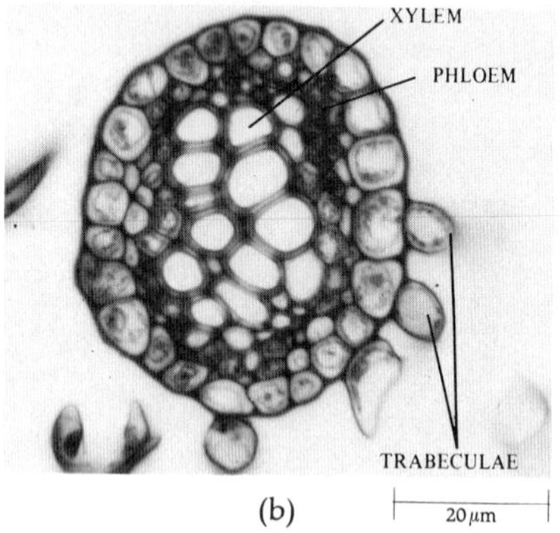

(b)

During development of the female gametophyte (megametophyte), the megaspore wall ruptures, and the gametophyte protrudes through the rupture to the outside. This is the portion of the female gametophyte in which the archegonia develop. Although it has been reported that the female gametophytes sometimes develop chloroplasts, it is likely that the gametophytes derive their nutrition largely from food stored within the megaspores.

Water is required for the sperms to swim to the archegonia and fertilize the eggs. Commonly this occurs after the gametophytes have been shed from the strobilus. During development of the embryos of both *Lycopodium* and *Selaginella*, a structure called a suspensor is formed. Although inactive in *Lycopodium* and some species of *Selaginella*, in other *Selaginella* species it serves to thrust the developing embryo deep within the nutrient-rich tissue of the female gametophyte. Gradually, the developing sporophyte emerges from the gametophyte (Figure 15–16b) and becomes an independent plant.

The life cycle of *Selaginella* is summarized in Figure 15–17.

15–16
Selaginella. (a) *Female (♀) and male (♂) gametophytes. The much larger female gametophyte has been removed from the megaspore, a portion of the wall of which can be seen below.* (b) *Young sporophyte emerging from female gametophyte. Note many microspores on leaf and root of sporophyte.*

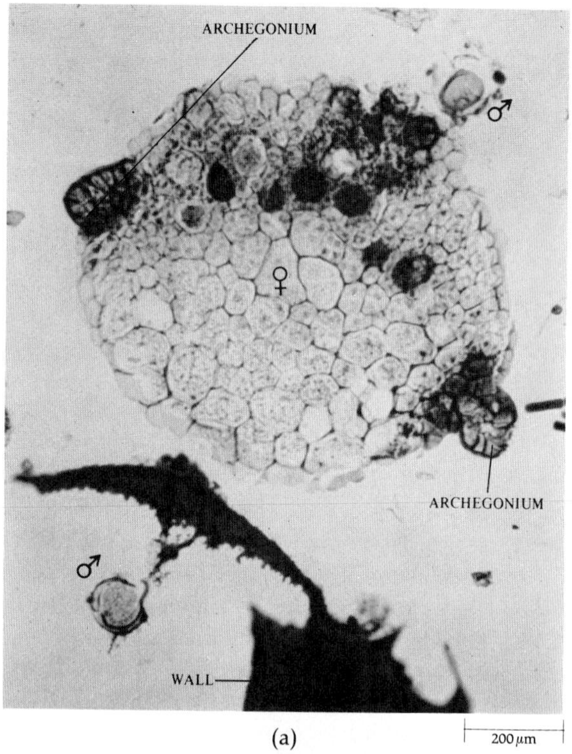

(a)

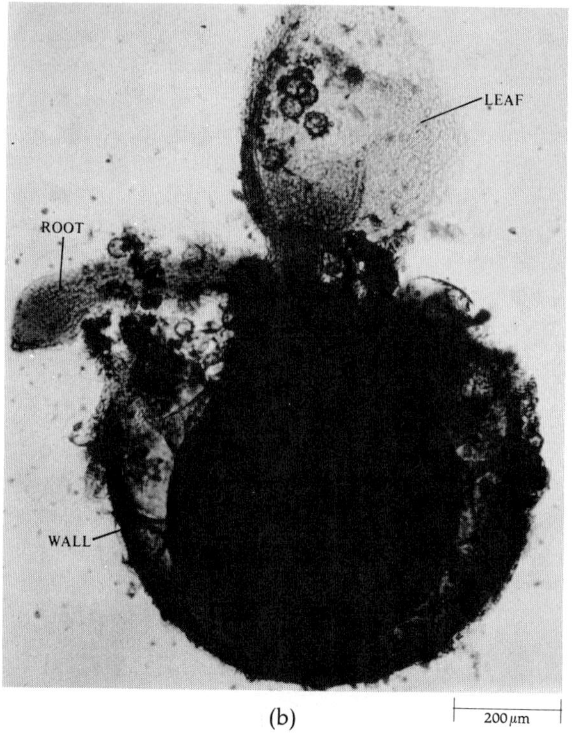

(b)

15–17
The life cycle of Selaginella, *which is heterosporous. Two kinds of sporangia are borne on the sporophyte and give rise to two morphologically distinct kinds of gametophyte, both much smaller than the sporophyte. As in the seed plants, the young sporophyte is enclosed by the tissues of the megagametophyte, and the major source of food for the developing embryo is material stored in the megaspore. There is no dormant period in the development of the embryo of* Selaginella *as there is in many seed plants, however.*

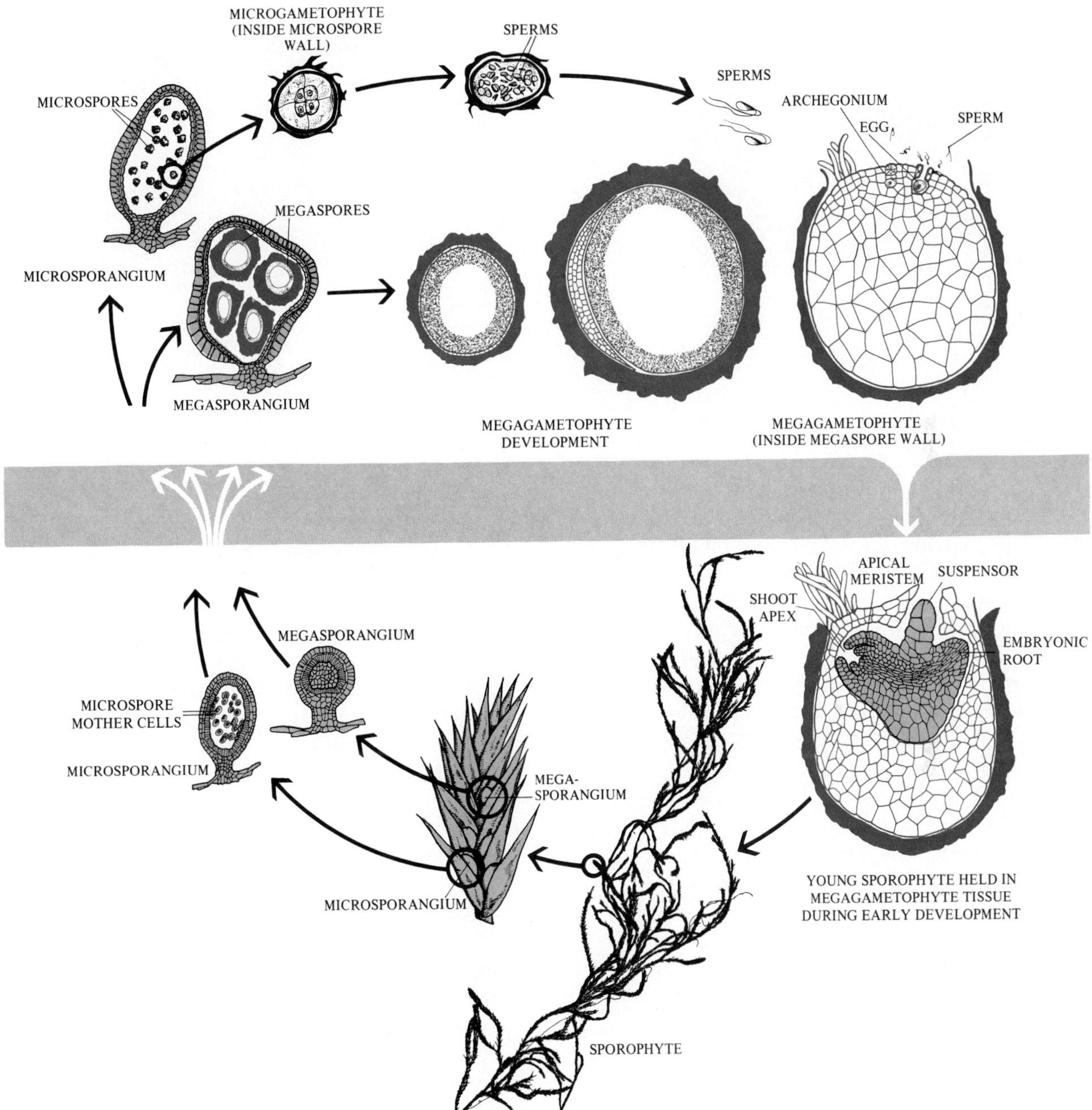

15–18
Isoetes muricata. *Habit, showing quill-like leaves, stem, and roots.*

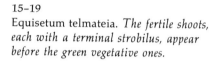

15–19
Equisetum telmateia. *The fertile shoots, each with a terminal strobilus, appear before the green vegetative ones.*

Isoetes

A very distinctive member of the Lycophytina is *Isoetes*, the quillwort. The sporophyte of this aquatic or amphibious plant consists of a short, fleshy undergound stem bearing quill-like leaves on its upper surface and roots on its lower portion (Figure 15–18). Like *Selaginella*, *Isoetes* is heterosporous. The megasporangia are borne at the base of certain leaves, and the microsporangia on leaves nearer the center of the plant in a similar position.

Although the leaves of *Isoetes* are relatively long, they are considered to be microphylls—they are univeined, and their leaf traces are not associated with leaf gaps. *Isoetes* is one of the few living seedless vascular plants with a vascular cambium and secondary tissues.

SUBDIVISION SPHENOPHYTINA

Like the Lycophytina, the Sphenophytina, or horsetails, extend back to the Devonian period. The sphenophytes reached their maximum abundance and diversity late in the Paleozoic era, about 300 million years ago. During the upper Devonian and Carboniferous periods, they were represented by trees that reached over 2 decimeters in diameter and 15 meters or more in height. Today the Sphenophytina are represented by a single herbaceous genus, *Equisetum*, with 15 species.

Equisetum is widespread in distribution and is often found in moist or damp places, by streams, or along the edge of woods (Figure 15–19). The horsetails are easily recognized because of their conspicuously jointed stems and rough texture. The small, scalelike leaves, which by definition are microphylls, are whorled at the nodes. When present, the branches arise laterally at the nodes and alternate with the leaves. The internodes (the portions of the stems between successive nodes) are ribbed, and the ribs are tough and strengthened with siliceous deposits in the epidermal cells. The unbranched species have been used in cleaning pots and pans, particularly in colonial and frontier times, and earned the name "scouring rushes." The roots are adventitious, arising at the nodes of the rhizomes.

The aerial stems of *Equisetum* arise from branching underground rhizomes and, although they may die back during unfavorable seasons, the rhizomes are perennial. Anatomically, the structure of the aerial stem is quite complex (Figure 15–20). At maturity, the internodes contain a hollow pith, surrounded by a ring of smaller canals. Each of these smaller canals is associated with a discrete strand of primary xylem and primary phloem (Figure 15–21).

Equisetum is homosporous. The sporangia are borne in groups of 5 to 10 on the margins of umbrellalike structures known as *sporangiophores* (sporangia-bearing branches), which are clustered into strobili at the apex of the stem (Figure 15–22). The fertile stems of some species

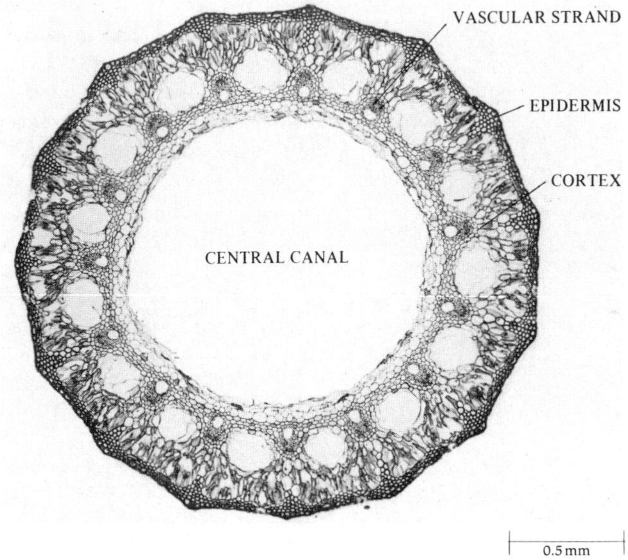

15–20
Equisetum. *Transverse section of stem, showing mature tissues.*

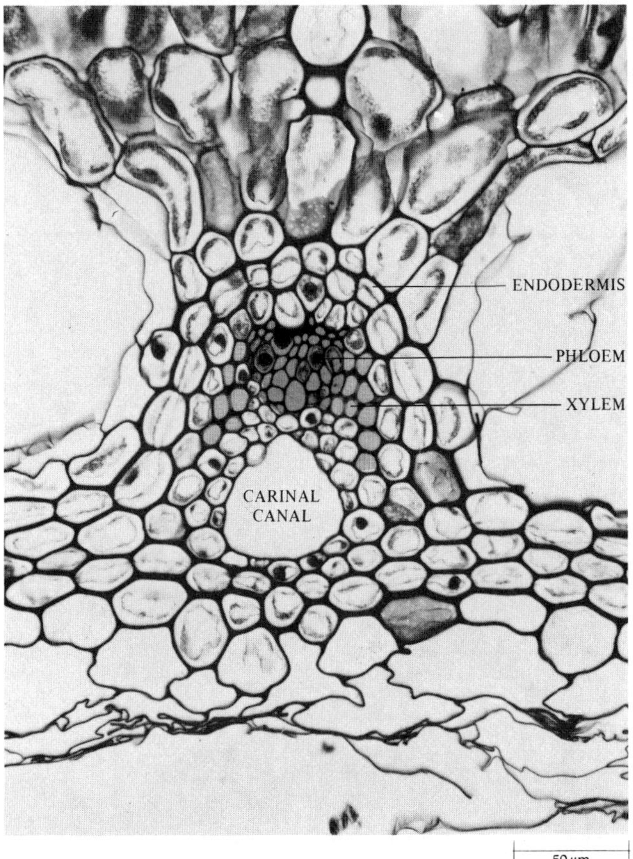

15–21
Equisetum. *Detail of vascular strand, showing xylem and phloem.*

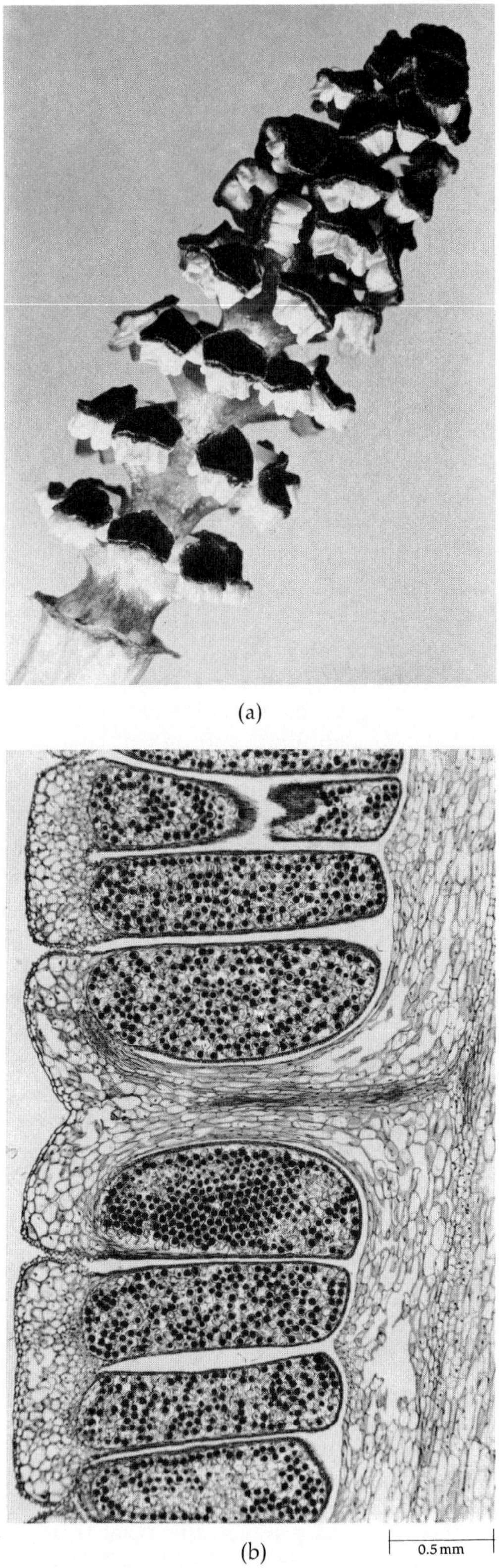

15–22
Equisetum arvense. (a) *Strobilus, showing sporangiophores with sporangia.* (b) *Longitudinal section of strobilus, showing sporangiophores and immature sporangia.*

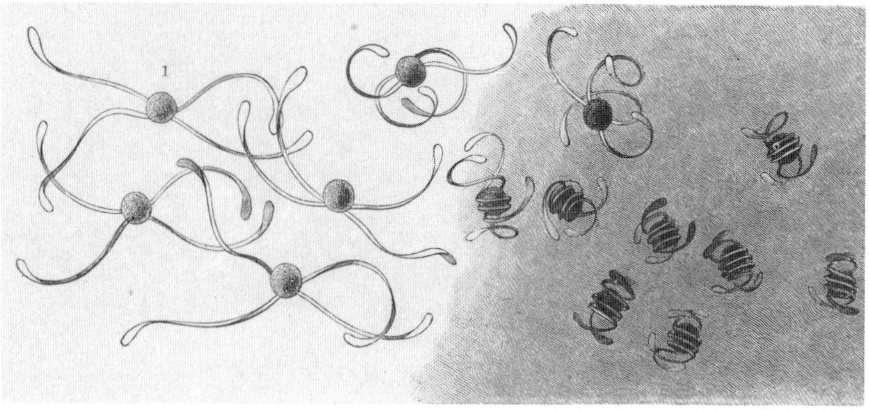

15-23
Spores of Equisetum telmateia *in dry* (left) *and moist* (right) *conditions. If a spore falls on the ground too dry to permit germination, its elaters remain outspread and it is carried off again by the wind.*

do not contain chlorophyll, and these stems are sharply distinct from the vegetative ones, often appearing before them early in spring. In other species, the strobili are borne at the tips of normal vegetative stems. When the spores are mature, the sporangia contract and split along their inner surface releasing numerous spores. Elaters, which are differentiated from the outer layer of the spore wall, coil when moist and uncoil when dry, thus presumably playing a role in spore dispersal (Figure 15-23).

The gametophytes of *Equisetum* are green and free-living, most being about the size of a pinhead. The gametophytes are of two types: either strictly male or bisexual. In the bisexual gametophytes, the archegonia are produced before the antheridia (Figure 15-24). The sperms are multiflagellated and require water to swim to the eggs. The eggs of several archegonia may be fertilized and develop into embryos, or young sporophytes, on a single gametophyte.

15-24
Equisetum. *Gametophyte, showing sex organs and rhizoids.*

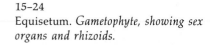

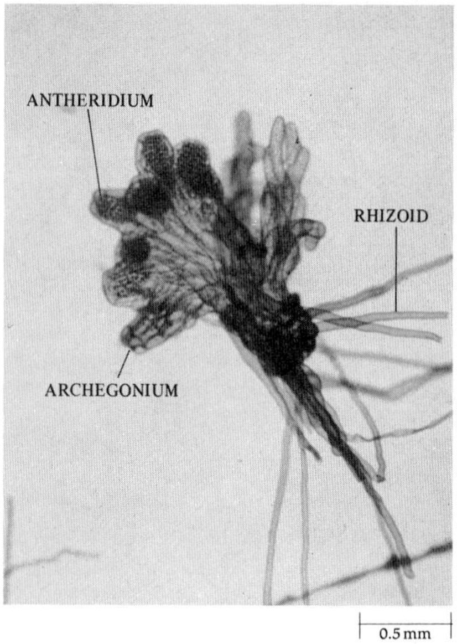

SUBDIVISION FILICOPHYTINA

The ferns are unique among seedless vascular plants in their possession of megaphylls. This characteristic has been used by some botanists to place the ferns in the same subdivision as the seed plants, which also have megaphylls. Ferns have been relatively abundant in the fossil record from the Carboniferous period to the present but are not known from the Devonian. About two-thirds of the approximately 12,000 living species are found in tropical regions, with the other third in temperate regions of the globe, including desert areas.

In both form and habitat, ferns exhibit great diversity (Figure 15-25). Some ferns are quite unfernlike in appearance; for example, the very small aquatic *Azolla* have leaves about 1 centimeter long (Figure 15-25d). At the other extreme in size are tree ferns (Figure 15-25b and 15-28), such as those of the genus *Cyathea,* some of

(a)

(b)

(c)

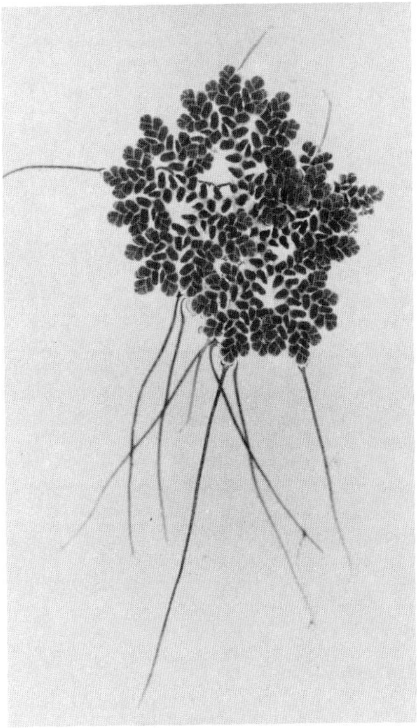

(d)

(e)

15–25
Ferns. (a) *Cinnamon fern,* Osmunda cinnamomea. (b) *Tree fern,* Cyathea arborea, *from the West Indies.* (c) *Hart's-tongue fern,* Phyllitis scolopendrium, *growing on mossy limestone at Owen Sound, Ontario.* (d) Azolla caroliniana, *a heterosporous water fern.* (e) Marsilea, *another heterosporous water fern.*

15–26
Adiantum. *Transverse section of rhizome with siphonostele. Note wide leaf gap.*

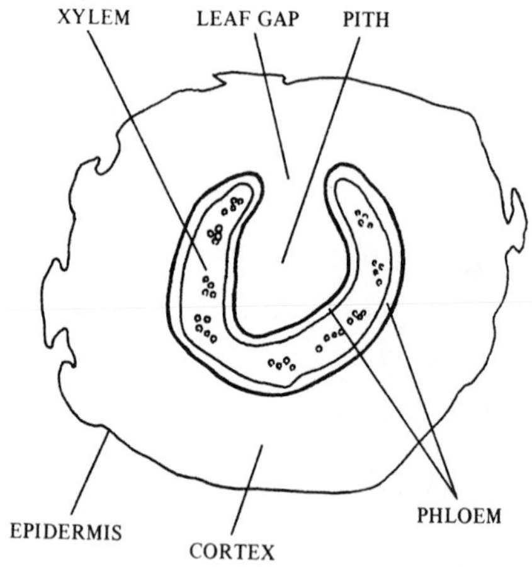

which have been recorded to reach heights of more than 24 meters and with leaves 5 meters or more in length. Although the trunks of such tree ferns may be 3 decimeters or more in diameter, their tissues are entirely primary in origin. Only *Botrychium,* the relatively small "grape fern" (Figure 15–29d), is known to have a vascular cambium.

Most garden and woodland ferns of temperate regions consist of fleshy, underground siphonostelic rhizomes (Figures 15–26 and 15–27), which produce new sets of leaves each year. The roots are adventitious, arising near the bases of the leaves from the rhizome. The leaf, or *frond,* is the conspicuous part of the sporophyte. Commonly the fronds are compound; that is, the lamina is divided into leaflets, or *pinnae,* which are attached to the *rachis,* an extension of the leaf stalk, or petiole. In nearly all ferns (bracken is a common exception), the young leaves are coiled in the bud and commonly are referred to as "fiddleheads." This type of leaf development is known as *circinate vernation* (Figure 15–28). It results from more rapid growth on the lower than the upper surface of the leaf early in development, mediated by the hormone auxin produced by the young pinnae on the inner side of the fiddlehead.

All but a few genera of ferns are homosporous. The sporangia are variously placed—on the lower surface of the leaves, on specially modified leaves, or on separate stalks. The sporangia commonly occur in clusters called *sori.* In many genera, the sori are covered by specialized outgrowths of the leaf, the *indusia,* which may shrivel when the sporangia are ripe (Figures 15–29 and 15–30).

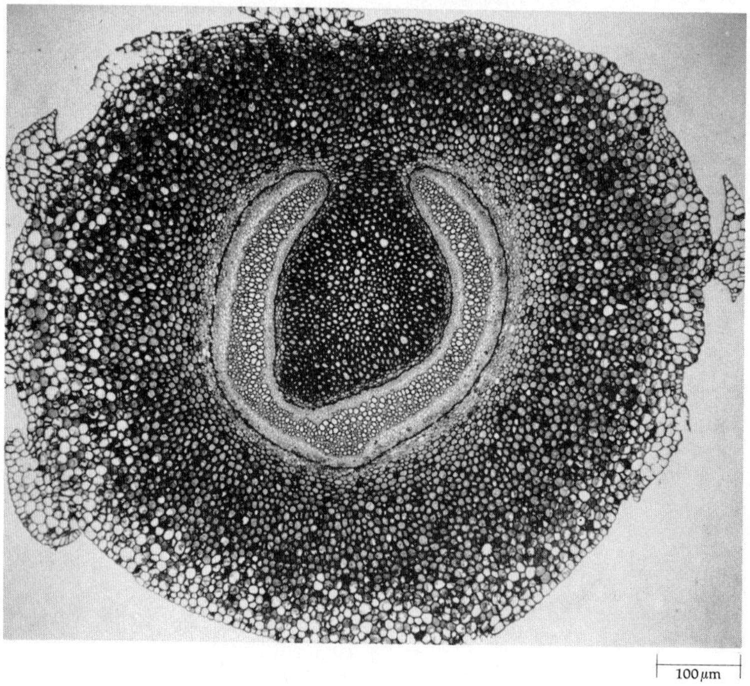

15-27
Dicksonia. *Transverse section of part of vascular region of rhizome. The phloem is composed mainly of sieve elements; the xylem is composed entirely of tracheids (see Chapter 19).*

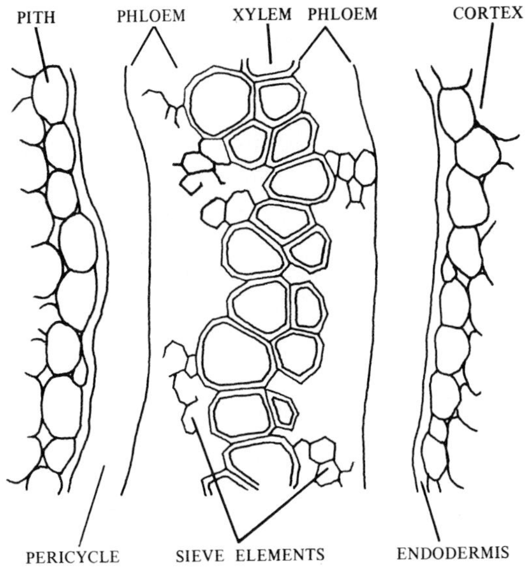

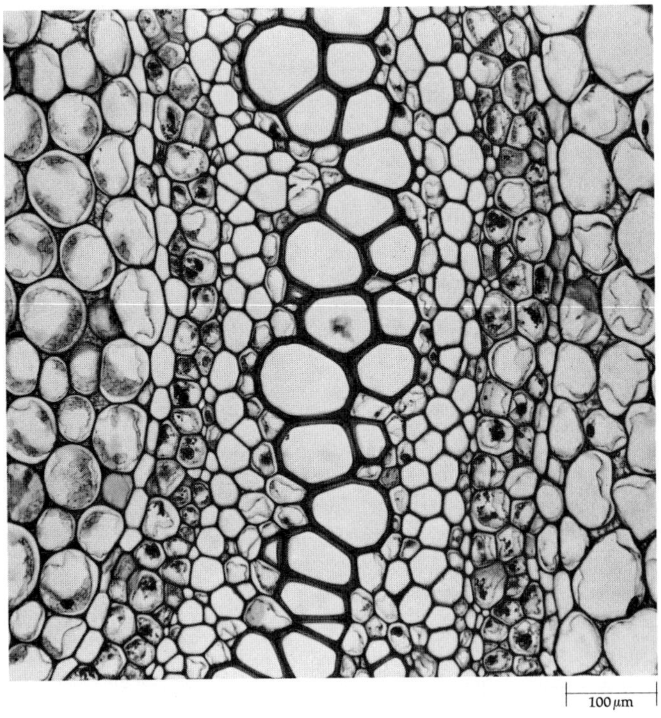

15-28
"Fiddleheads" of Cyathea australis, *a tree fern. Years ago, the young coiled leaves of ferns were served as a delicacy in the best hotels in New England and New York.*

15–29
Sori are clusters of sporangia on the undersides of fern leaves. (a) *In the common polypody* (Polypodium virginianum) *and other ferns of this genus, the sori are bare.* (b) *In* Pellaea glabella, *shown here, as well as in the bracken fern* (Pteridium aquilinum) *and the maidenhair ferns* (Adiantum), *the sori are located along the margins of the leaf blades, which are rolled back over them.* (c) *In the evergreen wood fern* (Dryopteris marginalis), *the sori, which are also located near the margins of the leaf blades, are completely covered by kidney-shaped indusia.* (d) *In* Botrychium virginianum, *the rattlesnake fern, the sporangia, which are globular, are borne on stalks that differ greatly from the leaves. Many sporangia have specialized dehiscence mechanisms and walls that are only a single layer of cells thick. However, in some more primitive groups of ferns, such as that to which* Botrychium *belongs, the sporangia split along a simple line of thin-walled cells and have walls that are several layers of cells thick.*

15–30
Cyrtomium falcatum, *a homosporous fern. Transverse section of leaf, showing sorus on lower surface. Sporangia are in different stages of development and are protected by umbrellalike indusium.*

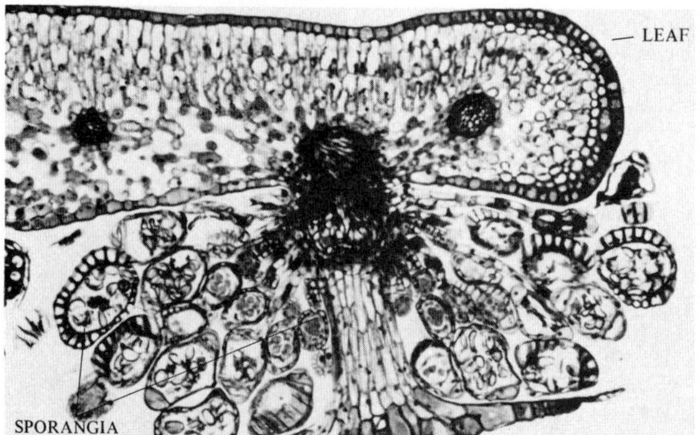

15-31
Male (left) and female (right) gametophytes of the fern Platyzoma microphyllum *from northern Australia. This species of* Platyzoma *is unique in that it produces spores of two size classes. Spores of intermediate size are also formed.*

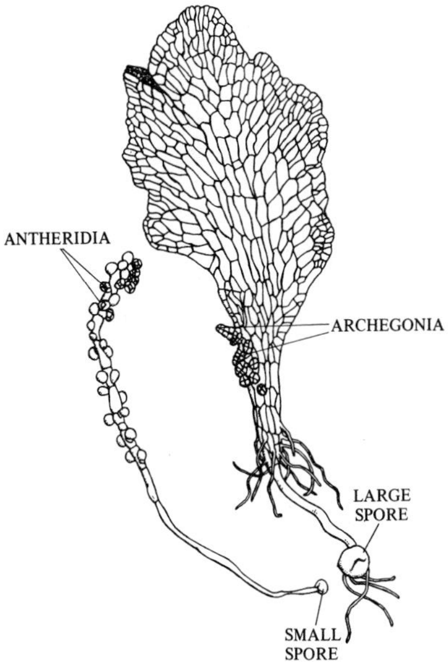

15-32
Osmunda. *Gametangia.* (a) *Antheridia. The sperms are nearly mature in the two middle antheridia.* (b) *Archegonium.*

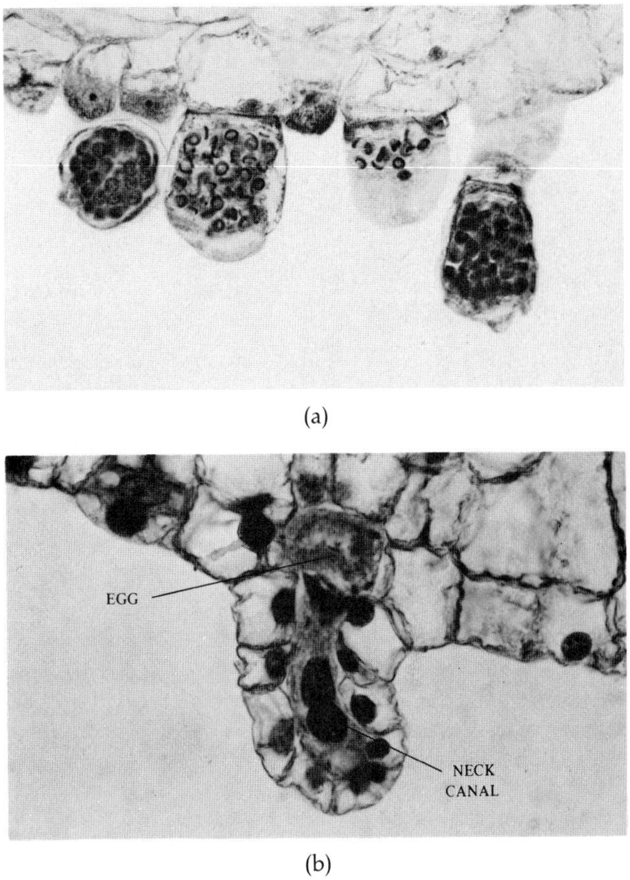

15-33
Mature sperms of the fern Marsilea. *Each sperm possesses more than 100 flagella.*

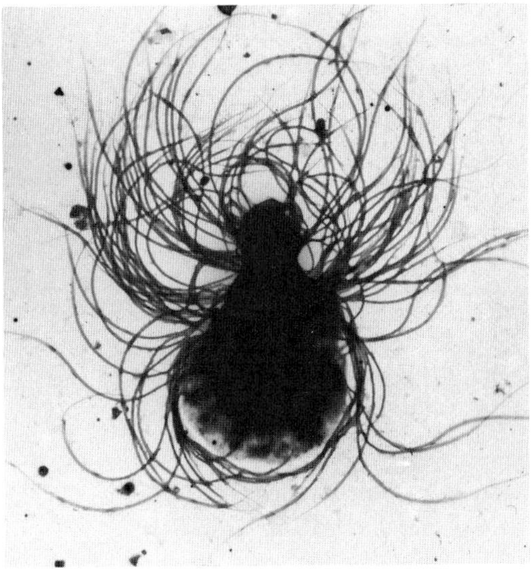

At that time the mature spores—the result of meiosis in the spore mother cells—are ejected through a crack in the so-called lip cells of the sporangium. The sporangia are stalked, and each contains a special layer of unevenly thick-walled cells called an annulus. Contraction of the annulus causes tearing of the lip cells. Sudden expansion of the annulus then results in a catapultlike discharge of the spores.

Heterospory in ferns is restricted to two specialized groups of water ferns, although a number of extinct ferns were heterosporous. The living genus *Platyzoma*, a peculiar fern native to northern Australia, represents a stage intermediate to the beginning of heterospory (Figure 15-31).

The spores of most homosporous ferns give rise to green, free-living bisexual gametophytes. The gametophyte begins development as a small, pale green, alga-like chain of cells called the germ filament or protonema. It then develops, under the influence of blue light, into a flat, heart-shaped membranous structure, the prothallus, with numerous rhizoids on its lower, or ventral, surface. Both antheridia and archegonia develop on the lower surface of the prothallus (Figure 15-32). Water is required for the multiflagellated sperms to swim to the eggs (Figure 15-33).

Early in development, the embryo or young sporophyte receives its nutrition from the gametophyte via a foot (Figure 15-34). However, development is rapid, and the sporophyte soon becomes an independent plant. The gametophyte then disintegrates.

The fern life cycle is summarized in Figure 15-35.

15–34
Young fern sporophyte about to emerge from gametophyte.

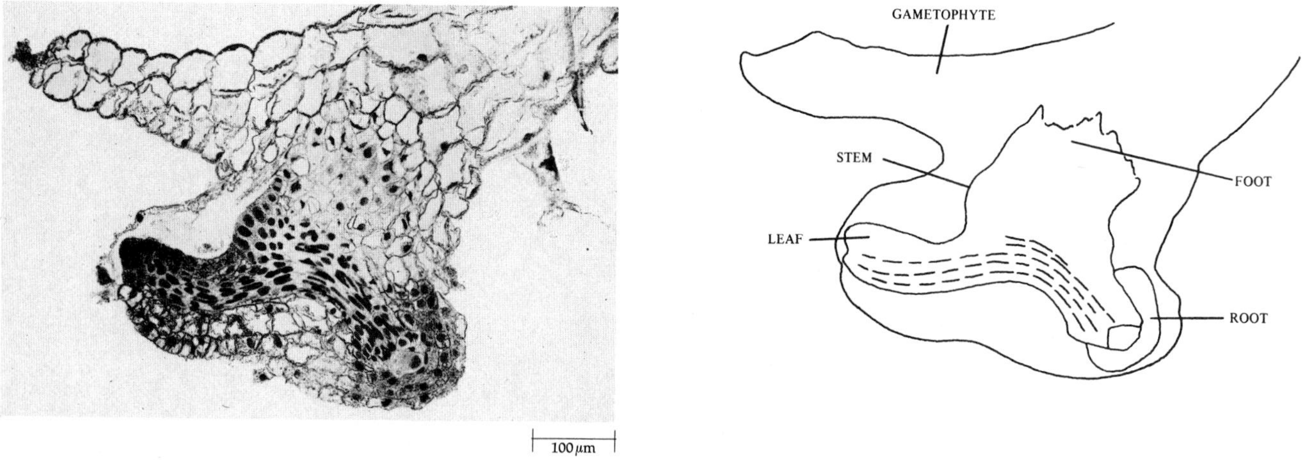

15–35
Life cycle of a homosporous fern. Following meiosis, spores are produced in the sporangia and shed. The gametophytes are green and nutritionally independent in most species. Many are only one layer of cells thick and somewhat heart-shaped with an apical notch; others are thicker and may be more irregular in form. From the lower surface of the gametophyte, specialized cellular filaments known as rhizoids extend downward into the substrate.

On the lower surface of the gametophyte are borne the flask-shaped archegonia, with their swollen lower portions sunken into the gametophyte tissue. The neck is composed of several to many tiers of cells. Antheridia are also borne on the lower surface of the gametophyte, and have a sterile jacket one cell layer thick. Within they bear numerous spirally coiled, multiflagellated sperms. When the sperms are mature and there is an adequate supply of water, the antheridia burst to release them, and they swim into the archegonium. Apparently they are attracted by malic acid or other substances released by the archegonium.

In the lower portion of the archegonium, the egg is fertilized and a zygote is formed, which begins to divide immediately. The young embryo grows and differentiates directly into the adult sporophyte, obtaining its nutrition from the gametophyte for a time but soon achieving a level of photosynthesis sufficient to maintain itself. After the young sporophyte becomes rooted in the soil, the gametophyte disintegrates.

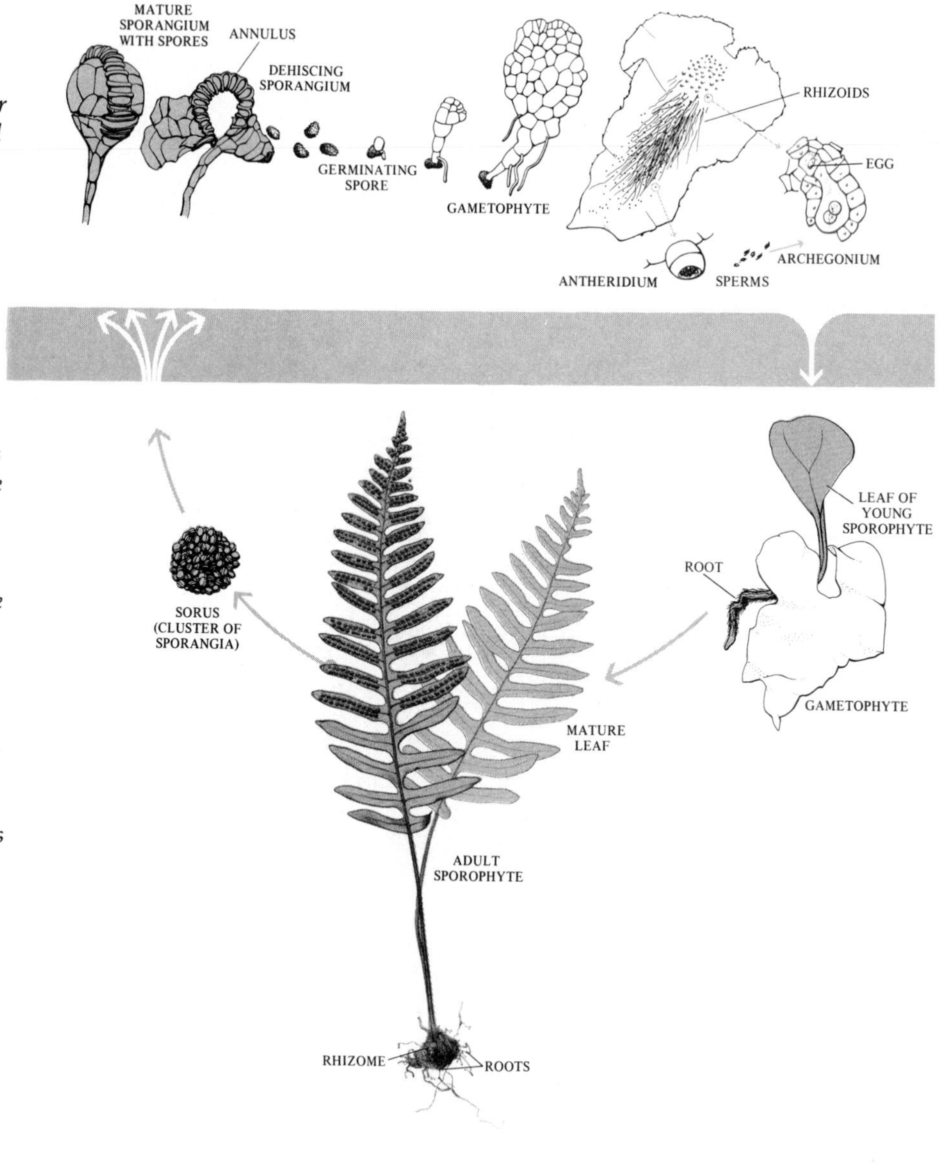

SUMMARY

The living seedless vascular plants are classified in four subdivisions: the Psilophytina *(Psilotum* and *Tmesipteris),* the Lycophytina *(Lycopodium, Selaginella,* and *Isoetes),* the Sphenophytina *(Equisetum),* and the Filicophytina (ferns). Most of the seedless and nonflowering vascular plants that reproduce via spores, the vascular cryptogams, are homosporous. Heterospory is exhibited by *Selaginella, Isoetes,* and some water ferns.

The life cycles of the vascular cryptogams are essentially similar to one another: a heteromorphic alternation of generations in which the sporophyte is dominant and free-living. The gametophytes of the homosporous species are bisexual, producing both antheridia and archegonia, and independent of the sporophyte for their nutrition. Those of heterosporous species are much reduced in size, unisexual, and dependent upon stored food derived from the sporophyte for their nutrition. All of the vascular cryptogams have motile sperms, and water is necessary for them to swim to the eggs.

The Psilophytina differ from other living vascular cryptogams in their lack of leaves (with the possible exception of *Tmesipteris*) and roots. The Lycophytina are characterized by the presence of microphylls. Although by definition microphylls, the leaves of *Equisetum* may be much-reduced megaphylls in an evolutionary sense. The most distinctive characteristic of the ferns is the presence of megaphylls, which at times has been considered sufficient cause to place the ferns in the same subdivision as the seed plants, which also have megaphylls.

All four groups of seedless vascular plants extend back to Devonian times, but only the ferns are represented by many species at present.

CHAPTER 16

The Seed Plants

16–1
One of the most interesting of all gymnosperm groups is the class Pteridospermae, the "seed ferns." The remnants of these bizarre plants are common in rocks of Carboniferous age and have been well known to paleobotanists for a century or more. Their vegetative parts are so fernlike that for many years they were grouped with the ferns. In 1904, however, two British workers, F. W. Oliver and D. H. Scott, demonstrated that these plants bore seeds and were, therefore, gymnosperms. This drawing is a reconstruction of the Carboniferous seed fern Medullosa noei. *The plant is about 5 meters tall.*

As mentioned in Chapter 14, the seed constitutes one of the most dramatic innovations to arise during the evolution of vascular plants, and it seems to be one of the factors responsible for the dominance of seed plants in today's flora. The reason is simple: The seed has survival value.

The oldest known seeds are from the late Devonian Period, some 350 million years ago. During the next 50 million years, a wide array of seed types evolved, including some that were attached to fernlike fronds (Figure 16–1). This group of extinct gymnosperms belongs to the class Pteridospermae, which means literally "seed ferns." Their resemblance to ferns has led to speculation that they and the ferns evolved from a common ancestor.

Within the seeds of modern gymnosperms and angiosperms, the megagametophyte is held within a fleshy covering known as the *nucellus*, which is the morphological equivalent of the megasporangium. This, in turn, is enveloped by one or more additional layers that make up the *integument*, which completely encloses the megasporangium except for an opening at the apex called the *micropyle*. It is the integument that develops into the *seed coat*. In most modern seed plants, the young sporophyte, or embryo, develops within the seed before dispersal; in ancient seed plants, the seeds were generally shed before the embryo developed. The oldest known embryo-containing seeds were reported in 1973 from the Lower Permian period, some 270 million years ago, in west Texas. Perhaps the development of an embryo before dispersal gives a seed a better chance of survival in cold and harsh conditions, and certainly the Permian was a period of climatic extremes.

In addition to the megaspore or embryo and the seed coat, all seeds contain stored food. The megasporangium plus its integument or integuments is called an *ovule*, and a seed may be defined as a mature ovule.

The evolution of the seed provided valuable means of adaptation to life on land; within the seed coat, the embryo can often remain dormant until conditions are favorable for germination and so survive drought, frost, or other unfavorable environmental circumstances. The supply of nutrients in the stored food generally is more than adequate to nourish the young plant until it becomes an independent entity.

The seed plants, all of which possess megaphylls, belong to the subdivision Spermatophytina, which is divided into five classes, the Cycadinae, the Ginkgoinae, the Coniferinae, the Gnetinae, and the Angiospermae.

THE GYMNOSPERMS

The gymnosperms include four classes with living representatives: Cycadinae (cycads), Ginkgoinae (maidenhair tree), Coniferinae (conifers), and Gnetinae (vessel-containing gymnosperms). The name gymnosperm, which means literally "naked seed," points to one of the principal characteristics of all members of this group of vascular plants: Their ovules and seeds are borne exposed on the surface of sporophylls or analogous structures. The gymnosperms are best understood as a number of parallel lines of seed-bearing plants that do not have the special characteristics of angiosperms. The classes of gymnosperms probably represent the achievement of a particular stage in evolutionary progression by various groups of ferns.

With few exceptions, the female gametophytes of gymnosperms produce several archegonia each. As a result, more than one egg may be fertilized, and hence several embryos may begin to develop within a single ovule. This phenomenon is known as polyembryony. In most cases, only one embryo survives, so that relatively few fully developed seeds contain more than one embryo.

In the lower vascular plants, water is required for the motile, flagellated sperms to reach and fertilize the eggs. In the gymnosperms, water is not required as a medium of transport of the sperms to the eggs. Instead, the partly developed male gametophyte, the *pollen grain*, is transferred bodily to the vicinity of a female gametophyte by the process of *pollination*, which is usually accomplished by wind. After pollination, the male gametophyte produces a tubular outgrowth, the *pollen tube*. In the conifers and Gnetales, the sperms are nonmotile, and the pollen tubes convey them directly to the archegonium. In the cycads and *Ginkgo*, the sperms are flagellated, and the pollen tube apparently functions largely as a haustorial structure, analogous to the haustoria of parasitic fungi, which, you may remember (page 212), penetrate the cells of other organisms and absorb nourishment directly from them. The pollen tube may grow for several months in the tissue of the nucellus, or megasporangium, before reaching a cavity above the female gametophyte. At that time, the pollen tube bursts and liberates its two sperms into the cavity. The sperms then swim to an archegonium, and one fertilizes the egg. With the development of sperm-conveying pollen tubes, the evolving vascular plants were no longer dependent upon water to assure fertilization.

The Conifers

By far the largest and most significant of gymnosperm classes today, the Coniferinae comprise some 50 genera and about 550 species. The tallest vascular plants, such as the redwood *Sequoia sempervirens* of coastal California and southwestern Oregon, are found in this group. These trees attain heights of up to 117 meters and diameters in excess of 11 meters. The conifers—which also include pines, firs, and spruces, for example—are of great commercial value, their stately forests providing the wealth of vast regions of the North Temperate zone. During the early Tertiary period, about 60 million years ago, some genera were more widespread than they are now and dominated huge expanses on all the northern continents.

The history of the conifers extends back at least to Late Carboniferous times, some 290 million years ago. Their leaves have many drought-resistant features, and perhaps the origin of conifers ought to be sought during the relatively dry Permian period, when increasing worldwide aridity must have provided a powerful evolutionary stimulus.

The Pines

The pines (genus *Pinus*) include among them perhaps the most familiar of all gymnosperms (Figure 16-2), because they dominate broad stretches of North America and

16–2
Longleaf pine, Pinus palustris, *in North Carolina.*

(a)

(b)

16–3
(a) *Seedling of longleaf pine,* Pinus palustris, *in Georgia, showing the juvenile leaves (long singly borne needles) and the first mature leaves borne in this species, in fascicles of 3s.* (b) *Seedling of pinyon pine,* Pinus edulis, *showing juvenile leaves and young tap root system; the mature leaves of this species are borne in fascicles of two.*

Eurasia and are widely cultivated even in the Southern Hemisphere. There are some 90 species of pines, all of which are characterized by an arrangement of the leaves that is unique among living conifers. The conspicuous leaves of pines are needlelike. In the seedlings, they are borne singly and are spirally arranged on the stems (Figure 16–3). After a year or two of growth, a pine begins to produce its leaves in bundles, each bundle containing a characteristic number of long, needlelike leaves—ranging from one to eight, depending on the species of pine. These bundles, wrapped at the base by a series of short, scalelike leaves, are actually short shoots

(a)

(b)

16–4
(a) *Bristlecone pine,* Pinus longaeva, *at Bryce Canyon, Utah. Branch showing clusters of mature needles in 5s and a mature ovulate cone.* (b) *Branch of red pine,* Pinus resinosa, *with young ovulate cones. Note bundles, or fascicles, of needlelike leaves characteristic of the adult plant.*

in which the activity of the apical meristem becomes suspended (Figure 16–4). Thus a bundle of needles in a pine is morphologically a determinate (not continuing indefinitely) branch system. Under unusual circumstances, the apical meristem within a bundle of needles in a pine may be reactivated and grow into a new shoot with indeterminate growth, or sometimes may even produce roots and grow into an entire pine tree (Figure 16–5).

The leaves of pines, like those of many other conifers, are impressively suited for growth under arid conditions (Figure 16–6). The epidermis is covered with a thick cuticle, beneath which are one or more layers of thick-walled cells, the hypodermis. The stomata are sunken below the surface of the leaf. The mesophyll, the ground tissue of the leaf, is compact and penetrated by conspicuous resin ducts. There are one or two veins in the center of the leaf. The veins are embedded in transfusion tissue. The transfusion tissue, a tissue composed of living parenchyma cells and nonliving tracheids similar to the water-conducting tracheids of the xylem, is delimited from the mesophyll by an endodermis.

In the pines and other conifers, secondary growth begins early and leads to the formation internally of

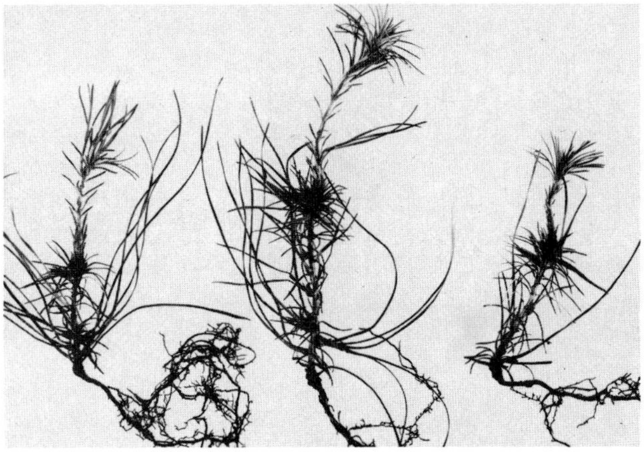

16–5
Three one-year-old pines (Pinus radiata) *grown from rooted needle bundles. These experiments, carried out by Jochen Kummerow of San Diego State University, emphasize that a pine needle bundle is actually a short shoot in which the activity of the apical meristem is suspended and can be regenerated.*

16–6
Pinus. *Transverse section of needle, showing mature tissues.*

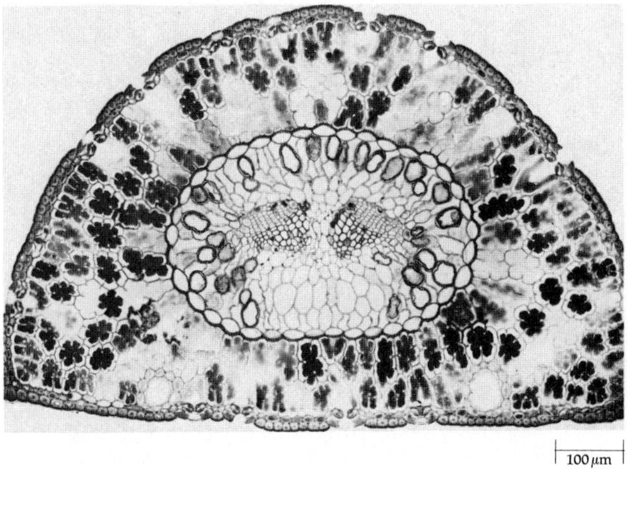

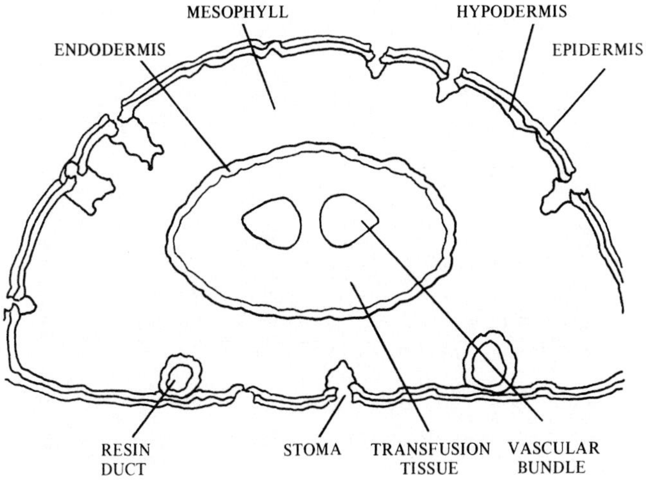

substantial amounts of secondary xylem. Secondary phloem is produced externally. The xylem consists primarily of tracheids and the phloem of sieve cells (typical food-conducting cells of gymnosperms and seedless vascular plants). Both tissues are traversed radially by narrow rays. With the initiation of secondary growth, the epidermis is eventually replaced with a periderm, which has its origin in the outer layer of cortical cells. As secondary growth continues, subsequent periderms are produced deeper in the bark.

Reproduction in Pine. The microsporangia and megasporangia in pine and other conifers are borne in separate cones, usually on the same plant. Ordinarily the microsporangiate cones are borne on the lower branches, the megasporangiate or ovulate cones on the higher branches of the tree.

Microsporangiate cones in the pines are relatively small, usually 1 to 2 centimeters in length (Figure 16–7). The microsporophylls (Figure 16–8) are spirally arranged and more or less membranous in texture. Each bears two microsporangia. Each young microsporangium contains many microsporocytes or microspore mother cells, which in early spring undergo meiosis to give rise to four haploid microspores each. Each microspore develops into a winged pollen grain, consisting of two *prothallial cells*, a *generative cell*, and a *tube cell* (Figure 16–9). This four-celled pollen grain is an immature male gametophyte. It is at this stage that the pollen is shed in enormous quantities and carried by the wind to the ovulate cones.

Ovulate cones of pines are much larger than the pollen-bearing cones and considerably more complex in structure (Figure 16–10). The cone scales, which bear the ovules, are not megasporophylls, but entire modified branch systems of determinate growth, properly known as *seed-scale complexes*. Each seed-scale complex consists of an ovuliferous scale, which bears two ovules on its upper surface, and a subtending sterile bract (Figure 16–11). The scales are arranged spirally around the axis of the cone. Each ovule contains a multicellular nucellus (the megasporangium) surrounded by a massive integument with an opening, the micropyle, facing the cone axis. Each megasporangium contains a single megasporocyte or megaspore mother cell, which ultimately undergoes meiosis to give rise to a linear series of four megaspores. However, only one of these megaspores is functional. The three nearest the micropyle soon degenerate.

16–7
Jack pine, Pinus banksiana. *Microsporangiate cones shedding pollen.*

16–8
Pinus. *Longitudinal view of portion of microsporangiate cone, showing microsporophylls and microsporangia with mature pollen grains.*

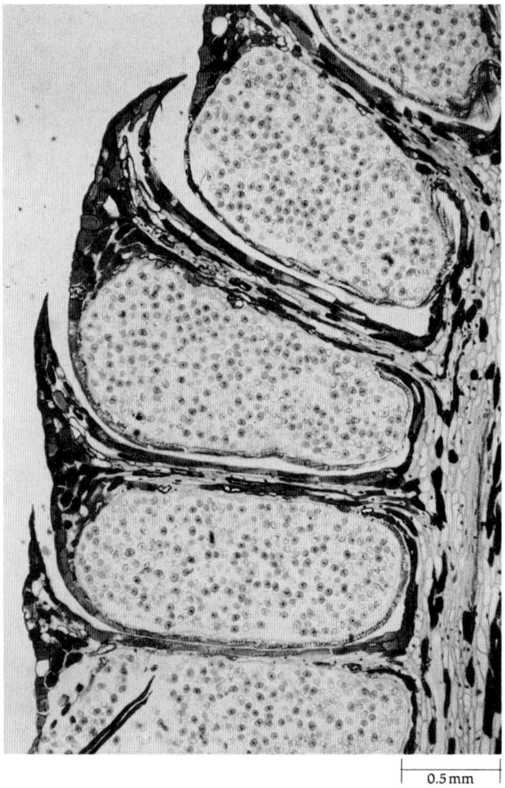

16-9
Pinus. *Pollen grains with enclosed immature male gametophytes.* (a) *The four-celled male gametophyte consists of two prothallial cells, a relatively small generative cell, and a relatively large tube cell.* (b) *Somewhat older pollen grain than that shown in* (a). *Here the prothallial cells, which have no apparent function, have degenerated.*

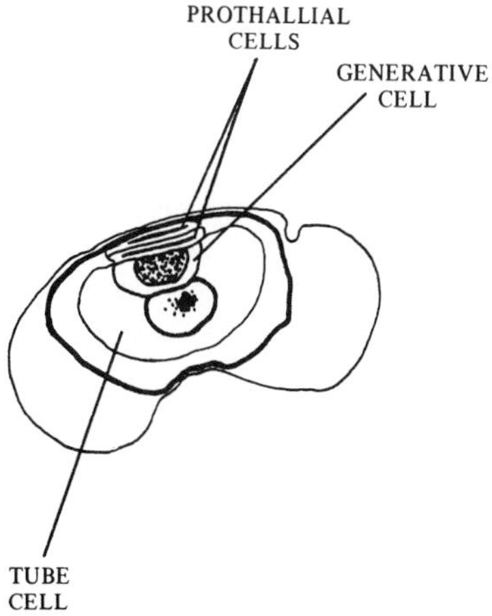

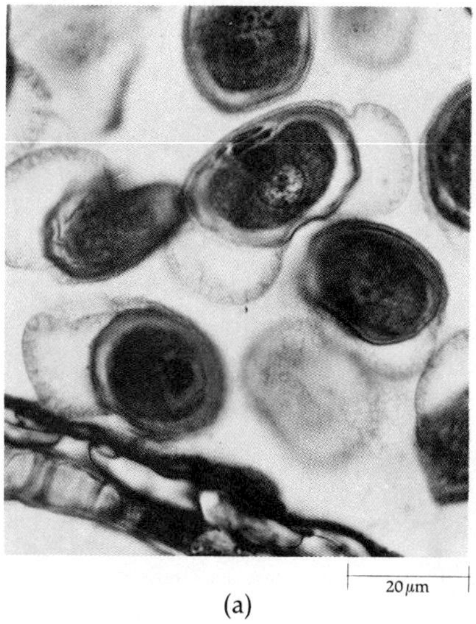

(a)

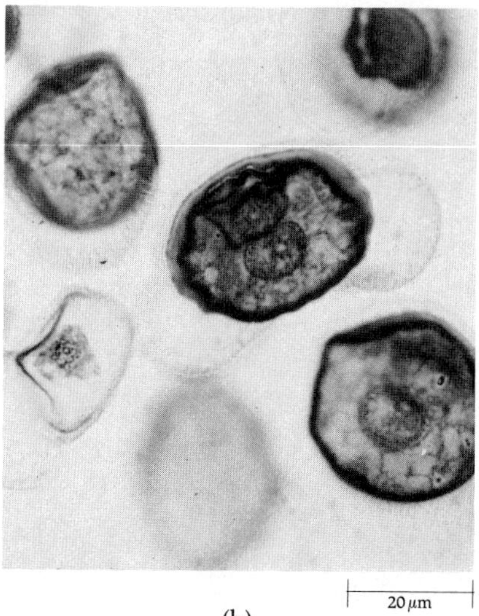

(b)

16-10
Some mature ovulate pine cones, showing relative sizes. (a) *Digger pine,* Pinus sabiniana. (b) *A pinyon pine,* Pinus edulis, *in surface and side views. The edible seeds of this and other pinyon pines are called "pine nuts."* (c) *Sugar pine,* Pinus lambertiana. (d) *Western yellow pine,* Pinus ponderosa. (e) *Eastern white pine,* Pinus strobus. (f) *Red pine,* Pinus resinosa.

(a)

(b)

(c)

(d)

(e)

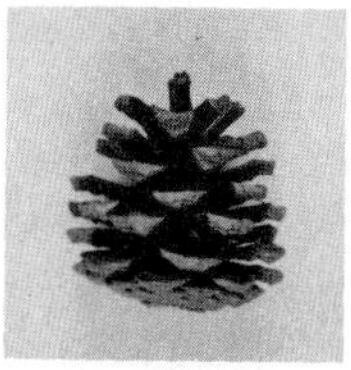

(f)

16–11
Pinus. *Longitudinal view of portion of young female cone, or strobilus, showing its complex structure. Note the megasporocyte (megaspore mother cell) surrounded by the nucellus.*

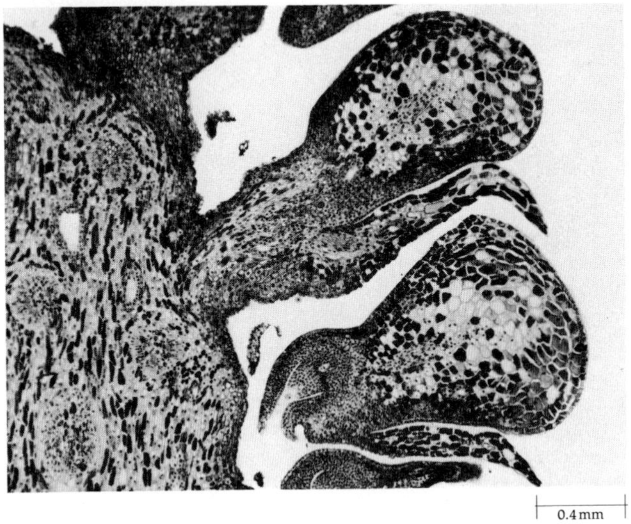

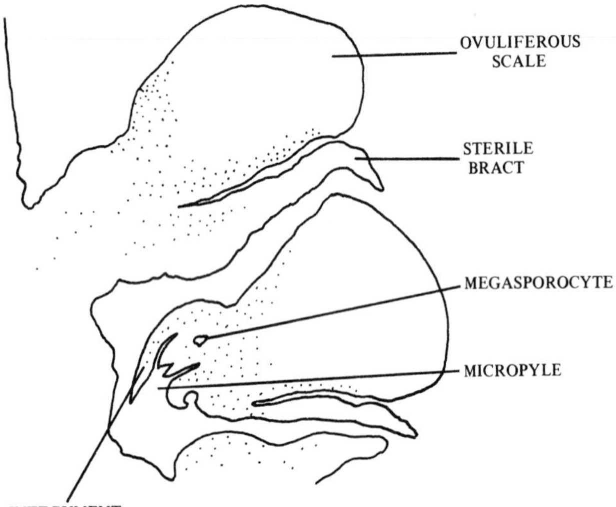

Pollination in pines occurs in spring, the pollen adhering to a drop of sticky fluid about the micropyle. At this stage, the scales of the ovulate cone are widely separated. As the micropylar fluid evaporates, the pollen grain is drawn into the space between the micropyle and nucellus, or megasporangium. Following pollination, the scales grow together and afford a higher degree of protection for developing ovules. Shortly after the pollen grain comes in contact with the nucellus, it germinates, forming a pollen tube. At this time, meiosis has not yet occurred in the megasporangium. About a month after pollination, the four megaspores are produced, only one of which develops into a megagametophyte. Development of the female gametophyte is very sluggish; often it does not begin until some 6 months after pollination and may then require another 6 months for completion. In the early stages of megagametophyte development, mitosis proceeds without immediate cell-wall formation. About 13 months after pollination, when the female gametophyte may contain some 2000 free nuclei, cell-wall formation begins. Approximately 15 months after pollination, archegonia, usually two or three in number, differentiate at the micropylar end of the megagametophyte (Figure 16–12). The stage is now set for fertilization.

16–12
Pinus. *Longitudinal section, showing distal portion of ovule prior to fertilization.*

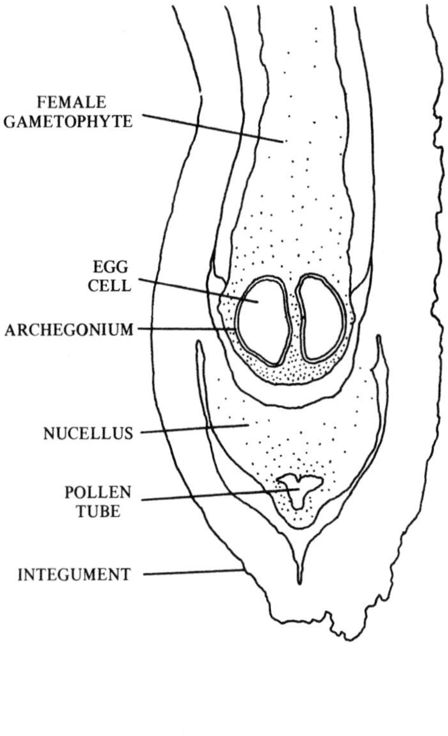

When we left the pollen grain a year earlier, it had begun to germinate, slowly digesting its way through the tissues of the nucellus on its way toward the developing female gametophyte. About 12 months after pollination, the generative cell of the four-celled male gametophyte undergoes division to give rise to a *stalk cell* and a *body cell.* Subsequently, before the pollen tube reaches the female gametophyte, the body cell divides to produce two sperms. The male gametophyte, or germinating pollen grain, is now mature. Note that an antheridium is not formed by the male gametophyte.

Some 15 months after pollination, the pollen tube reaches the egg cell of an archegonium, where it discharges much of its cytoplasm and both of its sperms into the cytoplasm of the egg (Figure 16–13). One sperm nucleus unites with the egg nucleus, the other degenerates. Commonly, the eggs of all archegonia are fertilized and begin to develop into embryos (polyembryony). However, only one embryo generally develops fully.

During early embryogeny, four tiers of cells are produced near the lower end of the archegonium. Each of the four cells of the lowermost tier (i.e., the tier farthest from the micropylar end of the ovule) begins to form an embryo, while the four cells of the tier next to the bottom, the suspensor cells, elongate greatly and force the

16–13
Pinus. *Fertilization: union of small sperm nucleus with large egg nucleus. Second sperm nucleus (above) is nonfunctional and will disintegrate.*

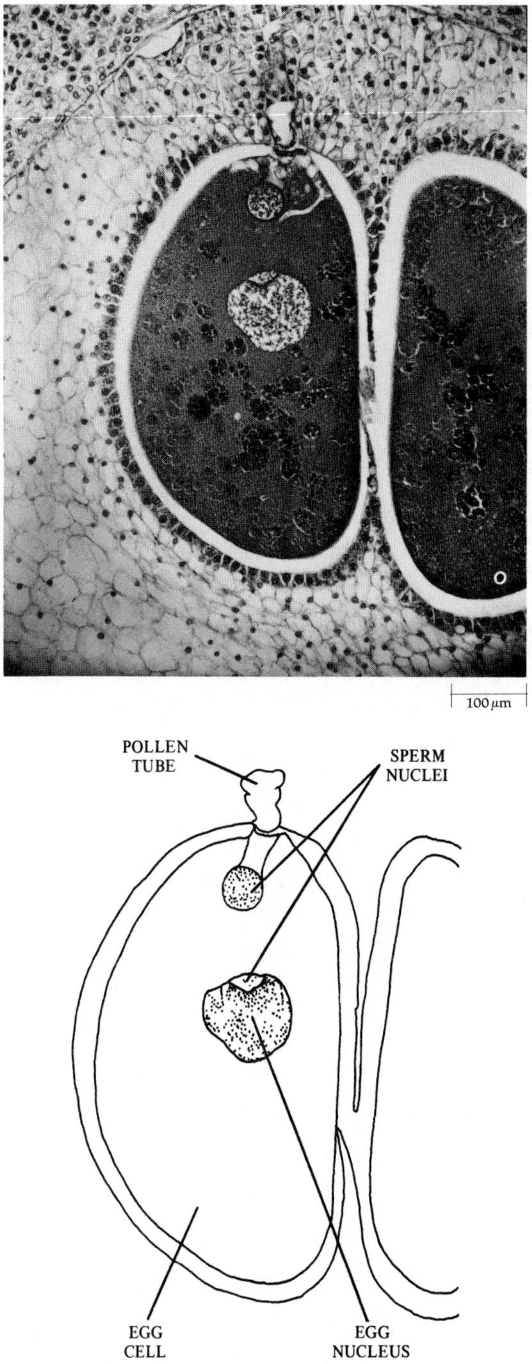

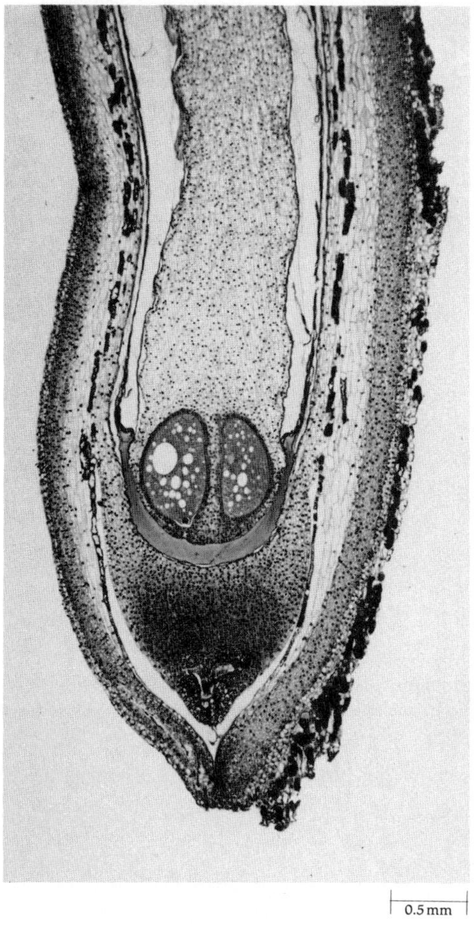

developing embryos into the female gametophyte. Thus, for a second time in the pine life cycle polyembryony is exhibited. However, generally only one of the developing embryos develops fully. During embryogeny the integument develops into a seed coat.

The seed of a pine and other conifers is a remarkable structure, for it consists of a combination of two sporophytic generations, the seed coat and embryo, and one gametophytic generation, which serves as reserve food, or nutritive tissue (Figure 16–14). The seed coat and embryo are diploid, and the female gametophyte is haploid. The embryo consists of a hypocotyl-root axis, with a root cap and apical meristem at one end and an apical meristem and several (generally eight) cotyledons (seed leaves) at the other. The integument consists of three layers, of which the middle layer becomes hard and serves as the seed coat.

The seeds of pines are often shed from the cones during the autumn of the second year following the initial appearance of the cones and the occurrence of pollination. At maturity, the cone scales separate; the winged seeds flutter through the air and are sometimes carried considerable distances by the wind. In some species of pines, the scales do not separate until the cones are subjected to extreme heat. The closed cones often remain on the branches of the trees for years, with the wood of the growing trunks and limbs swelling out around them. If a fire sweeps through the groves and kills the parent trees, the cones open and release the seed crop accumulated over many years, thus reestablishing the species in the recently burned soil.

The pine life cycle is summarized in Figure 16–15.

16–14
Pinus. *Longitudinal section of seed.*

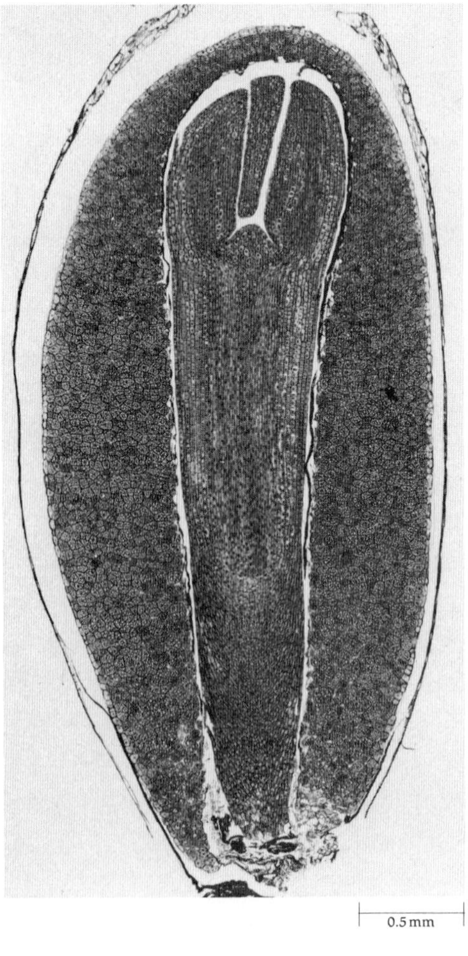

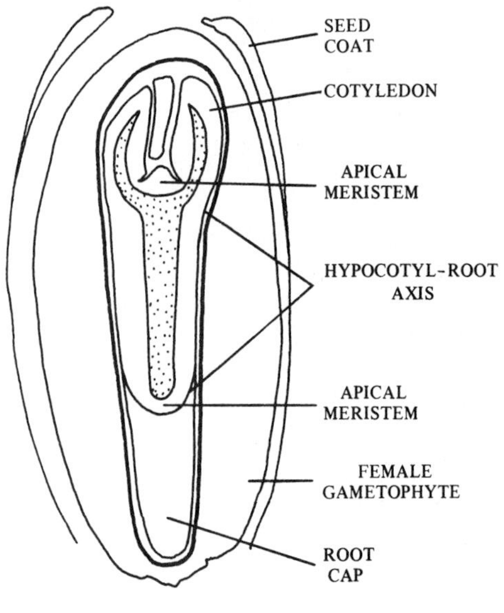

16–15
Life cycle of a pine. The gametophytes are much reduced and nutritionally dependent upon the sporophyte. They are enclosed within the tissue of the sporophyte. The ovule that encloses the megagametophyte matures after fertilization and becomes the seed. The elaborate suspensor, which is characteristic of the pines, disintegrates by the time the embryo is fully developed.

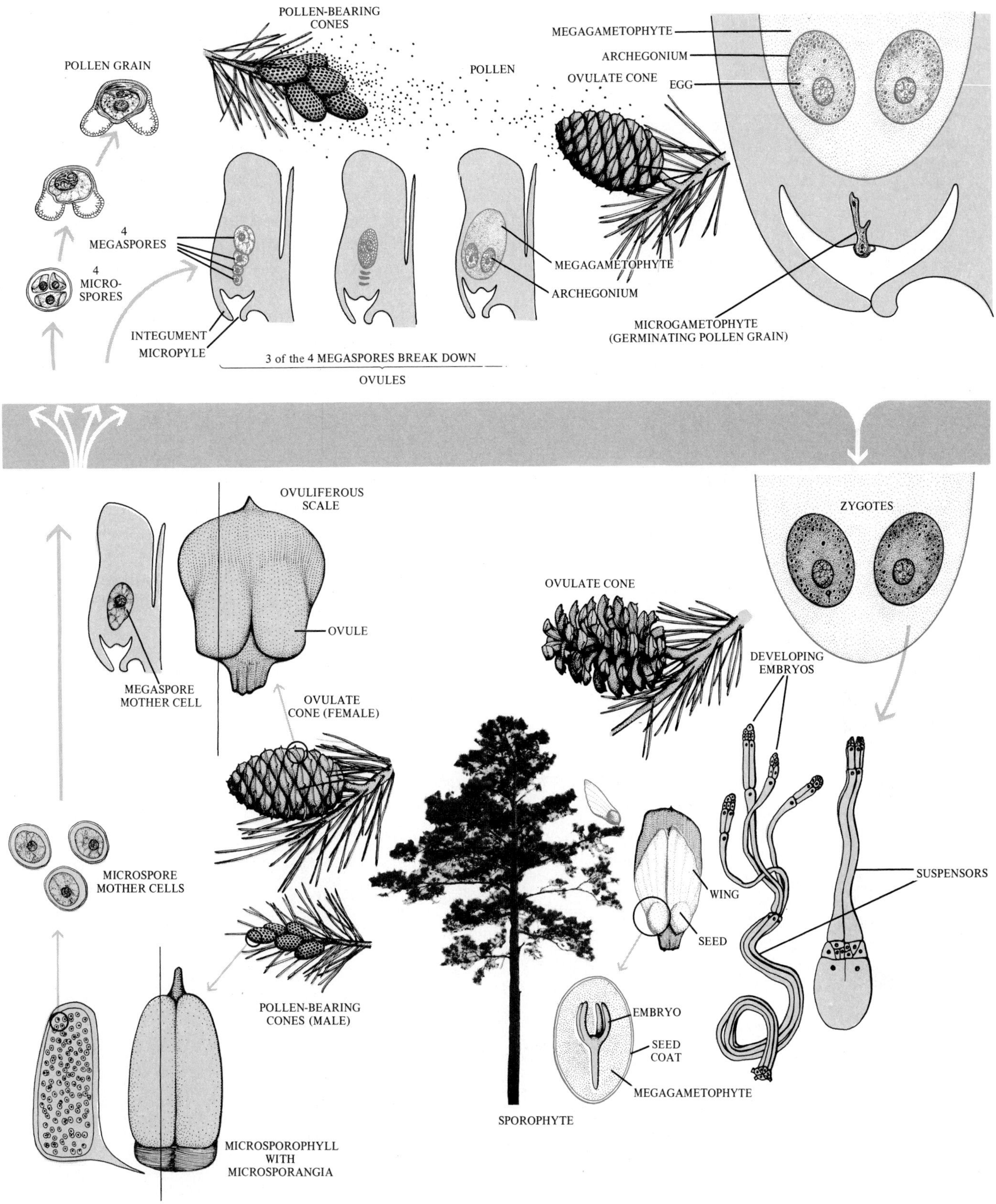

16–16
Ovulate cone of Abies magnifica, *red fir. The largest of the firs, this species grows in the mountains of California and southern Oregon. It sometimes reaches a height of 60 meters. The upright cones, which are 15 to 25 centimeters long, do not fall to the ground whole, as they do in pines, but shatter and completely fall apart, scattering winged seeds.*

Other Conifers

Although other conifers (Figures 16–16—16–21) lack the needle clusters of pines and may also differ in a number of relatively minor details of their reproductive systems, the group is a fairly homogeneous one. Among its important representatives are the firs *(Abies)*, spruces *(Picea)*, hemlocks *(Tsuga)*, Douglas fir *(Pseudotsuga)*, cypresses *(Cupressus)*, and junipers *(Juniperus)*. In the yews (family Taxaceae), the ovules are not borne in cones but are solitary and surrounded by a fleshy, cuplike structure, the *aril*. A number of other genera of conifers are characteristic of the Southern Hemisphere. Some of them, like the monkey-puzzle tree *(Araucaria)*, are frequently cultivated.

One of the most interesting groups of conifers is the family Taxodiaceae, which includes the tallest vascular plant, *Sequoia sempervirens* (see Figure 29–1). The famous "big tree," *Sequoiadendron giganteum* (see Figure 27–4), which forms spectacular, widely scattered groves along the west slope of the Sierra Nevada of California, also belongs to this group, as do the bald cypresses, *Taxodium*, of the southeastern United States and Mexico. All these genera were more widespread in the Tertiary period than they are at present (Figure 16–22).

Another genus that was abundant in the Tertiary in both Eurasia and North America was *Metasequoia*, the dawn redwood (Figure 16–23). Indeed, *Metasequoia* was the most abundant conifer in western North America and the American Arctic from the Upper Cretaceous period to the Miocene (in other words, through most of the Tertiary period, up to about 25 million years ago).

16–17
American larch (Larix laricina). *The larch genus occurs all around the Northern Hemisphere, with this species, which is also called tamarack, ranging from Labrador to Alaska southward, mainly in swamps, to West Virginia, the upper Midwest, and northern British Columbia. Leaves of larch are needlelike as in the pines. They are borne singly on short branch shoots and are spirally arranged. Unlike most conifers, the larches are deciduous, that is, they shed their leaves at the end of each growing season.*

16-18
The Gowen cypress (Cupressus goveniana). *These small trees—only about 6 meters in height at maturity—belong to an extremely local species found only near Monterey, California.*

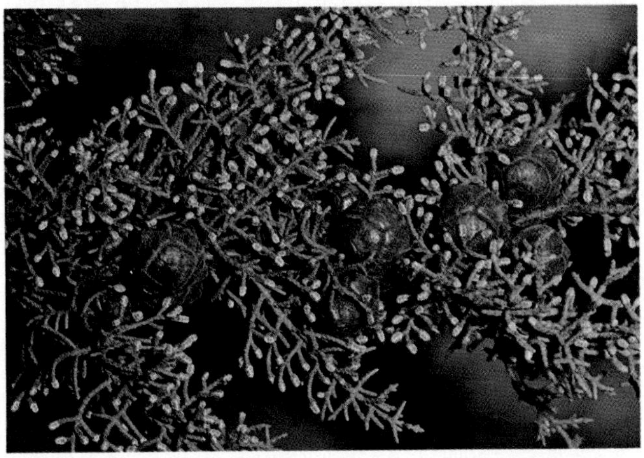

16-19
The western juniper (Juniperus occidentalis) *has round ovulate cones like those of the cypresses, but the scales are fleshy and fused together. These juniper "berries" give gin its distinctive taste and aroma.*

16-20
In conifers of the yew family (Taxaceae), the seeds are surrounded by a fleshy cup, the aril, which attracts birds and other animals that eat the "fruits" and spread the seeds. (a) *Species of the genus* Taxus, *the yews, which occur around the northern hemisphere, produce red, flashy ovulate structures.* (b) *Sporophylls and sporangia of the pollen-bearing cones of the same species; they are found on separate individuals.*

(a)

(b)

16–21
The bald cypress (Taxodium distichum) *is a deciduous member of the redwood family which grows in swamps in the southern United States.*

16–22
Geographic distribution of living and fossil sequoias.

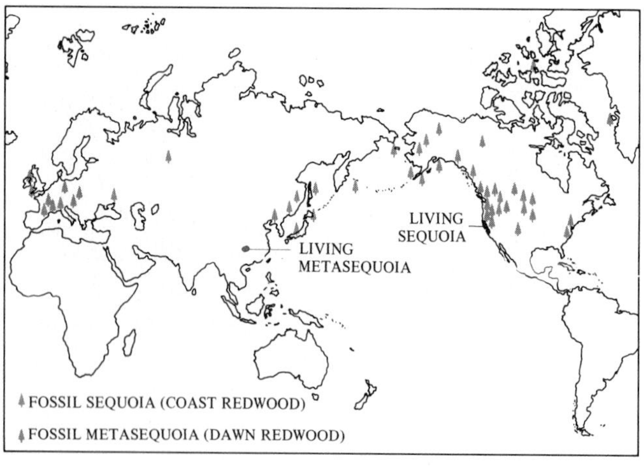

The genus *Metasequoia* was first described from fossil material by the Japanese paleobotanist Shigeru Miki in 1941. Three years later the Chinese forester Tsang Wang, from the Central Bureau of Forest Research of China, visited the village of Mo-tao-chi in the remote Szechuan province and discovered a huge tree of a sort he had never seen before. The natives of the area had built a temple around the tree's base. Tsang collected specimens of the tree's needles and cones, and study of these revealed that the fossil, *Metasequoia,* had "come to life." In 1948, the American paleobotanist Ralph Chaney of the University of California led an expedition down the Yangtze River and across three mountain ranges to valleys where a thousand dawn redwoods were growing, the last remnant of the once great *Metasequoia* forest. Now thousands of seeds have been distributed, largely through the efforts of the late Elmer Drew Merrill of the

16–23
(a) *The dawn redwood* (Metasequoia glyptostroboides) *at Mo-tao-chi in central China. The building at the base is a temple.* (b) *Fossil branchlet of* Metasequoia, *about 50 million years old.* [b, *Carolina Biological Supply Co.*]

(a)

(b)

Arnold Arboretum in Boston, and this "living fossil" can be seen growing in parks and gardens all over the world.

Other Living Gymnosperms

Cycads

The other classes of living gymnosperms are remarkably diverse and scarcely resemble one another at all. Among them are the cycads, class Cycadinae, which are a series of palmlike plants found mainly in the tropical and subtropical portions of the world. These bizarre plants were so numerous in Mesozoic times that this era is often called the "Age of Cycads and Dinosaurs." There are 9 genera and about 100 species of living cycads. *Zamia*, which occurs commonly in sandy woods in Florida, is an example (Figure 16–24).

Most cycads are good-sized plants, and some reach 18 meters or more in height. Many have a distinct trunk that is densely covered with the bases of shed leaves. The functional leaves characteristically occur in a cluster at the top of the stem, giving the cycads an aspect similar to that of the palms (indeed, a common name for the cycads is "sago palms"). Unlike the palms, however, the cycads exhibit true, if sluggish, secondary growth from a vascular cambium; the central portion of their trunks is occupied by a great mass of pith. The reproductive units of cycads are more or less reduced leaves with attached sporangia that are loosely or tightly clustered into cone-like structures near the apex of the plant. In cycads, pollen and seed cones are borne on different plants (Figure 16–25).

16–24
An ovulate plant of Zamia integrifolia, *the only species of cycad native to the United States. It was used by the Seminole Indians for making bread.*

16–25
Encephalartos altensteinii, *a cycad native to South Africa.* (a) *A microsporangiate plant with male cones.* (b) *An ovulate plant with ovulate cones. The top of one ovulate cone has been removed to reveal seeds on upper surfaces of megasporophylls.*

(a)

(b)

16–26
(a) Ginkgo biloba. (b) Ginkgo *leaves and fleshy seed.*

(a)

(b)

Ginkgo

The maidenhair tree, *Ginkgo biloba,* is easily recognized by its fan-shaped leaves with their openly branched, forking (dichotomous) pattern of veins. The leaves on the numerous spur shoots are more or less entire, whereas on the long shoots and in seedlings they are deeply lobed. Unlike most gymnosperms, *Ginkgo* is deciduous, its leaves turning a beautiful golden color before falling in autumn (Figure 16–26).

Ginkgo is the sole living survivor of an evolutionary line that probably extends back to the late Paleozoic era and was common during much of the Mesozoic era. The class Ginkgoinae was once widespread but seems to have been represented by relatively few species. There may indeed no longer be any truly wild trees of the living species, but *Ginkgo biloba* has long been cultivated in the temple grounds of China and Japan and has been an important feature of the gardens of the temperate regions of the world for more than 150 years. It is especially resistant to air pollution and so is commonly cultivated in cities.

Like the cycads, the *Ginkgo* bears the ovules and microsporangia on different individuals. The ovules of *Ginkgo* are borne in pairs on the end of short stalks and ripen to produce seeds in autumn. In *Ginkgo,* fertilization within the ovules may not occur until after the ovules have been shed from the tree. Embryos are formed during the later stages of maturation of the seeds, which occur on the ground. The seeds have a rancid odor as a result of the butyric acid in their fleshy coats, and for this reason only male trees are usually cultivated on streets or in parks or gardens. The microsporophylls are clustered in conelike structures, each microsporophyll bearing two microsporangia.

Gnetum, Ephedra, and *Welwitschia*

The class Gnetinae is a small group (about 70 species) of gymnosperms consisting of three genera: *Gnetum, Ephedra,* and *Welwitschia.* Although placed in the same class as a matter of convenience, each genus probably should be placed in a separate class, for each differs greatly from the others both structurally and reproductively.

Gnetum, a genus of about 30 species, consists of tropical trees and vines with large leathery leaves that closely resemble those of dicotyledons (Figure 16–27). It is found throughout the moist tropics of the world.

Most of the approximately 35 species of *Ephedra* are profusely branched shrubs with inconspicuous small, scalelike leaves (Figure 16–28). With its small leaves and apparently jointed stems, *Ephedra* superficially resembles *Equisetum. Ephedra* occurs in subarid and semihumid regions all over the world.

Welwitschia is probably the most bizarre of vascular plants (Figure 16–29). Most of the plant is buried in

16–27
The large leathery leaves of Gnetum, *one of the three living genera of Gnetinae, resemble those of some angiosperms. The species of* Gnetum *grow as shrubs or woody vines in tropical or subtropical forests* (a). (b) *Seeds of* Gnetum cuspidatum, *a climber, on a tree in Borneo.*

(a)

(b)

16–28
Ephedra *is the largest genus of Gnetinae.* (a) Ephedra californica, *a densely branched shrub that, like other members of the genus, has scalelike leaves.* (b) *Fleshy seed-bearing structures of* Ephedra procera.

(a)

(b)

16–29
Welwitschia mirabilis. *The plant produces just two long leaves during its lifetime, which may last a century. The leaves, which have a basal meristem, keep growing, break off at the distal ends and split lengthwise. Its root and huge, carrot-shaped stem, which projects slightly above ground, store water. The cone-bearing branches are formed on the inside of the rim. The specimen shown here is a seed-producing plant.*

sandy soil. The exposed part consists of a massive woody, concave disk that produces only two strap-shaped leaves. *Welwitschia* grows in desert areas of southwestern Africa.

The Gnetinae have many angiospermlike features (such as the similarity of strobili to inflorescences of angiosperms, the presence of vessels in the xylem, and the lack of archegonia in *Gnetum* and *Welwitschia*). As a consequence, some botanists have considered them as possible connecting links between the gymnosperms and angiosperms. However, today they are generally regarded as specialized end points of gymnosperm evolution.

THE ANGIOSPERMS

The flowering plants, class Angiospermae, comprise about 235,000 species, by far the largest number of species of any plant group. In their vegetative structures, they are enormously diverse. In size, they range from species of *Eucalyptus*, trees well over 100 meters in height (Figure 16–30), to some duckweeds of the genus *Wolffia*, which are simple floating plants scarcely 1 millimeter in length (Figure 16–31). Some are vines that climb high into the canopy of the tropical rain forest, and others are epiphytes that grow in the canopy. Many angiosperms, such as cacti, are adapted to grow in extremely arid regions of the world. Although the vast majority of flowering plants are free-living, both parasitic and saprophytic forms exist (Figure 16–32).

The Angiosperms are divided into two large subclasses: the Monocotyledoneae (monocots) with about 65,000 species, and the Dicotyledoneae (dicots) with about 170,000 species. The similarities between these two groups are far greater than the differences; nonetheless the two subclasses are clearly recognizable natural units. Among the monocots are such familiar plants as grasses, lilies, irises, orchids, cattails, and palms (Figure 16–33). The dicots include almost all familiar trees and shrubs, other than conifers, and almost all annual herbs, except for grasses, plus many other plants (Figure 16–34). The major differences between the monocots and dicots are summarized in Table 16–1.

In Section 5, we shall consider in some detail the structure and development of the plant body, that is, of the sporophyte, of flowering plants. The remainder of this chapter will be concerned with the principal characteristic of angiosperms, the flower, and with reproduction in flowering plants.

Table 16–1 *Summary of Main Differences Between Monocots and Dicots*

CHARACTERISTIC	DICOTS	MONOCOTS
Flower parts	In 4s or 5s (usually)	In 3s (usually)
Pollen	Basically tricolpate (having three furrows or pores)	Basically monocolpate (having one furrow or pore)
Cotyledons	Two	One
Leaf venation	Usually netlike	Usually parallel
Primary vascular bundles in stem	In a ring	Scattered
True secondary growth, with vascular cambium	Commonly present	Absent

16–30
Eucalyptus regnans, *the largest living angiosperm, a dicot. This specimen, which is 19 meters in girth, was photographed in Tasmania.*

16–31
A tiny duckweed, Wolffia microscopica, *the smallest of the angiosperms. These float submerged in water. The accompanying drawing diagrams the plant body and the extremely simple flower. The duckweeds are monocots.*

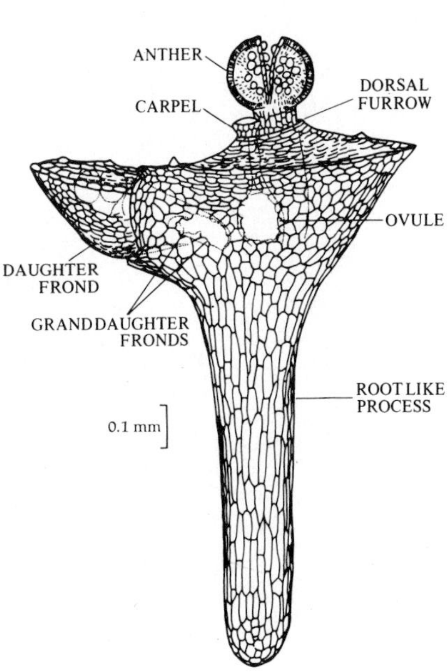

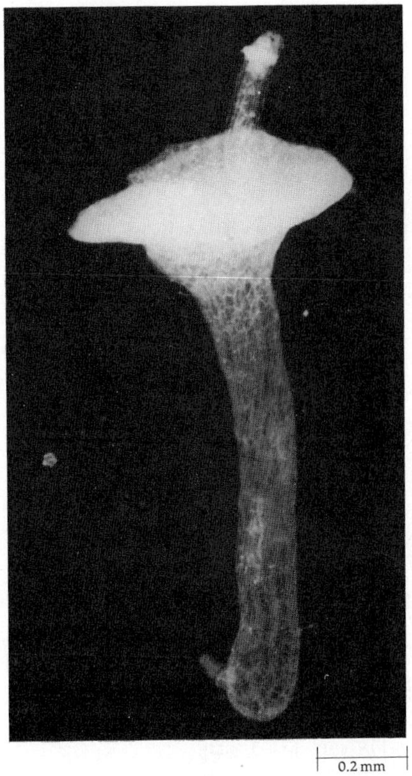

16–32
Nongreen angiosperms. (a) *Dodder* (Cuscuta cuspidata), *which has only small amounts of chlorophyll and is a parasite on other angiosperms. Plants of* Cuscuta, *which are bright orange or yellow, obtain most of their carbohydrate supply from their host plant.* Cuscuta *is a member of the morning glory family (Convolvulaceae).* (b) *Dodder on* Atriplex. (c) *Indian pipe* (Monotropa uniflora), *was long considered saprophytic—depending on decaying organic matter for its nutrition. Recently it has been shown that these "saprophytes" have obligate relationships with mycorrhizal fungi that are attached to a second angiosperm, in this case a green, actively photosynthetic one. The fungus forms a bridge that transfers carbohydrates actively from the photosynthetic plant to the Indian pipe.* Monotropa *entirely lacks chlorophyll. Both dodder and Indian pipe are dicots.*

(a)

(b)

(c)

(a)

(b)

(c)

16–33
Monocots. (a) *A member of the palm family,* Phoenix reclinata, *along the Zambezi River near Victoria Falls in Africa. The date palm belongs to the same genus.* (b) *An epiphytic orchid* (Barkeria lindleyana) *in Costa Rica.* (c) *Rice,* Oryza sativa, *a grass.*

(a)

(b)

(c)

16–34
Dicots. (a) *Butter-and-eggs* (Linaria vulgaris), *a plant of Eurasian origin that is now widespread as a weed along roadsides in North America and elsewhere. The flower is bilaterally symmetrical and has a long, nectar-filled spur.* (b) *Giant saguaro cactus* (Carnegiea gigantea). *The cacti, of which there are about 2000 species, are almost exclusively a New World family. The thick fleshy stems, which conserve water, have taken over the function of photosynthesis.* (c) *Round-lobed hepatica,* Hepatica americana, *which flowers in deciduous woodlands in the early spring. The flowers have no petals but 6 to 10 sepals and numerous spirally arranged stamens and carpels.*

16–35
The parts of a dicot flower. The carpel consists of the stigma, style, and ovary.

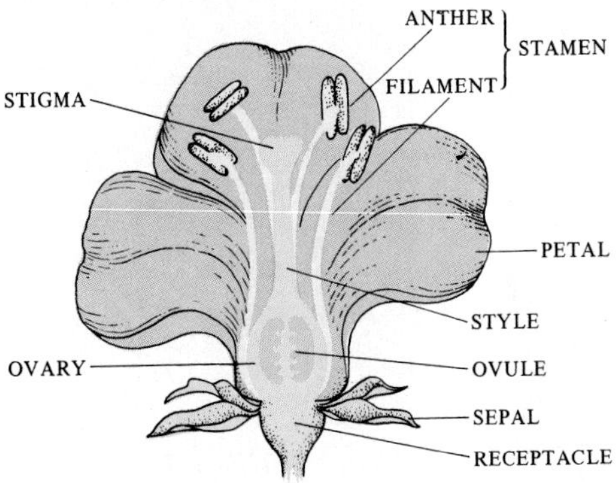

16–36
Diagrams of some of the common types of inflorescences found in the angiosperms.

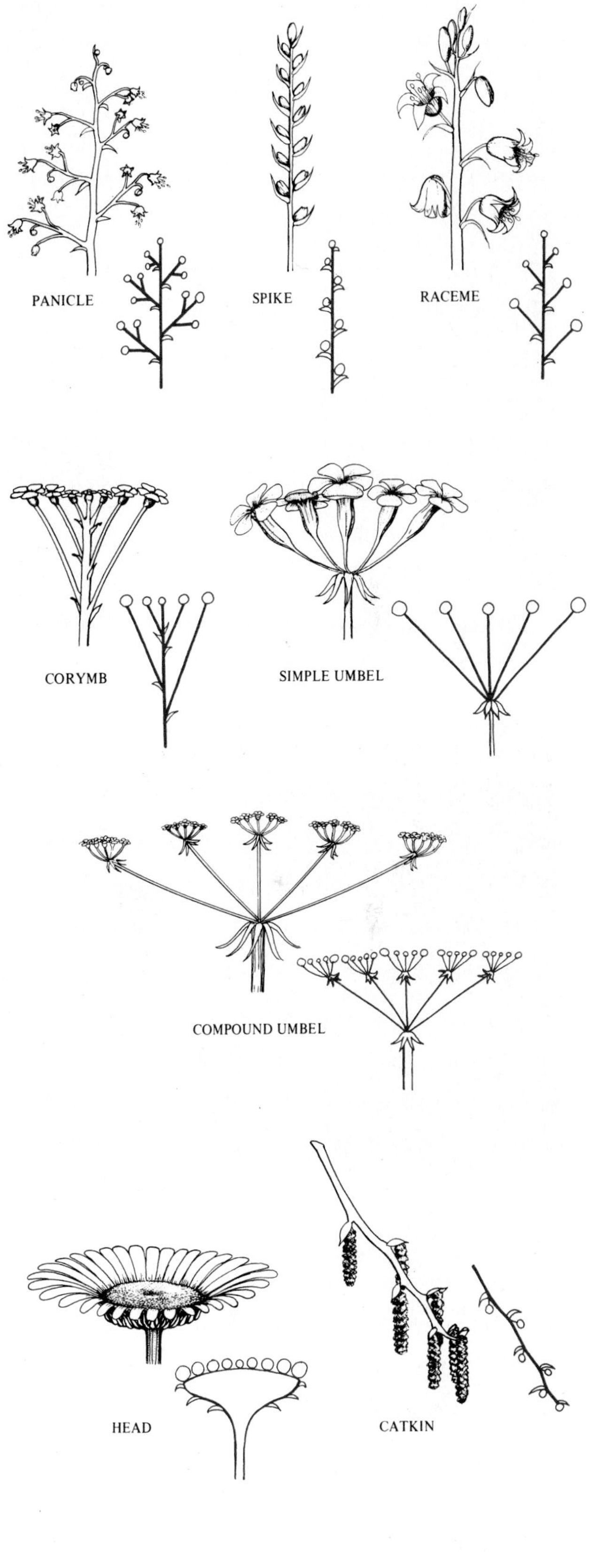

The Flower

The outstanding characteristic of angiosperms is the flower, which is a determinate shoot that bears sporophylls (Figure 16–35). The name angiosperm is derived from the Greek *angeion,* for vessel or receptacle, and *sperma,* for seed. Probably the most distinctive structure of the flower is the carpel, the vessel or container enclosing the ovules, which develop, after fertilization, into the seeds.

Flowers may be clustered in various ways into aggregations called *inflorescences* (Figures 16–36 and 16–37). The stalk of an inflorescence or of a solitary flower is known as a *peduncle.* The stalk of an individual flower in an inflorescence is called a *pedicel.* The portion of the pedicel to which the floral parts are attached is termed the *receptacle.* Like any other shoot tip, the receptacle consists of nodes and internodes. In the flower, the internodes are very short and, consequently, the nodes are very close together.

Most flowers contain two sets of sterile appendages, the *sepals* and *petals,* which are attached to the receptacle below the fertile parts of the flower, the *stamens* and *carpels.* The sepals occur below the petals, and the stamens below the carpels. Collectively, the sepals form the *calyx* and the petals the *corolla.* Together, the calyx and corolla constitute the *perianth* ("around the flower"). The sepals and petals are essentially leaflike in structure. Commonly the sepals are green and the petals brightly colored, although in many flowers both parts are similar in color (Figure 16–38).

The stamens—collectively the *androecium* ("house of man")—are microsporophylls. In the vast majority of living angiosperms, the stamen consists of a slender stalk, or *filament,* on top of which is borne a two-lobed *anther* containing four microsporangia, the pollen sacs.

The carpels—collectively the *gynoecium* ("house of woman")—are megasporophylls that are folded lengthwise, enclosing one or more ovules. A given flower may contain one or more carpels. If more than one carpel is

(a)

(b)

(c)

(d)

(e)

16–37
Inflorescences of (a) *shootingstar* (Dodecatheon pauciflorum), (b) *goldenrod* (Solidago altissima), (c) *lupine* (Lupinus diffusus), (d) *common plantain* (Plantago lanceolata), (e) *water hemlock* (Cicuta maculata). *Using Figure 16–36 as a key, can you identify the types shown here?*

16-38
Lilium superbum. *View of tiger lily flower. In some flowers, such as lily, the sepals and petals are not distinct from one another and the individual members of the perianth are called tepals.*

16-39
Three placentation types. (a) *Parietal.* (b) *Axile.* (c) *Free central.*

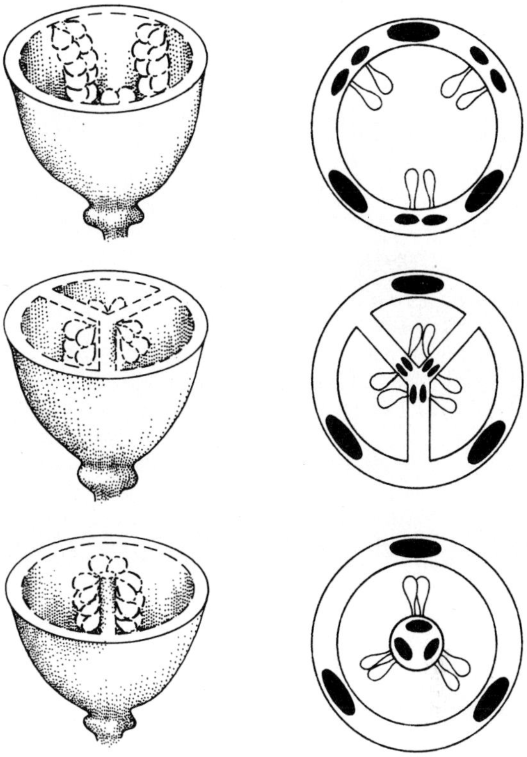

present, they may be separate or fused together, in part or entirely. Sometimes the individual carpel or the group of fused carpels is called a pistil. The word "pistil" comes from the same root as "pestle," the instrument pharmacists use for powdering substances within a mortar, which has a similar shape.

In most flowers, the individual carpels or group of fused carpels are differentiated into a lower part, the *ovary,* that encloses the ovules, and an upper part, the *stigma,* that receives the pollen. In many flowers a more or less elongated structure, the *style,* connects the stigma with the ovary. If the carpels are fused, there may be a common style or stigma, or each carpel may retain separate ones. The common ovary of such fused carpels are generally but not always partitioned into two or more locules, chambers of the ovary in which the ovules occur.

The portion of the ovary to which the ovules are attached is called the *placenta*. The arrangement of the placentae and, consequently, the ovules, varies among different flowers (Figure 16-39). In some flowers, the placentation is *parietal;* that is, the ovules are borne on the ovary wall or on extensions of it. In other flowers, the ovules are borne on a central column of tissue in a partitioned ovary with as many locules as there are carpels. This is *axile* placentation. In still others, the placentation is *free central,* the ovules being borne upon a central column of tissue not connected by partitions with the ovary wall. And finally, in some flowers a single ovule occurs at the very base of a unilocular ovary. This is *basal* placentation.

Given the basic structure of the flower, many variations may exist. The majority of flowers contain both stamens and carpels. Such flowers are said to be *perfect*. If either stamens or carpels are missing, the flower is said to be *imperfect,* and the flowers are either *staminate* or *carpellate* (Figure 16-40). If both staminate and carpellate flowers occur on the same plant, as in corn or oaks, the species is said to be *monoecious* ("in one house"). If they are found on separate plants, the species is said to be *dioecious*, as in American holly or willows.

Any one of the floral parts—sepals, petals, stamens, or carpels—may normally be lacking from a flower. Flowers with all four floral parts are called *complete* flowers. If any part is lacking, the flower is said to be *incomplete*. Thus, an imperfect flower is also incomplete, but not all incomplete flowers are imperfect.

16–40
Staminate and carpellate flowers of scrub oak (Quercus dumosa). *The oaks are monoecious, with the staminate and the acorn-shaped carpellate flowers borne on the same tree.*

16–41
The flowers of (a) *cherry* (Prunus pensylvanica) *and* (b) *apple* (Malus sylvestris). *The cherry flower exhibits perigyny, its sepals, petals, and stamens being attached to a cup-shaped extension of the receptacle, whereas the apple flower exhibits epigyny, the sepals, petals, and stamens apparently arising from the top of the ovary.*

(a)

(b)

The arrangement of the floral parts may be either spiral on a more or less elongated receptacle, or similar parts—such as the petals—may be located at one level in whorls. The parts may be fused to other members of the same whorl (*coalescence*) or to members of other whorls (*adnation*). An example of adnation is the fusion of stamens to the corolla, which is fairly common. When the floral parts of the same whorl are not joined, the prefix *apo-* (separate) or *poly-* may be used to describe the condition (for example, aposepalous or polysepalous, separate sepals). When the parts are coalesced, either *syn-* or *sym-* is used (for example, synsepaly, sympetaly, synandry, syncarpy).

In addition to the floral parts being either spiral or whorled in arrangement, the level of insertion of the sepals, petals, and stamens on the floral axis varies in relation to the level of the ovary or ovaries. If the sepals, petals, and stamens are attached to the receptacle below the ovary, the ovary is said to be *superior,* and the flower is said to be *hypogynous.* In some flowers with superior ovaries, the petals and stamens are attached to the margin of a cup-shaped extension of the receptacle (the hypanthium). Such flowers are said to be *perigynous* (Figure 16–41a). In still other flowers the sepals, petals, and stamens apparently grow from the top of the ovary, which is *inferior.* Such flowers are said to be *epigynous* (Figure 16–41b).

Finally, with regard to variation in floral structure, mention should be made of symmetry. In some flowers, the corolla is made up of petals of similar shape that radiate from the center of the flower and are equidistant from each other. Such flowers are said to be *actinomorphic,* regular, or radially symmetrical. In other flowers, one or more members of at least one whorl are of different form from other members of the same whorl. These flowers are said to be *zygomorphic,* irregular, or bilaterally symmetrical.

The Angiosperm Life Cycle

The gametophytes of angiosperms are highly reduced—more so than those of any other heterosporous plants, including the gymnosperms. The mature microgametophyte consists of only three cells, and the mature megagametophyte, held for its entire existence within the tissues of the sporophyte, in most instances consists of only seven. Both antheridia and archegonia are lacking. Pollination is indirect: pollen is deposited on the stigma, after which the pollen tube conveys two nonmotile sperms to the female gametophyte. After fertilization, the ovule develops into a seed, which is enclosed in the ovary. Concomitantly, the ovary develops into a fruit.

Microsporogenesis and Microgametogenesis

Microsporogenesis is the formation of microspores within the microsporangia, or pollen sacs, of the anther. Microgametogenesis is the development of the microspore into the microgametophyte, or pollen grain.

Initially, the anther consists of a mass of cells without visible differentiation, with the exception of a partly differentiated epidermis. Eventually, within this mass of cells four more or less distinct groups of cells appear, two in each lobe of the two-lobed anther. As each of the four groups continues to divide and enlarge, the peripheral cells of each group show evidence of being sterile, whereas those in the center show subsequent evidence of being fertile (Figure 16–42). The sterile cells develop into the wall of the pollen sac, including nutritive cells, which supply food to the developing microspores. The nutritive cells constitute the *tapetum*. The fertile, or sporogenous, cells become microsporocytes, or microspore mother cells, which divide meiotically, each diploid microsporocyte giving rise to four haploid microspores (Figure 16–43). Microsporogenesis is completed with formation of the single-celled microspores.

During meiosis, nuclear division may be followed immediately by cell-wall formation, or the four microspore protoplasts may be walled off simultaneously. The first condition is common in monocots, the second in dicots. The entire tetrad and the individual grains within it are at first enclosed by callose walls which are not penetrated by plasmodesmata. Following this period, the major features of the pollen grains are established (Figure 16–44). With secretion of the enzyme callase, the individual grains are released by rapid dissolution of the callose walls, and the pollen grains develop a resistant outer wall, the *exine*, and a cellulosic inner wall, the *intine*. The exine is composed of a very resistant polymer know as *sporopollenin*, which apparently is derived partly from the tapetum and partly from the spores following the breakdown of the callose walls. The intine, which is composed of cellulose and pectin, is laid down by the protoplast of the spores.

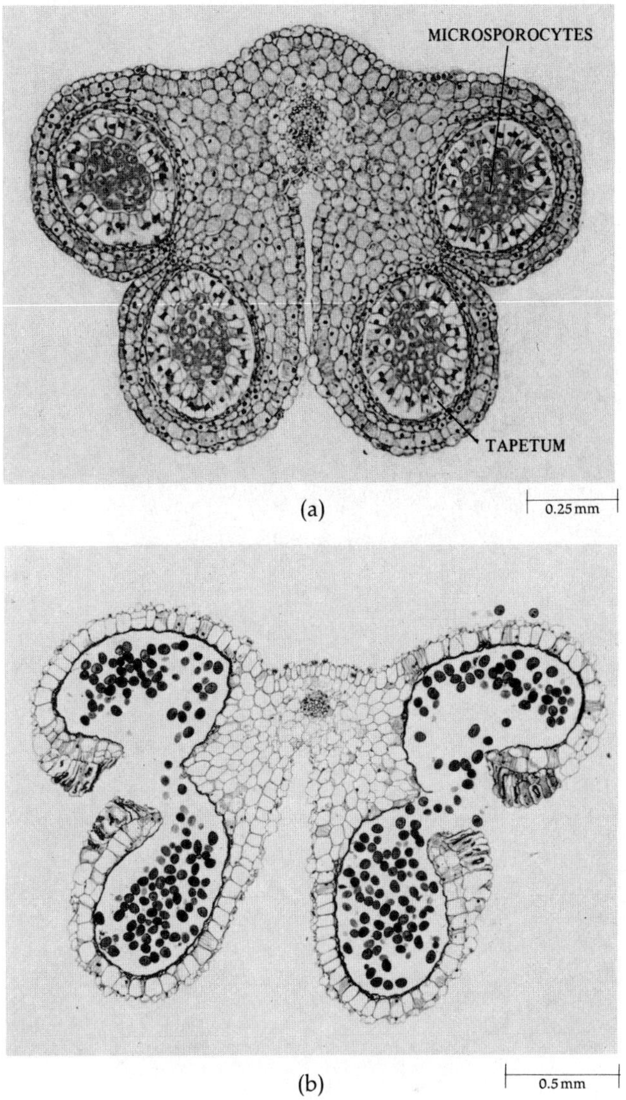

16–42
Transverse section of lily (Lilium) *anthers.* (a) *Immature anther, showing the four pollen sacs containing microsporocytes, or microspore mother cells, surrounded by tapetum.* (b) *Mature, dehiscing anther with pollen grains. Partitions between adjacent pollen sacs break down prior to dehiscence.*

16–43
Spore tetrad in Lilium longiflorum *before* (a) *and after* (b) *the formation of individual spore walls. In* (b), *a spore newly released from a tetrad lies near the bottom right of the micrograph.*

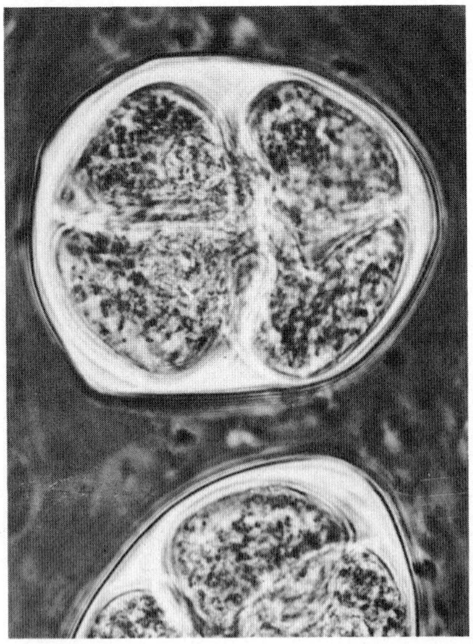

(a) 20 μm

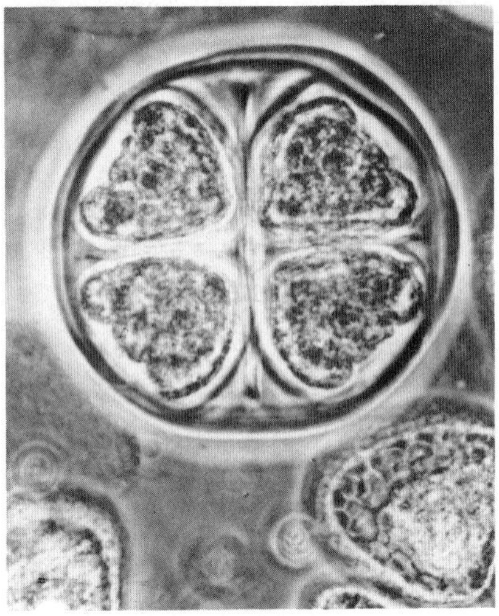

(b)

16–44
The wall of the pollen grain serves to protect the male gametophyte on its often hazardous journey between the anther and the stigma. The outer layer, or exine, is composed chiefly of a substance known as sporopollenin, which appears to be a polymer composed chiefly of fatty acids. The exine, which is remarkably tough and resistant, is often elaborately sculptured. The pollen grains of different species are distinctly different. (a) *A horse chestnut,* Aesculus hippocastanum. (b) *A lily,* Lilium longiflorum. (c) *Chrysanthemum* (Chrysanthemum mycosis). *Spiny pollen grains such as these are common among Asteraceae.* (d) *A morning glory,* Ipomoea purpurea.

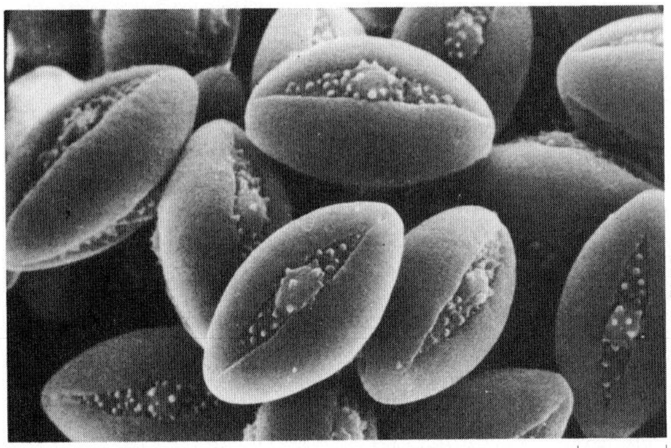

(a) 10 μm

Pollen grains, like spores, vary considerably in size and shape, ranging from less than 20 nanometers to over 250 nanometers in diameter. They also differ in the number and arrangement of the pores through which the pollen tubes ultimately grow. Nearly all families, many genera, and a fair number of species of flowering plants can be identified from their pollen grains alone. In contrast with the larger portions of plants, such as leaves, flowers, and fruits, the pollen grains, because of the chemical nature of the exine, are extremely well represented in the fossil record. Thus pollen provides a valuable index to the kinds of plants and the nature of the climate that prevailed in the past.

Microgametogenesis in angiosperms is monotonously uniform and begins when the uninucleate microspore divides mitotically to form two cells within the original spore wall. One cell is called the *tube cell,* the other the *generative cell* (Figure 16–45). In many species the microgametophyte is in this two-celled stage at the time the pollen grains are liberated by dehiscence of the anther. In others, the generative nucleus divides prior to release of the pollen grains, giving rise to two male gametes, or sperms (Figure 16–46).

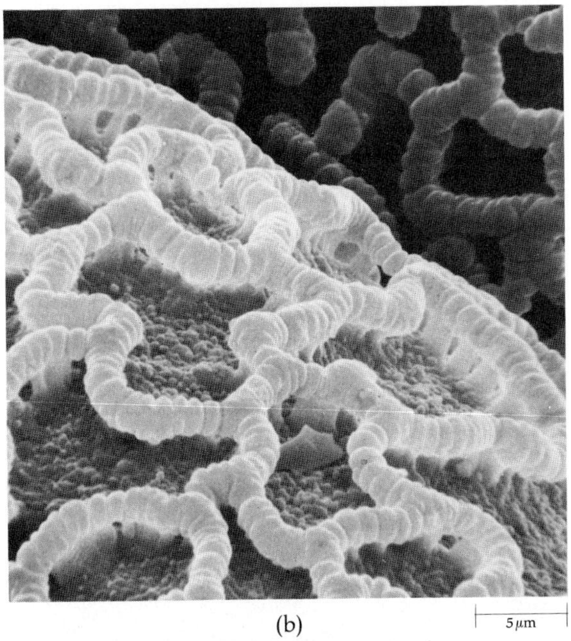

(b)

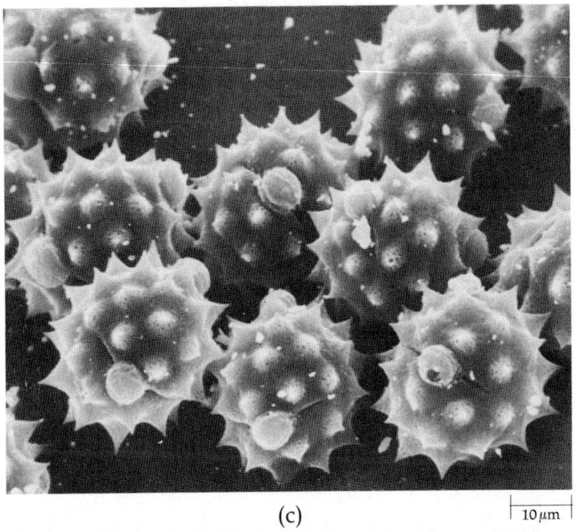

(c)

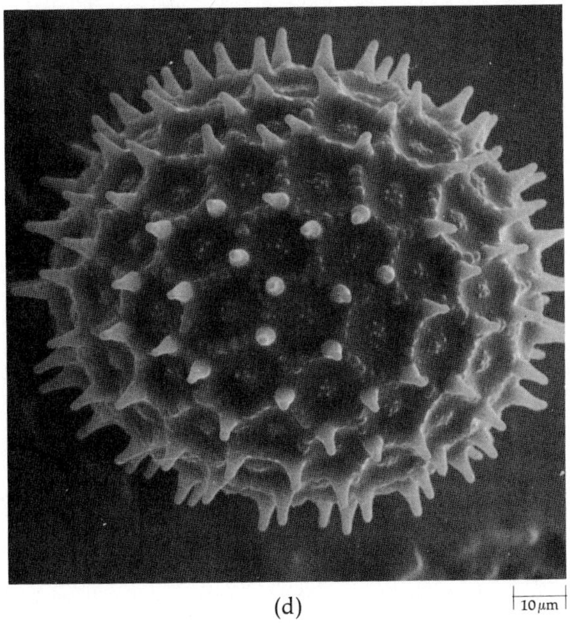

(d)

16–45
Mature pollen grain with two-celled male gametophyte in Lilium. *The generative cell will divide mitotically to give rise to two sperms; the larger tube cell will form the pollen tube.*

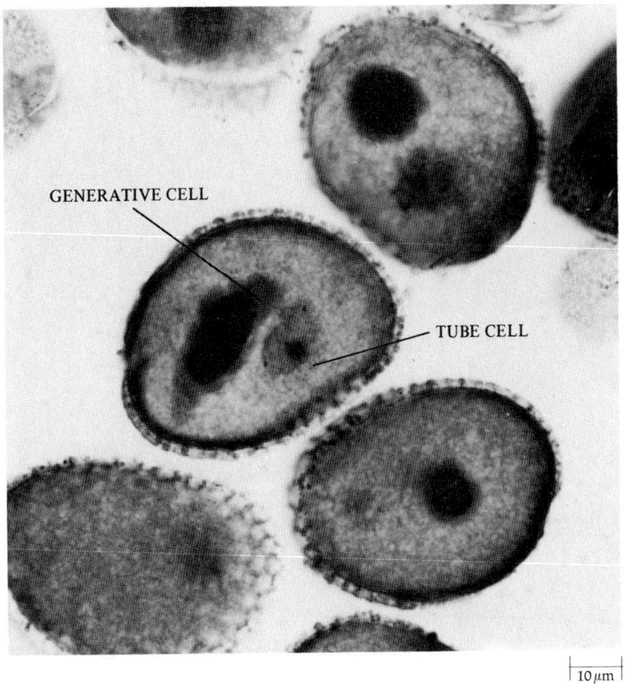

16–46
Mature male gametophyte (microgametophyte) in Polygonatum. *Both sperms and tube nucleus can be seen in the pollen tube.*

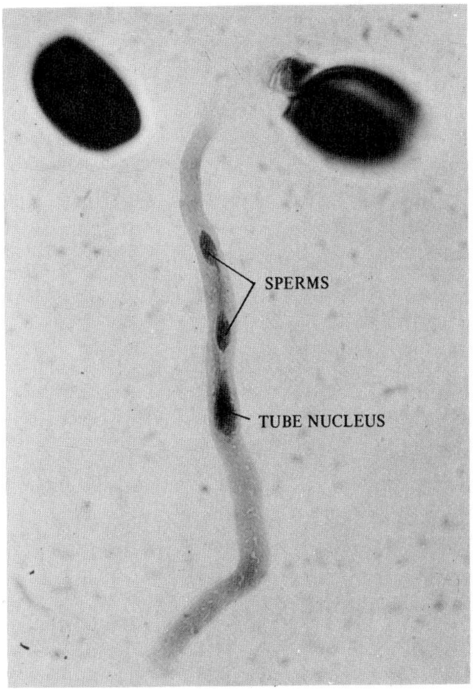

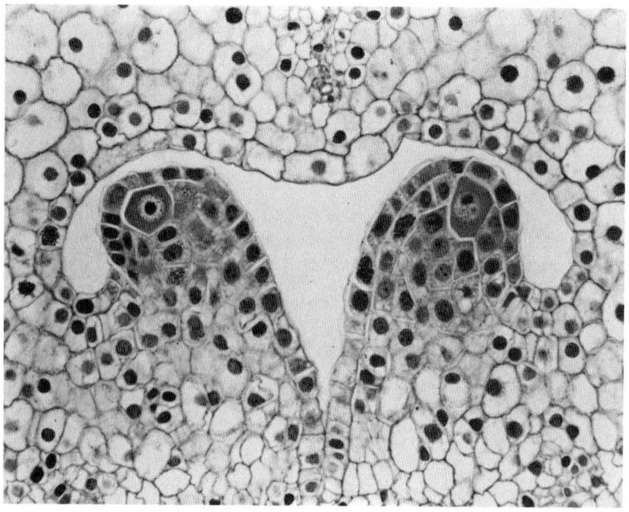

(a) 30 μm

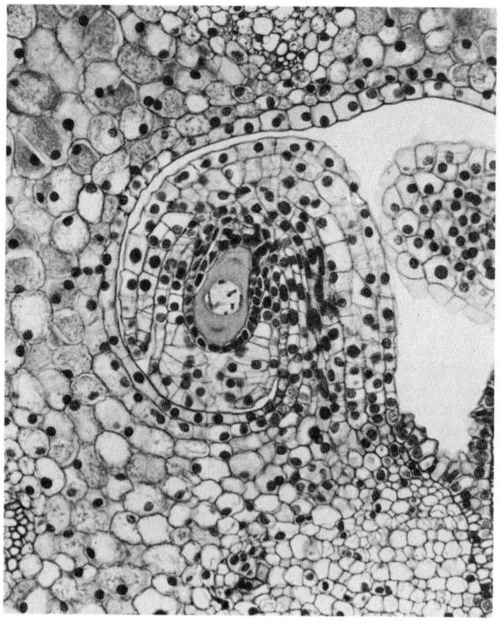

(b) 100 μm

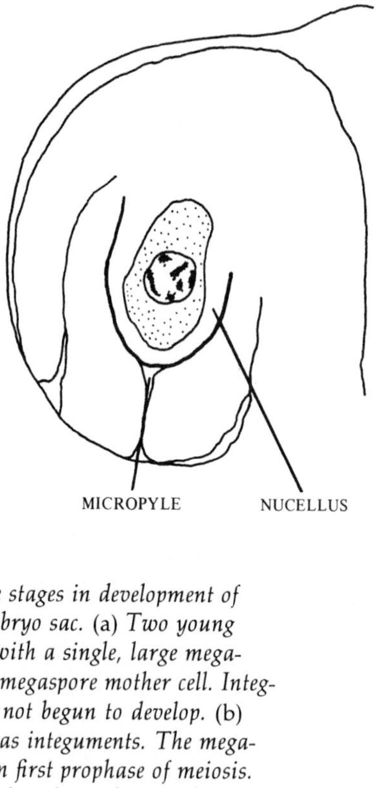

16–47
Lilium. *Some stages in development of ovule and embryo sac.* (a) *Two young ovules, each with a single, large megasporocyte, or megaspore mother cell. Integuments have not begun to develop.* (b) *Ovule now has integuments. The megasporocyte is in first prophase of meiosis.* (c) *Ovule with eight-nucleate embryo sac (only six of the nuclei are shown here). The polar nuclei have not yet migrated to the center of the sac.*

Megasporogenesis and Megagametogenesis

Megasporogenesis is the process of megaspore formation within the nucellus (megasporangium). Megagametogenesis involves development of the megaspore into the megagametophyte.

Structurally, the ovule is relatively complex, consisting of a stalk, the *funiculus,* which bears the nucellus, enclosed by one or two integuments (Figure 16–47). There are from one to many ovules, depending on the species, attached to the placenta of the ovary. Initially the developing ovule is entirely nucellus, but soon it develops one or two enveloping layers, the integument(s), with a small opening, the micropyle, at one end.

Early in the development of the ovule, a single megasporocyte, or megaspore mother cell, arises in the nucellus (Figure 16–47). The diploid megaspore mother cell divides meiotically to form four haploid megaspores, which are generally arranged in a linear tetrad. With this, megasporogenesis is completed. Of the four megaspores, three usually disintegrate, the one farthest from the micropyle surviving and developing into the megagametophyte.

The functional megaspore soon begins to enlarge at the expense of the nucellus, and its nucleus divides mitotically. Each of these nuclei divides mitotically, followed by yet another mitotic division of the resultant nuclei. At the end of the third mitotic division, the eight nuclei are arranged in two groups of four, one group near the micropylar end of the megagametophyte and the other at the opposite (*chalazal*) end. One nucleus from each group migrates into the center of the eight-nucleate cell; these two nuclei are then known as the *polar nuclei.* The three nuclei at the micropylar end of the megagametophyte form cell walls. These cells become organized as the *egg apparatus,* consisting of an *egg cell* and two

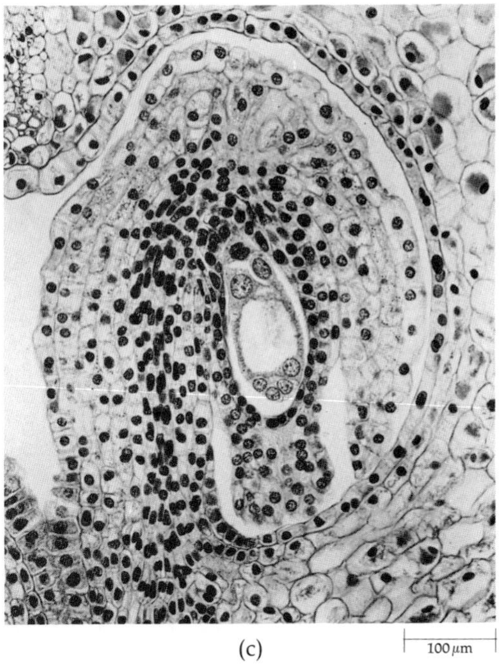

(c)

cellular *synergids*. Cell-wall formation also occurs at the opposite end to form the so-called *antipodal cells*. The *central cell* remains binucleate. This eight-nucleate, seven-celled structure is the mature female gametophyte, or *embryo sac* (Figure 16–47). Different patterns of embryo sac development occur in about a third of the species of angiosperms that have been investigated.

Pollination and Fertilization

With dehiscence of the anthers, the pollen grains are transferred to the stigmas by a variety of vectors (see Chapter 17). Once in contact with the stigma, the pollen grains germinate and form a pollen tube. If the generative cell has not already divided, it soon does so to form the two sperms. The germinated pollen grain, with its two sperms and tube nucleus, constitutes the mature microgametophyte.

The stigma and style are modified both structurally and physiologically to facilitate germination of the pollen grain and growth of the pollen tube. The surface of the stigma is glandular and excretes a solution that becomes sugary. This glandular tissue (stigmatic tissue) is continuous through the style to the ovary, and sometimes even to the funiculus of the ovule. Some styles contain open canals, and the pollen tubes grow either among or on the surfaces of cells lining the canals. Other styles are solid, and the pollen tubes grow between the glandular cells on their way to the ovules.

Commonly the pollen tube enters the ovule through the micropyle and penetrates one of the synergids, which often begins to disintegrate first and, in any case, is destroyed in the process. The sperms and vegetative nucleus are then released into the synergid through a subterminal pore that develops in the pollen tube. Ultimately, one sperm enters the egg cell, and the other the central cell, where it unites with the two polar nuclei (Figure 16–48). You may recall that in the gymnosperms, only one of the two sperms is functional; one unites with the egg and the other degenerates. The involvement of both sperms in angiosperms—the union of one sperm with the egg and of the other with the polar nuclei—is called *double fertilization* and represents one of the principal characteristics of the group. (As previously noted, true fertilization or syngamy involves only the union of gametes, in this case, of sperm and egg.) With union of sperm and egg, a diploid zygote is formed. Union of the other sperm with the two polar nuclei, called *triple fusion*, results in formation of a triploid *primary endosperm nucleus*.

In about a quarter of the species of angiosperms examined, there are more than two polar nuclei and the endosperm is then not triploid; in a few, there is only one, and the endosperm is then diploid and of the same genetic constitution as the embryo. The number of polar nuclei ranges from 1 to 14. Consequently, the primary endosperm nucleus ranges from $2n$ to $15n$.

16–48
Lilium. *Double fertilization. Union of sperm and egg nuclei, "true" fertilization, can be seen in lower half of the micrograph; triple fusion of the other sperm nucleus and two polar nuclei is taking place above.*

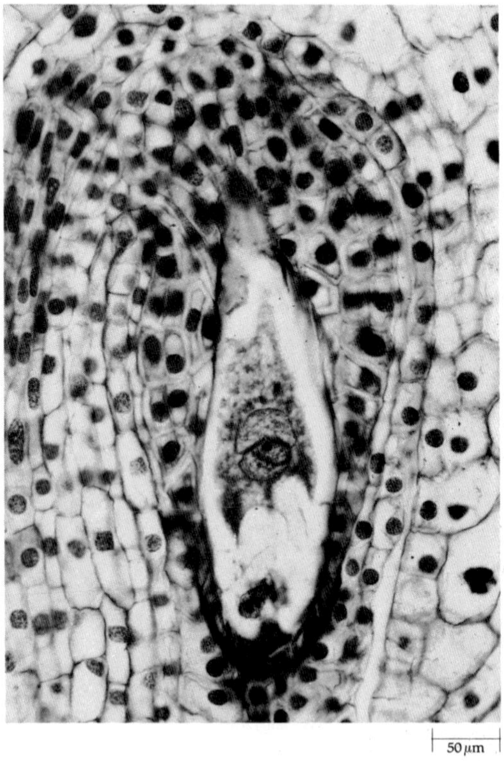

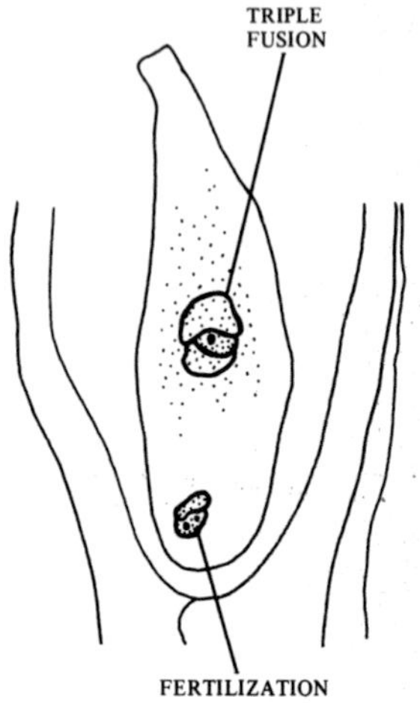

Development of Seed and Fruit

With double fertilization, a number of processes are initiated: the primary endosperm nucleus divides to form *endosperm;* the zygote develops into an embryo; the integuments develop into a seed coat; and the ovary wall and related structures develop into a fruit.

In contrast to the embryogeny of the majority of gymnosperms, which begins with a free nuclear stage, embryogeny in angiosperms is similar to that in lower vascular plants in that the first nuclear division of the zygote is accompanied by cell-wall formation. In the early stages of development, the embryos of dicotyledons and monocotyledons undergo similar sequences of cell divisions, both becoming spherical bodies. It is with the formation of the cotyledon(s) that a distinction first appears between dicot and monocot embryos, for whereas the dicot embryo begins to develop two cotyledons, the monocot embryo remains cylindrical. Embryogeny in angiosperms is presented in detail in Section 5.

Endosperm formation begins with the mitotic division of the primary endosperm nucleus and usually is initiated prior to division of the zygote. In some species a variable number of free nuclear divisions precede cell-wall formation. This type of endosperm formation is known as the *nuclear type* (Figure 16–49a and b). In other species, initial and subsequent mitoses are followed by cytokinesis. This is known as the *cellular type* of endosperm formation (Figure 16–49c). Although endosperm development may take place in a variety of ways, the function of the resulting tissue remains the same: to provide essential food materials for the developing embryo and, in many cases, the young seedling. In some angiosperms, especially dicots, the endosperm is completely digested by the developing embryo. The embryos of such seeds commonly develop fleshy food-storing cotyledons. In other angiosperms, especially monocots, variable amounts of endosperm are present in the seed and are utilized by the embryo when it resumes growth at the time of germination.

Angiosperm seeds differ from those of gymnosperms in the origin of their stored food. In gymnosperms the stored food is provided by the female gametophyte, in angiosperms by endosperm. Endosperm is neither gametophytic nor sporophytic tissue.

Concomitantly, with development of the ovule into a seed, the ovary—and sometimes other portions of the flower or inflorescence—develops into a fruit. As the ovary develops into a fruit, its wall, the *pericarp,* often thickens and becomes differentiated into distinct layers, the exocarp (outer layer), mesocarp (middle), and endocarp (inner), or into exocarp and endocarp only. These layers are generally more conspicuous in fleshy fruits than in dry ones. Fruits are considered in greater detail in Chapter 17.

The angiosperm life cycle is summarized in Figure 16–50 on page 356.

16–49
The growth of the embryo is preceded by the growth of the triploid (3n) *endosperm tissue, which nourishes the young plant. In nuclear endosperms* (a), *no cell walls are formed in the endosperm in the earlier stages and the endosperm is relatively liquid. Cell walls are formed later in some species* (b). *In cellular endosperms* (c), *each nuclear division is followed by the formation of a cell wall.*

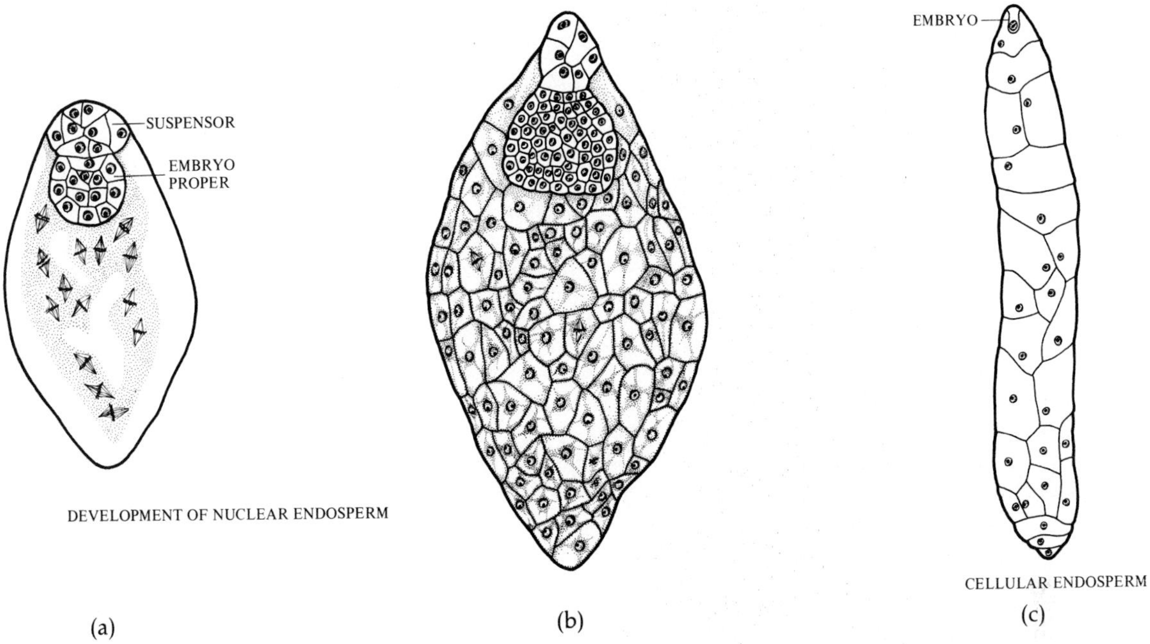

SUMMARY

The subdivision Spermatophytina, the seed plants, consists of the gymnosperms and angiosperms. In addition to seeds, all members of this subdivision bear megaphylls. The prerequisites of the seed habit include: heterospory; retention of a single megaspore within the megasporangium; development of the embryo, or young sporophyte, within the megagametophyte (within the megaspore, within the megasporangium); and integuments. All seeds consist of a seed coat, which is derived from the integument(s), an embryo, and stored food. In gymnosperms the stored food is provided by the haploid female gametophyte, in angiosperms by generally triploid endosperm. The importance of the seed resides in its great survival value. The oldest known seeds occur in strata of the Upper Devonian Epoch.

The living gymnosperms occur in four classes: Cycadinae, Ginkgoinae, Coniferinae, and Gnetinae. The angiosperms, or flowering plants, are divided into two large subclasses, the Monocotyledoneae and the Dicotyledoneae.

The life cycles of gymnosperms and angiosperms are essentially similar: a heteromorphic alternation of generations with large, independent sporophytes and greatly reduced gametophytes. In gymnosperms the ovules (megasporangia plus integuments) are borne exposed on the surfaces of the megasporophylls or analogous structures, whereas in the angiosperms the ovules are borne within the megasporophylls (carpels). The distinctive reproductive structure of angiosperms, the flower, is characterized by the presence of carpels.

The gametophytes of angiosperms are more greatly reduced than those of gymnosperms. At maturity the female gametophyte of most gymnosperms is a multicellular structure with several archegonia. In angiosperms the female gametophyte, commonly called an embryo sac, is most often a seven-celled, eight-nucleate structure. Archegonia are lacking, and the egg cell is associated with two synergid cells. The egg cell and two synergids together are called the egg apparatus.

The male gametophytes of both gymnosperms and angiosperms develop as pollen grains. Antheridia are lacking in both groups of seed plants. In gymnosperms the male gametes, or sperms, arise directly from the body cell, whereas in angiosperms, they arise directly from the generative cell. The germinated pollen grain, with its two sperms, is the mature male gametophyte. Except for the Cycadinae and Ginkgoinae, which have flagellated sperms, the sperms of seed plants are nonmotile and are conveyed to the egg by the pollen tube.

In seed plants, water is not necessary for the sperms to reach the eggs. Rather, the sperms are conveyed to the eggs by a combination of pollination and pollen tube formation. Pollination in gymnosperms is the transfer of pollen from microsporangium to megasporangium. In

16–50
The life of an angiosperm begins with the seed (a) *which consists of an embryo, stored food (endosperm in the seed shown here), and a protective seed coat. The sporophyte develops and eventually flowers* (b). *Within the anther of the flower, microspore mother cells* (c) *develop which divide meiotically, each giving rise to four haploid microspores. Each microspore divides once to form a tube cell and a generative cell. This two-celled structure is the immature microgametophyte, or pollen grain* (d). *Either before or during germination* (h) *the generative cell divides to form two sperms which are conveyed to the egg apparatus by the pollen tube. The germinated pollen grain, with its tube nucleus and two sperms, constitutes the mature male gametophyte.*

Within the ovule, a single megaspore mother cell develops. The megaspore mother cell (e) *gives rise to four megaspores, three of which disintegrate* (f). *The fourth develops into the female megagametophyte* (g), *which at maturity* (i) *is a seven-celled, eight-nucleate structure. The megametophyte is conventionally know as the embryo sac. The pollen germinates on the stigma, producing a pollen tube* (h), *which grows down the style into the ovary* (i). *One sperm nucleus from the pollen tube fuses with the egg to produce the zygote. A second sperm nucleus fuses with the two polar nuclei of the megagametophyte to produce the triploid* (3n) *primary endosperm nucleus. This phenomenon of double fertilization is found only among the angiosperms. The embryo develops within the embryo sac as the integuments of the ovule develop into a seed coat. Eventually, the seed is shed from the ovary.*

The flower shown in this life cycle is the Bermuda buttercup (Oxalis pes-caprae), *which was introduced from South Africa and has become common in many other parts of the world.*

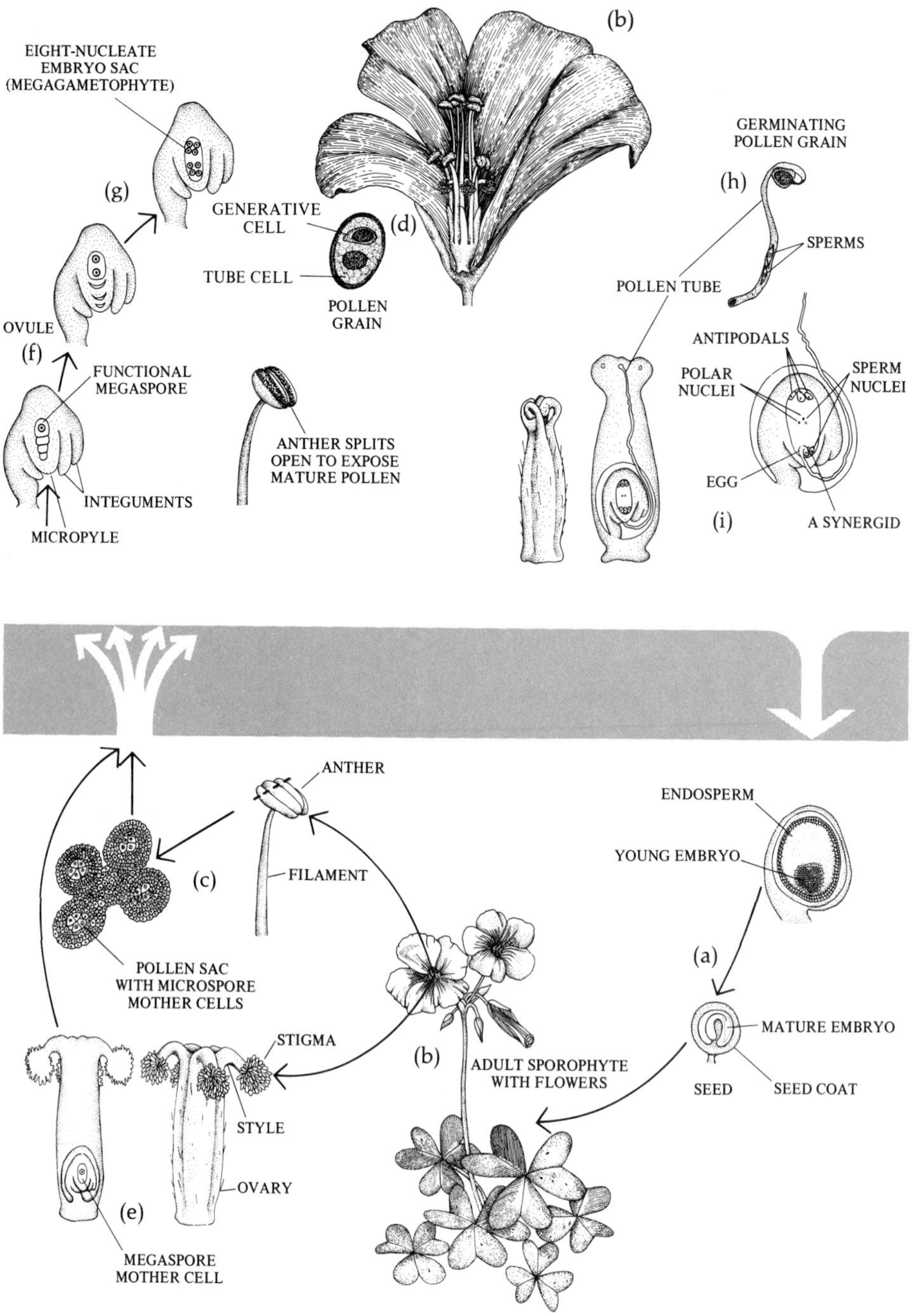

angiosperms, pollination is the transfer of pollen from anther to stigma.

In gymnosperms, one sperm of the male gametophyte (germinated pollen grain) unites with the egg of an archegonium. The second sperm has no apparent function, and it degenerates. In angiosperms, both sperms are functional: One unites with the egg (true fertilization, or syngamy), the other with the two polar nuclei, resulting in formation of a $2n$ zygote and a $3n$ primary endosperm nucleus, respectively. This phenomenon, which represents one of the principal characteristics of angiosperms, is called double fertilization.

With fertilization in gymnosperms, the ovules develop into seeds. In angiosperms, the ovules develop into seeds and the ovaries (and sometimes associated floral parts) into fruits, which enclose the seeds.

SUGGESTIONS FOR FURTHER READING

BIERHORST, D. W.: *Morphology of Vascular Plants,* The Macmillan Company, New York, 1971.

An encyclopedic and magnificently illustrated treatise on the morphology of the vascular plants by one of their leading contemporary students.

FOSTER, A. S., and E. M. GIFFORD, JR.: *Comparative Morphology of Vascular Plants,* 2nd ed., W. H. Freeman and Company, San Francisco, 1974.

A well-organized and general account of the vascular plants, containing much summary and interpretative material.

CHAPTER 17

Evolution of the Flowering Plants

17-1
Longhorn beetle visiting the flower head of a composite. The evolution of the flowering plants is, in large part, the story of increasingly specialized relationships between flowers and their insect pollinators.

The angiosperms make up much of the visible world of modern plants. Trees, shrubs, lawns, gardens, fields of wheat and corn, wildflowers, the fruits and vegetables on the grocery shelves, the bright splashes of color in the florists' windows, the geranium on the fire escape, duckweed and pond lilies, eel and turtle grass, the saguaro cacti and the prickly pears—wherever you are, flowering plants are there too.

In a letter to a friend, Charles Darwin referred to the apparent sudden appearance of this great group of plants as "an abominable mystery." In the fossil strata, one first finds layer after layer of arborescent (treelike) lycopods (pages 308–315) and sphenophytes (pages 316–318), together with primitive gymnosperms—the lush vegetation of which the Carboniferous forests were formed; then ferns, cycadophytes (page 334), seed ferns, ginkgos, and conifers; and finally, during the first half of the Cretaceous period, the angiosperms. During this period, the flowering plants appear, gradually assume worldwide dominance in the vegetation, and, by some 75 million years ago, come to be represented by many modern families and some modern genera (Figure 17-2).

Why did the angiosperms rise to world dominance and then continue to diversify to such a spectacular extent? In this chapter, we shall attempt to answer these two questions, grouping our discussion around three topics—the origin of the angiosperms, the evolution of the flower, and the role of certain chemical substances in angiosperm evolution.

THE ORIGIN OF THE ANGIOSPERMS

It is now generally agreed that the angiosperms evolved from some primitive gymnosperm, probably a shrub. No likely candidates are known from the Cretaceous period, but there are a number of gymnosperms represented earlier in the Mesozoic and in the Paleozoic that display certain combinations of angiospermlike traits. This in itself suggests an earlier origin for the angiosperms than that which the fossil record can document.

(a)

(b)

(c)

17–2
Gymnosperm (a) *and angiosperm* [(b) *and* (c)] *fossils from Upper Cretaceous deposits found in Wyoming.* (a) *Twig and individual cone scales of* Araucarites longifolia, *an extinct species of conifer belonging to the Araucaria family that is at present restricted to the Southern Hemisphere.* (b) *Leaf of a fan palm,* Sabalites montana, *an extinct species distantly related to the palmettos of the southeastern United States.* (c) *Leaf of an extinct species of* Viburnum, V. marginatum, *related to the widespread shrubs generally known as arrow-woods.*

The first fossil remains that can definitely be assigned to the angiosperms date from the early Cretaceous, some 125 million years ago. They consist of pollen closely similar to but distinguishable from the spores of ferns and the pollen of gymnosperms, with a single germ pore. It is likely that the angiosperms that existed more than 125 million years ago had pollen that cannot be distinguished as that of this group. By 120 million years ago, the three-pored angiosperm pollen characteristic of all but the most primitive dicots had appeared in the fossil record, and it appears reasonable to assume that the ancestors of some existing major groups of angiosperms were in existence by that time. By about 100 million years ago, they were more abundant than other plants all over the world.

University of California paleobotanist Daniel Axelrod has suggested that the early evolution of the angiosperms may have taken place away from the lowland basins where fossil deposits are normally found. In the hills and uplands of the tropics the angiosperms could have radiated into a variety of forms without much chance of their remains being preserved in the fossil record. When they invaded the lowlands, about 120 million years ago, they became dominant and literally took over the leading role in the world's vegetation.

Great strides have been made in the earth sciences in the past decade, and we now understand a great deal about the past positions of the continents. About 125 million years ago, when angiosperm pollen first appears in the fossil record, Africa and South America were directly linked with one another and with Antarctica, India, and Australia in a great southern supercontinent

17-3
Relationship between the southern lands at the time of first appearance of the angiosperms in the fossil record. In the Middle Cretaceous, 110 ± 10 million years ago, South America was directly connected with Africa, Madagascar, and India, and via Antarctica with Australia.

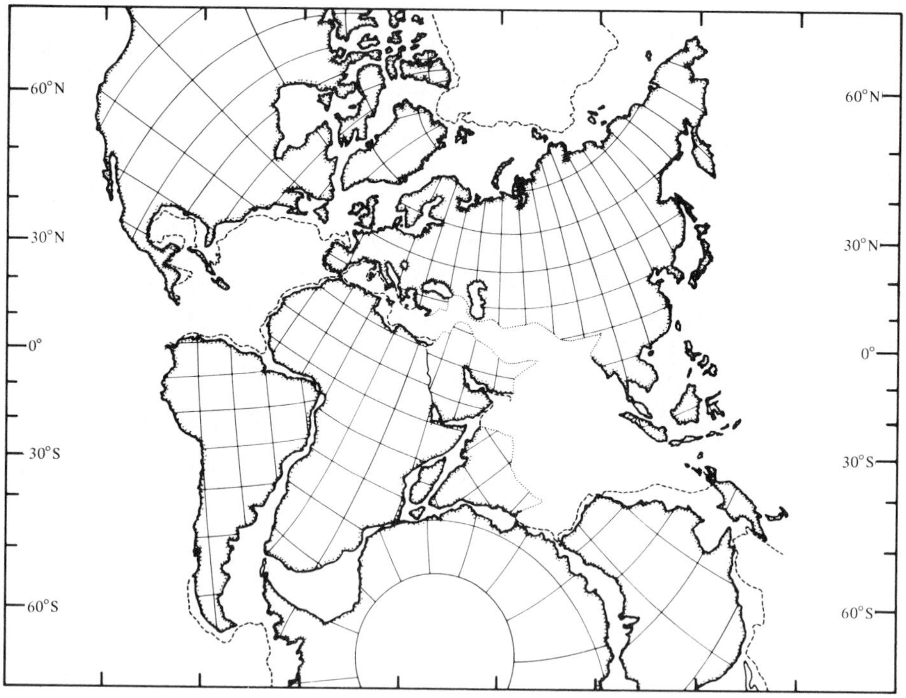

called Gondwanaland (Figure 17-3). Africa and South America began to separate, forming the South Atlantic Ocean at about this time, but they did not come completely apart in tropical regions until about 90 million years ago. India began to move northward at about the same time and collided with Asia about 45 million years ago, thrusting up the Himalayas in the process. Australia began to separate from Antarctica about 55 million years ago, but their separation did not become complete until about 40 million years ago.

Within the central regions of West Gondwanaland, formed by what are now the continents of South America and Africa, the sort of arid-to-subhumid habitats envisioned by Axelrod and others as an important early site of angiosperm evolution would have been frequent. With the final separation of these two continents at about the time the angiosperms became abundant in the fossil record on a worldwide basis, the world climate had changed greatly, especially in these regions, which had become milder, with fewer extremes. These changes are thought to be related to the expansion and evolutionary success of the early angiosperms.

When India and Australia were in the far south, they were covered with cool temperate vegetation. Certain groups of plants and animals are common to southern South America and southeastern Australia-Tasmania at the present time, having achieved their ranges by migration across Antarctica long before the onset of full glaciation there, which occurred about 20 million years ago (Figure 17-4). As Australia moved northward during the past 55 million years, it reached the great zone of aridity flanking the tropics, and the sorts of arid plant communities that are now so widespread there expanded greatly. As it neared Asia, the northern edge of Australia reached truly tropical climates and was invaded by the plants and animals of tropical Asia. The plants and animals characteristic of Asia and Australia have remained largely distinct geographically, however, and the line separating the areas where each is predominant is known as "Wallace's line," after Alfred Russel Wallace, an early naturalist-explorer in these regions. (Wallace is also well remembered as the young man who proposed the theory of evolution by natural selection simultaneously with Darwin.) The original plants and animals of cool-temperate Australia have survived in the southeast corner, in Tasmania, and in New Zealand, which parted company with Australia-Antarctica about 80 million years ago and moved northeast.

The original flora and fauna of India did not fare so well, and there are few remnants surviving. India moved much farther than Australia, crossing the south arid zone, the torrid tropics, and the north arid zone in the process. In consequence of this, nearly all of its original plants and animals became extinct and were eventually replaced by desert, tropical, and mountain flora and fauna from Eurasia.

Very early in the history of the flowering plants, different lines evolved a number of adaptations that made them particularly resistant to drought and cold. Among these were tough leaves, often reduced in size; vessel

(a)

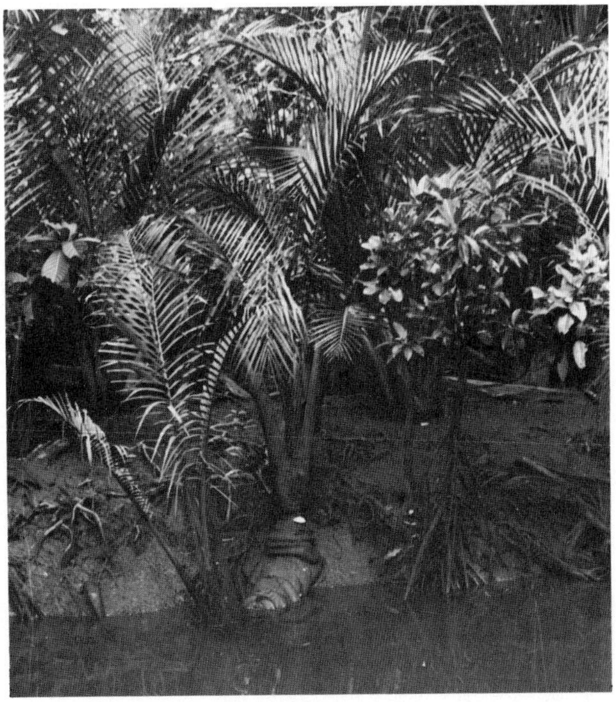

(b)

17–4
(a) *This silver beech* (Nothofagus menziesii), *photographed in Upper Caples Valley, Southland, New Zealand, is a relic of the Antarcto-Tertiary Geoflora forest that extended from what is now southern South America across Antarctica to Australia and New Zealand. These areas remained in proximity into Middle Eocene time, about 45 million years ago.* (b) *The palm* Nypa, *shown growing along the edges of a tidal swamp on Truk in Micronesia, with coconut palms* (Cocos nucifera) *in the background. In Eocene time this palm was distributed widely through the warmer parts of the world, when Africa and South America were in closer contact. It is now confined to Southeast Asia and some Pacific Islands.*

elements (water-conducting cells); and a tough, resistant seed that protects the young embryo from drying out. Many groups of angiosperms became deciduous; that is, they shed their leaves at particular times of year when vegetative growth is not possible. This happened first in tropical areas with periodic drought. The deciduous plants then spread to the north, where parts of the year were so cold that no water was available for growth. The sort of pollination system found in angiosperms, in which free water is not required, is likewise highly favored in dry habitats. The features that make many angiosperms particularly resistant to drought and cold are not found in all of the flowering plants, nor are they restricted to the angiosperms, but they have certainly played a major role in the diversification of the group.

ANGIOSPERMS IN THE TERTIARY PERIOD

In mid-Cretaceous time, some 110 million years ago, the climate of the earth was warmer and more uniform than it is at present. By the end of the Cretaceous period, much of the land was carpeted with a rich and highly developed flora of flowering plants—magnolia and its relatives, maple, grape, dogwood, ebony, ash, willow, birch, oak, figs, breadfruit, and palms. Many plants that we now consider exclusively tropical or subtropical were found far north of their present areas of distribution.

At the start of the Tertiary period, some 65 million years ago, three major assemblages of plants, or *geofloras,* comprised the vegetation of North America (Figures 17–5 and 17–6). The three are known as the Neotropical-Tertiary, Arcto-Tertiary, and Madro-Tertiary Geofloras.

1. The Neotropical-Tertiary Geoflora inhabited the southern half of the continent and consisted of an assemblage of broad-leaved, evergreen angiosperms like those characteristic of the tropical rain forest and savannas at the present time (Figure 17–5a).
2. The Arcto-Tertiary Geoflora occupied the northern portions of North America and was a mixed forest of conifers and hardwood, deciduous angiosperm trees somewhat similar to those now found in the southern Appalachians and portions of China and Japan (Figure 17–5b).
3. Between the Neotropical-Tertiary and Arcto-Tertiary Geofloras, centering in the southern Rocky Mountains and adjacent Mexico, a third major group of plants was making its first appearance—the Madro-Tertiary Geoflora. Characteristic of the western part of the continent, this geoflora was dominated by drought-resistant shrubs and trees with relatively small, tough leaves. The live oak and chaparral communities of California are derivatives of this group (Figure 17–5c), as is much of the other vegetation of the dry mountains of the southwestern United States and adjacent Mexico.

(a)

(b)

(c)

17–5
The vegetation of North America in the Tertiary period was made up of three major assemblages of plants, each of which has its modern counterpart. (a) *A Venezuelan rainforest which resembles the Neotropical-Tertiary Geoflora.* (b) *The Joyce Kilmer Memorial Forest in North Carolina, an Appalachian hardwood forest, a remnant of the Arcto-Tertiary Geoflora.* (c) *The east side of the Napa Valley, California, showing live oak and chaparral, a derivative of the Madro-Tertiary Geoflora. Grapes and fruit trees are being cultivated in the foreground.*

17-6
The history of vegetation types in North America, as illustrated by maps prepared by Daniel Axelrod. The changing outlines of the different vegetation types indicate the major trends that have taken place over the last 60 million years: contraction of the tropics, diversification of the products of the Arcto-Tertiary Geoflora, and the expansion of the Madro-Tertiary Geoflora and of grassland communities, with the recent appearance of the desert as a plant association of regional extent.

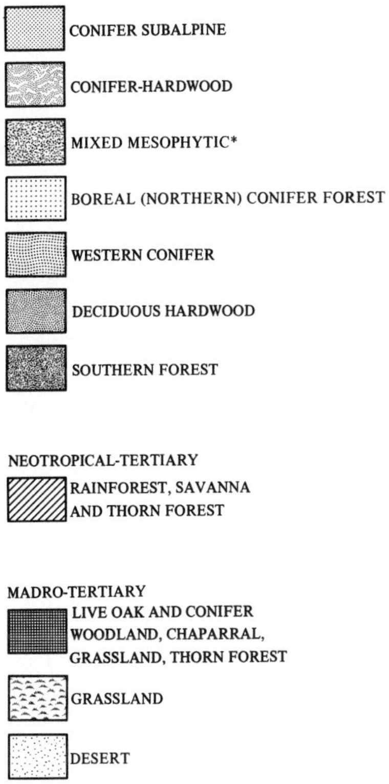

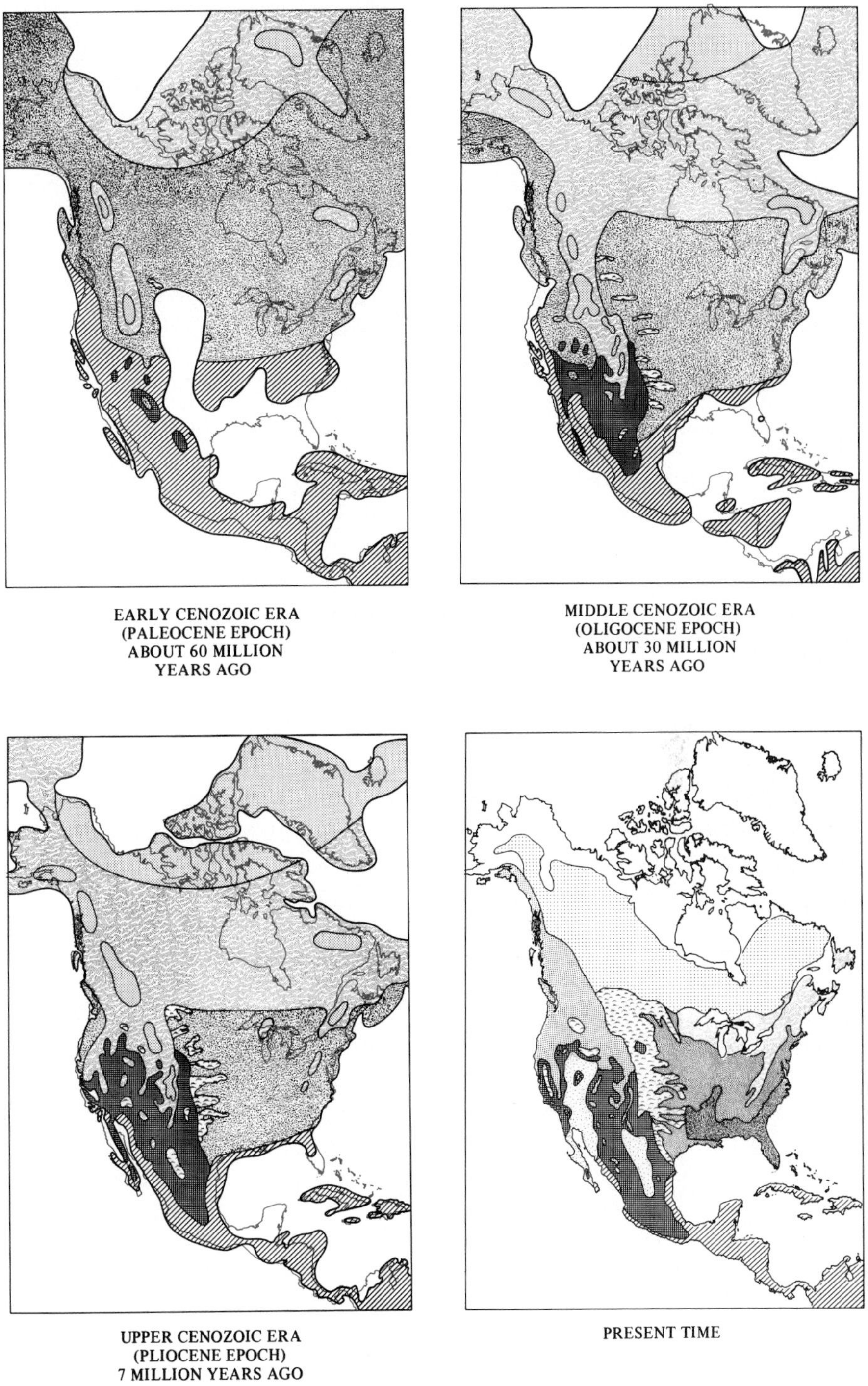

*MESOPHYTES ARE PLANTS THAT GROW IN AN ENVIRONMENT HAVING A MODERATE AMOUNT OF MOISTURE.

We can infer the conditions under which each geoflora lived from what we know of the conditions best suited to related, existing assemblages of plants. Each geoflora shows changes in its boundaries and composition during the course of the Tertiary period. We can trace several prominent trends.

The history of the Neotropical-Tertiary Geoflora has largely been one of progressive contraction. At the start of the Tertiary period, assemblages of plants that were tropical to subtropical in character extended well to the north of the present boundary between the United States and Canada. Now such plants are confined to a much narrower belt from southern Mexico south into South America. During the long-term retreat of these tropical and subtropical communities, there have been abundant opportunities for evolution and the production of new kinds of plants and newly constituted communities at the margins of the tropics, always an evolutionarily strategic area.

Remnants of the Arcto-Tertiary Geoflora—that rich assemblage of conifers and hardwoods that characterized most of the northern half of the continent during the first half of the Tertiary period—are found in the mountains of the southeastern United States, as well as in central China and Japan. In these areas a relatively mild climate and abundant year-round rainfall provide conditions similar to those that prevailed over most of North America some 40 million years ago. Such plants as redwoods, maples, elms, walnuts, butternuts, magnolias, firs, pines, sumac, willows, alders, and birches were common in the Arcto-Tertiary Geoflora. The largely coniferous forests that cover vast areas in Canada and Alaska are derivatives of the Arcto-Tertiary Geoflora. However, most of the hardwood and many of the coniferous species have been eliminated by the progressive development of the colder climates that now characterize these regions. In some parts of the western United States, notably California, a very rich coniferous forest, comprising many species, flourishes at present. Most of the hardwoods that were earlier associated with these conifers were eliminated in the latter part of the Tertiary period as summer rainfall in the western United States fell to lower and lower levels. In addition, away from the coast, the climate of this area has become progressively colder and drier, particularly during the last 10 million years. This has resulted in the elimination of many Arcto-Tertiary species in the area of the Rocky Mountains and the Great Basin (an area comprising most of Nevada and portions of Utah, California, Idaho, Wyoming, and Oregon), a trend that was hastened following the uplift of the Sierra Nevada–Cascade axis about 1 million years ago.

These same trends of climatic deterioration led to the elimination of many species of the Madro-Tertiary Geoflora in the area of the southern Rockies and the Great Basin. In the United States, they now can be found only in areas with a relatively mild climate in the southwestern states. Certain mountain ranges in northern Mexico where there is summer rainfall, such as the Sierra Madre Occidental, contain plants similar to those found over much of the western United States during Tertiary times. It is from these mountain ranges that this geoflora derives its name.

Throughout the history of the angiosperms, new evolutionary lines have been produced as the world climate changed. Both in the origin of the angiosperms and in their subsequent diversification and reassortment into modern communities, the challenge of aridity has been met repeatedly. Drought has been a powerful selective force, with relatively few organisms able to live in the more arid communities. The structure of those angiosperms that are able to grow in arid lands is often profoundly different from that of their ancestors.

EVOLUTION OF THE FLOWER

The Parts of the Flower

The single characteristic that sets the angiosperms off from all other groups is the flower. All that we surmise about the evolution of the flower is based on a comparative study of modern forms, because flowers, being delicate, are rarely found in the fossil record, and if they are found are usually poorly preserved. We have noted previously that the flower is a determinate shoot bearing various leaflike appendages, and now we can consider in more detail how the various parts of the flower are related to leaves.

The Carpel

The carpel in its most primitive form is a folded blade. As you can see in Figure 17-7, in some generalized carpels there is no localized area for the entrapment of pollen grains. Both margins of the folded blade are covered with stigmatic hairs. The folded carpel surrounds the ovules, which are located on its inner surface. The carpels are closed in all the living angiosperms, although the different taxa show different stages in the closure process. Primitive angiosperms have large stigmatic surfaces located directly on the "unsealed" carpels. More specialized forms, including nearly all of those that exist today, have a much smaller stigma held well above the carpel on a style.

The ovules, which were probably in rows along and near the edges of the inner surface of the carpel in the original angiosperms, became arranged in various ways in more specialized carpels. There are many ovules in primitive angiosperms, and few in advanced species. The gynoecium of the primitive angiosperm flower contained a number of separate carpels, which have been reduced in number, fused together, or both, during the course of evolutionary specialization (Figure 17-8).

17–7
Examples of changes that have taken place during the evolution of the angiosperm carpel. The least modified carpel is conduplicate (folded together lengthwise). The folded blade encloses a number of ovules attached to the inner surface. The margins are not sealed and there is no localized stigmatic surface, although stigmatic hairs are extensively developed on the inner surfaces. In the carpels derived from this primitive one, the margins of the blade are more completely sealed, and the pollen-trapping stigmatic surface is localized in a crest.

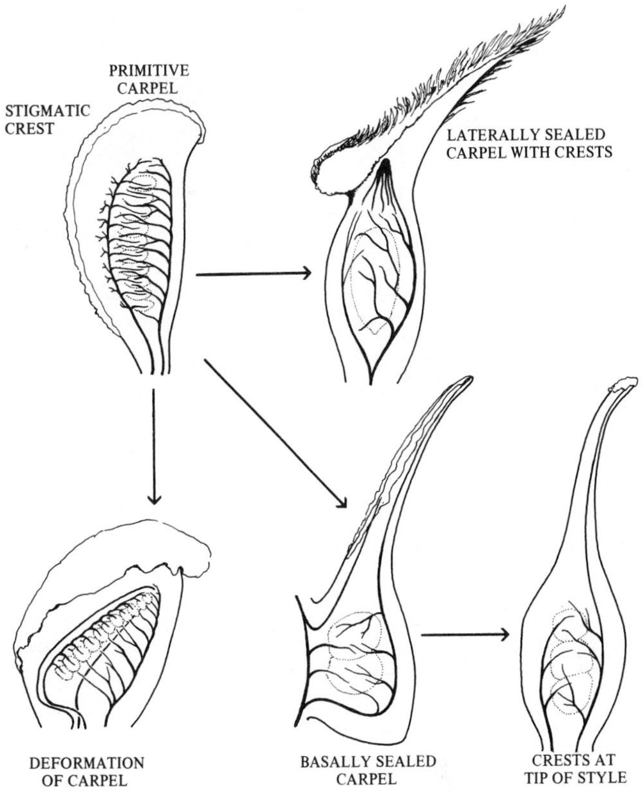

17–8
Presumed evolutionary development of the gynoecium.

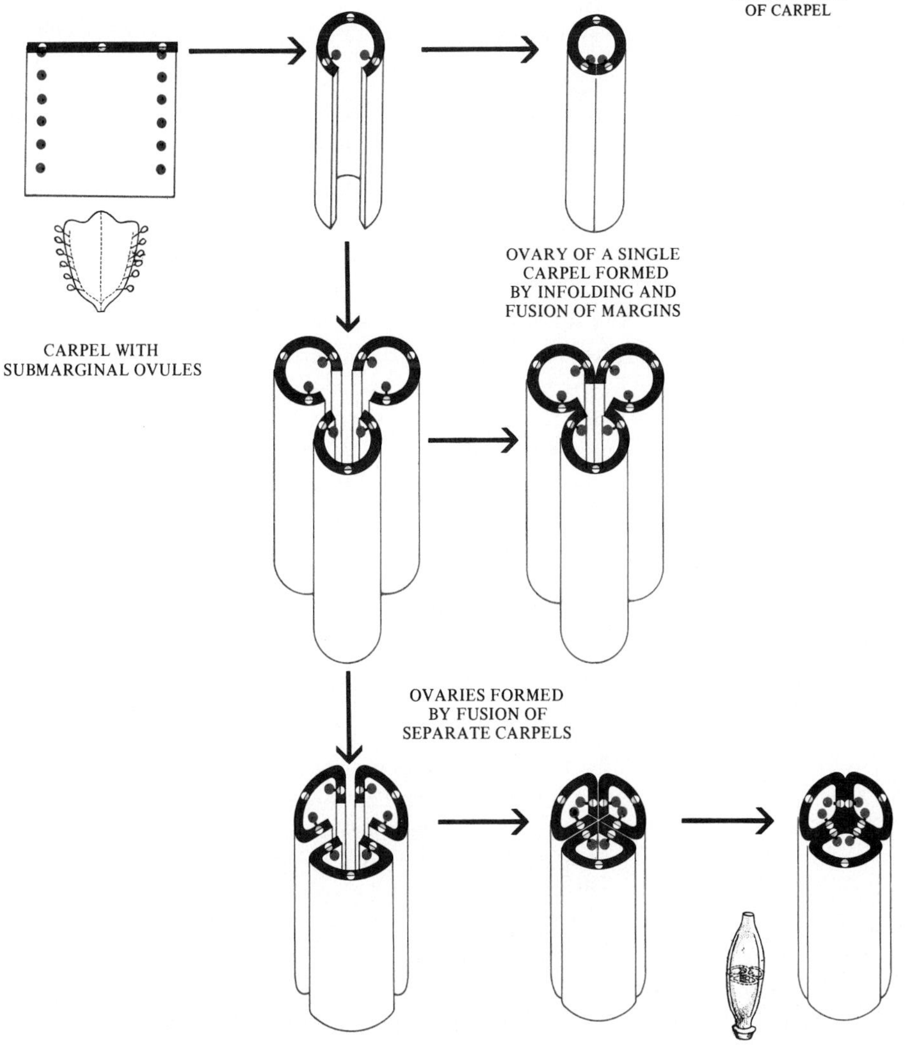

The Androecium

Although the stamens of modern plants rarely resemble leaves, it is possible to find leaflike stamens among the magnolias and their relatives. Like these, the primitive stamen may have been blade-shaped, with sporangia near the center of the blade (Figure 17-9). According to one theory, the blade became differentiated into a slender stalk (the filament), with the sporangia near its apex. An alternative theory, which is also attractive, holds that stamens were derived from slender branch systems bearing terminal sporangia that gradually became fused and, in some instances, leaflike.

In some specialized flowers, the stamens, like the carpels, became fused. Stamens may be fused with one another into columnar structures, as in the melon, pea, and sunflower families; or they may be fused with the corolla, as in the phlox, snapdragon, and mint families. In some evolutionarily advanced flowers, stamens have become secondarily sterile, that is, they have lost their sporangia and have become modified into such structures as nectaries, glands that secrete nectar, a sugary fluid that attracts pollinators and provides food for them. As we shall see, stamens have also played a part in the evolution of petals.

The Perianth

The perianth consists of the sepals and the petals. The sepals of most flowers are green and photosynthetic. Supplied by more than one vascular strand, they resemble the leaves of the plant on which they occur. They are considered to have been derived directly from foliage leaves.

In a few families—for example, the water lilies—petals seem to have been derived from sepals. In most angiosperms, however, the petals appear to be stamens that have lost their sporangia and have become specially modified for their new role—that of drawing attention to the flower. Most petals have just one vascular strand, as do stamens, but sepals, like leaves, normally have three or more.

Petal fusion has occurred in the evolution of many groups of angiosperms, resulting in the familiar tubular corolla that is characteristic of many families. When a tubular corolla is present, the stamens often fuse with and appear to arise from it. In a number of evolutionarily advanced families, the sepals are likewise fused into a tube.

Evolutionary Trends Among Flowers

A modern flower that may resemble the primitive flower in general structure is the magnolia (Figure 17-10). The magnolia flower has numerous carpels, stamens, and perianth parts, all of which are well separated from one another, and the spiral arrangement of these parts on the tip of the stalk, or receptacle, is still clearly evident. The very elongate receptacle and its conelike nature are, however, clearly secondary specializations within the magnolia family not shared by other primitive angiosperms.

By comparing such a primitive flower with specialized ones, we can trace four main trends in flower evolution. Examples are given in Figures 17-11 and 17-12.

1. From flowers with many parts, indefinite in number, flowers have evolved that have few parts, definite in number.
2. The number of types of appendages has been reduced from four in the primitive flower to three, two, or sometimes even just one in the more advanced flowers. The shoot has become shortened so that the original spiral arrangement of parts is no longer evident. Floral parts have become fused.
3. The ovary has become inferior rather than superior.
4. The regularity (radial symmetry) of the primitive flower has given way to irregularity (bilateral symmetry) in the more advanced flowers.

17-9
Stamens of primitive angiosperms. In these woody plants, the anthers are borne on the upper or lower surface of a leaflike microsporophyll. Himantandra *and* Degeneria *are here viewed from below, and* Austrobaileya *and* Magnolia *from above. In most modern angiosperms, the amount of sterile tissue is much less in relation to the sporangia, and the anther is at the end of a usually slender filament. Such stamens are not at all leaflike. These differences are difficult to explain and have led to the hypothesis that leaflike stamens such as these may actually have been derived from branch systems bearing terminal sporangia that gradually became fused and leaflike.*

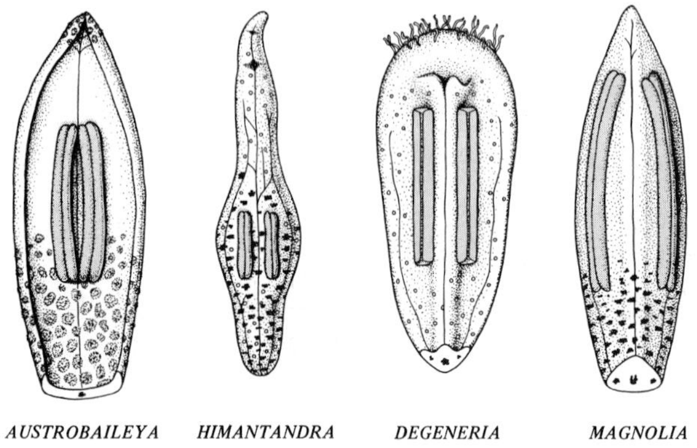

17-10
Southern Magnolia (Magnolia grandiflora). *The cone-shaped receptacle is made up of spirally arranged carpels from which emerge curved styles. Below the styles in the first two photographs are the cream-white stamens.* (a) *Unopened bud slit open to reveal the receptive stigmas; the anthers have not as yet shed their pollen.* (b) *Floral axis of a second-day flower with stigmas no longer receptive and stamens shedding pollen.* (c) *Fruit, showing carpels and bright red seeds, each exserted on a slender stalk.*

(a)

(b)

(c)

17-11
Types of flowers in three common families of dicots, showing changes in the position of the ovary. Many Rosaceae have superior ovaries with the flower parts attached below the ovaries; their flowers are perigynous. The Apiaceae and Onagraceae have inferior ovaries; that is, the flower parts are attached above them. The flowers of these two families are epigynous.

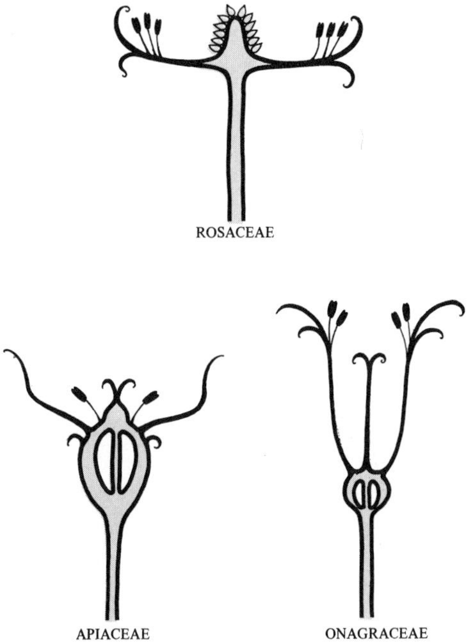

17-12
Examples of specialized trends in flowers. (a) *Wintergreen,* Chimaphila umbellata. *The sepals (not visible in this photograph) and petals are reduced to five each, the stamens to ten, and the five carpels fused into a compound gynoecium, or pistil, with a single stigma.* (b) *Sacred lotus,* Nelumbium nelumbo. *The undifferentiated sepals and petals (which are called tepals) and stamens are numerous and spirally arranged, but the carpels are fused into a compound gynoecium.* (c) *Baby blue eyes,* Nemophila menziesii. *The sepals (not visible), petals, and carpels are each fused and the stamens are reduced to five and arise on the corolla.* (d) *Flower of cultivated cotton* (Gossypium), *showing the column of stamens fused around the style that is characteristic of the family Malvaceae.*

(a)

(b)

(c)

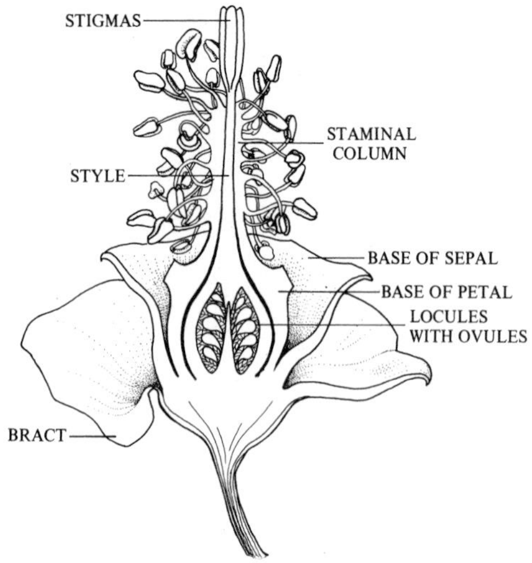

(d)

Examples of Specialized Families

Among the most evolutionarily specialized of the flowers are the Asteraceae (Compositae), dicots, and the Orchidaceae, monocots. These are the two largest families of angiosperms in terms of the number of included species.

Asteraceae

In the Asteraceae (Compositae), the epigynous flowers are relatively small and closely bunched together into a head. In each of the tiny flowers, there is an inferior ovary composed of two fused carpels with a single ovule in one locule (Figure 17-13).

In flowers of Asteraceae, the stamens are reduced to five in number and are usually fused to one another (coalesced) and to the corolla (adnate). The petals, also five in number, are fused to one another and, of course, also to the ovary, and the sepals are absent or reduced to a series of bristles or scales known as the pappus. As in the familiar dandelion, the pappus often serves as an agent of dispersal by wind; in other Asteraceae, as in beggar-ticks *(Bidens),* it may be barbed and serve for attachment to animals. In many Asteraceae, each head includes two types of flowers—disk flowers, which make up the central portion of the aggregate, and ray flowers, which are arranged on the outer periphery. In the ray flowers, which are sometimes completely sterile and often carpellate (pistillate), the fused corolla forms a long strap-shaped "petal" in species such as the sunflower, daisy, and black-eyed Susan.

17–13
(a) *Diagram of the organization of the head of a composite (family Asteraceae). The individual flowers are subordinated to the overall effect of the head, which acts as a large single flower in attracting insects.* (b)–(d) *Composites,* Arctotis acaulis; *Canada thistle,* Cirsium arvense; *and* Cosmos sulphureus.

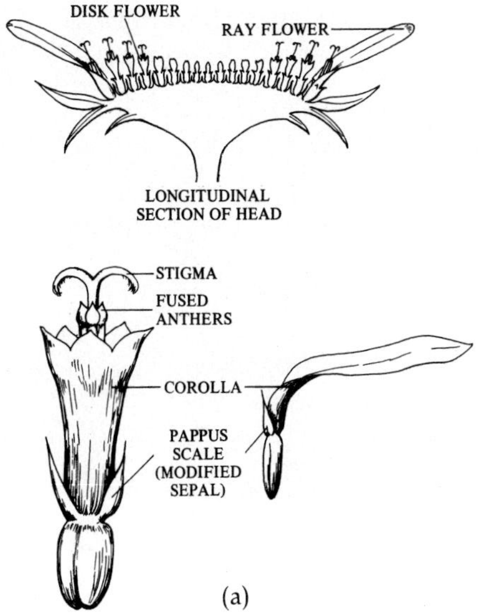

(a)

(b)

(c)

(d)

In general, the head in the composites has the appearance of a large single flower. Unlike many single flowers, however, the head matures over a period of days, with the individual flowers opening serially in a centripetal spiral. As a consequence, the ovules in a given head may be fertilized by a number of different pollen donors. The success of this plan as an evolutionary strategy is attested to by the great abundance of the members of Asteraceae, which, with some 13,000 species, is the second largest family of flowering plants.

Orchidaceae

Another successful flower plan is that of the orchids (family Orchidaceae, the largest of the plant families), which, unlike the composites, are monocots. There are about 17,000 species of orchids, but most of the species are tropical, and only some 140 occur north of Mexico. In the orchids, the three carpels are fused, and, as in the composites, the ovary is inferior. Unlike the composites, however, each ovary contains many thousands of minute ovules; consequently, each pollination event may result in the production of a very large number of seeds. Usually only one stamen is present (in the lady-slipper orchids, there are two), and this is characteristically fused with the style and stigma into a single complex structure, the column. The entire contents of the anther are held and distributed as a unit, the pollinium. The three petals are modified so that the two lateral ones form wings and the third forms a cuplike lip that is often very large and showy. The sepals, also three in number, are often colored and petal-like in appearance. The flower is always irregular (Figure 17–14).

The Agents of Evolution

Land plants, unlike most animals, cannot go from place to place to find food or shelter and cannot move about to seek their mates. In general, they must satisfy these needs passively. The flowering plants, however, evolved a set of features that allows them to "move" in seeking a mate; this set of features is embodied in the flower. By attracting insects and other animals and by directing their activities so that a high frequency of cross-fertilization of the plants will result, the angiosperms have, in a sense, transcended their rooted condition and become just as motile as the higher animals in this one respect. How did this all come about?

The earliest seed-bearing plants—various groups of gymnosperms—were pollinated passively, by the action of the wind. The ovules, which were borne on the leaves or within the cones, exuded drops of sticky sap from their micropyles, just as in the modern conifers. These drops caught the pollen grains and drew them to the micropyle. Insects, probably beetles, feeding on the sap and resin of stems and on leaves, must have come across the protein-rich pollen grains and the sticky droplets from the ovules. Once they began returning regularly to these new-found sources of food supply, they began inadvertently to carry pollen to the ovules. For some plants, even this desultory pollination by wandering insects must have been more effective than wind pollination alone.

The more attractive the plants were to the beetles, the more frequently they would be visited, and the more seeds they would produce. Any chance mutations that made the visits more frequent or more efficient offered an immediate selective advantage. Several evolutionary developments are seen as a direct consequence of pollination by insects. For example, plants that had flowers that provided special sources of food for their pollinators had a selective advantage. In addition to edible flower parts, pollen, and sticky fluid around the ovules, plants evolved that had specialized glands known as nectaries in their flowers. These glands secrete nectar, a nutritious sugary fluid that is attractive to insects.

Attraction of the insects to the flowers raised a new problem, that of protecting the ovule from predatory insects. The evolution by natural selection of the closed carpel was probably a direct consequence of this. Further changes in the shape of the flower, such as the development of the inferior ovary, may also have served to protect the ovules from being eaten.

Another development was the bisexual flower. The presence of both carpels and stamens in a single flower

17–14
(a) *An orchid* (Cattleya superba), *representing the most specialized of the monocot families.* (b) *A comparison of the parts of an orchid flower, shown on the left, with those of a radially symmetrical flower, shown on the right. The lip is a modified petal which serves as a landing platform for insects.*

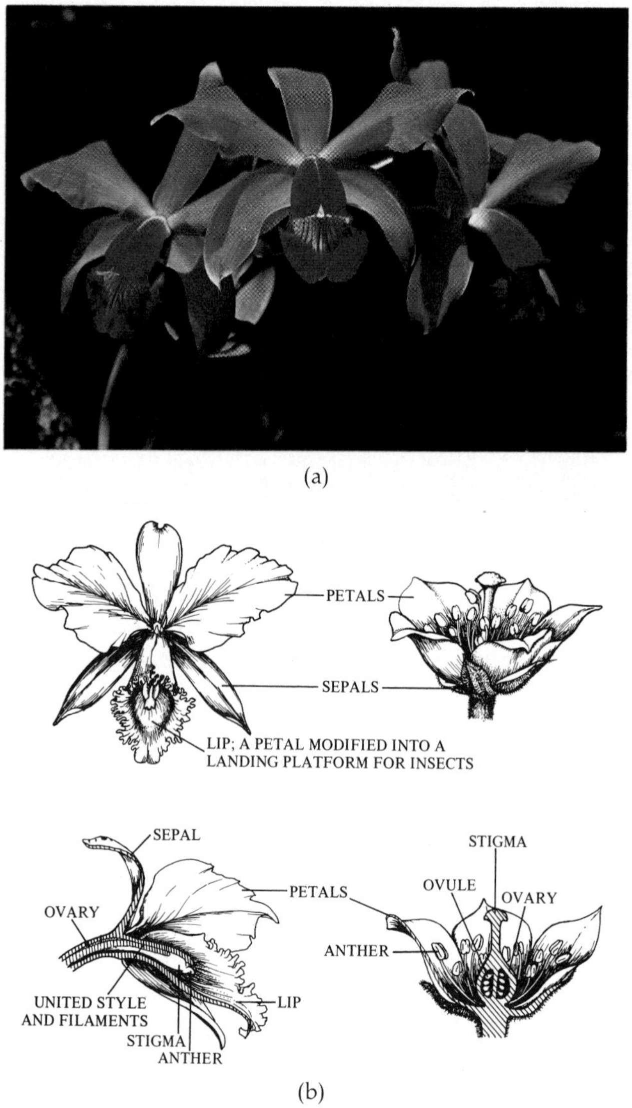

(in contrast, for instance, to the microsporangiate and ovulate cones of conifers) serves to make each visit by the pollinator more effective, because the pollinator can both pick up and deliver at each stop. This in turn led to the early development of genetic self-incompatibility (the inability of an organism to fertilize itself). Genetic self-incompatibility naturally promotes cross-pollination (outcrossing).

Sexual reproduction produces variability in natural populations, thereby allowing them to adapt to changing environments. This source of variability is lost to plants that regularly self-pollinate. Early in the history of flowering plants, mechanisms evolved that make cross-pollination mandatory, even when the flowers are bisexual or when both staminate and ovulate flowers are present on the same plant. The scanning electron micrographs below, from the laboratory of H. Roggen in the Netherlands, reveal how such genetically determined self-incompatibility systems work in Brussels sprouts, Brassica oleracea var. gemnifera.

The stigma of the Brassica flower is covered with protuberances called papillae. A layer of wax covers each papilla. "Compatible" (cross-pollinated) pollen grains stick to this wax layer. The outer wall of a pollen grain, the exine, penetrates the wax layer. When the exine comes into direct contact with the cuticle of the papilla, the papilla collapses, the pollen grain germinates, and the pollen tube digests its way through the cuticle and papilla wall on its way to the ovary.

When pollen grains and stigmata from the same plant come into contact, these "incompatible" pollen grains do not stick to and penetrate the wax layer, and germination does not take place. The mechanism seems to depend on proteins in the exine that carry out "recognition" reactions with the waxy layer similar to the immunological responses in animals. While it is the sporophyte that determines incompatibility in Brassica—the exine is derived from the sporophyte—in other plants it is the intine, and thus the gametophyte, that controls these reactions.

An unpollinated stigma showing numerous papillae (a). Stigma papillae and incompatible pollen grains one hour (b) and 6 hours (c) after pollination; the waxy layer is visible and the pollen grains have not germinated. (d) Half an hour after a compatible cross-pollination. The papillae have collapsed, probably discharging water, and the pollen grains will soon germinate.

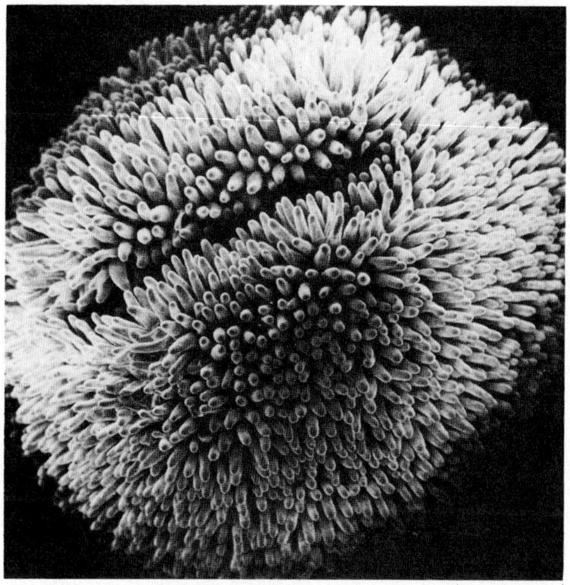

(a) 40 μm

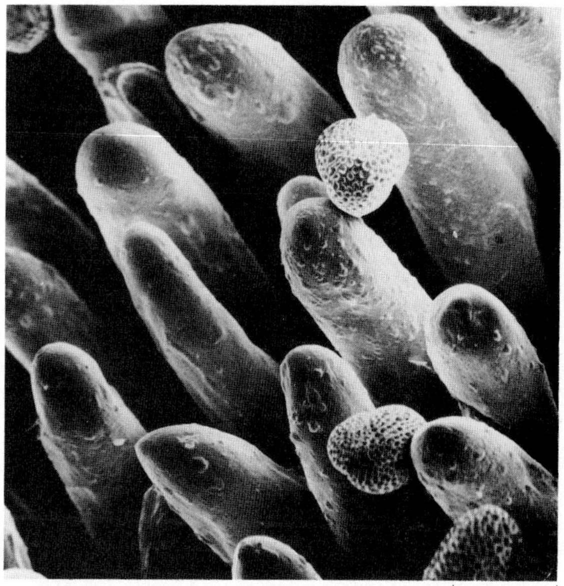

(c) 2 μm

(b) 3 μm

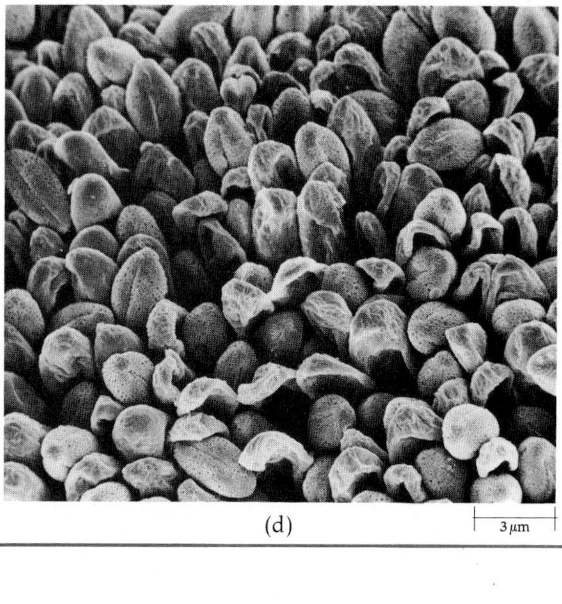

(d) 3 μm

By the beginning of the Cenozoic era, bees, wasps, butterflies, and moths had made their debut in evolutionary history. The rise and diversification of these long-tongued insects, for which flowers are often the only source of nutrition, was a direct result of angiosperm evolution. In turn, the insects profoundly influenced the evolutionary course of the angiosperms and contributed greatly to their diversification. Some of the insects visit a very restricted range of plants for food, and a given species of plant is almost never completely dependent upon only one kind of agent for pollination. Most flowers are visited by more than one type of insect, and, in many cases, even the more casual visitors serve to transfer pollen. Conversely, an insect is almost never dependent upon just one type of flower, and even the ones seemingly the most restricted to a single species do move from one species to another as various kinds come into bloom.

If a given plant species is visited by a relatively narrow spectrum of plant visitors, it tends to become specialized in terms of the characteristics of these visitors. Many of the modifications that evolved in the original primitive flower were special adaptations to encourage constancy. There were two types of changes. The first made a given flower more clearly distinguishable from other flowers. The highly distinctive colors, odors, and shapes, all of which make the flowers so pleasurable to human observers, are "brand names" for the guidance of their pollinators, and flowers thus progressively became more diversified in color, scent, and form (Figure 17–15). The second type of change involved structural features developed to exclude some pollinators, the chief one being the corolla tube. The nectar is hidden below the base of this tube where it can be reached only by long-tongued visitors and not by short-tongued and often indiscriminate ones.

More efficient pollination mechanisms resulted in additional structural changes, such as the fusion of flower parts. The fusion of carpels made it possible for the delivery of a single load of pollen to result in the fertilization of a large number of ovules simultaneously, forming many seeds in contrast to the single seed formed as the result of a wind-blown grain of pollen. (Wind-blown pollen grains are dispersed individually, but a single insect often carries a considerable load of pollen grains on its body.)

The development of the tubular corolla and the fusion of the floral parts occurred independently in members of many diverse families of plants as a result of these evolutionary pressures. Thus, as the flowers diverged in some characteristics, they converged in others. By the early Tertiary period, a course of mutual evolution involving flowers with fused petals, and long-tongued flower-constant insects—chiefly bees—was well established.

In the pages that follow, we shall describe some of the special adaptations of the flower that came about in response to the attentions of specific pollinators. Each must have begun by chance; that is, it was a fortuitous preadaptation on the part of the flower or of its agent. This chance variation gave both flower and pollinator a slight advantage over competing species, and so groups of individuals in which this originally fortuitous variation was somehow enhanced tended to leave more offspring than other populations.

Beetle-Pollinated Flowers

A number of modern species of angiosperms are pollinated solely or chiefly by beetles (Figure 17–16). The flowers of one distinct type are large and borne singly, like those of the magnolia, the lily, the California poppy, and the wild rose; the flowers of another type are small

17–15
"Honey guides" in the foxglove (Digitalis purpurea).

and aggregated in an inflorescence, such as those of the dogwood, elder, and spiraea. Members of some 16 families of beetles are frequent visitors to flowers, although, as a rule, they derive most of their nourishment from other sources, such as sap, fruit, leaves, dung, and carrion. In beetles, the sense of smell is much more highly developed than the visual sense, and beetle flowers are often white or dull in color and often have strong odors (Figure 17-17). These odors are usually fruity, spicy, or like certain fermenation odors and are thus distinct from the sweeter odors of flowers pollinated by bees, moths, and butterflies. Some beetle flowers secrete nectar; in others, the beetles chew directly on the petals or on the food bodies (pads or clusters of cells on the surfaces of various floral parts) and eat the pollen. Most beetle flowers have the ovules well buried beneath the floral chamber, out of reach of the chewing jaws of the beetles.

17-16
A pollen-eating beetle (family Oedemeridae) at the open, bowl-shaped flowers of round-leaved hepatica (Hepatica americana).

17-17
(a) *Western skunk cabbage* (Lysichiton americanum) *is pollinated by small, actively flying beetles (Staphylinidae). The beetles are attracted by its very strong odor and probably guided by the bright yellow modified leaf (spathe) that enfolds the inflorescence. Other species of this plant family (Araceae) have inflorescences with odors resembling dead fish or carrion, and some of these are pollinated by carrion flies. One such plant is shown in Figure 7-18 (page 138) and another, skunk cabbage* (Symplocarpus foetidus), *in* (b). *A number of plants in other families are likewise pollinated by carrion flies and have similar odors; an example is the foul-smelling milkweed,* Stapelia nudiflora (c) *and its relatives, which are almost entirely native to Africa.*

(a)

(c)

(b)

17-18
Bees have become as highly specialized as the flowers with which they have coevolved. Their mouth parts have become fused into a sucking tube containing a tongue. The first segment of each of the three pairs of legs has a patch of bristles on its inner surface. Those of the first and second pairs are pollen brushes that gather the pollen that sticks to the bee's hairy body. On the third pair of legs, the bristles form a pollen comb that collects pollen from the brushes and the abdomen. From the comb, the pollen is forced up into the pollen basket from the upper segment of the third pair of legs. Show here is a honey bee (Apis mellifera) *foraging in a flower of* Salvia. *Note the anthers depositing pollen on the top of the bee's thorax. The stigma will curve down later and assume the same position as the anthers, thus picking up pollen deposited on the bee's thorax earlier.*

Bee-Pollinated Flowers

Bees are the most important group of flower-visiting animals. Both males and females live on nectar, and the females also collect pollen to feed the larvae. Bees have mouth parts, body hairs, and other appendages specially fitted for collecting and carrying these food materials (Figure 17-18). As Karl von Frisch and other investigators of insect behavior have shown, bees can quickly learn to recognize colors, odors, and outlines. The bee's color spectrum is somewhat different from ours; it can see ultraviolet, which is invisible to us, but it cannot distinguish red, which appears black to a bee.

Bees have developed a high degree of constancy to particular kinds of flowers. Such constancy increases the efficiency of the bee, and in bees with narrowly restricted foraging habits, there are often conspicuous morphological and physiological adaptations related to the characteristics of the plant of choice. When constant to this degree, bees exert a powerful evolutionary force for specialization in the plants they visit. There are some 20,000 known species of bees, nearly all of which visit flowers for food.

Bee flowers—that is, flowers that coevolved with bees—have showy, brightly colored petals, which are usually blue or yellow. They often have a distinctive pattern by which the bee can quickly recognize them. This pattern may include a "honey guide," special markings that indicate the position of the nectar (Figure 17-15). Bee flowers are never pure red, and, as special photographic techniques have shown, they often have distinctive markings traced out in ultraviolet (see Essay on page 112).

In bee flowers, the nectary is characteristically situated at the base of the corolla tube and is usually set below the surface, accessible only to a specialized sucking organ. This is, of course, to the bee's advantage as well as to the flower's. Bee flowers characteristically are provided with a "landing platform" of some sort.

Some of the evolutionarily more advanced flowers—orchids in particular—have developed complex passageways and traps that force the bee to follow a particular route in and out of the flower, ensuring that both anther and stigma touch the bee's body at a particular point and in the right sequence.

An even more bizarre pollination strategy has been adopted by orchids of the genus *Ophrys.* The flower resembles a female bee, wasp, or fly (Figure 17-19). The males of these insects emerge early in the spring, before the females of their own species, and the orchids bloom at the same time. The males attempt to copulate with the orchid flower. As a result, they carry away a load of pollen and transfer it to the next flower that they visit. This pollination technique is known as pseudocopulation.

A whole additional spectrum of flowers with different characteristics is pollinated by flies of various kinds, including mosquitoes. These insects neither gather pollen nor provision larval cells. Examples of pollination by flies are shown in Figure 17-20.

17–19
A male bee of the species Eucera longicornis *attempting to copulate with the labellum (the median membrane of the corolla) of the bee orchid,* Ophrys apifera, *in Sweden. The pollinia (masses of agglutinated pollen grains) are being loosened by the pressing of the front side of the bee's head against the upper part of the flower. Flowers of* Ophrys *produce no nectar, but attract the male bees by means of specific odors which differ from species to species. A few other orchids, belonging to diverse genera, are also pollinated in this way.*

(a)

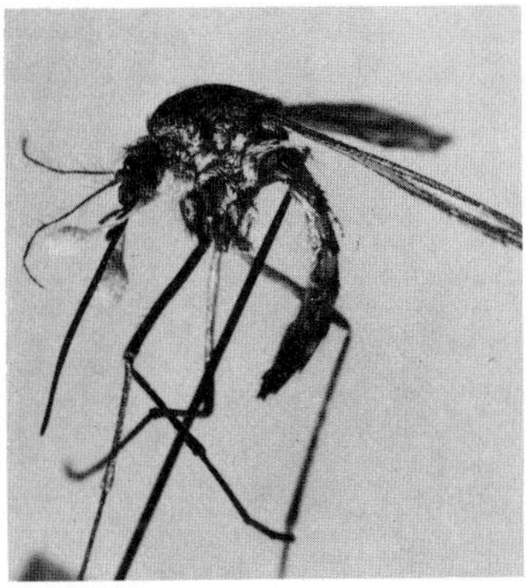

(b)

17–20
Pollination by flies. Small-flowered orchids (Habenaria elegans), *in which the flowers are white or green and relatively inconspicuous* (a), *are visited and pollinated by mosquitoes in north temperate and Arctic regions. The mosquitoes obtain nectar from the flowers.* (b) *A female mosquito of the genus* Aedes, *with an orchid pollinium attached to its head.* (c) *A flower fly of the family Syrphidae at flowers of the lily,* Zigadenus fremontii. *Note the conspicuous yellow nectaries.*

(c)

Flowers Pollinated by Moths and Butterflies

Flowers that have coevolved with diurnal (active during the day rather than at night) moths and butterflies are similar in many respects to bee flowers because these insects are guided to flowers by a combination of sight and smell similar to that of the bees (Figure 17–21). At least some species of butterfly, however, may be able to see red as well as blue and yellow, and some butterfly flowers are red and orange. Most moths are night fliers, and the typical moth flower—as seen in many species of tobacco *(Nicotiana),* for example—is white and has a heavy fragrance, a sweet penetrating odor that is emitted only after sunset. Other moth flowers, although not white, display colors that stand out against a dark background in the evening (for example, the yellow evening primrose, *Oenothera hookeri,* or the pink *Amaryllis belladonna*).

The nectary of moth and butterfly flowers is found at the base of a long slender tubular corolla or a spur where it is accessible only to the long tongues of moths and butterflies. There is a close correspondence between the length of the tongue in a species of moth or butterfly and the length of the corolla tube or spur in some of the species of flowers that it visits. Moths do not usually enter flowers as bees do, but hover above them inserting their long tongues into the floral tube. Consequently, moth flowers do not have the landing platforms, traps, and elaborate internal machinery seen in some of the bee flowers. One of the most specialized moth–flower relationships is shown in Figure 17–22.

17–21
(a) *A skipper* (Poanes hobomok) *visiting a flower of a wild geranium* (Geranium maculatum). (b) *Painted lady* (Cynthia annabella), *a butterfly. Their long sucking tongues, which are coiled up at rest and extended when feeding, vary in length from a few millimeters in some of the smaller moths, 1 to 2 centimeters in many butterflies, 2 to 8 centimeters in some hawk moths of the north temperate zones, and 10 to 25 centimeters in tropical hawk moths.*

(a)

(b)

Bird-Pollinated Flowers

Some birds regularly visit flowers to feed on nectar, floral parts, and flower-inhabiting insects; many of these birds also serve as pollinators. In North and South America, the chief bird pollinators are hummingbirds (Figure 17–23); in other parts of the world, flowers are visited regulary by representatives of other specialized bird families (Figure 17–24). Pure red flowers are inconspicuous to insects and are thus often pollinated by birds; in Europe, where there are no bird pollinators, there are no native pure red flowers.

Although bird flowers have a copious thin nectar (some even drip with nectar when the pollen is ripe), they usually have little odor, which is a corollary of the fact that the sense of smell is feebly developed in birds. Birds have keen vision, however (with a color sense much like our own), and most bird flowers are colorful, with red and yellow being the most common colors.

Bird-pollinated flowers include the red columbine, fuchsia, passion flower, eucalyptus, hibiscus, and many members of the cactus, banana, and orchid families. They are large or are parts of large inflorescences, features that can be correlated with their importance as visual stimuli and their role in holding large amounts of nectar.

17–22
Yucca moth (Tegeticula yucasella) *scraping pollen from a yucca flower. The female moth visits the creamy white flowers by night and gathers pollen, which it rolls into a tight little ball and carries in its specialized mouth parts to another flower. In the second flower, it pierces the ovary wall with its long ovipositor and lays a batch of eggs among the ovules. It then packs the sticky mass of pollen through the openings of the stigma. Moth larvae and seeds develop simultaneously, with the larvae feeding on the developing yucca seeds. When the larvae are fully developed, they gnaw their way through the ovary wall and lower themselves to the ground, where they pupate until the yuccas bloom again. An estimated 20 percent of the seeds are eaten. Few pollination relationships are as specialized as this one.*

17–23
Male white-necked jacobin hummingbird, Florisuga mellivora. *The flower being pollinated is an abutilon.*

17–24
Male marico sunbird (Cinnyris mariquensis) *perching and feeding on the red flowers of an aloe in South Africa.*

That bird and other animal pollinators usually restrict their visits to the flowers of a particular plant species is only one factor promoting outcrossing. For outcrossing to result, it is also necessary that the pollinator not confine its visits to a single flower or to the flowers of a single plant. When flowers are visited regularly by large animals with a high rate of energy expenditure, such as birds, sphinx moths (hawkmoths), or bats, they must produce large amounts of nectar to support the metabolic requirements of these animals and keep them coming back. On the other hand, if abundant supply of this nectar is available to animals with a lower rate of energy expenditure, such as small bees or beetles, these will tend to remain at a single flower and, being satisfied there, will not move on to other plants and thus bring about outcrossing. As a consequence, in the course of evolution, flowers that are regularly pollinated by animals operating at a high rate of energy consumption, such as hummingbirds, have tended to enclose their nectar in tubes or otherwise make it unavailable to smaller animals of lower energy consumption. Similarly, red color is a signal to birds, but not to insects; red is outside the insects' visual spectrum and blends with the background. Birds, like ourselves, do not respond very strongly to odor clues. Odorless, red flowers then are inconspicuous to, and do not tend to attract visits by, insects, an adaptation that is advantageous in view of their copious production of nectar.

Bat-Pollinated Flowers

Flower-visiting bats are found in both the Old World and the New World tropics. Bats that derive all or most of their nourishment from flowers have slender and elongated muzzles and long extensible tongues, sometimes with a brushlike tip, and their front teeth are often reduced in size or missing altogether.

Bat flowers are similar in many respects to bird flowers, being large, strong flowers with copious nectar (Figure 17-25). Because bats feed at night, bat flowers are usually dull in color and open only at night. They are often tubular, or protect their nectar in other ways. Bats are attracted to the flowers largely through their sense of smell, and bat-pollinated flowers characteristically have very strong fermenting or fruitlike odors. Bats fly from tree to tree, often in squeaking flocks, lap nectar, eat pollen and other flower parts, and carry pollen on their fur.

Wind-Pollinated Flowers

At about the turn of the century, it was thought by leading botanists that wind-pollinated flowers were the most primitive of the angiosperm flowers and that all others had been derived from these. The conifers, thought by some scientists at the time to be direct ancestors of the angiosperms, have small, colorless, odorless, unisexual clusters of cones and are pollinated by the wind. Similarly, wind-pollinated flowers, of which there are large numbers, have dull colors, are relatively odorless, and do not have nectar; the petals are small or absent; and the sexes are often separated. However, studies of other characteristics of these wind-pollinated angiosperms, in particular the wood (in which the vessel elements are often specialized), have convinced plant anatomists that wind-pollinated angiosperms evolved not from the conifers but from earlier insect-pollinated plants. According to present evidence, the wind-pollinated angiosperms originated independently from different ancestral stocks. They are best represented in temperate regions and are relatively rare in the tropics. In temperate climates, many trees of the same species, for example, are found close together, and the dispersal of pollen by wind can take place readily in early spring, often when the trees are leafless.

Wind-pollinated flowers usually have their stamens well exposed, where the pollen can be caught by the wind. In some, the anthers are suspended from long filaments hanging free from the flower (Figures 17-26 and 17-27). The abundant pollen grains, which are generally smooth and small, do not adhere to one another like the pollen grains of insect-pollinated species. The large stigmas are characteristically exposed, and they often have branches or feathery outgrowths adapted for intercepting pollen grains. Most wind-pollinated plants

17-25
A bat (Leptonycteris sanborni) *feeding on and pollinating* Agave palmeri *(century plant) in Arizona.*

17–26
Unlike most angiosperms, the grasses are wind-pollinated. (a) *Corn* (Zea mays) *has staminate inflorescences (tassels) at the top of the stem, and ovulate inflorescenses with long exserted stigmas, the silk, below.* (b) *Grasses characteristically have enlarged, feathery stigmas to receive the wind-blown pollen, which is shed by the hanging anthers, as seen here in* Bromus inermis. (c) *Scanning electron micrograph of a pollen grain of corn* (Zea mays), *showing the smooth pollen wall, found in most wind-pollinated plants, and a single aperture, characteristic of the monocots.*

(a)

(b)

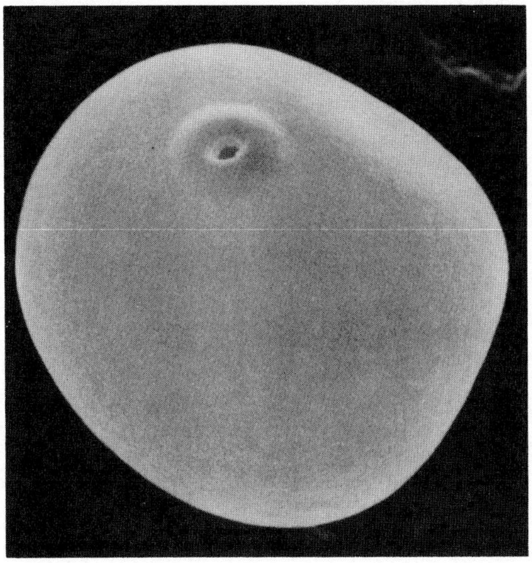

(c)

17–27
Grass flowers (florets) usually develop in clusters. (a) *As a cluster matures, a single pair of dry, chaffy bracts—the glumes—separate a little, exposing the elongating spikelet, with from one to many florets (depending on the type of grass) attached to a central axis, or rachilla. Each floret* (b) *is surrounded by two distinctive bracts of its own, the palea and the lemma. These are forced apart, exposing the inner parts of the flower* (c) *by the swelling of the lodicules, small, rounded bodies at the base of the carpel, and spread wide when the grass is in flower. The stamens, usually three in number, have slender filaments and long anthers; and the stigmas are typically long and feathery and so are efficient at intercepting the wind-dispersed pollen.*

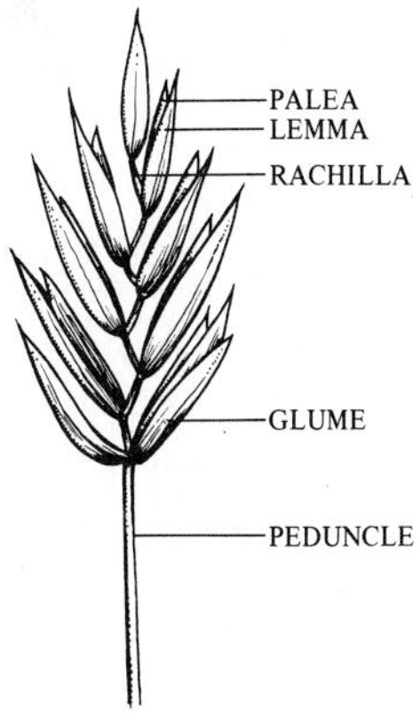

(a)

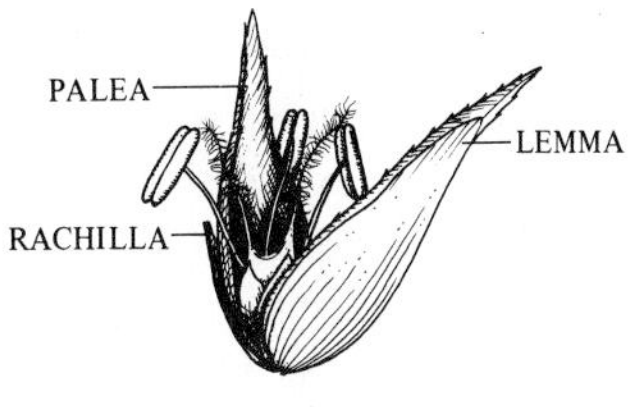

(b)

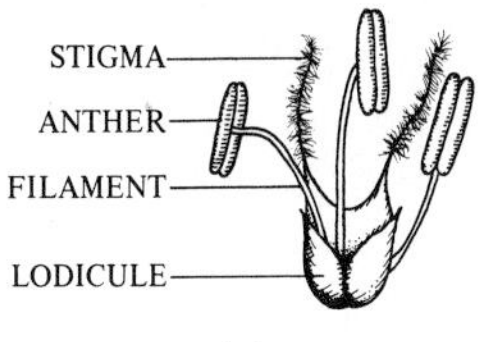

(c)

have ovaries with single ovules (and hence single-seeded fruit), because each pollination consists of the meeting of one pollen grain with one stigma and leads to the fertilization of one ovule for each flower. Thus, the oak flower produces a single acorn and the grass flower one grain, but plants with very small flowers tend to have, in compensation, multiple inflorescences.

Wind-pollinated angiosperms do not depend on insects to transport their pollen from place to place and therefore do not offer nutritious rewards to visitors. Wind pollination is very inefficient, however, and is successful only in situations where a large number of individuals grow close together. Nearly all of the pollen falls to the ground within a hundred meters of the parent plant, and if the individuals are widely scattered, the chance of a pollen grain finding a receptive stigma is very low. Many wind-pollinated plants are dioecious, like cottonwoods, or genetically self-incompatible, like oaks; for them, outcrossing is obligatory. In the tropical forest, where populations of plants are very spread out, wind pollination characterizes very few plants. Some representatives of wind-pollinated groups of plants that enter the tropics from temperate regions become insect-pollinated in the tropics. Insects, orienting to particular kinds of flowers, can carry pollen for distances of up to 25 kilometers or more, even if there are no individual plants in the intervening area.

17–28
Most common species of trees in temperate regions are wind-pollinated. The staminate flowers of the turkey oak (Quercus laevis) *hang in catkins, flexible, thin tassels several inches long. These catkins are whipped by passing breezes and the pollen, when ripe, is scattered and blown about by the wind.*

Flower Colors

Surprisingly, the wide array of colors that we see in the flowers of the angiosperms is produced by a very small number of pigments. Many red, orange, or yellow flowers owe their color to the presence of carotenoids similar to those that occur in the leaves. The most important pigments in floral coloration, however, are the *flavonoids*, compounds in which two six-carbon rings are linked by a three-carbon unit. Flavonoids probably occur in all angiosperms and are more sporadically distributed among the members of other groups of vascular plants, with a few reports even from other photosynthesizing organisms and animals. In the leaves, they block far ultraviolet radiation, highly destructive to nucleic acids and proteins, and usually selectively admit light of blue-green and red wavelengths, important in photosynthesis. Many other biological functions have also been proved or suggested for the flavonoids.

One major class of flavonoids that is important in flowers are the anthocyanins (Figure 17–29). Most red and blue plant pigments are anthocyanins, which are water-soluble and found in vacuoles, unlike the carotenoids which are fat-soluble and found in plastids. Anthocyanins change color depending on the acidity of the solution in which they are found; thus cyanidin is red in acid solution, violet in neutral solution, and blue in alkaline solution. Another group of flavonoids, very common in leaves and also in many flowers, are the flavonols. These are often colorless but may contribute to the ivory or white hues of certain flowers.

In the goosefoot, cactus, nightshade, and portulaca families and their relatives, the reddish pigments are not anthocyanins nor even flavonoids but a group of more complex aromatic compounds known as betacyanins. The red flowers of *Bougainvillea* and the red color of beets are due to betacyanins. No anthocyanins occur in these plants, which are shown quite clearly by their biochemistry to be close relatives and in turn are related to other groups such as cacti.

For all flowering plants, different mixtures of flavonoids and carotenoids, as well as changes in cellular pH and in the structural and therefore reflective properties of the flower parts, produce the characteristic pigmentation. Fall coloring of leaves comes about when large quantities of colorless flavonols are converted into anthocyanins as chlorophyll breaks down. In the all-yellow flower of *Calylophus lavandulifolius* (see the Essay on page 112), the ultraviolet reflective outer portion is colored by carotenoids, whereas the ultraviolet nonreflective inner portion is yellow to our eye because of the presence of a yellow chalcone, one of the flavonoids. To a bee or other insect, the outer portion of the flower would appear to be a mixture of yellow and ultraviolet, a color called "bee's

17-29
Three anthocyanin pigments, the basic pigments on which flower colors in many angiosperms depend: pelargonidin (red); cyanidin (violet); and delphinidin (blue). Related compounds known as flavonols are yellow or ivory, and the carotenoids are red, orange, or yellow. Mixtures of these pigments, together with changes in cellular pH, produce the entire range of flower color in the angiosperms. The changes in flower color that often follow fertilization provide "signals" to pollinators, telling them which flowers have opened recently and are more apt to provide food.

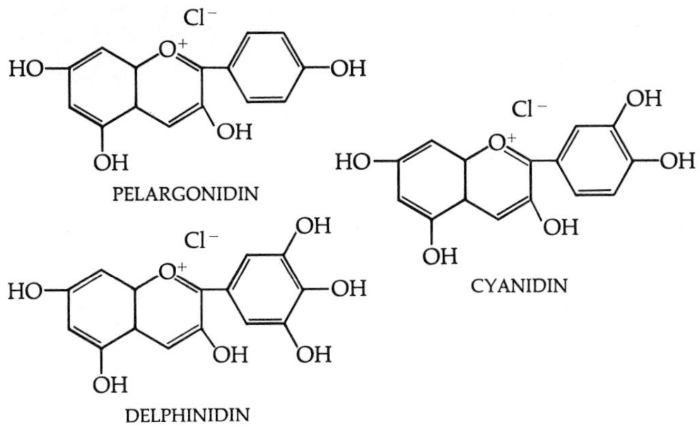

17-30
Fruit of the coconut (Cocus nucifera), *a drupe. Only the hard inner shell and its contents are present at maturity, the husk having rotted away. The coconut milk is nuclear endosperm. It acquires cell walls by the time of germination. The entire fruit floats easily in the sea, and coconuts have been widely dispersed by this means.*

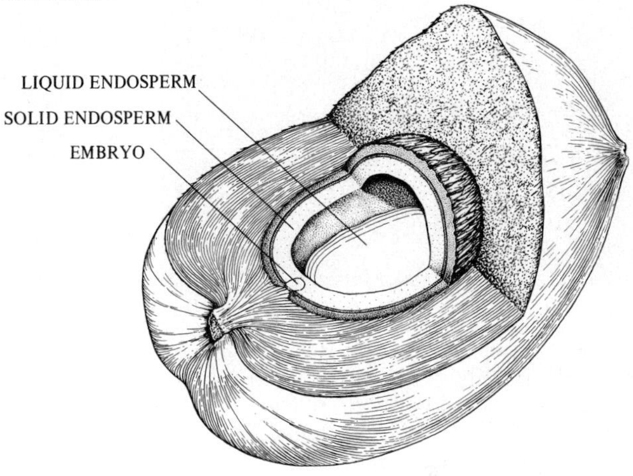

purple," whereas the nonreflective inner portion would appear pure yellow. Most, but not all, ultraviolet reflectivity in flowers is related to the presence of carotenoids, and thus ultraviolet patterns are more common in yellow flowers than in any other group.

EVOLUTION OF FRUITS

The fruit is a mature ovary, which may or may not retain some additional floral parts. A fruit in which such additional parts are retained is known as an accessory fruit.

Fruits are generally classified as *simple*, *multiple*, or *aggregate* according to the arrangement of the carpels from which the fruit developed. Simple fruits develop from one carpel or several united carpels. Aggregate fruits, such as magnolia, raspberry, and strawberry, consist of a number of separate carpels of one gynoecium. Multiple fruits consist of the gynoecia of more than one flower. The pineapple, for example, is a multiple fruit consisting of an inflorescence with many previously separate ovaries fused on the axis on which pineapple flowers were borne (the other flower parts are squeezed between the expanding ovaries.)

Simple fruits are by far the most diverse of the three groups. When ripe, they may be soft and fleshy, dry and woody, or papery. There are three main types of fleshy fruit—the *berry*, the *drupe*, and the *pome*. In the berry, examples of which are tomatoes, dates, and grapes, there are one to several carpels, each of which is usually many-seeded. The inner layer of the fruit coat is fleshy. In the drupe, there are also one to several carpels, but each usually contains only a single seed. The inner coat of the fruit is stony and usually tightly adherent to the seed. The coconut is a drupe whose outer covering is fibrous rather than fleshy, but in temperate regions we usually see only the seed with the adherent stony inner coat of the fruit (Figure 17-30). Other familiar drupes are the peach, cherry, olive, and plum. A highly specialized sort of fleshy fruit is the pome, which is characteristic of one subfamily of the rose family. The pome is derived from a compound inferior ovary in which the fleshy portion comes largely from the enlarged base of the perianth. Apples and pears are pomes.

Dry simple fruits are classified as *dehiscent* or *indehiscent* (Figures 17-31—17-33). In dehiscent fruits, the tissues of the mature ovary wall (the pericarp) break open, freeing the seeds. In indehiscent fruits, the seeds remain in the fruit after the fruit is shed from the parent plant.

There are several sorts of dehiscent simple dry fruits. The *follicle* is derived from a single carpel that splits

(a)

(b)

17–31
Bursting follicle of (a) *milkweed* (Asclepias) *and capsule of* (b) *cottonwood* (Populus). *In both, the seeds have tufts of light silky hair which aid in their dispersal.*

17–32
Seed discharge in dwarf mistletoe (Arceuthobium), *a parasitic angiosperm that is the most serious cause of loss to forest productivity in the western United States. Very high hydrostatic pressure that builds up in the fruit blasts the seeds up to 15 meters laterally. They have an initial velocity of about 100 kilometers per hour, and presumably this is one of the ways they are spread from tree to tree, although they are also sticky and carried by birds.*

down one side at maturity, as in columbines and milkweeds. In the pea family, the characteristic fruit is a *legume.* Legumes resemble follicles but split along both sides. In the mustard family (Brassicaceae), the fruit is called a *silique* and is formed of two fused carpels. At maturity, two halves split off, leaving the seeds attached to a persistent central portion. The most common sort of dehiscent simple dry fruit is the *capsule,* which is formed from a compound ovary in plants with either a superior or an inferior ovary. Capsules shed their seeds in a variety of ways. In the poppy family, Papaveraceae, the seeds are often shed when the capsule splits longitudinally, but in some members of this family they are shed through holes near the top of the capsule.

Indehiscent simple dry fruits are found in a great variety of plant families. Most common is the *achene,* a small single-seeded fruit in which the seed lies free in the cavity except for its attachment by the funiculus. Achenes are characteristic of buttercups and buckwheat. Winged achenes, such as those found in the elm and ash, are termed *samaras*. In the grass family (Poaceae), the achenelike fruit *(caryopsis)* is derived from a compound ovary and the seed coat is firmly united to the fruit wall. In the Asteraceae, the achenelike fruit is derived from a compound, inferior ovary and is called a *cypsela*. Acorns and hazelnuts are examples of *nuts,* which resemble achenes but have a stony coat and are derived from a compound ovary. Finally, in the parsley family (Apiaceae) and the maples, as well as a number of other unrelated groups, the fruit is a *schizocarp*. The schizocarp is derived from a compound ovary but splits at maturity into a number of one-seeded portions.

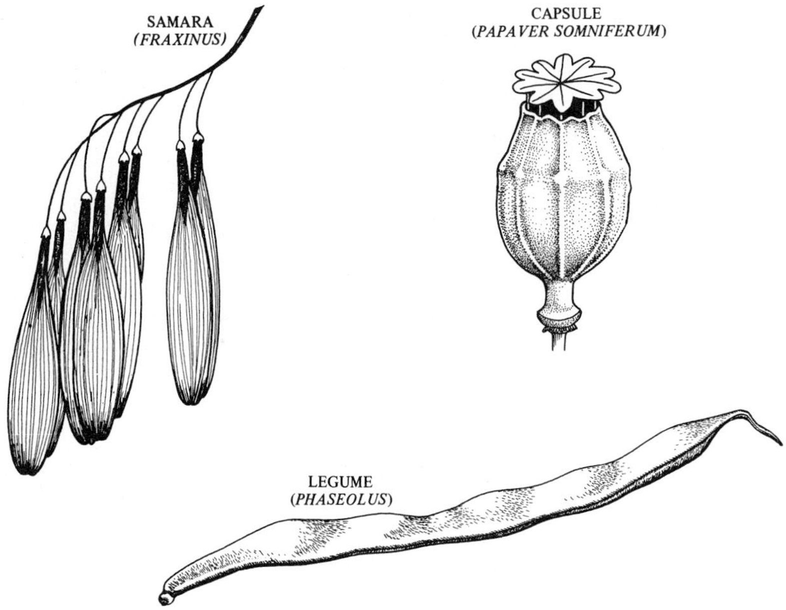

17–33
The samara, a winged fruit characteristic of the ash, as shown here, and the elm (Ulmus) *is indehiscent, retaining its single seed at maturity. Capsules and legumes are common types of dehiscent fruits—fruits that shed their seeds at maturity. In the poppy capsule shown here, the ripe seeds are shaken out through pores in the dry fruit, like salt from a saltshaker.*

Dispersal of Fruits

Just as flowers may be classified according to their pollinators, fruits may be grouped according to their dispersal agents.

Wind-Borne Fruits

Some plants have extremely light fruits or seeds and can be dispersed by wind because of this lightness. The dustlike seeds of all members of the orchid family, for example, are wind-borne. Other fruits have wings, usually formed from perianth parts. In the maple, for example, which has a gynoecium composed of two fused carpels, each carpel develops a long wing (Figure 17–34). The two carpels separate and fall when mature. Many of the Asteraceae—the dandelion, for example—develop a plumelike pappus, which aids in keeping the light fruits aloft (Figure 17–34). In some plants, it is the seed itself rather than the fruit that carries the wing or plume; the familiar butter-and-eggs *(Linaria)* has a winged seed, and fireweed *(Epilobium)* and milkweed *(Asclepias)* have plumed seeds. In the willows and the populars *(Salicaceae)*, the seed coat is covered with wooly hairs. In the tumbleweeds *(Salsola kali)*, the whole plant or the fruiting portion of it is blown by the wind and scatters seeds as it goes (Figure 17–35).

Other plants shoot their seeds aloft. In touch-me-nots *(Impatiens)*, the valves of the capsules separate suddenly, throwing seeds for some distance. In the witch hazel *(Hamamelis)*, the endocarp contracts as the fruit dries, discharging the seeds sometimes as far as 15 meters.

17–34
Wind-dispersed fruits. In maples (Acer), *each half of the schizocarp (a two-part indehiscent fruit that splits at maturity) has a long wing. The fruits of the dandelion* (Taraxacum officinale) *and some other composites have a modified calyx, the pappus, which may be adherent to the mature achene and form a plumelike structure that aids in wind dispersal.*

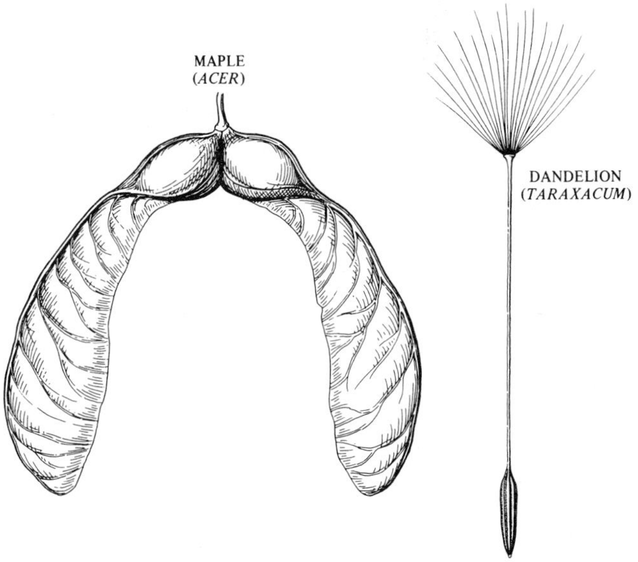

17–35
In the tumbleweed (Salsola kali), *the plant breaks off and is blown across open country, scattering the seeds.*

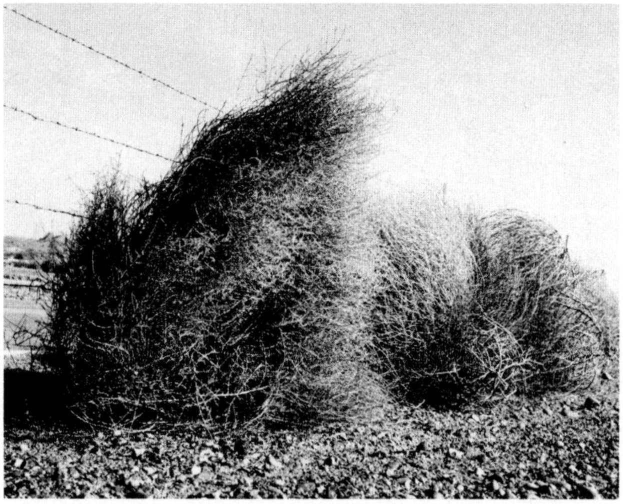

17–36
The seeds of fleshy fruits are usually dispersed by vertebrates that eat the fruits and deposit the seeds in their feces. The tree sparrow (Spizella arborea) *shown here is eating the bright red fruits of winterberry* (Ilex montana), *a member of the holly family.*

Water-Borne Fruits

The fruits and seeds of many plants, especially those growing in or near water sources, are adapted for floating, either because air is trapped in some part of the fruit or because the fruit contains buoyant tissue. Some fruits are especially adapted for dispersal by ocean currents; notable among these is the coconut, which is why almost every Pacific atoll is quick to acquire its own coconut tree. Rain is also a common means of fruit and seed dispersal and is particularly important for plants that live on hillsides or mountain slopes.

Animal-Borne Fruits

The evolution of sweet and often highly colored fleshy fruits was also accompanied by coevolution involving animals and flowering plants. The majority of fruits in which the pericarp is fleshy, such as cherries, raspberries, and grapes, for example, are eaten by vertebrates. When such fruits are eaten by birds or mammals, the seeds that lie within them are spread by being passed unharmed through the digestive tract (Figure 17–36).

When fleshy fruits ripen, they undergo a series of characteristic changes, mediated by the growth regulator ethylene, which will be discussed in Chapter 23. Among these are a rise in sugar content, a general softening of the fruit through the breakdown of pectic substances, and often a change in color from inconspicuous, leaflike green to very conspicuous bright red, yellow, blue, or black. The seeds of some plants, especially tropical ones, often have fleshy appendages, or arils, that have the bright colors characteristic of fleshy fruits and, like them, are aided in dispersal by vertebrates. In this way, fruit in which the fruit wall is dry, such as a capsule, can achieve the dispersal advantages exhibited by fleshy ones.

Unripe fruits are often green or colored in such a way that they are hidden among the green leaves and somewhat concealed from birds, mammals, and insects. They may actually be disagreeable to the taste—as in the acidic unripe fruits of cherries *(Prunus)*, for example—thereby discouraging animals from eating them before the seeds within are ripe. The changes in color that accompany ripening are the plant's "signal" that the fruit is ready to be eaten and that the seeds are ripe and ready for dispersal.

It is no coincidence that red is such a prominent color among ripe fruits. The fruit is thereby still concealed from insects, which are too small to disperse the large seeds of fleshy fruits effectively. As we have seen, pure red is invisible to insects and other invertebrates but is very conspicuous to birds and mammals (including ourselves).

A number of other angiosperms have fruits or seeds that are dispersed by adhering to fur or feathers (Figure 17–37). These have prickles, hooks, barbs, spines, hairs,

17-37
Two fruits with bristly spines that are dispersed by adhering to fur or feathers. In burdock, the entire head is dispersed. In Acaena, *the spines are on the outside of the floral cup, which, at maturity, grows up around the carpels.*

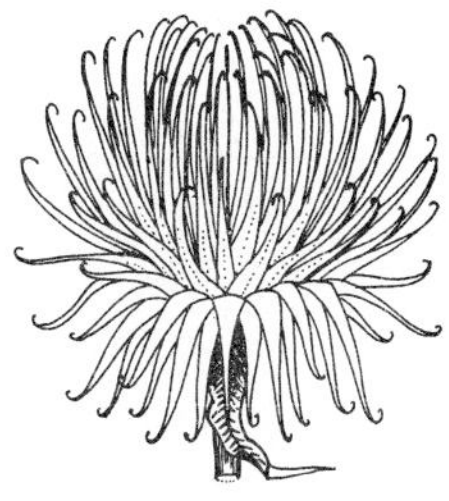

BURDOCK
(*ARCTIUM MINUS*)

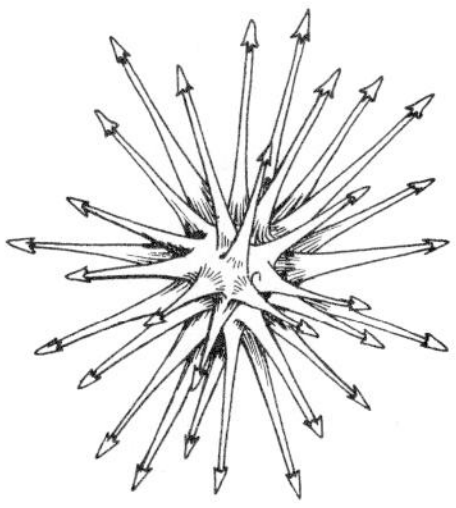

ACAENA

or sticky coverings and so are transported, often for great distances, by animals.

BIOCHEMICAL COEVOLUTION

Another important factor in the evolution of angiosperms has to do with the so-called secondary plant substances. Once thought of as waste products of the plants, these include an array of chemically unrelated compounds such as alkaloids, quinones, essential oils (including terpenoids), glycosides (including cyanogenic substances and saponins), flavonoids, and even raphides (needlelike crystals of calcium oxalate). These compounds generally characterize whole families or groups of families of flowering plants, although they sometimes occur in different families that are more or less unrelated in other respects (Figure 17-38).

These chemicals appear to play a major role in nature in restricting the palatability of the plants in which they occur (Figure 17-39). When a given family of plants is characterized by a distinctive group of secondary plant substances, it is apt to be eaten by insects belonging to certain families and only by those particular insects.

17-38
Secondary plant substances: sinigrin, from black mustard, Brassica nigra; *calactin, a cardiac glycoside, from the milkweed* Asclepias curassavica; *nicotine, from tobacco,* Nicotiana tabacum, *a member of the nightshade family (Solanaceae); caffeine, from* Coffea arabica *of the madder family; and theobromine, a prominent alkaloid in coffea, tea* (Thea sinensis), *and cocoa* (Theobroma cacao). *All but sinigrin are alkaloids, a diverse class of nitrogen-containing ring compounds that are physiologically active in vertebrates. Various other, often related, compounds are found in other members of the same plant families. They are often useful in plant classification.*

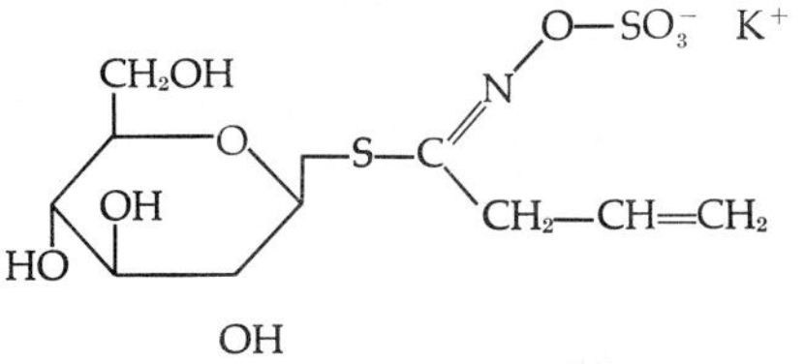

SINIGRIN

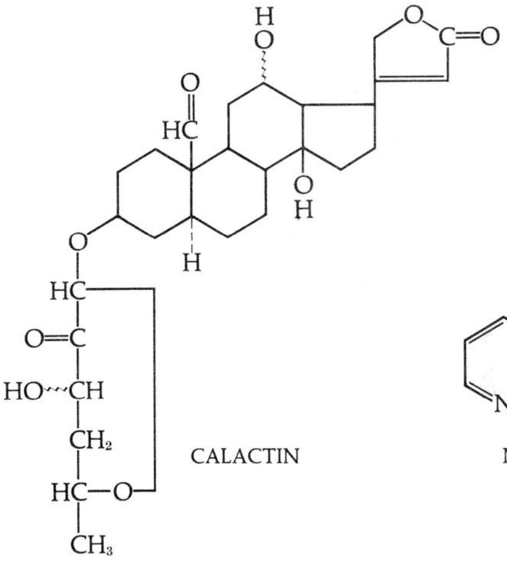

CALACTIN

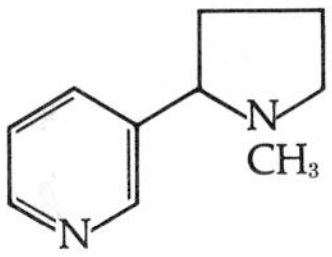

NICOTINE

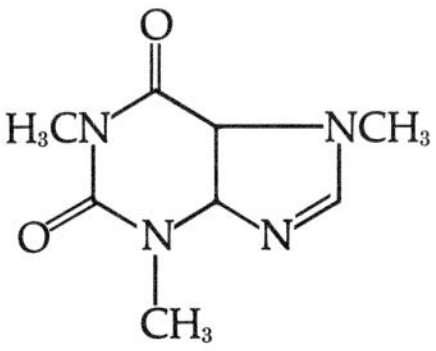

CAFFEINE

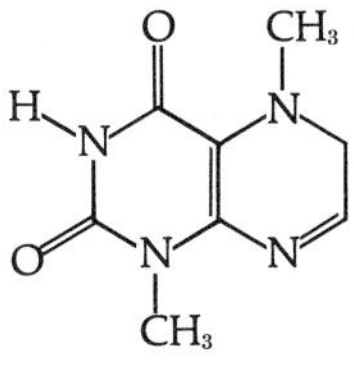

THEOBROMINE

17-39
Poison ivy (Toxicodendron radicans) *produces a "secondary plant substance," 3-pentadecanedienyl catechol, that produces an irritating and persistent rash on the skin of humans. The ability to produce this alcohol presumably evolved under the selective pressure exerted by herbivores. Fortunately, the plant is easily identifiable by its characteristic and unusual compound leaves with their three leaflets.*

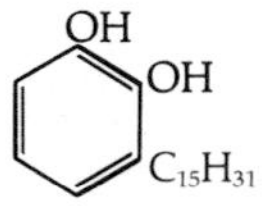

Thus, the mustard family, Brassicaceae (Cruciferae), is characterized by the presence of mustard oil glycosides and associated enzymes that break down these glycosides to release the pungent odors we associate with cabbage, horseradish, and mustard. Certain groups of moths, true bugs, and beetles feed only on the leaves of plants of this family, and the butterfly subfamily Pierinae (including the cabbage butterflies and orange-tips) also feed only on these plants. On the other hand, other families of insects belonging to these same groups ignore plants of the mustard family and will not feed on them even if starved. For the narrowly restricted feeders, the same chemicals that act as deterrents to most groups of plant feeders often act as feeding stimuli. For example, certain moth larvae that feed on cabbage will extrude their mouth parts and go through their characteristic feeding behavior when presented with agar or filter paper treated with juices pressed from these plants.

Thus, it is clear that the ability to manufacture these chemicals and retain them in their tissues is an important evolutionary step for the plants concerned and gives them biochemical protection from the predations of most herbivores. The evolution of the Brassicaceae doubtless took place in part behind just such a biochemical shield. From the standpoint of the insects, such protected plants, because they are not already heavily utilized by other plant feeders, represent an unexploited food source for any group of insects that can tolerate or break down the poisons manufactured by the plant. The main evolution of the butterfly group Pierinae probably took place after their ancestors had acquired the ability to feed on plants of the mustard family by breaking down these molecules.

It is interesting to note in passing that those insects which are narrowly restricted in their plant feeding habits to groups of plants with secondary plant substances are often brightly colored, a signal to *their* predators that they carry the noxious chemicals in their bodies and have hence protected themselves. If we consider, for example, the assemblage of insects found feeding on a milkweed plant on a summer day, we think of bright green chrysomelid beetles, bright red cerambycid beetles and true bugs, and orange and black monarch butterflies, among others. Milkweeds (family Asclepiadaceae) are richly endowed with alkaloids and cardiac glycosides, heart poisons that have potent effects in vertebrates, the main potential predators of these insects. If a bird ingests a monarch butterfly, severe vomiting and distress will follow, and the orange-and-black pattern will be avoided in the future. Other insects, such as in this case the viceroy butterfly, have evolved similar patterns and escape predation by capitalizing upon their resemblance to the poisonous monarch: a phenomenon known as *mimicry*, ultimately dependent upon the plant's chemical defenses. Various drugs and psychedelic chemicals (Figures 17-40 and 17-41), such as the active

17–40
Hallucinogenic and medicinal compounds. Mescaline, from the peyote cactus (Lophophora williamsii), *used ceremonially by many Indian groups of northern Mexico and the southwestern United States. Tetrahydrocannabinol (THC), the most important active molecule in marijuana* (Cannabis sativa). *Quinine, a valuable drug used in the treatment and prevention of malaria, derived from tropical trees and shrubs of the genus* Cinchona *(Rubiaceae). All of these substances presumably serve to protect the plants in which they occur from the depredations of insects, but are physiologically active in vertebrates, including man.*

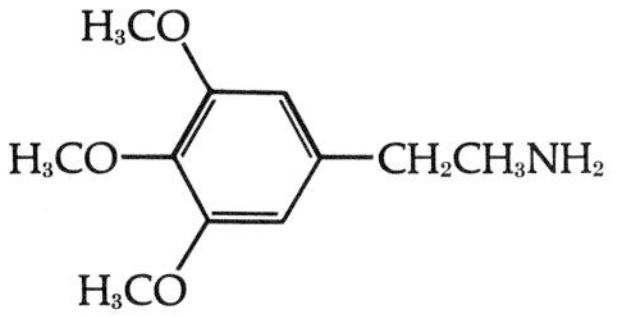
MESCALINE

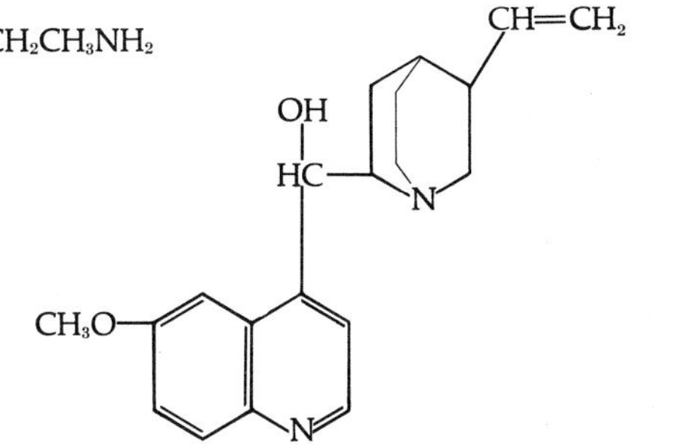
QUININE

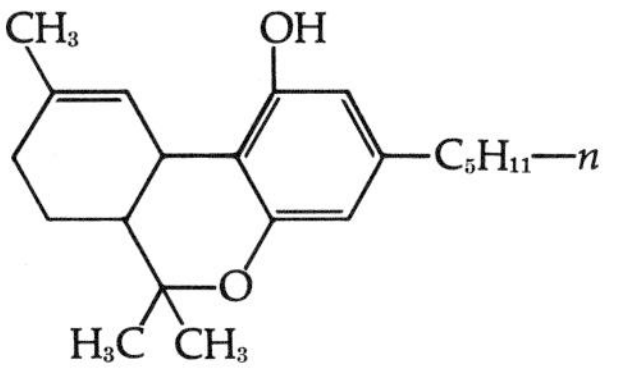
Δ[1] TETRAHYDROCANNABINOL

17–41
(a) *Peyote cactus* (Lophophora williamsii). (b) *A group of 13 mescal eaters, members of the Comanche tribe.*

(a)

(b)

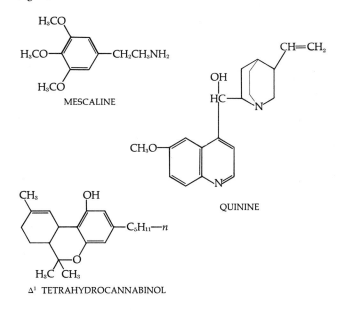

ingredients in marijuana *(Cannabis sativa)* and the opium poppy, are likewise secondary plant compounds that in nature play a role in discouraging the attacks of herbivores.

Still more sophisticated systems are known. When the leaves of potato or tomato plants are wounded, as by the Colorado potato beetle, proteinase inhibitors, which interfere with the digestive enzymes in the gut of the beetles, rapidly increase in concentration in the tissues of the plant that have been exposed to air. Other plants manufacture molecules that resemble the hormones of insects or other predators and thus interfere with their normal growth and development. One of the most useful of these is a complex molecule called diosgenin, obtained mainly from wild yams in Mexico, but with smaller amounts obtained in India and China. Diosgenin is only two simple chemical steps away from 16-dehydropregnenolone (16D), the main active ingredient in oral contraceptives, and wild yams have been the best source for 16D to date. Unfortunately, they grow very slowly, and the supply in the wild shows signs of exhaustion. The Soviets and Ecuadorians, among others, have recently been experimenting with species of *Solanum,* the potato and nightshade genus, which contains solasodine, a molecule that can be converted to 16D in two steps. Some of these plants may be developed into commercial crops.

As we have already seen, pollination systems have developed a particular coevolutionary pattern in which all the possible pollination types have evolved not once, but usually several or many times, in each group. The resulting array of forms gives the angiosperms an almost incomprehensible variety. In the case of biochemical relationships, however, the evolutionary steps appear to have been large and definitive, and whole families of plants can be characterized biochemically and associated with major groups of plant-eating insects. These biochemical relationships may have played a role in the early success of the angiosperms.

The main conclusion we can draw from the story of angiosperm evolution is that several interconnected forces combined to give the first angiosperms their evolutionary impetus. Such is almost always the case in evolution, where a single factor is probably never selected for in isolation, and it is genetic correlations, in the end, that play a dominant role. It is interesting to note that each of the three main factors that we have postulated as playing a role in the early success of the angiosperms has also been influential in determining the rich diversification of this group of vascular plants that is dominant in our modern world.

SUMMARY

The angiosperms, the dominant group of vascular plants at the present day, appear in the fossil record about 125 million years ago, in the lower Cretaceous period. The group became dominant all over the world 75 to 80 million years ago, in the upper Cretaceous, when many modern families and some genera can be recognized. The pollen of primitive angiosperms may be indistinguishable from gymnosperm pollen or fern spores, so that it is difficult to be certain of the presence of angiosperms more than 125 million years ago, but the group is doubtless somewhat older.

Angiosperms may have evolved in the semiarid uplands of West Gondwanaland, a supercontinent that existed until about 125 million years ago and was formed by the union of South America and Africa. By the time these continents finally separated about 90 million years ago, their climates had changed considerably, and the angiosperms were nearing world dominance.

Possible reasons for their success include conspicuous adaptations for drought resistance, including the evolution of the deciduous habit for angiosperms which must undergo periods unfavorable to active growth each year. The conducting systems of angiosperms are also notably efficient, most having vessels.

The flower is the most conspicuous feature of the angiosperms and has played a very significant role in their evolution. The carpel is a leaflike structure that has undergone infolding to enclose the ovules (which contain the megasporangia) and subsequent differentiation into a basal, swollen ovary, a stalklike style, and a stigma that is receptive to pollen. Similarly, the stamen has evolved from leaflike precursors or slender, branching systems with terminal sporangia to become specialized into the slender structure characteristic of most living angiosperms. Sepals are specialized leaves that protect the flower in bud, and petals in most subdivisions are sterilized stamens that have assumed a function in attracting insects. Some petals, however, are derived from sepals. The spiral arrangement of the flower parts of primitive angiosperms, coupled with their possession of numerous free parts in each whorl, has given way to a whorled arrangement in most modern forms with a definite number of parts which are often fused with one another or with the other whorls of the flower.

Examples of specialized families are Asteraceae (Compositae), in which numerous highly specialized flowers are aggregated into a head, the head itself being the attractive unit for insects; and Orchidaceae, in which bizarre elaboration of the flower parts has resulted in a highly irregular flower with the most specialized pollination systems known.

Pollination by insects is basic in the angiosperms, and the first pollinating agents were probably beetles or similar insects. The closing of the carpel may have been a device to protect the ovules from being eaten in the course of such activities. More specialized groups of insects evolved later in the history of the angiosperms, and wasps, flies, butterflies, and moths have each left their mark on the morphology of certain angiosperm flowers. The bees are, however, the most specialized of flower-visiting insects as well as the most constant and have probably had the greatest effect on the evolution of angiosperm flowers. Each group of flower-visiting animals is associated with a particular group of floral characteristics related to the visual and olfactory senses of the animals they attract. Some angiosperms have become wind-pollinated, shedding copious quantities of small, nonsticky pollen and having well-developed, often feathery stigmas that are efficient in collecting such pollen from the air.

Flowers that are regularly visited by and pollinated by animals with high energy requirements, such as hummingbirds and bats, must produce large amounts of nectar. They must then protect and conceal these sources of nectar from other potential visitors of low energy requirements, which might satiate themselves at a single flower or the flowers of a single plant and then not move on to another plant and effect cross-pollination. Wind pollination is inefficient, and the individuals must grow close together in large colonies for the system to work, whereas insects, birds, or bats can carry pollen great distances with precision.

Fruits are as diverse as the flowers from which they are derived and can be classified either morphologically, anatomically, or in terms of their methods of dispersal. They are basically mature ovaries, but, if additional floral parts are retained, the fruits are said to be accessory. Simple fruits are derived from one carpel or a group of united carpels, aggregate fruits from the free carpels of one flower, and multiple fruits from the fused carpels of several or many flowers. Dehiscent fruits split open to release the seeds, and indehiscent ones do not.

Wind-borne fruits or seeds are light and often have wings or tufts of trichomes that aid in their dispersal. Some plants shoot their seeds off explosively. Some seeds or fruits are borne by water, in which case they have resistant coats. Others are disseminated by animals, particularly vertebrates, and have evolved fleshy coverings that are tasty and often conspicuous for this purpose. Others adhere to the coats of mammals or birds and are distributed in this manner.

A third agency in the evolutionary success and diversification of the angiosperms has been biochemical coevolution. Certain groups of angiosperms have evolved various "secondary plant substances," such as alkaloids, which protect them from most herbivores. Certain herbivores, however, normally those with narrow feeding habits, are found associated with these plants, from which their competitors are excluded. This pattern is a sure indication of a stepwise pattern of coevolutionary interaction, and it appears likely that the early angiosperms also may have been protected by their ability to produce some chemicals that functioned as poisons for herbivores.

SUGGESTIONS FOR FURTHER READING

ALSTON, R. E., and B. L. TURNER: *Biochemical Systematics*, Prentice-Hall, Inc., Englewood Cliffs, N.J., 1963.

Although increasingly out of date, this book still provides in summary form a useful outline of the principal classes of secondary plant substances.

FAEGRI, K., and L. VAN DER PIJL: *The Principles of Pollination Ecology*, 2nd ed., Pergamon Press, Oxford and New York, 1971.

A rigorous and scholarly examination of the field on a worldwide basis.

HEISER, CHARLES B., JR.: *Seed to Civilization, The Story of Man's Food*, W. H. Freeman and Co., San Francisco, 1973.

Broadly based and useful review of the evolution of domesticated plants and animals.

PIJL, L. VAN DER: *Principles of Dispersal in Higher Plants*, 2nd ed., Springer-Verlag, Berlin, 1972.

A brief and rather technical but informative book on all matters of seed and fruit dispersal.

PROCTOR, M., and P. YEO: *The Pollination of Flowers*, Collins, St. James Place, London, 1973.

Outstanding and beautifully illustrated introduction to all aspects of pollination biology.

RADFORD, ALBERT E., *et al.*: *Vascular Plant Systematics*, Harper & Row, Publishers, New York, 1974.

An invaluable source book, including glossaries, techniques, bibliographies, indices, and useful discussions of all the various aspects of plant systematics—the scientific study of the kinds and diversity of plants, and of the relationships among them.

STEBBINS, G. LEDYARD: *Flowering Plants. Evolution Above the Species Level*, The Bellknap Press of Harvard University Press, Cambridge, Mass., 1974.

A scholarly and fascinating treatise on the origin and subsequent diversification of the angiosperms by one of the masters of evolutionary theory.

SECTION 5 The Plant Body of Angiosperms: Its Structure and Development

CHAPTER 18

Early Development of the Plant Body

In the previous section, we traced the long evolutionary development of the angiosperm from its presumed ancestor, a single-celled alga floating just beneath the surface of the sunlit water, through a series of early vascular plants whose forked axes were the forerunners of the leaves and roots characteristic of the higher vascular plants. We noted that, as a result of their common evolutionary origin, these various plant parts, although superficially different, have many features in common.

In this section, we shall be concerned primarily with the end result of this long evolutionary history, the flowering plant. This chapter thus begins where the story of the angiosperm life cycle ended in Section 4, with the seed—which consists, you will remember, of a seed coat, stored food, and an embryo. We begin with the formation of the embryo because it is through this process—known as embryogeny—that the vegetative parts of the plant—root, stem, and leaf—have their origin, and the organization of tissues is initiated. With this background, we shall be prepared to consider further the development, structure, and relationships of root, stem, and leaf, which are the subjects of the remaining chapters of the section.

THE MATURE EMBRYO AND SEED

The mature embryo of flowering plants consists of a stemlike axis bearing either one or two cotyledons (Figures 18-2 and 18-3). The cotyledons—sometimes referred to as the seed leaves—are the first leaves, or foliar structures, of the young sporophyte. As the names dicotyledon and monocotyledon imply, the embryos of dicotyledons have two cotyledons and those of monocotyledons have only one.

At either end of the embryo axis are found the apical meristems of the shoot and root. You may recall that apical meristems are found at the tips of all shoots and roots of the plant body, and that meristems are composed of meristematic cells—cells that are physiologically young

18-1
Acorns germinating on the forest floor. The first structure to emerge from the germinating seed is the root, which will anchor the young plant to the soil and absorb water essential for growth of the developing seedling.

18–2
Seeds and stages in germination of some common dicotyledons. (a) *The garden bean* (Phaseolus vulgaris). *Seed shown open and from external edge view.* (b) *Castor bean* (Ricinus communis). *Seed open, showing both flat and edge views of embryo.* (c) *Pea* (Pisum sativum). *External view of seed only.*

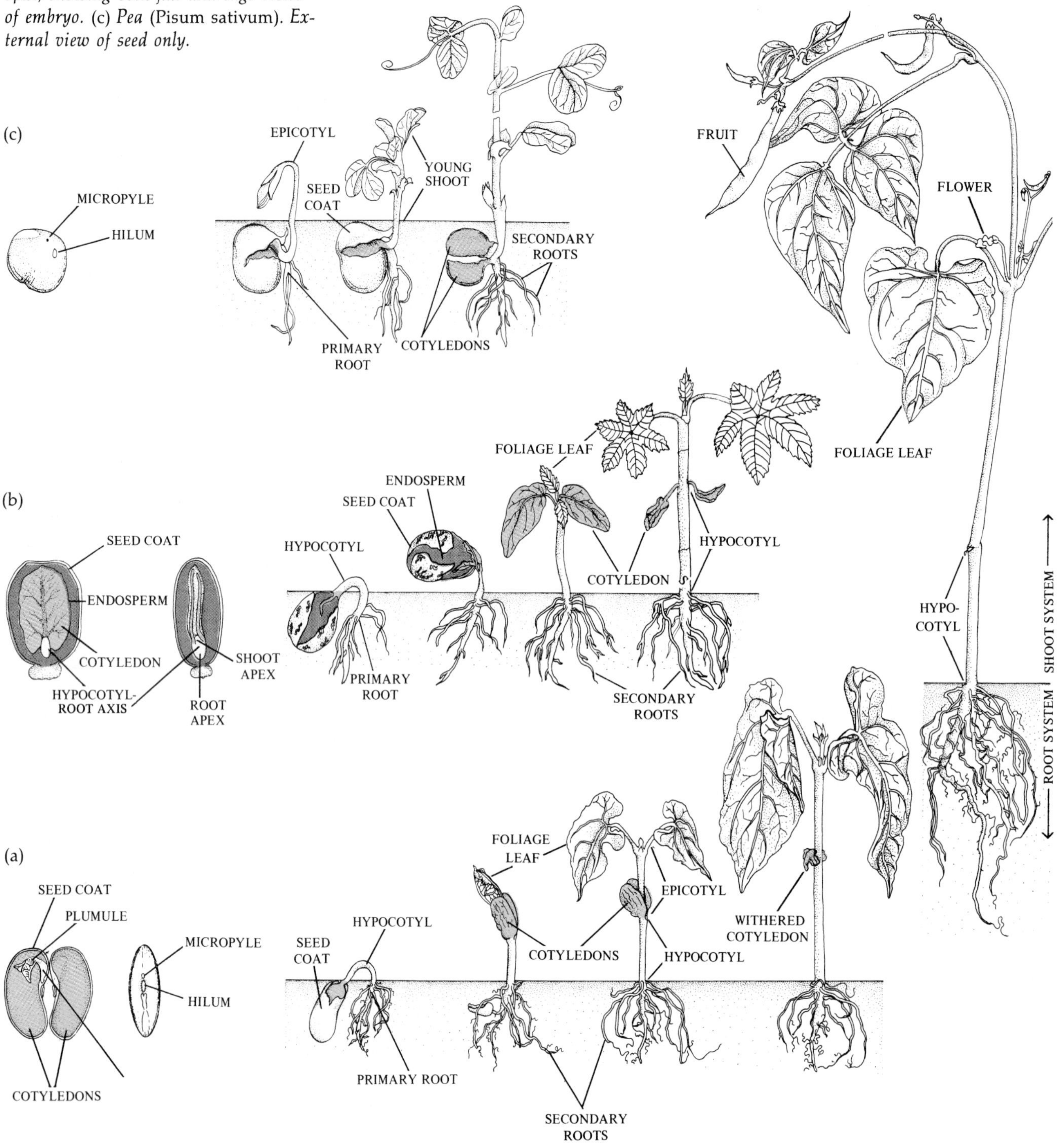

18-3
Seeds and stages in germination of some common monocotyledons (a) *corn* (Zea mays) *and* (b) *onion* (Allium cepa). *Both seeds shown in longitudinal section.*

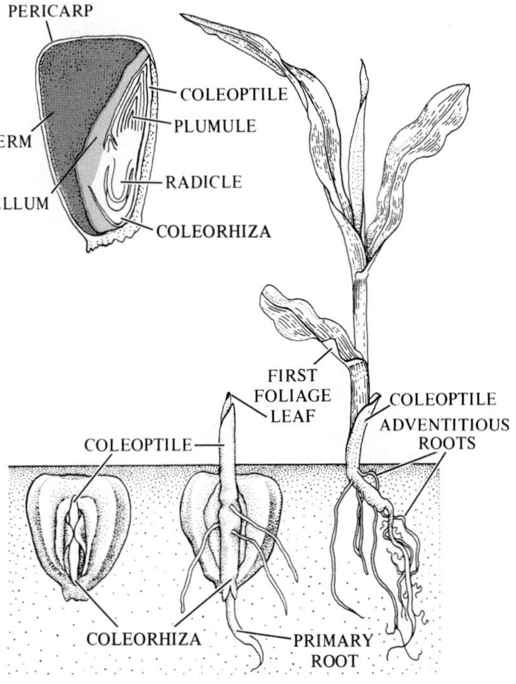

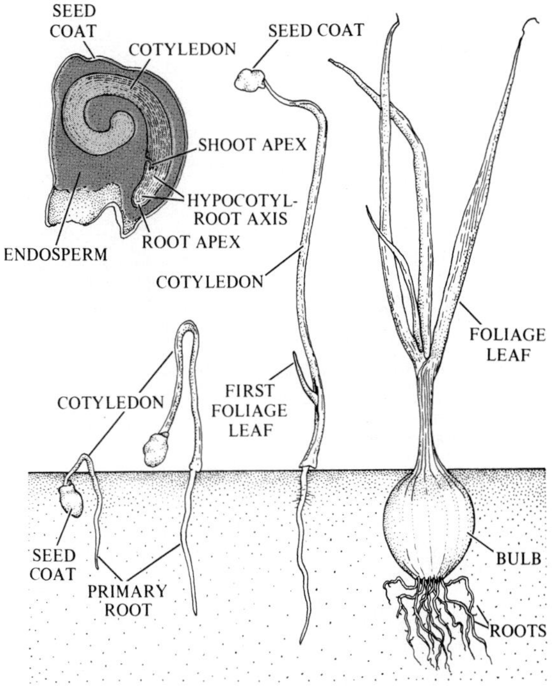

and capable of repeated division. (Apical meristems will be considered in more detail in Chapters 19, 20, and 21.) In the embryo, the apical meristem of the shoot terminates the axis above *(epi)* the cotyledons. In some embryos, the *epicotyl* consists of little more than the apical meristem (Figures 18-2b and 18-3b), whereas in others it bears one or more young leaves (Figures 18-2a and 18-3a). The epicotyl, together with its young leaves, is called a *plumule*, although some botanists use the terms epicotyl and plumule interchangeably.

The portion of the embryo axis between the root tip and cotyledons is referred to as the *hypocotyl* (*hypo*, below). In some plants, the lower end of the embryo axis has distinct root characteristics and is called the *radicle* (Figure 18-4). In many plants, however, the lower end of the axis consists of little more than an apical meristem covered by a root cap. If a radicle cannot be distinguished in the embryo, the embryo axis below the cotyledons may be called the hypocotyl-root axis.

During our discussion of the development of the angiosperm seed in Chapter 16, we noted that the endosperm is sometimes completely digested by the developing embryo and that the embryos of such seeds develop fleshy food-storing cotyledons. The cotyledons of most dicotyledonous embryos are fleshy and occupy the largest volume of the seed. Familiar examples of seeds lacking endosperm are sunflower, walnut, pea, and bean (Figures 18-2a and 18-2c). In dicots with large amounts of endosperm, the cotyledons are thin and membranous (Figure 18-2b) and serve to absorb the stored food from the endosperm.

In the monocotyledons, the single cotyledon, although somewhat fleshy, usually performs an absorbing rather than a food-storing function (Figure 18-3). Embedded in endosperm, the cotyledon absorbs food digested by enzymatic activity. The digested food is then moved by way of the cotyledon to the growing regions of the embryo. Among the most highly differentiated of monocot embryos are those of grasses (Figures 18-3a and 18-4). When fully formed, the grass embryo possesses a massive cotyledon, the *scutellum*, which is closely appressed to the endosperm. The scutellum, like the cotyledons of other monocots, functions in the absorption of food stored in the endosperm. The scutellum is attached to one side of the axis of the embryo, which has a radicle at its lower end and a plumule at its upper end. Both the radicle and the plumule are enclosed by sheathlike, protective structures, called the *coleorhiza* and the *coleoptile*, respectively (Figure 18-4).

All seeds are enclosed by a *seed coat*, which develops from the integuments of the ovule. The seed coat is usually much thinner than the integuments were originally. The thin, dry seed coat may have a papery texture, but in many seeds it is very hard and highly impermeable to water. The micropyle is often visible on the seed coat as a small pore. Commonly, the micropyle is asso-

18-4
Longitudinal section of mature embryo of wheat (Triticum aestivum).

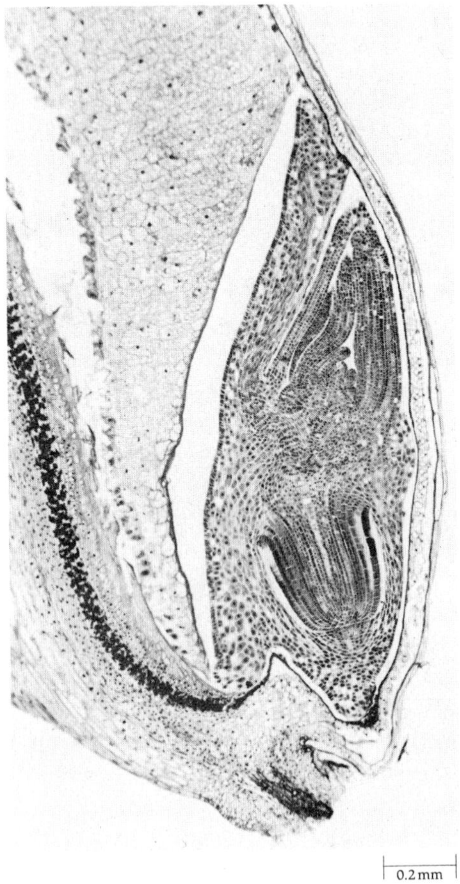

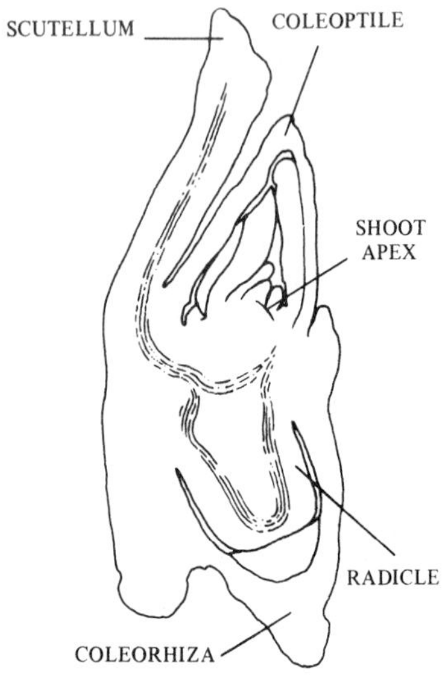

WHEAT

Like all grasses, wheat (Triticum aestivum) *is a monocot, and its fruit, the grain, or kernel, is one-seeded. More than 80 percent of the bulk of the wheat kernel and 70 to 75 percent of its protein is in the endosperm. White flour is made from the endosperm. The embryo (wheat germ) forms about 3 percent of the kernel; it is usually removed as wheat is processed because it contains oil, which makes the grain more likely to spoil. The bran is the seed coat plus the aleurone layer (outer part of endosperm); it constitutes about 14 percent of the kernel. The bran is also removed when wheat is milled to make white flour. Actually, the bran somewhat decreases the nutritional value of the wheat kernel. Because it is mostly cellulose, it cannot be digested by man and tends to speed the passage of food through the intestinal tract, resulting in a lower absorption. The wheat germ and the bran are sometimes used for human consumption but more often are fed to livestock.*

Wheat is about 9 to 14 percent protein, most of which, as we noted, is contained in the endosperm. Its protein value is diminished, however, by its lack of certain essential amino acids, notably lysine. Most of the vitamins are in the bran and wheat germ.

---

ciated with a scar, the *hilum*, which is left on the seed after the seed separates from its stalk, or *funiculus* (Figures 18-2a and 18-2c).

FORMATION OF THE EMBRYO

The early stages of embryo development are essentially similar in dicotyledons and monocotyledons (Figures 18-5 and 18-6). Formation of the embryo begins with the division of the fertilized egg, or zygote, within the embryo sac of the ovule. In most flowering plants, the first division of the zygote is transverse, or nearly so, with regard to the long axis of the zygote (Figures 18-5a and 18-6a). With this division, the polarity of the embryo is established: Of the two cells formed by division of the zygote, one will develop into the upper part of the embryo and the other (the one nearer the micropyle) into the lower part. In other words, the embryo has a shoot pole and a root pole. In some angiosperms polarity is already established in the egg cell, where most of the cytoplasm is aggregated near the upper end of the cell.

Polarity is a universal property of all higher organisms. The term arises by analogy with a magnet, which has plus and minus poles, and when one speaks of something having polarity, one means simply that whatever is being discussed—a plant, an animal, an organ, or a molecule—has one end that is different from the other end. Polarity in plant stems is a familiar phenomenon. In plants that are propagated by stem cuttings, for example, roots will form at the lower end of the stem,

18–5
Some early stages in development of the embryo of shepherd's purse (Capsella bursa-pastoris), *a dicot.* (a) *Two-celled stage.* (b) *Four-celled stage.* (c) *Beginning of development of embryo proper as distinct from suspensor.* (d) *Globular embryo proper has a protoderm, the future epidermis.*

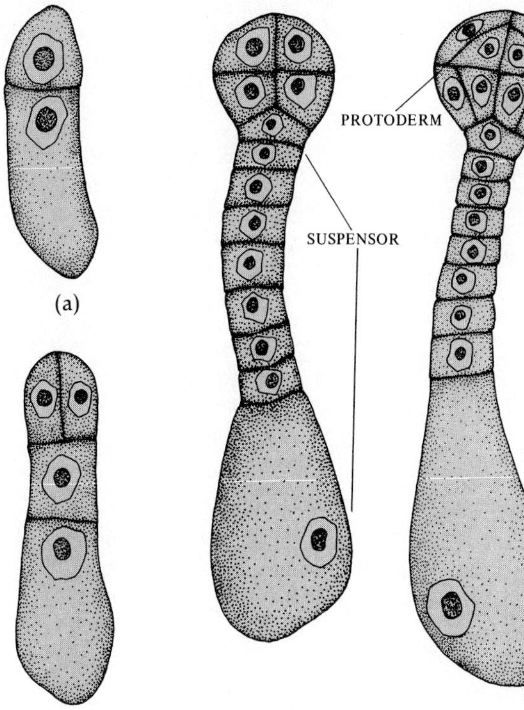

18–6
Some early stages in development of the embryo of the woodrush (Luzula forsteri), *a monocot.* (a) *Two-celled stage.* (b) *Four-celled stage.* (c) *Embryo proper, with protoderm, distinct from suspensor.* (d) *Root apex developing at lower end of embryo proper.* (e) *Apical meristem developing in depression, or notch (arrow), at base of the single cotyledon.*

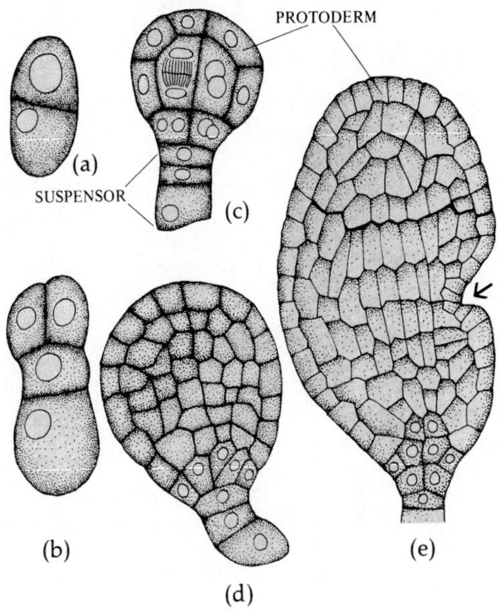

and leaves and buds at what was previously the upper end. The establishment of polarity is an essential first step in the development of all higher organisms, for it fixes the structural axis of the organism, the "backbone" on which the lateral appendages will be arranged.

Following the first division of the zygote, subsequent divisions may be transverse or vertical and in various combinations. Through an orderly progression of divisions, the embryo eventually differentiates into a nearly spherical structure, the embryo proper, and a stalklike structure, the *suspensor* (Figures 18–5 through 18–8). Before this stage is reached, the developing embryo is often referred to as the *proembryo.*

Suspensors were mentioned in Chapters 15 and 16, in relation to the embryos of *Lycopodium, Selaginella,* and *Pinus,* as structures that push the developing embryos into nutritive tissues. Until recently, it was believed that the suspensors of angiosperm embryos played a similarly limited role, merely pushing the developing embryo into the endosperm. It now appears that the suspensors of angiosperms are also actively involved in absorption of nutrients from the endosperm. In addition, in some embryos, proteinaceous substances manufactured in the suspensor apparently are utilized by the embryo proper during periods of rapid growth.

When first formed, the embryo proper consists of a mass of relatively undifferentiated cells. Soon, however, changes in the internal structure of the embryo result in the initiation of the tissue systems of the plant. The future epidermis, the *protoderm,* is formed by periclinal divisions beneath the surface of the embryo. (Periclinal divisions are those in which the cell plates that form between the two new cells are parallel with the surface of the plant part in which they occur.) In addition, differences in the degree of vacuolation and density of the cells within the embryo result in the initiation of the *procambium* and *ground meristem.* The highly vacuolated, less dense ground meristem gives rise to the *ground tissue,* which surrounds the less vacuolated and denser procambium, the precursor of the vascular tissues, xylem and phloem. The protoderm, ground meristem, and procambium—the so-called *primary meristems*—are continuous between the cotyledons and the axis of the embyro (Figures 18–7d and 18–8c).

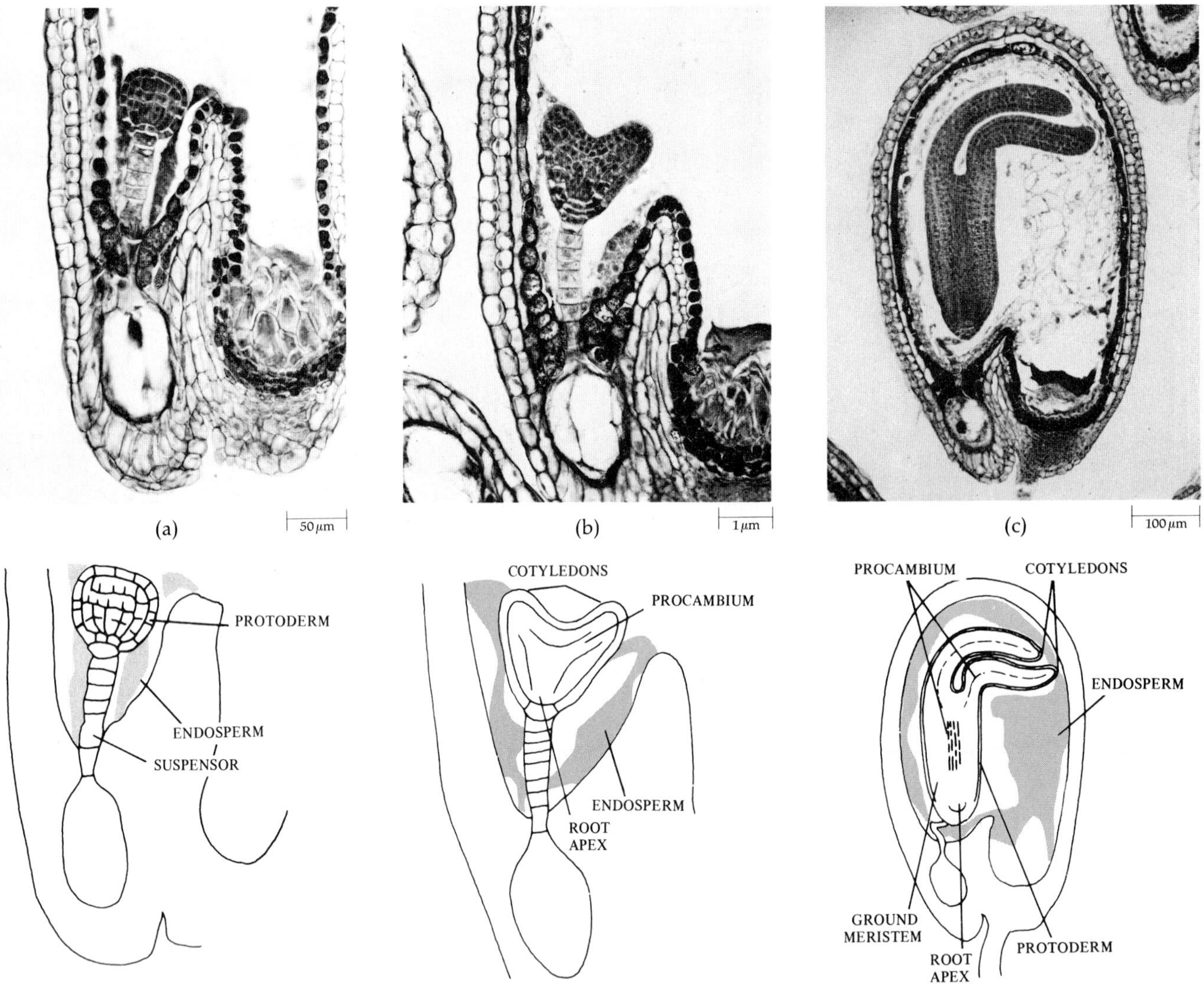

18–7
Some stages in development of the embryo of shepherd's purse (Capsella bursa-pastoris), *a dicot.* (a) *Embryo proper is globular and has a protoderm. Large cell, below, is basal cell of suspensor.* (b) *Embryo at heart-shaped (emergence of cotyledons) stage.* (c) *Embryo at torpedo stage. In* Capsella *the embryos curve.* (d) *Mature embryo.*

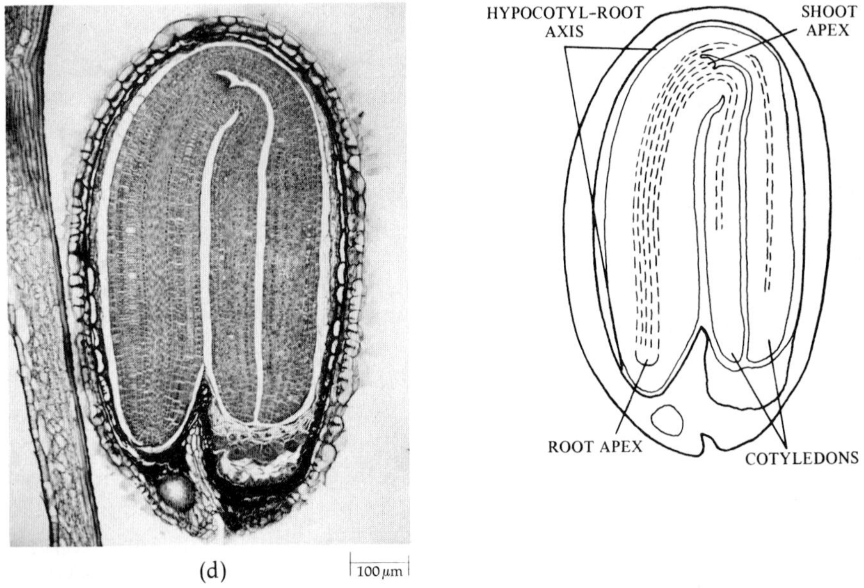

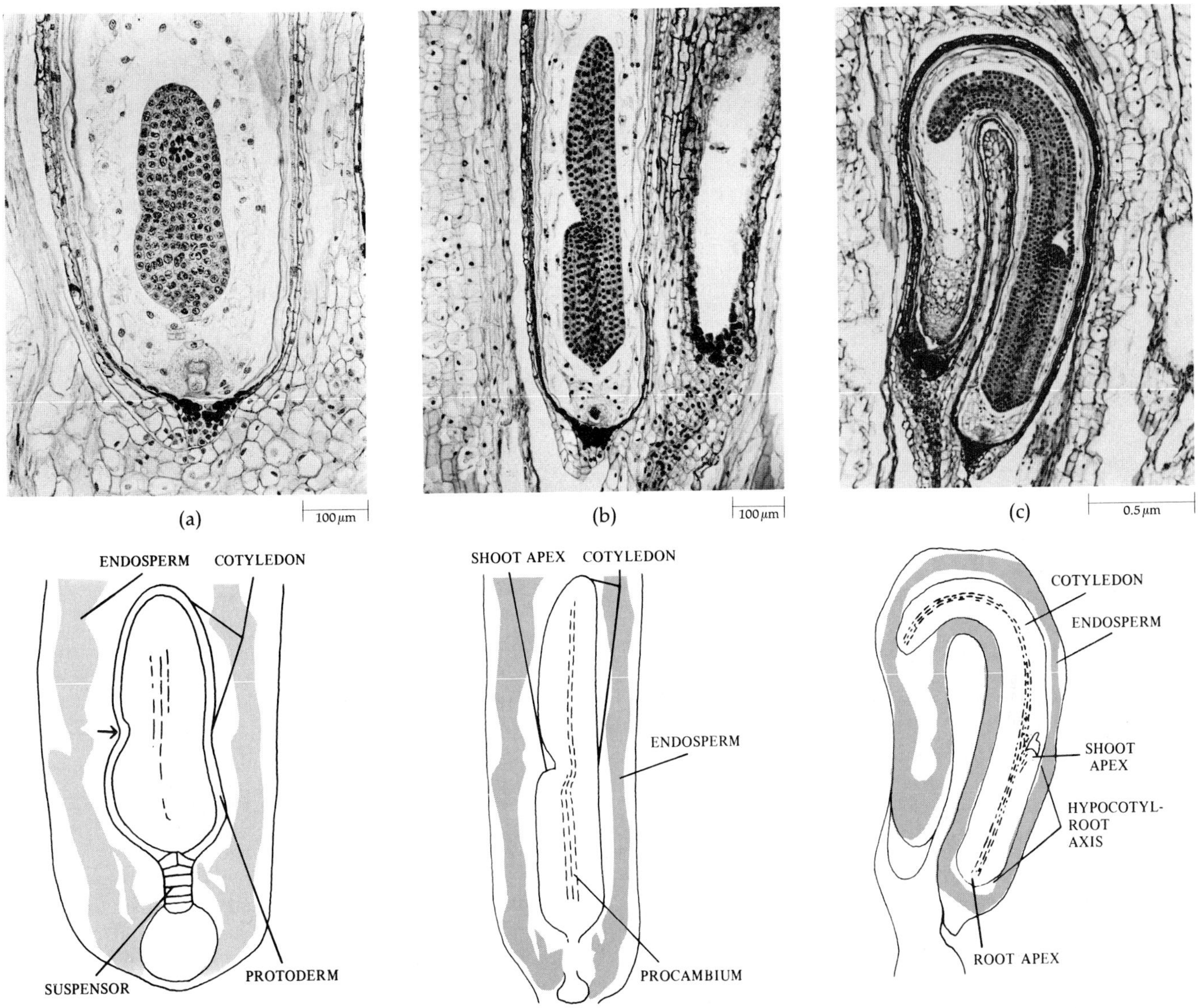

18–8
Some stages in development of the embryo of arrowroot (Sagittaria), *a monocot.* (a) *Depression, or notch (arrow), is beginning to form at base of emerging cotyledon.* (b) *Elongating embryo; apical meristem developing in notch; procambium now apparent.* (c) *Curving cotyledon; embryo approaching maturity.*

Development of the cotyledons may begin either during or after initiation of the primary meristems. (The stage of embryo development preceding cotyledon development is often referred to as the spherical or globular stage.) With the initiation of the cotyledons, further development of the embryo differs in dicots and monocots. The spherical embryo in dicots gradually assumes a bilobed shape (Figure 18–7b). This stage of the development of the embryo in dicots is often called the heart-shaped stage. Inasmuch as the embryos of monocots form only one cotyledon, they do not have a heart-shaped stage. Instead, their embryos become cylindrical in shape (Figure 18–8). As embryo development continues, the cotyledons and axes elongate (the so-called torpedo stage of embryo development), and the primary meristems extend along with them. During elongation, the embryo may remain straight or become curved. The single cotyledon of the monocot often becomes so large in comparison with the rest of the embryo that it is the dominating structure (see Figure 18–3b, page 395).

During the early stages of embryogeny, cell division takes place throughout the young sporophyte. However, as the embryo develops, the addition of new cells is gradually restricted to the *apical meristems* of the shoot and root. In dicotyledons, the apical meristem of the shoot arises between the two cotyledons (Figure 18–7). In monocotyledons, it arises on one side of the cotyledon and is completely surrounded by a sheathlike extension from the base of the cotyledon (Figure 18–8). The apical meristems of shoot and root are of great importance: These tissues are the source of virtually all of the new cells responsible for the development of the seedling and the adult plant from the embryo.

REQUIREMENTS FOR GERMINATION

The growth of the embryo is usually arrested while the seed matures and is dispersed. Resumption of growth of the embryo, or *germination* of the seed, is dependent upon many factors, both external and internal. Among the external, or environmental, factors, three are especially important: (1) water, (2) oxygen, and (3) temperature. Of these, water is the most critical.

Most mature seeds are extremely dry, normally containing only 5 to 20 percent of their total weight as water. Thus, germination is not possible until the seed imbibes the water required for metabolic activities. As the seed imbibes this considerable amount of water, it swells and much pressure may develop within the seed.

During the early stages of germination, respiration may be entirely anaerobic, but as soon as the seed coat is ruptured, the seed switches to aerobic respiration, which requires oxygen (Chapter 5). If the soil is waterlogged, the amount of oxygen available to the seed may be inadequate for aerobic respiration to take place, and the seed will fail to germinate.

Although many seeds will germinate over a fairly wide range of temperatures, they usually will not germinate either below or above a certain temperature specific for the species. The minimum temperature for most species is 0–5°C; the upper, 45–48°C; with an optimum of 25–30°C. At either extreme of temperature, the percentage of germination is usually very low.

Even when external conditions are favorable, some seeds will fail to germinate. Such seeds are said to be *dormant* (Chapter 24). The two most common causes of dormancy in seeds are (1) the impermeability of the seed coat to water and, sometimes, to oxygen, and (2) the physiological immaturity of the embryo. Some physiologically immature seeds must undergo a complex series of enzymatic and biochemical changes before they will germinate. Collectively these changes are called *after-ripening*. In temperate regions, after-ripening is triggered by the low temperatures of winter. Thus, the necessity for after-ripening helps prevent germination of the seed during the inclement period of winter when it would be unlikely to survive.

Dormancy is of great survival value to the plant or species. As in the example of after-ripening, it is a method of ensuring that conditions will be favorable for growth of the seedling when germination occurs. Some seeds must pass through the intestines of birds or mammals before they will germinate, resulting in wider dispersal of the species. Some seeds of desert species will germinate only when inhibitors in their coats are leached away by rainfall; this adaptation ensures that the seed will germinate only in those rare intervals when sufficient water is present for the seedling to mature. Similarly, others must be cracked mechanically, as by rushing water in a gravelly stream bed. Still other seeds lie dormant until the intense heat of a fire cracks the seed coat, thus promoting survival of the species in areas frequently swept by fires (see page 334).

FROM EMBRYO TO ADULT PLANT

When germination takes place, the first structure to emerge from most seeds is the radicle, or embryonic root; this order of events is a reflection of the primary needs of germinating seeds for water and for anchorage of the developing seedling. This first root, which is called the *primary root,* develops *lateral,* or *secondary roots,* and these, in turn, may give rise to more lateral roots. In this manner, a much-branched root system develops. Commonly the primary root in monocots is short-lived, and the root system of the adult plant develops from roots that are adventitious (arising not in its usual place) in origin. In the monocots, these adventitious roots arise at the nodes (the parts of the stem at which the leaves are attached) of the stems, and then produce lateral roots.

The way in which the shoot emerges from the seed during germination varies from species to species. For example, after the root emerges from the bean *(Phaseolus vulgaris)* seed, the hypocotyl elongates and becomes bent in the process (Figure 18–2a). Thus, the delicate shoot tip is not pushed through the soil and so is protected from injury. When the bend, or hook, as it is called, reaches the soil surface, it straightens out and pulls the cotyledons and plumule up into the air.

During germination and subsequent development of the seedling, the food stored in the cotyledons is digested and transported to the growing parts of the young plant. The cotyledons gradually decrease in size, wither, and eventually drop off. By this time, the seedling has become established; that is, it is no longer dependent upon the stored food of the seed for its nourishment; it is a photosynthesizing, autotrophic organism.

The germination of the castor bean *(Ricinus communis)* seed (Figure 18–2b) is essentially similar to that of the bean, except that in the castor bean the stored food resides in endosperm. As the hook straightens out, the endosperm and often the seed coat are carried upward, along with the cotyledons and plumule. During this period, the digested foods of the endosperm are absorbed by the cotyledons and transported to the growing parts of the seedling. In both the bean and castor bean, the cotyledons become green upon exposure to light, but they do not play an important photosynthetic function.

In the pea *(Pisum sativum)* it is the epicotyl that elongates and forms the hook. As the epicotyl straightens out, the plumule is raised above the soil surface. The cotyledons remain in the soil (Figure 18-2c), where they eventually decompose.

In the large majority of monocot seeds, the stored food is found in the endosperm. In relatively simple monocot seeds, such as that of the onion *(Allium cepa)*, it is the single tubular cotyledon that emerges from the seed and forms the hook (Figure 18-3b). When the cotyledon straightens, it carries the seed coat and enclosed endosperm upward. Throughout this period, and for some time afterward, the embryo obtains much of its nourishment from the endosperm by way of the cotyledon. Furthermore, the green cotyledon in onion functions as a photosynthetic leaf, and in this way it also contributes significantly to the food supply of the developing seedling. Soon the plumule, enclosed within the sheathlike base of the cotyledon, elongates and emerges from the cotyledon, which has provided it with protection.

Our last example of seedling development is provided by corn *(Zea mays)*, which, you will recall, has a highly differentiated embryo (Figure 18-3a). Both radicle and plumule are enclosed in sheathlike structures, the coleorhiza and coleoptile, respectively. The coleorhiza is the first structure to grow through the pericarp of the corn grain. It is then followed by the radicle, or primary root, which elongates very rapidly and quickly penetrates the coleorhiza. After the primary root emerges, the coleoptile is pushed upward by elongation of the first *internode*. (Internodes are the parts of the stem between two successive nodes.) When the base of the coleoptile reaches the soil surface, its edges spread apart at the tip, and the first leaves of the plumule begin to emerge. In addition to the primary root, two or more adventitious roots, which arise from the cotyledonary node, grow through the pericarp and then bend downward.

Regardless of the manner in which the shoot emerges from the seed, the activity of the apical meristem of the shoot results in the formation of an orderly sequence of leaves, nodes, and internodes. Apical meristems, which develop in the axils (the upper angles between leaves and stems) of the leaves, produce axillary or lateral shoots and these, in turn, may form additional axillary shoots.

The period from germination to the time the seedling becomes established as an independent organism constitutes the most crucial phase in the life history of the plant. It is at that time when the plant is most susceptible to injury by a wide range of insect pests and parasitic fungi, and when water stress can very rapidly prove fatal.

This type of growth of the root system and the shoot system is called vegetative growth. Eventually, one or more of the vegetative apical meristems of the shoot is changed into a reproductive apical meristem, that is, a meristem that develops into a flower or inflorescence (Chapter 21). When the flowers are formed, the plant is prepared to repeat the life cycle.

SUMMARY

The seeds of flowering plants consist of an embryo, a seed coat, and stored food. When fully formed, the embryo consists basically of a hypocotyl-root axis bearing either one (monocots) or two (dicots) cotyledons and an apical meristem at shoot and root apices. The cotyledons of most dicots are fleshy and contain the stored food of the seed. In other dicots and in most monocots, the stored food resides in the endosperm, and the cotyledons function as absorptive structures, absorbing digested food from the endosperm. The digested food is then transported to growing regions of the embryo.

During embryo development, the shoot and root of the young plant are initiated as one continuous structure beginning with the zygote. Polarity is established early in embryogeny, the portion of the embryo nearest the micropyle becoming the root pole and the portion farthest from the micropyle, the shoot pole. Eventually, the embryo differentiates into an embryo proper and a suspensor. At first an undifferentiated mass of cells, the embryo proper initiates the so-called primary meristems—protoderm, ground meristem, and procambium—the precursors of the epidermis, ground tissue, and vascular tissues, respectively. Either before or during development of the primary meristem, the embryo proper begins initiation of the cotyledons, the first leaves of the plant.

As the embryo develops, the addition of new cells is gradually restricted to certain parts of the plant body, the apical meristems, which arise at the shoot and root ends of the embryo axis. After dispersal of the seed, resumption of growth of the embryo—germination of the seed—is dependent upon environmental factors, including water, oxygen, and temperature. Many seeds must pass through a period of dormancy before they are able to germinate. All must imbibe water. The root is the first structure to emerge from most germinating seeds.

Following a period of vegetative growth, one or more apical meristems of the shoot is changed into a reproductive apical meristem, which develops into a flower or inflorescence.

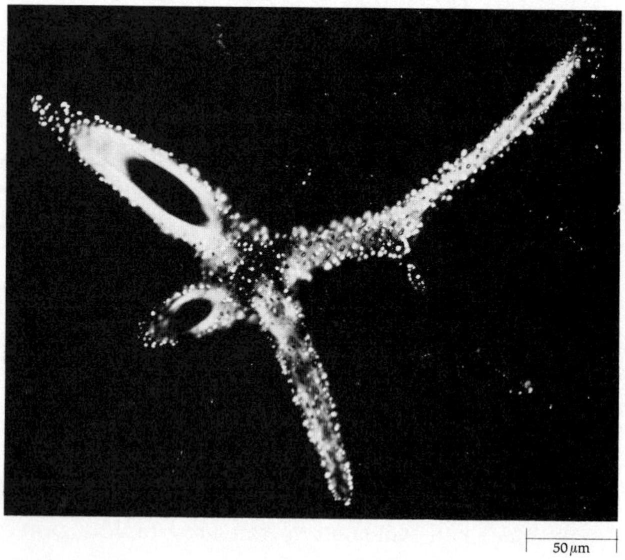

19–1
Branched sclereid from the leaf of the water lily (Nymphaea odorata) *seen in polarized light. Numerous small angular crystals are embedded in the sclereid wall. (Sclereid cells are described on page 405.)*

CHAPTER 19

Cells and Tissues of the Plant Body

Following the development of the embryo, the formation of new cells, tissues, and organs becomes restricted almost entirely to the meristems—the perpetually young, or embryonic, tissues concerned with growth. In Chapter 14 we noted that there are two main types of meristems, apical meristems and lateral meristems. The apical meristems are involved primarily with extension of the plant body and occur at the tips of shoots and roots. This type of growth is called *primary growth*. During primary growth, primary tissues are formed; the part of the plant body composed of these tissues is called the primary plant body.

The lateral meristems, the *vascular cambium* and the *cork cambium*, produce the secondary tissues, which make up the secondary plant body. You may recall (Chapter 14) that the vascular cambium produces secondary xylem and secondary phloem and that the cork cambium produces mostly cork. Detailed accounts of primary and secondary growth are presented in the three chapters that follow this one.

In both the apical and the lateral meristems, certain cells are able to divide repeatedly and in such a way that after each division one of the sister cells remains in the meristem while the other moves into the plant body. The self-perpetuating cells are called *initials*, and their sister cells are called *derivatives*. The initials perpetuate the meristem. It is important to note that the derivatives commonly divide one or more times before they begin to differentiate into specific types of cells. Consequently, the meristem generally is considered as consisting of the initials and their immediate derivatives.

Cell division is not limited to the apical and lateral meristems. For example, the protoderm, procambium, and ground meristem, which are partly differentiated tissues, are called *primary meristems* because (1) they give rise to the primary tissues and (2) many of their cells often divide several times before they begin to differentiate into specific cell types. With the exception of their initiation in the embryo proper during embryogeny, the primary meristems are derived from the apical meristems.

Growth of the plant body involves both cell division and cell enlargement. As one progresses from younger to older meristematic tissues, the overall size of the cells increases, and eventually the main factor involved in the increase in size of the particular region of the root, stem, or leaf is cell enlargement. *Differentiation*—the process by which cells become different from one another and from the meristematic cells from which they have their origin—often begins while the cell is still enlarging. At maturity, when differentiation is complete, some cells are living and others are dead. Among these living and dead cells are many different cell types (Figure 19-2). How cells that have a common origin come to be so different is generally considered one of the crucial questions of modern biology. Some factors involved in the control of cellular differentiation are discussed in Chapter 23.

In the plant body, basic tissue patterns are established by early meristematic activity. The shape of the plant and organization of its tissues are influenced greatly by both cell division and cell enlargement. This acquiring of a particular shape or form is known as *morphogenesis* (*morphe*, meaning "form," and *genere*, "to create").

THE TISSUE SYSTEMS

Botanists have long recognized that the principal tissues of vascular plants are grouped or organized into larger units in all parts of the plant. These groups of tissues, of which there are three, are known as the tissue systems, and their presence in root, stem, and leaf reveals both the basic similarity of the plant organs and the continuity of the plant body. The three tissue systems are: (1) the *fundamental* or *ground tissue system*, (2) the *vascular tissue system*, and (3) the *dermal tissue system*.

The ground, or fundamental, system consists of the so-called ground tissues: *parenchyma*, *collenchyma*, and *sclerenchyma*. Parenchyma is by far the most common of ground tissues. The vascular tissue system consists of the two conducting tissues, *xylem* and *phloem*. The dermal tissue system is represented by the *epidermis*, the outer protective covering of the primary plant body, and later by the *periderm*, in the secondary plant body.

THE TISSUES AND THEIR COMPONENT CELLS

Tissues may be defined as groups of cells that are distinct structurally or functionally or both. Tissues composed of only one type of cell are termed *simple tissues*, and those composed of two or more types of cells, *complex tissues*. Parenchyma, collenchyma, and sclerenchyma are simple tissues; xylem, phloem, and epidermis are complex tissues.

Parenchyma

Parenchyma tissue, the tissue from which all other tissues have evolved, is composed of parenchyma cells. In the primary plant body, parenchyma cells commonly occur as continuous masses in the cortex of stems (Figure

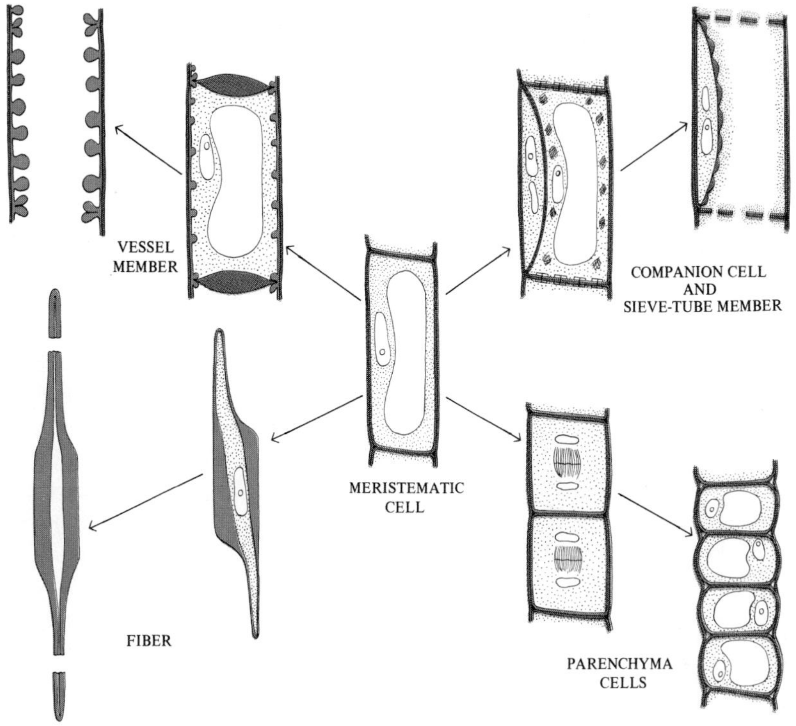

19-2
Diagram illustrating some of the cell types that may originate from a single kind of meristematic cell in the procambium or vascular cambium.

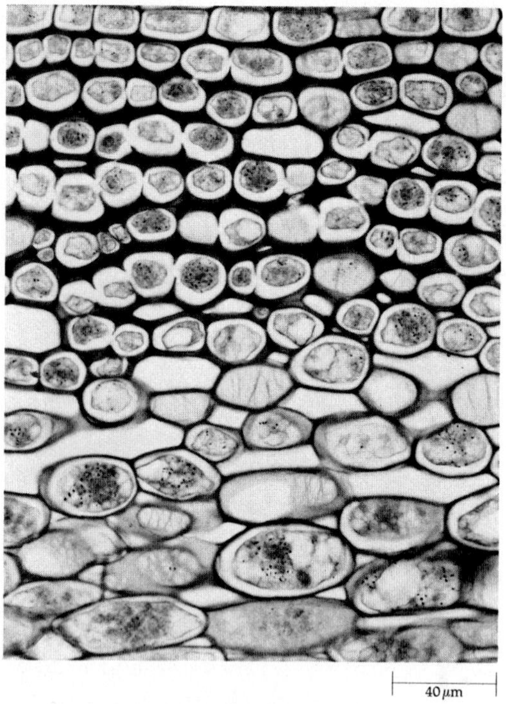

19–3
Transverse section showing collenchyma cells with unevenly thickened walls (above) and parenchyma cells (below) in the cortex of an elderberry (Sambucus canadensis) *stem. Clear areas between cells are intercellular spaces. In a few parenchyma cells, a meshwork of lines is visible. These are parenchyma cell walls. The light areas within the meshwork are primary pit-fields (page 35).*

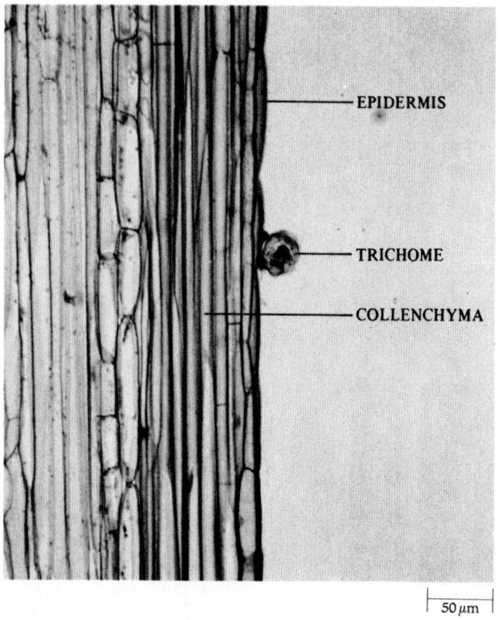

19–4
Longitudinal view of elongate collenchyma cells from the stem of a squash (Cucurbita maxima). *Trichomes are described on page 414.*

19–3) and roots, in the pith of stems, in leaf mesophyll, and in the flesh of fruits. Parenchyma cells also occur as vertical strands of cells in the primary and secondary vascular tissues and as rays in the secondary vascular tissues (Chapter 22).

Characteristically alive at maturity, parenchyma cells are capable of cell division and, although their walls are commonly primary, some parenchyma cells also have secondary walls. Because of their ability to divide, parenchyma cells play an important role in wound healing and regeneration. It is these cells that initiate adventitious structures such as adventitious roots on stem cuttings. Parenchyma cells are concerned with such activities as photosynthesis, storage, and secretion—activities dependent upon living protoplasm. In addition, parenchyma cells may play a role in the movement of water and the transport of food substances in plants.

During the past few years, considerable attention has been given to a special kind of parenchyma cell containing ingrowths of the cell wall which often greatly increase the surface area of the plasma membrane. These cells, called *transfer cells* by J. S. Pate and B. E. S. Gunning, are believed to play an important role in the transfer of solutes over short distances. Although such cells have long been known to exist in various plant parts, botanists have only recently realized that they are exceedingly common and probably have a common function. Transfer cells occur in association with the xylem and phloem of fine and minor veins in cotyledons and leaves of many herbaceous dicotyledons, and with the xylem and phloem of leaf traces at the nodes in both dicotyledons and monocotyledons. In addition, they are found in various tissues of reproductive structures (e.g., placentae, embryo sacs, endosperm) and in various glandular structures (e.g., nectaries, salt glands, the glands of carnivorous plants). Each of these locations is a potential site of intensive short-distance solute transfer. Whether the transfer cells actually function in the exchange of solutes remains to be determined experimentally.

Collenchyma

Collenchyma tissue is composed of collenchyma cells, which, like parenchyma cells, are living at maturity (Figure 19–3). Collenchyma tissue commonly occurs in discrete strands or as continuous cylinders beneath the epidermis in stems and petioles, and it borders the veins in dicot leaves. The generally elongated collenchyma cells (Figure 19–4) contain unevenly thickened, nonlignified primary walls, making them especially well adapted for the support of young, growing organs (see the description of the primary wall on page 34). The name collenchyma derives from the Greek word *colla,* meaning "glue," which refers to their characteristic thick, glistening walls in fresh tissue (Figure 19–5). Being primary in nature, the walls of collenchyma cells are readily stretched and offer relatively little resistance to elonga-

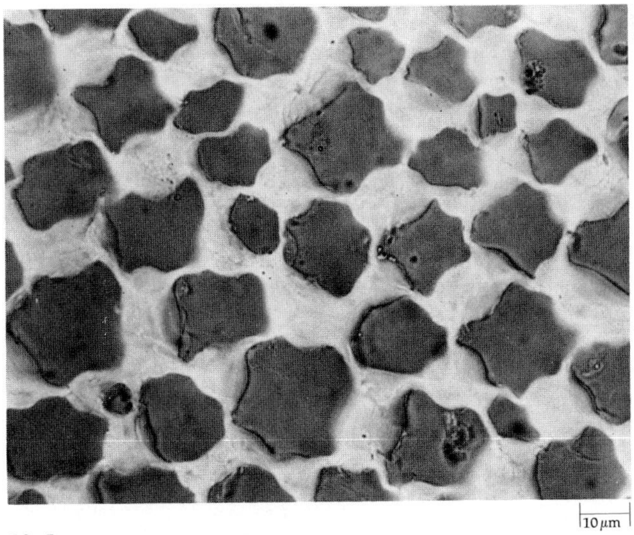

19–5
Transverse section of collenchyma tissue from petiole of celery (Apium graveolens). *In fresh tissue such as this the unevenly thickened collenchyma cell walls have a pearly or glistening appearance. The light background is the cell wall.*

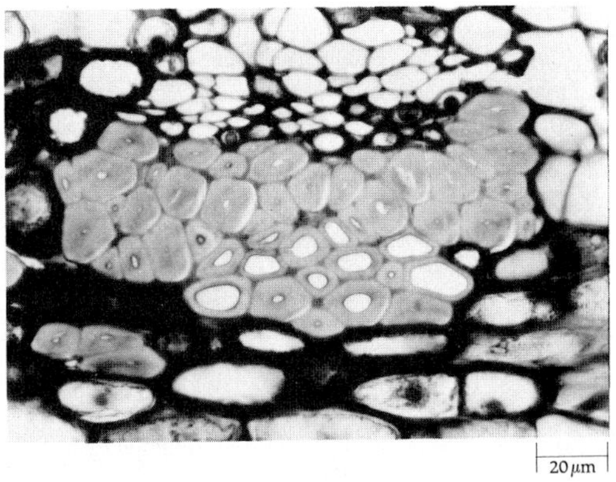

19–6
Primary phloem fibers (grey area) in cross-section from the stem of a linden, or basswood (Tilia americana). *Such thick-walled fibers contain relatively few and inconspicuous pits.*

tion of the plant part in which they are found. In addition, because collenchyma cells are living at maturity, they can continue to develop thick, flexible walls while still elongating.

Sclerenchyma

Sclerenchyma cells may develop in any or all parts of the primary and secondary plant body; they often lack protoplasts at maturity. The term sclerenchyma is derived from the Greek *skleros,* meaning "hard," and the principal characteristic of sclerenchyma cells is their possession of thick, often lignified secondary walls. Because of these walls, sclerenchyma cells are important strengthening and supporting elements in plant parts that have ceased elongating (see the description of the secondary wall on page 35).

Two types of sclerenchyma cells are recognized: *fibers* and *sclereids*. Fibers are generally long, slender cells, and they commonly occur in strands or bundles (Figure 19–6). The so-called bast fibers—for example, hemp, jute, and flax—are derived from dicotyledons. Other economically important fibers—such as Manila hemp—are present in the leaves of monocots. Sclereids are variable in shape and they are often branched (Figures 19–1 and 19–7), but compared with most fibers, sclereids are relatively short cells. Sclereids may occur singly or in aggregates throughout the ground tissue. They make up the seed coats of seeds, the shells of nuts, the endocarp, or stone, of stone fruits, and they give the pear fruit (Figure 19–8) its characteristic gritty texture.

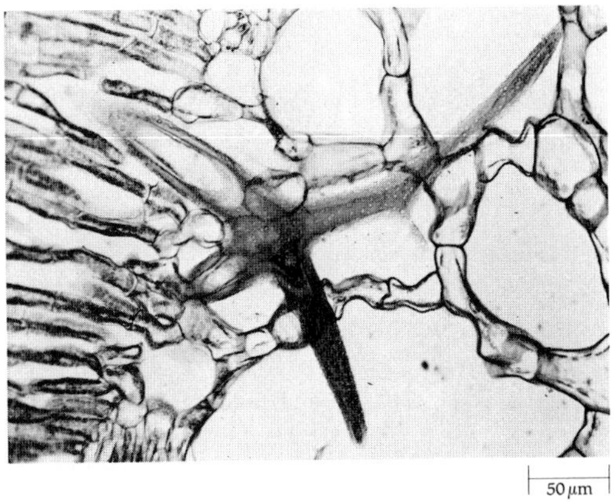

19–7
Branched sclereid from the leaf of the water lily (Nymphaea odorata), *seen in ordinary light. See Figure 19–1 for the same sclereid seen in polarized light.*

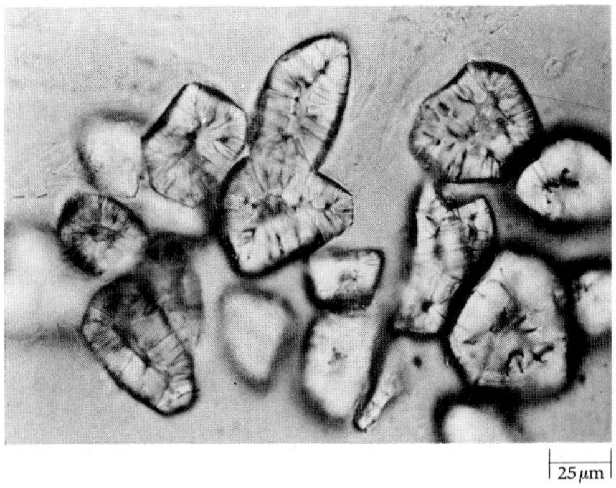

19–8
Stone cells (sclereids) from the flesh of the pear (Pyrus communis) *fruit. The secondary walls contain numerous conspicuous simple pits, with many branches.*

19–9
Transverse section of a vascular bundle from the stem of a squash (Cucurbita maxima). *Phloem occurs on both sides of the xylem, and a vascular cambium has developed between the external phloem and the xylem.*

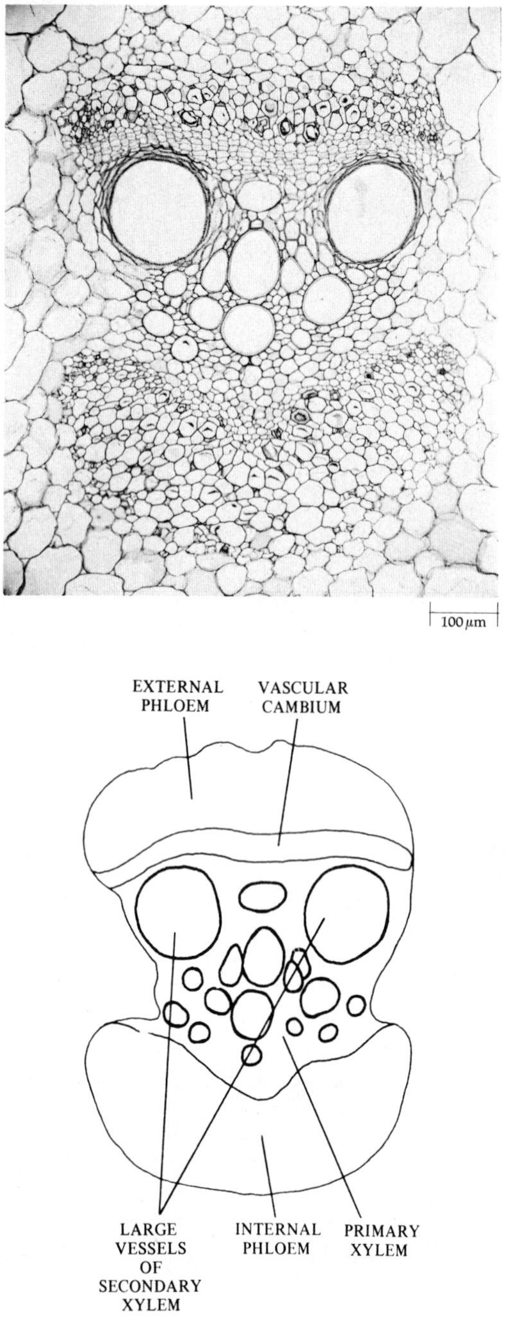

Xylem

Xylem is the principal water-conducting tissue of vascular plants. It is also involved in the conduction of minerals, in food storage, and in support. Together with the phloem, the xylem forms a continuous system of vascular tissue throughout the plant body (Figure 19–9). Xylem may be primary or secondary in origin. The primary xylem is derived from the procambium and the secondary xylem from the vascular cambium (Chapter 22).

The principal conducting cells of the xylem are the tracheary elements, of which there are two types, the *tracheids* and the *vessel members* (or *vessel elements*). Both tracheids and vessel members are elongated cells that have secondary walls and lack protoplasts at maturity (Figure 19–10). Both tracheids and vessel members may have pits in their walls. Besides the presence of pits, vessel members contain perforations (Figure 19–11), which are areas lacking both primary and secondary walls. The perforations are literally holes in the wall. Perforations generally occur on the end walls, but they may also be found on the lateral walls. The part of the wall bearing the perforation is called the *perforation plate.* Vessel members are joined into long, continuous columns, or tubes, called *vessels* (Figure 19–12). The pits of the long, tapering tracheids are concentrated on the overlapping ends of the cells.

The term *tracheary element* dates back to the seventeenth century and the Italian physician Marcello Malpighi, one of the founders of plant anatomy. Malpighi was especially interested in finding similarities between plants and animals. During his examination of xylem, he observed air bubbles escaping from a vessel with spiral wall thickenings. Immediately he compared the vessel with the tracheae, or air ducts, of insects and later applied the same term to the vessels of the xylem. That term has been used ever since for the water-conducting cells of the xylem.

Vessel members generally are thought of as being more efficient conductors of water than the tracheids, because water can flow relatively unimpeded from vessel member to vessel member through the perforations. When flowing through tracheids, water must pass through the membranes of the pit-pairs (Chapter 1, page 35). The pit membranes probably offer relatively little resistance to the flow of water, because during the final stages of differentiation they are partially hydrolyzed and only a highly permeable meshwork of cellulose microfibrils is left.

In the secondary xylem and late-formed primary xylem, or metaxylem, the pitted secondary walls of the tracheids and vessel members cover the entire primary wall, except at the pits and at the perforations of the vessel members (Figure 19–10). Consequently, these walls are rigid and cannot be stretched. During the

19–10
Cell types in the secondary xylem of an oak (Quercus), *as illustrated from macerated tissue (tissue that has been broken down into its constituent cells).* (a) *and* (b) *Wide vessel members.* (c) *Narrow vessel member.* (d) *Tracheid.* (e) *Fiber-tracheid.* (f) *Libriform fiber (the longest fiber in oak wood) and parenchyma cells. The spotted appearance of these cells is due to pits in the walls; pits are absent in* (f).

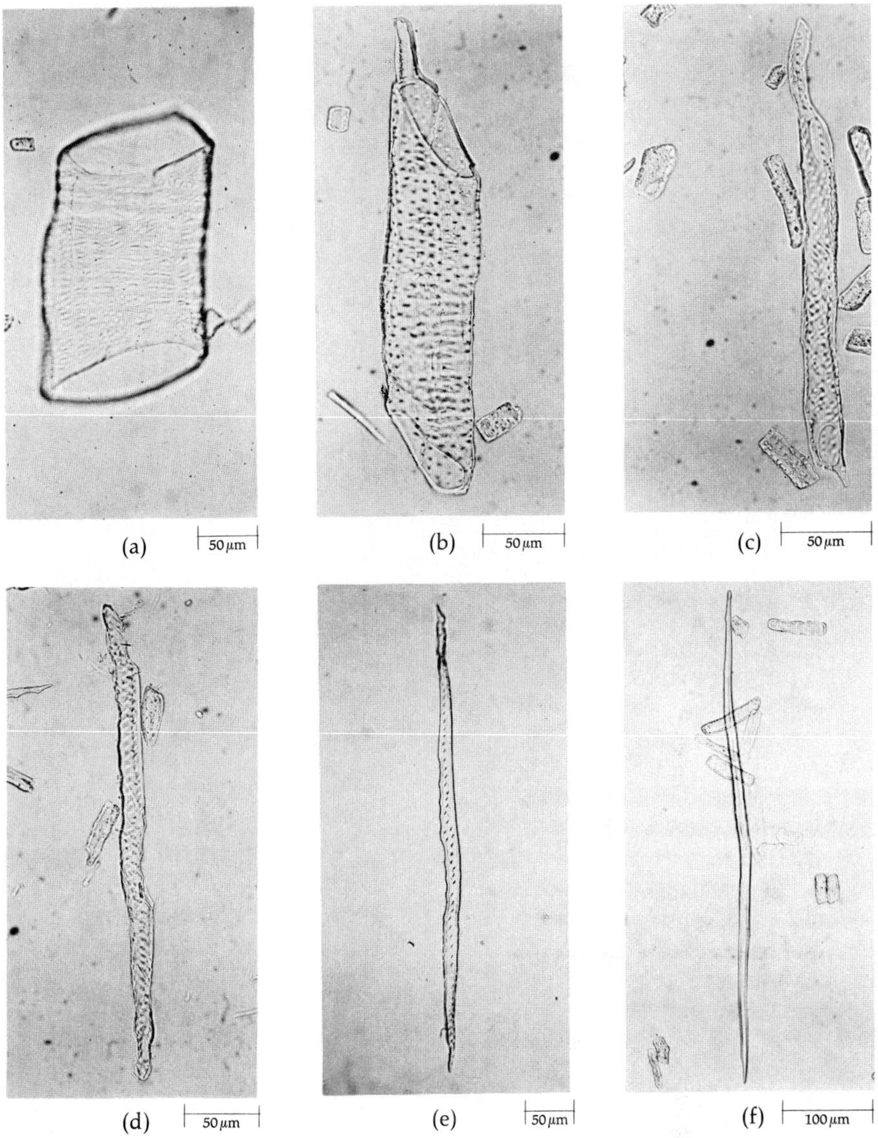

19–11
Perforated end walls of vessel members. These end walls are not perpendicular to the length of the vessel members but are at an oblique angle which makes the side walls of the vessel members appear to be wedge-shaped. The light areas within the thin grey lines of (a) *are perforations. The entire network of grey and dark lines surrounding the perforations is the perforation plate. The perforation plate in* (a) *is called scalariform (ladderlike) and in* (b) *it is called simple, due to the single, large circular perforation.* (a) *is from the tulip tree* (Liriodendron tulipifera); (b) *is from a linden* (Tilia americana).

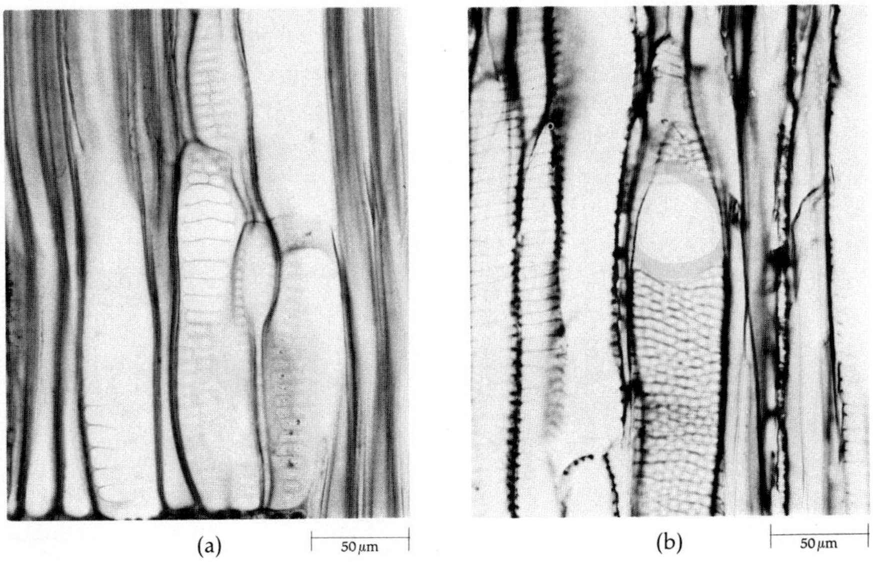

19-12
Scanning electron micrographs of vessel members from secondary xylem of (a) *red oak* (Quercus rubra) *and* (b) *squash* (Cucurbita maxima). (a) *Shows external view of parts of three vessel members. Notice rims between vessel members. In* (b), *parts of two vessel members are cut lengthwise so that half the cylinder is seen from the inside. The arrow points to the boundary between the two elements. Notice the numerous pits in the walls of vessel members in* (a) *and* (b).

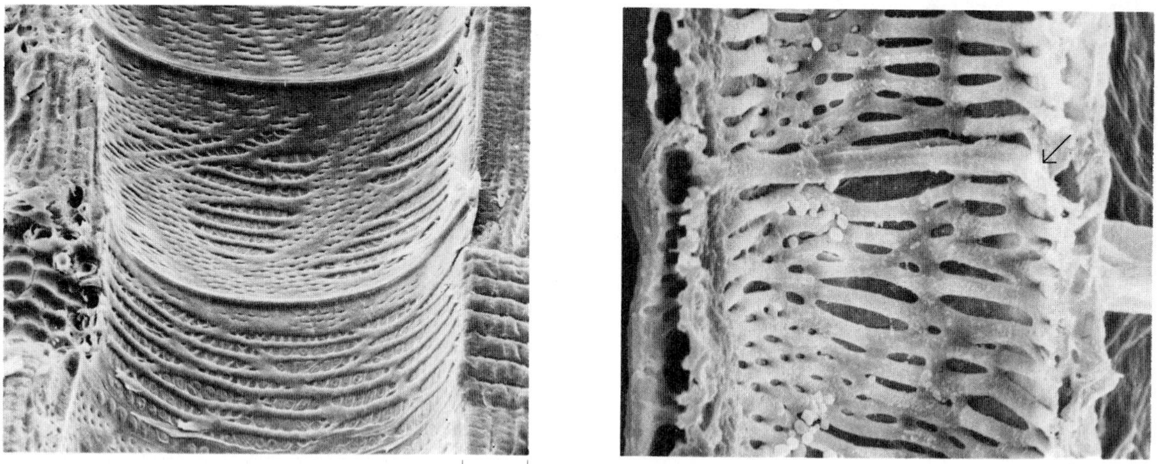

19-13
Parts of tracheary elements from the first-formed primary xylem (the protoxylem) of the castor bean (Ricinus communis). (a) *Annular (6 ringlike shapes at left) and helical wall thickenings in partly extended elements.* (b) *Double helical thickenings in elements that have been extended. The element on the left has been greatly extended, and the coils of the helices have been pulled far apart.*

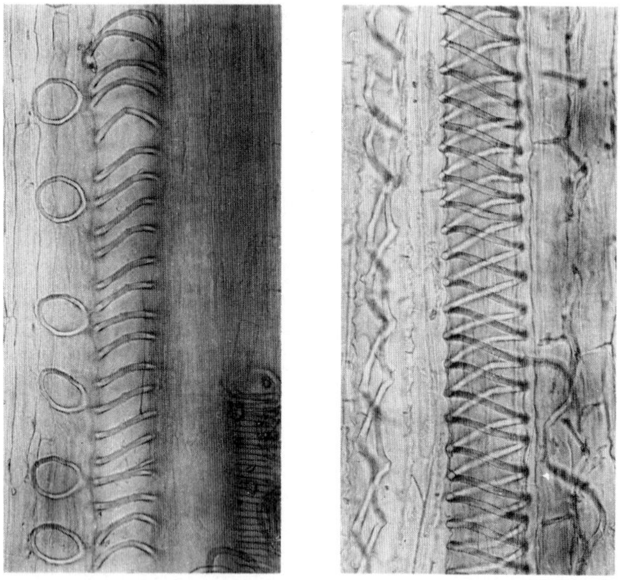

period of elongation or expansion of roots, stems, and leaves, many of the first-formed tracheary elements of the early primary xylem, or protoxylem, have their secondary walls deposited in the form of rings or spirals (Figure 19-13). These thickenings, which may be annular (ring-shaped) or helical (spiral), make it possible for such tracheary elements to be stretched or extended during overall elongation of the organ. The annular and helical thickenings are frequently destroyed during elongation. In the primary xylem, the nature of the wall thickening is greatly influenced by the amount of elongation of the organ. If little elongation occurs, pitted elements rather than extensible types appear. On the contrary, if much elongation takes place, many elements with annular and helical thickenings will be present.

The tracheid is the only type of water-conducting cell found in most lower vascular plants and gymnosperms; the great majority of angiosperms contain vessel members in addition to tracheids in their xylem. Extensive comparative studies of the tracheary elements in a wide range of vascular plants have clearly established that the tracheid is a more primitive cell than the vessel member and that vessel members evolved from tracheids independently in several groups of vascular plants.

In addition to tracheids and vessel members, xylem contains parenchyma cells that store various substances. The xylem parenchyma cells commonly occur in vertical strands. In the secondary xylem, they also occur in horizontal strands called *rays.* Fibers (Figure 19-10) and sclereids also occur in xylem. Many xylem fibers are living at maturity and serve a dual function of storage and support.

Phloem

Phloem is the principal food-conducting tissue in vascular plants (Figure 19-9). The phloem may be primary or secondary in origin and, as with primary xylem, the first-formed primary phloem, the protophloem, is frequently stretched and destroyed during elongation of the organ.

The principal conducting cells of the phloem are the *sieve elements*, of which there are two types, the *sieve cells* (Figures 19-14 and 19-15) and the *sieve-tube members* (Figures 19-16, 19-17, and 19-18). The term "sieve" refers to the clusters of pores, the *sieve areas*, through which the protoplasts of adjacent sieve elements are interconnected. In sieve cells, the pores are narrow and the sieve areas are rather uniform in structure on all walls. Most of the sieve areas are concentrated on the overlapping ends of the long, slender cells (Figure 19-14). In the sieve-tube members, the sieve areas on some walls have larger pores than those on other walls of the same cell. The part of the wall bearing the sieve areas with larger pores is called a *sieve plate* (Figures 19-17 and 19-18). Although sieve plates may occur on any wall, they generally are located on the end walls. Sieve-tube members occur end-on-end in longitudinal series called *sieve tubes*. Thus, the principal distinction between sieve cells and sieve-tube members is the presence of sieve plates in sieve-tube members and their absence in sieve cells.

Considerably less is known about the evolution of sieve elements than of tracheary elements. Sieve cells are the only type of food-conducting cell in most lower vascular plants and gymnosperms, whereas sieve-tube members occur in angiosperms, which lack sieve cells. Sieve-tube members have been reported in certain parts of the plant body of some species of *Equisetum* and, more recently, in a homosporous fern. Although the Gnetinae contain vessel members, they lack sieve-tube members.

The walls of sieve elements are composed chiefly of cellulose and generally are described as primary. In cut sections of phloem tissue, the pores of the sieve areas and sieve plates are generally lined or occluded by a wall substance called *callose* (Figure 19-17), which is a polysaccharide composed of spirally wound chains of glucose residues. The presence of callose in the pores of conducting sieve elements long puzzled botanists; it seemed illogical for the pores to contain any substance that would impede the movement of other substances from cell to cell. It is now known that most, if not all, of the callose found in the pores of conducting sieve elements is deposited there in reaction to wounding.

Unlike tracheary elements (tracheids and vessel members), sieve elements (sieve cells and sieve-tube members) have living protoplasts at maturity (Figure 19-18). The protoplasts of mature sieve elements are unique among the living cells of the plant in that they either lack nuclei entirely or contain only remnants of

19–14
Longitudinal (radial) view of secondary phloem of yew (Taxus canadensis), *showing vertically oriented sieve cells, strands of parenchyma cells, and fibers. Parts of two horizontally oriented rays can be seen traversing the vertically oriented cells.*

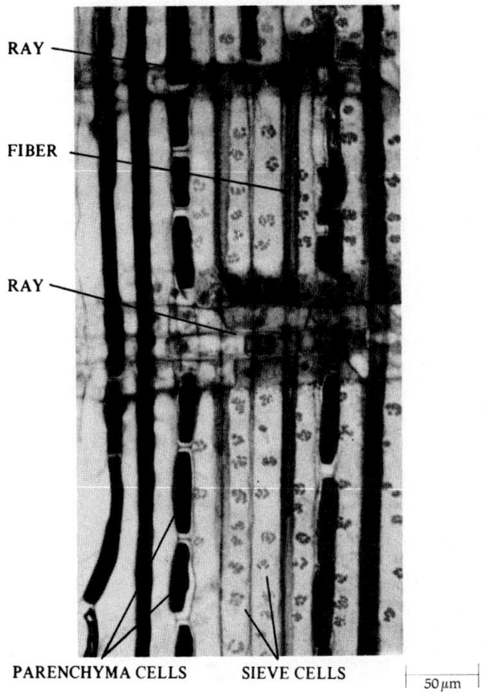

19–15
Detail of portion of the secondary phloem of yew (from Figure 19–14), showing sieve areas (arrows) on walls of sieve cells, and albuminous cells (see page 412), which here constitute the top row of cells of the ray.

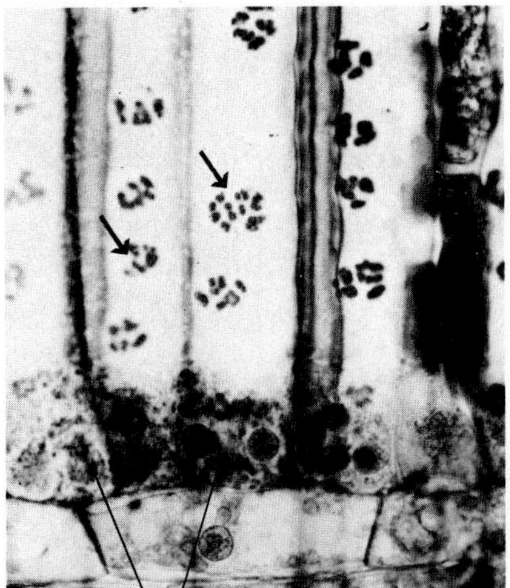

19–16
Longitudinal (radial) view of secondary phloem of linden (Tilia americana), *showing sieve-tube members and conspicuous groups of thick-walled fibers.*

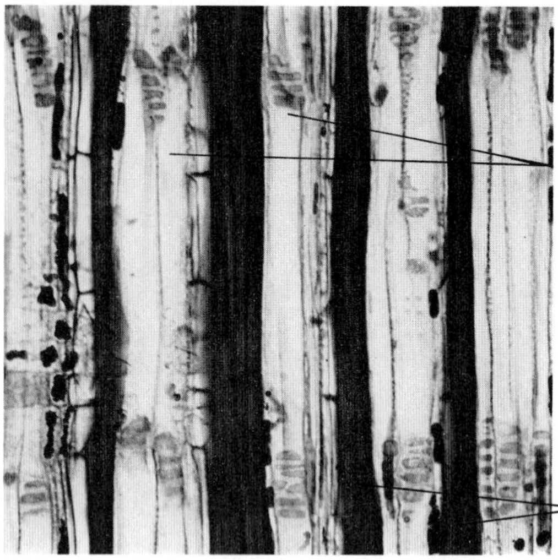

19–17
Detail of a portion of the secondary phloem of Tilia americana *(from Figure 19–16), showing a compound sieve plate, which is a sieve plate consisting of two or more sieve areas. The sieve areas are the gray oblong shapes in the center of the micrograph. Each sieve area is composed of pores bordered by cylinders of callose. The pores are visible in four of the sieve areas shown here. The arrows point to very small sieve areas on the lateral walls of sieve-tube members. The pores of these lateral sieve areas are too small to be seen in this micrograph.*

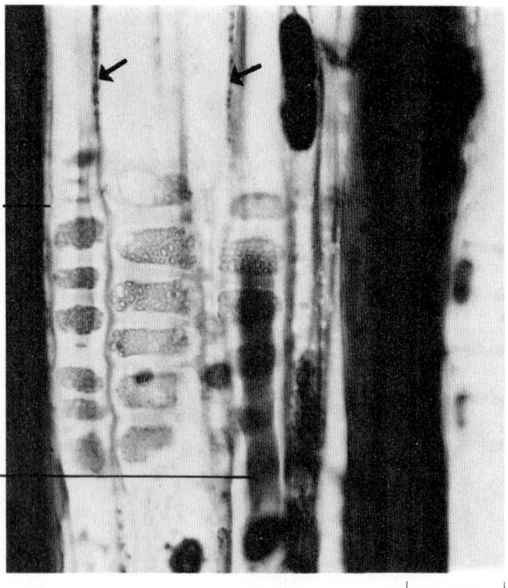

19–18
The phloem of the squash (Cucurbita maxima) *stem.* (a)–(c), *electron micrographs;* (d)–(f) *photomicrographs.* (a) *Longitudinal view of parts of two mature sieve-tube members and a sieve plate, showing distribution of P-protein (arrows) along the wall.* (b) *Face view of simple sieve plate (one sieve area per plate) between two mature sieve-tube members. The sieve-plate pores are wide open.* (c) *Transverse section of mature sieve-tube member, showing distribution of P-protein (arrows) along the wall.* (d) *Transverse section, showing two immature sieve-tube members. Slime bodies, or P-protein bodies, can be seen in the sieve-tube member on the left, an immature sieve plate in the one to the right, above. The small, dense cells are companion cells.* (e) *Transverse section, showing some mature sieve-tube members. A slime plug can be seen in the sieve-tube member on left, a mature sieve plate in one on right. Small, dense cells are companion cells.* (f) *Longitudinal section, showing mature and immature sieve-tube members. Arrows point to P-protein bodies in immature cells.*

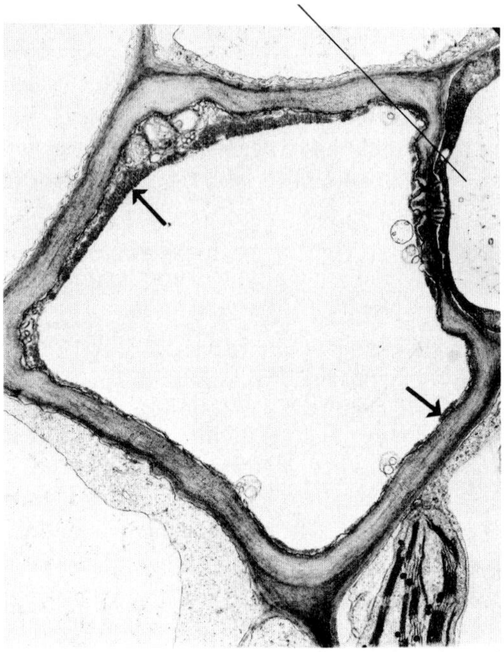

(c)

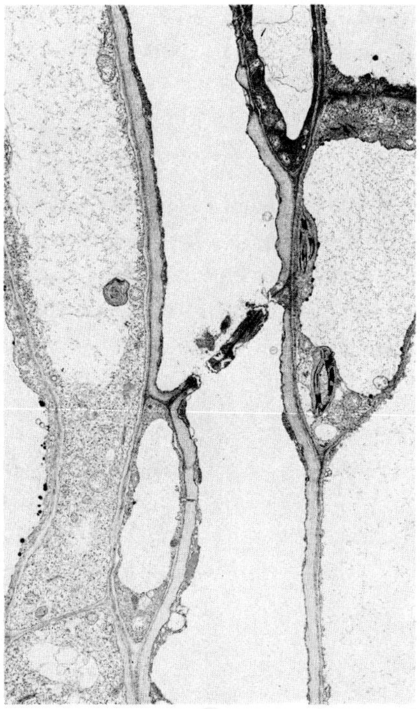

(a) 5 μm

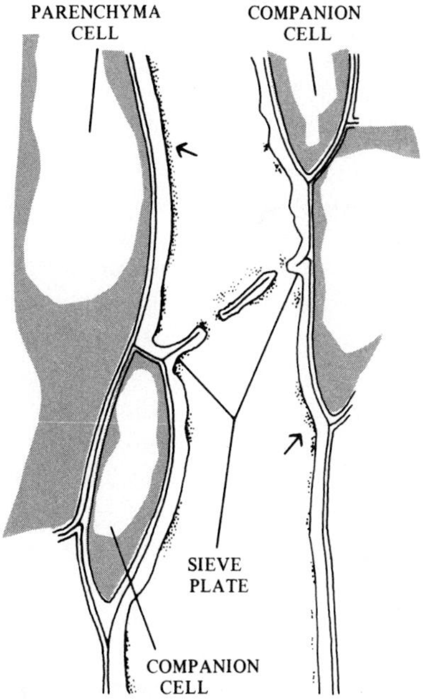

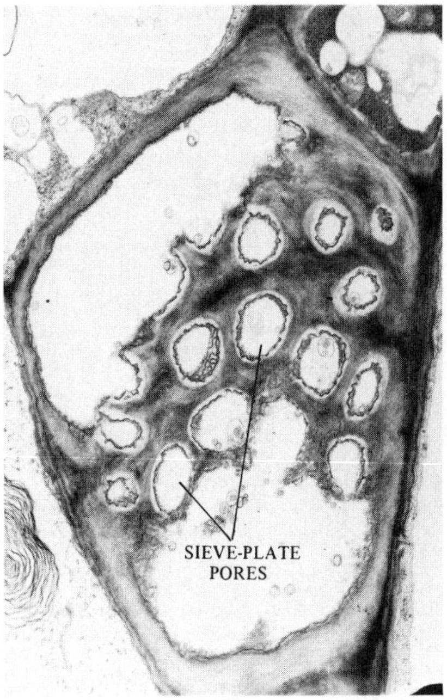

(b) 2 μm

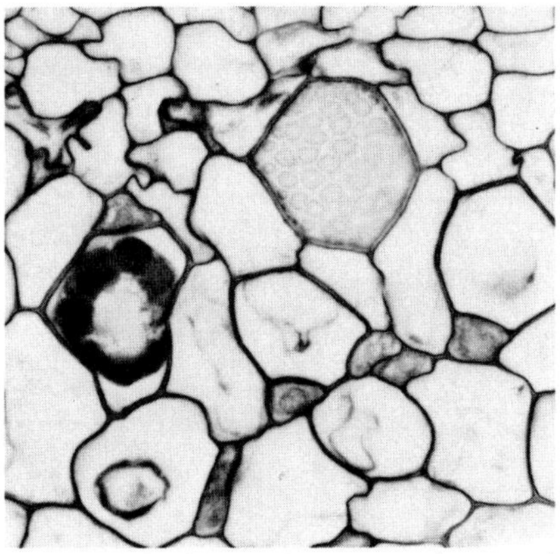

(d) 20 μm

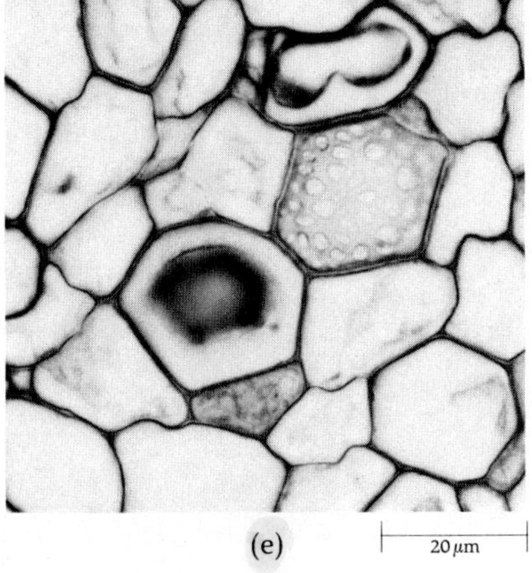

(e) 20 μm

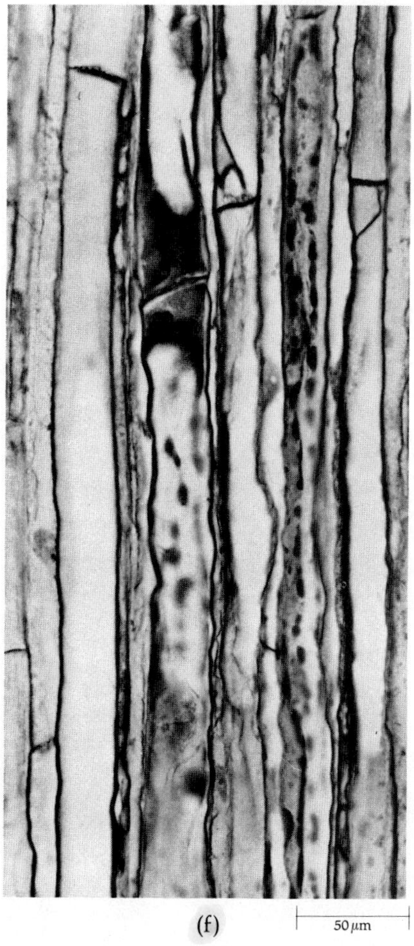

(f) 50 μm

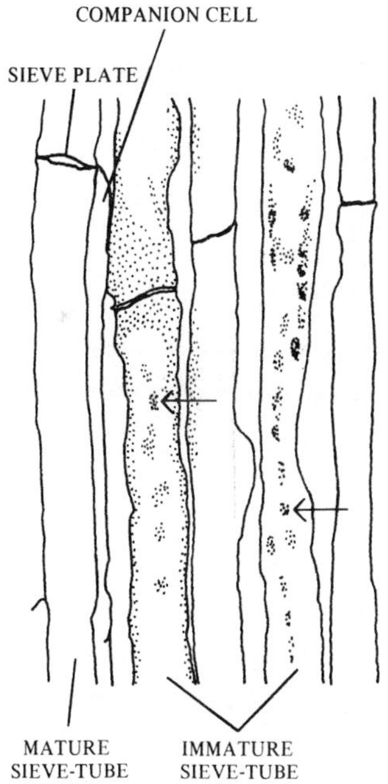

them. In addition, most mature sieve elements lack a clear boundary between the cytoplasm and the vacuoles. When young, the sieve element contains several or many vacuoles, each delimited from the cytoplasm by a tonoplast, or vacuolar membrane. In the final stages of differentiation, the tonoplasts disappear and the distinction between cytoplasmic and vacuolar contents no longer exists. The mature sieve-element protoplast is delimited from the cell wall by a differentially permeable membrane, the plasma membrane, and contains a network of endoplasmic reticulum, which is closely appressed to the plasma membrane along the wall. Plastids and mitochondria are also present, but ribosomes, Golgi bodies, and microtubules are lacking.

The protoplasts of the sieve-tube members of most angiosperms are characterized by the presence of a proteinaceous substance known as *slime* or *P-protein*. P-protein has its origin in the young sieve-tube member in the form of discrete bodies, called slime bodies or P-protein bodies (Figure 19–18). During the late stages of differentiation, the P-protein bodies elongate and disperse. In cut sections of phloem tissue, "slime plugs" of P-protein are usually found near the sieve plates. Like callose, slime plugs are not found in undisturbed cells and so are considered to be the result of the surging of the contents of the sieve-tube members that occurs when the tissues and sieve tubes are severed from the plant. Recent studies indicate that in normal mature sieve-tube members, P-protein is distributed along the walls and is continuous from one sieve-tube member to the next through the pores of the sieve areas and sieve plates. The sieve-plate pores are lined with P-protein, but not occluded with it.

Sieve-tube members are characteristically associated with specialized parenchymatic cells called *companion cells* (Figure 19–18d and e), which contain all of the components commonly found in living plant cells, including a nucleus. The sieve-tube member and its associated companion cells are closely related developmentally—they are derived from the same mother cell—and they have numerous connections with one another. Functionally the companion cells are very important, for they are largely responsible for the active secretion of substances into and their removal from the sieve-tube members. This subject will be discussed further in Chapter 25 when possible mechanisms of phloem transport in vascular plants are considered.

Sieve cells of gymnosperms are characteristically associated with specialized parenchyma cells called *albuminous cells* (Figure 19–15). Although generally not derived from the same mother cell as their associated sieve cells, the albuminous cells are believed to perform the same roles as companion cells. Like the companion cell, the albuminous cell contains a nucleus, in addition to other cytoplasmic components characteristic of living cells.

The sieve elements in most species apparently are short-lived and die in less than a year from their origin. This is not true, however, of all sieve elements. In the secondary phloem of the basswood, or linden tree (*Tilia americana*), some sieve elements remain alive and presumably function as conducting elements for 5 to 10 years. Sieve elements are known to remain alive for many years in perennial monocots. In certain palms, some sieve elements at the base of the main stem may be more than a century old. When the sieve elements die, their associated companion cells or albuminous cells also die, which is one more indication of the intimate relation that exists between sieve elements and their companion cells or albuminous cells.

Other parenchyma cells occur in the primary and secondary phloem (Figures 19–14 through 19–18). They are largely concerned with the storage of various substances. Fibers (Figures 19–14 through 19–16) and sclereids may also be present.

Epidermis

The epidermis is the outermost layer of cells on the primary plant body, and it constitutes the dermal tissue system of floral parts, fruits, and seeds, and of stems and roots until they undergo considerable secondary growth. Both functionally and structurally the epidermal cells are quite variable. In addition to the ordinary epidermal cells, which form the bulk of the epidermis, the epidermis may contain *stomata* (Figures 19–19 , 19–20, and 19–21), many types of appendages, or *trichomes* (Figure 19–22), and other kinds of cells specialized for specific functions.

19–19
Surface view of lower epidermis of red oak (Quercus rubra) *leaf, taken with the scanning electron microscope. Numerous stomata and rods of wax can be seen here.*

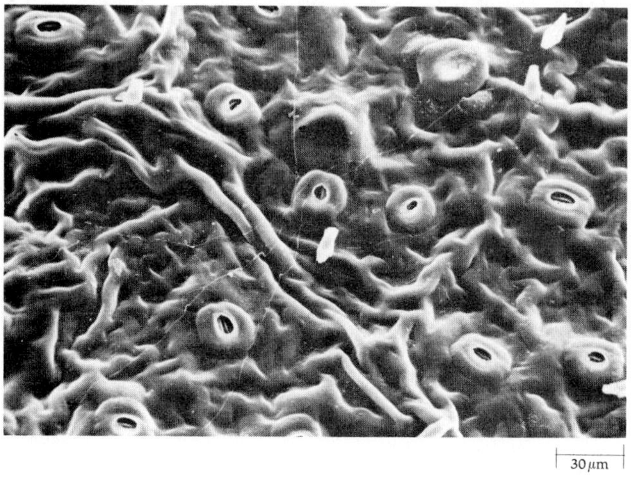

19–20
Electron micrographs of corn (Zea mays) stomata. (a) *Surface view of immature guard cells (center) and associated accessory, or subsidiary, cells (one on either side of the guard cells).* (b) *Transverse section through mature, thick-walled guard cells, each of which is attached to an accessory cell.*

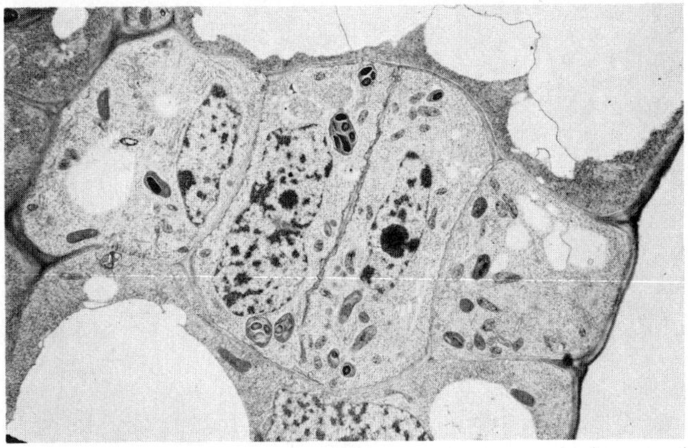

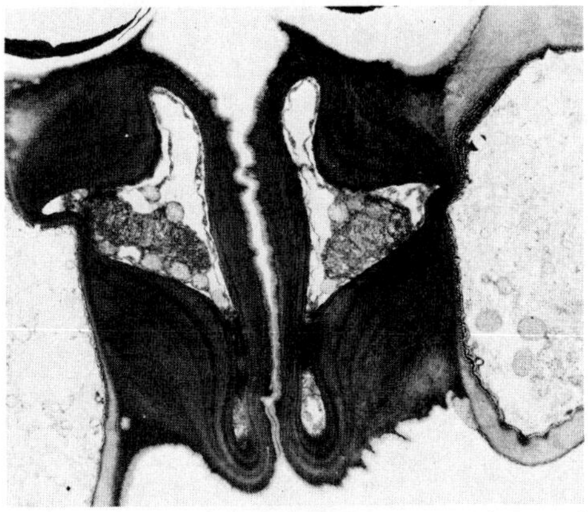

19–21
Surface (a) *and sectional* (b) *views of developing stomata.* (c) *Diagram of a mature stoma, showing its relationship to the epidermis and underlying cells.*

The guard cells originate from unequal division of a protodermal cell. The smaller of the two cells is called a stoma- or guard-cell mother cell, and it is the stoma-mother cell that divides to give rise directly to the two guard cells (a *and* b). *After the guard cells are formed, the intercellular substance in the median part of their common wall swells and then dissolves to form the pore. During this process the guard cells develop unevenly thickened walls, a structural feature generally considered to be of great importance in the mechanism responsible for opening and closing of the stoma. The walls next to the pores are generally thicker than those adjacent to other epidermal cells. While the stomata are being formed, they may be raised or lowered below the surface of the epidermis. Often there is a large air space, or substomatal chamber, just behind the stoma. Unlike ordinary epidermal cells, guard cells contain chloroplasts.*

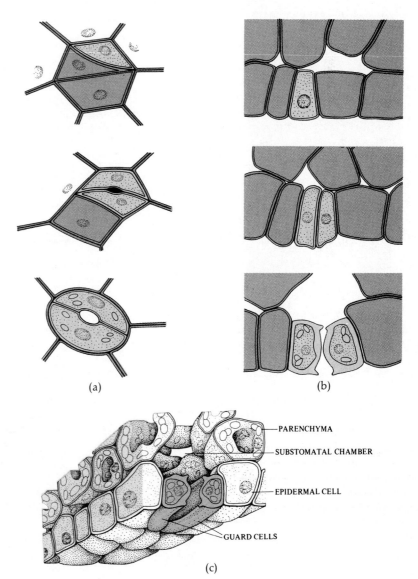

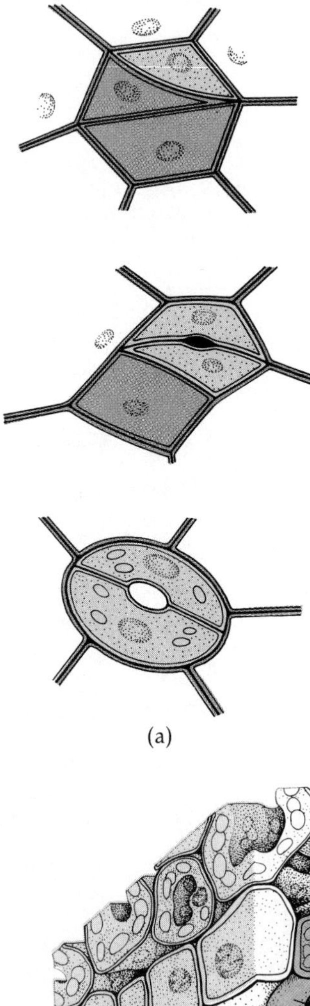

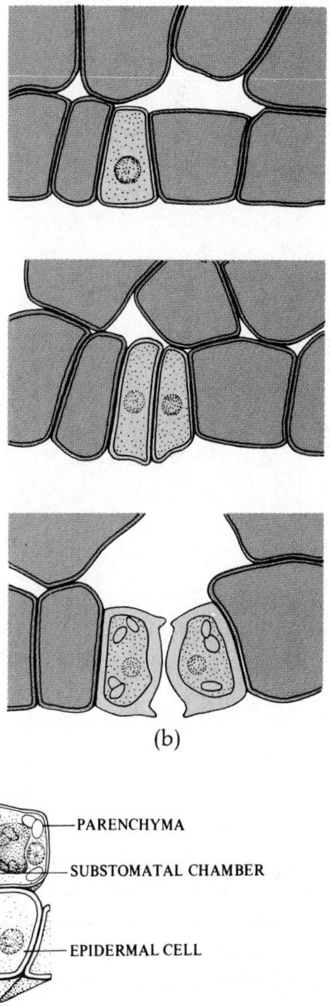

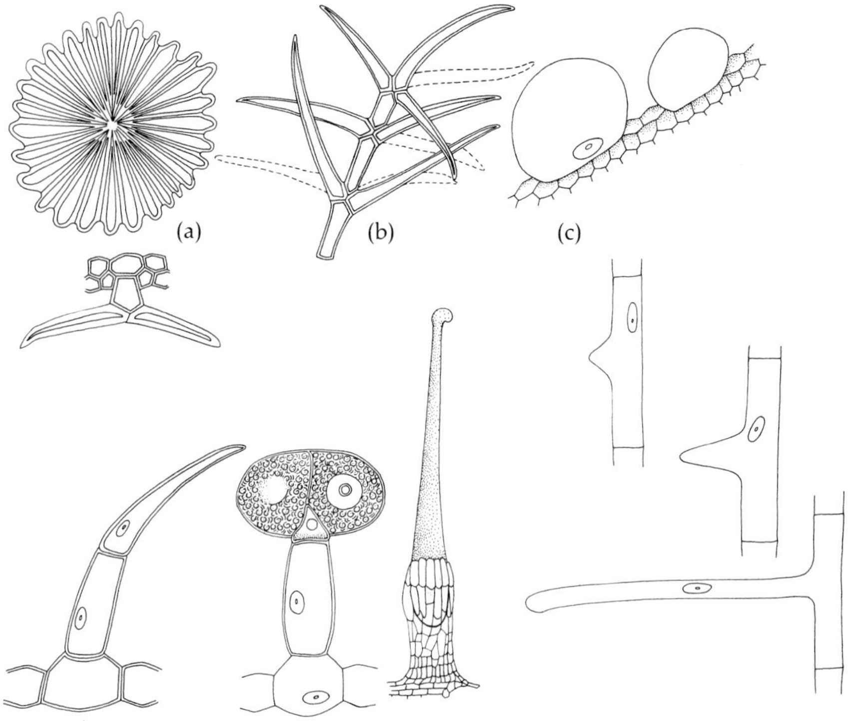

19-22
Trichomes. (a) *Surface (above) and sectional (below) views of the scale or peltate hair of the olive* (Olea europaea) *leaf.* (b) *Dendroid (treelike) hair from the leaf of the plane tree* (Platanus orientalis). (c) *Water vesicle of the "ice plant"* (Mesembryanthemum crystallinum). (d) *Short, unbranched hair from tomato* (Lycopersicum esculentum) *stem.* (e) *Glandular hair from tomato stem.* (f) *Stinging hair of* Urtica. *The stinging hair consists of a long needlelike part and a broad base surrounded by other epidermal cells. When the hair is touched, the tip breaks off and poisonous cell contents (histamine and acetylcholine) are injected into the skin.* (g) *Some stages of root hair development in corn* (Zea mays).

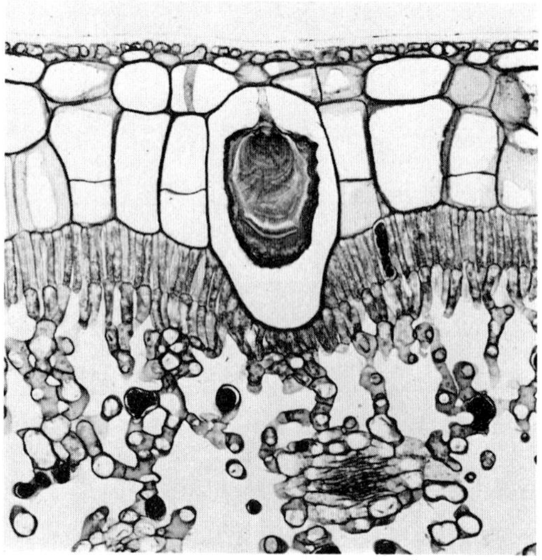

19-23
Transverse section showing upper portion of leaf blade of the rubber plant (Ficus elastica). *Notice the thick cuticle covering the multiple epidermis of mostly large cells. The club-shaped structure in the largest epidermal cell consists mostly of calcium carbonate deposited on a cellulose stalk. The narrow, elongate cells below the large, clear epidermal cells are palisade parenchyma.*

In most plants, the epidermis is only one layer of cells in thickness. However, in some plants, divisions in the protoderm of the leaf are parallel with the surface (periclinal divisions) and an epidermis with several layers (a multiple epidermis) is formed. Multiple epidermises are found in the leaves of such familiar house plants as the rubber plant (*Ficus elastica;* Figure 19-23) and *Peperomia* (Figure 19-24). The multiple epidermis is believed to serve as a water-storage tissue.

The main mass of epidermal cells is closely knit and affords considerable mechanical protection to the plant part. In order to minimize water loss, the walls of the epidermal cells of the aerial parts contain cutin and are covered with a cuticle (Chapter 1, page 34). The cuticle may also be covered with wax, either in smooth sheets or as rods or filaments extending upward from the surface (Figure 19-19; see also Figure 2-12). It is this wax that is responsible for the whitish or bluish "bloom" on leaves.

Interspersed among the flat, tightly packed epidermal cells are specialized cells filled with chloroplasts. These are the *guard cells* (Figures 19-20 and 19-21), which regulate the small openings, or stomata, in the leaf or young stem. (At one time the term stoma—the singular of stomata—was used to refer exclusively to the pore encompassed by a pair of guard cells. Today it is commonly used to refer to the entire stomatal apparatus: the guard cells plus pore between them.) In order for photosynthesis to take place in the chloroplasts of the leaf or stem, carbon dioxide must be permitted to diffuse into the leaf or stem and oxygen must be allowed to pass out.

19-24
Transverse section of leaf blade of Peperomia. *The very thick multiple epidermis, visible on its upper surface, presumably functions as a water-storage tissue.*

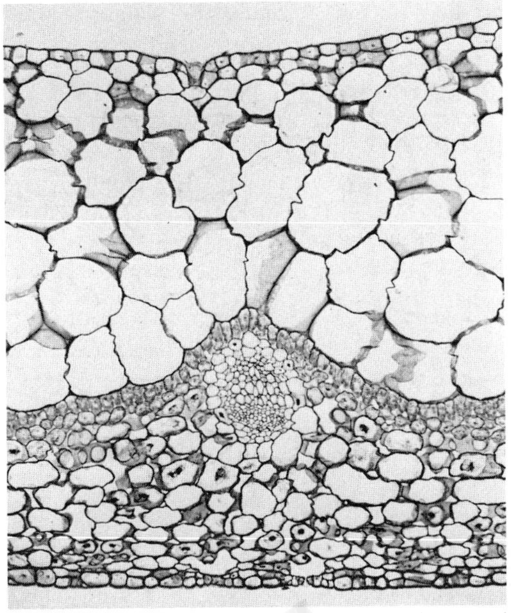

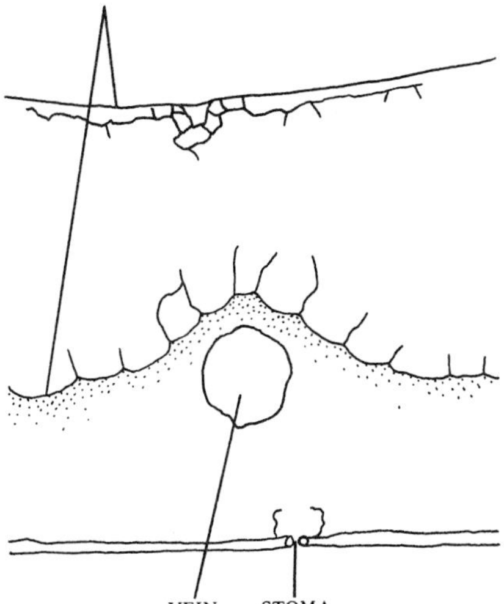

But in the course of this exchange of gases, water is lost by evaporation from the moist interior surfaces of the leaf or stem. If the guard cells are turgid (bulging with water), they pull apart, permitting the stomata to open and gases to be exchanged freely. In response to a decrease in moisture, the guard cells relax, closing the stomata and conserving the plant's water supply. Some factors involved in stomatal opening and closing are discussed in Chapter 25. Stomata often are associated with other epidermal cells that differ in shape from the ordinary epidermal cells. Such cells are termed subsidiary or accessory cells (Figure 19-20).

Periderm

In stems and roots having secondary growth, the epidermis commonly is replaced by a periderm. The periderm consists largely of *cork tissue,* in addition to the *cork cambium* and *phelloderm.* The origin of the cork cambium is variable, depending on the species and plant part. The periderm will be considered in detail in Chapter 22.

SUMMARY

A summary of tissues and cell types is found in Tables 19-1 and 19-2.

Table 19-1 *Summary of Tissues and Cell Types*

TISSUE	CELL TYPES
1. Epidermis	Generally parenchyma cells, including guard cells and trichomes; sclerenchyma cells
2. Periderm	Generally parenchyma cells; sclerenchyma cells
3. Xylem	Tracheids; vessel members; sclerenchyma cells; parenchyma cells
4. Phloem	Sieve cells or sieve-tube members; albuminous cells or companion cells; parenchyma cells; sclerenchyma cells
5. Parenchyma	Parenchyma cells
6. Collenchyma	Collenchyma cells
7. Sclerenchyma	Fibers or sclereids (in general, either may be called a sclerenchyma cell)

Table 19-2 *Summary of Cell Types*

CELL TYPE	CHARACTERISTICS	LOCATION	FUNCTION
Parenchyma	Shape: commonly polyhedral (many-sided); variable Cell wall: primary or primary and secondary; may be lignified, suberized, or cutinized Living at maturity	Just about anywhere in plant, as parenchyma tissue in cortex, as pith and pith rays, or in xylem and phloem	Such metabolic processes as respiration, digestion, and photosynthesis; storage and conduction; wound healing and regeneration
Collenchyma	Shape: elongated Cell wall: primary only—*never* lignified Living at maturity	On the periphery (beneath the epidermis) in young elongating stems; sometimes as a cylinder of tissue or only in patches; in ribs; along the veins in some leaves	Support in primary plant body
Fibers	Shape: generally very long Cell wall: primary and thick secondary—often lignified Often but not always dead at maturity	Sometimes in cortex of stems, most often associated with xylem and phloem; in leaves of monocotyledons	Support
Sclereids	Shape: variable; generally shorter than fibers Cell wall: primary and thick secondary—generally lignified May be living or dead in maturity.	Just about anywhere in plant	Mechanical, protective
Tracheid	Shape: elongated and tapering Cell wall: primary and secondary; lignified Contains pits, but no perforations Dead at maturity	Xylem	Chief water-conducting element in gymnosperms and lower vascular plants—also found in angiosperms
Vessel member (or vessel element)	Shape: elongated, but generally not so long as tracheids Cell wall: primary and secondary; lignified; in addition to pits, contains perforations; several vessel members in series (end-on-end) constitute a vessel Dead at maturity	Xylem	Chief water-conducting element in angiosperms

CELL TYPE	CHARACTERISTICS	LOCATION	FUNCTION
Sieve cell	Shape: elongated and tapering Cell wall: primary in most species; with sieve areas (sieve areas are wall areas with pores through which the protoplasts of adjoining cells are connected); callose often associated with wall and pores Living at maturity; either lacks a nucleus at maturity, or contains remnants of the nucleus; lacks distinction between vacuole and cytoplasmic contents	Phloem	Chief food-conducting element in gymnosperms and lower vascular plants.
Albuminous cell	Shape: generally elongated Cell wall: primary Living at maturity; associated with sieve cell, but generally not derived from same mother cell as sieve cell; has numerous connections with sieve cell	Phloem	Believed to play a role in movement of food into and out of the sieve cell
Sieve-tube member	Shape: elongated Cell wall: primary, with sieve areas; sieve areas on end wall with much larger pores than those on side walls; this wall part (commonly end wall) termed a "sieve plate"; callose often associated with wall and pores Living at maturity; either lacks nucleus at maturity, or contains only remnants of nucleus; contains proteinaceous substance known as "slime" or "P-protein"; several sieve-tube members in vertical series constitute a sieve tube	Phloem	Chief food-conducting element in angiosperms
Companion cell	Shape: variable, generally elongated Cell wall: primary Living at maturity; closely associated with sieve-tube members; derived from same mother cell as sieve-tube member; has numerous connections with sieve-tube member	Phloem	Believed to play a role in the movement of food into and out of the sieve-tube member

CHAPTER 20

The Root: Primary Structure and Development

In most vascular plants, the roots constitute the underground portion of the sporophyte and are specialized for anchorage and absorption. Two other functions associated with roots are storage and conduction. Most roots are important storage organs and some, such as those of the carrot, sugar beet, and sweet potato, are specifically adapted for the storage of food. Foods manufactured in above-ground, photosynthesizing portions of the plant body move through the phloem to the storage tissues of the root. Much of this food may be used eventually by the root itself, but very often the stored food is digested and transported back through the phloem to the above-ground parts. In biennials (plants that complete their life cycle over a 2-year period) such as the sugar beet, large food reserves accumulate in the storage regions of the root during the first year and then are used during the second year to produce flowers, fruits, and seeds. Water and minerals absorbed by the roots move through the xylem to the aerial parts of the plant.

ROOT SYSTEMS AND THEIR EXTENT

The first root of the plant originates in the embryo and usually is called the *primary root* (Chapter 18). In gymnosperms and dicotyledons this root, which is also called a *taproot,* grows directly downward, giving rise to *secondary roots,* also called *branch roots* or *lateral roots,* along the way. The older roots are found nearer the neck of the root (where the root and stem meet) and the younger ones nearer the root tip. This type of root system—that is, one that develops from a taproot and its branches—is called a *taproot system* (Figure 20–1a). In monocotyledons, the primary root is usually short-lived and the root system develops from adventitious roots that arise from the stem. These adventitious roots and their branch, or lateral, roots give rise to a *fibrous root system,* in which no one root is more prominent than the others (Figure 20–1b). Taproot systems generally penetrate deeper into the soil than fibrous root systems. The shallowness of fibrous root systems and the tenaciousness with which they

(a)

(b)

(c)

20–1
Types of root systems. (a) *Taproot system of dandelion* (Taraxacum officinale). (b) *Fibrous root system of a grass.* (c) *Prop roots of corn* (Zea mays), *a type of adventitious root (see page 426).*

cling to soil particles make them especially well suited as ground cover for the prevention of soil erosion.

The extent of a root system—that is, the depth to which it penetrates the soil and spreads laterally—is dependent upon several factors, including soil moisture, soil temperature, and the composition of the soil. The bulk of most feeding roots (roots actively engaged in the uptake of water and minerals) occurs in the upper meter of soil, and most of the feeding roots of most trees occur in the upper 15 centimeters of soil, the part of the soil normally richest in organic matter. Some trees, such as the spruces, beeches, and poplars, rarely produce deep taproots, whereas others, such as the oaks and many pines, commonly produce relatively deep taproots, making such trees rather difficult to transplant. The deepest known tree root was that of a pine growing on a highly porous sandy soil; it penetrated the soil about 6.5 meters. Commonly, the lateral spread of tree roots is greater than the spread of the crown of the tree. The root systems of corn plants *(Zea mays)* often reach a depth of about 1.5 meters and a lateral spread of about a meter on all sides of the plant.The roots of alfalfa *(Medicago sativa)* may extend to depths of up to 6 meters or more.

As a plant grows, it needs to maintain a balance between the total food-manufacturing (photosynthesizing) surface and the total water- and mineral-absorbing surface. In a young plant becoming established (Chapter 18), the total water- and mineral-absorbing surface usually far exceeds the photosynthesizing surface. However, the proportion of the plant surface below ground to that above ground changes with age. For example, in trees the shoot-root ratio tends to increase with age. The balance between shoot and root is invariably disturbed when plants are transplanted. Most of the fine, feeding roots generally are left behind when the plant is removed from the soil; cutting back the shoot system helps to reestablish a balance between it and the root system. Fungi and insects that attack the shoots or roots of plants often cause an imbalance in the shoot-root ratio.

One of the most detailed studies conducted on the extent of the root and shoot systems of any one plant was made by H. J. Dittmer on a 4-month-old rye plant *(Secale cereale)*. Dittmer found the total surface area of the root system, including root hairs, to be 639 square meters, or 130 times the surface area of the shoot system. What is even more amazing is the fact that the roots occupied only about 6 liters of soil.

GROWTH AND ORIGIN OF THE PRIMARY TISSUES

Apparently the growth of many roots is an almost continuous process that stops only under adverse conditions such as drought and low temperatures. During their growth through the soil, roots follow the path of

20–2
Longitudinal sections of onion (Allium cepa) *root tip.* (a) *The primary meristems can be distinguished very close to the apical meristem.* (b) *Detail of apical meristem. Compare the organization of this apical meristem with that of corn* (Zea mays), *Figure 20–3.*

(a)

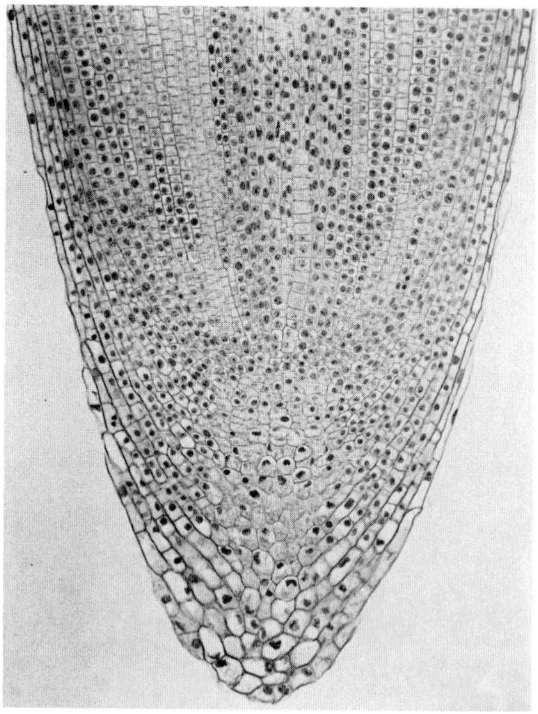

(b)

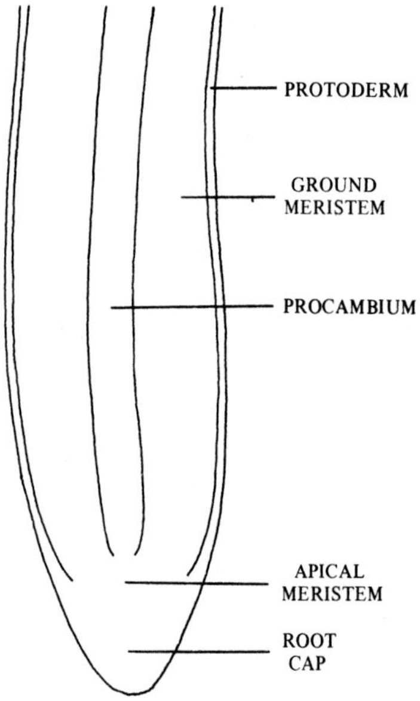

least resistance and frequently follow spaces left by earlier roots that have died and rotted.

The tip of the root is covered by a <u>*root cap*</u> (Figures 20–2, 20–3, and 20–4), a thimblelike mass of cells that protects the apical meristem behind it and aids the root in its penetration of the soil. As the root grows longer and the root cap is pushed forward, the cells on the periphery of the root cap are sloughed off. These sloughed-off cells form a slimy covering around the root and lubricate its passage through the soil. As quickly as root cap cells are sloughed off, new ones are added by the apical meristem.

The slimy substance of the root cap is a highly hydrated polysaccharide, probably a pectic substance, which is secreted by the outer root cap cells. Studies by several botanists, including D. H. Northcote, J. D. Pickett-Heaps, and D. J. Morré and his co-workers, have revealed that the slimy substance accumulates in Golgi vesicles. The Golgi vesicles then fuse with the plasma membrane and the slime is released into the wall. Eventually it passes to the outside, where it forms droplets.

Some studies suggest that the root cap plays a role in controlling the response of the root to gravity (geotropism). When placed in a horizontal position, a root normally curves downward, in response to gravity. However, if the root cap is removed, the root may not curve downward until a new root cap is formed—although root growth continues. It has been suggested that the starch-containing plastids (amyloplasts) in the median cells of the root cap act as statoliths, or gravity sensors. When a root is placed in a horizontal position, the plastids, which

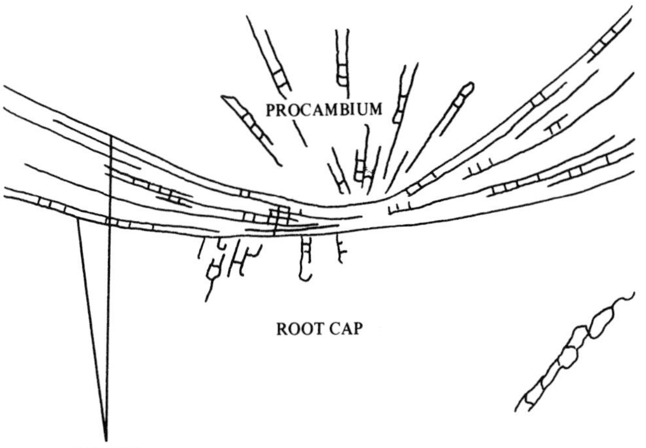

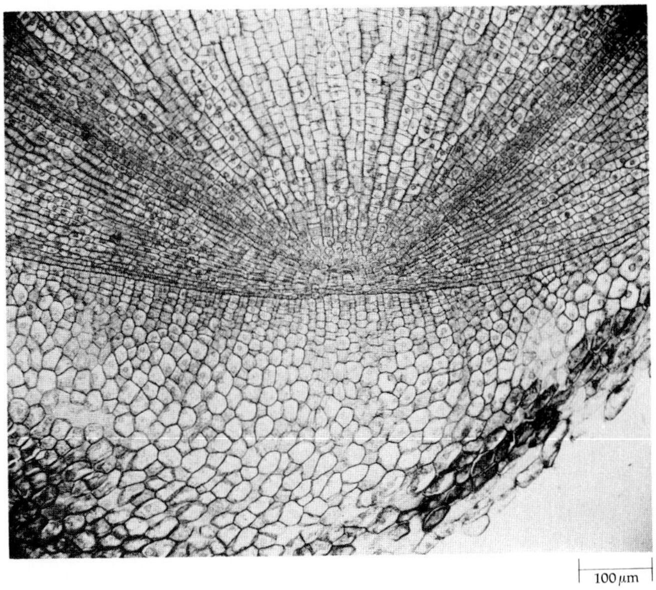

20–3
Apical meristem of corn (Zea mays) *root. Notice the three distinct layers of initials. The lower layer gives rise to the root cap, below; the middle layer to the ground meristem or cortex; and the upper to the procambium or vascular cylinder.*

were sedimented on the horizontal walls of the vertically growing roots, slide downward and come to rest against what were previously vertically oriented walls. After several hours, the root curves downward and the plastids return to their previous position. A recent statistical study failed, however, to reveal a direct link between amyloplast movement and the geotropic response, casting serious doubt on the role of amyloplasts in the response of roots to gravity. Both the endoplasmic reticulum and dictyosomes (Golgi bodies) of root cap cells also have been implicated in the geotropic response. Nevertheless, how any of these cellular components might signal the growing portion of the root of changes in orientation has not been explained.

Growth Regions of the Root

The apical meristem of the root is composed of relatively small (10 to 20 micrometers in diameter), many-sided cells—the initials and their immediate derivatives (Chapter 19)—with dense cytoplasm and large nuclei (Figures 20-2 and 20-3). The organization and number of initials in the apical meristems of roots are variable.

Two main types of apical organization are found in the roots of seed plants. In one, all regions emerge from a common group of initials (Figure 20-2). In the other, the root cap, the vascular cylinder of xylem and phloem, and the cortex can be traced to independent layers of initials (Figure 20-3). In the second type of organization, the epidermis has a common origin with either the root cap or the cortex.

20–4
Diagram illustrating early stages in primary development in a root tip.

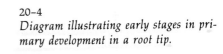

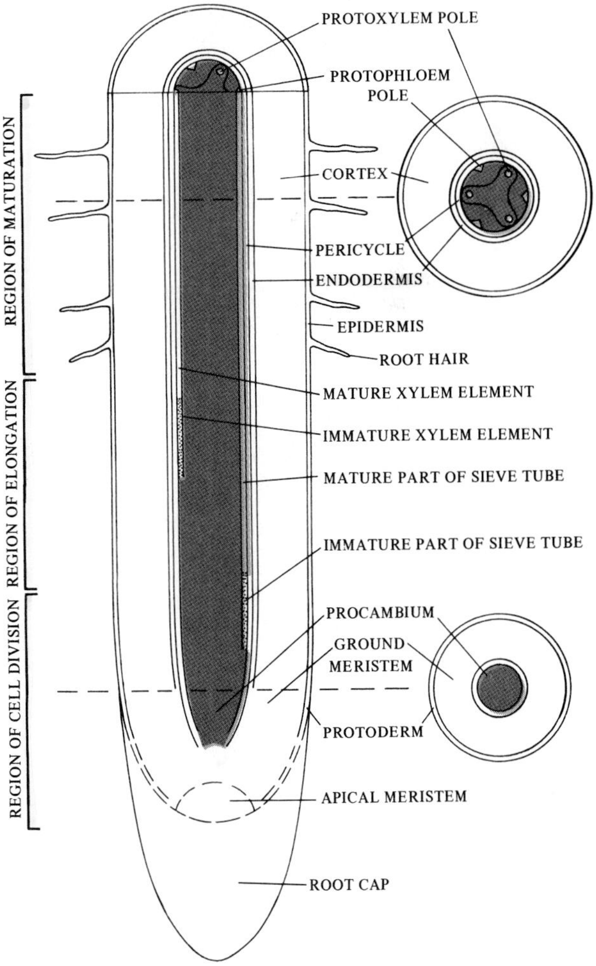

20–5
Apical meristem of a root tip of corn (Zea mays), *showing the quiescent center. To prepare this autoradiograph, the root tip was supplied for one day with a building block of the nucleotide of thymine labeled with radioactive tritium (^3H). In the rapidly dividing cells around the quiescent center, the radioactive material was incorporated quickly into the nuclear DNA and left its marks (the dark grains) on this autoradiograph.*

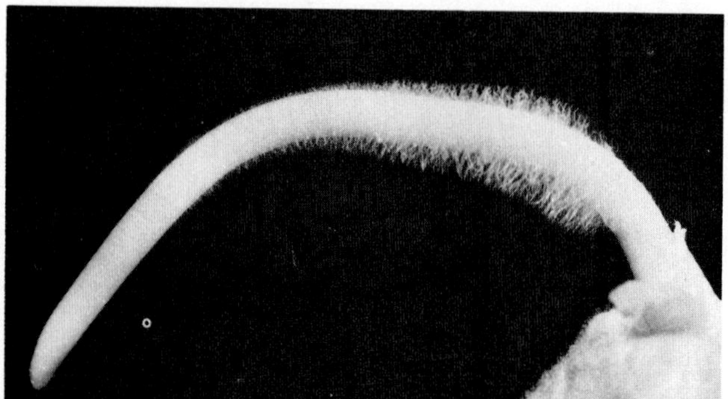

20–6
Rootlet of corn (Zea mays) *showing root hairs. Root hairs may be as much as 1.30 centimeters long and may attain their full size within hours. Each hair is comparatively short-lived, but the formation of root hairs continues as long as the root is growing.*

Although the initial region in the apical meristem of the root was once considered to be a region of active cell division, studies on the apical meristems of many roots indicate that this region is relatively inactive (Figure 20–5) and that most cell division takes place a short distance beyond the relatively quiescent initials. This relatively inactive region is known as the *quiescent center.* The distance beyond the apical meristem at which most cell division takes place varies from species to species and also within the same species, depending on the age of the root. The apical meristem and nearby portion of root in which cell division takes place are together referred to as the *region of cell division* (Figure 20–4).

Behind the region of cell division, but not sharply delimited from it, is the *region of elongation,* which usually measures only a few millimeters in length (Figure 20–4). The elongation of cells in this region results in most of the increase in length of the root. Beyond this region the root does not increase in length. Thus, growth in length of the root occurs near the root tip and results in a very limited portion of the root constantly being pushed through the soil.

Following the region of elongation is the *region of maturation,* in which most of the cells of the primary tissues mature (Figure 20–4). Root hairs are also produced in this region, and sometimes this part of the root is called the root-hair zone (Figure 20–6). It is important to note that there is a gradual transition from one region of the root to the other. The regions we have just described are not sharply delimited from one another. Some cells begin to elongate and to differentiate in the region of cell division, whereas others reach maturity in the region of elongation. For example, the first-formed elements of the phloem and xylem mature in the region of elongation and are often stretched and destroyed during elongation of the root.

The protoderm, procambium, and ground meristem can be distinguished very close to the apical meristem (Figures 20–2 and 20–4). These are the primary meristems that differentiate into the epidermis, the primary vascular tissues, and the cortex, respectively (Chapter 18).

PRIMARY STRUCTURE

The internal structure of the root is usually relatively simple compared with that of the stem. This is due in large part to the absence of leaves in the root and the corresponding absence of nodes and internodes. Thus, from one level of the root to another the arrangement of tissues shows very little difference.

The three tissue systems of the root in the primary state of growth can be readily distinguished in both transverse and longitudinal sections (Figures 20–2, 20–7, and 20–8). Clearly separated from one another are the

20–7
Transverse sections of buttercup (Ranunculus) *root.* (a) *Overall view of mature root.* (b) *Detail of immature vascular cylinder. Notice intercellular spaces among the cortical cells.* (c) *Detail of mature vascular cylinder. Notice the numerous starch grains in the cortical cells. A small amount of secondary growth has taken place between the phloem strands and the xylem.*

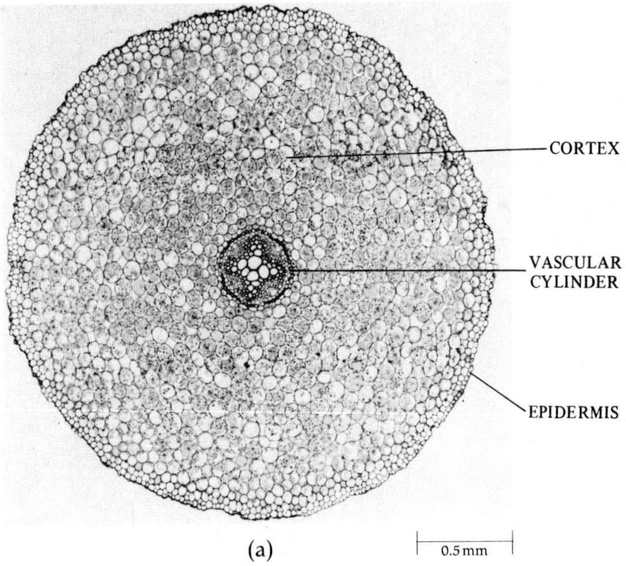

(a)

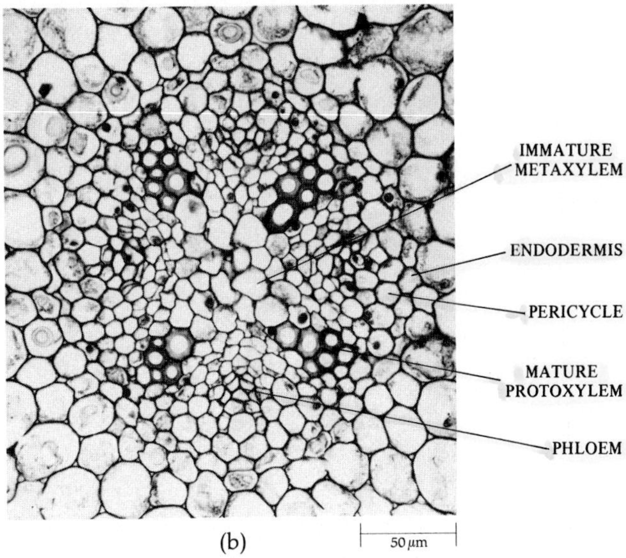

(b)

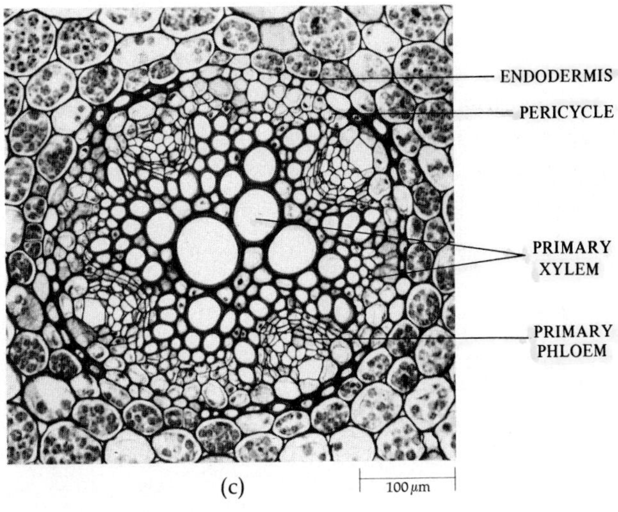

(c)

20-8
Transverse sections of corn (Zea mays) *root.* (a) *Overall view of mature root. Notice numerous root hairs of epidermis. Part of a branch root can be seen at lower right. The vascular cylinder, with its pith, is quite distinct.* (b) *Detail of portion of immature vascular cylinder.* (c) *Detail of portion of mature vascular cylinder.*

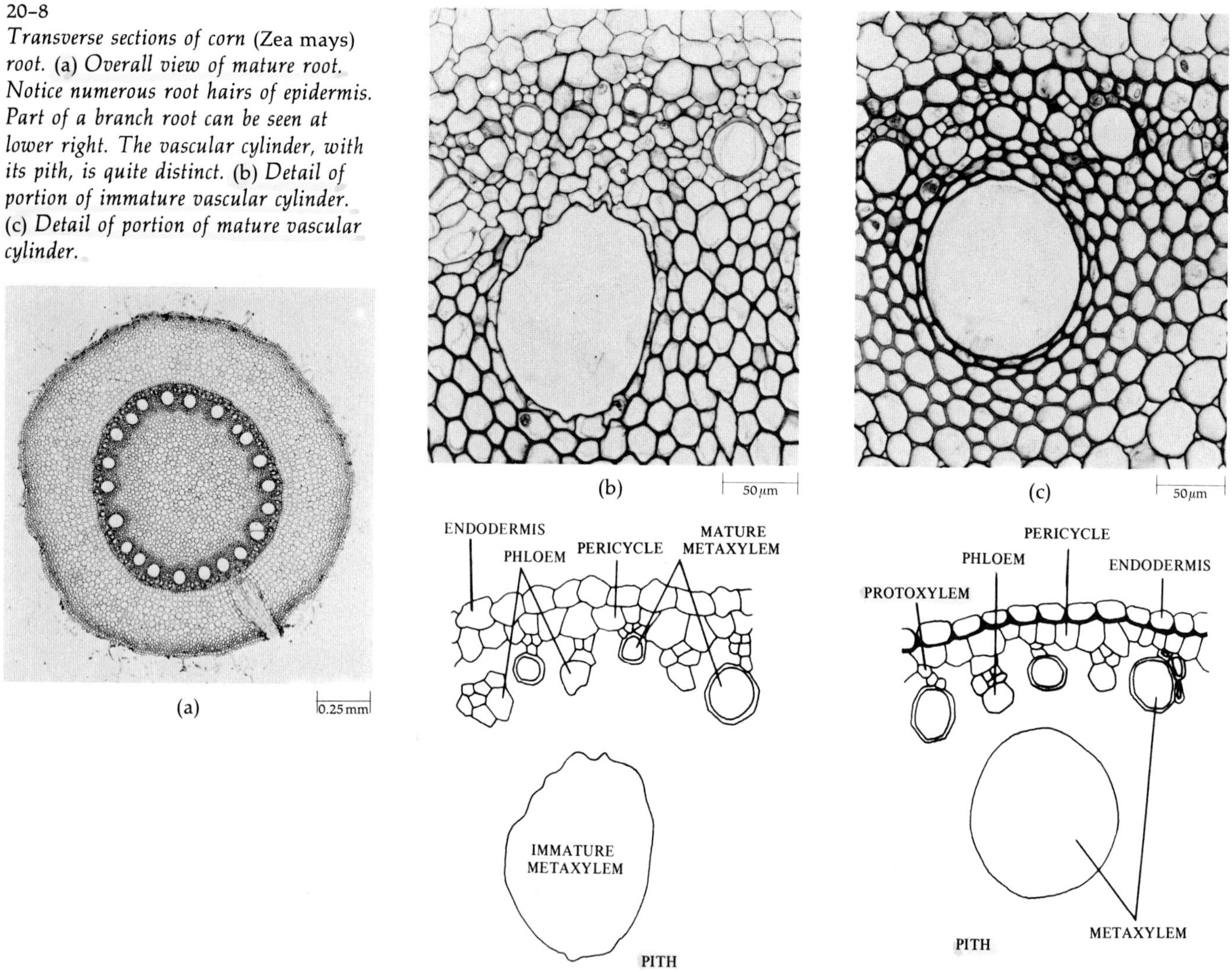

epidermis (dermal tissue system), the cortex (ground tissue system), and the vascular tissues (vascular tissue system). In most roots, the vascular tissues form a solid cylinder (Figure 20-7), but in some they form a hollow cylinder around a pith (Figure 20-8).

Epidermis

The function of the epidermis in young roots is to absorb water and minerals, and this function is facilitated by root hairs, which are tubular extensions of the epidermal cells (Figure 19-22, p. 414), that greatly increase the absorbing surface of the root.* In his study of the 4-month-old rye plant, Dittmer estimated that the plant contained approximately 14 billion root hairs, with an absorbing surface of 401 square meters. Placed end on end, the root hairs would have extended 10,628 kilometers.

Root hairs are relatively short-lived and are confined largely to the region of maturation of the root. The production of new root hairs occurs just beyond the region of elongation (Figures 20-4 and 20-6), at about the same rate as that at which the older root hairs, at the upper end of the root hair zone, are dying off. As the tip of the root penetrates the soil, new root hairs are produced immediately behind it, providing the root with new surfaces capable of absorbing new supplies of water and minerals. (For a discussion of root absorption, see pages 546-547). Obviously, it is the new and growing roots—the so-called

* See pages 240-242 in Chapter 11 for a discussion of the role of mycorrhizae in absorption.

feeding or feeder roots—that are primarily involved in the absorption of water and minerals. For this reason great care must be taken to remove as much soil as possible along with the root system when transplanting. If the plant is simply "torn" from the soil, most of the feeder roots will be left behind and the plant will probably not survive.

The epidermal cells of the root, including those with root hairs, are parenchyma cells, which are closely packed together and usually lack a cuticle. If a cuticle were present, it would impede the movement of substances into the root.

Cortex

As seen in transverse sections (Figure 20-7), the cortex occupies by far the greatest area of the primary body of most roots. The cells of the cortex store starch and other substances, but commonly lack chloroplasts. Roots that undergo secondary growth—which include those of gymnosperms and most dicots—shed their cortex early (Chapter 22). In such roots, the cortical cells remain parenchymatous. In monocots, the cortex is retained for the life of the root, and many of the cortical cells develop secondary walls and become lignified. Regardless of the degree of differentiation, the cortical tissue contains numerous intercellular spaces—air spaces essential for aeration of the cells of the root (Figures 20-7 and 20-8). The cortical cells have numerous contacts with one another, and their protoplasts are connected by plasmodesmata. Thus substances moving across the cortex may move from cell to cell by way of the protoplasts and plasmodesmata or by way of the cell walls.

Unlike the rest of the cortex, the innermost layer of the cortex is compactly arranged and lacks air spaces. This layer, the *endodermis* (Figures 20-7 and 20-8), is characterized by the presence of *Casparian strips* (Figures 20-9 and 20-10) on its anticlinal walls (that is, the walls perpendicular to the surface of the root). The Casparian strip is a bandlike portion of the primary wall that is impregnated with a fatty substance called suberin and is sometimes lignified. The protoplasts of endodermal cells are firmly attached to the Casparian strips and adhere tenaciously to them (Figure 20-9). Inasmuch as the endodermis is compact and the Casparian strips are impermeable to water, all substances entering and leaving the vascular cylinder by way of the endodermis must pass through the living protoplasts of the endodermal cells. A recent study by G. Nagahashi and co-workers clearly demonstrated the effectiveness of the Casparian strip as a barrier to movement across the walls of the endodermis in corn *(Zea mays)* roots. Roots were made to absorb the element lanthanum, a positively charged ion that cannot penetrate cell membranes. When examined with the electron microscope, the lanthanum was found only in the cell walls of the cortex. It was abruptly and

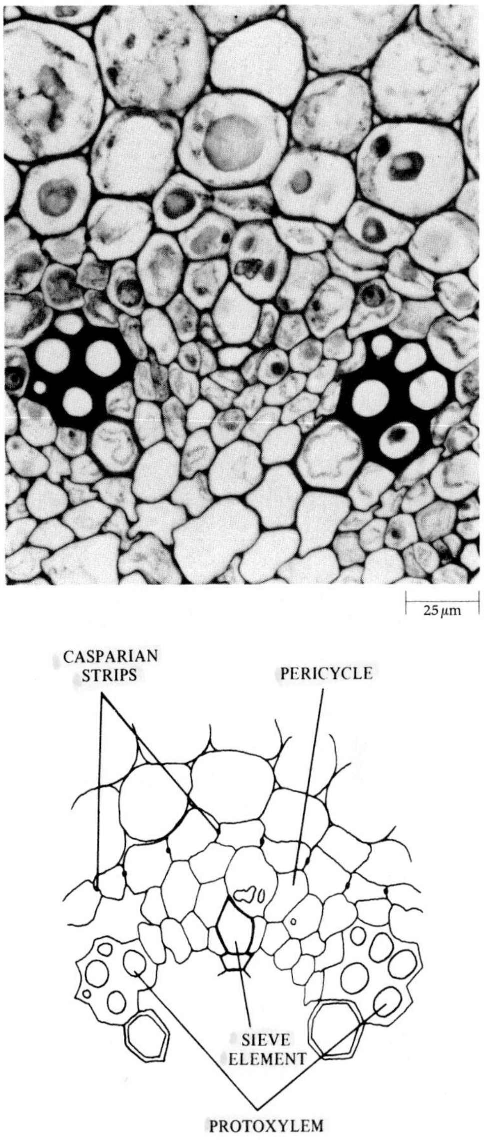

20-9
High-magnification view of portion of immature root of buttercup (Ranunculus), *showing Casparian strips in endodermal cells. Notice that the plasmolyzed protoplasts of the endodermal cells cling to the strips.*

20–10
Three-dimensional diagram showing Casparian strips (in color) on anticlinal walls (walls at right angles to surface of the root) of some endodermal cells. The Casparian strip—a bandlike portion of the wall impregnated with suberin—ensures that dissolved materials passing between the cortex and vascular cylinder of the root must pass through the protoplasts of the compactly arranged endodermal cells.

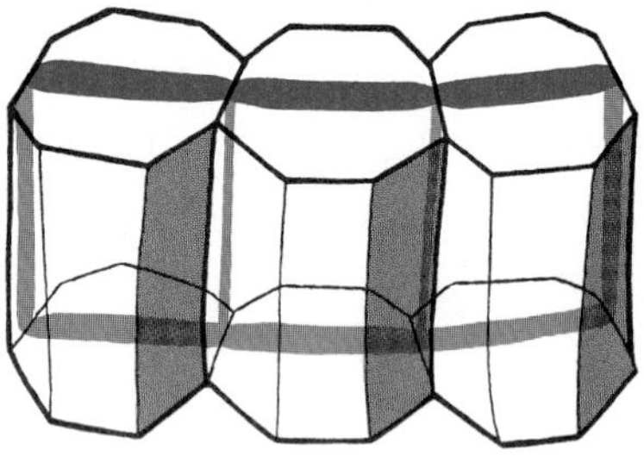

completely stopped in its progress across the root by the Casparian strip. Thus the endodermis clearly exercises some control over the movement of substances into and out of the vascular cylinder of the root.

As we mentioned previously, in roots that undergo secondary growth, the cortex and its endodermis are shed early. In roots in which the cortex is retained, a layer of suberin is eventually deposited internally over all wall surfaces of the endodermis. This is then followed by the deposition of secondary walls, which may become lignified. These changes in the endodermis begin opposite the phloem strands and spread toward the protoxylem poles (Figure 20–7). Opposite the protoxylem, many of the endodermal cells do not undergo such modifications, but remain thin-walled and retain their Casparian strips. Such cells are called *passage cells.*

Vascular Cylinder

The vascular cylinder of the root consists of the vascular tissues and one or more layers of cells, the *pericycle,* that completely surrounds the vascular tissues (Figures 20–7 and 20–8). In the young root, the pericycle is composed of parenchyma cells with primary walls, but as the root ages, the cells of the pericycle may develop secondary walls (Figure 20–8). The pericycle plays several important roles. In most seed plants, lateral roots arise in the pericycle. In plants undergoing secondary growth, the pericycle contributes to the vascular cambium and generally gives rise to the first cork cambium. And pericycle often proliferates, that is, gives rise to more pericycle. The last three functions of the pericycle will be discussed in Chapter 22.

The center of the vascular cylinder of most roots is occupied by a solid core of primary xylem from which ridgelike projections extend toward the pericycle (Figure 20–7). Nestled between the ridges of xylem are the strands of primary phloem, which alternate with the ridges of xylem. Obviously, the vascular cylinder of the root is a protostele (Chapter 14).

The number of ridges of primary xylem varies from species to species, and sometimes along the axis of a given root. If two ridges are present, the root is said to be *diarch,* if three are present, *triarch* (Figure 20–7), four, *tetrarch,* and so on; and if many are present, *polyarch* (Figure 20–8). The first *(proto)* xylem elements to mature in roots are located next to the pericycle, and the tips of the ridges commonly are referred to as *protoxylem poles* (Figures 20–7 and 20–8). The metaxylem *(meta,* after) occupies the inner portions of the ridges and the center of the vascular cylinder and matures after the protoxylem. The roots of some monocotyledons (for example, corn) have a pith (Figure 20–8), which is interpreted by some botanists as potential vascular tissue.

ORIGIN OF BRANCH ROOTS

In most seed plants, branch roots arise in the pericycle. (As noted previously, the terms branch root, lateral root, and secondary root are used synonymously.) Because of their origin deep from within the parent root, branch roots are said to be endogenous (originating within an organ (Figure 20–11).

Divisions in the pericycle that initiate branch roots occur some distance beyond the region of elongation. As the young branch root, or *root primordium,* increases in size, it pushes its way through the cortex (Figure 20–11), possibly secreting enzymes that digest some of the cortical tissue lying in its path. While still very young, the primordium develops a root cap and apical meristem, and the primary meristems appear. Initially the vascular cylinders of branch root and parent root are not connected with one another. The two vascular cylinders are joined later, when derivatives of intervening pericycle cells differentiate as xylem and phloem.

AERIAL ROOTS

Aerial roots are adventitious roots produced from above-ground structures. The aerial roots of some plants serve as *prop roots* for support, as in corn. When they come in contact with the soil, they branch and function also in the absorption of water and minerals. Prop roots are produced from the stems and branches of many tropical trees, such as the red mangrove *(Rhizophora mangle),* the banyan tree *(Ficus bengalensis),* and some of the palms. Other aerial roots, as in the English ivy *(Hedera helix),*

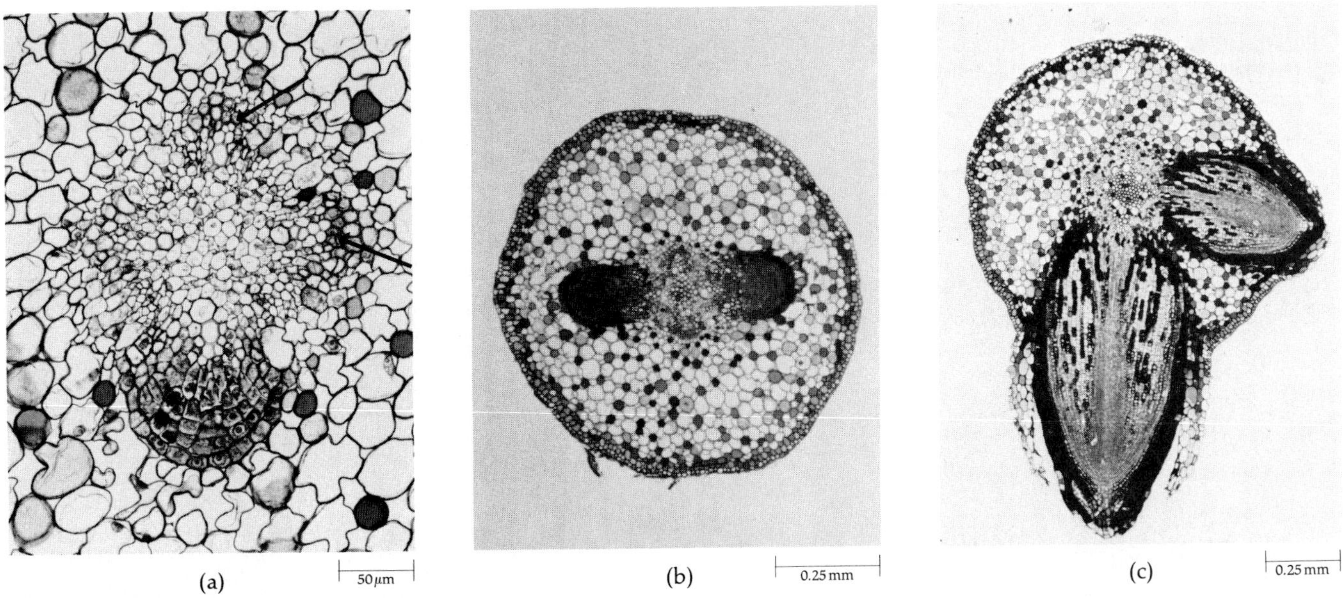

20–11
Three stages in the origin of branch roots in a willow (Salix). (a) *One root primordium is present (below) and two others are being initiated, above and to the right, in the region of the pericycle (arrows). The vascular cylinder is still very young.* (b) *Two root primordia penetrating the cortex.* (c) *One lateral root has reached the outside, the other nearly has.*

cling to the surface of objects and provide support for the climbing stem.

Roots require oxygen for respiration, which is why most plants cannot live in soil in which there is not adequate drainage and which consequently lacks air spaces.

Some trees that grow in swampy habitats develop roots that grow out of the water and serve not only to anchor but to aerate the plant. For example, the root system of the black mangrove, *Avicennia nitida* develops negatively geotropic extensions called *pneumatophores,* which grow upward and out of the mud and so provide adequate aeration (Figure 20–12). The "knees" of the bald cypress (*Taxodium distichum;* Figure 16–21, page 338) have had a similar function attributed to them, but this is in doubt.

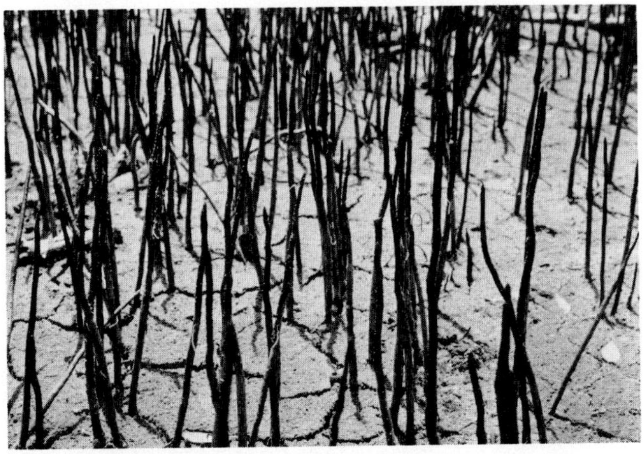

20–12
Pneumatophores (air roots) of the black mangrove (Avicennia nitida).

Special Adaptations

Many special adaptations are found among epiphytes, plants that grow on other plants but are not parasitic on them. The epidermis of the orchid root, for example, is several layers thick and, in some species, is the only photosynthetic organ of the plant (Figure 20–13). Special structures in the epidermis apparently provide for gas exchange at times when the epidermis is saturated with water. The epidermis also may function in the absorption of water (although this has been questioned).

Among epiphytes, *Dischidia rafflesiana,* the "flower pot plant," has a very unusual modification. Some of its leaves are flattened, succulent structures, but others form tubes—the "flower pots"—that collect debris and rain runoff (Figure 20–14). Ant colonies live in the "pots" and add to the nitrogen supply of the plant. Roots,

20–13
(a) *Aerial roots of an orchid* (Cattleya). (b) *Transverse section of an orchid root, showing the multiple epidermis or velamen.*

(a)

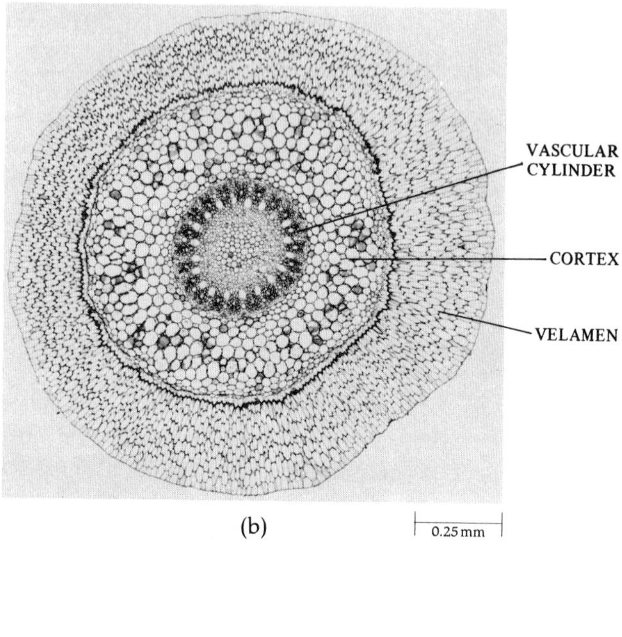

(b)

20–14
Dischidia rafflesiana, *the "flower pot plant."* (a) *Tubular leaf, or "pot," which collects debris and rain runoff.* (b) *Leaf cut open to show roots that have grown down into it from node.*

(a)

(b)

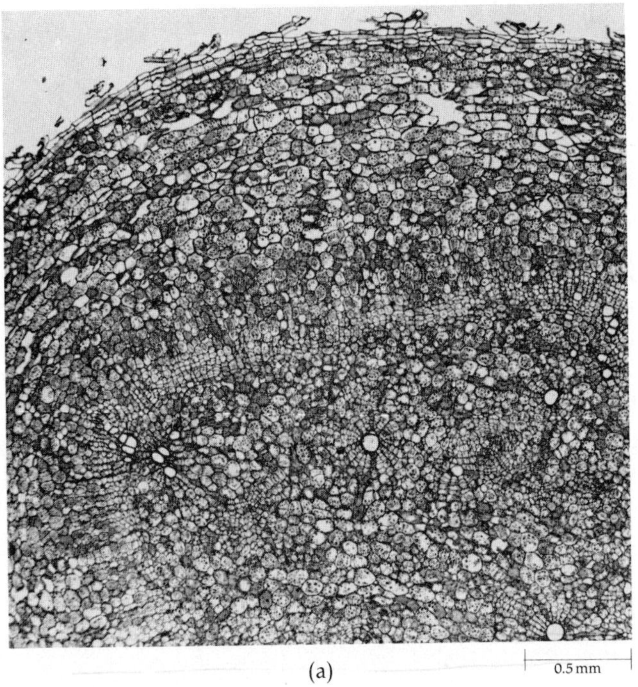

(a)

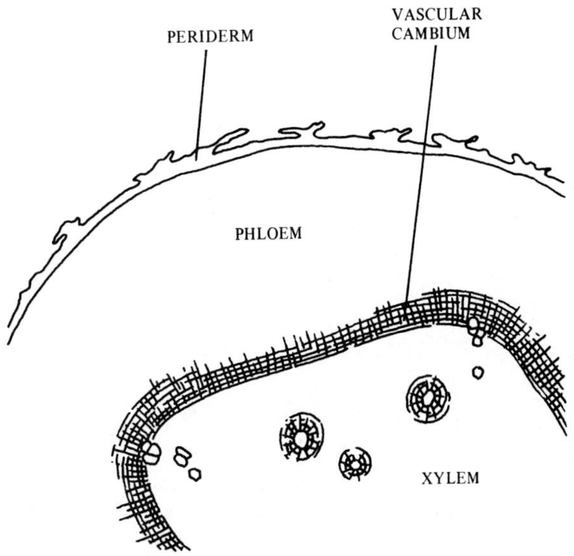

formed at the node above the modified leaf, grow downward and into the pot, from which they absorb water and minerals.

ADAPTATIONS FOR FOOD STORAGE

Most roots are storage organs, and in some plants the roots are specialized for this function. Such roots become fleshy, owing to an abundance of storage parenchyma, which is permeated by vascular tissue. The development of some storage roots, such as the carrot *(Daucus carota),* is essentially similar to that of "nonfleshy" roots, except for a predominance of parenchyma cells in their secondary xylem and phloem. The root of the sweet potato *(Ipomoea batatas)* develops in a manner similar to the carrot; however, in the sweet potato, additional vascular cambium cells develop within the secondary xylem around individual vessels or groups of vessels (Figure 20–15). While producing a few tracheary elements toward the vessels and a few sieve tubes away from them, these additional cambia also produce many storage parenchyma cells in both directions. In the beet *(Beta),* most of the increase in thickness results from the development of supernumerary (extra) cambia around the original vascular cambium (Figure 20–16). These concentric layers of cambia, which superficially resemble growth rings in woody roots and stems, produce parenchyma-dominated xylem and phloem toward the inside and outside, respectively. The upper portion of most fleshy roots actually develops from the hypocotyl.

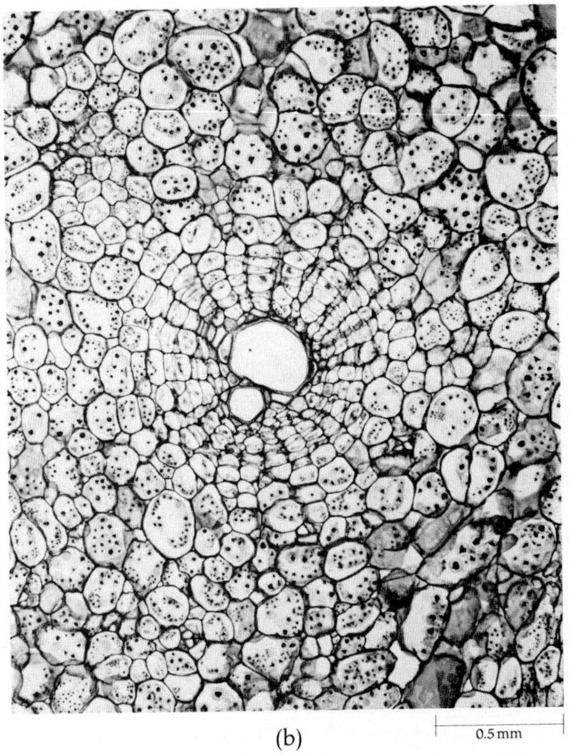

(b)

20–15
Transverse sections of a sweet potato (Ipomoea batatas) *root.* (a) *Overall view, showing periderm, phloem, vascular cambium, and xylem.* (b) *Detail of xylem, showing cambium around vessels.*

20–16
Transverse section of sugar beet (Beta vulgaris) *root, showing supernumerary cambia (arrows).*

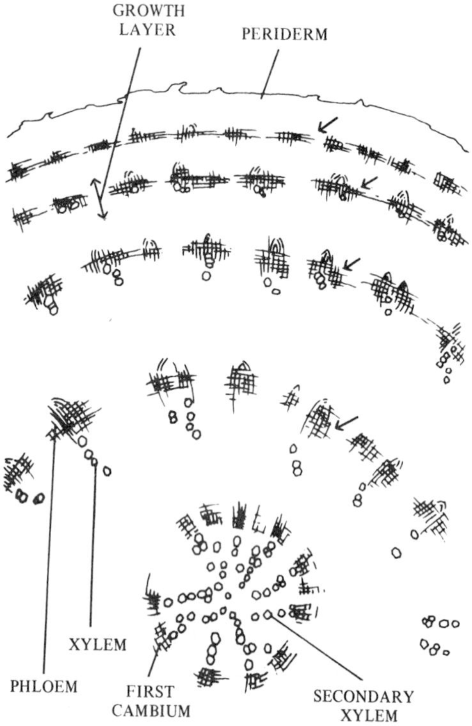

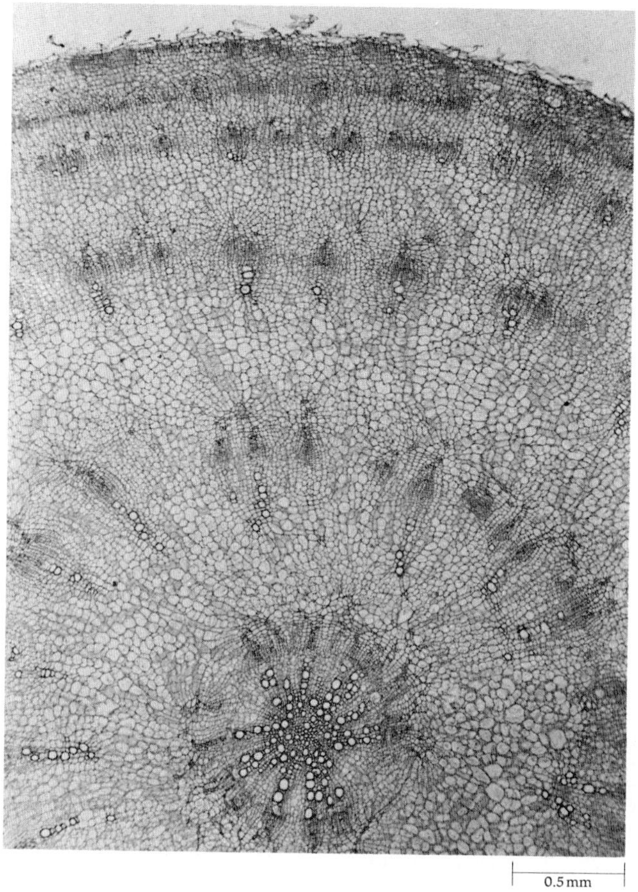

SUMMARY

Roots are organs specialized for anchorage, absorption, storage, and conduction. Gymnosperms and dicots commonly produce taproot systems, monocots commonly produce fibrous root systems. The extent of the root system is dependent upon several factors, but the bulk of most feeding roots are found in the upper meter of soil.

The apical meristems of most roots contain a quiescent center; most meristematic activity, or cell division, occurs a short distance from the apical initials. During primary growth, the apical meristem gives rise to the three primary meristems, protoderm, ground meristem, and procambium, and the latter differentiate into epidermis, cortex, and vascular cylinder, respectively. In addition, the apical meristem produces the root cap, which serves to protect the meristem and aid the root in its penetration of the soil.

Many epidermal cells of the root develop root hairs, which greatly increase the absorbing surface of the root. With the exception of the endodermis, the cortex contains numerous intercellular spaces. The compactly arranged endodermal cells contain Casparian strips on their anticlinal walls. Consequently, all substances moving between the cortex and vascular cylinder must pass through the protoplasts of the endodermal cells. The vascular cylinder consists of pericycle and the primary vascular tissues, which are completely surrounded by the pericycle. The primary xylem occupies the center of the vascular cylinder and has radiating ridges that alternate with strands of primary phloem.

Branch roots originate in the pericycle and push their way to the outside through the cortex and epidermis.

Aerial roots are adventitious roots that may serve as prop roots and, in some trees that live in swampy habitats, provide aeration.

Some roots, such as those of carrots, sweet potatoes, and beets, are specialized for storage. Such fleshy roots contain an abundance of storage parenchyma permeated by vascular tissue.

CHAPTER 21

The Shoot: Primary Structure and Development

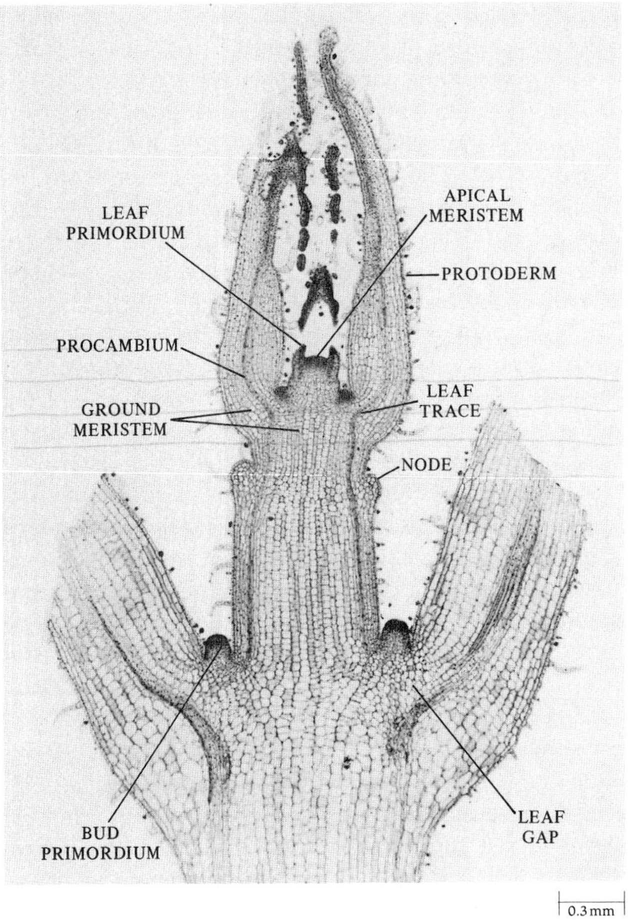

21–1
Longitudinal section of shoot tip of Coleus blumei, *a dicot. The leaves in* Coleus *are arranged opposite one another at the nodes, each successive pair at right angles to the previous pair; thus the leaves of the labeled node are at right angles to the plane of section.*

The shoot, which consists of the stem and its leaves, is initiated during development of the embryo (Chapter 18), where it is represented by the plumule. The plumule may be thought of as the first bud, consisting of a stem (the epicotyl), one or more rudimentary leaves (*leaf primordia*), and an apical meristem. With resumption of growth of the embryo during germination of the seed, new leaves develop from the flanks of the apical meristem, and the axis elongates and differentiates into nodes and internodes. Gradually *bud primordia* form in the axils of the leaves (Figure 21–1), and sooner or later they follow a sequence of growth and differentiation more or less similar to that of the first bud. This pattern is repeated many times as the shoot system of the plant is produced.

Often the terminal apical meristem of a shoot inhibits development of lateral buds, a phenomenon known as *apical dominance* (see page 487). As the distance between shoot apex and lateral buds increases, the retarding influence of the shoot apex is lessened and the lateral buds go forward with their development. Thus, pinching off the shoot apices, a common practice of home gardeners, results in fuller and bushier plants.

The two principal functions associated with stems are conduction and support. Substances manufactured in the leaves are transported through the stems by way of the phloem to sites of utilization of those substances, including growing leaves, stems, and roots and developing flowers, seeds, and fruits. Much of the food material is stored in parenchyma cells of roots, seeds, and fruits, but stems are also important storage organs and some, such as the white potato, are specifically adapted for storage. The principal photosynthetic organs of the plant, the leaves, are supported by the stems, which place the leaves in favorable positions for light essential to photosynthesis. In addition, most of the plant's loss of water vapor takes place through the leaves (Chapter 25).

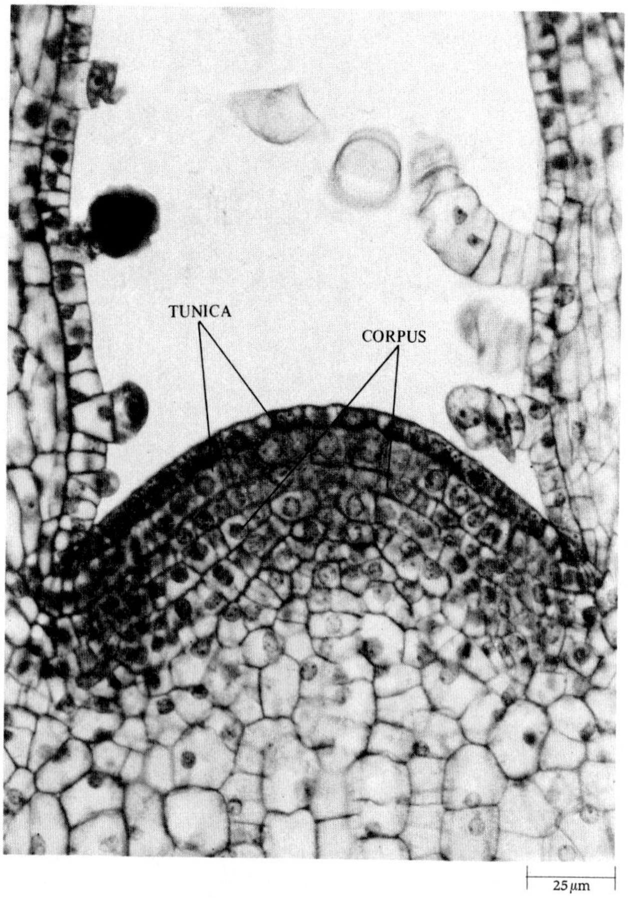

21–2
Detail of Coleus blumei *shoot apex, showing tunica-corpus organization.*

GROWTH AND ORIGIN OF THE PRIMARY TISSUES OF THE STEM

The organization of the apical meristem of the shoot is more complex than that of the root. In addition to adding cells to the primary plant body, the apical meristem of the shoot is involved with the formation of leaf primordia and often with bud primordia (Figure 21–1), which form lateral branches. The apical meristem of the shoot also differs from that of the root in its absence of a protective covering comparable to the root cap.

The vegetative shoot apices of most flowering plants have what is termed a *tunica-corpus* type of organization (Figure 21–2). The tunica consists of one or more peripheral layers of cells that contribute primarily to surface growth. The corpus, which consists of a mass or body of cells enclosed by the tunica layers, adds volume or bulk to the developing shoot. Presumably, each layer of tunica and the corpus has its own initials. The outer tunica layer always gives rise to the protoderm, whereas the procambium and ground meristem are derived from either the corpus or from a combination of the remaining tunica layers and the corpus.

A quiescent center somewhat similar to that of root apical meristems has been reported for the apical meristems of some shoots. This quiescent region in shoots has been termed a méristème d'attente (waiting meristem) by the French cytologists who advocate its presence. These reports of a *méristème d'attente* in shoot tips have stimulated much research, which has generally tended to support the concept of a quiescent center in the shoot apex. Apparently the méristème d'attente plays little or no role in development of the vegetative apex (see Chapter 18, page 401), but at the onset of floral development gives rise to most of the reproductive structures, as we shall see later in this chapter.

Although the primary tissues of the stem pass through periods of growth similar to those of the root, the stem cannot be divided along its axis into regions of cell division, elongation, and maturation in the same manner as roots. When active, the apical meristem of the shoot gives rise to leaf primordia in such close succession that nodes and internodes cannot at first be distinguished. Gradually, growth occurs between the levels of leaf attachment, the elongated parts of the stem become recognizable as internodes, and the portions of the stem at which the leaves are attached become recognizable as nodes (Figure 21–3). Thus, increase in length of the stem occurs largely by internodal elongation.

Commonly, the meristematic activity causing the elongation of the internode is more intense at the base of the developing internodes than elsewhere. If elongation of the internode takes place over a prolonged period, the meristematic region at the base of the internode may be called an *intercalary meristem* (a meristematic region between two more highly differentiated regions). Some elements of the primary xylem and primary phloem dif-

(a) (b) (c)

(d)

21–3
Stages in growth of the terminal bud of the shagbark hickory (Carya ovata). (a) *The young shoot is tightly packed in the bud and protected by bud scales.* (b) *and* (c) *With expansion of the bud, the scales open and fold back.* (d) *Internodal elongation is underway, and the nodes are beginning to separate from one another. This bud is a mixed bud; that is, it contains both leaves and flowers.*

ferentiate within the intercalary meristem and connect the more highly differentiated regions of the stem above and below the meristem.

As in the root, the apical meristem of the shoot gives rise to the primary meristems—protoderm, ground meristem, and procambium (Figure 21–1)—and they in turn develop into epidermis, ground tissue, and primary vascular tissues, respectively.

PRIMARY STRUCTURE

Considerable variation exists in the primary structure of stems of seed plants, but three basic types of organization can be recognized: (1) In some conifers and dicotyledons, the narrow, elongated procambial cells—and consequently the primary vascular tissues that develop from them—appear as a more or less continuous hollow cylinder within the ground tissue (Figure 21–4). The outer region of ground tissue is called the *cortex*, and the inner region the *pith*. (2) In other conifers and dicotyledons, the primary vascular tissues develop as a cylinder of interconnected strands separated from one another by ground tissue (Figure 21–5). The ground tissue separating the interconnected procambial strands, and later the mature *vascular bundles*, is continuous with the cortex and pith, and is referred to as *interfascicular* (between the bundles) parenchyma. The interfascicular regions are often called *pith rays*. (3) In the stems of most monocotyledons and of some herbaceous (nonwoody) dicotyledons, the arrangement of the procambial strands and vascular bundles is more complex. The vascular tissues do not appear as a single ring of bundles between a cortex and a pith, but commonly develop in more than one ring or as an anastomosing (interconnecting and branched) system of strands scattered throughout the ground tissue, which often cannot be distinguished as cortex and pith (Figure 21–6).

21–4
(a) *Transverse section of linden* (Tilia americana) *stem in primary state of growth.* (b) *Detail of portion of* Tilia *stem. See also description in text.*

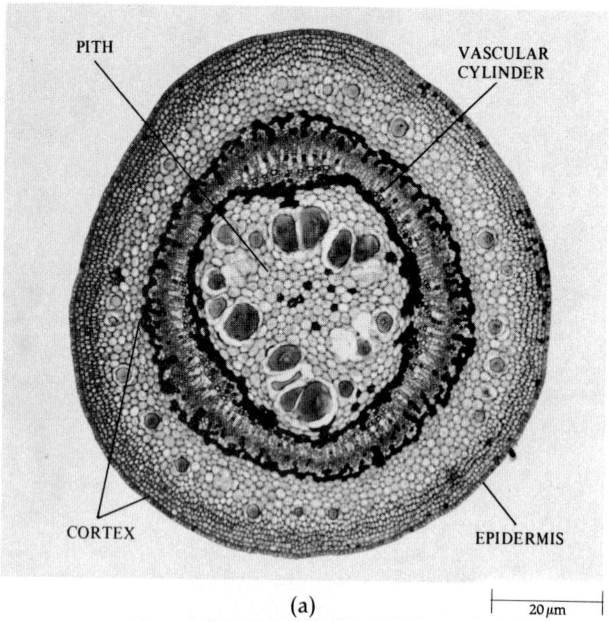

(a)

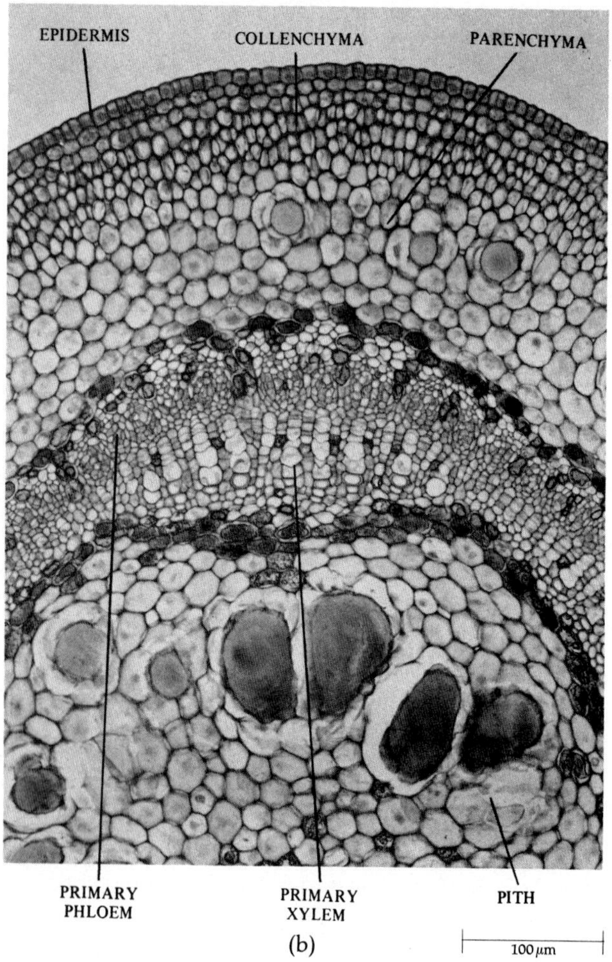

(b)

21–5
(a) *Transverse section of alfalfa* (Medicago sativa) *stem, an herbaceous dicot with discrete vascular bundles.* (b) *Detail of portion of alfalfa stem.*

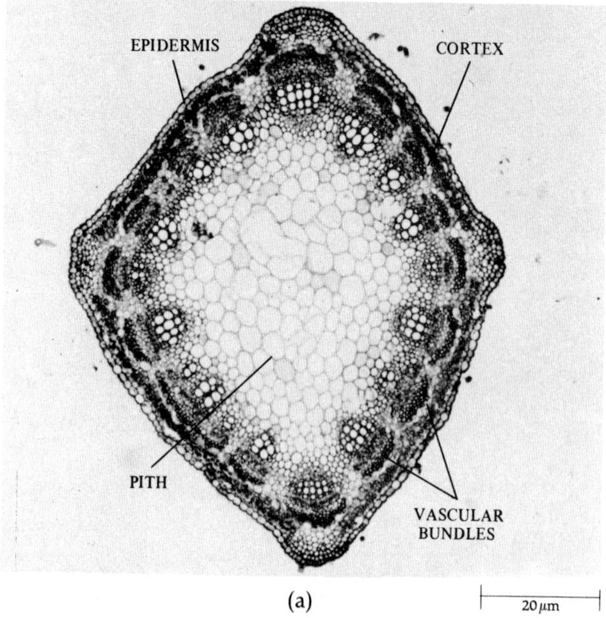

(a)

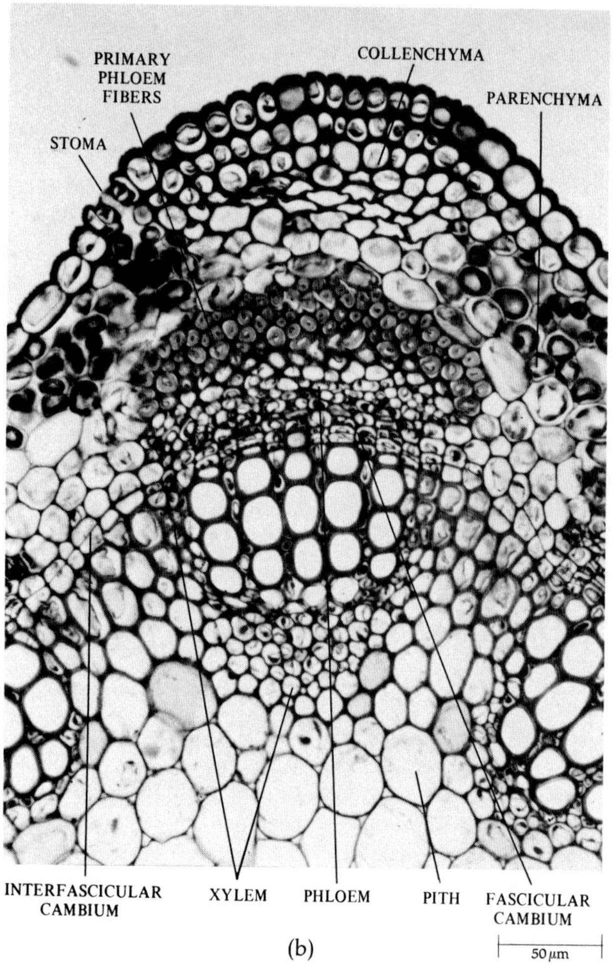

(b)

21-6
Transverse sections of the stem of corn (Zea mays). (a) *Internodal region, showing numerous vascular bundles scattered throughout the ground tissue.* (b) *Nodal region of young stem, showing horizontal procambial strands that interconnect with vertical bundles* [(a), *Carolina Biological Supply Co.*]

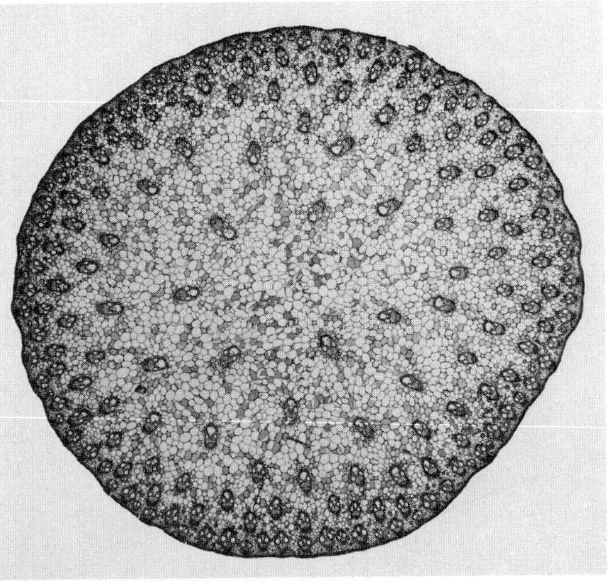

(a)

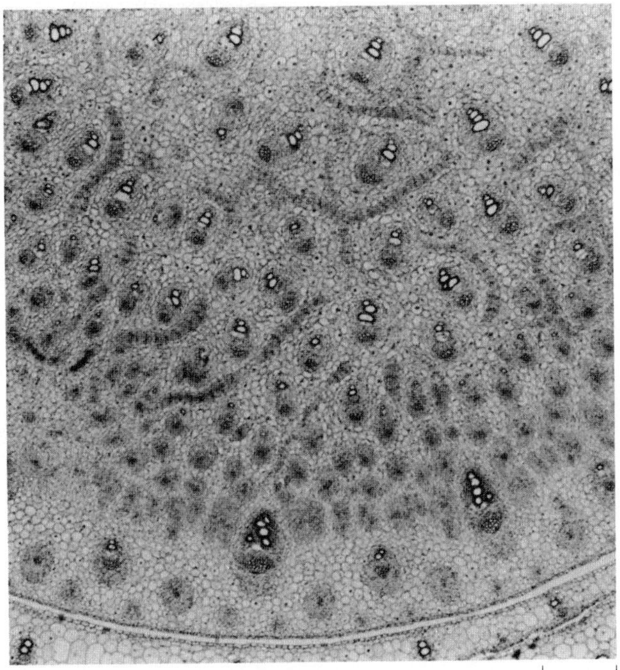

(b)

50 μm

The *Tilia* Stem

The stem of the basswood or linden *(Tilia)* exemplifies the first type of organization (Figure 21-4). As in most stems, the epidermis is a single layer of cells covered by a cuticle (see page 414). The stem epidermis generally contains far fewer stomata than the leaf epidermis.

The cortex of the *Tilia* stem consists of collenchyma and parenchyma cells (Chapter 19). The several layers of collenchyma cells, which provide support to the young stem, form a continuous cylinder beneath the epidermis. The rest of the cortex consists of parenchyma, which eventually contains chloroplasts. The innermost layer of cortical cells, which are dark in color, sharply delimits the cortex from the cylinder of procambium and primary vascular tissues.

In the great majority of stems, the primary phloem develops from the outer cells of the procambium and the primary xylem from the inner ones. However, not all of the procambial cells differentiate into primary xylem and primary phloem. A single layer of cells between the xylem and phloem remains meristematic and becomes the vascular cambium. *Tilia* is also an example of a woody stem—a stem that produces much secondary xylem (Chapter 22).

The inner boundary of the primary xylem in *Tilia* is sharply delimited by one or two layers of pith cells which are dark in appearance. The pith is composed primarily of parenchyma cells and contains numerous large canals containing mucilage (a slimy carbohydrate). Similar canals are formed in the cortex. As the cortical and pith cells increase in size, numerous intercellular spaces develop among them; these air spaces are essential for interchange of gases with the atmosphere. The cortical and pith parenchyma store various substances.

The *Sambucus* Stem

In the *Sambucus canadensis,* or elderberry, stem, the procambial strands and primary vascular bundles form a system of interconnected strands around the pith. The epidermis, cortex, and pith are essentially similar to those of *Tilia,* and the remainder of our discussion of the elderberry stem will be used to explain in more detail the development of the primary vascular tissues of stems (Figure 21-7).

In Figure 21-7a three procambial strands can be seen in which the primary vascular tissues have just begun to differentiate. The strand on the left is somewhat older than the two on the right and contains at least one mature sieve element and one mature tracheary element. Note that the first mature sieve element appears in the outer part of the procambial strand (next to the cortex), and the first mature tracheary element in the inner part (next to the pith) of the procambial strand. If you compare Figure 21-7a with 21-7c, you will see that the more recently formed sieve elements appear closer to the

21-7
Transverse sections of elderberry (Sambucus canadensis) *stem in primary state of growth.* (a) *Very young stem, showing protoderm, ground meristem and discrete procambial strands. Procambial strand on left contains one mature sieve element (arrow, above) and one mature xylem element (arrow, below).* (b) *Primary tissues further along in development.* (c) *Stem near completion of primary growth. Fascicular and interfascicular cambia are not yet formed.*

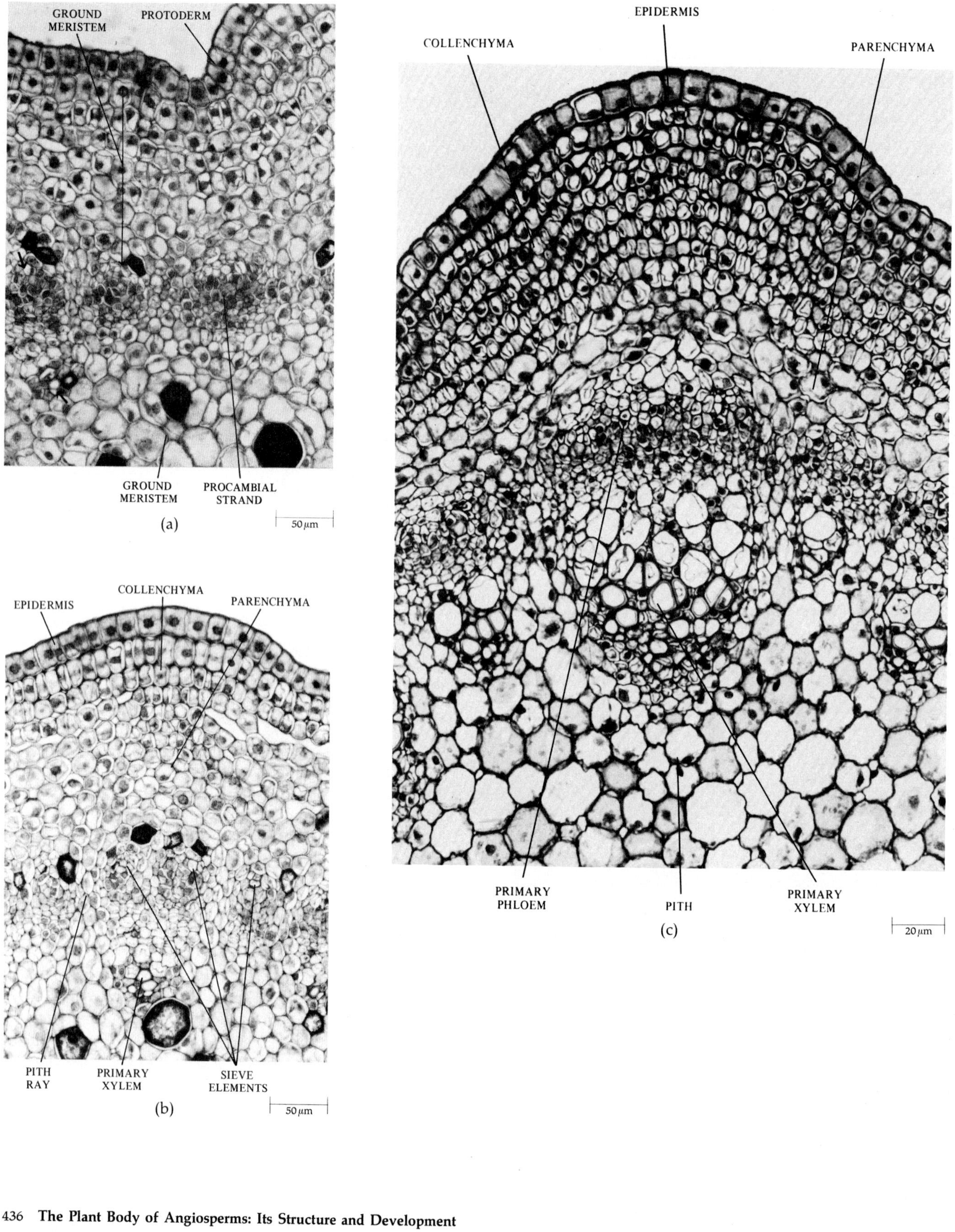

center of the stem and that the xylem differentiates in the opposite direction.

The first-formed primary xylem and primary phloem elements (protoxylem and protophloem, respectively) are stretched during elongation of the internode and are frequently destroyed. In addition, in both *Tilia* and *Sambucus* stems, fibers develop in the primary phloem after internodal elongation is completed (see Figure 22–8, page 464). These fibers are called *primary phloem fibers* (Figure 19–6, page 405).

Like the stems of *Tilia*, those of *Sambucus* become woody. In *Tilia*, almost all the vascular cambium originates from procambial cells between the primary xylem and primary phloem, because the interfascicular regions are very narrow. In *Sambucus*, the interfascicular regions are relatively wide. Consequently, in *Sambucus*, a substantial portion of the vascular cambium develops from the interfascicular parenchyma between the bundles. The part of the vascular cambium derived from the procambial cells of the vascular bundles or larger segments of the primary vascular system is called *fascicular cambium*, and that derived from the interfascicular parenchyma is called the *interfascicular cambium*. Together, the fascicular and interfascicular cambia form a continuous cylinder between the xylem and phloem. (For a discussion of secondary growth see Chapter 22.)

The *Medicago* and *Ranunculus* Stems

The structure and development of the primary tissues of the *Medicago sativa*, or alfalfa, stem are essentially similar to those of *Sambucus* and other woody dicotyledons. The stems of *Medicago*, however, are *herbaceous*, or nonwoody.

Medicago is an example of an herbaceous dicotyledon that exhibits some secondary growth (Figure 21–5). The vascular bundles are separated from one another by wide interfascicular regions and surround a large pith. The vascular cambium is partly fascicular and partly interfascicular in origin, but secondary vascular tissues are formed only in the bundles. The interfascicular cambium generally produces only sclerenchyma cells on the xylem side.

The extreme herbaceous stem of *Ranunculus*, the buttercup, lacks a vascular cambium, and its vascular bundles resemble those of many monocotyledons. Vascular bundles such as those of *Ranunculus* (Figure 21–8) and of monocotyledons (Figure 21–6), in which all the procambial cells differentiate and the potential for further growth is lost within the bundle, are said to be closed. Vascular bundles which give rise to a cambium are said to be open.

The *Zea* Stem

The *Zea mays*, or corn, stem exemplifies the stems of monocots in which the vascular bundles form an anastomosing system of strands scattered throughout the ground tissue (Figure 21–6). As in most monocots, the vascular bundles of corn are closed.

Figure 21–9 shows three stages in the development of a corn vascular bundle. As in the bundles of dicot stems, the phloem develops from the outer cells of the procambial strand, the xylem from the inner ones. Also, differentiation of the phloem and xylem are in opposite directions, as seen in transverse sections: The phloem differentiates from outside to inside, and the xylem from inside to outside. The first-formed phloem and xylem elements (protophloem and protoxylem) commonly are stretched and destroyed during elongation of the internode. This results in the formation of a very large air

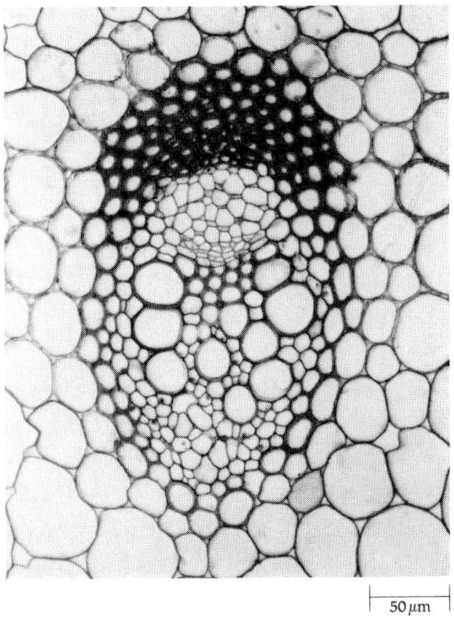

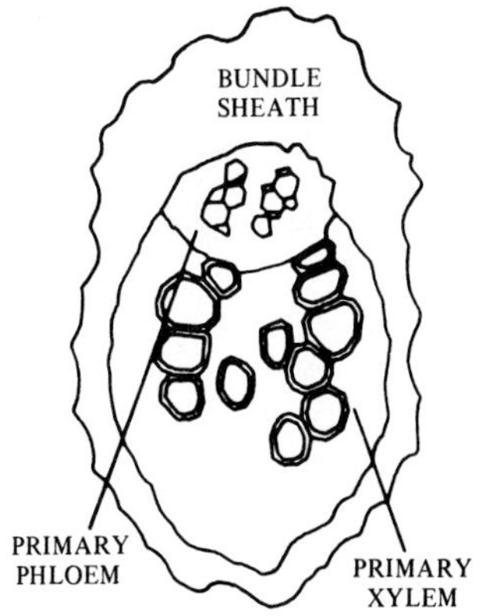

21–8
Transverse section of vascular bundle of buttercup (Ranunculus), *an herbaceous dicot. The bundles of buttercup are closed, all of the procambial cells becoming mature cells. The primary phloem and primary xylem are surrounded by a bundle sheath of thick-walled sclerenchyma cells.*

21–9
*Three stages in the differentiation of the vascular bundles of corn (*Zea mays*), from transverse sections of the stem.* (a) *The protophloem elements and two protoxylem elements are mature.* (b) *The protophloem sieve elements are now crushed, and much of the metaphloem is mature. Three protoxylem members are now mature, and the two metaxylem vessel members are almost fully expanded.* (c) *Mature vascular bundle surrounded by sheath of thick-walled sclerenchyma cells. The metaphloem is composed entirely of sieve elements and companion cells. The portion of the vascular bundle once occupied by the protoxylem elements is now a large air space. Notice wall thickenings of destroyed protoxylem elements bordering the space.*

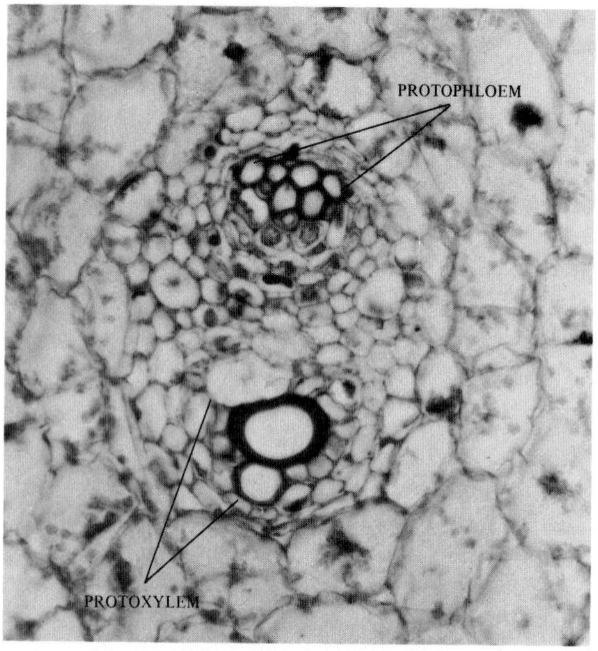

(a)

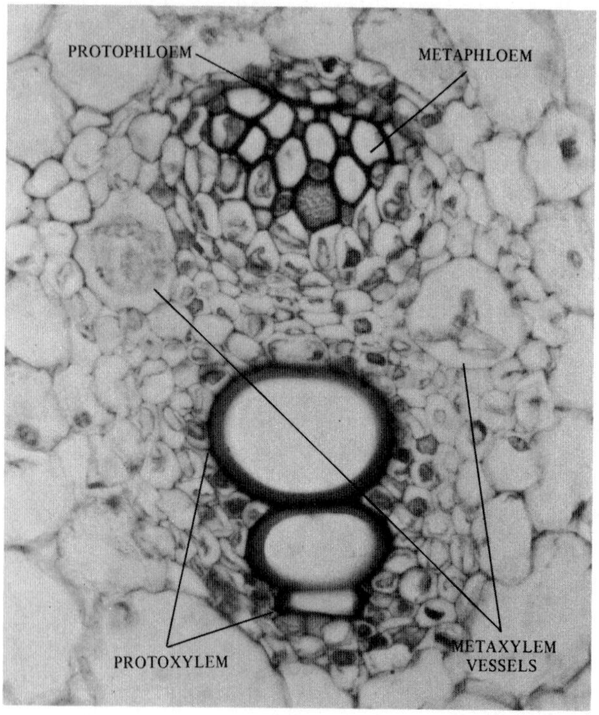

(b)

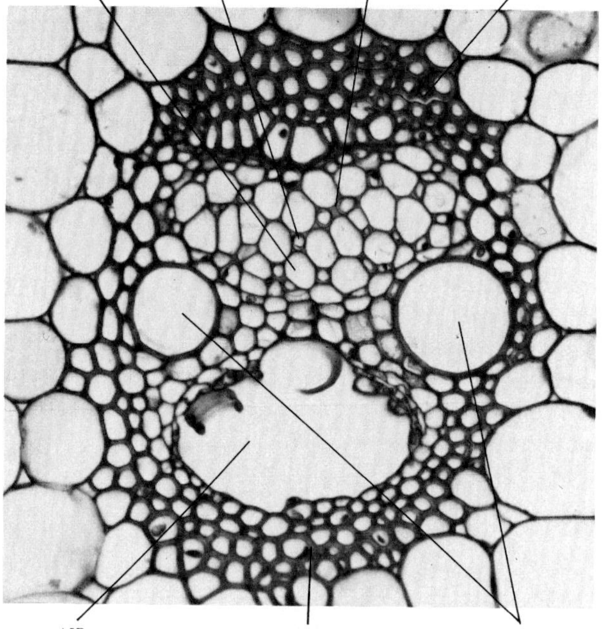

(c)

space on the xylem side of the bundle (Figure 21–9c). The mature vascular bundle contains two very large vessel members (the metaxylem vessel members), and phloem (metaphloem) tissue is composed of sieve-tube members and companion cells. The entire bundle is enclosed in a sheath of sclerenchyma cells (Figure 21–9c).

THE LEAF

The leaves of seed plants are megaphylls (pages 300–301) and, in their development and arrangement, have a profound influence on the structure of the stem. This is particularly true of those stems in which the primary vascular bundles are arranged as a cylinder of interconnected vascular strands around a pith (Figure 21–10), such as in *Sambucus* and *Medicago*.

The procambial strands of the stem arise behind the apical meristem just below the developing leaf primordia and sometimes are present below the sites of origin of the leaves even before the primordia have begun to

21–10
Diagrams of the primary vascular system of a stem in which discrete bundles form a cylinder of interconnected strands around a pith. (a) *Vascular system spread out in one plane.* (b) *Three-dimensional diagram showing the arrangement of the bundles within the stem. Notice the numerous leaf traces diverging outward from the cylinder and the relation of the traces to the vascular bundles of the stem.*

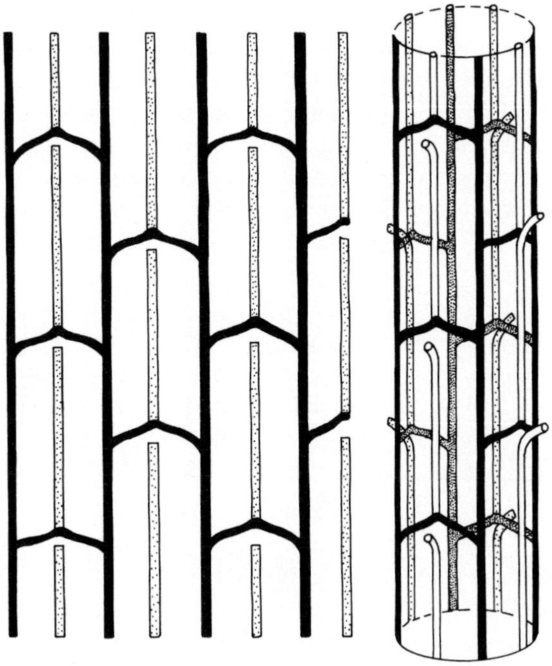

21–11
Diagrams of longitudinal (a) *and transverse* (b) *sections of part of the stem of tobacco* (Nicotiana tabacum), *illustrating the relationship of the vascular systems of the leaf and the stem. In addition to leaf traces, branch traces occur at the nodes and are often closely associated with the leaf traces. In tobacco, two branch traces extend from the vascular system of the stem to the bud, and the branch gap is continuous with the leaf gap.*

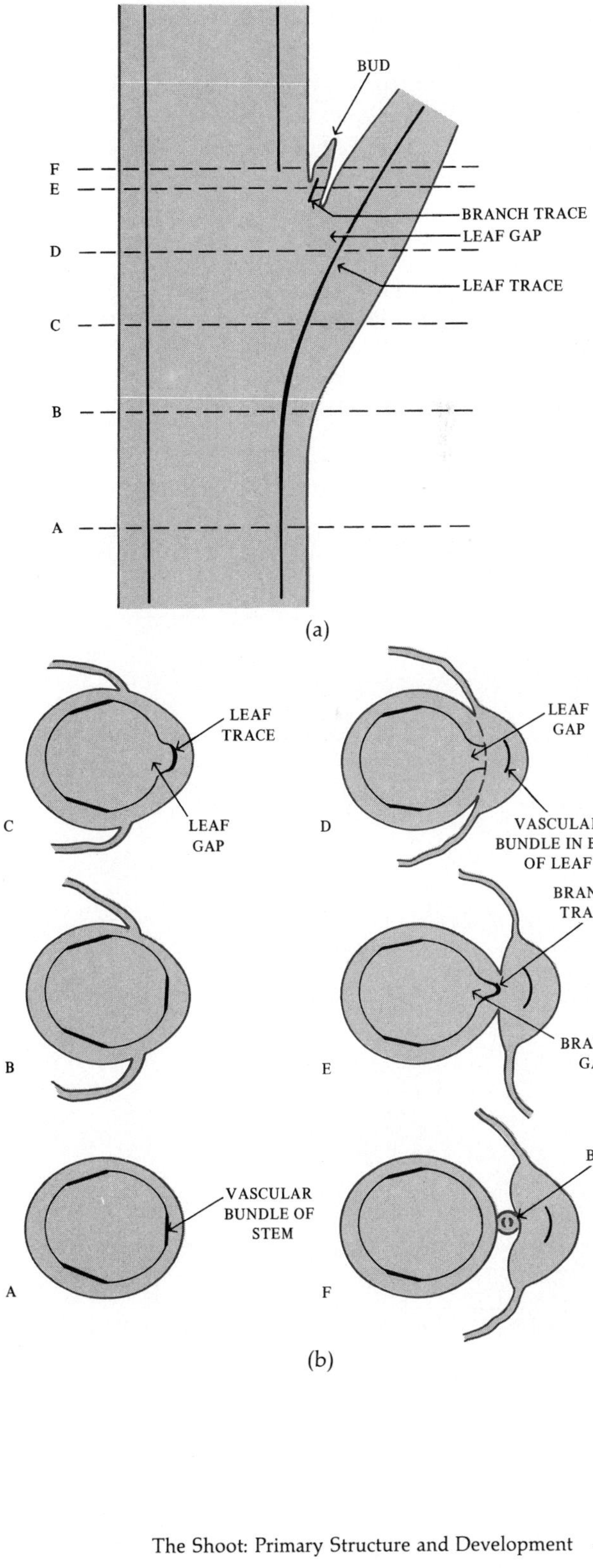

develop. Thus, the position of the leaves determines, in large part, the pattern of the vascular system in the stem. The intimate association of leaf and stem is clearly apparent if one follows the vascular bundles downward from the leaf and into the stem (Figure 21–11). As the bundles enter the stem, they traverse the cortex and then merge with other parts of the vascular system, or vascular cylinder, of the stem. The part of the vascular bundle extending from the base of the leaf to its connection with a vascular bundle in the stem is the *leaf trace*, and the wide gap or region of ground tissue found in the vascular system where the leaf trace is bent away from the vascular cylinder is the *leaf gap*. If the vascular bundle continuous with the leaf trace is followed downward into the stem, it will be found to be interconnected with other vascular bundles (Figures 21–10 and 21–11), each of which is associated with a leaf trace at some level of the stem. Thus the primary vascular system of the shoot may be considered a unit structure composed of leaf traces and their associated and interconnected vascular tissues in the stem.

DEVELOPMENT OF THE LEAF

The leaf primordium is initiated by periclinal divisions of subsurface layers of cells along the periphery of the apical meristem. (Note that periclinal divisions are divisions in which the newly-formed cell plates between dividing cells are parallel with the surface of the plant part concerned—in this case, the apical meristem. Anticlinal divisions are divisions in which the newly-formed cell plates are perpendicular to the surface.) Depending on the number of tunica layers present, these divisions may involve only tunica or corpus, or both tunica and corpus layers. The outer tunica layer divides only anticlinally, contributing to the increase in surface. As a result of the periclinal divisions, a bulge, or *leaf buttress*, is formed (Figure 21–12a). Either before or during buttress formation, a procambial strand appears beneath the developing leaf primordium.

With continued growth, the leaf buttress develops into an erect peglike structure, the leaf primordium (Figure 21–12b). In dicotyledons, this structure soon develops localized regions of meristematic activity on approximately opposite sides of its axis. These regions, which will initiate formation of the blade of the leaf, are called *marginal meristems* (Figure 21–13). The marginal meristems give rise to the *protoderm* and *ground meristem* of the blade. The procambium of the blade arises through localized divisions in the ground meristem. Expansion and increase in length of the leaf occur largely by *intercalary growth*, that is, by divisions throughout the blade. Differences in rates of cell division and cell enlargement in the various layers of the blade result in the formation of numerous intercellular spaces and in the form of mesophyll characteristic of the leaf. Compared

21–12
Some early stages of leaf development in Coleus blumei, *as seen in longitudinal sections of the shoot tip. The leaves in* Coleus *occur in pairs, opposite one another at the nodes (see Figure 21–1).* (a) *Two small bulges, or leaf buttresses, can be seen opposite one another on the flanks of the apical meristem. In addition, a bud primordium can be seen arising in the axil of each of the two young leaves, below.* (b) *Two erect peglike leaf primordia have developed from the leaf buttresses. Notice the procambial strands extending upward into the leaf primordia. The bud primordia, below, are further along in development than those in*

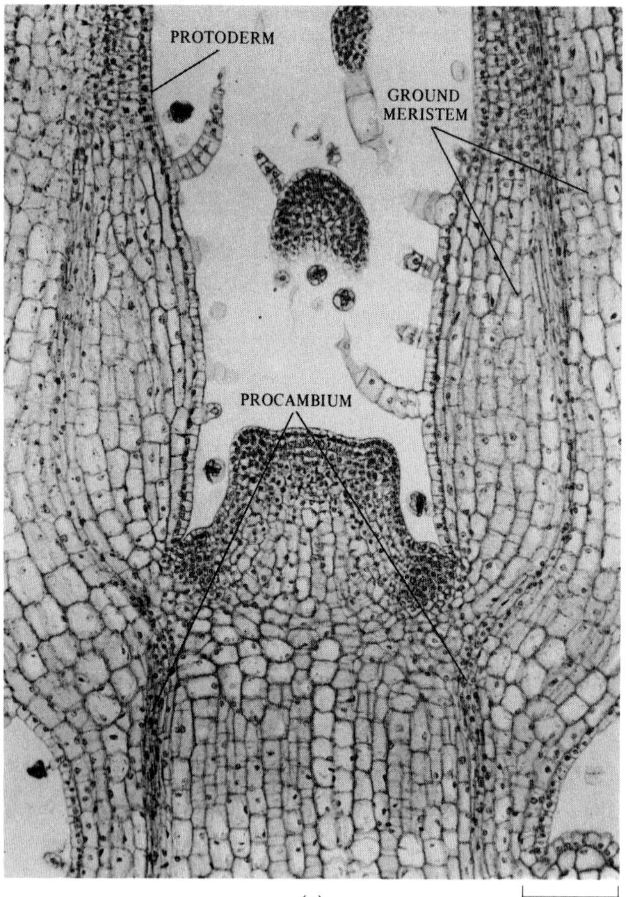

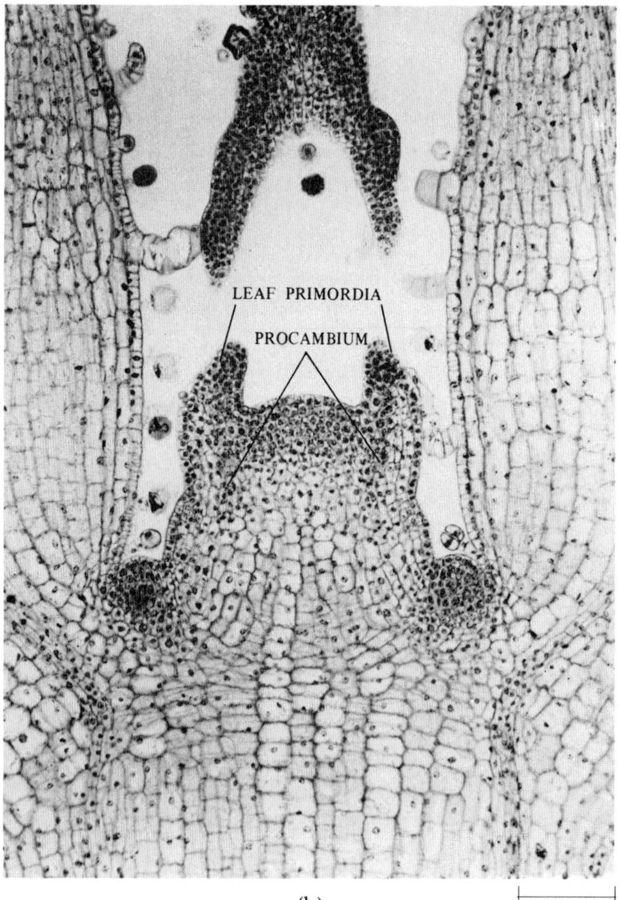

with growth of the stem, the growth of most leaves is of short duration. Whereas the unlimited or prolonged growth of the vegetative apical meristems is described as being indeterminate, the restricted type of growth exhibited by the leaf and by floral apices is said to be determinate.

As the peglike leaf primordium elongates, the procambial strand beneath it develops upward into it (Figure 21–12) and is continuous with the coarse veins emanating from the main vein, or midvein, of the blade. The smaller veins of the leaf are initiated at the tip of the leaf. They develop from the top to the bottom of the leaf in continuity with the coarser veins. Thus, the tip of the leaf is the first part to have a complete system of veins. This course of development reflects the overall maturation of the leaf, which is from the tip to the base of the leaf.

(a). (c) *As the leaf primordia elongate, the procambial strands, which are continuous with the vascular bundles in the stem, continue to develop into the leaves.*

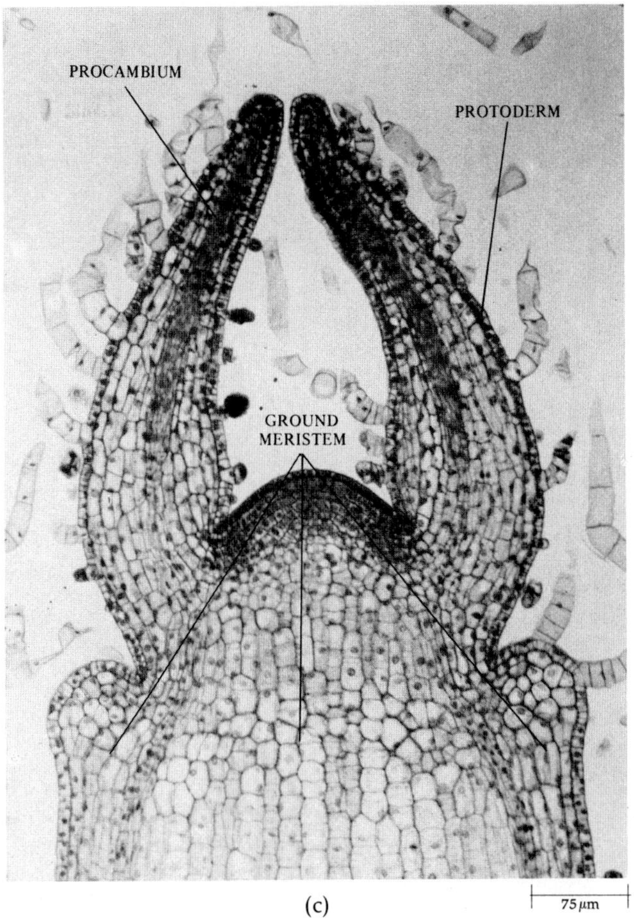

21–13
Transverse section of developing leaves of tobacco (Nicotiana tabacum) *grouped around the shoot tip, sectioned below the apical meristem. The younger leaves are nearer the axis. A leaf primordium at first lacks differentiation into midrib and blade. Some early stages in development of blade and midrib can be seen here.*

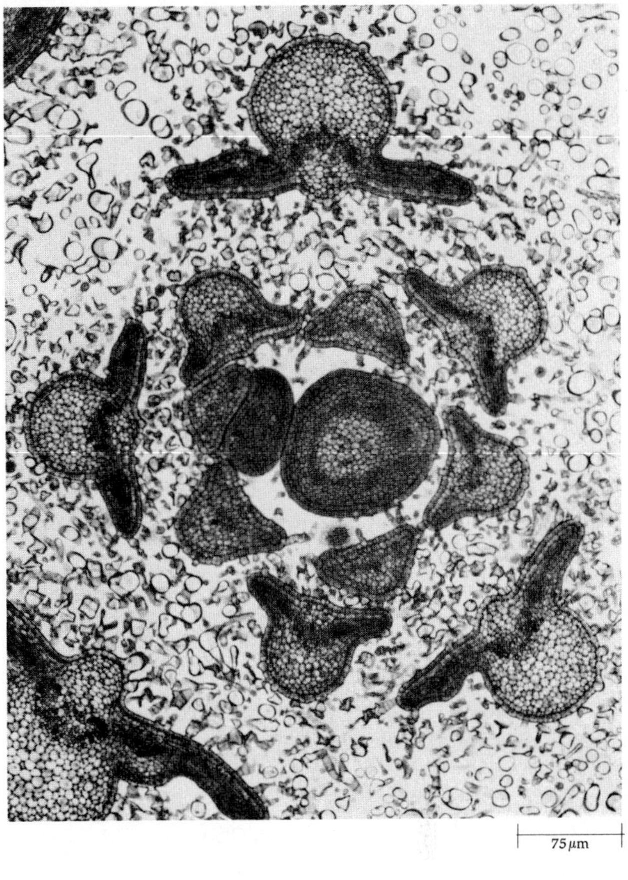

21-14
Some examples of simple leaves. (a) *Mulberry* (Morus alba). (b) *Culver's-root* (Veronicastrum virginicum). (c) *Sugar maple* (Acer saccharum). (d) *Silver maple* (Acer saccharinum). (e) *Red oak* (Quercus rubra). *Note the spiral (alternate) arrangement of the leaves in mulberry, and the whorled arrangement of those in Culver's-root. Leaf arrangement in the maples is opposite and in oak it is spiral, although only single leaves of these trees are shown here.*

(a)

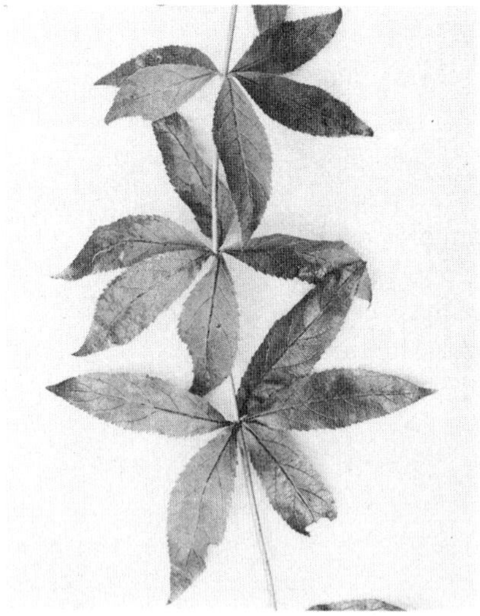

(b)

(c)

(d)

(e)

MORPHOLOGY OF THE LEAF

Leaves vary greatly in form and in internal structure. In dicots, the leaf commonly consists of an expanded portion, the blade, or *lamina,* and a stalklike portion, the *petiole* (Figure 21-14). Small scalelike or leaflike structures called *stipules* develop at the base of some leaves (Figure 21-15). Many leaves lack petioles and are said to be *sessile* (Figure 21-16). In most monocots and certain dicots, the base of the leaf is expanded into a *sheath,* which encircles the stem. In some grasses, the sheath extends the length of an internode. The arrangement of leaves on the stem may be *spiral* (alternate), *opposite* (in pairs), or *whorled* (three or more at a node).

The leaves of dicotyledons are either simple or compound. In simple leaves, the blades are not divided into distinct parts, although they may be deeply lobed (Figure 21-14). The blades of compound leaves are divided into leaflets, each usually with its own small stalk, or petiole (petiolule). Two types of compound leaves can be distinguished: pinnately compound leaves and palmately compound leaves (Figure 21-17). In pinnately compound leaves, the leaflets arise from either side of an axis, the *rachis,* like the pinnae of a feather. (The rachis is an extension of the petiole.) The leaflets of a palmately compound leaf diverge from the tip of the petiole and a rachis is lacking.

21–15
Pinnately compound leaf of pea (Pisum sativum). *Notice the stipules at the base of the leaf and the slender tendrils at the tip of the leaf. In the pea leaf, the stipules are larger than the leaflets.*

21–16
Sessile leaves—leaves without a petiole—are often found among dicots, but are particularly characteristic of the grasses and other monocots, such as sorghum (Sorghum eulgare).

21–17
Some examples of compound leaves: (a) *palmately compound,* (b)–(e) *pinnately compound.* (a) *Red buckeye* (Aesculus pavia). (b) *Shagbark hickory* (Carya ovata). (c) *Green ash* (Fraxinus pennsylvanica *var.* subintegerrima) (d) *Black locust* (Robinia pseudo-acacia). (e) *Honey locust* (Gleditsia triacanthos). *In honey locust, each leaflet is subdivided into leaflets.*

(a)

(b)

(c)

(d)

(e)

21–18
Sections of lilac (Syringa vulgaris) *leaf. Views* (b), (c), *and* (d) *show enlargements of portions of* (a), *which is a section cut approximately parallel with the leaf surface. The section cuts deeper into the leaf in the lower part of the micrograph. Thus part of the upper epidermis can be seen in the light area at the top of the micrograph and part of the lower epidermis at the bottom. Notice the greater number of stomata in the lower epidermis. The venation in lilac is netted.* (b) *Section through palisade parenchyma.* (c) *Section through spongy parenchyma.* (d) *Section through lower epidermis, showing two trichomes and several stomata, in addition to many ordinary epidermal cells.* (e) *Transverse section.*

STRUCTURE OF THE LEAF

Variations in the structure of angiosperm leaves are to a great extent related to the habitat and are often used to characterize the so-called ecological types of plants: mesophytes (plants that grow where it is neither too wet nor too dry), hydrophytes (plants that grow wholly or partly submerged in water), and xerophytes (plants that grow in dry, or arid, habitats). Such distinctions are not sharp, however, and leaves often exhibit a combination of features characteristic of different ecological types. Regardless of their shapes and sizes, all leaves are composed of the same tissues: *epidermis, mesophyll,* and *vascular bundles* or *veins.*

Epidermis

The ordinary epidermal cells of the leaf, like those of the stem, are compactly arranged and covered with a cuticle that reduces water loss (see Chapter 19, page 414). Stomata may occur on both sides of the leaf, but commonly are more numerous on the lower surface (Figure 21–18). In leaves of hydrophytes that float on the surface of the water, stomata may occur in the upper epidermis only (Figure 21–19); the immersed leaves usually lack stomata entirely. The leaves of xerophytes generally contain

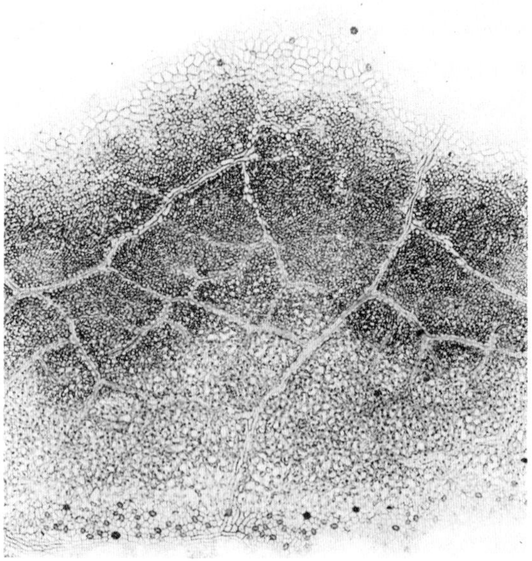

(a)

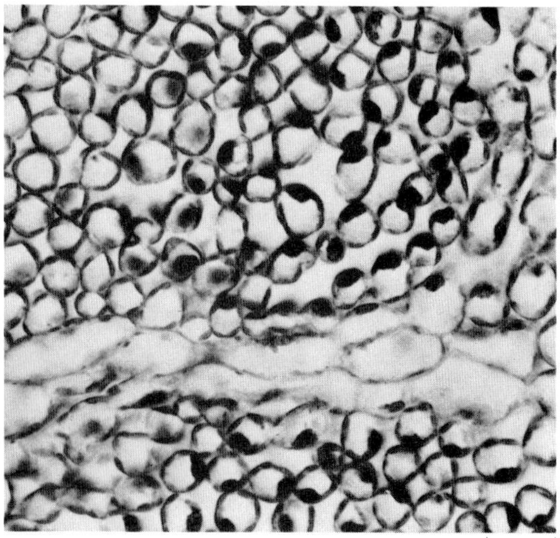

(b)

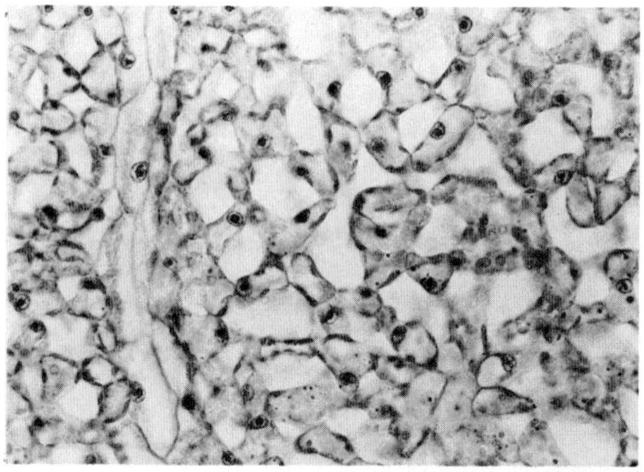

(c)

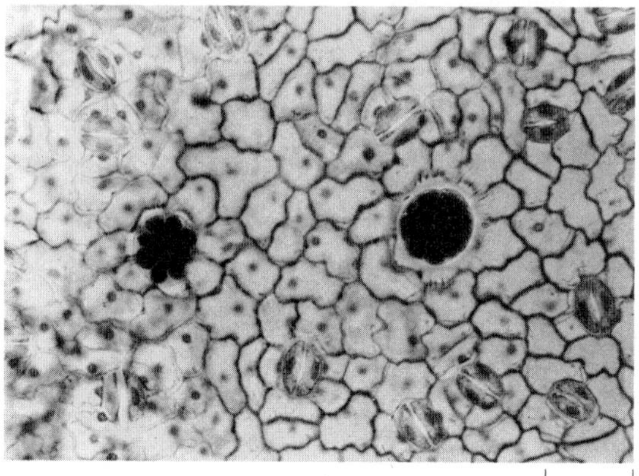

(d)

greater numbers of stomata than those of other plants. Presumably these numerous stomata permit a higher rate of gas exchange under conditions of favorable water supply. In many xerophytes, the stomata are sunken in depressions on the lower surface of the leaf (Figure 21–20). The depressions may also contain many epidermal hairs. Epidermal hairs, or trichomes (page 414), may occur on either or both surfaces of a leaf. Thick coats of epidermal hairs may retard water loss from leaves.

In the leaves of dicotyledons, the stomata are scattered and randomly arranged, and their development is mixed—that is, mature and immature stomata occur side by side. In monocotyledons, the stomata are arranged in rows parallel with the long axis of the leaf. Their development begins at the tips of the leaves and progresses downward.

Mesophyll

The mesophyll—the ground tissue of the leaf—is specialized for photosynthesis. It contains a large system of intercellular spaces, which are connected with the outer atmosphere through the stomata. The intercellular spaces facilitate rapid gas exchange, an important factor

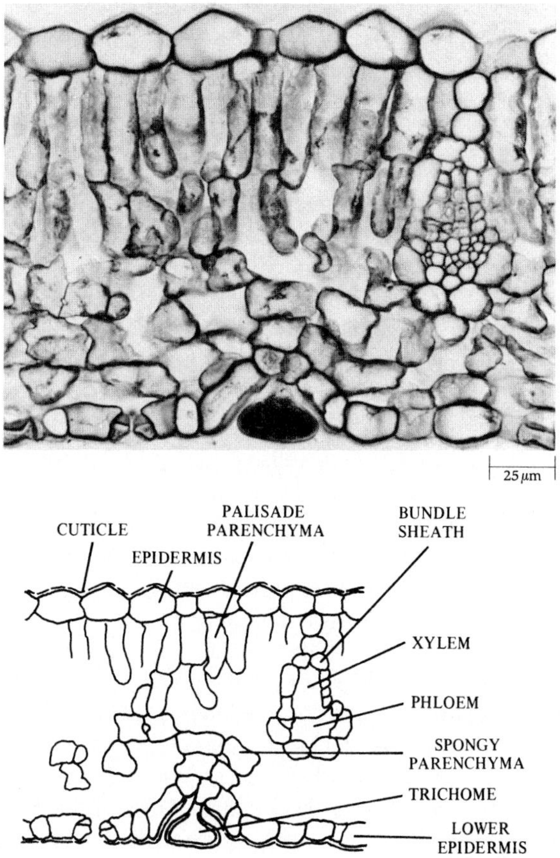

(e)

21–19
Transverse section of the water lily (Nymphaea odorata) *leaf. The* Nymphaea *leaf floats on the surface of the water and has stomata in the upper epidermis only. As is typical of hydrophytes, the vascular tissue in the* Nymphaea *leaf is much reduced, especially the xylem. The palisade parenchyma consists of three or four layers of cells above the spongy parenchyma with its very large intercellular spaces. The branched, thick-walled cell in the center of the leaf is a sclereid.*

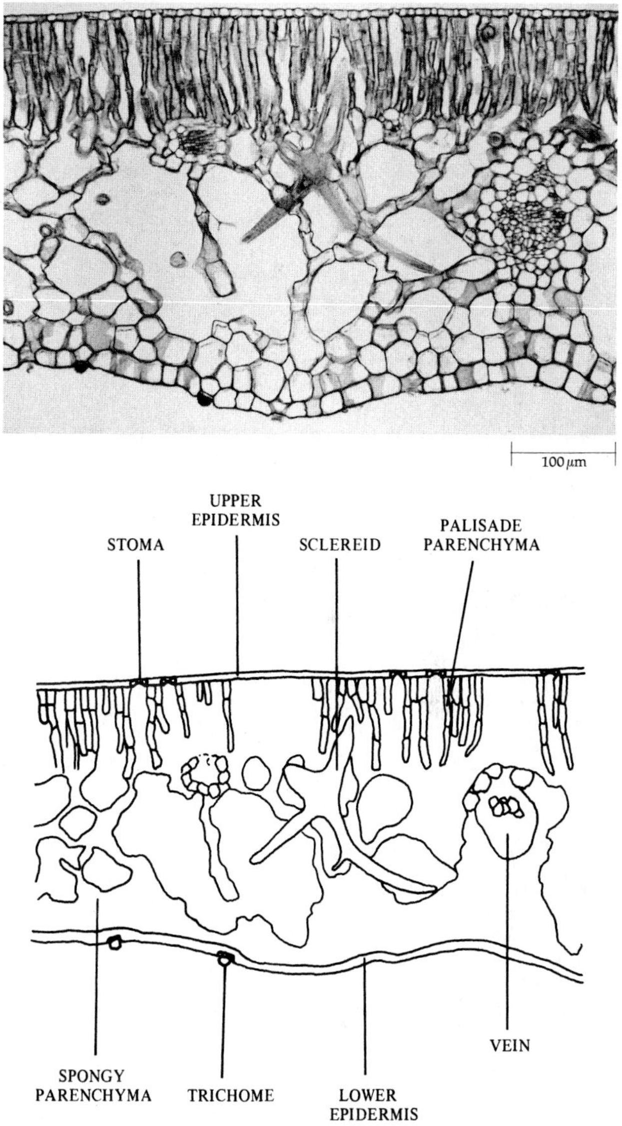

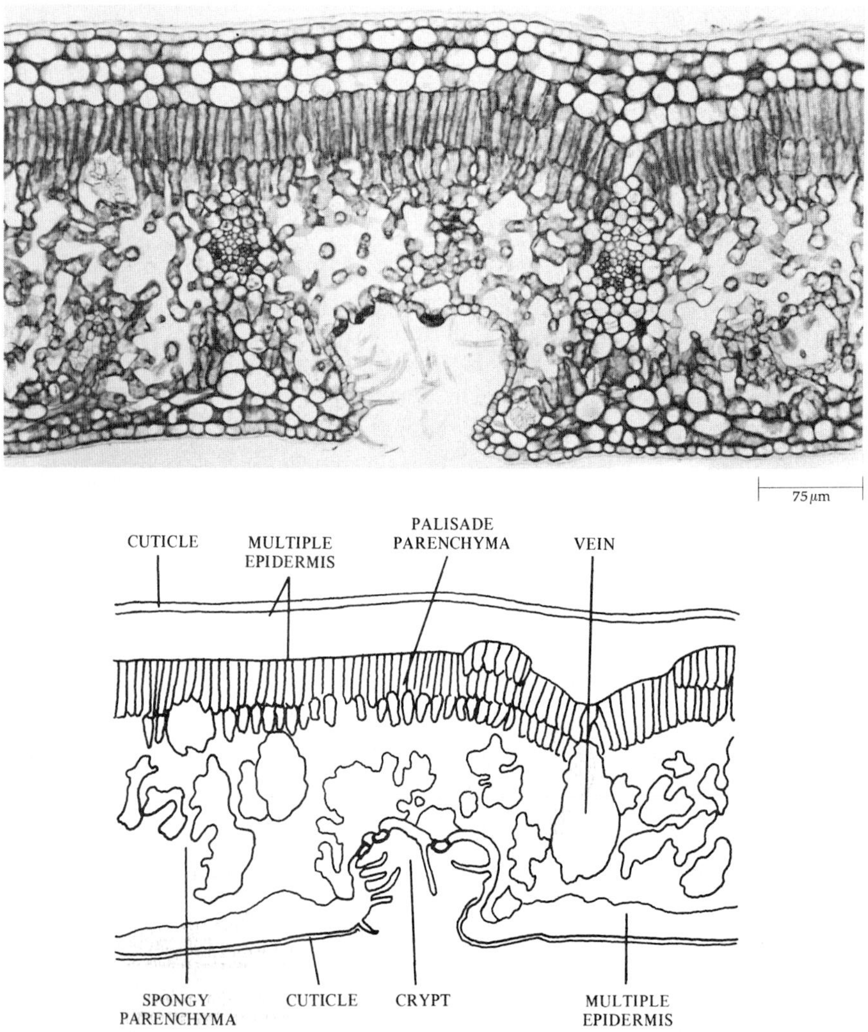

21-20
Transverse section of oleander (Nerium oleander) *leaf. Oleander is a xerophyte, and this is reflected in the structure of the leaf. Notice the very thick cuticle covering the multiple (several layered) epidermis on the upper and lower surfaces of the leaf. The stomata and epidermal hairs are restricted to invaginated portions of the lower epidermis called stomatal crypts.*

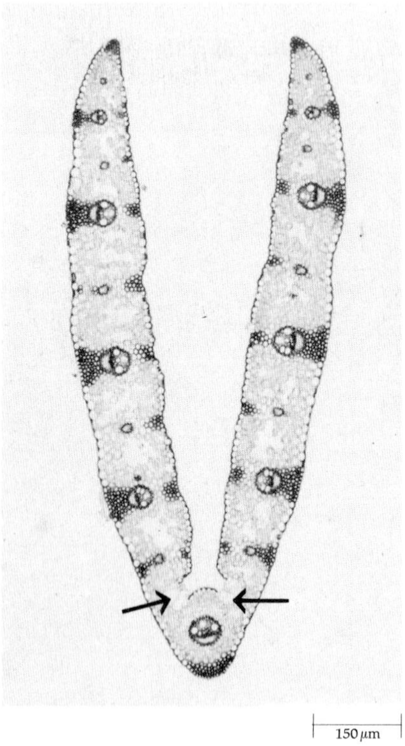

21-21
Transverse section of annual blue grass (Poa annua) *leaf. In the grass leaf, the mesophyll is not differentiated as palisade and spongy parenchyma. Notice the sclerenchyma cells (the dark cells bordering the veins and extending to the epidermis). The epidermis of the grass leaf contains bulliform cells, which are larger than the other epidermal cells. The bulliform cells are thought to play a part in the rolling or unrolling, folding or unfolding of grass leaves. In the* Poa *leaf shown here, the bulliform cells (arrows) are partly collapsed and the leaf is folded. An increase in turgor in the bulliform cells would presumably cause the leaf to unfold, although studies indicate that other tissues are also involved in this phenomenon.*

in photosynthetic efficiency. In mesophytes, the mesophyll is differentiated into *palisade parenchyma* and *spongy parenchyma*. The cells of the palisade tissue are columnar in shape, with their long axes oriented at right angles to the epidermis (Figures 21–18b and 21–18e), and the spongy parenchyma cells are irregular in shape (Figures 21–18c and 21–18e). Although the palisade parenchyma appears more compact than the spongy, most of the vertical walls of the palisade cells is exposed to intercellular space, and in some leaves the palisade surface may be two to four times greater than the spongy. Chloroplasts are also more numerous in palisade than in spongy cells. Thus, most of the photosynthesis in the leaf apparently takes place in the palisade parenchyma.

Commonly the palisade parenchyma is located on the upper side of the leaf, and the spongy on the lower (Figure 21–18). In leaves of xerophytes, palisade parenchyma often occurs on both sides of the leaf. And in some plants, such as corn and other grasses (Figure 21–21), the mesophyll cells are of more or less similar shape and a distinction between spongy and palisade parenchyma does not exist.

PLANTS AND AIR POLLUTION

The plant leaf, like the human lung, can function only when it is exchanging gases with the surrounding air. As a consequence, the leaf, like the lung, is an organ exceedingly susceptible to air pollution.

Air pollution has many forms. Some pollutants are particulate. The particles may be organic, such as those present in the smoke produced by burning fossil fuels, leaves, and garbage, or inorganic, such as the lead compounds released in the combustion of leaded gasoline. As a major component of smog, these particles reduce the amount of sunlight reaching the earth's surface. Such particles also have direct ill effects on plants. They may clog stomata and prevent them from functioning or they may, in particular the metallic particles, act as plant poisons.

Fluorides, which enter the air as waste products from the manufacture of phosphates, steel, aluminum, and other industrial products, act as cumulative poisons, entering the leaf through the stomata and causing collapse of leaf tissue, apparently by inhibiting enzymes concerned with cellulose synthesis. Thousands of acres of Florida citrus groves have been damaged by fluorides discharged from phosphate fertilizer plants.

When ores containing sulfur are processed, sulfur dioxide is produced:

$$2CuS + 3O_2 \longrightarrow 2CuO + 2SO_2$$

(Sulfur dioxide has a disagreeable pungent-sweet taste; it is unusual among air pollutants in that it can be tasted at lower concentrations than it can be smelled.) Sulfur oxides are also produced by the burning of fossil fuels containing sulfur. In moist air, sulfur oxides react with water to form droplets of sulfuric acid, a strong corrosive acid. In several parts of the United States, virtual deserts have been created by the emission of sulfur dioxide combined with metals. As long ago as 1905, air pollution controls were instituted in areas surrounding copper smelters in Tennessee, but the surrounding area, once covered by luxuriant forest, remains barren to this day. Not only was all the vegetation killed, but the acid leached the soil of nutrients. Similarly, in a copper-smelting area of Sacramento Valley, California, all vegetation was killed over an area of 260 square kilometers and growth severely affected over a further 320 square kilometers.

The most familiar type of air pollution to Californians is photochemical smog, which is produced by sunlight acting on automobile exhausts. The Los Angeles Basin offers an ideal setting for the formation of photochemical smog because the life of Los Angeles is heavily geared to the use of the automobile, and the mountains to the north and east prevent the dispersal of the reactants. Not only are many species of plants unable to survive in the city itself (as is true in many other major cities of the world), but smog moving out of the Los Angeles Basin is damaging agricultural crops and killing pine forests in mountains as far as 160 kilometers away. One of the principal ingredients of photochemical smog is nitrogen dioxide, NO_2, *which is produced by any combustion process that occurs in air (dry air is 77 percent nitrogen) and so is present in automobile exhausts. Under the influence of light,* NO_2 *is split into NO and atomic oxygen. The latter is extremely reactive and forms ozone,* O_3, *by reaction with normal (molecular) oxygen. (Similar reactions, powered by ultraviolet light, take place at the outer layers of the atmosphere, producing the ozone shield described on page 2. Ozone can also be produced by an electric spark, and is the source of the "clean" smell after a lightning storm.) Ozone is a highly toxic substance. In plants, it damages the thin-walled palisade cells, apparently affecting the permeability of the membranes of both the cells and their chloroplasts. Another component of photochemical smog is PAN (peroxyacetyl nitrate,* $C_2H_3O_5N$). *It is several times more toxic than ozone but is normally present in much lower concentrations. Photosynthesis is reduced 66 percent by photochemical smog concentrations of 0.25 part per million.*

Vascular Bundles

The mesophyll of the leaf is thoroughly permeated by a system of vascular bundles or veins, which is connected with the vascular system of the stem. In most dicotyledons, the veins are arranged in a branching pattern, with successively smaller veins branching from somewhat larger ones. This type of vein arrangement is known as *netted venation* (Figure 21-22). Often the largest vein extends along the long axis of the leaf as a midvein which, with its associated ground tissue, comprises the so-called midrib of such leaves. By contrast, most monocot leaves have many veins of fairly similar size that are oriented parallel to one another along the leaf—*parallel venation.* In these parallel-veined leaves, the longitudinal veins are interconnected by much smaller veins, to form a complex network (Figure 21-23).

The veins contain xylem and phloem, which generally are entirely primary in origin. (The midvein and sometimes the coarser veins undergo secondary growth in some dicot leaves.) The vein-endings in dicot leaves often contain only tracheids, although both xylem and phloem elements may extend to the ends of the vein. Commonly the xylem occurs on the upper side of the leaf and the phloem on the lower side (Figure 21-18e).

The vascular tissues of the veins are rarely exposed to intercellular spaces of the mesophyll. The large veins are surrounded by parenchyma cells that contain few chloroplasts, whereas the small veins are enclosed by one or more layers of compactly arranged cells that form a *bundle sheath* (Figures 21-18 through 21-21). The cells of the bundle sheath often resemble the mesophyll cells in which the small veins are located. The bundle sheaths extend to the ends of the veins, assuring that no part of

21–22
Leaf of tulip tree (Liriodendron tulipifera), *cleared to show the veins, at three successive magnifications. No mesophyll cell of the leaf is far from a vein. Water and dissolved minerals are carried to the leaf through the xylem, whereas organic molecules produced in the leaf are carried out of the leaf through the phloem.*

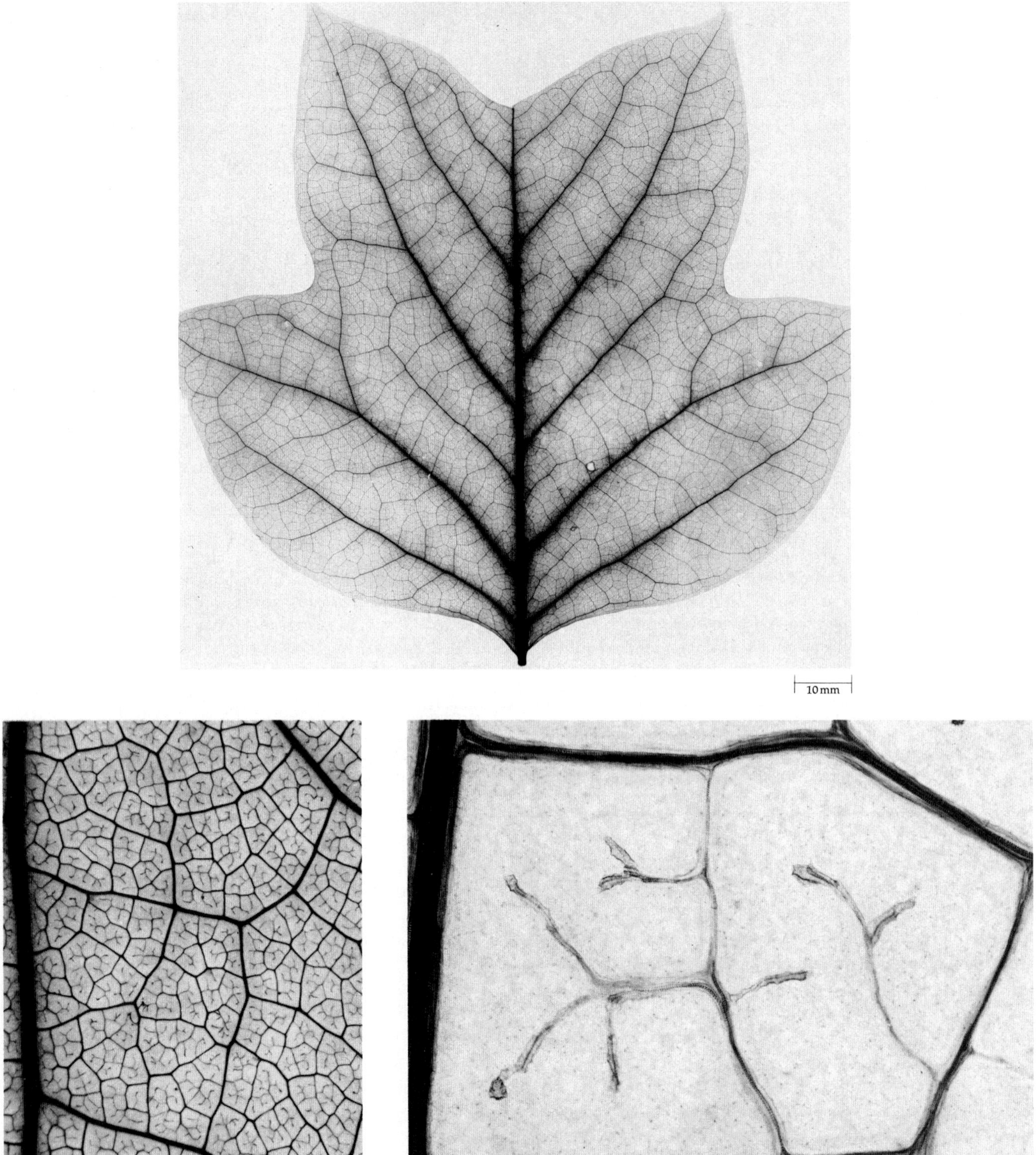

21–23
(a) *The basal leaves of this lady's slipper orchid* (Cypripedium acaule) *show the characteristic parallel venation of a monocot. It is widespread in the eastern United States.* (b) *A portion of a cleared leaf of the orchid* Cochleanthus.

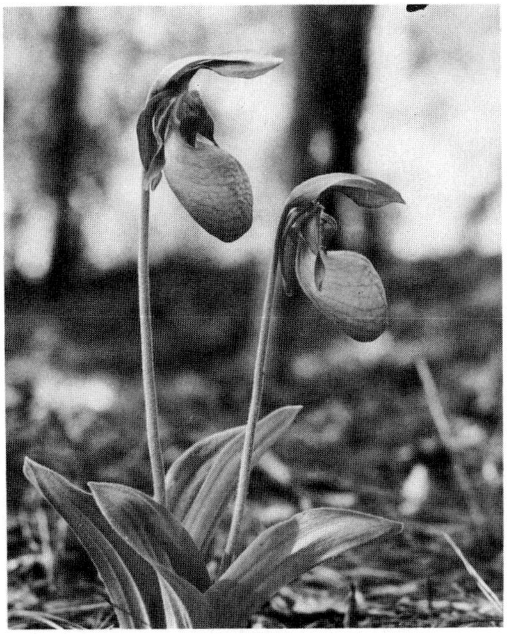

(a)

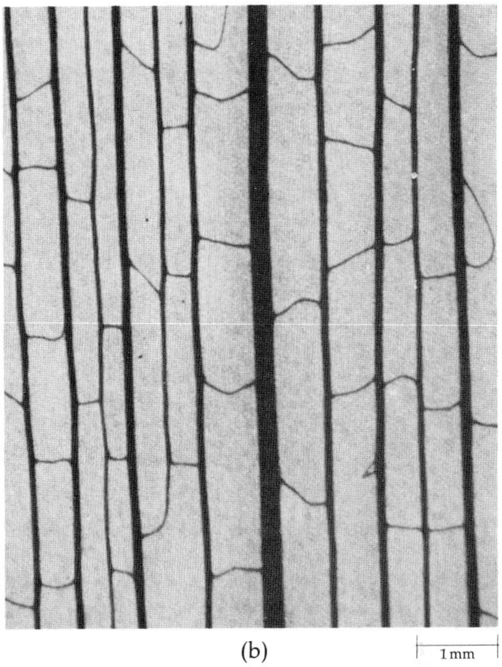

(b)

the vascular tissue is exposed to air of the intercellular spaces and that all substances entering and leaving the vascular tissues must pass through the sheath. Thus the bundle sheath performs a function analogous to that of the endodermis of the root (Chapter 20).

In many leaves, the bundle sheaths are connected with either or both upper and lower epidermis by cells resembling the sheath cells (Figures 21–18e and 21–21). Such extensions of the sheaths are called *bundle-sheath extensions.* Besides offering mechanical support to the leaf, in dicotyledons they apparently conduct water to the epidermis.

Because of its compact structure and its cuticle, the epidermis provides considerable strength to the leaf. In addition, the larger veins of dicot leaves are often bordered by collenchyma cells, which provide support to the leaf. In monocot leaves, the veins may be bordered by fibers. Collenchyma cells and fibers may also be found along the leaf margins of dicot and monocot leaves, respectively.

LEAF ABSCISSION

In many plants, the normal separation of the leaf from the stem—the process of *abscission*—is preceded by certain structural and chemical changes near the base of the petiole, which result in the formation of an *abscission zone* (Figure 21–24). In woody dicotyledons, two layers can be

21–24
Abscission zone in maple (Acer *sp.*) *leaf, as seen in longitudinal section through the base of the petiole.*

distinguished in the abscission zone: an abscission or separation layer, and a protective layer. The separation layer consists of relatively short cells with poorly developed wall thickenings, which make it structurally weak. A protective layer is formed by the deposition of suberin in the cell walls and intercellular spaces beneath the separation layer. After the leaf falls, the protective layer is recognized as a *leaf scar* on the stem (Figure 22–12, page 468).

TRANSITION BETWEEN VASCULAR SYSTEMS OF ROOT AND SHOOT

In previous chapters, we have seen that the distinction between plant organs is based primarily on the relative distribution of the vascular and ground tissues. For example, in dicot roots the vascular tissues generally form a solid cylinder surrounded by the cortex. In addition, the strands of primary phloem alternate with the radiating ridges of primary xylem. In contrast, in the stem the vascular tissues often form a cylinder of interconnected strands around a pith, with the phloem on the outside of the vascular bundles and the xylem on the inside. In addition, the pattern of the vascular system in the stem is determined, to a great extent, by the position of the leaves. Obviously, somewhere in the primary plant body a change must take place from the type of structure found in the root to that found in the shoot. This change is a gradual one, and the region of the plant axis through which it occurs is called the *transition region*.

In Chapter 18, we saw that the shoot and root are initiated as one continuous structure during development of the embryo. Consequently, it is in the axis of the embryo or young seedling where vascular transition is accomplished. This transition is initiated during appearance of the procambial system in the embryo and is completed with differentiation of the variously distributed procambial tissues in the seedling.

The structure of the transition region can be very complex, and much variation exists in the transition region among plants. In most gymnosperms and dicotyledons, vascular transition occurs between the root and cotyledons. Figure 21–25 depicts a type of transition region common among dicotyledons. Note the diarch (having two protoxylem poles) structure of the root; the branching and reorientation of the xylem and the phloem, which in the upper part of the axis results in the formation of a pith; and the traces of the first leaves of the epicotyl.

21–25
The transition region—the connection between root and cotyledons—in the seedling of a dicotyledon with a diarch root. In the root, the primary vascular system is represented by a single cylinder of vascular tissue. In the hypocotyl-root axis, the vascular system branches and diverges into the cotyledons, and the xylem and phloem become reoriented along the hypocotyl-root axis.

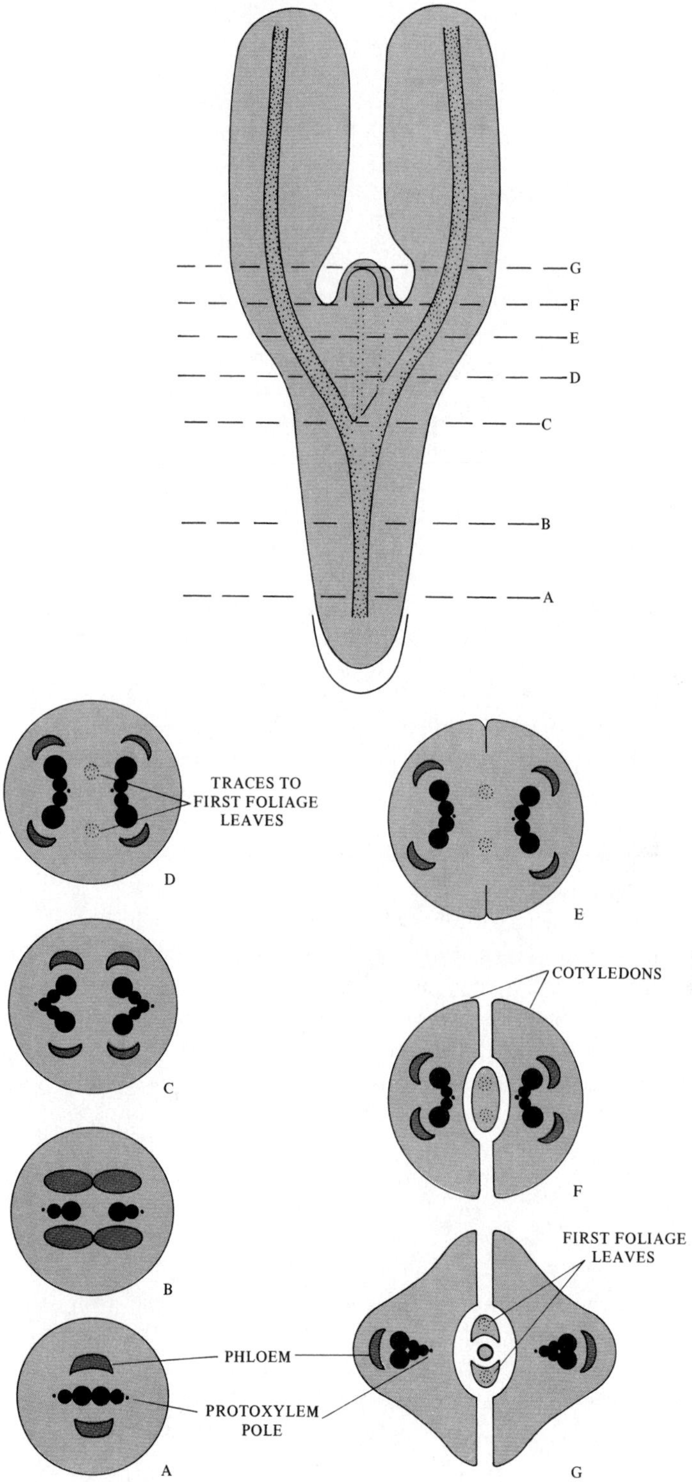

DEVELOPMENT OF THE FLOWER

The development of the flower or inflorescence terminates the meristematic activity of the vegetative shoot apex.

At flowering the vegetative shoot apex is transformed into a reproductive apex. Consequently, flowering is considered by many botanists as a stage in the development of the shoot apex. Various environmental factors not yet clearly understood by botanists are involved in the induction to flowering (see Chapter 18, page 400).

The transition from a vegetative to a floral apex is often preceded by an elongation of internodes and early development of lateral buds below the shoot apex. The apex itself undergoes marked increase in mitotic activity, accompanied by changes in dimensions and organization: From a relatively small apex with a tunica-corpus type of organization, the apex becomes broad and dome-like. It is at this time that the méristème d'attente (page 432) presumably becomes actively involved in formation of the reproductive structures.

The initiation and early stages of development of the sepals, petals, stamens, and carpels are quite similar to those of leaves. Commonly, the initiation of the floral parts begins with the sepals, followed by the petals, then the stamens, and finally the carpels (Figures 21–26 and 21–27). This usual order of appearance of the floral parts may be modified in certain flowers, but the floral parts always have the same relative spatial relation to one another (Figures 21–28 and 21–29).

The basic structure of the flower and some of its variations were discussed in Chapter 16.

21–26
Development of the tomato (Lycopersicum esculentum) *flower, as seen in longitudinal sections.* (a) *Part of a young inflorescence; flowers with sepal primordia.* (b) *Part of a young inflorescence; flowers with sepal, petal, and stamen primordia.*

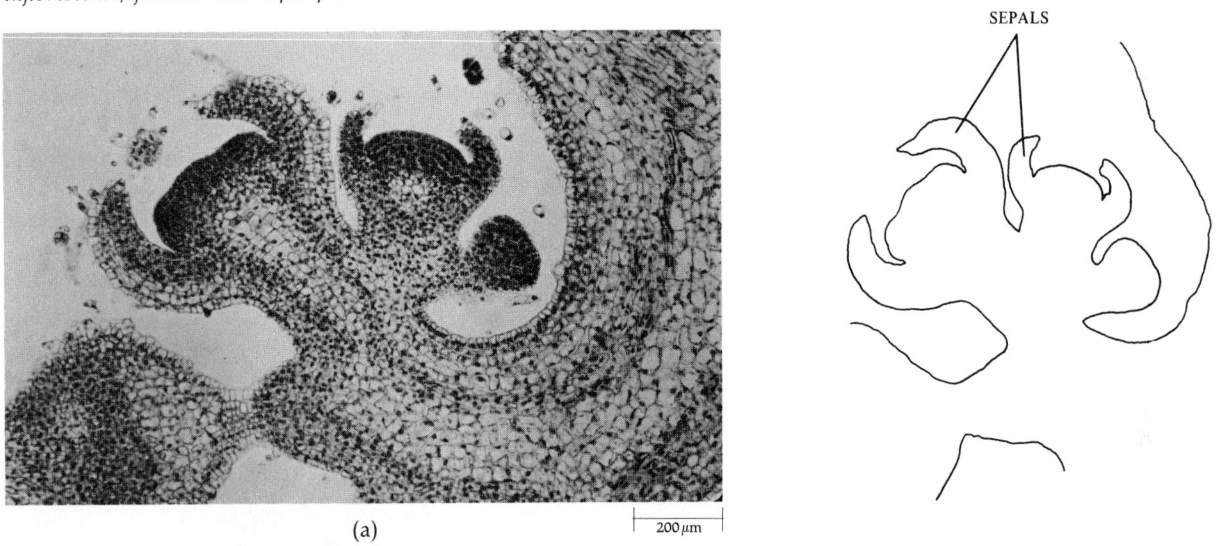

(a)

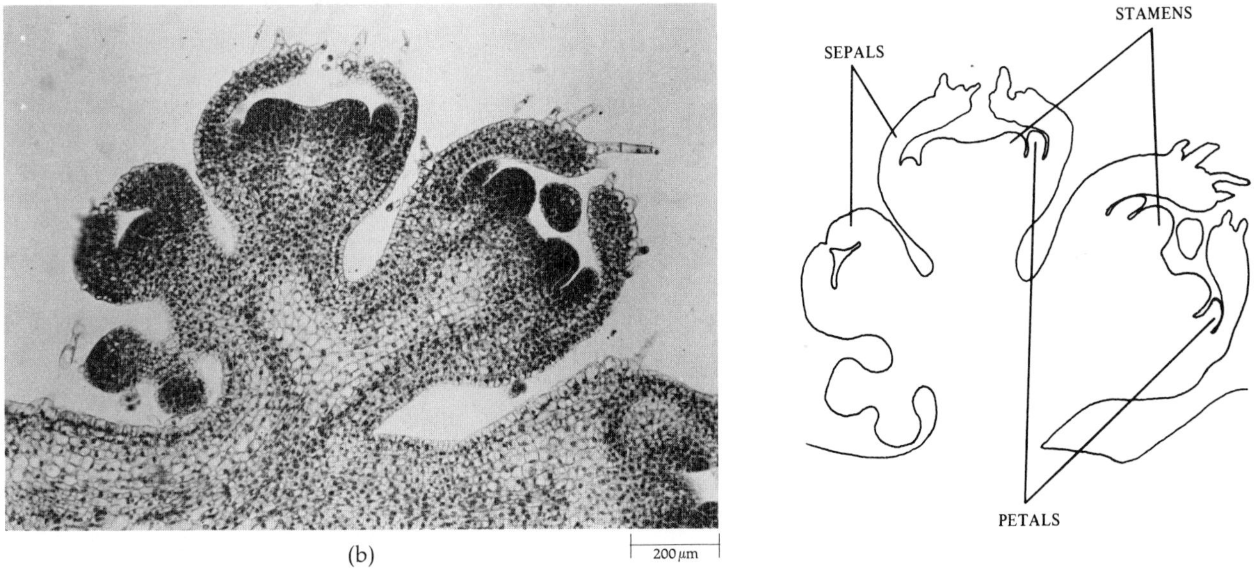

(b)

21-27
Development of the tomato (Lycopersicum esculentum) *flower, as seen in longitudinal sections (continued).* (a) *Flower cut in plane showing sepal, petal, stamen, and carpel primordia.* (b) *Flower cut in plane showing sepal, petal, stamen,*

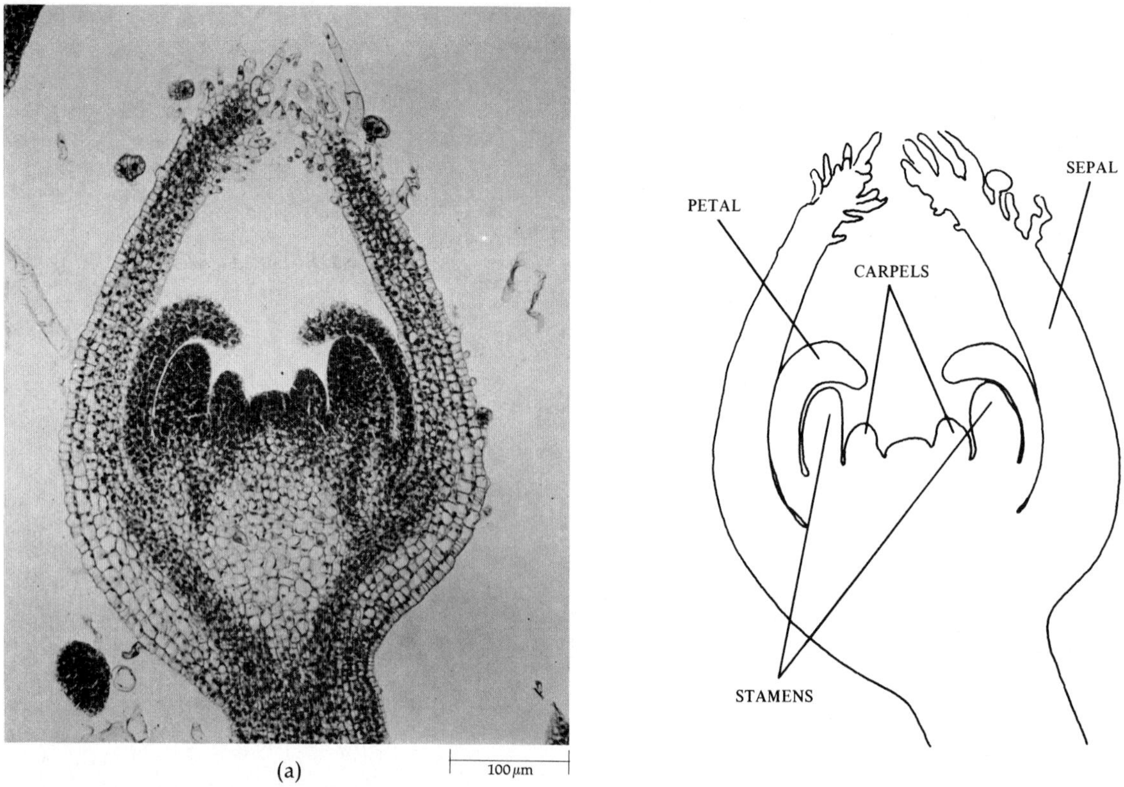

(a)

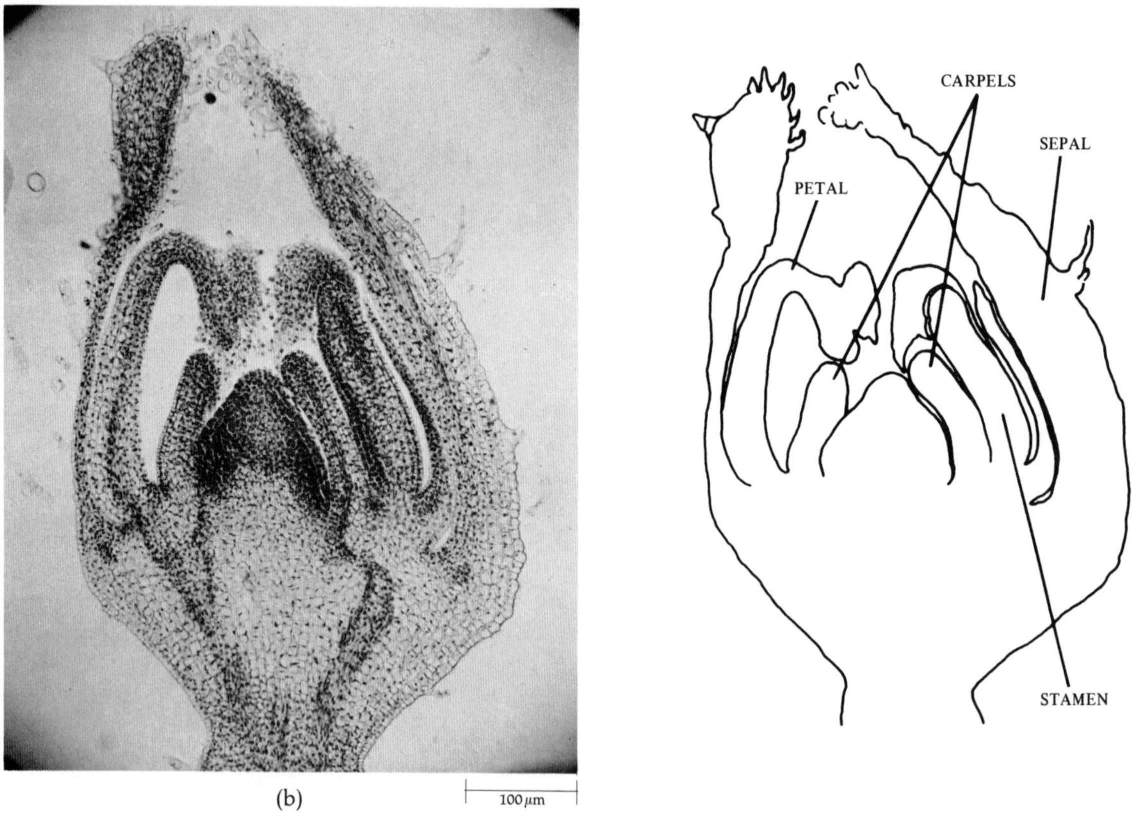

(b)

21–27 (cont'd)
and carpel primordia, but not cut through stamen on left, and barely through petal on right. (c) *Flower with developing ovules and style.* (d) *Fully-formed flower. Notice that the ovary is superior.*

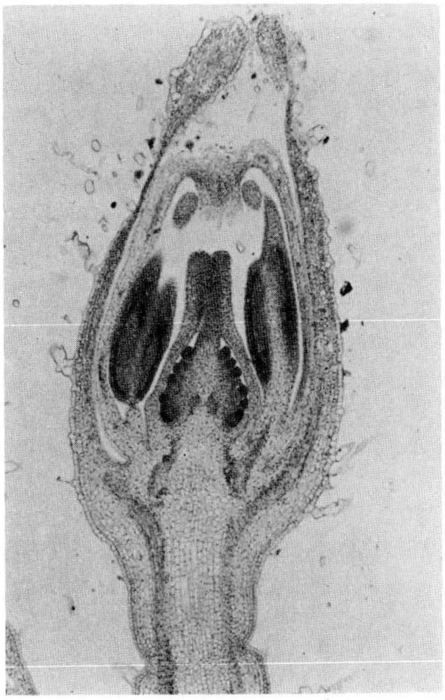

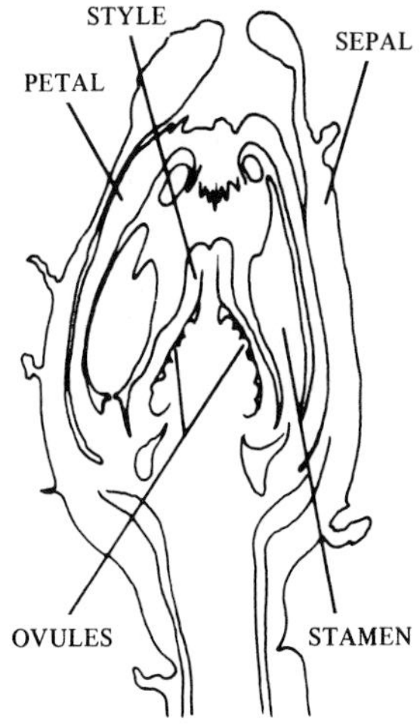

(c) 100 μm

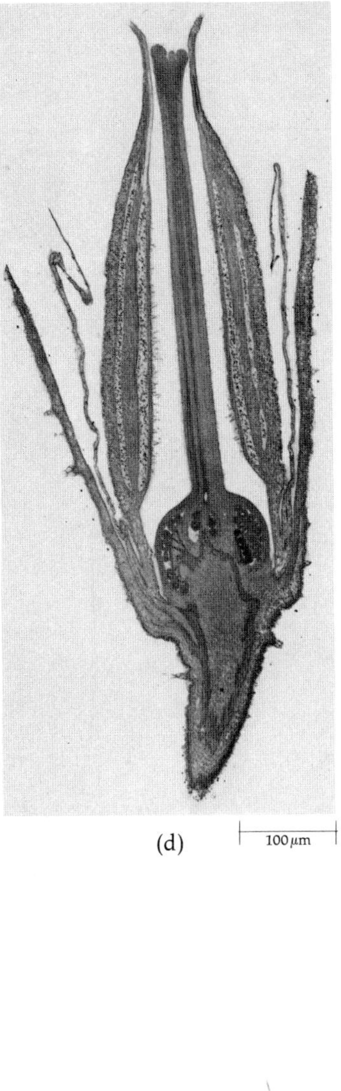

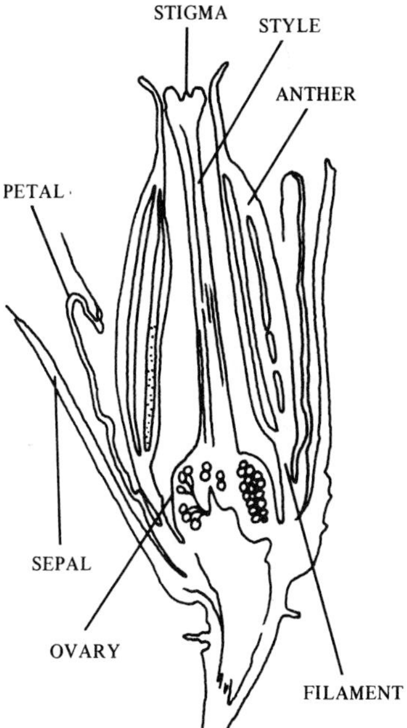

(d) 100 μm

21–28
Longitudinal section of young inflorescence (head) of fleabane (Erigeron), *a member of the Asteraceae. The inflorescence contains numerous flower primordia. The older flower primordia are on the periphery of the head, the younger in the center.*

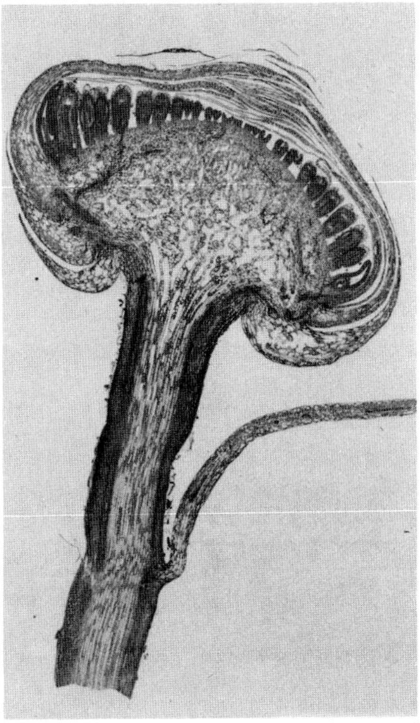

0.5 mm

21-29
Some stages in development of the fleabane (Erigeron) *flower, as seen in longitudinal sections.* (a) *Flower primordium with developing corolla.* (b) *Initiation of stamens, which are adnate to the corolla.* (c) *Flower with developing corolla, stamens, and carpels.* (d) *and* (e) *Flowers with developing ovules and styles. The two carpels overarch the cavity containing the ovule and then become prolonged into a solid style with a two-parted stigma. The calyx, or pappus, is initiated at about the same time as the anthers, but develops slowly. Notice that the fleabane flower has an inferior ovary.*

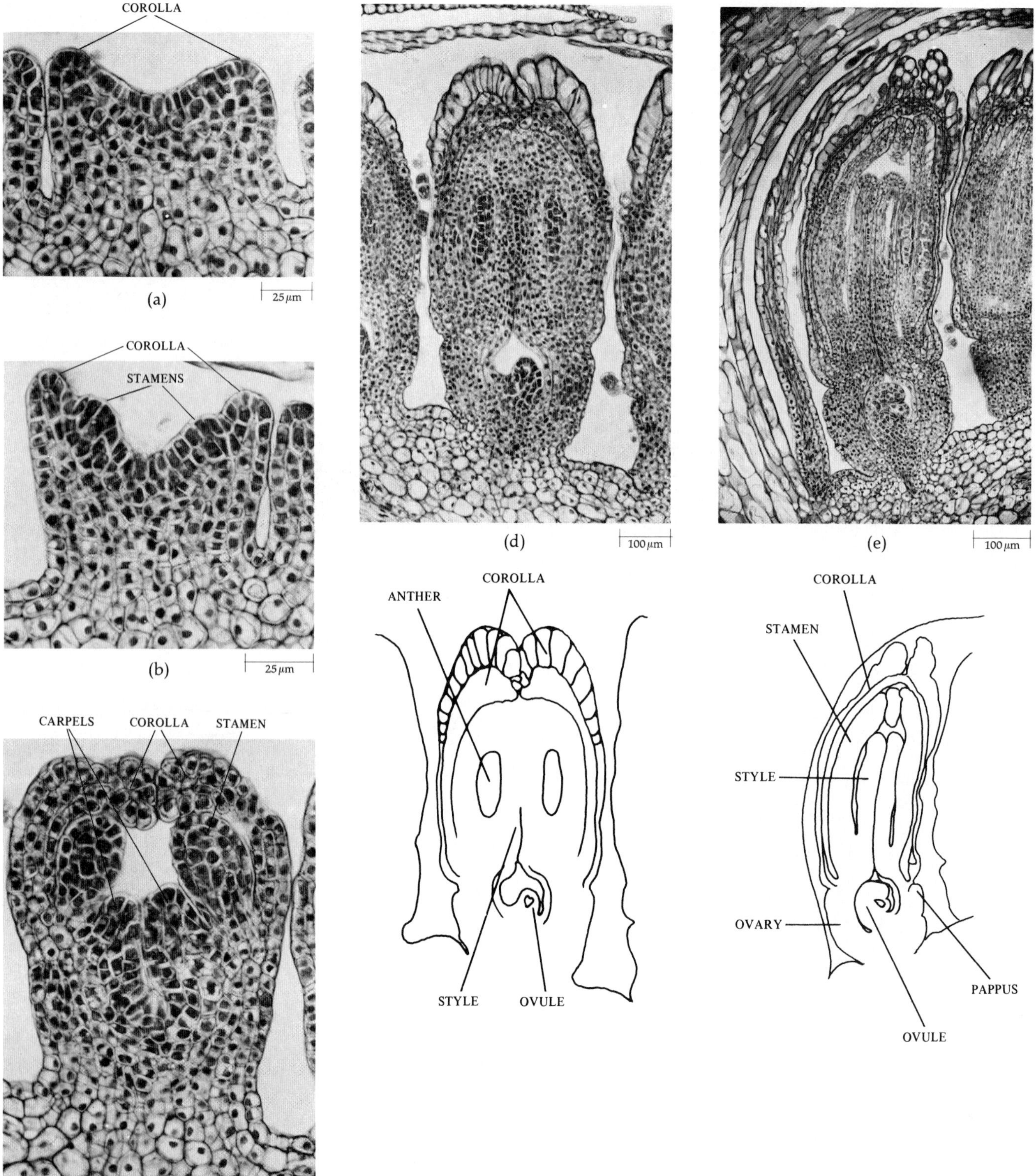

21–30
The tendrils of grape (Vitis) *are modified stems.*

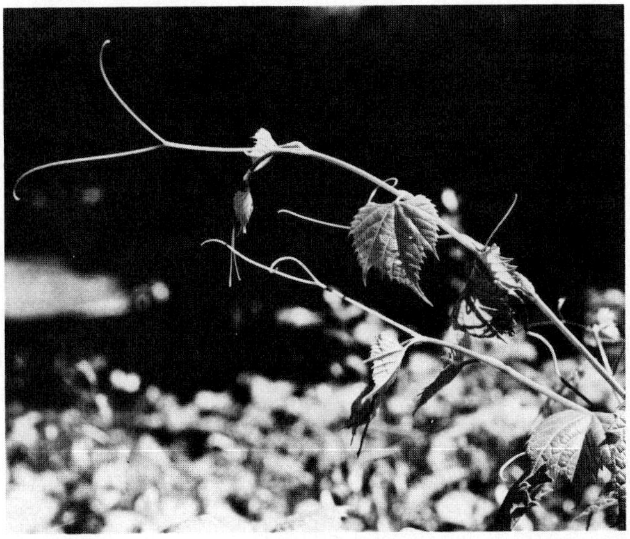

STEM AND LEAF MODIFICATIONS

The stems of some climbing plants coil themselves around the structure on which they are growing. Others produce modified branches called tendrils. In the English ivy the tendrils produce enlarged, cuplike structures, called holdfasts, at their tips. The tendrils of grape *(Vitis)* (Figure 21–30) and of Virginia creeper *(Parthenocissus quinquefolia)* also are modified stems that coil around the supporting structure. In grape, the tendrils sometimes produce small leaves or flowers.

Most tendrils are leaf modifications. In legumes, such as the garden pea *(Pisum sativum),* the tendrils constitute the terminal part of the pinnately compound leaf (Figure 21–15). Not all legumes form tendrils. One of these, the peanut, or goober *(Arachis hypogaea),* has another interesting adaptation. After fertilization takes place, the stamens and corolla of the flower fall off and the internode between ovary and receptacle begins to elongate. With continued elongation, the stalk bends downward and buries the developing fruit several centimeters into the ground, where it ripens. If the ovary is not buried, it withers and fails to mature.

Branches that assume the form of and closely resemble foliage leaves are called cladophylls. The filmy, leaflike branches of asparagus *(Asparagus officinalis)* are familiar examples of cladophylls (Figure 21–31). The thick and fleshy aerial shoots ("spears") of asparagus are the edible portion of the plant. The scales found on the spears are true leaves. If the asparagus plants are allowed to continue growing, cladophylls will develop in the axils of the minute, inconspicuous scales and will function as photosynthetic organs.

In some plants, the leaves are modified as spines, which are hard and dry and nonphotosynthetic (Figure 21–32). [The terms spine and thorn are often used in-

21–31
The filmy branches of the common edible asparagus (Asparagus officinalis) *resemble leaves. Such modified stems are called cladophylls.*

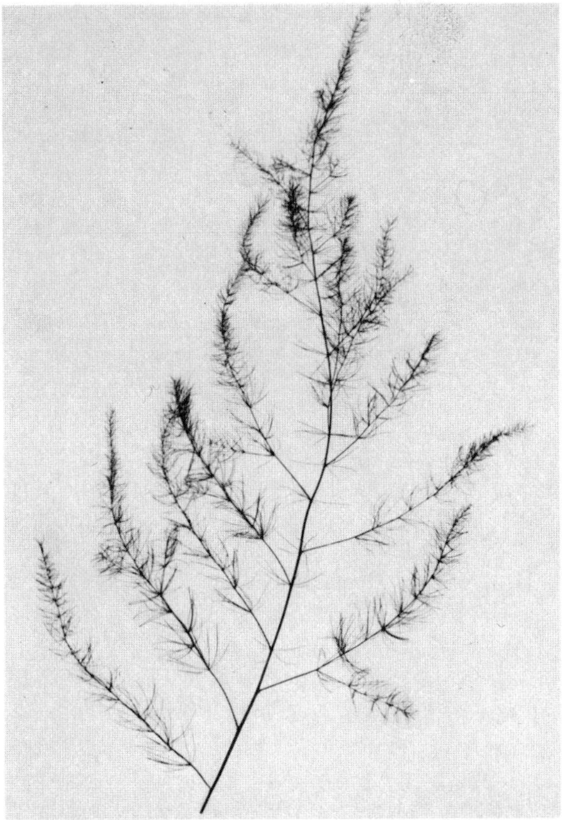

21–32
(a) *Spines, as in this cereus cactus, are modified leaves.* (b) *Thorns are modified branches. In this photograph of the hawthorn,* Crataegus, *you can see that the thorns arise in the axils of the leaves.*

(a)

(b)

terchangeably. However, thorns technically are modified branches that arise in the axils of leaves (Figure 21–32).] In some cacti, the branches resemble leaves (Figure 21–33).

Among the most spectacular of modified or specialized leaves are those of the carnivorous plants, such as the pitcher plant, sundew, and Venus flytrap, which capture insects and then digest them with enzymes secreted by the plant. The available nutrients then are absorbed by the plant (Figure 26–14, page 550).

Food Storage

Stems, like roots, serve food storage functions. Probably the most familiar type of specialized storage stem is the *tuber,* as exemplified by the Irish, or white, potato *(Solanum tuberosum).* In the white potato, tubers arise at the tips of *stolons* (underground branches of the aerial stem) of plants grown from seed. However, when segments (cuttings) of tubers are used for propagation, the tubers arise at the ends of long thin *rhizomes,* or underground stems (Figure 21–34). Except for vascular tissue, almost the entire mass of the tuber inside the "skin" (periderm) is storage parenchyma. The so-called eyes of the white potato are depressions containing groups of buds. The depression is the axil of a scalelike leaf.

A *bulb* is a large bud consisting of a small, conical stem with numerous modified leaves attached to it. The leaves are scalelike and contain thickened bases where food is stored. Adventitious roots arise from the bottom of the stem. Familiar examples of plants with bulbs are the onion (Figure 21–35a) and lily.

21–33
The branches of the spineless cactus Epiphyllum *resemble leaves.*

Although superficially similar to bulbs, corms consist primarily of stem tissue. Their leaves commonly are thin and much smaller than those of bulbs; consequently, the stored food of the corm is found in the fleshy stem. Such well-known plants as gladiolus (Figure 21–35b), crocus, and cyclamen produce corms.

Kohlrabi *(Brassica oleracea* var. *caulo-rapa)* is an example of an edible plant with a fleshy storage stem. The short, thick stem stands above the ground and bears several leaves with very broad bases (Figure 21–35c). The common cabbage *(Brassica oleracea* var. *capitata)* is closely related to kohlrabi. The so-called "head" of cabbage consists of a short stem bearing numerous thick, overlapping leaves. In addition to a terminal bud, several well-developed axillary buds may be found within the head.

The leaf stalks, or petioles, of some plants become quite thick and fleshy. Celery *(Apium graveolens)* and rhubarb *(Rheum rhaponticum)* are two familiar examples.

Water Storage: Succulency

Succulent plants are plants that have juicy tissues, that is, tissues specialized for the storage of water. Most of these plants, such as the cacti of the American deserts, the *Euphorbia* of similar appearance of the African deserts, and the century plant *(Agave)*, normally grow in arid regions, where the ability to store water is necessary for their survival. The green, fleshy stems of the cacti serve both as photosynthetic and storage organs. The water-storing tissue consists of large, thin-walled parenchyma cells that lack chloroplasts.

21–34
The Irish, or white, potato (Solanum tuberosum). (a) *Plant with potatoes, or tubers, attached to a rhizome (underground stem).* (b) *Detail of rhizome with potatoes. Notice the two, very young tubers arising from the rhizome above the larger potatoes.*

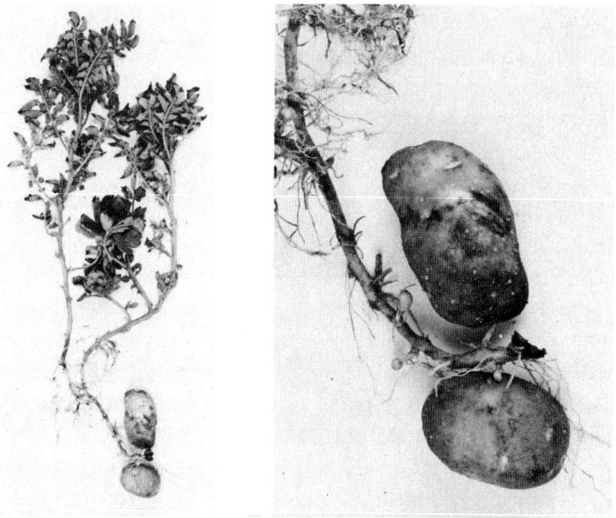

(a) (b)

21–35
Examples of modified leaves or stems. (a) *An onion* (Allium cepa) *bulb, consisting of a conical stem with scalelike, food-containing leaves attached to it. The leaves are the part of the onion we eat; the stem cannot be seen here.* (b) *A gladiolus* (Gladiolus grandiflora) *corm, a fleshy stem with small, thin leaves.* (c) *The fleshy storage stem of kohlrabi* (Brassica oleracea *var.* caulo-rapa).

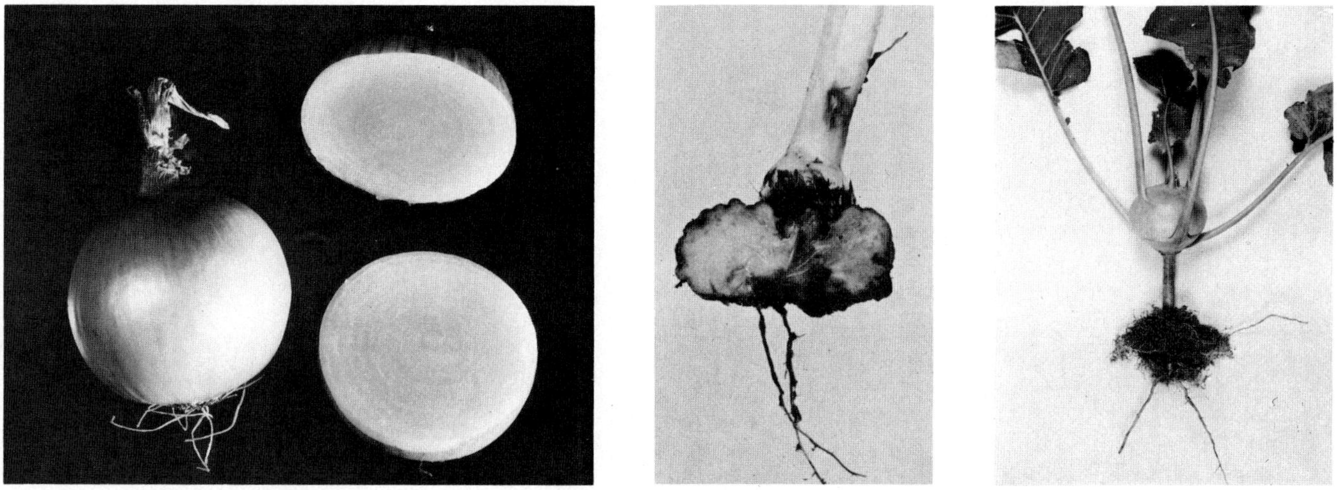

(a) (b) (c)

In the century plant, the leaves are succulent. As in succulent stems, nonphotosynthetic parenchyma cells of the ground tissue constitute the water-storing tissue. Other examples of plants with succulent leaves are the "ice plant" *(Mesembrysanthemum crystallinum),* the stonecrops *(Sedum),* and some species of *Peperomia.* In the ice plant, large epidermal cells called water vesicles, which superficially resemble beads of ice, serve for water storage (see Figure 19-22c, page 414). The water-storing cells of the *Peperomia* leaf are part of a multiple (several layered) epidermis derived by anticlinal divisions of the protoderm (see Figure 19-24, page 415).

SUMMARY

The vegetative shoot apices of most flowering plants have a tunica-corpus type of organization consisting of one or more peripheral layers of cells (the tunica) and an interior (the corpus). Although the primary tissues of the stem pass through periods of growth similar to those of the root, the stem cannot be divided into regions of cell division, elongation, and maturation in the same manner as roots are. The stem increases in length largely by internodal elongation.

As in the root, the apical meristem of the shoot gives rise to protoderm, ground meristem, and procambium, which develop into the primary tissues. Three basic variations exist in stems with regard to the relative distribution of ground and primary vascular tissues: The primary tissues may develop (1) as a continuous hollow cylinder, (2) a cylinder of interconnected strands, or (3) an anastomosing (that is, interconnecting and branched) system of strands scattered throughout the ground tissue. Regardless of the type of organization, the phloem is commonly located outside the xylem.

Leaves have their origin on the sides of the apical meristem, and their position on the stem determines the patterns of the vascular system in the stem. Leaves are determinate in growth; that is, their development is of relatively short duration. Vegetative shoot apices exhibit unlimited or indeterminate growth.

In dicotyledons, most leaves consist of a blade and petiole. The blade of some leaves is divided into leaflets. Stomata are commonly more numerous on the lower than the upper surface of the leaf. The ground tissue, or mesophyll, of the leaf is specialized as a photosynthetic tissue and, in mesophytes, is differentiated into palisade parenchyma and spongy parenchyma. The mesophyll is thoroughly permeated by air spaces and by veins, which are composed of xylem and phloem surrounded by a parenchymatous bundle sheath. The xylem commonly occurs on the upper side of the vein, the phloem on the lower. In many plants, leaf abscission is preceded by formation of an abscission zone at the base of the petiole.

The change in the type of structure found in the root to that in the shoot takes place in a region of the plant axis of the embryo and young seedling called the transition region.

At flowering, the vegetative shoot apex is directly transformed into a reproductive apex.

Stem and leaf modifications include tendrils, found in many climbing plants. Some tendrils are leaf modifications; others are modified stems. Cacti and other plants have spines, which are modified leaves. Thorns resemble spines, but are modified branches.

Stems, like roots, may serve food-storage functions. Examples of fleshy stems are tubers, bulbs, and corms. Water-storing plants are known as succulents. The water-storing tissue of succulent plants is made up of large parenchyma cells. Stems or leaves or both may be succulent.

CHAPTER 22

Secondary Growth

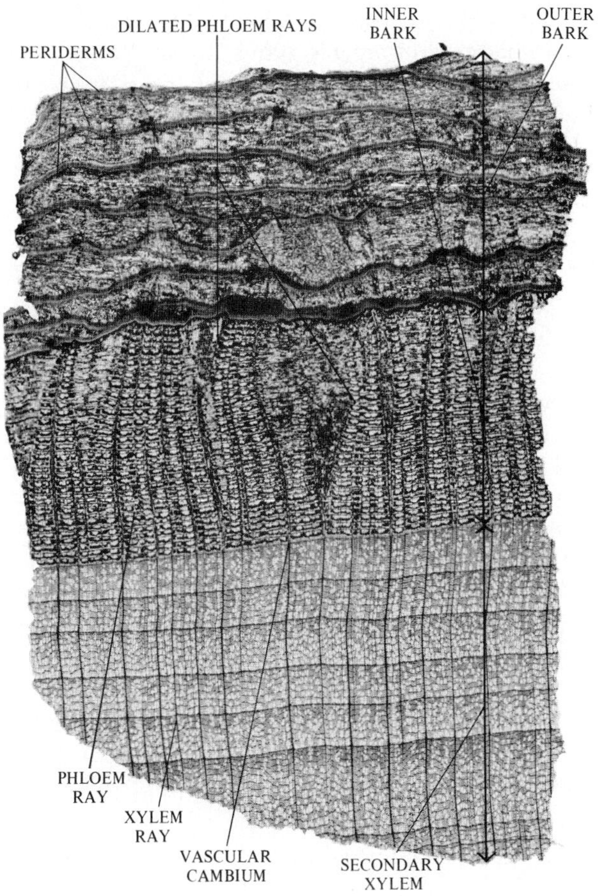

22–1
Transverse section of the bark and some secondary xylem from an old stem of linden (Tilia americana).

In many plants—most monocotyledons and the extreme herbaceous dicotyledons such as *Ranunculus*—growth in a given part of the plant body ceases with maturation of the primary tissues. At the other extreme are the gymnosperms and woody dicotyledons, in which the roots and stems continue to increase in girth in regions that are no longer elongating. This increase in thickness or girth of the plant body—secondary growth—results from activity of the two *lateral meristems*, the *vascular cambium* and the *cork cambium*.

Herbs, or herbaceous plants, are plants with shoots that undergo little or no secondary growth. In temperate regions either the shoot or the entire plant lives for only one season, depending on the species. Woody plants—trees and shrubs—live for several or many years. Each year new primary growth is resumed, and additional secondary tissues are added to the older plant parts through reactivation of the lateral meristems. Although most monocots lack secondary growth, some (such as the palms) may develop thick stems by primary growth alone.

Plants are often classified according to their seasonal growth cycles as annuals, biennials, or perennials. In the *annuals*, which include many of our weeds, wild flowers, garden flowers, and vegetables, the entire cycle from seed to vegetative plant to flowering plant to seed again takes place within a single season—which may be only a few weeks in length. In the annuals, only the dormant seed bridges the gap between one season and the next. In *biennials*, two seasons are needed from seed germination to seed formation. The first season of growth ends with the formation of a root, a short stem, and a rosette of leaves near the soil surface. In the second growing season, flowering, fruiting, seed formation, and death occur, completing the life cycle. In temperate regions, annuals and biennials seldom become woody, although both their stems and roots may undergo some secondary growth.

Perennials are plants in which the vegetative structures live year after year. The herbaceous perennials pass unfavorable seasons as dormant underground roots, rhizomes, bulbs, or tubers. Others, the woody perennials, which include vines, shrubs, and trees, survive above ground but usually stop growing during the unfavorable seasons. Woody perennials flower only when they become adult plants, which may take many years. The horse chestnut, *Aesculus hippocastanum,* for instance, does not flower until it is about 25 years old. *Puya raimondii,* a large (up to 10 meters high) relative of the pineapple that is found in the Andes, takes about 150 years to flower. Many woody plants are deciduous, losing all their leaves at the same time and developing new leaves from buds when the season again becomes favorable for growth. In evergreen trees and shrubs, leaves are also lost and replaced, but not simultaneously.

THE VASCULAR CAMBIUM

Unlike the many-sided initials of the apical meristems, which contain dense cytoplasm and large nuclei, the meristematic cells of the vascular cambium are highly vacuolated. They exist in two forms: as vertically elongated *fusiform initials,* and as horizontally elongated or squarish *ray initials.* The fusiform initials are much longer than wide and appear flattened or brickshaped in transverse section (Figure 22–2). In white pine, *Pinus strobus,* the fusiform initials average 3.2 millimeters long; in the apple, *Malus sylvestris,* 0.53 millimeters (Figure 22–3); and in the black locust, *Robinia pseudo-acacia,* 0.17 millimeters (Figure 22–4).

It is through periclinal divisions of the cambial initials and their derivatives that secondary xylem and secondary phloem are produced. In other words, the cell plate that forms between the dividing cambial initials (the dashed lines in Figure 22–2) is parallel to the surface of the stem. If the derivative of a cambial initial is divided off toward the outside of the stem, it eventually becomes a phloem cell, and if it divides off toward the

22–2
Growth in the vascular cambium. The vascular cambium is made up of two types of cells, ray initials and fusiform initials. The fusiform initials divide along the long axis of the cell parallel with the surface of the root or stem. One daughter cell remains meristematic (the initial), and the other (the derivative of the initial) eventually develops into a cell or cells in the vascular tissue. The derivatives of the initials commonly divide one or more times before differentiating into elements of the secondary xylem or secondary phloem. Cells that divide toward the inner surface of the vascular cambium become xylem elements, and those that divide toward the outer surface become phloem elements. The ray initials divide to form vascular rays, which lie at right angles to the long axis of the root or stem.

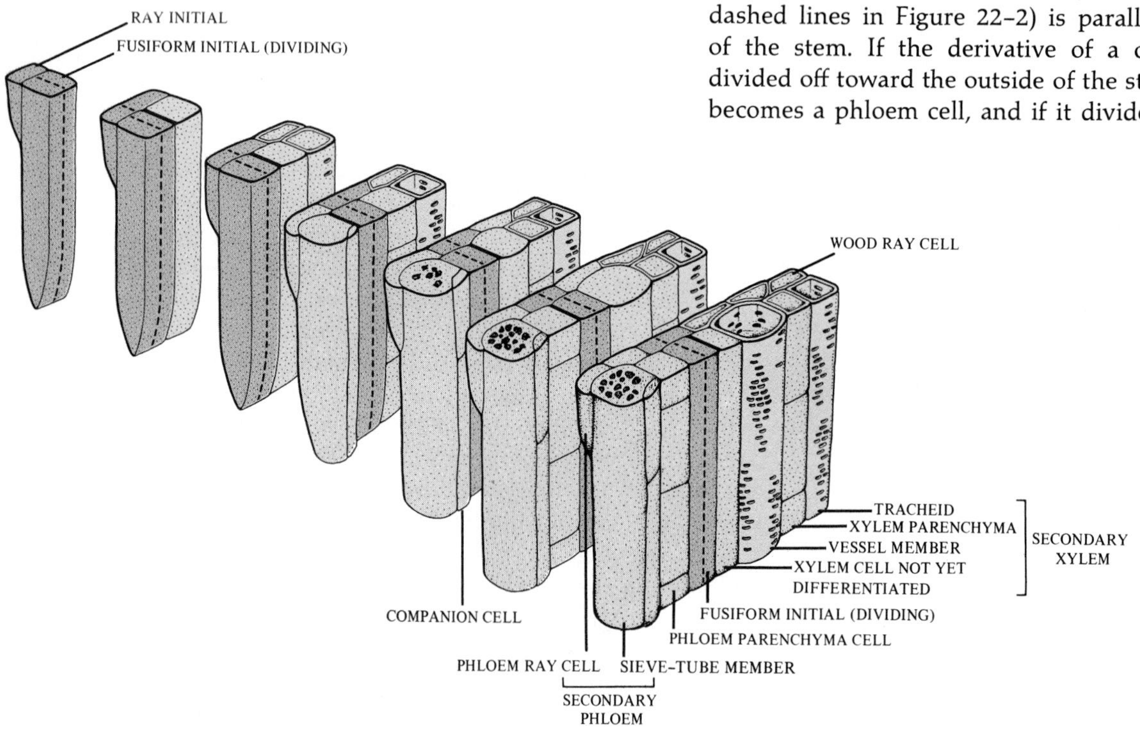

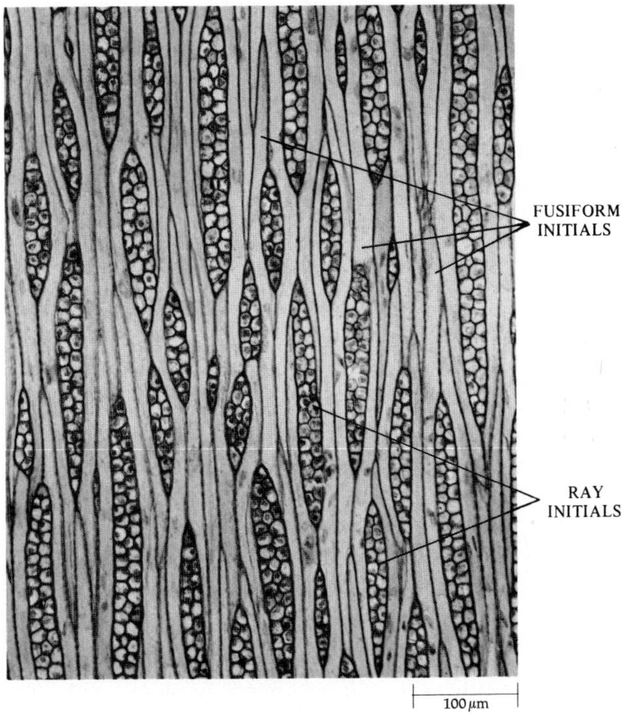

22–3
Vascular cambium of the apple (Malus sylvestris) *tree, tangential view. Tangential sections are cut at right angles to the rays, and so we see the rays here in cross-section.*

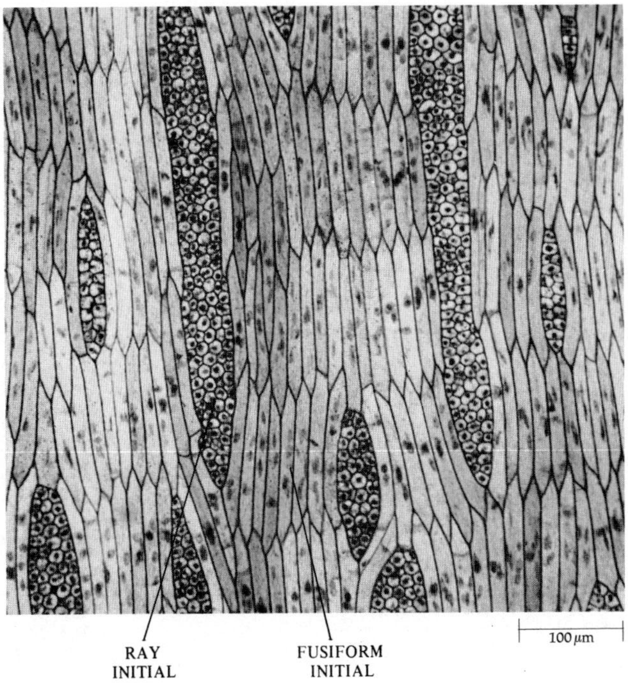

22–4
Vascular cambium of the black locust (Robinia pseudo-acacia) *tree, tangential view.*

inside, it becomes a xylem cell. In this manner a long, continuous radial file, or row, of cells is formed, extending from the cambial initial outwards to the phloem and inwards to the xylem (Figure 22–2).

The xylem and phloem cells produced by the fusiform initials have their long axes oriented vertically and make up what is known as the *axial system* of the secondary vascular tissues. The ray initials produce horizontally oriented *ray cells,* which form the *vascular rays* or *radial system* (Figure 22–2). The rays are composed largely of parenchyma cells and are variable in length. Water passes from the secondary xylem to the vascular cambium and the secondary phloem through the vascular rays, and nutrients move from the secondary phloem to the cambium and to the living cells of the secondary xylem through the same cells. The rays also serve as storage centers for such substances as starch and lipids.

In a restricted sense, the term vascular cambium is used to refer only to the cambial initials, of which there is one per radial file. However, sometimes it is difficult, if not impossible, to distinguish between the initials and their immediate derivatives, which may remain meristematic for a period of time. Even in the winter condition, when the cambium is inactive, or dormant, several layers of undifferentiated cells of similar appearance may occur between xylem and phloem. Consequently, some botanists use the term vascular cambium to refer to the initials and their immediate derivatives, which are indistinguishable from the initials. Others refer to this region as the *cambial zone.*

As the vascular cambium adds cells to the secondary xylem and the core of xylem increases in width, the cambium is displaced outward. In order to accommodate to this, the vascular cambium undergoes an increase in circumference. The increase in circumference of the cambium is accomplished by anticlinal divisions of the initials. Along with an increase in the number of fusiform initials, new ray initials and rays are added, so that a fairly constant ratio of rays to fusiform cells is maintained in the secondary vascular tissues. Obviously, the developmental changes that take place in the cambium are exceedingly complex and must be closely correlated.

In temperate regions, the vascular cambium is dormant during winter and becomes reactivated in the spring. New increments, or growth layers, of secondary xylem and secondary phloem are laid down during the growing season. Reactivation of the vascular cambium is triggered by the expansion of the buds and resumption of their growth. Apparently, the hormone auxin (Chapter 23), produced by the developing shoots, moves downward in the stems and stimulates resumption of cambial activity. Other factors are also involved in cambial reactivation and in continued normal growth of the cambium.

22-5
Comparison of primary and secondary structure in root and stem of a woody dicot. (a) *Root and stem at completion of primary growth;* (b) *origin of vascular cambium;* (c) *after formation of some secondary xylem and secondary phloem in root and stem, and periderm formation in root;* (d) *at end of first year's growth, showing effect of secondary growth—including periderm formation—on the primary plant body.*

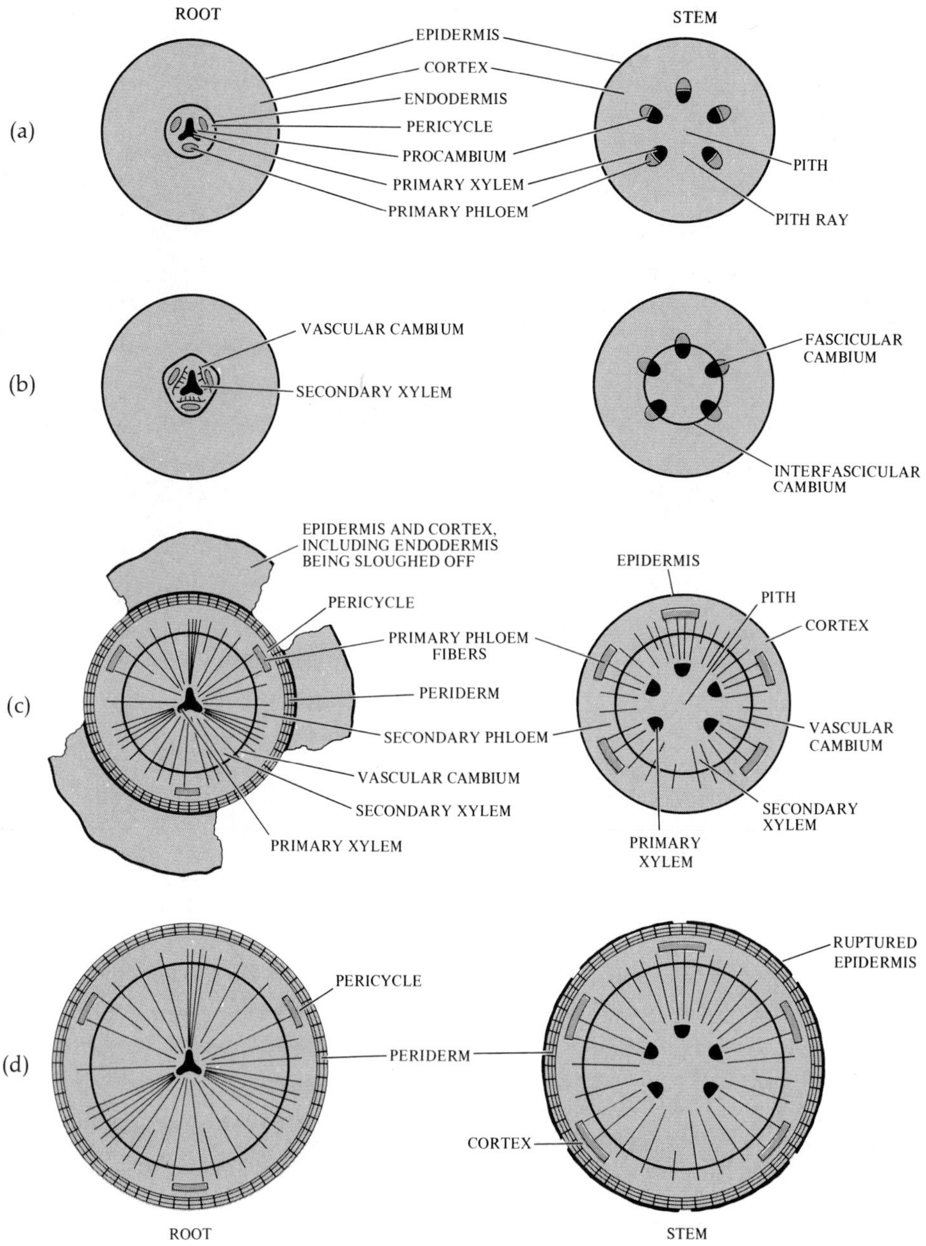

EFFECT OF SECONDARY XYLEM AND PHLOEM FORMATION ON THE PRIMARY PLANT BODY

Root

In roots, the vascular cambium is initiated by procambial cells that remain meristematic between the primary xylem and primary phloem. Thus, depending on the number of phloem strands present in the root, two or more independent regions of cambial activity are initiated more or less simultaneously (Figure 22-5). Soon afterward, the pericyclic cells opposite the protoxylem poles divide periclinally, and the inner sister cells contribute to the vascular cambium. Now the cambium completely surrounds the core of xylem.

As soon as it is formed, the vascular cambium opposite the phloem strands begins to produce secondary xylem, and in the process the strands of primary phloem are displaced from their positions between the ridges of primary xylem. By the time the cambium opposite the protoxylem poles is actively dividing, the cambium is circular in outline and the primary phloem has been separated from the primary xylem (Figure 22-5).

By repeated divisions toward the inside and outside, secondary xylem and secondary phloem, respectively, are added to the root (Figures 22-5 and 22-6). In some roots the vascular cambium derived from the pericycle forms wide rays, whereas narrower rays are produced in other parts of the secondary vascular tissues.

With increase in width of the secondary xylem and

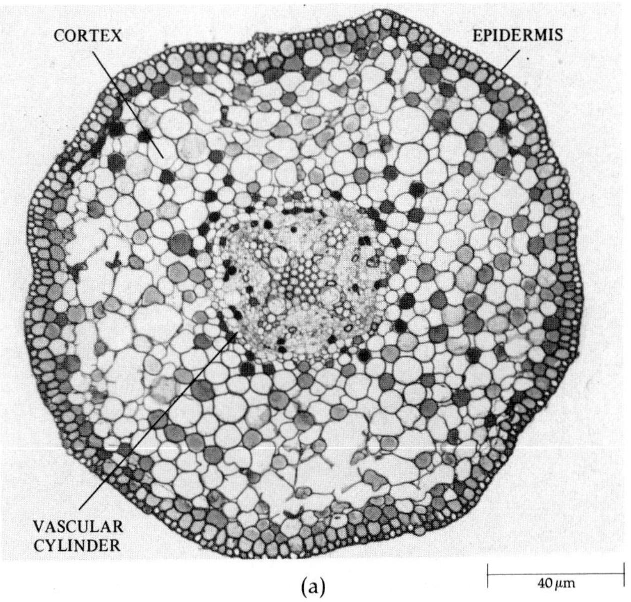

(a)

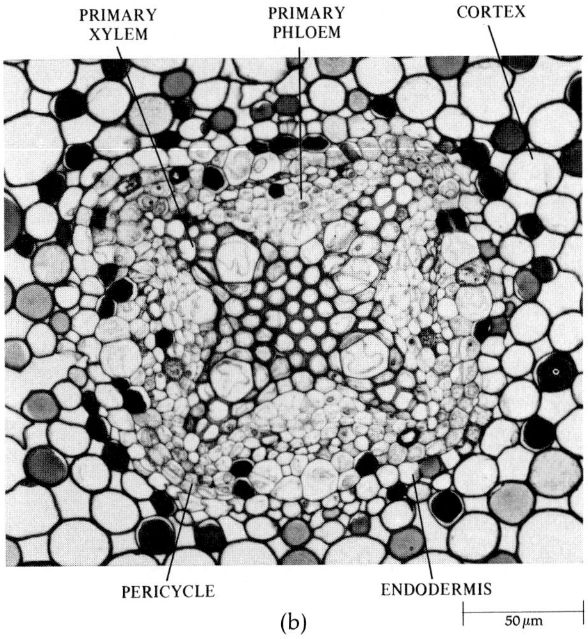

(b)

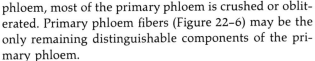

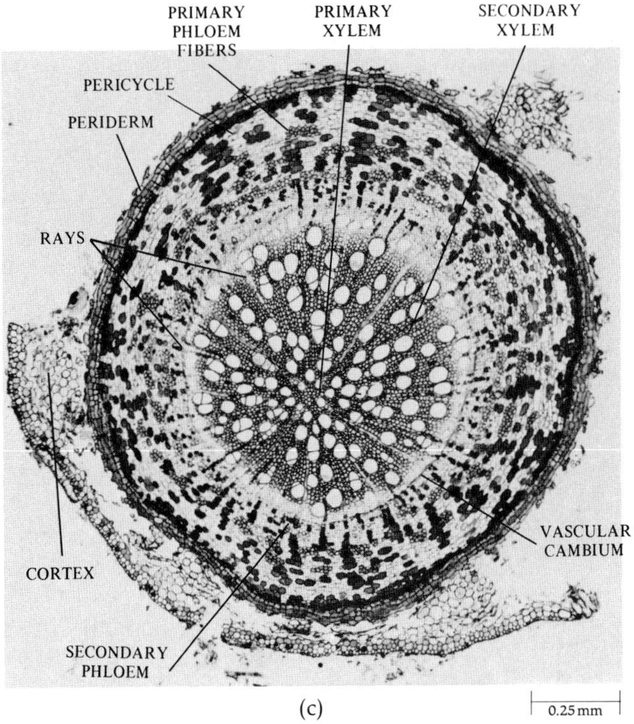

(c)

22–6
Transverse sections of the willow (Salix) *root, which becomes woody.* (a) *Overall view of root near completion of primary growth.* (b) *Detail of primary vascular cylinder.* (c) *Overall view of root at end of first year's growth, showing effect of secondary growth on primary plant body.*

phloem, most of the primary phloem is crushed or obliterated. Primary phloem fibers (Figure 22–6) may be the only remaining distinguishable components of the primary phloem.

Stem

As we mentioned previously, the vascular cambium of the stem arises from the procambium that remains undifferentiated between the primary xylem and primary phloem and from parenchyma of the interfascicular regions (page 443). The part of the cambium arising within the vascular bundles is called fascicular cambium and that arising in the interfascicular regions, or pith rays, interfascicular cambium. The vascular cambium of the stem, unlike that of the root, is essentially circular in outline from its inception (Figure 22–5).

In woody stems, the production of secondary xylem and secondary phloem results in formation of a cylinder of secondary vascular tissues, with the rays extending radially through the cylinder (Figure 22–5). Commonly, much more secondary xylem than phloem is produced any given year. This is also true in the root. And, as in the root, with secondary growth the primary phloem is pushed outward and its thin-walled cells are destroyed. Only the thick-walled primary phloem fibers remain intact (Figure 22–8).

Figures 22–7 and 22–8 show the elderberry, *Sambucus canadensis,* stem in two stages of secondary growth. (See Chapter 21 for primary growth of the *Sambucus* stem.)

22–7
Transverse section of elderberry (Sambucus canadensis) *stem in which a small amount of secondary growth has taken place. A cork cambium has not yet been formed.*

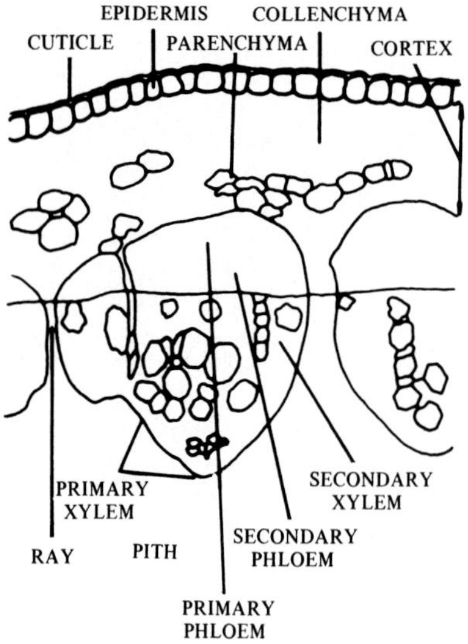

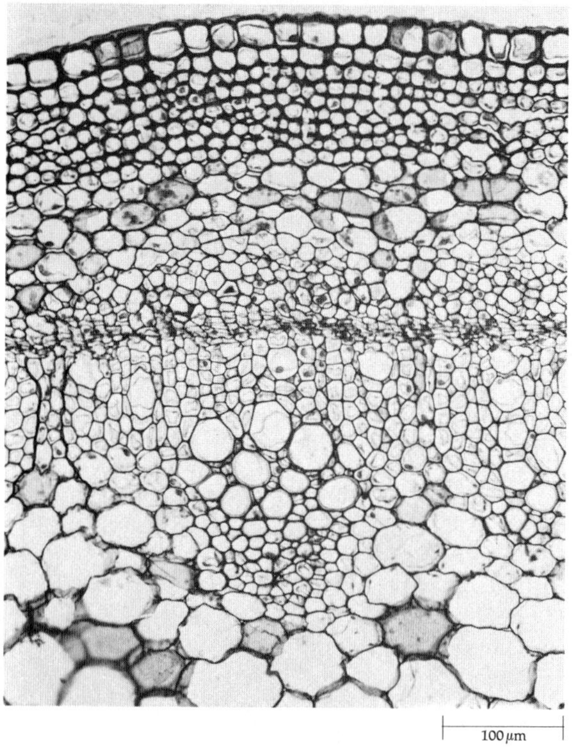

22–8
Transverse section of elderberry (Sambucus canadensis) *stem at end of first year's growth.*

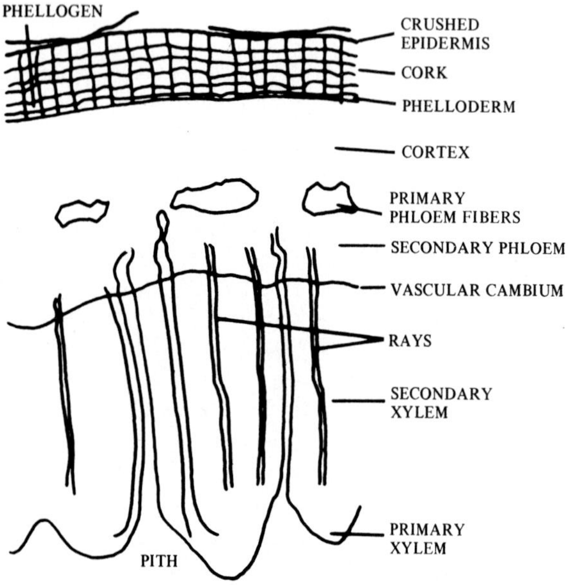

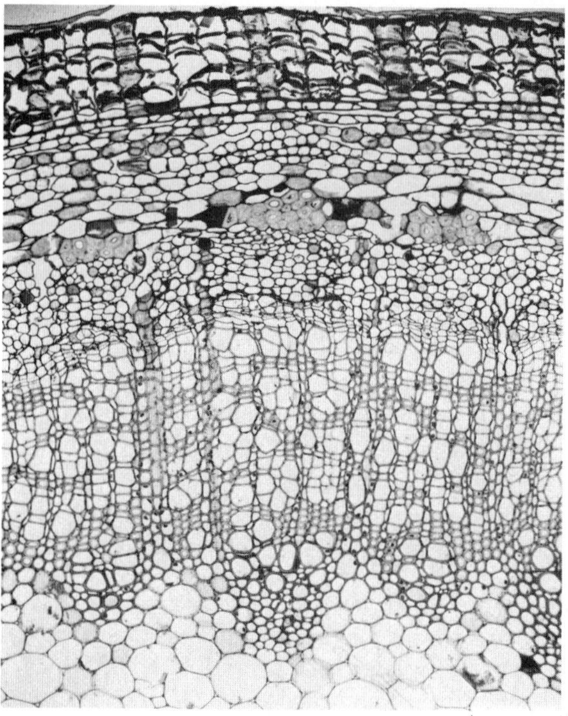

22-9
Transverse sections of linden (Tilia americana) *tree stems.* (a) *One-year-old stem.* (b) *Two-year-old stem.* (c) *Three-year-old stem. Numbers indicate growth rings in secondary xylem.*

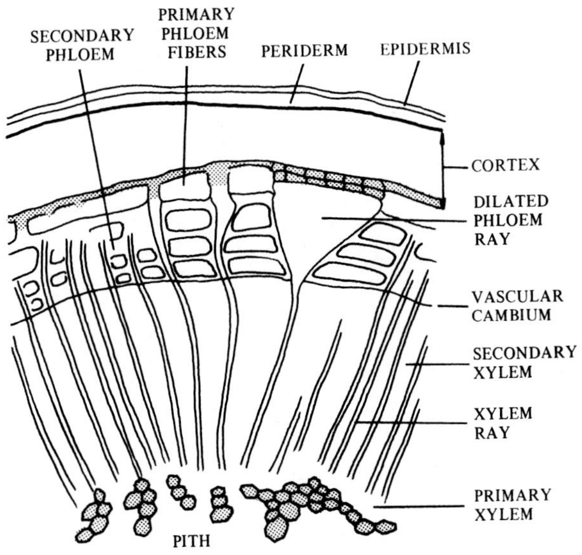

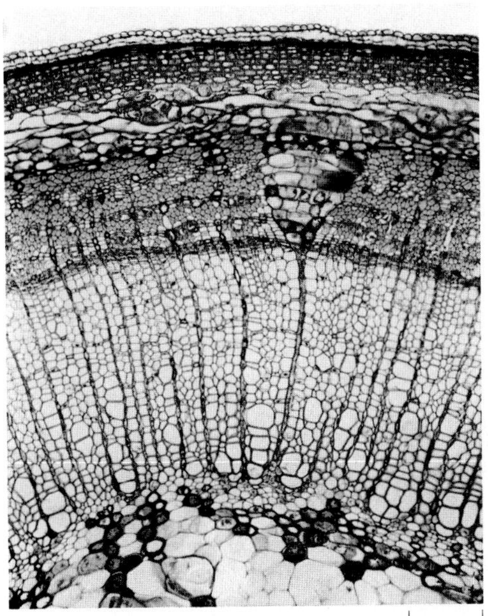

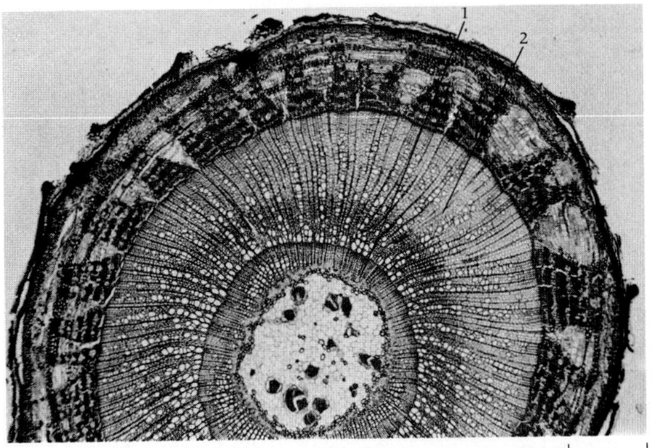

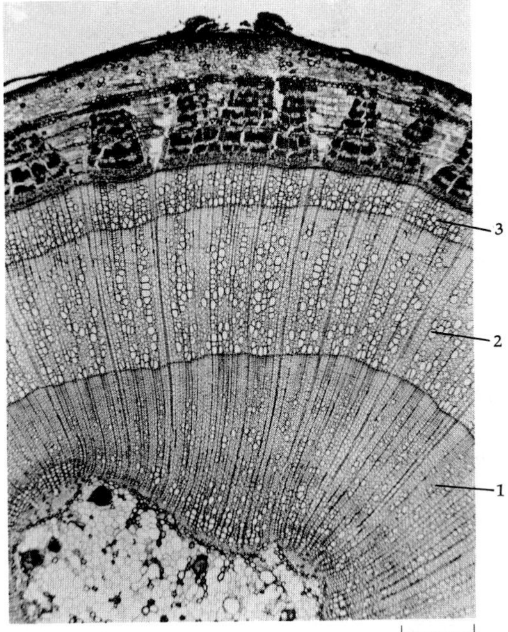

Only a small amount of secondary xylem and secondary phloem has been produced in the stem of Figure 22-7. Figure 22-8 shows a stem at the end of the first-year's growth. Note that considerably more secondary xylem than secondary phloem has been formed. The thick-walled cells outside the secondary phloem are primary phloem fibers.

Figure 22-9 shows one-, two-, and three-year-old stems of linden, *Tilia americana.* In Chapter 21, the stem of *Tilia* was given as an example of one in which the primary tissues arise as an almost continuous hollow cylinder; thus most of the vascular cambium in the *Tilia* stem is fascicular in origin. Some of the rays in the secondary phloem of *Tilia* become very wide as the stem increases in girth. This is one way in which the tissues outside the vascular cambium keep up with the increase in girth of the core of xylem.

The vascular cambia and secondary tissues of root and stem are continuous with one another. Hence there is no transition region in the secondary plant body.

The Periderm and Cork

In most woody roots and stems, cork formation usually follows the initiation of secondary xylem and secondary phloem production, and the cork tissue replaces the epidermis as the protective covering on the portion of the plant part concerned. Cork, or *phellem* as it is technically termed, is formed by a cork cambium, or *phellogen,* which may also form *phelloderm* ("cork skin"). The cork is formed toward the outer surface, and the phelloderm toward the inner surface of the cork cambium (Figures

22–10
Some stages of periderm and lenticel development in elderberry (Sambucus canadensis), *as seen in transverse sections.* (a) *Newly formed periderm beneath epidermis; collenchyma and parenchyma of cortex.* (b) *Initiation of lenticel; collenchyma and parenchyma of cortex beneath developing lenticel.* (c) *Well-developed lenticel. The phelloderm in* Sambucus *generally consists of a single layer of cells.*

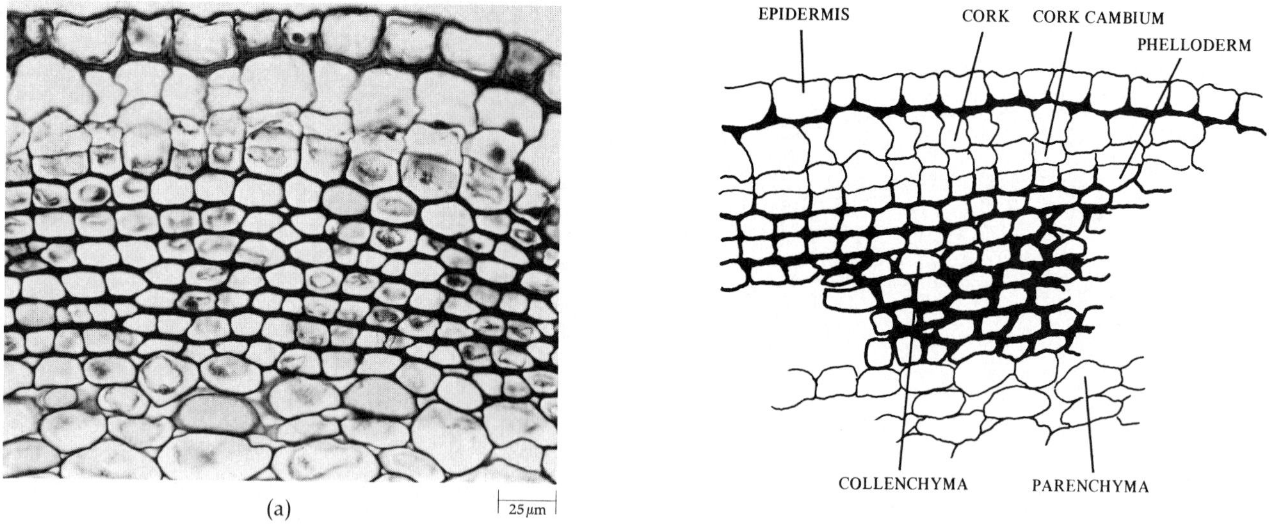

(a)

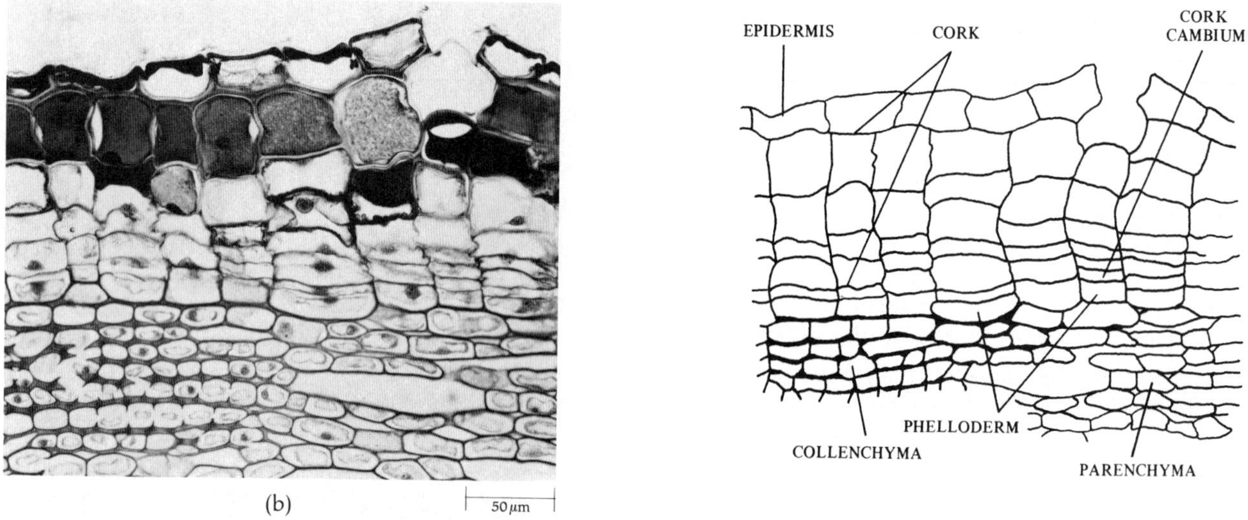

(b)

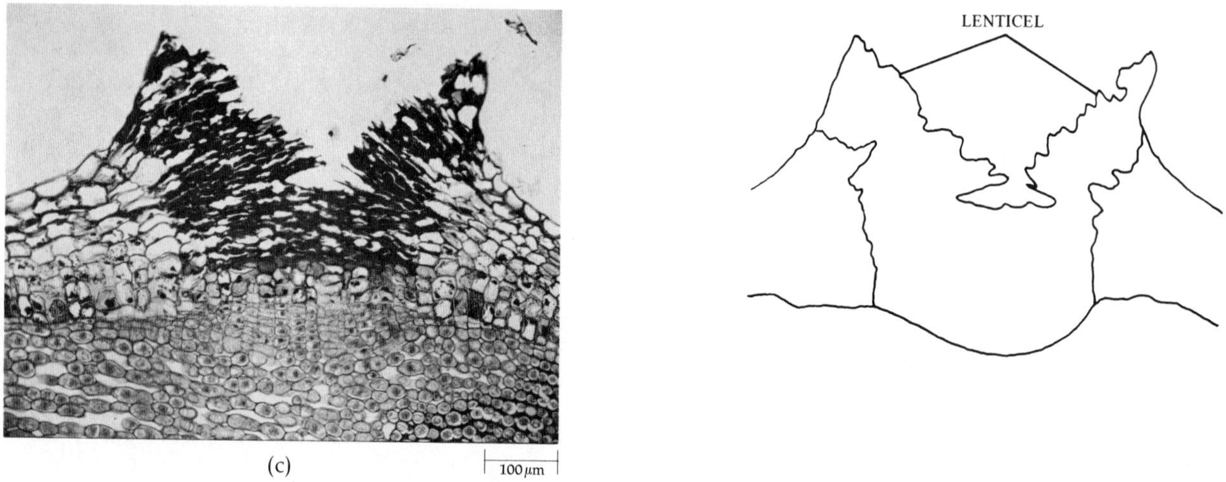

(c)

22–11
Lentícel of Aristolochia, *a vine, stem, as seen in transverse section. Unlike* Sambucus, *the phelloderm of* Aristolochia *consists of several layers of cells.*

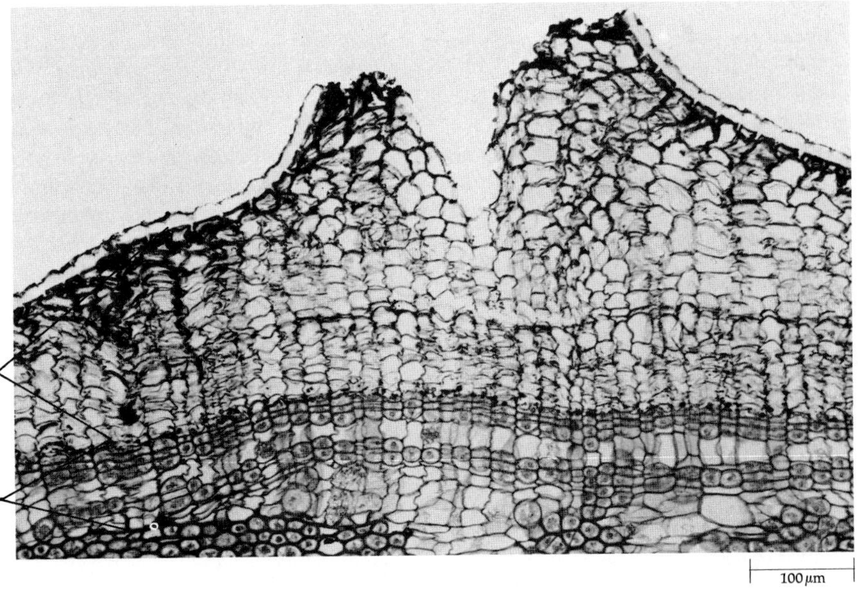

22–10 and 22–11). Together, these three tissues—cork, cork cambium, and phelloderm—make up the *periderm.*

In most dicots and gymnosperms, the first periderm commonly appears during the first year of growth of the root or stem, in those portions that are no longer elongating. In stems, the first cork cambium most commonly originates in a layer of cortical cells immediately below the epidermis (Figures 22–5 and 22–10), although in many species it originates in the epidermis. In roots, the first cork cambium arises through periclinal division of pericycle cells, the outer sister cells combining to form a complete cylinder of cork cambium. Afterwards, the remaining cells of the pericycle may proliferate below the periderm and give rise to a tissue that resembles a cortex (Figures 22–5 and 22–6).

Repeated divisions of the cork cambium result in the formation of radial rows of compactly arranged cells, most of which are cork cells (Figures 22–10 and 22–11). During differentiation of the cork cells, their inner walls are lined by a relatively thick layer of a fatty substance, suberin, which makes the tissue highly impermeable to water and gases. The walls of the cork cells also may become lignified. At maturity, the cork cells are dead.

The cells of the phelloderm are living at maturity, lack suberin, and resemble cortical parenchyma cells. The phelloderm cells may be distinguished from cortical cells by their inner position in the radial rows of other periderm cells (Figure 22–11).

With the formation of the first periderm in the root, the cortex—including the endodermis—and epidermis are isolated from the rest of the root, eventually die, and are sloughed off. Because the first periderm of the stem usually arises immediately below the epidermis, the cortex of the stem is not sloughed off during the first year (Figures 22–5 and 22–6), although the epidermis dries up and peels off.

At the end of the first year's growth, the following tissues are present in a woody root (from outside to inside): remnants of the epidermis and cortex, periderm, pericycle, primary phloem (fibers and crushed soft-walled cells), secondary phloem, vascular cambium, secondary xylem, and primary xylem; in the stem: remnants of the epidermis, periderm, cortex, primary phloem (fibers and crushed soft-walled cells), secondary phloem, vascular cambium, secondary xylem, primary xylem, and pith (Figure 22–5).

The Lenticels

In the preceding discussion, we noted that the suberin-containing cork cells are compactly arranged and, as a tissue, present an impermeable barrier to water and gases. However, the inner tissues of the stem, like all metabolically active tissues, need to exchange gases with the surrounding air. In stems and roots containing periderms, this necessary gas exchange is accomplished through *lenticels* (Figures 22–10 and 22–11), portions of the periderm in which the phellogen (cork cambium) is more active than elsewhere, resulting in the formation of a tissue with numerous intercellular spaces. In addition, the phellogen itself contains intercellular spaces in the region of the lenticels.

Lenticels begin to form during development of the first periderm (Figure 22–10) and, in the stem, generally appear below a stoma or group of stomata. On the surface of the stem or root, the lenticels appear as raised circular, oval, or elongated areas (Figures 22–12 and 22–17a). Lenticels are also formed on some fruits—for instance, the small dots on the surface of apples and pears are lenticels. As the roots and stems grow older, lenticels continue to develop at the bottom of cracks in the bark in newly formed periderms.

22–12
External features of woody stems. Examination of the twigs of deciduous woody plants reveals many important developmental and structural features of the stem. The most conspicuous structures on the twigs are the buds. Buds occur at the tips of the twigs, the terminal buds, and in the axils of the leaves, the lateral, or axillary, buds. In addition, accessory buds occur in some species. Commonly occurring in pairs, the accessory buds are located one each on either side of an axillary bud. In some species, the accessory buds do not develop if their associated axillary bud undergoes normal development. In still others, the accessory buds give rise to flowers and the axillary bud to a leafy shoot.

After the leaves fall, leaf scars, with their included bundle scars, can be seen beneath the axillary buds. The protective layer of the abscission zone produces the leaf scar. The bundle scars are the severed ends of vascular bundles that extended from the leaf traces into the petiole of the leaf, prior to abscission.

Groups of terminal-bud-scale scars reveal the locations of previous terminal buds, and until they are obscured by secondary growth these groups of scars may be used to determine the age of portions of the stem. The portion of stem between two groups of terminal-bud-scale scars represents one year's growth. The lenticels appear as slightly raised areas on the stem.

(a) *Green ash* (Fraxinus pennsylvanica *var.* subintegerrima). (b) *White oak* (Quercus alba). (c) *Linden* (Tilia americana). (d) *Boxelder* (Acer negundo). (e) *American elm* (Ulmus americana). (f) *Horse-chestnut* (Aesculus hippocastanum). (g) *Butternut* (Juglans cinerea). (h) *Black locust* (Robinia pseudo-acacia).

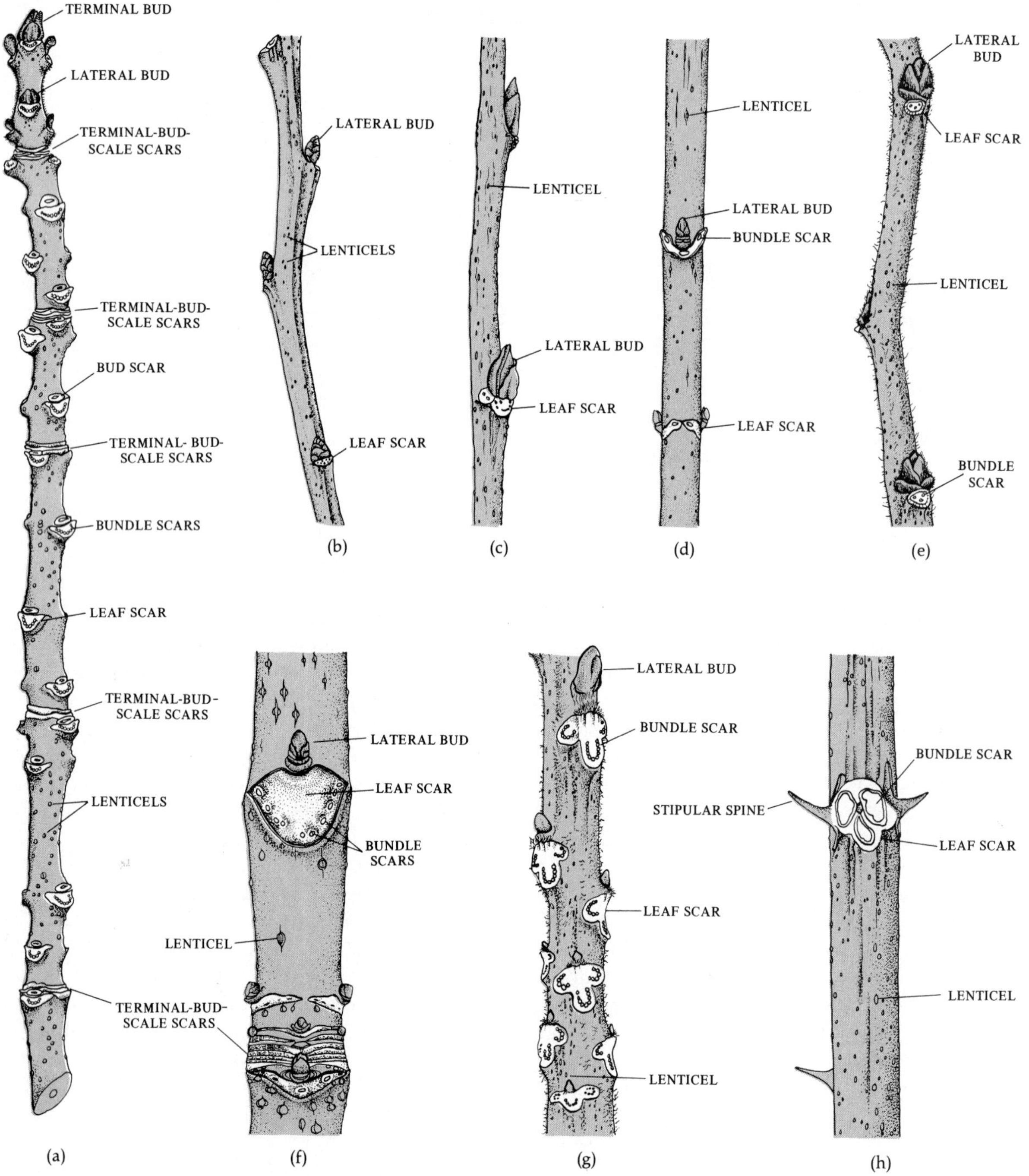

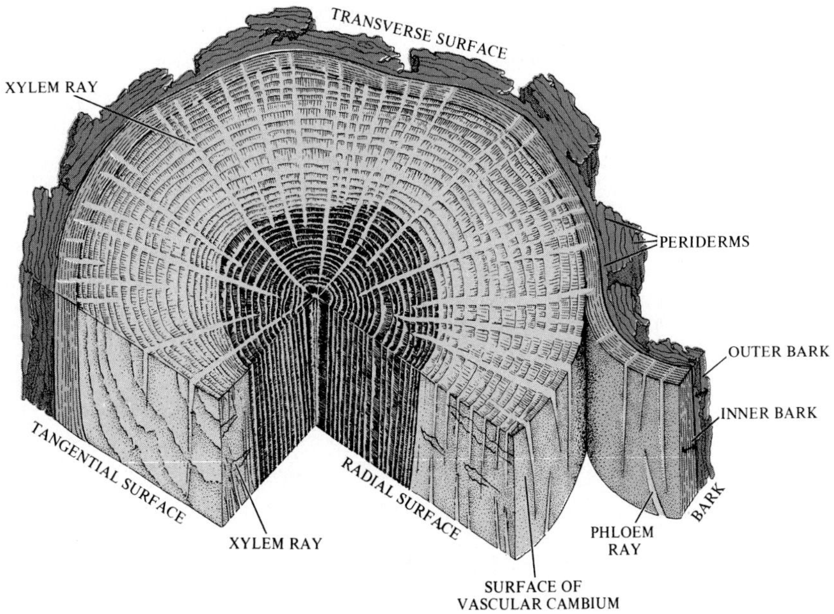

22–13
Diagram of part of a red oak (Quercus rubra) *stem, showing transverse, tangential, and radial surfaces. The dark area in the center of the stem is heartwood. The lighter part of the wood is sapwood.*

The Bark

The terms periderm, cork, and bark are often and unnecessarily confused with one another. As we have just learned, cork is one of three parts of the periderm, a secondary tissue that replaces the epidermis in most woody roots and stems. The term *bark* refers to all tissues outside the vascular cambium, including the periderm when present (Figures 22–1 and 22–13). When the vascular cambium first appears and secondary phloem has not yet been formed, the bark consists entirely of primary tissues. At the end of the first year's growth, the bark includes any primary tissues still present, the secondary phloem, the periderm, and any dead tissues remaining outside the periderm.

Each growing season, the vascular cambium adds secondary phloem to the bark and secondary xylem, or wood, to the core of the stem or root. As mentioned, usually less secondary phloem is produced by the vascular cambium than secondary xylem. In addition, the soft-walled cells (sieve elements and various kinds of parenchymatic elements) of the old secondary phloem commonly are crushed (Figures 22–14 through 22–16). Eventually the old secondary phloem is separated from the rest of the phloem by newly formed periderms. As a result, considerably less secondary phloem accumulates in the stem or root than secondary xylem, which continues to accumulate year after year.

As the stem or root increases in girth, considerable stress is placed on the older tissues of the bark. In some plants, tearing of these tissues results in the formation of large air spaces. In many plants, the parenchyma cells of the axial system and rays divide and enlarge, and in this way the old secondary phloem keeps up for a while with the increase in circumference of the plant part. Earlier we noted that certain rays in the *Tilia* stem become very wide as the stem increases in girth; such rays are called dilated rays.

22–14
Transverse section of the bark of the black locust (Robinia pseudo-acacia) *stem, consisting mostly of nonfunctional phloem. For detail of functional phloem and some nonfunctional phloem see Figure 22–15.*

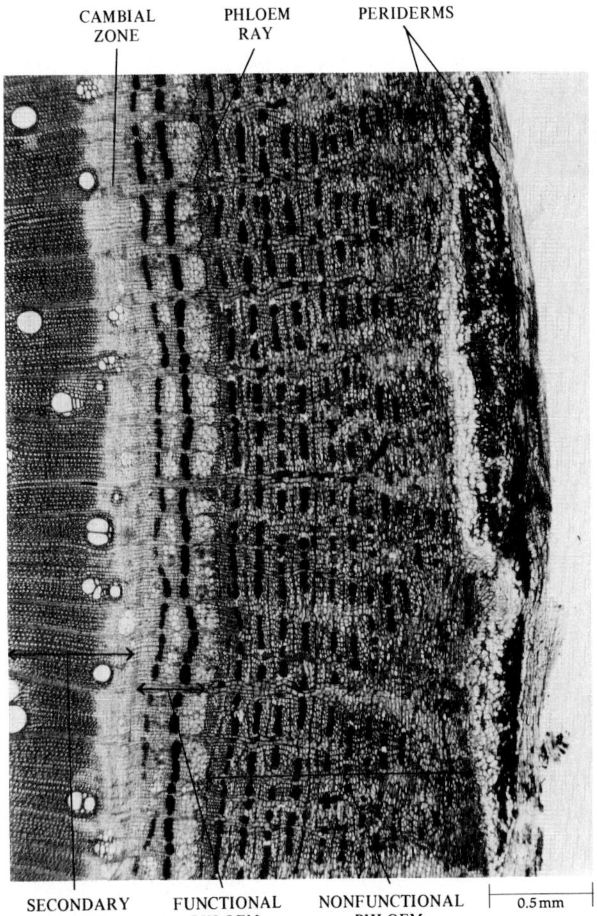

22–15
Transverse section of secondary phloem of the black locust (Robinia pseudo-acacia) *tree, showing mostly functional phloem. Sieve elements (arrows) of the nonfunctional phloem have collapsed.*

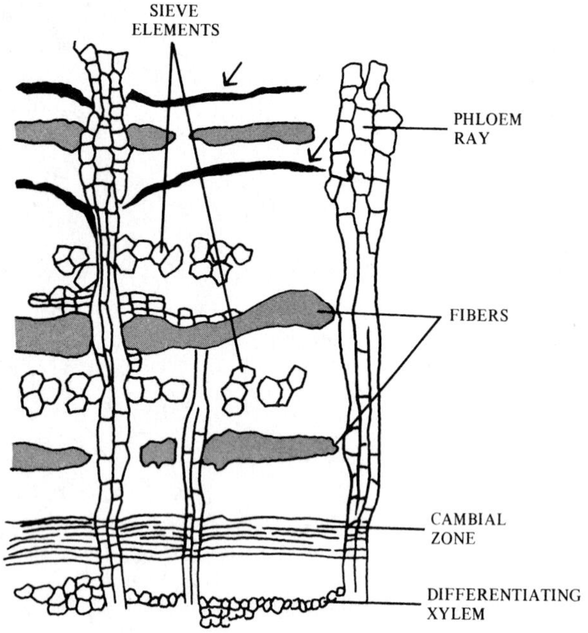

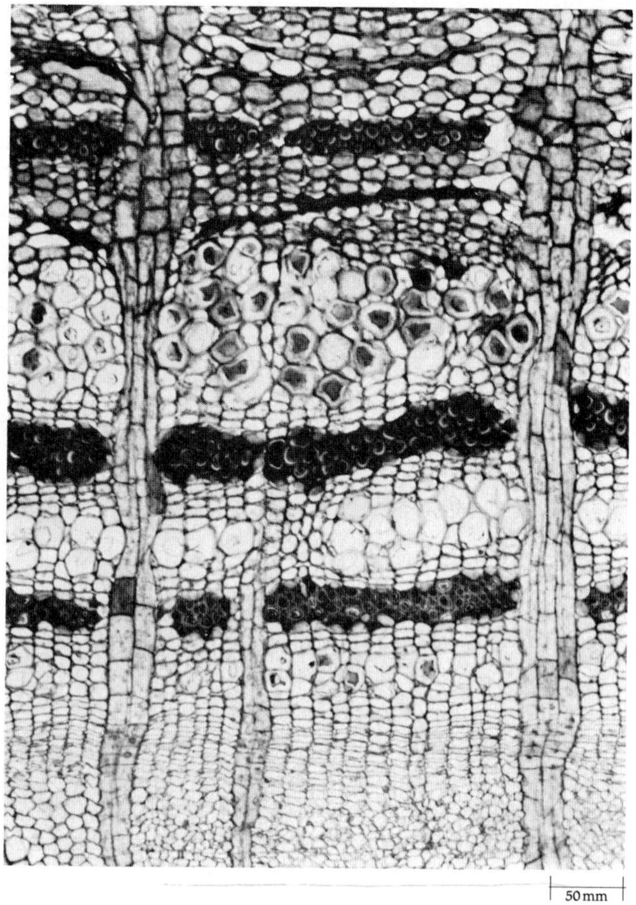

22–16
Radial section of the bark of the black locust (Robinia pseudo-acacia) *tree. Most of the section consists of nonfunctional phloem, in which the sieve elements are collapsed (arrows).*

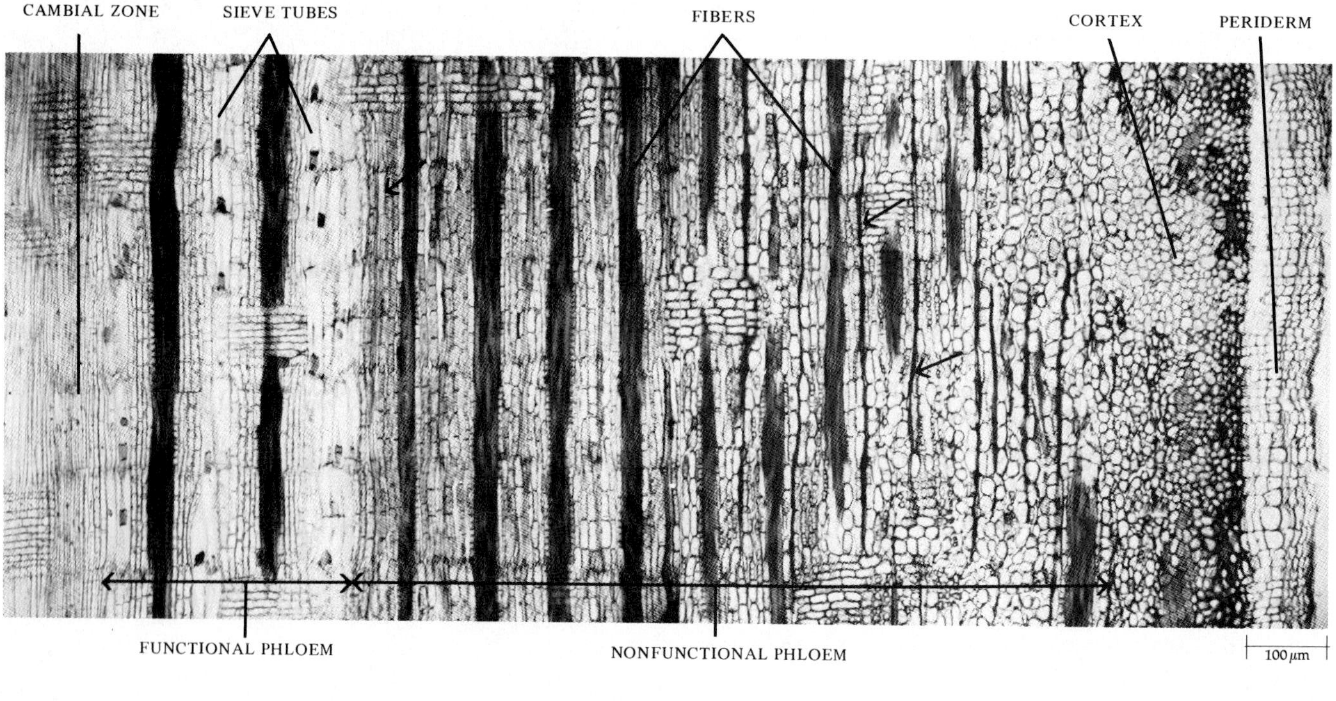

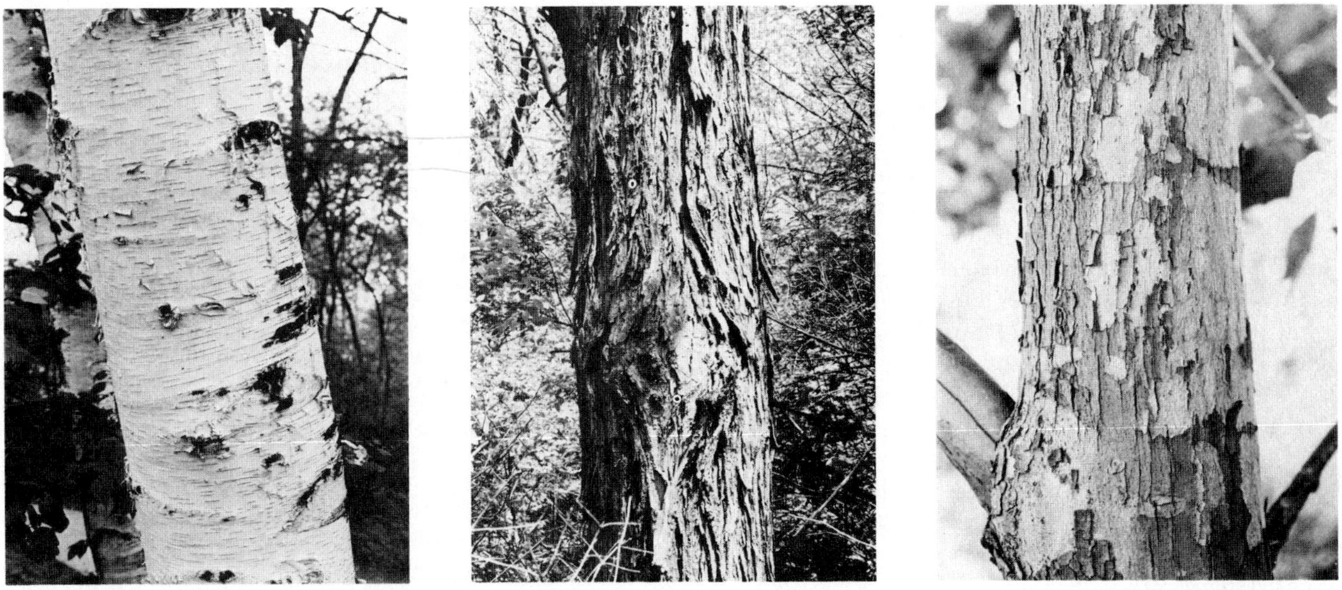

(a) (b) (c)

(d)

22–17
Bark of four species of trees. (a) *Papery, peeling bark of the paper birch* (Betula papyrifera). *Elongated areas on surface of bark are lenticels.* (b) *Shaggy bark of the shagbark hickory* (Carya ovata). (c) *Scaly bark of the plane tree* (Platanus occidentalis). (d) *Deeply furrowed bark of the black oak* (Quercus velutina).

The first-formed periderm may keep up with the increase in girth of the root or stem for several years, with the cork cambium exhibiting periods of activity and inactivity that may or may not correspond to the periods of activity of the vascular cambium. In the stems of apple *(Malus sylvestris)* and pear *(Pyrus communis)* trees, the first cork cambium may remain active for about 20 years. In most woody roots and stems, additional periderms are formed as the axis increases in circumference. After the first periderm, subsequently formed periderms originate deeper and deeper in the bark (Figures 22–1 and 22–13) from parenchyma cells of the phloem no longer actively engaged in the transport, or translocation, of food substances.

All of the tissues outside the innermost cork cambium—all of the periderms, together with any cortical and phloem tissues included among them—make up the outer bark (Figures 22–1 and 22–13). You will recall that with maturation of the suberin-containing cork cells, the tissues outside them are separated from the supply of water and nutrients. Hence the outer bark consists entirely of dead tissues. The living part of the bark inside the innermost cork cambium is called the inner bark (Figures 22–1 and 22–13).

The manner in which new periderms are formed and the kinds of tissues isolated by them have a marked influence on the appearance of the outer surface of the bark (Figure 22–17). In some barks the newly formed periderms develop as discontinuous overlapping layers and result in formation of a scaly bark, or scale bark (Figures 22–1 and 22–13). Scale barks are found on relatively young stems of pine *(Pinus)* and pear *(Pyrus communis)* trees, among others. In other barks, the newly formed periderms arise as more or less continuous, concentric rings around the axis and result in formation of a ring bark. Grape *(Vitis)* and honeysuckle *(Lonicera)* are examples of plants with ring barks, which are less common than scale barks. The barks of many plants are intermediate between ring and scale barks.

Commercial cork is obtained from the bark of the cork oak, *Quercus suber*, which is native to the Mediterranean region. The first cork cambium of this tree has its origin in the epidermis, and the cork produced by it is of little commercial value. When the tree is about 20 years old, the original periderm is removed, and a new cork cambium is formed in the cortex, just a few millimeters below the site of the first one. The cork produced by the new cork cambium accumulates very rapidly and after about 10 years is thick enough to be stripped off the tree. Once again a new cork cambium arises beneath the previous one, and after about another 10 years the cork can be stripped again. This procedure may be repeated at about 10-year intervals until the tree is 150 or more years old. The spots and long dark streaks seen on the surfaces of commercial cork are lenticels.

In most woody roots and stems, very little secondary phloem actually is concerned with the conduction of food. In most species, only the current year's growth increment, or growth ring, of secondary phloem is active in the long-distance transport of food through the stem. This is because the sieve elements are short-lived (Chapter 19), most dying by the end of the same year in which they are derived from the vascular cambium. In some plants, such as black locust *(Robinia pseudo-acacia)*, the sieve elements collapse and are crushed relatively soon after they die (Figures 22–14 through 22–16).

The part of the inner phloem actively engaged in the transport of food substances is called *functional phloem*. Although the sieve elements outside the functional phloem are dead, the phloem parenchyma cells (axial parenchyma) and parenchyma cells of the rays may remain alive and function as storage cells for many years. That part of the inner phloem is called *nonfunctional phloem*. Only the outer bark is composed entirely of dead tissue.

THE WOOD: SECONDARY XYLEM

Apart from the use of various plant tissues as food for man, no single plant tissue has played a more indispensible role to man's survival throughout recorded history than wood, or secondary xylem. Commonly, woods are classified as either *hardwoods* or *softwoods*. The so-called hardwoods are dicot woods, and the softwoods, conifer woods. As we shall see, the two kinds of woods have basic structural differences, but the terms hardwood and softwood do not accurately express the degree of density or hardness of the wood. For example, one of the lightest and softest of woods is balsa *(Ochroma lagopus)*, a tropical dicot. By contrast, the wood of some conifers, such as hemlock *(Tsuga)*, are harder than some softwoods.

Conifer Wood

The structure of conifer wood is relatively simple compared with that of most dicots. The principal features of conifer wood is its lack of vessels (Chapter 19) and its relatively small amount of axial or wood parenchyma. Long, tapering tracheids constitute the dominant cell type in the axial system. In some genera, such as *Pinus*, the only parenchyma cells of the axial system are those associated with *resin ducts*. The resin ducts are relatively large intercellular spaces lined with thin-walled parenchyma cells, which secrete resin into the duct. In *Pinus*, resin ducts occur in both the axial system and the rays (Figures 22–18 and 22–19).

22–18
Block diagram of the secondary xylem of white pine (Pinus strobus). *With the exception of the parenchyma cells associated with the resin ducts, the axial system consists entirely of tracheids. The rays are only one cell wide, except for those containing resin ducts. Early wood and late wood are described on page 477.*

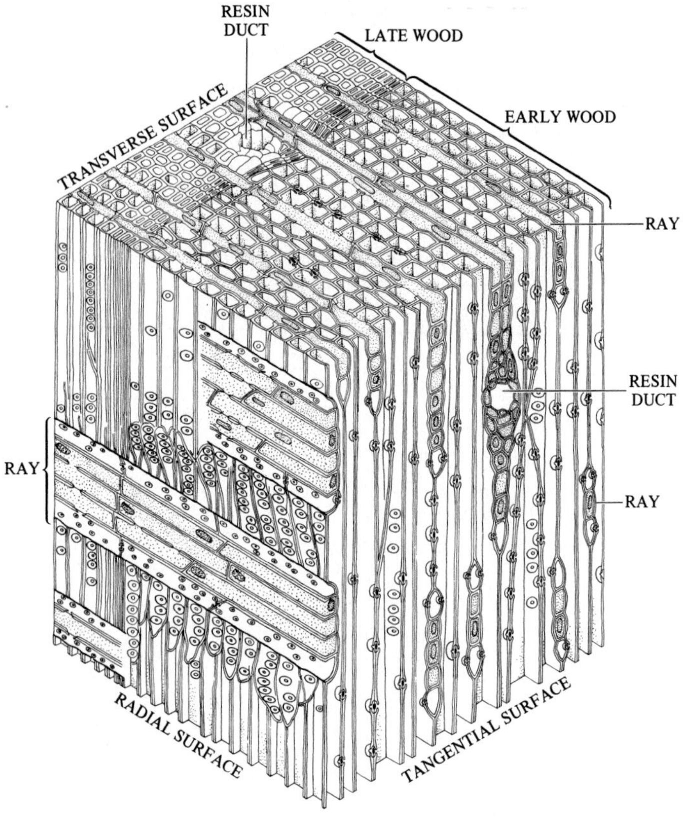

22–19
Wood of white pine (Pinus strobus), *a conifer, in* (a) *transverse,* (b) *radial, and* (c) *tangential sections.*

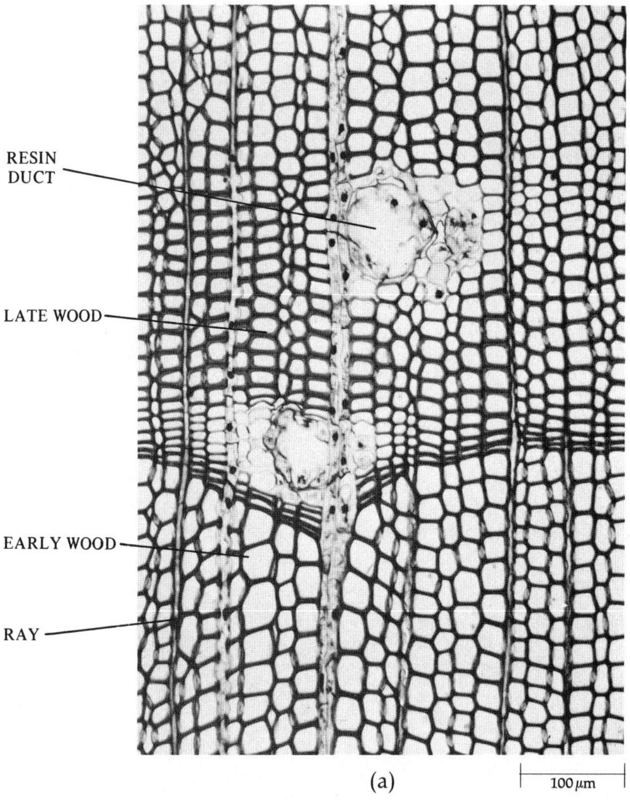

(a)

The tracheids of conifers are characterized by large, circular bordered pits, which are most abundant on the ends of the cells where they overlap with other tracheids (Figures 22–18 through 22–20). The pit-pairs (Chapter 1, page 35) between conifer tracheids are unique in their presence of tori (singular: torus). The torus is a thickened central portion of the pit-membrane (Figure 22–21) and is slightly larger than the openings or apertures in the pit-borders (Figure 22–20). The pit-membrane is flexible, and under certain conditions, the torus may block one of the apertures and prevent the movement of water or gases through the pit-pair (Figure 22–20).

Figure 22–18 is a three-dimensional diagram of the wood of white pine *(Pinus strobus)* based on the three wood sections shown in Figure 22–19. In sections cut at right angles to the long axis of the root or stem—transverse or cross sections—the tracheids appear angular or squarish and the rays can be seen on end traversing the wood (Figure 22–19a). There are two kinds of longitudinal sections, radial and tangential. Radial sections are cut parallel with the rays and, in such sections, the rays appear as sheets of cells oriented at right angles to the vertically elongated tracheids of the axial system (Figures 22–19b and 22–20d). Tangential sections are cut at right angles to the rays and reveal the width and height of the rays. In *Pinus* the rays are one cell wide, except for those containing resin ducts (Figure 22–19c). Details of white pine wood are shown in Figure 22–20.

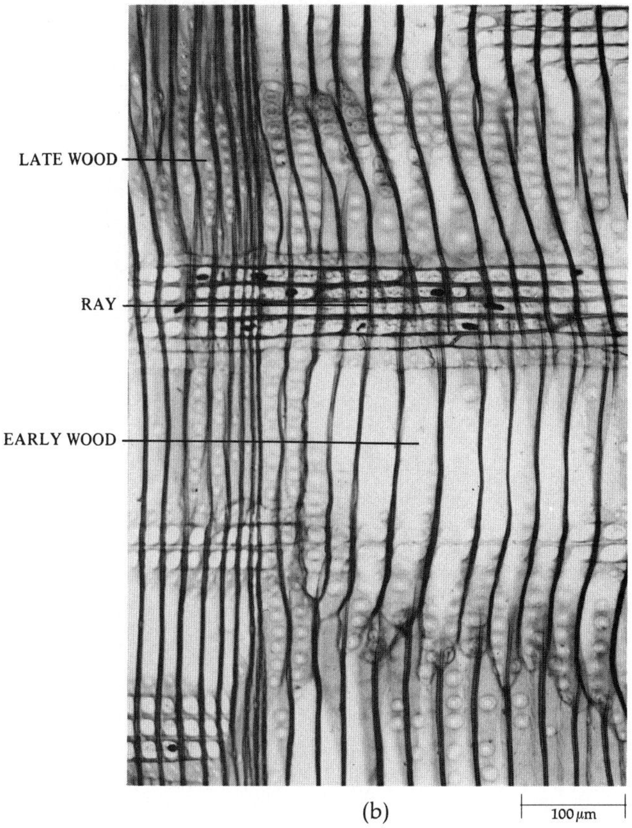

(b)

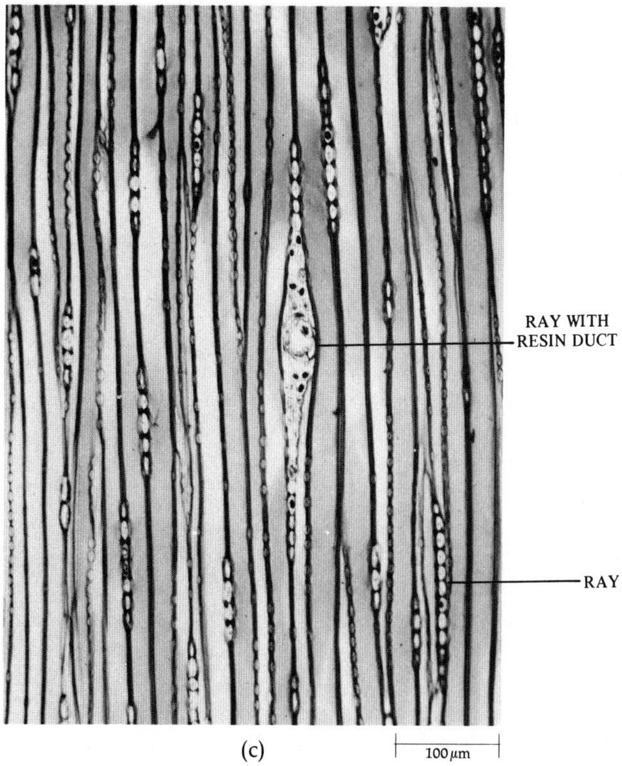

(c)

22–20
Details of white pine (Pinus strobus) *wood.* (a) *Transverse section, showing bordered pit-pairs of tracheids.* (b) *Radial section, showing face-view of bordered pit-pairs in walls of tracheids.* (c) *Tangential section, showing bordered pit-pairs of tracheids.* (d) *Radial section, showing ray. The rays of pine and other conifers are composed of ray tracheids and ray parenchyma cells. Notice bordered pits of ray tracheids.*

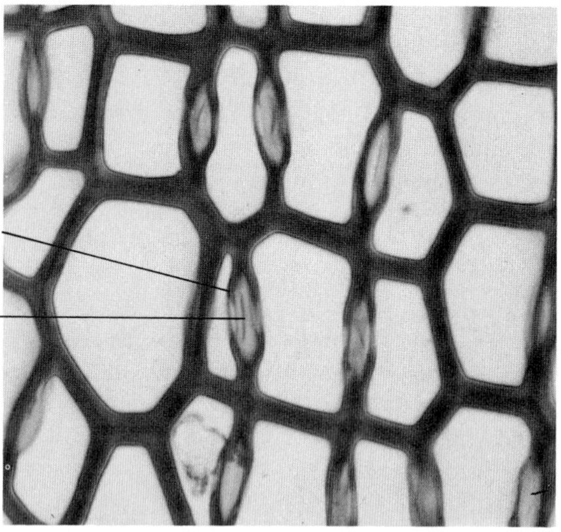

(a)

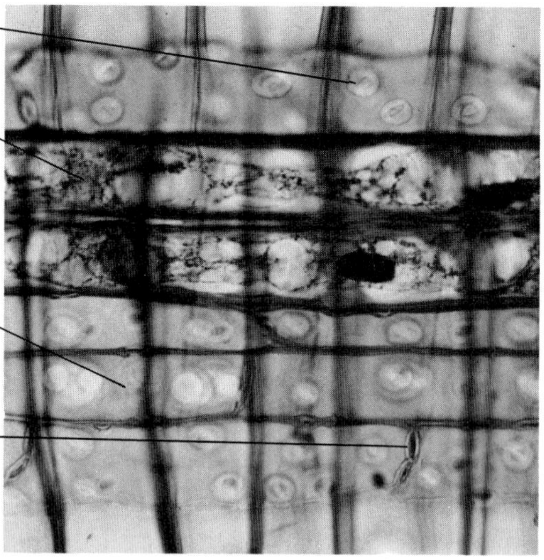

(d)

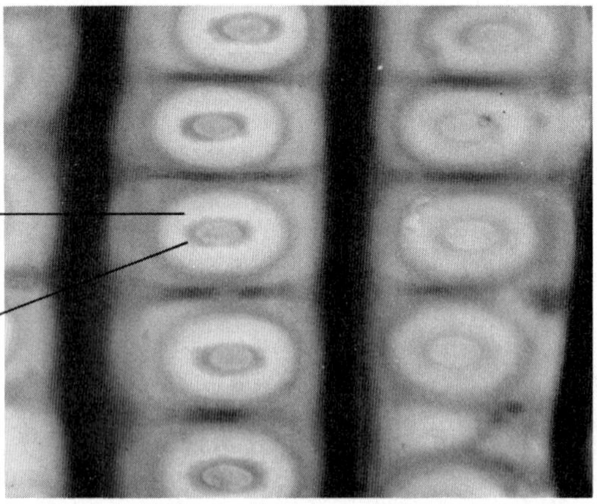

(b)

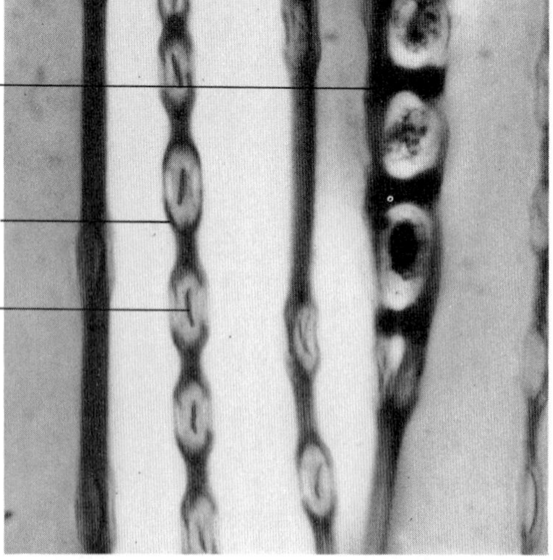

(c)

22–21
Scanning electron micrograph of pit membrane of bordered pit-pair in white pine (Pinus strobus) *tracheid. The thickened part of the membrane is the torus. The part of the membrane surrounding the torus is called the margo.*

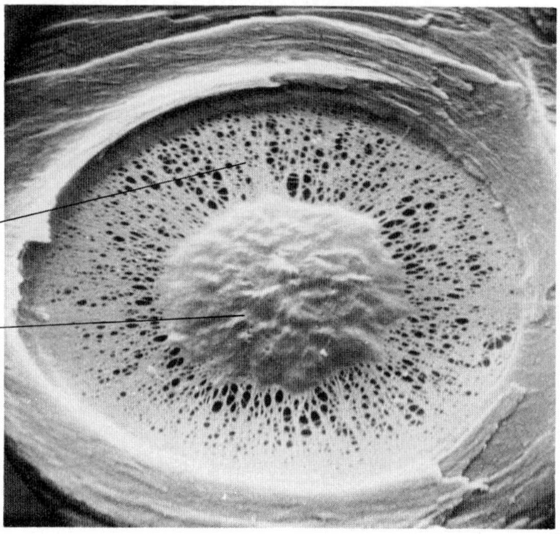

Dicotyledonous Wood

Wood structure in dicotyledons is much more varied than in conifers, owing in part to a greater number of cell types in the axial system, including vessel members, tracheids, several types of fibers, and parenchyma cells (Figure 22–22; also see Figure 19–10, page 407). It is the presence of vessel members, in particular, that distinguishes dicot woods from conifer woods.

The rays of dicot woods are often considerably larger than those of conifer woods. In conifer wood, the rays are predominately one cell wide, and most range from 1 to 20 cells high. The rays of dicot woods range from one to many cells wide and from one to several hundred cells high. In some dicot woods, such as oak, the large rays can be seen with the unaided eye (Figure 22–13). The large rays of the red oak wood illustrated in Figure 22–22c are 12 to 30 cells wide and hundreds of cells high. Besides the large rays, oak wood has numerous rays only one cell wide (Figure 22–22c). In red oak, the rays average about 21 percent of the volume of the wood. Overall, the rays of hardwoods average about 17 percent of the volume of the wood. The average ray volume for conifers is 7.8 percent.

22–22
Wood of red oak (Quercus rubra), *a dicot, in* (a) *transverse,* (b) *radial, and* (c) *tangential sections.*

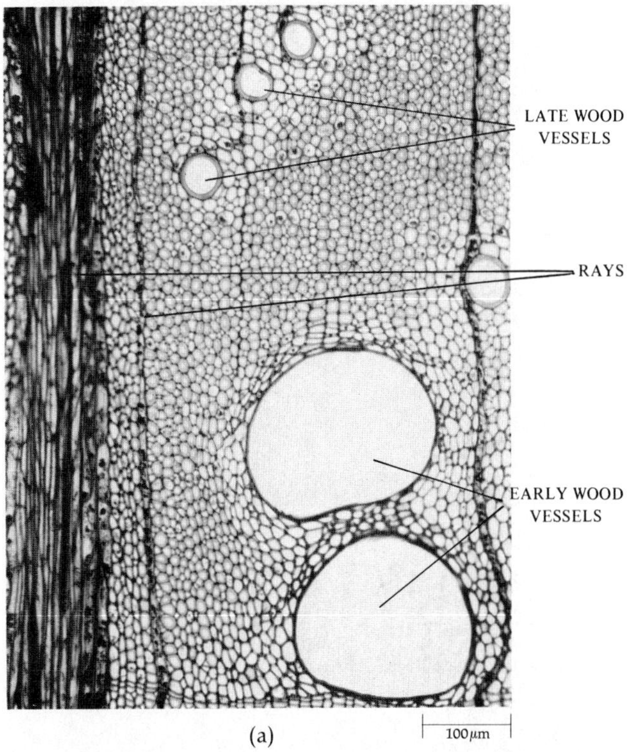

(a)

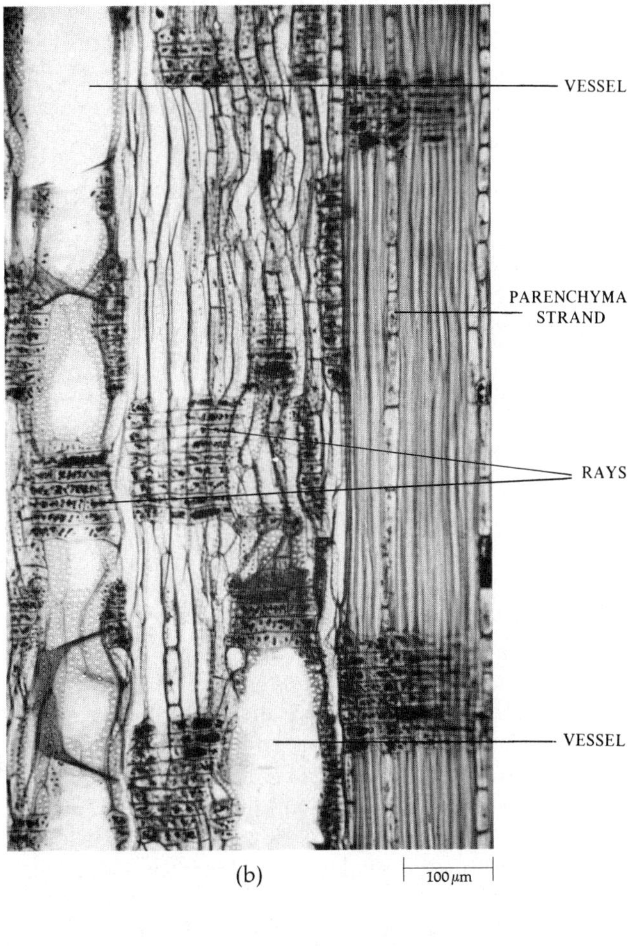

(b)

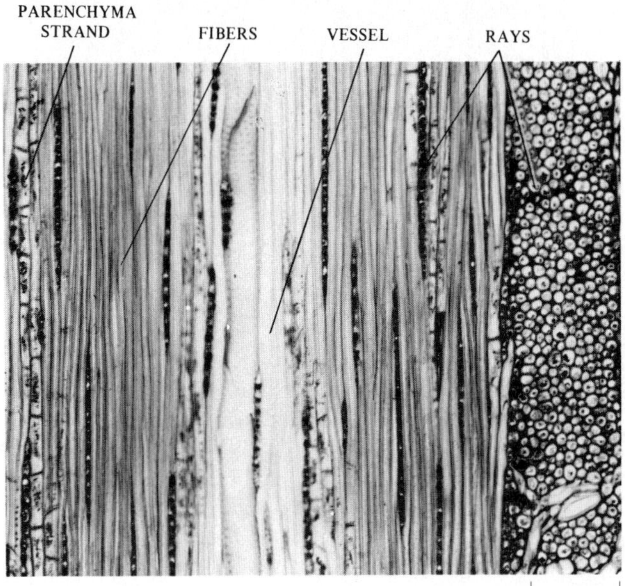

(c)

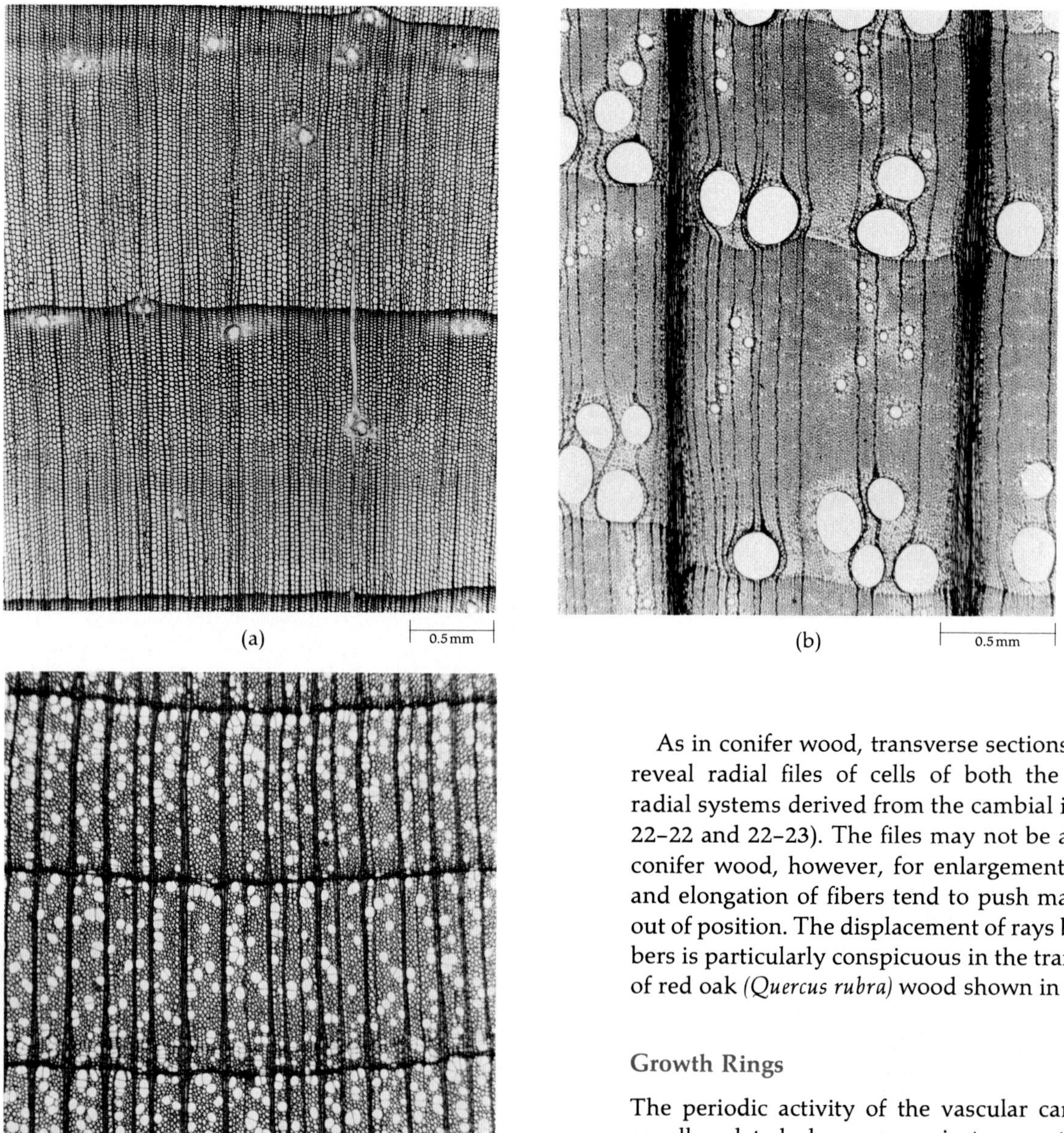

22–23
Transverse sections of wood, showing growth layers. (a) *White pine* (Pinus strobus), *a conifer. Conifer woods lack vessels; that is, they are nonporous. The round, clear areas found primarily in the late wood are resin ducts.* (b) *Red oak* (Quercus rubra). *The large vessels of ring-porous wood such as red oak are found in the early wood. The dark vertical lines are rays.* (c) *Tulip tree* (Liriodendron tulipifera), *a diffuse-porous wood.*

As in conifer wood, transverse sections of dicot wood reveal radial files of cells of both the axial and the radial systems derived from the cambial initials (Figures 22–22 and 22–23). The files may not be as orderly as in conifer wood, however, for enlargement of the vessels and elongation of fibers tend to push many of the cells out of position. The displacement of rays by vessel members is particularly conspicuous in the transverse section of red oak *(Quercus rubra)* wood shown in Figure 22–23b.

Growth Rings

The periodic activity of the vascular cambium—a seasonally related phenomenon in temperate zones—is responsible for the production of *growth increments,* or *growth rings,* in both secondary xylem and secondary phloem, although in the phloem the increments are not always readily discernible. If a growth layer represents one season's growth, it is called an *annual ring.* Abrupt changes in available water and other environmental factors may be responsible for the production of more than one growth ring in a given year; such rings are called false annual rings. Thus the age of a given portion of the old woody stem can be estimated by counting growth rings, but the estimates may be inaccurate if false annual rings are included.

The widths of growth rings may vary greatly from year to year, and they are affected by such environmental factors as light, temperature, rainfall, available soil water, and length of the growing season. The width of a growth ring, it has been found, is a fairly accurate index of the rainfall of a particular year. Under favorable con-

(a)

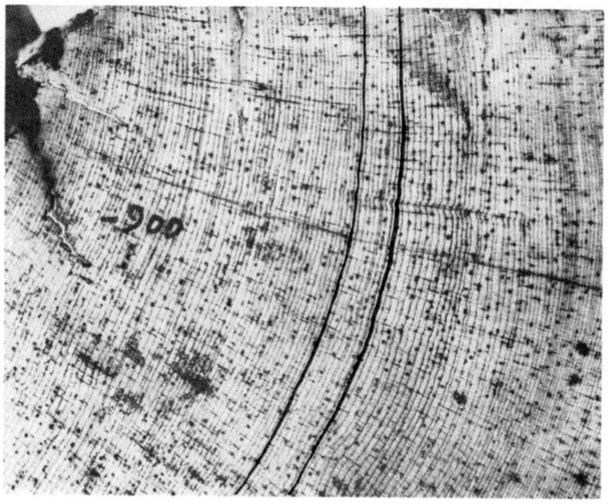

(b)

22-24
(a) *Bristlecone pine* (Pinus longaeva) *from the White Mountains of east-central California. These pines, which grow near the timberline of the mountains, are the oldest living trees; one reached an age of 4900 years.* (b) *A transverse section of the wood, showing the variation in width of annual rings. This section begins approximately 2900 years ago, with the two colored rings being those of 860* B.C. *and 850* B.C. *The overlapping patterns of rings in dead trees have made possible records of relative precipitation extending back some 8200 years.*

ditions—that is, during periods of adequate or abundant rainfall—the growth rings are wide; under unfavorable conditions, they are narrow. In semiarid regions, where there is very little rain, the tree is a sensitive rain gauge. An excellent example of this is the bristlecone pine *(Pinus longaeva)* of the White Mountains of California (Figure 22-24). Each growth ring is different from every other, and a study of the rings tells a story that dates back thousands of years. The oldest-known living specimen of bristlecone pine is 4900 years old, but dendrochronologists—scientists who conduct historical research through the growth rings of trees—have been able to match samples of wood from living and dead trees, and in this way they have built up a continuous series of rings dating back 8200 or more years. The widths of the growth rings of bristlecone pines at the higher elevations (upper tree line) have been found to be closely related to temperature changes, and a record of average ring width in these trees provides a valuable guide to past temperatures and climatic conditions. For example, in the White Mountains of California, the summers were relatively warm from 3500 to 1300 B.C., and the tree line was about 150 meters above its present level. Summers were cool from 1300 to 200 B. C.

The structural basis for the visibility of growth layers in the wood is the difference in density of the wood produced early in the growing season and that produced later (Figures 22-19, 22-22, and 22-23). The early wood is less dense (with larger cells and proportionally thinner walls) than the late wood (with narrower cells and proportionally thicker walls). In a given growth layer, the change from early to late wood may be very gradual and almost imperceptible. However, where the late wood of one growth layer abuts on the early wood of a newer growth layer, the change is abrupt.

In dicot woods, size differences of the vessels, or pores, in early and late woods may be quite marked, the pores of the early wood being distinctly larger than in the late wood. (The term pore is used by the wood anatomist for a vessel in cross section.) Such woods are termed ring-porous woods (Figures 22-22a and 22-23b). In other dicot woods, the pores are fairly uniform in distribution and size throughout the growth layer. These woods are called diffuse-porous woods (Figure 22-23c). In ring-porous woods, almost all the water is conducted in the outermost growth layer, and at speeds about 10 times greater than in diffuse-porous woods.

Sapwood and Heartwood

As the wood grows older and no longer serves as a conducting tissue, its parenchyma cells eventually die. Before this happens, however, the wood often undergoes visible changes, which involve the loss of reserve food substances and the infiltration of the wood by various substances (such as oils, gums, resins, and tannin),

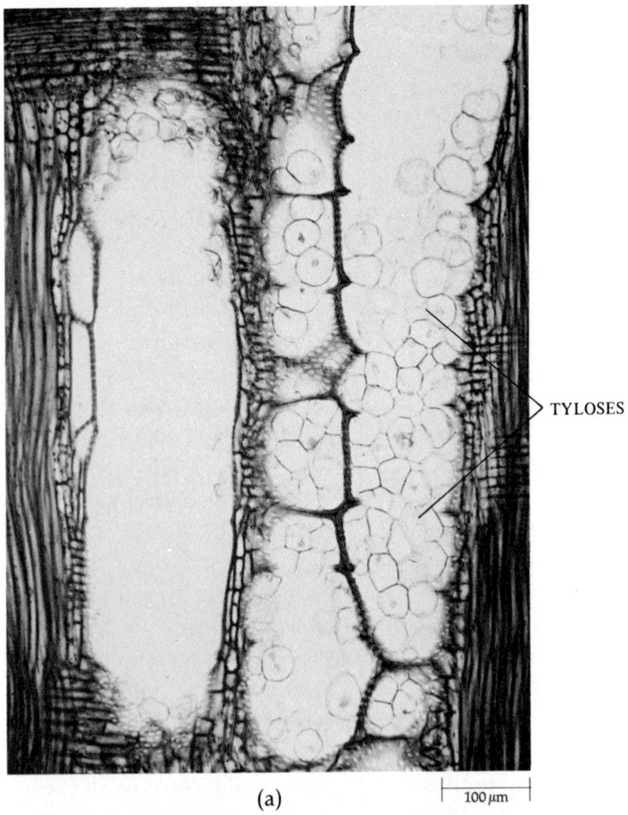

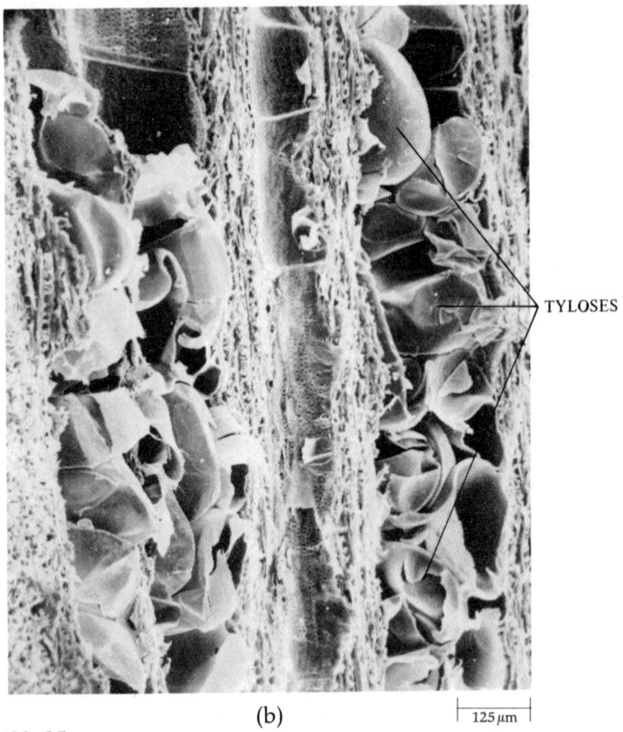

22–25
Tyloses, balloonlike outgrowths of parenchyma cells, which partially or completely block the lumen of the vessel. (a) *In vessels of black locust* (Robinia pseudo-acacia). (b) *In vessels of white oak* (Quercus alba), *as seen with the scanning electron microscope.*

which color it and sometimes make it aromatic. This often darker, nonconducting wood is called *heartwood*, and the generally lighter conducting wood, *sapwood* (Figure 22–13). In many woods, tyloses are formed in the vessels when they become nonfunctional (Figure 22–25). These are outgrowths from ray or axial parenchyma cells through the pit cavity in the vessel wall. Tyloses may completely occlude the lumen (space bounded by the cell wall) of the vessel. Tyloses often are induced to form prematurely or unnaturally by plant pathogens and result in death of the plant. Many of the so-called "wilting diseases" exert their effects in this way.

The proportion of sapwood to heartwood and the degree of visible difference between them varies greatly from species to species. Some trees, such as maple *(Acer)*, birch *(Betula)*, and ash *(Fraxinus)* have thick sapwoods; whereas others, such as locust *(Robinia)*, catalpa *(Catalpa)*, and yew *(Taxus)*, have thin sapwoods. Still other trees have no clear distinction between sapwood and heartwood *(Populus, Salix, Abies)*.

SUMMARY

Secondary growth (the increase in girth in regions that are no longer elongating) occurs in all gymnosperms and in most dicotyledons and involves activity of the two lateral meristems, the vascular cambium and the cork cambium, or phellogen. Herbs may undergo little or no secondary growth, whereas woody plants—trees and shrubs—may continue to increase in thickness for many years. Figure 22–26 presents summaries of root and stem development of a woody plant, beginning with the apical meristem and ending with the secondary tissues produced during the first year's growth.

The vascular cambium contains two types of initials, fusiform initials and ray initials. Through periclinal divisions, the fusiform initials give rise to the components of the axial system and ray initials to those (ray cells) of the radial system or rays. Increase in circumference of the cambium is accomplished by anticlinal division of the initials.

The first cork cambium in most stems originates in a layer of cells immediately below the epidermis. In the root, the first cork cambium arises in the pericycle. The cork cambium produces cork, or phellem, to the outside, and phelloderm to the inside. Together, the cork cambium, cork, and phelloderm comprise the periderm.

With formation of a periderm, lenticels develop in the periderm to assure aeration of the plant part concerned. The bark consists of all tissues outside the vascular cambium. In old roots and stems, most of the phloem comprising the bark is nonfunctional. Sieve elements are short-lived and, generally, only the present year's growth increment contains conducting, or functional,

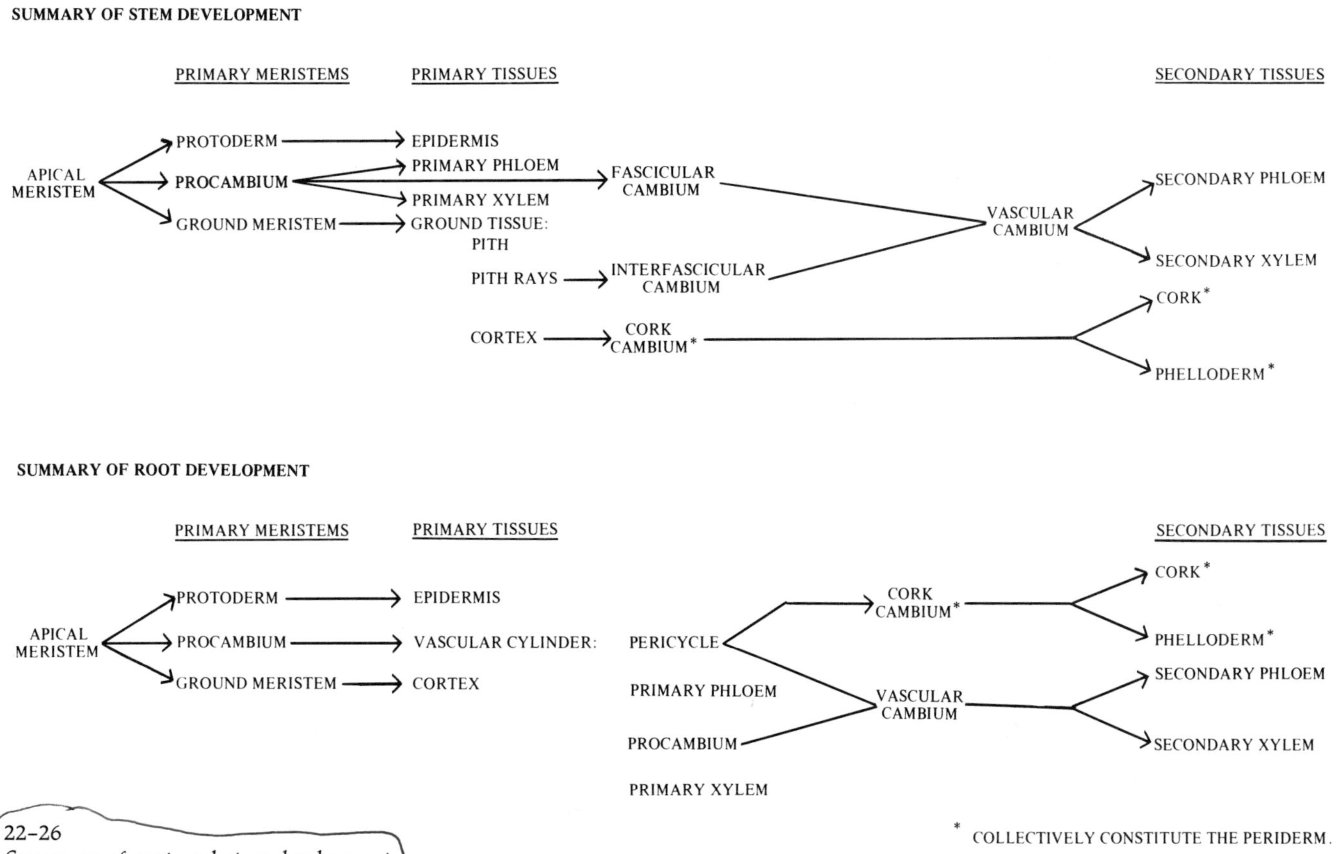

22–26
Summary of root and stem development in a woody dicot during first year's growth.

sieve elements. After the first periderm, subsequently formed periderms originate deeper and deeper in the bark from parenchyma cells of nonfunctional phloem.

Woods are classified as either softwoods or hardwoods. All so-called softwoods are conifers and all so-called hardwoods, dicotyledons. Compared with dicot woods, those of conifers are simple, consisting of tracheids and parenchyma cells. Some contain resin ducts. Dicot woods may contain a combination of all of the following cell types: vessel members, tracheids, several types of fibers, and parenchyma cells. Growth layers that correspond to yearly increments of growth are called annual rings. The difference in density between the late wood of one growth increment and the early wood of the following increment makes it possible to distinguish the growth layers. In some plants, the nonconducting heartwood is visibly distinct from the actively conducting sapwood.

SUGGESTIONS FOR FURTHER READING

CUTTER, E. G.: *Plant Anatomy: Experiment and Interpretation,* Part I: *Cells and Tissues,* Addison-Wesley Publishing Co., Inc., Reading, Mass., 1969.*

An introduction to plant cells and tissues, with emphasis on experimental work.

CUTTER, E. G.: *Plant Anatomy: Experiment and Interpretation,* Part II: *Organs,* Addison-Wesley. Publishing Co., Inc., Reading, Mass., 1971.*

An introduction to the organs of the plant and possible causes underlying their development.

EPSTEIN, E.: "Roots," *Scientific American* 228(5): 48–58, 1973.

A discussion of root structure and function, including some new ideas about the mechanism of ion uptake by the root.

* Available in paperback.

ESAU, KATHERINE: *Plant Anatomy,* 2nd ed., John Wiley & Sons, Inc., New York, 1965.

The standard work in the field, this well-illustrated book considers all aspects of plant anatomy.

______: *Anatomy of Seed Plants,* 2nd ed., John Wiley & Sons, Inc., New York, 1976.

A shorter book than the preceding, this is also an excellent textbook and reference.

FAHN, A.: *Plant Anatomy,* 2nd ed., Pergamon Press, Elmsford, N.Y., 1974.*

A well-illustrated, up-to-date textbook considering all aspects of plant anatomy.

O'BRIEN, T. P., and M. E. MCCULLY: *Plant Structure and Development,* The Macmillan Company, New York, 1969.

A pictorial and physiological approach to plant structure and development.

RAY, P. M.: *The Living Plant,* 2nd ed., Holt, Rinehart & Winston, Inc., New York, 1972.*

A short, readable account of plant growth and development by one of the leading workers in the field.

STEEVES, T. A., and I. M. SUSSEX: *Patterns in Plant Development,* Prentice-Hall, Inc., Englewood Cliffs, N.J., 1972.

A structural approach to plant development, with emphasis on experimental and analytical data.

WARDLAW, C. W.: *Morphogenesis in Plants. A Contemporary Study, 2nd ed.,* Methuen & Co. Ltd., London, 1968.

A fascinating book, aptly described by its title, by one of the acknowledged masters of the field.

ZIMMERMANN, M. H., and C. L. BROWN: *Trees: Structure and Function,* Springer-Verlag, New York, 1971.

An up-to-date discussion of how trees work, with emphasis on function as it relates to structure.

* Available in paperback.

SECTION 6 Growth Regulation and Growth Responses

CHAPTER 23

Regulating Growth and Development: The Plant Hormones

A plant, in order to grow, needs light, carbon dioxide, water, and minerals, including nitrogen from the soil. From these things, it makes more of its own substance, turning simple materials into the complex organic substances of which living things are composed. As we saw in the previous section, the plant does far more than simply increase its mass and volume as it grows. It differentiates, develops, and takes shape, forming a variety of cells, tissues, and organs. How can a single cell, the fertilized egg, be the source of the myriad tissues and organs that make up the extraordinary individual known as a "normal" plant? Many of the details of how these processes are regulated are not known, but it has become clear that normal development depends on the interplay of a number of internal and external factors. The principal internal factors that regulate plant growth and development are the subject of this chapter. Chapter 24 will take up some of the external factors—light, temperature, day length, gravity, and so on—that affect plant growth.

Plant hormones play a major role in regulating growth. Hormones are organic substances that are produced in one tissue and transported to another, where they cause physiological responses. They are active in very small quantities. In the shoot of a pineapple plant, for example, only 6 micrograms of auxin, a common plant hormone, are found per kilogram of plant material. One enterprising plant physiologist calculated that the weight of the hormone in relation to that of the shoot is comparable to the weight of a needle in a 20 metric-ton truckload of hay.

The term "hormone" comes from the Greek word meaning "to excite." It is now clear, however, that many hormones have inhibitory influences. So, rather than thinking of hormones as stimulators, it is perhaps more useful to consider them as chemical messengers. But this term, too, needs qualification. As we shall see, the response to the particular "message" depends not only on its content but on how it is "read" by the recipient.

AUXIN

Some of the first recorded experiments on growth-regulating substances were performed by Charles Darwin and his son Francis and reported in *The Power of Movement in Plants,* published in 1881. The Darwins first made systematic observations of the bending toward light *(phototropism)* of seedlings of canary grass *(Phalaris canariensis)* and of oats *(Avena sativa).* They then showed that if they covered the upper portion of the shoot of the seedling with a cylinder of metal foil or a hollow tube of glass blackened with India ink and exposed the plant to a lateral light (that is, from the side), the characteristic bending in the lower portion of the shoot did not occur. If, however, the tips were enclosed in transparent glass tubes, bending occurred normally (Figure 23–1). "We must therefore conclude," they stated, "that when seedlings are freely exposed to a lateral light some influence is transmitted from the upper to the lower part, causing the latter to bend."

In 1926, the Dutch plant physiologist Frits W. Went succeeded in separating this "influence" from the plants that produced it. Went cut off the coleoptile tips from a number of oat seedlings and placed them for about an hour on a slice of agar with their cut surfaces in contact with the agar. (The coleoptile is the sheathlike, pointed structure covering the shoot of grass seedlings; morphologically it is a portion of cotyledon. Agar is a gelatinous substance derived from certain red algae.) He then cut the agar into small blocks and placed the blocks on one side of the stumps of the decapitated plants, which were kept in the dark during the entire experiment. Within one hour, he observed a distinct bending *away* from the side on which the agar block was placed. Agar blocks that had not been exposed to a coleoptile tip produced either no bending or a slight bending toward the side on which the block had been placed. Agar blocks that had been exposed to a section of coleoptile from lower on the shoot also produced no physiological effect (Figure 23–2).

By these experiments, Went showed that the coleoptile tip exerted its effects by means of a chemical stimulus rather than a physical stimulus, such as an electrical one. This chemical stimulus came to be known as *auxin,* a term coined by Went from the Greek word *auxein,* "to increase."

The only naturally occurring auxin is indoleacetic acid, abbreviated IAA. As you can see (Figure 23–3), IAA closely resembles the amino acid tryptophan (Figure 2–15, page 56). Tryptophan is probably the precursor from which IAA is formed in the living plant. Auxin is produced in the coleoptile tips of the grasses and in the apical meristems of shoots and roots. It is probably abundant in embryos and is also found in young leaves and in fruits.

Auxin is actively transported from the tips of the shoots toward the base of the plant. The movement usually occurs in the tissue at large rather than in the "pipelines" of the xylem and phloem. The transport process is presumed to involve an interaction between IAA and the plasma membranes of the cells.

Auxin has a variety of effects, which differ from time to time, from species to species, and most particularly, from tissue to tissue. Like many other physiologically active compounds, auxin is toxic at high concentrations. The weedkiller 2,4–D is a synthetic auxin, one of many that have been manufactured and that have a variety of applications.

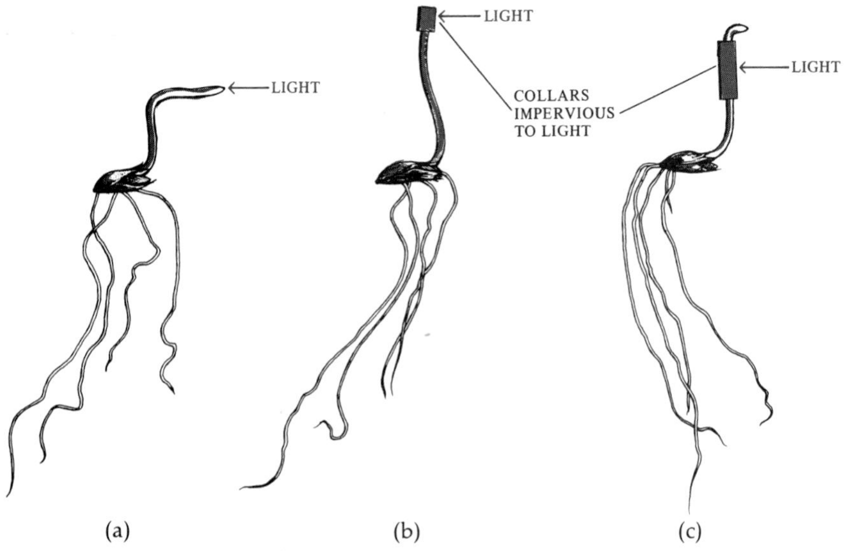

23–1
The Darwins' experiment. Seedlings normally bend toward the light (a). *When the tip of a seedling was covered by a light-proof collar (but not when it was covered by a transparent one), this bending did not occur* (b). *When the collar was placed below the tip* (c), *the characteristic light response took place. From these experiments, the Darwins concluded that, in response to light, an "influence" that causes bending was transmitted from the tip of the seedling to the area below the tip, where bending normally occurs.*

23–2
Went's experiment. (a) *Went removed the coleoptile tips from seedlings and placed the tips on agar.* (b) *The agar was then cut into small blocks and placed on one side of the decapitated shoots of the seedlings.* (c) *The seedlings, which were kept in the dark during the entire experiment, always bent away from the side on which the agar block was placed. From this, Went concluded that the "influence" that caused the seedling to bend was chemical and that it accumulated on the side away from the light. The degree to which a seedling bends under the experimental conditions provides the common bioassay for auxin. A bioassay is a method for quantitatively determining the concentration of a substance by its effect on the growth of a suitable organism under controlled conditions.*

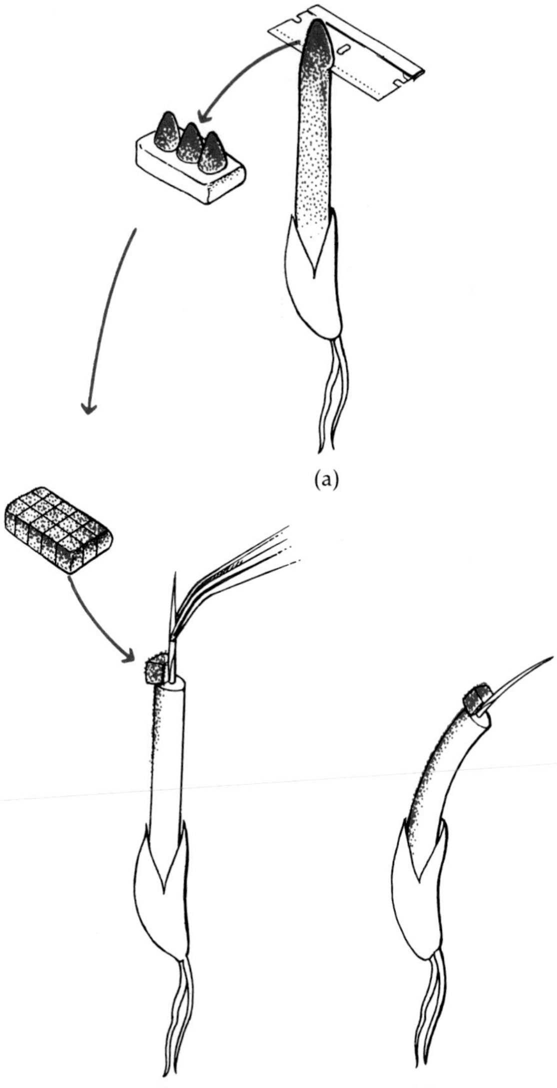

23–3
Auxins. (a) *Indoleacetic acid (IAA) is the only known naturally occurring auxin.* (b) *Dichlorophenoxyacetic acid (2,4–D), a synthetic, is a widely used herbicide.* (c) *Naphthalenacetic acid, another synthetic auxin, is commonly employed to induce the formation of adventitious roots in cuttings and to reduce fruit drop in commercial crops. The synthetic auxins, unlike IAA, are not readily broken down by natural plant enzymes and microbes and so are better suited for commercial purposes.*

(a)

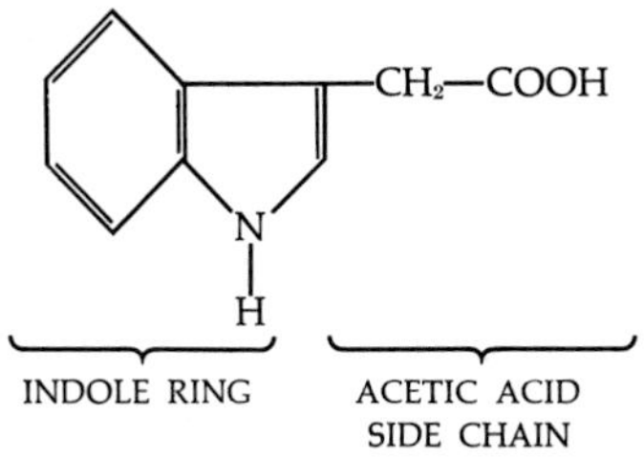

INDOLEACETIC ACID (IAA)

(b)

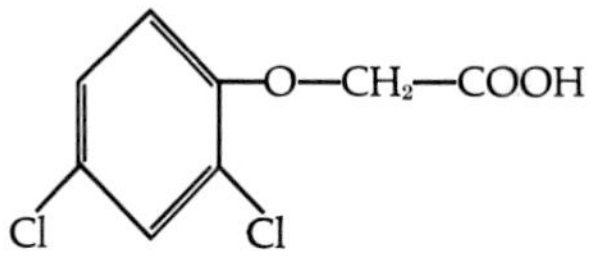

2,4-DICHLOROPHENOXYACETIC ACID (2,4-D)

(c)

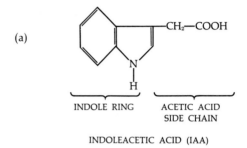

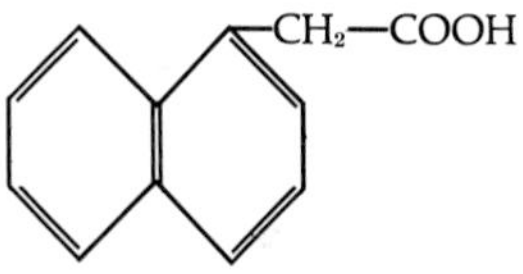

ALPHA-NAPHTHALENACETIC ACID

Auxin and the Elongation of the Shoot

Auxin causes the shoot to elongate. Auxin is produced at the shoot apex, primarily by rapidly growing leaf primordia and leaf blades, and migrates toward the base, causing the cells below the apex to expand. It is this expansion (rather than cell division) that is primarily responsible for the increase in the size of the plant. If the shoot tip is cut off, elongation stops. If a suitable amount of IAA—either in an agar block or in a paste—is applied to the cut surface, growth resumes. When comparable quantities of auxin are applied to an intact plant, however, with care taken to assure that the molecules actually penetrate the cuticle and gain access to the cells, only a small enhancement of growth typically occurs. This indicates that the internal auxin supply is the optimal amount for shoot elongation.

Auxin and Root Growth

Auxin, in very small amounts, may stimulate the growth of roots. In somewhat larger amounts, however, it clearly inhibits the growth of the main roots, although it may promote the initiation of new branch roots. It stimulates the development of adventitious roots, and some synthetic auxins are used commercially to stimulate root formation in cuttings (Figure 23-4).

Auxin and Abscission

Auxin is produced in young leaves, but it does not appear to have direct effects on the rate of leaf growth. It does, however, affect leaf drop. As leaves grow older, first certain reusable ions and molecules are returned to the stem, among them magnesium ions, amino acids (derived from proteins), and sugars (some derived from starch). Next, in some plants at least, enzymes break up the middle lamellas in a special layer of weak, thin-walled parenchyma cells, the separation layer of the abscission zone (Figure 21-24), across the base of the petiole. Beneath the separation layer a protective layer composed of heavily suberized cells is formed, further isolating the leaf from the main body of the plant before it drops. Eventually, the leaf is held to the plant only by a few strands of vascular tissue, and these are ultimately broken because of the weight of the leaf or its fluttering in the wind. Abscission has been correlated with a diminished production of auxin in the leaf, among other things, and under many circumstances can be prevented by the application of auxin.

Abscission layers also form in the stems of fruits such as apples and oranges, permitting the fruit to drop as it ripens. Commercial fruit growers often spray their orchards with synthetic auxins in order to keep the ripe fruit on trees until it can be harvested.

23-4
The holly cuttings in the upper row were treated with auxin 21 days before the picture was taken. The cuttings in the lower row were not. Note the growth of adventitious roots on the plants in the upper row.

Auxin and Fruit Growth

Auxin promotes the growth of fruit. Ordinarily, if the flower is not pollinated and fertilization does not take place, the fruit will not develop. In some plants, fertilization of one egg cell is sufficient for normal fruit development, but in others, such as apples or melons, which have many seeds, several must be fertilized for the ovary wall to mature and become fleshy. By treating the female flower parts of certain species with auxin, it is possible to produce parthenocarpic fruit (from *parthenos*, meaning "maiden" or "virgin"), which is fruit produced without fertilization, such as seedless tomatoes, cucumbers, and eggplants.

Apparently, developing seeds are the source of auxin. In the strawberry, if the seeds are removed during the fruit's development, the strawberry stops growing altogether. If a narrow ring of seeds is left, the fruit (actually the fleshy receptacle) forms a bulging girdle of growth in the area of the seeds. If auxin is applied to the denuded receptacle, growth proceeds normally (Figure 23-5).

Auxin and Cell Differentiation

Auxin influences the differentiation of the vascular tissue in the elongating shoot. If one cuts a wedge out of a stem of *Coleus* in such a way as to sever and remove portions of the vascular bundles, new vascular tissues will be formed from cells in the pith and will connect

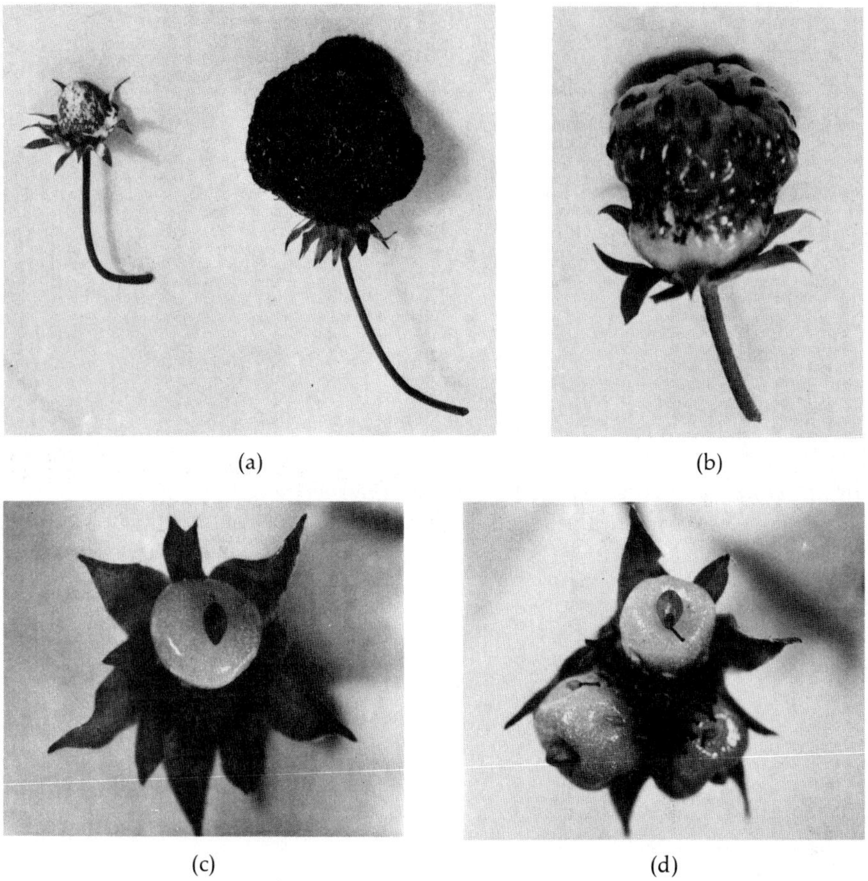

23–5
(a) *Normal strawberry (right) and strawberry from which all the seeds (enclosed in the fruits, which are achenes) were removed (left).* (b) *Strawberry in which three horizontal rows of seeds were left.* (c) *Growth in strawberry induced by one developing seed.* (d) *Growth induced by three developing seeds. If a paste containing auxin is applied to the strawberry from which the seeds have been removed, the strawberry grows normally.*

with the bundles in the uncut regions. If one takes away the leaves and buds above the excision, the formation of new cells is delayed. If one adds IAA to the petiole of the cut leaf just above the excision, formation of vascular tissue resumes. Auxin similarly plays an important role in the joining of vascular traces from developing leaves to the bundles in the stem.

Similar effects are seen in calluses. A callus is a mass of undifferentiated cells that forms when a plant is wounded or when isolated cells are grown in tissue culture. If one takes a callus of lilac pith tissue and grafts a bud on it, vascular tissue is induced in the callus. Similarly, if the callus is grown in a medium containing auxin and sugar (because the callus does not contain photosynthetic cells), vascular tissues form. And this next discovery, made by R. H. Wetmore and his co-workers, is curious. By adjusting the amount of sugar in the medium, one can induce formation of xylem alone, xylem and phloem, or phloem alone. A low concentration of sucrose (1.5 to 2.5 percent) favors xylem, 4 percent favors phloem, in between produces both. This raises a point that needs emphasis: None of the growth-regulatory substances ever acts alone. Their effects are usually produced in concert with other internal factors (such as available sugar), both known and unknown, that are at work in the plant body.

Auxin and the Vascular Cambium

In woody plants, auxin promotes the growth of the cambium. When the meristematic region of the shoot begins to grow in the spring, auxin moving down from the shoot tip stimulates cambial cells to divide, forming secondary phloem and secondary xylem. Again, these effects are modulated by other growth-regulating substances in the plant body. For example, experiments with externally applied IAA and gibberellic acid (see page 488) indicate that, in the intact plant, interactions between auxins and gibberellins determine the relative rates of secondary phloem and secondary xylem production.

Auxin and Lateral Buds

Auxin can also act as an inhibitor. It inhibits the growth of lateral buds, accounting in some species for the phenomenon of apical dominance, mentioned in the preceding section. If the apical meristem is removed, the lateral buds begin to grow. If, however, auxin is applied immediately to the cut surface, the growth of the buds is then inhibited. In a potato plant, which as we have seen is actually a modified stem, the eyes are buds. Treatment of potatoes with synthetic auxins inhibits bud growth and permits the potatoes to be stored for longer periods. Although auxin inhibits lateral buds in most plants tested, it does not do so in all. Apical dominance in some plants may be regulated by competition for ions and nutrients.

How Does Auxin Control Elongation?

Auxin increases the plasticity of the cell wall. When the cell wall softens, the cell enlarges owing to the water pressure of its contents. As the water pressure is reduced by expansion of the cell, the plant cell takes up more water, and the cell thus continues to enlarge until it encounters sufficient resistance from the wall. (The factors involved in the uptake of water will be discussed in Section 7.)

The softening of the cell wall is brought about by a complex series of interactions that are not yet fully understood. During continued growth, auxin stimulates specific RNA and protein biosyntheses, and under the control of this newly forming protein, certain of the old bonds holding the cell wall together are broken as new carbohydrate material is incorporated into the structure. However, application of auxin to coleoptiles can result in increased growth within 3 minutes—too short a time for new protein synthesis to be initiated. Moreover, over short periods, auxin can soften cell walls even when protein synthesis is inhibited. Further, it is known that auxin binds to isolated fragments of plasma membrane. Therefore, it is believed that the very first step of auxin action is to bind to the membrane and cause some critical change in it, but details of this primary action have yet to be worked out.

THE GIBBERELLINS

In the same year (1926) that Went first performed his experiments with blocks of agar, E. Kurosawa of Japan was studying a disease of rice plants called "foolish seedling disease," in which the plants grew rapidly, were spindly, pale-colored, and sickly, and tended to fall over. The cause of these symptoms, Kurosawa discovered, was a chemical produced by a fungus, *Gibberella fujikuroi*, parasitic on seedlings. The substance found in the rice plants was named *gibberellin.*

Gibberellin was isolated and identified chemically by biochemists in Japan in the 1930s, but for several decades it attracted little interest. Then, in 1956, the first successful isolation of gibberellin from a plant rather than a fungus (the seed of the bean *Phaseolus vulgaris*) was made. Since that time, gibberellins have been isolated from many species of plants, and it is now generally believed that they probably occur in all plants. They are present in varying amounts in all parts of the plant, but the highest concentrations are found in immature seeds. More than 40 gibberellins now have been isolated from plant tissues and identified chemically. (Gibberellins are also known as gibberellic acids, often abbreviated GA, with subscript numbers to distinguish them.) They vary slightly in structure (see Figure 23-6) and also in activity. The best studied of the group is GA_3, which is also produced by Kurosawa's fungus.

The gibberellins have dramatic effects on stem elongation in intact plants. A marked increase in the growth of the shoot is the most general response seen in higher plants; often the stems become long and thin and the leaves pale in color. The gibberellins stimulate both cell division and cell elongation and affect leaves as well as stems.

Gibberellins and Dwarf Mutants

The most remarkable results are seen when gibberellins are applied to certain plants that are single-gene dwarf

23-6
Three of the more than 40 gibberellins that have been isolated from natural sources. Gibberellic acid (GA_3) is the most abundant in fungi and the most biologically active in many tests. The minor structural differences that distinguish the other two gibberellins are indicated by arrows.

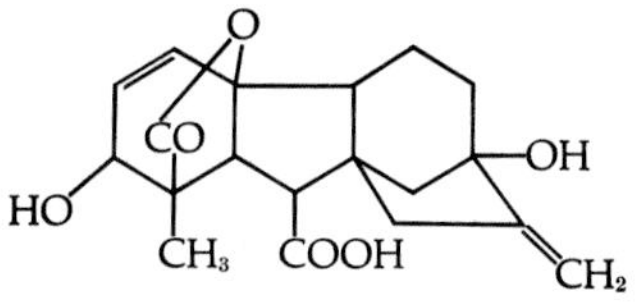

GIBBERELLIC ACID (GA_3)

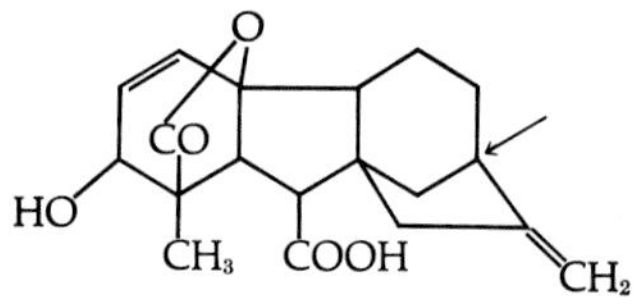

GA_7

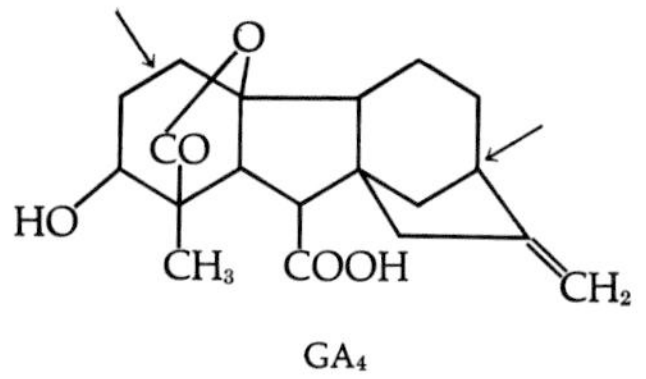

GA_4

mutants (Figure 23-7). Under gibberellin treatment, such plants become indistinguishable from normal tall plants. This striking effect leads to the speculation that the result of the mutation, in biochemical terms, is a loss of the plant's ability to synthesize its own gibberellins. One bioassay for these hormones is their effect on dwarf plants, particularly dwarf corn, an effect that cannot be duplicated by auxin or any of the other known hormones.

Gibberellins and Flowering

Some plants, such as mustard *(Brassica juncea)* or the biennial henbane *(Hyoscyamus niger)*, form rosettes before flowering. (In a rosette, leaves develop but the internodes between them do not elongate.) In these plants, flowering can be induced by exposure to long days, to cold (as in the biennials), or to both. Following the appropriate exposures, the stems elongate—a phenomenon known as bolting—and the plants flower (see Figure 23-8). Application of gibberellin to some of these rosette plants causes bolting and flowering without appropriate cold or long-day exposures.

Gibberellins and Juvenility

The juvenile stages of some plants are different from their adult stages (Figure 23-9). Ivy offers a very familiar example. If sufficiently adult plants of ivy are growing on a building or a wall near you, compare the upper branches, with the lower ones. The form of the leaf is different. Also, the behavior is different. The juvenile branch roots readily; the adult one does not. The adult branch flowers; the juvenile one does not.

If you take an adult branch and nip off the apical meristem, the axillary buds will develop and form new adult branches. If you apply gibberellin to such a bud, however, it will grow into a typical juvenile branch.

23-7
The dwarf plant on the right was given gibberellin, with the one on the left serving as a control. The plants are contender beans, a variety of the common bean (Phaseolus vulgaris).

23-8
Bolting in cabbage (Brassica oleracea) *produced by gibberellin treatment. The plant on the right was treated once a week for 8 weeks.*

(a)

(b)

23–9
Juvenile (a) *and mature* (b) *leaves of* Eucalyptus globulus, *showing the great differences that can occur within a given species. The juvenile leaves are softer and opposite to one another. Their only layer of palisade parenchyma is just below their upper epidermis. The mature leaves are hard, spirally arranged, and hang vertically. In the mature leaves, both surfaces are equally exposed to the light, and there are two layers of palisade parenchyma. In some species, gibberellins promote the maintenance of juvenile characteristics.*

Gibberellins and Pollen and Fruit Development

Gibberellins have been shown to stimulate pollen germination and the growth of pollen tubes in a number of genera, including lilies, lobelias, petunias, and peas. Like auxin, gibberellins can cause the development of parthenocarpic fruits, including apples, currants, cucumbers, and eggplant. In some fruits, such as the mandarin orange, the almond, and the peach, the gibberellins have been effective where auxin was not.

Auxins and gibberellins together, in some instances, produce fruit more than twice as large as those obtainable by the application of either one alone.

Gibberellins and Seeds

The seeds of most plants require a period of dormancy before they will germinate. In certain plants, dormancy usually cannot be broken except by an exposure to cold or to light. In many species, including lettuce, tobacco, and wild oats, gibberellins will substitute for the dormancy-breaking cold or light requirement and promote the growth of the embryo and the emergence of the seedling.

How Do Gibberellins Work?

The most important studies on the mechanism of gibberellin action were carried out simultaneously by investigators in Japan, Australia, and the United States. These studies, which trace the sequence of events in the germination of a barley seed and the early growth of its embryo, show the key role played by gibberellin in this sequence.

In barley and other grass seeds, there is a specialized layer of cells, the aleurone layer, just inside the seed coat. These cells are rich in protein. When the seeds begin to germinate—triggered by the imbibition of water—the embryo releases gibberellins. In response to the gibberellins, the aleurone cells synthesize hydrolytic enzymes, the principal one of which is alpha-amylase, the enzyme that breaks down starch into sugar (Figure 23–10). The enzymes digest the stored food reserves of the starchy endosperm, which are released in the form of sugars and amino acids, absorbed by the scutellum, and then transported to the growing regions of the embryo (Figure 23–11). In this way, the embryo calls forth the substances needed for its growth at the moment it requires them.

23-10
The release of sugar from endosperm induced by gibberellin treatment. These data show that sugars are produced only when the aleurone layer is present. It is, in fact, the aleurone layer that is the source of the enzyme, alpha-amylase, which digests the starches stored in the endosperm.

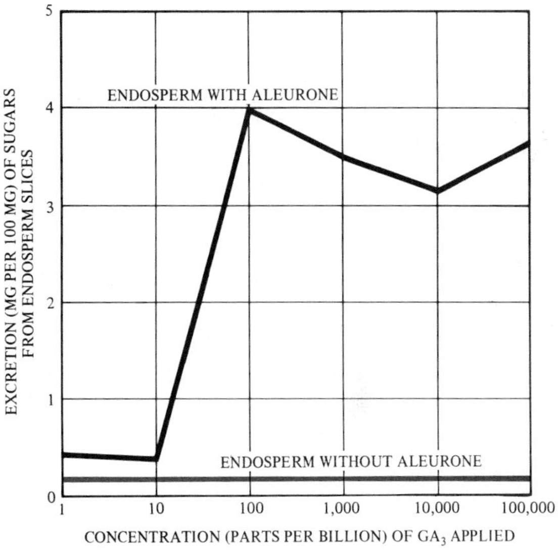

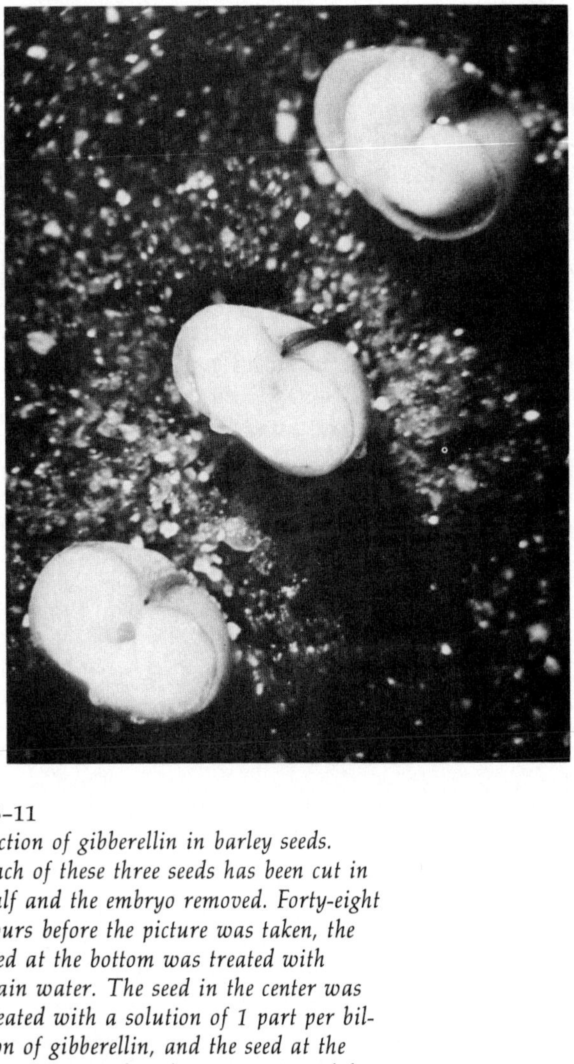

23-11
Action of gibberellin in barley seeds. Each of these three seeds has been cut in half and the embryo removed. Forty-eight hours before the picture was taken, the seed at the bottom was treated with plain water. The seed in the center was treated with a solution of 1 part per billion of gibberellin, and the seed at the top was treated with 100 parts per billion of gibberellin. As you can see, digestion of the starchy storage tissue has begun to take place in the treated seeds.

The investigators believe that gibberellins activate certain genes, causing the synthesis of specific messenger RNA molecules, which, in turn, direct the synthesis of the enzymes. It has not been proved, however, that gibberellins act directly on the gene, although it has been shown that both RNA and protein synthesis take place and are necessary for the appearance of the enzymes. Whatever the details of the mechanism of gibberellin action in aleurone cells, it is clear that the aleurone layer cells constitute a highly differentiated tissue poised to respond to the demands—mediated by the gibberellins—of the growing embryo. This is one of the best-described examples of how hormones integrate the biochemistry and physiology of the different tissues of a whole plant. It is not known whether the way in which gibberellin works in these seeds is related to its effects on other plant organs.

CYTOKININS

In 1941, Johannes van Overbeek, a Dutch plant physiologist, now at Texas A&M University, found that coconut milk (which is a liquid endosperm) contained a potent growth factor different from anything known at that time. This factor, or factors, greatly accelerated the development of plant embryos and promoted the growth of isolated tissues and cells in the test tube. Van Overbeek's discovery had two effects: It gave impetus to studies of isolated plant tissues and organs, and it launched the search for another major group of plant-growth regulators.

The basic medium for tissue culture in plants contains sugar, vitamins, and various salts. Folke Skoog and his co-workers at the University of Wisconsin showed that a stem segment of the tobacco plant *(Nicotiana tabacum)* grew initially in such a culture medium, but that its growth soon slowed or stopped. Apparently, some growth stimulus originally present in the tobacco stem became exhausted. The addition of IAA had no effect. However, when coconut milk was added to the medium, the cells began to divide and growth resumed.

While studying the differentiation of plants, particularly the carrot, in tissue culture, F. C. Steward and his group at Cornell University observed that whatever was present in the endosperm of coconut milk was able to induce previously differentiated, mature carrot cells to divide again. Similar effects were seen in a variety of isolated tissues. This group of growth regulators became known as the *cytokinins,* from cytokinesis, a term for cell division.

Skoog and his co-workers set out to identify this new growth factor. Coconut milk is a rich and complex mixture of growth-promoting substances, and after many years of effort, although they succeeded in purifying one growth factor a thousandfold and in identifying it as a purine, they could not isolate it. Finally, they reluctantly decided that it would be impossible to find the needle in this particular chemical haystack. So, changing course, they tested a variety of purine-containing substances—largely nucleic acids—in the hope of finding a new source of the material.

Pursuing this new course, Carlos O. Miller, then a postdoctoral student with Skoog, searched for bottles with a nucleic acid label. He tested one marked "Herring Sperm DNA" and found that it made tobacco cells divide. More herring sperm DNA was ordered. To the consternation of the investigators, these fresh preparations did not work. As a last resort, they tried another old sample. This one, too, was active. Apparently, the factor was to be found only as a breakdown product of DNA. Subsequently, they found that a variety of preparations of nucleic acids that had aged, or had been aged artificially, as by heating, contained material that produced the cell-division response.

Subsequently, Skoog and co-workers succeeded in isolating the growth factor from one of these DNA preparations and identifying it chemically. They called this substance kinetin. Kinetin, as you can see in Figure 23–12, resembles the purine adenine, the clue that led to its discovery. Kinetin, which probably does not exist at all in plants in nature, has a relatively simple structure, and biochemists were soon able to synthesize a number of other, related compounds that behaved liked cytokinins. Eventually, a natural cytokinin was isolated from kernels of corn *(Zea mays).* Called zeatin, it is the most active of the cytokinins.

Cytokinins have now been found in about 40 different species of higher plants, largely in actively dividing tissues, including seeds, fruits, and roots. They have also been found in bleeding sap, the sap that drips out of cracks and pruning cuts of many types of plants.

Auxin-Cytokinin Interactions

Studies of interactions involving auxin and cytokinins are helping physiologists understand how plant hormones work to produce the total growth pattern of the plant. Apparently, the undifferentiated plant cell has two courses open to it: Either it can enlarge, divide, enlarge, and divide again, or it can elongate without cell division. The cell that divides repeatedly remains essentially undifferentiated, or embryonic, whereas the elongating cell tends to differentiate, or become specialized. In studies of tobacco stem tissues, the addition of IAA to the tissue culture produced rapid cell expansion, so that giant cells were formed. Kinetin alone had little or no effect. IAA plus kinetin resulted in rapid cell division, so that large numbers of relatively tiny cells were formed. In other

23–12
Note the resemblances between the purine adenine and these four cytokinins. Kinetin and 6-benzylamino purine (BAP) are commonly used synthetic cytokinins. Zeatin and i^6 Ade have been isolated from plant material.

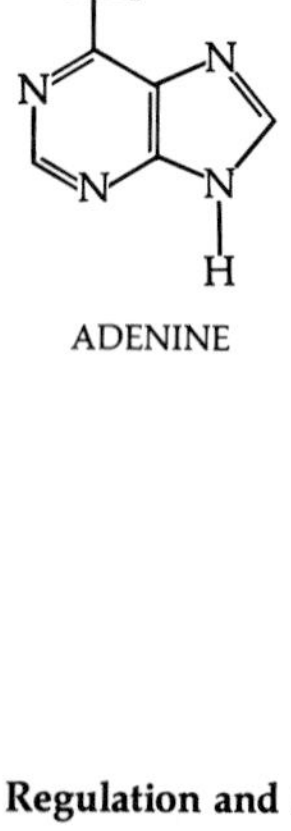
ADENINE

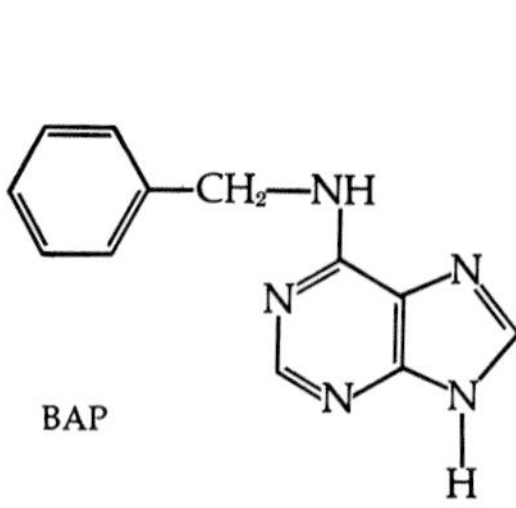
KINETIN

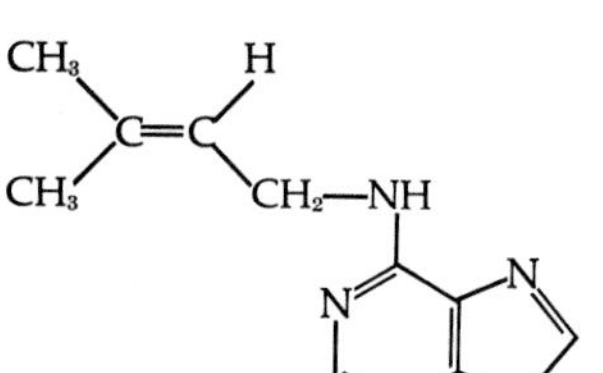
ZEATIN

BAP

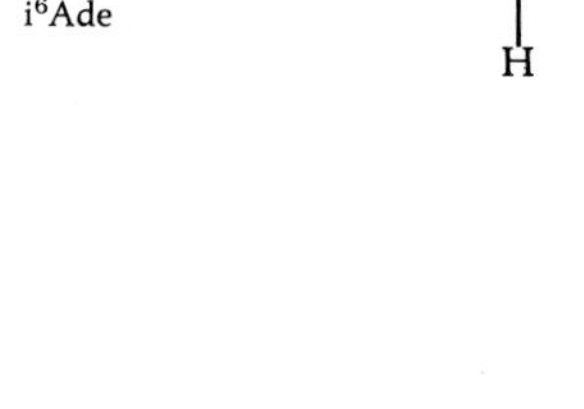
i^6Ade

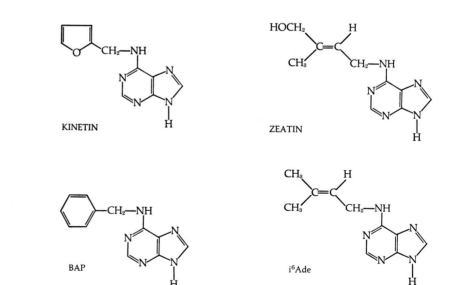

23-13
Auxin-cytokinin interactions. Kinetin alone has little effect on the growth of tobacco callus in tissue culture. IAA alone, regardless of the concentration used, causes the culture to grow to a weight of about 10 grams. When both hormones are present, growth is greatly increased. Notice, however, that when optimum concentrations are exceeded, growth responses cease.

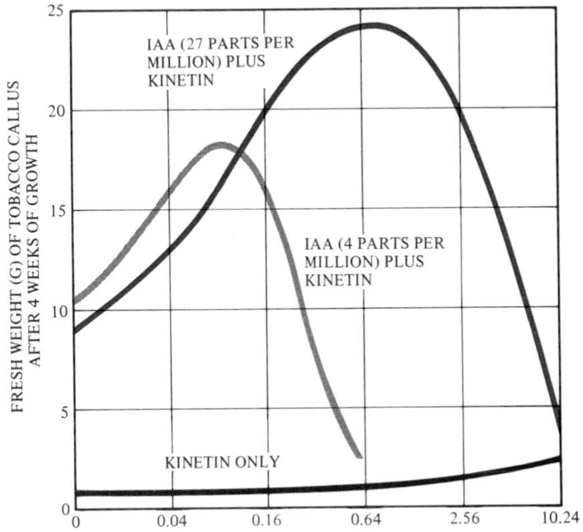

23-14
Two small buds formed on geranium (Pelargonium) *callus tissue cultured on a medium containing auxin and cytokinin.*

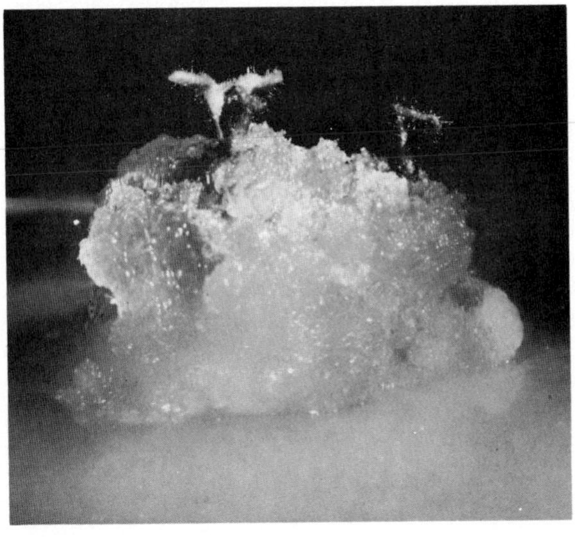

words, the addition of the kinetin together with IAA (although not kinetin alone) switched the cells to a meristematic course.

In the presence of a high concentration of auxin, callus tissue frequently gives rise to organized roots. In tobacco pith callus, whether roots or buds form is dependent on the relative concentration of auxins and kinetin. With higher concentrations of auxin, roots form; with higher concentrations of kinetin, buds form; when the two are present at roughly equal concentrations, the callus continues to produce undifferentiated cells (Figure 23-13). Although these effects have not been observed in tissues of plant types other than tobacco, they do serve to suggest how hormonal interactions may be responsible for differentiation.

In another tissue culture study, in which tuber tissue of the Jerusalem artichoke was used, it was shown that a third substance, the calcium ion, can modify the action of the auxin-cytokinin combination. In this study, IAA plus low concentrations of kinetin was shown to favor cell enlargement, but as Ca^{2+} was added to the culture, there was a steady shift in the growth pattern from cell enlargement to cell division. High concentrations of calcium prevent the cell wall from expanding, and at such concentrations the cell switches course and divides. Thus, not only do hormones modify the effects of hormones, but these combined effects are, in turn, modified by nonhormonal factors, such as calcium.

Effects of Cytokinins on Buds and Leaves

Cytokinins have been shown in some experiments to have effects that oppose the effects of auxin. Application of cytokinins to lateral buds will sometimes cause them to develop even in the presence of auxin, thus modifying apical dominance.

Cytokinins and Seed Growth

The seeds of certain varieties of lettuce that ordinarily require light for germination will germinate even in the dark if treated with kinetin solutions. Cytokinins are not naturally present in the dry seed, however; they do not appear until the seed starts to grow, indicating that they are a result rather than a cause of germination. According to present concepts, cytokinins (and probably auxin also) are released by the endosperm as a consequence of its breakdown by the hydrolytic enzymes under the influence of gibberellin.

Cytokinins and Leaf Senescence

In most species of plants, the leaves begin to turn yellow as soon as they are removed from the plant. This yellowing, which is due to a loss of chlorophyll, can be prevented by cytokinin. Leaves of the cocklebur plant,

MERISTEM CULTURE

As long ago as the 1930s, laboratory botanists developed techniques for tissue culture—the growing of plant cells in test tubes. Tiny fragments of meristem were implanted under sterile conditions in a medium employing minerals and various combinations of organic compounds. Under these conditions, the meristematic cells proliferated to form uniform undifferentiated tissue masses. Somewhat more recently, it was discovered that by adjusting the hormone balance in the medium, it was possible to make these cells differentiate and grow into mature plants (see Figure 23–14). Because of the special skills and equipment needed for tissue culture, the use of these techniques has been confined largely to research laboratories. However, meristem culture has now found two wide practical applications. First, it has been discovered that in virus-infected plants, the meristem is often free of virus. Thus, by growing new plants from meristems, it is possible to develop virus-free strains of plants such as chrysanthemums and carnations, in which particular desirable characteristics and virus infection had previously been inextricably combined. Second, it has been found that it is practical to culture comparatively large quantities of undifferentiated tissue of meristematic origin by transferring cells to subcultures every few weeks. Hundreds or, if desired, thousands of subcultures can be produced in this way in a relatively short time and in a small space. Then, by altering the hormone balance, each of these can be turned into small but perfect plantlets. This technique has already been employed in the culture of exotic orchids and is being extended to other hard-to-grow varieties.

for example, when excised and floated on plain water, turn yellow in about 10 days. If a small amount of kinetin is present in the water, much of the chlorophyll and the fresh appearance of the leaf are maintained. If excised leaves are spotted with kinetin-containing solutions, the spots remain green while the rest of the leaf yellows. Such studies, which also have been carried out on radishes and other plants, lead to the hypothesis that senescence in leaves, and probably in other plant parts as well, results from the progressive "turning off" of segments of DNA, with a consequent loss of messenger RNA production and protein synthesis. The cytokinins, it is proposed, prevent the DNA from being turned off and so promote continued enzyme synthesis and the continued production of other compounds such as chlorophyll.

However, if a cytokinin-spotted leaf contains radioactive amino acids, labeled with ^{14}C, it can be shown that the amino acids migrate from other parts of the leaf to the cytokinin-treated areas. The current hypothesis does not account for this migration.

How Do Cytokinins Work?

Ever since the original isolation of the cytokinins from preparations of nucleic acids, plant physiologists have suspected that the hormones might in some way be involved with the nucleic acids. Transfer RNA molecules, as you may recall, contain a number of unusual bases (see Figure 3–12, page 75). In some types of transfer RNA, the natural cytokinin i^6Ade (^{6}N-isopentenyladenine), itself an unusual base, is incorporated into the molecule. For example, i^6Ade is found in the serine and tyrosine tRNA molecules, in which it is located immediately adjacent to the anticodon. It is still not known, however, if its presence or its position in tRNA is related to its activity in promoting cell division.

ABSCISIC ACID

At certain times, the survival of the plant depends on its ability to restrain its growth or its reproductive activities. Following the early discovery of the growth-promoting hormones, plant physiologists began to speculate that regulatory hormones with inhibitory actions would also be found. Finally, in 1949, it was discovered that the dormant buds of ash and potatoes contained large amounts of inhibitors. These inhibitors blocked the effects induced by IAA in the *Avena* coleoptile. When dormancy in the buds was broken, the inhibitor contents declined. These inhibitors became known as *dormins.*

During the 1960s, several investigators reported the discovery in leaves and fruits of a substance capable of accelerating abscission. One of these, called abscisin, was identified chemically. In 1965, one of the dormins was also identified chemically, and the two, the abscisin and the dormin, were found to be identical. The compound is now known as *abscisic acid* (Figure 23–15).

Abscisic acid is collected largely from the ovary bases of fruits. The fruit of the cotton plant has proved to be a particularly rich source. The largest amounts of abscisic acids are found at the time of fruit drop.

Application of abscisic acid to vegetable buds changes them to winter buds by converting the leaf primordia into bud scales. Its inhibitory effects on buds can be overcome by gibberellin. The appearance of alpha-amylase, induced by gibberellin in the barley seed, is inhibited by abscisic acid, which seems to depress protein production in general. Auxin, on the other hand, seems to act both by interacting with the plasma membrane and by accelerating the production of specific proteins. It therefore seems to be antagonistic to abscisic acid in its action. There may even be more direct interactions between abscisic acid and gibberellin in plant growth.

23–15
Abscisic acid, an inhibitor that blocks the action of the growth-promoting hormones.

ABSCISIC ACID (DORMIN)

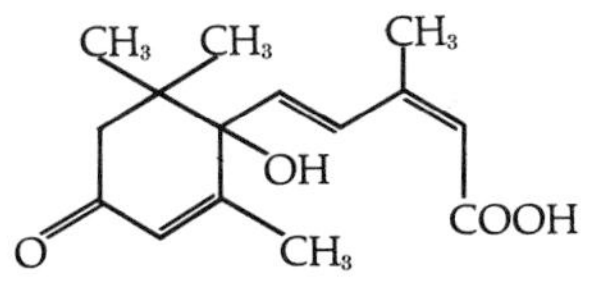

23–16
Ethylene, a simple hydrocarbon involved in the ripening of fruit.

$CH_2{=}CH_2$

ETHYLENE

If a drop of abscisic acid is spotted on a leaf, the treated areas yellow rapidly, even though the rest of the leaf stays green, an effect that is opposite to that of the cytokinins. Whether this is a direct or indirect action is not known at present.

ETHYLENE

In recent years, increasing attention has been paid to other natural substances that accelerate abscission. One of these is *ethylene* (Figure 23–16), a hydrocarbon that affects fruit ripening. In leaves, ethylene presumably triggers the enzymes that bring about the changes associated with abscission. Moreover, ethylene is produced when an unknown senescence factor, different from abscisic acid, is released from the cells of the leaf after protein synthesis in them is depressed by abscisic acid.

Ripening in fruit involves a number of changes. In fleshy fruits, the chlorophyll is degraded, and other pigments may form, changing the fruit color. Simultaneously, the fleshy part of the fruit softens. This is a result of the enzymatic digestion of pectin, the principal component of the middle lamella. When the middle lamella is weakened, cells are able to slip past one another. During this same period, starches and organic acids or, as in the case of the avocado, oils, are metabolized into sugars. As a consequence of these changes, fruits become conspicuous and palatable and thus attractive to animals that eat the fruit and so scatter the seed.

During the time of ripening of many fruits, there is a large increase in cellular respiration, evidenced by an increased uptake of oxygen. This phase is known as the climacteric. The relationship between the climacteric and the other events of fruit ripening is not known, but the ripening of fruits can be suppressed by suppressing the intensity of the climacteric. For example, cold suppresses it, and in some fruits, cold stops the climacteric permanently. Fruits can be stored for very long periods of time in a vacuum; under such conditions, the amount of available oxygen is minimal, which suppresses respiration; and ethylene, which speeds the onset of the climacteric, is held at low levels. After the climacteric, senescence sets in, and the fruit becomes susceptible to invasions by fungi and other microorganisms.

In the early 1900s, many fruit growers made a practice of improving the color and increasing the sweetness of citrus fruits by "curing" them in a room with a kerosene stove. (Long before this, the Chinese used to ripen fruits in rooms where incense was being burned.) It was long believed that it was the heat that ripened the fruits. Ambitious fruit growers, who installed more modern equipment, found to their sorrow that this was not the case. As experiments showed, it is actually the incomplete combustion products of the kerosene that are responsible. The most active gas was identified as ethylene (Figure 23–16). As little as 1 part per million of ethylene in the air will speed the onset of the climacteric.

As early as 1910, it was reported that gases emanating from oranges hastened the ripening of bananas, but it was not until almost 25 years later that ethylene was identified as a natural product of numerous fruits and plant tissues. The amounts produced by plants are very small, and new and extremely sensitive assay methods had to be developed before it could be proved that ethylene production began *before* the climacteric, even though the largest amounts coincide with the climacteric. When this was established, ethylene became generally accepted as a natural plant-growth regulator. It has now been found in fruits (in all the types tested), flowers, leaves, leafy stems, and roots of many different plant species and also in some species of fungi.

Auxin, at certain concentrations, causes a burst of ethylene production in some parts of some plants. It is believed that some of the effects on fruits and flowers once attributed to auxin are related to auxin's effects on ethylene production.

In addition to its effects on fruit ripening, ethylene causes leaves to abscise, chlorophyll to blanche, flowers to fade, and the petioles of seedlings to grow more rapidly on the upper side and therefore curve downward. This effect, which is known as epinasty, is so specific that it is used as a test for ethylene. Ethylene is also responsible for a host of other effects that may or may not have anything to do with the growth of the plant under normal conditions. In some plants, ethylene has been shown to cause a lateral expansion (in contrast to the usual elongation) of the cell, apparently by changing the orientation of the microfibrils. The way in which ethylene affects the ripening of fruit is not known, but a variety of hypotheses are currently under investigation, and this, too, is an area in which new data may be expected very soon.

SUMMARY

Hormones are important regulators of growth in both higher animals and higher plants. A hormone is a chemical produced in certain tissues of the organism and transported to other tissues of the organism, where it causes a physiological response. It is active in extremely small amounts.

Naturally occurring auxin is a hormone that is produced in the tips of coleoptiles and in apical meristems of shoots and probably of roots. In coleoptiles and shoots, it travels unidirectionally toward the base of the plant. It causes lengthening of the shoot and the coleoptile, chiefly by promoting cell elongation. Studies indicate that its effect on cell elongation is achieved in some indirect way by a relaxation of the cellulose fibrils of the cell wall, permitting the cell to expand. It also plays a role in differentiation of vascular tissue. Auxin initiates cell division in the vascular cambium. It often inhibits growth in lateral buds, thus maintaining apical dominance. The same quantity of auxin that promotes growth in the stem inhibits growth in the main root system. Auxin promotes the initiation of branch roots and adventitious roots, however. It retards abscission in leaves and fruits. In fruits, auxin produced by seeds or the pollen tube stimulates growth of the ovary wall. Its capacity to produce such varied effects is believed to result from the different responses of the various target tissues.

The gibberellins were first isolated from a parasitic fungus that causes abnormal growth in rice seedlings. They were subsequently found to be natural growth hormones present in higher plants. The most dramatic effects of gibberellins are seen in dwarf plants, in which application of gibberellins restores normal growth, and in plants with a rosette form of growth, in which gibberellins cause bolting. Gibberellins cause seed germination in grasses. In the barley seed, the embryo releases gibberellins, which cause the aleurone layer of the endosperm to produce several enzymes, including alpha-amylase. Alpha-amylase breaks down the starch stored in the endosperm, releasing sugar. The sugar nourishes the embryo and promotes the germination of the seed.

The cytokinins, a third class of growth hormone, were first discovered as a consequence of their capacity to promote cell division in cultures of plant tissues. They are chemically related to components of nucleic acids. Cytokinins plus auxin cause cell division in plant tissue culture. In tobacco pith cultures, a high concentration of auxin promotes root formation while a high concentration of cytokinin promotes bud formation. In intact plants, cytokinin promotes the growth of lateral buds, opposing the effects of auxin. Cytokinins prevent senescence in leaves by stimulating protein synthesis.

Abscisic acid is a growth-inhibiting hormone that has been found in dormant buds and in fruits, with a maximum amount present just before the fruit drops. Abscisic acid has opposed the effects of all three types of growth hormones in various laboratory tests.

Ethylene is a gas produced by the incomplete combustion of hydrocarbons. It is also a natural plant growth regulator. It produces a number of physiological effects, including ripening in fruit.

Exposing a plant tissue to a hormone has been compared to putting a dime in a vending machine. You may get your morning newspaper, a candy bar, or a record on the jukebox. It depends not so much on the dime as on the machine in which you put it. Similarly, the effects of plant hormones depend largely on the target tissues and the chemical environment in which these tissues find themselves.

CHAPTER 24

External Factors and Plant Growth

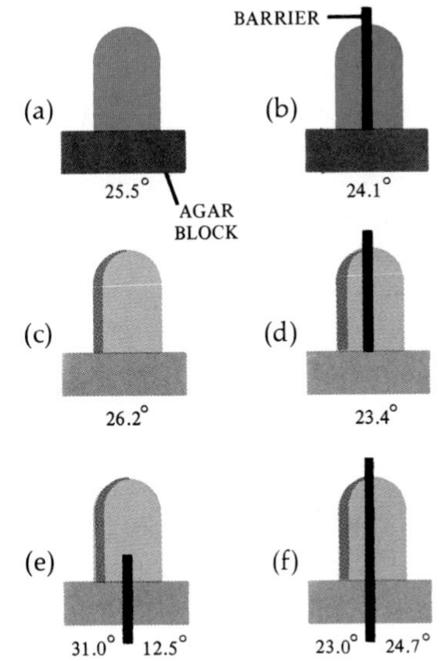

24–1
Phototropism and auxin. Experiments (a) *and* (b), *performed in the dark, show that splitting the tip of the coleoptile and inserting a barrier does not significantly affect the total amount of auxin diffused from the tip. The amount of auxin produced is shown by the numbers below the agar block. These numbers indicate the degree of curvature produced in an amputated coleoptile by the amount of hormone that has diffused into the agar.*

The other experiments were performed with light coming from the right, as indicated by the shading. A comparison of (c) *and* (d) *with* (a) *and* (b) *shows that auxin production is not dependent upon light. (The slight differences in curvature shown are not significant.) But if a barrier is inserted in the agar block* (e), *it is discovered that light causes the displacement of the auxin away from the light. Finally, experiment* (f) *shows that it is displacement that has occurred, not different rates of production in the light and dark sides, because when displacement is prevented with a barrier, auxin production is not significantly different in the two sides.*

Living things must regulate their activities in accordance with the world around them. Animals, being mobile, can change their circumstances to some extent—foraging for food, courting a mate, and seeking or even making shelters in bad weather. A higher plant, however, once it sends down its first root, is immobilized and unable to modify the environment in which it must grow and reproduce. For this reason, the higher plants are characterized by remarkable abilities to adjust to and even anticipate changes in their environment. These special adaptations are manifested chiefly in changing patterns of growth.

THE TROPISMS

Perhaps the most familiar interaction between plants and the external world is the bending toward light of the growing tips of plants, studied by the Darwins some 90 years ago. This *phototropism,* it is now known, is caused by the elongation, under the influence of auxin, of the cells on the shaded side of the tip (Figure 24–1). Why do these cells elongate more than those on the lighted side? Three possible answers to this question have been suggested: (1) Light decreases the sensitivity of the cells to auxin. (2) Light destroys auxin. (3) Light drives auxin to the shaded side of the growing tip. To choose among these hypotheses, recent experiments, based on earlier work by Frits Went, have been carried out by Winslow Briggs and his co-workers. These investigators showed first that the same total amount of auxin is diffused from the tip in the light as in the dark. Following the exposure to light, however, the amount diffused from the shaded side is greater than the amount diffused from the lighted side. If the tip is split and a barrier, such as a thin piece of glass, is placed between the two halves, the differential distribution of auxin does not occur. In other words, Briggs clearly demonstrated that auxin (or perhaps a precursor) migrates from the light side to the dark side and that the turning of the shoot is a response to this

(a)

(b)

24–2
Leaves of the wood sorrel (Oxalis), *day* (a) *and night* (b). *One hypothesis concerning the function of such "sleep" movements is that they protect the leaves from absorbing moonlight on bright nights, thus protecting the photoperiodic phenomena discussed later in the chapter. Another, proposed by Darwin almost 100 years ago, is that the folding protects against heat loss from the leaves by night. (Richard F. Trump, OMIKRON.)*

unequal distribution. Recent experiments using IAA (indoleacetic acid) labeled with ^{14}C have shown clearly that it is the auxin that migrates. It has also been shown that only light of wavelengths less than 500 nanometers induces this migration.

Another familiar tropism is *geotropism,* a response to gravity manifested especially conspicuously by seedlings. If a seedling is placed on its side, its root will grow downward and its shoot will grow upward. In shoots that for any reason are oriented horizontally, differences both in gibberellin and in auxin concentration develop between the upper and lower sides. Collectively, these cause the lower side of the shoot to elongate more than the upper side, and the shoot to grow upward. When it becomes vertical, the lateral asymmetries in concentration disappear, and growth continues in the upright direction.

Asymmetries of growth-regulating agents are less well understood in roots, but it is known that these are such as to cause the upper side of the root to outgrow the lower side so that the root bends downward. It is not understood how plants receive the gravitational stimulus. Some work indicates that heavy intracellular particles, such as starch-containing plastids, may be involved, but no one has suggested a plausible way for them to participate in the sensory process (see page 420).

CIRCADIAN RHYTHMS

It is a common observation that some plants open their flowers in the morning and close them at dusk or spread their leaves in the sunlight and fold them toward the stem at night (Figure 24–2). As long ago as 1729, the French scientist Jean-Jacques de Mairan noticed that these diurnal movements continue even when the plants are kept in dim light (Figure 24–3). More recent studies have shown that less evident activities, such as photosynthesis, auxin production, and the rate of cell division, also have regular daily rhythms, which continue even when all environmental conditions are kept constant. These regular, approximately 24-hour cycles have come to be called *circadian rhythms,* from the Latin *circa,* meaning "about," and *dies,* "day." Circadian rhythms, which have been found to exist almost universally throughout the plant and animal kingdoms, appear to be absent in the prokaryotes (bacteria and blue-green algae) only.

Are the Rhythms Endogenous?

Are these rhythms actually internal—that is, caused by factors entirely within the organism—or is the organism keeping itself in tune with some external factor? For a number of years, biologists debated whether it might not be some environmental force, such as cosmic rays, the magnetic field of the earth, or the earth's rotation, that

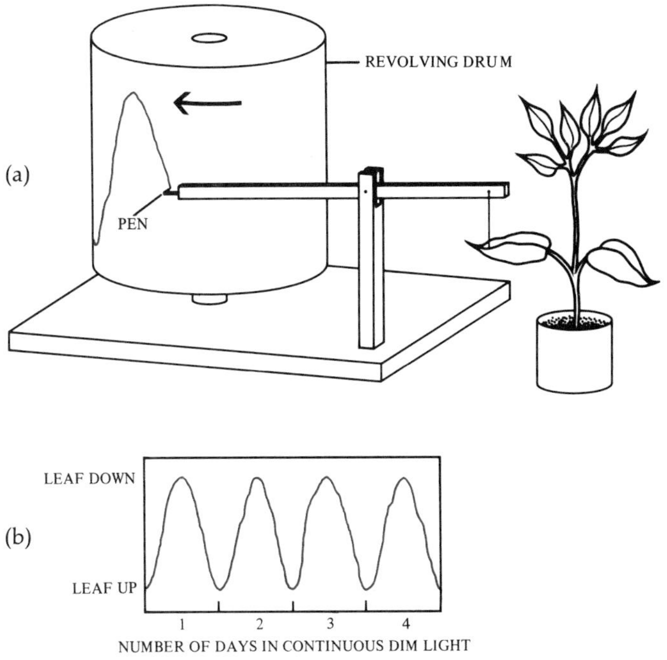

24–3
In many plants, the leaves move outward perpendicular to the sun's rays during the day and upward to the stem at night. These "sleep movements" can be recorded on a revolving drum using a delicately balanced pen-and-lever system attached to a leaf by a fine thread (a). *Many plants, such as the bean* (Phaseolus vulgaris) *shown here, will continue to exhibit these movements for several days even when kept in continuous dim light. A recording of this circadian rhythm is shown in* (b).

was setting the rhythms. In this regard, it was, of course, theoretically important that the rhythms studied were circadian (that is, the period of a rhythm is *about* 24 hours) and that there were slight individual differences among organisms.

Attempts to settle this recurrent controversy have led to countless experiments under an extraordinary variety of conditions. Organisms have been taken down in salt mines, shipped to the South Pole, flown halfway around the world in airplanes, and, most recently, orbited in satellites. Although there is still a vocal minority that believes that circadian rhythms are under the influence of a subtle geophysical factor, most workers now agree that the rhythms are endogenous—that is, they are controlled by internal mechanisms in the organism. The internal timing device is often referred to as a *biological clock.*

Setting the Clock

Under constant environmental conditions, the period of a circadian rhythm is *free-running*—that is, its natural period (usually between 21 and 27 hours) does not have to be reset at each cycle. Although circadian rhythms originate within the organisms themselves, the environment acts as a synchronizing agent, a factor that is important to the survival of individuals and species. In fact, the environment is responsible for keeping a circadian rhythm in step with the daily 24-hour light-dark cycle. If a circadian rhythm of a plant were greater or less than 24 hours, the rhythm would soon get out of step with the 24-hour light-dark cycle. Then, if only the circadian rhythm were followed, a phenomenon such as flowering, which generally takes place in the light period, would occur at a different time each day, including during the dark period. It is necessary, therefore, for the plant to become resynchronized—that is, to become *entrained* to the 24-hour day.

Entrainment is the process by which a periodic repetition of light and dark, or another external cycle, causes a circadian rhythm to remain synchronized with the same cycle as the entraining factor. Light and temperature cycles are the principal factors in entrainment (Figure 24–4).

Another feature of interest to biological clock watchers is that these rhythms do not automatically speed up as the temperature rises. (The reason one might expect them to is that biochemical activities—and the clock must have a biochemical basis—take place more rapidly at high temperatures than at low ones.) Some clocks run slightly faster as the temperature rises, but others go more slowly, and many are almost unchanged. Therefore, clocks must contain within their workings some sort of compensatory mechanism, a feedback system that adjusts them to temperature changes. Such a feature would be very important, of course, in plants and in those animals that cannot regulate their internal temperatures.

Some recent evidence suggests that the plasma membranes of cells may be the key to the biological clock, regulating as they do the passage of ions into and out of cells. These movements of ions are probably mediated by proteins in the membrane (see page 62), but their activity would probably depend in turn on the fluidity of

24-4
(a) *In* Gonyaulax polyedra, *a single-celled marine alga, three separate functions follow separate circadian rhythms, two of which are shown in the graph: (1) bioluminescence, which reaches a peak in the middle of the night; (2) photosynthesis, which reaches a peak in the middle of the day; and (3) cell division, which is restricted to the hours just before dawn. If* Gonyaulax *is kept in continuous dim light, these three functions continue to occur with the same rhythm for days and even weeks, long after a number of cell divisions have taken place.* (b) *The rhythm of bioluminescence in* Gonyaulax, *like most circadian rhythms, can be altered by modifying the cycles of illumination. For example, if the investigators expose cultures of the cells to alternating light-dark periods of 6 or 7 hours each, the rhythmic function will become entrained to the imposed cycle. In continuous dim light, however, the organism will return to its original rhythm of about 24 hours.* (c) *Photomicrograph of* Gongaulax polyedra.

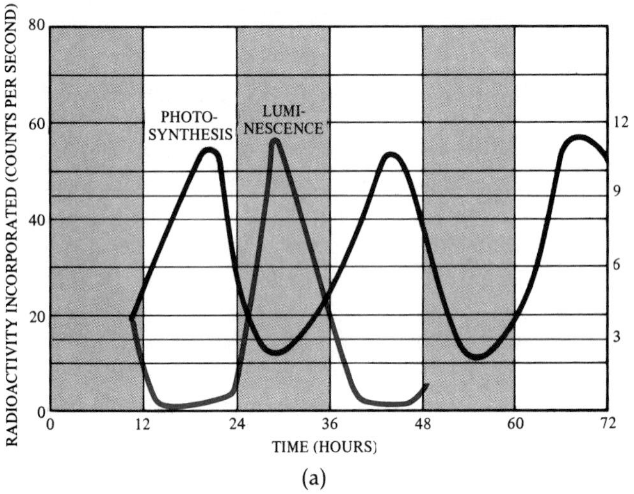

(a)

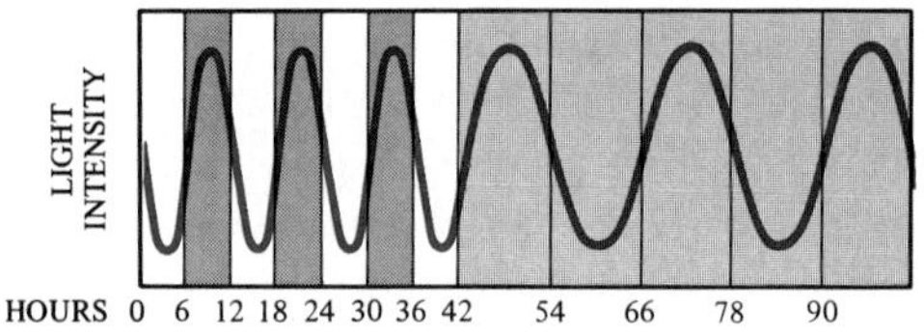

(b)

lipids in the membrane. Lipids in membranes are not sensitive to changes in temperature and could well provide the basis for the temperature insensitivity of biological clocks.

Some biological rhythms regulate the interaction between organisms. For example, some plants secrete nectar at certain specific times of the day. As a result, bees—which have their own biological clocks—become accustomed to visiting these flowers at these times, thereby ensuring maximum rewards for the bees and cross-pollination for the flowers.

For most organisms, however, the use of the biological clock for such special purposes is probably a secondary one. The primary usefulness of the clock is to enable the plant or animal to recognize the changing seasons of the year by accurately measuring changing day length. In this way, the organism is prepared for coming changes in the environment, regulating its growth, reproduction, and other activities accordingly.

PHOTOPERIODISM

Fifty years ago a mutant appeared in a field of tobacco plants *(Nicotiana tabacum)* growing near Washington, D.C. The new variety had unusually large leaves and stood over 3 meters tall. As the season progressed, the regular plants flowered, but Maryland Mammoth, as it came to be known, just grew bigger and bigger. Investigators from the Department of Agriculture, W. W. Garner and H. A. Allard, took cuttings from the Mammoth and put them in the greenhouse, where they would be safe from frost. These cuttings flowered in December, although by then they were only 1.5 meters tall, half the size of their parent. New Maryland Mammoths grew from their seed, and these, too, did not flower until almost winter.

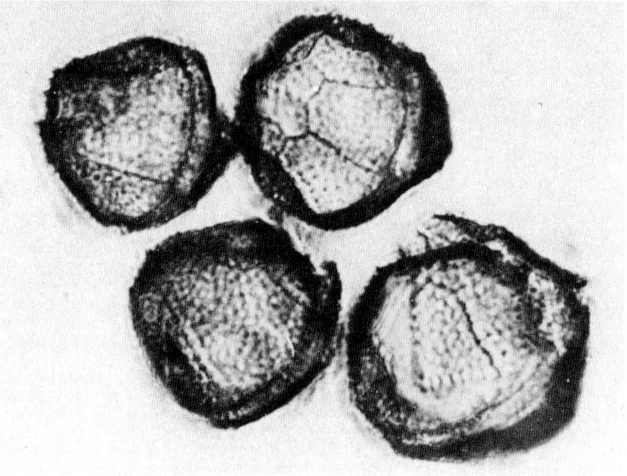

(c)

Coincidentally, Garner and Allard were carrying out experiments with the Biloxi variety of soybean *(Glycine max).* Agriculturalists were interested in spacing out the soybean harvest by making successive sowings of seeds at two-week intervals from early May through June. But spacing out the planting had no effect; all the plants, no matter when the seeds were sown, came into flower at the same time—in September.

The investigators started growing these two kinds of plants—Maryland Mammoth tobacco and Biloxi soybeans—under a wide variety of controlled conditions of temperature, moisture, nutrition, and light. They eventually found that the critical factor in both species was the length of day. Neither plant would flower unless the day length was shorter than a critical number of hours. Consequently, soybeans, no matter when they were planted, all flowered as soon as the days became short enough, which was in September, and the Maryland Mammoth, no matter how tall it grew, would not flower until December, when the days had become even shorter.

Garner and Allard called this phenomenon *photoperiodism.* Photoperiodism is a biological response to a change in the proportions of light and dark in a 24-hour daily cycle. Photoperiodism has now been shown to initiate mating and other activities in animals as diverse as codling moths, spruce budworms, aphids, and potato worms, as well as many species of fish, birds, and mammals.

Long-Day Plants and Short-Day Plants

Garner and Allard went on to test and confirm their discovery with many other species of plants. Following this single lead, they were able to answer a host of questions that had long troubled both professional botanists and amateur gardeners. Why, for example, is there no ragweed in northern Maine? *Answer:* Because ragweed starts making flowers when the day is about $14\frac{1}{2}$ hours long. The long summer days do not shorten to $14\frac{1}{2}$ hours in northern Maine until August, and then there is not time for ragweed seed to develop before the frost. Why does spinach not grow in the tropics? *Answer:* Because spinach needs 14 hours of light a day for a period of at least two weeks in order to flower, and this never happens in the tropics. As you can see, the discovery of the photoperiodic control of flowering was not only interesting but of great practical importance.

The investigators found that plants are of three general types, which they called *short-day, long-day,* and *day-neutral.* Short-day plants flower in early spring or fall; they must have a light period shorter than a critical length. For instance, the common cocklebur *(Xanthium strumarium)* is induced to flower by 16 hours or less of light (Figures 24-5 and 24-6). Other short-day plants are some chrysanthemums, poinsettias, strawberries, and primroses.

Long-day plants, which flower chiefly in the summer, will flower only if the light periods are longer than a critical length. Spinach, some potatoes, some wheat varieties, henbane, and lettuce are examples of long-day plants.

Cocklebur and spinach will both bloom if exposed to 14 hours of daylight, yet one is designated as short-day and one as long-day. The important factor is not the absolute length of the photoperiod but rather whether it is longer or shorter than a particular critical interval. Day-neutral plants flower without respect to day length.

Within individual species of plants that cover a large north–south range, different photoperiodic ecotypes have often been observed. Thus, in many prairie grasses, species that may occur from southern Canada to Texas, northern ecotypes flower before southern ones when they are grown together in a common environment. Different populations are precisely adjusted to the demands of the photoperiodic regime where they occur.

The photoperiodic response can be remarkably pre-

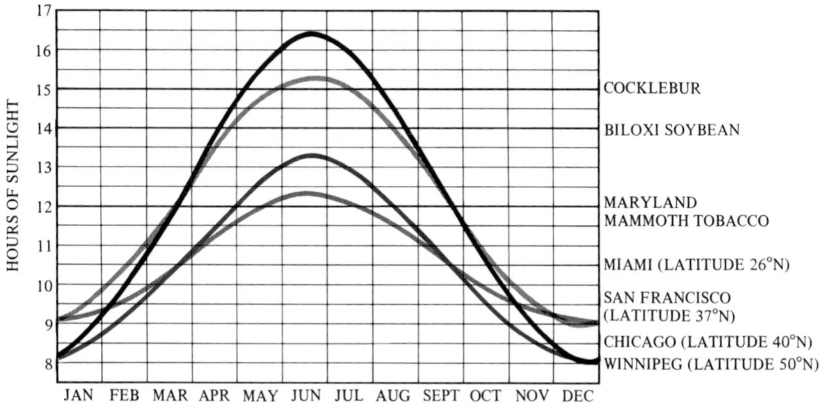

24-5
Length of day determines when plants flower. The four curves depict the annual change of day length in four North American cities at four different latitudes. The heavy horizontal lines indicate the flowering requirements of three different short-day plants. The cocklebur, for instance, requires 16 hours or less *of light. In Miami, it can flower as soon as it matures, but in Winnipeg, the buds do not appear until early August, so late that the frost will probably kill the plants before the seed is set. (Adapted from "The Control of Flowering" by Aubrey W. Naylor. Copyright © May 1952 by Scientific American, Inc. All rights reserved.)*

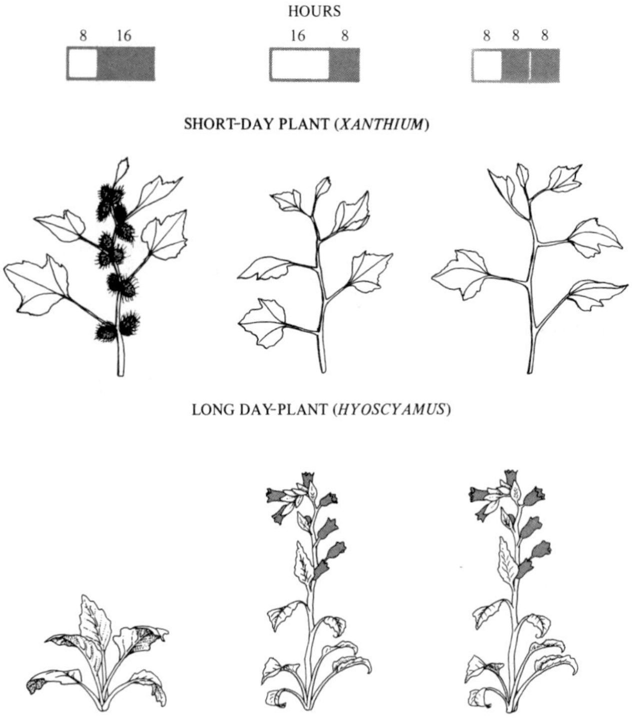

24–6
Photoperiodism in flowering plants. Short-day plants flower when the photoperiod is less than *some critical value.* Xanthium *requires 16 hours* or less *of light to flower. If the period of darkness is interrupted even briefly, flowering will not occur.* Hyoscyamus *requires about 10 hours (depending on temperature) or more to flower. However, if the dark periods are interrupted by a flash of light,* Hyoscyamus *will flower even on a short-day period. Bars above indicate duration of light and dark periods in a 24-hour day.*

cise. At 22.5°C, the long-day plant *Hyoscyamus niger* (henbane) will flower when exposed to photoperiods of 10 hours and 20 minutes (see Figure 24–6). But with a photoperiod of 10 hours, it will not.

The response varies with different species. Some plants require only a single exposure to the appropriate day–night cycle, whereas others require several weeks of exposure. In many plants, there is a correlation between the number of induction cycles and the rapidity of flowering or the number of flowers formed. Some plants have to reach a certain degree of maturity before they will flower (although others will respond to the appropriate photoperiod when they are seedlings). Some plants, when they get older, will eventually flower even if not exposed to the appropriate photoperiod, although they will flower much earlier with the proper exposure.

Environmental conditions also affect photoperiodic behavior. At 28.5°C, henbane requires $11\frac{1}{2}$ hours light, for instance, whereas at 15.5°C, it requires only $8\frac{1}{2}$ hours.

Measuring the Dark

In 1938, another pair of investigators, Karl C. Hamner and James Bonner, began a study of photoperiodism, using the cocklebur as their experimental tool. As we mentioned previously, the cocklebur is a short-day plant, requiring 16 hours *or less* of light per 24-hour cycle to flower. It is particularly useful for experimental purposes because a single exposure under laboratory conditions to a short-day cycle will induce flowering two weeks later, even if the plant is immediately returned to long-day conditions. The cocklebur can withstand a great deal of rough treatment, surviving even if its leaves are removed. Hamner and Bonner showed that it is the leaf blade of the cocklebur that perceives the photoperiod. A completely defoliated plant cannot be induced to flower. But if as little as one-eighth of a fully expanded leaf is left on the stem, the single short-day exposure induces flowering.

In the course of these studies, in which they tested a variety of experimental conditions, Hamner and Bonner made a crucial and totally unexpected discovery. If the period of darkness is interrupted by as little as a one-minute exposure to a 25-watt bulb, flowering does not occur. Interruption of the light period by darkness has no effect on flowering whatsoever. Subsequent experiments with other short-day plants showed that they, too, required periods not of uninterrupted light but of uninterrupted darkness.

On the basis of the findings of Garner and Allard, commercial growers of chrysanthemums had found that they could hold back blooming in the short-day plants by extending the daylight with artificial light. Now, on the basis of the new experiments by Hamner and Bonner, they were able to economize on their electric bill and still have flowers for late-season football games.

What about long-day plants? They also measure darkness. A long-day plant that will flower if it is kept in a laboratory in which there is light for 16 hours and dark for 8 hours will also flower on 8 hours of light and 16 hours of dark if the dark is interrupted by even a brief exposure of light.

Chemical Basis of Photoperiodism

The next important clue to the response of plants to light came from a team of research workers at the U.S. Department of Agriculture Research Station in Beltsville, Maryland. This team, whose members have included Sterling B. Hendricks, Harry A. Borthwick, and M. W. Parker among others, has been making contributions in this area of research for more than three decades.

Hamner and Bonner had shown that if the dark period is interrupted by a single flash of light from an ordinary bulb, the cocklebur will not flower. The Beltsville group, following this lead, began to experiment with light of different wavelengths, varying the intensity and duration of the flash. They found that red light at about 660 nanometers (orange-red) was most effective at preventing flowering in the cocklebur and other short-day plants. It was also the most effective, they found, at promoting flowering in long-day plants. The light exposure could thus be compared to an electric switch; the same switch can be used to turn a motor on and to turn it off.

The Beltsville group found their next clue in the report of an earlier study performed with lettuce seeds *(Lactuca sativa).* Lettuce seeds germinate only if they are exposed to light. (This requirement is true of many small seeds, which need to germinate in loose soil and near the surface in order for the seedlings to be sure of breaking through.) The earlier workers, in studying the light requirement of lettuce seeds, had shown that red light stimulated germination and that light of a slightly longer wavelength (far-red) inhibited germination even more effectively than did no illumination at all. The Beltsville group found that when red light was followed by far-red light, the seeds did not germinate. The red light most effective at inducing germination in seeds was light of the same wavelength as that involved in the flower response—about 660 nanometers. Furthermore, they found that the light most effective in inhibiting the effect produced in seeds by red light was light of a wavelength of 730 nanometers. The series of flashes could be repeated over and over; the number of flashes did not matter, but the nature of the final one did. If the series ended with a red flash, the great majority of the seeds germinated. If it ended with a far-red flash, the great majority did not (Figure 24-7).

Far-red light was then tried on short-day and long-day plants, with the same on-off effect. Far-red light alone, when given during the dark period, had no effect. But a flash of far-red light immediately following a flash of red light canceled the effects of the red light.

Discovery of Phytochrome

Here is how the Beltsville group and others have interpreted these results. The plant contains a pigment that exists in two different forms: P_{660} and P_{730} (Figure 24-8). P_{660} absorbs red light and is converted to P_{730}. The conversion of P_{660} to P_{730} takes place in the daylight. or in incandescent light; in both kinds of light, red wavelengths predominate over far-red. P_{730} absorbs far-red light, which converts it back to P_{660}. In nature, the P_{730}-to-P_{660} conversion takes place slowly in the dark (Figure 24-9). P_{730} is the active form. In short-day plants, P_{730} inhibits flowering under conditions in which flowering would otherwise occur. In long-day plants, P_{730}

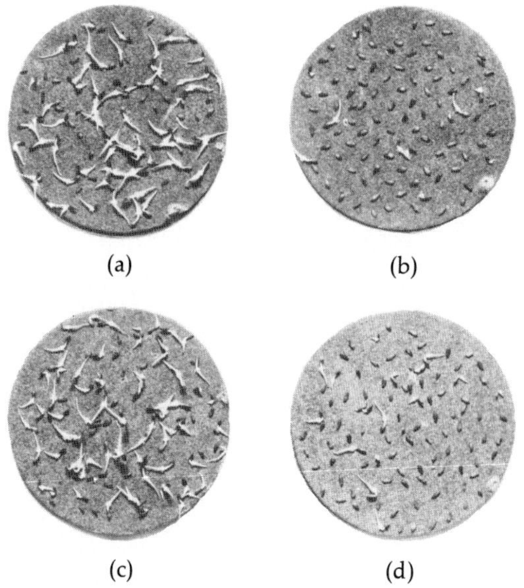

24-7
Light and the germination of lettuce seeds. The seeds in (a) *were exposed briefly to red light, those in* (b) *to red–far-red, in* (c) *to red–far-red–red and in* (d) *to red–far-red–red–far-red. All were then incubated on wet blotters at 20°C. As you can see, whether or not the seeds germinated depended largely on which wavelength they were exposed to last.*

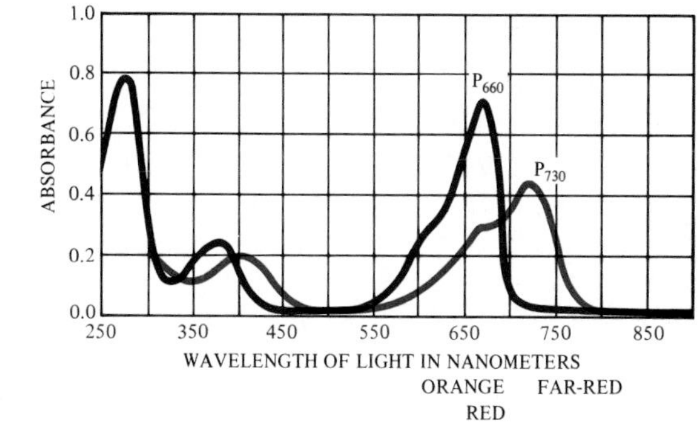

24-8
Absorption spectra of the two forms of phytochrome, P_{660} *and* P_{730}. *This shift in absorption spectra made it possible to isolate the pigment.*

24-9
P_{660} changes to P_{730} when exposed to red light. P_{730} reverts to P_{660} when exposed to far-red light. In darkness, P_{730} reverts to P_{660}.

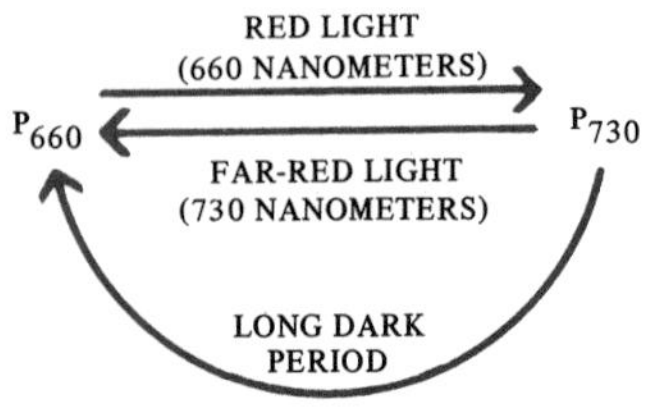

promotes flowering under appropriate conditions. This hypothesis has been conclusively borne out by subsequent work.

In 1959, Borthwick and co-workers gave this pigment the name of *phytochrome* and in the same year presented conclusive physical evidence that it existed.

Isolation of Phytochrome

Phytochrome is present in plants in very small amounts. To detect it, a spectrophotometer was needed that was sensitive to extremely small changes in light absorbency. (Large changes in light absorbency can be detected by the eye, of course, as changes in color.) Such a spectrophotometer was introduced some seven years after the existence of phytochrome was proposed, and it was used first to detect and subsequently to isolate the pigment.

Tissues with substantial amounts of chlorophyll are opaque to the measuring beams of the spectrophotometer. Furthermore, chlorophyll, like phytochrome, absorbs light of about 660 nanometers. For this reason, dark-grown seedlings (in which chlorophyll has not yet developed) were chosen as the source from which to isolate phytochrome. The pigment proved to be blue in color (why might you expect this color?) and to show the characteristic of red–far-red conversion in the test tube by reversibly changing color slightly in response to red or far-red light.

There are two portions to the phytochrome molecule: a light-absorbing portion and a large protein portion. The light-absorbing portion is a phycobilin, much like certain accessory pigments found in blue-green and red algae. Its exact structure is not known.

The way in which phytochrome works is also not yet established. One early hypothesis was that the molecule acted like an enzyme, playing a key role at some metabolic cross road. This has not been supported by more recent evidence. A second hypothesis, supported by extensive studies, is that phytochrome functions through gene activation. A third recent suggestion is that phytochrome acts in some way to affect membrane permeability. One indication of a change in membrane permeability is a change in electric potential of plasma membranes following the phytochrome response. This change is presumably caused by a flow of ions across the membrane, analogous to the action potential generated in a nerve axon of an animal. There is evidence that the active form of phytochrome may bind to special receptors on the plasma membrane, thus permitting or prohibiting the entry of substances involved in the regulation of cellular activities.

Other Phytochrome Responses

Phytochrome has now been shown to be involved in a number of other responses of plants. The germination of many seeds, for instance, occurs in the dark. In the seedlings, the stem elongates rapidly, pushing the shoot (or, in most monocots, the cotyledon) up through the dark soil layers. During this stage of growth, there is essentially no enlargement of the leaves; such enlargement would interfere with the passage of the shoot through the soil. Soil is not necessary for this growth pattern; any seedling grown in the dark will be elongated and spindly, with small leaves. It will also be almost colorless, because plastids do not turn green until they are exposed to light. Such a seedling is said to be etiolated (Figure 24–10).

When the seedling tip emerges into the light, the etiolated growth gives way to normal plant growth. In dicots, the plumular hook unbends, the stem growth rate may slow down somewhat, and leaf growth begins. In grasses, the mesocotyl (the part of the embryo axis between scutellum and coleoptile) growth stops, the stem begins to elongate, and the leaves open.

A dark-grown bean seedling that receives one minute of red light a day, for instance, will show these light effects beginning on the fourth day. But if the exposure to red light is followed by a one-minute exposure to far-red, none of the changes usually produced by the red light will appear (Figure 24–11). Similarly, in the seedlings of grains, termination of mesocotyl growth is triggered by exposure to red light, and the effect of red light is canceled by far-red.

A recent report from England suggests that an important function of phytochrome in plants growing in a natural environment is the detection of shading by other plants. Radiation below 700 nanometers is almost completely reflected or absorbed by vegetation, whereas that between 700 and 800 nanometers (in the far-red range) is largely transmitted. Thus, as a response to this shift in the normal P_{660}–P_{730} equilibrium, it is suggested, plants show increased internodal elongation.

Red–far-red reactions are involved in anthocyanin formation in the apple, turnip, and cabbage; in the germination of seeds; in changes in the chloroplasts and other plastids; and in a tremendous variety of other plant responses.

24–10
Dark-grown seedlings, such as the ones on the right, are thin and pale with longer internodes and smaller leaves than normal seedlings, such as those on the left. This group of characteristics, known as etiolation, has survival value for the seedling because it increases its chances of reaching light before its stored energy supplies are used up. (Richard F. Trump, OMIKRON.)

24–11
All three bean plants received 8 hours of daylight each day. The center plant was exposed to 5 minutes of elongation-promoting far-red light during the dark period. The plant on the right received the same 5-minute exposure to far-red light followed by a 5-minute exposure to red light, which counteracted the effect of the far-red.

Phytochrome and Photoperiodism

When the existence of phytochrome was first demonstrated, its discoverers hypothesized that the behavior of phytochrome might explain photoperiodism, that is, that the red–far-red conversion might be the time-measuring mechanism, the biological clock. According to this hypothesis, in short-day plants, P_{730}, which inhibits flowering in these plants, would accumulate in the light and revert back to the inactive form, P_{660}, in the dark. When the nights were long enough, all (or a critical amount) of P_{730} would be inactivated, and flowering would no longer be inhibited. Long-day plants, on the other hand, would require short nights, during which the P_{730} would not be completely destroyed; if the night were short enough, enough P_{730} would remain at the end of it to promote flowering.

How would you test this hypothesis? One way that the investigators tested it was by measuring the actual amount of time required for all the P_{730} to revert to P_{660} in the dark. (This is a fairly simple test to perform, using the special spectrophotometer.) The dark conversion in all the plants studied was found to take only about three hours. What happens during the rest of the dark period? Remember that in some plants, such as *Hyoscyamus*, as little as 20 minutes can make the difference. What happens during that last important 20 minutes, which in some plants is decisive in determining whether flowering occurs?

Or try another test. In a long-day plant, such as henbane, P_{730} promotes flowering. P_{730} is inactivated by far-red light. P_{730} is also inactivated by darkness. If P_{730} is inactivated, a long-day plant does not flower. So what would you expect if you exposed henbane to a dose of far-red light plus an exact borderline period of darkness? What actually does happen is that the henbane flowers.

On the basis of these and other observations, it is now generally agreed that the time-measuring phenomenon of photoperiodism is not controlled by the interconversion of P_{660} and P_{730}. There seem to be two variables involved in the phytochrome response. One is the amount of P_{730}. The second is evidently determined by the circadian rhythm of the plant. Both must be in appropriate condition for the response to occur. "P_{730}," in the words of Bruce G. Cumming, who worked on this problem with the Beltsville group, "is a sort of master gate which can control things passing through it but does not determine their arrival."

HORMONAL CONTROL OF FLOWERING

You will remember that Hamner and Bonner in their early cocklebur experiments showed that the leaf "perceived" the light, which caused the bud to flower. Apparently something is transmitted from leaf to bud that has profound effects on growth and development. This hypothetical substance has been termed the flowering hormone, or more precisely the flower-evocating factor.

The earliest experiments on the flower-evocating factor were carried out independently in several laboratories in the 1930s. Some of the first, those of M. H. Chailakhyan in Russia, carried out just a few years before the first cocklebur studies, are representative. Using the short-day plant *Chrysanthemum indicum,* Chailakhyan showed that if the upper portion of the plant was defoliated and the leaves on the lower part exposed to a short-day induction period, the plant would flower. If, however, the upper, defoliated part was kept on short days and the lower, leafy part on long days, no flowering occurred. He interpreted these results as indicating that the leaves formed a hormone that moved to the apex and initiated flowering. He named this hypothetical hormone florigen, the "flower maker."

Subsequent experiments showed that the flowering response will not take place if the leaf is removed immediately after photoinduction. But if the leaf is left on the plant for a few hours after the induction cycle is complete, it can then be removed without affecting flowering. The flowering hormone can pass through a graft from a photoinduced plant to a noninduced plant. Unlike auxin, however, which can pass through agar or other nonliving tissue, as Went showed, florigen can travel from one plant tissue to another only if there are anatomical connections of living tissue between them. If a branch is girdled, that is, if the "bark" (the tissues—including cortex and phloem—outside the vascular cambium) is removed, florigen movement ceases. On the basis of these data, it was concluded that florigen moves by way of the phloem system, the means by which most organic substances are transported.

In some plants—the Biloxi soybean is an example—leaves must be removed from the grafted receptor plant or it will not flower. This observation suggests that in noninduced plants, the leaves may produce an inhibitor. In fact, some investigators have concluded on the basis of such evidence that there is no substance that initiates flowering but rather a substance that inhibits flowering unless it is removed by the proper conditions. Strong evidence now suggests that both inhibitors and promoters are involved in the control of flowering.

What would you expect to happen if a long-day plant was grafted to a short-day plant and the short-day plant was then induced to flower? Grafts can be carried out successfully only if the two plants are related genetically. Fortunately for the purposes of this experiment, some species that are very different in their photoperiodic responses are so related. In fact, Maryland Mammoth tobacco (short-day) and black henbane (long-day) are members of the same family (Solanaceae). When the two are grafted together, the henbane will flower if both are kept under a short-day period. Similarly, henbane kept under a long-day period will induce flowering in the tobacco plant kept under the same conditions. Day-neutral plants may also serve as donors or receivers of the flowering stimulus when grafted to long-day or short-day plants. So the flowering hormone, if it exists, is the same, or at least physiologically equivalent, in all three groups.

As you can see, evidence for the existence of florigen is very compelling, although no actual chemical substance has yet been isolated for characterization.

DORMANCY

Plants do not grow at the same rate all of the time. During unfavorable seasons, they rest, limiting their growth or ceasing to grow altogether. This ability to rest enables plants to survive periods of water scarcity or low temperature.

Dormancy is a special condition of rest. After periods of ordinary rest, growth resumes when the temperature becomes milder or when water or any other limiting factor becomes available again. A dormant bud or embryo, however, can only be "activated" by certain, often quite precise, environmental cues. This adaptation is of great survival importance to the plant. For example, we know that plants bud and bloom and seeds germinate in the spring—but how do they recognize spring? If warm weather alone were enough, in many years all the plants would bud and all the seedlings start to grow during Indian summer, only to be destroyed by the winter frost, or during any one of the warm spells that often punctuate the winter season. The dormant seed or bud does not respond to these apparently favorable conditions because of endogenous inhibitors which must first be removed or neutralized before the rest period is terminated. In contrast to this reluctance to grow too rapidly, developed over the course of centuries by natural selection, commercial seeds are artificially selected for their readiness to germinate promptly when they are exposed to favorable conditions, a trait that would be a great hazard for wild seeds.

Dormancy in Seeds

Almost all seeds growing in areas with marked seasonal temperature variations require a period of cold prior to germination. Many seeds require drying before they germinate. This requirement prevents their germinating within the moist fruit of the parent plant. Some seeds, as we have seen in the case of lettuce, require exposure to light, but others are inhibited by light. Some seeds will not germinate in nature until they have become abraded, as by soil action. Such abrasion wears away the seed coat, permitting water or oxygen to enter the seed and, in some cases, removing the source of inhibitors. The seeds of some desert species germinate only when sufficient rain has fallen to leach away inhibitory chemicals in the seed coat; the amount of rainfall necessary to wash off these inhibitors is directly related to the supply of water the plant needs to complete a hasty cycle from seed to flower to seed again.

Some seeds may survive a long time in the dormant condition, enabling them to exist for many years, decades, and even centuries under favorable conditions. In 1879, seeds of 20 species of common Michigan weeds were stored for an experiment designed to continue 160 years. At the last sampling, the seeds of three species were still viable. Although this demonstration of endurance is impressive, it does not approach the record for seeds of the sacred lotus *(Nelumbo nucifera)* found by a Japanese botanist in a peat deposit in Manchuria. Radiocarbon dating showed the seeds to be some 2000 years old, but when the seed coats were filed to permit water to enter, every single one germinated.

In 1967, even this record was broken with seeds of the arctic tundra lupine *(Lupinus arcticus)*. These seeds were found in a frozen lemming burrow in the Yukon with animal remains estimated by carbon dating to be at least 10,000 years old. Their cold requirement having been adequately fulfilled, a sample of seeds germinated within 48 hours (Figure 24-12).

During recent years, plant scientists have become increasingly interested in the factors involved in maintaining seed viability. Various enzyme systems may fail progressively in the stored seed, eventually leading to a complete loss of viability. Under what circumstances might viability best be prolonged? Such questions are relevant to the worldwide interest in developing seed banks with the goal of preserving the genetic characteristics of the original wild-type varieties of various crop plants for use in future breeding programs. The need for such seed banks arises from the progressive replacement of older varieties with newer ones and the elimination of the replaced varieties, chiefly through the destruction of their habitats caused by the rapidly increasing human population.

24-12
Lupinus arcticus *grown from a seed at least 10,000 years old. The seeds were found by a Yukon mining engineer in lemming burrows deeply buried in permanently frozen silt of the Pleistocene age.*

Dormancy in Buds

As with seeds, buds of many species require cold for breaking dormancy. You may know this from your own experience. If branches of flowering trees and shrubs are cut and brought inside in the fall, they do not flower, but the same branches left out until late winter or early spring will bloom in the warmer temperature indoors. Similarly, bulbs such as those of tulips, hyacinths, narcissus, and jonquils can be "forced"—that is, made to bloom inside in the winter—but only if they have previously been outside or in a cold place. (Such bulbs, as we noted in Chapter 21, are actually large buds in which the leaves are modified for storage.) Deciduous fruit trees, such as the apple, chestnut, and peach, cannot be grown in climates where the winters are not cold.

Investigations carried out under controlled conditions confirm these suggestions that cold is required for the breaking of dormancy in many species. Most varieties of peach, for example, must remain for 600 to 900 hours at temperatures below 4°C before they will respond to the activating influence of warmer temperatures and longer days. Some plants will respond to a brief exposure to freezing temperatures; if one bud of a greenhouse-cultivated lilac bush is briefly exposed to freezing temperatures, that bud and that alone will break into bloom soon after. Cold is not required to break dormancy in all cases, however. In the potato, for instance, in which the "eyes" are dormant buds, at least two months of dry storage is the chief requirement; temperature is not a factor. In many plants, including particularly trees, the photoperiodic response breaks winter dormancy, with the dormant buds being the receptor organs. Photoperiodism often regulates the onset of dormancy as well, presumably through some hormonal mechanism.

Application of gibberellins may sometimes break dormancy. For instance, gibberellin treatment of a peach bud may induce development after the bud has been kept for 164 hours below 8°C. Does this mean that under normal conditions, increase in gibberellin terminates the dormancy? Not necessarily. Dormancy may be a state of balance between growth inhibitors and growth stimulators. Addition of any growth stimulator (or removal of inhibitors) may alter the balance so that growth begins.

As you can see even from this brief review of a complex subject, there does not seem to be any common mechanism by which dormancy is induced or broken. This fact, although it considerably multiplies the problems of the plant physiologist, is in keeping with our sense of evolutionary history. Dormancy became advantageous to plants only comparatively recently, when the seed plants began to spread into a variety of different ecological domains. Presumably dormancy evolved independently in many separate groups of plants, each of which, among the survivors, found its solution separately and often in a different way.

COLD AND THE FLOWERING RESPONSE

Cold also may affect the flowering response. For example, if winter rye *(Secale cereale)* is planted in the autumn, it germinates during the winter and flowers the following summer seven weeks after growth begins. If it is planted in the spring, it does not flower for 14 weeks, and it remains vegetative through most of the growing season. The German plant physiologist Gustav Gassner discovered in 1915 that he could influence the flowering of winter rye and other cereal plants by controlling the temperature of the germinating seeds. He found that if the seeds of the winter strain are kept at near-freezing (1°C) temperatures during germination, the winter rye, even when planted in late spring, will flower the same summer it is planted. This procedure, which was later adopted commercially, came to be known as vernalization, from *vernus,* the Latin word for "spring."

Even after vernalization, the plant must be subjected to a suitable photoperiod, usually long days. The vernalized winter rye behaves like a typical long-day plant, flowering in response to the long days of summer. A similar example is seen with henbane. Besides the annual strain of *Hyoscyamus,* which we have discussed previously, there is a biennial strain. The vegetative rosette, which culminates the first year's growth of the biennial strain, flowers only if it is exposed to cold. After cold exposure, it becomes a typical long-day plant, with the same photoperiodic response as that of the annual strain.

As the above examples indicate, in some plants the cold treatment affects the photoperiodic response. Spinach, ordinarily a long-day plant, does not usually flower until the days are 14 hours in length. If the spinach seeds are cold-treated, however, they will flower when the days are only 8 hours long. Similarly, cold treatments of the clover *Trifolium subterranum* can completely remove its dependency on day length for flowering.

In the biennial variety of *Hyoscyamus* and in most biennial rosette long-day plants, gibberellin treatment can substitute for cold. If gibberellin is applied, such plants will elongate rapidly and then flower. Application of gibberellin to short-day plants or to nonrosette long-day plants has little effect on (or inhibits) the flowering re-

24–13
Relationship between day length and the developmental cycle of plants in the temperate zone.

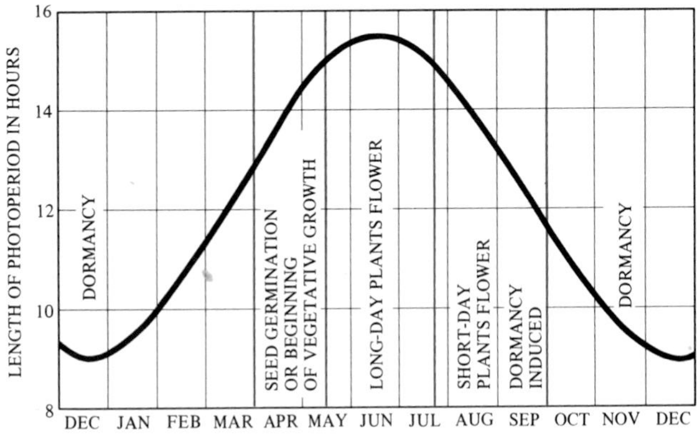

sponse. However, if you inhibit gibberellin synthesis when the plant is being exposed to the appropriate inductive cycle, the plant will not flower unless its gibberellin is replaced. In other words, in these plants, gibberellin and something else must be required.

TOUCH RESPONSES IN PLANTS

Plants also respond to touch. One of the most common examples is seen in tendrils, which are modified leaves in some species of plants, and modified stems in others. The tendrils wrap around any object with which they come in contact (see Figure 24-14) and so enable the plant to cling and climb. The response can be rapid; a tendril may wrap around a support in less than a minute. Cells touching the support shorten slightly and those on the other side elongate. There is some evidence that auxin plays a role in this response.

A more spectacular touch response is seen in the sensitive plant, *Mimosa pudica,* in which the leaflets and sometimes entire leaves droop suddenly when touched (Figure 24-15). This response is a result of a sudden change in turgor pressure at the base of leaflets and leaf, brought about, apparently, by the propagation of an electrical impulse along the plasma membranes (see Essay on page 547). There is some controversy about its survival value to the plant. *Mimosa pudica* often grows in dry, exposed areas where it may be subjected to drying winds; strong winds may shake the leaves enough to make them fold up, so conserving water. Another sug-

24-14
Tendrils of Smilax *(greenbrier). Twisting is caused by varying growth rates on different sides of the tendril.*

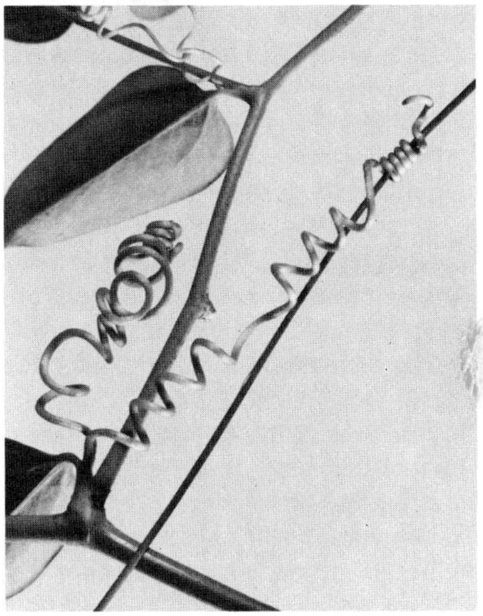

24-15
Sensitive plant (Mimosa pudica).
(a) *Normal position of leaves and leaflets.* (b, c) *Successive responses to touch.*

(a)

(b)

(c)

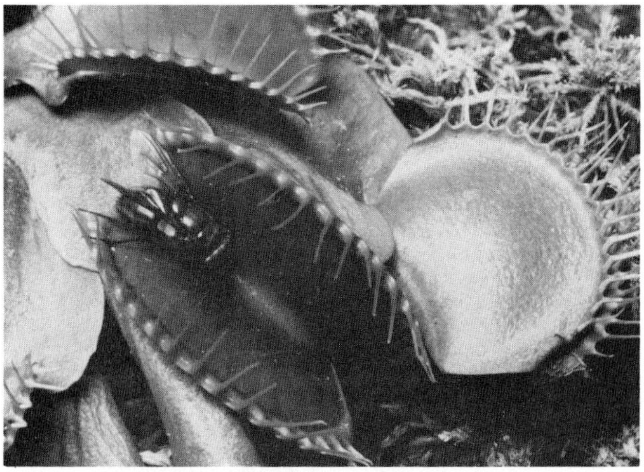

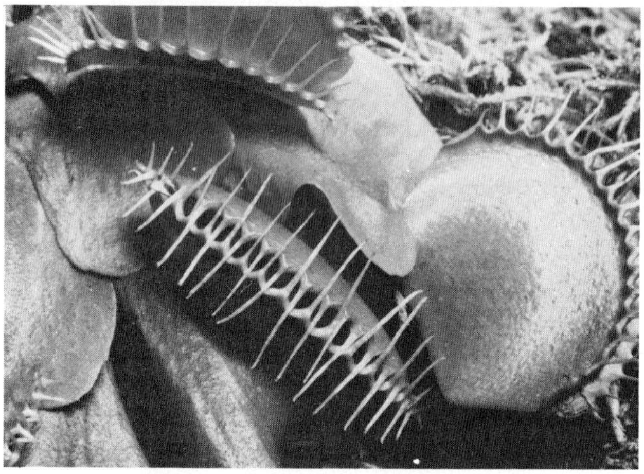

24–16
Touch responses in the Venus flytrap
(Dionaea muscipila).

gestion is that the wilting response makes the plant unattractive to large herbivores. Finally, it is conceivable that the folding response startles insects; there have been claims that other "nonsensitive" species of *Mimosa* growing near *Mimosa pudica* show more evidence of attack by chewing insects.

The triggering of turgor changes by touch is also involved in the capture of prey by the carnivorous Venus flytrap *(Dionaea muscipila).* The leaves of the Venus flytrap are hinged in the middle, and each leaf half is equipped with three sensitive hairs. When an insect walks on one of these leaves, attracted by the nectar on the leaf surface, it brushes against the hairs, triggering the traplike closing of the leaf. The toothed edges mesh, the leaf halves gradually squeeze closed, and the insect is pressed against digestive glands on the inner surface of the trap.

The trapping mechanism is so specialized that it can distinguish between living prey and inanimate objects, such as pebbles and small sticks, that fall on the leaf by chance: The leaf will not close unless two of its hairs are touched in succession or one hair is touched twice.

IN CONCLUSION

Each developmental event in the life of a plant is under the control not of any single factor but of a complex variety of factors. These factors and the environment interact. They may enhance, modify, or neutralize one another. In the words of William S. Hillman of the Brookhaven National Laboratory, "After all, if plants were as simple as the physiologist might wish, there would be nothing left to do."

SUMMARY

Plants possess a variety of adaptations for detecting and responding to alterations in their environment. Phototropism, the turning of the growing shoot toward light, is an example of such an adaptation. The differential growth of the seedling is caused by lateral migration of the growth hormone auxin under the influence of light. Geotropism is caused by the movement of auxin to the lower surface of a horizontal shoot, the accumulated auxin causing the shoot to bend upward and the root, which is inhibited by auxin, to bend downward.

Circadian rhythms are cycles of activity in an organism that recur at intervals of approximately 24 hours under constant environmental conditions. They are probably endogenous, caused not by an external factor such as alternating light and darkness or the earth's rotation but by some internal timing mechanism within the organism. Such a timing mechanism, the chemical and physical nature of which is unknown, is called a biological clock. The possession of a biological clock makes it possible for an organism to perceive changes in external diurnal cycles, such as the lengthening and shortening of days as the seasons progress. In this way, activities such as dormancy, leaf abscission, and flowering can be brought into synchrony with the external environment.

Photoperiodism is the response of organisms to changing 24-hour cycles of light and darkness. Such responses control the onset of flowering in many plants. Some plants will flower only when the periods of light exceed a critical length. Such plants are known as long-day plants. Other plants, short-day plants, flower only when the periods of light are less than some critical period. Day-neutral plants flower regardless of photoperiods. Factors such as temperature and the age of the plant may affect the photoperiodic response. Interruption of the dark phase of the photoperiod, even by a brief flash of light, can serve to reverse the photoperiodic effects.

Phytochrome, a pigment commonly present in small amounts in higher-plant tissues, is sensitive to the transitions between light and darkness. The pigment can exist in two forms, P_{660} and P_{730}. P_{660} absorbs red light of a wavelength of 660 nanometers and is thereby converted to P_{730}. P_{730} is converted to P_{660} over a period of hours in the dark; it can also be converted to P_{660} by exposure to far-red light. P_{730} is the active form of the pigment; it promotes flowering in long-day plants and inhibits flowering in short-day plants. P_{730} also is responsible for changes that take place in seedlings as they penetrate the soil to the light, for germination of seeds, and for development of anthocyanins, which cause the red and purple colors of apples, turnips, and cabbages. Its mechanism of action is under active study but is still unknown.

In both long-day and short-day plants, the photoperiod is perceived in the leaves but the response takes place in the bud. Apparently, a chemical stimulus that causes the bud to flower is transmitted from leaf to bud. This chemical stimulus, or flower-evocating factor, has been given the name of florigen, but it has not yet been isolated or identified. Experiments have indicated that the chemical travels through the plant by way of the phloem system and that its structure and function are similar in long-day, short-day, and day-neutral plants.

Alternation of periods of growth and rest permits the plant to survive water shortages and extremes of hot or cold. Dormancy is a special type of rest in which the plant or plant tissues such as seeds or buds do not begin to grow again without special environmental cues. The requirement for such cues, which include cold exposure, dryness, and a suitable photoperiod, prevents the tissue from breaking dormancy during superficially favorable conditions, such as Indian summer or within the succulent fruit of the parent plant. There is apparently no uniform mechanism common to all plant groups for the induction and breaking of dormancy. Vernalization refers to the promotion of flowering winter strains by keeping seeds at low temperatures. Hormones, cold, and light interact to modify plant responses.

Some species of plants respond to touch. Examples include the winding of tendrils, the collapse of the leaves of the sensitive plant, the triggering of the carnivorous Venus flytrap, and the infoldings of the tentacles of the sundew. At least some of these responses have been shown to be mediated by electrical signals.

SUGGESTIONS FOR FURTHER READING

GALSTON, A. W.: *The Green Plant,* Prentice-Hall, Inc., Englewood Cliffs, N.J., 1968.*

A convenient and concise summary of plant growth and development.

HILLMAN, WILLIAM S.: *The Physiology of Flowering,* Holt, Rinehart & Winston, Inc., New York, 1962.

A clear and thoughtful review of research in this very active field of investigation, now somewhat out of date.

LEOPOLD, A. C., and KRIEDEMAN: *Plant Growth and Development,* McGraw-Hill Book Company, New York, 1975.

An important research work that provides much material of interest in connection with this and following chapters.

RAY, P. M.: *The Living Plant,* 2nd ed., Holt, Rinehart & Winston, Inc., New York, 1972.*

A short, readable account of plant growth and development which has proved extremely useful for students.

SALISBURY, FRANK B.: *The Biology of Flowering,* Natural History Press, New York, 1971.

A semipopular account of research in the field, concentrating mainly on the short-day plant Xanthium.

SALISBURY, FRANK B., and CLEON ROSS: *Plant Physiology,* Wadsworth Publishing Co., Inc., Belmont, Calif., 1969.

A detailed and useful review of the entire subject.

STEEVES, T. A., and I. M. SUSSEX: *Patterns in Plant Development,* Prentice-Hall, Inc., Englewood Cliffs, N.J., 1972.

An exceptionally well-written and well-illustrated text which should be read by all students interested in plant morphogenesis.

STEWARD, F. C.: *Growth and Organization in Plants,* Addison-Wesley Publishing Co., Inc., Reading, Mass., 1968.

A treatment of the problems of growth and development which synthesizes all aspects of knowledge of plants—their physiology, biochemistry, and morphology.

———: *Plants at Work,* Addison-Wesley Publishing Co., Inc., Reading, Mass., 1964.

A comprehensive introductory study of plants which should be useful to all students of biology.

TORREY, J. G.: *Development in Flowering Plants,* The Macmillan Company, New York, 1967.*

The morphogenetic approach to plant growth and development is here presented in a brief and readable book.

* Available in paperback.

SECTION 7 Water and Soil Relationships

CHAPTER 25

The Movement of Water and Solutes in Plants

Three-quarters of the surface of the earth is covered by water. In fact, if the earth's surface were absolutely smooth, all the land would be 2.5 kilometers under water. Hidden reservoirs of water lie beneath the soil, and water vapor clings to the surface of the land and the sea. Water makes up more than half of all living tissue and more than 90 percent of most plant tissues. In short, water is a very common liquid, by far the most common. But do not mistake "common" for "ordinary"; water is not in the least an ordinary liquid. It is, in fact, quite extraordinary in comparison with other liquids.

Before we examine the relationship of water and plants, we shall discuss some of the properties of water that form the bases of this relationship and some of the principles of water movement.

SOME CHARACTERISTICS OF WATER

Surface Tension

Look, for example, at water dripping from a faucet. Each drop clings to the rim and dangles for a moment by a thread of water; then, just as it breaks loose, tugged by gravity, its outer surface is drawn taut, enclosing the entire sphere as it falls free (Figure 25-2). Take a needle or a razor blade and place it gently on the surface of a glass of water. It will float. Metal is denser by far than water, yet despite the learned Archimedes, metal objects float on water. Look at a pond in spring or summer; you will see water striders and other insects walking on its surface as if it were a taut elastic mat (Figure 25-3).

The phenomena just described are all the result of *surface tension*. Surface tension is caused by the cohesiveness of water molecules. The reason for this extraordinary cohesiveness lies in the structure of the molecule (Figure 25-4). The water molecule, you will recall, is a tetrahedron; that is, it is four-cornered, with two positive corners and two negative corners. As a consequence, each water molecule tends to form hydrogen bonds with

25-1
Water droplets clinging to grass blades.

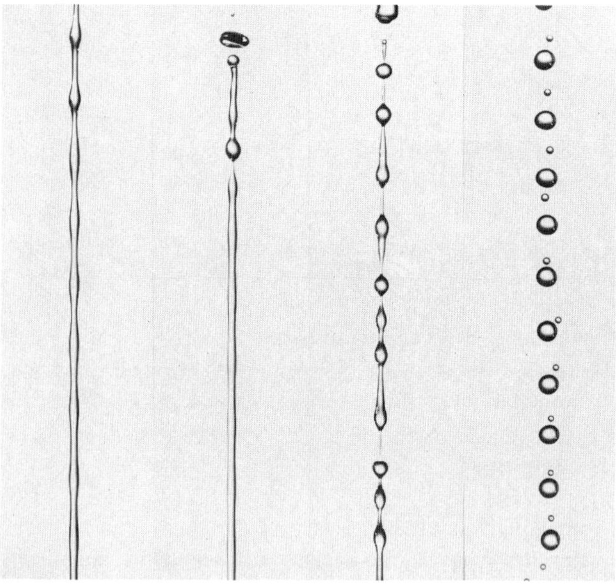

25–2
The formation of drops of water.

25–3
Spider's legs and the margin of a leaf dimple the surface of a pond.

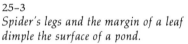

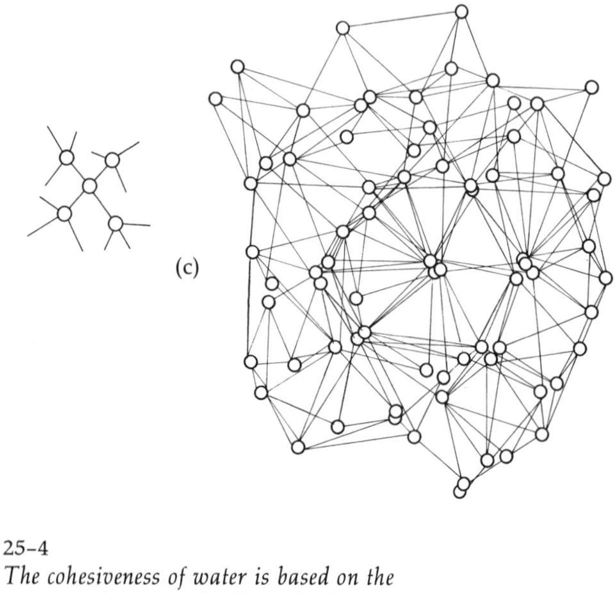

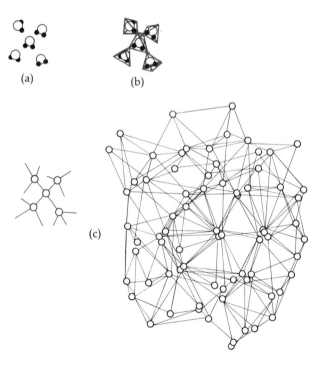

25–4
The cohesiveness of water is based on the capacity of water molecules to bond with one another. Each water molecule (a) *can join four others to form a complex network* (b). *The diagram* (c) *is liquid water as seen by a computer.*

four other molecules. Each individual bond is fragile and short-lived, but the sum total of all the bonds holds water together with tremendous strength.

Cohesion and Adhesion

The same intramolecular attractions that result in high surface tension also result in a tremendous cohesiveness of water. Pure water—that is, water that has no microscopic bits of dirt or bubbles of air to interrupt the intricate, ever-changing patterns of bonding between the water molecules—can be subjected to tremendous tugging forces before the molecules of water are pulled apart. Of course, because a given volume of water will not maintain any shape unless it is confined in some way, and because no pull can effectively be exerted on it unless it is in some kind of container with walls to which the water will adhere, it is important to note that molecules of water are strongly attracted to many substances besides other water molecules. In other words, water both coheres and adheres strongly. (*Adhesion* is the holding together of unlike substances, and *cohesion* is the holding together of like substances.)

If you hold two dry glass slides together and dip one corner in water, the interaction of cohesion and adhesion can cause water to spread between the two slides. This is *capillary action.* Capillary action can similarly cause liquids to rise in very thin glass tubes, to creep up a piece of blotting paper, or to move slowly through the micropores of the soil.

Imbibition

The adhesion of water molecules is caused by the difference in charge between one end of the molecule and the other. As a consequence of this difference in charge, water molecules can cling either to positively charged or to negatively charged surfaces. Many biological macromolecules, such as cellulose, tend to develop charges when they are wet and so to attract water molecules. The adherence of water molecules is responsible for another biologically important phenomenon called *imbibition* or, sometimes, hydration.

Imbibition ("drinking up") is the movement of water molecules into substances such as wood or gelatin, which swell or increase in volume as a result of adhesion between them and the water molecules. The pressures developed by imbibition can be astonishingly large. It is said that stone for the ancient Egyptian pyramids was quarried by driving wooden pegs into holes drilled in the rock face and then soaking the pegs with water. As the wood swelled, a force sufficient to break the slab of stone was created. In living plants, imbibition is seen particularly in seeds, which may increase to many times their original size as a result.

Water and Heat

The tendency of water molecules to cohere gives water three other properties of great biological significance. These are (1) high specific heat, (2) high latent heat of vaporization, and (3) high latent heat of melting.

Let us look first at the high specific heat of water. *Specific heat* refers to the amount of heat a substance requires for a given increase in temperature. One calorie of heat, you will recall, is required to raise the temperature of 1 gram of water 1°C. The specific heat of water is about twice the specific heat of oil or alcohol; in other words, 0.5 calorie will raise 1 gram of oil or alcohol 1°C (see Table 25-1). It is four times the specific heat of air or aluminum and nine times that of iron. Only liquid ammonia has a higher specific heat. In other words, it is difficult to raise the temperature of water.

Why should this be so? Temperature is a measure of the rapidity of movement—the kinetic energy—of molecules. In order for the kinetic energy of molecules of water to increase sufficiently for the temperature to rise 1°C, it is necessary to rupture a number of the hydrogen bonds holding the molecules together. If you put an iron skillet over a gas flame, the skillet will soon be red hot as a result of the transfer to the metal of the heat energy produced by the combustion of gas. If you put a pot of water of about the same size or weight over the same flame, it takes much longer to come to a boil. Much of the heat energy produced by the gas flame is "used up" in breaking hydrogen bonds holding the water molecules together.

What does this high specific heat mean in biological terms? If you have ever spent a day and a night on the desert, you will have had a glimpse of what it means. On the desert one can plunge from searing daytime temperatures to temperatures near freezing at night. On the rest of the surface of the earth, the rising and setting of the sun have much less effect on temperature because of the moderating influence of water. For this reason, large bodies of water are remarkably stable in their temperature; the temperature of ocean depths rarely varies more than a few degrees. Land-dwelling plants and animals that live near the ocean or other large bodies of water are also protected to some extent against violent temperature fluctuations by the high specific heat of water.

Heat of Vaporization

Vaporization of a substance—or, as we commonly refer to it, evaporation—is the change from liquid to gas (Table 25-2). It comes about because the rapidly moving molecules of a liquid break loose from the surface and enter the air. The hotter the liquid is, the more rapid the

Table 25-1 *Comparative Specific Heats (the Quantity of Heat Required to Raise the Temperature of 1 Gram Through 1°C)*

	CALORIES
Water	1.00
Lead	0.03
Iron	0.10
Salt (NaCl)	0.21
Glass	0.20
Sugar (sucrose)	0.30
Liquid ammonia	1.23
Chloroform	0.24
Ethyl alcohol	0.60

Table 25-2 *Comparative Latent Heats of Vaporization (the Quantity of Heat Required to Convert 1 Gram of Liquid to 1 Gram of Gas)*

	CALORIES
Water (at 0°C)	596
Water (at 100°C)	540
Ammonia	295
Chlorine	67.4
Hydrofluoric acid	360
Nitric acid	115
Carbon dioxide	72.2
Ethyl alcohol	236.5
Ether	90.4

movement of its molecules and, hence, the more rapid the rate of evaporation—but so long as a liquid is exposed to air that is below 100 percent in humidity, evaporation will take place and will continue to take place right down to the last drop.

In order for a water molecule to break loose from its neighboring molecules—that is, to vaporize—the hydrogen bonds have to be broken. This requires heat energy. In fact, more than 500 calories are needed to change 1 gram of liquid water into vapor (the exact amount depends on the temperature of the water). In other words, the same amount of heat would raise the temperature of 500 grams of water 1°C. As a consequence, when water evaporates, as from the surface of your skin or a leaf, it absorbs a great deal of heat from the immediate environment. Thus, evaporation has a cooling effect. Evaporation of water from a leaf is an important cooling mechanism in plant life. Table 25–2 shows the amount of heat necessary to change a gram of liquid to a gram of gas for a number of common substances.

High Latent Heat of Melting

The latent heat of melting is defined as the number of calories required to convert 1 gram of solid at the freezing point to 1 gram of liquid *at the same temperature.* As you can see in Table 25–3, the same quantity of heat must be employed to melt 1 gram of ice as to raise the temperature of the resulting ice water to 80°C. Conversely, when water turns back to ice, 80 calories of heat are given up to the surroundings. This is sometimes referred to as heat of fusion. This giving up and absorbing of heat as water freezes and ice melts acts as a sort of buffer system or thermostat. When a body of water becomes cooled to its freezing point, so long as water and ice exist in contact, additional heat will tend only to melt the ice and cooling will tend to freeze the water. As a consequence, a body of water at the freezing point will remain at about the same temperature despite a fairly wide range of changes in air temperature. This too, of course, serves to stabilize the temperature of the surrounding air and land.

Table 25–3 *Comparative Latent Heats of Melting (the Quantity of Heat Required to Convert 1 Gram of Solid to 1 Gram of Liquid)*

	CALORIES
Water (at 0°C)	79.7
Methyl alcohol	16
Benzene	30
Acetic acid	45

Water and Ice

Water becomes less dense when it freezes. Liquid water, like other liquids, contracts as it gets colder. At 4°C it has its maximum density. Then when it falls below 4°C, it expands. This, of course, is quite within the range of common experience: We know that ice cubes float at the top of a water glass and that icebergs cruise the surface of the oceans. But have you thought about what would happen if water behaved like an "ordinary" liquid and just got denser and denser (as ammonia does) as it froze? Not only would ice cubes and icebergs plummet to the bottom, but ponds and lakes and perhaps even oceans would begin to freeze from the bottom up. Spring and summer might stop the freezing process, but laboratory experiments have shown that if ice is held to the bottom of even a relatively shallow tank, water can be boiled on the top without melting the ice. Thus, if water did not expand when it froze, it would continue to freeze from the bottom up, year after year, and never melt again. But as it is, the water at 4°C sinks down, displacing the colder water, which remains at the top. And so life goes on, nurtured by the remarkable properties of that extraordinary liquid that has been its cradle since the beginning of biological time.

PRINCIPLES OF WATER MOVEMENT

The movement of water, whether in living systems or the abiotic world, is governed by three principles: mass flow (which we shall discuss first because it is simpler and more familiar); diffusion; and osmosis.

Mass Flow

Mass flow is the overall movement of water (or some other liquid). It occurs in response to differences in the potential energy of water, usually referred to as *water potential.*

A simple example of water that has potential energy is water at the top of a hill or waterfall. As this water runs downhill, its potential energy can be converted to mechanical energy by a watermill or to electrical energy by a hydroelectric turbine (Figure 25–5).

Pressure is another source of water potential. If we put water into a rubber bulb and squeeze the bulb, this water, like the water at the top of a hill, also has water potential, and it will move to an area of less water potential. Can we make the water that is running downhill run uphill by means of pressure? Obviously we can, but only so long as the water potential produced by the pressure exceeds the water potential produced by gravity. Water moves from an area where water potential is greater to an area where water potential is less, regardless of the reason for the water potential.

The concept of water potential is a useful one because it enables physiologists to predict the way in which

25–5
Water at the top of a falls, like a boulder on a hilltop, has potential energy. The movement of water from one energy level to another, as from the top of the falls to the bottom, is an example of mass flow.

water will move under various combinations of circumstances. Measurements of water potential are usually made in terms of the pressure required to stop the movement of water—that is, the hydrostatic (water-stopping) pressure—under the particular circumstances. This pressure is usually expressed in bars. (A bar is a unit of pressure equal to the average pressure of the air at sea level—about 70.3 grams per square centimeter.)

25–6
Diagram of the diffusion process. Diffusion is the random movement of molecules from a more concentrated to a less concentrated area. Notice that as one type of molecule (indicated by color) diffuses to the right, the other diffuses in the opposite direction. The result will be an even distribution of both types of molecules. Can you see why the net movement of molecules will slow down as equilibrium is reached?

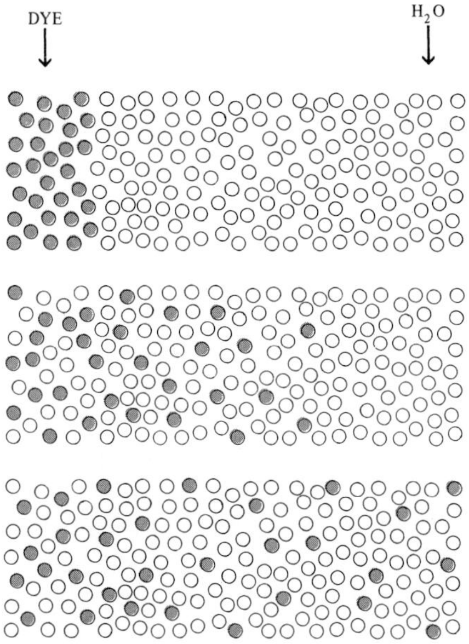

Diffusion

Diffusion is a familiar phenomenon. If you sprinkle a few drops of perfume in one corner of a room, the scent will eventually permeate the entire room even if the air is still. If you put a few drops of dye in one end of a glass tank full of water, the dye molecules will slowly distribute themselves evenly throughout the tank. (The process may take a day or more, depending on the size of the tank.) Why do the dye molecules move apart?

One possible explanation is that the molecules are behaving like members of a crowded population who, in order to find "elbow room," move out of the city into the suburbs. This is not true, however, of diffusion processes. If you could observe the individual dye molecules in the tank (see Figure 25–6), you would see that each one of them moves individually and at random; looking at any single molecule—at either its rate of motion or its direction of motion—gives you no clue at all as to where the molecule is located with respect to the others. So how do the molecules get from one side of the tank to the other? In your imagination, take a thin cross section of the tank, running from top to bottom. Dye molecules will move in and out of the section, some moving in one direction, some moving in the other. But you will see more dye molecules moving from the side of greater concentration. Why? Simply because there are more dye molecules at that end of the tank. Because there are more dye molecules on the left, more dye molecules, moving at random, will move to the right, even though there is an equal probability that any one molecule of dye will move from right to left. Consequently, the overall (net) movement of dye molecules will be from left to right. Similarly, if you could see the movement of the individual water molecules in the tank, you would see their overall movement is from right to left.

What happens when all the molecules are distributed evenly throughout the tank? The even distribution does not affect the behavior of the molecules as individuals; they still move at random. And, because the movements are random, just as many molecules go to the left as to the right. But because there are now as many molecules of dye and as many molecules of water on one side of the tank as on the other, there is no overall direction of motion. Under such circumstances, the net transfer of the molecules is zero, and the system is said to be in a state of *equilibrium*. There is, however, just as much overall molecular motion as before, provided that the temperature has not changed.

Diffusion thus may be defined as the dispersion of substances by the independent movement of their ions or molecules which tends to equalize their concentrations throughout a system. Note that the direction of diffusion is from a region of greater concentration to one of lesser concentration of the diffusing substance. Clearly, then, one important factor governing the rate of diffusion is the magnitude of the difference in concentration from one part of a system to another. A second factor is the distance over which the difference in concentration is distributed. Combining these two factors, one may speak of the concentration gradient, or the concentration difference per unit distance. Substances that are moving from a region of greater to one of lesser concentration are said to be moving *along the gradient*, whereas substances moving in the opposite direction are moving *against the gradient*. (As we shall see, movement against the gradient requires an input of energy.)

Notice that, in our imaginary tank, there are two gradients; the dye molecules are moving along one of them, and the water molecules are moving along the other in the opposite direction. When the molecules have reached a state of equal distribution—that is, when there are no more gradients—they continue to move; but as we noted previously, there is no net movement in either direction.

Factors Affecting the Rate of Diffusion

Diffusion is not an effective means of transferring large amounts of material across great distances. For example, it has been calculated that 940 days would be required for 1 milligram of sucrose to diffuse 1 meter through a water-filled tube with a cross section of 1 square centimeter, starting from a sheet of solid sucrose at one end with crystalline sugar as a source. Such a long period of time would be required because even a large concentration difference distributed over a large distance does not result in a steep concentration gradient except near the source of diffusing matter at the beginning of the diffusion period.

In contrast, large gradients are readily built up over short distances, so that diffusion over short distances can be very fast. In simple cases, the time required for diffusion is proportional to the square of the distance covered. For example, diffusion over a distance of 1 micrometer is 10^8 times faster than diffusion over a distance of 1 centimeter. Thus, movement by diffusion between and within cells is relatively rapid.

Of course, the size of the diffusing molecules is also a factor in determining the rate of diffusion. Small molecules or ions diffuse much more rapidly than large ones. If two particles are of the same size but have different densities, the heavier particle will diffuse more slowly than the lighter one.

The greater the area across which a given concentration occurs, the greater the number of molecules that will diffuse per unit time.

An increase in temperature also increases the rate of diffusion of a substance by increasing the kinetic energy of its molecules.

A system in which diffusion is occurring is a system that possesses water potential. Because water, like other substances, will move from a region of greater concentration to a region of lesser concentration, the area of the tank in which there is pure water has a greater water potential than the area containing water plus dye or some other dissolved substance (solute).

The essential characteristics of diffusion are (1) that each molecule moves independently of the others and (2) that these movements are random. The net result of diffusion is that the diffusing substance becomes evenly distributed.

Cells and Diffusion

Most organic molecules cannot freely diffuse through the lipid barrier of the cell membrane. Carbon dioxide and oxygen, however, which are soluble in lipids, move freely through the membrane. Water also moves in and out freely. Water is not soluble in lipids (which is, of course, just another way of saying that lipids are not soluble in water), and so the fact that water moves freely has led biologists to postulate the presence of small pores in the membrane, which permit the passage of water molecules (and also of some small ions). Within a cell, materials may be produced at one place and used at another place. Thus, a concentration gradient is established between the two areas, and the material diffuses down the gradient from the site of production to the area of use.

As we noted previously, diffusion is essentially a slow process, except over very short distances. It is efficient only if the concentration gradient is steep, the surface area is relatively large, and the volume relatively small. For instance, the rapid spread of a scent, such as perfume, through the air is due not primarily to diffusion but rather to the circulation of air currents. Similarly, in many cells, the transport of materials is speeded by active streaming of the cytoplasm (see page 20). Also, cells hasten diffusion by their own metabolic activities. For example, oxygen is used up within the cell almost as rapidly as it enters, thereby keeping the gradient steep. Carbon dioxide is produced by the cell, and so a gradient from inside to outside is maintained.

Osmosis

Water moves freely through membranes. However, the membranes, while permitting the passage of water, block the passage of most materials dissolved in it. (These substances are the *solutes*, and the water is the *solvent* of the solution.) Such a membrane is known as a differentially permeable membrane, and the diffusion of water (or any solvent) through such a membrane is known as *osmosis*. In the absence of other factors that influence water potential (such as pressure), the movement or diffusion of the water in osmosis will typically be from a region of lesser solute concentration (and therefore of greater water concentration) into a region of greater solute concentration (lower water concentration). The presence of solute decreases the water potential and so causes the movement of water from a region of greater to a region of lesser water potential.

If water is separated from a solution by a differentially permeable membrane, as shown in Figure 25–7, the water will move across the membrane and cause the solution to rise in the tube until equilibrium is reached, that is, until the water potential on both sides of the membrane is equal. If enough pressure were applied from the upper part of the tube, it would be possible to prevent movement of water into it. The pressure that must be applied to the solution to stop the movement of water is called the *osmotic pressure*.* The term osmotic pressure is also used to express the amount of reduction of water potential in a solution caused by the solute. An increase in solute concentration increases the osmotic pressure and decreases the water potential of a solution.

The word *isotonic* was coined to describe solutions that have equal osmotic pressures. No net movement of water will occur across a differentially permeable membrane separating two solutions that are isotonic to each other, unless pressure is exerted on one side. In comparing solutions of different solute concentrations, a solution that has less solute and therefore a lower osmotic pressure is said to be *hypotonic*, and one that has more solute and a higher osmotic pressure is said to be *hypertonic*.

* There is an increasing tendency among botanists working with plant-water relationships to substitute the term osmotic potential or solute potential for osmotic pressure. Numerically the terms are equal, but osmotic (solute) potential carries a negative (minus) sign.

25–7
Osmosis and osmotic pressure. (a) *The tube contains a solution and the beaker contains distilled water.* (b) *The differentially permeable membrane permits the passage of water but not of solute. The movement of water into the solution causes the solution to rise in the tube until the osmotic pressure resulting from the tendency of water to move into a region of lower water concentration is counterbalanced by the height,* h, *of the column of solution.* (c) *The force that must be applied to the piston to oppose the rise in the tube of the solution is a measurement of the osmotic pressure. It is proportional to* h.

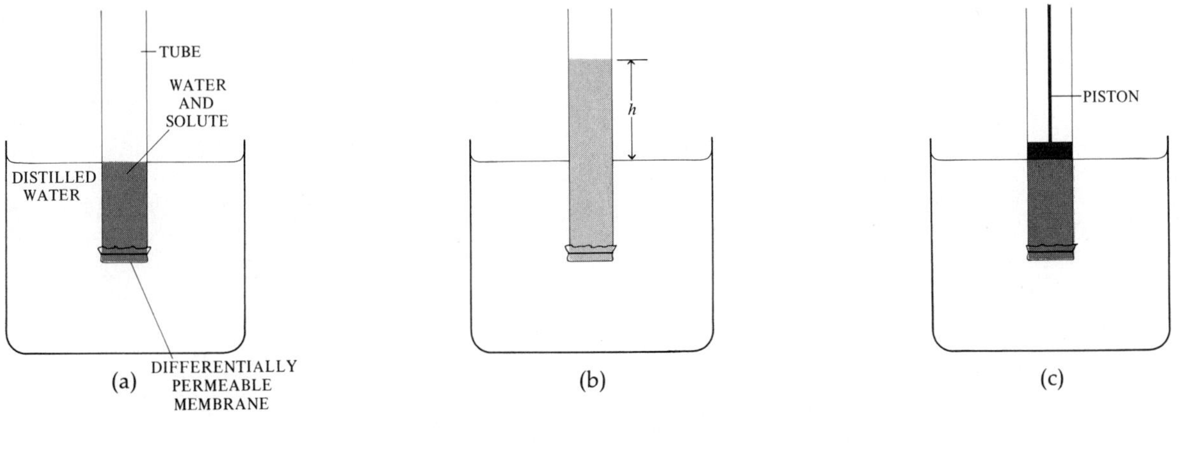

Osmosis and Living Organisms

The movement of water across the cell membrane from a hypotonic to a hypertonic solution causes some crucial problems for living systems. These problems vary according to whether the cell or organism is hypotonic, isotonic, or hypertonic in relation to its evironment. One-celled organisms that live in salt water, for example, are usually isotonic with the medium they inhabit, which is one way of solving the problem. Similarly, the cells of higher animals are isotonic with the blood and lymph that constitute the watery medium in which they live.

Many types of cells live in a hypotonic environment. In all single-celled organisms that live in fresh water, such as *Euglena* (page 274), the interior of the cell is hypertonic to the surrounding water; consequently, water tends to move into the cell by osmosis. If too much water were to move into the cell, it could dilute the cell contents to the point of interfering with function and could even eventually rupture the plasma membrane. This is prevented by a specialized organelle known as a contractile vacuole, which collects water from various parts of the cell body and pumps it out with rhythmic contractions.

Turgor

If a plant cell is placed in a hypotonic solution, the protoplast expands and the plasma membrane stretches and exerts pressure against the wall. However, the plant cell will not rupture because it is restrained by the relatively tough cell wall.

Plant cells tend to concentrate relatively strong solutions of salts within their vacuoles, and in addition they may accumulate sugars, organic acids, and amino acids as well. As a result, plant cells tend constantly to absorb water by osmosis and to build up hydrostatic pressures within them. This pressure against the cell wall keeps the cell stiff, or *turgid.* Consequently, the hydrostatic pressure in plant cells is commonly referred to as *turgor pressure.* Turgor pressure may be defined as the pressure that develops in a plant cell as a result of osmosis (and/or imbibition). Equal to and opposing the turgor pressure is the inwardly directed pressure of the cell wall, called the wall pressure.

Turgor in the plant is especially important in the support of nonwoody plant parts. In Chapter 1 we noted that some 90 percent of the growth of a plant cell is a direct result of water uptake. Auxin presumably contributes to this water uptake by relaxing the cell wall and so decreasing the resistance the wall exerts to turgor pressure.

Turgor is maintained by most plant cells because they are generally in a hypotonic medium. If, however, a turgid plant cell is placed in a hypertonic solution, water will leave the cell by osmosis, and the vacuole and protoplast will shrink, causing the plasma membrane to pull away from the cell wall (Figure 25–8). This phenomenon is known as *plasmolysis.* Figure 25–9 shows some cells of an *Elodea* leaf before and after plasmolysis. Note in both Figures 25–8 and 25–9 that although the plasma membrane and the tonoplast, the membrane surrounding the vacuole, are permeable only to water, the cell walls allow both solutes and water to pass freely through them.

25–8
Plasmolysis in a leaf epidermal cell. Under normal conditions, the protoplasm fills the space within the cell walls (a). *When the cell is placed in a solution of 0.22 M (molar) sucrose, water passes out of it and the plasma membrane contracts slightly* (b). *When immersed in a stronger (more concentrated) solution, the cell loses large amounts of water and contracts still further* (c). *However, the process may be reversed if the cell is then transferred to pure water.*

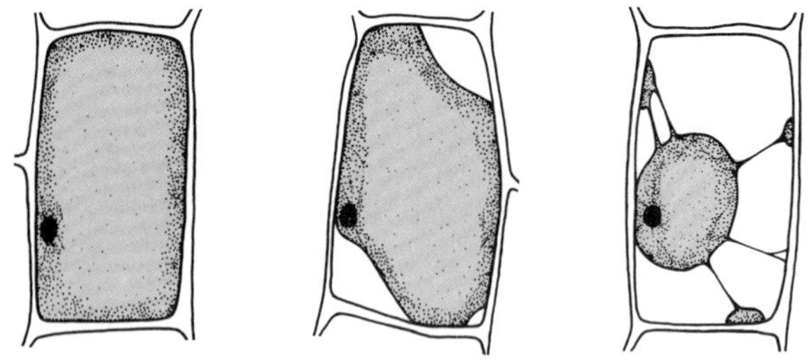

PLANTS AND WATER

In the preceding discussion, we may seem to have digressed from the subject of plant physiology. This is not, in fact, the case. These properties of water we have been considering have been among the most powerful of all forces dictating the evolutionary history of living systems, in general, and, because of their dependence on underground water, of land plants in particular. We are now going to look first at the movement of water and solutes through the plant body from the ground to the atmosphere and, toward the end of the chapter, at the movement of solutes and water from the sites of photosynthesis to the nonphotosynthetic parts of the plant body. We shall begin the former discussion with a description of transpiration because, as you will see, this process is a major determining factor in the movement of water through the plant body.

Two striking examples of the magnitude of transpiration are provided by a corn plant and a deciduous forest. A single corn plant weighing 0.4 kilograms (dry weight) at harvest time has absorbed, transported, and transpired 130 to 180 liters (or kilograms) of water (Figure 25–10). It has been estimated that the water transpired by an average corn crop during the growing season is equivalent to about 38 centimeters of rainfall. A deciduous forest consisting largely of oak trees may transpire the equivalent of 43 to 56 centimeters of rainfall in one year.

25–9
Elodea *leaf cells. Before* (a) *and after* (b) *being placed in a 0.5* M *sucrose solution. The cells in* (b) *are plasmolyzed.*

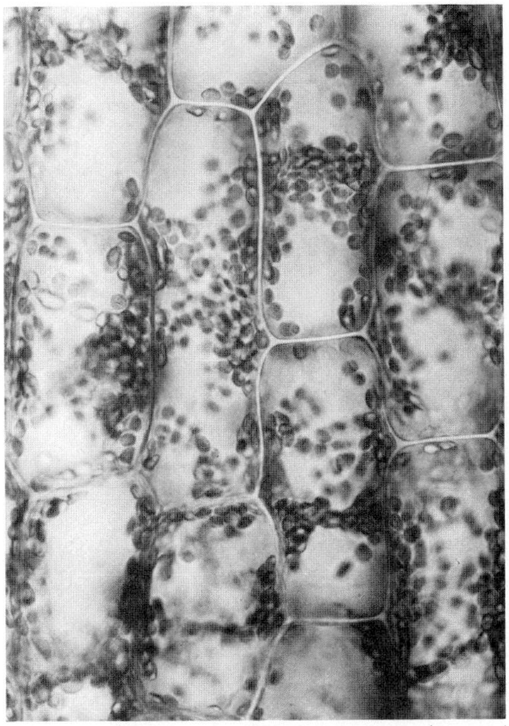

(a) 25 μm

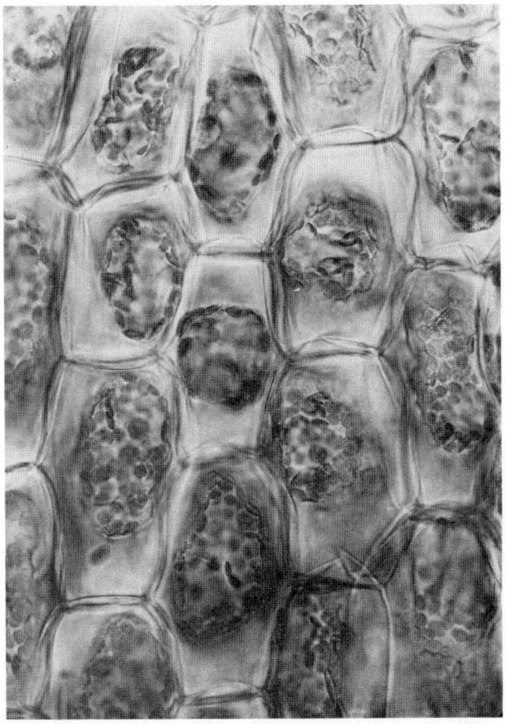

(b) 25 μm

25–10
Most of the water taken up by plants is transpired—that is, lost by evaporation from the plant body. Only small fractions are retained. The example shown here is corn (Zea mays), *which transpires (clear area) over 98 percent of the water it absorbs.*

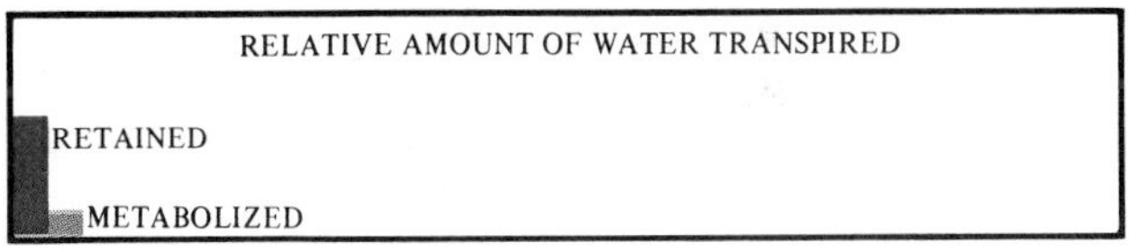

THE MOVEMENT OF WATER THROUGH THE XYLEM

Transpiration

In the early eighteenth century, Stephen Hales noted that plants "imbibe" a much greater amount of water than animals. He calculated that one sunflower plant, bulk for bulk, "imbibes" and "perspires" 17 times more water than a man every 24 hours. Indeed, the quantity of water absorbed by any plant is enormous—far greater than that used by any animal of comparable weight. An animal uses less water because a great deal of its water is recirculated through its body over and over again, in the form (in vertebrates) of blood plasma and other fluids. In plants, more than 90 percent of the water taken in by the roots is given off into the air as water vapor (Table 25-4). This process is known as *transpiration*, which is defined as the loss of water vapor by any part of the plant body.

Why do plants lose such large quantities of water to transpiration? To answer this question, let us look again at the structure of the leaf. The chief function of the leaf is photosynthesis, which is the source of all the food for the entire plant body. The necessary energy for photosynthesis, as you know, comes from sunlight. Therefore, for maximum photosynthesis, a plant must spread a maximum surface to the sunlight. But sunlight is only one of the requirements for photosynthesis; the chloroplast also needs carbon dioxide. Under most circumstances, carbon dioxide is readily available in the air surrounding the plant, but in order for carbon dioxide to enter the plant cell, which it does by diffusion, it must go into solution, because cell membranes are almost impervious to gaseous carbon dioxide. Therefore, there must be contact with a moist cell surface. But wherever water is exposed to air, evaporation occurs.

Plants, as we shall see, have developed a number of special adaptations for limiting evaporation, but all of these cut down the supply of carbon dioxide. In other words, photosynthesis and water loss by transpiration are inextricably bound together in the life of the green plant.

At the end of this discussion of water transport, we shall discuss adaptations that help plants to regulate the movement of water and gases. For the moment, however, let us continue to focus on the principles of movement of water through the plant body.

Absorption by Roots

The root system serves to anchor the plant in the soil and, above all, to meet the tremendous water requirements of the leaves. Almost all the water that a plant takes from the soil enters through the younger parts of the root. Absorption takes place directly through the epidermis of the root, largely in the region of root hairs. Several millimeters above the root tip the root hairs provide an enormous area for absorption (Figure 25-11; Table 25-5). From the root hairs, the water moves through the cortex, the endodermis (the inner layer of cortical cells), and the pericycle, and into the primary xylem. (For a review of root structure, see Chapter 20.) Once in the conducting elements of the xylem, the water moves upward through the root and stem and into the leaves.

During periods of rapid transpiration, water may be removed from around the root hairs so quickly that the soil becomes depleted; water will then move from some distance away toward the root hairs through fine pores in the soil. By and large, however, the roots come into contact with water by growing. Under normal conditions, roots of apple trees grow an average of about 3 to 9 millimeters a day; roots of prairie grasses may grow more than 13 millimeters a day; and the main roots of corn plants average 52 to 63 millimeters a day. A four-month-old rye plant has over 10,000 kilometers of roots and many billions of root hairs.

Table 25-4 *Water Loss by Transpiration per Single Plant during the Growing Season*

KIND OF PLANT	WATER LOSS BY TRANSPIRATION (LITERS)
Cowpea *(Vigna sinensis)*	49
Irish potato *(Solanum tuberosum)*	95
Winter wheat *(Triticum aestivum)*	95
Tomato *(Lycopersicum esculentum)*	125
Corn *(Zea mays)*	206

After J. F. Ferry, *Fundamentals of Plant Physiology*, The Macmillan Company, New York, 1959.

Table 25-5 *Density of Root Hairs in Three Species of Plants*

PLANT	NO. OF ROOT HAIRS/SQ CM OF ROOT SURFACE
Loblolly pine *(Pinus taeda)*	217
Black locust *(Robinia pseudo-acacia)*	520
Winter rye *(Secale cereale)*	2,500

After J.F. Ferry, *Fundamentals of Plant Physiology*, The Macmillan Company, New York, 1959.

25–11
(a) *Primary root of radish seedling* (Raphanus sativus), *showing root hairs.* (b) *Root hair surrounded by soil particles with water adhering to them.*

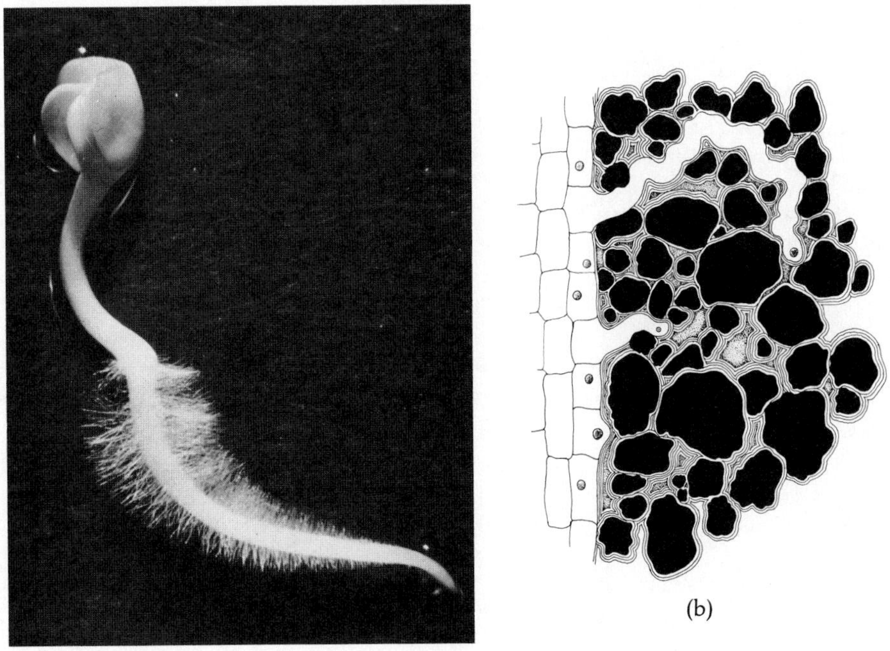

(a) (b)

25–12
A demonstration, using radioactive potassium, that xylem is the channel for the upward movement of water and minerals. Wax paper was inserted between the xylem and the phloem to prevent the lateral transport of the isotope. The relative amounts of radioactive potassium in each of the segments (S) is shown in (b).

(a)

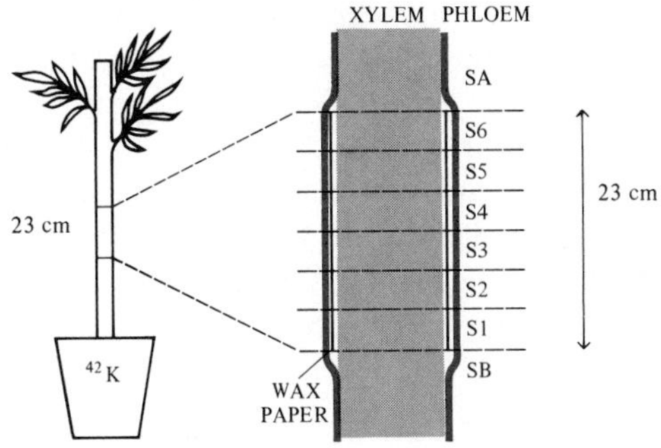

Stripped branch (xylem and phloem separated by wax paper)

(b)

		ppm ^{42}K in phloem	ppm ^{42}K in xylem
Above strip	SA	53	47
Stripped section	S6	11.6	119
	S5	0.9	122
	S4	0.7	112
	S3	0.3	98
	S2	0.3	108
	S1	20	113
Below strip	SB	84	58

The roots of some species of plants grow downward for great distances; a deep narrow root system is often found in plants growing on beach dunes, for example. In most kinds of plants, however, the greatest extension of the roots is lateral (page 418). Consequently, if a tree is watered at its base, most of the water will not be absorbed. Generally speaking, you can expect to find the largest area of new root growth just under the outermost leaves.

The transpiration stream, in addition to keeping the leaves of the plant provided with water, distributes mineral ions to the shoot as well (Figure 25–12). After ions are absorbed by the outer cells of the root, they are transferred through the cortical cells and finally released into the xylem. When transpiration is occurring, the ions are carried rapidly throughout the plant.

Root Pressure

When transpiration is very slow or absent, as at night, the root cells may still secrete ions into the xylem. Because the vascular tissue of the root is surrounded by the endodermis, a layer of tightly packed cells with radial and transverse walls that are impermeable to water and ions, ions do not tend to leak out of the xylem. Therefore, the water potential of the xylem becomes more negative, and water moves into the xylem by osmosis through the surrounding cells. In this manner a positive pressure, called *root pressure,* is created, and it forces both water and dissolved ions up the xylem (Figure 25–13).

You have undoubtedly observed the effects of root pressure. In the early morning you are likely to see dewlike droplets of water at the tips of grass leaves. These

25–13
(a) *Demonstration of root pressure in the cut stump of a plant. Uptake of water by the plant roots causes the mercury to rise in the column. Pressures of 3 to 5 bars have been demonstrated by this method.* (b) *Guttation droplets on the edge of a wild strawberry* (Fragaria virginiana) *leaf.*

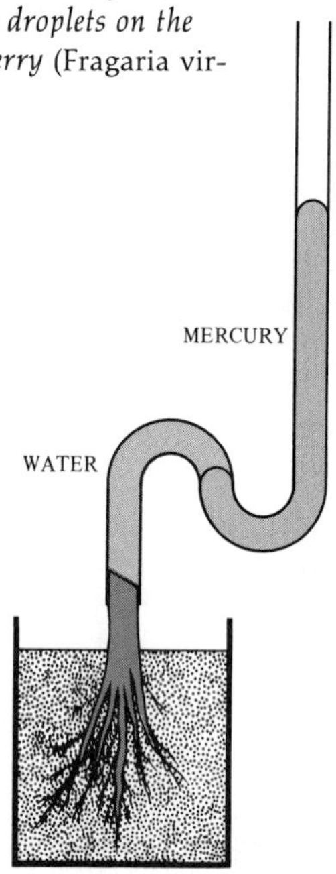

(a)

(b)

droplets are not dew, which is water that has condensed from the air, but come from within the leaf by a process known as guttation. They exude not through the stomata but through special openings called hydathodes, which occur at the tips and margins of leaves (Figure 25–13). The water of guttation is literally forced out of the leaves by root pressure.

It is to be emphasized that root pressure is least effective during the day, when the movement of water through the plant is the fastest. Root pressure never becomes large enough to move water to the top of a tall tree. Moreover, many plants, such as pines, develop no root pressure at all. Thus, root pressure is perhaps to be regarded in part as a by-product of the mechanism of pumping ions into the xylem and perhaps in part as a subsidiary means of moving water into the shoot under special conditions.

Water Transport

We have seen that water enters the plant by the roots and is given off, in large quantities, by the leaf. So we are faced with the unavoidable question of how the water gets from one place to another, often over large vertical distances. This question has intrigued many generations of botanists.

The general pathway that the water follows in its ascent has been clearly identified. You can trace this pathway yourself in a simple experiment. Put a cut stem in water colored with any harmless dye (preferably, cut the stem under the water to prevent air from entering the conducting elements of the xylem) and then trace the path of the liquid into the leaves. You will find that the stain quite clearly delineates the conducting elements of the xylem. More recent experiments using radioactive isotopes placed in water given to the plant confirm that the water does indeed travel by way of the vessel elements (or tracheids) in the xylem. You will notice in the experiment shown in Figure 25–12 that care must be taken to separate the xylem from the phloem. Earlier experiments in which this separation was not made produced ambiguous results, because apparently there is a great deal of lateral movement from the xylem into the phloem, presumably across the rays. (This lateral movement, however, as the experiment shows, is not necessary for the overall movement of water and minerals from soil to leaf.)

So we know the path that water takes, but how does it move? Logic suggests two possibilities: It can be pushed from the bottom or pulled from the top. (A third possibility, involving active pumps, "hearts," along the way has been proposed from time to time but is no longer considered seriously by botanists.) We have already ruled out the first of these possibilities, however. Root pressure, as we noted previously, does not exist in all plants and, in those plants in which it is present, is not

25-14
Measurement in ash trees shows that a rise in water uptake follows a rise in transpiration. These data suggest that as water is lost, forces for its uptake are acquired. (G stands for grams.)

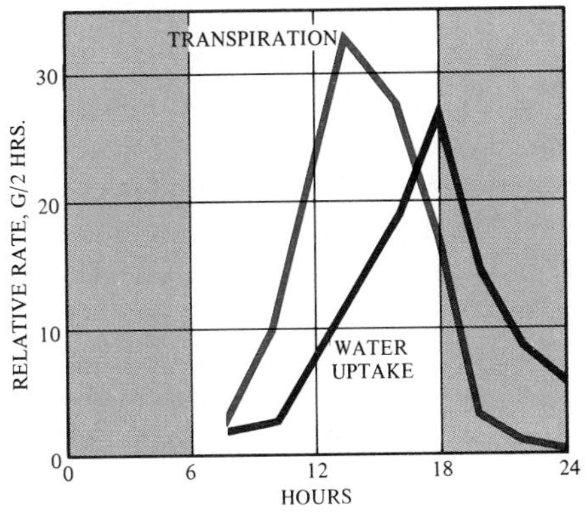

25-15
A simple physical system that demonstrates the cohesion-adhesion-tension theory. A porous clay pot is filled with water and attached to the end of a long, narrow glass tube also filled with water. The water-filled tube is placed with its lower end below the surface of a volume of mercury contained in a beaker. As water evaporates from the pores in the pot, it is replaced by water "pulled up" through the narrow glass tube in a continuous column. As the water evaporates, mercury rises in the tube to replace it. Similarly, it has been demonstrated, transpiration from plant leaves can result in sufficient water loss to create a similar negative pressure.

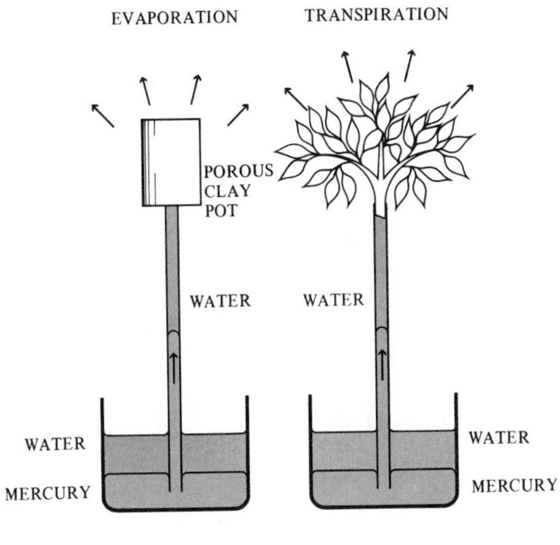

sufficient to move water to the top of a tall tree. Moreover, the simple experiment just cited involving the cut stem rules out root pressure as a crucial factor. So we are left with the hypothesis that water is pulled up through the plant body. And, as we shall see, this hypothesis is correct according to all present evidence.

The Cohesion-Adhesion-Tension Mechanism

When the cells of the leaf lose water to the air during transpiration, their ions and molecules become more concentrated—that is, the water potential of the cells becomes increasingly more negative (Figure 25-14). Because the intracellular water is in direct connection, through the water-permeable plasma membranes, with the water in the walls, a pull is exerted on this water when water evaporates from the walls, increasing its negative water potential. The water of the wall is in connection with the water of the xylem, so the negative potential is transmitted to the contents of the xylem as well, causing a buildup of tension there. Because of the extraordinary cohesiveness of water, this tension is transmitted all the way down the stem to the roots, so that water is withdrawn from the roots, pulled up the xylem, and distributed to the cells which are losing water to the atmosphere. However, this loss makes the water potential of the roots more negative, thus increasing their ability to extract water from the soil.

This theory of water movement is known as the cohesion-tension theory, because it depends on the great cohesiveness of water, which permits it to withstand tension (Figure 25-15). However, the theory might better be known as the cohesion-adhesion-tension theory, because adhesion of the water molecules to the walls of the tracheids and vessels of the xylem and to the cell walls of the leaf and root cells is as important as cohesion and tension for the rise of water. The cell walls along which the water moves have evolved as a very effective water-attracting surface, taking maximal advantage of water's adhesiveness, and thus providing a situation in which cohesiveness is readily expressed. Moreover, the filtering of water by the cortex of the root doubtless contributes to the cleanliness and hence cohesiveness of the water that enters the xylem, and the small size of the conduits, or capillaries, through which the water moves also contributes to the reliability of the system.

There is no doubt that the tensile strength of water is great enough to prevent the pulling apart of water molecules under the tension required to move water up the xylem of tall trees. For example, it has been demonstrated that a column of water in a fine capillary tube is capable of withstanding a tension of −264 bars, whereas the estimated tension required to move water to the top of a giant redwood (*Sequoia sempervirens*) is only about −20 bars.

How can the cohesion-adhesion-tension theory be

25–16
A method for measuring the water tension in the xylem. A branch whose xylem tension the investigator wishes to measure is cut off and placed in a "pressure bomb." When the branch is cut, some of the xylem sap—which was under tension before the branch was cut—recedes into the xylem below the cut surface. Pressure is raised in the bomb until sap emerges from the cut end of the stem. Presuming that equal pressure is required to force the sap in either direction, the positive pressure needed to force out the sap is ideally equal to the tension that existed in the branch before it was severed.

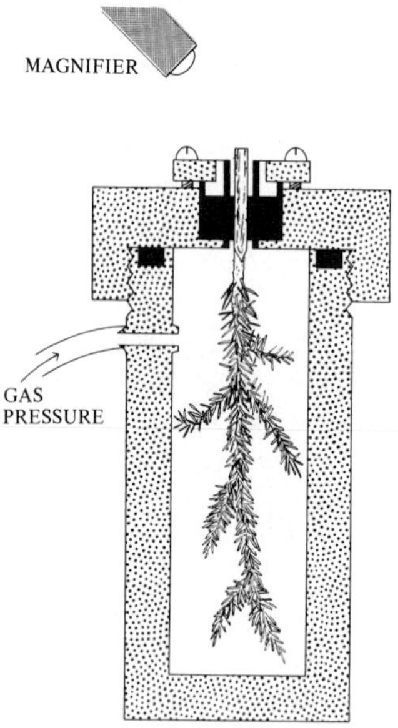

tested? One way to test this theory directly is by measuring the tension of water within a xylem vessel. When a twig is cut from a transpiring tree, the columns of water in the xylem vessels snap back. By mounting the twig in a pressure chamber as shown in Figure 25–16, it is possible to apply pressure to the leaves until the curved upper surface of the water columns appear (when examined with a microscope) at the cut surface of the twig. The magnitude of the pressure required to return the water to the cut surface is equal to the magnitude of the tension under which the water existed in the twig before its excision. Results obtained by this method are entirely consistent with predictions of the cohesion-adhesion-tension theory.

A second set of data that is in accord with the cohesion-adhesion-tension theory indicates that the movement of water begins at the top of the tree. The velocity of sap flow in various parts of the tree has been measured by an ingenious method involving a small heating element to warm the xylem contents for a few seconds and a sensitive thermocouple to detect the moment at which the heated xylem sap moves past (see Figure 25–17). As shown in the graph, in the morning the sap begins to flow first in the twigs as tension arises close to the leaves and later in the trunk. In the evening, the flow diminishes first in the twigs, as water loss from the leaves diminishes, and later in the trunk.

A third set of supporting data comes from measurements of minute changes in the diameter of the tree trunk (Figure 25–18). The shrinking of the trunk occurs because of the negative pressures in the water passages of the xylem. The water molecules clinging to the sides of the vessel pull them inward. When transpiration begins in the morning, first the upper part of the stem shrinks, as water is pulled out of the xylem before it can be replenished from the roots; then the lower part shrinks. Later in the day, as the transpiration rate decreases, the upper trunk expands, then the lower.

Note that the energy for the evaporation of water molecules—and thus for the movement of water and minerals through the plant body—is supplied not by the plant but directly by the sun. Note also that the movement is possible because of the extraordinary properties of water to which the plant is so exquisitely adapted.

25–17
Method for measuring velocity of sap flow. A small heating element inserted into the xylem heats the ascending sap for a few minutes. A thermocouple above the heating element records the passing wave of heat. The experimenter times the interval between these two events. As shown in the graph, in the morning the sap begins to increase its velocity of flow first in the twigs (upper curve) and then in the trunk (lower curve). In the evening, velocity diminishes first in the twigs and then in the trunk. (Adapted from "How Sap Moves in Trees" by Martin H. Zimmermann. Copyright © March 1963 by Scientific American, Inc. All rights reserved.)

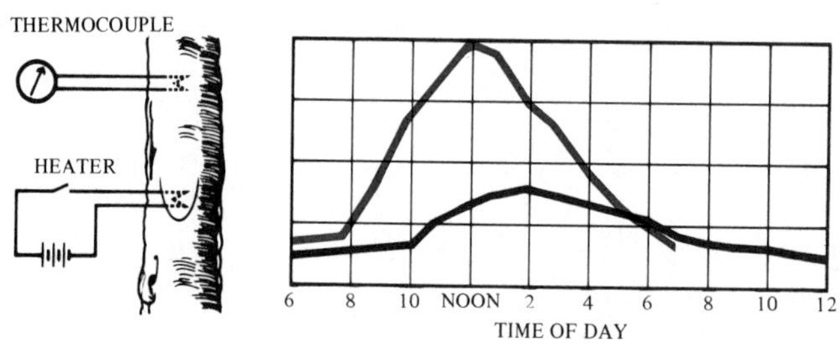

INCREASE IN DIAMETER

WEDNESDAY THURSDAY FRIDAY SATURDAY

Regulation of Transpiration

As we noted previously, transpiration is extremely costly to the plant, especially when the water supply is limited, and the intake of carbon dioxide is absolutely essential for photosynthesis. These two processes are not separable, at least not by any evolutionary solution yet devised. However, a number of special adaptations exist that minimize water loss while optimizing the gain of carbon dioxide.

The Cuticle and the Stomata

Leaves are covered by a cuticle that makes the surface of the leaf largely impervious both to water and to carbon dioxide. (See page 444 for a discussion of leaf structure.) A small fraction of the water transpired by plants is lost through this protective outer coating, and another small fraction is lost through the lenticels in the bark. By far the largest amount of water transpired by a higher plant is lost through the stomata (Figure 25-19). Stomatal transpiration involves two processes: (1) evaporation of water from cell wall surfaces bordering the intercellular spaces, or air spaces, of the mesophyll tissue, and (2) diffusion of the water vapor from the intercellular spaces into the atmosphere by way of the stomata.

Stomata, as we have noted previously, are small openings in the epidermis that are opened and closed by changes in turgor of the two surrounding guard cells (see Figure 25-20). Stomata are also found on young stems, but they are far more abundant on leaves. The number of stomata may be quite large; for example, there are 12,000 stomata per square centimeter of leaf surface in tobacco leaves. The stomata lead into a honeycomb of air spaces within the leaf that surround the thin-walled mesophyll cells. The air in these spaces, which make up 15 to 40 percent of the total volume of the leaf, is saturated with water vapor that has evaporated from the damp surfaces of the mesophyll cells. Although the stomatal openings take up only about 1 percent of the total leaf surface, more than 90 percent of the water transpired by the plant is lost through the stomata. The rest is lost through the cuticle.

Closing of stomata not only prevents the loss of water vapor from the leaf but also, of course, prevents the entry of carbon dioxide into the leaf. A certain amount of carbon dioxide, however, is produced by the plant during respiration, and as long as light is available this carbon dioxide can be used to sustain a very low level of photosynthesis even when the stomata are closed. Photosynthesis in plants that live in very dry climates is often kept at a low level because of the necessity for water retention, but many of these plants—the C_4 plants—have evolved an efficient method of utilizing the low levels of carbon dioxide available (see page 117).

The Mechanism of Stomatal Movements

As illustrated in Figure 25-20, stomata open when the guard cells are more turgid than the surrounding cells and close when the guard cells are less turgid. Turgor is maintained or lost due to the passive osmotic movement of water in or out of the cells along a gradient of solute concentration.

Energy is expended, however, in the preferential accumulation of solute that creates the gradient. Preferential loss of solute may, of course, occur because of differentially increased permeability of the cell membrane, but some believe that an active mechanism may be involved here, too.

The major solute responsible for the gradients is a potassium ion (K^+). Several techniques for estimating potassium levels inside single guard cells show that potassium levels rise inside the guard cells when the stomata open and drop when the stomata close. The surrounding cells provide a great reservoir of potassium ions, so that whereas in some plants the K^+ content of the surrounding cells may change in dramatic opposition to that of the guard cells, in others it may remain almost the same. In either situation, the gradient of potassium between the guard cells and surrounding cells changes significantly, accompanied by the osmotic flow of water and resultant turgor changes illustrated in Figure 25-20.

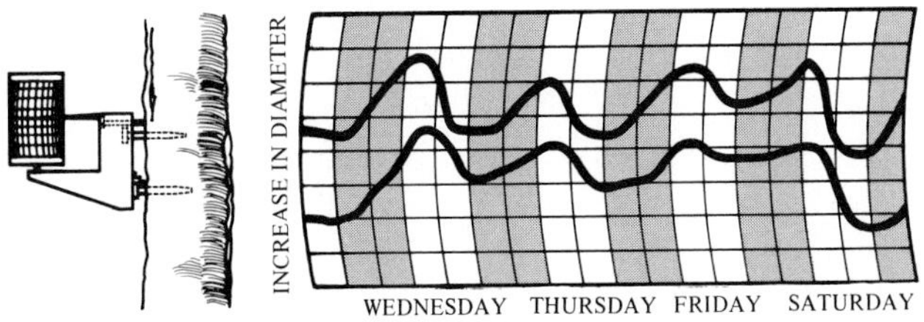

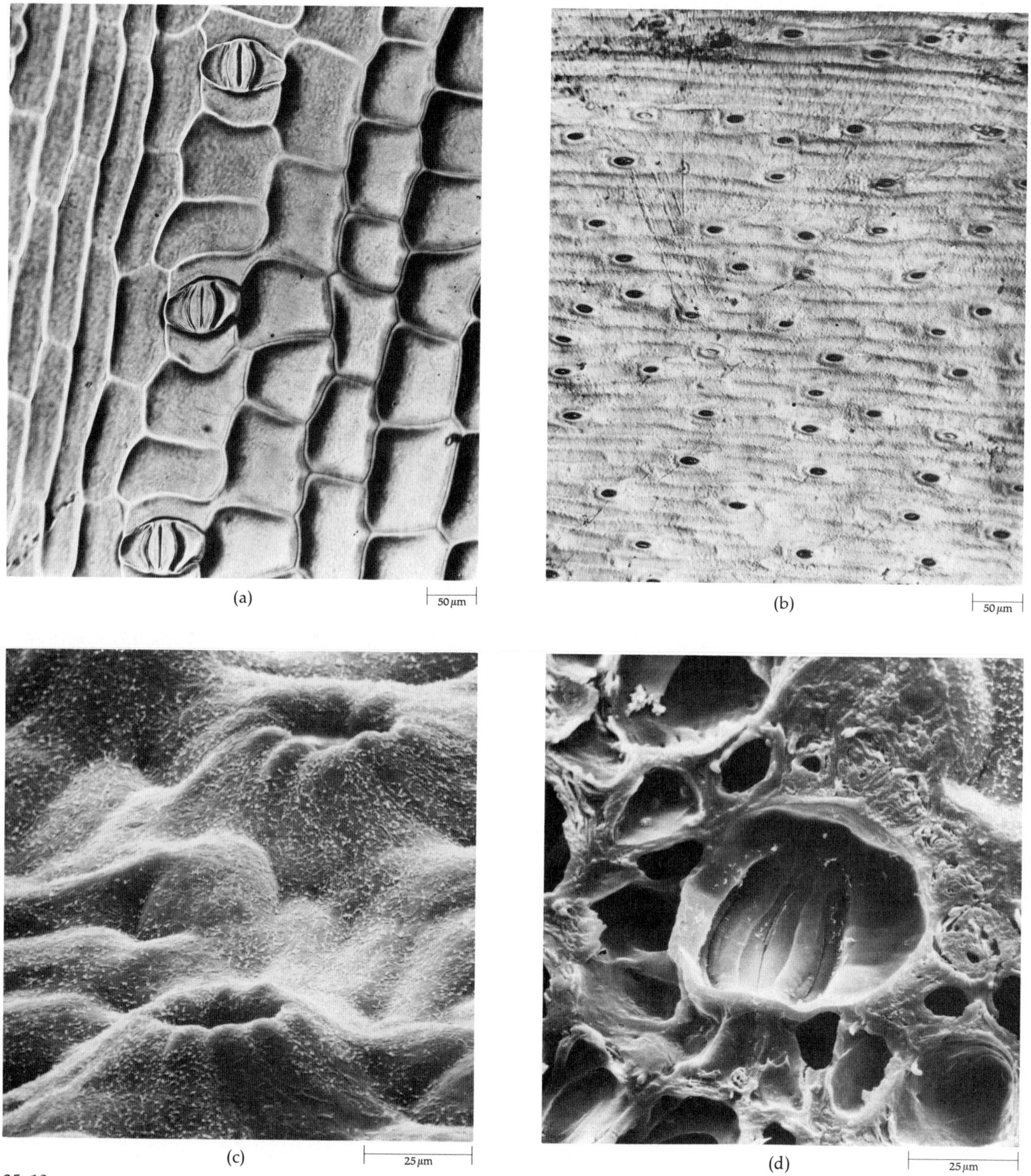

25–19
Surface views of stomata. (a) *Lower leaf surface of the spiderwort,* Tradescantia virginiana. *Note open stoma above, and the accessory cells on either side (right and left) of the guard cells.* (b) *Lower leaf surface of the bluebell,* Endymion nonscriptus. *These reflected-light micrographs were made from a replica of the leaf surface made with dental plastic.* (c) *Lower surface of the leaf of* Cycas revoluta. *The stomata are sunken beneath the raised structures shown here. Each structure is composed of nine epidermal cells. Cuticular wax is dispersed over the leaf surface.* (d) *Portion of a* Cycas *leaf showing substomatal chamber and lower surface of stoma.*

25-20
Mechanism of stomatal movements. Each stoma is flanked by two guard cells that open the stoma when they are turgid and collapse and close it when they lose turgor. In many species, the guard cells have thickened walls adjacent to the stomatal opening. As turgor pressure increases, the thinner parts of the cell wall are stretched more than the thicker parts, causing the cells to bow out and the stoma to open. A surface view is shown in (a) *and a cross section in* (b).

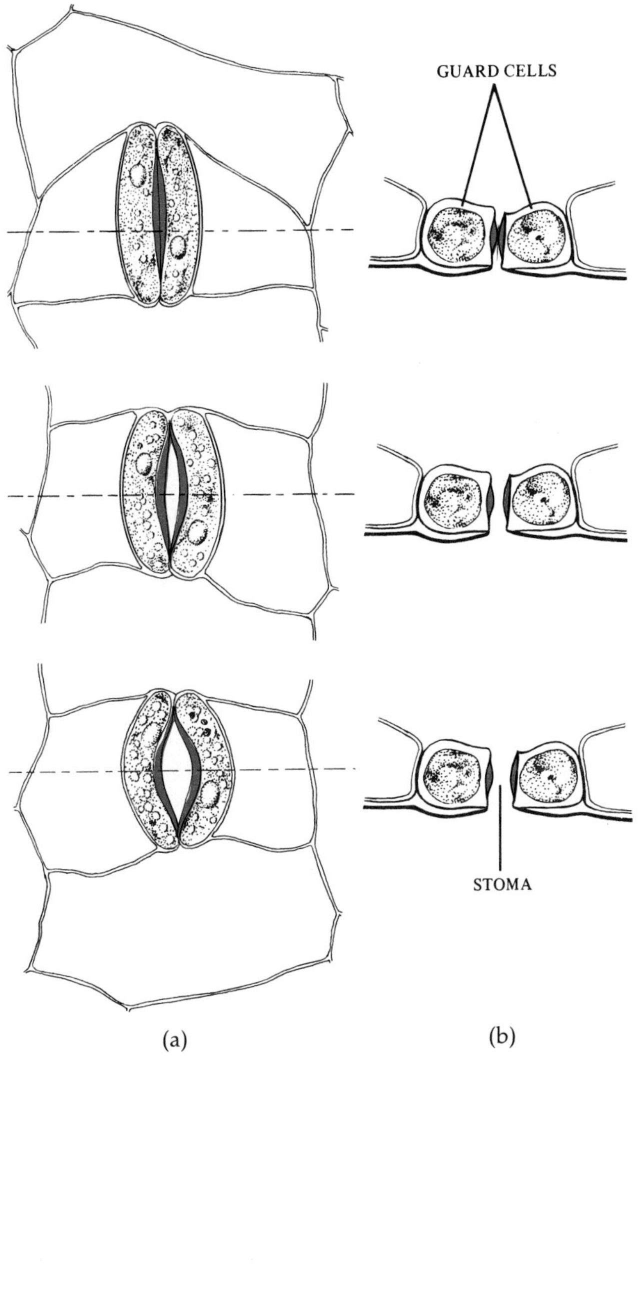

Factors Affecting Stomatal Movements

A number of environmental factors affect stomatal opening and closing, water loss being the major influence. When the turgor of a leaf drops below a certain critical point, which varies with different species, the stomatal aperture becomes smaller. The effect of water loss overrides other factors affecting the stomata, but stomatal changes can occur independent of overall water gain or loss by the plant. The most conspicuous example is found in the many species in which the stomata close regularly in the evening and open in the morning, even though there may be no changes in water available to the plant.

Factors other than water loss that affect the stomata include carbon dioxide concentration, light, and temperature. In most species, an increase in carbon dioxide concentration in the intercellular spaces causes the stomata to close. Conversely, in some species, exposure of the leaf to carbon dioxide-free air may cause the stomata to open. As long as the water content of the leaf is sufficient, both of these effects can occur either in light or in darkness.

Similarly, in most species, when a turgid leaf that has been kept in darkness for a few hours is exposed to light, the stomata open. When the light is switched off, the stomata close. There is a short time lag before opening, but the closing may be detected within 15 seconds. The action of light is primarily mediated through the photosynthetic utilization of carbon dioxide.

Within normal ranges of temperature (10 to 25°C), changes in temperature have little effect on stomatal behavior, but temperatures higher than 30 to 35°C can cause stomatal closure. The closing can be prevented by holding the plant in carbon dioxide-free air, however, which suggests that temperature changes work primarily by affecting the concentration of carbon dioxide in the leaf. An increase in temperature results in an increase in respiration and a concomitant increase in the concentration of intercellular carbon dioxide, which may be the cause of stomatal closure. Many plants in hot climates close their stomata regularly at midday, apparently because of the effect of temperature on carbon dioxide accumulation and because of dehydration of leaves as the water loss of transpiration exceeds the water uptake by absorption.

Although the stomata of most plants are open during the day and closed at night, this is not true of all plants. A wide variety of succulents—including cacti, the pineapple, and members of the stonecrop family, Crassulaceae, among others—open their stomata at night, when conditions are least favorable to transpiration. The Crassulacean acid metabolism (CAM) characteristic of such plants has a pathway for carbon flow not substantially different from that of C_4 plants, as we discussed briefly in Chapter 6. At night, when their stomata are open, the CAM plants take in carbon dioxide and convert it to malic and isocitric acids. During the day, when their stomata are closed, the carbon dioxide is released from these organic acids and used immediately in photosynthesis.

Other Factors Affecting the Rate of Transpiration

Although stomatal opening and closing are the major factors affecting the rate of transpiration, there are a number of other factors both in the environment and in the plant itself. One of the most important of these is temperature. The rate of water evaporation doubles for every temperature rise of about 10°C. However, because evaporation cools the leaf surface, its temperature does not rise as rapidly as that of the surrounding air. As we noted previously, stomata close when temperatures exceed 30 to 35°C.

Humidity is also important. Water is lost much more slowly into air already laden with water vapor. Leaves of plants growing in shady forests, where the humidity is generally high, typically spread large luxuriant leaf surfaces, because their chief "problem" is getting enough light, not losing water. In contrast, plants of grasslands or other exposed areas often have narrow leaves and relatively little total leaf surface. They get all the light they can use but are constantly in danger of excess water loss.

Air currents also affect the rate of transpiration. A breeze cools your skin on a hot day because it blows away the water vapor that has accumulated near the skin surface and so accelerates the rate of evaporation of water from your skin surface. Similarly, wind blows away the water vapor from leaf surfaces. Sometimes, if the air is very humid, wind may actually decrease the transpiration rate by cooling the leaf, but a dry breeze will greatly increase evaporation. Leaves of plants that grow in exposed, windy areas are often hairy; these hairs are believed to protect the leaf surface from wind action and so slow the rate of transpiration by stabilizing the boundary layer of air over the leaf.

Coastal ecotypes of various plant species often have leaves that are relatively thick and broad, whereas those found in the interior tend to have leaves that are more highly dissected, thin, and narrow. Under conditions of high solar radiation, such as those apt to be found away from the coast, the chief limiting factor for photosynthesis is the availability of carbon dioxide. Its availability to the cells of the leaf is limited to some extent by the resistance of the mesophyll. The average distance to chloroplast-bearing cells is decreased in thin, broad, dissected leaves. In addition, such leaves have very thin boundary layers, and a small difference in temperature is enough to bring about an effective convection, so that small or dissected leaves can remain near the ambient temperature in hot weather. On the other hand, large flat surfaces, such as the stems of beavertail cactus (*Opuntia*), may rise to temperatures as much as 16°C above the ambient and would therefore be severely damaged if not protected by a thick, strongly protective cuticle. These are examples of the ways in which leaf shape may be adaptive in particular situations.

TRANSLOCATION AND THE PHLOEM

The discovery by William Harvey (1578–1657) of the circulation of blood of animals stimulated the search for the analogous pathways and pumping mechanisms in plants. During the last half of the seventeenth century, two physicians, Marcello Malpighi of Italy and Nehemiah Grew of England, working independently, concluded that there were two conducting streams in plants, an ascending stream located in the wood and a descending stream located in the bark. In 1679, Malpighi postulated that the ascending stream of "raw sap" from the soil was modified in the leaves by the action of sunlight and, thereby, converted into an "elaborated sap," which fed other parts of the plant.

In 1686, Malpighi provided evidence for the existence of a descending stream in the bark. He showed that when a tree was "ringed"—that is, when a complete girdle of bark was removed—the tissue below the ring did not receive nourishment (Figure 25–21) and the tissue above the ring became swollen—presumably by accumulation of the "elaborated sap." He further demonstrated that the swelling phenomenon did not occur in the winter months, when the tree lacked leaves.

Grew's (1682) observations of the so-called bleeding of plants led him to conclude that sap moved upward in the xylem. This phenomenon—the exudation of water or sap from the wood of cut plants—is caused by root pressure, the same force responsible for guttation.

During the next century, great strides were made toward understanding the movement of water and dissolved minerals in the xylem, and by the end of the nineteenth century the cohesion-tension theory for the rise of water in tall plants was proposed by Dixon and Joly (1894). Unfortunately, much less progress was made in the understanding of food transport in the phloem, although many fairly accurate descriptions of phloem structure were available.

Why the great delay in progress in the understanding of phloem transport? In both the eighteenth and nineteenth centuries, one of the most eminent and influential investigators of the time discounted all evidence pointing to a descending sap stream in the phloem. In the eighteenth century it was Stephen Hales, and in the nineteenth century Julius von Sachs. Hales demonstrated the force of suction in wood and roots and the presence of root pressure in vines (Figure 25–22). He was also among the earliest investigators to measure tran-

25–21
Portion of white pine (Pinus strobus) *stem that had been girdled by porcupines. Note the differences in development of the stem immediately above and below the girdle. Food manufactured by leaves of the whorl of branches below the girdle stimulated growth in thickness of the stem below the girdle and prevented it from dying.*

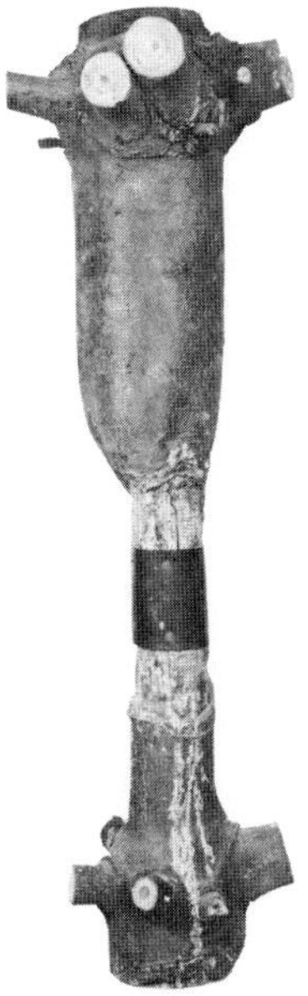

spiration. However, Hales denied the existence of a descending sap stream, admitting only that sap might sink in the wood at night because of lowering temperatures. Sachs believed that food substances were translocated in the xylem. He also believed that they moved by diffusion and, based on this belief, he constructed an elaborate theory of water movement. Of course, both investigators were wrong about the phloem and its role in food, or assimilate, transport. Nevertheless, their ideas influenced thinking for several decades. (The term "assimilate" is commonly used by phloem workers to refer to the food substances transported in the phloem.)

A second deterrent to progress in the understanding of phloem transport was the lack of convincing evidence that the phloem—and, more specifically, the sieve elements—formed the pathway of assimilate transport in plants. The sieve element was discovered by Theodor Hartig in 1837, and it was he who suggested that the sieve element was the principal food-conducting conduit in the phloem. In 1860, Hartig also discovered that when the bark of trees is cut with a sharp knife deep enough to reach the active or functional phloem (page 472), sap will exude from it. Hartig immediately suggested that this exudation of sap was related to the descending sap stream and, with the aid of a microscope, he proved that the exudate came from the phloem. However, Sachs dismissed Hartig's discovery with a few unkind remarks, and it was not until the late 1920s and 1930s that the role of phloem was properly understood and widely accepted.

Movement of Substances in the Phloem

Much valuable information on movement of substances in the phloem has come from studies utilizing aphids, small insects that suck the juices of plants. Most spe-

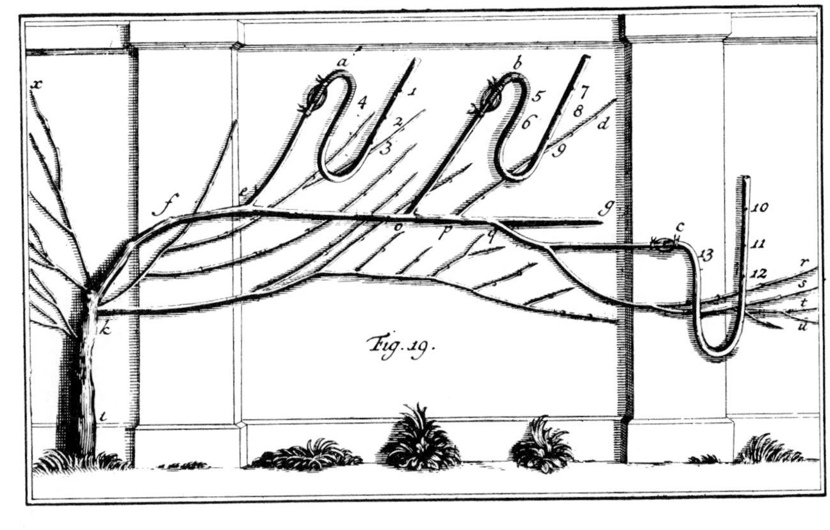

25–22
Stephen Hales, the first to measure blood pressure in animals, also measured the ascent of sap in plants. The drawing is from his Statical Essays *of 1726. "Experiment XXXVII. April 4th, I fixed three mercurial gages, (Fig. 19.)* a, b, c, *to a vine, on a south-east aspect, which was 50 feet [15 meters] long, from the root to the end* r u. . . . *The branches to which* a *and* c *were fixed were thriving shoots two years old, but the branch* b *was much older. When I first fixed them, the mercury was pushed up by the force of the sap, in all the gages down the legs 4, 5, 13, so as to rise nine inches [22.8 centimeters] higher in the other legs." Hales noted that the mercury rose still higher next morning and reached a maximum height of 53, 66, and 66 centimeters respectively in a, b, and c. He also noted diurnal fluctuation in this pressure.*

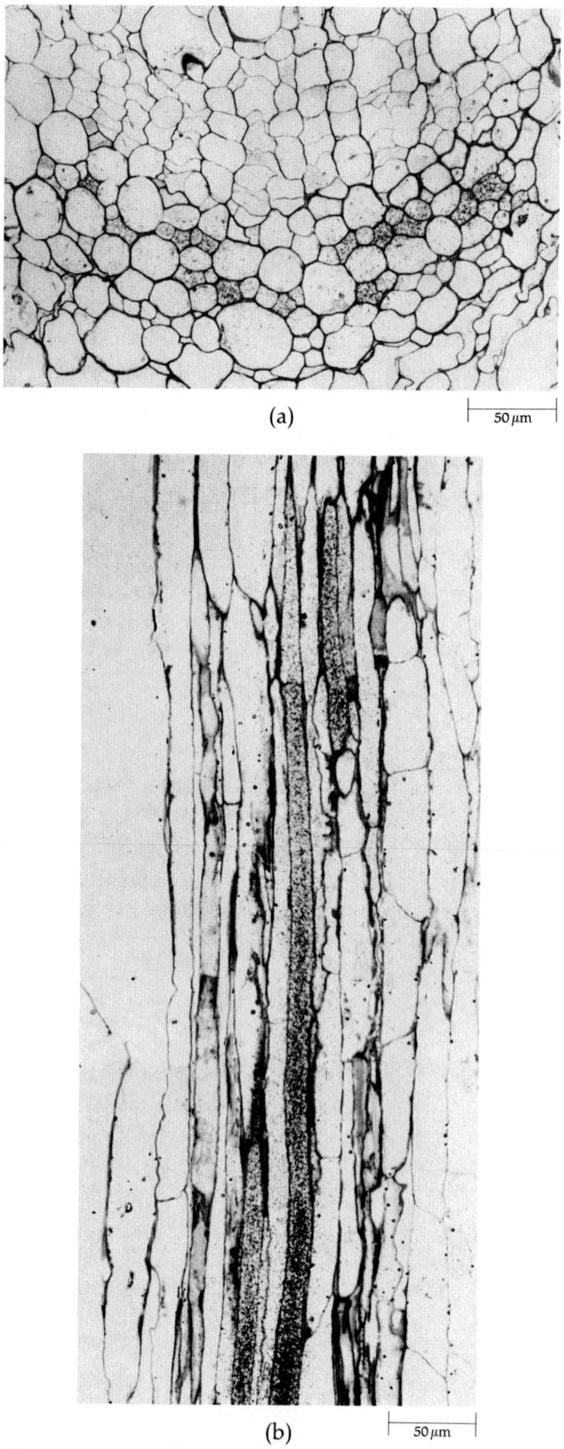

25-23
Historadiographs of (a) *transverse and* (b) *longitudinal sections of a vascular bundle from a stem of the broad bean* (Vicia faba). *A leaf of this stem was exposed to* $^{14}CO_2$ *for 35 minutes, and the photographic film covering the sections was exposed to radiation from the* ^{14}C *for 32 days. The labeled contents are found almost entirely in sieve elements.*

cies of aphids are phloem feeders. When these aphids insert their modified mouth parts, or stylets, into a stem or leaf, they extend them until the tips of the stylets puncture a conducting sieve tube. The turgor pressure of the sieve tube then forces the sieve-tube sap through the aphid's digestive tract and out its posterior end as droplets of "honeydew." If feeding aphids are anesthetized and severed from the stylets, exudation often continues from the cut stylets for many hours. The sieve-tube exudate can then be collected from the cut ends of the stylets with a micropipette and analyzed. Analyses of exudates obtained in this manner reveal sieve-tube sap to contain 10 to 25 percent dry matter, 90 percent or more of which is sugar, mainly sucrose in most plants. Low concentrations (less than 1 percent) of amino acids and other nitrogen-containing substances are also present.

The use of radioactive tracers has also provided an enormous amount of valuable information on phloem transport. Before such tracers were available, it was necessary to cut into the intact plant to introduce dyes and other substances in an attempt to study certain transport phenomena. However, when the high hydrostatic pressures of the sieve tubes are released at the time the sieve tubes are severed, the contents of the sieve elements surge toward the cut surfaces, greatly disturbing the system. In Chapter 19 we noted that this phenomenon is responsible for the formation of slime (P-protein) plugs in injured sieve elements. With the use of radioactive tracers, it is possible to experiment with entire plants and thus to obtain a fairly good understanding of normal transport phenomena. Results of experiments with radioactive assimilates (for example, ^{14}C-labeled sucrose) early confirmed the movement of such substances in the phloem and, more recently, in the sieve tubes of the phloem (Figure 25-23).

Data obtained from studies utilizing aphids and radioactive tracers indicate that the rates of longitudinal movement of assimilates in the phloem are far in excess of the normal rate of diffusion of sucrose in water. For example, in one series of experiments utilizing severed aphid stylets, it was estimated that at the sites of the stylet tips, the sap was moving at a rate of about 100 centimeters per hour.

The Pressure-Flow Mechanism

The first person in this century to give serious attention to Hartig's ideas on assimilate movement in the phloem was E. Münch, who in 1927 proposed the *pressure-flow hypothesis.* Since it was first proposed, this hypothesis has been modified, but its major features remain unchanged. According to the pressure-flow hypothesis, assimilates move through the sieve tubes along concentration gradients between the *sources* of assimilates and the sites of utilization, or *sinks,* of those assimilates. Sources are places where food substances are available for transport, such as photosynthesizing leaves or storage regions. Examples of sinks are growing and differentiating tissue regions and storage regions.

If we take a photosynthesizing leaf as the source and a growing root tip as the sink, the pressure-flow mechanism would operate as follows. Sugar manufactured in mesophyll cells of the leaf is actively secreted or "pumped" into the sieve tubes of the veins by neighboring parenchymatic cells, such as companion cells. This decreases the water potential in the sieve tube and causes water to move into the sieve tube from the xylem. With the movement of water into the sieve tube, the sugar is carried passively to the growing root tip, where the sugar is removed from the sieve tube through the expenditure of energy by neighboring parenchymatic cells. This results in increased water potential in the sieve tube and the eventual movement of water out of the tube at the sink. Note that the role of the sieve tube is a relatively passive one. Although energy is expended at source and sink, it is expended by the parenchyma cells, not the sieve tube. The pressure-flow mechanism is an example of "mass flow," the solute or sugar molecules moving along with the water through the sieve tube. What is the driving force for the movement from source to sink in the pressure-flow mechanism? Although it has often been stated that the driving force is the turgor, or hydrostatic, pressure gradient along the direction of flow, the pressure-flow mechanism does not depend on the presence of such a pressure gradient, but rather on the differences in the water potential on either side of the differentially permeable membranes (the plasma membranes) at and between the sources and sinks. These differences in water potential result in osmosis and provide the driving force for the pressure-flow mechanism (Figure 25–24).

Since Münch first proposed the pressure-flow hypothesis in 1927, several interesting alternative proposals have been made to explain the mechanism of assimilate movement in the phloem. All of these alternative hypotheses call upon the sieve elements to play an active role in the movement of assimilates, in contrast to the pressure-flow mechanism, which casts the sieve element in a passive role. At present, however, the pressure-flow mechanism is the most widely accepted explanation for the movement of substances in the phloem.

25–24
Diagram of pressure-flow mechanism. Sugar (colored arrows) enters a sieve tube of a leaf by active transport. As a consequence of the increased concentration of sugar, water (black arrows) enters the sieve tube by osmosis. Sugar is removed in the root by an active transport process and the concentration falls. As a consequence of the lowered sugar concentration, water leaves the sieve tube by osmosis. Because of the active secretion of sugar into the sieve tube at the source and its active absorption from the sieve tube at the sink, a flow of sugar solution takes place between source and sink.

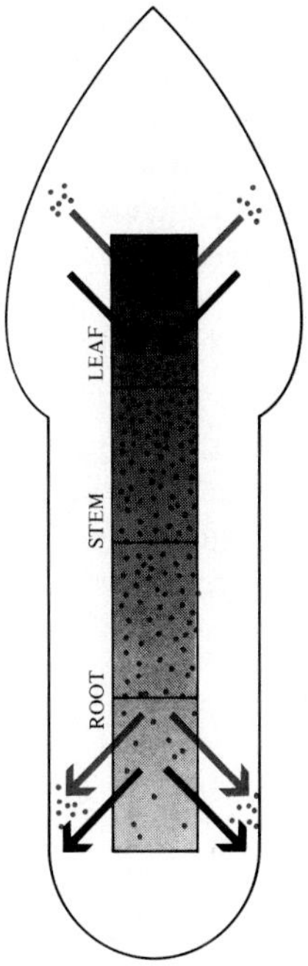

SUMMARY

Water has a number of unusual properties that are important in its biological roles. These include a high surface tension, strong adhesion and cohesion, high specific heat, a high heat of vaporization, high latent heat of melting, and the fact that water becomes less dense as its temperature falls below 4°C. These properties result from the molecular structure of water, its consequent polarity, and the formation of hydrogen bonds between adjacent water molecules.

Principles governing the movement of water include mass flow, diffusion, and osmosis. Mass flow is the overall, unidirectional movement of molecules of a liquid, typically as a result of gravity or pressure. Diffusion is the dispersion of substances by the independent movement of their ions and molecules, which tends to equalize their concentrations throughout a system. Osmosis is the movement of water (or any solvent) across a differentially permeable membrane that does not permit the passage of the solute. The movement of water always occurs from an area of greater water potential to an area of lesser water potential. Factors affecting water potential are the presence of a solute (which decreases water potential) and pressure.

The diffusion of water through differentially permeable membranes may result in the development of hydrostatic pressures. Hydrostatic pressure in plant cells is commonly referred to as turgor pressure. Opposing the turgor pressure with a force of equal magnitude is the wall pressure of the cell wall.

In plants, more than 90 percent of the water taken in by the roots is given off into the air as water vapor. This loss of water vapor by plant parts is termed transpiration, and most of the water transpired by higher plants is lost through stomata of the leaves.

Absorption of water takes place largely through the root hairs, which provide an enormous surface area for water uptake. In some plants when the roots absorb water from the soil and transport it into the xylem, the water within the xylem builds up positive pressure, called root pressure. This osmotic uptake depends on the transport of mineral ions from the soil into the xylem by the living cells of the root.

From the roots, water makes its way to the leaves by means of the xylem. The current and widely accepted theory of water movement to the top of tall plants through the xylem is the cohesion-adhesion-tension theory. According to this theory, water within the vessels is under tension because the water molecules cling together in continuous columns pulled by evaporation from above. It has been demonstrated that water has sufficient tensile strength to withstand such tension. Other supporting evidence includes observations that water in the xylem is under sufficient tension, that water movement in trees begins in the topmost branches, and that the tree trunks shrink slightly when water movement begins.

The rate of transpiration is affected by such factors as carbon dioxide concentration in the intercellular spaces (and, conversely, the exposure of the leaf to carbon dioxide from the air), light, temperature, atmospheric humidity, air currents, and availability of soil water. Most of these factors have an effect on the behavior of stomata. Stomatal opening and closing are controlled by changes in turgor of the guard cells. The stomata are open when the guard cells are turgid and closed when they are flaccid.

Research on the movement of substances in the phloem has been greatly aided by the use of aphids and radioactive tracers. Analyses of sieve-tube sap reveal that it contains sugar, mainly sucrose, and small quantities of nitrogenous substances. Rates of longitudinal movement of substances in the phloem are far in excess of the normal rate of diffusion of sucrose in water—for example, 100 centimeters per hour.

According to the pressure-flow hypothesis, assimilates move from source to sink along a concentration gradient. Sugars are actively secreted into and absorbed from the sieve tube by parenchyma cells at source and sink, respectively, resulting in mass flow of solution in the sieve tube. The role of the sieve tube is a passive one.

CHAPTER 26

Soils and Plant Nutrition

COMPOSITION OF THE EARTH

The earth is composed of about 100 different elements; oxygen and silicon are by far the most abundant of these, followed by aluminum and iron. Elements are found in the earth in the form of *minerals*. A mineral is a naturally occurring inorganic substance, usually solid, with a definite chemical composition. Some minerals, such as diamond, sulfur, and copper, consist of single elements. Others, such as quartz (SiO_2) and calcite ($CaCO_3$), are compounds.

Rocks are mixtures of minerals. At or just under the land surface of the earth is granite, which is an igneous rock (from the Latin *ignis*, fire), that is, rock formed directly from molten material. Just below the granite is a layer of heavy igneous rock, basalt, which contains more iron compounds than granite. The bottom of the ocean basin is composed of basalt covered by a thin layer of sediment; no granite is present.

THE SOIL

Formation of Soil

Rocks are broken down by weathering processes, such as freezing and thawing or heating and cooling, which cause substances in the rocks to expand and contract, thus splitting the rocks apart. Movement by water and wind exerts a scouring action that breaks the fragmented rock into smaller particles, often carrying the fragments great distances. Water enters between the particles, and soluble materials dissolve in the water. Water in combination with carbon dioxide from the air forms a mild acid which dissolves substances that will not dissolve in the water alone. Chemical reactions begin to take place that contribute to the disintegration of the rock. Soon, if other conditions such as light and temperature permit, bacteria, fungi, algae, lichens, and bryophytes and other small plants begin to gain a foothold. Growing roots also split rocks, and their disintegrating bodies and those of the animals associated with them add to the accumulating material. Finally, the larger plants move in, anchoring the soil in place with their root systems, and a new community has begun.

Composition of Soil

Soil is the end product of physical, chemical, and biological phenomena and is that portion of the earth's crust in which plants grow. It is composed of a mixture of inorganic materials—the fragments of weathered rock—with organic ones. Some of the organic materials in soil, such as roots and small animals, are living, whereas others are dead and often disintegrating. The spaces between the components of the soil may be occupied by either air or water. Soil is, therefore, a complex mosaic made up of a number of interrelated materials.

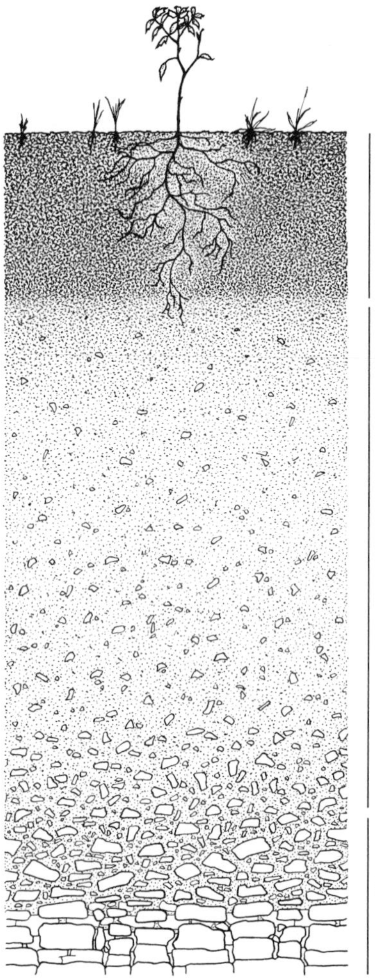

26–1
Soil typically has three layers: the A horizon is the zone of maximum organic accumulation; the B horizon, or subsoil, consists of weathered soil formed either from the C layer below or from some other source and transported to its present site by wind or water; and the C layer, or substratum, is relatively unweathered rock.

As a consequence of the processes involved in its formation, soil often has three layers (Figure 26–1). The first, or A horizon, contains the greatest portion of the soil's organic material, both living and dead. Here are found large amounts of decaying leaves, dead plant parts, insects and other small arthropods, earthworms, protozoa, nematodes, and decomposing fungi and bacteria. In the A horizon of the soil, the inorganic material is usually more extensively broken up and weathered than it is at lower depths. The B horizon contains much less organic material and is in general far less weathered than the A horizon, but it may contain an accumulation of materials that were formed in the A horizon and carried down with rainwater. Finally, the C horizon, which underlies the true soil, is composed of relatively unweathered rock.

The fragments of rock in the soil are of many different sizes. The following classification is used for the smaller sizes:

Classification	*Diameter of fragment* (micrometers)
Coarse sand	200–2000
Fine Sand	20–200
Silt	2–20
Clay	Less than 2

Most soils contain a mixture of particles of different sizes, and soils are named according to the proportions of different particles in the mixture. Soils are called sands if sand particles make up 70 percent or more of materials by weight. Clays are soils in which at least 35 percent of the particles are clay. The loam group, which contains many subdivisions, is more difficult to define. An ideal loam is a mixture of all three particle types with about an equal balance of heavy and light particles (Figure 26–2). Most agricultural soils are loams.

SOILS AND WATER

If the fragments that make up a soil are large, the pores and spaces between them will be large; water will drain through the soil rapidly, and relatively little will be available for plant growth in the A and B horizons. Because of their finer pores, clay soils are able to hold a much greater amount of water against the action of gravity, which tends to pull the water below the level at which it is available for plant growth. Thus, clay soils may retain three to six times as much water as a compa-

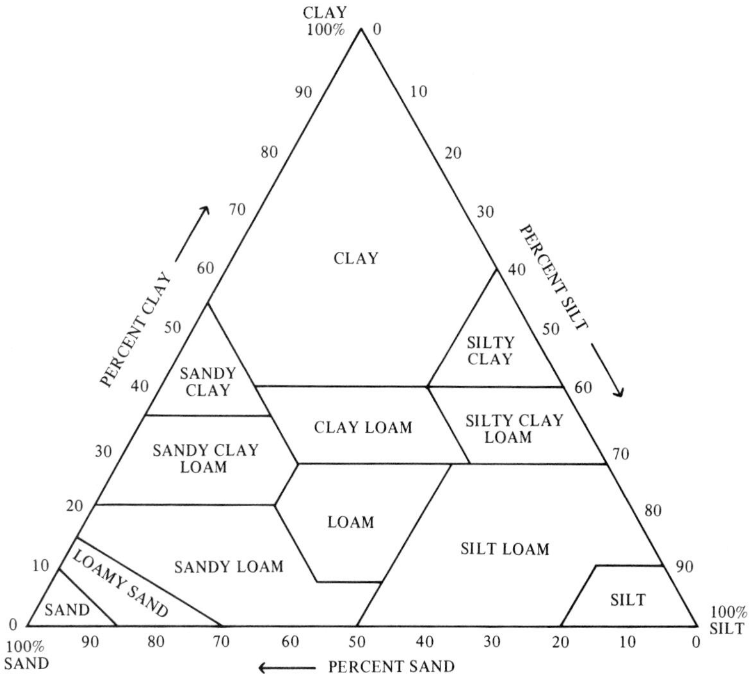

26–2
A diagram for determining soil type. Soils contain mixtures of three types of particles—sand, clay, and silt—and are named according to the proportion of each type by weight. To use this diagram for soil analysis, you would first locate the points corresponding to the percentages of silt and clay present in the sample on the silt and clay lines inward from these points, with the silt line parallel to the clay side of the triangle and the clay line parallel to the sand side. The name of the compartment in which two lines intersect is the class name of the soil in question.

rable volume of sand. The amount of water that a soil holds against the action of gravity is defined as its *field capacity*.

Pure clay soils are not very suitable for plant growth, however, because they are generally packed too tightly for normal root growth. Without the oxygen supplied by the air in the larger spaces in soil, roots cannot grow, because respiration cannot take place. The balance between the availability of oxygen and the availability of water is one factor that determines how good a certain kind of soil is for plant growth.

If a plant is allowed to grow indefinitely in a sample of soil and if no water is added, the plant eventually will not be able to obtain adequate water from the soil. At this point, the remaining water will be held in very small pores and in thin films around soil particles. The moisture content of a soil at the point at which a plant can no longer remove appreciable quantities of water from it is defined as the permanent wilting percentage of that soil. The permanent wilting percentage varies with the plant being tested. Many plants adapted to arid areas can extract more water from a soil that has reached the permanent wilting percentage for other species.

In general, the field capacity is much higher in fine-grained soils, and clays hold the greatest amount of moisture. To make this moisture available for plant growth, however, the soils must contain enough larger particles to be sufficiently aerated, which is why the loams are generally the best soils for plant growth.

26–3
A soil sample from McClain County, Illinois. The black soil of the corn belt is among the richest and most fertile in the world.

The roots of plants have evolved in relation to their three primary functions: anchoring the plant body in the soil, obtaining the minerals that the plant requires, and absorbing water as the roots grow through the soil. In the following pages, we shall explore further the nature of the substances that the plant extracts from the soil through its roots and the ways in which the plant is dependent upon them.

26–4
The fibrous root systems of grasses cling to the prairie soil, binding and anchoring it in place.

MINERAL ELEMENTS

The mineral elements required are usually separated into two categories—the *macronutrients,* which are generally expressed in percentage of dry weight of the plant and which range from about $\frac{1}{2}$ of 1 percent to 3 or 4 percent, and the *micronutrients,* of which as little as a few parts per million may be sufficient. Determination of many of the elements is generally made by burning the plant completely, which permits the carbon, hydrogen, oxygen, nitrogen, and sulfur to escape as gases, and analyzing the ash. Proportions of each element vary in different species and in the same species grown under different conditions. Also, the ash will often contain elements such as silicon, which are present in the soil and are taken up by plants but which are not generally required for plant growth.

THE WATER CYCLE

The earth's supply of water is stable and is used over and over again. Most of the water (98 percent) is present in oceans, lakes, and streams. Of the remaining 2 percent, some is frozen in polar ice and glaciers, some is found in the soil, some is in the atmosphere as water vapor, and some is in the bodies of living organisms.

Sunshine evaporates water from the oceans, lakes, and streams, from the moist soil surfaces, and from the bodies of living organisms, drawing the water up into the atmosphere, from which it again falls as rain. This constant movement of water from the earth into the atmosphere and back again is known as the water cycle. *The water cycle is driven by solar energy.*

Some of the water that falls on the land percolates down through the soil until it reaches a zone of saturation. In the zone of saturation, all holes and cracks in the rock are filled with water. Below the zone of saturation is solid rock through which the water cannot penetrate. The upper surface of this zone of saturation is known as the water table.

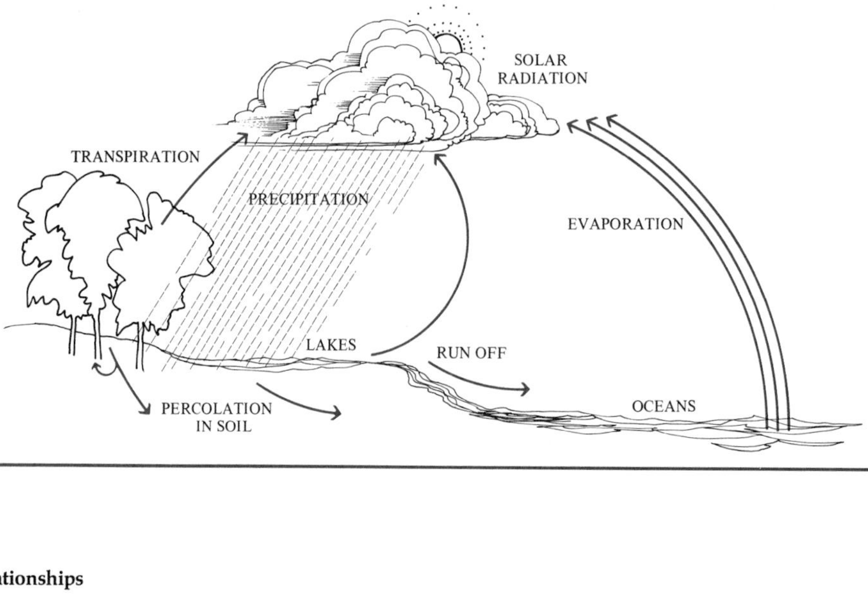

Table 26-1 lists the mineral elements required by plants, the form in which they usually are absorbed, and some of the uses plants make of them. You might anticipate that organisms make use of what is most readily available, as indeed they seem to have done to a considerable extent when life originated from elements in the gases of the primitive atmosphere. But the table reveals some findings that you might not expect. Sodium, for instance, which is one of the most abundant of the elements, is apparently not required at all by most plants. The fact that plants evolved with no functions requiring sodium is even more unusual when you consider that sodium is vital to animals. In the seas, where both plant and animal life seem

Table 26-1 *A Summary of Mineral Elements Required by Plants*

ELEMENT	FORM IN WHICH ABSORBED	APPROXIMATE CONCENTRATION IN WHOLE PLANT (AS % OF DRY WEIGHT)	SOME FUNCTIONS
Macronutrients			
Nitrogen	NO_3^- (or NH_4^+)	1–3%	Amino acids, proteins, nucleotides, nucleic acids, chlorophyll, and coenzymes.
Potassium	K^+	0.3–6%	Enzymes, amino acids, and protein synthesis. Activator of many enzymes. Opening and closing of stomata.
Calcium	Ca^{2+}	0.1–3.5%	Calcium of cell walls. Enzyme cofactor. Cell permeability.
Phosphorus	H_2PO_4 or HPO_4^{2-}	0.05–1.0%	Formation of "high-energy" phosphate compounds (ATP and ADP). Nucleic acids. Phosphorylation of sugars. Several essential coenzymes. Phospholipids.
Magnesium	Mg^{2+}	0.05–0.7%	Part of the chlorophyll molecule. Activator of many enzymes.
Sulfur	SO_4^{2-}	0.05–1.5%	Some amino acids and proteins. Coenzyme A.
Micronutrients			
Iron	Fe^{2+}, Fe^{3+}	10–1500 parts per million (ppm)	Chlorophyll synthesis, cytochromes, and ferredoxin.
Chlorine	Cl^-	100–10,000 ppm	Osmosis and ionic balance; probably essential in photosynthesis in the reactions in which oxygen is produced.
Copper	Cu^{2+}	2–75 ppm	Activator of some enzymes.
Manganese	Mn^{2+}	5–1500 ppm	Activator of some enzymes.
Zinc	Zn^{2+}	3–150 ppm	Activator of many enzymes.
Molybdenum	MoO_4^{2-}	0.1–5.0 ppm	Nitrogen metabolism.
Boron	BO^{3-} or $B_4O_7^{2-}$ (borate or tetraborate)	2–75 ppm	Influences Ca^{2+} utilization. Functions unknown.
Elements Essential to Some Plants or Organisms			
Cobalt	Co^{2+}	Trace	Required by nitrogen-fixing microorganisms.
Sodium	Na^+	Trace	Osmotic and ionic balance, probably for many plants not essential. Required by some desert and salt-marsh species and may be required by all plants that utilize C_4 photosynthesis.

to have had their origins, sodium is the most abundant mineral element and is far more readily available than potassium, which it closely resembles in its essential properties. Similarly, although silicon and aluminum are almost always present in large amounts in soils, few plants require silicon and none requires aluminum. On the other hand, all plants need molybdenum, which is relatively rare.

Certain large land areas are deficient in one or more of the elements usually necessary for plant growth. The plant species and ecotypes that are native to these areas have altered nutritional requirements and can tolerate low levels of the elements in question. Thus a soil's deficiency of a particular element, such as copper, can be determined only in relation to a particular plant; many native species grow in soils where crop plants may fail. The composition of soils is one of the bases for the differentiation of plant populations in nature, as we discuss in Chapters 8 and 29.

Functions of Minerals in Plants

Plants require mineral elements for many different functions. For example, mineral ions affect osmosis (page 52) and so regulate water balance. Because in many plants several mineral ions can serve interchangeably in this role, this requirement is described as non-specific. On the other hand, a mineral element may be a functional part of an essential biological molecule, and in this case the requirement is highly specific. An example is the presence of magnesium in the structure of chlorophyll. Some minerals are required constituents of cell membranes, and some control their permeability. Others are indispensable components of a variety of enzyme systems that catalyze biological reactions in the cell. Still others provide a proper ionic environment in which biological reactions can occur.

Because mineral elements fill such basic needs and are involved in such fundamental processes, the effects of mineral deficiencies are typically very widespread, affecting a number of structures and functions in the plant body.

Catalysts

A key role of the mineral elements is their participation with enzymes in catalyzing reactions in the cell. In some cases, they are an essential structural part (a "prosthetic group") of the enzyme. In others, they serve as activators or regulators of certain enzymes. Potassium, for instance, which affects several enzyme systems, is believed to regulate the conformation of some proteins. Changing the shape of an enzyme could, for example, expose or obstruct active sites.

26–5
(a) *Plants of the mustard family, such as the wintercress* (Barbarea vulgaris) *shown here, use sulfur in the synthesis of the mustard oils which give the plants both their name and their characteristic sharp taste.* (b) *Horsetails* (Equisetum) *incorporate silicon into their cell walls, making them indigestible to most herbivores but useful, at least in colonial America, for scouring pots and pans.*

(a)

(b)

Electron Transport

As we have seen, many of the biochemical activities of cells, including photosynthesis and respiration, are oxidation–reduction reactions. In such reactions, electrons are often transferred to or from an electron acceptor. Among the important electron acceptors are the cytochromes and ferredoxin, which contain iron.

Structural Components

Some mineral elements serve as structural components of cells. Calcium combines with pectic acid in the middle lamella of the plant cell wall. Phosphorus occurs in the sugar phosphate backbone of the DNA helix, in RNA, and in the phospholipids of the cell membrane. Nitrogen is an essential component of amino acids and porphyrins. Sulfur is found in two amino acids, forming an important structural element in proteins (page 56).

Osmosis

The movement of water in and out of plant cells, as we saw in the last chapter, is largely dependent on the concentration of solute in the cells and in the surrounding medium. The uptake of ions by the plant cells thus may result in the entry of water into the cell. The resultant hydrostatic pressure from within the cell produces expansion of the immature cell, which is the chief component in cellular growth and provides for turgor in the mature cell. This is another example of conversion of energy from one form to another by a living system; the chemical (ATP) energy expended in the active uptake of ions by the plant cell is translated into the physical energy of water movement.

Effects on Cell Permeability

Ions have a direct effect on the permeability of cell membranes to minerals. In general, ions with a charge of +1 tend to increase the permeability of membranes, whereas ions with a charge of +2 apparently tend to reduce it. If, for example, you put a red beet cell in solution and increase the amount of potassium (K^+), the red pigments will leak out. If you add calcium (Ca^{2+}), the leakage will stop. One ion may reduce the uptake of another by competing with it. Nitrates and phosphates often increase uptake of minerals by generally enhancing cellular metabolism and, thereby, speeding active transport.

Calcium also has a direct effect on the physical properties of the membrane. When there is a calcium deficiency, membranes seem to lose their integrity and the ions begin to leak out.

Elements Essential for Some Plants

Some elements are required only by some plants or only by plants under certain conditions. It has recently been suggested that all C_4 plants need sodium. Some salt-marsh species (Figure 26–6) not only can tolerate levels of sodium that would be toxic to other plants, but may actually require sodium. Many other plants that grow in harsh habitats, such as in the desert, may require minute quantities of sodium. Cobalt is needed by nitrogen-fixing microorganisms, some of which may exist in symbiotic association with and provide a nitrogen source for some plants; however, the cobalt is not needed directly by the plant, which may also have other sources of nitrogen.

Nonessential Elements Taken Up by Plants

In addition to the elements they require for growth, plants may absorb elements they do not require if they are present in the soil. Sodium is found in the ash of almost all higher plants, although it has been proved to be essential only to certain species. Aluminum, also common, is usually found in traces in most plants, particularly plants that grow in acid soils in which the proportion of soluble aluminum is usually relatively high. The addition of soluble aluminum to soil in which *Hydrangea macrophylla* is growing has a surprising effect: The flowers change from pink to blue.

26–6
A salt marsh. Smooth cordgrass (Spartina alterniflora) *in the foreground and, at slightly higher elevations away from the water's edge, black needlerush* (Juncus roemerianus).

Some mineral elements are poisonous. Copper and boron, for example, are highly toxic to plants except in very dilute concentrations, although plants cannot grow without them. Some elements can be accumulated by plants without injury but may be poisonous to animals grazing on the plants.

It has recently been shown that plutonium—formerly considered to be immobile in the soil—is, in fact, absorbed by and concentrated in the roots of plants. This finding is particularly significant because radioactive plutonium is a fuel for nuclear power plants, and small amounts of plutonium are discharged from some nuclear facilities during routine operation. The radioactive plutonium can then become concentrated in the tissues of animals that eat the plants and can lead to radiation damage, including cancer, in these tissues. Interestingly, as the concentrations in soil increase, the relative amounts in the plants decrease, indicating that toxic effects of the plutonium inhibit uptake.

Cation Exchange

The minerals taken in through the roots of plants have their source in the parent rock. As the rock weathers, the minerals enter the solution as ions. Most metals form positively charged ions, that is, cations, for example, Ca^{2+}, K^+, and Na^+. Clay particles provide a reservoir of such cations for the plant because at various points on their crystalline lattice there is an excess of negative charge, where cations can be bound and held against leaching from the soil.

The cations bound in this way to the clay particles can be replaced by other cations (*cation exchange*) and then released into the soil solution, where they become available for plant growth. This is why clay particles are an essential element in all "good" soils.

The principal negatively charged ions—anions—found in soil are NO_3^-, SO_4^{2-}, HCO_3^-, and OH^-. Anions are leached out of the soil more rapidly than cations, because they do not cling to clay particles. An exception is the phosphate (PO_4^{3-}), which is held to the soil particles in various ways.

The acidity or alkalinity of soil has a great deal to do with the availability of its minerals for plant growth. Soils vary widely in pH, and many plants have a narrow range of tolerance on this scale. In alkaline soils, many cations are precipitated (removed from solution), and such elements as iron, manganese, copper, and zinc may thereby become unavailable for plants.

MINERAL CYCLES

The bulk of the organic matter of soil is made up of dead leaves and other dead plant material, together with the decomposing bodies of animals. This organic debris is mixed with the inorganic particles of the soil, and in this mixture live astonishing numbers of small organisms that spend all or a part of their lives beneath the soil surface. A single teaspoon of soil may contain 5 billion bacteria, 20 million small filamentous fungi, and 1 million protozoa. The soil animals and microorganisms break down the organic matter, releasing its mineral elements so that they can be reutilized by the plants (Figure 26-8). In this way, except for the minerals that are leached out of the soil and run off eventually to the ocean, substances that are taken from the soil are constantly returned to it, with both macronutrients and micronutrients constantly recycled through plant and animal bodies, returned to soil, broken down, and taken up into plants again. Each element has a different cycle, and this cycle is often extremely complex, involving many different organisms and different enzyme systems. The end results are the same, however: A significant amount of the element is constantly returned to the soil. The phosphorus cycle is an example.

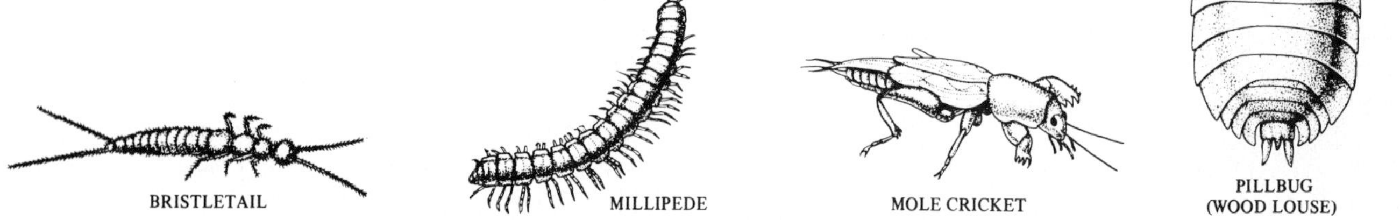

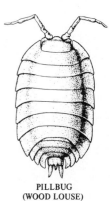

26-7
Soil animals. Plants share the soil with a vast number of other living organisms. The four shown here are common in the soils of temperate deciduous forests and play an important role in breaking down dead organic matter—plants and animals.

26–8
Among the largest of the soil organisms are the earthworms. Earthworms make burrows in the earth by passing soil and organic matter through their digestive tracts and depositing them in the form of castings. In a single year, their combined activities may produce as much as 500 metric tons of castings per hectare.

The Phosphorus Cycle

In the phosphorus cycle, the main reservoir of the element being cycled is in the solid state (Figure 26–9). Except for carbon, hydrogen, oxygen, and nitrogen, all of the elements required by plants are parts of similar though not identical cycles.

Phosphorus is directly involved in the transfer of energy in biological systems, as well as in other crucial processes that occur within cells. No living organism can exist without phosphorus; yet of all the elements that cycle from a pool that exists in solid form, phosphorus is the most likely to be limiting. In Australia, for example, where the rocks are extremely weathered and many kinds of soil are poor in phosphates, the distribution and limits of plant communities are often determined by the levels of phosphate in the soil.

Many kinds of rock contain phosphorus. When such rocks are eroded by water, minute amounts of phosphorus dissolve and become available to plants. However, most of this phosphorus, present as dissolved ions, is washed into streams and rivers, eventually reaching the sea. Very little of it ever moves back to the land, and so the availability of phosphorus to organisms depends on its being continually dissolved from rocks. Over large areas of land, however, man has increased the rate of erosion substantially; the soils that contain phosphorus are rapidly being eroded away. To counter this loss, we obtain fertilizer by mining phosphate rock and by harvesting guano, thick deposits of sea bird feces, rich in phosphorus derived from the fish upon which the birds feed. On a worldwide scale, it is probable that phosphorus is being lost to the sea more rapidly than it is being returned. Eventually, the deposits of phosphate rock may be exhausted. It will then be necessary to find ways of returning phosphorus from the deep sea basins into which rivers have gradually deposited it.

The essential role of mycorrhizae in phosphorus and other mineral cycles has been discussed on page 240.

26–9
The phosphorus cycle.

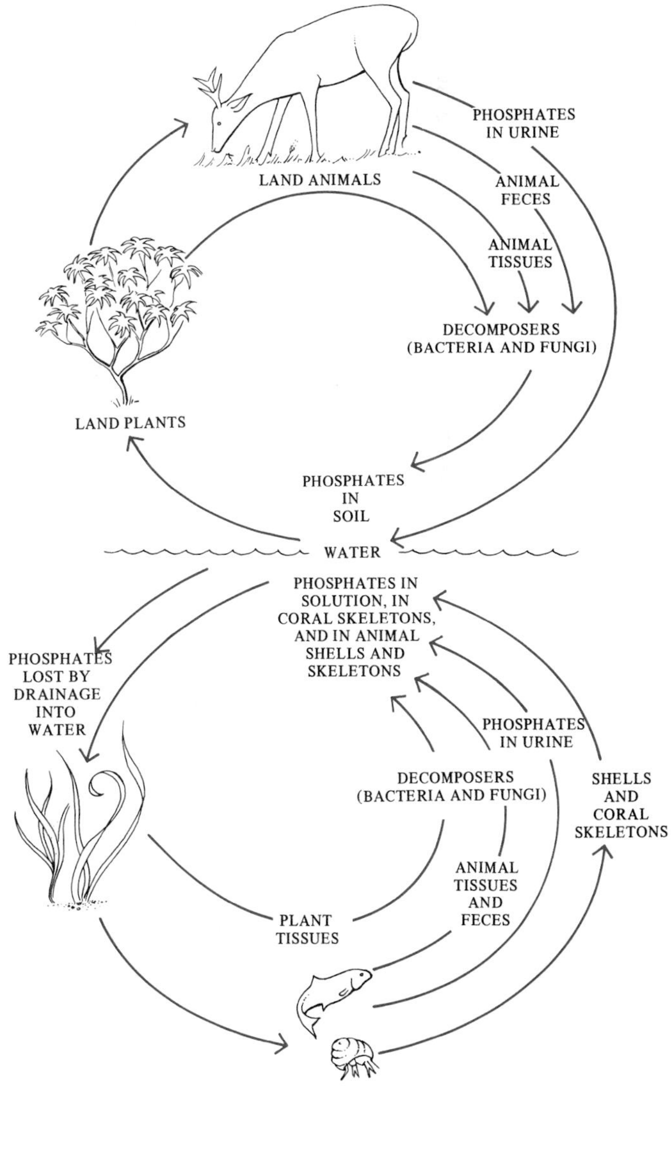

THE UPTAKE OF MINERAL ELEMENTS

The mineral composition of an organism is far different from the mineral composition of the medium in which it grows. An example is given in Table 26–2 for the cells of *Nitella clavata*, a fresh-water alga. Note that the algal cells have concentrations of potassium and chloride ions about 100 times greater than that of the external medium. The vacuoles of rutabaga *(Brassica napo-brassica)* cells may contain 10,000 times more potassium than the external solution. Although potassium is one of the mineral elements required by plants in the largest quantities, it is often deficient in soils. In arid and semiarid regions, moreover, sodium is usually much more abundant than potassium, and plants must selectively concentrate potassium without at the same time taking in large quantities of sodium. Similarly, the data of Table 26–2 indicate that not all ions are concentrated to the same extent as are potassium and chloride.

Because substances do not diffuse against a concentration gradient (see page 534), it is clear that energy must be spent by the cell in order for it to accumulate selected molecules and ions. The energy-requiring movements of substances into cells against a concentration gradient are known as *active transport.*

There is extensive evidence that active transport is involved in the accumulation of minerals by plants. If roots are deprived of oxygen, as in waterlogged soil, they are unable to oxidize carbohydrates completely for energy; under such conditions, the uptake of ions from the soil is drastically impaired. If a plant is deprived of light, the plant will cease to absorb salts and finally will release them back into the soil water (Figure 26–10).

Ion Pumps

One of the ways in which roots move ions such as potassium against a concentration gradient is by the action of carrier proteins which form part of the plasma membrane (see page 62). Such a protein combines with the ion and then, utilizing the energy of ATP, carries it to the inner side of the membrane and releases it. Because the plasma membrane is partially impermeable to ions such as potassium, the continued action of the "pump proteins" builds up a high internal concentration of the transported ions.

The same pump that carries potassium ions inward may move hydrogen ions outward. The ability to extrude hydrogen ions is essential, because it prevents the cell from becoming too acidic.

The active transport of ions across the plasma membrane may result in a difference of electrical charge on either side of the membrane. In the nerve cells of animals, stimulation of the plasma membrane results in an abrupt change in the permeability of the membranes to ions. The rapid movement of ions across the membrane produces an electrical charge, the action potential, which is the basis for transmission of nerve impulses. Similar events appear to be involved in certain plant responses.

Entry into the Plant

As shown in Figure 26–11, ions move through the roots into the xylem, from which they travel to other plant tissues. They may diffuse freely, passing through the cell walls (but not the protoplasts) until they reach the endodermis, or they may enter the cells by active transport and travel through the cytoplasm.

Table 26–2 *Analysis of the Vacuolar Sap of* Nitella clavata *and of the Pond Water in Which It Was Growing*

ION	SAP CONCENTRATION, MILLIEQUIVALENTS PER LITER*	POND WATER CONCENTRATION, MILLIEQUIVALENTS PER LITER*
Ca^{2+}	13.0	1.3
Mg^{2+}	10.8	3.0
Na^{+}	49.9	1.2
K^{+}	49.3	0.51
Sum of cations	123.0	
Cl^{-}	101.1	1.0
SO_4^{2-}	13.0	0.67
$H_2PO_4^{2-}$	1.7	0.008
Sum of anions	115.8	

* A milliequivalent of an ion is one-thousandth its gram ionic weight divided by its valence.

26–10
The rate of phosphate absorption by corn (Zea mays) plants fell to zero after 4 days of continuous darkness. It began to rise again when the plants were reilluminated. These and other data indicate that salt uptake in plants is an energy-requiring process.

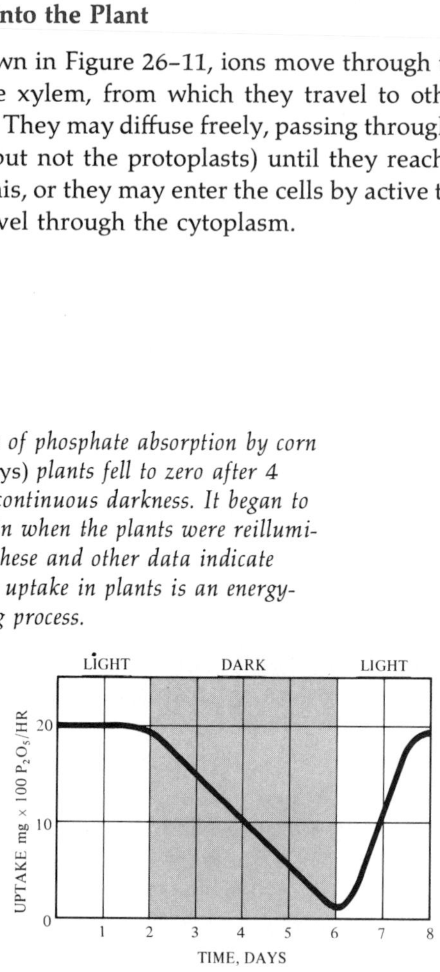

ACTION POTENTIALS IN *Drosera* TENTACLES

Studies have been carried out by Stephen E. Williams and Barbara G. Pickard at Washington University in St. Louis on leaf movements in a sundew (Drosera intermedia). *The plant has a rosette of 8 to 15 club-shaped leaves, each about 2 centimeters long (see Figure 26–14b on page 550). The upper surface of the leaves is covered with hairlike structures, the tentacles, each about 3 millimeters long, enlarged at the tip. A sticky droplet surrounding each tentacle tip attracts and catches insects, whereupon the tentacles secrete enzymes that digest their prey. If an insect is trapped on the outer tentacles of a leaf, the tentacles bend in, rapidly carrying the insect to the center. When the insect is trapped in the middle of the leaf or carried there by the infolding tentacles, the outer tentacles slowly fold in around it. Microelectrodes placed in the tentacles have shown that the rapid response of the outer tentacles of* Drosera *results from an abrupt change in the electrical potential across the plasma membrane, similar to the action potential of the nerve impulse in the nerve axons of animals. The stimulus is received in the head of the tentacle and transmitted down the stalk, which responds by bending. By contrast, it appears that the slower movements of the leaf—it eventually curves around to envelope the insect—are stimulated by a chemical agent that is transmitted when an inner tentacle is provided with an insect.*

The investigators predict that futher studies will show that higher plants also use electrical signals in coordinating a variety of functions.

(*a*) *Drosera* tentacle, showing placement of electrodes. (*b*) Action potentials measured from the stalk. They were produced by positioning a fruitfly so that its feet stroked the head of a tentacle on the margin of the leaf. The recording ended when the contact was broken as a result of the bending of the tentacle.

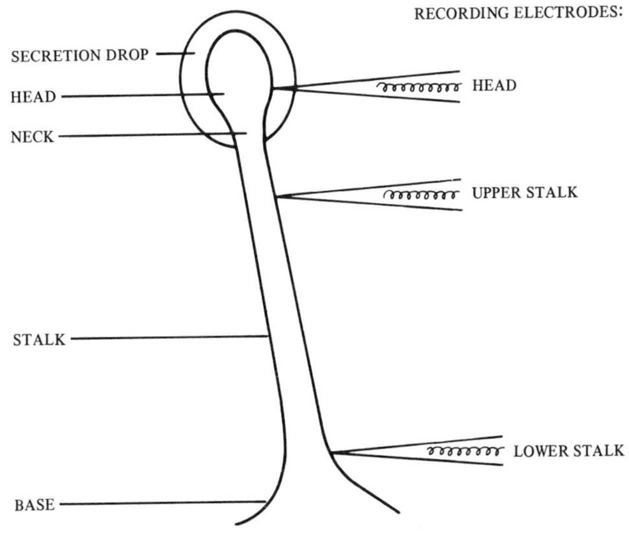

(a)

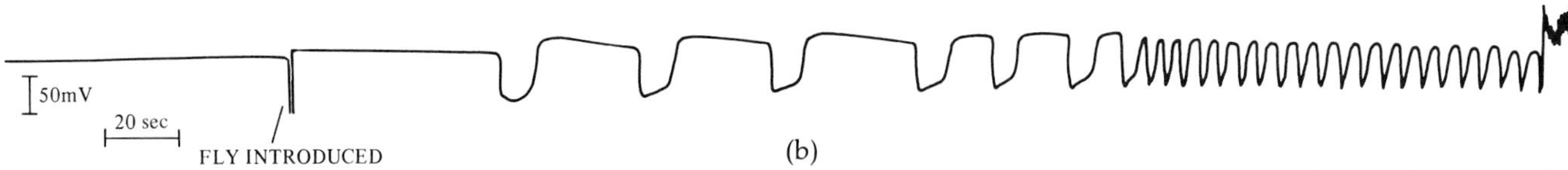

(b)

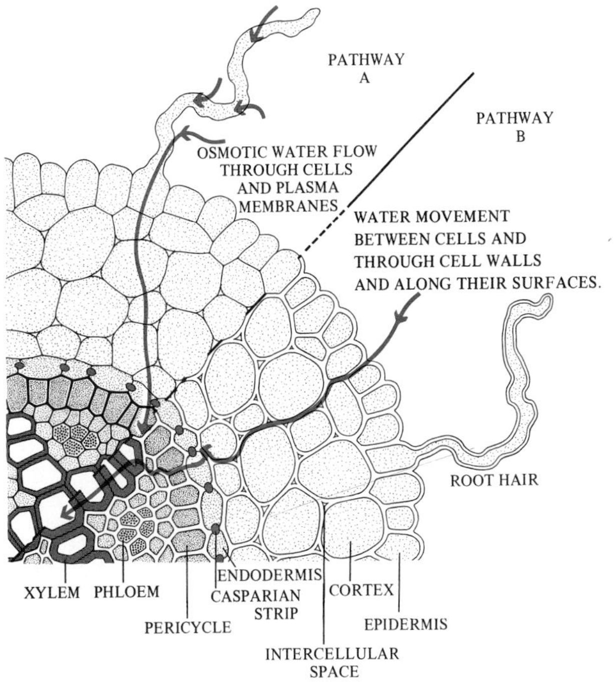

26–11
Diagrammatic cross section of a root, showing the pathways of uptake of water and minerals. Along pathway A, water moves by osmosis and salts by active transport across the plasma membranes of living cells and from cell to cell by way of the plasmodesmata in their walls. Along pathway B, the water flows through the cell walls and along their surfaces, and the solutes flow with the water or move by diffusion. Notice how the Casparian strip blocks off pathway B all around the vascular cylinder of the root. In order to get by the Casparian strip, the solutes must pass through the cytoplasm of the cells of the endodermis.

Some minerals are recirculated through the plant. Tracer studies using radioactive phosphorus, for example, have shown that phosphorus circulates more or less continuously in the plant body by way of the xylem and phloem. Minerals differ in mobility. Iron is immobile, whereas magnesium is mobile. For instance, in a magnesium-deficient plant, the magnesium moves from the old leaves to the younger ones, leaving the old leaves deficient in chlorophyll. In the iron-deficient plant, however, the young leaves suffer more.

Minerals can be given to plants by spraying them on the leaves, even though they do not usually enter the plant body by this route. Foliage sprays have proved a relatively inexpensive means of supplying crop plants with needed iron, for example.

Interaction Between Ions

Ions in solution affect the absorption of one another in a variety of ways. Cations are absorbed more readily in the presence of a readily absorbable anion, and vice versa, because of the requirement for electrical neutrality. When two or more ions with the same charge are present, either antagonistic or synergistic effects may be observed. Chemically related ions tend to interfere with one another. Chloride uptake, for example, is decreased in the presence of bromide or iodide. Sulfate and selenate have been shown to compete with one another for uptake in a variety of plants, as do arsenate and phosphate. Similarly, potassium, rubidium, and cesium appear to compete with one another for the same uptake sites in barley roots.

Not all ions compete on equal terms, however; potassium absorption is not inhibited by the presence of sodium, for instance, as much as sodium uptake is reduced by the presence of potassium. In a potassium-deficient medium, however, the uptake of sodium is greatly increased, indicating that it is competition by the potassium that inhibits sodium uptake.

Some ions, such as nitrate, phosphate, and sulphate, actually often stimulate the absorption of other ions. This is probably the result of a general enhancement of the metabolism of the plant.

NITROGEN AND THE NITROGEN CYCLE

We saw in the preceding discussion that plants require a number of elements in addition to those they obtain directly from the atmosphere (carbon and oxygen in the form of carbon dioxide) and from soil water (hydrogen and oxygen). All but one of these elements are derived from the weathering of rocks and come to the plants from the soil. The exception is nitrogen, which makes up 78 percent of the earth's atmosphere. Although the ultimate source of nitrogen is also the rock of the earth's surface, it enters the soil—and, through the soil, the plants that grow on it—indirectly by way of the atmosphere. Most living things, however, cannot use elemental atmospheric nitrogen to make proteins and other organic substances. (Unlike carbon and oxygen, nitrogen is very unreactive chemically, and only certain bacteria and blue-green algae have the highly specialized capacities for assimilating nitrogen from the atmosphere and converting it into a form that can be used by cells. This process—nitrogen fixation—is discussed on pages 553–557.) A shortage of usable nitrogen is often the major limiting factor in plant growth.

The processes by which nitrogen is circulated through plants and the soil by the action of living organisms is known as the nitrogen cycle (Figure 26–12).

Ammonification

Much of the nitrogen found in the soil reaches it from dead organic materials where it exists in the form of complex organic compounds such as proteins, amino acids, nucleic acids, and nucleotides. However, these nitrogenous compounds are usually rapidly decomposed into simple compounds by soil-dwelling organisms. The

26–12
The nitrogen cycle.

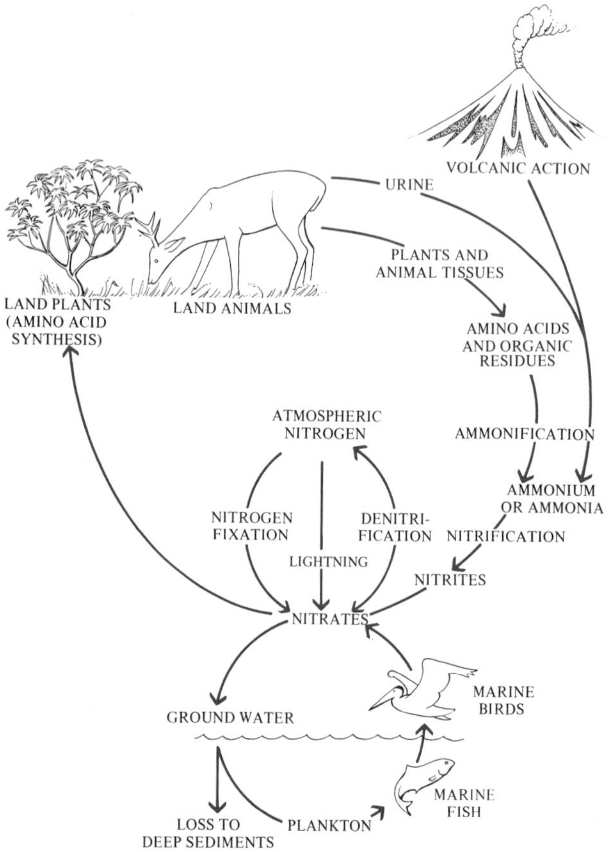

saprobic bacteria and various species of fungi are mainly responsible for the decomposition of dead organic materials. These microorganisms use the protein and amino acids as a source for their own needed proteins and release the excess nitrogen in the form of ammonium (NH_4^+). This process is known as *ammonification*. The nitrogen may be given off as ammonia gas (NH_3), but this usually occurs only during the decomposition of large amounts of nitrogen-rich material, as in a manure pile or a compost heap. Usually the ammonia produced by ammonification is dissolved in the soil water, where it combines with protons to form the ammonium ion.

Nitrification

Several species of bacteria common in soils are able to oxidize ammonia or ammonium (Figure 26–13). The oxidation of ammonia, known as *nitrification,* is an energy-yielding process, and the energy released in the process is used by these bacteria to reduce carbon dioxide in much the same way the photosynthetic autotrophs use light energy for the reduction of carbon dioxide. Such organisms are known as *chemosynthetic autotrophs* (as distinct from photosynthetic autotrophs, the plants and the algae). The chemosynthetic nitrifying bacteria *Nitrosomonas* and *Nitrosococcus* oxidize ammonia to nitrite (NO_2^-):

$$2NH_3 + 3O_2 \rightarrow 2NO_2^- + 2H^+ + 2H_2O$$

Nitrite is toxic to higher plants, but it rarely accumulates in the soil. *Nitrobacter,* another genus of bacteria, oxidizes the nitrite to form nitrate (NO_3^-), again with a release of energy:

$$2NO_2^- + O_2 \rightarrow 2NO_3^-$$

Nitrate is the form in which almost all nitrogen moves from the soil into the roots.

A few species of plants are able to use animal proteins as a nitrogen source. These carnivorous plants (Figure 26–14) have special adaptations used to lure and trap small animals. They then digest them, absorbing the nitrogenous and other organic compounds and minerals, such as potassium and phosphate, that they contain. Most of the carnivores of the plant world are found in bogs, which are usually strongly acid and therefore not favorable for the growth of nitrifying bacteria.

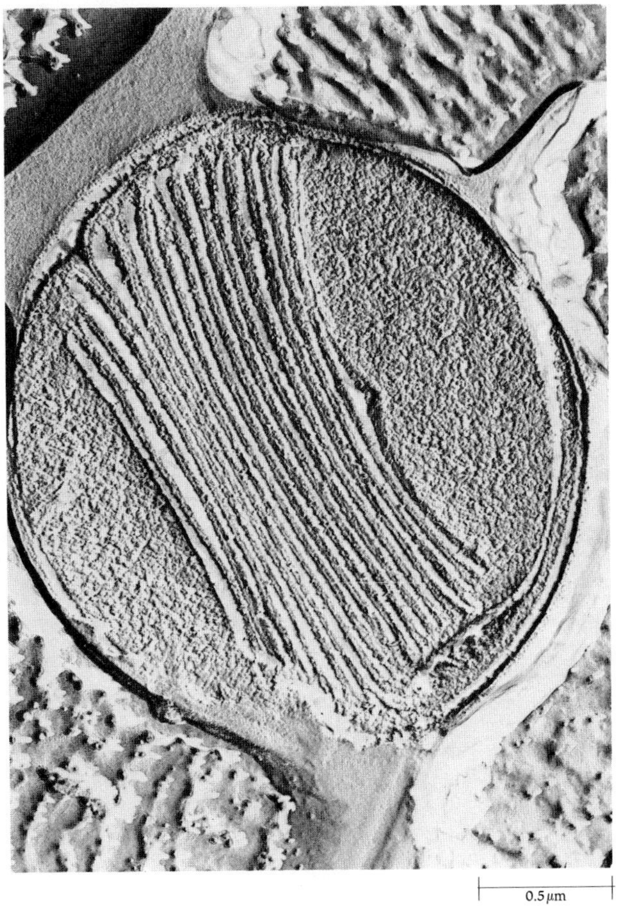

26–13
Marine nitrifying bacterium (Nitrosocystis oceanus), *prepared by the freeze-etch technique of electron microscopy.*

26-14
Carnivorous plants can obtain minerals, including nitrogen, from their prey. (a) *Yellow pitcher plants* (Sarracenia flava) *attract insects into their flowerlike tubular leaves by means of nectar. The lip of the "pitcher" of this species is extremely smooth, and many insects that enter cannot escape. Once in the leaf, the insects slide or drop into a foul-smelling broth of rainwater, digestive enzymes, and decomposing bacteria at the pitcher's base.* (b) *The sundew* (Drosera intermedia) *is a tiny plant, often only an inch or two across, with club-shaped tentacles on the upper surface of its leaf. These tentacles secrete a clear sticky liquid which attracts insects. When an insect is caught in the secretion, the tentacles bend inward until the leaf finally curves around the insect. The tentacles also secrete digestive enzymes.* (c) *The Venus flytrap* (Dionaea muscipula) *has leaves like bear traps that snap closed around insects seeking the nectar on the leaf surface. The closing of the trap is triggered by touching one or two of the three trigger hairs in the middle of each leaf lobe. Venus flytraps are found in nature only on the coastal plain of North and South Carolina, usually on the edges of wet depressions and pools.* (d) *The common bladderwort* (Utricularia vulgaris) *is a free-floating aquatic plant. The traps are tiny, flattened, pear-shaped bladders. Each bladder has a mouth guarded by a hanging door. The tripping mechanism consists of four stiff bristles near the lower free edge of the door. When a small animal brushes against these bristles, the hairs distort the lower edge of the door, causing it to spring open. Water then rushes into the bladder, carrying the animal with it, and the door closes behind. The animals decay by bacterial action, and released minerals and organic compounds are taken up by the cellular walls of the trap. The undigested exoskeletons remain within the bladders.* [(d), Carolina Biological Supply Co.]

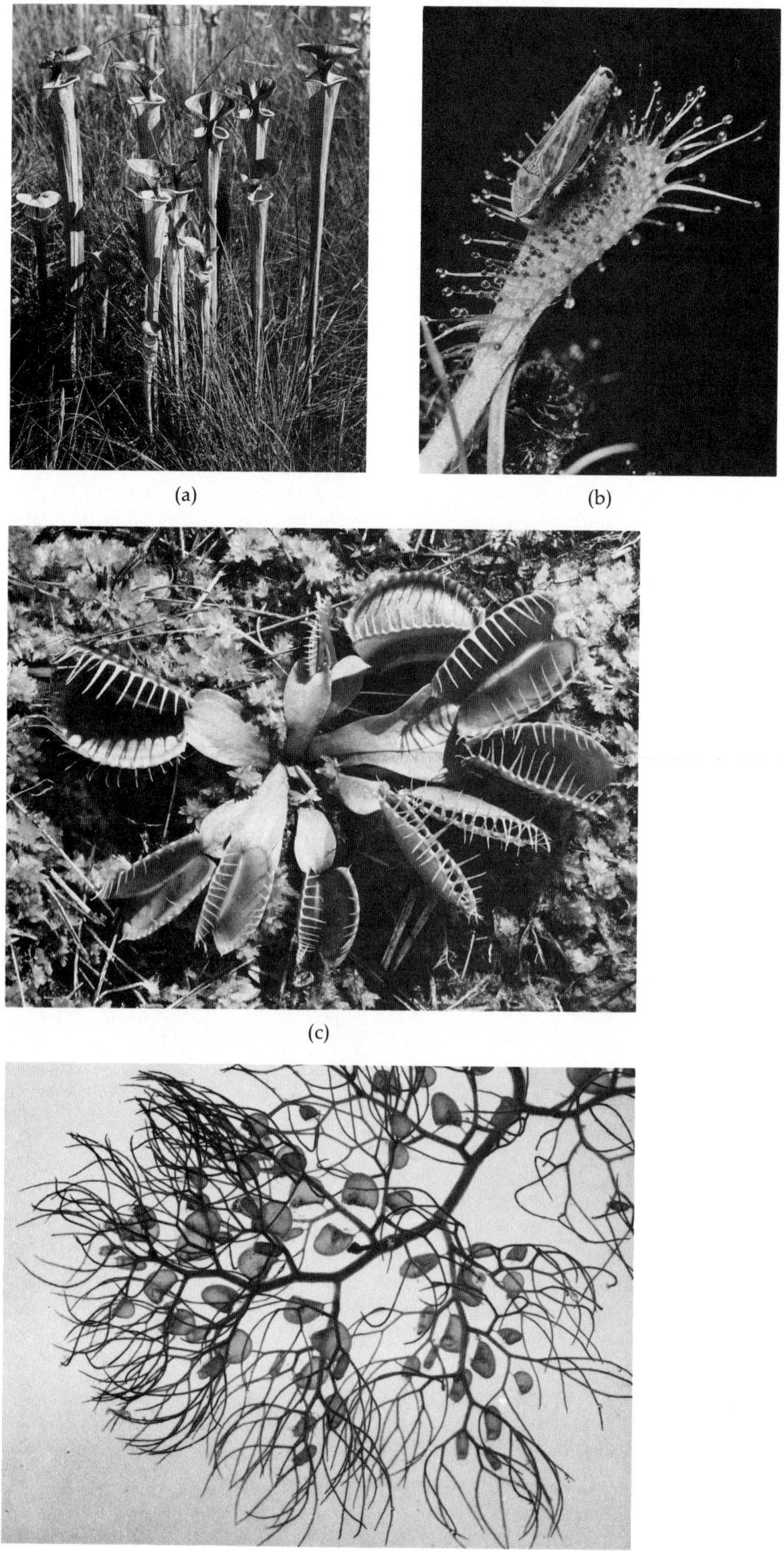

(a) (b) (c) (d)

Assimilation of Nitrogen

Once the nitrate is within the cell, it is reduced back to ammonium. This reduction process requires energy, in contrast to the nitrification process which involves oxidation (of NH_4^+) and which releases energy. The ammonium ions formed by the reduction process are transferred to carbon-containing compounds to produce amino acids and other nitrogen-containing organic compounds. This process is known as *amination*. The incorporation of nitrogen into organic compounds takes place largely in young, growing root cells. The initial stages in the metabolism of nitrogen appear to occur right in the root; almost all the nitrogen ascending the stem in the xylem is already in the form of organic molecules, largely amino acids.

Formation of Amino Acids

Amino acids are formed from ammonium ions and keto acids. The keto acids are usually products of the breakdown of sugar. Figure 26–15 shows the overall reaction by which a keto acid is combined with ammonium to form an amino acid. The major amino acid that is formed in this fashion is glutamic acid. Glutamic acid provides the chief carrier of nitrogen through the plant body.

From the amino acid produced by the amination of a keto acid, other amino acids are formed by *transamination*. Transamination is the transfer of the amino group ($-NH_2$) from an amino acid to a keto acid to produce another amino acid. Figure 26–16 shows the formation of alanine by the transfer of the amino group from glutamic acid to pyruvic acid (see page 101).

26–15
Plants incorporate nitrogen into organic compounds. This process is known as amination. A major pathway for the incorporation of nitrogen is the combination of alpha-ketoglutaric acid with ammonia (NH_3) to form the amino acid glutamic acid. NAD_{red} provides energy and hydrogen ions for the second step of this reaction. Alpha-ketoglutaric acid is one of the intermediates in the Krebs cycle.

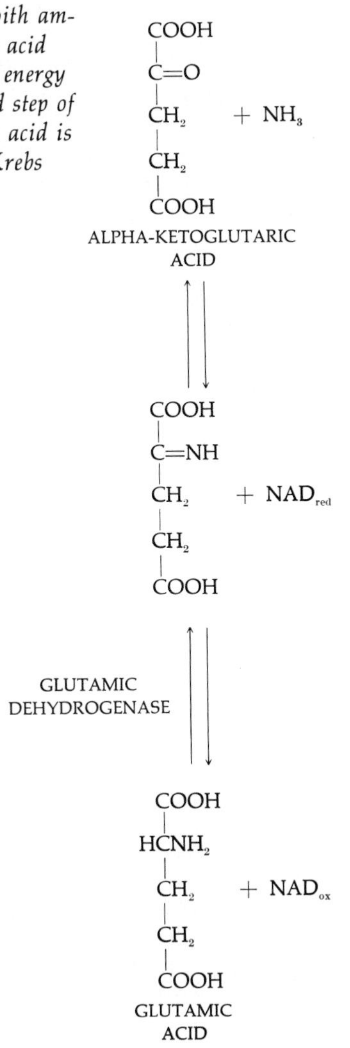

26–16
Transamination is the process by which the amino group of an amino acid is transferred to an alpha-keto acid to form a new amino acid.

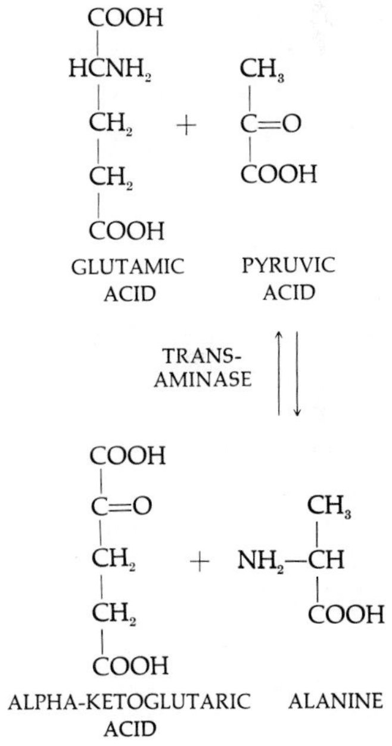

Plants, either by amination or transamination, can make all the amino acids they require from inorganic nitrogen. Animals can make only about 8 of the 20 required amino acids and must get the others in their diet. Thus, the animal world is completely dependent on the plant world for its proteins, as it is for its carbohydrates.

Other Nitrogen-Containing Compounds

Other important organic nitrogen-containing compounds include the nucleotides, such as ATP, ADP, NAD, and NADP; chlorophyll and other similar organic molecules with porphyrin ring structures; and the nucleic acids, DNA and RNA. Many of the vitamins, such as the vitamin B group, contain nitrogen. These, like the amino acids, can be synthesized from inorganic nitrogen by plants but must be obtained from plants by animals.

Another major difference between the plants and animals in their management of nitrogen lies in the capacity of plants to recycle ammonia or ammonium. Animals cannot recycle these breakdown products of nitrogen metabolism to any important extent, and as a consequence, nitrogenous compounds are constantly being excreted by animals in urine, feces, sweat, and even tears. Although some nitrogen is lost from plants in fallen leaves, only small amounts of nitrogen are excreted from the plant body. In fact, excretion of large amounts of waste materials of any sort is typical only of animals.

Nitrogen Loss

As we have seen, the nitrogen-containing compounds of green plants are returned to the soil with the death of the plants (or of the animals that have eaten the plants) and are reprocessed by soil organisms and microorganisms, taken up by the plant roots in the form of nitrate dissolved in the soil water, and reconverted to organic compounds. In the course of this cycle, a certain amount of nitrogen is always "lost," in the sense that it becomes unavailable to the plants.

A main source of nitrogen loss is the removal of plants from the soil. Soils under cultivation often show a steady decline of nitrogen content. Nitrogen may also be lost when topsoil is carried off by soil erosion or when ground cover is destroyed by fire. Nitrogen is also removed by leaching; nitrates and nitrites, both of which are anions, are particularly susceptible to being washed away by water percolating down through the soil. In some soils, denitrifying bacteria break down nitrates and release nitrogen gas into the air. This process, which provides the bacteria with needed oxygen for the respiration of carbon compounds, is expensive in terms of energy requirements (that is, O_2 can be reduced more readily than NO_3^-) and is carried out extensively only in

COMPOST

Composting, a practice as old as agriculture itself, has recently been attracting increased interest as a means of disposing of organic wastes by converting them to fertilizer. The starting product is any collection of organic matter—leaves, garbage, animal manure, straw, lawn clippings, sewage sludge, sawdust—and the population of bacteria and other microorganisms normally present. The only other requirements are oxygen and moisture. Grinding of the organic matter is not essential, but it provides greater surface area for microbial attack and so speeds the process.

In a compost heap, microbial growth accelerates rapidly, generating heat, much of which is conserved in the pile because the outer layers of organic matter act as an insulator. In a large pile (2 meters × 2 meters × 1.5 meters, for instance), the interior temperature rises to 70°C (158°F); in small piles, it usually reaches 40°C. As the temperature rises, the population of decomposers changes, with thermophilic and thermotolerant forms replacing the organisms originally present. As the previous forms die, their organic matter also becomes part of the product. A useful side effect of the temperature increase is that most of the common pathogenic bacteria present in sewer sludge are destroyed, as are also cysts, eggs, and other immature forms of plant and animal parasites.

Changes in pH also occur. The initial pH is usually slightly acid (about pH 6), as is the liquid portion of most plant material. During the early stages of decomposition, the production of organic acids causes a further acidification to about pH 4.5 to 5.0. However, as the temperature rises, the pH also increases, leveling off at slightly alkaline values (pH 7.5 to 8.5).

An important factor in composting (as in any biological growth process), is the ratio of carbon to nitrogen. About 30 to 1 (by weight) is optimal. If the carbon ratio is higher, microbial growth slows. If the nitrogen ratio is higher, some escapes as ammonia. Adding limestone (calcium carbonate) to the pile, although a common practice, increases the nitrogen loss.

Studies with municipal compost piles at Berkeley, California, demonstrated that if large piles were kept moist and aerated, composting could be completed in as little as 2 weeks. Three months or more during the winter is a more usual schedule. If compost is added to the soil before the composting process is complete, it may actually rob the soil of nitrogen.

Because it greatly reduces the bulk of plant wastes, composting can be a very useful means of waste disposal. In Scarsdale, New York, for example, leaves composted in a municipal site were reduced to one-fifth their original volume. At the same time, they formed a useful soil conditioner, improving aeration and water-holding capacities. Chemical analyses indicate, however, that the value of compost as fertilizer is limited. A rich compost commonly contains, in dry weight, about 1.5 to 3.5 percent nitrogen, 0.5 to 1.0 percent phosphorus, and 1.0 to 2.0 percent potassium, far less than a commercial fertilizer. It is also, in terms of manpower, far less efficient to prepare. However, as commercial fertilizers are becoming more expensive and less available and our waters are becoming increasingly polluted with fertilizer run-off and organic wastes, composting, at least under certain circumstances, is coming to seem an increasingly attractive alternative.

soils where oxygen is deficient, that is, in soils that are badly drained and hence poorly aerated.

Sometimes a high proportion of the nitrogen present in the soil is unavailable to plants. This immobilization comes about when an excess of carbon is present. When organic substances rich in carbon but poor in nitrogen—straw is a good example—are abundant in the soil, the microorganisms attacking these substances will need more nitrogen than they contain in order to make full use of the carbon present. As a consequence, they will use not only the nitrogen present in the straw or similar material but also all the nitrogen salts available in the soil. Eventually, this imbalance tends to right itself as the carbon is given off as carbon dioxide by microbial respiration and the ratio of nitrogen to carbon in the soil increases.

Nitrogen Fixation

As you can see, if the nitrogen that is removed from the soil were not steadily replaced, virtually all life on this planet would finally flicker out. The nitrogen is replenished in the soil by *nitrogen fixation.*

Nitrogen fixation is the process by which gaseous nitrogen from the air is incorporated into organic nitrogen-containing compounds and, thereby, brought into the nitrogen cycle. Nitrogen fixation, which can be carried out to a significant extent only by certain bacteria and blue-green algae, is a process on which all living organisms are now dependent, just as all organisms are ultimately dependent on photosynthesis for energy.

One to two hundred million metric tons of nitrogen are added to the earth's surface each year by biological systems. Man produces another 28 million metric tons, much of which is used for fertilizers, but this is done at high energetic cost in terms of fossil fuels. The total amount of energy required for the production of ammonium fertilizers at present is estimated at the equivalent of 2 million barrels of oil per day. In fact, it is estimated that the costs of nitrogen fertilization are just reaching the point of diminishing returns. The traditional crops in areas such as India do not achieve significantly increased yield with nitrogen fertilization, having low nitrogen requirements, but they are now being replaced with "miracle grains" and other crops that do yield better with heavy nitrogen fertilization—just at a time when such treatment threatens to become prohibitively expensive.

Of the various classes of nitrogen-fixing organisms, the symbiotic bacteria are by far the most important in terms of total amounts of nitrogen fixed. The most common of the nitrogen-fixing symbiotic bacteria is *Rhizobium,* a type of bacterium that invades the roots of leguminous plants (angiosperms of the family Fabaceae), such as clover, peas, beans, vetches, and alfalfa (Figure 26–17).

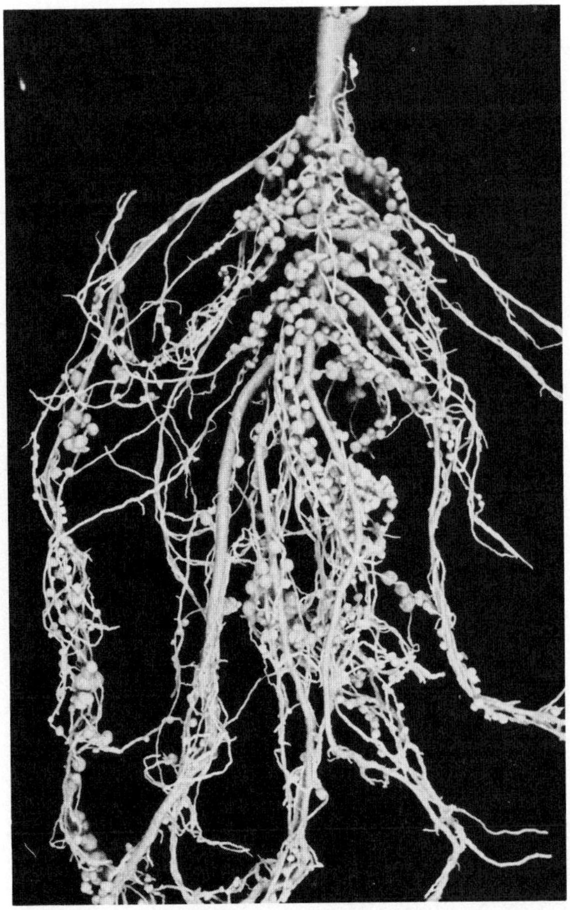

(a)

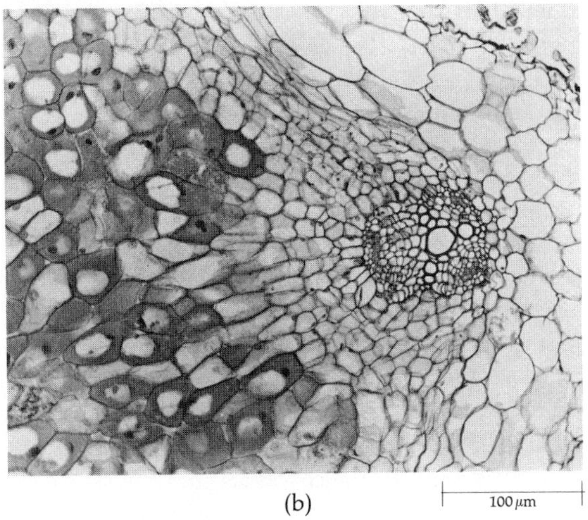

(b)

26–17
(a) *Nitrogen-fixing nodules on the roots of birdfoot trefoil* (Lotus corniculatus), *a legume. These nodules are the result of a symbiotic relationship between a soil bacterium* (Rhizobium) *and plant root cells.*
(b) *Cross-section of nodules from the root of a pea* (Pisum sativum). *The root structure can be seen at the right.*

26–18
Sorghum plants are being shredded and returned to the soil, in order to add organic matter to the soil and recycle the nutrients removed from it by the growing plants.

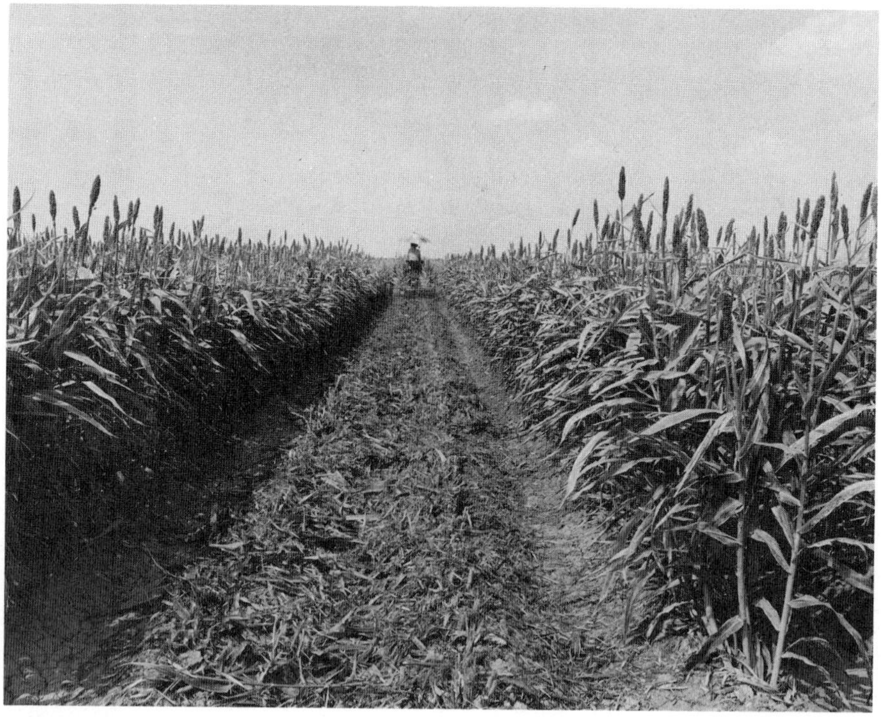

The beneficial effects on the soil of growing leguminous plants are so obvious that they have been recognized for hundreds of years. Theophrastus, who lived in the third century B.C., wrote that the Greeks used crops of broad beans to enrich the soils. Where leguminous plants are grown, some of the "extra" nitrogen may be released into the soil, where it becomes available for other plants. In modern agriculture, it is common practice to rotate a nonleguminous crop, such as corn, with a leguminous one, such as alfalfa. The leguminous plants are then either harvested for hay, leaving behind the nitrogen-rich roots or, better still, plowed back into the field (Figure 26–18). A good crop of alfalfa that is plowed back into the soil may add as much as 450 kilograms of nitrogen to the soil per hectare. Applications of the trace elements cobalt and molybdenum, required by the symbiotic bacteria, greatly enhance the nitrogen yield if these elements are present in limiting quantities, as over much of Australia.

Rhizobium

Rhizobia enter the roots of the leguminous plants, usually by way of the root-hair tips, when the plants are still seedlings (Figure 26–19). The way in which they penetrate the epidermis is a matter of lively speculation. One theory holds that they enter through a physical break in the growing root tip; no such break has ever been found, however. Other evidence suggests that in response to materials exuded by the bacterial cells, the legume releases an enzyme that apparently helps to loosen the fibrils in the walls of the root hairs through which the rhizobia enter. But no cellulase or other such enzyme has been isolated. Another possibility is that the bacteria build themselves right into the primary wall of the growing cell of the host. One clearly visible effect of the rhizobia is that they induce curling in the root hairs. All rhizobia curl the root hairs of all legumes, even those in which they do not establish an infection. This curling, it has been suggested, is the result of the action of hormones produced by the bacteria.

Once inside, the bacterial cell moves through what is known, because of its appearance, as an infection thread (Figure 26–20). The infection thread is a tube or tunnel of cellulose produced by the plant cell. Thus, the bacteria at this stage are not actually in contact with the protoplasm, but are separated from it by the cellulose wall that the plant cells have been induced to manufacture. Once started, the infection thread branches, becoming sparsely populated with the slowly multiplying bacteria. These threads typically pass close to the cell nuclei and seem to cause their degeneration. The infection threads pass from one cell to another, apparently through the primary pit-fields in the cell wall, and finally invade the cortex. Soon there is a proliferation of the surrounding cortical cells, presumably owing to the effects of a zeatin-like cytokinin released during the growth of *Rhizobium.* Finally, a vesicle forms in the thread, near a cell nucleus. The vesicle breaks open, and the bacteria enter the cy-

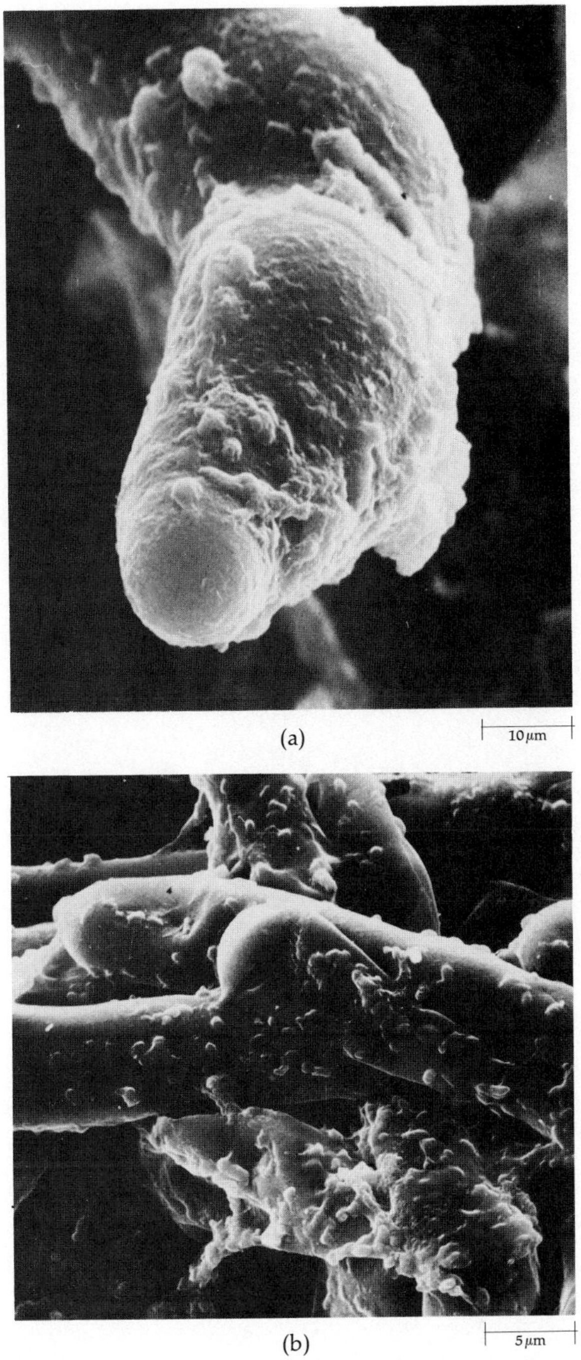

26–19
(a) *Tip of root hair of a cluster clover* (Trifolium glomeratum) *seedling, with several rhizobia and some soil particles.* (b) *Surfaces of root hairs, showing rhizobia and other soil bacteria adhering to surface.*

toplasm of the host cell. Soon after their release, the bacteria begin to grow, increasing in size some tenfold.

In cell cultures of nonleguminous plants, *Rhizobium* can fix nitrogen under anaerobic conditions if the plant is provided a pentose sugar and a dicarboxylic acid, both common plant constituents. These simple requirements suggest that the function of the root nodule in legumes may merely be to provide a convenient anaerobic site and that agriculturally important associations with nonleguminous plants may be formed under field conditions much more easily than formerly thought. The enzyme necessary for the fixation of atmospheric nitrogen is produced only by the bacterium and seems to utilize molybdenum as a binding site for N_2. Recently, it has been possible to form symbiotic associations with nonleguminous plants such as grasses and rapeseed *(Brassica napus)*. It has also been possible to transfer genetic factors into *Rhizobium*, an achievement that widens the scope of possible experimentation considerably. In addition, strains of *Escherichia coli* and other related bacteria have been produced into which the genes for nitrogen fixation have been transferred. Because of its high economic importance, the process of nitrogen fixation currently is a very active field of investigation.

The legumes are by far the largest group of plants that enter into this nitrogen-fixing partnership with symbiotic bacteria. In the legumes, the bacterial partner is invariably *Rhizobium*, a genus that rarely has been reported to form symbiotic associations with other plants. There are, however, at least a dozen genera of angiosperms other than legumes that have also established a nitrogen-fixing partnership with symbiotic bacteria, mostly actinomycetes. Associations with blue-green algae are likewise common. In general, they occur in the roots.

One of the most recently described associations is that between the Brazilian grass *Digitaria decumbens* and the bacterium *Spirillum lipoferum*, which occurs in the inner cortex of the root, but also fixes nitrogen when grown in culture with appropriate nutrients. Associations between *Spirillum lipoferum* and corn *(Zea mays)* roots have been produced recently in Brazil, and one strain of corn achieved yields of fixed nitrogen comparable with that of *Rhizobium* in nodules on the roots of soybeans *(Glycine max)*. The economic potential of this discovery is likely to be enormous.

Free-Living Nitrogen-Fixing Microorganisms

Nonsymbiotic bacteria of the genera *Azotobacter* and *Clostridium* are both able to fix nitrogen. *Azotobacter* is aerobic, and *Clostridium* anaerobic; both are common saprophytic soil bacteria. It is estimated that they probably add about 7 kilograms of nitrogen to a hectare of soil per year. Another important group includes many photosynthetic bacteria. Free-living blue-green algae also play an important role in the fixation of nitrogen. They

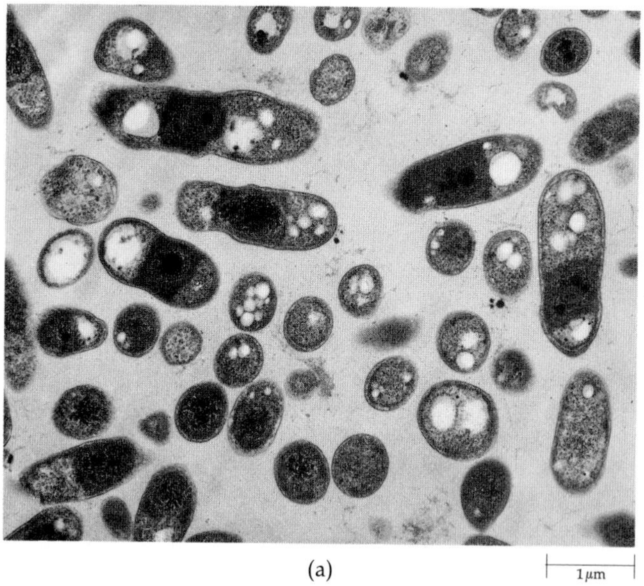

(a)

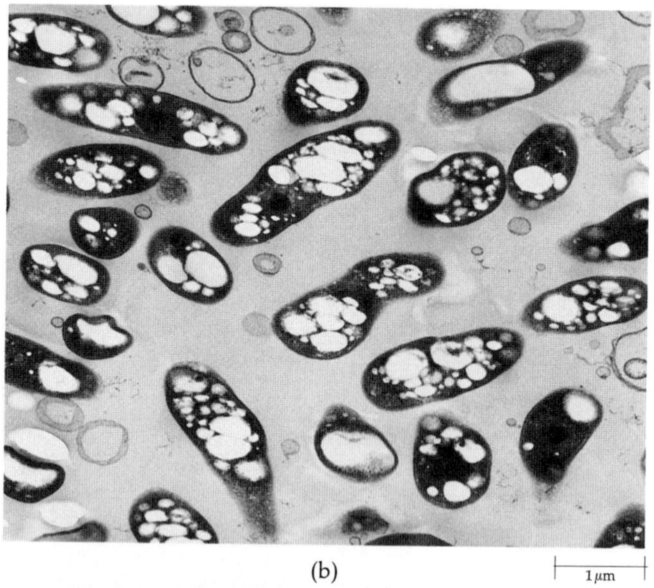

(b)

26–20
Rhizobium. (a) *Bacteria in their free-living form.* (b) *The bacteroid form which they assume when they enter the root cells.* (c) *Cross-section of an infected nodule.* (d) *Infection thread (which, as you can see, is passing near the cell nucleus, at the righthand side of the micrograph).*

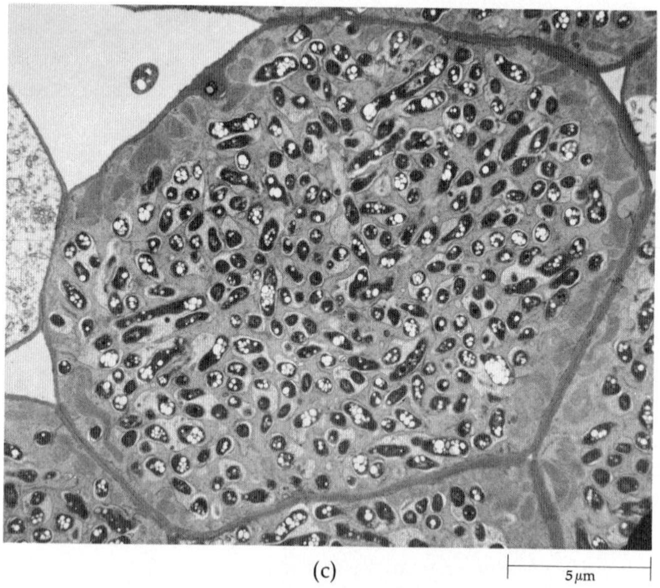

(c)

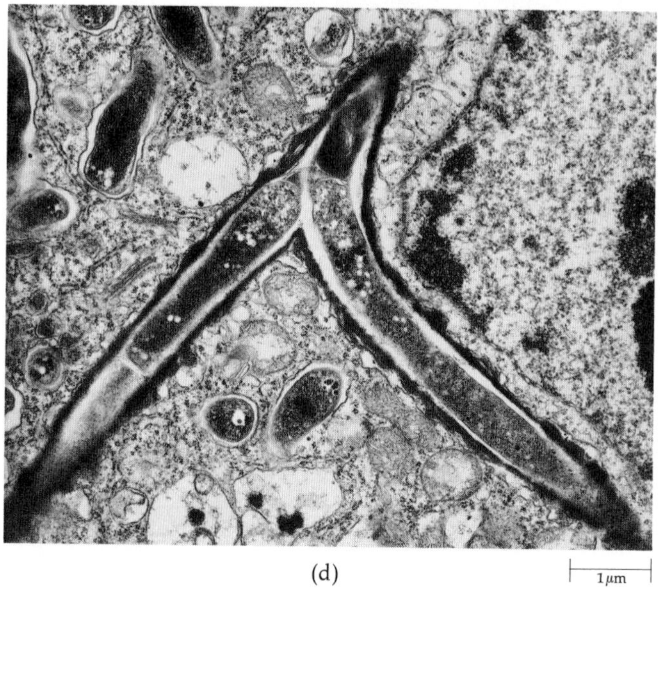

(d)

are crucial in the cultivation of rice, which is a staple diet for more than half of the world's population. The blue-green algae may also play an important ecological role in nitrogen fixation in the ocean.

The distinction between free-living and symbiotic nitrogen fixation may not be as great as has been traditionally thought. Some microbes regularly occur in soil around the roots of certain plants which exude carbohydrates, consuming these compounds while contributing nitrogen to the plants indirectly. Symbiotic associations between normally free-living bacteria, such as *Azotobacter*, and the cells of higher plants in tissue culture have been forced by growing them on an artificial medium lacking nitrogen.

Other Sources of Nitrogen

Nitrogen gas may be oxidized by lightning and brought to the earth in rainfall, and rainwater can sometimes bring down ammonia which has escaped into the atmosphere. Measurements at an experimental station in England over a 5-year period showed that the rainwater brought down 7.1 kilograms of nitrogen per hectare per year.

SOILS AND AGRICULTURE

The elements essential for plant growth constitute only a small proportion of even the most fertile soil, as can be seen in Table 26–3. The most common minerals in soils are derivatives of aluminum and silicon, but these, although they play a role in maintaining the essential elements, do not contribute directly to plant nutrition.

In a natural situation, the elements present in the soil recirculate and so become available again for plant growth. As discussed previously, negatively charged clay particles are able to bind such positively charged ions as, for example, Ca^{2+}, Mg^{2+}, and K^{+}. The ions are removed from the particles by the roots of the plant either directly or after they pass into the soil solution. In general, the cations that are required by plants are present in large amounts in fertile soils, and the amounts removed by single crops are small. However, when a series of crops is grown on a particular field and the nutrients are continuously removed from the cycle as the crops are harvested, some of these cations (commonly potassium) may become depleted to such an extent that fertilizers containing the missing element must be added.

Nitrogen and phosphorus are also apt to be limiting under agricultural conditions, largely as a result of crop removal. Thus nitrogen, phosphorus, and potassium are the three elements that are commonly included in commercial fertilizers. Fertilizers are usually labeled with a formula that indicates the percentage of each of these elements. A 10-5-5 fertilizer, for example, is one that contains 10 percent nitrogen, 5 percent phosphoric acid, and 5 percent potassium.

Table 26–3 *Concentrations and Amounts of the Essential Elements in Representative Agricultural Soils*

ESSENTIAL ELEMENT	PERCENTAGE IN SOIL
Iron	3.5
Potassium	1.5
Calcium	0.5
Magnesium	0.4
Nitrogen	0.1
Phosphorus	0.06
Sulfur	0.05
Manganese	0.05
Boron	0.002
Zinc	0.001
Copper	0.0005
Molybdenum	0.0001

Other essential elements are sometimes limiting in soils on which crops are grown, even though required in very small amounts.

SUMMARY

The earth is made up of about 90 naturally occurring elements, the most common of which are oxygen, silicon, aluminum, and iron. Elements are found in the earth in the form of minerals. A mineral is a naturally occurring inorganic substance with a definite chemical composition. Rocks are mixtures of minerals. The soil is formed by the weathering of rocks and consists of an inorganic and an organic portion. The A horizon of soil contains the bulk of the organic matter, both living and dead, and the most highly weathered mineral particles. In the lower B horizon occur weathered fragments but little organic matter. The C horizon is composed of relatively unweathered rock.

A total of 16 elements is known to be required by higher plants for normal growth. Of these, carbon, hydrogen, and oxygen are derived from air and water. The rest are absorbed by roots in the form of ions. These 13 elements are sometimes categorized as macronutrients and micronutrients. The macronutrients are nitrogen, potassium, calcium, phosphorus, magnesium, and sulfur. The micronutrients are iron, chlorine, copper, manganese, zinc, molybdenum, and boron.

Sodium is required by some plants and cobalt is indirectly essential to others.

Mineral elements play a number of important roles in

cells. They regulate osmosis and affect cell permeability. Some also serve as electron acceptors, as structural components of cells, and as accessories for catalysts or as actual structural components of enzymes.

Minerals become available to plants in the soil solution in the form of ions. Plants employ metabolic energy to concentrate the ions they require. Some of the ions are taken up by active transport processes, whereas others apparently flow in passively due to the electrochemical gradients created by the actively moving ions and their pumps.

In a natural community, elements are taken from the soil by plants and then returned to it when the plants or the animals that have eaten them die. Mycorrhizal associations between fungi and the roots of plants are important in the functioning of this system and are also important in directly mediating the uptake of ions.

Under agricultural conditions, nitrogen, phosphorus, and potassium are most apt to be limiting to plant growth, and these are the elements commonly added to soil in fertilizers.

The circulation of nitrogen through the soil, through the bodies of plants and animals, and back to the soil again is known as the nitrogen cycle. It involves several stages. Nitrogen reaches the soil in the form of organic material of plant and animal origin. These substances are decomposed by soil organisms. Ammonification, the release of ammonium (NH_4^+) from nitrogen-containing compounds, is carried out by soil bacteria and fungi. Nitrification is the oxidation of ammonium to form nitrites and nitrates; these steps are carried out by two different types of bacteria. Nitrogen enters plants almost entirely in the form of nitrates. Within the plants, nitrates are reduced to ammonium. Amino acids are formed from the combination of ammonium with a keto acid (amination) or the transfer of an amino group ($-NH_2$) from an amino acid to a keto acid to produce another amino acid (transamination). These organic compounds are eventually returned to the soil, completing the nitrogen cycle.

Nitrogen is lost from the soil by removal of crops from the soil, erosion, fire, leaching, and the action of denitrifying bacteria. Nitrogen is added to the soil by nitrogen fixation, which is the incorporation of elemental nitrogen into organic components. Biological nitrogen fixation is carried out entirely by microorganisms. These include bacteria *(Rhizobium)*, which are symbionts of leguminous plants, and free-living soil bacteria and blue-green algae. In agriculture, plants are removed from the soil. As a consequence, nitrogen and other elements are not recycled, as they are in nature, and so, often must be replenished in either an organic or inorganic form.

SUGGESTIONS FOR FURTHER READING

BUCKMAN, HARRY O., and KYLE C. BRADY: *The Nature and Properties of Soils*, 7th ed. The Macmillan Company, New York, 1969.

The most comprehensive elementary text on soil science.

BUNTING, BRIAN T.: *The Geography of Soils*, Aldine Publishing Co., Chicago, 1965.*

A brief description of the main soil groups of the world and their influence on plant distribution.

CRAFTS, A. S., and C. E. CRISP: *Phloem Transport in Plants*, W. H. Freeman and Co., San Francisco, 1971.

A definitive text on phloem structure and function. Various experimental methods employed in phloem research and the influence of environmental factors on translocation are discussed. Five hypotheses for the phloem-transport mechanism are reviewed, and strong evidence in support of pressure flow is presented.

EPSTEIN, EMANUEL: *Mineral Nutrition of Plants: Principles and Perspectives*, John Wiley & Sons, Inc., New York, 1972.

A thorough coverage of the whole field of mineral nutrition, well illustrated and written, by one of the leading students of the subject.

FARB, PETER: *Living Earth*, Pyramid Publications, Inc., New York, 1969.*

A writer of many popular science books and an amateur naturalist, Peter Farb eloquently describes the teeming life that exists below the surface of the ground. A book to be read for pleasure.

International Symposium on Nitrogen Fixation, Washington State University Press, Pullman, Washington, 1975.

A collection of outstanding papers on all aspects of this very active field of research.

MEYER, B. S., D. B. ANDERSON, R. H. BOHNING, and D. G. FRATIANNE: *Introduction to Plant Physiology*, Van Nostrand Reinhold Company, New York, 1973.

A text designed for the student with training only in general biology and general chemistry. The presentation is concerned with the whole plant, with ecological overtones. Biochemical and cellular levels are introduced insofar as they contribute to the understanding of a plant as a coordinated unit.

PEEL, A. J.: *Transport of Nutrients in Plants*, John Wiley & Sons, Inc., New York, 1974.

* Available in paperback.

A concise, up-to-date account of long-distance transport physiology. A balanced view of transport in phloem and xylem is presented. A good text for the student who is unfamiliar with transport processes.

POSTGATE, J.R. (ed.): *The Chemistry and Biochemistry of Nitrogen Fixation,* Plenum Publishing Co., London and New York, 1971.

The standard modern reference on this important subject.

STEWART, W. D. P.: *Nitrogen Fixation in Plants,* The Athone Press, University of London, 1966.

In this concise volume, the author presents a balanced view of the expanding field of research in nitrogen fixation.

SUTCLIFFE, JAMES: *Plants and Water* (Crane-Russale Co.) St. Martin's Press, Inc., New York, 1968.*

An excellent short account of water relationships, including a review of experimental work in this area.

VIORST, JUDITH: *The Changing Earth,* Bantam Books, Inc., New York, 1967.*

A well-written introduction for the layman to the science of modern geology.

WARDLAW, I.F.: "Phloem Transport: Physical, Chemical, or Impossible," *Ann. Rev. Plant Physiol.* 25:515–539, 1974.

An assessment of the metabolic control of phloem transport and of the water relations of the sieve elements. Ultrastructure and source-sink relationships are discussed in relation to the mechanism of translocation.

ZIMMERMANN, M. H., and C. L. BROWN: *Trees: Structure and Function,* Springer-Verlag, New York, 1971.

This book is devoted to those aspects of tree physiology that are peculiar to tall woody plants. The emphasis is on function and includes material not found in general physiology texts.

* Available in paperback.

SECTION 8 Ecology

CHAPTER 27

The Biosphere

Ecology is the study of the interactions of organisms with one another and with their environment. It is one way of viewing and dealing with the diversity of life. Ecology attempts to tell us why particular plants and animals can be found living in one area and not in others, why there are so many of one sort and so few of another, and what changes we may expect the interactions among them to produce in a particular area in the future.

In this chapter, we first examine some of the major factors, such as light, temperature and rainfall, that affect conditions in the biosphere—the zone of air, land, and water where life exists. We then survey the general patterns of life in the sea, in fresh water, and on the land, concluding with descriptions of the principal terrestrial *biomes,* those large complexes of communities of living organisms that are characterized by distinctive vegetation and climate—deserts or grasslands, for example.

In Chapter 28, we examine the structure of ecological systems, or ecosystems. If we study a particular community—a patch of woodland, a pasture, a pond, or a coral reef—we begin to see that none of the organisms living in it exists in isolation; rather each is involved in a number of relationships, both with other organisms and with factors in the nonliving environment. These interactions have two consequences: (1) a flow of energy through photosynthetic autotrophs to heterotrophs, which eat either the autotrophs or other heterotrophs, and (2) a cycling of inorganic materials, which move from the nonliving environment through the bodies of living organisms and back to the environment. Such a combination of living (biotic) and nonliving (abiotic) elements—through which energy flows and essential elements, such as nitrogen and phosphorus, recycle—is known as an *ecosystem.*

In Chapter 29 we examine communities and populations still more closely. A *community* consists of all the plants, animals, and other organisms that live in a particular area. A *population* is a group of organisms of the same species occupying a particular area. Thus, com-

27–1
A redwood community along the coast of northern California and southern Oregon.

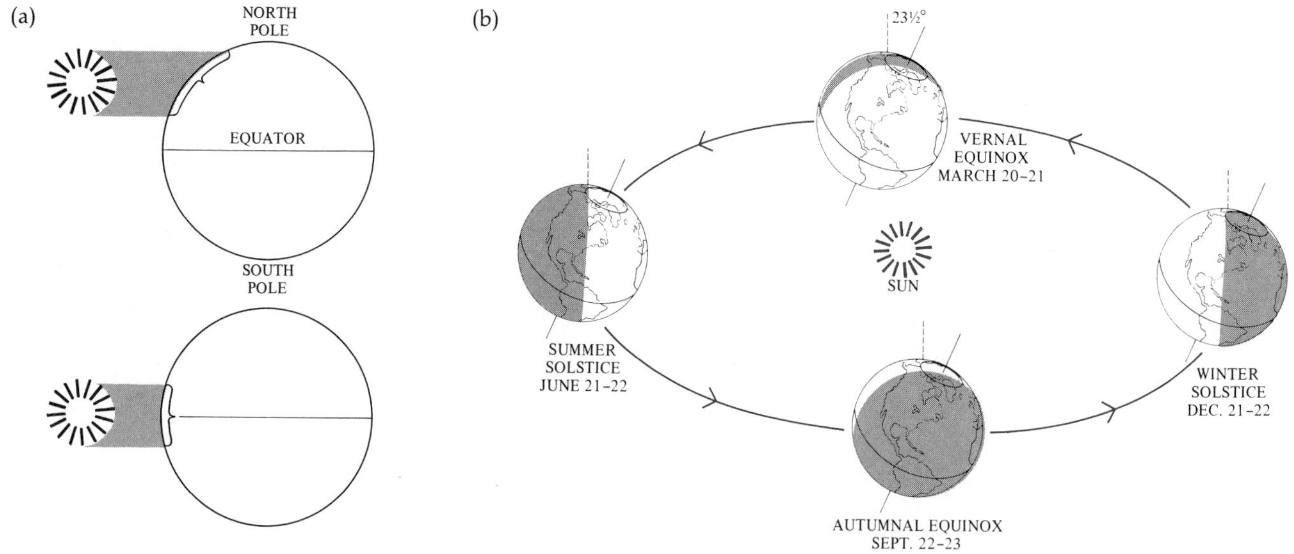

27–2
(a) *A beam of solar energy striking the earth near one of the poles is spread over a wider area of the earth's surface than is a similar beam striking the earth near the equator.* (b) *In the Northern and Southern Hemispheres, temperatures change in an annual cycle because the earth is slightly tilted on its axis in relation to its pathway around the sun. In winter, the Northern Hemisphere tilts away from the sun, which decreases the angle at which the sun's rays strike the surface and also decreases the duration of daylight, both of which result in lower temperatures. In the summer, the Northern Hemisphere tilts toward the sun. Note that the polar region of the Northern Hemisphere is continuously dark during the winter and continuously light during the summer.*

munities are made up of populations, and, as we shall see, interactions among populations within a community are major factors in determining the numbers and kinds of organisms to be found there.

THE EARTH IN THE SOLAR SYSTEM

Although the solar radiation reaching the outside of the earth's atmosphere is relatively constant, the amount that reaches the surface of the earth varies greatly from place to place and from time to time. In determining the amount of energy that reaches a given point on the earth's surface, both the annual cycle of the earth around the sun and its daily rotation around its own axis are of importance. The daily cycle keeps the temperature at a given latitude relatively constant. The annual cycle causes seasonal changes, because the earth's axis is not perpendicular to its plane of revolution around the sun but is inclined at about 23.5°. The angle and direction of this inclination are maintained as the earth travels around the sun, causing its position relative to the sun to change during its annual passage (Figure 27–2). As a result, the angle of the sun and the length of the daily photoperiod change markedly from season to season. In areas distant from the equator, an organism has to survive in a progression of changing environmental conditions. The many adaptations to the problems of survival in the face of environmental change have a great deal to do with the patterns of diversity we have observed among organisms.

THE CIRCULATION OF THE ATMOSPHERE

The oceans and the atmosphere together are components of a single dynamic system that serves to transport some of the heat that forms near the equator to higher latitudes closer to the poles. Because cold air is denser than warm air, global differences in temperature produce pressure gradients that give rise to air motion. Near the equator, the air is heated, causing it to rise and flow toward the poles. There it cools, but because of the rotation of the earth, the air does not travel all the way to the poles before descending. It tends instead to cool and sink at about 30°N and 30°S latitude, moving back toward the equator. Such masses of moving air are called cells of circulation. These forces also set into motion the associated cells of circulation and wind patterns shown in Figure 27–3 and result ultimately in the patterns of oceanic circulation shown in Figure 27–6 (page 567).

In addition, there are three major air currents moving around the world. Between about N and S 30° latitude is a broad belt of "trade winds," which blow from the northeast in the Northern Hemisphere and from the southeast in the Southern Hemisphere. Between 30° and 60° N and S are belts of winds flowing from west to east—the "prevailing westerlies." Finally, weaker winds,

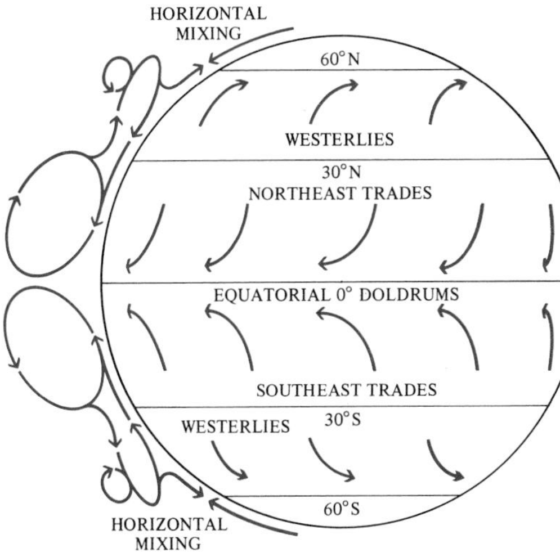

27–3
Cells of air circulation resulting in rising and cooling air and ample precipitation near the equator are modified by the rotation of the earth to produce this overall pattern of movement. In the regions of sinking and warming air at about N and S 30° latitude are found most of the world's great deserts.

(a)

(b)

27–4
(a) *Forest on the moist western slope of the Sierra Nevada in California. The tree in the center is an individual big tree* (Sequoiadendron giganteum). (b) *Dry slopes with sagebrush* (Artemisia tridentata) *and Joshua trees* (Yucca brevifolia) *east of the same range at about the same latitude. The major pattern of air circulation is from west to east. As air rises on the western slope of the mountains, it is cooled and abundant precipitation results. When the dry air descends on the eastern slope of the mountains, it is warmed and its moisture-holding capacity is increased.*

blowing from east to west, occur above 60° N and below 60° S.

Warm air has a greater water-holding capacity than cold air, and precipitation is in general greatest near the equator, where warm air is continually rising and being cooled. For the same reason, precipitation is also high near N and S 60° latitude. Precipitation is low where the air movement is predominantly downward, as at about N and S 30° latitude. Here are found the great semiarid and subhumid zones that flank the tropics on both sides. In these latitudes are found most of the great deserts of the world. Other deserts, such as the Gobi Desert of Asia, are found at the centers of huge continental areas in which the air, traveling from the oceans, has lost most of its moisture. The descending air is also very dry at the poles, so that these areas, in terms of precipitation, are also semiarid.

In the Sierra Nevada of California, the western slopes are densely forested, the eastern ones dry and desertlike (Figure 27–4). Any elevated area that lies in the path of the prevailing winds will receive more precipitation on its windward than on its leeward side. Thus, in zones of prevailing westerlies, north–south mountain ranges such as the Sierra Nevada are moister on their western flanks, where moisture-laden air rises, expands, and becomes cooler, thus losing its capacity for holding water. When the air passes over the crest of the range, however, it begins to fall and becomes warmer. It is then able to hold more moisture, and rainfall is rare (Figure 27–5).

27-5
Mean annual rainfall (vertical columns) in relation to altitude at a series of stations across the Coast Range and Sierra Nevada in California. The prevailing wind is from the Pacific Ocean at the west (left). Note the abrupt drop in precipitation east of the mountains.

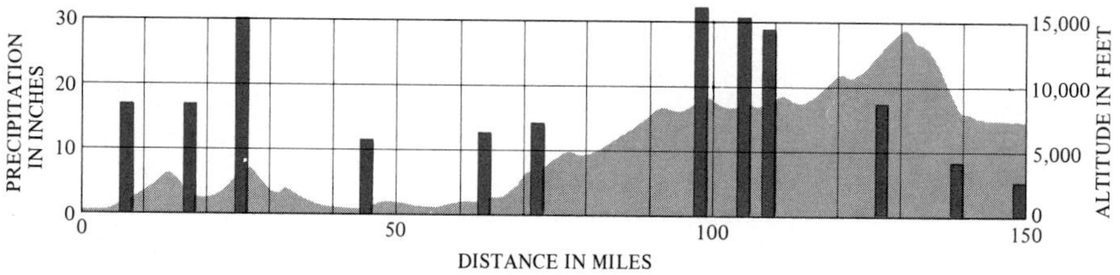

THE OCEANS

The major movements of water in the sea are governed by principles similar to those that control circulation in the atmosphere. There are two important differences, however. First, in the oceans not only heat but also salinity and wind patterns are important in controlling the density of the water and its movement. Second, oceanic currents are often deflected by the continents that bound the oceans.

Wind is the most important factor in producing currents in the ocean. Huge patterns of circulation, the surface gyrals, move around the great subtropical zones of high pressure at about N and S 30° latitude. They move clockwise in the Northern Hemisphere and counterclockwise in the Southern Hemisphere. Not only do these major currents have important effects on marine organisms, but they also profoundly modify the climates of the bordering continents. In the North Atlantic, for example, the Gulf Stream is deflected away from the shore of North America in the vicinity of Cape Hatteras, and its warming effects are felt on the western shores of Europe, which are much more temperate than those of eastern North America at comparable latitudes (Figure 27–6). Similar effects are produced by the circulation of oceanic currents in the North Pacific.

LIFE IN THE OCEAN

The sea covers almost three-quarters of the surface of the earth. Life extends to its deepest portions, but photosynthesizing organisms are restricted to the upper, lighted zones. The sea has an average depth of more than 3 kilometers and, except for a very small fraction of its surface, is dark and cold. Most of it, therefore, is inhabited only by bacteria, fungi, and animals.

The sea absorbs light readily. Even in clear water, less than 40 percent of the sunlight reaches a depth of 1 meter, and less than 1 percent of the sunlight that reaches the surface penetrates below 50 meters. The red, orange, and yellow wavelengths of light are absorbed first, so that only the shorter, blue and green ones penetrate deeply. Thus, below depths of a few meters only those photosynthesizing organisms that can make use of the available short wavelengths of light can grow.

Water is much denser than air and moves much more slowly. As a result, oxygen is not distributed as evenly in the sea as in the atmosphere, and in some locations it is present only in short supply. Temperature also has an effect on the availability of oxygen; in warm regions, the solubility of oxygen is low, and the organisms that require it may not be able to survive. In contrast, there is almost always an abundant supply of carbon dioxide available for photosynthesis. In addition to affecting the solubility and availability of gases, temperature affects the rates of photosynthesis and respiration of algae and plants directly. For this reason, it is probably the primary factor limiting the distribution of the various kinds of marine algae and plants latitudinally.

Other factors limit the growth and distribution of marine algae. Algae of the coastal shelf must be securely anchored if they are to withstand the constant action of the waves. On sandy or gravelly beaches, therefore, seaweeds are almost invariably absent. Except in the very calmest waters, only rocky shores are abundantly populated with seaweeds (Figure 27–7).

In the sea, the distribution of minerals is more constant than on land, with adjacent areas not sharply differentiated from one another. Nevertheless, minerals are often limiting in particular areas of the sea, such as tropical waters, and are then responsible for limiting plankton density. Following a bloom of plankton, a marked depletion of most nutrients can be detected. In areas of upwelling, plankton populations (and populations of the animals that depend on them) are often large and stable. Such areas of upwelling along coasts make up only about 0.1 percent of the total area of the oceans, yet are estimated to contain more than half of the world's fish catch. Such areas occur mainly along the western shores of the continents in the zone of the prevailing westerlies, when polar winds enter these areas.

The open ocean, away from the continental shelf, and the surface layers of the water above the shelf are inhabited by free-floating, or planktonic, organisms. The total biomass of the phytoplankton—free-floating microscopic algae—is several times greater than that of the attached seaweeds.

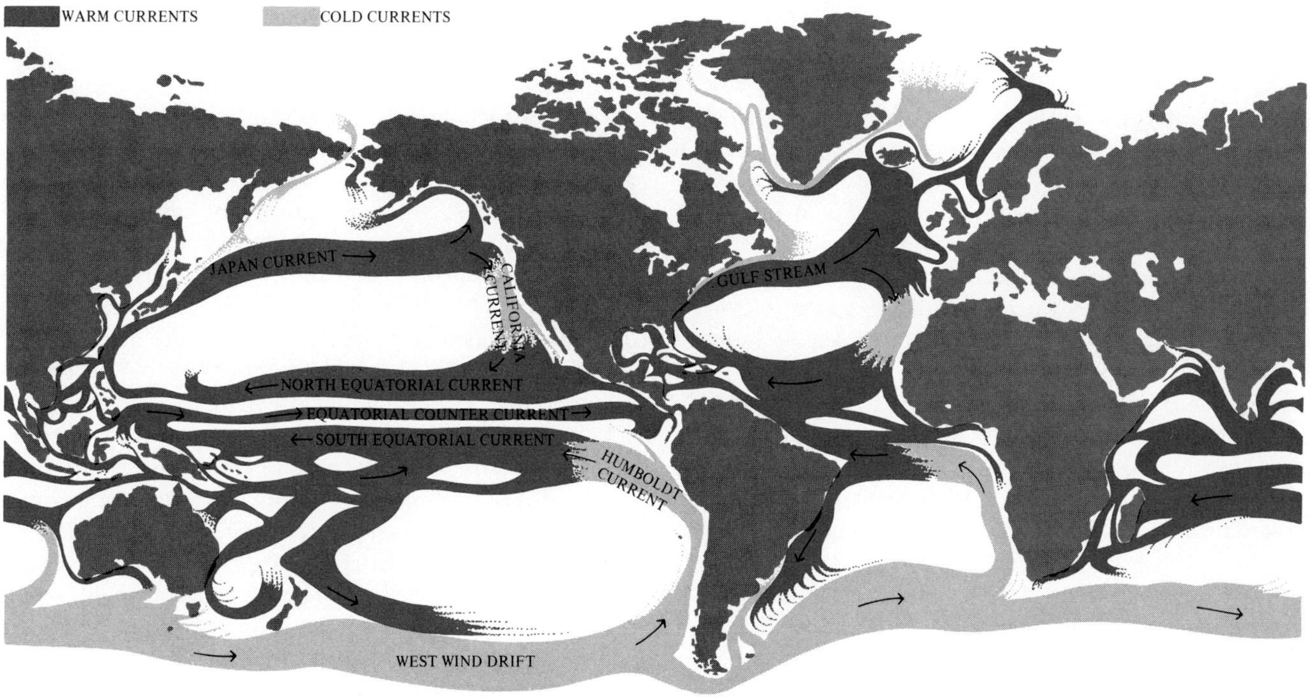

27–6
The major currents of the ocean have profound effects on climate. Because of the warming effects of the Gulf Stream, western Europe is milder in temperature than is eastern North America at similar latitudes. The eastern coast of South America is warmed by water from the equator, whereas the Humboldt current brings cooler weather to the western coast of South America. Where winds constantly move surface water away from coastal slopes, cold water rich in nutrients is brought to the surface in upwellings such as those responsible for the rich fishing industry off the coast of Peru.

27–7
Where the sea meets the edge of the continents, many kinds of multicellular algae occur on the rocks. These, at Fort Fisher, North Carolina, are covered with a green alga (Enteromorpha).

Kinds of Marine Algae

In the sea, the plants are represented only by a few genera, all of them monocots. Dicots, gymnosperms, ferns, and bryophytes are totally absent. Marine bacteria are abundant and diverse, and they extend to the lowest depths. Fungi, mainly Ascomycetes and Fungi Imperfecti, are relatively common on decaying organic material of all sorts. They occur at all depths but are rarely conspicuous, although they may play a major role in the decomposition and eventual detachment of large seaweeds.

The algae are the predominant group of photosynthesizing organisms in the sea. The blue-green algae, which are relatively common but generally inconspicuous in temperate waters, often form readily visible tufts or filamentous masses in the tropics. On the whole, however, they are less common and less well represented in the sea than they are in terrestrial or freshwater habitats. Green algae are also of relatively minor importance as compared with their rich diversity in fresh water, but some species and genera, such as *Ulva* and *Enteromorpha* (Figure 27-7), are widespread and often locally abundant. The marine green algae are best developed in the tropics, but are frequent in marine habitats everywhere.

On the Atlantic and Pacific shores of the United States, the brown algae are the most familiar. There are approximately 1500 species of this group. They can be found in the drift on beaches, in tide pools, around pilings, and in offshore kelp beds along many temperate coasts. The brown algae are relatively poorly represented in the tropics.

In tropical waters and at greater depths (down to 30 meters or so), the red algae predominate. With about 4000 species, they are the largest and most diverse group of seaweeds. Like the brown algae, they are almost exclusively marine. Red algae occur at the greatest depths inhabited by photosynthesizing organisms, where their accessory pigments give them a tremendous advantage in utilizing short-wavelength light, but they also range far into the intertidal zones where they may only occasionally be wet by salt spray.

Other groups of algae that occur in the sea are microscopic organisms generally confined to the plankton. Whereas only about 1 percent of the surface of the sea lies over a bottom that is shallow enough for the relatively large attached seaweeds to grow, the entire surface layer of the water is filled with microscopic floating organisms.

Diatoms are ubiquitous, and they can be found in every conceivable marine habitat—on the bottom in shallow water, on floating or attached objects, and in the plankton. Other members of the Chrysophyta and also the Pyrrophyta are abundant in the plankton, but the diatoms and dinoflagellates normally predominate.

The relatively large members of the phytoplankton, which are still small enough to be invisible to the naked eye except in mass, are vastly outnumbered by others so small that they pass through ordinary plankton nets. How small they are is emphasized by the diameter of mesh openings in such nets—0.040 to 0.076 millimeter. Some of these organisms are very small diatoms and dinoflagellates, but other groups of uncertain affinity, such as the coccolithophorids, are often abundant. Some of these organisms, such as *Micromonas pusilla*, are only 0.001 to 0.003 millimeter in length. These exceedingly small organisms may occur in such large numbers that their contributions to the productivity of the sea are sometimes vastly greater than those of the larger plankton.

On certain occasions, particularly when drastic shifts in nutrient level occur, phytoplankton in the sea may become so abundant as to color the water reddish, brownish, or yellowish. Ordinarily, such colored water is accompanied by brilliant luminescence at night, owing largely to the bioluminescent dinoflagellate *Gonyaulax*. Blooms of this sort occur as the result of sharp increases in some species of diatoms or dinoflagellates, in which the number of cells may reach 2 million per liter or more. They often result in mass mortality for other marine organisms, either as a result of the depletion of oxygen or minerals, or because of the production of toxins. Dinoflagellates of the genera *Gymnodinium* and *Gonyaulax*, for example, which produce a water-soluble nerve toxin, can become abundant in areas of upwelling, where a large supply of nutrients is coupled with sufficiently high temperatures for rapid reproduction. It is hypothesized that the mass mortality of marine algae and animals following such blooms may have contributed in the past to the formation of petroleum deposits.

FRESHWATER BIOMES

There are many similarities between life in the ocean and life in fresh water, but also a few striking differences. The freshwater habitat is a very limited one, with inland lakes covering some 1.8 percent of the world's surface and running water some 0.3 percent. Freshwater habitats are stikingly discontinuous and intimately interconnected with terrestrial ones. However, they provide relatively constant conditions, and thus freshwater organisms with reasonably efficient means of motility often have very wide distributions.

In general, limnologists—scientists who study natural fresh waters in all their aspects—classify freshwater habitats into two groups: standing water and running water. In running-water communities, current plays a dominant role in determining the distribution of organisms (Figure 27-8). These are not complete ecosystems, in the sense that the communities of organisms living in them do not produce all their own food. Life in streams

27-8
Castle Creek in Aspen, Colorado is a swift, constantly flowing mountain stream. There is little rooted vegetation in the stream itself but there is an abundance of small animals.

depends on the continual supply of organic material from adjacent terrestrial habitats or, in some cases, from quieter waters.

Because of the continual agitation of waters in streams and rivers, the supply of oxygen is more constant than it is in still bodies of fresh water, and the temperature is more uniform throughout. Firm attachment to the substrate is an important factor for plants living in streams, as it is for plants living along the shores of the ocean.

Ponds and Lakes

In ponds and lakes, conditions are more stable. Lakes and large ponds contain three distinct ecological zones: (1) the *littoral zone*, along the shore, in which rooted vegetation lives; (2) the *limnetic zone* of open water, in which planktonic forms occur, and which extends down to the limit of light penetration; and (3) the deep-water *profundal zone*, below the limit of effective light penetration. Photosynthesizing organisms are confined to the first two zones, but their remains regularly sink into the profundal zone, which is present only in large lakes. Bacteria and fungi play important roles in the depths, the profundal zone.

In the large lakes of temperate regions, there is a pattern of thermal stratification similar to that found in the sea (Figure 27-9). During the summer, the upper layers of the lake are heated and stirred by the wind. At the bottom of the lake is a zone of dense water at about 4°C. Fresh water is most dense at this temperature and becomes less dense either upon heating or cooling. The transition between the upper, warm layer of water and the lower, cold one is called the *thermocline*, defined in fresh water as the region in which the temperature changes 1°C or more per meter of depth. The thermocline often lies below the lower limits for effective photosynthesis, and the supply of oxygen below it often becomes seriously depleted during the summer.

In autumn, as the surface water is cooled and becomes denser, it sinks, replenishing the supply of oxygen in the lower depths. At the same time, the often nutrient-rich bottom waters are circulated throughout the lake. If the surface water freezes in the winter, the resulting blanket of ice and snow may cut off photosynthesis in the lake, causing serious oxygen depletion. In spring, the ice melts, the water temperature rises to 4°C, and the water reaches its maximum density, sinking to the bottom again and redistributing oxygen and nutrients.

Because of the continually changing conditions, a large lake in temperate regions never reaches a stable ecological condition. Instead, a succession of organisms replace one another seasonally. In the spring and autumn, when oxygen and dissolved minerals are redistributed by the overturn of the water, there is often a spectacular bloom of plankton (Figure 27-10; see also Figure 27-11).

27–9
Seasonal cycle of temperature changes in a temperate-zone lake. Water increases in density as it cools, reaching a maximum density at 4°C. In the summer (a), *the top layer of water, called the epilimnion, becomes warmer than the lower layers. The middle layer, in which there is an abrupt drop in temperature, is the thermocline. This stratification cuts off the supply of oxygen below the thermocline, producing summer stagnation. In the fall, the temperature of the epilimnion drops until it is the same as that of the hypolimnion* (b). *Then, aided by the fall winds, the water of the lake begins to circulate and oxygen is returned to the depths* (c). *As the surface water cools below 4°C, it expands, becomes lighter, remains on the surface and, in many areas, freezes. Then winter stratification occurs* (d). *In spring* (e), *as ice melts and the water on the surface warms to 4°C, it sinks to the bottom, producing the spring overturn, after which the water again circulates freely* (f).

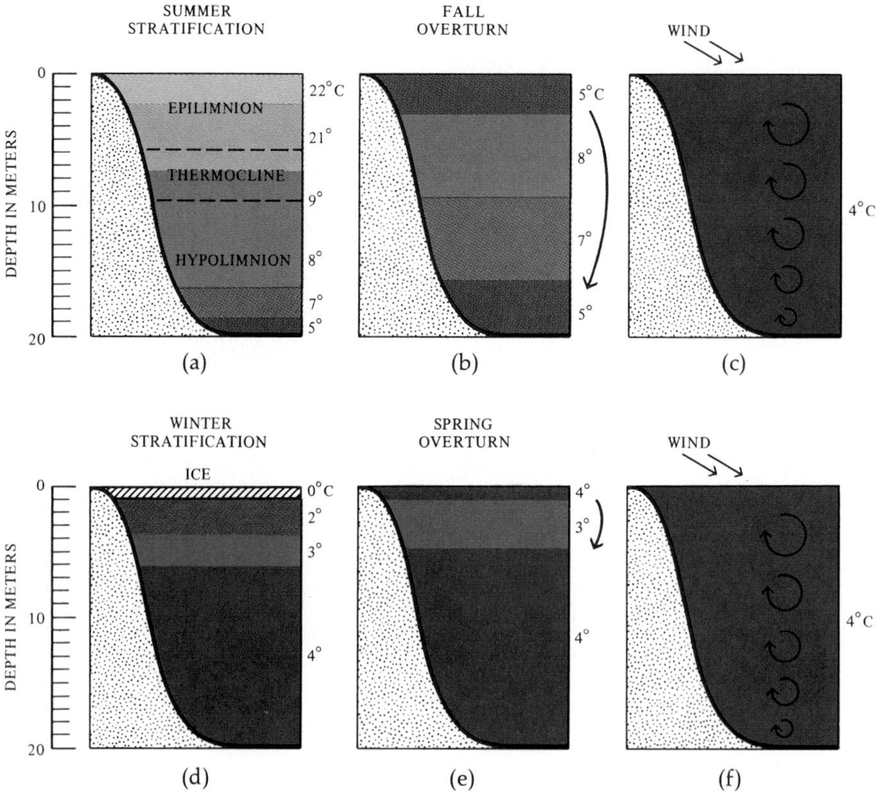

27–10
Freshwater plankton. In the diatom Fragilaria, *the individuals are linked into chains. The large animal near the center is a rotifer, and the star-shaped diatoms are* Asterionella.

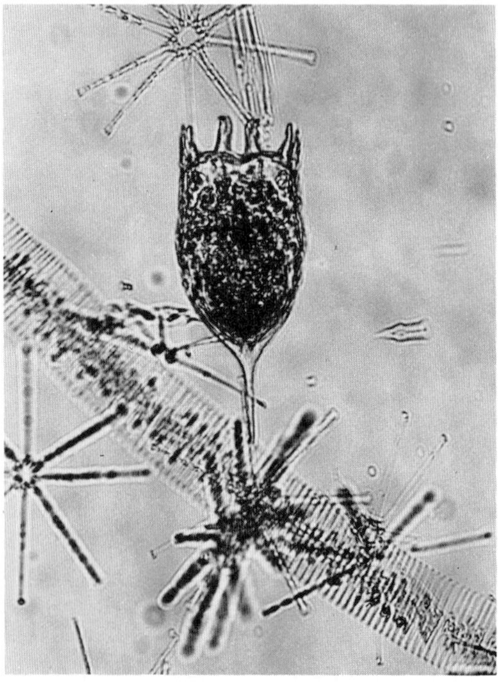

27–11
A seasonally dry stream in northern California, with an abundant bloom of green algae. The tall trees are Eucalyptus, *introduced from Australia.*

27-12
This freshwater marsh is dominated by cattails (Typha angustifolia *and* Typha latifolia). *The water lilies are* Nymphaea odorata. *Aquatic succession will gradually return the ditch to marsh and then to dry land.*

The edge of a lake or pond is its most productive zone and is normally dominated by angiosperms which are rooted to the bottom. Many of these, like the cattails and rushes, stand above the water, and others, like the water lilies and certain pond weeds, are rooted on the bottom but spread their leaves on the water's surface (Figure 27-12). The duckweeds, minute seed plants, float on the top of the water or just under the surface. Certain pond weeds, belonging to several families of monocots, are entirely submerged.

Unlike many organisms found in fresh water, the freshwater seed plants are clearly derived from terrestrial ancestors. In their new habitat, they have become modified in various ways—such as the reduction of vascular strands and development of buoyant stems or floaters, underwater pollination, and thin, often much-dissected leaves.

Not only do these angiosperms make an important contribution to the productivity of ponds and lakes, they also provide habitats for many species of plants and animals. Every protruding object tends to be densely covered with organisms of many different groups, especially in the richer, relatively shallow portions of the ponds.

The algae found in freshwater habitats are presumably derived from marine ancestors. Some of the fungi, on the other hand, appear to have originated as freshwater groups; no major group of the fungi appears to be linked with the sea in its early evolutionary history. The few freshwater bryophytes undoubtedly had terrestrial ancestors.

LIFE ON THE LAND

Land plants face a variety of problems. They are subjected to periodic drought and to rapid diurnal and seasonal changes in temperature. They must survive during unfavorable seasons, they must often grow on substrates of varying mineral composition, and they are all subject to the action of gravity (which affects them much more directly than it does water plants). Except at high elevations, however, oxygen is more uniformly distributed than in the sea, and carbon dioxide is normally readily available.

The earth's land areas are discontinuous, and this discontinuity has an important effect on the distribution of organisms. In addition, the radical changes that often occur from place to place on the land mean that the distribution of any particular kind of land organism is apt to be much more limited than would be the distribution of an organism of similar size and motility in the sea.

Local Modifications in Climate

On the land, plants and plant communities vary as the land varies. For instance, the mean atmospheric temperature decreases about 0.5°C for each degree of latitude. Increases in elevation produce a similar effect, and in general, a change in mean atmospheric temperature corresponding to 1 degree in latitude occurs with a rise of about 100 meters in elevation. This has important con-

27-13
We meet a similar series of plant communities whether we travel north for hundreds of kilometers or ascend a mountain. This particular kind of relationship between altitude and latitude was first pointed out by Alexander von Humboldt (see Essay on page 601).

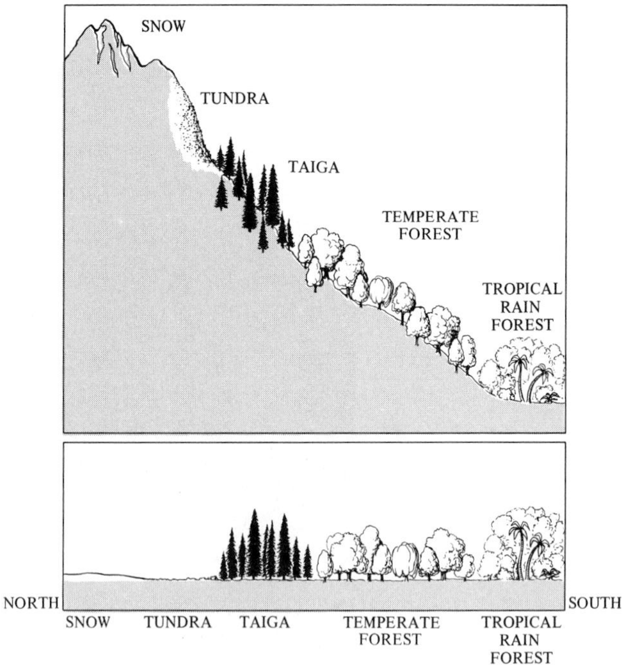

27-14
In high mountains in the tropics, temperatures do not change seasonally but freezing may occur every night of the year. These unusual conditions have resulted in the evolution of plants with bizarre growth form, like these giant senecios (family Asteraceae) photographed on Mt. Kenya in east Africa. In such plants, the apical meristem is protected from freezing by the tightly infolded leaves, thick fleshy construction, and matted, reflective covering of silvery hairs.

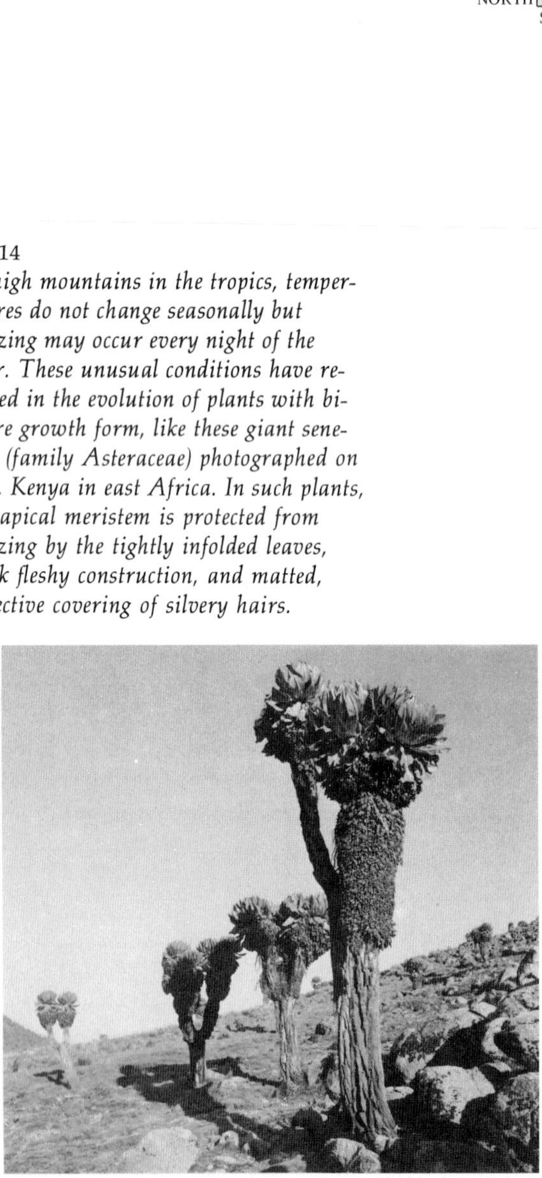

sequences for the distribution of land organisms. For example, plants and animals of almost arctic characteristics may approach or even reach the equator, particularly in north–south mountain ranges (Figure 27–13).

There are, however, important differences between high-latitude and high-altitude habitats. In the mountains the air is clearer and the solar radiation is more intense. Most of the water vapor in the earth's atmosphere, which plays the major role in preventing heat from radiating away from the earth at night, occurs below 2000 meters. Consequently, nights in the mountains are often much cooler than those at lower elevations in the same latitude (Figure 27–14). Moreover, there is pronounced seasonal variation both in day length and in temperature near the poles, and relatively little variation near the equator. Those organisms that do range toward the equator in the mountains must make physiological adjustments for the differences between these environments.

There are often pronounced temperature variations from slope to slope of a particular mountain. In the middle latitudes of the Northern Hemisphere, for example, sunlight reaches only the south and west slopes of mountains in the winter, and such slopes are often drier than the north and east slopes of the same mountains. Further, in summer, the afternoon sun is much hotter than the morning sun owing to the decrease in the water vapor in the atmosphere. Because of the interaction of these two factors, the driest slope of a Northern-Hemi-

27-15
Krummholz (crooked wood), a type of stunted forest characteristic of alpine regions, on lee side of ridge, northeast end of Cove Mountain, Manti National Forest, Utah. The dwarfed trees are Engelmann spruce, Picea engelmannii *(the taller trees), and limber pine,* Pinus flexilis, *at the very limits of their distribution. This photograph illustrates the line between two sharply demarcated biomes.*

sphere mountain is usually the southwest. (Which would you expect to be the driest slope of a mountain in the Southern Hemisphere?) On low hills in an otherwise barren Arctic tundra, clusters of small trees and shrubs may occur on the southern slope. For the same reasons, "moss" grows, or is better developed, on the northeast side of trees in the Northern Hemisphere. Ant hills in comparable regions are often built with a longer slope on the southern side, which provides an efficient temperature-regulating mechanism.

TERRESTRIAL BIOMES

A *biome* can be defined as a climatically controlled group of plants and animals with a unique appearance that is distributed over a wide geographical area. Terrestrial biomes can be classified in various ways, but the seven categories recognized in this book provide the basis for a minimum account of the vegetation of the world.

Tropical Rain Forest

More species of plants and animals live in the tropical rain forest than in all the rest of the biomes of the world put together. Neither water nor temperature are limiting during any part of the year. Although there are many species, there are few individuals per species; a species may be represented only once per hectare. Not only is there a large number of organisms in the tropical rain forest, but their interrelationships are more complex than those found between plants and animals in any other biome.

Little light penetrates to the floor of the tropical rain forest, and the rainfall is generally between 200 and 400 centimeters per year. There is little accumulation of organic debris; decomposers rapidly break down leaves, stems, and the bodies of animals that fall from the canopy. Although there may be notable variation in precipitation from month to month, there is no pronounced dry season.

In the tropical rain forest, plants in general have not evolved particular mechanisms that would permit them to survive unfavorable seasons of drought or cold. Almost all the plants are woody, and woody vines are abundant. There is a large flora of *epiphytes,* which grow on the branches of other plants in the illuminated zone far above the forest floor. Epiphytes, which have no direct contact with the forest floor, obtain water and minerals from the humid air of the canopy. With the epiphytes and climbing vines, many kinds of animals have moved into the treetops; and that is the area of the tropical rain forest in which animal life is most abundant and diverse.

Because so little light reaches the forest floor, there are very few herbaceous plants, and those that do occur are mostly epiphytes or grow in clearings. Many plants in the tropical rain forest are trees; they are larger than those usually found in temperate forests, often reaching 40 to 60 meters in height. Further, the trees are diverse, and there are seldom fewer than 40 species per hectare. This stands in sharp contrast with temperate forests where there are rarely more than a few species per hectare. The trees of the tropical rain forest are remarkably homogeneous in appearance. Generally, they branch only near the crown. Because their roots are usually shallow, they often have buttresses at the base of the trunk to provide a firm, broad anchorage. Their leaves are medium-sized, leathery, and dark green; their bark is thin and smooth; and their flowers are generally inconspicuous, and greenish or whitish in color. Such forests often form several layers of foliage, the lower layers being produced by the seedling trees of the taller species and some species that are lower-growing.

There are three major areas of the world in which the tropical rain forest is well developed. The largest is that in the Amazon basin of South America, with outliers in coastal Brazil, Central America (Figure 27–16), and eastern Mexico, as well as some of the islands in the West Indies. In Africa, there is a large area of rain forest in the basin of the Congo, with an extension along the west coast of Liberia. The third area of rain forest extends from Ceylon and eastern India to Thailand, the Philippines, the large islands of Malaysia, and a narrow strip along the northeast coast of Queensland, Australia.

The tropical rain forest now forms about half the forested area of the earth, but it is in the process of being systematically destroyed. The rapidly expanding human population in the tropics has made the traditional tropical agricultural practice of clearing and short-term cultivation immensely destructive because it is now carried out on such a wide scale.

27–16
Map showing the distribution of rain forests in Central America.

One reason for this is the nature of tropical soils. These soils are conditioned by high and constant temperatures and abundant rainfall and are relatively infertile. Such soils are known as *latosols*—red clays largely leached of their nutrients. Tree roots never reach very deep into laterites, although the processes leading to the formation of these soils may extend to depths greater than 15 meters. When clearing is carried out on laterites, the leaching process accelerates greatly and the soils either erode rapidly and spectacularly or form thick, impenetrable crusts on which no cultivation is possible. Despite this, the tropical forests of the world are being cut and burned at an ever-increasing rate to produce fields that become completely useless to agriculture within a few years. It is estimated that by the end of the present century, all the tropical rain forests may have disappeared, with the exception of a few small reserves and those in relatively inaccessible areas.

Savanna

Savannas are transitional areas between the evergreen tropical rain forest and the deserts (Figure 27–18a). They usually have much lower annual rainfall than the tropical forest—frequently in the range of 90 to 150 centimeters a year. There is also a wider fluctuation in average monthly temperatures, owing to the seasonal drought and sparse covering of vegetation.

Many plant communities in the tropics are characterized by a marked period of drought each year. One of the most widely distributed of the tropical plant communities is the tropical grassland, which has widely scattered trees that are generally deciduous, losing their leaves in the dry season. Such regions cover large areas of East Africa (Figure 27–18b), but they are also found on the margins of rain forests everywhere. In Southeast Asia, there are extensive areas of monsoon forest. Here there is high precipitation during portions of the year, when the moisture-bearing monsoon is blowing off the ocean, but there also exists a well-defined dry season during which the trees lose their leaves. Thorn forest occurs under drier conditions and is characterized by dense, spiny, low trees. In drier regions, the thorn forests grade into the desert.

The existence of thorn forest and the open grassy plains of central Africa depends to a large extent on periodic burning. The winter is hot and long, and the natives frequently burn the plain to produce young grass for the game and for their own herds.

In savanna communities, because of the scattered distribution of trees, the forest floor is generally well illuminated, and perennial herbs (mostly grasses) are common. Bulbous plants, which are able to withstand periodic burning, are abundant. Because of the dense

27–17
Rain forest. (a) *A thick growth of epiphytes.* (b) *Huge buttressed tree in the forest, with a* Philodendron *growing up its trunk.* (c) *Stinkhorn fungus,* Dictyophora duplicata. (d) *Staghorn fern,* Platycerium, *an epiphyte.* (e) *Fruits on the forest floor.*

27–18
(a) *The savanna communities of North America.* (b) *Zebras* (Equus burchelli) *and Coke's hartebeest* (Alcelaphus buselaphus cokii) *grazing in the savanna near Nairobi, Kenya. The trees and bushes are mixed* Acacia *species.*

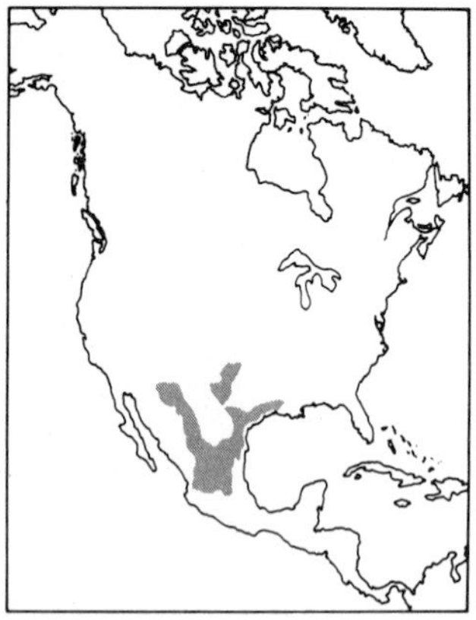

(a)

(b)

cover of perennial herbs made possible by the abundant, seasonal rainfall, there are few annual herbs. Epiphytes are also rare. The trees found in savanna communities often have thick bark. They are well branched but are seldom more than 15 meters tall. Almost all of them are deciduous; they lose their leaves at the start of the dry season and flower in a leafless condition. Their leaves are smaller than those characteristic of the evergreen trees of the rain forest and so lose less water by transpiration.

Desert

The great deserts of the world are all located in the zones of atmospheric high pressure that flank the tropics at about N and S 30° latitude, and they extend poleward in the interior of the large continents. Many deserts are characterized by less than 10 centimeters of rain per year. In the Atacama Desert of coastal Peru and northern Chile, the average rainfall is less than 2 centimeters per year. Extensive deserts are located in North Africa and in the southern part of the continent, in the Near East, in western North and South America, and in Australia. The Sahara Desert, which extends all the way from the Atlantic coast of Africa to Arabia, is the largest in the world. Australia's desert, which covers some 44 percent of the continent, is next in size. Less than 5 percent of North America is desert (Figure 27–19).

Desert regions are characterized by very high temperatures, and summer temperatures of over 36°C are com-

27–19
Only a small proportion of North America is desert.

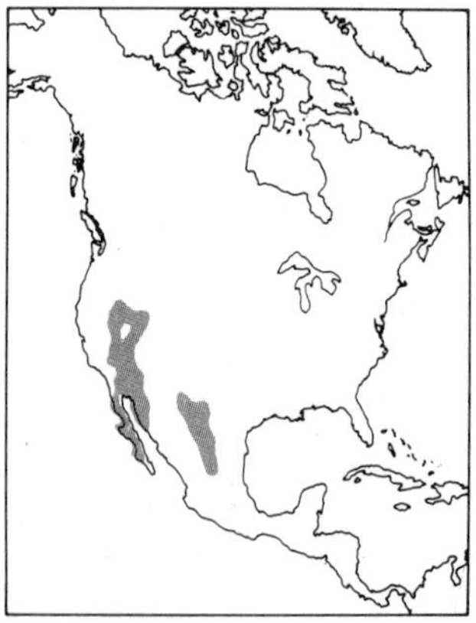

(a)

(b)

mon. On the other hand, because of the sparse plant cover, there is little water vapor in the atmosphere over deserts, and the consequent rapid radiation of heat at night creates strong differences in temperature in the course of a single day.

The annual distribution of rainfall in desert areas generally reflects that of the bordering areas. On their equatorial side, it rains in the summer; on their poleward side, it rains in the winter. Between the two, as in the lowlands of Arizona, there may be two annual peaks of precipitation. As a result, in such an area there are generally two periods of active plant growth—one in the winter, and one in the summer—and different plants are active in each. In general, the patterns of activity of desert plants reflect their respective origins—the present deserts of the world are of relatively recent origin, and their plants and animals have been selected from those of the bordering areas.

Annual plants are most important, both in number and kind, in the desert and semiarid regions of the world (Figure 27-20c). Because of the erratic supply of water, perennial herbs do not succeed well in such regions, and there is no dense covering of perennials, which would inhibit the growth of annuals. Because of their very rapid growth, annuals can spring up in the open areas during those limited periods when water is available. The seeds of these annuals can survive in the soil through long periods of drought—sometimes extending over many years—and then germinate rapidly.

(c)

27-20
Desert plants. (a) *Carpet of annual plants flowering in the Mojave Desert of California.* (b) *Washington palms* (Washingtonia filifera) *in Palm Canyon, California.* (c) *Cholla* (Opuntia)—*saguaro* (Carnegeia gigantea) *community in southern Arizona.*

27-21
The short-grass prairies of North America.

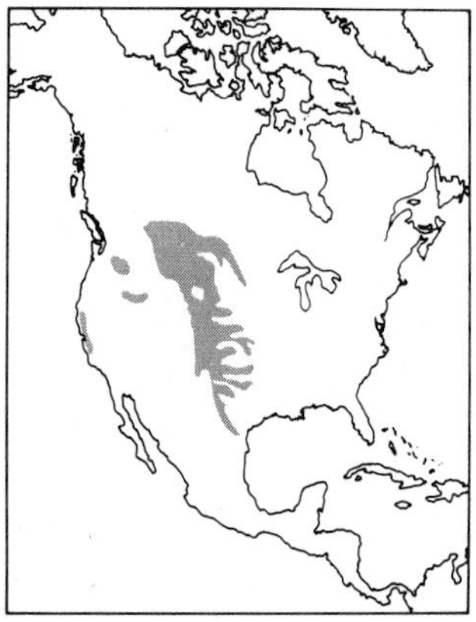

27-22
The prairie soils were once so bound together with the roots of grasses that they could not be cultivated until special plowing instruments were developed. But once the plants are removed by overgrazing or careless cultivating, prairie soil rapidly deteriorates and is carried away by the wind. This photograph, taken in Texas County, Oklahoma, in 1937, vividly recalls the "dust bowl" conditions that led to migrations from the central United States. John Steinbeck's novel, The Grapes of Wrath, *was based on the experiences of these migrants.*

Those relatively few perennial plants that do occur in the desert are generally bulbous and dormant for much of the year. Most of the taller plants are either succulent—for example, the cacti, euphorbias, and other characteristic desert plants—or have small leaves that are either leathery or are shed during unfavorable seasons (Figure 27-20b). Usually the leaves have a thicker cuticle and fewer stomata than those of plants in less arid regions. Photosynthesis often takes place in the stems as well as in the leaves during dry periods. Many of the succulent plants have adopted CAM photosynthesis (page 117) and absorb carbon dioxide only at night. Temperatures for photosynthetic maxima are often much higher for desert plants than for other species, and in many the orientation of the leaves minimizes heat absorption. Woody plants either have wide-ranging roots that effectively absorb the periodic rainfall (like the majority of shrubs) or are restricted to washes and arroyos where the moisture is concentrated (like most trees in these regions).

Grasslands

The grasslands include a wide variety of plant communities; some are related to savannas, others to deserts, and still others to temperate deciduous forests. The major grasslands of the world often grade into deserts toward the equator as the amount of precipitation decreases. The richer grasslands have as much as 100 centimeters of precipitation per year and intergrade into temperate deciduous forest where the moisture supply is more abundant.

Grasslands form the zone that lies between deserts and temperate woodland, occurring where the amount of rainfall is intermediate between that characteristic of those two biomes. Grasslands have often been changed into either woods, when disturbed, or deserts, when man has failed to recognize the existence of this delicate ecological balance. Such failures were the major cause of the "dust bowl" disasters of the central United States in the 1930s (Figure 27-22).

Grasslands have traditionally been heavily exploited for agricultural purposes, both as pasture and, when cleared, for farms. Much effort has been spent in learning how to convert other biomes into grassland and, once the conversion has been made, learning how to maintain them in this condition. The most productive soils for agriculture occur in areas formerly occupied by tall-grass prairies.

Grasslands generally occur over large areas in the interior portions of the continents. In North America, there is a transition from the more desertlike, western, short-grass prairie (the Great Plains), through the moister, richer, tall-grass prairie (the corn belt), to the eastern temperate deciduous woodland. Grasslands become drier and drier at increasing distances from the Atlantic

(a)

(b)

27–23
(a) *A June day on a wheat field in North Dakota. The cottonwood* (Populus) *grove by the prairie creek is characteristic of this biome. A thunderstorm is gathering on the horizon.* (b) *An overgrazed prairie.*

Ocean and the Gulf of Mexico, which are the major sources of moisture-bearing winds in the eastern half of the continent (Figure 27–23).

Perennial bunchgrasses and sod-forming grasses are dominant, but other perennial herbs are common. Although the growth of grassland plants may be somewhat seasonal, there is little room for the development of annual herbs, and these are essentially absent in this biome. Occasionally, annuals and weeds from other areas become established in disturbed areas, as, for example, around the burrows of animals, the sites of man's habitations, and along roadsides.

The great grasslands of the world are inhabited by herds of grazing mammals that are associated with a complex of large predators. Many of these have been hunted almost to the point of extinction and survive mainly in refuges, having given way to herds of domestic animals and cultivated fields.

Temperate Deciduous Forest

Temperate deciduous forest is almost absent in the Southern Hemisphere but is represented in all the major land masses of the north (Figure 27–24). This type of forest reaches its best development in areas with warm summers and relatively cold winters (Figure 27–25). Annual precipitation generally ranges from about 75 to 250 centimeters. The deciduous (leaf-dropping) habit may be related to the unavailability of water during much of the winter.

27–24
The deciduous forests of North America.

(a)

(b)

27–25
The deciduous forest in the southern Appalachians is one of the richest temperate-zone forests in the world, containing many kinds of hardwoods. (a) *Fog over the Great Smokies.* (b) *Autumn in the mountains of North Carolina.*

(a)

(b)

27–26
(a) *Forest of slash pine* (Pinus elliottii) *in Florida. Plant associations of this sort were derived from the richer ones of the Arcto-Tertiary geoflora (see page 362) during the past 20 million years.* (b) *Western coniferous forest in the Cascade Mountains of Washington, dominated by Douglas fir* (Pseudotsuga menziesii). *During the course of the Tertiary, most characteristic eastern hardwoods were eliminated from the western United States as temperatures became colder and precipitation less.*

27-27
(a) *Among the common harbingers of spring is the white-flowered bloodroot* (Sanguinaria canadensis). *The bloodroot derives its name from the orange juice that flows from a broken stem.*
(b) *The marsh marigold* (Caltha palustris) *forms colonies in wet woods throughout temperate North America and Eurasia.*

(a)

(b)

The ecology of the temperate deciduous forest is somewhat different from that of evergreen forests (Figure 27-26). In winter, the trees are leafless and tree activity is greatly reduced. In spring, a variety of herbaceous plants burst forth in profusion on the temporarily well-illuminated forest floor (Figure 27-27). In the time before leaves appear on the trees, reducing light on the forest floor, most of the early spring herbs have been able to set seed. Very few annual plants occur in the deciduous forest, probably because they lack the storage organs that enable the perennials to grow very rapidly during the relatively short time that conditions are favorable for them.

One of the outstanding characteristics of the temperate deciduous forest is the similarity of the plants found in its three main Northern-Hemisphere regions. The various deserts of the world, for example, are generally inhabited by very different groups of plants which have converged in their ecological characteristics. The plants of the temperate deciduous forest, however, are a more uniform group. For example, the deciduous forests of Japan resemble those of eastern North America much more closely than either resemble the forests of western North America. Most of the deciduous trees and their associated herbs were eliminated in the western United States during the later half of the Tertiary and during the Pleistocene era as the amount of summer rainfall was greatly reduced. In their place now is the magnificent coniferous forest of the western mountains, where grow such trees as the coastal redwood, *Sequoia sempervirens;* the big tree, *Sequoiadendron giganteum;* the Douglas fir, *Pseudotsuga menziesii;* and the sugar pine, *Pinus lambertiana.*

A local but distinctive plant association has evolved in areas of mediterranean climate that are characterized by cool, moist winters and hot, dry summers. Such climates are found along the shores of the Mediterranean, over a large part of California and into some adjacent states, in the coastal portions of central Chile, in southern Africa, and along portions of the coast of southern and southwestern Australia. The plants in these areas, often evergreen or summer deciduous trees and shrubs, have relatively short growing seasons in fall and spring, being limited by low temperatures in winter and by drought in summer. In such a climate, precipitation and the need for water are out of phase, the luxuriant growth of spring being followed by the drought of summer.

Each of the areas of mediterranean climate is isolated from the others, and each has evolved its own distinctive assemblage of plants and animals. The degree of ecological convergence has been high, however, and the chaparral of California is closely similar in appearance to the matorral of Chile or the maquis of the Mediterranean, even though the plants are essentially unrelated. Seasonal drought enhances the importance of edaphic (soil related) and biotic variation, and small differences in precipitation often have profound effects on the species in the area. Thus, these areas often have high proportions of extremely local species of plants and animals (Figure 27-28).

27–28
Chaparral in the mountains near Los Angeles, California. This sort of plant association, consisting of broad-leaved, drought-resistant, and, often, spiny evergreen shrubs, occurs only in limited areas. Like the deserts, these areas originated independently in different parts of the world. The very similar appearing shrubs found in each area were derived from completely different ancestral stocks and have become similar in appearance as they responded over millions of years to the similar mediterranean-type climates characterized by a long, hot, dry summer and a moist, cool winter during which most of the growth occurs. Similar plant associations occur in central Chile (where they are known as matorral), around the Mediterranean (where they are often called maquis), in South Africa, and western Australia.

Taiga

This northern coniferous forest, or snow forest, is characterized by severe winters and a persistent winter cover of snow. The climates that produce such biomes develop only in the interior of large continental masses at the appropriate latitudes and are thus virtually absent from the Southern Hemisphere. Owing to the influence of the prevailing westerlies between N 40° and 50° latitude, the western portions of North America and Eurasia are characterized by milder climates than the eastern portions. Consequently, taiga is found in a narrower band in the west and farther north (Figure 27–29); the band is narrower because its spread northward is ultimately limited by Arctic conditions.

In the taiga, most of the precipitation falls in the summer; the cold winter air in these regions has a very low moisture content. The rate of evaporation is low, and lakes, bogs, and marshes are frequent (Figure 27–30).

"Taiga" is a Russian word, and this biome extends over vast areas of the Soviet Union and also in North America. It is flanked on the south either by deciduous forest or grassland, depending upon the precipitation. Similar plant communities extend southward in the mountains. A number of genera of coniferous trees, such as spruce *(Picea)*, hemlock *(Tsuga)*, fir *(Abies)*, and pines *(Pinus)* (in relatively warm, dry areas) are common, with a lesser representation of willows *(Salix)* and birches *(Betula)*, particularly in moist places. Perennial herbs are common, and there are a few shrubs. Annual plants are virtually absent.

27–29
Taiga of North America.

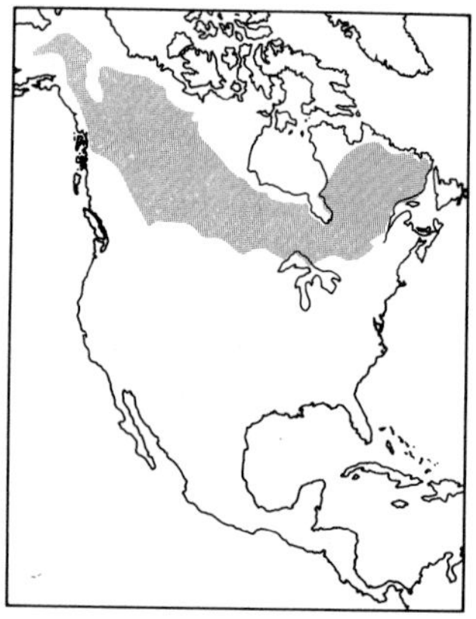

27–30
The taiga, which covers hundreds of thousands of square kilometers in the cooler part of the north temperate zone, is dominated by white spruce (Picea glauca) *and tamarack* (Larix laricina), *which become smaller as they extend northward. This photo was taken in North Manitoba, Canada.*

At its northern limits, taiga grades unevenly into tundra. In both these biomes, there is an abundant supply of light during the relatively short growing season. Cultivated plants may grow very rapidly, attaining large size in a very short period of time.

Tundra

This treeless region, which extends to the farthest limits of plant growth, is an enormous biome occupying one-fifth of the earth's land surface. It is best developed in the Northern Hemisphere and is mostly found north of the Arctic Circle, although it extends farther south along the eastern sides of the continents (Figures 27–31 and 27–32). The tundra essentially comprises one huge band across Eurasia and North America, with more or less separated outposts southward in the mountains (Figure 27–33). Some species of plants that occur in this biome have wide circumpolar ranges.

In general, permanent ice, or permafrost, exists within less than a meter of the surface, and, because the water cannot percolate through the ice, ground conditions in the tundra are usually moist. The precipitation is usually less than 25 centimeters per year, but much of this water is held in the surface layers of the soil and accumulated vegetation. Nitrogen is generally present in very short supply. Evaporation is slow owing to the low moisture-holding capacity of the air at low temperatures, but some areas are so dry as to constitute true polar deserts.

27–31
Tundra of North America.

27–32
(a) *Tundra in Mt. McKinley National Park, Alaska, in autumn. Although precipitation is limited, tundra soils are usually moist. This is because the permanent ice a short distance below the surface traps the water at the surface and because the evaporation rates in the cool northern climates are low.* (b) *Crustose lichens on a rock in the tundra of interior Alaska.*

(a)

(b)

27–33
Tundra in Alpine Meadows, Logan Pass in Utah. Such mountain tundra is comparable in many respects with that found miles to the north in the arctic. Here, however, forested slopes are found within a hundred meters or so. There is much more snow than in arctic areas, and the relatively long, cold nights of the alpine summer do not occur in the arctic, where there is continuous daylight during the growing season.

There are approximately 235,000 species of flowering plants in existence today. About a third (85,000) of these are native to temperate regions, the balance are found in the tropics. A vast number of tropical plants are in danger of extinction in the wild within the next hundred years because the human populations of most tropical countries continue to double every 20 to 25 years and the forests are being cleared for wood and marginal cultivation. About 24 percent of the Amazon forest had been cleared by 1975, and at least 25 million additional acres are being cleared each year. We know so little of the plants of the tropics that many have not even been given a scientific name. Whatever samples of these plants are preserved may well be all we shall be able to pass on to our descendents in the 21st century and beyond.

In temperate regions, about 5 percent of the 85,000 native species are in current danger of extinction. Habitat destruction is only one problem. Overgrazing by domestic animals, use of fertilizers and herbicides, introduction of foreign plants without their natural controls, and destruction of insect, bird, and and bat pollinators all can endanger plants.

Of the approximately 20,000 species, subspecies, and varieties of native higher plants of the continental United States, at least ten percent warrant concern, according to a study completed in 1974 by the Smithsonian Institution. About 100 species are recently (within the last 200 years) extinct or presumed extinct, about 750 endangered (currently in danger of extinction throughout all or a significant portion of their range), and more than 1200 threatened (likely to become endangered within the foreseeable future).

The Tiburon mariposa lily *(Calochortus tiburonensis)* is restricted to a single hilltop on the Tiburon Peninsula which extends into the northern portion of San Francisco Bay. Discovered for the first time about 1970, this lily provides an excellent example of the extreme restriction of plant species which is characteristic of regions with a mediterranean climate. *(Sherwin Carlquist)*

Extinct in the wild by 1806, *Franklinia alatamaha*, a handsome tree which belongs to the camellia family, was discovered in 1765 on two acres near the Alatamaha River in Georgia and named in honor of Benjamin Franklin.

Hawaii's *Rollandia angustifolia*, of the lobelia family, is pollinated by birds whose curved bills match the curve of the flowers. More than half of the approximately 2200 species of native plants in the Hawaiian Islands are endangered or recently extinct. The plants of islands all over the world are in particular danger of extinction, having evolved in isolation and having few natural defenses against aliens.

For plant growth to occur, the mean temperature must be above freezing for at least 1 month of the year. The growing season in many areas of tundra is less than 2 months. A wide variety of perennials occur, but there are almost no annuals and few woody plants. Vegetative propagation is characteristic of many of the perennials, and this may be correlated with the uncertainties of setting seed in the brief Arctic summer. Large woody plants are absent, probably because of the low temperature and permafrost, which permits only shallow root penetration; most of the volume of tundra vegetation is underground, as much as 96 to 98 percent in some plant communities.

A COMPARISON OF LIFE ON LAND AND IN THE SEA

Although the sea occupies approximately 71 percent of the earth's surface, only about 16 percent of the species of animals and perhaps 4.5 percent of the species of photosynthesizing organisms (plants and algae) occur there. This is true even though there are more major groups of organisms in the sea than on the land. The land is inhabited by many types of organisms belonging to relatively few major divisions.

It would appear, therefore, that terrestrial organisms occupy narrower and more sharply defined habitats than marine ones. When moisture is limiting, the effects of all other ecological variables that serve to prevent the spread of organisms seem to be enhanced. Further, terrestrial habitats, being sharply delimited, are so diverse that the specific characteristics required for the efficient occupation of a given habitat by a particular ecotype seem to preclude its penetration into another, slightly different habitat. A few groups of organisms—notably insects and angiosperms—have produced a great many species on land.

The oceans have been inhabited by living things for far greater periods of time and have been the scene of the evolution of most major groups. Only a few of these seem to have been able to meet the requirements for a terrestrial existence. Once these groups diversified and came to occupy most of the available habitats on land, there was even less opportunity for additional marine groups to spread into this new environment. The ancestors of the terrestrial vertebrates, the insects, and the vascular plants doubtless replaced other, early terrestrial organisms that proved less successful than themselves.

SUMMARY

Ecology is the study of the interactions of organisms with one another and with their environment. As a science, it seeks to discover how an organism affects and is affected by its environment and to define how these interactions determine the kinds and number of organisms found in a particular place and time.

The populations of plants and animals in an area constitute a community. Together with the nonliving factors in their environment, they form an ecosystem. Major groupings of ecosystems can be called biomes.

The environment has played a crucial role in establishing the diversity of organisms. Nearly all energy used in biological systems comes ultimately from the sun, and the amount of solar radiation that reaches different portions of the earth varies greatly from time to time and from place to place. Organisms living away from the equator have to adjust to a highly variable set of conditions.

Climatic differences are also produced by the patterns of circulation of the atmosphere and its water content, as well as those of the ocean. In the ocean, the major patterns of warm and cold currents profoundly modify the distribution of life, not only in the ocean itself but also on the adjacent continents.

In the ocean, photosynthesizing organisms occur only in the uppermost zones, rarely below 30 meters in depth. Most of the photosynthesis in the sea takes place in the plankton, where diatoms and other golden-brown algae predominate, together with dinoflagellates.

In freshwater habitats, there are distinct differences between the communities of running water and those of standing water. The former depend on a continual influx of organic material from neighboring terrestrial communities, and the organisms that occur in running water need efficient anchoring mechanisms. In ponds and lakes, rooted plants occur around the margins and often emerge above the surface of the water, and planktonic forms occur in the illuminated zone of the center. The spatial and temporal characteristics of lake life are closely related to a complex pattern of thermal stratification and seasonal vertical circulation of lake water.

Terrestrial biomes can be divided into seven major groups. Richest is the tropical rain forest, where neither water nor low temperatures are limiting. Here the trees are evergreen and characterized by large leathery leaves. There are many vines and epiphytes, but a poorly developed layer of herbs on the forest floor. Tropical soils often are red clays known as latosols, which may erode or solidify and promptly become rocklike when the forest is cleared.

Those tropical communities characterized by a seasonal drought are termed savannas. Away from the equator, savanna intergrades into desert, which is characterized by very low precipitation and high daytime temperatures.

The major grasslands of the world lie between deserts and deciduous woodland. They are a fertile region for agriculture, and in the past they often supported large herds of grazing mammals now seriously depleted by the activities of man.

In the temperate deciduous forest, most of the trees lose their leaves during the cold, often snowy winters when moisture may be unavailable for growth. Communities of similar origin, now less rich in kinds of woody plants, are the coniferous forests of western North America and the evergreen and summer deciduous scrub communities found in those widely scattered areas with a summer-dry mediterranean climate.

The taiga is a vast northern coniferous forest that extends in an unbroken band across Eurasia and North America. At its northern limits, taiga intergrades into tundra, a treeless region that also extends around the Northern Hemisphere, mostly above the Arctic Circle, in a band broken only by water.

Comparing life on land with that in the sea, we find that only about 16 percent of animal species and perhaps 4.5 percent of the species of photosynthesizing organisms (plants and algae) are marine, even though the sea occupies about 71 percent of the earth's surface. The relative scarcity of marine species appears to be a reflection of the much less sharply defined habitats in the sea. Yet, more major groups are found in the sea than on land, probably because they evolved there. Only a few have been able to send successful colonists onto the land, but several of these—notably the insects and the flowering plants—have attained a truly spectacular level of diversity.

CHAPTER 28

The Dynamics of the Ecosystem

28–1
A stream is a heterotrophic ecosystem, because its communities of organisms do not produce all their own food. They depend on a constant supply of organic material from adjacent terrestrial areas.

The intricate and ever-changing relationships that exist among the vast multitude of living things can be given a unifying perspective by viewing them in the context of an ecosystem. In an ecosystem, organisms and their nonliving environments are regarded either as temporary carriers of the relatively fixed supply of nutrients that continually recycle through the ecosystem, or as waystations in the regulated transfer and eventual dissipation of the energy originally stored in organic molecules by the process of photosynthesis. In this chapter, we shall consider some of the dynamics of ecosystems: how they are organized, the efficiency with which they make use of the available nutrients, and how they achieved maximum stability through successional development.

THE CYCLING OF NUTRIENTS

In terms of its nutrient supply, an ecosystem is more or less self-sustaining, and one of the most important reasons for this autonomy is the continuous cycling of chemical elements between its organisms and the environment. The paths of some of these elements, known as biogeochemical cycles, were discussed in Chapter 26. Ideally, nothing is removed from the ecosystem and nothing is used up, so that the pool of nutrients is continually renewed and continually available for the growth of organisms. The rate of flow from the nonliving pool to the organisms and back again, the amount of available material in the nonliving pool, and the form of this pool all differ from element to element.

Recycling in a Forest Ecosystem

As we noted in the preceding chapter, most of the nutrients of the tropical rain forest are present in the bodies of the plants, animals, and other organisms that make up the rain forest community, and little is present in the soil. Studies of a deciduous forest ecosystem have shown that the plant life of this community also plays a major role in retaining the nutrient elements. The studies were made in the Hubbard Brook Experimental Forest in the White Mountain National Forest in New Hampshire (Figure 28-2). The investigators first established a procedure for determining the mineral budget—input and output, "profit" and "loss"—of areas in the forest. By analyzing the content of rain and snow, they were able to estimate input, and by constructing concrete weirs that channeled the water flowing out of selected areas, they were able to calculate output. (A particular advantage of the site is that bedrock is present just below the soil surface so that little material leaches downward.) They discovered, first, that the natural forest was extremely efficient in conserving its mineral elements. For example, the annual net loss of calcium from the ecosystem was 9.2 kilograms per hectare. This represents only about 0.3 percent of the calcium in the system. In the case of nitrogen, the ecosystem was actually accumulating this element at a rate of about 2 kilograms per hectare per year. There was a similar, though somewhat smaller, net gain of potassium.

In the winter of 1965–1966, all of the trees, saplings, and shrubs in one 15.6-hectare area of the forest were cut down. No organic materials were removed, however, and the soil was undisturbed. During the following summer, the area was sprayed with a herbicide to inhibit regrowth. During the four months from June through September 1966, the runoff of water from the area was 4 times higher than in previous years. Net losses of calcium were 10 times higher than in the undisturbed forest, for example, and potassium 21 times higher. The most severe disturbance was seen in the nitrogen cycle. The tissues of dead plants and animals continued to be decomposed to ammonia or ammonium, which then were acted upon by nitrifying bacteria to produce nitrates, the form in which nitrogen is usually assimilated by higher plants. However, no higher plants were present, and the nitrate, a negatively charged ion, was not held in the soil. Net losses of nitrate nitrogen averaged 120 kilograms per hectare per year from 1966 to 1968. As a side effect, the stream draining the area became polluted with algal blooms, and its increase in nitrate concentrations came to exceed the levels established by the U.S. Public Health Service for drinking water.

28-2
Weir at Hubbard Brook Experimental Forest in New Hampshire. Water from each of six experimental ecosystems was channeled through a weir, such as this one built where the water leaves the watershed, and was analyzed for chemical elements. The watershed behind the weir in this photograph has been stripped of vegetation. The experiments showed that such deforestation greatly increased the loss of elements from the system.

TROPHIC LEVELS

Any ecosystem includes, in addition to its abiotic, or nonliving, components, at least two biotic elements: autotrophs and heterotrophs. As we have seen, autotrophs are mainly photosynthesizing organisms, which are able to use light energy to manufacture their own food. Heterotrophs cannot manufacture their own food and use the organic molecules made by the autotrophs as a food source. Several feeding levels are recognized among the heterotrophs. First, there are the *primary consumers*, or herbivores (animals that eat plants). Second, there are the *secondary consumers*, or carnivores and parasites, which feed on the primary consumers. Finally, there are the *decomposers*. All these levels are represented in any fairly complicated ecosystem.

The levels we have just discussed are called *trophic* levels (feeding levels) and describe one sort of structure that is apparent in an ecosystem. In a given ecosystem, organisms from each of these levels make up what is called a *food chain* (Figures 28-3 and 28-4). The relationships between the organisms involved in such a food chain regulate the flow of energy through the ecosystem, and the length and complexity of such food chains vary a great deal. Usually an organism has more than one source of food and is itself preyed upon by more than one kind of organism. Under these circumstances, it is more appropriate to speak of a *food web* (Figure 28-5). The complexity of such relationships has important implications for the overall properties of the ecosystem, as we shall see later in this chapter.

28–3
A marine food chain. (a) *Krill* (Euphausia superba) *spilling out of the opened stomach of a 22-meter blue whale* (Balaenoptera musculus) *on the deck of a factory ship in the Antarctic. This whale's stomach was estimated to contain about a metric ton of krill. The two fish in the center of the pile of krill were also feeding on the krill. The handle of the flensing knife is 1.5 meters long.* (b) *Krill, marine crustaceans that are abundant in Antarctic waters and that provide the chief source of food for the blue whale.* (c) *Marine plankton, the food of krill. This sample includes four species of diatoms, two species of dinoflagellates, and a copepod crustacean.*

(a)

(b)

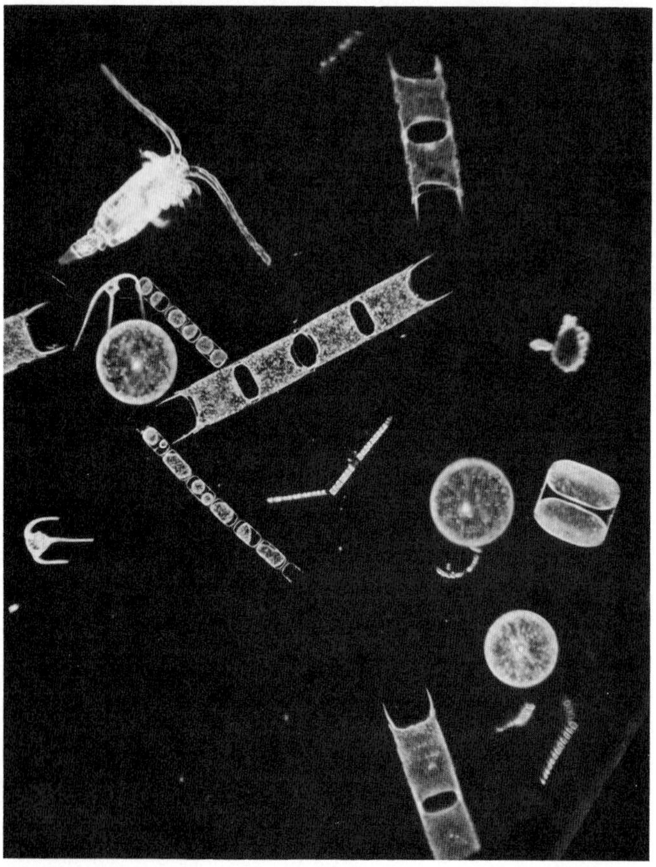

(c)

28–4
A terrestrial food chain. A three-toed box turtle feeds on a snail that has fed on the mushrooms that are decomposing organic matter in the soil.

28–5
The feeding relationships of the adult herring (Clupea harengus) *in the North Atlantic, illustrating the complex interrelationships of a food web.*

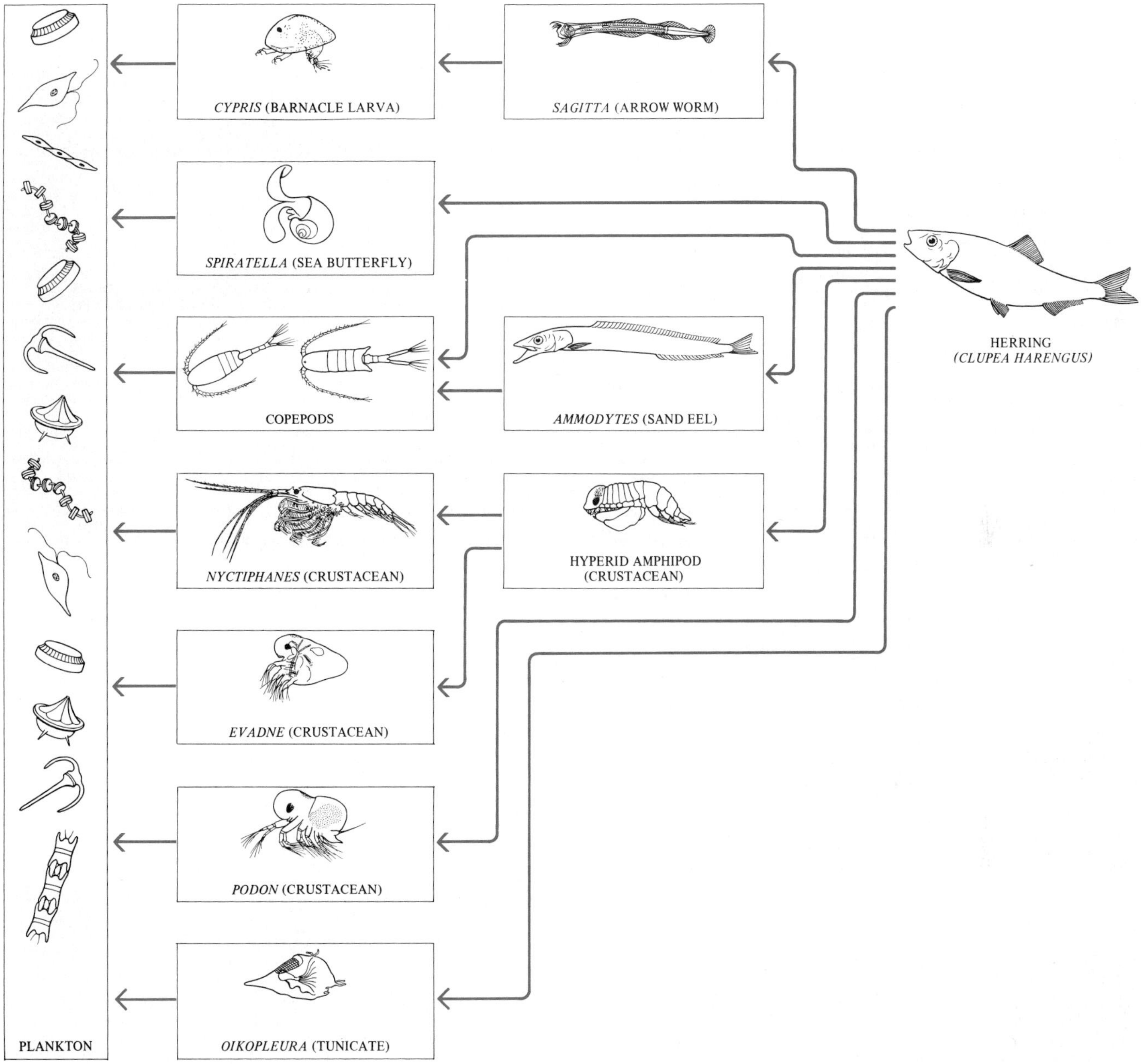

The Flow of Energy

In an ecosystem, the flow of energy starts with photosynthesis and the production of carbohydrate molecules. Looking at this process in another way, when a photon of light strikes a chlorophyll molecule, it raises an electron to a new, higher-energy orbit, from which the electron drops back, usually almost instantly, to the ground state. Photosynthetic systems use various chemical strategies to delay the electron's return to the ground state. The energy gained thereby is used to manufacture carbohydrate molecules. Thereafter, there is a slow release of the energy incorporated into the carbohydrate molecules; this release is mediated by the ATP-ADP system.

Energy does not cycle in the ecosystem; it enters and is gradually dissipated. A large proportion is lost at each step of a food chain, and we shall consider these steps sequentially.

To begin with, a very large amount of biomass is produced each year. (Biomass is a convenient shorthand term meaning organic matter; it includes woody parts of trees, stored food, bones, and so on.) Current estimates are that some 110 to 120 billion metric tons of organic material are produced on land each year and about twice that much in the sea. Despite this enormous total figure, however, photosynthesizing organisms are not really very efficient in converting the sun's energy into organic compounds. Generally, only about 1 percent of the light that falls on a plant is actually utilized. For very productive stands of vegetation, 1 to 3 percent of the annual incident radiation may be converted to chemical energy.

When the organic material produced by plants is consumed by a herbivore, energy is released. Most is lost as heat. A fraction is converted to animal tissue. In general, only about 10 percent of the usable energy of the plant becomes usable energy at the next level. Approximately the same relationship is true at each succeeding level. Thus, if an average of 1500 calories of light energy per square meter of land surface is utilized by plants per day, about 15 calories are converted to plant material. Of these, about 1.5 calories are incorporated into the bodies of the herbivores that eat the plants and about 0.15 calories into the bodies of the carnivores that prey on the herbivores.

To give a concrete example, Lamont Cole of Cornell University has calculated that in Cayuga Lake, for each 1000 calories of light energy utilized by algae, about 150 calories are reconstituted as small aquatic animals and 30 calories as smelt. If we were to eat these smelt, we would gain about 6 calories from each original 1000 contained in the algae. But if trout eat the smelt and we then eat the trout, we gain only about 1.2 calories from the original 1000 calories. From this example it is clear that there is more energy available to us if we eat smelt rather than the trout that feed on the smelt; yet trout are considered a delicacy, smelt a coarse fish. Under conditions of starvation, humans must turn to an all-plant diet, not being able to afford the 10-fold loss in energy that occurs when these plants are fed to animals. For us to make the maximum use of the solar energy trapped by plants, we must become mainly herbivores.

Food chains are generally limited to three or four links; the amount of food remaining at the end of a long food chain is so small that few organisms can be supported by it. Body size also plays a role in the structure of food chains. For example, an animal constituting one link generally has to be large enough to capture prey on the next lower link of the food chain.

Owing to the relationships just discussed, the total biomass at successive trophic levels in an ecosystem decreases sharply, setting up the sort of relationship described by the expression "pyramid of mass" (Figure 28–6). If energy is measured, it follows the same rapid decrease characteristic of mass; there is far less energy in the bodies of all the predators present in a given community, for example, than in the plants. In general, there are also far more individuals at the lower levels than at the higher levels, a "pyramid of numbers." From this it also follows that if we divide all the organisms in an ecosystem into size classes, the small animals will be far more numerous than the large ones. We saw earlier that despite the origin of eukaryotic organisms about 1.3 billion years ago, multicellularity did not come into existence until about 700 million years ago. Thereafter, the green algae, brown algae, red algae, and multicellular animals appear in the fossil record during the relatively short interval, geologically speaking, of perhaps 50 million years. Their evolution may have been triggered by the sudden proliferation of complex food webs. Among the structures evolved by organisms at that time, prob-

28–6
Pyramids of mass (a), *energy* (b), *and numbers of organisms* (c) *in various communities. A relatively small amount of mass or of energy is transferred to each successively higher level.*

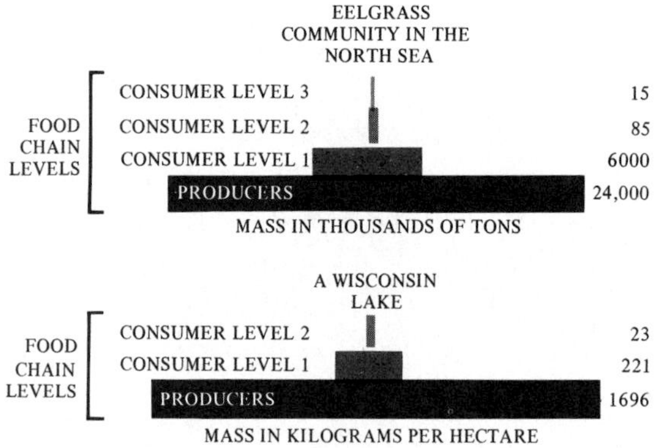

(a)

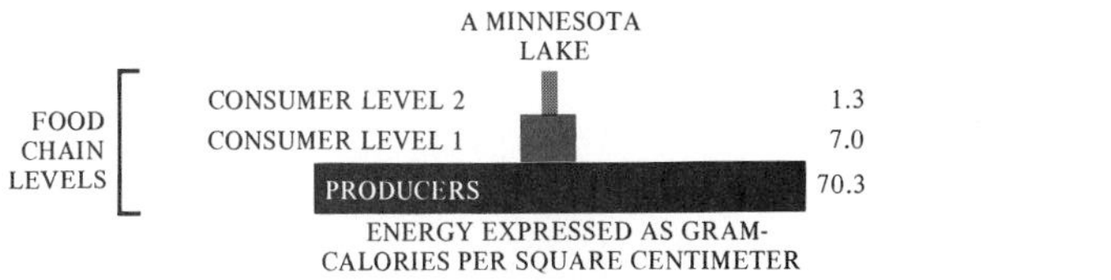

(b)

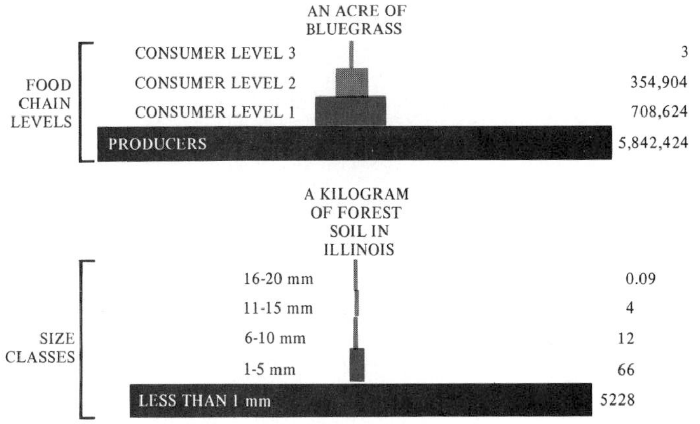

(c)

ably in relation to competition with other newly evolved forms of life, were skeletons and other supporting structures, which made the groups in which they occur more likely to appear in the fossil record.

THE DEVELOPMENT OF ECOSYSTEMS

Succession

Some plant communities remain the same year after year, whereas others change rapidly. A cleared wood lot is rapidly colonized by the trees of the vicinity, and pasture gives way to forest. Similar series of events happen in many naturally open areas, such as lakes, meadows, or rocky hillsides. All of these are examples of *succession*, a process that is continuous and world-wide in scope. Ecosystems tend to become more and more complex over time.

Succession occurs at a variable rate in all temporarily open areas. Some ponds, for example, fill with aquatic plant remains and debris, emergent vegetation builds soil, the site is taken over by meadow, moisture-loving shrubs may come in, and finally the forest characteristic of the region develops in the meadow that was formerly a pond (Figure 28–7). On dry rocks, lichens will grow. The lichens break down the rocks directly because of the chemicals they secrete, and accumulate soil around their bases; they then give way to mosses, and finally to flowering plants (Figure 28–8). The roots of the flowering plants will probe cracks in the rock, breaking them further, and eventually, perhaps after many centuries, the rock will be reduced to soil and the soil will be occupied by the forest or other vegetational types characteristic of the region.

(a)

(b)

(c)

28–7
(a) *Emerging vegetation grows along the edge of a pond.* (b) *Plants with surface-floating leaves grow across the surface of a pond, choking out bottom-dwelling plants.* (c) *Marsh grasses, sedges, and cattails growing on an old pond bed.*

28–8
An early stage of succession on a rocky slope. Lichens have begun to accumulate soil, and a small fern has sprung up in a small crevice.

(a)

(b)

28–9
(a) *Seedling trees of balsam fir* (Abies balsamea) *growing under and replacing quaking aspen* (Populus tremuloides) *in northern Minnesota—a stage in forest succession leading to a climax community of white spruce and balsam fir.* (b) *Seedling of a red maple* (Acer rubrum) *rising above needles of white pine* (Pinus strobus). *Mature white pines filter the light so that their own seedlings cannot survive and only those tolerant of shade, such as those of the maple and oak, can gain a foothold.*

The Climax Community

The various stages of succession an ecosystem undergoes finally produce a mature ecosystem, or *climax community*, the nature of which varies according to the climate of the area. A climax community is relatively self-sustaining; in it, organisms have their most intricate relationship. Each climax community constitutes part of a perfectly self-contained ecosystem driven by energy from the sun (Figure 28–9).

Some aspects of succession and climax have great significance for man. For example, when Western European settlers first came to California in large numbers, they found a magnificent forest of sugar pine along much of the length of the Sierra Nevada. Later, although conservationists tried to preserve some of these forests in national parks and forests, many of the stands of pines were eventually replaced by other trees, such as white fir, *Abies concolor,* and incense cedar, *Calocedrus decurrens.* Why did this change take place?

28–10
(a) *When fire sweeps through a forest, secondary succession—with regeneration from nearby unburned stands of vegetation—is initiated. Some plants produce sprouts from the base, others seed abundantly on the burn. In one group of pines, the closed-cone pines, the cones do not open to release their seeds until they have been burned.* (b) *Sugar pines* (Pinus lambertiana) *in the southern Sierra Nevada of California. With the control of forest fires, these trees are being replaced by others such as the incense cedar* (Calocedrus decurrens), *the large tree on the right.*

(a)

(b)

The answer is that the sugar pine was a member of a climax community before the advent of a large human population. Thereafter, the forest that contained it was no longer a climax community for the region. The variable was fire, which was greatly reduced after the influx of human inhabitants. Without periodic, lightning-set fires of low intensity raging through the groves, a thick growth of brush and smaller trees arises and prevents reproduction of the sugar pines. Only a system of controlled burning can now preserve the few remaining groves of sugar pine (Figure 28–10).

Recolonization

When man alters a landscape, changes begin to take place. Yucca Flat in southern Nevada has been the site of a number of nuclear detonations with an energy yield of about 40 kilotons at a height of about 100 meters above the ground. Such a blast eliminates all of the above-ground portions of the desert scrub vegetation to a distance of about a kilometer from ground zero. Within this area various perennial herbs and bunchgrasses started to appear in the first few years after the cessation of testing (1958), and one native annual herb, *Mentzelia albicaulis,* extended its range to within 150 meters of ground zero. Interestingly, tumbleweed, *Salsola kali,* an introduced annual from central Eurasia, has occupied the entire area of the blast in profusion from the first growing season onward. Successional processes may gradually restore the original vegetation to the area if there is sufficient time.

28–11
Old field in Wake County, North Carolina, with a young stand of loblolly pine (Pinus taeda). *The plow furrows of the abandoned field are still visible.*

28–12
Sand dune near Lake Michigan with beach grass in the foreground. Later successional stages can be seen on the more remote dunes.

In old fields that have been abandoned, denuded sand dunes, and on the streets of the ghost towns of the American West, succession is taking place and tending to produce ecosystems that more and more closely resemble those of adjacent, less-disturbed areas (Figures 28–11 and 28–12). Following natural disasters, recolonization produces similar changes. For example, in August, 1883, a violent explosion destroyed half of the island of Krakatau, in the Java Straits about 40 kilometers from Java, and covered the remaining half beneath a layer of pumice and ash more than 30 meters thick. The neighboring islands of Verlaten and Lang were also buried, and the entire assemblage of plants and animals on these islands was wiped out at a single stroke. Soon afterwards, however, the recolonization of the island began, and the expected number (based on the number originally occupying the area) of about 30 species of land and freshwater birds was reached in about 30 years. Recolonization by plants seems to have been even more rapid, with a total of over 270 species being recorded for Krakatau by 1934.

Volcano Paricutín, in the Mexican state of Michoacán, erupted violently on February 20, 1943. During the period of the eruption, all vegetation was destroyed over an area of some 13,000 hectares. After the eruption ceased on March 4, 1952, lichens, algae, and mosses were growing on the lava flows within 3 years; ferns a year later; and flowering plants appeared on the rim of the crater within 5 years. It is likely that complete weathering of the volcanic rock and reforestation may take several centuries or more. Figure 28–13 shows another example of the results of a volcanic eruption.

28–13
Yapoah Crater, a volcanic cinder cone east of the Cascade Mountains in central Oregon. Succession leading to the establishment of a climax forest on such a cone may take centuries and may often be interrupted by further volcanic activity long before it is complete.

Structural Changes During Succession

A number of important changes in the structure of an ecosystem take place during the course of succession. First, the kinds of plants and animals change continuously. Under dry conditions, as on a bare, rocky hillside, the first organisms to become established are those that are able to stand up under all environmental conditions. These include the lichens, such plants as mosses, and such animals as mites, arachnids, and small insects. Appropriately, these are called "pioneer organisms," and they are said to form a "pioneer community." In general, they are relatively simple organisms, and they form short food chains—ones with few links. They depend on disturbance for their continued existence in the area. They are not present in the intermediate or final stages of succession. Often the plants, animals, and other organisms characteristic of a particular successional stage are present only at this stage.

Second, the amount of organic matter in the community that is incorporated in living organisms or the remains of living organisms increases during the course of succession. Larger and more complex organisms appear and food chains lengthen. Some of the new organisms persist for a long time in the community, and others are relatively short-lived. The kinds of interrelationships between organisms that will be discussed in the next chapter become more and more complex.

Third, the overall diversity of species, particularly of heterotrophic groups of organisms, tends to increase with succession. The relationships between these are more and more difficult to discern, and the feeding range of each species is apt to become narrower and narrower as the community approaches its mature condition.

Finally, autotrophic organisms such as plants occupy any site more rapidly than do heterotrophic organisms. Therefore, in any successional series, the maximum production for the site is reached relatively early. Afterwards, respiration (for the entire ecosystem) continues to increase, and the net production for the ecosystem decreases as the proportion of the heterotrophs and the complexity of the food chains increases.

SUMMARY

The ecosystem is the highest level of biological integration; it is a self-sustaining system driven by energy from the sun, in which the regulated cycling of essential materials takes place.

An ecosystem consists of nonliving elements and at least two kinds of living ones—autotrophs and heterotrophs. Among the latter are the primary consumers, or herbivores; the secondary consumers, or carnivores and parasites; and finally, there are the decomposers. The organisms found at these levels are members of food chains or food webs.

Energy flows through an ecosystem in an orderly manner, with about 1 percent of the incident light being, converted into chemical energy by green plants. When these plants are consumed, about 10 percent of their potential energy reaches the next trophic level, and a similar efficiency characterizes transfers further up the food chain. The amounts of energy remaining after several transfers are so small that food chains rarely exceed three or four links in length.

When an area is denuded by artificial or natural means, or a new area, such as a lava bed, appears, succession takes place. The kinds of plants and animals change continuously, some being characteristic only of early, others only of middle, stages of succession. The amount of organic material that is incorporated into organisms increases during succession, as the biomass of the community grows rapidly. The diversity of species in the ecosystem increases greatly, and the relationships between them become more and more complex. Production (related to the representation of photosynthetic organisms) reaches a peak early, and then respiration increases rapidly in the later stages of succession. Eventually, succession results in the production of a climax community, which will undergo major change only if there are major environmental changes.

CHAPTER 29

The Integration of the Community

By the middle of the nineteenth century, European botanists were describing the repeatable associations of plants they found around them. The perceptive Austrian botanist Anton Kerner von Marilaun, in his classic work on the plant life of the Danube basin, stated in 1863 that "Every plant has its place, its time, its function, and its meaning—In every zone the plants are gathered into definite groups which appear either as developing or as finished communities but never transgress the orderly structure and correct composition of their kind." The work of Kerner and his European successors gave rise to the concept of the *community* as a unit of organization with evolutionary importance. They felt that the repeatable associations of plants that were observed functioned almost as "superorganisms," which had, in effect, a life of their own to which their individual components were subordinated.

COMMUNITY STRUCTURE

Certain kinds of plants occur repeatedly together in similar places, such as in marshes, around small ponds, or on rocky cliffs. In its simplest sense, "community" merely means the populations of plants and animals that live together in such a specific area, and no one really objects to this definition. Yet we have to find out how a given community can be defined and delimited.

The Redwood Forest: A Plant Community

The redwood forest of coastal California and southern Oregon is an example of a plant community (Figure 29–1). In this community, the redwoods themselves create and profoundly modify the habitat for all other species. (An organism's habitat is simply its natural environment, the place where it is usually found.) Not only are the redwoods responsible for gradients of shade and moisture, but their rotting leaves and branches condition the

(a)

(b)

(c)

(d)

29–1
The redwood community. (a) *One of the principal ecological factors determining the range of the redwood* (Sequoia sempervirens) *is fog, which ensures an ample supply of water during the otherwise dry summer. Dripping from the leaves of the gigantic trees, this fog also provides moisture for the plants associated with the redwoods* (b). *At the foot of the redwood trees characteristically grow swordfern* (Polystichum munitum) (c) *and redwood sorrel* (Oxalis oregana) (d). *They are members of the "redwood community," even though their distribution is not linked with that of the redwood, as can be seen in Figure 29–2. Rather they occur together where their requirements overlap with those of the redwood. When they are associated, they form a "redwood community"; when they are not, they do not. We call this particular association of plants the redwood community, rather than the swordfern community, because the redwoods so thoroughly dominate it. In fact, it can be defined only in terms of the redwoods themselves and does not have any other sort of unity.*

ALEXANDER VON HUMBOLDT

Alexander von Humboldt (1769–1859) was perhaps the greatest scientific traveler who ever lived. Humboldt ranged widely across the trackless interior of Latin America around the start of the nineteenth century and climbed some of its highest mountains. Exploring the region between Ecuador and central Mexico, Humboldt was the first to recognize the incredible diversity of tropical life and, consequently, the first to realize just how many species of plants and animals there must be in the world.

In his travels, Humboldt was impressed with the fact that plants tended to occur in repeatable groups, or communities, and that whenever there were similar conditions—climatic, edaphic, and biotic—similar groupings of plants appeared. He also discovered a second major ecological principle, the relationship between altitude and latitude. Climbing a mountain in the tropics, he found, was analogous to passing farther north (or south), away from the equator. Humboldt illustrated this point with a famous diagram of the zones of vegetation on Mt. Chimborazo in Ecuador.

On his return from Latin America in 1804, Humboldt visited the United States for eight weeks. He spent three of these weeks as Thomas Jefferson's guest at Monticello, talking over many matters of mutual interest, and it is thought that Humboldt's enthusiasm for exploring America encouraged Jefferson's own great scheme for the exploration of the western United States. Thus, it is fitting that Humboldt's name is commemorated in the names of several counties, mountain ranges, and rivers in the American West. After returning to Berlin, Humboldt lived for more than half a century, one of the greatest writers and scientists of his era, dying in his 90th year.

soil, that is, change its characteristics, so that only certain plants and animals can live in it. These redwood-dominated plants and animals form associations that appear much the same, whether they occur in southern Oregon or in central California. But when we analyze the situation in detail, some startling differences become apparent.

Herbert L. Mason of the University of California plotted the distribution of the species associated with the redwood and found that not a single one even approximated a complete coincidence of geographical area with that occupied by the redwood (Figure 29–2). Most of the associated plants ranged widely beyond the area of the redwood. A few species were found only within a portion of the redwood's range. Thus, the unity of the redwood community is based solely upon one species—the redwood itself, *Sequoia sempervirens.* This community comprises a number of coincidently distributed species whose genetic variability has been such that they all evolved ecotypes that could grow in this particular environment. (Some of the very different ecological requirements of the associated species are shown graphically in Figure 29–3.)

If we add the historical dimension, the picture becomes even more complex. A number of different evolutionary lines have evolved, diverged, come together, and run parallel to one another, finally producing the assemblage of species that occur together in the redwood forest today. Some of the evolutionary lines, like that including the redwood itself, have changed very slowly; others have been modified rapidly. Thus, if we consider the plants that were associated with the redwood in the early Tertiary period, we find that they were very different from those associated with it in the middle Tertiary, and different again from those that grow in these forests at present.

In summary, we can view the redwood community either as it exists at the present day or in its historical dimension. In either case, the only property that can be used to *define* the community is the occurrence of the redwood itself. So defined, the term "redwood community" is a precise concept. Defined as an aggregation of particular species in particular geographical areas, however, the concept becomes equivocal and imprecise.

29–2
Geographical distributions of "members of the redwood community."

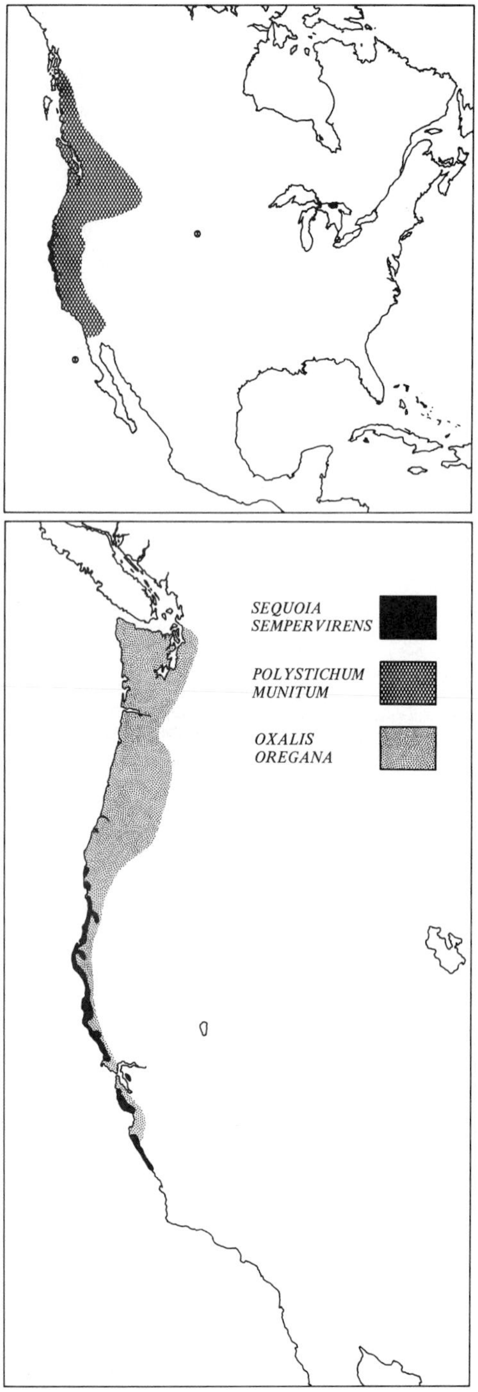

29–3
There are many kinds of environmental gradients that determine whether species are able to coexist with the redwood (Sequoia sempervirens), *and that may play a role in determining the rest of their range. In each of the three graphs presented here, the amount of ground each plant species covers is taken as an index of its response to an environmental gradient. In* (a) *the gradient is expressed in terms of minimum available moisture, which is related to the moisture-retaining capacity of the soil. The Pacific madrone* (Arbutus mensiesii) *and the bigleaf maple are compared with the redwood. In* (b), *the gradient is expressed in terms of available light, which, in the redwood community, is determined to a great extent by the shade cast by the redwoods.* Asarum caudatum *(wild ginger) grows well under redwoods,* Eschscholzia californica *(California poppy) not at all. In* (c), *both* Arctostaphylos glandulosa *(manzanita) and* Corylus cornuta *(hazel nut) can grow with the redwood in this range of calcium concentration. However, manzanita is able to grow better where concentrations of calcium are less.*

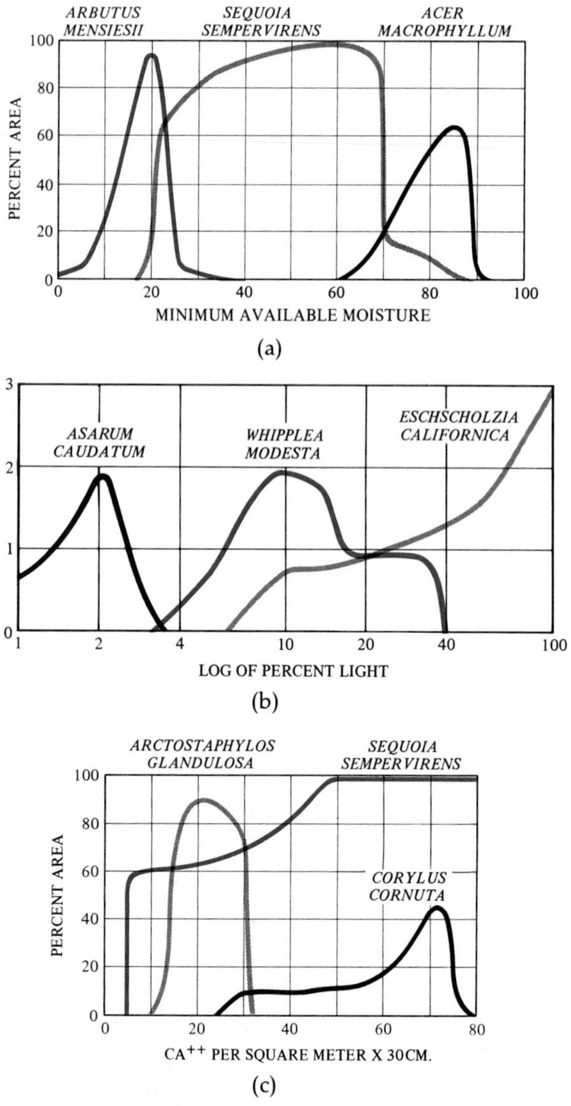

COMMUNITY ANALYSIS

The difficulty in defining a particular community increases when that community is not dominated by a species as conspicuous as the redwood. Two major attempts have been made to determine whether species populations occur as groups that are discontinuous with others and recognizable as a series of discrete classes.

1. One approach examined samples of vegetation along an environmental gradient such as altitude or moisture. Thus, Robert H. Whittaker, in his detailed study of the vegetation of the Great Smoky Mountains, considered the representation of species along a moisture gradient at different elevations. One of his graphs is shown in Figure 29-4. In addition, for habitats of equivalent moisture, Whittaker analyzed the way in which species composition changed with altitude. An example of this sort of analysis is presented in Figure 29-5. From information like this, Whittaker concluded that species populations form a shifting series of combinations along various environmental gradients. He interpreted the plant associations he studied as forming a "complex and largely continuous pattern."
2. Another approach to the problem of vegetational analysis was pioneered by the late J. T. Curtis and his school of plant ecologists at the University of Wisconsin. Plant associations were studied in terms of the proportionate representation of different species in a large number of subsamples. Forest trees were scored in each sample in terms of an index value based on the density, size, and distribution of the species concerned.

   In the upland forest of southern Wisconsin (Figure 29-6), there are four common species of trees: black oak *(Quercus velutina)*, white oak *(Quercus alba)*, red oak *(Quercus rubra)*, and sugar maple *(Acer saccharum)*. Stands in which black oak is dominant were found to contain lesser amounts of white oak, red oak, and sugar maple, in that order, and similar relationships were derived for stands in which each of the other species is dominant. No matter how the data were analyzed, it was not possible to determine natural, repeatable groupings of these species. The order of species is called a vegetational continuum, and this continuum can be divided into groupings only arbitrarily. It represents a shifting series of combinations of species in a linear pattern or gradient.

As stated by Robert P. MacIntosh of the University of Notre Dame, "A community is thus visualized as a continuous, orderly pattern of species individually distributed in time and space and most effectively considered in terms of gradients, orders, and probabilities. Species

29-4
Percentage of forest made up by each of four different species of trees—yellow birch (Betula alleghaniensis), *flowering dogwood* (Cornus florida), *chestnut oak* (Quercus prinus), *and Virginia pine* (Pinus virginiana)*—along a moisture gradient in the Great Smoky Mountains.*

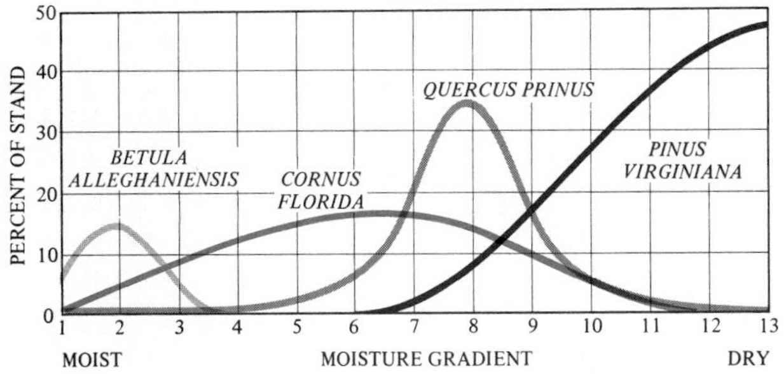

29-5
Percentage of forest made up by each of three species of pines (Pinus)*—Virginia* (virginiana), *table-mountain* (pungens), *and pitch pine* (rigida)*—and two of oak* (Quercus)*—scarlet* (marilandica) *and blackjack* (coccinea)*—growing in relatively dry sites along an elevation gradient in the Great Smoky Mountains.*

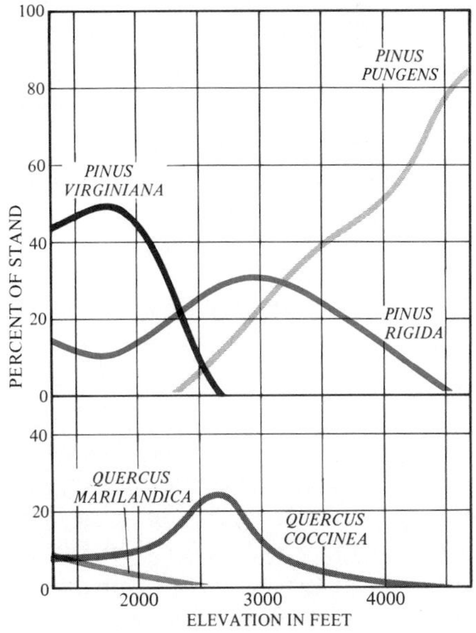

29-6
Upland forest in southern Wisconsin.

having ecological amplitudes over similar sectors of the gradient have greater probabilities of being represented in frequent combinations in nature." In the present decade, modern mathematical analysis, with the aid of high-speed digital computers, has been applied to the analysis of plant communities. This will doubtless provide a key to the study of communities in the future.

FACTORS LIMITING PLANT DISTRIBUTION

Climatic Factors

Plant distribution is primarily controlled by the distribution of climatic factors such as moisture, light, and temperature. In determining the limits of distribution, the extremes are more significant than the mean values. For example, redwoods were planted to the north of their range, where they thrived until 1932, when an abnormally cold winter caused the freezing of the young trees. Doubtless the northern limits of the redwood are determined in part by the exceptional occurrence of very cold winters which prevent the establishment of individuals north of the main range of the species.

It should be pointed out, however, that such abnormal weather conditions present exceptional evolutionary challenges. If a few redwoods were more resistant to the cold than others and could propagate themselves following the elimination of most individuals of the species, a new ecotype might become established that could extend the range of the redwood farther north. In fact, the trees in the northern groves are doubtlessly very different from those in the south. Plants that occur along an environmental gradient vary genetically and physiologically also, even though they may belong to the same species, and are likely to respond very differently to a given environmental stress.

Edaphic Factors

Differences in soils—which are called *edaphic differences*—are very important in controlling plant distribution.

The mineral content of soils, which we considered in Chapter 26, is a major factor in determining patterns of plant growth. In dry areas, where soil building proceeds relatively slowly, differences in the parent rock are

29-7
The ways in which plants interact to form communities varies with the genetic constitution of the individual populations. In the prairies of the central United States, three species of grasses regularly grow together to form a community. These are Andropogon scoparius *(left column),* Panicum virgatum *(center column), and* Andropogon gerardi *(right column). All of the plants seen in these photographs were grown in central Texas. The three strains at the top are from Devils Lake, North Dakota; the three in the center from Lincoln, Nebraska; and the three in the lower row from Austin, Texas. The most efficient use of energy input is shown by the local ecotypes from central Texas, the least efficient by those from farthest away. Although these three grasses appear to form a repeatable community that has wide geographical extent, the way they adjust to the environment and to each other's presence obviously differs greatly from place to place. Each of the three strains of each species is genetically very different from all of the others.*

reflected more or less directly in the composition of the soil. Here soil types may differ sharply within a short distance, and each may be characterized by peculiar and often restricted species of plants.

Thus, in contrast to growth factors such as sunlight and water, which generally affect large areas of plant life, soil differences, owing to differences in parent rock, may be extremely localized with sharp lines of demarcation. In fact, plants are sometimes used by geologists as indicators of mineral deposits. *Merceya latifolia* (copper moss) thrives only in soils with a high copper concentration. Iron ore deposits may cause stunting, cobalt can cause white spots on leaves, and unusually large amounts of molybdenum can make leaves turn yellow-orange.

A group of soil-rock systems known as the serpentines has been studied fairly extensively. The serpentines occur in many parts of the world, including Europe and Asia as well as North America, and although the plants that grow on serpentine soils differ in the different areas, they share three characteristics in common. (1) Serpentine soils are generally sterile and unproductive as farm or timber lands, (2) among the sparse growth on the serpentine is found a concentration of unusual flora, and (3) these unusual plants are often absent from neighboring areas.

Serpentine soils, which are composed essentially of magnesium, iron, and silicate, are often reddish brown or gray at the surface but change to yellowish or greenish at the lower levels. They often have a spotted or mottled appearance, resembling reptilian skin (hence their name). Compared with other soils, serpentine is high in magnesium, and often in chromium and nickel, and low in molybdenum, calcium, and nitrates.

For many years, physiologists were inclined to theorize that common plants would not grow on serpentine soils, because they could not tolerate the high levels of magnesium that the serpentine varieties could. But this did not explain why the serpentine varieties were not found growing in neighboring areas.

Studies with *Phacelia californica* have answered these questions, at least as far as that particular species goes. This plant, which is a perennial with small blue flowers, is common in California and is found on both serpentine and nonserpentine soil. The ecotype that grows on the nonserpentine soil is much stronger and more vigorous than the type growing on the nearby serpentine. A laboratory investigation was undertaken in which

29-8
(a) *A large digger pine* (Pinus sabiniana) *in the center of the picture stands on the dividing line between brush-covered serpentine terrain (left), and the nonserpentine hillsides covered with oak and grass.* (b) Helianthus annuus, *the common sunflower, growing on a series of soil lots artificially altered to vary in calcium content from A, in which there is a 5 percent saturation of the soil with calcium, to G, in which there is an 80 percent saturation.* (c) Helianthus bolanderi *subsp.* exilis, *a sunflower that is characteristically found on serpentine soils, growing on a similar series of soil preparations. As you can see, the serpentine-inhabiting species variety has a very low requirement for calcium compared with the closely related common species, which is one of the reasons that it is able to grow successfully on the low-calcium serpentine soil. However, on normal soils, the common sunflower grows more vigorously and would crowd out the serpentine variety. In this way, differences in soils control plant distribution.*

(a)

(b)

(c)

growth from seeds of both ecotypes was studied. On the serpentine soil, only the serpentine ecotype grew. On serpentine soil plus nitrates and phosphates, growth of the serpentine ecotype was more vigorous, but the nonserpentine ecotype still would not grow. If calcium was added, however, the nonserpentine grew; the serpentine also grew, but no better than in the nontreated soil. On nonserpentine soil, both plants germinated and grew.

In short, the critical factor is the calcium content. The serpentine ecotype can "make do" with the limited amounts of calcium in the serpentine soil and, in fact, cannot use more when more is available. The nonserpentine ecotype, with a higher calcium requirement, can grow only in the presence of additional calcium. Why then, do we find serpentine ecotypes only on serpentine soil? The answer to this question can be found in the greater vigor of the nonserpentine plants. On nonserpentine soils, the serpentine ecotypes cannot compete with the more common and more vigorous ecotypes and are quickly crowded out. Thus, they have gained an exclusive claim to the serpentine soil—but only at a price (Figure 29-8).

Biotic Factors

The genetic variability of a particular plant population sets certain limits on that population in terms of its climatic, edaphic, and biotic limits. The range of tolerance of a given kind of plant differs at different stages in its life cycle, and the stage at which it is least tolerant will, of

(a)

(b)

29–9
Factors that limit plant distribution can apply to one part of the life cycle, not to others. For example, shoestring ferns (Vittaria) *cannot usually produce sporophytes in the United States, although once the gametophytes become established, they persist and reproduce vegetatively, as shown in the photograph of this clump of gametophytes from sandstone cliffs in Tallulah Gorge, Georgia* (a). *Although gametophytes of* Vittaria *occur in suitable situations north to Virginia and Ohio, sporophytes are formed in the United States only in Florida and Hawaii. Spores taken from* Vittaria lineata *plants in Florida gave rise to the cultivated gametophytes shown in* (b). *On their margins can be seen the characteristic gemmae, strings of cells that break off to produce new plants asexually.*

course, be critical (Figure 29–9). For example, a drought of a week's duration might kill the seedlings of a certain species of plant but leave the adult individuals of the same species unaffected. If such droughts occurred periodically, the species might not be able to reproduce and would thus eventually be eliminated. In California orchards, flowers are not fertilized and therefore fruits do not develop when the blooming period is marked by rains or heavy fog, although the same degree of moisture may actually stimulate the growth of all other portions of the plant. This may be because of destruction of the pollen or because of limitation of insect activity, but, in any case, the blooming period is critical in the production of fruit.

The bristlecone pines, *Pinus longaeva,* which are found principally in the White Mountains of eastern California, include what appear to be the world's oldest living trees. A number of the individuals are more than 4000 years old. In the groves of these trees there are few seedlings, and the individuals found there are relics of earlier, more favorable periods when establishment was possible. The very long life span of this pine apparently allows it to survive over relatively long unfavorable periods, from which it is occasionally "rescued" by the occurrence of conditions under which seedling establishment is possible.

One additional example from the same mountain range will illustrate some of the ways in which different kinds of factors can interact in determining the range of plant species. In the White Mountains of California, two closely related species of daisy, *Erigeron clokeyi* and *E. pygmaeus,* were studied by Harold Mooney of Stanford University. These plants occur at elevations of approximately 3000 to 4100 meters. They are similar in growth form and general appearance. The species occur both on soils derived from dolomite, which are very light in color, and on those derived from sandstone, which are

29–10
(a) Erigeron clokeyi *(left) and* E. pygmaeus *(right), two closely related species of composite (Asteraceae) that grow together in the White Mountains of California and Nevada.* (b) *Relative abundance of* E. pygmaeus *and* E. clokeyi *on sandstone (black line) and on dolomite (colored line) along an elevational transect in the White Mountains. At 3475 meters, for example (arrow), of the total number of plants of the two species counted on sandstone, about 40 percent were* E. clokeyi *and 60 percent* E. pygmaeus. *On dolomite at the same elevation, only 25 percent of the plants found were* E. clokeyi.

(a)

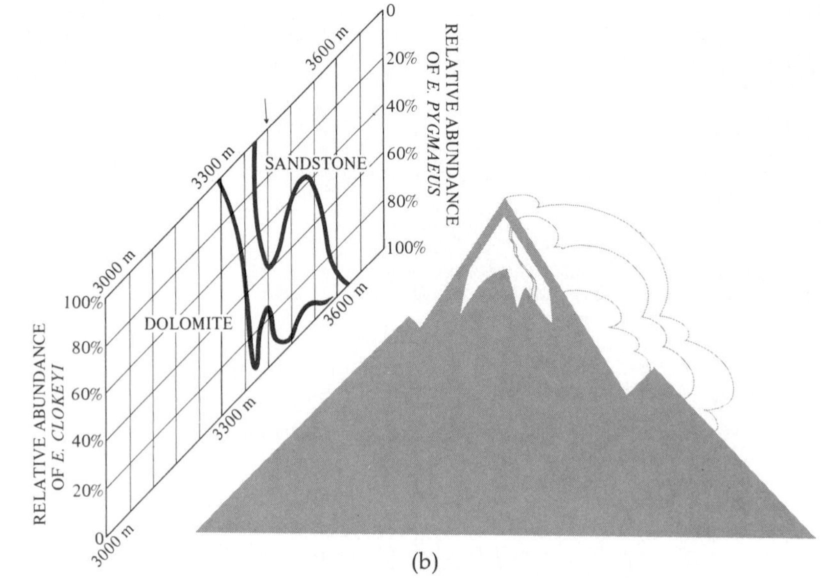

(b)

dark. Yet as shown in Figure 29–10, their abundance on these two different series of soils differs greatly at different elevations. Individuals of *E. clokeyi* on dolomite are abundant below 3400 meters and rare above this elevation. On sandstone, plants of *E. clokeyi* remain abundant to elevations of almost 3600 meters. What is the basis for this difference?

Temperatures of the dark sandstone soils are considerably warmer than those of the light dolomite ones. Plants of *E. clokeyi* on the colder dolomite soils near the upper limits of their range lag 2 to 3 weeks behind those on adjacent sandstone soils in their development. Differences of this magnitude are critical, because the White Mountain area has such a short growing season. Warm temperatures favorable for seed survival and seedling establishment in *E. clokeyi* are highly unfavorable for *E. pygmaeus.* At the lower limits of the range, periodic droughts are frequent, and *E. clokeyi* survives much better under such a regime than does *E. pygmaeus.* Thus, temperature is the primary limiting factor on the differential distribution of the two species at their upper limits; moisture, at their lower limits.

Interactions in Communities

The different kinds of interactions possible between organisms are virtually without limit, and many of them are of extreme importance in determining the distribution and abundance of the species concerned. We shall group the sorts of interactions we wish to discuss under three headings: mutualism, competition, and plant-herbivore interactions.

Mutualism

As we have seen previously, *mutualism* is a form of symbiosis in which the growth and survival of both interacting populations are benefited. Neither population can survive without the other, at least in nature. The formation of lichens (discussed in Chapter 11) is a familiar example. Another is the relationship between legumes and the nitrogen-fixing bacteria that occur on their roots. Also, some of the closely linked pollination relationships discussed in Chapter 17, such as that between the yucca moth and the yucca plant, might be described as mutualism.

One of the most interesting and significant examples of mutualism in the plant world, one that has received a great deal of attention in recent years, concerns an interaction between seed plants and fungi. Indeed, as we pointed out in Chapter 11, the roots of many, if not most, seed plants are closely associated with fungal mycorrhizae, without which normal growth of the seed plants is impossible. These associations may have played the crucial role in the first invasion of the land by plants.

The fungi that form mycorrhizal associations include a

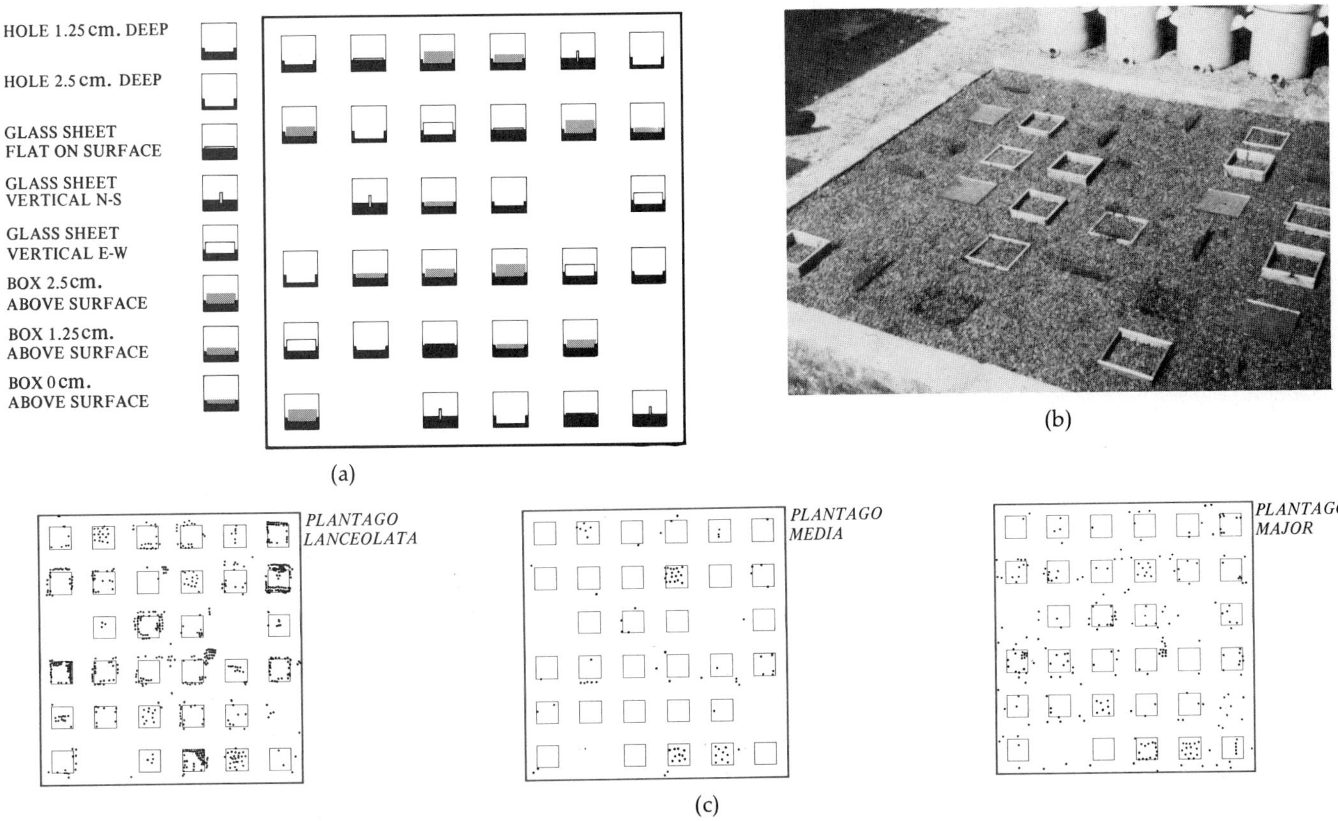

29–11
Why do different plant species grow in slightly different habitats in nature, and how can they coexist? One of the important levels of control is at the stage of seedling germination. In this experiment, a variety of artificial seed beds were prepared as described in the key (a). *As you can see in* (b) *and* (c), *the germination response of three species of plantain,* Plantago, *was very different. The experimental plot is shown in the photograph. Differences of this sort would result in different distributions for related plant species in nature. They appear to be utilizing the resources in different ways.*

great many Basidiomycetes, a few Ascomycetes, and, in endomycorrhizal associations, a number of Oomycetes. Many of the fungi are found only in mycorrhizal association, and some seem to be highly specific. The basidiomycete *Boletus elegans*, for example, is known to associate only with the larch, *Larix*. Others, for instance, *Cenococcum grandiforme*, have been recorded in association with more than a dozen genera of forest trees.

The more we learn of mycorrhizal relationships, the more important they seem to be to the higher plants. In many higher plants, uninfected individuals are rarely encountered under natural conditions, even though growth may be possible without fungi if other conditions are narrowly regulated. Most woody plants are dual organisms in the same sense that lichens are dual organisms, even if the relationship is not so obvious above ground. As University of Wisconsin soil scientist S. A. Wilde has stated, "A tree removed from the soil is only a part of the whole plant, a part surgically separated from its . . . absorptive and digestive organ." For most plants, it is the mycorrhizal fungi that play this role.

Competition

Under experimental conditions, when two kinds of plants are grown together for a long enough time in a simple environment, one is always eliminated. Under natural conditions, two species cannot coexist indefi-

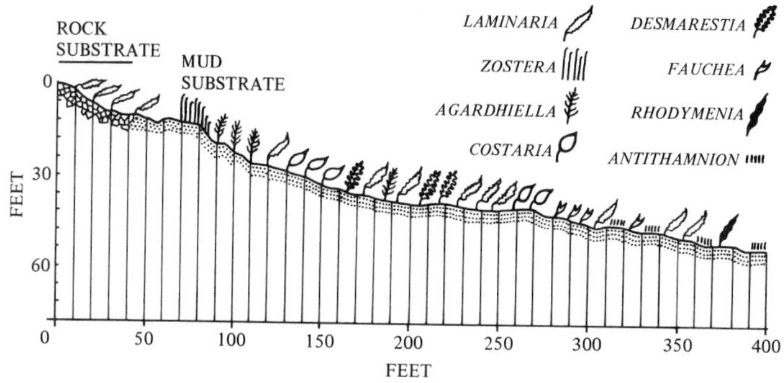

29–12
Distribution of subtidal marine vegetation along a transect near Friday Harbor, Washington. Overlapping distributions and the gradual replacement of one kind of alga by another can be seen. Marine algae are found most abundantly on solid rock where they can be firmly anchored. The larger algae are supported by gas-filled floats. In the strong currents, they form overlapping layers of vegetation, greatly increasing the shade on certain areas of the bottom. The patterns observed in these marine communities are very similar to those discussed by Whittaker for plant communities in the Great Smoky Mountains (Figure 29–4).

nitely in the same habitat. This is a simplified statement of what Garrett Hardin has aptly called the competitive exclusion principle. If they are growing together and utilizing any of the same essential resources, the individuals of one or both species will be smaller, or fewer in number, than they would be if they were growing alone. Ecologists group interactions of this sort under the general heading *competition.* If the environment is complex, as in nature, the organisms may use it in different ways, subdivide it, and then continue to coexist indefinitely.

In a bog, mosses of the genus *Sphagnum* often appear to form a continuous cover, and several species are usually involved. How can these species continue to coexist? When the situation is examined in more detail, it is found that there are semiaquatic species growing along the bottoms of the ditches and wet hollows; other species growing in drier places on the sides of the hummocks, which they help to form; and still other species growing only in the driest situations on the tops of the hummocks, where they are eventually succeeded by one or more species of flowering plants. Thus, although all the species of *Sphagnum* coexist in the sense that they are all present in the same bog, actually they occupy different habitats and they continually replace one another as the characteristics of each microhabitat change.

If the populations of coexisting species are kept at low levels, they may not eliminate one another. This effect was seen in England during the present century, when a severe epidemic of myxomatosis drastically reduced the population of rabbits. Formerly, the grassland on chalk soils was kept closely cropped by the rabbits, and a diverse assemblage of flowering plants was able to grow in this habitat. After the decline in numbers of rabbits, the grass cover of the chalk soils became deeper and more dense and many of the formerly abundant species of plants became rare (Figure 29–13). Similar effects are often seen when comparing grazed versus ungrazed pasture or grassland.

(a)

(b)

29–13
Horn Heath, an area of chalk grassland in England, before (a) *and after* (b) *the elimination of rabbits by myxomatosis. The first photograph was taken on September 13, 1954, the second on August 31, 1957.*

Unlike animals, green plants are dependent upon a single mechanism for the conversion of energy—photosynthesis. Competition in plants is manifested largely in terms of the "struggle for light," and plants that grow in the shade of others have evolved mechanisms for carrying on photosynthesis at low light intensities (Figure 29–14). Variations in plant height, arrangement of leaves, and shape of crown seem to be significant factors in the subdivision of the community's environment, whether it is low grassland or tall forest. In the tropical rain forest, for example, there are several layers of foliage, each characterized by plants in which the leaves have particular characteristics. Thus, as we have seen, the leaves of the trees in the upper strata tend to be dark green and leathery, often with abruptly pointed tips ("drip tips"). The leaves of plants growing in the dim light of the forest floor, on the other hand, tend to be larger and less leathery than leaves of the upper strata, but still have drip tips.

The importance of leaf adaptation in the struggle for light can be seen more simply in the relationship between two species of clover, *Trifolium repens* and *T. fragiferum.* These clovers were sown at two densities—3.9 and 6.9 plants per square decimeter, respectively. As they developed, the amount of light reaching ground level declined until it was about 3 percent of daylight. When the two species were grown mixed, a dense canopy of leaves of *Trifolium repens* developed first. Later, this was overtopped by the leaves of *T. fragiferum,* which had longer petioles and simply formed a layer over the top of that formed by the shorter-petioled leaves of *T. repens.* Thus, whereas *T. repens* was initially successful in mixed culture, *T. fragiferum* ultimately tended to overtake and surpass it.

Most competitive situations, however, are much more complicated. There are a number of different ways of expressing the relative success of two species growing together. Figure 29–15, for example, shows the variations in performance that occurred with changes in density of planting of a particular species of corn. With increasing density, the dry weight of the shoots of corn increased within the limits of the experiment. The dry weight of the shoots less ears increased even more rapidly than that of the shoots alone, but the dry weight of the ears decreased markedly when there were more than 8 plants per square meter. Which measures of performance would most adequately express success of the plants?

Population ecologists consider that two types of selection are operating in natural situations. In what is called *r* selection, rapid breeding and the production of many seeds is called for; in *K* selection, maximum adjustment to the environment with a relatively lower reproductive potential is needed. Plants in which *K* selection predominates are perennials of relatively closed communities; those in which *r* selection predominates are annuals or short-lived perennials that occur in fluctuating, open communities, such as disturbed ground.

29–14
Competition in plants as shown by the growth of asexually reproducing strains of floating duckweeds, Lemna gibba *and* L. polyrrhiza, *in pure and in mixed cultures.* L. gibba *replaces* L. polyrrhiza *in mixed cultures because* L. gibba *has air-filled sacs in the plant body which float it like little balloons to form a mass over the surface of the other species, shading it from the light.*

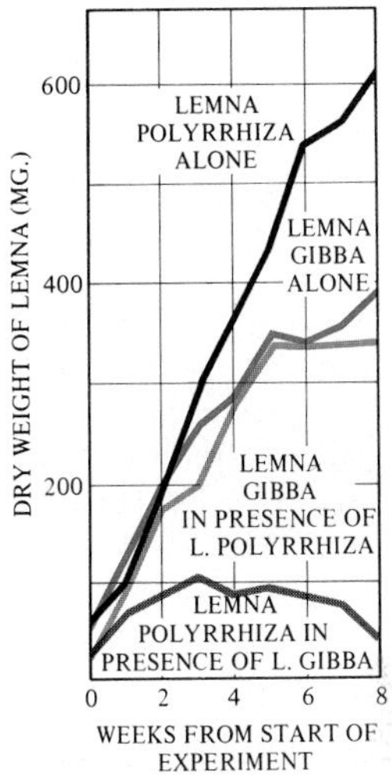

29–15
The influence of plant density on yield in corn (Zea mays) *as measured in three different ways. Thus, if ear weight is the farmer's goal, an optimum density is about 7 or 8 plants per square meter.*

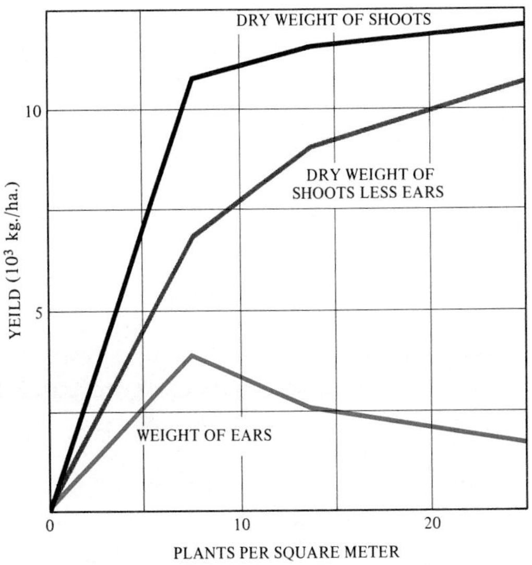

Many organisms produce chemical substances that either inhibit their own growth, resulting in increased spacing of individuals, or the growth of other species. For example, the fungus *Penicillium chrysogenum,* which grows on organic substrates such as seeds, produces significant quantities of penicillin in nature. Eventually, however, it is often replaced by bacteria, such as *Bacillus cereus,* that produce penicillinases (enzymes that break down penicillin).

More complex relationships are evident in the higher plants. In coastal California, for example, a bare zone normally occurs between shrub and grass communities. The feeding activity of rodents, rabbits, and birds, which find shelter in the shrubs, is concentrated in this zone; if this activity is prevented by means of wire-mesh exclosures, annual herbs grow vigorously within what normally would be a bare zone. In addition, plants such as *Salvia leucophylla,* purple sage, produce volatile terpenes that inhibit the establishment of seedlings of many species of plants in their vicinity. Such plants will not grow in the bare zone even if exclosures are constructed, but often only at the border between the bare zone and the fully developed grassland, where toxic substances are less abundant and the grassland is still somewhat open (Figure 29-16).

Such chemical inhibition of one species of plant by another is called *allelopathy.* Toxic chemicals often play an important role in structuring communities.

29-16
Shrubs of purple sage, Salvia leucophylla, *produce volatile terpenes which inhibit the growth of other plants in their vicinity. Here, near Santa Barbara, California, they are surrounded first by a completely bare zone and then by a zone of inhibited grassland inhabited by stunted annual herbs, to which certain species may be restricted.*

Plant-Herbivore Interactions

Vast areas of the Australian continent were at one time covered with spiny clumps of prickly-pear cactus, *Opuntia,* a plant that was introduced from Latin America. Fertile lands became useless for grazing and the economy was severely threatened over great stretches of the interior. Today, the cactus has all but been eliminated by a cactus moth discovered in South America and deliberately introduced into Australia. The moth, once abundant, now can scarcely be found, even by a careful inspection of the remaining clumps of cactus, yet there is no doubt that it continues to exert a controlling influence over the populations of the plant in Australia (Figure 29-17).

In central Colorado, a seemingly insignificant herbivore, the small butterfly *Glaucopsyche lygdamus,* which has a wing length of about 14 millimeters, lays its eggs on the inflorescences of a common lupine, *Lupinus amplus.* The eggs are laid when the plants are in bud, and the larvae feed on portions of the flowers, but rarely on the ovary itself. It would seem that this butterfly could have little effect on the reproductive potential of the plant. In a census of a population near Gothic, Colorado, however, this assumption was disproved. A sample of 41 control inflorescenses, on which no eggs had been laid, had the potential of producing 967 mature flowers and actually produced 693, or about 72 percent of the potential. A second sample of 51 inflorescences, on which eggs were deposited, had the potential for producing 1433 mature flowers, but actually produced only 533, or only 37 percent of its potential. And of these 533 flowers, an additional 138 were so badly damaged that they probably would have dropped off without producing seed, so that a more realistic estimate of realized potential in the group on which *Glaucopsyche* had deposited its eggs would be about 28 percent.

Thus, although the effects of herbivores on plants may not be obvious, they are clearly profound. As discussed in Chapter 17, these interactions have led to the production by plants of a wide variety of chemical defenses—in the form of those biochemicals once regarded as "secondary plant substances." The ability of plants to produce such toxic chemicals and retain them in their tissues obviously gives them a tremendous competitive advantage. This advantage is analogous to that which is achieved by the production of thorns or tough, leathery leaves.

The protection of plants by their production of toxic

(a) (b)

29–17
(a) *Dense prickly pear cactus* (Opuntia inermis) *in* Casuarina *scrub, Queensland, Australia, in October, 1926, and* (b) *the same area in October, 1929, after the cacti were destroyed by the deliberately introduced South American moth,* Cactoblastis cactorum. *The moth* (c) *was successfully introduced starting in May, 1925 and destroyed the cacti over 120 million hectares of rangeland, bringing it under control in another 12 million hectares.*

(c)

molecules has interesting economic implications. Normally, plants of the squash family (Cucurbitaceae) produce bitter terpenes in their fruits and leaves, and these substances protect them from the attacks of most herbivores. Under cultivation, however, these substances have been bred out in order to improve the flavor of the fruits, and the plants have become palatable again to herbivores. Special steps must be taken to protect them—for example, heavy spraying with insecticides. Cultivated watermelon also can be attacked by a much wider range of insects than its wild relative, and these must be discouraged if the crop is to be grown successfully.

Pollination relationships are, of course, a specialized form of plant-herbivore interaction in which a particular portion of the plant is eaten and pollination takes place. Here the emphasis is on attracting the herbivore, and we have seen in Chapter 17 how this has produced the great diversity of angiosperm flowers. This is a different sort of selective race—one in which both plants and insects become better adapted to one another and increasingly specialized systems occur.

The secondary plant substances that animals ingest may in turn play a role in the ecological relationships of the animals. Some insects store these poisons within their tissues and are then protected from their own predators by them (Figure 29–18). Many sex attractants in insects (and other animals) are derived from the plants on which the insects feed. The insects concentrate these and then use them to attract the opposite sex of the same species.

The most intricate coevolutionary systems involving plant-herbivore interaction occur, not surprisingly, in the tropics. Here we shall describe a single example to give some idea of the complexity that can be involved in such a relationship. Trees and shrubs of the genus *Acacia* occur widely in the tropical and subtropical regions of the world. In Africa and tropical America, where grazing mammals are abundant, the acacia species are protected by thorns. In Australia, where plants of this group are even more important, there are no large grazing mammals, and most of the species of acacia lack thorns. This is a result of selective pressure exerted by herbivores on a particular group of plants.

(a)

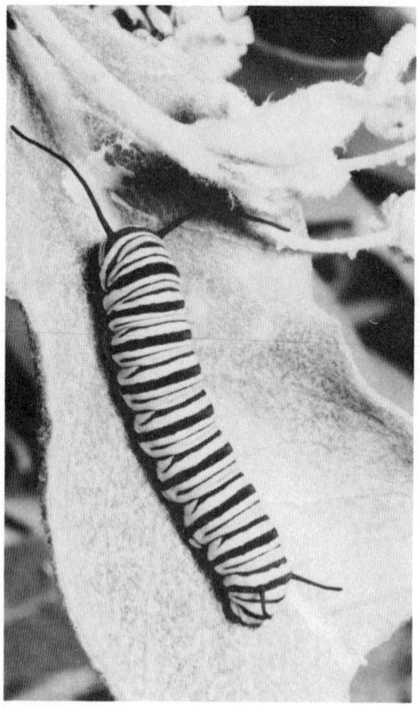

(b)

29–18
(a) *The monarch butterfly* (Danaus plexippus) *obtains cardiac glycosides from the plants of the milkweed family (Asclepiadaceae) upon which the larvae* (b) *feed. As a result, it is unpalatable to birds and other vertebrates. It "advertises" this fact by its bright orange-and-black adult coloration and striking larvae, which are conspicuously banded with white, yellow, and black. Even the eggs of the monarch, which are bright yellow and highly conspicuous, receive enough cardiac glycosides from the adult that lays them to be protected. Other insects that feed on plants of this family also tend to be brightly colored, presumably for similar reasons.*

On some of the African species of *Acacia,* ants of the genus *Crematogaster* live in the galls that appear, apparently spontaneously, on the stipular thorns. Each colony of ants inhabits a series of galls on one or more trees. The ants obtain food from nectar-secreting glands on the leaves of the acacias and eat caterpillars and other herbivores that occur on the trees. Both the ant and the acacia clearly benefit from the association between them.

This sort of system has been developed to an extraordinary degree in the lowlands of Mexico and Central America. The relationship between the bull's-horn acacia, *Acacia cornigera,* and an ant inhabitant, *Pseudomyrmex ferruginea,* which was studied in the Mexican state of Veracruz and has been described by Daniel Janzen of the University of Michigan, will be used as the basis for our discussion here (Figure 29–19).

The bull's-horn acacia has a pair of greatly swollen thorns more than 2 centimeters long at the base of each leaf. Nectaries are borne on the petioles, and small nutritive organs known as Beltian bodies are borne at the tip of each leaflet. The ants live in the thorns and obtain sugars from the nectaries and oils and proteins by eating the Beltian bodies. The acacia grows extremely rapidly and is particularly frequent in cut-over or disturbed areas, where the competition for light is intense.

Since Thomas Belt, in his book, *The Naturalist in Nicaragua* (1874), first described the relationship between *Pseudomyrmex* and the swollen-thorn acacias, a controversy has raged about whether the presence of the ants benefited the acacia plants or not. This question was finally and definitively solved in 1963–1964 by Janzen. He found that the worker ants, which swarm over the surface of the plant, bite and sting animals of all sizes that contact the plant, thus protecting it from the activities of herbivores and ensuring a home for the ants. Moreover, whenever the branches of other plants touch an inhabited acacia tree, the ants girdle and remove them, thus producing a tunnel to the light through the rapidly growing tropical vegetation. When Janzen removed the ants from the plants artificially—by poisoning them or clipping off the portions of the plant that had ants in them—or when the acacia was naturally unoccupied, growth was extremely slow and the plant usually died after a few months as a result of insect damage and shading by other species. Plants inhabited by ants grew very rapidly, soon reaching 6 meters or more in height and overtopping the other second-growth vegetation. Therefore, it is clear that the ant-acacia system is just as much a biological entity as, for example, a lichen. One element cannot survive without the other in the community of which it is a part.

There are many other kinds of relationships that link organisms that occur together. For example, trees in a forest, as well as lower herbs, are often joined by their roots in a phenomenon known as root grafting. Nu-

(a)

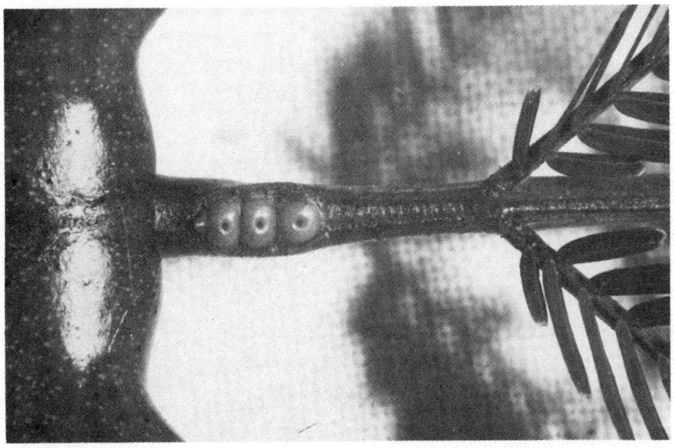

(b)

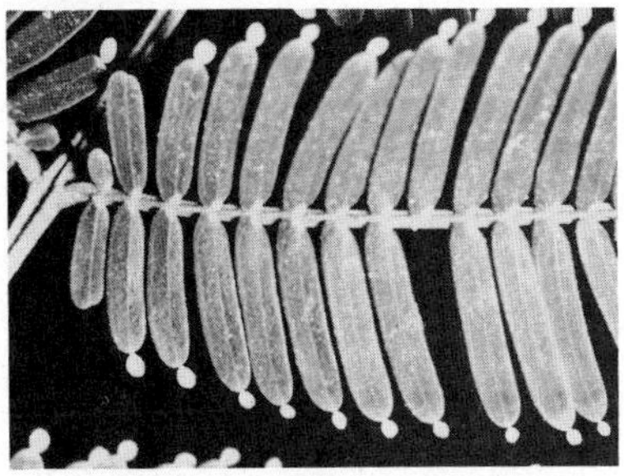

(c)

(d)

(e)

(f)

29–19
Ants and acacia. Seedling of bull's-horn acacia, Acacia cornigera (a), *with young queen ants of* Pseudomyrmex ferruginea *which live in the hollow thorns (being modified stipules, the acacia "thorns" are technically spines). One of the queens will found a colony that will later patrol the surface of the seedling.* Acacia cornigera, *showing the nectaries at the base of the petioles of the compound leaves* (b). *The pear-shaped Beltian bodies at the ends of the leaflets* (c) *of* Acacia collinsii. (d) *Ants* (Pseudomyrmex ferruginea) *attacking the tendrils of a vine growing on* A. cornigera. *Obtaining all of their food from the plant, the ants in turn girdle all plants that come into contact with it and kill most of the other insects that attempt to feed on it.* (e) *Regeneration and growth of* A. cornigera *in the lowlands of Veracruz, Mexico. At the left is shown regeneration in stumps that are occupied by ants and at the right, in an unoccupied stump. In* (f) *a single individual is seen overtopping the dense second-growth vegetation of the tropical lowlands. If it were not occupied by ants, it would have died as a small seedling, overtopped by the vegetation and devoured by insects.*

trients present in one plant can thus be transferred to another in complex and unexpected patterns; and the survival of one species in an area may depend on the presence of another with which it forms root grafts. Stumps of trees may live indefinitely, even though they have no photosynthetic surface of their own, because their roots are linked with those of other individuals in the forest from which they obtain nutrients.

Viewed as a whole, the relationships within a community are incredibly complex. Plants that occur together affect one another in an endless variety of ways, a few of which we are just beginning to understand. It may be that it is not the individual in a species that is important in survival, but that the linked group of individuals and the kinds of interactions that they have with one another determine the success of a particular species at a particular place.

Any organism's ability to survive at a given place, at a given point in time, is conditioned by dozens if not hundreds of complex relationships that we perceive only dimly. The organisms have evolved as ecological entities, each playing a highly specified role in its community. When taken from this community and viewed in isolation, they have no more meaning than would a single human being separated from all his fellows and standing alone.

SUMMARY

The community is an important, highly integrated unit of nature. It is formed because of the overlapping tolerance limits of its members, none of which entirely coincide. When it is dominated by one very prominent kind of plant, such as the redwood, it can be recognized easily and defined by the presence of that plant. When there is no one dominant species, a community can only be defined statistically.

Plant distributions are limited by the potential physiological variability of the species and the extent of its range, given the sort of habitats in which it occurs and their distribution in the particular geographical area. Climatic, edaphic, and biotic factors are all important in determining distributions, and they interact in complex ways.

Some of the kinds of relationships that occur in communities can be grouped under three main headings: mutualism, competition, and plant-herbivore relationships. In mutualism, both interacting populations are benefited. Examples are the formation of lichens; the growth of nitrogen-fixing bacteria in nodules on the roots of legumes; the formation of mycorrhizal associations between fungi and the roots of higher plants; and closely linked pollination-flower relationships.

Competitive interactions are found between most plants that grow together. One of the most important sorts of interaction concerns competition for light. Plants have evolved various strategies—tall stems, long petioles, and broad leaves—that enable them to win in this competition, and other strategies—large amount of photosynthetic surface, low light saturation values for photosynthesis, and accessory pigments—that enable them to survive at low light intensities. They have also evolved chemicals that limit or enhance the success of other plants in their vicinity, and such allelopathic relationships are of great importance in determining the structure of communities.

Herbivores control the reproductive potential of plants by destroying their photosynthetic surface or by eating their reproductive structures directly. Plants counter these attacks by the evolution of spines, tough leaves, or, most importantly, chemical defenses. When an insect has overcome these chemical defenses, it not only has a new and often largely untapped food resource at its disposal, it also may utilize the toxic substances to gain a degree of protection from its own predators. Pollination relationships are a special case of plant-herbivore interactions in which the emphasis is on attraction, not repulsion. Some plant-herbivore interactions involve a high degree of mutualism; the obligate interaction between the bull's-horn acacia and its associated ants in eastern Mexico is such a system.

SUGGESTIONS FOR FURTHER READING

ANDREWARTHA, H. G.: *Introduction to the Study of Animal Populations*, University of Chicago Press, Chicago, 1961.

This brief volume presents many of the most important principles of ecology as they apply both to plant and to animal populations.

BILLINGS, W. D.: *Plants and the Ecosystem*, Wadsworth Publishing Co., Inc., Belmont, Calif., 1966.

This slim paperback presents an excellent summary of the field as it is approached by the plant ecologist interested especially in the physiology of individual organisms.

BRAUN, E. LUCY: *The Deciduous Forests of Eastern North America*, McGraw Hill Book Company, New York, 1950.

A detailed and well-documented study of one of the richest temperate biomes.

BROCK, THOMAS D.: *Principles of Microbial Ecology*, Prentice-Hall, Inc., Englewood Cliffs, N.J., 1966.

An excellent and imaginative account of the field, written in a lively style by one of its outstanding workers.

CAIN, STANLEY A., and G. M. DEO CASTRO: *Manual of Vegetation Analysis*, Harper & Row, Publishers, Inc., New York, 1959.

Describes modern methods for studying the plant community.

COLINVAUX, PAUL: *Introduction to Ecology*, John Wiley & Sons, Inc., New York, 1973.

A modern synthetic text in ecology from a very broad point of view, emphasizing the interrelationships between ecology and evolution.

CURTIS, J. T.: *The Vegetation of Wisconsin*, University of Wisconsin Press, Madison, Wisc., 1960.

An outstanding study of the vegetation of the state by an ecologist whose influence on the field has been extensive. Vegetation is treated as a continuum which can be analyzed.

DAUBENMIRE, R. R.: *Plants and Environment*, 2nd ed., John Wiley & Sons, Inc., New York, 1959.

A standard text in the field by a community-oriented ecologist.

ELTON, CHARLES S.: *The Ecology of Invasions by Animals and Plants*, John Wiley & Sons, Inc., New York, 1958.

Excellent summary of this important aspect of ecology, with interesting chapters on conservation and biological, rather than chemical, control of insect pests.

EMLEN, J. MERRITT: *Ecology: An Evolutionary Approach*, Addison-Wesley Publishing Co., Menlo Park, Calif., 1973.

A theoretical consideration of ecology leading to the construction of models, which are then applied imaginatively in field situations.

GLEASON, HENRY A., and ARTHUR CRONQUIST: *The Natural Geography of Plants*, Columbia University Press, New York, 1964.

Written for the layman, this handsome volume presents a readable and accurate survey of the principles of plant ecology and plant geography in nontechnical terms.

HANSON, HERBERT C., and ETHAN D. CHURCHILL: *The Plant Community*, Litton Educational Publishing Corp., New York, 1961.

A modern treatment of the "classical" view of the plant community, dealing with it as a "superorganism."

KREBS, CHARLES J.: *Ecology. The Experimental Analysis of Distribution and Abundance*, Harper & Row, Publishers, Inc., New York, 1972.

Aptly described by its subtitle, this outstanding book provides a sound basis for understanding ecology.

ODUM, EUGENE P.: *Fundamentals of Ecology*, 2nd ed., W. B. Saunders Co., Philadelphia, 1959.

The standard text in the field, written by one of the leaders of American ecology; very heavily slanted towards animal examples.

RICHARDS, PAUL W.: *The Tropical Rain Forest*, Cambridge University Press, Cambridge, England, 1952.

The standard modern reference on this complex and rapidly disappearing biome.

TRESHOW, MICHAEL: *Environment and Plant Response*, McGraw-Hill Book Company, New York, 1970.

This short book reviews the responses of plants to climatic extremes, mineral element balance, and pollution of various types, and provides extensive and useful bibliographies on these subjects.

WATT, K. E. F.: *Ecology and Resource Management*, McGraw-Hill Book Company, New York, 1968.

A clearly written mathematical approach to the principles of ecology which constitutes a valuable introduction to this vital and rapidly expanding field.

APPENDIX A

Fundamentals of Chemistry

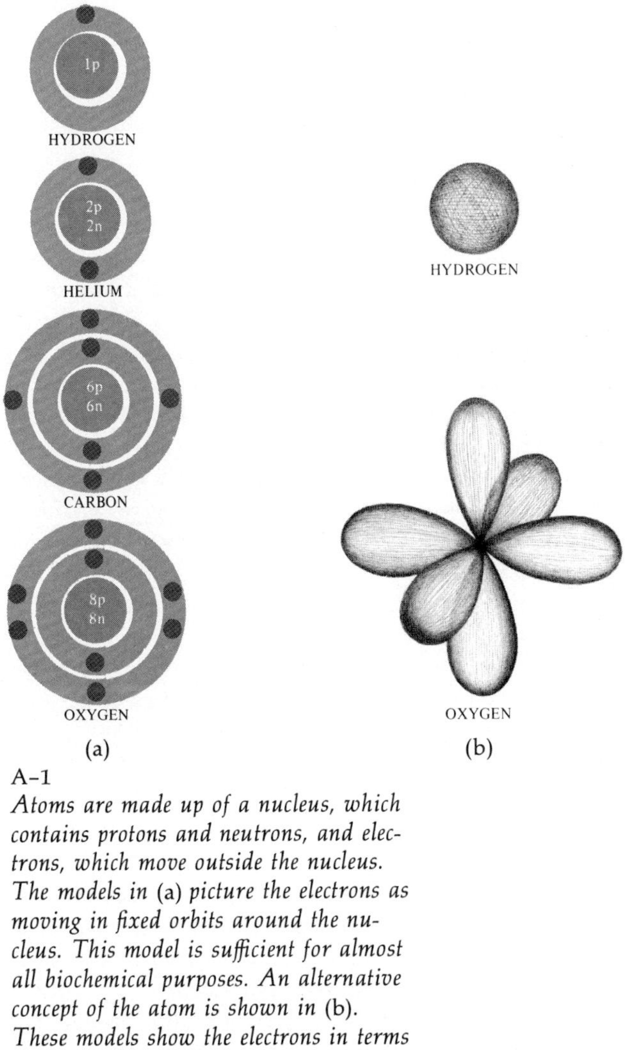

A–1
Atoms are made up of a nucleus, which contains protons and neutrons, and electrons, which move outside the nucleus. The models in (a) *picture the electrons as moving in fixed orbits around the nucleus. This model is sufficient for almost all biochemical purposes. An alternative concept of the atom is shown in* (b). *These models show the electrons in terms of their wave functions, or orbitals. The sphere represents the space in which the hydrogen electron is usually found. The three dumbbell-shaped orbitals of the oxygen atom are the areas in which the electrons of the outer shell are most likely to be found.*

THE STRUCTURE OF ATOMS

All matter is composed of atoms, which are the smallest complete units of elements. Thus, the element hydrogen, for example, is made up of hydrogen atoms; one hydrogen atom is represented by the letter H, which also stands for the element as a whole. The parts of the atom—that is, subatomic particles such as the proton, neutron, and electron—have properties that are quite different from those of the atom itself. The nucleus, or center, of an atom is formed by protons and neutrons, and electrons occur outside of this nucleus. The atomic weight, or mass, of an atom is determined by the number of protons and neutrons present (the assigned weight of each proton and neutron is 1), because electrons are virtually weightless. The atomic number of an atom represents the number of protons in its nucleus.

Protons have a positive electrical charge. Neutrons, as might be inferred from their name, have no charge. Therefore, the nucleus of any atom has a net positive charge, and the strength of this charge can be determined from the atomic number. This net positive charge of the nucleus is offset by electrons, negative particles traveling at a high velocity that move about the nucleus at varying distances from it in layers, or shells. The negative charge of one electron is sufficient to balance out the positive charge of one proton. Therefore, because the number of electrons surrounding the nucleus equals the number of protons within it, an atom in a normal state is electrically neutral. Thus, the atomic number equals the sum of all electrons in the shells surrounding the nucleus of the atom in its electrically stable state. An atom that lacks one of its orbital electrons has a charge of $+1$; one that lacks two electrons has a charge of $+2$; and an atom that has gained an electron has a charge of -1.

Figure A–1 shows two ways of representing atoms. To the left are four examples of the model first proposed in 1913 by Niels Böhr. In this model, negatively charged electrons are pictured as moving in planetary orbits around the nucleus. The relative sizes of electrons and nucleus are not meant to be represented accurately. As we have said, electrons are much, much smaller and lighter than atomic nuclei. When you weigh yourself on a scale, about 50 grams of your total weight is made up of electrons; all the rest is protons and neutrons. If a picture of the hydrogen atom were drawn to scale and the proton that is in the nucleus of the atom were depicted as being 1 centimeter in diameter, the electron, in proportion, would be about half a kilometer away. Most of the atom is empty space.

The simple Böhr, or planetary, model that we follow in this text has been replaced for some purposes by the representations shown at the right in Figure A–1. Here the positions of the electrons are represented in terms of their so-called wave functions, or orbitals. An orbital is the region in which we are likely to find a given electron most of the time. The electron of a single hydrogen atom, for example, will most probably be found somewhere in the sphere shown. The regions in which there is a high probability (95 percent) of finding the electrons of the outer shell of the oxygen atom are three dumbbell-shaped orbitals that are perpendicular to each other.

Isotopes

The number of protons in an atom determines its chemical character. Chemical elements consist of atoms of one particular kind. Hydrogen consists of atoms with one proton each; helium consists of atoms with two protons. The carbon atom has six protons; nitrogen, seven; oxygen, eight; and so on up to an as yet unnamed element which has 104. Sometimes an atom loses a proton. This can be caused by natural radiations, or it can be induced artificially in atomic reactors. When an atom loses or gains a proton, it changes its chemical identity.

Although atoms of the same element always contain the same number of protons, they vary in their neutron content and, therefore, in their atomic weights. Atoms that have the same atomic number but different atomic weights are known as *isotopes.* For almost all biological purposes, isotopes of a given atom can be regarded as the same element, because they have the same chemical properties. In nature, the most common isotope of hydrogen is ^{1}H, which contains one proton in the nucleus and no neutrons. Another isotope of hydrogen is deuterium, or heavy hydrogen (^{2}H), which contains one proton and one neutron. A third hydrogen isotope, tritium (^{3}H), contains one proton and two neutrons. Tritium is unstable, however, which means that it tends to emit high-energy particles from its nucleus. Such unstable forms are known as *radioactive isotopes.* Both deuterium and tritium have nearly the same chemical properties as the more common isotope of hydrogen (^{1}H), and either can substitute for it in chemical reactions. If an atom gains a proton, however, rather than a neutron, it is no longer hydrogen but helium. The fusion of hydrogen nuclei (protons) to form helium is the source of energy at the heart of the sun and also provides the terrible destructive force of the hydrogen bomb.

Many naturally occurring isotopes are radioactive. All the heavier elements—atoms that have 84 or more protons in their nucleus—are unstable and, therefore, radioactive. All radioactive elements emit nuclear particles at a fixed rate; they are said to undergo radioactive "decay" to another element. This rate is measured in terms of half-life: The half-life of a radioactive element is defined as the time in which half the atoms in a sample of the element lose their radioactivity and become stable. Because the half-life of an element is constant, it is possible to calculate the fraction of decay that will take place for a given isotope in a given period of time.

Half-lives vary widely, depending on the isotope. The radioactive nitrogen isotope ^{13}N has a half-life of 10 minutes, and tritium has a half-life of $12\frac{1}{4}$ years. The most common isotope of uranium (^{238}U) has a half-life of $4\frac{1}{2}$ billion years. The uranium atom undergoes a series of decays, eventually being transformed to an isotope of lead (^{206}Pb).

Isotopes have a number of important uses in biological research. One of these is in dating the age of fossils and of the rocks in which fossils are found. The proportion of ^{238}U to ^{206}Pb in a given rock sample, for example, is a good indication of how long ago that rock was formed. (The lead formed as a result of the decay of uranium is not the same as the lead commonly present in the original rock, ^{204}Pb.) A second use of isotopes is as radioactive tracers. The use of radioactive carbon dioxide ($^{14}CO_2$) has played an important role in enabling plant physiologists to trace the path of carbon in photosynthesis, as described in Chapter 6. A third use is autoradiography, a technique in which a sample of material containing a radioactive isotope is placed on a photographic film. Energy emitted from the isotope leaves traces on the film and so reveals the exact location of the isotope within the specimen; Figure 20–5, page 422, is an example of an autoradiograph.

Electronic Structure

Positive charges and negative charges attract one another, and thus an atomic nucleus tends to attract as many electrons as there are protons in the nucleus. Each shell surrounding the nucleus can, however, accommodate only a fixed number of electrons, and this also determines the electron pattern in a given atom. The first shell can hold two electrons; and the second, eight. Every outer shell similarly can hold no more than eight, although if the fourth shell is filled, the third shell can hold as many as eighteen. Atoms with as many as seven shells are known, but only atoms with a large number of protons, such as uranium, with an atomic number of 92, can

Table A–1 *Some Elements Present in Living Matter*

ELEMENT	SYMBOL	ATOMIC NUMBER	NUMBER OF ELECTRONS SHELL 1	SHELL 2	SHELL 3	SHELL 4
Hydrogen	H	1	1	0	0	0
Carbon	C	6	2	4	0	0
Nitrogen	N	7	2	5	0	0
Oxygen	O	8	2	6	0	0
Sodium	Na	11	2	8	1	0
Magnesium	Mg	12	2	8	2	0
Silicon	Si	14	2	8	4	0
Phosphorus	P	15	2	8	5	0
Sulfur	S	16	2	8	6	0
Chlorine	Cl	17	2	8	7	0
Potassium	K	19	2	8	8	1
Calcium	Ca	20	2	8	8	2
Iron	Fe	26	2	8	8	8

attract enough electrons to fill so many shells. As you can see in Table A–1, most of the elements involved in life processes are elements with relatively few protons, and so for our purposes, we need only be concerned with the first two or three shells.

Atoms tend to stabilize or complete their shells. Oxygen, for example, requires eight electrons to achieve a neutral charge. Two of these electrons are in the inner shell and six are in the outer shell. Thus, oxygen needs to gain two electrons to stabilize its outer shell at eight. (The gain of electrons is called *reduction*, and the loss of electrons is *oxidation*.) Sodium has an atomic number of 11 and so, as you can predict, has two electrons in its inner shell, eight in its second shell, and one in its third shell. As a consequence, the sodium atom has a tendency to lose this odd electron—or become oxidized—achieving a stable electron configuration of eight in what then becomes its outer shell. The loss of this electron gives sodium a net positive charge of 1. In chemical symbols, sodium is Na and sodium that has been oxidized is Na^+. Magnesium, with an atomic number of 12, needs to lose two electrons in order to gain a stable outer shell. It then becomes Mg^{2+}. Chlorine, which has an atomic number of 17, needs to gain one electron in order to stabilize its outer shell. In gaining an electron, it acquires a negative charge and becomes Cl^-.

The number of electrons an atom or molecule must gain or lose to attain a stable configuration in its outer shell is known as its *valence* (Figure A–2). Thus, using the examples above, magnesium has a valence of +2 because it needs to lose two electrons, and chlorine has a valence of −1 because it needs to gain one. In general, atoms with less than four electrons in the outer shell

A–2
Valence. The atoms shown here are electrically neutral. To gain a full outer shell of eight electrons, the oxygen atom needs to gain two electrons. It would then have a total of ten electrons, each with a negative charge, and eight protons, each with a positive charge. It is therefore said to have a valence of −2. The sodium atom, which needs to lose one electron to stabilize its outer shell, has a valence of +1.

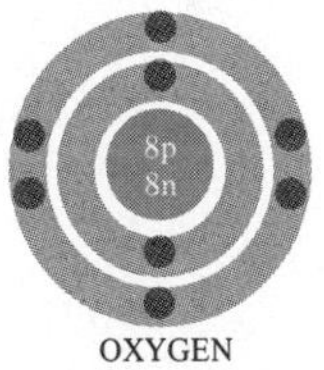

OXYGEN

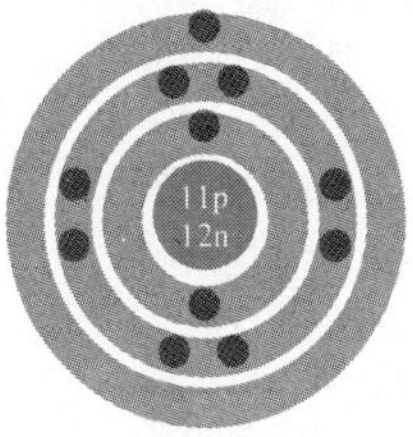

SODIUM

tend to lose electrons, or become oxidized, thereby achieving a net positive charge; those with more than four electrons tend to gain electrons, or become reduced.

Some elements may have more than one valence. Such elements are heavier atoms that have three or more filled shells and in which an electron may move from one shell to another. Iron, for instance, can have a charge of +2 or a charge of +3, a property that is important in its biological activities.

Electrically charged atoms or groups of atoms are known as *ions.* They are also sometimes called *electrolytes,* because solutions that contain ions can conduct an electric current, which is carried by the charged particles. Many of the inorganic chemical components of living organisms are found in the form of ions in the interior fluids of the cells or in extracellular fluids.

MOLECULES

Atoms may be held to one another by forces known as *chemical bonds,* and an assembly of atoms held together by chemical bonds is called a *molecule.* A molecule is made of two or more atoms. These atoms may be the same, as in the case of the oxygen molecule, which is composed of two atoms of oxygen, or they may be different. A molecule made of two or more kinds of atoms is known as a *compound.*

The formula of a chemical compound consists of the symbols of the chemical elements composing it along with subscript figures showing how many atoms of that element are present if there is more than one.

Chemical Bonds

There are two general types of chemical bonds that bind atoms into molecules—ionic bonds and covalent bonds.

Ionic Bonds

The most common type of bond in inorganic molecules is the ionic bond. Ionic bonds are formed between positively charged and negatively charged atoms or groups of atoms. Sodium ions and chlorine ions, for example, are mutually attracted to one another, producing sodium chloride, NaCl (table salt). Ionic bonds may be formed between atoms that are already ionized, or the ions may be produced by electron exchanges between atoms. Sodium may lose an electron to chlorine, leaving sodium with a valence of +1 and chlorine with a valence of −1 (Figure A–3a). The two oppositely charged particles then attract one another. In this process, you will note, sodium becomes oxidized and chlorine reduced. Magnesium tends to lose two electrons; since it then has a charge of +2, it has enough attractive force to hold two chlorine ions. Sodium chloride and other ionic salts crystallize in a lattice made up of alternating electropositive and electronegative ions, which gives the crystal great stability. (See Figure A–3b). Crystalline salts dissolve in water when water molecules get between the oppositely charged ions and separate them.

Covalent Bonds

Covalent bonds figure prominently in organic chemistry, which is essentially the chemistry of hydrogen, carbon, nitrogen, and oxygen (the first four elements listed in Table A–1). This type of bond results from the sharing of a pair of electrons. For example, each hydrogen atom has one proton and one electron, and the hydrogen atom needs one more electron to complete its outer shell. Each oxygen atom has eight protons and eight electrons. Two of the electrons are in the first shell and six in the second shell, so oxygen needs two more electrons to stabilize its outer shell. If two hydrogen atoms share their electrons with one oxygen atom, the requirements of all three are satisfied, as shown in Figure A–4. A nitrogen atom has seven protons and so an uncharged atom of nitrogen has seven electrons, two in the inner shell and five in the outer shell. Nitrogen needs three electrons to fill its outer shell; hence, it combines with three hydrogen atoms to make NH_3, which is ammonia.

Notice that there is no sharp dividing line between ionic and covalent bonds. In most ionic bonds, there is some sharing of the electron or electrons donated from one atom to another; and many covalent bonds have some ionic (or polarized) character.

Carbon-Atom Combinations

A carbon atom has six protons and six electrons, two in the inner shell and four in the outer shell. As a consequence, carbon does not tend either to gain or to lose electrons but usually forms covalent bonds. Because it is neither strongly electropositive nor strongly electronegative, it can combine both with electronegative elements, such as oxygen, nitrogen, phosphorus, and sulfur, and with the electropositive element hydrogen. Also, it can form four covalent bonds simultaneously. For these reasons, a tremendous variety of compounds can be built up from carbon atoms. Figure A–5 shows a few of the combinations that carbon atoms make with hydrogen atoms to form methane, ethane, propane, and butane; these and other molecules that consist only of hydrogen and carbon are called hydrocarbons. Furthermore, carbon atoms can combine with each other to form very long chains or rings, which are the backbones of most organic molecules.

The only possible rival for the biological role of carbon would be silicon, which has an atomic number of 14 and, thus, four electrons in its outer shell. However, it is

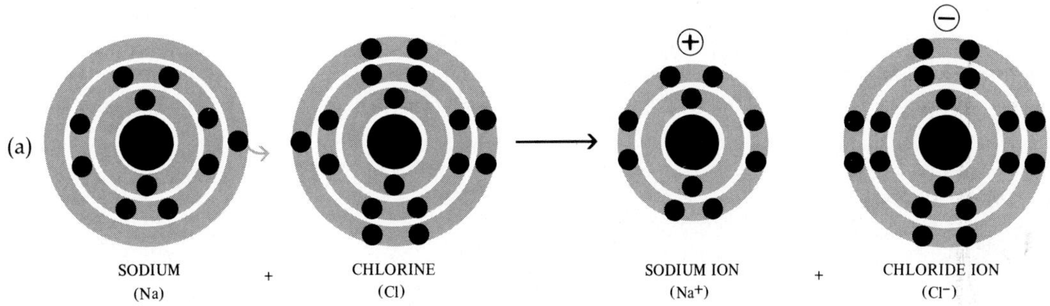

A–3

Ionic bonds. (a) *When sodium and chlorine interact, sodium loses the single electron in its outer shell to chlorine so that each contains eight electrons in its outer shell. Sodium thus acquires a positive charge and chlorine a negative charge.*

(b) *The oppositely charged ions attract each other strongly. Crystalline NaCl is a latticework of alternating* Na^+ *and* Cl^- *ions held together by their opposite charges.*

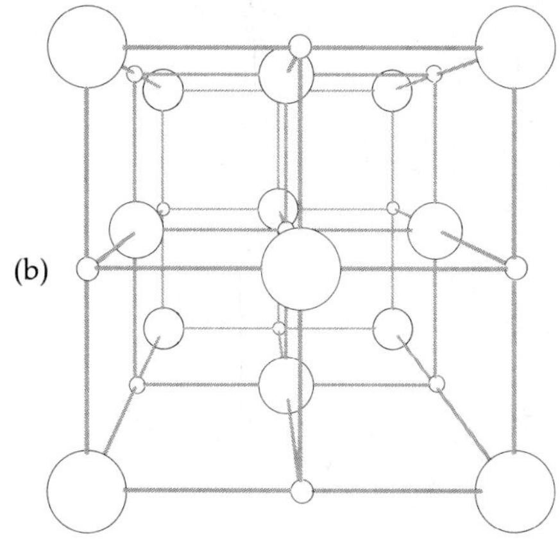

A–4

Covalent bonds result from the sharing of a pair of electrons. In (a), *one oxygen atom shares a pair of electrons with each of two hydrogen atoms to form water. In this way, all three atoms complete their outer shells. In* (b), *three hydrogen atoms and one nitrogen atom combine to form ammonia.*

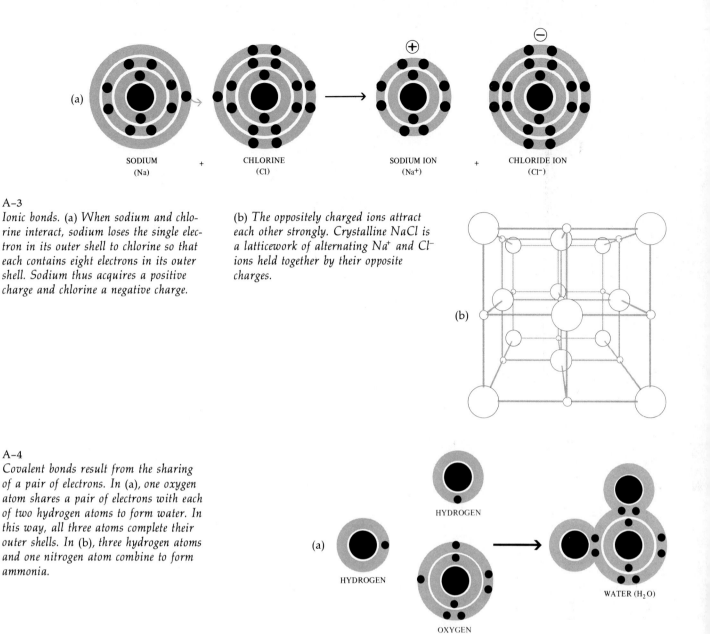

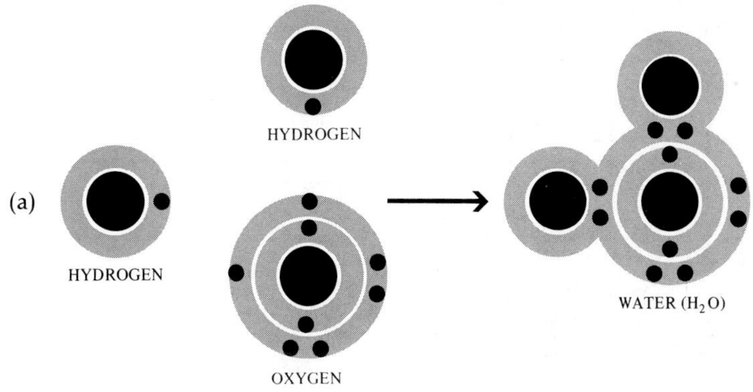

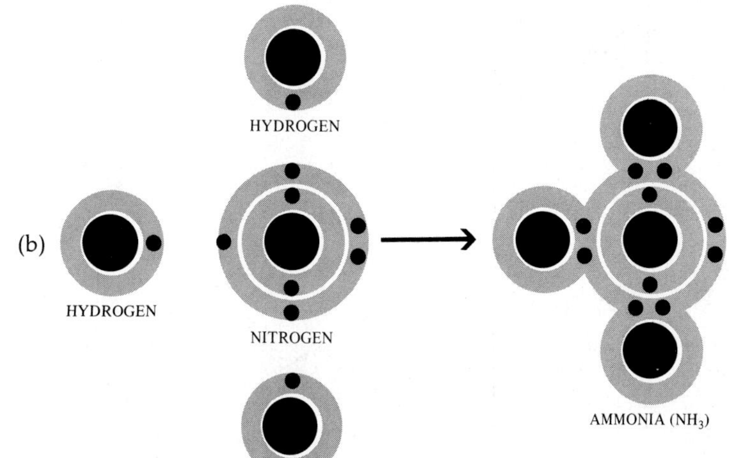

A-5
The carbon atom, with four electrons in its outer shell, can form four covalent bonds simultaneously. This property is responsible for carbon's crucial role in organic chemistry. (a) *For example, one carbon atom can combine with four hydrogen atoms to form methane gas.* (b) *Three ways of representing methane three-dimensionally.* (c) *Two-dimension representations of methane and three other hydrocarbons.*

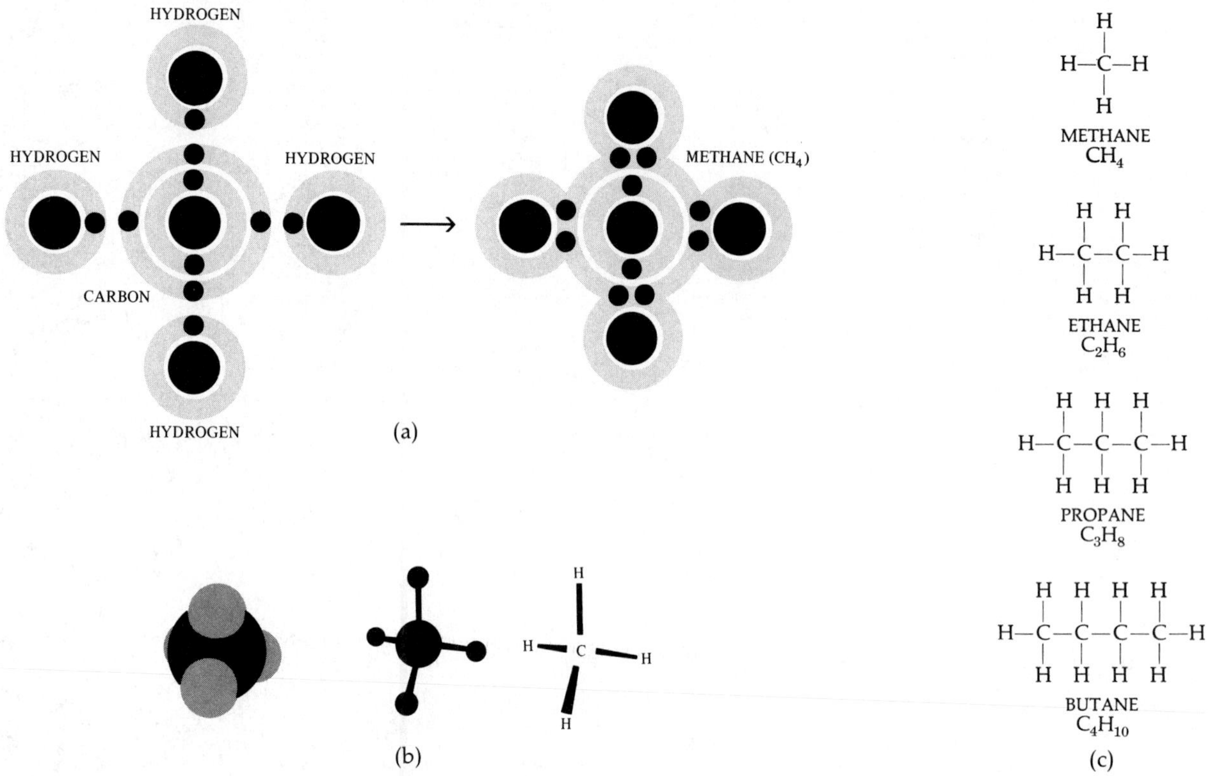

just a little bit larger than carbon and so cannot form double bonds. Consequently, it cannot form a gas with oxygen. For this reason, it could not substitute for carbon in living systems as we know them, because the flow of energy through the living world is dependent upon the capacity of plants and animals to exchange carbon dioxide with the atmosphere.

Functional Groups

Sometimes clusters of atoms joined by covalent bonds tend to move or react together as a group, as if they were a single atom. One class of such atomic clusters is represented by certain complex ions. Water, for example, has a slight tendency to dissociate, forming the hydroxyl ions OH^- and hydrogen ions H^+; the latter combine with water molecules to form hydronium ions H_3O^+. A second class is represented by organic radicals or functional groups, which play important roles as building units in organic compounds. Figure A-6 shows some functional groups that occur repeatedly in organic molecules.

A-6
Some functional groups that play important roles in organic compounds.

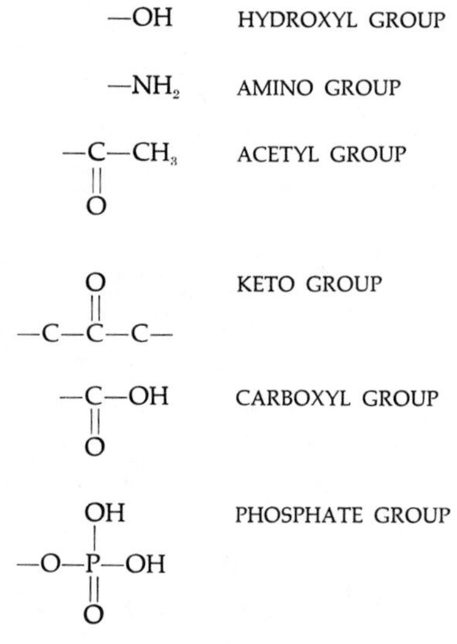

Atomic and Molecular Weights

The atomic weight of an element is its weight relative to that of the common isotope of carbon (^{12}C) with an atomic weight of 12 (6 protons and 6 neutrons). Hypothetically, an element exactly twice as heavy as carbon would have an atomic weight of 24; an element one-half as heavy, 6. Because atoms are much too small to be weighed individually, they are measured in amounts called *gram atoms.* One gram atom of an element is that amount of the element that equals, in grams, its atomic weight. For example, 1 gram atom of carbon is that amount of carbon that weighs 12 grams, and 1 gram atom of hydrogen weighs about 1 gram (actually 1.008 grams because of the presence of isotopes in any naturally occurring sample). *Molecular weight* is the sum of the atomic weights of all the atoms in a molecule. A *gram molecule* is the amount of a substance that equals, in grams, its molecular weight. One gram molecule of oxygen gas (O_2) weighs about 32 grams; 1 gram molecule of hydrogen gas (H_2) weighs abut 2 grams; and 1 gram molecule of CO_2 weighs about 44 grams.

The gram molecule is called a *mole.* The number of particles in 1 mole of any substance (whether atoms, ions, or molecules) is always the same: 6.022×10^{23}. For example, 1 mole of water (H_2O) contains 6.022×10^{23} water molecules, and 1 mole of glucose ($C_6H_{12}O_6$) contains 6.022×10^{23} glucose molecules. This constant is known as the *Avogadro number.* The mole is useful for defining quantities of substances involved in chemical reactions.

In order to make water, for instance, one would combine 2 moles of hydrogen (4 grams) with 1 mole of oxygen (32 grams); in other words, four hydrogen atoms for every two oxygen atoms. Two moles of water would be produced, each weighing 18 grams. Similarly, to make table salt (NaCl), one would combine 1 gram atom of sodium (about 23 grams) and 1 gram atom of chlorine (about 35.5 grams).

ACIDS AND BASES

Acids taste sour, like sour milk, citrus fruits, and vinegar. Bases taste flat, like milk of magnesia, and feel slippery and soapy in solution.

To define "acid" and "base" in chemical terms, it is easiest to begin by looking at water. Water consists of two atoms of hydrogen and one of oxygen held together by covalent bonds. Water molecules also have a slight tendency to ionize, separating into H^+ and OH^- ions.* In any given volume of pure water, a very small but constant number of water molecules will be dissociated into ions. The number is constant because the tendency of water to dissociate is exactly offset by the tendency of the ions to reunite; thus, even as some are ionizing, an equal number of others are forming covalent bonds.

In pure water, the number of H^+ (or H_3O^+) ions exactly equals the number of OH^- ions. This is necessarily the case, because neither ion can be formed without the other when only H_2O molecules are present. A solution acquires the properties we recognize as acid when the concentration of H^+ ions exceeds the number of OH^- ions; conversely, a solution is basic when the OH^- ions exceed the H^+ ions. There is always an inverse relationship between the concentration of H^+ and OH^- ions; when H^+ is high, OH^- is low, and vice versa. This is because the product of their concentrations is a constant. Thus,

$$[H^+]\,[OH^-] = 1 \times 10^{-14} \text{ mole at } 25°C$$

We are now in a position to define our terms chemically:

1. An *acid* is a substance that donates H^+ ions to a solution without donating OH^- ions. Because an H^+ ion is really a proton, acids can also be defined as *proton donors.*
2. A *base* is a substance that decreases the number of H^+ ions, or protons. More specifically, a base is a *proton acceptor.* The OH^- ion is a base because it can accept a proton and thus be neutralized:

$$H^+ + OH^- \rightarrow H_2O$$

3. In any acid–base reaction, there is always a proton donor and a proton acceptor.

Strong and Weak Acids and Bases

Hydrochloric acid (HCl) is an example of a common acid. It is a strong acid, meaning that it tends to be almost completely ionized. Sodium hydroxide (NaOH) is a common strong base; in solution, it exists entirely as Na^+ and OH^-. Weak acids and weak bases are those that ionize only slightly. Compounds that contain the carboxyl group (—COOH) are often weak acids because the hydrogen atom may partially dissociate from the carboxyl group to yield a proton:

$$R\text{—}COOH \rightleftharpoons R\text{—}COO^- + H^+$$

In this equation, R represents any chemical structure that may be attached to a carboxyl group.

Compounds that contain the amino group (—NH_2) act as weak bases, because (—NH_2) group has a weak tendency to accept hydrogen ions, thereby forming $NH_3{}^+$:

$$R\text{—}NH_2 + H^+ \rightleftharpoons R\text{—}NH_3{}^+$$

* As previously mentioned, the H ions tend to combine with other H_2O molecules to produce the ion hydronium (H_3O^+), but we shall omit the extra molecule from our consideration in this discussion.

Because of the strong tendency of H^+ and OH^- to combine and the weak tendency of water to ionize, the product of their concentrations is a very small and constant number, as we noted previously; therefore the concentration of H^+ ions will always decrease as the concentration of OH^- increases, and vice versa. If HCl is added to a solution in which NaOH is present, the following reaction will take place:

$$Na^+ + OH^- + H^+ + Cl^- \rightarrow H_2O + Na^+ + Cl^-$$

If the acid and base are added in equivalent amounts, the solution will once more be neutral. This is another characteristic of acids and bases—that is, added together, they produce a salt and water.

The pH Scale

Chemists define degrees of acidity by means of the pH scale. (In the expression "pH," the p stands for "power" and the H stands for the hydrogen ion.)

In a liter of pure water, $\frac{1}{10,000,000}$ mole of hydrogen ions can be detected. For convenience, this is written in terms of a power, 10^{-7}, and in terms of the pH scale, it is referred to simply as pH 7 (see Table A-2). At pH 7, the concentrations of free H^+ and OH^- ions are exactly the same, and thus pure water is "neutral." Any pH below 7 is acidic, and any pH above 7 is basic. The lower the pH number, the higher the concentration of hydrogen ions. Thus pH 2 means 10^{-2} mole of hydrogen ions per liter of water, or $\frac{1}{100}$ mole per liter—which is, of course, a much larger figure than $\frac{1}{10,000,000}$.

Lemon juice has a pH of about 2, as do the stomach contents of man and other animals. Orange juice has a pH of about 3; human blood has a pH of 7.4. The best soil pH for most plants is about 6.4, but alkaline soils have a pH between 7 and 9, and peat bogs may have a pH as low as 3.

CHEMICAL REACTIONS

Chemical compounds, as we have seen, are formed by bonds that result from the giving over or the sharing of electrons. All chemical reactions involve the breaking of such bonds and the formation of new bonds.

Types of Reactions

All chemical reactions can be classified into a few general types. One type can be represented by the formula A + B → AB. An example of this sort of reaction is the combination of hydrogen gas H_2 with oxygen gas O_2 to produce water H_2O. However, we know that each molecule of water contains two atoms of hydrogen and one of oxygen. We can, therefore, write the following balanced equation:

$$2H_2 + O_2 \rightarrow 2H_2O$$

One of the most important reactions in biology is the

Table A-2 *The pH Scale*

	Concentration of H^+ ions (moles per liter)		pH	Concentrations of OH^- ions (moles per liter)	
Acidic	1.0	10^0	0	10^{-14}	
	0.1	10^{-1}	1	10^{-13}	
	0.01	10^{-2}	2	10^{-12}	
	0.001	10^{-3}	3	10^{-11}	
	0.0001	10^{-4}	4	10^{-10}	
	0.00001	10^{-5}	5	10^{-9}	
	0.000001	10^{-6}	6	10^{-8}	
Neutral	0.0000001	10^{-7}	7	10^{-7}	
Basic		10^{-8}	8	10^{-6}	0.000001
		10^{-9}	9	10^{-5}	0.00001
		10^{-10}	10	10^{-4}	0.0001
		10^{-11}	11	10^{-3}	0.001
		10^{-12}	12	10^{-2}	0.01
		10^{-13}	13	10^{-1}	0.1
		10^{-14}	14	10^0	1.0

oxidation of glucose, $C_6H_{12}O_6$, to form carbon dioxide, CO_2, and water, H_2O. It is easily demonstrated that the equation for this reaction balances if we end with six molecules of carbon dioxide and six molecules of water for every molecule of glucose:

$$C_6H_{12}O_6 + 6O_2 \rightarrow 6CO_2 + 6H_2O$$

We may also write this equation as:

$$(CH_2O) + O_2 \rightarrow CO_2 + H_2O$$

By enclosing the (CH_2O) in parentheses, we indicate that the unit may be repeated a number of times to form a complex molecule; in any case, however, the equation still balances.

The Energy Factor

Although the equations written above are balanced in terms of their chemical components, one crucial factor is missing from each of them. This factor is energy. As we have seen, forces of electrical attraction between atoms or groups of atoms produce the chemical bonds that hold a compound together. Depending on the strength of these forces, each compound has a greater or lesser energy content, or bonding energy. All chemical reactions involve rearrangement of bonds and, therefore, changes in energy. In general, chemical reactions proceed in such a way as to result in a loss of energy to the surroundings. The end products of chemical reactions that proceed spontaneously contain less energy than that which was in the starting molecules. The combination of hydrogen and oxygen to make water liberates energy and will occur spontaneously. It is possible, however, to put energy into a reaction. In the laboratory this is usually done by heating. The separation of water into hydrogen and oxygen can be produced in the laboratory by running an electric current through water. In nature the energy for this reaction, which takes place during photosynthesis, comes directly from the sun. The oxidation of glucose or any other carbon compound to carbon dioxide and water liberates energy. We know this from everyday experience. If we touch a match to many carbon compounds, such as natural gas, petroleum, or coal, we start a reaction in which large amounts of energy are liberated in the form of heat. In the cell, sugars are oxidized to CO_2 and H_2O, and the energy thus released is used to perform a wide variety of cellular activities.

The heat changes taking place during chemical reactions are measured in terms of calories, and 1 calorie is the amount of heat necessary to raise the temperature of 1 gram of water 1°C.

Endergonic and Exergonic Reactions

A reaction that requires energy is known as *endergonic* (from the Greek *endon*, meaning "within," and *ergon*, meaning "work"), because energy must be put into it. A reaction that liberates energy is an *exergonic* reaction. An endergonic reaction can be thought of as an "uphill" reaction and an exergonic reaction as a "downhill" reaction. Only exergonic (downhill) reactions can proceed spontaneously. Endergonic reactions do not occur by themselves; they must be coupled to some downhill process in such a way that the energy released by the downhill process can be used for the uphill process. Thus, the energy released in the downhill process must be greater than that required for the uphill process. In living systems, the uphill and downhill movements are often accomplished in very small stages, so that large amounts of energy are not required or released all at once.

Even in an exergonic reaction, a certain amount of energy is needed to start the reaction; this is known as the energy of activation. The application of a small flame to a piece of paper provides the energy of activation for the rapid oxidation of the cellulose of the paper, and in chemical laboratories, energy of activation is usually supplied by heat. Heat serves to increase the motion of the molecules, which results in an increase in the frequency of reaction. (Generally, the rate of a chemical reaction doubles for each 10°C increase in temperature.) Once an exergonic reaction begins, it often becomes self-sustaining by the heat energy it produces, as in the example of the burning piece of paper.

APPENDIX B

Metric Table

	FUNDAMENTAL UNIT	QUANTITY	NUMERICAL VALUE	SYMBOL	ENGLISH EQUIVALENT
Area		hectare	10,000 m^2	ha	2.471 acres
Length	meter			m	39.37 inches
		kilometer	1000 (10^3) m	km	.62137 miles
		centimeter	.01 (10^{-2}) m	cm	.3937 inches
		millimeter	.001 (10^{-3}) m	mm	
		micrometer	.000001 (10^{-6}) m	μm	
		nanometer (millimicron)	.000000001 (10^{-9}) m	nm (mμ)	
		angstrom	.0000000001 (10^{-10}) m	Å	
Mass	gram			g	0.03527 ounces
		kilogram	1000 g	kg	2.2 pounds
		milligram	.001 g	mg	
		microgram	.000001 g	μg	
Time	second			sec	
		millisecond	.001 sec	msec	
		microsecond	.000001 sec	μsec	
Volume (solids)	cubic meter			m^3	35.314 cubic feet
		cubic centimeter	.000001 m^3	cm^3	.061 cubic inches
		cubic millimeter	.000000001 m^3	mm^3	
Volume (liquids)	liter			l	1.06 quarts
		milliliter	.001 liter	ml	
		microliter	.000001 liter	μl	

Temperature Conversion Scale

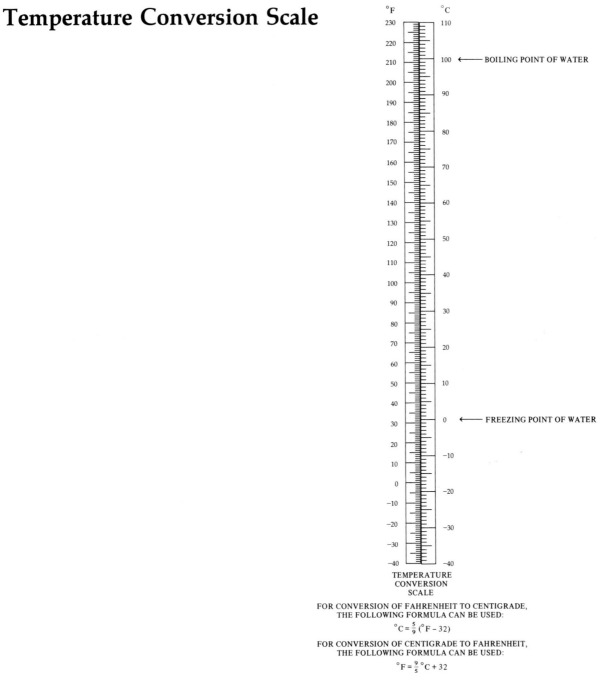

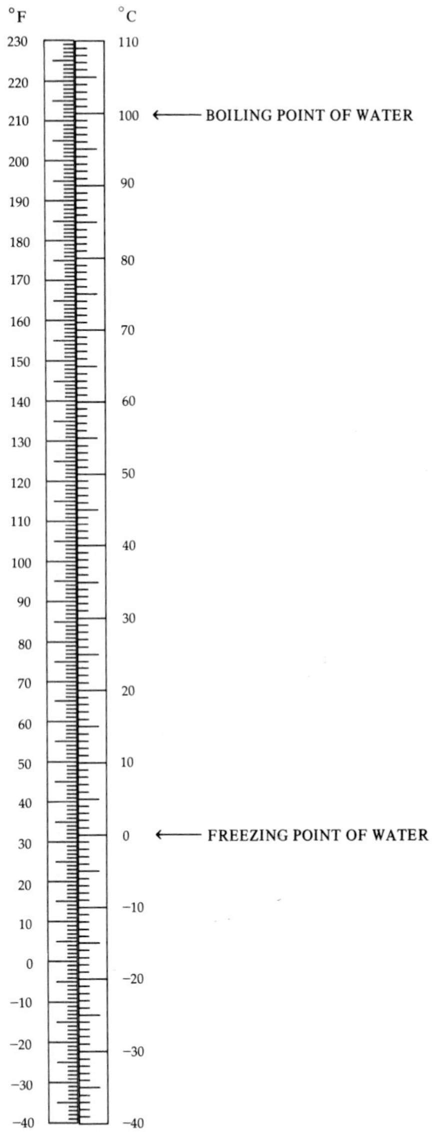

TEMPERATURE CONVERSION SCALE

FOR CONVERSION OF FAHRENHEIT TO CENTIGRADE, THE FOLLOWING FORMULA CAN BE USED:

$$°C = \frac{5}{9}(°F - 32)$$

FOR CONVERSION OF CENTIGRADE TO FAHRENHEIT, THE FOLLOWING FORMULA CAN BE USED:

$$°F = \frac{9}{5}°C + 32$$

APPENDIX C

Classification of Organisms

There are several alternative ways to classify organisms. The one presented here follows the overall scheme described at the end of Chapter 9 in which organisms are divided into five major groups, or kingdoms: Monera, Protista, Animalia, Fungi, and Plantae.

The chief taxonomic categories are kingdom, division (phylum), class, order, family, genus, species. This classification, outlined below, includes most of the divisions of Protista that include autotrophic members and the slime molds and all the recognized divisions of Fungi and Plantae. Certain subdivisions and classes given prominence in this book are also included, but the listing is far from complete. The number of species given for each group is the estimated number of living species described and named.

KINGDOM MONERA

Prokaryotic cells that lack a nuclear envelope, plastids and mitochondria, and 9-plus-2 flagella. Monera are unicellular but sometimes aggregate into filaments or other superficially multicellular bodies. Their predominant mode of nutrition is absorption, but some groups are photosynthetic or chemosynthetic. Reproduction is primarily asexual, by fission or budding, but portions of DNA molecules may also be exchanged between cells under certain circumstances. They are motile by simple flagella, gliding, or nonmotile.

About 1600 species of bacteria and bacterialike organisms are recognized at present, but doubtless thousands more await discovery. The recognition of species is not comparable with that in eukaryotes, and is based largely upon metabolic features. One group, the class Rickettsiae—very small bacterialike organisms—occurs widely as parasites in arthropods, and may contain tens or even hundreds of thousands of species, depending upon the criteria used; they have not been included in the estimate given. No satisfactory classification of the Monera has yet been proposed. One of the included groups is the Cyanobacteria, the blue-green algae, with motility by gliding and photosynthesis based on chlorophyll *a*. Although some 7500 species have been described, a more reasonable estimate puts the total number of these specialized bacteria at about 200 nonsymbiotic distinct species.

KINGDOM FUNGI

DIVISION

DIVISION MYCOTA: Eukaryotic unicellular or multinucleate organisms in which the nuclei occur in a basically continuous mycelium; this mycelium becomes septate in certain groups and at certain stages of the life cycle. They are heterotrophic, with nutrition by absorption. Reproductive cycles typically include both sexual and asexual phases. All are provisionally included in a single division, Mycota. There are some 100,000 valid species of fungi to which names have been given, but at least twice this many more probably await discovery, and many of the described ones will be eventually found to have been named two or more times; this is particularly so for fungi that may be classified both as Ascomycetes and as members of the Fungi Imperfecti.

CLASS

Class Chytridiomycetes. Aquatic fungi with motile cells characteristic of certain stages in their life cycle. These motile cells have a single posterior flagellum of the whiplash type. Their cell walls are composed of chitin, but cellulose may be present. About 750 species.

Class Oomycetes. Mostly aquatic fungi with motile cells characteristic of certain stages of the life cycle, their cell walls are composed of glucose polymers including cellulose. The flagella are two in number, one tinsel and one whiplash. There are about 475 known species.

Class Zygomycetes. Terrestrial fungi with the hyphae septate only during the formation of reproductive bodies; chitin predominant in the cell walls. The class includes about 600 described species.

Class Ascomycetes. Terrestrial and aquatic fungi with the hyphae septate but the septa perforated; complete septa cut off the reproductive bodies, such as spores or gametangia. Chitin is predominant in the cell walls. Sexual reproduction involves the formation of a characteristic cell, the ascus, in which meiosis takes place and within which ascospores are formed. The hyphae in many Ascomycetes are packed together into complex "fruiting bodies" known as ascocarps. Yeasts are unicellular Ascomycetes that reproduce asexually by budding. There are about 30,000 species, in addition to some 25,000 described species of Fungi Imperfecti, in which sexual stages do not occur or are not known.

Lichens. The lichens are Ascomycetes that have obligate symbiotic relationships with unicellular algae that multiply within their densely packed hyphae. There are about 25,000 described species.

Class Basidiomycetes. Terrestrial fungi with the hyphae septate but the septa perforated; complete septa cut off reproductive bodies, such as spores. Chitin is predominant in the cell walls. Sexual reproduction involves formation of basidia, in which meiosis takes place and on which the basidiospores are borne. Basidiomycetes are dikaryotic during most of their life cycle, and there is often complex differentiation of "tissues" within their basidiocarps. There are some 25,000 described species.

KINGDOM PROTISTA

Eukaryotic unicellular or multicellular organisms. Their modes of nutrition include ingestion, photosynthesis, and sometimes absorption. True sexuality is present in some divisions. They move by 9-plus-2 flagella or are nonmotile. Plants and animals are specialized multicellular groups derived from Protista. The photosynthetic divisions treated in this book are:

DIVISION

DIVISION CHLOROPHYTA: Green algae. Unicellular or multicellular organisms charcterized by chlorophylls *a* and *b* and various carotenoids. The carbohydrate food reserve is starch. Motile cells have two whiplash flagella at the apical end. True multicellular genera do not exhibit complex patterns of differentiation. Multicellularity has arisen at least three times, and quite possibly more often. There are about 7000 known species, and possibly many more.

DIVISION PHAEOPHYTA: Brown algae. Multicellular marine algae characterized by the presence of chlorophylls *a* and *c* with fucoxanthin. The carbohydrate food reserve is laminarin. Motile cells are biflagellated, with one forward flagellum of the tinsel type and one trailing one of the whiplash type. A considerable amount of differentiation is found in some of the kelps, with specialized conducting cells for transporting photosynthate to the dimly lighted regions of the body present in some genera. There is, however, no differentiation into roots, leaves, and stems, as in the vascular plants. There are about 1500 species.

DIVISION RHODOPHYTA: Red algae. Primarily marine algae characterized by the presence of chlorophyll *a* and phycobilins. Their carbohydrate reserve is Floridean starch. No motile cells are present at any stage in the complex life cycle. The vegetative body is built up of closely packed filaments in a gelatinous matrix and is not differentiated into roots, leaves, and stems. It lacks specialized conducting cells. There are some 4000 species, at least some of which have chlorophyll *d*.

DIVISION CHRYSOPHYTA: Golden-brown algae and diatoms. Autotrophic organisms with chlorophylls *a* and *c* and the accessory pigment fucoxanthin. Food stored as the carbohydrate leucosin or as large oil droplets. Cell walls consisting mainly of pectic compounds, sometimes heavily impregnated with siliceous materials. Some 6000 to 10,000 living species.

CLASS

Class Bacillariophyceae. Diatoms. Chrysophyta with double siliceous shells, the two halves of which fit together like a pillbox. They are sometimes motile by the secretion of mucilage fibrils along a specialized groove, the raphe. There are many extinct and 5000 to 9000 living species.

Class Chrysophyceae. Golden-brown algae. A diverse group of organisms including flagellate, amoeboid, and nonmotile forms, some naked and others with a cell wall that may be ornamented with siliceous scales. At least 1100 species.

DIVISION

DIVISION XANTHOPHYTA: Autotrophic unicellular, amoeboid, and colonial organisms with chlorophyll *a* and carotenoids. Food stored as leucosin. Cell walls consisting primarily of cellulose and pectin. About 450 species.

DIVISION PYRROPHYTA: Autotrophic organisms with chlorophylls *a* and *c*. Food is stored as starch. Cell walls contain cellulose. This division contains some 1100 species, mostly biflagellated organisms, of which the great majority belong to the following class:

CLASS

Class Dinophyceae. The dinoflagellates. Pyrrophyta with lateral flagella, one of which beats in a groove that encircles the organism. They probably have no form of sexual reproduction, and their mitosis is unlike that in any other organism. There are more than 1000 species.

DIVISION

DIVISION EUGLENOPHYTA: The euglenoids. Autotrophic (or sometimes derived heterotrophic) organisms with chlorophylls *a* and *b*. They store food as paramylon, an unusual carbohydrate. Euglenoids usually have a single apical flagellum of the tinsel variety and a contractile vacuole. The flexible pellicle is rich in proteins. Sexual reproduction is unknown. Euglenoids occur mostly in fresh water. There are more than 800 species.

DIVISION GYMNOMYCOTA: The slime molds. Heterotrophic amoeboid organisms that form sporangia, but lack a cell wall for most of their life cycle. Predominant mode of nutrition is by ingestion. There are two principal classes:

CLASS

Class Myxomycetes. The plasmodial slime molds. Slime molds with a multinucleate plasmodium that creeps along as a mass and eventually differentiates into sporangia, each of which is multinucleate and eventually gives rise to many spores. About 450 species.

Class Acrasiomycetes. Cellular slime molds. Slime molds in which there are separate amoebas that eventually swarm together to form a mass but retain their identity within this mass, which eventually differentiates into a compound sporangium. Seven genera and about 26 species.

KINGDOM PLANTAE	Autotrophic green plants, advanced tissue differentiation, diploid phase (sporophyte) includes an embryo, haploid phase (gametophyte) produces gametes by mitosis, primarily terrestrial.
DIVISION	DIVISION BRYOPHYTA: Mosses, hornworts, and liverworts. Multicellular plants with photosynthetic pigments and food reserves similar to those of the green algae. They have multicellular gametangia with a sterile jacket layer. The sperms are biflagellated. Gametophytes and sporophytes both exhibit complex multicellular patterns of development, but conducting tissues are usually absent and not well differentiated when present. Most of the photosynthesis in these primarily terrestrial plants is carried out by the gametophyte, upon which the sporophyte is at least initially dependent. There are some 24,000 species.
CLASS	*Class Hepaticae.* The liverworts. The gametophytes are thallose or leafy, the rhizoids are single-celled, and the sporophytes, which lack stomata, are relatively simple in construction. There are about 9000 species.
	Class Antherocerotae. The hornworts. The gametophytes are thallose. The sporophyte grows from a basal intercalary meristem for as long as conditions are favorable. Stomata are present on the sporophyte. There are about 100 species.
	Class Musci. The mosses. The gametophytes are leafy. Sporophytes have complex patterns of dehiscence. Rhizoids are multicellular. Stomata are present on the sporophyte. There are about 14,500 species.
DIVISION	DIVISION TRACHEOPHYTA: The vascular plants. Terrestrial plants with complex differentiation of organs into roots, leaves, and stems. The vascular plants have well-developed strands of conducting tissue in which water and organic materials are transported. The main trends of evolution in the vascular plants involve a progressive reduction of the gametophyte, which ranges from green, free-living forms in some groups to heterotrophic forms that are more or less enclosed by sporophytic tissue in others; the loss of multicellular gametangia and motile sperms; and the evolution of the seed. The division includes the following subdivisions with living representatives:
SUBDIVISION	SUBDIVISION PSILOPHYTINA: Whisk ferns. Homosporous vascular plants with or without microphylls and extremely simple sporophytes; no differentiation between root and shoot. Sperms motile. Two genera and several species.

AGGREGATE FRUIT: A fruit developing from the several separate carpels of a single flower.

ALBUMINOUS CELL: Certain ray and axial parenchyma cells in gymnosperm phloem that are closely associated with the sieve cells both morphologically and physiologically.

ALEURONE (ȧ·lū′rōn) [Gk. *aleuron*, flour]: Proteinaceous material, usually in the form of small granules, occurring in the outermost cell layer of the endosperm of wheat and other grains.

ALGA, *pl.* ALGAE (ăl′gȧ, al′je); A photosynthetic eukaryotic organism lacking multicellular sex organs (except for the charophytes); the blue-green algae are photosynthetic prokaryotes.

ALKALI (ăl′kȧ·lī) [Arabic *alqili*, the ashes of the plant saltwort]: A substance with marked basic properties.

ALKALINE (ăl′kȧ·lĭn): Pertaining to substances that release hydroxyl (OH^-) ions in water, having a pH greater than 7.

ALKALOIDS (ăl′kȧ·loids): Nitrogen-containing organic compounds produced by plants and physiologically active in vertebrates. Many have a bitter taste and some are poisonous, for example, nicotine, morphine, quinine, caffeine, strychnine.

ALLELE (allelomorph) (ă·lēl′) (ă·lē·lo·môrf) [Gk. *allelon*, of one another, + *morphe*, form]: One of the two or more alternative states of a gene that occupy the same position (locus) on homologous chromosomes. Alleles are separated from each other at meiosis.

ALLOPOLYPLOID (ăl·o·pŏl′ĭ·ploid) [Gk. *allos*, different, + *poly*, many, + *ploos*, fold]: A polyploid in which one or more sets of chromosomes come from different species or widely different strains.

ALTERNATE: Referring to bud or leaf arrangement in which there is one bud or one leaf at a node.

ALTERNATION OF GENERATIONS: A reproductive cycle in which a haploid ($1n$) phase, the gametophyte, gives rise to gametes which, after fusion to form a zygote, germinate to produce a diploid ($2n$) phase, the sporophyte. Spores produced by meiotic division from the sporophyte give rise to new gametophytes, completing the cycle.

AMINO ACIDS (ă·mē′no) [Gk. *Ammon*, referring to the Egyptian sun god, near whose temple ammonium salts were first prepared from camel dung]: Nitrogen-containing organic molecules, the "building stones" or units from which protein molecules are built.

AMMONIFICATION (ă·mōn′ĭ·fĭ·kā′shŭn): Decomposition of amino acids and other nitrogen-containing organic compounds, resulting in the production of ammonia (NH_3) and ammonium ($NH_{4}+$).

AMOEBOID (ȧ·mē′boid) [Gk. *amoibe*, change]: Moving or eating by means of pseudopodia (temporary cytoplasmic protrusions from the cell body).

AMPHI- [Gk. *amphi*, on both sides]. Prefix meaning "on both sides," "both," or "of both kinds."

AMYLASE (ăm′ĭ·lās): An enzyme that hydrolyzes starch.

AN- [Gk. *an-*, not]: Prefix, equivalent to "a-", meaning "not"; used before a vowel or "h."

ANAEROBIC (ăn·ā ĕr·ō′bĭk) [Gk. *a-*, without, + *aer*, air, + *bios*, life]: Applied to cells that can live without oxygen; strict anaerobes cannot live in the presence of oxygen.

ANALOGOUS (ȧ·năl′o·gŭs) [Gk. *analogos*, proportionate]: Applied to structures similar in function but different in evolutionary origin.

ANAPHASE (ăn′ȧ·fāz) [Gk. *ana*, away, + *phasis*, form]: A stage in mitosis in which the chromatids of each chromosome separate and move to opposite poles; similar stages in meiosis in which chromatids or paired chromosomes move apart.

ANATOMY (ȧ′nat·ă·mē): In botany, the area that deals with the internal structure of organisms.

ANDRO- [Gk. *andros*, man]: Prefix meaning "male."

ANDROECIUM (ăn·drē′shĭ·ŭm) [Gk. *andros*, man, + *oikos*, house]: Stamens collectively.

ANGIOSPERM (ăn′jĭ·o·spûrm′) [Gk. *angion*, a vessel, + *sperma*, a seed]: Literally a seed borne in a vessel (carpel), thus one of a group of plants whose seeds are borne within a matured ovary (fruit).

ANGSTROM (ăng′strŭm) [after A. J. Ångstrom, a Swedish physicist, 1814–1874]: A unit of length equal to 0.0001 of micrometer; abbreviation Å.

ANION (ăn′ī′ŏn) [Gk. to go up]: Any negative ion.

ANISOGAMY (ăn′i·sog′ȧ·mĭ) [Gk. *aniso*, dissimilar, unequal, + *gamos*, marriage]: The condition of having dissimilar motile gametes.

ANNUAL [L. *annulus*, year]: A plant in which the entire life cycle is completed in a single growing season.

ANNUAL RING: In wood, term applies to one growth layer formed during one year; *see also* false annual ring *and* growth layer.

ANNULUS (ăn′u·lŭs) [L. ring]: In ferns, a row of specialized cells in a sporangium; in gill fungi, a ring on the stalk, the remnant of the inner veil.

ANTERIOR: Situated before or toward the front.

ANTHER (ăn′thẽr) [Gk. *anthos*, flower]: The pollen-bearing portion of a stamen.

ANTHERIDIOPHORE [Gk. *anthos*, flower, + *phoros*, bearing]: In some liverworts, a stalk that bears antheridia.

ANTHERIDIUM (ăn′thẽr·ĭd′ĭ·ŭm): Multicellular male gametangium of plants other than seed plants.

ANTHOCYANIN (ăn′tho·sĭ′ȧ·nĭn) [Gk. *anthos*, flower, + *kyanos*, dark blue]: A water-soluble blue, purple, or red pigment occurring in the cell sap.

ANTIBIOTIC (ăn′tĭ·bĭ·ŏt′ĭk) [Gk. *anti*, against or opposite, + *biotikos*, pertaining to life]: Natural organic substance that retards or prevents the growth of organisms, generally used to designate substances formed by microorganisms that prevent growth of other microorganisms.

ANTICLINAL: Perpendicular to the surface.

ANTIPODALS (ăn·tĭp′o·däl) [Gk. *anti*, against + *podus*, foot]: Three (sometimes more) cells of the mature embryo sac located at the end opposite from the micropyle.

APICAL DOMINANCE (ăp′ĭ·kăl; ā′pĭ·kăl): Influence exerted by a terminal bud in suppressing the growth of lateral buds.

APICAL MERISTEM (ăp′ĭ·kăl mĕr′ĭ·stĕm):

The growing point, composed of meristematic tissue, at the tip of the root or stem in vascular plants.

APOMIXIS (ăp′o•mĭk′sĭs) [Gk. *apo*, separate, away from, + *mixis*, act of mingling]: Reproduction without meiosis or syngamy.

APOTHECIUM (ăp′o•thē′shĭ•ŭm) [Gk. *apotheke*, a storehouse]: A cup-shaped or saucer-shaped open ascocarp.

ARCH-, ARCHEO- [Gk. *arche*, *archos*, beginning]: Prefix meaning "first," "main," "earliest."

ARCHEGONIOPHORE [Gk. *archegonos*, the first of a race, + *phoros*, bearing]: In some liverworts, a stalk that bears archegonia.

ARCHEGONIUM (är′ke•gō′nĭ•ŭm) [Gk. *archegonos*, the first of a race]: Multicellular female sex organ, containing an egg; found in the Bryophyta and some vascular plants.

ARIL (ăr′ĭl) [L. *arillus*, raisin, grape, seed]: An accessory seed covering, often formed by an outgrowth at the base of the ovule, often brightly colored, which may aid in dispersal by attracting animals that eat it and, in the process, carry the seed away from the parent plant.

ARTIFACT (är′tĭ•făkt) [L. *ars*, art, + *facere*, to make]: A product that is due to an extraneous, especially human, agency and that does not occur in nature.

ASCOCARP (ăs′ko•kärp) [Gk. *askos*, a bag, + *karpos*, fruit]: A fruiting body of the Ascomycetes, generally either an open cup, a vessel, or a closed sphere lined with specialized cells called asci, in which nuclear fusion and meiosis occur.

ASCOGENOUS HYPHAE [Gk. *asco*, a bladder, + *genous*, producing]: Hyphae containing paired haploid male and female nuclei; develop from an ascogonium, and eventually give rise to asci.

ASCOGONIUM (ăs′ko•go′nĭ•ŭm) [Gk. *asco*, a bladder, + *gigesthai*, to be born]: The oogonium or female gametangium of the ascomycetes.

ASCOSPORE (ăs′ko•spōr): A fungus spore produced within an ascus.

ASCUS, *pl.* ASCI (ăs′kŭs, ăs′ī): A specialized cell, characteristic of the Ascomycetes, in which two haploid nuclei fuse to produce a zygote that immediately divides by meiosis; at maturity, an ascus contains ascospores.

ASEPTATE [Gk. *an-*, not, + L. *septum*, fence]: A term used to describe algal or fungal filaments lacking crosswalls; other term, nonseptate.

ASEXUAL REPRODUCTION: Any reproductive process, such as fission or budding, that does not involve the union of gametes.

ATOM (ăt′ŭm) [Gk. *atomos*, indivisible]: The smallest unit into which a chemical element can be divided and still retain its characteristic properties.

ATOMIC NUCLEUS: The central core of an atom, containing protons and neutrons, around which electrons orbit.

ATOMIC NUMBER: The number of protons in the nucleus of an atom.

ATOMIC WEIGHT: The weight of a representative atom of an element relative to the weight of an atom of carbon ^{12}C, which is assigned the integral value of 12.

ATP: *See* adenosine triphosphate.

AUTO- [fr. Gk. *autos*, self, same]: Prefix, meaning "same" or "self-same."

AUTOECIOUS (ô•tē′shŭs) [Gk. *auto*, self, + *oikia*, dwelling]: As in some rust fungi, completing the life cycle on a single species of host plant.

AUTOPOLYPLOID (ô′to•pŏl′ĭ•ploid): A polyploid in which the chromosomes all come from the same species, usually resulting from the doubling of chromosomes in a single individual.

AUTORADIOGRAPH, RADIOAUTOGRAPH (ô′to•rā′dĭ•o•grăf): A photographic print made by a radioactive substance acting upon a sensitive photographic film.

AUTOTROPH (ô′to•trŏf) [Gk. *autos*, self, + *trophos*, feeder]: An organism that is able to synthesize the nutritive substances it requires from inorganic substances, in contrast to heterotroph.

AUXIN (ôk′sĭn) [Gk. *auxein*, to increase]: A plant growth-regulating substance regulating cell elongation among other effects.

AXIAL SYSTEM: In secondary xylem and secondary phloem, term applied collectively to cells derived from fusiform cambial initials. The long axes of these cells are oriented parallel with the main axis of the root or stem. Also called longitudinal system and vertical system.

AXIL (ăk′sĭl) [Gk. *axilla*, armpit]: The upper angle between a twig or leaf and the stem from which it grows.

AXILLARY (ăk′sĭ•lĕr′ĭ): Term applied to buds or branches occurring in the axil of a leaf.

BACILLUS (bȧ•sĭl′ŭs) [L. dim of *baculum*, rod]: A rod-shaped bacterium.

BACKCROSS: The crossing of a hybrid with one of its parents or with a genetically equivalent organism; a cross between an individual whose genes are to be tested and one that is homozygous for all recessive genes involved in the experiment.

BACTERIOPHAGE (băk•tēr′ĭ•o•fāj) [Gk. *bakterion*, little rod, + *phagein*, to eat]: A virus that parasitizes a bacterial cell.

BACTERIUM (băk•tir′ĭ•ŭm) [Gk. *bakterion*, dim, of *baktron*, staff]: A unicellular prokaryotic organism lacking chlorophyll *a*.

BARK: A nontechnical term applied to all tissues outside the vascular cambium; *see* also inner bark and outer bark.

BASAL BODY (băs′ăl): A cytoplasmic organelle that organizes cilia or flagella, identical in form and structure to the centriole.

BASE: A substance that, on dissociation, releases hydroxyl (OH^-) ions, but not hydrogen (H^+) ions, having a pH of more than 7; *see* acid.

BASE-PAIRING RULE: The requirement that adenine must always pair with thymine (or uracil) and guanine with cytosine in a nucleic acid double helix.

BASIDIOCARP (bȧ•sĭd′i•o•kärp′) [L. *basidium*, a little pedestal, + *carpus*, a fruit]: A "fruiting" body of the Basidiomycetes; familiar examples are mushrooms, puffballs, and shelf fungi.

BASIDIOSPORE (bȧ•sĭd′i•o•spōr′): A spore of the Basidiomycetes produced within and borne on a basidium following nuclear fusion and meiosis.

BASIDIUM, *pl.* BASIDIA (bȧ•sĭd′ĭ•ŭm): A specialized reproductive cell of the Basidiomycetes, often club-shaped, in which nuclear fusion and meiosis occur.

BERRY: A simple fleshy fruit, which includes a fleshy ovary wall and one or more carpels and seeds; examples are grapes, tomatoes, and bananas.

BI- [L. *bis*, twice, double, two]: Prefix

meaning "two," "twice," "having two points."

BIENNIAL (bī·ĕn'ī·ăl): A plant that normally requires two growing seasons to complete the life cycle. Only vegetative growth occurs the first year, often resulting in the formation of an over-wintering rosette; the flowering and fruiting occur in the second year.

BILATERALLY SYMMETRICAL: *See* zygomorphic.

BIOMASS: Total dry weight of all organisms in a particular habitat or area.

BIOME (bī'ōm): A complex of communities of very wide extent, characterized by distinctive vegetation and climate; for example, all grassland areas collectively form the grassland biome.

BIOSPHERE (bī'o·sfēr'): The zone of air, land, and water at the surface of the earth occupied by living things.

BIOTIC (bī·ŏt'ik) [Gk. *bioticos*]: Relating to life.

BISEXUAL FLOWER: One that has at least one functional stamen and carpel.

BIVALENT (bī·vā'lĕnt) [L. *bis*, twice, + *valens*, pres. part. *valere*, to have power, to be strong]: A pair of synapsed chromosomes.

BLADE: The broad, expanded part of a leaf; also called lamina.

BODY CELL: The cell of the male gametophyte, or pollen grain, of gymnosperms, which divides mitotically to form two sperms.

BORDERED PIT: A pit in which the secondary wall overarches the pit membrane.

BRACT: A modified, usually reduced leaflike structure.

BRANCH ROOT: A root that arises from another, older root; also called lateral root or secondary root, if the older root is the primary root.

BRYOPHYTES (brī'o·fīt): Nonvascular terrestrial green plants, for example, mosses and liverworts.

BUD: (1) An embryonic shoot; (2) vegetative outgrowth of yeasts and some bacteria.

BULB: An underground storage organ, composed chiefly of an enlarged and fleshy leaf base.

BUNDLE SCAR: Scar or mark left on leaf scar by vascular bundles broken at the time of leaf fall, or abscission.

BUNDLE SHEATH: Layer or layers of cells surrounding a vascular bundle; may consist of parenchyma or sclerenchyma cells, or both.

BUNDLE-SHEATH EXTENSION: A group of cells extending from a bundle sheath of a vein, located in the leaf mesophyll, to either or both upper and lower epidermis; may consist of parenchyma, collenchyma, or sclerenchyma.

CALLOSE (kăl'ōs) [L. *callosus*, callous]: A complex branched carbohydrate, which is a common wall constituent associated with the sieve areas of sieve elements; may develop in reaction to injury in sieve elements and parenchyma cells.

CALLUS (kăl'ŭs) [L. hard skin]: Undifferentiated tissue; a term used in tissue culture, and also in grafting and wound healing.

CALORIE [L. *calor*, heat]: The amount of energy in the form of heat required to raise the temperature of one gram of water one degree centigrade (1°C); in making metabolic measurements, the kilocalorie (kcal) is generally used. A kilocalorie, or Calorie, is the amount of heat required to raise the temperature of one kilogram of water one degree centigrade.

CALVIN CYCLE: The series of enzymatically mediated photosynthetic reactions during which carbon dioxide is reduced to 3-phosphoglyceraldehyde and the carbon dioxide acceptor, ribulose-1,5-diphosphate, is regenerated. For every six molecules of carbon dioxide entering the cycle, a net gain of one molecule of 3-phosphoglyceraldehyde results.

CALYPTRA (kȧ·lĭp'trȧ) [Gk. *kalyptra*, covering for head]: The hood or cap of some moss species, partially or entirely covering the capsule, and formed from the expanded archegonial wall.

CALYX (kā'lĭks) [Gk. *kalyx*, a husk, cup]: Sepals collectively; the outermost flower whorl.

CAMBIAL ZONE: A region of thin-walled, undifferentiated meristematic cells between the secondary xylem and secondary phloem; consists of cambial initials and their recent derivatives.

CAMBIUM (kăm'bĭ·ŭm) [L. *cambiare*, to exchange]: A meristem that gives rise to parallel rows of cells; commonly applied to the vascular cambium and the cork cambium, or phellogen.

CAPSULE (kăp'sūl): (1) A dehiscent, dry fruit that develops from two or more carpels; (2) a slimy layer around the cells of certain bacteria; (3) the sporangium of Bryophyta.

CARBOHYDRATE (kär'bo·hī'drāt) [L. *carbo*, ember, + *hydor*, water]: An organic compound consisting of a chain of carbon atoms to which hydrogen and oxygen are attached in a 2:1 ratio; includes sugars, starch, glycogen, cellulose, etc.

CARBON CYCLE: Worldwide circulation and utilization of carbon atoms.

CARNIVOROUS (kär·nĭv'o·rŭs): Feeding upon animals; opposed to herbivorous. The term also refers to plants that are able to utilize proteins obtained from trapped animals, chiefly insects.

CAROTENE (kăr'o·tēn) [L. *carota*, carrot]: A yellow or orange pigment found in chloroplasts and chromoplasts of plants.

CAROTENOIDS (kă·rŏt'e·noids): A class of fat-soluble pigments that includes the carotenes (yellows and oranges) and the xanthophylls (yellow); found in chloroplasts and chromoplasts of plants.

CARPEL (kär'pĕl) [Gk. *karpos*, fruit]: A leaflike organ in angiosperms that encloses one or more ovules; one of the members of the gynoecium.

CARPOGONIUM (kär'po·gō'nĭ·ŭm) [Gk. *karpos*, fruit, *gonos*, offspring]: In red algae, the female gametangium.

CARPOSPORANGIUM (kär'po·spo·răn'jĭ·ŭm) [Gk. *karpos*, fruit, + *spora*, seed, + *angeion*, vessel]: In red algae, a carpospore-containing cell.

CARPOSPORE (kär'po·spōr) [Gk. *karpos*, fruit, + *spora*, seed]: In red algae, the single diploid protoplast found within a carposporangium.

CARYOPSIS (kăr'ĭ·ŏp'sĭs) [Gk. *karyon*, a nut, + *opsis*, appearance]: Simple, dry, one-seeded indehiscent fruit, with pericarp firmly united all around the seed coat; a grain characteristic of the grasses (family Poaceae).

CASPARIAN STRIP [Robert Caspary, a German botanist]: A bandlike region of primary wall containing suberin and lignin; found in anticlinal—radial and transverse—walls of endodermal cells.

CATALYST (kăt'ȧ·lĭst) [Gk. *katalysis*, dissolution]: A substance that accelerates the rate of a chemical reaction but that is not used up in the reaction; enzymes are catalysts.

CATION (kăt'ī'ŏn) [Gk. *kata,* downward + *ion,* going]: Any positive ion.

CATKIN [obs. D. *katteken,* kitten]: A spikelike inflorescence of unisexual flowers; found only in woody plants.

CELL [L. *cella,* small room]: The structural unit of living organisms; in plants, consists of cell wall and protoplast.

CELL DIVISION: The division of the cytoplasm into two equal parts, brought about in higher plants by the formation of a cell plate, which grows outward to the margins of the cell; also referred to as cytokinesis.

CELL PLATE: Structure that forms at the equator of the spindle in dividing cells of plants and a few green algae during early telophase; the predecessor of the middle lamella.

CELL SAP: The fluid contents of the vacuole.

CELLULASE (sĕl'u•lās): An enzyme that hydrolyzes cellulose.

CELLULOSE (sĕl'u•lōs): A carbohydrate, the chief component of the cell wall in most plants; an insoluble complex carbohydrate formed of microfibrils of glucose molecules.

CELL WALL: The rigid outermost layer of the cells found in plants, some protista, and prokaryotes.

CENTRIC DIATOM [Gk. *kentron,* center]: A radially symmetrical diatom.

CENTRIOLE (sĕn•trĭ•ōl) [Gk. *kentron,* center, + L. *-ole,* little one]: A cytoplasmic organelle generally found in animal cells and flagellated cells in other groups; usually outside of the nuclear membrane, which doubles before mitosis; the two centrioles then move apart and organize the spindle apparatus.

CENTROMERE (sĕn'tro•mēr) [Gk. *kentron,* center, + *meros,* a part]: That portion of the chromosome to which the spindle fiber is attached; also called the kinetochore.

CHALAZA [Gk. *chalaza,* small tubercle]: The region of an ovule or seed where the funiculus unites with the integument(s) and nucellus.

CHEMICAL POTENTIAL: The activity or free energy of a substance, dependent upon the rate of motion of the average molecule and the concentration of the molecules.

CHEMOAUTOTROPHIC (kĕm'o•ô•to•trŏf'ĭk): Organisms (bacteria) able to manufacture their own basic foods with chemical energy; *see* autotroph.

CHIASMA (kī•ăz'mȧ) [Gk. *chiasma,* a cross]: The x-shaped figure formed by the meeting of two nonsister chromatids of homologous chromosomes; probably the site of crossing-over.

CHITIN (kī'tīn) [Gk. *chiton,* a coat of mail]: A tough, resistant polysaccharide forming the cell walls of certain fungi, the exoskeleton of arthropods, and the epidermal cuticle or other surface structures of many other invertebrates.

CHLOR- [Gk. *chloros,* green]: Prefix meaning "green."

CHLORENCHYMA (klo•rĕng'kĭ•mȧ): General term applied to parenchyma cells that contain chloroplasts.

CHLOROPHYLL (klō'ro•fĭl) [Gk. *chloros,* green, + *phyllon,* leaf]; The green pigment of plant cells, necessary for photosynthesis; also found in some protista and some prokaryotes.

CHLOROPLAST: A membrane-bounded organelle in algal and green plant cells in which chlorophylls are contained; site of photosynthesis.

CHLOROSIS (klo•rō'sĭs): Loss or reduced development of chlorophyll.

CHROMA [Gk. *chroma,* color]: Prefix meaning "color."

CHROMATID (krō'mȧ•tĭd) [Gk. *chroma,* color, + L. *-id,* daughters of]: One of the two daughter strands of a duplicated chromosome which are joined by a single centromere.

CHROMATIN (krō'mȧ•tīn): The deeply staining nucleoprotein complex of the chromosomes.

CHROMATOPHORE [Gk. *chroma,* color, + *phorus,* a bearer]: In some bacteria, a discrete vesicle delimited by a single membrane and containing photosynthetic pigments.

CHROMOPLAST: A plastid containing pigments other than chlorophyll, usually yellow and orange carotenoid pigments.

CHROMOSOME (krō'mo•sōm) [Gk. *chroma,* color, + *soma,* body]: The body in the cell nucleus containing genes in a linear order; threads or rods of chromatin that appear during mitosis and meiosis and that bear the genes.

CILIUM, *pl.* CILIA (sĭl'ĭ•ŭm) [L. eyelash]: A short, hairlike structure present on the surface of certain specialized cell types; has a highly characteristic internal structure of two inner fibrils surrounded by nine pairs of outer fibrils; usually numerous and arranged in rows.

CIRCADIAN RHYTHMS (sēr'kăh•de•ăn) [L. *circa,* about, + *dies,* day]: Regular rhythms of growth and activity, which occur on a close to 24-hour basis.

CIRCINATE VERNATION [L. *circinare,* to make round, + *vernare,* to flourish, be verdant]: As in ferns, coiled arrangement of leaves and leaflets in the bud; it uncoils gradually as the leaf develops further.

CISTERNA, *pl.* CISTERNAE [L. *cistern,* a reservoir]: A flattened or saclike portion of the endoplasmic reticulum or a Golgi body, or dictyosome.

CLADOPHYLL ('klad•ē•fyl) [Gk. *klados,* shoot, + *phyllon,* leaf]: A branch resembling a foliage leaf.

CLAMP CONNECTION: In the Basidiomycetes, a lateral connection between adjacent cells of a dikaryotic hypha; ensures that each cell of the hypha will contain two dissimilar nuclei.

CLASS: A taxonomic category between division (or phylum) and order.

CLEISTOTHECIUM (klīs'to•thē'shĭ•ŭm) [Gk. *kleistos,* closed, + *thekion,* small receptacle]: A closed, spherical ascocarp.

CLIMAX COMMUNITY [Gk. *klima,* ladder]: Final or stable community in successional series that is more or less in equilibrium with existing environmental conditions.

CLINE (klīn) [Gk. *klin-,* stem of *klinein,* to slope]: Gradation of differences within a species over a geographic area.

CLONE [Gk. *klon,* twig]: A population of individuals descended by mitotic division from a single ancestor.

CLOSED BUNDLE: A vascular bundle in which a cambium does not develop.

COALESCENCE [L. *coalescere,* to grow together]: The union of floral parts of the same whorl, as petals to petals.

COCCUS, *pl.* COCCI (kŏk'ŭs) [Gk. *kokkos,* a berry]: A spherical bacterium.

CODON (kō'dŏn): Sequence of three adjacent nucleotides that form the code for a single amino acid.

COENOCYTIC (se•nō•sĭ'tic) [Gk. *koinos,* shared in common, + *kytos,* a hollow vessel]: A term used to describe an organism or part of an organism that is multinucleate, the nuclei not separated by walls; siphonaceous; siphonous.

COENZYME (kō•ĕn'zīm): An organic molecule which plays an accessory role in enzyme-catalyzed processes, often by acting as a donor or acceptor of a substance involved in the reaction; ATP and NAD are common coenzymes.

COHESION (lo-hē'zhŭn) [L. *cohaerere,* to stick together]: Union or holding together of parts of the same materials.

COLEOPTILE (kō'le•op'till) [Gk. *koleos,* sheath, + *ptilon,* feather]: The sheath enclosing the apical meristem and leaf primordia of the grass embryo. Often interpreted as the first leaf.

COLEORHIZA (kō'le•o•rī'za) [Gk. *koleos,* sheath, + *rhiza,* root]: The sheath enclosing the radicle in the grass embryo.

COLLENCHYMA (kō•lĕng'kĭ•ma) [Gk. *kolla,* glue]: A supporting tissue composed of collenchyma cells; often found in regions of primary growth in stems and in some leaves.

COLLENCHYMA CELL: Elongated living cell with irregularly thickened primary cell wall.

COLLOID (kŏl'oid): A permanent suspension of fine particles.

COMMUNITY: All the organisms inhabiting a common environment and interacting with one another.

COMPANION CELL: A specialized parenchyma cell associated with the sieve-tube members in angiosperm phloem and arising from the same mother cell as the sieve-tube member.

COMPETITION: The effect of a common demand by two or more organisms on a limited supply of food, water, light, minerals, etc.

COMPLETE FLOWER: A flower having four whorls of floral parts: sepals, petals, stamens, and carpels.

COMPOUND: A combination of atoms in definite ratios, held together by chemical bonds.

COMPOUND LEAF: A leaf whose blade is divided in several distinct leaflets.

CONCENTRATION GRADIENT: The concentration difference of a substance per unit distance.

CONE: Reproductive structure consisting of a number of sporophylls more or less compactly grouped on a central axis; a strobilus.

CONIDIOPHORE (ko•nĭd'ĭ•o•fōr): Hypha on which one or more conidia are produced.

CONIDIUM, *pl.* CONIDIA (ko•nĭd'ĭ•ŭm) [Gk. *konis,* dust]: An asexual fungus spore not contained within a sporangium. It may be produced singly or in chains. Most conidia are multinucleate.

CONIFER (kō'nĭ•fēr) [Gk. *kanōs,* cone, + L. *ferre,* to carry]: A cone-bearing tree.

CONJUGATION: The process in bacteria, protozoa, and certain algae and fungi by which two individuals transfer nuclear material.

CONJUGATION TUBE: As in the green alga, *Spirogyra,* a tube through which a gamete, or gametes, may move to unite with other gametes.

CONNATE (kōn'āt): Said of similar parts that are united or fused, as petals fused in a corolla tube.

CONVERGENT EVOLUTION [L. *convergere,* to turn together]: The independent development of similar structures in forms of life that are unrelated or only distantly related; often found in organisms living in similar environments; also called parallel evolution.

CORK, OR PHELLEM: A secondary tissue produced by a cork cambium; made up of polygonal cells, nonliving at maturity, with walls infiltrated with suberin, a waxy or fatty material resistant to the passage of gases and water vapor; outer part of the periderm.

CORK CAMBIUM, OR PHELLOGEN: The lateral meristem that forms the periderm, producing cork (phellem) toward the surface (outside) of the plant and phelloderm toward the inside; common in stems and roots of gymnosperms and dicots.

CORM: A short, thickened, underground stem, upright in position, in which food is accumulated, usually in the form of starch.

COROLLA (ko•rŏl'ȧ) [L. *corolla,* dim. of *corona,* a wreath, crown]: Petals, collectively; usually the conspicuously colored flower whorl.

COROLLA TUBE: A tubelike structure resulting from the fusion of the petals along their edges.

CORTEX [L. *cortic-, cortex,* bark, cork]: Ground-tissue region of a stem or root bounded externally by the epidermis and internally by the vascular system; a primary tissue region.

COTYLEDON (kŏt'ĭ•lē'dŭn) [Gk. *kotyledon,* cup-shaped hollow]: Seed leaf: generally stores food in dicotyledons; and generally absorbs food in the monocotyledons.

COVALENT BOND: A chemical bond formed between atoms as a result of the sharing of two electrons.

CRISTAE (krĭs'tē) [L. *crista,* crest]: A crest or ridge, used here to designate the infoldings of the inner mitochondrial membrane.

CROP ROTATION: The practice of growing different crops in regular succession to aid in the control of insects and diseases, to increase soil fertility, and to decrease erosion.

CROSS-POLLINATION: The transfer of pollen from the anther of one plant to the stigma of a flower of another plant.

CROSS SECTION: *See* transverse section.

CROSSING-OVER: The exchange of corresponding segments of genetic material between chromatids of homologous chromosomes at meiosis.

CRYPTOGAM [Gk. *kryptos,* hidden, + *gamete,* wife]: An archaic term for all organisms except the flowering plants (phanerogams), animals, and heterotrophic Protista.

CUTICLE (kū'tĭk'l) [L. *cuticula,* dim. of *cutis,* the skin]: Waxy or fatty layer on outer wall of epidermal cells.

CUTIN (kū'tĭn) [L. *cutis,* skin]: Fatty substance deposited on outer surface of plant cell walls exposed to air; forms a layer known as the cuticle.

CYCLOSIS (sī•klō'sis) [Gk. *kyklosis*, circulation]: The streaming of cytoplasm within a cell.

-CYTE, CYTO- [Gk. *kytōs*, vessel, container]: Suffix or prefix meaning "pertaining to cell."

CYTOKINESIS: *See* cell division.

CYTOKININ (sī'to•kĭ•nĭn) [Gk. *kytōs*, hollow vessel, + *kinesis*, motion]: Class of plant growth substance promoting cell division, among other effects.

CYTOLOGY: The study of the cell.

CYTOPLASM (sī'to•plăz'm): Term commonly used to refer to the protoplasm of the cell exclusive of the nucleus.

CYTOPLASMIC GROUND SUBSTANCE: The least differentiated part of the cytoplasm, as seen with the electron microscope, and the part surrounding the nucleus and various organelles; also called hyaloplasm.

CYTOSINE (sī'to•sēn): A pyrimidine base found in the nucleic acids DNA and RNA.

DE- [L. *de-*, away from, down, off]: Prefix meaning "down," "away from" or "off," for example, dehydration, removal of water.

DECIDUOUS (de•sĭd'u•ŭs) [L. *decidere*, to fall off]: Shedding leaves at a certain season.

DECOMPOSERS: Organisms (bacteria, fungi) in an ecosystem that break down organic material into smaller molecules so that they can recirculate.

DEHISCENCE (de•hĭs'ĕns) [L. *de*, down, off, from, + *hiscere*, to gape, split open]: The opening of an anther, fruit, or other structure, permitting the escape of reproductive bodies contained within.

DENITRIFICATION (dē•nī'trĭ•fĭ•kā'shŭn): The process by which nitrogen is released from the soil by the action of denitrifying bacteria.

DEOXYRIBONUCLEIC ACID (DNA) (dĕ•ŏk'-sĭ•rī'bo•nu•klē'ik): Carrier of genetic information in cells, composed of chains of phosphate, sugar molecules (deoxyribose), and purines and pyrimidines; capable of self-replication as well as of determining RNA synthesis.

DERMAL TISSUE SYSTEM: The epidermis or the periderm.

DETERMINATE GROWTH: Growth of limited duration, as is characteristic of floral meristems and of leaves.

DEUTERIUM (dū'tĕr'ĭ•ŭm): Or heavy hydrogen; a hydrogen atom, the nucleus of which contains one proton and one neutron. The common nucleus of hydrogen consists of only one proton.

DICHOTOMY (dī•kŏt'o•mī): The division or forking of an axis into two branches.

DICOTYLEDON (dī•kŏt'ĭ•lē'dŭn): One of the two classes of angiosperms, Dicotyledoneae, often abbreviated as dicot; a plant whose embryo has two cotyledons.

DICTYOSOME: *See* Golgi body.

DIFFERENTIALLY PERMEABLE MEMBRANE: A membrane through which substances diffuse at different rates.

DIFFERENTIATION (dĭf'ĕr•ĕn'shĭ•ā'shŭn): A process by which a relatively unspecialized cell undergoes a progressive change to a more specialized cell; the specialization of cells and tissues for particular functions during development.

DIFFUSE-POROUS WOOD: A wood in which the pores, or vessels, are fairly uniformly distributed throughout the growth layers or in which the size of pores changes only slightly from early wood to late wood.

DIFFUSION (dĭ•fū'zhŭn) [L. *diffundere*, to pour out]: The movement of suspended or dissolved particles from a more concentrated to a less concentrated region as a result of the random movement of individual molecules; the process tends to distribute them uniformly throughout a medium.

DIGESTION: The conversion of complex, usually insoluble foods into simple, usually soluble forms by means of enzymatic action.

DIKARYON (dī•kăr'ē•ŏn) [Gk. *di*, two, + *karyon*, a nut]: In fungi, mycelium with paired nuclei, each usually derived from a different parent.

DIKARYOTIC (dī•kăr'ĭ•ŏ'tĭc) [Gk. *di*, two + *karyon*, nut, kernel]: In fungi, having pairs of nuclei within cells or compartments.

DIMORPHISM (d ī•môr'fiz'm) [Gk. *di*, two, + *morphe*, form]: The condition of having two distinct forms, such as sterile and fertile leaves in ferns, or sterile and fertile shoots in horsetails.

DIOECIOUS (dī•ē'shŭs) [Gk. *di*, two + *oikos*, house]: Unisexual; having the male and female (or staminate and ovulate) elements on different individuals of the same species.

DIPLOID (dĭp'loid): Having two sets of chromosomes; the $2n$ number characteristic of the sporophyte generation.

DISK FLOWERS: The actinomorphic, tubular flowers that compose the central part of a head of flowers in most Asteraceae; contrasted with the flattened, zygomorphic ray-shaped flowers (ray flowers) on the margins of the head.

DISTAL: Situated away from or far from point of reference (usually the main part of body); opposite to proximal.

DIVISION: One of the major kinds of groups used by botanists in classifying organisms; the taxonomic category between kingdom and class. *See also* phylum.

DNA: *See* deoxyribonucleic acid.

DOMINANT: Applied to a gene that exerts its full phenotypic effect regardless of its allelic partner; a gene that masks the effect of its allele.

DORMANCY: A period of inactivity in bulbs, buds, seeds, and other plant organs, during which growth ceases and is resumed only if certain requirements, as of moisture, temperature or day length, have been fulfilled.

DOUBLE FERTILIZATION: The fusion of the egg and sperm (resulting in a $2n$ fertilized egg, the zygote) and the simultaneous fusion of the second male gamete with the polar nuclei (resulting in a $3n$ primary endosperm nucleus); a unique characteristic of all angiosperms.

DOUBLING RATE: The length of time required for a population of a given size to double in number.

DRUPE (droop) [Gk. *dryppa*, overripe olive]: A simple, fleshy fruit, derived from a single carpel, usually one-seeded, in which the inner fruit coat is adherent to the seed.

DRUSE: A compound, more or less spherical crystal with many component crystals projecting from its surface; composed of calcium oxalate.

EARLY WOOD: The first-formed wood of a growth layer. It contains larger cells and is less dense than the subsequently formed late wood. Replaces the term spring wood.

ECO- [Gk. *oikos*, house, home]: Prefix, meaning "house" or "home."

ECOLOGY (ē•kŏl'ō•ji): The study of the relationships between organisms and their environment.

ECOSYSTEM (ē'ko•sĭs'tĕm): An interacting system of living organisms and their nonliving environment.

ECOTYPE (ē'co•tīp) [Gk. *oiko-*, from *oikos*, house, + L. *typus*, a figure, image]: Locally adapted variant of an organism.

ECTOPLAST: *See* plasma membrane.

EDAPHIC (e•dăf'ĭk) [Gk. *edaphos*, bottom, ground, soil]: Pertaining to soil conditions that influence plant growth.

EGG: A nonmotile female gamete.

EGG APPARATUS: The egg cell and synergids located at the micropylar end of the female gametophyte, or embryo sac, of angiosperms.

ELATER (ĕl'ȧ•tẽr) [Gk. *elater*, driver]: (1) An elongated, spindle-shaped, sterile cell in the sporangium of a liverwort sporophyte (aiding in spore dispersal); (2) clubbed, hygroscopic bands attached to the spores of the horsetails.

ELECTROLYTE (e•lĕk'tro•līt) [Gk. *elektron*, amber, + *lytos*, soluble]: A substance that dissociates into ions in aqueous solution and so makes possible the conduction of an electric current through the solution.

ELECTRON: A subatomic particle with a negative electric charge equal in magnitude to the positive charge of the proton, but with a mass of $\frac{1}{1837}$ of that of the proton. Electrons surround the atom's positively charged nucleus and determine the atom's chemical properties.

ELECTRON-DENSE: In electron microscopy, not permitting the passage of electrons and so appearing dark in the micrograph.

ELEMENT: A substance composed of only one kind of atom; one of about 100 distinct natural or man-made types of matter that, singly or in combination, compose all materials of the universe.

EMBRYO [Gk. *en*, in, + *bryein*, to swell]: A young sporophytic plant, before the start of a period of rapid growth (germination in seed plants).

EMBRYOGENY: The formation of the embryo.

EMBRYO SAC: The female gametophyte of angiosperms, generally an eight-nucleate, seven-celled structure. The seven cells are the egg cell, two synergids and three antipodals (each with a single nucleus), and the central cell (with two nuclei).

ENDERGONIC (ĕn'dẽr•gŏ•nĭk): Energy-requiring, as in a chemical reaction; applied to an "uphill" process.

ENDO- [Gk. *endon*, within]: Prefix meaning "within."

ENDOCARP (ĕn'dō•kärp) [Gk. *endon*, within, + *karpos*, fruit]: The innermost layer of the mature ovary wall, or pericarp.

ENDODERMIS (ĕn'do•dûr'mĭs) [Gk. *endon*, within, + *derma*, skin]: A single layer of cells forming a sheath around the vascular region in roots and some stems; characterized by a Casparian strip within radial and transverse walls. In roots and stems of seed plants, the endodermis is the innermost layer of the cortex.

ENDOGENOUS [Gk. *endon*, within, + *genos*, race, kind]: Arising from deep-seated tissues, as lateral roots.

ENDOPLASMIC RETICULUM (ĕn•do•plăz'mĭk re•tĭk'u•lŭm): An extensive system of double membranes present in most cells, dividing the cytoplasm into compartments and channels, often coated with ribosomes (rough endoplasmic reticulum).

ENDOSPERM [Gk. *endon*, within, + *sperma*, seed]: A tissue containing stored food that develops from the union of a male nucleus and polar nuclei of central cell, found only in angiosperms, which is digested by the growing sporophyte either before or after the maturation of the seed.

ENZYME [Gk. *en*, in, + *zyme*, yeast]: A protein of complex chemical constitution produced in living cells that, even in very low concentration, speeds the rate of (catalyzes) a chemical reaction.

EPI- [Gk. *epi*, upon]: Prefix meaning "upon" or "above."

EPICOTYL (ĕp'ĭ•kŏt'ĭl): The upper portion of the axis of embryo or seedling, above the cotyledons.

EPIDERMIS (ĕp'ĭ•dûr'mĭs): The outermost layer of cells of the leaf and of young stems and roots; primary in origin.

EPIGYNY [Gk. *epi*, upon, + *gyne*, woman]: Floral organization in which the sepals, petals, and stamens apparently grow from the top of the ovary.

EPIPHYTE (ĕp'ĭ•fīt): An organism that grows upon another plant but is not parasitic upon it.

EPISTATIC (ĕp' ĭ•stăt'ĭk) [NL., fr. Gk. *epistasis*, a stopping, scum]: Term used to describe a gene the action of which determines whether or not the effects of another gene will occur.

ETHYLENE (ĕth'ĭlēn): A simple hydrocarbon that is a plant growth substance involved in the ripening of fruit.

ETIOLATION (ē'tĭ•o•lā'shŭn) [Fr. *étioler*, to blanch): A condition involving increased stem elongation, poor leaf development, and lack of chlorophyll found in plants growing in the dark or with a greatly reduced amount of light.

EUKARYOTIC (u•kă•rē•äd'ik) [Gk. *eu*, good, + *karyon*, nut, kernel]: Applied to organisms having membrane-bound nuclei; contrasts with prokaryotic.

EUSTELE: A stele in which the primary vascular tissues are arranged in discrete strands around a pith; typical of gymnosperms and angiosperms.

EXERGONIC (ĕk'sẽr•gŏ•nĭk) [L. *ex*, out, + Gk. *ergon*, work]: Energy-producing, as in a chemical reaction; applied to a "downhill" process.

EXINE [L. *exterus*, outside]: The outer wall layer of a spore or pollen grain.

EXOCARP (ĕk'sō•kärp) [Gk. *exo*, without, + *karpos*, fruit]: The outermost layer of the mature ovary wall, or pericarp.

EYESPOT: Also stigma; a small, pigmented structure in flagellated unicellular organisms that may be sensitive to light.

F_1: First filial generation in a cross between any two parents; F_2 and F_3 are the second and third generations.

FALSE ANNUAL RING: In wood, one of more than one growth layer formed during a single growing season, as seen in transverse section.

FAMILY: A taxonomic group between order and genus; the suffix to the family name in animals and some protista is -idae; in other organisms, including plants and fungi, it is -aceae.

FASCICLE (făs'ĭ•k'l) [L. *fasciculus*, a small bundle]: A bundle of pine leaves or other needle leaves; a vascular bundle (obsolete).

FASCICULAR CAMBIUM: The vascular cambium originating within a vascular bundle, or fascicle.

FATS: Organic compounds containing carbon, hydrogen, and oxygen, as in carbohydrates. The proportion of oxygen to carbon is much less in fats than it is in carbohydrates. Fats in the liquid state are called oils.

FERREDOXIN: An electron-transferring protein of high iron content; some are involved in photosynthesis.

FERTILIZATION: The fusion of two gametes, especially of their nuclei, to form a diploid zygote.

FIBER: An elongated, tapering, generally thick-walled sclerenchyma cell of vascular plants. Its walls may or may not be lignified; it may or may not have a living protoplast at maturity.

FIBRIL (fī'brĭl): Submicroscopic threads, composed of cellulose molecules, that constitute the form in which cellulose occurs in the cell wall.

FIELD CAPACITY: The amount of water a particular soil will hold against the action of gravity.

FIELD PERCENTAGE: The normal upper limit of the available capillary water.

FILAMENT (fĭl'ă•mĕnt): (1) the stalk of a stamen; (2) a term used to describe the threadlike bodies of certain algae or fungi.

FISSION (fĭsh'ŭn): Asexual reproduction involving the division of a single-celled individual into two new single-celled individuals of equal size.

FITNESS: The reproductive capacity of an organism.

FLAGELLUM, *pl.* FLAGELLA (flȧ•jĕl'ŭm) [L. *flagellum*, whip]: A fine, long thread, composed of protoplasm, protruding from a cell body; longer than a cilium, but having the same internal structure; capable of a vibratory motion; used in locomotion and feeding; common in algae, fungi, and motile gametes.

FLAVOPROTEIN: A dehydrogenase that contains a flavin and often a metal and plays a major role in oxidation.

FLORAL TUBE: A cup or tube formed by the fusion of the basal parts of the sepals, petals, and stamens, often in plants that have an inferior ovary.

FLORET (flō'rĕt) [Fr. *fleurette*, dim. of *fleur*, flower]: One of the small flowers that make up the composite inflorescence or the spike of the grasses.

FLORIDEAN STARCH: The reserve carbohydrate of the red algae.

FLOWER: The reproductive structure of angiosperms; a complete flower includes calyx, corolla, androecium (stamens), and gynoecium (carpels), but all contain at least one stamen or one carpel.

FOLLICLE (fŏl'ĭ•k'l) [L. *folliculus*, small ball]: A dry, dehiscent fruit derived from a single carpel and opening along one side.

FOOD CHAIN, FOOD WEB: A chain of organisms existing in any natural community such that each link in the chain feeds on the one below and is eaten by the one above; there are seldom more than six links in a chain, with autotrophs on the bottom and the largest carnivores at the top.

FOSSIL [L. *fossilis*, dug up]: The remains, impressions, or traces of an organism that has been preserved in the earth's crust.

FP: *See* flavoprotein.

FREE ENERGY: Energy available to do work.

FROND: The leaf of a fern; any large, divided leaf.

FRUIT: In angiosperms, a matured, ripened ovary (or group of ovaries), containing the seeds; also applied informally, for example, *fruiting* body, to reproductive structures of other groups of plants, together with any adjacent parts that may be fused with it at maturity.

FRUSTULE [L. *frustulum*, a small piece]: The siliceous cell wall of a diatom.

FUCOXANTHIN (fū'ko•zăn'thĭn) [Gk. *phykos*, seaweed, + *xanthos*, yellowish-brown]: A brown pigment found in brown algae and certain groups of protista.

FUNDAMENTAL TISSUE SYSTEM: *See* ground tissue system.

FUNICULUS [L. *funiculus*, dim. of *funis*, rope or small cord]: The stalk of the ovule.

FUSIFORM INITIALS [L. *fusus*, spindle]: The vertically elongated cells in the vascular cambium that give rise to the cells of the axial system in the secondary xylem and secondary phloem.

GAMETANGIUM (găm'e•tăn'jĭ•ŭm) [Gk. *gamein*, to marry, + L. *tangere*, to touch]: General term applied to any cell or organ in which gametes are formed.

GAMETE (gȧ•mēt') [Gk. *gamete*, wife]: The mature functional haploid reproductive cell whose nucleus fuses with that of another gamete of opposite sex (fertilization) with the resulting diploid cell (zygote) developing into a new individual.

GAMETOPHORE [Gk. *gamein*, to marry, + *phoros*, bearing]: In the bryophytes, a fertile stalk that bears gametangia.

GAMETOPHYTE (gȧ•mē'•to•fīt): In plants having alternation of generations, the haploid ($1n$), gamete-producing phase.

GEL: A mixture of semisolid or solid constituency containing a large proportion of liquid trapped in its solid component.

GEMMA, *pl.* GEMMAE (jĕm'ă) [L. *gemma*, bud]: A small mass of vegetative tissue; an outgrowth of the thallus, for example, in liverworts or certain fungi. It can develop into an entire new plant.

GENE: A unit of heredity which is a portion of the DNA of a chromosome.

GENE FREQUENCY: The incidence (relative occurrence) of a particular allele in a population.

GENE POOL: All the alleles of all the genes in a population.

GENERATIVE CELL: (1) In many gymnosperms, the cell of the male gametophyte that divides to form the stalk and body cells; (2) in angiosperms, the cell of the male gametophyte that divides to form two sperms.

GENE RECOMBINATION: The appearance of gene combinations in the progeny different from the combinations present in the parents, as a result of the sexual process.

GENETIC CODE: The three-symbol system of base-pair sequences in DNA; referred to as a code because it determines the amino acid sequence in the enzymes and other protein components synthesized by the organism.

GENOTYPE (jēn'o•tīp): The genetic constitution, latent or expressed, of an organism, as contrasted with the phenotype; the sum total of all the genes present in an individual.

GENUS, *pl.* GENERA (jēn'ŭs) The taxonomic group between family and species; includes one or more species that have certain characteristics in common.

GEOFLORA (je•o•flō•ra): A biome having great extent in both space and time.

GEOTROPISM (je•ŏt'•ro•pĭz'm) [Gk. *ge*, earth, + *tropes*, turning]: Direction of growth induced by gravity.

GERMINATION [L. *germinare*, to sprout]: The beginning or resumption of growth by a spore, seed, bud, or other structure.

GIBBERELLINS (jĭb•ĕ•rĕ'lĭn) [fr. NL *Gibberella*, genus of fungi]: A group of growth hormones, the best known effect of which is to increase the elongation of stems in a number of higher plants.

GILL: The plates on the underside of the cap in the gill fungi class (Basidiomycetes).

GIRDLING: The removal from a woody stem of a ring of bark extending inward to the cambium; also called ringing.

GLUCOSE (glōō'kōs): A common six-carbon sugar ($C_6H_{12}O_6$); the most common monosaccharide.

GLYCOGEN [Gk. *glykys*, sweet, + *gen*, of a kind]: A carbohydrate similar to starch, the reserve food of the blue-green algae; also known as cyanophycean starch.

GLYCOLYSIS (glī•kŏl'ĭ•sĭs): A process in which sugar is changed anaerobically to pyruvic acid with the liberation of a small amount of useful energy.

GLYOXYSOME: A microbody containing enzymes necessary for the conversion of fats into carbohydrates during germination in many seeds.

GOLGI BODY (gôl'je): An organelle present in the cells of eukaryotes consisting of a group of flat, disk-shaped sacs that are often branched into tubules at the margins; they seem to function as collecting and packaging centers for substances the cells manufacture; often called dictyosomes in plants. The term Golgi apparatus is used to refer collectively to all of the Golgi bodies of a given cell.

GRAFTING: A union of different individuals in which a portion, the *scion*, is inserted on a root or stem, the *stock*.

GRAIN: *See* caryopsis.

GRANA: Structures within chloroplasts, seen as green granules with the light microscope and as a series of stacked thylakoids with the electron microscope. The grana contain the chlorophylls and carotenoids and are the actual site of the light reactions of photosynthesis.

GROUND MERISTEM [Gk. *meristos*, divisible]: A primary meristem, or meristematic tissue, which gives rise to the ground tissues.

GROUND TISSUE SYSTEM: All tissues other than the epidermis (or periderm) and the vascular tissues; also called fundamental tissue system.

GROWTH LAYER: A layer of growth in secondary xylem or secondary phloem; may be an annual ring or a false annual ring.

GROWTH RING: A growth layer in secondary xylem or secondary phloem, as seen in transverse section; may be called growth increment, especially where seen in other than transverse section.

GUANINE (gŭ'a•nēn): A purine base found in DNA and RNA.

GUARD CELLS: Specialized epidermal cells surrounding a pore, or stoma; changes in turgor of a pair of guard cells causes opening and closing of the pore.

GUTTATION (gŭ•tā'shŭn) [L. *gutta*, a drop]: The exudation of liquid water from leaves.

GYMNOSPERM (jĭm'no•spûrm) [Gk. *gumnos*, naked, + *sperma*, seed]: A seed plant with seeds not enclosed in an ovary; the conifers are the most familiar group.

GYNOECIUM (jĭ•nē'sĭ•ŭm) [Gk. *gyne*, woman, + *oikos*, house]: The aggregate of carpels in the flower of a seed plant.

HABIT [L. *habitus*, to have]: Characteristic form or bodily appearance of an organism.

HABITAT [L. it inhabits, fr. *habitare*]: The natural environment of an organism; the place where it is usually found.

HAPLOID (hăp'loid) [Gk. *haploos*, single]: The state in which each chromosome is represented only once ($1n$).

HARDWOOD: A name commonly applied to the wood of a dicot tree.

HARDY-WEINBERG LAW: The mathematical expression of the relationship between relative frequencies of two or more alleles in a population; it demonstrates that the frequencies of dominant and recessive genes tend to remain constant.

HAUSTORIUM, *pl.* HAUSTORIA (hôs•tō'rĭ•ŭm) [M.L. *haustrum*, pump]: A projection of fungal hypha, stem, or plant part that acts as a penetrating and absorbing organ.

HEARTWOOD: Nonliving and commonly dark-colored wood in which no water transport occurs; it is surrounded by sapwood.

HEMICELLULOSE (hĕm'ĭ•sĕl'u•lōs): Polysaccharide resembling cellulose but more soluble and less ordered; found particularly in cell walls.

HERB (ûrb; hûrb) [L. *herba*, grass]: A nonwoody seed plant with a relatively short-lived aerial portion.

HERBACEOUS (hûr•bā'shŭs): A term referring to any nonwoody plant.

HERBARIUM (hûr•bâr'ĭ•ŭm) [L. *herba*, grass]: A collection of dried and pressed plant specimens.

HERBIVORE (hûr'bĭ•vōr): A plant eater.

HERMAPHRODITE (hûr•măf'ro•dīt) [fr. Gk. for Hermes and Aphrodite]: An organism possessing both male and female reproductive organs.

HETERO- [Gk. *heteros*, other, different]: Prefix meaning "other" or "different."

HETEROCYST (hĕt'ĕr•ō•sĭst) [Gk. *heteros*, different, + *cystis*, a bag]: A large, transparent, thick-walled cell in the filaments of certain blue-green algae.

HETEROECIOUS (hĕt'ĕr·ē'shŭs) [Gk. *heteros*, different, + *oikos*, house]: As in some rust fungi, requiring two different host species to complete the life cycle.

HETEROGAMY (hĕt'ĕr·ŏg'ȧ·mĭ) [Gk. *heteros*, different, + *gamos*, union or reproduction]: Reproduction involving two types of gametes.

HETEROKARYOTIC (hĕt'ĕr·o·kar'ĭ·ō'tĭc) [Gk. *heteros*, other + *karyon*, nut, kernel]: In fungi, having two or more genetically distinct types of nuclei within the same mycelium.

HETEROMORPHIC [Gk. *heteros*, different, + *morphe*, form]: A term used to describe a life history in which the haploid and diploid generations are dissimilar in form.

HETEROSIS (hĕt'ĕr·ō·sĭs) [Gk. *heterosis*, alteration]: Hybrid vigor, the superiority of the hybrid over either parent in any measurable character.

HETEROSPOROUS (hĕt'ĕr·ŏs'po·rŭs): Having spores of two kinds, usually designated as microspores and megaspores.

HETEROTHALLIC (hĕt'ĕr·ō·thăl'ĭk) [Gk. *heteros*, different, + *thallus*]: A term used to describe a species, the individuals of which are self-sterile or self-incompatible; two compatible strains or individuals are required for sexual reproduction.

HETEROTROPH (hĕt'ĕr·o·trŏf') [Gk. *heteros*, other, + *trophos*, feeder]: An organism that cannot manufacture organic compounds and so must feed on organic food materials that have originated in other plants and animals, in contrast with the autotrophs.

HETEROZYGOUS (hĕt·ĕr·o·zī'gŭs): Having two different alleles at the same locus on homologous chromosomes.

HILUM (hīlŭm) [L. *hilum*, a trifle]: (1) Scar left on seed after separation of seed from funiculus; (2) the part of a starch grain around which the starch is laid down in more or less concentric layers.

HOLDFAST: (1) Basal part of an algal thallus that attaches it to a solid object; may be unicellular or composed of a mass of tissue; (2) cuplike structures at the tips of some tendrils, by means of which they become attached.

HOMEO-, HOMO-, HOMOLO- [Gk. *homos*, same, similar]: Prefix meaning "similar" or "same."

HOMEOSTASIS (hō'me·ō·stā'sĭs) [Gk. *homos*, similar, + *stasis*, standing]: The maintaining of a relatively stable internal physiological environment or equilibrium in an organism, population, or ecosystem.

HOMOKARYOTIC (hō'mō·kăr'ĭ·ō'tĭc) [L. *homo*, man + Gk. *karyon*, nut, kernel]: In fungi, having nuclei with the same genetic makeup within a mycelium.

HOMOLOGOUS CHROMOSOMES: Chromosomes that associate in pairs in the first stage of meiosis; each member of the pair is derived from a different parent.

HOMOLOGY (ho·mŏl'o·jĭ) [Gk. *homologia*, agreement]: A condition indicative of the same phylogenetic, or evolutionary, origin, but not necessarily the same structure and/or function.

HOMOSPOROUS (ho·mŏs'po·rŭs) Having but one kind of spore.

HOMOTHALLIC (hō'mō·thăl'ĭk) [Gk. *homos*, same, + *thallus*]: A term used to describe a species, the individuals of which are self-fertile.

HOMOZYGOUS (hō'mo·zī'gŭs): Having identical alleles at the same locus on homologous chromosomes.

HORMONE [Gk. *hormaein*, to excite]: A chemical substance produced usually in minute amounts in one part of an organism, from which it is transported to another part of that organism on which it has a specific effect.

HOST: An organism on or in which a parasite lives.

HUMUS: Decomposing organic matter in the soil.

HYALOPLASM: *See* cytoplasmic ground substance.

HYBRID: Offspring of two parents that differ in one or more heritable characteristics; offspring of two different varieties or of two different species.

HYBRIDIZATION (hī·brĭd·ĭ·zā'shŭn): The formation of offspring between unlike parents.

HYBRID VIGOR: *See* heterosis.

HYDROGEN BOND: A weak bond between a hydrogen atom attached to one oxygen or nitrogen atom and another oxygen or nitrogen atom.

HYDROLYSIS (hī·drŏl'ĭ·sĭs) [Gk. *hydor*, water, + *lysis*, loosening]: Splitting of one molecule into two by addition of H^+ and OH^- ions of water.

HYDROPHYTE [Gk. *hydro*, water, + *phyton*, a plant]: A plant that grows wholly or partly submerged in water.

HYMENIUM [Gk. *hymen*, a membrane]: The layer of asci on an ascocarp, or of basidia on a basidiocarp, plus any associated sterile hyphae.

HYPER- [Gk. above, over]: Prefix meaning "above" or "over."

HYPERTONIC (hī'pēr·tŏn'ĭc): Having a concentration high enough to gain water across a membrane from another solution.

HYPHA, *pl.* **HYPHAE** [Gk. *hyphe*, web]: A single tubular filament of a fungus; the hyphae together comprise the mycelium.

HYPO- [Gk. less than]: Prefix meaning "under" or "less."

HYPOCOTYL (hī'po·kŏt'ĭl): The portion of an embryo or seedling situated between the cotyledons and the radicle.

HYPOCOTYL-ROOT AXIS: The embryo axis below the cotyledon or cotyledons consisting of the hypocotyl and the apical meristem of the root or the radicle.

HYPODERMIS [Gk. *hypo*, under, + *derma*, skin]: A layer or layers of cells beneath the epidermis, which are distinct from the underlying cortical or mesophyll cells.

HYPOGYNY [Gk. *hypo*, under, + *gyne*, female]: Floral organization in which the sepals, petals, and stamens are attached to the receptacle below the ovary.

HYPOTHESIS [Gk. *hypo*, under, + *tithenai*, to put]: A temporary working explanation or supposition based on accumulated facts and suggesting some general principle or relation of cause and effect; a postulated solution to a scientific problem that must be tested by experimentation and, if not validated, discarded.

HYPOTONIC (hi'po·tŏn'ĭc): Having a concentration low enough to lose water across a membrane to another solution.

IAA: *See* indoleacetic acid.

IMBIBITION (ĭm·bĭ·bĭsh'ŭn): Adsorption

of water and swelling of colloidal materials because of the adsorption of water molecules onto the internal surfaces of the materials.

IMPERFECT FLOWER: A flower lacking either stamens or carpels.

IMPERFECT FUNGI: Fungi reproducing only by asexual means, or in which the sexual cycle has not been observed. Most are Ascomycetes.

INBREEDING: The breeding of closely related plants or animals. In plants, it is usually brought about by repeated self-pollination.

INCOMPLETE DOMINANCE: The condition that results when two different alleles together produce an effect intermediate between the effects of these same genes in the homozygous condition.

INCOMPLETE FLOWER: A flower lacking one or more of the four kinds of floral parts: sepals, petals, stamens, or carpels.

INDEHISCENT (ĭn'de•hĭs'ĕnt): Remaining closed at maturity, as are many fruits (samaras, for example).

INDEPENDENT ASSORTMENT: *See* Mendel's second law.

INDETERMINATE GROWTH: Unrestricted or unlimited growth, as with a vegetative apical meristem that produces an unrestricted number of lateral organs indefinitely.

INDOLEACETIC ACID: The only known naturally occurring auxin; a growth regulator.

INDUSIUM, *pl.* INDUSIA (ĭn•dū'zi•ŭm) [L. *indusium*, a woman's undergarment]: Membranous growth of the epidermis of a fern leaf that covers a sorus.

INFERIOR OVARY: An ovary that is more or less united with the calyx, from the summit of which the other floral whorls appear to arise.

INFLORESCENCE (ĭn'flo•rĕs'ĕns): A flower cluster, with a definite arrangement of flowers.

INITIAL: (1) In meristem, a cell that remains within the meristem indefinitely and at the same time, by division, adds cells to the plant body; (2) of a cell, or element, a meristematic cell that eventually differentiates into a mature, more specialized cell, or element.

INNER BARK: In older trees, the living part of the bark. The bark inside the innermost periderm.

INTEGUMENT (ĭn•tĕg'u•mĕnt): Outermost layer or layers of tissue enveloping the nucellus of the ovule; develops into the seed coat.

INTER- [L. between]: Prefix meaning "between," "in between," or "in the midst of."

INTERCALARY (ĭn•tûr'kă•lĕr'ĭ) [L. *intercalare*, to insert]: Descriptive of meristematic tissue or growth not restricted to the apex of an organ; i.e., growth in the region of the nodes.

INTERCELLULAR SPACES: Spaces, of varying origin, among cells within a tissue.

INTERFASCICULAR CAMBIUM: The vascular cambium arising between the vascular bundles, or fascicles, from interfascicular parenchyma.

INTERFASCICULAR REGION: Tissue region between vascular bundles in stem; also called pith ray or medullary ray.

INTERNODE: The region of a stem between two successive nodes.

INTERPHASE: The state between two mitotic or meiotic cycles.

INTINE [L. *intus*, within]: The inner wall layer of a spore or pollen grain.

INTRA- [L. within]: Prefix meaning "within," for example, intracellular, "within cells."

INVOLUCRE (ĭn'vo•lū•kĕr) [L. *involucrum*, wrapper]: A whorl or rosette of bracts surrounding an inflorescence.

ION: An atom or molecule that has lost or gained one or more electrons. By this process, known as ionization, it becomes electrically charged.

IRREGULAR FLOWER: A flower in which one or more members of at least one whorl are of different form from other members of the same whorl.

ISO- [Gk. *isos*, equal]: Prefix meaning "equal"; like "homo-."

ISOGAMY (ī•sŏg'ă•mĭ): A type of sexual reproduction in algae and fungi in which the gametes (or gametangia) are alike in size.

ISOMER (ī'so•mĕr) [Gk. *isos*, equal, + *meros*, part]: One of a group of compounds identical in atomic composition but differing in structural arrangement, for example, glucose and fructose ($C_6H_{12}O_6$).

ISOMORPHIC [Gk. *isos*, equal, + *morphe*, form]: A term used to describe a life history in which the haploid and diploid generations are similar in form.

ISOTONIC (ī'sō•tŏnĭc): Having the same osmotic concentration.

ISOTOPE (ī'so•tōp): One of several possible forms of a chemical element, differing from other forms in the number of neutrons in the atomic nucleus, but not in chemical properties.

KARYOGAMY [Gk. *karyon*, nut, + *gamos*, marriage]: The union of two nuclei following plasmogamy.

KELP [M.E. *culp*, seaweed]: A common name for any of the large brown algae.

KINETIN (kĭ•nē'tĭn): [Gk. *kinetikos*, causing motion]: A purine that probably does not occur in nature but that acts as a cytokinin in plants.

KINETOCHORE: *See* centromere.

KINGDOM: The chief taxonomic category, for example, Monera or Plantae.

KREBS CYCLE: The series of reactions that results in the oxidation of pyruvic acid to hydrogen atoms, electrons, and carbon dioxide. The electrons, passed along electron-carrier molecules, then go through the oxidative phosphorylation and terminal oxidation processes.

LAMELLA (lă•mĕl'ă) [L., thin metal plate]: Layer of cellular membranes, particularly photosynthetic, chlorophyll-containing membranes.

LAMINA: *See* blade.

LAMINARIN [L. *lamina*, thin plate]: The principal storage product of the brown algae; a polymer of glucose and mannitol.

LATERAL MERISTEMS: Meristems that give rise to secondary tissue; the vascular cambium and cork cambium.

LATERAL ROOT: *See* branch root.

LATE WOOD: The wood formed in the later part of a growth layer. It contains smaller cells and is more dense than the early wood. Replaces the term summer wood.

LATOSOL (lat'ə,sôl) [L. *later*, brick, tile]: A red, iron-containing soil found in the tropics.

LEACHING: The downward movement and drainage of minerals from the soil by percolating water.

LEAF BUTTRESS: A lateral protrusion below the apical meristem, it represents the initial stage in the development of a leaf primordium.

LEAF GAP: Region of parenchyma tissue in the primary vascular cylinder above the point of departure of the leaf trace or traces.

LEAFLET: One of the parts of a compound leaf.

LEAF PRIMORDIUM [L. *primordium*, a beginning]: A lateral outgrowth from the apical meristem that will become a leaf.

LEAF SCAR: A scar left on the twig when the leaf falls.

LEAF TRACE: That part of the vascular bundle extending from the base of the leaf to its connection with a vascular bundle in the stem.

LEGUME [Fr. a vegetable]: (1) A member of the Fabaceae, the pea or bean family; (2) a type of dry fruit developed from one carpel and opening along two sides.

LENTICELS (len'tĭ•sĕls) [L. *lenticella*, a small lentil]: Spongy areas in the cork surfaces of stem, roots, and other plant parts that allow interchange of gases between internal tissues and the atmosphere through the periderm.

LEUCOPLAST (lū'ko•plăst) [Gk. *leuko*, white, + *plasein*, to form]: A colorless plastid, commonly the center of starch formation.

LIANA (le•ä'nȧ) [Fr. *liane*, from *lier*, to bind]: A large, woody vine that climbs upon other plants.

LICHEN (lī'kĕn) [Gk. *leichen*, thallus plants growing on rocks and trees]: An Ascomycete fungus which characteristically incorporates living algal cells, forming a composite organism.

LIFE CYCLE: The entire sequence of phases in the growth and development of any organism from time of zygote formation to gamete formation.

LIGNIN: A polymer of phenylpropanoid units, and one of the most important constituents of the secondary wall, although not all secondary walls contain lignin.

LIGULE [L. *ligula*, dimin. of *lingua*, tongue]: A minute outgrowth or appendage at the base of grass leaves and leaves of certain Lycophytina.

LINKAGE: The tendency for certain genes to be inherited together owing to the fact that they are located on the same chromosome.

LIPID (lip'pĭd) [Gk. *lipos*, fat]: A large variety of organic fat or fatlike compounds; including fats, oils, steroids, phospholipids, and carotenes.

LOCULE (lŏk'ūl) [L. *loculus*, dim. of *locus*, place]: A cavity within a sporangium or a cavity of the ovary in which ovules occur.

LOCUS (lō'kŭs): The position on a chromosome occupied by a particular gene.

LUMEN [L. *lumen*, light, an opening for light]: The space bounded by the plant cell wall.

LYSIS (lī'sĭs) [Gk. *lysis*, a loosening]: A process of disintegration or cell destruction.

LYSOGENIC BACTERIA (lī•so•jĕn'ĭk) [Gk. *lysis*, a loosening]: Bacteria-carrying viruses (phages) that eventually break loose from the bacterial chromosome and set up an active cycle of infection, producing lysis in their bacterial hosts.

LYSOSOME [Gk. *lysis*, loosening, + *soma*, body]: An organelle, bounded by a single membrane, and containing acid hydrolytic enzymes capable of breaking down proteins and other complex macromolecules.

MACROFIBRIL: An aggregation of microfibrils, visible with the light microscope.

MACROMOLECULE [Gk. *makros*, large]: A molecule of very high molecular weight; refers specifically to proteins, nucleic acids, polysaccharides, and complexes of these.

MALTASE (môl'tās): An enzyme that hydrolyzes maltose to glucose.

MALTOSE: Malt sugar.

MARGINAL MERISTEM: The meristem located along the margin of a leaf primordium and forming the blade.

MATING TYPE: A strain of organisms incapable of sexual reproduction with one another but capable of such reproduction with members of other strains of the same organism.

MEGA- [Gk. *megas*, great, large]: Prefix meaning "large."

MEGAGAMETOPHYTE [Gk. *megas*, large, + *gamos*, marriage, + *phyton*, plant]: In heterosporous plants, the female gametophyte; located within the ovule of seed plants.

MEGAPHYLL (mĕg'ȧ•fĭl) [Gk. *mega-*, *meg-*, from *megas*, *megalou*, great, mighty + *phyllon*, leaf]: A generally large leaf with several to many veins; its leaf trace is associated with a leaf gap. Contrasts with microphyll.

MEGASPORANGIUM (mĕg'ȧ•spo•răn'jĭ•ŭm) [Gk. *megas*, large sporangium]: Sporangium in which megaspores are produced.

MEGASPORE: In heterosporous plants, a haploid ($1n$) spore that develops into a female gametophyte; in some groups, megaspores are larger than microspores.

MEGASPORE MOTHER CELL: A diploid cell in which meiosis will occur, resulting in the production of four megaspores; also called megasporocyte.

MEGASPOROCYTE: *See* megaspore mother cell.

MEGASPOROPHYLL (mĕg'ȧ•spō'ro•fĭl) [Gk. *megos*, large spore, + *phyton*, leaf]: A leaf or leaflike structure bearing megasporangia.

MEIOSIS (mī•ō'sĭs) [Gk. *meioun*, to make smaller]: The two successive nuclear divisions in which the chromosome number is reduced from diploid ($2n$) to haploid ($1n$) and segregation of the genes occurs; gametes or, as in higher plants, spores may be produced as a result of meiosis.

MENDEL'S FIRST LAW: The factors for a pair of alternate characteristics are separate and only one may be carried in a particular gamete (genetic segregation).

MENDEL'S SECOND LAW: The inheritance of one pair of characteristics is independent of the simultaneous inheritance of other traits, such characteristics "assorting independently" as though there were no others present (later modified by the discovery of linkage).

MERISTEM (mĕr'ĭ•stĕm) [Gk. *merizein*, to divide]: The undifferentiated plant tissue from which new cells arise.

MESO- [Gk. *mesos*, middle]: Prefix meaning "middle."

MESOCARP (mĕs'ō•kärp) [Gk. *mesos*, middle, + *karpos*, fruit]: The middle layer of the mature ovary wall, or pericarp, between the exocarp and endocarp.

MESOPHYLL: The photosynthetic ground tissue (parenchyma) of a leaf, located between the layers of epidermis.

MESOPHYTE [Gk. *mesos*, middle, + *phyton*, a plant]: Plant that grows where it is neither too wet nor too dry.

MESSENGER RNA (mRNA): The RNA that carries genetic information from the gene to the ribosome, where it determines the order of the amino acids in a polypeptide.

METABOLISM (mĕ•tăb'o•lĭz'm) [Gk. *metabole*, change]: The sum of all chemical processes occurring within a living cell or organism.

METAPHASE (mĕt'ȧ•fāz) [Gk. *meta*, middle, + *phasis*, form]: Stage of mitosis or meiosis during which the chromosomes lie in the central plane of the spindle.

METAXYLEM [Gk. *meta*, after, beyond]: The part of the primary xylem that differentiates after the protoxylem. It reaches maturity after the portion of the plant part in which it is located has finished elongating.

MICRO- [Gk. *mikros*, small]: Prefix meaning "small."

MICROBODY: An organelle bounded by a single membrane and containing a variety of enzymes; generally associated with one or two cisternae of endoplasmic reticulum. Peroxisomes and glyoxysomes are kinds of microbodies.

MICROFIBRIL (mī•kro•fī'brĭl): Exceedingly small fiber, visible only with the electron microscope.

MICROGAMETOPHYTE [Gk. *mikros*, small, + *gamos*, marriage, + *phyton*, plant]: In heterosporous plants, the male gametophyte.

MICROMETER: A unit of microscopic measurement convenient for describing cellular dimensions; 1/1,000 of a millimeter, or 1/25,000 inch; symbol μm.

MICROPHYLL (mī'kro•fĭl) [Gk. *mikro-*, *mikr-*, from *mikros*, small, short + *phyllon*, leaf]: A small leaf with one vein and one leaf trace not associated with a leaf gap; contrasts with megaphyll.

MICROPYLE (mī'kro•pīl): In the ovules of seed plants, the opening in the integuments through which the pollen tube usually enters.

MICROSPORANGIUM (mī'kro•spo•răn'jĭ•ŭm): A sporangium within which microspores are formed.

MICROSPORE: A spore that develops into a male gametophyte.

MICROSPORE MOTHER CELL: A cell in which meiosis will occur, resulting in four microspores; in seed plants, often called a pollen mother cell. Also called microsporocyte.

MICROSPOROCYTE: *See* microspore mother cell.

MICROSPOROPHYLL (mī'kro•spo'ro•fĭl): A leaflike organ bearing one or more microsporangia.

MICROTUBULE [Gk. *mikros*, little, small, + L. *tubulus*, little pipe]: Narrow (about 25 nanometers in diameter), elongate, nonmembranous tubule of indefinite length occurring in the cytoplasm of many eukaryotic cells and flagella.

MIDDLE LAMELLA: The layer of intercellular material, rich in pectic compounds, cementing together the primary walls of adjacent cells.

MILLIMICRON: *See* nanometer.

MINERAL [ML. *minera*, ore, mine]: A chemical element or compound occurring naturally as a result of inorganic processes.

MITOCHONDRION, *pl.* MITOCHONDRIA (mī'to•kŏn'drĭ•ŏn): A double-membrane-bound organelle found in eukaryotic cells that contains the enzymes of the Krebs cycle and the electron-transport chain; the major source of ATP in nonphotosynthetic cells.

MITOSIS (mītō'sĭs) [Gk. *mitos*, thread]: A process during which the duplicated chromosomes divide longitudinally and the daughter chromosomes then separate to form two genetically identical daughter nuclei; usually accompanied by cytokinesis.

MOLE: The name for a gram molecule. The number of particles in 1 mole of any substance is always the same: 6.022×10^{23}.

MOLECULAR WEIGHT: The relative weight of a molecule when the weight of the common carbon atom is taken as 12; the sum of the relative weights of the atoms in a molecule.

MOLECULE (mŏl'e•kūl) [L. *moles*, mass]: Smallest possible unit of a compound, consisting of two or more atoms.

MONO- [Gk. *monos*, single]: Prefix meaning "one," "single."

MONOCOTYLEDON (mŏn'o•kŏt'ĭ•lē'dŭn): A member of one of the two great classes of angiosperms, Monocotyledoneae; abbreviated as monocot. A plant whose embryo has one cotyledon.

MONOECIOUS (mo•nē'shŭs) [Gk. *monos*, single, + *oikos*, house]: Having the anthers and carpels produced in separate flowers but borne on the same individual.

MONOKARYOTIC (mō'nō•kăr'ĭ•ō'tĭc) [Gk. *monos*, one, single + *karyon*, nut, kernel]: In fungi, having a single haploid nucleus within one cell or compartment.

MONOSACCHARIDE (mŏn'o•săk'ȧ•rīd) [Gk. *monos*, single, + *sakcharon*, sugar]: A simple sugar, such as five- and six-carbon sugars.

-MORPH, MORPH- [Gk. *morphe*, form]: Suffix or prefix meaning "form."

MORPHOGENESIS (môr'fo•jĕn'e•sĭs): The development of form.

MORPHOLOGY (môr•fŏl'o•jĭ) [Gk. *morphe*, form, + *logos*, discourse]: The study of form and its development.

MULTIPLE EPIDERMIS: A tissue composed of several cell layers derived from the protoderm; only the outer layer assumes characteristics of a typical epidermis.

MULTIPLE FRUIT: A cluster of matured ovaries produced by a cluster of flowers, as a pineapple.

MUTAGEN (mū'ta•jĭn) [L. *mutare*, to change, + Gk. *genaio*, to produce]: An agent that increases the mutation rate.

MUTANT (mū'tănt): A mutated gene or an organism carrying a gene that has undergone a mutation.

MUTATION (mu•tā'shŭn): An inheritable change of a gene from one allelic form to another.

MUTUALISM (mū'tu•ăl•ĭz'm): The living together of two or more organisms in an association that is mutually advantageous.

MYC-, MYCO- [Gk. *mykes,* fungus]: Prefix meaning "pertaining to fungi."

MYCELIUM (mī•sē'lĭ•ŭm) [Gk.*mykes,* fungus]: The mass of hyphae forming the body of a fungus.

MYCOLOGY (mī•kŏl'o•jĭ): The study of fungi.

MYCOPLASMAS [Gk. *mykes,* fungus, + *plasma,* something molded]: The smallest of known prokaryotic organisms.

MYCORRHIZA, *pl.* MYCORRHIZAE (mī'ko•rī'*za*): The combination of the hyphae of certain fungi with the root of a vascular plant.

NAD: *See* nicotinamide adenine dinucleotide.

NADP: *See* nicotinamide adenine dinucleotide phosphate.

NANNOPLANKTON (năn'o•plăngk'tŏn) [NL. from Gk. *nano-,* from *nanos,* dwarf + *plankton,* neut. of *planktos,* wandering]: Plankton with dimensions of less than 70 to 75 micrometers.

NANOMETER (NM): A millionth of a meter; one twenty-five millionth of an inch; formerly termed millimicron.

NATURAL SELECTION: The differential reproduction of genotypes.

NECTARY [GK. *nektar,* the drink of the gods]: In angiosperms, a gland that secretes a sugary fluid which pollinators utilize as food.

NETTED VENATION: The arrangement of veins in the leaf blade resembles a net; characteristic of dicot leaves. Also called reticulate venation.

NEUTRON (nū'trŏn) [L. *neuter,* neither]: An uncharged particle with a mass slightly greater than that of a proton, found in the atomic nucleus of all elements except hydrogen, in which the nucleus consists of a single proton.

NICHE (nĭch): The role played by a particular species in its ecosystem.

NICOTINAMIDE ADENINE DINUCLEOTIDE (NAD) (nĭk'o•tĭn'*ă*•mīd ăd'e•nēn dī•nū'kle•o•tīd): A coenzyme which functions as an electron acceptor.

NICOTINAMIDE ADENINE DINUCLEOTIDE PHOSPHATE (NADP) (nĭk'o•tĭn'*ă*•mīd ăd'e•nēn dī•nū'kle•o•tīd): A coenzyme which functions as an electron acceptor; similar in structure to NAD except with an extra phosphate.

NITRIFICATION (nī'trĭ•fĭ•kā'shŭn): The conversion of ammonium or ammonia to nitrate, a process carried out by specific bacteria and fungi.

NITROGEN BASE: A nitrogen-containing molecule having basic properties (tendency to acquire an H atom); a purine or pyrimidine; one of the building blocks of nucleic acids.

NITROGEN CYCLE: Worldwide circulation of nitrogen atoms in which certain microorganisms take up atmospheric nitrogen and convert it into other forms which may be assimilated into the bodies of other organisms; excretion, burning, and bacterial and fungal action in dead organisms return nitrogen atoms to the atmosphere.

NITROGEN FIXATION: Incorporation of atmospheric nitrogen into nitrogen compounds available to green plants, a process that can be carried out only by certain microorganisms, or by higher plants in symbiotic association with microorganisms.

NITROGEN-FIXING BACTERIA: Soil bacteria that convert atmospheric nitrogen into nitrogen compounds.

NODE [L. *nodus,* knot]: The part of a stem where one or more leaves are attached; *see* internode.

NODULES (nŏd'ūl): Enlargements or swellings on the roots of legumes and certain other plants inhabited by nitrogen-fixing bacteria.

NONSEPTATE: *See* aseptate.

NUCELLUS (nu•sĕl'ūs) [L. *nucella,* a small nut]: Tissue composing the chief part of the young ovule, in which the embryo sac develops; equivalent to a megasporangium.

NUCLEAR ENVELOPE: The double membrane surrounding the nucleus of a cell.

NUCLEIC ACID (nu•klē'ĭk): An organic acid consisting of joined nucleotide complexes; the two types are deoxyribonucleic acid (DNA) and ribonucleic acid (RNA).

NUCLEOLAR ORGANIZER: A special area on certain chromosomes associated with the formation of the nucleolus.

NUCLEOLUS (nu•klē'o•lŭs) [L. *nucleolus,* a small nucleus]: A spherical body composed chiefly of RNA and protein found in the nucleus of eukaryotic cells; site of production of ribosomes.

NUCLEOTIDE (nū'kle•o•tīd): A single unit of nucleic acid composed of phosphate, five-carbon sugar (either ribose or deoxyribose), and a purine or a pyrimidine.

NUCLEUS (nū'kle•ŭs): (1) A specialized body within the eukaryotic cell bounded by a double membrane and containing the chromosomes; (2) the central part of the atom.

NUT [L. *nux,* nut]: A dry, indehiscent, hard, one-seeded fruit, generally produced from a gynoecium of more than one fused carpel.

OBLIGATE ANAEROBE (ŏb'lĭ•gat ăn•ā'ĕr•ōb): An organism that is metabolically active only in the absence of oxygen.

-OID [Gk. like, resembling]: Suffix meaning "like or similar to."

ONTOGENY (ŏn•tŏj'e•nĭ) [Gk. *on,* being, + *genesis,* origin]: The development, or life history, of an individual organism or of part of it.

OO- [Gk. *oion,* egg]: Prefix meaning "egg."

OOGAMY (o•ŏg'ȧ•mĭ): Kind of sexual reproduction in which one of the gametes (the egg) is large and nonmotile, whereas the other gamete (the sperm) is smaller and motile.

OOGONIUM (ō'o•gō'nĭ•ŭm): In certain algae and fungi, a unicellular female sex organ that contains one or several eggs.

OOSPORE (ō'ō•spōr) [Gk. *oion,* egg, + *spora,* seed]: The thick-walled zygote characteristic of the Oomycetes, a class of fungi.

OPEN BUNDLE: A vascular bundle in which

a vascular cambium develops.

OPERATOR GENE: The site at which the protein product of a regulator gene works.

OPERCULUM (o•pûr'ku•lŭm) [L. *operculum*, lid]: In mosses, the lid of the sporangium.

OPERON: A group of adjacent genes that are under the control of a single operator gene.

OPPOSITE: Term applied to buds or leaves occurring in pairs at a node.

ORDER: A category of classification above family and below class; composed of one or more families. The suffix to the ordinal name is *-ales*.

ORGAN: A structure composed of different tissues, such as root, stem, leaf, or flower parts.

ORGANELLE (ôr'găn•ĕl'): A formed body in the cytoplasm of a cell.

ORGANIC: Pertaining to organisms or living things generally, to compounds formed by living organisms, and to the chemistry of compounds containing carbon.

ORGANISM (ôr'găn•iz'm): Any individual living creature, either unicellular or multicellular.

OSMOMETER [Gk. *osmos*, pushing, + *meter*, measure]: A device used to measure the rate of osmosis.

OSMOSIS (ŏs•mō'sĭs) [Gk. *osmos*, impulse or thrust]: The diffusion of water, or any solvent, across a differentially permeable membrane; in the absence of other forces, movement of water during osmosis will always be from a region of greater chemical potential of water to one of lesser chemical potential.

OSMOTIC POTENTIAL: *See* solute potential.

OSMOTIC PRESSURE: The potential pressure that can be developed by a solution separated from pure water by a differentially permeable membrane; it is an index of the solute concentration of the solution.

OUTER BARK: In older trees, the dead part of the bark; the innermost periderm and all tissues outside it. Also called rhytidome.

OVARY (ō'vȧ•rĭ) [L. *ovum*, an egg]: An enlarged basal portion of a carpel or of a gynoecium composed of fused carpels; the ovary becomes the fruit.

OVULE (ō'vūl) [L. *ovulum*, a little egg]: A structure in seed plants containing the female gametophyte with egg cell, all being surrounded by the nucellus and one or two integuments; when mature, the ovule becomes a seed.

OVULIFEROUS SCALE [ovule, + L. *ferre*, to bear, + scale]: In certain conifers, the appendage or scalelike shoot to which the ovule is attached.

OXIDANT: The molecule that accepts the electron in an oxidation-reduction reaction; the oxidant becomes reduced.

OXIDATION: Loss of an electron by an atom. Oxidation and reduction (gain of an electron) take place simultaneously, because an electron that is lost by one atom is accepted by another. Oxidation-reduction reactions are an important means of energy transfer within living systems.

OXIDATIVE PHOSPHORYLATION (ŏk'sĭ•dā'tĭv fŏs'fo•rĭl•ā'shŭn): The formation of ATP from ADP and inorganic phosphate that takes place in the electron-transport chain of the mitochondrion.

PAIRING OF CHROMOSOMES: Side-by-side association of homologous chromosomes.

PALEOBOTANY (pā'le•o•bŏt'ȧ•nĭ) [Gk. *palaios*, old]: The study of fossil plants.

PALISADE PARENCHYMA (păl'ĭ•sād' pȧ•rĕng'kĭ•mȧ): A leaf tissue composed of columnar chloroplast-bearing parenchyma cells with their long axes at right angles to the leaf surface.

PANICLE (păn'ĭ•k'l) [L. *panicula*, tuft]: An inflorescence, the main axis of which is branched, and whose branches bear loose flower clusters.

PARA- [Gk. at the side of, beside]: Prefix, meaning "beside."

PARADERMAL SECTION [Gk. *para*, beside, + *derma*, skin]: Section cut parallel with the surface of a flat structure, such as a leaf.

PARALLEL EVOLUTION: *See* convergent evolution.

PARALLEL VENATION: The pattern of venation in which the principal veins of the leaf are parallel or nearly so, characteristic of monocots.

PARAPHYSIS, *pl.* PARAPHYSES [Gk. *para.*, beside, + *physis*, growth]: As in certain fungi, a sterile hypha growing among reproductive cells in the fruiting body.

PARASEXUAL CYCLE (păr'ȧ•sĕk'shoo•ăl): The fusion and segregation of heterokaryotic haploid nuclei in certain fungi to produce recombinant nuclei.

PARASITE (păr'ȧ•sīt): An organism that lives on or in an organism of a different species and derives nutrients from it.

PARENCHYMA (pȧ•rĕng'kĭ•mȧ) [Gk. *paren*, beside, + *en*, in, + *chein*, to pour]: A tissue composed of parenchyma cells.

PARENCHYMA CELL: Living, generally thin-walled cell, concerned with one or more of the many physiological activities in plants; varies in size and form. The most common cell type in the plant.

PARTHENOCARPY (pär'the•no•kär'pĭ) [Gk. *parthenos*, virgin, + *karpos*, fruit]: The development of fruit without fertilization; parthenocarpic fruits are usually seedless.

PASSAGE CELL: Endodermal cell of root that retains thin wall and Casparian strip when other associated endodermal cells develop thick secondary walls.

PATHOGEN (păth'o•jĕn) [Gk. *pathos*, suffering, + *genesis*, beginning]: An organism which causes a disease.

PATHOLOGY [Gk. *pathos*, suffering, + *logos*, account]: The study of plant or animal diseases, their effects on the organism, and their treatment.

PECTIN (pĕk'tĭn): A complex organic compound present in the intercellular layer and primary wall of plant cell walls; the basis of fruit jellies.

PEDICEL (pĕd'ĭ•sĕl) [L. *pediculus*, a little foot]: The stem of an individual flower.

PEDUNCLE (pe•dŭng'k'l): The stem of an inflorescence.

PENNATE DIATOM [L. *pennate*, feathered]: A bilaterally symmetrical diatom.

PEPTIDE (pĕp'tīd): Two or more amino acids linked by peptide bonds.

PEPTIDE BOND: The type of bond formed when two amino acid units are joined end to end; the acid group (COOH) of one amino acid is attached to the basic group (NH_2^-) of the next and a molecule of water (H_2O) is lost.

PERENNIAL (pĕr·ĕn'ĭ·ŭl) [L. *per*, through, + *annus*, a year]: A plant that persists from year to year and usually produces reproductive structures in two or more different years.

PERFECT FLOWER: A flower having both stamens and carpels; hermaphroditic flower.

PERFECT STAGE: Phase of the life history of a fungus that includes sexual fusion and the spores associated with such fusions.

PERFORATION PLATE: Part of the wall of a vessel member that is perforated.

PERI- [Gk. around]: Prefix meaning "around."

PERIANTH (pĕr'ĭ·ănth) [Gk. *peri*, around, + *anthos*, flower]: The petals and sepals taken together.

PERICARP (pĕr'ĭ·kärp) [Gk. *peri*, around, + *karpos*, fruit]: Fruit wall, developed from the ovary wall.

PERICLINAL: Parallel with the surface.

PERICYCLE (pĕr'ĭ·sī·k'l) [Gk. *peri*, around, + *kykos*, circle]: Tissue, generally of root, bounded externally by the endodermis and internally by the phloem.

PERIDERM (pĕr'ĭ·dûrm) [Gk. *peri*, all around, about + *derma*, skin]: Outer protective tissue that replaces epidermis when it is destroyed in secondary growth; includes cork, cork cambium, and phelloderm.

PERIGYNY [Gk. *peri*, about, + *gyne*, a female]: Floral organization in which the petals and stamens are attached to the margin of a cup-shaped extension of the receptacle. Superficially, the sepals, petals, and stamens appear to be attached to the ovary.

PERISTOME (pĕr'ĭ·stōm) [Gk. *peri*, around, + *stoma*, a mouth]: In mosses, a fringe of teeth about the opening of the sporangium.

PERITHECIUM (pĕr'ĭ·thē'shĭ·ŭm): A spherical or flask-shaped ascocarp.

PERMANENT WILTING PERCENTAGE: The amount of water a given soil holds as unavailable to a particular plant being tested.

PERMEABLE [L. *permeare*, to pass through]: Usually applied to membranes through which substances may diffuse.

PEROXISOME: A microbody that plays an important role in glycolic acid metabolism associated with photosynthesis.

PETAL: A flower part, usually conspicuously colored; one of the units of the corolla.

PETIOLE (pĕt'ĭ·ōl): The stalk of the leaf.

pH: A symbol denoting the relative concentration of hydrogen ions in a solution; pH values run from 0 to 14, and the lower the value the more acid a solution, that is, the more hydrogen ions it contains; pH 7 is neutral, less than 7 acid, more than 7 alkaline.

PHAGE: *See* bacteriophage.

PHELLEM: *See* cork.

PHELLODERM (fĕl'o·dûrm) [Gk. *phellos*, cork, + *derma*, skin]: A tissue formed inwardly by the cork cambium, opposite the cork; inner part of the periderm.

PHELLOGEN (fĕl'lo·jĕn) [Gk. *phellos*, cork, + *genesis*, birth]; *see* cork cambium.

PHENOTYPE (fē'no·tīp): The physical appearance of an organism resulting from interaction between its genetic constitution (genotype) and the environment.

PHLOEM (flō'ĕm) [Gk. *phloos*, bark]: Food-conducting tissue basically composed of sieve elements, various kinds of parenchyma cells, fibers, and sclereids.

PHOSPHATE (fŏs'fāt): A compound of phosphorus; in general, phosphoric acid.

PHOSPHORYLATION (fŏs'fo·rĭl·ā'shŭn): A reaction in which phosphate is added to a compound, e.g., the formation of ATP from ADP and inorganic phosphate.

PHOTO-, -PHOTIC [Gk. *photos*, light]: Prefix or suffix meaning "light."

PHOTOPERIODISM (fō'to·pḛr'ĭ·ŭd·ĭz'm): Response to duration and timing of day and night; a mechanism evolved by organisms for measuring seasonal time.

PHOTORESPIRATION: The light-dependent production of glycolic acid in chloroplasts and its subsequent oxidation in peroxisomes.

PHOTOSYNTHESIS (fō'to·sĭn'the·sĭs) [Gk. *photos*, light, + *syn*, together, + *tithenai*, to place]: The conversion of light energy to chemical energy; the production of carbohydrate from carbon dioxide in the presence of chlorophyll, using light energy.

PHOTOTROPISM (fō'to·trō'pĭz'm) [Gk. *photos*, light, + *trope*, turning]: Growth movement in which the direction of the light is the determining factor, as the growth of a plant toward a light source; turning or bending response to light.

PHRAGMOPLAST: A spindle-shaped system of fibrils, which arises between two daughter nuclei at telophase and within which the cell plate is formed during cell division, or cytokinesis. The fibrils of the phragmoplast are composed of microtubules.

PHYCOBILINS (fī·ko·bī'lĭn): A group of water-soluble accessory pigments, including phycocyanins and phycoerythrins, which occur in certain groups of algae.

PHYCOLOGY (fī·kŏl'o·jĭ) [Gk. *phykos*, seaweed]: The study of algae.

PHYLLO-, PHYLL- [Gk. *phyllon*, leaf]: Prefix meaning "leaf."

PHYLLODE (fĭl'ōd): A flat, expanded petiole replacing the blade of a leaf in photosynthetic function.

PHYLOGENY (fī·lŏj'e·nĭ) [Gk. *phylon*, race, tribe]: Evolutionary relationships among organisms; developmental history of a group of organisms.

PHYLUM (fī'lŭm) [NL. fr. Gk. *phylon*, race, tribe]: One of the major kinds of groups used by zoologists in classifying organisms; the category between class and kingdom. *See also* division.

PHYSIOLOGY (fĭz·ĭ·ŏl'o·jĭ): The study of the activities and processes of living organisms.

PHYTO-, -PHYTE [Gk. *phyton*, plant]: Prefix or suffix meaning "plant."

PHYTOCHROME (fī'to·krom): A phycobilinlike pigment found in cytoplasm of green plants that is associated with the absorption of light; photoreceptor for red–far-red light; involved with a number of timing processes, such as flowering, dormancy, leaf formation, and seed germination.

PHYTOPLANKTON (fī'to·plăngk'tŏn): Autotrophic plankton.

PIGMENT: Substance that absorbs light, often selectively.

PILEUS [L. *pileus*, a cap]: The caplike part of the mushroom basidiocarp and of certain ascocarps.

PINNA, *pl.* PINNAE (pĭn'ȧ) [L. *pinna*, feather]: A primary division, or leaflet, of a compound leaf or frond. Pinnae may be divided into pinnules.

PINOCYTOSIS (pin·ō·sī'tō·sȧs) [Gk. *pinein*, to drink, + *kytos*, hollow vessel]: The process by which invaginations of the plasma membrane pinch off in the cytoplasm, their contents then become incorporated into the protoplasm.

PISTIL [L. *pistillum*, pestle]: Central organ of flowers typically consisting of ovary, style, and stigma; a pistil may consist of one or more fused carpels.

PISTILLATE (pĭs'tĭ·lat): Pertaining to a flower with one or more carpels but no functional stamens; carpellate.

PIT: A recess or cavity in a cell wall where the secondary wall does not form.

PITH: The ground tissue occupying the center of the stem or root within the vascular cylinder; it usually consists of parenchyma.

PITH RAY: *See* interfascicular region.

PLACENTA, *pl.* PLACENTAE (plȧ·sĕn'tȧ) [L. *placenta*, a cake]: The part of the ovary wall to which the ovules or seeds are attached.

PLACENTATION [L. *placenta*, a cake, + *-tion*, state of]: The manner of ovule attachment within the ovary.

PLANKTON (plăngk'tŏn) [Gk. *planktos*, wandering]: Free-floating, mostly microscopic, aquatic organisms.

PLAQUE: Clear area in a sheet of cells resulting from the killing or lysis of contiguous cells by virus multiplication.

-PLASMA, PLASMO-, -PLAST [Gk. *plasma*, form, mold]: Prefix or suffix meaning "formed," or "molded": for example, protoplasm, "first-molded" (living matter); chloroplast, "green-formed" (body).

PLASMA MEMBRANE OR PLASMALEMMA: Outer boundary of the protoplast, next to the cell wall; consists of a single three-ply membrane. Also called cell membrane and ectoplast.

PLASMODESMA, *pl.* PLASMODESMATA (plăs'-mo·dĕz·mȧ): The minute cytoplasmic threads that extend through openings in cell walls and connect the protoplasts of adjacent living cells.

PLASMODIUM (plăz·mo'dĭ·ŭm): Stage in life cycle of Myxomycetes; a multinucleate mass of protoplasm surrounded by a membrane.

PLASMOGAMY [Gk. *plasma*, something formed, + *gamos*, marriage]: Union of the protoplasts of sex cells, or gametes, not accompanied by union of their nuclei.

PLASMOLYSIS (plăz·mŏl'ĭ·sĭs) [Gk. *plasma*, form, + *lysis*, a loosening]: The separation of the cytoplasm from the cell wall due to removal of water from the protoplast by osmosis.

PLASTID (plăs'tĭd): Organelle in the cells of certain groups of eukaryotes that is the site of such activities as food manufacture and storage; plastids are bounded by a double membrane.

PLEIOTROPISM (plī·o·tro·pĭz'm): The capacity of a gene to affect a number of different characteristics.

PLUMULE (ploo͞'mūl) [L. *plumula*, a small feather]: The first bud of an embryo; the portion of the young shoot above the cotyledons.

PNEUMATOPHORES ('n(y)ümәd·ә,fō(ә)r) [Gk. *pneuma*, breath, + *phore*, fr. *pherein*, to carry]: Negatively geotropic extensions of the root systems of some trees growing in swampy habitats. They grow upward and out of the water to assure adequate aeration.

POLARITY (po·lăr'ĭ·tĭ): A characteristic resulting from differentiation between two ends of a living system; the condition of a substance, such as a magnet, that exhibits opposite or contrasted properties in opposite or contrasted parts.

POLAR NUCLEI: Two nuclei (usually), one derived from each end (pole) of the embryo sac, which become centrally located. They fuse with a male nucleus to form the primary ($3n$) endosperm nucleus.

POLLEN [L. *pollen*, fine dust]: A collective term for pollen grains.

POLLEN GRAIN: A microspore containing a mature or immature microgametophyte (male gemetophyte).

POLLEN MOTHER CELL: *See* microspore mother cell.

POLLEN SAC: A cavity in the anther that contains the pollen grains.

POLLEN TUBE: A tube formed following germination of the pollen grain which carries the male gametes into the ovule.

POLLINATION (pŏl'ĭ·nā'shŭn): The transfer of pollen from where it was formed (e.g., the anther) to a receptive surface.

POLY- [Gk. *polys*, many]: Prefix meaning "many."

POLYEMBRYONY: Having more than one embryo within the developing seed.

POLYMER (pŏl'ĭ·mĕr): A large molecule composed of many like molecular subunits.

POLYMERIZATION (pŏl'ĭ·mĕr·ĭ·zā'shŭn): The chemical union of monomers such as glucose or nucleotides to form polymers such as starch or nucleic acid.

POLYNUCLEOTIDES (pŏl'ĭ·nū'kle·o·tīds): Long-chain molecules composed of units (monomers) called nucleotides; DNA is a polynucleotide.

POLYPEPTIDE (pŏl'ĭ·pĕp'tīd): Numerous amino acids linked together by peptide bonds.

POLYPLOID (pŏl'ĭ·ploid) [Gk. *polys*, many, + *ploos*, fold]: Referring to a plant, tissue, or cell with more than two complete sets of chromosomes.

POLYRIBOSOME: An aggregation of ribosomes; ribosomes actively involved in protein synthesis; also called polysome.

POLYSACCHARIDE (pŏl'ĭ·săk'ȧ·rīd): A carbohydrate composed of many monosaccharide units joined in a long chain, for example, glycogen, starch, and cellulose.

POME (pōm) [Fr. *pomme*, apple]: A simple fleshy fruit, the outer portion of which is formed by the floral parts that surround the ovary and expand with the growing fruit; found only in one subfamily of the Rosaceae (apples, pears, quince, *Cotoneaster*, *Pyracantha*, etc.)

POPULATION: Any group of individuals, usually of a single species.

P-PROTEIN: Phloem-protein; a proteinaceous substance found in cells of angiosperm phloem, especially in sievetube members; also called slime.

PREPROPHASE BAND: A ringlike band of microtubules that outlines the equatorial plane of the future mitotic spindle in many cells. May play a role in positioning of the nucleus before mitosis begins.

PRIMARY ENDOSPERM NUCLEUS: The result of the fusion of a sperm nucleus and the two polar nuclei.

PRIMARY GROWTH: In plants, growth originating in the apical meristems of shoots and roots, as contrasted with secondary growth, which originates in the vascular and cork cambiums.

PRIMARY MERISTEM OR MERISTEMATIC TISSUE (mĕr′ĭ•ste•măt′ĭk): A tissue derived from the apical meristem; of three kinds (1) protoderm, (2) procambium, and (3) ground meristem.

PRIMARY PIT-FIELD: Thin area in a primary cell wall through which plasmodesmata pass, although plasmodesmata may occur elsewhere in the wall as well.

PRIMARY PLANT BODY: The part of the plant body arising from the apical meristems and their derivative meristematic tissues; composed entirely of primary tissues.

PRIMARY ROOT: The first root of the plant, developing in continuation of the root tip or radicle of the embryo; in gymnosperms and dicots it becomes the taproot.

PRIMARY TISSUES: Cells derived from the apical meristems and primary meristematic tissues of root and shoot; opposed to secondary tissues derived from a cambium. Primary growth results in an increase in length.

PRIMARY WALL: The wall layer deposited during the period of cell expansion.

PRIMORDIUM, *pl.* PRIMORDIA [L. *primus*, first, + *ordiri*, to begin to weave]: A cell or organ in its earliest stage of differentiation; e.g., a fiber primordium, leaf primordium.

PROCAMBIUM (pro•kăm′bĭ•ŭm) [L. *pro*, before, + cambium]: A primary meristematic tissue that gives rise to primary vascular tissues.

PROEMBRYO: Embryo in early stages of development, before embryo proper and suspensor become distinct.

PROKARYOTIC (prō′ka•rē•ăd′ĭk): Lacking a membrane-bound nucleus, plastids, Golgi apparatus, and mitochondria, as in bacteria.

PROMOTER: A specific site at the beginning of an operon at which a polymerase can start its synthesis.

PROPHAGE (prō′fāj′): Noninfectious phage units linked with the bacterial chromosome which multiply with the growing and dividing bacteria but do not bring about lysis of the bacteria. Prophage is a stage in the life cycle of a temperate phage.

PROPHASE (prō′fāz′) [Gk. *pro*, before, + *phasis*, form]: An early stage in nuclear division, characterized by the shortening and thickening of the chromosomes and their movement to the metaphase plate.

PROPLASTID (pro′plăs•tĭd): A minute self-reproducing body in the cytoplasm from which a plastid develops.

PROP ROOTS: Adventitious roots arising from the stem above soil level and helping to support the plant; common in many monocots, for example, corn (*Zea mays*).

PROTEASE (prō′te•ās): An enzyme that digests protein by hydrolysis of peptide bonds.

PROTEIN [Gk. *proteios*, primary]: A complex organic compound composed of many (about 100 or more) amino acids joined by peptide bonds.

PROTHALLIAL CELL (prŏ•thăl′ĭ•ăl) [Gk. *pro*, before, + *thallos*, sprout]: The sterile cell or cells found in the male gametophytes, or microgametophytes, of heterosporous plants; believed to be remnants of the vegetative tissue of the male gametophyte.

PROTHALLUS (pro•thăl′ŭs): In ferns and some other relatively unspecialized vascular plants, the more or less independent gametophyte; also called prothallium.

PROTO- [Gk. *protos*, first]: Prefix meaning "first," for example, Protozoa, "first animals."

PROTODERM (prō′to•dûrm) [Gk. *protos*, first, + *derma*, skin]: Primary meristematic tissue that gives rise to epidermis.

PROTON: A subatomic, or elementary, particle, with a single positive charge equal in magnitude to the charge of an electron, and a mass of 1; a component of every atomic nucleus. A proton can also be thought of as the nucleus of the lightest and most abundant hydrogen isotope.

PROTONEMA, *pl.* PROTONEMATA (prō′to•nē′mȧ) [Gk. *protos*, first, + *nema*, a thread]: Filamentous growth, an early stage in development of the gametophyte of mosses.

PROTOPLASM (prō′to•plăz′m): The living substance of all cells.

PROTOPLAST (prō′to•plăst): The entire contents, both protoplasmic and nonprotoplasmic, of a cell, not including the cell wall.

PROTOSTELE [Gk. *protos*, first, + stele, pillar]: The simplest type of stele, containing a solid column of vascular tissue.

PROTOXYLEM: The first part of the primary xylem, which matures during elongation of the plant part in which it is found.

PROXIMAL (prŏk′sĭ•măl) [L. *proximus*, near]: Situated near the point of reference, usually the main part of body or the point of attachment; opposite of distal.

PSEUDO- [Gk. *pseudes*, false]: Prefix meaning "false."

PSEUDOPLASMODIUM (sū′do•plăz•mō′-dĭ•ŭm): A multicellular mass of individual amoeboid cells, representing the aggregate phase in the cellular slime molds.

PTERIDOPHYTA (tĕr′ĭ•dŏf′ĭ•tȧ): A now-discarded unit of classification, including all vascular plants except the seed plants.

PUNNETT SQUARE: The checkerboard diagram used for analysis of gene distributions.

PURINE (pū′rēn): A nitrogenous base with a double-ring structure, such as adenine or guanine; one of the components of nucleic acids.

PYRAMID OF ENERGY: Energy relationships among various feeding levels involved in a particular food chain; autotrophs (at the base of the pyramid) represent the greatest amount of available energy; herbivores next; then primary carnivores; secondary carnivores; etc.

PYRENOID (pī•rē′noid) [Gk. *pyren*, the stone of a fruit, + L. *oides*, like]: A body found in the chloroplasts of certain algae and liverworts that seems to be associated with starch deposition.

PYRIMIDINE (pĭ•rĭm'ĭ•dēn): A nitrogenous base with a single-ring structure, such as cytosine, thymine, or uracil; one of the components of nucleic acids.

QUANTOSOME [L. *quantus*, how much, + Gk. *soma*, body]: Granules located on the inner surfaces of chloroplast lamellae; once believed to be a functional unit in photosynthesis.

QUIESCENT CENTER: The relatively inactive initial region in the apical meristems of roots.

RACEME (rȧ•sēm') [L. *racemus*, bunch of grapes]: An inflorescence in which the main axis is elongated but the flowers are borne on pedicels that are about equal in length.

RACHIS (rā'kĭs) [Gk. *rhachis*, a backbone]: Main axis of spike; axis of fern leaf (frond) from which pinnae arise; in compound leaves, the extension of the petiole corresponding to the midrib of an entire leaf.

RADIALLY SYMMETRICAL: *See* actinomorphic.

RADIAL SECTION: A longitudinal section cut parallel with a radius of a cylindrical body, such as a root or stem; in the case of secondary xylem, or wood, and secondary phloem, parallel with the rays.

RADIAL SYSTEM: In secondary xylem and secondary phloem, term applied to all the rays, the cells of which are derived from ray initials; also called horizontal system and ray system.

RADICLE [L. *radix*, root]: The embryonic root.

RADIOACTIVE ISOTOPE: An isotope with an unstable nucleus that stabilizes itself by emitting radiation.

RADIOISOTOPE (rā'dĭ•o•ī'so•tōp): An unstable isotope of an element that decays or disintegrates spontaneously, emitting radiation; radioactive isotope.

RAPHE (răf) [Gk. *raphe*, seam]: (1) Ridge on seeds, formed by the stalk of the ovule, in those seeds in which the stalk is sharply bent at the base of the ovule; (2) groove on the frustule of a diatom.

RAPHIDES (răf'ĭ•dēz) [Gk. *rhaphis*, a needle]: Bundles of fine, sharp, needlelike crystals of calcium oxalate found in some plants, such as Vitaceae and Onagraceae.

RAY FLOWERS: *See* disk flowers.

RAY INITIAL: An initial in the vascular cambium that gives rise to ray cells of secondary xylem and secondary phloem.

RECEPTACLE: That part of the axis of a flower stalk that bears the floral organs.

RECESSIVE: Describing a gene whose phenotypic expression is masked by a dominant allele; heterozygotes are phenotypically indistinguishable from dominant homozygotes.

RECOMBINATION: *See* gene recombination.

REDUCTANT: The molecule that donates the electron in an oxidation-reduction reaction. The reductant becomes oxidized.

REDUCTION [Fr. *reduction*, fr. L. *reductio*, a bringing back; originally "bringing back" a metal from its oxide; i.e., iron from iron rust or ore]: Gain of an electron by a compound that takes place simultaneously with oxidation (loss of an electron by an atom), because an electron that is lost by one atom is accepted by another.

REGULAR: *See* actinomorphic.

REGULATOR GENE: A gene that prevents or represses the activity of the structural genes in an operon.

REPLICATE: Produce a facsimile or a very close copy. Used to indicate the production of a second molecule of DNA exactly like the first molecule or of a sister chromatid.

RESIN DUCT: A tubelike intercellular space lined with resin-secreting cells (epithelial cells) and containing resin.

RESPIRATION: An intracellular process in which food is oxidized with release of energy. The complete breakdown of sugar or other organic compounds to carbon dioxide and water is termed aerobic respiration, although the earlier steps are anaerobic.

RETICULATE VENATION: *See* netted venation.

RHIZOBIA (rī•zō'bĭ•ȧ) [Gk. *rhiza*, root + *bios*, life]: Bacteria of the genus *Rhizobium* which may be involved with leguminous plants in a symbiotic relationship as a result of which nitrogen fixation occurs.

RHIZOIDS (rī'zoid) [Gk. *rhiza*, root]: (1) Branched rootlike extensions of fungi and algae that absorb water, food, and nutrients; (2) root-hair-like structures in liverworts, mosses, and some vascular plants, usually borne by the gametophyte generation. True roots are found only in vascular plants.

RHIZOME (rī'zōm) [Gk. *rhizoma*, mass of roots]: A more or less horizontal underground stem.

RIBONUCLEIC ACID (RNA) (rī'bo•nu•klē'ĭk): Type of nucleic acid formed on chromosomal DNA and involved in protein synthesis; composed of chains of phosphate, sugar molecules (ribose), and purines and pyrimidines; genetic material of many viruses.

RIBOSE (rī'bōs): A five-carbon sugar, a component of RNA.

RIBOSOME (rī'bo•sōm): A small particle composed of protein and RNA; the site of protein synthesis.

RING-POROUS WOOD: A wood in which the pores, or vessels, of the early wood are distinctly larger than those of the late wood, forming a well-defined ring in cross sections of the wood.

RNA: *See* ribonucleic acid.

ROOT [A.S. *rot*]: The descending axis of a plant, normally below ground, and serving to anchor the plant and to absorb and conduct water and minerals.

ROOT CAP: A thimblelike mass of cells covering and protecting the growing tip of a root.

ROOT HAIRS: Tubular outgrowths of epidermal cells of the root in the zone of maturation.

ROOT PRESSURE: The pressure developed in roots as the result of osmosis which causes guttation from leaves and exudation from cut stumps.

RUNNER: *See* stolon.

SAMARA (săm'ȧ•rȧ): Simple, dry, one- or two-seeded indehiscent fruit with pericarp-bearing, winglike outgrowths.

SAP: (1) A name applied to the fluid contents of the xylem or the sieve elements of the phloem; (2) the fluid contents of the vacuole are called cell sap.

SAPROBE [GK. *sapros*, rotten]: An organism that secures its food directly from nonliving organic matter.

SAPWOOD: Outer part of the wood of stem or trunk, usually distinguished from the heartwood by its lighter color, in which active conduction of water takes place.

SCHIZO- [Gk. *schizein*, to split]: Prefix meaning "split."

SCHIZOCARP (skĭz'o•kärp): Dry fruit with two or more united carpels which split apart at maturity.

SCLEREID (sklĕr'e•ĭd) [Gk. *skleros*, hard]: A sclerenchyma cell with a thick, lignified secondary wall having many pits. Variable in form, but typically not very long; may or may not be living at maturity.

SCLERENCHYMA (skle•rĕng'kĭ•má) [Gk. *skleros*, hard + L. *enchyma*, infusion]: A supporting tissue composed of sclerenchyma cells, including fibers and sclereids.

SCLERENCHYMA CELL: Cell of variable form and size with more or less thick, often lignified, secondary walls; may or may not be living at maturity; includes fibers and sclereids.

SCUTELLUM (sku•tĕl'ŭm) [L. *scutella*, a small shield]: Single cotyledon of grass embryo, specialized for absorption of the endosperm.

SECONDARY GROWTH: In plants, growth derived from secondary or lateral meristem, the vascular and cork cambiums; secondary growth results in an increase in girth.

SECONDARY PLANT BODY: The part of the plant body produced by the vascular cambium and the cork cambium. Consists of secondary xylem, secondary phloem, and periderm.

SECONDARY ROOT: *See* branch root.

SECONDARY TISSUES: Tissues produced by the vascular cambium and cork cambium.

SECONDARY WALL: Innermost layer of the cell wall, formed in certain cells after cell elongation has ceased; secondary walls have a highly organized microfibrillar structure.

SEED: A structure formed by the maturation of the ovule of seed plants following fertilization. In conifers it consists of seed coat, embryo, and female gametophyte (n) as storage tissue. Some angiosperm seeds, when mature, are composed only of seed coat and embryo; others also still contain endosperm ($3n$), a storage tissue.

SEED COAT: The outer layer of the seed, developed from the integuments of the ovule.

SEEDLING: A young sporophyte developing from a germinating seed.

SEGREGATION: The separation of the chromosomes (and genes) from different parents at meiosis.

SEMIPERMEABLE MEMBRANE: A membrane that is permeable to water but differentially permeable to solutes.

SEPAL (sē'păl) [M.L. *sepalum*, a covering]: One of the outermost flower structures which usually enclose the other flower parts in the bud; a unit of the calyx.

SEPTATE (sĕp'tāt) [L. *septum*, fence]: Divided by cross walls into cells or compartments.

SESSILE (sĕs'ĭl) [L. *sessilis*, low, dwarfed, fr. *sedere*, to sit]: Sitting; referring to a leaf lacking a petiole or to a flower or fruit lacking a pedicel.

SETA (sē'tȧ) [L. *seta*, bristle]: In bryophytes, the stalk that supports the capsule, if present.

SEXUAL REPRODUCTION: The fusion of gametes followed by meiosis and recombination at some point in the life cycle.

SHEATH: (1) The base of a leaf that wraps around the stem, as in grasses; (2) a tissue layer surrounding another tissue, as a bundle sheath.

SHOOT: The above-ground portions, such as the stem and leaves, of a vascular plant.

SHRUB: A perennial woody plant of relatively low stature, typically with several stems arising from or near the ground.

SIEVE AREA: A portion of sieve-element wall containing clusters of pores through which the protoplasts of adjacent sieve elements are interconnected.

SIEVE CELL: A long, slender sieve element with relatively unspecialized sieve areas and with tapering end walls that lack sieve plates, found in the phloem of gymnosperms and lower vascular plants.

SIEVE ELEMENT: The cell of the phloem concerned with the long-distance transport of food substances. Classified into sieve cells and sieve-tube members.

SIEVE PLATE: The part of the wall of sieve-tube members bearing one or more highly differentiated sieve areas.

SIEVE TUBE: A series of sieve-tube members arranged end-on-end and interconnected by sieve plates.

SIEVE-TUBE MEMBER: One of the component cells of a sieve tube. Found primarily in flowering plants, and typically associated with a companion cell. Also called sieve-tube element.

SILIQUE (sĭ•lēk') [L. *siliqua*, pod]: The fruit characteristic of Brassicaceae (mustards); two-celled, the valves splitting from the bottom and leaving the placentae with the false partition stretched between.

SIMPLE FRUIT: A fruit derived from one carpel or several united carpels.

SIMPLE LEAF: An undivided leaf; opposed to compound.

SIMPLE PIT: Pit not surrounded by an overarching border of secondary wall; opposed to bordered.

SIPHONACEOUS, SIPHONOUS (sī'fŏn•ā'shŭs, sī'fŏn•ŭs) [F. *siphon*, fr. L. *sipho, -onis*, fr. Gk. *siphon*, a siphon, tube, pipe]: In algae, multinucleate, without crosswalls; coenocytic.

SIPHONOSTELE [Gk. *siphon*, pipe, + *stele*, pillar]: A type of stele containing a hollow cylinder of vascular tissue surrounding a pith.

SLIME: *See* P-protein.

SOFTWOOD: A name commonly applied to the wood of a conifer tree.

SOLUTE (sŏl'ŭt): A dissolved substance.

SOLUTE POTENTIAL: The change in free energy or chemical potential of water produced by solutes; carries a negative (minus) sign. Also called osmotic potential.

SOLUTION: Usually liquid, in which the molecules of the dissolved substance—e.g., sugar—the solute, are dispersed between the molecules of the solvent; e.g., water.

SOMATIC CELLS (so•măt'ĭk) [Gk. *soma*,

body]: The differentiated, usually diploid ($2n$) cells composing body tissues of multicellular plants and animals.

SOREDIUM, *pl.* SOREDIA (so•rē′dĭ•ŭm) [Gk. *soros*, heap]: A special, reproductive unit of lichens consisting of a few algal cells surrounded by fungal hyphae.

SORUS, *pl.* SORI (sō′rŭs) [Gk. *soros*, heap]: A group or cluster of sporangia.

SPECIALIZED: (1) Of organisms, having special adaptations to a particular habitat or mode of life; (2) of cells, having particular functions.

SPECIES, *pl.* SPECIES [L. kind, sort]: A kind of organism; species are designated by binomial names written in italics.

SPECIFICITY (spĕs′ĭ•fĭs′ĭ•tĭ): Uniqueness, as in proteins in given organisms and of enzymes in given reactions.

SPERM: A mature male sex cell or gamete, usually motile and smaller than the female gamete.

SPERMAGONIUM, *pl.* SPERMAGONIA (spûr′-mȧ•gō′nĭ•ŭm) [Gk. *sperma*, sperm, + *gonos*, offspring]: In the rust fungi, the structure that produces spermatia.

SPERMATANGIUM, *pl.* SPERMATANGIA (spûr′-mȧ•tăn′jĭ•ŭm) [Gk. *sperma*, sperm, + L. *tangere*, to touch]: In the red algae, the cell that produces spermatia.

SPERMATIUM, *pl.* SPERMATIA [Gk. *sperma*, sperm]: In the red algae and some fungi, a minute nonmotile male gamete.

SPERMATOPHYTE (spûr′mȧ•to•fīt′) [Gk. *sperma*, seed, + *phyton*, plant]: A seed plant.

SPHEROSOME: Spherical structures in the cytoplasm of plant cells, many of which contain mostly lipids and apparently are centers of lipid synthesis and accumulation.

SPIKE [L. *spica*, ear of grain]: An inflorescence in which the main axis is elongated and the flowers are sessile.

SPIKELET (spīk′lĕt [L. *spica*, ear of grain, + dim. ending *-let*]: The unit of inflorescence in grasses; a small group of grass flowers.

SPINDLE FIBERS: A group of microtubules that extend from the centromeres of the chromosomes to the poles of the spindle or from pole to pole in a dividing cell.

SPINE: A hard, sharp-pointed structure; usually a modified leaf, or part of a leaf.

SPONGY PARENCHYMA (pȧ•rĕng′kĭ•mȧ): A leaf tissue composed of loosely arranged, chloroplast-bearing cells.

SPORANGIOPHORE (spo•răn′jĭ•o•fōr′) [Gk. *phore*, fr. *phorein*, go bear]: A branch bearing one or more sporangia.

SPORANGIUM, *pl.* SPORANGIA (spo•răn′-jĭ•ŭm): [Gk. *spora*, seed, + *angeion*, a vessel]: A hollow unicellular or multicellular structure in which spores are produced.

SPORE: A reproductive cell, usually unicellular, capable of developing into an adult without fusion with another cell.

SPORE MOTHER CELL: A diploid ($2n$) cell that undergoes meiosis and produces (usually) four haploid cells (spores) or four haploid nuclei.

SPOROPHYLL (spō′ro•fĭl): A modified leaf or leaflike organ that bears sporangia; applied to the stamens and carpels of angiosperms, fertile fronds of ferns, etc.

SPOROPHYTE (spō′ro•fīt): The spore-producing, diploid ($2n$), phase in the life cycle of a plant having alternation of generations.

SPOROPOLLENIN: The very tough substance (a cyclic alcohol) of which the exine, or outer wall, of spores and pollen grains is composed.

STALK CELL: One of two cells produced by division of the generative cell in developing pollen grains of gymnosperms; it is a sterile cell and eventually degenerates.

STAMEN (stā′mĕn) [L. *stamen*, thread]: The part of the flower producing the pollen, composed (usually) of anther and filament; collectively, the stamens make up the androecium.

STAMINATE (stăm′ĭ•nat): Pertaining to a flower having stamens but no functional carpels.

STARCH [M.E. *sterchen*, to stiffen]: A complex insoluble carbohydrate, the chief food storage substance of plants, which is composed of several hundred glucose units ($C_6H_{10}O_5$) and which is readily broken down enzymatically into these separate units.

STELE (stē′le) [Gk. *stele*, a post]: The central cylinder, inside the cortex, of roots and stems of vascular plants.

STEM: The part of the axis of vascular plants that is above ground, as well as anatomically similar portions below ground (rhizomes, corms, etc.)

STERIGMA, *pl.* STERIGMATA [Gk. *sterigma*, a prop]: A small, slender protuberance that bears a basidiospore; it is borne by a basidium.

STIGMA: (1) The region of a carpel serving as a receptive surface for pollen grains and on which they germinate; (2) eyespot of algae.

STIPE (stīp): A supporting stalk, such as the stalk of a gill fungus or the leaf stalk of a fern.

STIPULE (stĭp′ŭl): A leaflike appendage on either side of the basal part of a leaf of some species of plants.

STOLON (stō′lŏn) [L. *stolo*, shoot]: A stem that grows horizontally along the ground surface and may form adventitious roots, as runners in the strawberry.

STOMA, *pl.* STOMATA (stō′mȧ) [Gk. *stoma*, mouth]: A minute opening bordered by guard cells in the epidermis of leaves and stems through which gases pass. Also used to refer to the entire stomatal apparatus: the guard cells plus their included pore.

STROBILUS (strŏb′ĭ•lŭs) [Gk. *strobilos*, a cone]: A number of modified leaves (sporophylls) or ovule-bearing scales grouped terminally on a stem; a cone.

STROMA [Gk. *stroma*, anything spread out]: The ground substance of plastids.

STRUCTURAL GENE: One of the genes of an operon responsible for the production of its characteristic product.

STYLE [Gk. *stylos*, a column]: Slender column of tissue which arises from the top of the ovary and through which the pollen tube grows.

SUB-, SUS- [L. under, below]: Prefix meaning "under" or "below," for example, subepidermal, "underneath the epidermis."

SUBERIN (sū′bĕr•ĭn) [L. *suber*, the cork oak]: Fatty material found in the cell walls of cork tissue and in the Casparian strip of the endodermis.

SUBSIDIARY CELL: An epidermal cell morphologically distinct from other epidermal cells and associated with a pair of guard cells; also called accessory cell.

SUBSPECIES: A subdivision of a species.

SUBSTRATE [L. *substratus*, strewn under]: The foundation to which an organism is attached; substance acted on by an enzyme.

SUCCESSION: In ecology, the slow, orderly progression of changes in community composition during development of vegetation in any area, from initial colonization to the attainment of the climax typical of a particular geographic area.

SUCCULENT: A plant with fleshy, water-storing stems or leaves.

SUCKER: A sprout produced by the roots of some plants and that gives rise to a new plant.

SUCRASE (sū'krās): An enzyme that hydrolyzes sucrose into glucose and fructose; also called invertase.

SUCROSE (sū'krōs): Cane sugar; a disaccharide (glucose plus fructose) found in many plants; the primary form in which sugar produced by photosynthesis is translocated.

SUPERIOR OVARY: An ovary free and separate from the calyx.

SUSPENSION: A heterogeneous dispersion in which the dispersed phase consists of solid particles sufficiently large that they will settle out of the fluid dispersion medium under the influence of gravity.

SUSPENSOR (sŭs•pĕn'sẽr): A structure in the embryo of many vascular plants that pushes the terminal part of the embryo into the endosperm.

SYMBIOSIS (sĭm'bī•ō'sĭs) [Gk. *syn*, with, + *bios*, life]: The living together in close association of two or more dissimilar organisms. Includes parasitism (in which the association is harmful to one of the organisms) and mutualism (in which the association is advantageous to both.)

SYN- [Gk. *syn*, together]: Prefix, meaning "together"; like "sym-".

SYNAPSIS (sĭ'năp'sĭs): The pairing of homologous chromosomes that occurs prior to the first meiotic division; crossing over occurs during synapsis.

SYNERGIDS (sĭ•nûr'jĭd): Two ephemeral cells lying close to the egg in the mature embryo sac of the ovule of flowering plants.

SYSTEMATICS: Scientific study of the kinds and diversity of organisms and of the relationships between them.

SYNTHESIS: The formation of a more complex substance from simpler ones.

TANGENTIAL SECTION (tăn•jĕn'shăl): A longitudinal section cut at right angles to a radius of a cylindrical structure, such as a root or stem; in the case of secondary xylem, or wood, and secondary phloem, at right angles to the rays.

TAPETUM (tȧ•pē'tŭm) [Gk. *tapes*, a carpet]: Nutritive tissue in the sporangium, particularly an anther.

TAPROOT: The primary root of a plant formed in direct continuation with the root tip or radicle of the embryo; forms a stout, tapering main root from which arise smaller, lateral branches.

TAXON: Any one of the categories such as species, class, order, or division, into which living organisms are classified.

TAXONOMY (tăks•ŏn'o•mĭ) [Gk. *taxis*, arrangement, + *nomos*, law]: The science of the classification of organisms.

TELIOSPORE (tē'lĭ•ȯ•spōr'): In the rust fungi, a thick-walled spore in which karyogamy and meiosis occur, and from which basidia develop.

TELIUM, *pl.* TELIA [Gk. *telos*, completion]: In the rust fungi, the structure that produces teliospores.

TELOPHASE (tĕl'o•fāz) [Gk. *telos*, end, + phase]: The last stage in mitosis and meiosis, during which the chromosomes become reorganized into two new nuclei.

TEMPERATE PHAGE: A bacterial virus that may remain latent in its host bacterial cell. In this latent (prophage) state, it is associated with the bacterial chromosome and is replicated with it.

TEMPLATE (tĕm'plĭt): A pattern or mold guiding the formation of a negative or complement: a term applied especially to DNA duplication, which is explained in terms of a template hypothesis.

TENDRIL (tĕn'drĭl) [L. *tendere*, to stretch out, to extend]: A slender coiling structure—usually a modified leaf or part of a leaf—that aids in support of the stems.

TEPAL (tĕp'ăl): Unit structure of a perianth that is not differentiated into sepals and petals.

TEST CROSS: A cross of a dominant with a homozygous recessive; used to determine whether the dominant is homozygous or heterozygous.

TETRAD (tĕt'răd): A group of four spores formed from a spore mother cell by meiosis.

TETRAPLOID (tĕt'rȧ•ploid) [Gk. *tetra*, four, + *ploos*, fold]: Twice the usual, diploid, number of chromosomes (that is, $4n$).

TETRASPORANGIUM, *pl.* TETRASPORANGIA [Gk. *tetra*, four, + *spora*, seed, + *angeion*, vessel]: In certain red algae, a sporangium in which meiosis occurs, resulting in the production of tetraspores.

TETRASPORE [Gk. *tetra*, four, + spores]: In certain red algae, the four spores formed by meiotic division, in the tetrasporangium, of a spore mother cell.

TETRASPOROPHYTE [Gk. *tetra*, four, + *spora*, seed, + *phyton*, plant]: In certain red algae, a diploid plant that produces tetrasporangia.

THALLOPHYTE (thăl'o•fīt): A term previously used to designate fungi and algae collectively, now largely abandoned.

THALLUS (thăl'ŭs) [Gk. *thallos*, a sprout]: A type of plant body that is undifferentiated into root, stem or leaf.

THEORY [Gk. *theorein*, to look at]: A generalization based on observation and experiments conducted to test the validity of a hypothesis and found to support the hypothesis.

THERMODYNAMICS [Gk. *therme*, heat, + *dynamis*, power]: The study of energy, using heat as the most convenient form of measurement of energy. The first law of thermodynamics states that in all processes, the total energy of the universe remains constant. The second law of thermodynamics states that the entropy or degree of randomness tends to increase.

THORN: A hard, woody, pointed branch.

THYLAKOID [Gk. *thylakos*, sac, + *oides*, like]: Unit of the chloroplast, a saclike membranous structure; stacks of thylakoids form the grana.

THYMINE (thī'mēn): A pyrimidine occurring in DNA but not in RNA.

TISSUE [L. *texere*, to weave]: A group of

similar cells organized into a structural and functional unit.

TISSUE CULTURE: A technique for maintaining fragments of plant or animal tissue alive in a medium after removal from the organism.

TISSUE SYSTEM: A tissue or group of tissues organized into a structural and functional unit in a plant or plant organ. There are three tissue systems: dermal, vascular, and fundamental, or ground.

TONOPLAST [Gk. *tonos*, stretching tension, + *plastos*, formed, molded]: The cytoplasmic membrane surrounding the vacuole in plant cells; also called vacuolar membrane.

TORUS, *pl.* TORI: The central thickened part of the pit-membrane in the bordered pits of conifers and some other gymnosperms.

TRACHEARY ELEMENT: The general term for a water-conducting cell in vascular plants; tracheids and vessel members.

TRACHEID (trā'ke•ĭd): An elongated, thick-walled conducting and supporting cell of xylem. It has tapering ends and pitted walls without perforations, as contrasted with a vessel member. Found in nearly all vascular plants.

TRANSDUCTION: The transfer of genetic material (DNA) from one bacterium to another by a temperate phage.

TRANSFER CELL: Specialized parenchyma cell with wall ingrowths that increase the surface of the plasma membrane; apparently functions in the short-distance transfer of solutes.

TRANSFER RNA (tRNA): Low-molecular-weight RNA that becomes attached to an amino acid and guides it to the correct position on the ribosome for protein synthesis; there is at least one tRNA molecule for each amino acid.

TRANSFORMATION: A genetic change produced by the incorporation into a cell of DNA from another cell.

TRANSITION REGION: The region in the primary plant body showing transitional characteristics between structures of root and shoot.

TRANSLOCATION: (1) In plants, the long-distance transport of water, minerals, or food; most often used to refer to food transport; (2) in genetics, the interchange of chromosome segments between non-homologous chromosomes.

TRANSPIRATION [Fr. *transpirer*, to perspire]: The loss of water vapor by plant parts; most transpiration occurs through stomata.

TRANSVERSE SECTION: A section cut perpendicular, or at right angles, to the longitudinal axis of a plant part.

TREE: A perennial woody plant generally with a single stem (trunk).

TRICHOGYNE [Gk. *trichos*, a hair, + *gyne*, female]: In the red algae and certain Ascomycetes and Basidiomycetes, a receptive protuberance of the female gametangium for spermatia.

TRICHOME [Gk. *tricho*, hair]: An outgrowth of the epidermis, such as a hair, scale, and water vesicle.

TRIOSE [Gk. *tries*, tree, + *ose*, suffix indicating a carbohydrate]: Any three-carbon sugar.

TRIPLE FUSION: In angiosperms, the fusion of the second male gamete, or sperm, with the polar nuclei, resulting in formation of a primary endosperm nucleus, which is most often triploid.

TRIPLOID (trĭp'loid) [Gk. *triploos*, triple]: Having three complete chromosome sets per cell ($3n$).

TRITIUM (trĭt'ĭ•ŭm): Radioactive isotope of hydrogen, ^3H. The nucleus of a tritium atom contains one proton and two neutrons, whereas the more common hydrogen nucleus consists only of a proton.

-TROPH, TROPHO- [Gk. *trophos*, feeder]: Suffix or prefix meaning "feeder," "feeding," for example, autotrophic, "self-nourishing."

TROPHIC LEVEL: A step in the movement of energy through an ecosystem.

TROPISM [Gk. *trope*, a turning]: Tropic movement; a response to an external stimulus in which the direction of the movement is usually determined by the direction from which the most intense stimulus comes.

TUBE CELL: In male gametophytes, or pollen grains, of seed plants, the cell that develops into the pollen tube.

TUBER [L. *tuber*, bump, swelling]: A much-enlarged, short, fleshy underground stem, such as that of the potato (*Solanum tuberosum*).

TUNICA-CORPUS: The organization of the shoot of most angiosperms and a few gymnosperms, consisting of one or more peripheral layers of cells (the tunica layers) and an interior (the corpus). The tunica layers undergo surface growth (by anticlinal divisions), and the corpus undergoes volume growth (by divisions in all planes).

TURGID (tûr'jĭd) [L. *turgidus*, swollen, inflated]: Swollen, distended, referring to a cell that is firm due to water uptake.

TURGOR PRESSURE [L. *turgor*, a swelling]: The pressure within the cell resulting from the movement of water into the cell.

TYLOSE, *pl.* TYLOSES [Gk. *tylos*, a lump or knot]: A balloonlike outgrowth from a ray or axial parenchyma cell through the pit in a vessel wall and into the lumen of the vessel.

UMBEL (ŭm'bĕl) [L. *umbella*, a sunshade]: An inflorescence, the individual pedicels of which all arise from the apex of the peduncle.

UNICELLULAR (ū'nĭ•cĕl'u•lĕr): Composed of a single cell.

UNISEXUAL: Usually applied to a flower lacking either stamens or carpels. A perianth may be present or absent.

URACIL (ū'ră•sĭl): A pyrimidine found in RNA but not in DNA.

UREDINIUM, *pl.* UREDINIA (ū'rē•dĭn•ĭ•ŭm) [L. *uredo*, a blight]: In rust fungi, the structure that produces uredospores.

UREDOSPORE [L. *uredo*, a blight, + spore]: In rust fungi, a reddish, binucleate spore produced in summer.

VACUOLAR MEMBRANE: *See* tonoplast.

VACUOLE [L. dimin, of *vacuus*, empty]: A space or cavity within the cytoplasm filled with a watery fluid, the cell sap; part of lysosomal compartment of the cell.

VALVE: One of the two halves of the diatom cell wall, or frustule.

VARIATION: The differences that occur within the offspring of a particular species.

VARIETY: A subdivision of a species.

vascular (văs'ku•lẽr) [L. *vasculum*, a small vessel]: Pertains to any plant tissue or region consisting of or giving rise to conducting tissue; e.g., xylem, phloem, vascular cambium.

vascular bundle: A strand of tissue containing primary xylem and primary phloem (and procambium if still present) and frequently enclosed by a bundle sheath of parenchyma or fibers.

vascular cambium: A cylindrical sheath of meristematic cells, the division of which produces secondary phloem and secondary xylem.

vascular rays: Ribbonlike sheets of parenchyma that extend radially through the wood, across the cambium, and into the secondary phloem. They are always produced by the vascular cambium.

vascular tissue system: All the vascular tissues in a plant or plant organ.

vegetative: Of, relating to, or involving propagation by asexual processes.

vegetative reproduction: (1) In seed plants, reproduction by means other than by seeds; (2) in other organisms, reproduction by vegetative spores, fragmentation, or division of the plant body. Unless a mutation occurs, each daughter cell individual is genetically identical with its parent.

vein: A vascular bundle forming a part of the framework of the conducting and supporting tissue of a leaf or other expanded organ.

velamen [L. *velumen*, fleece]: A multiple epidermis covering the aerial roots of some orchids and aroids; also occurs on some terrestrial roots.

venation: Arrangement of veins in leaf blade.

venter [L. *venter*, belly]: The enlarged basal portion of an archegonium containing the egg.

vernalization (vûr'năl•i•zā'shŭn) [L. *vernalis*, spring]: The induction of flowering by cold treatments.

vessel [L. *vasculum*, a small vessel]: A tubelike structure of the xylem composed of elongate cells (vessel members) placed end to end and connected by perforations. Its function is to conduct water and minerals through the plant body. Found in nearly all angiosperms and a few other vascular plants.

vessel member: One of the cells composing a vessel. Also called vessel element.

virus [L. slimy liquid, poison]: A submicroscopic, noncellular particle; composed of a nucleic acid core and a protein shell; viruses reproduce only within host cells.

volva (vŏl'vȧ) [L. *volva*, a wrapper]: A cuplike structure at the base of the stalk of certain mushrooms.

wall pressure: The pressure of the cell wall exerted against the turgid protoplast; opposite and equal to the turgor pressure.

water potential: The algebraic sum of the solute potential and the pressure potential, or wall pressure; the potential energy of water.

water vesicle: An enlarged epidermal cell in which water is stored; a type of trichome.

weed [AS. *weod*, used at least since 888 in its present meaning]: Generally a herbaceous plant not valued for use or beauty, growing wild and rank, and regarded as using ground or hindering the growth of superior vegetation.

whorl: A circle of leaves or of flower parts.

wild type: In genetics, the phenotype or genotype that is characteristic of the majority of individuals of a species in a natural environment.

wood: Technically, the secondary xylem.

xanthophyll (zăn'tho•fĭl) [Gk. *xanthos*, yellowish-brown, + *phyllon*, leaf]: A yellow chloroplast pigment; a member of the carotenoid group.

xerophyte [Gk. *xeros*, dry, + *phyton*, a plant]: A plant that grows in a dry, or arid, habitat.

xylem (zī'lĕm) [Gk. *xylon*, wood]: A complex vascular tissue through which most of the water and minerals of a plant are conducted; characterized by the presence of tracheary elements.

zeatin (zē'ă•tĭn): Plant hormone; a natural cytokinin isolated from corn.

zoosporangium [Gk. *zoon*, animal, + sporangium]: A sporangium bearing zoospores.

zoospore (zō'o•spōr): A motile spore, found among algae and fungi.

zygomorphic [Gk. *zygo*, yoke, pair, + *morphe*, form]: A type of flower capable of being divided into two symmetrical halves only by a single longitudinal plane passing through the axis; also called bilaterally symmetrical.

zygospore (zī'go•spōr): A thick-walled resistant spore developing from a zygote, resulting from the fusion of isogametes.

zygote (zī'gōt) [Gk. *zygotos*, paired together]: The diploid ($2n$) cell resulting from the fusion of male and female gametes.

Illustration Acknowledgments

I–1 Stephen B. Gough and William J. Woelkering

I–2 *(a)* Charlie Ott, National Audubon Society Collection/PR *(b)* John H. Gerard

I–3 E. S. Ross

I–5 E. M. Poirot

I–6 John H. Gerard

I–8 Gianni Tortoli, from Photo Researchers, Inc.

I–9 *(a)* Wide World Photos; *(b)* International Rice Research Institute

I–10 Grant Heilman Photography

1–1 Michael A. Walsh

Essay, page 16, Rare Book Division, The New York Public Library, Astor, Lenox, and Tilden Foundations

1–2 A. Ryter

1–3 J. F. M. Hoeniger, *Journal of General Microbiology,* **40**:29–42,1965.

1–4 George Palade

1–5 Michael A. Walsh

1–6 From *Cell Ultrastructure* by William A. Jensen and Roderic B. Park. © 1967 by Wadsworth Publishing Company, Inc., Belmont, California 90042. Reproduced by permission of the publisher.

1–7 L. A. Staehelin, *Zeitschrift Für Zellforschung und Mikroskopische Anatomie,* **74**:325–350, 1966.

1–9 Ray F. Evert

1–10 Ray F. Evert

1–11 Ray F. Evert

1–12 Eldon H. Newcomb

1–13 A. Berkaloff, J. Bourguet, P. Favard, and M. Guinnebault, *Introduction à la Biologie: Biologie et Physiologie Cellulaire,* Hermann Collection, Paris, 1967. (After William A. Jenson and Roderic B. Park, *Cell Ultrastructure,* Wadsworth Publishing Company, Inc., 1967.)

1–14 James Cronshaw

1–15 *(b), (c)* A. P. Kole, *Proc. Kon. Ned. Akad. v. Wet.,* Ser. C. **62**:404–408

1–17 Ursula Goodenough

1–18 *(a), (b)* L. K. Shumway

1–19 From Ursula Goodenough. (After D. Branton and R. B. Park, *Journal of Ultrastructure Research,* **19**:283–303, 1967.)

1–20 Roderic B. Park

1–21 Ray F. Evert

1–22 Michael A. Walsh

1–24 James Cronshaw

Essay, page 32, Patrick Echlin

1–25 *(b)* Myron C. Ledbetter

1–26 R. D. Preston

1–27 Ray F. Evert

1–29 Ray F. Evert

1–30 *(a)* and *(b)* Ray F. Evert

1–31 D. Branton, from *Cell Ultrastructure* by William A. Jenson and Roderic B. Park, Wadsworth Publishing Company, Inc., Belmont, California, 1967.

1–32 Donald E. Olins and Ada L. Olins, University of Tennessee—Oak Ridge Graduate School of Biomedical Sciences and The Oak Ridge National Laboratory

1–33 Michael A. Walsh

1–36 Photographs of onion cell by William Tai

1–37 Roland R. Dute

1–38 H. J. Wilson

1–39 General Biological Supply

1–40 James Cronshaw

2–1 After A. C. Leopold, *Plant Growth and Development,* McGraw-Hill Book Company, New York, 1964. (Adapted from Crafts, *et al.,* 1949.)

2–6 *(d)* M. C. Ledbetter; *(e), (f)* Ray F. Evert

2–12 B. E. Juniper

2–20 After Albert L. Lehninger, *Biochemistry,* 2nd ed., Worth Publishers, Inc., New York, 1975.

3–1 Drawing by B. O. Dodge, from E. L. Tatum

3–6 From *The Double Helix* by James D. Watson. Copyright © 1968 by James D. Watson. Reprinted by permission of Atheneum Publishers, New York.

3–7 A. C. Barrington-Brown

3–10 John Cairns

3–12 S. H. Kim, "Three-Dimensional Structure of Yeast Phenylalanine Transfer RNA: Folding of the Polynucleotyde Chain," *Science,* **179**:285–288, 19 January 1973. Copyright 1973 by the American Association for the Advancement of Science.

3–14 O. L. Miller, *et al.,* "Visualization of Bacterial Genes in Action," *Science,* **169**:392–395, 24 July 1970. Copyright 1970 by the American Association for the Advancement of Science.

3–17 Walter Fiers from "Molecular Biology's Flower Child," *Science News,* vol. 103, January 6, 1973.

3–20 After James D. Watson, *Molecular Biology of the Gene,* 2nd ed., W. A. Benjamin, Inc., Menlo Park, Calif., 1970.

3–21 Jack Griffith

3–24 O. L. Miller, Jr. and Barbara R. Beatty, Biology Division, Oak Ridge National Laboratory

4–1 Hale Observatories

4–2 The Bettmann Archive, Inc.

4–3 The Bettmann Archive, Inc.

4–4 The Bettmann Archive, Inc.

Essay, page 94, Howard Towner

4–5 After Lehninger, *Biochemistry,* 2nd ed., Worth Publishers, Inc., New York, 1975.

5–1 Michael A. Walsh.

Essay, page 105, Y. Haneda

6–1 L. K. Shumway.

Essay, page 113 *(a), (b)* Photographs by Howard Towner

Essay, page 122, Arnold Sparrow

7–1 The Bettmann Archive, Inc.

7–3 Peter B. Moens

7–4 Bernard John

7–5 G. Östergren

7–6 Photographs of crested wheat grass *(Agropyron cristatum)* by William Tai

7–7 Arnold Sparrow

Essay, page 130 *(top), (middle)* Ray F. Evert; *(bottom)* American Fern Society—Wherry Collection

7–10 Photograph courtesy of Professor J. William Schopf, Department of Geology, University of California, Los Angeles

7–11 After G. Ledyard Stebbins, "The Comparative Evolution of Genetic Systems," *The Evolution of Life,* edited by Sol Tax, The University of Chicago Press, Chicago, Ill., vol. 1, 1960.

7–14 After Karl von Frisch, *Biology,* Harper & Row Publishers, Inc., New York, 1964, translated by Jane Oppenheimer.

7–16 *(b)* Howard Towner

7–18 C. G. G. J. van Steenis

7–19 After E. D. Merrell, 1964

8–1 *(top)* Radio Times Hulton Picture Library; *(bottom)* The Bettmann Archive, Inc.

8–6 Mather and Harrison, *Heredity,* vol. 3, 1949.

8–7 Howard Towner; seeds supplied by Daniel Janzen

8–8 From Sherwin Carlquist, *Island Life,* The Natural History Press, Garden City, New York, 1965.

8–9 A. P. Nelson

8–10 Carnegie Institution of Washington, Publication 540. Berton Crandall, photographer

8–11 From Sherwin Carlquist, *Island Life,* The Natural History Press, Garden City, New York, 1965.

8–12 John A. Jump

8–14 Richard Straw

8–15 M. A. Nobs, Carnegie Institution of Washington, Publication 623, 1963

8–17 *(a), (b), (c)* C. J. Marchant; *(d)* Peter H. Raven

8–18 Professor Arne Strid, University of Copenhagen

9–1 Larry West

9–2 The Bettmann Archive, Inc.

9–3 *(a)* John H. Gerard; *(b)* Larry West; *(c)* E. S. Ross

9–4 Theodore Delevoryas, *Plant Diversification,* Holt, Rinehart and Winston, Inc., New York, 1966.

9–5 *(a)* Theodore Delevoryas, *Plant Diversification,* Holt, Rinehart and Winston, Inc., New York, 1966; *(b)* Charles S. Webber, Jepson Herbarium

9–6 *(a), (c)* Eric V. Gravé; *(b)* H. Forest

9–7 *(a)* Oxford Scientific Films, Bruce Coleman, Inc.; *(b)* Charlie Ott, National Audubon Society Collection/PR; *(c)* John Merkle, National Audubon Society Collection/PR; *(d)* © Alison Wilson; *(e)* © D. P. Wilson; *(f)* E. R. Degginger, Bruce Coleman, Inc.; *(g)* Eric V. Gravé; *(h)* E. S. Ross

9–8 *(a)* Lynn M. Stone, Bruce Coleman, Inc.; *(b)* E. R. Degginger, Bruce Coleman, Inc.; *(c)* David Overcash, Bruce Coleman, Inc.

9–9 *(a)* Irvin L. Oakes, National Audubon Society Collection/PR; *(b)* John M. Burnley, Bruce Coleman, Inc.; *(c)* Dennis Brokaw, National Audubon Society Collection/PR; *(d)* J. Shaw, Bruce Coleman, Inc.; *(e)* Jane Burton, Bruce Coleman, Inc.; *(f)* R. Carr, Bruce Coleman, Inc.; *(g)* Irvin L. Oakes, National Audubon Society Collection/PR; *(h)* Jack Dermid; *(i)* E. S. Ross; *(j)* Larry West

9–10 *(a)* Michael Jost; *(b)* L. A. Staehelin

10–1 Lee D. Simon

10–3 *(a)* J. W. Schopf and E. S. Barghoorn, *Science* **156**:508–512, 1967. Copyright 1967 by the American Association for the Advancement of Science; *(b)* E. S. Barghoorn and J. W. Schopf, *Science* **152**:758–763, 1966. Copyright 1966 by the American Association for the Advancement of Science.

10–4 S. M. Siegel, then of Union Carbide Corp. and B. J. Siegel of Yale University

10–5 *(a)* Hubert Lechevalier; *(b)* R. S. Wolfe; *(c)* Hans Reichenbach and Martin Dworkin; *(d)* National Medical Audiovisual Center

10–7 *(a)* J. F. M. Hoeniger, *Journal of General Microbiology,* **40**:29–42, 1965. *(b)* H. Stolp and M. P. Starr, *Antonie van Leeuwenhoek,* **29**:217–248, 1963.

10–8 A. Ryter

10–10 S. Conti

10–11 A. Berkaloff, J. Bourguet, P. Favard, and M. Guinnebault, *Introduction à la Biologie: Biologie et Physiologie Cellulaire,* Hermann Collection, Paris, 1967.

10–12 C. Robinow

10–13 National Medical Audiovisual Center

10–14 Charles C. Brinton, Jr.

10–15 Werner Braun, *Bacterial Genetics,* 2nd ed., W. B. Saunders Company, Philadelphia, 1965.

10–17 Werner Braun, *Bacterial Genetics,* 2nd ed., W. B. Saunders Company, Philadelphia, 1965.

10–18 J. Lederberg and E. M. Lederberg

10–20 Walter Reed Army Institute of Research

10–21 M. C. Pelczar and R. D. Reid, *Microbiology,* Copyright 1965 by McGraw-Hill Book Company. Used with permission of McGraw-Hill Book Company.

10–22 *(a)* Parke, Davis & Co.; *(b)* after R. Y. Stanier, Michael Doudoroff, and E. A. Adelberg, *The Microbial World,* 2nd ed. © 1963, pp. 658–659. Reprinted by permission of Prentice-Hall, Inc., Englewood Cliffs, New Jersey; *(d)* E. S. Ross

10–23 *(b), (c), (d)* Karl Maramorosch

10–24 W. A. Niering

10–26 Mercedes R. Edwards, *Journal of Phycology,* **4**:283–298, 1968; *(b)* Ray F. Evert; *(c)* Turtox/Cambosco; *(d)* Erling J. Ordal

10–28 Ray F. Evert

10–29 Photograph by Norma J. Lang, *Journal of Phycology,* **1**:127–134, 1965.

10–30 Norma J. Lang

10–31 After Wendell Stanley and Evan G. Valens, *Viruses and the Nature of Life,* E. P. Dutton & Co., New York, 1961.

10–32 J. D. Almeida and A. F. Howatson, *Journal of Cell Biology,* **16**:616, 1963.

10–33 *(a)* Lee D. Simon; *(b)* Constructed by Kim Allen

Essay, page 205 *(a)* Frederick A. Murphy, National Communicable Disease Center; *(b)* M. J. Pelczar and R. D. Reid, *Microbiology,* Copyright 1965 McGraw-Hill Book Company. Used with permission of McGraw-Hill Book Company.

10–34 *(a), (b)* Karl Maramorosch; *(c)* E. Shikata, *Journal of the National Cancer Institute,* **36**:January 1966; *(d)* E. Shikata and Karl Maramorosch, *Virology,* **32**:no. 3, July 1967. Copyright © Academic Press, Inc.

11–1 Jack Dermid

11–2 Ross E. Hutchins

11–3 *(a)* Eric V. Gravé; *(b)* Walter Dawn, National Audubon Society Collection /PR

11–4 Michael D. Coffey, Barry A. Palevitz, and Paul J. Allen, *The Canadian Journal of Botany,* **50**:231–240, 1972.

11–6 H. C. Hoch and J. E. Mitchell, *Phytopathology,* **62**:149–160,1972.

11–7 H. C. Hoch and D. P. Maxwell

11–11 *(a)* Alma W. Barksdale; *(b)* Alma W. Barksdale, *Mycologia,* **55**:493–501, 1963; *(c), (d)* Alma W. Barksdale; *(e)* Alma W. Barksdale, *Mycologia,* **55**:627–632, 1963.

11–12 John S. Niederhauser and William C. Cobb, "The Late Blight of Potatoes," *Scientific American,* May 1959.

11–13 Harold C. Bold, *Morphology of Plants,* 2nd ed., Harper & Row Publishers, Inc., New York, 1967, Figures 11–18A and 11–18B.

11–14 After Theodore Delevoryas, *Plant Diversification*, Holt, Rinehart and Winston, Inc., New York, 1966.

Essay, page 221, Drawing from A. H. Reginald Buller, *Researches on Fungi*, vol. 6, Longman Group Limited, Harlow, England; *(a)*, *(b)* Eric V. Gravé; *(c)* Robert M. Page

Essay, page 222 *(a)* U.S. Department of Agriculture

11–15 *(a)* Alvin E. Staffan; *(b)* John Shaw; *(c)* Agenzia Fotografica, Luisa Ricciarini—Milan

11–16 *(a)* C. Bracker; *(b)* Ray F. Evert; *(c)* J. Cooke, *The American Journal of Botany*, **56**:335–340, 1969.

11–17 Ray F. Evert

11–18 After Theodore Delevoryas, *Plant Diversification*, Holt, Rinehart and Winston, Inc., New York, 1966.

11–19 *(b)* Ray F. Evert

11–20 Agenzia Fotografica, Luisa Ricciarini—Milan

Essay, page 227, David Pramer, *Science*, **144**:382–388, 24 April 1964. Copyright 1964 by the American Association for the Advancement of Science.

11–21 After R. F. Scagel, *et al.*, *An Evolutionary Survey of the Plant Kingdom*, Wadsworth Publishing Company, Inc., Belmont, California, 1965.

11–22 James R. Leard, National Audubon Society Collection/PR

11–23 *(a)* Ross E. Hutchins; *(b)* R. Carr, Bruce Coleman, Inc.

11–24 *(a)* Ross E. Hutchins; *(b)* John R. Clawson, National Audubon Society Collection/PR; *(c)* Larry West

11–26 Ray F. Evert

11–27 Photo courtesy of Clifford Wetmore, University of Minnesota

11–29 *(a)*, *(b)* Jane Burton, Bruce Coleman, Inc.; *(c)* Agenzia Fotografica, Luisa Ricciarini—Milan

11–30 *(b)* C. Heintz and D. Niederpruem

11–31 Roger Moore

11–32 *(a)* Jane Burton, Bruce Coleman, Inc.; *(b)* J. Markham, Bruce Coleman, Inc.; *(c)* R. E. Pelham, Bruce Coleman, Inc.; *(d)* Charlie Ott, National Audubon Society Collection/PR

11–33 *(a)* Jeff Foott, Bruce Coleman, Inc.; *(b)* Richard Parker, National Audubon Society Collection/PR

11–34 Ray F. Evert

11–35 S. C. Bisserot, Bruce Coleman, Inc.

11–36 *(a)*, *(b)* R. Gordon Wasson

11–38 E. S. Ross

11–40 Ray F. Evert

11–41 F. W. Went

11–42 S. A. Wilde

11–43 John P. Limbach, Ripon Microslides, Inc.

11–44 Ray F. Evert

12–1 Jack Dermid

12–2 © Douglas P. Wilson

Essay, page 248 *(top)* Photograph by James Oschman; *(bottom)* Leonard Muscatine

12–8 R. W. Hoshaw

12–9 Ray F. Evert

12–10 Ray F. Evert

12–11 *(a)* Jack Dermid; *(b)* Ross E. Hutchins

12–13 Don Longanecker

12–14 Ray F. Evert

12–15 Ray F. Evert

12–16 Ray F. Evert

12–18 © M. Alison Wilson

12–19 R. F. Scagel, *et al.*, *An Evolutionary Survey of the Plant Kingdom*, Wadsworth Publishing Company, Inc., Belmont, California, 1965.

12–21 Gary L. Floyd, Kenneth D. Stewart, and Karl R. Mattox, *Journal of Phycology*, **7**:306–309, 1971.

12–22 Ray F. Evert

12–23 and 12–32 R. F. Scagel, *et al.*, *An Evolutionary Survey of the Plant Kingdom*, Wadsworth Publishing Company, Inc., Belmont, California, 1965.

12–24 *(a)* Photo courtesy of Laurie Stewart Radford, Chapel Hill, N.C., from R. D. Wood and K. Imahori, *A Revision of the Characeae*, vol. 2, 1964–5.

12–25 © Douglas P. Wilson

12–28 Bruce C. Parker

12–33 *(a)* Don Longanecker; *(b)* William Randolph Taylor, *Plants of Bikini and Other Northern Marshall Islands*, University of Michigan Press, Ann Arbor, Michigan. Copyright 1950 by The University of Michigan; *(c)* © Douglas P. Wilson; *(d)* Gilbert H. Smith, *Cryptogamic Botany*, vol. 2, 2nd ed. Copyright © 1955 by the McGraw-Hill Book Company. All rights reserved.

12–34 After F. T. Haxo and L. R. Blinks, *Journal of General Physiology*, **33**:408, 1956.

12–37 *(top)*, *(middle)* Franco Rossi; *(bottom)* G Dallas Hanna. Courtesy of California Academy of Sciences.

12–40 Ray F. Evert

12–42 *(a)* © Douglas P. Wilson; *(b)* Y. Haneda

Essay, pages 274–275, D. Kubai and H. Ris

12–43 Don Longanecker

Essay, page 276, Photographs by Akio Miura, Tokyo University of Fisheries

12–44 Ross E. Hutchins

12–45 *(a)*, *(b)*, *(c)* K. B. Raper; *(d)* J. T. Bonner

13–1 E. S. Ross

13–2 Ross E. Hutchins

13–3 Ray F. Evert

13–4 Ray F. Evert

13–5 Ray F. Evert

13–6 Gilbert H. Smith, *Cryptogamic Botany*, vol. 2, 2nd ed. Copyright © 1955 by the McGraw-Hill Book Company, Inc. All rights reserved.

13–7 Ray F. Evert

Essay, page 287, After C. T. Ingold, *Spore Discharge in Land Plants*, Clarendon Press, Oxford, 1939.

13–8 Photographs by Alvin Grove

13–9 Ray F. Evert

13–10 R. H. Hall

13–11 R. F. Scagel, *et al.*, *An Evolutionary Survey of the Plant Kingdom*, Wadsworth Publishing Company, Inc., Belmont, California, 1965.

13–13 Ray F. Evert

13–14 Turtox/Cambosco

13–15 Charles W. Mann, National Audubon Society Collection/PR

13–16 *(a)* H. A. Thornhill, National Audubon Society Collection/PR; *(b)* Ross E. Hutchins

13–17 Ray F. Evert

13–18 *(a)* Lewis E. Anderson; *(b)* Larry West; *(c)* Ross E. Hutchins

13–19 After C. T. Ingold, *Spore Discharge in Land Plants*, Clarendon Press, Oxford, 1939.

14–1 Field Museum of Natural History

14–2 *(a)* After J. Walton, *Fossil Plants*, Macmillan, Inc., New York, 1940; *(b)* From J. Walton, *Phytomorphology*, **14**:155–160, 1964; *(c)* After F. M. Hueber, *International Symposium on the Devonian System*, p. 817.

14–6 From Adriance S. Foster and Ernest M. Gifford, Jr., *Comparative Morphology of Vascular Plants*, W. H. Freeman and Company. Copyright © 1959.

14–7 Albert C. Long

14–8 John M. Pettitt and Charles B. Beck, *Science*, **156**:1727–1729, June 1967. Copyright 1967 by the American Association for the Advancement of Science.

15–1 J. Spurr

15–2 *(a)* Ray F. Evert; *(b)* Virgil N. Argo

15–3 Ray F. Evert

15–4 Ray F. Evert

15–6 Tamra Engelhorn

15–7 Ray F. Evert

15–8 Ray F. Evert

15–9 Ray F. Evert

15–10 *(a)* Ross E. Hutchins; *(b)* R. Carr, Bruce Coleman, Inc.

15–11 Ray F. Evert

15–13 Ray F. Evert

15–14 Ray F. Evert

15–15 Ray F. Evert

15–16 Ray F. Evert

15–18 Ray F. Evert

15–19 E. S. Ross

15–20 Ray F. Evert

15–21 Ray F. Evert

15–22 *(a)* Ross E. Hutchins; *(b)* Ray F. Evert

15–23 A. Kerner and F. W. Oliver, *The Natural History of Plants*, II, 1894.

15–24 Ray F. Evert

15–25 *(a)* Claude Schoepf; *(b)* W. H. Hodge; *(c)* American Fern Society—Wherry Collection; *(d)* Ray F. Evert; *(e)* E. S. Ross

15–26 Ray F. Evert

15–27 Ray F. Evert

15–28 Bruce Coleman, Inc.

15–29 Charles Neidorf

15–30 Ray F. Evert

15–32 Ray F. Evert

15–33 Ikuko Mikukami and Joseph Gall, *Journal of Cell Biology*, **29**:97–111, 1966.

15–34 Ray F. Evert

16–1 W. N. Stewart and T. Delevoryas, *Botanical Review*, **22(1)**:45–80, 1956.

16–2 Jack Dermid

16–3 *(a)* Jack Dermid; *(b)* Ray F. Evert

16–4 *(a)* R. and J. Spurr, Bruce Coleman, Inc.; *(b)* Ray F. Evert

16–5 J. Kummerow

16–6 Ray F. Evert

16–7 From *Patterns of Life: The Unseen World of Plants* by William M. Harlow. Copyright © 1966 by William M. Harlow.

16–8 Ray F. Evert

16–9 Ray F. Evert

16–10 Ray F. Evert

16–11 Ray F. Evert

16–12 Ray F. Evert

16–13 Ray F. Evert

16–14 Ray F. Evert

16–16 Reid Morgan, Jepson Herbarium

16–17 D. Hardley, Bruce Coleman, Inc.

16–18 Charles S. Webber, Jepson Herbarium

16–19 Charles S. Webber, Jepson Herbarium

16–20 Jane Burton, Bruce Coleman, Inc.

16–21 Gene Ahrens, Bruce Coleman, Inc.

16–23 *(a)* Photo courtesy of Save-the-Redwoods League, 114 Sansome St., Rm. 605, San Francisco, Calif. 94104

16–24 Kenneth W. Fink, National Audubon Society Collection/PR

16–25 Ray F. Evert

16–26 John H. Gerard

16–27 *(b)* Ding Hou

16–28 *(a)* Jack Dermid; *(b)* CCM: General Biological Supply Company, Chicago, Ill.

16–29 J. Rodin

16–30 Roy Hunold, National Audubon Society Collection/PR

16–31 From P. Maheshwari, *An Introduction to the Embryology of the Angiosperms.* Copyright 1950 by the McGraw-Hill Book Company. Used with permission of McGraw-Hill Book Company.

16–32 *(a)*, *(b)* E. S. Ross; *(c)* Craig Blacklock

16–33 *(a)*, *(c)* E. S. Ross; *(b)* N. Smythe, National Audubon Society Collection/PR

16–34 *(a)* John H. Gerard; *(b)* E. S. Ross; *(c)* Larry West

16–37 *(a)*, *(c)* Larry West; *(b)* John H. Gerard, National Audubon Society Collection/PR; *(d)* E. S. Ross; *(e)* John H. Gerard

16–38 David Overcash, Bruce Coleman, Inc.

16–40 E. S. Ross

16–41 *(a)* Michael P. Gadomski,National Audubon Society Collection/PR; *(b)* M. J. Manuel, National Audubon Society Collection/PR

16–42 Ray F. Evert

16–43 J. Heslop-Harrison

16–44 *(a)*, *(c)*, *(d)* Patrick Echlin; *(b)* J. Heslop-Harrison

16–45 Ray F. Evert

16–46 Ray F. Evert

16–47 Ray F. Evert

16–48 Ray F. Evert

16–49 Redrawn from *Morphology of the Angiosperms* by A. J. Eames. Copyright 1961 by McGraw-Hill Book Company. Used with permission of McGraw-Hill Book Company.

17–1 E. S. Ross

17–2 Erling Dorf

17–3 After McKenzie and Slater, 1971

17–4 Photo by J. H. Johns, A.R.P.S., courtesy of the New Zealand Forest Service

17–5 *(a)* Karl Weidmann, National Audubon Society Collection/PR; *(b)* Jack Dermid; *(c)* E. S. Ross

17–6 Daniel Axelrod

17–7 After Irving W. Bailey and B. G. L. Swamy with permission of *The American Journal of Botany;* and adapted from Irving W. Bailey and B. G. L. Swamy, "The Conduplicate Carpel of Dicotyledons and Its Initial Trends of Specialization," in *Contributions to Plant Anatomy* by Irving W. Bailey, The Ronald Press Company, New York, 1954.

17–8 After Lawrence, *Taxonomy of Vascular Plants*, The Macmillan Company, New York, 1951

17–9 After Canright, *The American Journal of Botany*

17–10 Agenzia Fotografica, Luisa Ricciarini—Milan

17–11 Porter, *Plant Taxonomy*

17–12 E. S. Ross

17–13 *(b)* E. S. Ross; *(c)* Marion Latch, Bruce Coleman, Inc.; *(d)* Charles E. Schmidt, Bruce Coleman, Inc.

17–14 *(a)* K. Weidmann, National Audubon Society Collection/PR

Essay, page 371, Dr. H. P. Roggen, Vossenlaun 3g, Bosch en Duin, The Netherlands

17–15 © John Markham, Photo Researchers, Inc.

17–16 Larry West

17–17 *(a)* S. C. Bisserot, Bruce Coleman, Inc.; *(b)* Larry West; *(c)* E. S. Ross

17–18 E. S. Ross

17–19 Peter H. Raven

17–20 *(a)*, *(c)* E. S. Ross; *(b)* Leonard B. Thien

17–21 *(a)* Larry West; *(b)* E. S. Ross

17–22 Ross E. Hutchins

17–23 Franz Lazi, Bruce Coleman, Inc.

17–24 C. Laubscher, Bruce Coleman, Inc.

17–25 Oxford Scientific Films, Bruce Coleman, Inc.

17–26 *(a)* E. S. Ross; *(b)* Ross E. Hutchins; *(c)* John F. Skvarla, University of Oklahoma

17–28 Jack Dermid

17–30 R. F. Scagel, *et al.*, *An Evolutionary Survey of the Plant Kingdom*, Wadsworth Publishing Company, Belmont, California, 1965.

17–31 *(a)* Charles W. Mann, National Audubon Society Collection/PR; *(b)* William Harlow, National Audubon Society Collection/PR

17–32 U.S. Forest Service

17–33 R. F. Scagel, *et al.*, *An Evolutionary Survey of the Plant Kingdom*, Wadsworth Publishing Company, Belmont, California, 1965

17–35 Jack Dermid

17–36 Karl H. Maslowski, National Audubon Society Collection/PR

17–37 After R. F. Scagel, *et al.*, *An Evolutionary Survey of the Plant Kingdom*, Wadsworth Publishing Company, Belmont, California, 1965.

17–39 Photograph by Hal H. Harrison, National Audubon Society Collection/PR

17–41 *(a)* Professor R. E. Schultes, Harvard Botanical Museum; *(b)* James Mooney, Smithsonian Office of Anthropology, Bureau of American Ethnology collection

18–1 Jack Dermid

18–4 Ray F. Evert

18–7 Ray F. Evert

18–8 Ray F. Evert

19–1 Ray F. Evert

19–3 Ray F. Evert

19–4 Ray F. Evert

19–5 Ray F. Evert

19–6 Ray F. Evert

19–7 Ray F. Evert

19–8 Ray F. Evert

19–9 Ray F. Evert

19–10 Ray F. Evert

19–11 Ray F. Evert

19–12 *(a)* J. Heslop-Harrison; *(b)* Irving B. Sachs, U.S. Forest Products Laboratory

19–13 Ray F. Evert

19–14 Ray F. Evert

19–15 Ray F. Evert

19–16 Ray F. Evert

19–17 Ray F. Evert

19–18 *(a)*, *(b)*, *(c)* Ray F. Evert, Walter Eschrich and Susan E. Eichhorn, *Planta* (Berl.) **109**:193–210, 1973; *(d)*, *(e)*, *(f)* Ray F. Evert

19–19 Irving B. Sachs, U.S. Forest Products Laboratory

19–20 Michael A. Walsh

19–21 *(c)* After Katherine Esau, *Plant Anatomy*, 2nd ed., John Wiley & Sons, Inc., New York, 1965.

19–23 Ray F. Evert

19–24 Ray F. Evert

20–1 *(a)* Lynwood M. Chace, National Audubon Society Collection/PR; *(b)*, *(c)* Jack Dermid

20–2 Ray F. Evert

20–3 Ray F. Evert

20–5 F. A. L. Clowes

20–6 Hugh Spencer, National Audubon Society Collection/PR

20–7 Ray F. Evert

20–8 Ray F. Evert

20–9 Ray F. Evert

20–11 Ray F. Evert

20–12 Ray F. Evert

20–13 Ray F. Evert

20–14 Ray F. Evert

20–15 Ray F. Evert

20–16 Ray F. Evert

21–1 Ray F. Evert

21–2 Ray F. Evert

21–3 Ray F. Evert

21–4 Ray F. Evert

21–5 Ray F. Evert

21–6 *(b)* Ray F. Evert

21–7 Ray F. Evert

21–8 Ray F. Evert

21–9 Ray F. Evert

21–12 Ray F. Evert

21–13 Ray F. Evert

21–14 Ray F. Evert

21–15 Ray F. Evert

21–16 Jesse Lunger, National Audubon Society Collection/PR

21–17 Ray F. Evert

21–18 Ray F. Evert

21–19 Ray F. Evert

21–20 Ray F. Evert

21–21 Ray F. Evert

21–22 Thomas Pray, *The American Journal of Botany*, **41**:659–670, October 1954.

21–23 *(a)* Jack Dermid; *(b)* Thomas Pray

21–24 Ray F. Evert

21–26 Ray F. Evert

21–27 Ray F. Evert

21–28 Ray F. Evert

21–29 Ray F. Evert

21–30 Ray F. Evert

21–31 Ray F. Evert

21–32 Ray F. Evert

21–33 Daniel H. Franck

21–34 Ray F. Evert

21–35 *(a)* Jack Dermid; *(b)*, *(c)* Ray F. Evert

22–1 Ray F. Evert

22–2 After Peter M. Ray, *The Living Plant*, Holt, Rinehart and Winston, Inc., New York, 1963

22–3 Ray F. Evert

22–4 Ray F. Evert

22–6 Ray F. Evert

22–7 Ray F. Evert

22–8 Ray F. Evert

22–9 Ray F. Evert

22–10 Ray F. Evert

22–11 Ray F. Evert

22–14 Ray F. Evert

22–15 Ray F. Evert

22–16 Ray F. Evert

22–17 Ray F. Evert

22–19 Ray F. Evert

22–20 Ray F. Evert

22–21 Irving B. Sachs, U.S. Forest Products Laboratory

22–22 Ray F. Evert

22–23 Ray F. Evert

22–24 *(a)* L. Ullberg; *(b)* C. W. Ferguson, Laboratory of Tree-Ring Research, University of Arizona

22–25 *(a)* Ray F. Evert; *(b)* Irving B. Sachs, U.S. Forest Products Laboratory

Essay, page 482, Jack Dermid

23–4 U.S. Department of Agriculture

23–5 J. P. Nitsch, *The American Journal of Botany*, **37**:3, March 1950.

23–7 S. H. Wittwer, courtesy of Michigan Agricultural Experiment Station

23–8 S. H. Wittwer, courtesy of Michigan Agricultural Experiment Station

23–9 Howard Towner

23–10 Johannes Van Overbeek, "Plant Hormones and Regulators," *Science*, **152**:2 May 1966. Copyright 1966 by the American Association for the Advancement of Science.

23–11 J. E. Varner

23–13 After Johannes Van Overbeek, "Plant Hormones and Regulators," *Science*, **152**:2, May 1966. Copyright 1966 by the American Association for the Advancement of Science.

23–14 H. R. Chen

24–1 After A. C. Leopold, *Plant Growth and Development*, McGraw-Hill Book Company, New York, 1964. (After Briggs, *et al.*, 1957.)

24–3 After A. W. Galston, *The Green Plant*. Reproduced by permission of Prentice-Hall, Inc., Englewood Cliffs, New Jersey, 1968.

24–4 *(a)* After Beatrice Sweeney, *Rhythmic Phenomena in Plants*, Academic Press, New York, 1969; *(c)* Alfred Loeblich III

24–6 After Peter M. Ray, *The Living Plant*, Holt, Rinehart and Winston, Inc., New York, 1963.

24–7 U.S. Department of Agriculture

24–11 U.S. Department of Agriculture

24–12 A. E. Porsild, C. R. Harrington, and G. A. Mulligan, *Science*, **158**:113–114, 5 October 1967. Copyright 1967 by the American Association for the Advancement of Science.

24–14 Jack Dermid

24–15 Jack Dermid

24–16 William Harlow

25–1 Larry West

25–2 Exxon Research and Engineering Co.

25–3 Jack Dermid

25–4 Drawings after James Sutcliffe, *Plants and Water*, Edward Arnold (Publishers) Ltd., London. (After Bernal, 1965.)

25–5 P. Berger, National Audubon Society Collection/PR

25–7 After Albert L. Lehninger, *Biochemistry*, 2nd ed., Worth Publishers, New York, 1975.

25–9 Ray F. Evert

25–10 Photograph by Jack Dermid; graph after James Sutcliffe, *Plants and Water*,

Edward Arnold (Publishers) Ltd., London. (Data from Miller, 1968.)

25–11 *(a)* Jack Dermid

25–12 After M. Richardson, *Translocation in Plants*, Edward Arnold (Publishers) Ltd., London. (After Stout and Hoagland, 1939.)

25–13 *(a)* After M. Richardson, *Translocation in Plants*, Edward Arnold (Publishers) Ltd., London. (After Stout and Hoagland, 1939.); *(b)* Larry West.

25–14 After A. C. Leopold, *Plant Growth and Development*, McGraw-Hill Book Company, New York, 1964. (After Kramer, 1937.)

25–15 After M. Richardson, *Translocation in Plants*, Edward Arnold (Publishers) Ltd., London.

25–16 P. F. Scholander, H. T. Hammel, E. D. Bradstreet, and E. A. Hemmingsen, *Science*, **148**:339–346, 16 April 1965. Copyright 1965 by the American Association for the Advancement of Science.

25–17 After Martin H. Zimmerman, *Scientific American*, **208**:132–142, March 1963.

25–18 After Martin H. Zimmerman, *Scientific American*, **208**:132–142, March 1963.

25–19 *(a)*, *(b)* Joan Sampson; *(c)*, *(d)* J. Heslop-Harrison

25–21 Theodore T. Kozlowski, *Trees Magazine*, March–April 1953.

25–22 *Endeavor*, **28**:no 103, January 1963.

25–23 Walter Eschrich and Eberhard Fritz

26–1 From James F. Bonner and Arthur W. Galston, *Principles of Plant Physiology*, W. H. Freeman and Company, San Francisco, California. Copyright © 1952. (After Lyon and Buckman, 1943.)

26–2 After James F. Ferry, *Fundamentals of Plant Physiology*, The Macmillan Company, New York, 1959. (After *Soils and Man*, Yearbook of Agriculture, U.S. Department of Agriculture, 1938.)

26–3 Standard Oil Company (N.J.)

26–4 Robert H. Wright, National Audubon Society Collection/PR

26–5 *(a)* John H. Gerard, National Audubon Society Collection/PR; *(b)* Ross E. Hutchins

26–6 Jack Dermid

26–8 U.S. Department of Agriculture

26–10 After James Sutcliffe, *Plants and Water*, Edward Arnold (Publishers) Ltd., London. (After Alberda, 1948.)

Essay, page 547, From "Receptor Potentials and Action Potentials in *Drosera* Tentacles," by Stephen E. Williams and Barbara G. Pickard, *Planta* (Berl.) **103**:193–221, 1972. © 1972 by Springer-Verlag.

26–11 After Peter M. Ray, *The Living Plant*, Holt, Rinehart and Winston, Inc., New York, 1963.

26–13 Stanley W. Watson

26–14 *(a)*, *(c)* Jack Dermid; *(b)* Ross E. Hutchins

26–17 *(a)* The Nitragin Company, Inc.; *(b)* Courtesy of Triarch Inc., Box 98, Ripon, Wisconsin 54971

26–18 U.S. Department of Agriculture

26–19 P. J. Dart

26–20 R. R. Herbert, R. D. Holsten, and R. W. F. Hardy, E. I. duPont de Nemours Co., Inc.

27–1 U.S. Forest Service

27–4 *(a)* U.S. Forest Service; *(b)* Edward S. Ross

27–7 Jack Dermid

27–8 Art Bilsten, National Audubon Society Collection/PR

27–10 Larry West

27–11 Edward S. Ross

27–12 Jack Dermid

27–14 © Sherwin Carlquist

27–15 U.S. Forest Service

27–17 *(a)*, *(b)* Karl Weidmann, National Audubon Society Collection/PR; *(c)*, *(d)*, *(e)* E. S. Ross

27–18 John S. Flannery, Bruce Coleman, Inc.

27–20 E. S. Ross

27–22 U.S. Department of Agriculture

27–23 *(a)* Patricia Caulfield; *(b)* George Porter, National Audubon Society Collection/PR

27–25 *(a)* Larry West; *(b)* Jack Dermid

27–26 *(a)* Jack Dermid; *(b)* K. Gunnar, Bruce Coleman, Inc.

27–27 *(a)* Jack Dermid; *(b)* Les Blacklock

27–28 Dennis Brokaw

27–30 Jen and Des Bartlett, Bruce Coleman, Inc.

27–32 *(a)* Jen and Des Bartlett, Bruce Coleman, Inc.; *(b)* Charlie Ott, National Audubon Society Collection/PR

27–33 Les Blacklock

Essay, page 585 *(a)* Edward S. Ayensu, National Museum of Natural History, Smithsonian Institution; *(b)* Ken Nagata, National Museum of Natural History, Smithsonian Institution

28–1 Les Blacklock

28–2 G. E. Likens

28–3 *(a)* © Robert Clarke; *(b)* G. Llano; *(c)* © Douglas P. Wilson

28–4 John H. Gerard, National Audubon Society Collection/PR

28–5 Redrawn from J. Phillipson, *Ecological Energetics*, Edward Arnold (Publishers) Ltd., London, Institute of Biology Studies in Biology, no. 1, 1966; based on A. C. Hardy, *Fish. Invest. Lond.* **II**:7:no 3, 1924, and Sir A. C. Hardy, The Open Sea: **II**, *Fish and Fisheries*, Collins Co., Ltd., London.

28–6 Redrawn from G. G. Simpson and W. S. Beck, *Life: An Introduction to Biology*, 2nd ed., Harcourt Brace Jovanovich, Inc., New York, 1965. (Adapted from Odum, 1949.)

28–7 Larry West

28–8 Larry West

28–9 Jack Dermid

28–10 *(a)* U.S. Forest Service; *(b)* Jack Dermid

28–11 Jack Dermid

28–12 Larry West

28–13 U.S. Forest Service

29–1 *(a)* Miller Redwood Company; *(b)* Moulin Studios; *(c)*, *(d)* Dennis Brokaw

Essay, page 601, The Bettmann Archive, Inc.

29–3 Redrawn from R. H. Waring and J. Major, *Ecological Monographs*, **34**:167–215, 1964, by permission of Duke University Press.

29–4 Redrawn from R. H. Whittaker, *Ecological Monographs*, **26**:16, 1956, by permission of Duke University Press.

29–5 Redrawn from R. H. Whittaker, *Ecological Monographs*, **26**:16, 1956, by permission of Duke University Press.

29–6 J. M. Conrader, National Audubon Society Collection/PR

29–7 Calvin McMillan, *Bioscience*, **19**:132, 1969.

29–8 R. B. Walker

29–9 Donald R. Farrar

29–10 H. A. Mooney, *Ecology*, **47**:950–951, 1966.

29–11 Redrawn from John L. Harper, J. T. Williams and G. R. Sager, *Journal of Ecology*, **53**:273–286, July 1965. Photograph, John L. Harper

29–12 Redrawn from Michael Neushul, *Ecology*, **48**:no. 1, Winter 1967.

29–13 A. S. Thomas, *Journal of Ecology*, **48**:287–306, 1960.

29–14 Redrawn from John L. Harper, *Symposia Soc. Exper. Biology*, **15**:25, 1961.

29–15 Redrawn from John L. Harper, *Symposia Soc. Exper. Biology*, **15**:25, 1961.

29–16 C. H. Muller, *Bulletin of the Torrey Botanical Club*, **93**:334, 1966.

29–17 John Mann, Australian Department of Lands

29–18 E. S. Ross

29–19 Daniel Janzen

Index